HIGH-FREQUENCY MAGNETIC COMPONENTS

HIGH-FREQUENCY MAGNETIC COMPONENTS

SECOND EDITION

Marian K. Kazimierczuk
Wright State University, Dayton, Ohio, USA

Registered office
John Wiley & Sons Ltd, The Atrium, Southern Gate, Chichester, West Sussex, PO19 8SQ, United Kingdom

For details of our global editorial offices, for customer services and for information about how to apply for permission to reuse the copyright material in this book please see our website at www.wiley.com.

Library of Congress Cataloging-in-Publication Data

Kazimierczuk, Marian.
High-frequency magnetic components / Marian K. Kazimierczuk. – Second edition.
pages cm
Includes bibliographical references and index.
ISBN 978-1-118-71779-0 (hardback)
1. Electromagnetic devices. 2. Magnetic devices. I. Title.
TK7872.M25K395 2009
621.3815 – dc23

2013026579

A catalogue record for this book is available from the British Library.
ISBN 9781118717790

Typeset in 9/11pt Times by Laserwords Private Limited, Chennai, India.

1 2014

To my Father

Contents

Preface **xvii**
About the Author **xix**
List of Symbols **xxi**

1 Fundamentals of Magnetic Devices 1

1.1 Introduction 1
1.2 Fields 2
1.3 Magnetic Relationships 2
1.3.1 Magnetomotive Force 2
1.3.2 Magnetic Field Intensity 3
1.3.3 Magnetic Flux 3
1.3.4 Magnetic Flux Density 4
1.3.5 Magnetic Flux Linkage 5
1.4 Magnetic Circuits 6
1.4.1 Reluctance 6
1.4.2 Magnetic KVL 8
1.4.3 Magnetic Flux Continuity 8
1.5 Magnetic Laws 9
1.5.1 Ampère's Law 9
1.5.2 Faraday's Law 13
1.5.3 Lenz's Law 15
1.5.4 Volt–Second Balance 16
1.5.5 Ohm's Law 16
1.5.6 Biot–Savart's Law 18
1.5.7 Maxwell's Equations 19
1.5.8 Maxwell's Equations for Good Conductors 24
1.5.9 Poynting's Vector 24
1.5.10 Joule's Law 26
1.6 Eddy Currents 29
1.7 Core Saturation 32
1.7.1 Core Saturation for Sinusoidal Inductor Voltage 34
1.7.2 Core Saturation for Square-Wave Inductor Voltage 36
1.7.3 Core Saturation for Rectangular Wave Inductor Voltage 38

1.8 Inductance 40
1.8.1 Definitions of Inductance 40
1.8.2 Inductance of Solenoid 45
1.8.3 Inductance of Inductor with Toroidal Core 47
1.8.4 Inductance of Inductor with Torus Core 49
1.8.5 Inductance of Inductor with Pot Core 49
1.8.6 Inductance Factor 49
1.9 Air Gap in Magnetic Core 51
1.9.1 Inductance 51
1.9.2 Magnetic Field in Air Gap 53
1.10 Fringing Flux 54
1.10.1 Fringing Flux Factor 54
1.10.2 Effect of Fringing Flux on Inductance for Round Air Gap 57
1.10.3 Effect of Fringing Flux on Inductance for Rectangular Air Gap 58
1.10.4 Method of Effective Air Gap Cross-Sectional Area 60
1.10.5 Method of Effective Length of Air Gap 60
1.10.6 Patridge's Fringing Factor 60
1.10.7 Distribution of Fringing Magnetic Field 62
1.11 Inductance of Strip Transmission Line 62
1.12 Inductance of Coaxial Cable 62
1.13 Inductance of Two-Wire Transmission Line 63
1.14 Magnetic Energy and Magnetic Energy Density 64
1.14.1 Magnetic Energy Density 64
1.14.2 Magnetic Energy Stored in Inductors with Ungapped Core 64
1.14.3 Magnetic Energy Stored in Inductors with Gapped Core 65
1.15 Self-Resonant Frequency 69
1.16 Quality Factor of Inductors 69
1.17 Classification of Power Losses in Magnetic Components 69
1.18 Noninductive Coils 71
1.19 Summary 71
1.20 References 74
1.21 Review Questions 76
1.22 Problems 78

2 Magnetic Cores 81

2.1 Introduction 81
2.2 Properties of Magnetic Materials 81
2.3 Magnetic Dipoles 83
2.4 Magnetic Domains 89
2.5 Curie Temperature 90
2.6 Magnetic Susceptibility and Permeability 91
2.7 Linear, Isotropic, and Homogeneous Magnetic Materials 93
2.8 Magnetic Materials 93
2.8.1 Ferromagnetic Materials 93
2.8.2 Antiferromagnetic Materials 94
2.8.3 Ferrimagnetic Materials 95
2.8.4 Diamagnetic Materials 95
2.8.5 Paramagnetic Materials 96
2.9 Hysteresis 96
2.10 Low-Frequency Core Permeability 98
2.11 Core Geometries 99
2.11.1 Toroidal Cores 99
2.11.2 CC and UU Cores 100

2.11.3 Pot Cores 100
2.11.4 PQ and RM Cores 101
2.11.5 EE and EDT Cores 102
2.11.6 Planar Cores 103
2.12 Ferromagnetic Core Materials 103
2.12.1 Iron Cores 104
2.12.2 Ferrosilicon Cores 104
2.12.3 Amorphous Alloy Cores 104
2.12.4 Nickel–Iron and Cobalt–Iron Cores 105
2.12.5 Ferrite Cores 105
2.12.6 Powder Cores 106
2.12.7 Nanocrystalline Cores 107
2.13 Superconductors 108
2.14 Hysteresis Loss 109
2.15 Eddy-Current Core Loss 113
2.15.1 General Expression for Eddy-Current Core Loss 113
2.15.2 Eddy-Current Core Loss for Sinusoidal Inductor Voltage 115
2.15.3 Eddy-Current Power Loss in Round Core for Sinusoidal Flux Density 117
2.15.4 Total Core Power Loss for Sinusoidal Inductor Voltage 118
2.15.5 Eddy-Current Core Loss for Square-Wave Inductor Voltage 122
2.15.6 Eddy-Current Core Loss for Rectangular Inductor Voltage 124
2.15.7 Eddy-Current Power Loss in Laminated Cores 128
2.15.8 Excess Core Loss 129
2.16 Steinmetz Empirical Equation for Total Core Loss 129
2.16.1 Losses of Ungapped Cores 129
2.16.2 Losses of Gapped Cores 134
2.17 Core Losses for Nonsinusoidal Inductor Current 135
2.18 Complex Permeability of Magnetic Materials 136
2.18.1 Series Complex Permeability 137
2.18.2 Loss Angle and Quality Factor 141
2.18.3 Complex Reluctance 144
2.18.4 Complex Inductance 145
2.18.5 Complex Impedance of Inductor 145
2.18.6 Approximation of Series Complex Permeability 146
2.18.7 Parallel Complex Permeability 148
2.18.8 Relationships Between Series and Parallel Complex Permeabilities 150
2.19 Cooling of Magnetic Cores 151
2.20 Summary 152
2.21 References 157
2.22 Review Questions 160
2.23 Problems 161

3 Skin Effect 163

3.1 Introduction 163
3.2 Resistivity of Conductors 164
3.2.1 Temperature Dependance of Resistivity 164
3.3 Skin Depth 166
3.4 AC-to-DC Winding Resistance Ratio 173
3.5 Skin Effect in Long Single Round Conductor 173
3.6 Current Density in Single Round Conductor 175
3.6.1 Bessel Differential Equation 175
3.6.2 Kelvin Functions 176

3.6.3 Approximations of Bessel's Equation Solution 177
3.6.4 Current Density $J(r)/J(0)$ 177
3.6.5 Current Density $J(r)/J(r_o)$ 178
3.6.6 Current Density $J(r)/J_{DC}$ 180
3.6.7 Approximation of Current Density in Round Conductor 183
3.6.8 Impedance of Round Conductor 184
3.6.9 Approximation of Resistance and Inductance of Round Conductor 189
3.6.10 Simplified Derivation of Round Wire Resistance 191
3.7 Magnetic Field Intensity for Round Wire 193
3.8 Other Methods of Determining the Round Wire Inductance 195
3.9 Power Loss Density in Round Conductor 200
3.10 Skin Effect in Single Rectangular Plate 204
3.10.1 Magnetic Field Intensity in Single Rectangular Plate 204
3.10.2 Current Density in Single Rectangular Plate 206
3.10.3 Power Loss in Single Rectangular Plate 208
3.10.4 Impedance of Single Rectangular Plate 211
3.11 Skin Effect in Rectangular Foil Conductor Placed Over Ideal Core 215
3.12 Summary 218
3.13 Appendix 220
3.13.1 Derivation of Bessel Equation for Long Round Wire 220
3.14 References 222
3.15 Review Questions 223
3.16 Problems 224

4 Proximity Effect 226

4.1 Introduction 226
4.2 Orthogonality of Skin and Proximity Effects 227
4.3 Proximity Effect in Two Parallel Round Conductors 227
4.4 Proximity Effect in Coaxial Cable 228
4.5 Proximity and Skin Effects in Two Parallel Plates 230
4.5.1 Magnetic Field in Two Parallel Plates 230
4.5.2 Current Density in Two Parallel Plates 231
4.5.3 Power Loss in Two Parallel Plates 235
4.5.4 Impedance of Each Plate 243
4.6 Antiproximity and Skin Effects in Two Parallel Plates 244
4.6.1 Magnetic Field in Two Parallel Plates 244
4.6.2 Current Density in Two Parallel Plates 247
4.6.3 Power Loss in Two Parallel Plates 248
4.7 Proximity Effect in Open-Circuit Conductor 249
4.8 Proximity Effect in Multiple-Layer Inductor 250
4.9 Self-Proximity Effect in Rectangular Conductors 256
4.10 Summary 259
4.11 Appendix 260
4.11.1 Derivation of Proximity Power Loss 260
4.12 References 261
4.13 Review Questions 263
4.14 Problems 263

5 Winding Resistance at High Frequencies 265

5.1 Introduction 265
5.2 Eddy Currents 265
5.3 Magnetic Field Intensity in Multilayer Foil Inductors 266

5.4 Current Density in Multilayer Foil Inductors 274
5.5 Winding Power Loss Density in Individual Foil Layers 278
5.6 Complex Winding Power in nth Layer 281
5.7 Winding Resistance of Individual Foil Layers 282
5.8 Orthogonality of Skin and Proximity for Individual Foil Layers 284
5.9 Optimum Thickness of Individual Foil Layers 286
5.10 Winding Inductance of Individual Layers 291
5.11 Power Loss in All Layers 292
5.12 Impedance of Foil Winding 293
5.13 Resistance of Foil Winding 294
5.14 Dowell's Equation 294
5.15 Approximation of Dowell's Equation 298
 5.15.1 Approximation of Dowell's Equation for Low and Medium Frequencies 298
 5.15.2 Approximation of Dowell's Equation for High Frequencies 299
5.16 Winding AC Resistance with Uniform Foil Thickness 300
 5.16.1 Optimum Uniform Foil Thickness of Inductor Winding for Sinusoidal Inductor Current 301
 5.16.2 Boundary Between Low and Medium Frequencies for Foil Windings 306
5.17 Transformation of Foil Conductor to Rectangular, Square, and Round Conductors 308
5.18 Winding AC Resistance of Rectangular Conductor 309
 5.18.1 Optimum Thickness of Rectangular Conductor for Sinusoidal Inductor Current 315
 5.18.2 Boundary Between Low and Medium Frequencies for Rectangular Wire Winding 318
5.19 Winding Resistance of Square Wire 318
 5.19.1 Winding AC Resistance of Square Conductor 320
 5.19.2 Optimization of Square Wire Winding at Fixed Pitch 321
 5.19.3 Optimization of Square Wire Winding at Fixed Porosity Factor 322
 5.19.4 Critical Thickness of Square Winding Resistance 324
 5.19.5 Boundary Between Low and Medium Frequencies for Square Wire Winding 325
5.20 Winding Resistance of Round Wire 326
 5.20.1 AC Resistance of Round Wire Winding 329
 5.20.2 Optimum Diameter of Round Wire at Fixed Pitch 331
 5.20.3 Optimum Diameter of Round Wire at Fixed Porosity Factor 332
 5.20.4 Critical Round Wire Diameter 334
 5.20.5 Boundary Between Low and Medium Frequencies for Round Wire Winding 335
5.21 Inductance 335
5.22 Solution for Round Conductor Winding in Cylindrical Coordinates 338
5.23 Litz Wire 338
 5.23.1 Litz-Wire Construction 338
 5.23.2 Model of Litz-Wire and Multistrand Wire Windings 339
 5.23.3 Litz-Wire Winding Resistance 341
 5.23.4 Optimum Strand Diameter at Fixed Porosity Factor 345
 5.23.5 Approximated Optimum Strand Diameter 346
 5.23.6 Optimum Strand Diameter at Variable Porosity Factor 348
 5.23.7 Boundary Between Low and Medium Frequencies for Litz-Wire Windings 349
 5.23.8 Approximation of Litz-Wire Winding Resistance for Low and Medium Frequencies 349

5.24 Winding Power Loss for Inductor Current with Harmonics 351
5.24.1 Copper Power Loss in PWM DC–DC Converters for Continuous Conduction Mode 353
5.24.2 Copper Power Loss in PWM DC–DC Converters for DCM 360
5.25 Winding Power Loss of Foil Inductors Conducting DC and Harmonic Currents 364
5.25.1 Optimum Foil Thickness of Inductors Conducting DC and Harmonic Currents 365
5.26 Winding Power Loss of Round Wire Inductors Conducting DC and Harmonic Currents 366
5.26.1 Optimum Diameter of Inductors Conducting DC and Harmonic Currents 367
5.27 Effective Winding Resistance for Nonsinusoidal Inductor Current 367
5.28 Thermal Effects on Winding Resistance 370
5.29 Thermal Model of Inductors 373
5.30 Summary 374
5.31 Appendix 375
5.31.1 Derivation of Dowell's Equation Approximation 375
5.32 References 377
5.33 Review Questions 381
5.34 Problems 381

6 Laminated Cores 383

6.1 Introduction 383
6.2 Low-Frequency Eddy-Current Laminated Core Loss 384
6.3 Comparison of Solid and Laminated Cores 389
6.4 Alternative Solution for Low-Frequency Eddy-Current Core Loss 389
6.4.1 Sinusoidal Inductor Voltage 391
6.4.2 Square-Wave Inductor Voltage 393
6.4.3 Rectangular Inductor Voltage 393
6.5 General Solution for Eddy-Current Laminated Core Loss 393
6.5.1 Magnetic Field Distribution at High Frequencies 393
6.5.2 Power Loss Density Distribution at High Frequencies 397
6.5.3 Lamination Impedance at High Frequencies 400
6.6 Summary 408
6.7 References 409
6.8 Review Questions 410
6.9 Problems 411

7 Transformers 412

7.1 Introduction 412
7.2 Transformer Construction 413
7.3 Ideal Transformer 413
7.4 Voltage Polarities and Current Directions in Transformers 416
7.5 Nonideal Transformers 417
7.6 Neumann's Formula for Mutual Inductance 422
7.7 Mutual Inductance 424
7.8 Magnetizing Inductance 425
7.9 Coupling Coefficient 427
7.10 Leakage Inductance 429
7.11 Dot Convention 432
7.12 Series-Aiding and Series-Opposing Connections 435

7.13 Equivalent T Network 435
7.14 Energy Stored in Coupled Inductors 436
7.15 High-Frequency Transformer Model 437
7.16 Stray Capacitances 438
7.17 Transformer Efficiency 438
7.18 Transformers with Gapped Cores 438
7.19 Multiple-Winding Transformers 439
7.20 Autotransformers 439
7.21 Measurements of Transformer Inductances 440
7.22 Noninterleaved Windings 442
7.23 Interleaved Windings 444
7.24 Wireless Energy Transfer 446
7.25 AC Current Transformers 446
7.25.1 Principle of Operation 446
7.25.2 Model of Current Transformer 447
7.25.3 Low-Frequency Response 448
7.25.4 High-Frequency Response 449
7.25.5 Maximum Power Transfer by Current Transformer 452
7.26 Saturable Reactors 454
7.27 Transformer Winding Power Losses with Harmonics 455
7.27.1 Winding Power Losses with Harmonics for CCM 455
7.27.2 Winding Power Losses with Harmonics for DCM 460
7.28 Thermal Model of Transformers 464
7.29 Summary 465
7.30 References 467
7.31 Review Questions 470
7.32 Problems 471

8 Integrated Inductors 472

8.1 Introduction 472
8.2 Skin Effect 472
8.3 Resistance of Rectangular Trace with Skin Effect 474
8.4 Inductance of Straight Rectangular Trace 477
8.5 Inductance of Rectangular Trace with Skin Effect 478
8.6 Construction of Integrated Inductors 480
8.7 Meander Inductors 481
8.8 Inductance of Straight Round Conductor 485
8.9 Inductance of Circular Round Wire Loop 486
8.10 Inductance of Two-Parallel Wire Loop 486
8.11 Inductance of Rectangle of Round Wire 486
8.12 Inductance of Polygon Round Wire Loop 486
8.13 Bondwire Inductors 487
8.14 Single-Turn Planar Inductor 488
8.15 Inductance of Planar Square Loop 490
8.16 Planar Spiral Inductors 490
8.16.1 Geometries of Planar Spiral Inductors 490
8.16.2 Inductance of Square Planar Inductors 493
8.16.3 Inductance of Hexagonal Spiral Inductors 502
8.16.4 Inductance of Octagonal Spiral Inductors 503
8.16.5 Inductance of Circular Spiral Inductors 504
8.17 Multimetal Spiral Inductors 505
8.18 Planar Transformers 506
8.19 MEMS Inductors 507

8.20 Inductance of Coaxial Cable 509
8.21 Inductance of Two-Wire Transmission Line 509
8.22 Eddy Currents in Integrated Inductors 509
8.23 Model of RF-Integrated Inductors 510
8.24 PCB Inductors 512
8.25 Summary 514
8.26 References 515
8.27 Review Questions 518
8.28 Problems 519

9 Self-Capacitance 520

9.1 Introduction 520
9.2 High-Frequency Inductor Model 520
9.3 Self-Capacitance Components 530
9.4 Capacitance of Parallel-Plate Capacitor 531
9.5 Self-Capacitance of Foil Winding Inductors 532
9.6 Capacitance of Two Parallel Round Conductors 533
9.6.1 Potential of Infinite Single Straight Round Conductor with Charge 533
9.6.2 Potential Between Two Infinite Parallel Straight Round Conductors with Nonuniform Charge Density 533
9.6.3 Capacitance of Two Parallel Wires with Nonuniform Charge Density 536
9.7 Capacitance of Round Conductor and Parallel Conducting Plane 539
9.8 Capacitance of Straight Parallel Wire Pair Over Ground 540
9.9 Capacitance Between Two Parallel Straight Round Conductors with Uniform Charge Density 540
9.10 Capacitance of Cylindrical Capacitor 542
9.11 Self-Capacitance of Single-Layer Inductors 542
9.12 Self-Capacitance of Multilayer Inductors 545
9.12.1 Exact Equation for Self-Capacitance of Multilayer Inductors 545
9.12.2 Approximate Equation for Turn-to-Turn Self-Capacitance of Multilayer Inductors 550
9.13 Self-Capacitance of Single-Layer Inductors 553
9.13.1 Exact Equation for Self-Capacitance of Single-Layer Inductors 553
9.13.2 Approximate Equation for Turn-to-Turn Self-Capacitance of Single-Layer Inductors 555
9.14 Δ-to-Y Transformation of Capacitors 557
9.15 Overall Self-Capacitance of Single-Layer Inductor with Core 557
9.16 Measurement of Self-Capacitance 559
9.17 Inductor Impedance 560
9.18 Summary 564
9.19 References 565
9.20 Review Questions 566
9.21 Problems 566

10 Design of Inductors 568

10.1 Introduction 568
10.2 Magnet Wire 569
10.3 Wire Insulation 572
10.4 Restrictions on Inductors 572
10.5 Window Utilization Factor 574
10.5.1 Wire Insulation Factor 574

10.5.2 Air and Wire Insulation Factor 575
10.5.3 Air Factor 576
10.5.4 Bobbin Factor 577
10.5.5 Edge Factor 578
10.5.6 Number of Turns 578
10.5.7 Window Utilization Factor 579
10.5.8 Window Utilization Factor for Foil Winding 581
10.6 Temperature Rise of Inductors 581
10.6.1 Expression for Temperature Rise of Inductors 581
10.6.2 Surface Area of Inductors with Toroidal Core 582
10.6.3 Surface Area of Inductors with Pot Core 583
10.6.4 Surface Area of Inductors with PQ Core 584
10.6.5 Surface Area of Inductors with EE Core 585
10.7 Mean Turn Length of Inductors 585
10.7.1 Mean Turn Length of Inductors with Toroidal Cores 585
10.7.2 Mean Turn Length of Inductors with PC and PQ Cores 586
10.7.3 Mean Turn Length of Inductors with EE Cores 586
10.8 Area Product Method 586
10.8.1 General Expression for Area Product 586
10.8.2 Area Product for Sinusoidal Inductor Voltage 587
10.9 Design of AC Inductors 590
10.9.1 Optimum Magnetic Flux Density 590
10.9.2 Examples of AC Inductor Designs 591
10.10 Inductor Design for Buck Converter in CCM 603
10.10.1 Derivation of Area Product A_p for Square-Wave Inductor Voltage 603
10.10.2 Inductor Design for Buck Converter in CCM Using Area Product A_p Method 603
10.11 Inductor Design for Buck Converter in DCM Using A_p Method 619
10.12 Core Geometry Coefficient K_g Method 654
10.12.1 General Expression for Core Geometry Coefficient K_g 654
10.12.2 AC Inductor with Sinusoidal Current and Voltage 655
10.12.3 Inductor for PWM Converter in CCM 656
10.12.4 Inductor for PWM Converter in DCM 656
10.13 Inductor Design for Buck Converter in CCM Using K_g Method 658
10.14 Inductor Design for Buck Converter in DCM Using K_g Method 660
10.15 Summary 663
10.16 References 664
10.17 Review Questions 666
10.18 Problems 666

11 Design of Transformers 668

11.1 Introduction 668
11.2 Area Product Method 668
11.2.1 Derivations of Core Area Product A_p 668
11.2.2 Core Window Area Allocation for Transformer Windings 670
11.3 Optimum Flux Density 673
11.4 Area Product A_p for Sinusoidal Voltages 674
11.5 Transformer Design for Flyback Converter in CCM 675
11.5.1 Practical Design Considerations of Transformers 675
11.5.2 Area Product A_p for Transformer Square Wave Voltages 675
11.5.3 Area Product A_p Method 676
11.6 Transformer Design for Flyback Converter in DCM 689

11.7 Geometrical Coefficient K_g Method 702
11.7.1 Derivation of Geometrical Coefficient K_g 702
11.7.2 K_g for Transformer with Sinusoidal Currents and Voltages 704
11.7.3 Transformer for PWM Converters in CCM 704
11.7.4 Transformer for PWM Converters in DCM 705
11.8 Transformer Design for Flyback Converter in CCM Using K_g Method 705
11.9 Transformer Design for Flyback Converter in DCM Using K_g Method 709
11.10 Summary 714
11.11 References 714
11.12 Review Questions 715
11.13 Problems 715

Appendix A Physical Constants **717**

Appendix B Maxwell's Equations **718**

Answers to Problems **719**

Index **725**

Preface

This book is about high-frequency magnetic power devices: high-frequency power inductors and high-frequency power transformers. It is intended as a textbook at the senior and graduate levels for students majoring in electrical engineering and as a reference for practicing engineers in the areas of power electronics and radiofrequency (RF) power amplifiers as well as other branches of physical sciences. Power electronics has evolved as a major enabling technology, which is used to efficiently convert energy from one form to another. The purpose of this book is to provide foundations for the analysis and design of high-frequency power magnetic devices: inductors and transformers. Magnetic components have a broad variety of applications across many diverse industries, such as energy conversion from one from to another (DC–DC, AC–DC, DC–AC, and AC–AC), switch-mode power supplies (SMPS), resonant inverters and converters, laptops, radio transmitters, uninterruptable power supplies (UPS), power factor correction (PFC), solar and wind renewable energy circuits, distributed energy generation systems (microgrids), automotive power electronics in hybrid and electric vehicles, battery chargers, wireless (or contactless) power transfer, energy harvesting, electrical machines, portable electronic devices, chokes, active power filters, electromagnetic interference (EMI)/radiofrequency interference (RFI) filters, RF noise suppressors, oscillators, energy storage systems, aviation power systems, induction heating, electronic ballasts, light-emitting diode (LED) lighting, magnetic sensors, ferrous metal detectors, fuel cell power supplies, medical equipment, implantable medical devices, and current measurement probes.

This book addresses the skin and proximity effects on winding and core losses in magnetic components at high frequencies. Magnetic components have often a large size and weight, are lossy, and have low-power density. Special topics in this book include optimization of the size of winding conductors, integrated inductors, analysis of the self-capacitance of inductors and transformers, temperature effects on the performance of magnetic components, and high-frequency physics-based models of inductors and transformers. The International System (SI) of Units are used in this book. All quantities are expressed in units of the meter–kilogram–second (MKS) system. The second edition of this book is a thoroughly revised and updated textbook and includes new research results and advances in magnetic device technology.

Introduction to Physical Constants and Maxwell's Equations is given in Appendices A and B, respectively.

This textbook assumes that the student is familiar with electromagnetic fields, general circuit analysis techniques, calculus, and vector algebra. These courses cover the fundamental laws of physics, such as Faraday's law, Ampère's law, Gauss's law, Lenz's law, Ohm's law, Joule's law, Poynting's theorem, and Maxwell's equations. There is sufficient material in this textbook for a one-semester course. Complete solutions for all problems are included in the *Solutions Manual*, which is available from the publisher for those instructors who adopt the textbook for their courses.

I am pleased to express my gratitude to Dr Nisha Kondrath for MATLAB® figures, Dr Dakshina Murthy-Bellur for his help in developing the design examples of inductors and transformers, Dr Rafal Wojda for his contributions to optimization of winding conductors and for MATLAB® figures, Dr Gregory Kozlowski for his help with Bessel functions and analysis of a single round conductor, and Dr Hiroo Sekiya for creative discussions. I am deeply indebted to the students, reviewers, scientists, industrial engineers, and other readers for valuable feedback, suggestions, and corrections. Most of these suggestions have been incorporated in the second edition of this book.

Throughout the entire course of this project, the support provided by John Wiley & Sons was excellent. I wish to express my sincere thanks to Laura Bell, Assistant Editor; Richard Davies, Senior Project Editor; and Peter Mitchell, Publisher. It has been a real pleasure working with them. Finally, my thanks goes to my family for their patience, understanding, and support throughout the endeavor.

The author would welcome and greatly appreciate suggestions and corrections from the readers for improvements in the technical content as well as the presentation style.

Marian K. Kazimierczuk

About the Author

Marian K. Kazimierczuk is a Robert J. Kegerreis, Distinguished Professor of Electrical Engineering at Wright State University, Dayton, Ohio, United States. He received the MS, PhD, and DSc degrees from Warsaw University of Technology, Department of Electronics, Warsaw, Poland. He is the author of six books, over 170 archival refereed journal papers, over 200 conference papers, and seven patents.

Professor Kazimierczuk is a Fellow of the IEEE. He served as the Chair of the Technical Committee of Power Systems and Power Electronics Circuits, IEEE Circuits and Systems Society. He served on the Technical Program Committees of the IEEE International Symposium on Circuits and Systems (ISCAS) and the IEEE Midwest Symposium on Circuits and Systems. He also served as an associate editor of the *IEEE Transactions on Circuits and Systems, Part I, Regular Papers*, *IEEE Transactions on Industrial Electronics*, *International Journal of Circuit Theory and Applications*, and *Journal of Circuits, Systems, and Computers*, and as a guest editor of the *IEEE Transactions on Power Electronics*. He was an IEEE Distinguished Lecturer.

Professor Kazimierczuk received Presidential Award for Outstanding Faculty Member at Wright State University in 1995. He was Brage Golding Distinguished Professor of Research at Wright State University in 1996–2000. He received the Trustees' Award from Wright State University for Faculty Excellence in 2004. He received the Outstanding Teaching Award from the American Society for Engineering Education (ASEE) in 2008. He was also honored with the Excellence in Research Award, Excellence in Teaching Awards, and Excellence in Professional Service Award in the College of Engineering and Computer Science, Wright State University.

His research interests are in power electronics, including pulse-width modulated (PWM) DC–DC power converters, resonant DC–DC power converters, modeling and controls, radio frequency (RF) high-efficiency power amplifiers and oscillators, high-frequency magnetic devices, semiconductor power devices, renewable energy sources, and evanescent microwave microscopy.

Professor Kazimierczuk is the author or co-author of *Resonant Power Converters* (Second Edition), *Pulse-Width Modulated DC-DC Power Converters*, *RF Power Amplifiers*, *Electronic Devices, A Design Approach*, and *Laboratory Manual to Accompany Electronic Devices, A Design Approach*.

List of Symbols

A Magnetic vector potential
A_c Cross-sectional area of core
A_{cell} Cross-sectional area of winding cell
A_{Cu} Cross-sectional area of copper
A_L Specific inductance of core, inductance factor
A_w Cross-sectional area of winding bare wire
A_{wo} Outer cross-sectional area of winding wire
A_{ws} Cross-sectional area of strand bare wire
A_{wso} Outer cross-sectional area of strand wire
A_p Area product of core
A_t Surface area of inductor or transformer
a Unity vector
a_n Unity vector normal to a surface
B Magnetic flux density
B_{DC} DC component of magnetic flux density
B_m Amplitude of the AC component of magnetic flux density
B_r Residual flux density, remnant magnetization, and remnance
B_{pk} Peak value of total magnetic flux density ($B_{pk} = B_{DC} + B_m$)
B_s Saturation flux density
BW Bandwidth
b Breadth of winding
b_b Breadth of bobbin winding window
C Capacitance
C_s Equivalent series-resonant capacitance
D Electric flux density, DC component of on-duty cycle of switch
D_{min} Minimum DC component of on-duty cycle of switch
D_{max} Maximum DC component of on-duty cycle of switch
D_i Inner diameter of toroidal core
D_o Outer diameter of toroidal core
D_w Dwell duty cycle
d_i Diameter of winding bare wire
d_o Outer diameter of insulated winding wire

d_s	Diameter of bare strand wire
d_{so}	Outer diameter of insulated strand wire
E	Electric field intensity
EMF	Electromotive force
F	Force
F_f	Fringing factor
F_g	Air gap factor
F_R	AC resistance factor
F_{Rh}	Harmonic AC resistance factor
F_{Rph}	Harmonic AC resistance factor of primary winding
F_{Rsh}	Harmonic AC resistance factor of secondary winding
f	Frequency
f_r	Self-resonant frequency of inductor
f_s	Switching frequency
f_H	Upper 3-dB frequency
f_L	Lower 3-dB frequency
H	Magnetic field intensity
H_c	Coercive force
h	Thickness of winding conductor
h_b	Height of winding window of bobbin
I_I	DC input current of converter
I_L	Average or DC current through inductor L
I_{Lmax}	Maximum current through inductor L
I_{Lrms}	rms current through inductor L
I_{Lpk}	Peak total current through inductor L
I_n	rms value of nth harmonic of inductor current
I_O	DC output current of converter
I_{rms}	rms value of current i
I_{Omax}	Maximum value of DC load current I_O
I_{Omin}	Minimum value of DC load current I_O
i	Current
i_L	Inductor current
i_{Lm}	Current through magnetizing inductance
i_p	Current through primary winding
i_s	Current through secondary winding
J	Conduction current density
J_D	Displacement current density
J_m	Amplitude of current density
J_{rms}	rms value of current density
K_a	Air factor
K_b	Bobbin factor
K_{ai}	Air and wire insulation factor
K_{ed}	Edge factor
K_g	Core geometry coefficient
K_f	Waveform factor
K_i	Wire insulation factor
K_u	Window utilization factor
k	Coupling coefficient, complex propagation constant
l_c	Magnetic path length (MPL)
l_g	Length of air gap
l_T	Mean turn length (MTL), length of turn
l_w	Length of winding wire
l_{wp}	Length of primary winding wire
l_{ws}	Length of secondary winding wire
L	Inductance

L_l	Leakage inductance
L_m	Magnetizing inductance of transformer
M	Mutual inductance
M_{VDC}	DC voltage transfer function of converter
m_o	Orbital magnetic moment
m_s	Spin magnetic moment
N	Number of turns
N_l	Number of layers
N_{lp}	Number of layers of primary winding
N_{ls}	Number of layers of secondary winding
N_p	Number of turns of primary winding
N_s	Number of turns of secondary winding
n	Transformer primary-to-secondary turns ratio
n	Unity vector normal to a surface
P	Power
P_C	Core loss
P_{cw}	Core and winding power loss
P_w	Winding power loss
P_{wDC}	Winding DC power loss
P_{wp}	Primary winding power loss
P_{wpDC}	Primary winding DC power loss
P_{wpDCs}	Primary winding DC power loss with strands
P_{ws}	Secondary winding power loss
P_{wsDC}	Secondary winding DC power loss
P_{wsDCs}	Secondary winding DC power loss with strands
P_O	Output power of converter or amplifier
P_v	Core power loss per unit volume
P_t	Total apparent power
p	Winding pitch
Q	Quality factor
Q_L	Loaded quality factor of resonant circuit at resonant frequency f_o
Q_{Lo}	Quality factor of inductor
Q_m	Quality factor of magnetic material
R	Resistance
R_{cs}	Core series equivalent resistance
R_L	Load resistance
R_{Lmax}	Maximum load resistance
R_{Lmin}	Minimum load resistance
R_w	Winding resistance
R_{wDC}	Winding DC resistance
R_{wpDC}	Primary winding DC resistance
R_{wpDCs}	Primary strand winding DC resistance
R_{wpDCs}	Primary strand winding DC resistance
R_{wsDCs}	Secondary strand winding DC resistance
r	Radius
r_L	Equivalent series resistance (ESR) of inductor
$\mathbf{S}$	Poynting vector
S_n	Number of strands
S_{np}	Number of strands in primary winding
S_{ns}	Number of strands in secondary winding
T	Switching period
T_c	Curie temperature
T_q	Torque
t	Time
V_c	Core volume

V_{rms}	rms value of voltage v
v	Voltage, velocity
v_d	Drift velocity
v_L	Voltage across inductance L
v_o	AC component of output voltage
W_a	Core window area
W_{ap}	Core window area of primary winding
W_{as}	Core window area of secondary winding
W_b	Bobbin cross-sectional area
W_m	Magnetic energy stored in inductor or transformer
w	Energy density
w_m	Magnetic energy density
Z	Impedance
X	Parallel reactance
x	Series reactance
$\mathcal{P}$	Permeance
$\mathcal{R}$	Reluctance
α	Regulation of transformer
γ	Complex propagation constant
Δi_L	Peak-to-peak value of inductor ripple current
ΔT	Temperature rise
δ	Skin depth
δ_c	Skin depth in core
δ_m	Loss angle of a magnetic material
δ_w	Skin depth in winding conductor
δ_{wn}	Skin depth in winding conductor at nth harmonic
λ	Flux linkage
λ_{11}	Flux linkage produced by current i_1
λ_{22}	Flux linkage produced by current i_2
λ_{21}	Flux linkage with winding 2 produced by current i_1
λ_{12}	Flux linkage with winding 1 produced by current i_2
ϵ_0	Permittivity of free space
ϵ_r	Relative permittivity
η	Efficiency, porosity factor
η_t	Efficiency of transformer
μ	Permeability
μ_0	Permeability of free space
μ_p	Parallel complex permeability
μ_r	Relative permeability
μ_{rc}	Relative core permeability
μ_{s}	Series complex permeability
$\mu_{s,}$	Real part of series complex permeability
μ_s	Imaginary part of series complex permeability
μ_{rs}	Series complex relative permeability
$\mu_{rs,,}$	Real part of series complex relative permeability
μ_{rs}	Imaginary part of series complex relative permeability
ϕ	Magnetic flux
ϕ_n	Phase angle of nth harmonic of inductor current
ξ	Voltage utilization factor
ρ	Resistivity
ρ_c	Resistivity of core material
ρ_{Cu}	Copper resistivity
ρ_w	Resistivity of winding conductor
σ	Conductivity
χ	Magnetic susceptibility

$\chi_{s'}$	Series complex magnetic susceptibility
$\chi_{s''}$	Real part of series complex magnetic susceptibility
χ_s	Imaginary part of series complex magnetic susceptibility
ψ	Surface power loss density
w	Angular frequency
w_r	Self-resonant angular frequency

1

Fundamentals of Magnetic Devices

1.1 Introduction

Many electronic circuits require the use of inductors and transformers [1–60]. These are usually the largest, heaviest, and most expensive components in a circuit. They are defined by their electromagnetic (EM) behavior. The main feature of an inductor is its ability to store magnetic energy in the form of a magnetic field. The important feature of a transformer is its ability to couple magnetic fluxes of different windings and transfer AC energy from the input to the output through the magnetic field. The amount of energy transferred is determined by the operating frequency, flux density, and temperature. Transformers are used to change the AC voltage and current levels as well as to provide DC isolation while transmitting AC signals. They can combine energy from many AC sources by the addition of the magnetic flux and deliver the energy from all the inputs to one or multiple outputs simultaneously. The magnetic components are very important in power electronics and other areas of electrical engineering. Power losses in inductors and transformers are due to DC current flow, AC current flow, and associated skin and proximity effects in windings, as well as due to eddy currents and hysteresis in magnetic cores. In addition, there are dielectric losses in materials used to insulate the core and the windings. Failure mechanisms in magnetic components are mostly due to excessive temperature rise. Therefore, these devices should satisfy both magnetic requirements and thermal limitations.

In this chapter, fundamental physical phenomena and fundamental physics laws of electromagnetism, quantities, and units of the magnetic theory are reviewed. Magnetic relationships are given and an equation for the inductance is derived. The nature is governed by a set of laws. A subset of these laws are the physics EM laws. The origin of the magnetic field is discussed. It is shown that moving charges are sources of the magnetic field. Hysteresis and eddy-current losses are studied. There are two kinds of eddy-current effects: skin effect and proximity effect. Both of these effects cause nonuniform distribution of the current density in conductors and increase the conductor AC resistance at high frequencies. A classification of winding and core losses is given.

High-Frequency Magnetic Components, Second Edition. Marian K. Kazimierczuk.

Companion Website: www.wiley.com/go/kazimierczuk_High2e

1.2 Fields

A *field* is defined as a spatial distribution of a quantity everywhere in a region. There are two categories of fields: scalar fields and vector fields. A *scalar field* is a set of scalars assigned at individual points in space. A scalar quantity has a magnitude only. Examples of scalar fields are time, temperature, humidity, pressure, mass, sound intensity, altitude of a terrain, energy, power density, electrical charge density, and electrical potential. The scalar field may be described by a real or a complex function. The intensity of a scalar field may be represented graphically by different colors or undirected field lines. A higher density of the field lines indicates a stronger field in the area.

A *vector field* is a set of vectors assigned at every point in space. A vector quantity has both magnitude and direction. Examples of vector fields are velocity $\mathbf{v}$, the Earth's gravitational force field $\mathbf{F}$, electric current density field $\mathbf{J}$, magnetic field intensity $\mathbf{H}$, and magnetic flux density $\mathbf{B}$. The vector field may be represented graphically by directed field lines. The density of field lines indicates the field intensity, and the direction of field lines indicates the direction of the vector at each point. In general, fields are functions of position and time, for example, $\rho_v(x, y, z, t)$. The rate of change of a scalar field with distance is a vector.

1.3 Magnetic Relationships

The magnetic field is characterized by magnetomotive force (MMF) $\mathcal{F}$, magnetic field intensity $\mathbf{H}$, magnetic flux density $\mathbf{B}$, magnetic flux ϕ, and magnetic flux linkage λ.

1.3.1 Magnetomotive Force

An inductor with N turns carrying an AC current i produces the *MMF*, which is also called the *magnetomotance*. The MMF is given by

$$\mathcal{F} = Ni \text{ (A} \cdot \text{turns)}. \tag{1.1}$$

Its descriptive unit is ampere-turns (A·t). However, the approved SI unit of the MMF is the ampere (A), where $1 \text{ A} = 6.25 \times 10^{18}$ electrons/s. The MMF is a *source* in magnetic circuits. The magnetic flux ϕ is forced to flow in a magnetic circuit by the MMF $\mathcal{F} = Ni$, driving a magnetic circuit. Every time another complete turn with the current i is added, the result of the integration increases by the current i.

The MMF between any two points P_1 and P_2 produced by a magnetic field $\mathbf{H}$ is determined by a line integral of the magnetic field intensity $\mathbf{H}$ present between these two points

$$\mathcal{F} = \int_{P_1}^{P_2} \mathbf{H} \cdot d\mathbf{l} = \int_{P_1}^{P_2} H \cos\theta dl, \tag{1.2}$$

where $d\mathbf{l}$ is the incremental vector at a point located on the path l and $\mathbf{H} \cdot d\mathbf{l} = (H \cos\theta)dl = H_l dl = H(dl \cos\theta)$. The MMF depends only on the endpoints, and it is independent of the path between points P_1 and P_2. Any path can be chosen. If the path is broken up into segments parallel and perpendicular to H, only parallel segments contribute to $\mathcal{F}$. The contributions from the perpendicular segments are zero.

For a uniform magnetic field and parallel to path l, the MMF is given by

$$\mathcal{F} = Hl. \tag{1.3}$$

Thus,

$$\mathcal{F} = Hl = Ni. \tag{1.4}$$

The MMF forces a magnetic flux ϕ to flow.

The MMF is analogous to the electromotive force (EMF) V. It is a potential difference between any two points P_1 and P_2. field **E** between any two points P_1 and P_2 is equal to the line integral of the electric field **E** between these two points along any path

$$V = V_{P2} - P_{P2} = \int_{P_1}^{P_2} \mathbf{E} \cdot d\mathbf{l} = \int_{P_1}^{P_2} E \cos \theta dl. \tag{1.5}$$

The result is independent of the integration path. For a uniform electric field E and parallel to path l, the EMF is

$$V = El. \tag{1.6}$$

The EMF forces a current $i = V/R$ to flow. It is the work per unit charge (J/C).

1.3.2 Magnetic Field Intensity

The *magnetic field intensity* (or *magnetic field strength*) is defined as the MMF $\mathcal{F}$ per unit length

$$H = \frac{\mathcal{F}}{l} = \frac{Ni}{l} = \left(\frac{N}{l}\right) i \text{ (A/m)}, \tag{1.7}$$

where l is the inductor length and N is the number of turns. Magnetic fields are produced by moving charges. Therefore, magnetic field intensity H is directly proportional to the amount of current i and the number of turns per unit length N/l. If a conductor conducts current i (which a moving charge), it produces a magnetic field H. Thus, the source of the magnetic field H is a conductor carrying a current i. The magnetic field intensity **H** is a vector field. It is described by a magnitude and a direction at any given point. The lines of magnetic field H always form closed loops. By Ampère's law, the magnetic field produced by a straight conductor carrying current i is given by

$$\mathbf{H}(r) = \frac{i}{2\pi r}\mathbf{a}_\phi. \tag{1.8}$$

The magnetic field intensity H is directly proportional to current i and inversely proportional to the radial distance from the conductor r. The Earth's magnetic field intensity is approximately 50 μT.

1.3.3 Magnetic Flux

The amount of the *magnetic flux* passing through an open surface S is determined by a surface integral of the magnetic flux density **B**

$$\phi = \int\int_S \mathbf{B} \cdot d\mathbf{S} = \int\int_S \mathbf{B} \cdot \mathbf{n} dS \text{ (Wb)}, \tag{1.9}$$

where **n** is the unit vector normal to the incremental surface area dS at a given position, $d\mathbf{S} = \mathbf{n}dS$ is the incremental surface vector normal to the local surface dS at a given position, and $d\phi = \mathbf{B} \cdot d\mathbf{S} = \mathbf{S} \cdot \mathbf{n}dS$. The magnetic flux is a scalar. The unit of the magnetic flux is Weber.

If the magnetic flux density B is uniform and forms an angle θ_B with the vector perpendicular to the surface S, the amount of the magnetic flux passing through the surface S is

$$\phi = \mathbf{B} \cdot \mathbf{S} = BS \cos \theta_B \text{ (Wb)}. \tag{1.10}$$

If the magnetic flux density B is uniform and perpendicular to the surface S, the angle between vectors **B** and $d\mathbf{S}$ is $\theta_B = 0^\circ$ and the amount of the magnetic flux passing through the surface S is

$$\phi = \mathbf{B} \cdot \mathbf{S} = BS \cos 0^\circ = BS \text{ (Wb)}. \tag{1.11}$$

If the magnetic flux density B is parallel to the surface S, the angle between vectors **B** and $d\mathbf{S}$ is $\theta_B = 90^\circ$ and the amount of the magnetic flux passing through the surface S is

$$\phi = SB \cos 90^\circ = 0. \tag{1.12}$$

For an inductor, the amount of the magnetic flux ϕ may be increased by increasing the surface area of a single turn A, the number of turns in the layer N_{tl}, and the number of layers N_l. Hence, $S = N_{tl}N_lA = NA$, where $N = N_{tl}N_l$ is the total number of turns.

The direction of a magnetic flux density B is determined by the right-hand rule (RHR). This rule states that if the fingers of the right hand encircle a coil in the direction of the current i, the thumb indicates the direction of the magnetic flux density B produced by the current i, or if the fingers of the right hand encircle a conductor in the direction of the magnetic flux density B, the thumb indicated the direction of the current i. The magnetic flux lines are always continuous and closed loops.

1.3.4 Magnetic Flux Density

The *magnetic flux density*, or *induction*, is the magnetic flux per unit area given by

$$B = \frac{\phi}{S} \text{ (T)}. \quad (1.13)$$

The unit of magnetic flux density B is Tesla. The magnetic flux density is a vector field and it can be represented by magnetic lines. The density of the magnetic lines indicates the magnetic flux density B, and the direction of the magnetic lines indicates the direction of the magnetic flux density at a given point. Every magnet has two poles: south and north. Magnetic monopoles do not exist. Magnetic lines always flow from south to north pole inside the magnet, and from north to south pole outside the magnet.

The relationship between the magnetic flux density B and the magnetic field intensity H is given by

$$B = \mu H = \mu_r\mu_0 H = \frac{\mu N i}{l_c} = \frac{\mu \mathcal{F}}{l_c} < B_s \text{ (T)}, \quad (1.14)$$

where the permeability of free space is

$$\mu_0 = 4\pi \times 10^{-7} \text{ (H/m)}, \quad (1.15)$$

$\mu = \mu_r\mu_0$ is the permeability, $\mu_r = \mu/\mu_0$ is the relative permeability (i.e., relative to that of free space), and l_c is the length of the core. Physical constants are given in Appendix A. For free space, insulators, and nonmagnetic materials, $\mu_r = 1$. For diamagnetics such as copper, lead, silver, and gold, $\mu_r \approx 1 - 10^{-5} \approx 1$. However, for ferromagnetic materials such as iron, cobalt, nickel, and their alloys, $\mu_r > 1$ and it can be as high as 100 000. The permeability is the measure of the ability of a material to conduct magnetic flux ϕ. It describes how easily a material can be magnetized. For a large value of μ_r, a small current i produces a large magnetic flux density B. The magnetic flux ϕ takes the path of the highest permeability.

The magnetic flux density field is a vector field. For example, the vector of the magnetic flux density produced by a straight conductor carrying current i is given by

$$\mathbf{B}(\mathbf{r}) = \mu\mathbf{H}(r) = \frac{\mu i}{2\pi r}\mathbf{a}_\phi. \quad (1.16)$$

For ferromagnetic materials, the relationship between B and H is nonlinear because the relative permeability μ_r depends on the magnetic field intensity H. Figure 1.1 shows simplified plots of the magnetic flux density B as a function of the magnetic field intensity H for air-core inductors (straight line) and for ferromagnetic core inductors. The straight line describes the air-core inductor and has a slope μ_0 for all values of H. These inductors are linear. The piecewise linear approximation corresponds to the ferromagnetic core inductors, where B_s is the saturation magnetic flux density and $H_s = B_s/(\mu_r\mu_0)$ is the magnetic field intensity corresponding to B_s. At low values of the magnetic flux density $B < B_s$, the relative permeability μ_r is high and the slope of the B–H curve $\mu_r\mu_0$ is also high. For $B > B_s$, the core saturates and $\mu_r = 1$, reducing the slope of the B–H curve to μ_0.

The total peak magnetic flux density B_{pk}, which in general consists of both the DC component B_{DC} and the amplitude of AC component B_m, should be lower than the saturation flux density B_s of a magnetic core at the highest operating temperature T_{max}

$$B_{pk} = B_{DC(max)} + B_{m(max)} \le B_s. \quad (1.17)$$

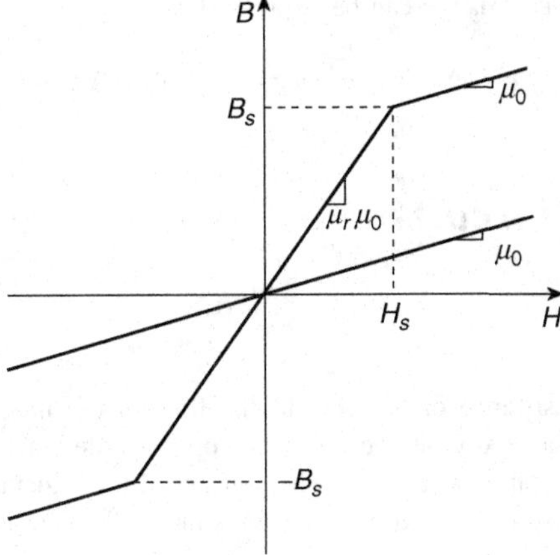

Figure 1.1 Simplified plots of magnetic flux density B as a function of magnetic field intensity H for air-core inductors (straight line) and ferromagnetic core inductors (piecewise linear approximation)

The DC component of the magnetic flux density B_{DC} is caused by the DC component of the inductor current I_L

$$B_{DC} = \frac{\mu_r \mu_0 N I_L}{l_c}. \tag{1.18}$$

The amplitude of the AC component of the magnetic flux density B_m corresponds to the amplitude of the AC component of the inductor current I_m

$$B_m = \frac{\mu_r \mu_0 N I_m}{l_c}. \tag{1.19}$$

Hence, the peak value of the magnetic flux density can be written as

$$B_{pk} = \frac{\mu_r \mu_0 N I_L}{l_c} + \frac{\mu_r \mu_0 N I_m}{l_c} = \frac{\mu_r \mu_0 N (I_L + I_m)}{l_c} = \frac{\mu_r \mu_0 N I_{Lpk}}{l_c} \leq B_s \tag{1.20}$$

where $I_{Lpk} = I_L + I_m$. The saturation flux density B_s decreases with temperature. For ferrites, B_s may decrease by a factor of 2 as the temperature increases from 20 °C to 90 °C. The amplitude of the magnetic flux density B_m is limited either by core saturation or by core losses.

1.3.5 Magnetic Flux Linkage

The *magnetic flux linkage* is the sum of the flux enclosed by each turn of the wire wound around the core

$$\lambda = N \int \int_S \mathbf{B} \cdot d\lambda S = \int v dt. \tag{1.21}$$

For the uniform magnetic flux density, the magnetic flux linkage is the magnetic flux linking N turns and is described by

$$\lambda = N\phi = NA_c B = A_{eff} B = NA_c \mu H = \frac{\mu A_c N^2 i}{l_c} = \frac{N^2}{\mathcal{R}} i = Li \ (\mathrm{V \cdot s}) \tag{1.22}$$

where $\mathcal{R}$ is the core reluctance and $A_{eff} = NA_c$ is the effective area through which the magnetic flux ϕ passes. Equation (1.22) is analogous to Ohm's law $v = Ri$ and the equation for the capacitor charge $Q = Cv$. The unit of the flux linkage is Wb·turn. For sinusoidal waveforms, the relationship among the amplitudes is

$$\lambda_m = N\phi_m = NA_c B_m = NA_c \mu H_m = \frac{\mu_r \mu_0 A_c N^2 I_m}{l_c}. \tag{1.23}$$

The change in the magnetic linkage can be expressed as

$$\Delta\lambda = \int_{t_1}^{t_2} v dt = \lambda(t_2) - \lambda(t_1). \tag{1.24}$$

1.4 Magnetic Circuits

1.4.1 Reluctance

The *reluctance* $\mathcal{R}$ is the resistance of the core to the flow of the magnetic flux ϕ. It opposes the magnetic flux flow, in the same way as the resistance opposes the electric current flow. An element of a magnetic circuit can be called a *reluctor*. The concept of the reluctance is illustrated in Fig. 1.2. The reluctance of a linear, isotropic, and homogeneous magnetic material is given by

$$\mathcal{R} = \frac{1}{\mathcal{P}} = \frac{l_c}{\mu A_c} = \frac{l_c}{\mu_0 \mu_r A_c} \text{ (A} \cdot \text{turns/Wb)} \quad \text{or} \quad \text{(turns/H)}, \tag{1.25}$$

where A_c is the cross-sectional area of the core (i.e., the area through which the magnetic flux flows) and l_c is the mean magnetic path length (MPL), which is the mean length of the closed path that the magnetic flux flows around a magnetic circuit. The reluctance is directly proportional to the length of the magnetic path l_c and is inversely proportional to the cross-sectional area A_c through which the magnetic flux ϕ flows. The *permeance* of a basic magnetic circuit element is

$$\mathcal{P} = \frac{1}{\mathcal{R}} = \frac{\mu A_c}{l_c} = \frac{\mu_0 \mu_r A_c}{l_c} \text{ (Wb/A} \cdot \text{turns)} \quad \text{or} \quad \text{(H/turns)}. \tag{1.26}$$

When the number of turns $N = 1$, $L = \mathcal{P}$. The reluctance is the magnetic resistance because it opposes the establishment and the flow of a magnetic flux ϕ in a medium. A poor conductor of the magnetic flux has a high reluctance and a low permeance. Magnetic Ohm's law is expressed as

$$\phi = \frac{\mathcal{F}}{\mathcal{R}} = \mathcal{P}\mathcal{F} = \frac{Ni}{\mathcal{R}} = \frac{\mu A_c Ni}{l_c} = \frac{\mu_{rc}\mu_0 A_c Ni}{l_c} \text{ (Wb)}. \tag{1.27}$$

Magnetic flux always takes the path with the highest permeability μ.

In general, the magnetic circuit is the space in which the magnetic flux flows around the coil(s). Figure 1.3 shows an example of a magnetic circuit. The reluctance in magnetic circuits is analogous to the resistance R in electric circuits. Likewise, the permeance in magnetic circuits is analogous to the conductance in electric circuits. Therefore, magnetic circuits described by the equation $\phi = \mathcal{F}/\mathcal{R}$ can be solved in a similar manner as electric circuits described by Ohm's law $I = V/R = GV = (\sigma A/l)V$, where ϕ, $\mathcal{F}$, $\mathcal{R}$, $\mathcal{P}$, B, λ, and μ, correspond to I, V, R, G, J, Q, and σ, respectively. For example, the reluctances can be connected in series or in parallel. In addition, the reluctance $\mathcal{R} = l_c/\mu A_c$ is analogous to the electric resistance $R = l/\sigma A$ and the magnetic flux density $B = \phi/A_c$ is analogous to the current density $J = I/A$. Table 1.1 lists analogous magnetic and electric quantities.

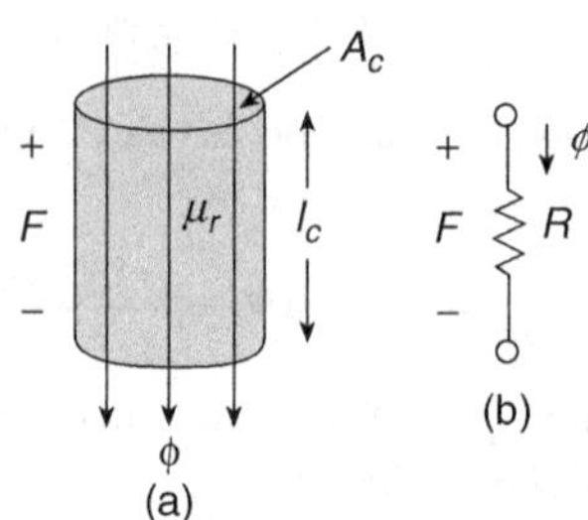

Figure 1.2 Reluctance. (a) Basic magnetic circuit element conducting magnetic flux ϕ. (b) Equivalent magnetic circuit

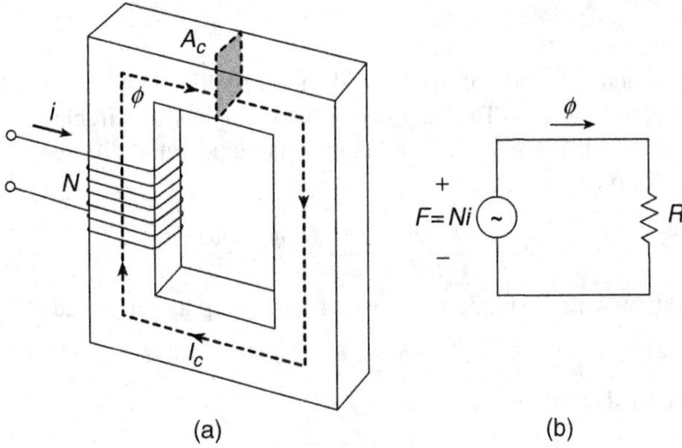

Figure 1.3 Magnetic circuit. (a) An inductor composed of a core and a winding. (b) Equivalent magnetic circuit

Table 1.1 Analogy between magnetic and electric quantities

Magnetic quantity	Electric quantity
$\mathcal{F} = Ni$	V
$\mathcal{F} = Hl$	$V = El$
ϕ	I
H	E
B	J
$\mathcal{R}$	R
$\mathcal{P}$	G
λ	Q
μ	ϵ
L	C
$\phi = \frac{\mathcal{F}}{\mathcal{R}}$	$I = \frac{V}{R}$
$B = \frac{\phi}{A}$	$J = \frac{I}{A}$
$H = \frac{\mathcal{F}}{l} = \frac{Ni}{l}$	$E = \frac{V}{l}$
$\mathcal{R} = \frac{l}{\mu A}$	$R = \frac{l}{\sigma A}$
$\mathbf{B} = \mu\mathbf{H}$	$\mathbf{D} = \epsilon\mathbf{E}$
$\lambda = Li$	$Q = Cv$
$i = \frac{d\lambda}{dt}$	$v = \frac{dQ}{dt}$
$v = L\frac{di}{dt}$	$i = C\frac{dv}{dt}$
$d\mathbf{B} = \frac{\mu(Id\mathbf{l}\times\mathbf{a}_R)}{4\pi R^2}$	$\mathbf{F} = \frac{Q_1Q_2\mathbf{a}_R}{2\pi\epsilon R^2}$
$w_m = \frac{1}{2}\mathbf{B}\cdot\mathbf{H}$	$w_e = \frac{1}{2}\mathbf{D}\cdot\mathbf{E}$
$w_m = \frac{1}{2}\mu H^2$	$w_e = \frac{1}{2}\epsilon E^2$
$W_m = \frac{1}{2}i\lambda$	$W_e = \frac{1}{2}vQ$
$W_m = \frac{1}{2}Li^2$	$W_e = \frac{1}{2}Cv^2$

1.4.2 Magnetic KVL

Physical structures, which are made of magnetic devices, such as inductors and transformers, can be analyzed just like electric circuits. The magnetic law, analogous to Kirchhoff's voltage law (KVL), states that the sum of the MMFs $\sum_{k=1}^{n} \mathcal{F}_k$ and the magnetic potential differences $\sum_{k=1}^{m} \mathcal{R}_k \phi_k$ around the closed magnetic loop is zero

$$\sum_{k=1}^{n} \mathcal{F}_k - \sum_{k=1}^{m} \mathcal{R}_k \phi_k = 0. \tag{1.28}$$

For instance, an inductor with a simple core having an air gap as illustrated in Fig. 1.4 is given by

$$Ni = \mathcal{F} = \mathcal{F}_c + \mathcal{F}_g = \phi(\mathcal{R}_c + \mathcal{R}_g), \tag{1.29}$$

where the reluctance of the core is

$$\mathcal{R}_c = \frac{l_c}{\mu_{rc}\mu_0 A_c} \tag{1.30}$$

the reluctance of the air gap is

$$\mathcal{R}_g = \frac{l_g}{\mu_0 A_c} \tag{1.31}$$

and it is assumed that $\phi_c = \phi_g = \phi$. This means that the fringing flux in neglected. If $\mu_r \gg 1$, the magnetic flux is confined to the magnetic material, reducing the leakage flux. The ratio of the air-gap reluctance to the core reluctance is

$$\frac{\mathcal{R}_g}{\mathcal{R}_c} = \mu_{rc}\frac{l_g}{l_c}. \tag{1.32}$$

The reluctance of the air gap $\mathcal{R}_g$ is much higher than the reluctance of the core $\mathcal{R}_c$ if $\mu_{rc} \gg l_c/l_g$.

The magnetic potential difference between points a and b is

$$\mathcal{F}_{ab} = \int_a^b \mathbf{H} \cdot d\mathbf{l} = \mathcal{R}_{ab}\phi, \tag{1.33}$$

where $\mathcal{R}_{ab}$ is the reluctance between points a and b.

1.4.3 Magnetic Flux Continuity

The continuity of the magnetic flux law states that the net magnetic flux through any closed surface is always zero

$$\phi = \oiint_S BdS = 0 \tag{1.34}$$

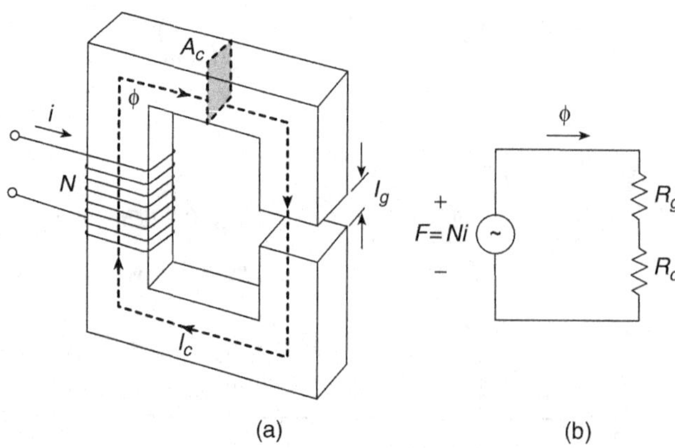

Figure 1.4 Magnetic circuit illustrating the magnetic KVL. (a) An inductor composed of a core with an air gap and a winding. (b) Equivalent magnetic circuit

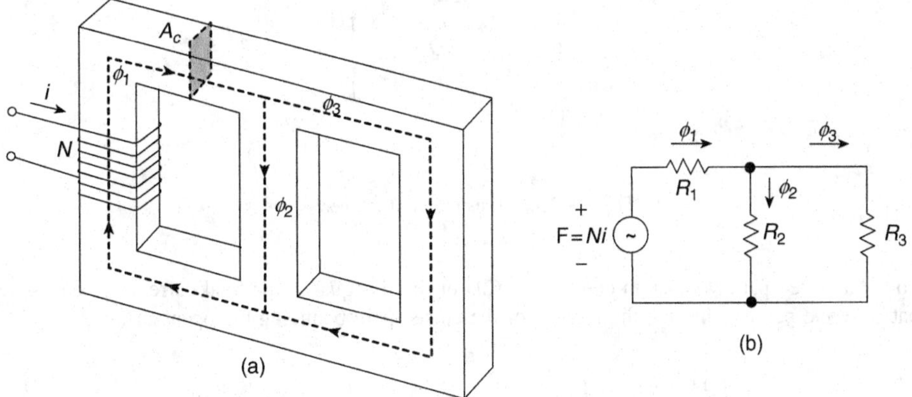

Figure 1.5 Magnetic circuit illustrating the continuity of the magnetic flux for EE core. (a) An inductor composed of a core and a winding. (b) Equivalent magnetic circuit

or the net magnetic flux entering and exiting the node is zero

$$\sum_{k=1}^{n} \phi_k = \sum_{k=1}^{n} S_k B_k = 0. \tag{1.35}$$

This law is analogous to Kirchhoff's current law (KCL) introduced by Gauss and can be called Kirchhoff's flux law (KFL). Figure 1.5 illustrates the continuity of the magnetic flux law. For example, when three core legs meet at a node,

$$\phi_1 = \phi_2 + \phi_3, \tag{1.36}$$

which can be expressed by

$$\frac{\mathcal{F}_1}{\mathcal{R}_1} = \frac{\mathcal{F}_2}{\mathcal{R}_2} + \frac{\mathcal{F}_3}{\mathcal{R}_3}. \tag{1.37}$$

If all the three legs of the core have windings, then we have

$$\frac{N_1 i_1}{\mathcal{R}_1} = \frac{N_2 i_2}{\mathcal{R}_2} + \frac{N_3 i_3}{\mathcal{R}_3}. \tag{1.38}$$

Usually, most of the magnetic flux is confined inside an inductor, for example, for an inductor with a toroidal core. The magnetic flux outside an inductor is called the *leakage flux*.

1.5 Magnetic Laws

1.5.1 Ampère's Law

Ampère[1] discovered the relationship between current and the magnetic field intensity. *Ampère's law* relates the magnetic field intensity H inside a closed loop to the current passing through the loop. A magnetic field can be produced by a current and a current can be produced by a magnetic field. Ampère's law is illustrated in Fig. 1.6. A magnetic field is present around a current-carrying conductor or conductors. The integral form of Ampère's circuital law, or simply Ampère's law, (1826) describes the relationship between the (conduction, convection, and/or displacement) current and the magnetic field produced by this current. It states that the closed line integral of the magnetic field intensity **H**

[1] André-Marie Ampère (1775–1836) was a French physicist and mathematician, who is the father of electrodynamics.

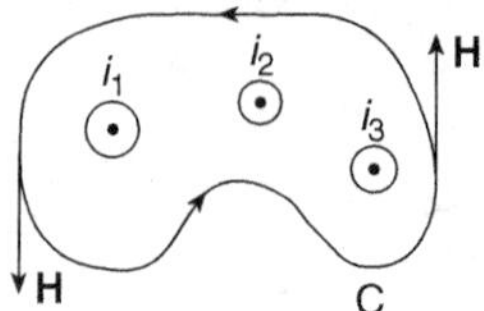

Figure 1.6 Illustration of Ampère's law

around a closed path (Amperian contour) C (2D or 3D) is equal to the total current i_{enc} enclosed by that path and passing through the interior of the closed path bounding the open surface S

$$\oint_C \mathbf{H} \cdot d\mathbf{l} = \int\int_S \mathbf{J} \cdot d\mathbf{S} = \sum_{n=1}^{N} i_n = i_1 + i_2 + \ldots + i_N = i_{enc}, \tag{1.39}$$

where $d\mathbf{l}$ is the vector length element pointing in the direction of the Amperian path C and $\mathbf{J}$ is the conduction (or drift) and convection current density. The current i_{enc} enclosed by the path C is given by the surface integral of the normal component $\mathbf{J}$ over the open surface S. The surface integral of the current density $\mathbf{J}$ is equal to the current I flowing through the surface S. In other words, the integrated magnetic field intensity around a closed loop C is equal to the electric current passing through the loop. The surface integral of $\mathbf{J}$ is the current flowing through the open surface S. The conduction current is caused by the movement of electrons originating from the outermost shells of atoms. When conduction current flows, the atoms of medium normally do not move. The convection current is caused by the movement of electrically charged medium.

For example, consider a long, straight, round conductor that carries current I. The line integral about a circular path of radius r centered on the axis of the round wire is equal to the product of the circumference and the magnetic field intensity H_ϕ

$$\oint_C \mathbf{H} \cdot d\mathbf{l} = 2\pi r H_\phi = I, \tag{1.40}$$

yielding the magnetic field intensity

$$H_\phi = \frac{I}{2\pi r}. \tag{1.41}$$

Thus, the magnetic field decreases in the radial direction away from the conductor.

For an inductor with N turns, Ampère's law is

$$\oint_C \mathbf{H} \cdot d\mathbf{l} = Ni. \tag{1.42}$$

Ampère's law in the discrete form can be expressed as

$$\sum_{k=1}^{n} H_k l_k = \sum_{k=1}^{m} N_k i_k. \tag{1.43}$$

For example, Ampère's law for an inductor with an air gap is given by

$$H_c l_c + H_g l_g = Ni. \tag{1.44}$$

If the current density J is uniform and perpendicular to the surface S,

$$HC = SJ. \tag{1.45}$$

The current density J in winding conductors of magnetic components used in power electronics is usually in the range 0.1–10 A/mm^2. The displacement current is neglected in (1.39). The generalized Ampère's law by adding the displacement current constitutes one of Maxwell's equations. This is known as Maxwell's correction to Ampère's law.

Ampère's law is useful when there is a high degree of symmetry in the arrangement of conductors and it can be easily applied in problems with symmetrical current distribution. For example, the magnetic field produced by an infinitely long wire conducting a current I outside the wire is

$$\mathbf{B} = \frac{I}{2\pi r}\mathbf{a}_\phi \text{ (A/m)}. \tag{1.46}$$

Ampère's law is a special case of Biot–Savart's law.

Example 1.1

An infinitely long round solid straight wire of radius r_o carries sinusoidal current $i = I_m \cos \omega t$ in steady state at low frequencies (with no skin effect). Determine the waveforms of the magnetic field intensity $H(r,t)$, magnetic flux density $B(r,t)$, and magnetic flux $\phi(r,t)$ inside and outside the wire.

Solution: At low frequencies, the skin effect can be neglected and the current is uniformly distributed over the cross section of the wire, as shown in Fig. 1.7. To determine the magnetic field intensity $H(r,t)$ everywhere, two Amperian contours C_1 and C_2 are required, one inside the conductor for $r \le r_o$ and the other outside the conductor for $r > r_o$.

The Magnetic Field Intensity Inside the Wire. The current in the conductor induces a concentric magnetic field intensity both inside and outside the conductor. The current density inside the conductor is uniform. The vector of the current density amplitude inside the conductor is assumed to be parallel to the conductor axis and is given by

$$\mathbf{J}_m = J_{mz}\mathbf{a}_z. \tag{1.47}$$

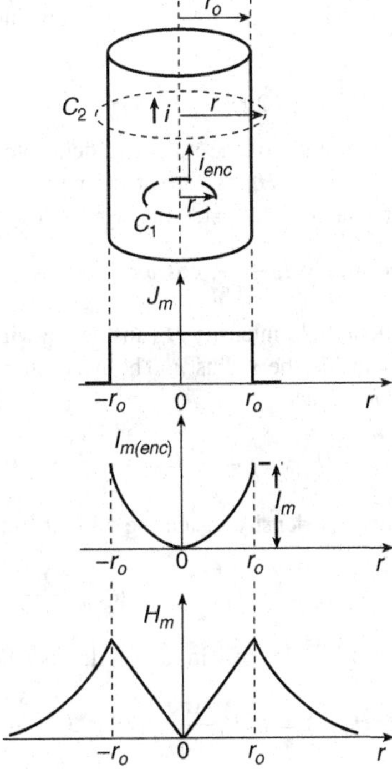

Figure 1.7 Cross section of an infinitely long round straight wire carrying a sinusoidal $i = I_m \cos \omega t$ and amplitudes of current density J_m, enclosed current $I_{m(enc)}$, and magnetic field intensity H_m as a function of the radial distance r from the wire center at low frequencies, that is, when the skin effect can be neglected ($\delta > r_o$)

Consider a radial contour C_1 inside the conductor. The current flowing through the area enclosed by the cylindrical shell of radius r at low frequencies is given by

$$i_{enc} = I_{m(enc)} \cos \omega t, \tag{1.48}$$

where $I_{m(enc)}$ is the amplitude of the current enclosed by the shell of radius r. Hence, the amplitude of the current density at a radius r is

$$J_m(r) = \frac{I_{m(enc)}}{\pi r^2} \quad \text{for} \quad 0 \le r \le r_o \tag{1.49}$$

and the amplitude of the current density at the wire surface $r = r_o$ is

$$J_m(r_o) = \frac{I_m}{\pi r_o^2}. \tag{1.50}$$

The current density is uniform at low frequencies (where the skin effect can be neglected), that is, $J_m(r) = J_m(r_o)$, yielding the amplitude of the enclosed current

$$I_{m(enc)} = I_m\left(\frac{\pi r^2}{\pi r_o^2}\right) = I_m\left(\frac{A_r}{A_{ro}}\right) = I_m\left(\frac{r}{r_o}\right)^2 \quad \text{for} \quad 0 \le r \le r_o, \tag{1.51}$$

where $A_r = \pi r^2$ and $A_{ro} = \pi r_o^2$. Figure 1.7 shows a plot of $I_{m(enc)}$ as a function of the radial distance from the conductor center r. The vector of the magnetic flux density is

$$\mathbf{H} = \mathbf{H}_\phi = H(r)\mathbf{a}_\phi. \tag{1.52}$$

From Ampère's law,

$$I_{m(enc)} = \oint_{C_1} \mathbf{H} \cdot d\mathbf{l} = H_m(r)\oint_{C_1} dl = 2\pi r H_m(r) \quad \text{for} \quad 0 \le r \le r_o, \tag{1.53}$$

where $C_1 = 2\pi r$ for $r \le r_o$. Equating the right-hand sides of (1.51) and (1.53), the amplitude of the magnetic field intensity inside the wire at low frequencies is obtained

$$H_m(r) = I_m\left(\frac{r}{r_o}\right)^2 \frac{1}{2\pi r} = I_m \frac{r}{2\pi r_o^2} \quad \text{for} \quad 0 \le r \le r_o. \tag{1.54}$$

Figure 1.7 shows a plot of the amplitude of the magnetic field intensity H_m as a function of r. The amplitude of the magnetic field intensity H_m is zero at the wire center because the enclosed current is zero. The waveform of the magnetic field inside the wire at low frequencies

$$H(r,t) = I_m \frac{r}{2\pi r_o^2} \cos \omega t \quad \text{for} \quad r \le r_o. \tag{1.55}$$

Thus, the amplitude of the magnetic field intensity H_m inside the wire at radius r is determined solely by the amplitude of the current inside the radius r. The maximum amplitude of the magnetic field intensity occurs on the conductor surface

$$H_{m(max)} = H_m(r_o) = \frac{I_m}{2\pi r_o}. \tag{1.56}$$

The amplitude of the magnetic flux density inside the wire at low frequencies is

$$B_m(r) = \mu_0 H_m(r) = \mu_0 I_m \left(\frac{r}{r_o}\right)^2 \frac{1}{2\pi r} = \mu_0 I_m \frac{r}{2\pi r_o^2} \quad \text{for} \quad 0 \le r \le r_o. \tag{1.57}$$

The amplitude of the magnetic flux inside the wire at low frequencies is

$$\phi_m(r) = AB_m(r) = \mu_0 A_r H_m(r) = \mu_0 I_m \frac{r(\pi r^2)}{2\pi r_o^2} = \mu_0 I_m \frac{r^3}{2r_o^2} \quad \text{for} \quad 0 \le r \le r_o. \tag{1.58}$$

The waveform of the magnetic flux is

$$\phi(x,t) = \phi \cos \omega t = \mu_0 I_m \frac{r^3}{2r_o^2} \cos \omega t \quad \text{for} \quad 0 \le r \le r_o. \tag{1.59}$$

The Magnetic Field Intensity Outside the Wire. Consider a radial contour C_2 outside the conductor. The entire current $i = I_m \cos \omega t$ is enclosed by a path of radius $r \geq r_o$. From Ampère's law, the amplitude of the entire current i is

$$I_m = \oint_{C_2} \mathbf{H} \cdot d\mathbf{l} = H_m(r) \oint_{C_2} dl = 2\pi r H_m(r) \quad \text{for} \quad r \geq r_o, \tag{1.60}$$

where $C_2 = 2\pi r$ with $r \geq r_o$. The amplitude of the near-magnetic field intensity outside the conductor at any frequency is given by the expression

$$H_m(r) = \frac{I_m}{2\pi r} \quad \text{for} \quad r \geq r_o \tag{1.61}$$

and the waveform of this field is

$$H(r,t) = \frac{I_m}{2\pi r} \cos \omega t \quad \text{for} \quad r \geq r_o. \tag{1.62}$$

The amplitude of the magnetic field intensity increases linearly with r inside the wire from 0 to $H_m(r_o) = I_m/(2\pi r_o)$ at low frequencies. The amplitude of the magnetic field intensity is inversely proportional to r outside the wire at any frequency.

The waveform of the magnetic flux density is

$$B(r,t) = \mu_0 H(r,t) = \frac{\mu_0 I_m}{2\pi r} \cos \omega t \quad \text{for} \quad r \geq r_o. \tag{1.63}$$

The waveform of the magnetic flux enclosed by a cylinder of radius $r > r_o$ is

$$\phi(x,t) = A_w B(r,t) = A_w \mu_0 H(r,t) = \frac{r_o^2 \mu_0 I_m}{2r} \cos \omega t \quad \text{for} \quad r \geq r_o. \tag{1.64}$$

Example 1.2

Toroidal Inductor. Consider an inductor with a toroidal core of inner radius a and outer radius b. Find the magnetic field inside the core and in the region exterior to the torus core.

Solution: Consider the circle C of radius $a \leq r \leq b$. The magnitude of the magnetic field is constant on this circle and is tangent to it. Therefore, $\mathbf{B} \cdot l\mathbf{l} = Bdl$. From the Ampère's law, the magnetic field density in a toroidal core (torus) is

$$\oint_C \mathbf{B} \cdot d\mathbf{l} = B \oint_C dl = B(2\pi r) = \mu_r \mu_0 NI \quad \text{for} \quad a \leq r \leq b \tag{1.65}$$

where r is the distance from the torus center to a point inside the torus. Hence,

$$B = \frac{\mu_r \mu_0 NI}{2\pi r} \quad \text{for} \quad a \leq r \leq b. \tag{1.66}$$

For an ideal toroid in which the turns are closely spaced, the external magnetic field is zero. For an Amperian contour with radius $r < a$, there is no current flowing through the contour surface, and therefore $\mathbf{H} = 0$ for $r < a$. For an Amperian contour C with radius $r > b$, the net current flowing through its surface is zero because an equal number of current paths cross the contour surface in both directions, and therefore $\mathbf{H} = 0$ for $r > b$.

1.5.2 Faraday's Law

A time-varying current produces a magnetic field, and a time-varying magnetic field can produce an electric current. In 1820, a Danish scientist Oersted[2] showed that a current-carrying conductor

[2]Hans Christian Oersted (1777–1851) was a Danish physicist and chemist, who discovered that an electric current produces a magnetic field. This discovery established the connection between electricity and magnetism, leading to the origination of science of electromagnetism.

produces a magnetic field, which can affect a compass magnetic needle. He connected electricity and magnetism. Ampère measured that this magnetic field intensity is linearly related to the current, which produces it. In 1831, the English experimentalist Michael Faraday[3] discovered that a current can be produced by an alternating magnetic field and that a time-varying magnetic field can induce a voltage, or an EMF, in an adjacent circuit. This voltage is proportional to the rate of change of magnetic flux linkage λ, or magnetic flux ϕ, or current i, producing the magnetic field.

Faraday's law (1831), also known as *Faraday's law of induction*, states that a time-varying magnetic flux $\phi(t)$ passing through a closed stationary loop, such as an inductor turn, generates a voltage $v(t)$ in the loop and for a linear inductor is expressed by

$$v(t) = \frac{d\lambda}{dt} = \frac{d(N\phi)}{dt} = N\frac{d\phi}{dt} = N\frac{d(AB)}{dt} = NA\frac{dB}{dt} = NA\mu\frac{dH}{dt} = \frac{\mu AN^2}{l}\frac{di}{dt}$$
$$= N\frac{d}{dt}\left(\frac{\mathcal{F}}{\mathcal{R}}\right) = N\frac{d}{dt}\left(\frac{Ni}{\mathcal{R}}\right) = \frac{N^2}{\mathcal{R}}\frac{di}{dt} = \mathcal{P}N^2\frac{di}{dt} = L\frac{di}{dt}. \tag{1.67}$$

This voltage, in turn, may produce a current $i(t)$. The voltage $v(t)$ is proportional to the rate of change of the magnetic linkage $d\lambda/dt$, or to the rate of change of the magnetic flux density dB/dt and the effective area NA through which the flux is passing. The inductance L relates the induced voltage $v(t)$ to the current $i(t)$. The voltage $v(t)$ across the terminals of an inductor L is proportional to the time rate of change of the current $i(t)$ in the inductor and the inductance L. If the inductor current is constant, the voltage across an ideal inductor is zero. The inductor behaves as a short circuit for DC current. The inductor current cannot change instantaneously. Figure 1.8 shows an equivalent circuit of an ideal inductor. The inductor is replaced by a dependent voltage source controlled by di/dt.

The voltage between the terminals of a single turn of an inductor is

$$v_T(t) = \frac{d\phi(t)}{dt}. \tag{1.68}$$

Hence, the total voltage across the inductor consisting of N identical turns is

$$v_L(t) = Nv_T(t) = N\frac{d\phi(t)}{dt} = \frac{d\lambda(t)}{dt}. \tag{1.69}$$

Since $v = Ldi/dt$,

$$di = \frac{1}{L}vdt \tag{1.70}$$

yielding the current in an inductor

$$i(t) = \int_0^t idt + i(0) = \frac{1}{L}\int_0^t vdt + i(0) = \frac{1}{\omega L}\int_0^{\omega t} vd(\omega t) + i(0). \tag{1.71}$$

For sinusoidal waveforms, the derivative d/dt can be replaced by $j\omega$ and differential equations may be replaced by algebraic equations. A phasor is a complex representation of the magnitude, phase, and space of a sinusoidal waveform. The phasor is not dependent on time. A graphical representation

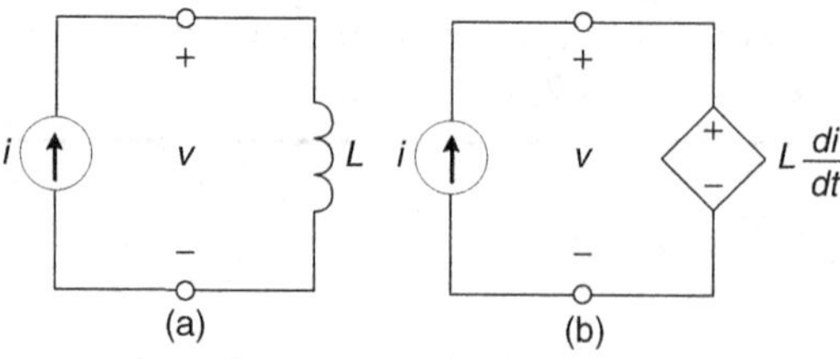

Figure 1.8 Equivalent circuit of an ideal inductor. (a) Inductor. (b) Equivalent circuit of an inductor in the form of dependent voltage source controlled by the rate of change of the inductor current di/dt

[3]Michael Faraday (1791–1867) was an English physicist and chemist, who discovered electromagnetic induction and invented the method of generating electricity.

of a phasor is known as a phasor diagram. Faraday's law in phasor form can be expressed as

$$\mathbf{V}_{Lm} = j\omega\boldsymbol{\lambda}_{\mathbf{m}} = j\omega L\mathbf{I}_{Lm} = \omega LI_{Lm}e^{j90^\circ}. \tag{1.72}$$

The sinusoidal inductor current legs the sinusoidal inductor voltage by 90°.

The impedance of a lossless inductive component in terms of phasors of sinusoidal inductor current $\mathbf{I}_{Lm}$ and voltage $\mathbf{V}_{Lm} = j\omega\boldsymbol{\lambda}_{\mathbf{m}}$ is

$$\mathbf{Z}_L = \frac{\mathbf{V}_m}{\mathbf{I}_{Lm}} = \frac{j\omega\lambda_m}{I_m} = j\omega L, \tag{1.73}$$

where $L = \lambda_m/I_m$. The impedance of lossy inductive components in terms of phasors is

$$\mathbf{Z}_L = \frac{\mathbf{V}_m}{\mathbf{I}_{Lm}} = R + j\omega L. \tag{1.74}$$

For nonlinear, time-varying inductors, the relationships are

$$\lambda(t) = L(i)i(t) \tag{1.75}$$

and

$$v(t) = \frac{d\lambda(t)}{dt} = L(i)\frac{di(t)}{dt} + i(t)\frac{dL(i)}{dt} = L(i)\frac{di(t)}{dt} + i(t)\frac{dL(i)}{di}\frac{di(t)}{dt}$$
$$= \left[L(i) + i(t)\frac{dL(i)}{di}\right]\frac{di(t)}{dt} = L_{eq}\frac{di(t)}{dt}, \tag{1.76}$$

where

$$L_{eq} = L(i) + i(t)\frac{dL(i)}{di}. \tag{1.77}$$

In summary, a time-varying electric current $i(t)$ produces magnetic fields $H(t)$, $\phi(t)$, and $\lambda(t)$ by Ampère's law. In turn, the magnetic field produces a voltage $v(t)$ by Faraday's law. This process can be reversed. A voltage $v(t)$ produces a magnetic fields $H(t)$, $\phi(t)$, and $\lambda(t)$, which produced electric current $i(t)$.

1.5.3 Lenz's Law

Lenz[4] discovered the relationship between the direction of the induced current and the change in the magnetic flux. *Lenz's law* (1834) states that the EMF $v(t) = -Nd\phi(t)/dt$ induced by an applied time-varying magnetic flux $\phi_a(t)$ has such a direction that induces current $i_E(t)$ in the closed loop, which in turn induces a magnetic flux $\phi_i(t)$ that tends to oppose the change in the applied flux $\phi_a(t)$, as illustrated in Fig. 1.9. If the applied magnetic flux $\phi_a(t)$ increases, the induced current $i_E(t)$

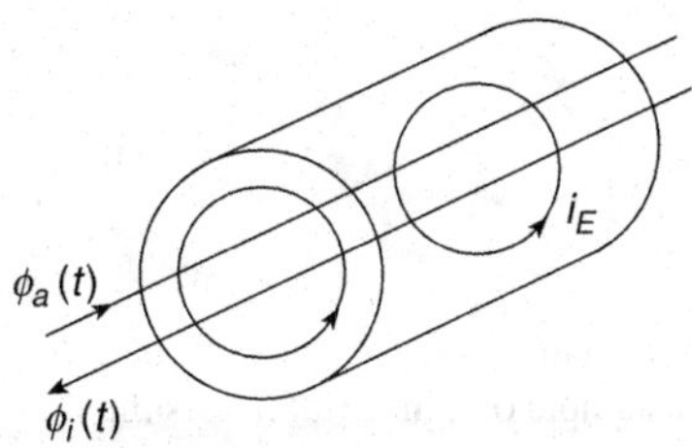

Figure 1.9 Illustration of Lenz's law generating eddy currents. The applied time-varying magnetic flux $\phi_a(t)$ induces eddy current $i_E(t)$, which in turn generates induced flux $\phi_i(t)$ that opposes changes in the applied flux $\phi_a(t)$

[4]Heinrich Friedrich Emil Lenz (1804–1865) was a Russian physicist of German ethnicity born in Estonia, who made a contribution to electromagnetism in the form of his law.

produces an opposing flux $\phi_{(t)}$. If the applied magnetic flux $\phi_a(t)$ decreases, the induced current $i_E(t)$ produces an aiding flux $\phi_i(t)$. The induced magnetic flux ϕ_i always opposes the inducing (applied) magnetic flux ϕ_a. If $\phi_a(t)$ increases, the induced current produces an opposing flux $\phi_i(t)$. If $\phi_a(t)$ decreases, the induced current produces an aiding magnetic flux $\phi_i(t)$. The direction of the induced current $i_E(t)$ with respect of the induced magnetic field $\phi_i(t)$ is determined by the RHR.

If a time-varying magnetic field is applied to a conducting loop (e.g., an inductor turn), a current is induced in such a direction as to oppose the change in the magnetic flux enclosed by the loop. The induced currents flowing in closed loops are called *eddy currents*. Eddy currents occur when a conductor is subjected to time-varying magnetic field(s). In accordance with Lenz's law, the eddy currents produce their own magnetic field(s) to oppose the original field.

The effects of eddy currents on winding conductors and magnetic cores are nonuniform current distribution, increased effective resistance, increased power loss, and reduced internal inductance. If the resistivity of a conductor was zero (as in a perfect conductor), eddy-current loops would be generated with such a magnitude and phase to exactly cancel the applied magnetic field. A perfect conductor would oppose any change in externally applied magnetic field. Circulating eddy currents would be induced to oppose any buildup of the magnetic field in the conductor. In general, nature opposes to everything we want to do.

1.5.4 Volt–Second Balance

Faraday's law is $v_L = d\lambda/dt$, yielding $d\lambda = v_L dt$. Hence,

$$\lambda(t) = \int d\lambda = \int v_L dt. \tag{1.78}$$

For periodic waveforms in steady state,

$$\int_0^T v_L(t)dt = \lambda(t)\Big|_0^T = \lambda(T) - \lambda(0) = 0. \tag{1.79}$$

This equation is called a *volt–second balance*, which states that the total area enclosed by the inductor voltage waveform v_L is zero for steady state. As a result, the area enclosed by the inductor voltage waveform v_L above zero must be equal to the area enclosed by the inductor voltage waveform v_L below zero for steady state. The volt–second balance can be expressed by

$$\int_0^{t_o} v_L(t)dt + \int_{t_o}^T v_L(t)dt = 0 \tag{1.80}$$

which gives

$$\int_0^{t_o} v_L(t)dt = -\int_{t_o}^T v_L(t)dt. \tag{1.81}$$

This can be written as $A^+ = A^-$.

1.5.5 Ohm's Law

Materials resist the flow of electric charge. The physical property of materials to resist current flow is known as *resistivity*. Therefore, a sample of a material resists the flow of electric current. This property is known as *resistance*. Ohm[5] discovered that the voltage across a resistor is directly proportional to its current and is constant, called resistance. Microscopic Ohm's law describes the relationship between the conduction current density **J** and the electric field intensity **E**. The conduction current is caused by the movement of electrons. Conductors exhibit the presence of many free (conduction or

[5] Georg Simon Ohm (1787–1854) was a German physicist and mathematician, who discovered the relationship between voltage and current for a resistor.

valence) electrons, from the outermost atom shells of a conducting medium. These free electrons are in random constant motion in different directions in a zigzag fashion due to thermal excitation. The average electron thermal energy per one degree of freedom is $E_{T1} = kT/2$ and the average thermal energy of an electron in three dimensions is $E_{T3} = 3E_{T1} = 3kT/2$. At the collision, the electron kinetic energy is equal to the thermal energy

$$\frac{1}{2}m_e v_{th}^2 = \frac{3}{2}kT, \tag{1.82}$$

where $k = 10^{-23}$ J/K is the Boltzmann's constant and $m_e = 9.1095 \times 10^{-31}$ kg is the rest mass of a free electron. The thermal velocity of electrons between collisions is

$$v_{th} = \sqrt{\frac{3kT}{m_e}} \approx 1.12238 \times 10^7 \text{ cm/s} = 112.238 \times 10^5 \text{ m/s} = 112.238 \text{ km/s}. \tag{1.83}$$

In good conductors, mobile free electrons drift through a lattice of positive ions encountering frequent collisions with the atomic lattice. If the electric field **E** in a conductor is zero, the net charge movement over a large volume (compared with atomic dimensions) is zero, resulting in zero net current. If an electric field **E** is applied to a conductor, a Coulomb's force **F** is exerted on an electron with charge $-q$

$$\mathbf{F} = -q\mathbf{E}. \tag{1.84}$$

According to Newton's second law, the acceleration of electrons between collisions is

$$a = \frac{F}{m_e} = \frac{-qE}{m_e}, \tag{1.85}$$

where $m_e = 9.11 \times 10^{-31}$ kg is the mass of electron. If the electric field intensity E is constant, then the average drift velocity of electrons increases linearly with time

$$v_d = at = -\frac{qEt}{m_e}. \tag{1.86}$$

The average drift velocity is directly proportional to the electric field intensity E for low values of E and saturates at high value of E. Electrons are involved in collisions with thermally vibrating lattice structure and other electrons. As the electron accelerates due to electric field, the velocity increases. When the electron collides with an atom, it loses most or all of its energy. Then, the electron begins to accelerate due to electric field E and gains energy until a new collision. The average position change x_{avg} of a group of N electrons in time interval Δt is called the *drift velocity*

$$v_d = \frac{x_{avg}}{\Delta t} = \frac{x_1 + x_2 + x_3 + \ldots + x_N}{N\Delta t} = \frac{v_1 + v_2 + v_3 + \ldots + v_N}{N}. \tag{1.87}$$

The drift velocity of electrons $\mathbf{v}_d$ has the opposite direction to that of the applied electric field **E**. By Newton's law, the average change in the momentum of a free electron is equal to the applied force

$$\mathbf{F} = \frac{m_e \mathbf{v}_d}{\tau_c}, \tag{1.88}$$

where the mean time between the successive collisions of electrons with atom lattice, called the *relaxation time*, is given by

$$\tau_c = \frac{l_n}{v_d} \tag{1.89}$$

in which l_n is length of the mean free path of electrons between collisions. Equating the right-hand sides of (1.84) and (1.88), we obtain

$$\frac{m_e \mathbf{v}_d}{\tau_c} = -q\mathbf{E} \tag{1.90}$$

yielding the average drift velocity of electrons

$$\mathbf{v}_d = -\frac{q\tau_c}{m_e}\mathbf{E} = -\mu_n \mathbf{E} \tag{1.91}$$

where the mobility of electrons in a conductor is

$$\mu_n = \frac{q\tau_c}{m_e} = \frac{ql_n}{m_e v_d}. \tag{1.92}$$

The volume charge density in a conductor is

$$\rho_v = -nq, \tag{1.93}$$

where n is the concentration of free (conduction or valence) electrons in a conductor, which is equal to the number of conduction electrons per unit volume of a conductor. The resulting flow of electrons is known as the conduction (or drift) current. The conduction (drift) current density, corresponding to the motion of charge forced by electric field E, is given by

$$\mathbf{J} = \frac{\mathbf{I}}{A} = \rho_v \mathbf{v}_d = -nq\mathbf{v}_d = -nq\mu_n \mathbf{E} = \frac{nq^2\tau_c}{m_e}\mathbf{E} = \sigma\mathbf{E} = \frac{\mathbf{E}}{\rho}, \tag{1.94}$$

where the conductivity of a conductor is

$$\sigma = nq\mu_n = nq\frac{q\tau_c}{m_e} = \frac{nq^2\tau_c}{m_e} = \frac{nq^2 v_d}{m_e l_n} \tag{1.95}$$

and the resistivity of a conductor is

$$\rho = \frac{1}{\sigma} = \frac{1}{nq\mu_n} = \frac{m_e}{nq^2\tau_c} = \frac{m_e l_n}{nq^2 v_d}. \tag{1.96}$$

Hence, the point (or microscopic) form of Ohm's law (1827) for conducting materials is

$$\mathbf{E} = \rho\mathbf{J} = \frac{\mathbf{J}}{\sigma}. \tag{1.97}$$

The typical value of mobility of electrons in copper is $\mu_n = 0.0032$ m^2/V·s. At $E = 1$ V/m, the average drift velocity of electrons in copper is $v_d = 0.32$ cm/s. The thermal velocity of electrons between collisions is $v_{th} = 1.12 \times 10^7$ cm/s. Due to collisions of electrons with atomic lattice and the resulting loss of energy, the velocity of individual electrons in the direction opposite to the electric field $\mathbf{E}$ is much lower than the thermal velocity. The average drift velocity is much lower than the thermal velocity by two orders on magnitude. The average time interval between collisions of electrons is called the *relaxation time* and its typical value for copper is $\tau_c = 3.64 \times 10^{-14}$ s $= 36.4$ fs. The convection current and the displacement current do not obey Ohm's law, whereas the conduction current does it.

To illustrate Ohm's law, consider a straight round conductor of radius r_o and resistivity ρ carrying a DC current I. The current is evenly distributed in the conductor. Thus, the current density is

$$\mathbf{J} = \frac{\mathbf{I}}{A_w} = \frac{I}{\pi r_o^2}\mathbf{a}_z. \tag{1.98}$$

According to Ohm's law, the electric field intensity in the conductor is

$$\mathbf{E} = \rho\mathbf{J} = \rho\frac{I}{\pi r_o^2}\mathbf{a}_z. \tag{1.99}$$

1.5.6 Biot–Savart's Law

Hans Oersted discovered in 1819 that *currents produce magnetic fields* that form closed loops around conductors (e.g., wires). Moving charges are sources of the magnetic field. Jean Biot and Félix Savart arrived in 1820 at a mathematical relationship between the magnetic field $\mathbf{H}$ at any point P of space and the current I that generates $\mathbf{H}$. Current I is a source of magnetic field intensity $\mathbf{H}$. The Biot–Savart's law allows us to calculate the differential magnetic field intensity $d\mathbf{H}$ produced by a small current element $Id\mathbf{l}$. Figure 1.10 illustrates the Biot–Savart's law. The differential form of the Biot–Savart's law is given by

$$d\mathbf{H} = \frac{I}{4\pi}\frac{d\mathbf{l} \times \mathbf{a}_R}{R^2} \tag{1.100}$$

where $d\mathbf{l}$ is the current element equal to a differential length of a conductor carrying electric current I and points in the direction of the current I, and $\mathbf{R} = R\mathbf{a}_R$ is the distance vector between $d\mathbf{l}$ and an observation point P with field H. The vector $d\mathbf{H}$ is perpendicular to both $d\mathbf{l}$ and to the unit vector $\mathbf{a}_R$

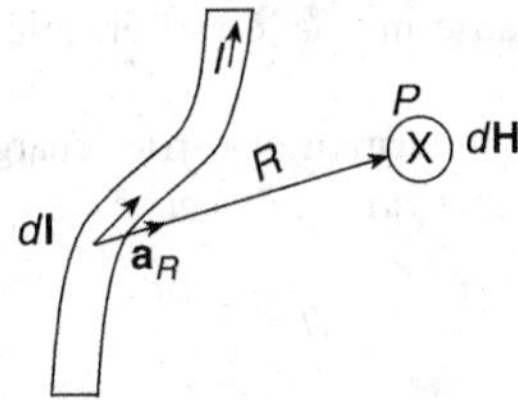

Figure 1.10 Magnetic field $d\mathbf{H}$ produced by a small current element $Id\mathbf{l}$

directed from $d\mathbf{l}$ to P. The magnitude of $d\mathbf{H}$ is inversely proportional to R^2, where R is the distance from $d\mathbf{l}$ to P. The magnitude of $d\mathbf{H}$ is proportional to $\sin\theta$, where θ is the angle between the vectors $d\mathbf{l}$ and $\mathbf{a}_R$. The Biot–Savart's law is analogous to Coulomb's law that relates the electric field E to an isolated point charge Q, which is a source of radial electric field $E = Q/(4\pi\epsilon R^2)$.

The total magnetic field $\mathbf{H}$ induced by a current I is given by the integral form of the Biot–Savart's law

$$\mathbf{H} = \frac{I}{4\pi}\int_l \frac{d\mathbf{l}\times\mathbf{a}_R}{R^2} \quad \text{(A/m)}. \tag{1.101}$$

The integral must be taken over the entire current distribution.

1.5.7 Maxwell's Equations

Maxwell[6] assembled the laws of Faraday, Ampère, and Gauss (for both electric and magnetic fields) into a set of four equations to produce a unified EM theory. Maxwell's equations (1865), together with the law of conservation of charge (the continuity equation), form a foundation of a unified and coherent theory of electricity and magnetism. They couple electric field **E**, magnetic field **H**, current density **J**, and charge density ρ_v. These equations provide the qualitative and quantitative description of static and dynamic EM fields. They can be used to explain and predict electromagnetic phenomena. In particular, they govern the behavior of EM waves.

Maxwell's equations in differential (point or microscopic) forms in the time domain at any point in space and at any time are given by

$$\nabla\times\mathcal{H} = \mathcal{J} + \frac{\partial\mathcal{D}}{\partial t} = \sigma\mathcal{E} + \epsilon\frac{\partial\mathcal{E}}{\partial t} \quad \text{(Ampere's law)}, \tag{1.102}$$

$$\nabla\times\mathcal{E} = -\frac{\partial\mathcal{B}}{\partial t} = -\mu\frac{\partial\mathcal{H}}{\partial t} \quad \text{(Faraday's law)}, \tag{1.103}$$

$$\nabla\cdot\mathcal{D} = \rho_v \quad \text{(Gauss's law)}, \tag{1.104}$$

and

$$\nabla\cdot\mathcal{B} = 0 \quad \text{(Gauss's magnetic law)}, \tag{1.105}$$

where $J_D = \partial\mathcal{D}/\partial t$ is the *displacement current density.* The conductive current density (corresponding to the motion of charge) $\mathcal{J}$ and the displacement current density J_D are sources of EM fields $\mathcal{H}$, $\mathcal{E}$, $\mathcal{B} = \mu\mathcal{H}$, and the volume charge density ρ_v is a source of the electric fields $\mathcal{E}$ and $\mathcal{D} = \epsilon\mathcal{E}$, where μ is permeability and ϵ is the permittivity of a material. Maxwell's equations include two Gauss's[7] laws. Gauss's law states that charge is a source of electric field. In contrast, Gauss's magnetic law states that magnetic field is sourceless (divergenceless), that is, there are no magnetic sources or sinks. This law also indicates that magnetic flux lines close upon themselves. Two Maxwell's equations

[6]James Clerk Maxwell (1831–1879) was a Scottish physicist and mathematician, who mathematically unified Faraday's, Ampère's, and Gauss's laws. "Maxwell's equations" are foundations of EM fields and waves.
[7]Karl Fredrich Gauss (1777–1855) was a German mathematician and physicist.

are partial differential equations because magnetic and electric fields, current, and charge may vary simultaneously with space and time.

Neglecting the generation and recombination of carrier charges like in semiconductors, the *continuity equation* or the *law of local conservation of electric charge* must be satisfied at all times

$$\nabla \cdot \mathcal{J} = -\frac{\partial \rho_v}{\partial t}. \tag{1.106}$$

This law states that the time rate of change of electric charge ρ_v is a source of electric current density field J. This means that the current density is continuous and charge can be neither created nor destroyed. It can only be transferred. The continuity equation is a point form of KCL known in circuit theory. The script letters are used to designate instantaneous field quantities, which are functions of position and time, for example, $\mathcal{E}(x, y, z, t)$. Maxwell's equations are the cornerstone of electrodynamics. A time-varying magnetic field is always accompanied by an electric field, and a time-varying electric field is always accompanied by a magnetic field. For example, a radio antenna generates radiofrequency (RF) waves that consist of both the electric and magnetic fields. The divergence of **B** equal to zero indicates that magnetic charges do not exist in the nature. It is a magnetic flux continuity law. Maxwell's equations also indicate that conductive and/or displacement current is a source of magnetic field, and charge is a source of electric field.

The *divergence* of the electric field intensity **E** at a point is the net outward electric field flow per unit volume over a closed incremental surface S and is defined as

$$\nabla \cdot \mathbf{E} = \lim_{\Delta V \to 0} \frac{\oint_S \mathbf{E} \cdot d\mathbf{S}}{\Delta V} = \frac{\rho_v}{\epsilon}. \tag{1.107}$$

where S is the closed surface, which encloses the volume V, and ϵ is the permittivity of a medium. The closed surface integral $\oint_S \mathbf{E} \cdot d\mathbf{S}$ is the flux of vector **E** outflowing from the volume V. In the limit, the volume V shrinks to a point. Electric fields **E** and $\mathbf{D} = \epsilon\mathbf{E}$ are *source fields* or *sink fields* because the divergence of these fields is not equal to zero ($\nabla \cdot \mathbf{E} \neq 0$ and $\nabla \cdot \mathbf{D} \neq 0$).

In general, a curl-free vector field is called *irrotational*, or a *conservative*, or a *potential* field. Electrostatic fields **D** and **E** are *irrotational* because their curl is equal to zero. If a scalar source (in the form of a charge) of the field **E** is present at a point P, then divergence of **E** is nonzero. Therefore, the vector field whose divergence is nonzero is called a source field. If $\nabla \cdot \mathbf{E} > 0$, the field is a source field. If $\nabla \cdot \mathbf{E} < 0$, the field is a sink field. If $\nabla \cdot \mathbf{E} = 0$, the field is sourceless. A positive charge Q is a source of an electric field **E**, and a negative charge Q is a sink of an electric field **E**.

The *curl* of the magnetic field density at a point is the circulation of **B** per unit area and is defined as

$$\nabla \times \mathbf{B} = \left[\lim_{\Delta S \to 0} \frac{\oint_C \mathbf{B} \cdot d\mathbf{l}}{\Delta S} \right]_{max} = \frac{\Delta I}{\Delta S}, \tag{1.108}$$

where the area ΔS of the contour C is oriented so that the circulation is maximum. In the limit as ΔS shrinks to zero around a point P, the curl of **B** is obtained. Magnetic fields **H** and $\mathbf{B} = \mu\mathbf{H}$ are *rotational* and *sourceless*. They are rotational because their curl is not equal to zero ($\nabla \times \mathbf{B} \neq 0$). They are sourceless because their divergence is equal to zero ($\nabla \cdot \mathbf{H} = 0$). It is worth noting that

$$\nabla \times \mathbf{H} = \mathbf{J} + \frac{\partial \mathbf{D}}{\partial t}. \tag{1.109}$$

The curl of **H** has a nonzero value whenever current is present.

The generalized Ampère's law given by (1.102) states that both conductive and displacement currents induce magnetic field. In other words, a time-varying electric field will give rise to a magnetic field, even in the absence of a conduction (drift) current flow. Maxwell added the displacement current to the Ampère's equation, making (1.109) general.

Gauss's law given by (1.104) states that the net outflow of the electric flux density at any point in space is equal to the charge density at that point. The electric flux starts from a charge and ends on a charge. This means that the electric field is a *divergent* field or *source* field. A positive divergence at a point indicates the presence of a positive charge at that point (i.e., a positive charge is a flux source). Conversely, the negative divergence at a point indicated the presence of a negative charge at that point (i.e., a negative charge is a flux sink).

Equation (1.105) states that the magnetic flux lines always form closed paths, that is, they close on themselves. This means that the magnetic field is a divergenceless field. This law implies that there is no isolated magnetic charges. The magnetic field is sourceless.

If $\mathbf{J} = 0$, Maxwell's equation in (1.102) becomes

$$\nabla \times \mathbf{H} = \frac{\partial \mathbf{D}}{\partial t} = \epsilon_r \epsilon_0 \frac{\partial \mathbf{E}}{\partial t}. \quad (1.110)$$

This equation states that a time-varying electric field induces a changing magnetic field without electric conduction and convection currents, and a changing magnetic field induces a changing electric field. There would be no radiation and propagation of EM waves without the displacement current. In particular, there would be no wireless communications.

Maxwell's equations in integral (or macroscopic) forms are as follows:

$$V = EMF = \oint_C \mathbf{E} \cdot d\mathbf{l} = -\int\int_S \frac{\partial \mathbf{B}}{\partial t} \cdot d\mathbf{S} = -\frac{d\phi}{dt} \quad \text{(Faraday's law)}, \quad (1.111)$$

and

$$\oint_C \mathbf{H} \cdot d\mathbf{l} = \int\int_S \mathbf{J} \cdot d\mathbf{S} + \int\int_S \frac{\partial \mathbf{D}}{\partial t} \cdot d\mathbf{S} = i_{enc} + \frac{d\phi_E}{dt} \quad \text{(Ampere–Maxwell's law)}, \quad (1.112)$$

$$\oiint_S \mathbf{D} \cdot d\mathbf{S} = \int\int\int_V \rho_v dV = Q_{enc} \quad \text{(Gauss's electric law)}, \quad (1.113)$$

$$\oiint_S \mathbf{B} \cdot d\mathbf{S} = 0 \quad \text{(Gauss's magnetic law)}. \quad (1.114)$$

The current density $\mathbf{J}$ may consist of a conduction (or drift) current $\mathbf{J}_c = \sigma \mathbf{E}$ caused by the presence of free electrons and an electric field $\mathbf{E}$ in a conducting medium, the diffusion current density $\mathbf{J}_{diff}$ caused by the gradient of charge carrier concentration, as well as a convection current density $\mathbf{J}_{conv} = \rho_v \mathbf{v}$ due to the motion of free-charge distribution (i.e., the movement of electrically charged medium).

Faraday's law of induction given by (1.111) describes the creation of an electric field by a changing magnetic flux. The EMF, which is equal to the line integral of the electric field $\mathbf{E}$ around any closed path C, is equal to the rate of change of magnetic flux through any surface area S bounded by that path. As a result, for instance, the current is induced in a conducting loop placed in a time-varying magnetic field.

The generalized Ampère's circuital law given by (1.112) describes how a magnetic field can be produced by both an electric current and/or a time-varying electric flux ϕ_E. It states that the line integral of magnetic field $\mathbf{H}$ around any closed path is the sum of the net current through that path and the rate of change of electric flux through any surface bounded by that path.

Gauss's law in the integral form for electric field given by (1.113) states that the total electric flux through any closed surface S is equal to the net charge Q inside that surface.

Gauss's law in the integral form for magnetic field given by (1.114) states that the net magnetic flux through any closed surface is always zero. This means that the number of magnetic field lines that enter a closed volume is equal to the number of magnetic field lines that leave that volume. Magnetic field lines are continuous with no starting or end points. There are no magnetic sources or sinks. A magnetic monopole does not exist. Equation (1.114) also means that there are no magnetic charges.

The phasor technique is a useful mathematical tool for solving problems in linear systems that involve periodic sinusoidal or periodic nonsinusoidal waveforms in steady state, where the amplitude A_m frequency ω and phase ϕ are time-invariant. In this case, complex algebra can be used as a mathematical tool. Periodic nonsinusoidal waveforms, such as a rectangular wave, can be expanded into a Fourier series of sinusoidal components, which is a superposition of harmonic sinusoids. If the excitation is a sinusoidal function of time, the steady-state waveforms described in the time domain can be represented by phasors (complex amplitudes), the trigonometric equations are replaced by algebraic equations, and linear integro-differential equations become linear algebraic equations with no sinusoidal functions, which are easy to solve. Differentiation in the time domain is equivalent to multiplication by $j\omega$ in the phasor domain, and integration in the time domain is equivalent to division by $j\omega$ in the phasor domain. The solutions in the phasor domain can be converted back into the time domain. The sinusoidal current $i(t) = I_m \cos(\omega t + \phi)$ can be represented as $i(t) = Re\{I_m e^{j(\omega t + \phi)}\} = Re\{I_m e^{j\phi} e^{j\omega t}\} = Re\{\mathbf{I}_m e^{j\omega t}\}$, where the complex amplitude $\mathbf{I}_m = I_m e^{j\phi}$ is called a phasor.

The electric field intensity for one-dimensional case in the time domain is given by

$$\mathcal{E}(x,t) = E_m(0)e^{-\frac{x}{\delta}} \cos\left(\omega t - \frac{x}{\delta} + \phi_o\right) = Re\{\mathbf{E}(x)e^{j\omega t}\}, \tag{1.115}$$

where δ is the skin depth of a conductor and ϕ_o is the phase of the electric field. The phasor of the sinusoidal (harmonic) electric field intensity is

$$\mathbf{E}(x) = E_m(0)e^{-\frac{x}{\delta w}} e^{-j\frac{x}{\delta}} e^{j\phi_o}. \tag{1.116}$$

Similarly, the sinusoidal magnetic field intensity is

$$\mathcal{H}(x,t) = H_m(0)e^{-\frac{x}{\delta}} \cos\left(\omega t - \frac{x}{\delta} + \theta_o\right) = Re\{\mathbf{H}(x)e^{j\omega t}\}, \tag{1.117}$$

where θ_o is the phase of the magnetic field and the phasor of the magnetic field intensity is

$$\mathbf{H}(x) = H_m(0)e^{-\frac{x}{\delta}} e^{-j\frac{x}{\delta}} e^{j\theta_o}. \tag{1.118}$$

Substituting the electric and magnetic field intensities into Maxwell's equation in the time domain, we obtain

$$\nabla \times Re\{\mathbf{E}(x)e^{j\omega t}\} = -\frac{\partial}{\partial t} Re\{\mu\mathbf{H}(x)e^{j\omega t}\}, \tag{1.119}$$

which becomes

$$Re\{\nabla \times \mathbf{E}(x)e^{j\omega t}\} = Re\{-j\omega\mu\mathbf{H}(x)e^{j\omega t}\}. \tag{1.120}$$

Thus, $\frac{\partial}{\partial t}$ in Maxwell's equations in the time domain can be replaced by $j\omega$ to obtain Maxwell's equations for sinusoidal field waveforms in phasor forms

$$\nabla \times \mathbf{E} = -j\omega\mu\mathbf{H} = -j\omega\mathbf{B}, \tag{1.121}$$

$$\nabla \times \mathbf{H} = \mathbf{J} + j\omega\mathbf{D} = \mathbf{J} + j\omega\epsilon\mathbf{E} = \sigma\mathbf{E} + j\omega\epsilon\mathbf{E} = (\sigma + j\omega\epsilon)\mathbf{E}, \tag{1.122}$$

$$\nabla \cdot \mathbf{D} = \rho_v, \tag{1.123}$$

and

$$\nabla \cdot \mathbf{B} = 0. \tag{1.124}$$

The *constitutive equations* or *material equations* for linear and isotropic materials are

$$\mathbf{B} = \mu\mathbf{H} \tag{1.125}$$

$$\mathbf{D} = \epsilon\mathbf{E} \tag{1.126}$$

and

$$\mathbf{J} = \sigma\mathbf{E}, \tag{1.127}$$

where **D** is the electric flux density.

In general, the complex propagation constant is given by

$$\gamma = \sqrt{j\omega\mu(\sigma + j\omega\epsilon)} = \omega\sqrt{\frac{\epsilon\mu}{2}} \left[\sqrt{1 + \left(\frac{\sigma}{\omega\epsilon}\right)^2} + 1\right]^{\frac{1}{2}} + j\omega\sqrt{\frac{\epsilon\mu}{2}} \left[\sqrt{1 + \left(\frac{\sigma}{\omega\epsilon}\right)^2} - 1\right]^{\frac{1}{2}}$$

$$= \alpha + j\beta = \frac{1}{\delta} + j\beta, \tag{1.128}$$

where $\alpha = Re\{\gamma\}$ is the attenuation constant and $\beta = Im\{\gamma\}$ is the phase constant. The skin depth is

$$\delta = \frac{1}{\alpha} = \frac{1}{\omega\sqrt{\frac{\epsilon\mu}{2}}\left[\sqrt{1 + \left(\frac{\sigma}{\omega\epsilon}\right)^2} + 1\right]^{\frac{1}{2}}}. \tag{1.129}$$

Figure 1.11 shows a plot of skin depth δ as a function of frequency f for copper. The plot is made using MATLAB®. The quantities $\rho = 1/\sigma$, μ, and ϵ describe the electrical properties of materials. The quantities ω, ρ, μ, and ϵ determine whether a material behaves more like a conductor or more like a dielectric.

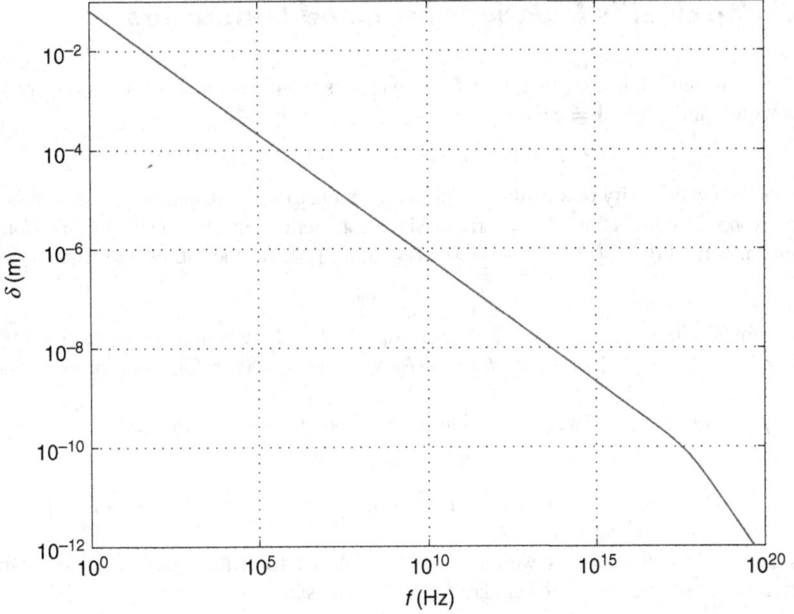

Figure 1.11 Skin depth δ as a function of frequency f

For good conductors, $\sigma \gg \omega\epsilon$, that is, $\sigma/(\omega\epsilon) \gg 1$ or $\omega\epsilon\rho \ll 1$, the complex propagation constant simplifies to the form

$$\gamma \approx \sqrt{\frac{\omega\mu\sigma}{2}} + j\sqrt{\frac{\omega\mu\sigma}{2}} = (1+j)\sqrt{\frac{\omega\mu\sigma}{2}} = \sqrt{j\omega\mu\sigma} = \alpha + j\beta, \tag{1.130}$$

where $(1+j)/\sqrt{2} = e^{45^\circ} = \sqrt{j}$. The skin depth for good conductors is

$$\delta = \sqrt{\frac{2}{\omega\mu\sigma}} = \sqrt{\frac{2\rho}{\omega\mu}} = \frac{1}{\sqrt{\pi\mu\sigma f}} = \sqrt{\frac{\rho}{\pi\mu f}}. \tag{1.131}$$

The wavelength for good conductors is

$$\lambda = \frac{2\pi}{\beta} = 2\sqrt{\frac{\pi}{\mu\sigma f}}. \tag{1.132}$$

The propagation speed or phase velocity for good conductors is

$$v_p = \lambda f = \frac{\omega}{\beta} = \sqrt{\frac{2\omega}{\mu\sigma}}. \tag{1.133}$$

Under the condition of $\omega\epsilon\rho \ll 1$, the system is magnetoquasistatic. This is the case if and only if the left-hand side of the inequality is no more than two orders of magnitude less than unity, that is, $\omega\epsilon\rho \ll 1/100$.

For copper windings, $\rho_{Cu} = 17.24$ nΩm at $T = 20\,^\circ$C and $\epsilon = \epsilon_0 = 10^{-9}/(36\pi) = 8.854 \times 10^{-12}$ F/m, the maximum frequency for magnetoquasistatic operation is

$$f_{max} = \frac{1}{2\pi \times 100 \times \epsilon_0 \rho_{Cu}} = \frac{36\pi}{2\pi \times 100 \times 10^{-9} \times 17.24 \times 10^{-9}} = 10.44 \times 10^{15}$$
$$= 10.44 \text{ PHz}. \tag{1.134}$$

For $\sigma/(\epsilon\omega) \ll 1$, the conductor becomes a dielectric. The skin depth is given by

$$\delta = \frac{1}{\omega\sqrt{\mu\epsilon}}. \tag{1.135}$$

A summary of Maxwell's equations is given in Appendix B.

1.5.8 Maxwell's Equations for Good Conductors

In general, Maxwell's equation in phasor form, which is the differential form of Ampère's equation, together with Ohm's law ($\mathbf{J} = \sigma\mathbf{E}$) is given by

$$\nabla \times H = \mathbf{J} + j\omega\mathbf{D} = \sigma\mathbf{E} + j\omega\epsilon\mathbf{E} = (\sigma + j\omega\epsilon)\mathbf{E}, \tag{1.136}$$

where σ is the conductivity of a medium. For good conductors, the displacement current is negligible in comparison with conduction (drift) current. Since conduction current density $\mathbf{J} = \sigma\mathbf{E}$ dominates the displacement current density $\mathbf{J}_d = j\omega\mathbf{D} = j\omega\epsilon\mathbf{E}$, that is, $\mathbf{J} \gg \mathbf{J}_\mathrm{D}$, the following inequality is satisfied

$$\sigma \gg \omega\epsilon, \tag{1.137}$$

which becomes $\sigma/(\omega\epsilon) \gg 1$ or $\omega\rho\epsilon \ll 1$. For copper, $J = J_D$ when $\sigma = \omega\epsilon_0$ at $f = 1/(2\pi\epsilon_0\rho) = 1.00441 \times 10^{18} = 1.0441$ EHz. Also, $J = 100J_D$ when $\sigma \geq 100\omega\epsilon$ for frequencies $f \leq 10^{16}$ Hz $=$ 10 PHz.

Since $J_D = j\omega D = 0$, Maxwell's equation for good conductors (which is Ampère's law) becomes

$$\nabla \times \mathbf{H} \approx \mathbf{J} = \sigma\mathbf{E}. \tag{1.138}$$

It states that the maximum circulation of $\mathbf{H}$ per unit area as the area shrinks to zero (called the curl of $\mathbf{H}$) is equal to the current density J.

For sinusoidal waveforms, Maxwell's equation in phasor form for good conductors, which is the differential (microscopic) form of Faraday's law, is expressed as

$$\nabla \times \mathbf{E} = -j\omega\mathbf{B} = -j\omega\mu\mathbf{H}. \tag{1.139}$$

Using Ohm's law $\mathbf{E} = \mathbf{J}/\sigma$, we obtain

$$\nabla \times \frac{\mathbf{J}}{\sigma} = -j\omega\mu\mathbf{H} \tag{1.140}$$

producing another form of Maxwell's equation

$$\nabla \times J = -j\omega\mu\sigma\mathbf{H}. \tag{1.141}$$

Assuming that σ and μ are homogeneous, taking the curl on both sides of the above equation and substituting into Maxwell's equation,

$$\nabla \times (\nabla \times \mathbf{J}) = -j\omega\mu\sigma\nabla \times \mathbf{H} = -j\omega\mu\sigma\mathbf{J}. \tag{1.142}$$

Expanding the left-hand side,

$$\nabla(\nabla \cdot \mathbf{J}) - \nabla^2\mathbf{J} = -j\omega\mu\sigma\mathbf{J}, \tag{1.143}$$

where the law of conservation of charge states that charge can be neither created nor destroyed and its point (microscopic) form is expressed by $\nabla \cdot \mathbf{J} = 0$. It is a point form of Kirchhoff's current law. The conduction (or drift) current density $\mathbf{J}$ in good conductors must satisfy the following second-order partial differential equation

$$\nabla^2\mathbf{J} = j\omega\mu\sigma\mathbf{J} = \gamma^2\mathbf{J}, \tag{1.144}$$

where $\gamma^2 = j\omega\mu\sigma$.

For good conductors,

$$\nabla \cdot (\nabla \times \mathbf{H}) = (\sigma + j\omega\epsilon)(\nabla \cdot \mathbf{E}) = 0. \tag{1.145}$$

Hence, Maxwell's equation for good conductors becomes

$$\nabla \cdot \mathbf{D} = \rho_v = 0. \tag{1.146}$$

1.5.9 Poynting's Vector

Poynting[8] developed the mathematical description of the magnitude and the direction of EM energy density transmission. The *instantaneous Poynting vector* (1883) at a given point describes the EM

[8]John Henry Poynting (1852–1914) was an English physicist (Maxwell's student), who described the magnitude and the direction of EM energy flow.

power flux surface density of EM wave

$$\mathcal{S} = \mathcal{E} \times \mathcal{H} \text{ (W/m}^2\text{)}. \tag{1.147}$$

The Poynting vector represents the density and the direction of power flow of electromagnetic fields at any point in space, that is, it is the rate at which energy flows through a unit surface area perpendicular to the direction of wave propagation. The direction of vector $\mathcal{S}$ is normal to both $\mathcal{E}$ and $\mathcal{H}$. The cross product $\mathbf{E} \times \mathbf{H}$ is *pointing* in the direction of power flow, that is, in the direction of wave propagation. The vector $\mathcal{S}$ represents an instantaneous surface power density. Since the unit of $\mathcal{E}$ is V/m and the unit of $\mathcal{H}$ is A/m, the unit of $\mathcal{S}$ is (V/m) × (A/m) = VA/m^2 = W/m^2.

For time-harmonic fields, the *complex Poynting vector* is

$$\mathbf{S}_c = \mathbf{E} \times \mathbf{H}^* \quad \text{(W/m}^2\text{)}. \tag{1.148}$$

The time-average power density, defined as the power density averaged over one period of the sinusoidal excitation, is given by the *time-average Poynting vector*

$$\mathbf{S}_{av} = \frac{1}{2} Re\{\mathbf{E} \times \mathbf{H}^*\} \quad \text{(W/m}^2\text{)}. \tag{1.149}$$

The amount of time-average power passing through a surface S is

$$P_{av} = \int\int_S \mathbf{S}_{av} \cdot d\mathbf{S} = \int\int_S \mathbf{S}_{av} \cdot \mathbf{n} dA = \frac{1}{2} Re \left\{ \oint_S (\mathbf{E} \times \mathbf{H}) \cdot d\mathbf{S} \right\} \text{ (W)}. \tag{1.150}$$

where $d\mathbf{S} = \mathbf{a}_n dA$, $\mathbf{a}_n$ is the unity vector normal to surface S, and dA is the differential surface. The surface integral of $\mathbf{S}_{av}$ describes the total power generated or dissipated inside the enclosed surface S.

For a linear, isotropic, and time-invariant medium of volume V enclosed in a closed surface S, the Poynting theorem relates the following energies: (i) the delivered energy,(ii) the dissipated energy, (iii) the magnetic stored energy, and (iv) the electric stored energy. This theorem describes the principle of conservation of energy. The integral form of the Poynting theorem is given by

$$\oint_S (\mathbf{E} \times \mathbf{H}) \cdot d\mathbf{S} = -\int\int\int_V \mathbf{J} \cdot \mathbf{E} dV - \frac{\partial}{\partial t} \int\int\int_V \left(\frac{1}{2}\mathbf{B} \cdot \mathbf{H} + \frac{1}{2}\mathbf{D} \cdot \mathbf{E} \right) dV$$

$$= -\int\int\int_V \mathbf{J} \cdot \mathbf{E} dV - \frac{\partial}{\partial t} \int\int\int_V \left(\frac{1}{2}\mu H^2 + \frac{1}{2}\epsilon E^2 \right) dV. \tag{1.151}$$

For sinusoidal field waveforms,

$$\oint_S (\mathbf{E} \times \mathbf{H}^*) \cdot d\mathbf{S} = -\int\int\int_V [\mathbf{E} \cdot \mathbf{J}^* + j\omega(\mathbf{H}^* \cdot \mathbf{B} + \mathbf{E}^* \cdot \mathbf{D})] dV$$

$$= -\frac{1}{2}\int\int\int_V \rho |J|^2 dV - j\omega \int\int\int_V \left(\frac{1}{2}\mu H^2 + \frac{1}{2}\epsilon E^2 \right) dV$$

$$= -\int\int\int_V p_D dV - \frac{\partial}{d\partial} \int\int\int_V (w_m + w_e) dV, \tag{1.152}$$

where the asterisk * in the phasor superscript indicates a complex conjugate quantity, $p_D = \frac{1}{2}\rho |J|^2$ is the ohmic power loss density (Joule's law), $w_m = \frac{1}{2}\mu |H|^2$ is the magnetic energy density stored in the magnetic field, and $w_e = \frac{1}{2}\epsilon |E|^2$ is the electric energy density stored in the electric field. The first term on the right-hand side of (1.152) represents the ohmic power dissipated as heat in the volume V (Joule's law) as a result of the flow of conduction current density $\mathbf{J} = \sigma \mathbf{E}$ due to the presence of the electric field $\mathbf{E}$ (Ohm's law). This power exits the volume V through its surface S. The second and third terms represent the time rate of change of the magnetic and electric energies stored in the magnetic and electric fields, respectively. The left-hand side of (1.152) describes the total power leaving the closed surface S. The Poynting theorem describes the principle of conservation of energy. It states that the total power flow out of a closed surface S at any time instant is equal to the sum of the ohmic power dissipated within the enclosed volume V and the rates of decrease of the stored magnetic and electric energies. If there are sources inside the volume V, the dot product $\mathbf{J} \cdot \mathbf{E}$ has the minus sign and represents the power density added to the volume V by these sources.

For steady state, the *complex power* flowing into a volume V surrounded by a closed surface S is given by

$$P = \frac{1}{2}\oint_S (\mathbf{E} \times \mathbf{H}^*) \cdot d\mathbf{S} = P_D + 2j\omega(W_m - W_e) \text{ (W)}, \tag{1.153}$$

where the time-average real power dissipated in the volume V is given by Joule's law as

$$P_D = \frac{1}{2}\int\int\int_V \mathbf{E} \cdot \mathbf{J}^* dV = \frac{1}{2}\int\int\int_V \rho|\mathbf{J}|^2 dV = \frac{1}{2}\int\int\int_V \sigma|\mathbf{E}|^2 dV. \tag{1.154}$$

If $W_m > W_c$, the device inside the volume V is inductive. If $W_c > W_c$, the device inside the volume V is capacitive. If $W_m = W_c$, the device inside the volume V operates at self-resonant frequency (SRF).

For harmonic fields, the instantaneous magnetic energy density in an isotropic medium is

$$w_m(t) = \frac{1}{2}Re\{\mathbf{B}_m e^{j\omega t}\} \cdot Re\{\mathbf{H}_m e^{j\omega t}\} = \frac{1}{2}B_m H_m \cos^2\omega t = \frac{B_m}{2\mu}\cos^2\omega t = \frac{1}{2}\mu H_m^2 \cos^2\omega t. \tag{1.155}$$

Hence, the time-average magnetic energy density is

$$w_{m(av)} = \frac{1}{2}Re\{\mathbf{H} \cdot \mathbf{B}^*\} = \frac{1}{4}H_m B_m = \frac{1}{4}\mu H_m^2 = \frac{B_m^2}{4\mu}. \tag{1.156}$$

1.5.10 Joule's Law

Joule's law (1841) states that the rate of heat dissipation in a conductor is proportional to the square of the current through it and the conductor resistance. The power dissipated in a conductor is $P = RI^2$ and the energy dissipated in a conductor during time interval Δt is $W = P\Delta t$. This law can be extended to distributed systems.

Let us consider the power dissipated in a conductor caused by the movement of electrons forced by electric field **E**. The charge density of free electrons is ρ_v. The electron charge in a small conductor volume ΔV is given by

$$q = \rho_v \Delta V. \tag{1.157}$$

The electric force exerted on the charge q by the electric field **E** is

$$\mathbf{F} = q\mathbf{E} = \mathbf{E}\rho_v \Delta V. \tag{1.158}$$

The incremental amount of energy (or work) ΔW done by the electric force **F** in moving the charge q by an incremental distance $\Delta \mathbf{l}$ is

$$\Delta W = \mathbf{F} \cdot \Delta \mathbf{l} = q\mathbf{E} \cdot \Delta \mathbf{l} = \mathbf{E} \cdot \Delta \mathbf{l} \rho_v \Delta V. \tag{1.159}$$

The power used to perform the work ΔW in time interval Δt is given by

$$\Delta P = \frac{\Delta W}{\Delta t} = \frac{\mathbf{F} \cdot \Delta \mathbf{l}}{\Delta t} = q\mathbf{F} \cdot \mathbf{v}_d = \mathbf{E} \cdot (\rho_v \mathbf{v}_d)\Delta V = \mathbf{E} \cdot \mathbf{J}\Delta V = \frac{\mathbf{E} \cdot \mathbf{E}}{\rho}\Delta V = \rho\mathbf{J} \cdot \mathbf{J}\Delta V, \tag{1.160}$$

where $\mathbf{v}_d = \Delta \mathbf{l}/\Delta t$ is the electron drift velocity. The power loss density describing the time rate at which energy is converted into heat per unit volume of a conductor is given by

$$p_D = \frac{\Delta P}{\Delta V} = \frac{\Delta W/\Delta t}{\Delta V} = \mathbf{E} \cdot \mathbf{J} = \frac{\mathbf{E} \cdot \mathbf{E}}{\rho} = \rho\mathbf{J} \cdot \mathbf{J} \text{ (W/m}^3\text{)}. \tag{1.161}$$

The power loss in the conductor converted into heat is

$$P = \int\int\int_V p_D dV = \int\int\int_V \mathbf{E} \cdot \mathbf{J} dV = \int\int\int_V \frac{\mathbf{E} \cdot \mathbf{E}}{\rho} dV = \int\int\int_V \rho\mathbf{J} \cdot \mathbf{J} dV \text{ (W)}. \tag{1.162}$$

For sinusoidal field waveforms, using Ohm's law, the power dissipated per unit volume is given by the point Joule's law

$$p_D = \mathbf{J} \cdot \mathbf{E}^* = \rho\mathbf{J} \cdot \mathbf{J}^* = \rho|J|^2 = \frac{\mathbf{E} \cdot \mathbf{E}^*}{\rho} = \frac{|E|^2}{\rho}. \tag{1.163}$$

The power dissipated in a conductor of volume V and resistivity ρ as thermal energy (i.e., heat) is given by the integral form of Joule's law

$$P_D = \int\int\int_V p_D dV = \int\int\int_V \mathbf{J}\cdot\mathbf{E}^* dV = \int\int\int_V \rho\mathbf{J}\cdot\mathbf{J}^* dV = \int\int\int_V \rho|J|^2 dV. \quad (1.164)$$

For a linear conductor carrying a multiple-harmonic inductor current waveform, the current density and electric fields can be expanded into Fourier series and the power loss density may be expressed by

$$p_D = \sum_{n=1}^{\infty} \mathbf{J}_n\cdot\mathbf{E}_n^* = \sum_{n=1}^{\infty} \rho\mathbf{J}_n\cdot\mathbf{J}_n^* = \sum_{n=1}^{\infty} \rho|J_n|^2 = \sum_{n=1}^{\infty} \frac{\mathbf{E}_n\cdot\mathbf{E}_n^*}{\rho} = \sum_{n=1}^{\infty} \frac{|E_n|^2}{\rho} \quad (1.165)$$

and the total power loss is

$$P_D = \sum_{n=1}^{\infty} \int\int\int_V p_{Dn} dV = \sum_{n=1}^{\infty} \int\int\int_V \mathbf{J}_n\cdot\mathbf{E}_n^* dV = \sum_{n=1}^{\infty} \int\int\int_V \rho\mathbf{J}_n\cdot\mathbf{J}_n^* dV$$
$$= \sum_{n=1}^{\infty} \int\int\int_V \rho|J_n|^2 dV = \sum_{n=1}^{\infty} \int\int\int_V \frac{\mathbf{E}_n\cdot\mathbf{E}_n^*}{\rho} = \sum_{n=1}^{\infty} \int\int\int_V \frac{|E_n|^2}{\rho}, \quad (1.166)$$

where J_n and E_n are the amplitudes of current density and electric field intensity at nth harmonic, respectively.

The current density in a conductor in the time domain in steady state for one-dimensional case is described by

$$Re\{\mathbf{J}(x)e^{j\omega t}\} = J(x,t) = J_m(0)e^{-\frac{x}{\delta_w}} \cos\left(\omega t - \frac{x}{\delta_w} + \phi_o\right), \quad (1.167)$$

where δ_w is the skin depth and ϕ_o is the initial phase. It is assumed that the current amplitude varies only in the x-direction. From Ohm's law,

$$E(x,t) = \rho J(x,t) = \rho J_m(0)e^{-\frac{x}{\delta_w}} \cos\left(\omega t - \frac{x}{\delta_w} + \phi_o\right), \quad (1.168)$$

where $E_m(0) = \rho J_m(0)$. Assuming that ρ is a real number, the phase shift between $J(x,t)$ and $E(x,t)$ is zero. The instantaneous power density at a point is given by

$$p(x,t) = J(x,t)E(x,t) = J_m(0)E_m(0)e^{-\frac{2x}{\delta_w}} \cos^2\left(\omega t - \frac{x}{\delta_w} + \phi_o\right)$$
$$= \rho J_m^2(0)e^{-\frac{2x}{\delta_w}} \cos^2\left(\omega t - \frac{x}{\delta_w} + \phi_o\right)$$
$$= \frac{J_m(0)E_m(0)}{2} e^{-\frac{2x}{\delta_w}} + \frac{J_m(0)E_m(0)}{2} e^{-\frac{2x}{\delta_w}} \cos 2\left(\omega t - \frac{x}{\delta_w} + \phi_o\right)$$
$$= \frac{\rho J_m^2(0)}{2} e^{-\frac{2x}{\delta_w}} + \frac{\rho J_m^2(0)}{2} e^{-\frac{2x}{\delta_w}} \cos 2\left(\omega t - \frac{x}{\delta_w} + \phi_o\right)$$
$$= p_D(x) + p_D(x) \cos 2\left(\omega t - \frac{x}{\delta_w} + \phi_o\right), \quad (1.169)$$

where $\cos^2 z = 1/2 + 1/2\cos 2z$. The first term in the above equation represents the time-average real power density dissipated in a conductor at a point, and the second term represents the AC component of the instantaneous real power density dissipated in a conductor as heat at a point. The time-average real power density dissipated in a conductor at a point is

$$p_D(x) = \frac{1}{T}\int_0^T p(x,t)dt = \frac{1}{2\pi}\int_0^{2\pi} p(x,\omega t)d(\omega t) = \frac{J_m(0)E_m(0)}{2} e^{-\frac{2x}{\delta_w}} = \frac{\rho J_m^2(0)}{2} e^{-\frac{2x}{\delta_w}}, \quad (1.170)$$

where T is the period. The total time-average power dissipated as heat in a conductor of volume V is

$$P_D = \int\int\int_V p_D(x)dV = \frac{1}{2}\int\int\int_V J_m(0)E_m(0)e^{-\frac{2x}{\delta_w}} dxdydz$$

$$= \frac{1}{2}\int\int\int_V \rho J_m^2(0)e^{-\frac{2x}{\delta_w}}dxdydz. \tag{1.171}$$

When EM fields are sinusoidal, phasors are described in space as follows: $\mathbf{H}(\mathbf{r}) = \mathbf{H}(x,y,z)$, $\mathbf{E}(\mathbf{r}) = \mathbf{E}(x,y,z)$, and $\mathbf{J}(\mathbf{r}) = \mathbf{J}(x,y,z)$. The instantaneous point (local) power density is

$$p(\mathbf{r},t) = Re\{\mathbf{J}(\mathbf{r},t\} \cdot Re\{\mathbf{E}(\mathbf{r},t\} = \frac{1}{4}[\mathbf{J}(\mathbf{r},t) + \mathbf{J}^*(\mathbf{r},t)][\mathbf{E}(\mathbf{r},t) + \mathbf{E}^*(\mathbf{r},t)]$$

$$= \frac{1}{4}[\mathbf{J}(\mathbf{r}) \cdot \mathbf{E}^*(\mathbf{r}) + \mathbf{J}(\mathbf{r}) \cdot \mathbf{E}(\mathbf{r})e^{2j\omega t} + \mathbf{J}^*(\mathbf{r}) \cdot \mathbf{E}(\mathbf{r}) + \mathbf{J}^*(\mathbf{r}) \cdot \mathbf{E}^*(\mathbf{r})e^{-2j\omega t}]$$

$$= \frac{1}{2}Re[\mathbf{J}(\mathbf{r}) \cdot \mathbf{E}^*(\mathbf{r}) + \mathbf{J}(\mathbf{r}) \cdot \mathbf{E}(\mathbf{r})e^{2j\omega t}]. \tag{1.172}$$

The time-average real power density dissipated in a conductor at a point $\mathbf{r}$ is

$$p_D(\mathbf{r}) = \frac{1}{T}\int_0^T p(\mathbf{r},t)dt = \frac{1}{2}Re[\mathbf{J}(\mathbf{r}) \cdot \mathbf{E}^*(\mathbf{r})]. \tag{1.173}$$

The time-average power dissipated as heat in the conductor of volume V is given by

$$P_D = \int\int\int_V p_D(\mathbf{r})dV = \frac{1}{2}Re\int\int\int_V \mathbf{J}(\mathbf{r}) \cdot \mathbf{E}^*(\mathbf{r})dV$$

$$= \frac{1}{2}\int\int\int_V \rho\mathbf{J}(\mathbf{r}) \cdot \mathbf{J}^*(\mathbf{r})dV = \frac{1}{2}\int\int\int_V \rho|\mathbf{J}(\mathbf{r})|^2 dV. \tag{1.174}$$

The current density in phasor form for one-dimensional case is given by

$$\mathbf{J}(x) = J_m(0)e^{-\frac{x}{\delta_w}}e^{-j\frac{x}{\delta_w}}e^{j\phi_o} = J_m(x)e^{j(\phi_o - \frac{x}{\delta_w})}, \tag{1.175}$$

where the amplitude is

$$J_m(x) = J_m(0)e^{-\frac{x}{\delta_w}}. \tag{1.176}$$

The time-average point power density for sinusoidal waveforms is given by point Joule's law in phasor form

$$P_D(x) = \frac{1}{2}Re(\mathbf{J} \cdot \mathbf{E}^*) = \frac{1}{2}\rho\mathbf{J} \cdot \mathbf{J}^* = \frac{1}{2}\rho|J(x)|^2 = \frac{1}{2}\rho J_m^2(0)e^{-\frac{2x}{\delta_w}}. \tag{1.177}$$

For periodic waveforms, the time-average real power dissipated in a conductor of volume V and resistivity ρ due to conversion of EM energy to thermal energy (heat) is given by Joule's law in phasor form

$$P_D = \frac{1}{2}Re\int\int\int_V \mathbf{J} \cdot \mathbf{E}^* dV = \frac{1}{2}\int\int\int_V \rho\mathbf{J} \cdot \mathbf{J}^* dV = \frac{1}{2}\int\int\int_V \rho|J|^2 dV, \tag{1.178}$$

where J and E are the amplitudes of the current density and the electric field intensity, respectively. The time-average power loss density P_v is defined as the total time-average power loss P_D per unit volume

$$P_v = \frac{P_D}{V}, \tag{1.179}$$

where V is the volume carrying the current.

Since $\mathbf{B} = \mu\mathbf{H}$, the point (local) magnetic energy density for sinusoidal waveforms is given by

$$w_m(x) = \frac{1}{2}\mathbf{B} \cdot \mathbf{H}^* = \frac{1}{2}\mu\mathbf{H} \cdot \mathbf{H}^* = \frac{1}{2}\mu|H(x)|^2. \tag{1.180}$$

The maximum magnetic energy stored in inductor L is given by

$$W_m = \int ivdt = \int iL\frac{di}{dt}dt = L\int_0^{I_m} idi = \frac{1}{2}LI_m^2 = \frac{1}{2}\frac{N^2}{\mathcal{R}}I_m^2 = \frac{\mathcal{F}_m^2}{2\mathcal{R}}$$

$$= \frac{1}{2}\left(\frac{\mu N^2 S}{l}\right)\left(\frac{Bl}{\mu N}\right)^2 = \frac{1}{2}\int\int\int_V \mu|H(x)|^2 dV, \tag{1.181}$$

where V is the volume of the interior of the inductor and $v = Ldi/dt$. The magnetic energy density w_m is defined as the magnetic energy W_m per unit volume

$$w_m = \frac{W_m}{V}. \tag{1.182}$$

The time-average local magnetic energy density is given by

$$w_m(x) = \frac{1}{2}\mathbf{B}\cdot\mathbf{H}^* = \frac{1}{2}\mu\mathbf{H}\cdot\mathbf{H}^* = \frac{1}{2}\mu|H(x)|^2 \text{ (J/m}^3\text{)}. \tag{1.183}$$

The total time-average magnetic energy is

$$W_m = \frac{1}{2}\int\int\int_V \mathbf{B(x)}\cdot\mathbf{H}^*(\mathrm{x})dV = \frac{1}{2}\int\int\int_V \mu|H(x)|^2 dV \text{ (J)}. \tag{1.184}$$

The time-average magnetic energy is the average energy per unit time over a period of time and is

$$W_m = \frac{1}{4}L|I|^2 = \frac{1}{4}\int\int\int_V \mathbf{B(x)}\cdot\mathbf{H}^*(\mathrm{x})dV = \frac{1}{4}\int\int\int_V \mu|H(x)|^2 dV \text{ (J)}. \tag{1.185}$$

1.6 Eddy Currents

Eddy currents were discovered by Foucault in 1851. Therefore, they are also called Foucault's currents. Figure 1.12 illustrates the eddy current $i_E(t)$ of density $J_e(t)$ induced by a time-varying magnetic field $H(t)$. Eddy currents circulate in closed paths. In a conductor, the induced magnetic field $H(t)$ may be caused by the Conductor's own AC current or by the AC current flowing in adjacent conductors.

In accordance with Faraday's law, a time-varying magnetic field $H(t) = B(t)/\mu$ induces an electric field E

$$\nabla \times \mathbf{E} = -\frac{\partial \mathbf{B}}{\partial t}. \tag{1.186}$$

According to Ohm's law, the electric field induces an eddy current

$$\mathbf{J}_e = \sigma\mathbf{E} = \frac{\mathbf{E}}{\rho}. \tag{1.187}$$

According to Ampère's law, eddy currents induce a magnetic field $H(t)$, as shown in Fig. 1.12. These currents are similar to a current flowing in turns of a multilayer solenoid, and therefore produce a magnetic field. According to Lenz's law, the induced magnetic field opposes the applied magnetic field.

The magnetic field induces an EMF $v(t) = d\phi/dt = A(dB/dt) = A\mu(dH/dt)$ in a conducting material of conductivity σ, which in turn produces eddy currents $i_E(t)$. The flow of eddy currents causes power losses in winding conductors and magnetic cores.

The applied time-varying magnetic field $H_a(t)$ induces the electric field $E(t)$, which induces eddy currents $i_E(t)$, and these currents generate a time-varying magnetic field $H(t)$ that opposes the original

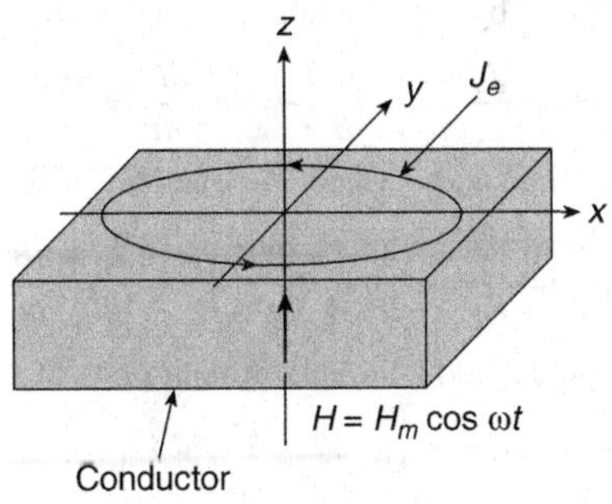

Figure 1.12 Eddy current

applied magnetic field $H_a(t)$, according to Lenz's law. The direction of the induced eddy currents $i_E(t)$ with respect to the induced magnetic field $H(t)$ is determined by the RHR, as shown in Fig. 1.12. The opposing magnetic flux can be found using Ampère's law

$$\nabla \times \mathbf{H} = \mathbf{J}_a + \mathbf{J}_e, \tag{1.188}$$

where $\mathbf{J}_a$ is the applied current density and $\mathbf{J}_e$ is the eddy current density. When the applied current $\mathbf{J}_a$ is zero and the magnetic field is generated by adjacent conductors, Ampère's law becomes

$$\nabla \times \mathbf{H} = \mathbf{J}_e. \tag{1.189}$$

The eddy-current density in a conductor of conductivity $\sigma = 1/\rho$ can be described by

$$\mathbf{J}_e = \sigma \mathbf{E} = \frac{\mathbf{E}}{\rho} \tag{1.190}$$

For sinusoidal waveforms, the phasor of the eddy-current density is given by

$$\mathbf{J}_e = -j\omega\sigma \mathbf{A}, \tag{1.191}$$

where $\mathbf{A}$ is the phasor of the magnetic vector potential. Eddy currents can be reduced using selecting high-resistivity materials (such as ferrites) or using thin plates, called laminations. These currents dissipate energy as heat in magnetic cores and winding conductors of inductors and transformers.

There are two effects associated with eddy currents: skin effect and proximity effect. The skin effect current density J_s is orthogonal to the proximity effect current density J_p. The winding power loss due to eddy currents is

$$P = \frac{1}{2\sigma} \int\int\int (J_s J_s^* + J_p J_p^*) dV. \tag{1.192}$$

Eddy currents are driven by a voltage induced in a conductor by the magnetic field. According to Faraday's law, the voltage induced in a conductor by the magnetic flux $\phi = AB = A\mu B$ is

$$v(t) = \frac{d\phi}{dt} = \frac{d(AB)}{dt} = A\frac{dB}{dt} = A\mu\frac{dH}{dt}. \tag{1.193}$$

Eddy currents flow in a plane perpendicular to the magnetic field density $\mathbf{B}$. From Ohm's law, the rms value of the eddy current in a conductor of resistance R is

$$I = \frac{V}{R}. \tag{1.194}$$

The power loss caused by sinusoidal eddy currents in a conductor of resistance R is

$$P_1 = \frac{V^2}{R} = \frac{A^2}{R}\left(\frac{dB}{dt}\right)^2. \tag{1.195}$$

Let us divide the conductor into two insulated parts so that the surface of one part $A_2 = A/2$. Hence, the resistance of one-half of the conductor is

$$R_2 = \frac{\rho l}{A_2} = \frac{2\rho l}{A} = 2R. \tag{1.196}$$

The voltage induced in a conductor by the magnetic flux $\phi = A_2 B$ is

$$V_2 = \frac{d\phi}{dt} = \frac{d(A_2 B)}{dt} = A_2\frac{dB}{dt} = \frac{A}{2}\frac{dB}{dt}. \tag{1.197}$$

The power loss caused by sinusoidal eddy currents in one-half of the conductor of resistance R_2 is

$$P_{2(1)} = \frac{V_2^2}{R_2} = \frac{1}{2R}\frac{A^2}{4}\left(\frac{dB}{dt}\right)^2 = \frac{A^2}{8R}\left(\frac{dB}{dt}\right)^2. \tag{1.198}$$

The power loss caused by sinusoidal eddy currents in both parts of the conductor is

$$P_2 = 2P_{2(1)} = \frac{A^2}{4R}\left(\frac{dB}{dt}\right)^2. \tag{1.199}$$

Let us divide the conductor into n parts so that $A_n = A/n$. Then, the power loss caused by sinusoidal eddy currents in all n parts of the conductor is

$$P_n = \frac{nA_n^2}{R_n}\left(\frac{dB}{dt}\right)^2 = \frac{n\left(\frac{A}{n}\right)^2}{nR}\left(\frac{dB}{dt}\right)^2 = \frac{A^2}{n^2R}\left(\frac{dB}{dt}\right)^2 = \frac{P_1}{n^2}. \tag{1.200}$$

Thus, the eddy-current power loss decreases by a factor of n^2 when the area of the conductor perpendicular to the magnetic flux is divided into n laminations, electrically insulated by the oxide. Laminations should be parallel to the magnetic field density **B**. Laminated magnetic cores are used to reduce eddy-current loss by reducing the magnitude of eddy currents.

Example 1.3

An infinitely long round solid straight wire of radius r_o carries sinusoidal current $i = I_m \cos \omega t$ in steady state at low frequencies (with no skin effect), as described in Example 1.1 and depicted in Fig. 1.7. Determine the eddy-current power loss, equivalent resistance, and optimum wire diameter.

Solution: The waveform of the induced voltage (EMF) is

$$V(x,t) = \frac{d\phi}{dt} = -\omega\mu_0 I_m \frac{r^3}{2r_o^2} \sin \omega t \quad \text{for} \quad 0 \le r \le r_o. \tag{1.201}$$

Hence, the amplitude of the induced voltage (EMF) inside the wire at low frequencies is

$$V_m(r) = \omega\phi_m(r) = \omega\mu_0 B_m(r) = \omega\mu_0 A H_m(r) = \omega\mu_0 I_m \frac{r^3}{2r_o^2} \quad \text{for} \quad 0 \le r \le r_o. \tag{1.202}$$

The cylindrical shell of radius r and thickness dr has a cross-sectional area $A_{sh} = l_w dr$ and the length of the current path $l_i = 2\pi r$. Hence, the resistance of the cylindrical shell is

$$R_{dr}(r) = \frac{\rho_w l_i}{A_{sh}} = \frac{2\pi r \rho_w}{l_w dr}. \tag{1.203}$$

The time-average power loss in the cylinder is

$$dP_e = \frac{V_m^2(r)}{2R_{dr}} = \left(\omega\mu_0 I_m \frac{r^3}{2r_o^2}\right)^2 \frac{l_w dr}{4\pi r \rho_w} = \frac{\omega^2\mu_0^2 I_m^2 l_w}{16\pi \rho_w r_o^4} r^5 dr \quad \text{for} \quad 0 \le r \le r_o. \tag{1.204}$$

The time-average eddy-current power loss inside the entire wire is

$$P_e = \int_0^{r_o} dP_e = \frac{\mu_0^2\omega^2 I_m^2 l_w}{16\pi \rho_w r_o^4}\int_0^{r_o} r^5 dr = \frac{\mu_0^2\omega^2 I_m^2 l_w r_o^2}{96\pi \rho_w} = \frac{\mu_0^2\omega^2 I_m^2 l_w d^2}{384\pi \rho_w}, \tag{1.205}$$

where $d = 2r_o$. The eddy-current power loss in terms of the equivalent eddy-current power loss resistance R_e is

$$P_e = \frac{1}{2}R_e I_m^2. \tag{1.206}$$

Thus,

$$\frac{1}{2}R_e I_m^2 = \frac{\mu_0^2\omega^2 I_m^2 l_w r_o^2}{96\pi \rho_w} = \frac{\mu_0^2\omega^2 I_m^2 l_w d^2}{384\pi \rho_w}. \tag{1.207}$$

Hence, the equivalent eddy-current resistance is

$$R_e = \frac{\mu_0^2\omega^2 l_w r_o^2}{48\pi \rho_w} = \frac{\mu_0^2\omega^2 l_w d^2}{192\pi \rho_w}. \tag{1.208}$$

The low-frequency resistance is

$$R_{wDC} = \frac{\rho_w l_w}{\pi r_o^2} = \frac{4\rho_w l_w}{\pi d^2}. \tag{1.209}$$

The total resistance of the wire is

$$R_w = R_{wDC} + R_e = \frac{\rho_w l_w}{\pi r_o^2} + \frac{\mu_0^2\omega^2 l_w r_o^2}{48\pi \rho_w} = \frac{4\rho_w l_w}{\pi d^2} + \frac{\mu_0^2\omega^2 l_w d^2}{192\pi \rho_w} = \frac{4\rho_w l_w}{\pi d^2}\left(1 + \frac{\mu_0^2\omega^2 d^4}{768\rho_w^2}\right)$$

$$= \frac{\rho_w l_w}{\pi}\left(\frac{1}{r_o^2} + \frac{\pi^2\mu_0^2\omega^2 r_o^2}{48\rho_w^2}\right) = \frac{4\rho_w l_w}{\pi}\left(\frac{1}{d^2} + \frac{\mu_0^2\omega^2 d^2}{768\rho_w^2}\right) = \frac{4\rho_w l_w}{\pi}\left(\frac{1}{d^2} + \frac{d^2}{192\delta_w^4}\right). \quad (1.210)$$

As the conductor diameter d is increased, the conductor DC resistance R_{DC} increases and the eddy-current resistance R_e decreases. Therefore, there is an optimum conductor diameter at which the total conductor resistance R_w takes on a minimum value. The derivative of the total wire resistance with respect to its diameter d is

$$\frac{dR_w}{dd} = \frac{4\rho_w l_w}{\pi}\left(-\frac{2}{d^3} + \frac{\mu_0^2\omega^2 d}{384\rho_w^2}\right) = \frac{4\rho_w l_w}{\pi}\left(-\frac{2}{d^3} + \frac{d}{96\delta_w^4}\right) = 0. \quad (1.211)$$

Hence, the optimum conductor diameter is

$$\frac{d_{opt}}{\delta_w} = \sqrt[4]{192} \approx 3.722. \quad (1.212)$$

The AC-to-DC resistance ratio is

$$F_R = \frac{R_w}{R_{wDC}} = 1 + \frac{\mu_0^2\omega^2 d^4}{96\pi\rho_w^2} = 1 + \frac{d^4}{192\delta_w^4}. \quad (1.213)$$

The AC-to-DC resistance ratio at $d = d_{opt}$ is

$$F_{Rv} = 1 + 1 = 2. \quad (1.214)$$

The minimum total resistance of the conductor is

$$R_{wmin} = F_{Rv} R_{wDC} = 2R_{wDC}. \quad (1.215)$$

1.7 Core Saturation

Many inductors are made up using magnetic cores. A magnetic core is a conductor of magnetic field H. For an inductor with a magnetic core of cross-sectional area A_c and a saturation magnetic flux density B_s, the magnetic flux at which the magnetic core begins to saturate is

$$\phi_s = A_c B_s, \quad (1.216)$$

resulting in the maximum value of the magnetic flux density

$$B_{pk} = \frac{\phi_{pk}}{A_c} = \frac{\phi_{DC(max)} + \phi_{AC(max)}}{A_c} < B_s. \quad (1.217)$$

The saturated magnetic flux density $B_s = \mu_0 H_s$ is nearly constant. Therefore, $v_L = NA_c dB(t)/dt \approx 0$ and the inductor behaves almost like a short circuit. To avoid core saturation, one has to reduce the maximum value of the magnetic flux ϕ_{pk} in the core or increase the core cross-sectional area A_c. A nonuniform magnetic flux distribution in ferrite cores creates localized magnetic saturation and hot spots.

The magnetic flux linkage at which the magnetic core begins to saturate is given by

$$\lambda_s = N\phi_s = NA_c B_s = LI_{m(max)}. \quad (1.218)$$

Thus,

$$N_{max} A_c B_{pk} = LI_{m(max)} \quad (1.219)$$

yielding the maximum number of turns

$$N_{max} = \frac{LI_{m(max)}}{A_c B_{max}}. \tag{1.220}$$

According to (1.7), the magnetic field intensity H is proportional to $\mathcal{F} = Ni$. Therefore, there is a maximum amplitude of the inductor current $I_{m(max)}$ at which the core saturates. Figure 1.13 shows plots of B as functions of H and i. The saturation flux density is

$$B_s = \mu H_s = \frac{\mu_{rc}\mu_0 N I_{m(max)}}{l_c}. \tag{1.221}$$

To avoid core saturation, the *ampere-turn limit* is given by

$$N_{max} I_{m(max)} = \frac{B_s l_c}{\mu_{rc}\mu_0} = B_s A_c \mathcal{R} = \frac{B_s l_c}{\mu_{rc}\mu_0}. \tag{1.222}$$

To avoid core saturation, one has to reduce the peak inductor current $I_{m(max)}$ or to reduce the number of turns N to satisfy the condition for all operating conditions

$$\frac{\mu_{rc}\mu_0 N I_m}{l_c} < B_s. \tag{1.223}$$

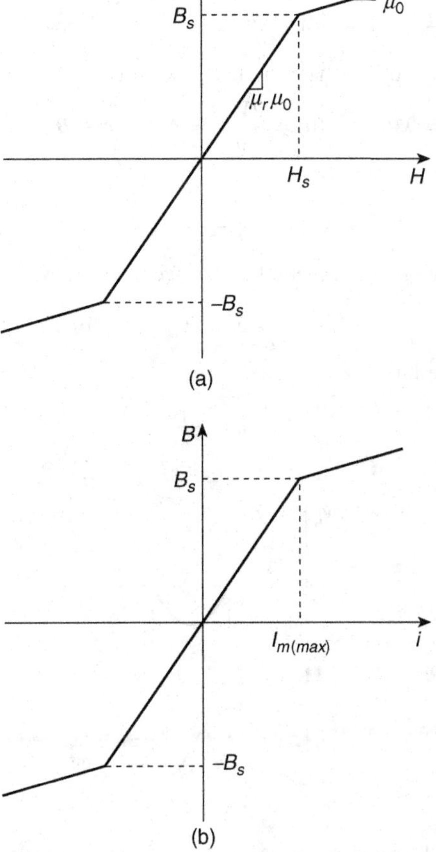

Figure 1.13 Magnetic flux density as functions of magnetic field intensity and inductor current. (a) Magnetic flux density B as a function of magnetic field intensity H. (b) Magnetic flux density B as a function of inductor current i at a fixed number of turns N

From Faraday's law, $d\lambda = v_L(t)dt$. Hence, the general relationship between the inductor voltage and the flux linkage is given by

$$\lambda(t) = \int_0^t v_L(t)dt + \lambda(0) = \frac{1}{\omega}\int_0^{\omega t} v_L(\omega t)d(\omega t) + \lambda(0). \tag{1.224}$$

For a transformer,

$$(N_1 i_1 + N_2 i_2 + \ldots)_{max} \leq B_s A_c \mathcal{R} = \frac{B_s l_c}{\mu_{rc}\mu_0}. \tag{1.225}$$

It is important to avoid both local and global core saturation.

1.7.1 Core Saturation for Sinusoidal Inductor Voltage

Consider an inductor with a magnetic core of saturation flux density B_s. Figure 1.14 shows sinusoidal waveforms of the inductor voltage v_L and the magnetic flux linkage λ. The DC components of these waveforms are assumed to be zero. The inductor voltage is given by

$$v_L = V_{Lm} \sin \omega t. \tag{1.226}$$

The magnetic flux linkage is

$$\lambda(t) = \frac{1}{\omega}\int_0^{\omega t} v_L(\omega t)d(\omega t) + \lambda(0) = \frac{1}{\omega}\int_0^{\omega t} V_{Lm} \sin \omega t d(\omega t) + \lambda(0)$$
$$= \frac{V_{Lm}}{\omega}(1 - \cos \omega t) + \lambda(0) = \frac{V_{Lm}}{\omega} - \frac{V_{Lm}}{\omega} \cos \omega t + \lambda(0). \tag{1.227}$$

Thus, the peak-to-peak value of the magnetic flux linkage is

$$\Delta\lambda = \lambda(\pi) - \lambda(0) = \frac{2V_{Lm}}{\omega} = N\phi = NA_c B_m < NA_c B_s. \tag{1.228}$$

The initial value of the flux linkage is

$$\lambda(0) = -\frac{\Delta\lambda}{2} = -\frac{V_{Lm}}{\omega}. \tag{1.229}$$

The steady-state waveform of the magnetic flux linkage is given by

$$\lambda(t) = -\frac{V_{Lm}}{\omega} \cos \omega t = -\lambda_m \cos \omega t, \tag{1.230}$$

where the amplitude of the flux linkage is

$$\lambda_m = \frac{V_{Lm}}{\omega}. \tag{1.231}$$

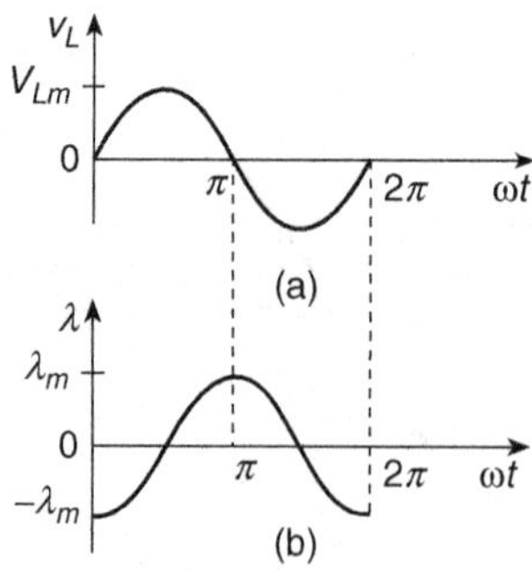

Figure 1.14 Waveforms of the square-wave inductor voltage and the corresponding magnetic flux linkage for sinusoidal inductor voltage. (a) Waveform of the inductor voltage v_L. (b) Waveform of the magnetic flux linkage λ

Thus, the amplitude of the magnetic flux linkage λ_m increases as the frequency f decreases. The minimum frequency f_{min} occurs when the amplitude of the magnetic flux linkage λ_m reaches the saturation value λ_s

$$\lambda_m = \lambda_s = \frac{V_{Lm(max)}}{\omega_{min}}. \tag{1.232}$$

The lowest frequency at which the inductor can operate without saturating the core is given by

$$f_{min} = \frac{V_{Lm(max)}}{2\pi\lambda_s} = \frac{V_{Lm(max)}}{2\pi N A_c B_s} = \frac{\sqrt{2}V_{Lrms(max)}}{2\pi N A_c B_s} = \frac{V_{Lrms(max)}}{K_f N A_c B_s} = \frac{V_{Lrms(max)}}{4.44 N A_c B_s}, \tag{1.233}$$

where the waveform factor for a sinusoidal inductor voltage is defined as the ratio of the rms value to the amplitude of the sinusoidal inductor voltage

$$K_f = \frac{V_{Lrms}}{V_{Lm}} = \frac{2\pi}{\sqrt{2}} = \pi\sqrt{2} = 4.44. \tag{1.234}$$

The minimum frequency f_{min} decreases, when N increases, A_c increases, B_s increases, and $V_{Lm(max)}$ decreases. As the temperature increases, B_s decreases. For ferrite cores, B_s may decrease by a factor of 2 as T increases from room temperature to 100 °C.

Another method to derive the minimum frequency is as follows. Assume that the initial condition is $\lambda(0) = -\lambda_s$. The magnetic flux linkage at core saturation is given by

$$\lambda_s = \frac{1}{\omega_{min}}\int_0^{\pi} v_L d(\omega t) + \lambda(0) = \frac{1}{\omega_{min}}\int_0^{\pi} V_{Lm}\sin\omega t d(\omega t) - \lambda_s = \frac{2V_{Lm}}{\omega_{min}} - \lambda_s, \tag{1.235}$$

resulting in

$$\lambda_s = \frac{\Delta\lambda_{max}}{2} = \frac{V_{Lm(max)}}{\omega_{min}}. \tag{1.236}$$

Hence, the lowest frequency at which the inductor can operate without saturating the core is given by

$$f_{min} = \frac{V_{Lm(max)}}{2\pi\lambda_s} = \frac{V_{Lm(max)}}{2\pi N A_c B_s}. \tag{1.237}$$

The maximum rms value of the sinusoidal voltage across an inductor is

$$V_{Lrms(max)} = \frac{\omega N A_c B_s}{\sqrt{2}} = \frac{2\pi f N A_c B_s}{\sqrt{2}} = \sqrt{2}\pi f N A_c B_s = K_f f N A_c B_s = 4.44 f N A_c B_s. \tag{1.238}$$

Example 1.4

A Philips 3F3 ferrite magnetic core material has the saturation flux density $B_s = 0.32$ T at temperature $T = 20$ °C. The core is expected to operate up to $T = 120$ °C. The core cross-sectional area is $A_c = 80 \times 10^{-6}$ m^2. The number of turns is $N = 10$. Find the maximum magnetic flux linkage and the minimum operating frequency for a sinusoidal voltage waveform with amplitude $V_{Lm} = 10$ V.

Solution: The saturation flux density of the ferrite magnetic core material is $B_s = 0.16$ T at temperature $T = 120$ °C. Hence, the saturation magnetic linkage is

$$\lambda_s = N\phi_s = N A_c B_s = 10 \times 80 \times 10^{-6} \times 0.16 = 128 \times 10^{-6} \text{ V} \cdot \text{s}. \tag{1.239}$$

The minimum frequency without core saturation is

$$f_{min} = \frac{V_{Lm(max)}}{2\pi\lambda_s} = \frac{10}{2\pi \times 128 \times 10^{-6}} = 12.434 \text{ kHz}. \tag{1.240}$$

1.7.2 Core Saturation for Square-Wave Inductor Voltage

If the inductor voltage waveform is a square wave $\pm V$, the magnetic flux linkage is a symmetrical triangular wave, as shown in Fig. 1.15. For the first half of the cycle,

$$v_L = V \quad \text{for} \quad 0 \le t \le \frac{T}{2} \tag{1.241}$$

and

$$\lambda(t) = \int_0^t v_L(t)dt + \lambda(0) = \int_0^t Vdt + \lambda(0) = Vt + \lambda(0) \quad \text{for} \quad 0 \le t \le \frac{T}{2}. \tag{1.242}$$

The flux linkage at $t = T/2$ is

$$\lambda\left(\frac{T}{2}\right) = \frac{VT}{2} + \lambda(0). \tag{1.243}$$

For the second half of the cycle,

$$v_L = -V \quad \text{for} \quad \frac{T}{2} \le t \le T \tag{1.244}$$

and

$$\lambda(t) = \int_{T/2}^t v_L(t)dt + \lambda\left(\frac{T}{2}\right) = \int_{T/2}^t (-V)dt + \lambda\left(\frac{T}{2}\right)$$
$$= -V\left(t - \frac{T}{2}\right) + \lambda\left(\frac{T}{2}\right) \quad \text{for} \quad \frac{T}{2} \le t \le T. \tag{1.245}$$

Hence, the peak-to-peak value of the magnetic flux linkage is

$$\Delta\lambda = \lambda\left(\frac{T}{2}\right) - \lambda(0) = \frac{VT}{2} + \lambda(0) - \lambda(0) = \frac{VT}{2} = \frac{V}{2f}, \tag{1.246}$$

$$-\lambda_m = \lambda(0) = -\frac{\Delta\lambda}{2} = -\frac{VT}{4} = -\frac{V}{4f}, \tag{1.247}$$

and

$$\lambda_m = \lambda\left(\frac{T}{2}\right) = \frac{\Delta\lambda}{2} = \frac{VT}{4} = \frac{V}{4f}. \tag{1.248}$$

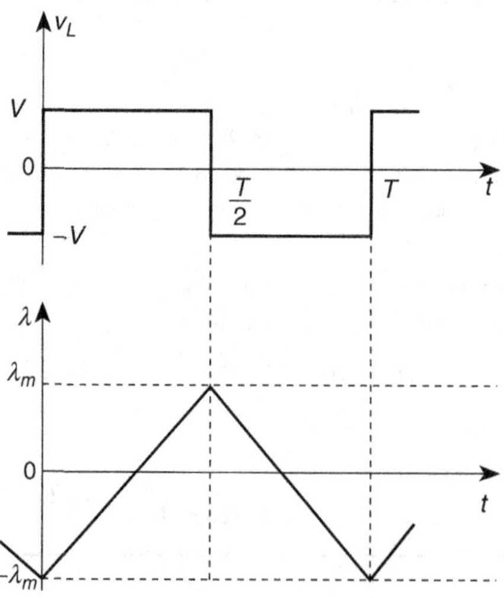

Figure 1.15 Waveforms of the square-wave inductor voltage and the corresponding magnetic flux linkage

The steady-state waveform of the magnetic linkage is

$$\lambda(t) = Vt - \frac{V}{4f} \quad \text{for} \quad 0 \le t \le \frac{T}{2} \tag{1.249}$$

and

$$\lambda(t) = -V\left(t - \frac{T}{2}\right) + \frac{V}{4f} \quad \text{for} \quad \frac{T}{2} \le t \le T. \tag{1.250}$$

The rms value of the square-wave inductor voltage is obtained as

$$V_{rms} = \sqrt{\frac{1}{T}\int_0^T v_L^2 dt} = \sqrt{\frac{1}{T}\int_0^T V^2 dt} = V. \tag{1.251}$$

For the core saturation,

$$\lambda_s = \frac{\Delta\lambda_m}{2} = \frac{V_{max}}{4f_{min}} = N\phi_s = NA_cB_s, \tag{1.252}$$

where $V_{max} = V_{Lrms(max)}$ for the square-wave inductor voltage. The minimum frequency at which the core can be operated without saturation is given by

$$f_{min} = \frac{V_{max}}{4\lambda_s} = \frac{V_{max}}{4NA_cB_s} = \frac{V_{max}}{K_fNA_cB_s}, \tag{1.253}$$

where the waveform factor of the square-wave inductor voltage is

$$K_f = 4. \tag{1.254}$$

The maximum peak voltage of the square-wave inductor voltage at the operating frequency f is

$$V_{max} = V_{rms} = 4fNA_cB_s. \tag{1.255}$$

In general, the minimum core cross-sectional area is given by

$$A_c = \frac{V_{Lrms}}{K_f f_{min} N B_{pk}} = \frac{V_{Lrms}}{K_f f_{min} N (B_{DC} + B_m)}, \tag{1.256}$$

where

$$B_{pk} = B_{DC} + B_m \le B_s \quad \text{for} \quad T \le T_{max} \tag{1.257}$$

and K_f is the waveform coefficient of the inductor voltage. The peak value of the flux density B_{pk} must be lower than B_s at the maximum operating temperature T_{max} to avoid the core saturation. The amplitude of the AC component of the flux density B_m must be limited to avoid core saturation or to reduce core loss. As the amplitude of the AC component of the flux density B_m increases, the core loss also increases.

The saturation flux density B_s limits the maximum amplitude of the magnetic field intensity

$$H_s = H_{m(max)} = \frac{B_s}{\mu_{rc}\mu_0} = \frac{NI_{Lm(max)}}{l_c}. \tag{1.258}$$

The maximum amplitude of the current in the winding at which the core saturates is

$$I_{SAT} = I_{m(max)} = \frac{l_cB_s}{\mu_{rc}\mu_0N}. \tag{1.259}$$

As the amplitude of the inductor current I_{Lm} increases, the amplitude of the magnetic field H_m also increases. To avoid core saturation,

$$NI_{Lm(max)} < \frac{B_sl_c}{\mu_{rc}\mu_0}. \tag{1.260}$$

When a core with an air gap is used, both amplitudes H_m and I_{Lm} can be increased to

$$H_{m(max)} = \frac{B_s}{\mu_{re}\mu_0} \tag{1.261}$$

and

$$I_{SAT} = I_{Lm(max)} = \frac{l_cB_s}{N\mu_{re}\mu_0}, \tag{1.262}$$

where μ_{re} is the effective relative permeability of a gapped core.

1.7.3 Core Saturation for Rectangular Wave Inductor Voltage

Consider the situation, where the inductor voltage waveform is a rectangular wave whose high level is V_H and low level is $-V_L$, as depicted in Fig. 1.16. The magnetic flux linkage is an asymmetrical triangular wave, as shown in Fig. 1.16. For the first part of the cycle,

$$v_L = V_H \quad \text{for} \quad 0 \leq t \leq DT \tag{1.263}$$

and

$$\lambda(t) = \int_0^t v_L(t)dt + \lambda(0) = \int_0^t V_H dt + \lambda(0) = V_H t + \lambda(0) \quad \text{for} \quad 0 \leq t \leq DT, \tag{1.264}$$

where D is the duty cycle. The flux linkage at $t = DT$ is given by

$$\lambda(DT) = V_H DT + \lambda(0). \tag{1.265}$$

For the second part of the cycle,

$$v_L = -V_L \quad \text{for} \quad DT \leq t \leq T \tag{1.266}$$

and

$$\begin{aligned}\lambda(t) &= \int_{DT}^t v_L(t)dt + \lambda(DT) = \int_{DT}^t (-V_L)dt + \lambda(DT)\\ &= -V_L(t - DT) + \lambda(DT) \quad \text{for} \quad DT \leq t \leq T.\end{aligned} \tag{1.267}$$

Hence, the peak-to-peak value of the magnetic flux linkage is

$$\Delta\lambda = \lambda(DT) - \lambda(0) = V_H DT + \lambda(0) - \lambda(0) = V_H DT = \frac{DV_H}{f}. \tag{1.268}$$

The rms value of the inductor voltage is

$$V_{Lrms} = \sqrt{\frac{1}{T}\int_0^T v_L^2 dt} = \sqrt{\frac{1}{T}\left(\int_0^{DT} V_H^2 dt + \int_{DT}^T V_L^2 dt\right)} = \sqrt{DV_H^2 + (1 - D)V_L^2}. \tag{1.269}$$

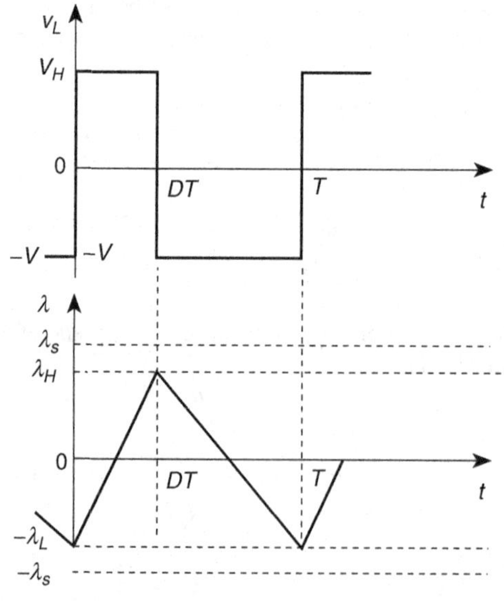

Figure 1.16 Waveforms of the rectangular inductor voltage and the corresponding magnetic flux linkage

Using the volt–second balance law, we obtain

$$V_H DT = V_L(1 - D)T, \tag{1.270}$$

yielding

$$\frac{V_L}{V_H} = \frac{D}{1 - D}. \tag{1.271}$$

Therefore,

$$V_{Lrms} = V_H \sqrt{\frac{D}{1 - D}} = V_L \sqrt{\frac{1 - D}{D}}. \tag{1.272}$$

The flux linkage at the beginning of the core saturation is

$$\lambda_s = \frac{\Delta\lambda_{max}}{2} = \frac{DV_H}{2f_{min}} = N\phi_s = NA_c B_s. \tag{1.273}$$

Hence, the minimum operating frequency is

$$f_{min} = \frac{DV_H}{2NA_c B_s} = \frac{V_{Lrms}}{NA_c B_s} \frac{\sqrt{D(1 - D)}}{2} = \frac{V_{Lrms}}{K_f NA_c B_s}, \tag{1.274}$$

where the waveform coefficient is

$$K_f = \frac{2}{\sqrt{D(1 - D)}}. \tag{1.275}$$

The minimum cross-sectional area is given by

$$A_c = \frac{V_{Lrms}}{K_{fmax} f B_s}. \tag{1.276}$$

Figure 1.17 shows a plot of K_f as a function of the duty cycle D. The minimum value of K_f occurs at $D = 0.5$. Figure 1.18 depicts $1/K_f$ as a function of D. The core cross-sectional area is proportional to $1/K_f$. The minimum value of the core cross-sectional area occurs at $D = 0.5$.

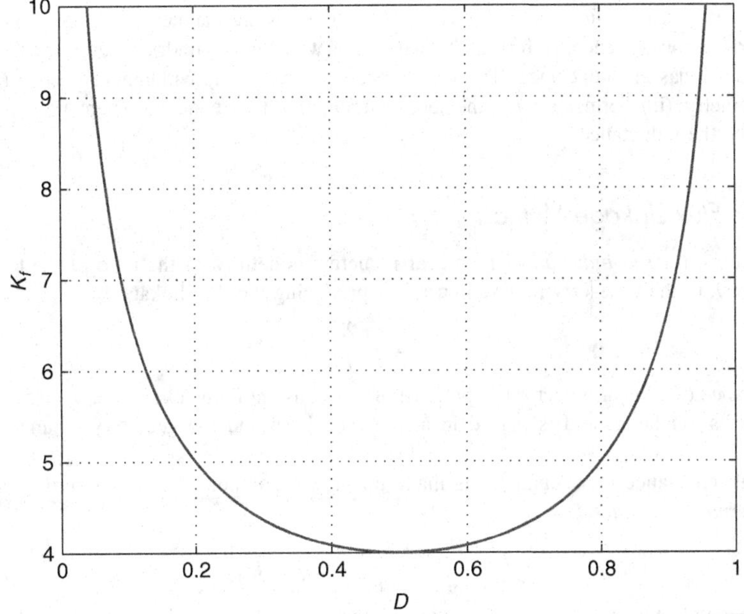

Figure 1.17 Waveform coefficient K_f as a function of duty cycle D

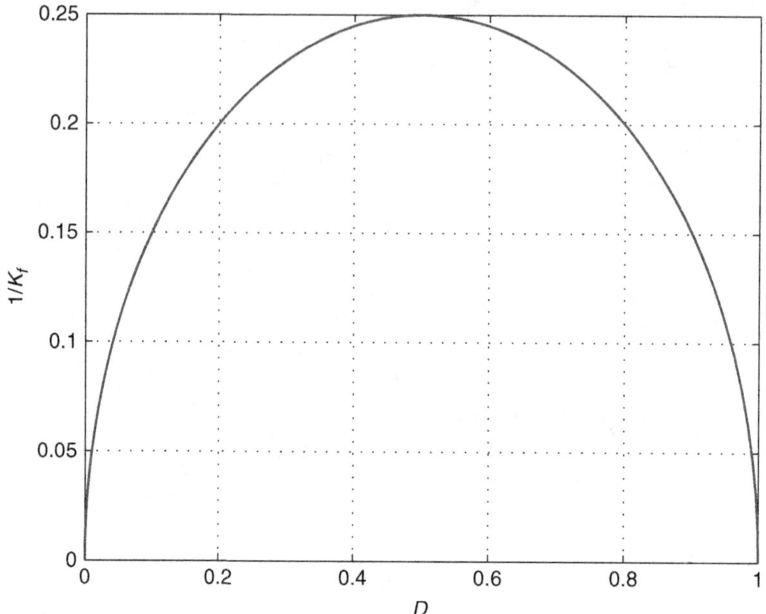

Figure 1.18 Coefficient $1/K_f$ as a function of duty cycle D

1.8 Inductance

1.8.1 Definitions of Inductance

An inductor is a two-terminal passive device that is able to store magnetic field and magnetic energy in this field. A coil is generally formed by winding a wire on a cylindrical former, called a bobbin. Any conductor has an inductance. The inductance depends on (i) winding geometry, (ii) core and bobbin geometry, (iii) permeability of the core material, and (iv) frequency. There are several methods to determine the inductance.

Magnetic Flux Linkage Method

The *inductance* (or *self-inductance*) for linear inductors is defined as the ratio of the total magnetic flux linkage λ to the time-varying (AC) current i producing the flux linkage

$$L = \frac{\lambda}{i}. \tag{1.277}$$

The inductance of a linear inductor is a proportionality constant in the expression $\lambda = Li$. An inductor is linear if its magnetic field is placed in a linear medium and the geometry of an inductor does not change.

The total inductance of an inductor is made up of two components: an *external inductance* L_{ext} and an *internal inductance* L_{int}

$$L = \frac{\lambda_{ext}}{i} + \frac{\lambda_{int}}{i} = L_{ext} + L_{int}. \tag{1.278}$$

The external inductance $L_{ext} = \lambda_{ext}/i$ is due to the external magnetic energy stored in the magnetic field outside the conductor. This inductance is usually independent of frequency. The internal

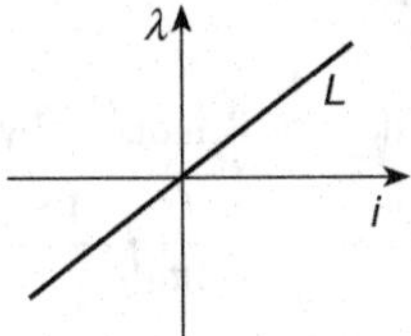

Figure 1.19 Magnetic flux linkage λ as a function of current i producing the flux linkage for linear inductors

inductance $L_{int} = \lambda_{int}/i$ is due to the magnetic energy stored in the internal magnetic field inside the conductor. This inductance depends on the frequency because the magnetic field intensity H distribution inside the conductor is a function of frequency due to skin effect. The internal inductance usually decreases with frequency.

A single conductor carrying an AC current i is linked by its own magnetic flux. For linear inductors, the flux linkage λ is proportional to the current i, resulting in $\lambda = Li$. The inductance L is the slope of the λ-i characteristic, as illustrated in Fig. 1.19. This characteristic is analogous to the resistor characteristic $v = Ri$ or the capacitor characteristic $Q = Cv$. A circuit that is designed to have a self-inductance is called an inductor. An inductor has a self-inductance of 1 H if a current of 1 A produces a flux linkage of 1 V·s (or 1 Wb·turn).

A change in the current flowing through the inductor produces an induced EMF, called an electromotance, or voltage

$$v = \oint_C \mathbf{E} \cdot d\mathbf{l} = \frac{d\lambda}{dt} = L\frac{di}{dt}. \tag{1.279}$$

An inductor has a self-inductance of 1 H if the change in the inductor current at a rate of 1 A/s produces a voltage difference between its terminals of 1 V. The inductance L is a function of the number of turns N, core permeability μ_{rc}, core geometry, and frequency f.

The inductance can be defined as

$$L = \frac{\lambda}{i} = \frac{N}{i}\int\int_S \mathbf{B} \cdot d\mathbf{S}. \tag{1.280}$$

The magnetic field produced by a current-carrying conductor links itself. The associated inductance is called a self-inductance. In some cases, the magnetic flux links only a part of the current and the inductance is defined as

$$L = \frac{1}{i}\int\int_S \frac{i_{enc}}{i} d\phi. \tag{1.281}$$

The voltage across the inductance is

$$v_L = \frac{d\lambda}{dt} = N\frac{d\phi}{dt} = N\frac{d\phi}{di_L}\frac{di_L}{dt} = L\frac{di_L}{dt}. \tag{1.282}$$

The self-inductance L relates the voltage induced in an inductor v_L to the time-varying current i_L flowing through the same inductor.

Reluctance Method

The inductance of an inductor can be determined using the core reluctance $\mathcal{R}$ or the core permeance $\mathcal{P}$

$$L = \frac{\lambda}{i} = \frac{N\phi}{i} = \frac{NA_cB}{i} = \frac{NA_c\mu H}{i} = \frac{NA_c\mu Ni}{l_c i} = \frac{A_c\mu}{l_c}N^2$$

$$= \frac{N^2}{\mathcal{R}} = \mathcal{P}N^2 = \frac{\mu_{rc}\mu_0 A_c N^2}{l_c}. \tag{1.283}$$

If $N = 1$, $L = \mathcal{P} = 1/\mathcal{R}$.

Biot–Savart's Law Method

Using (1.101), the total magnetic field density **B** induced by a current I is obtained

$$\mathbf{B} = \mu\mathbf{H} = \frac{I}{4\pi}\int_l \frac{d\mathbf{l} \times \mathbf{a}_R}{R^2}. \tag{1.284}$$

The magnetic flux inside the core with a cross-sectional area A_c is

$$\phi = A_c B = \frac{IA_c}{4\pi}\int_l \frac{d\mathbf{l} \times \mathbf{a}_R}{R^2}. \tag{1.285}$$

The magnetic flux linkage is

$$\lambda = N\phi = NA_c B = \frac{IA_c N}{4\pi}\int_l \frac{d\mathbf{l} \times \mathbf{a}_R}{R^2}. \tag{1.286}$$

Hence, the inductance is

$$L = \frac{\lambda}{I} = \frac{A_c N}{4\pi}\int_l \frac{d\mathbf{l} \times \mathbf{a}_R}{R^2}. \tag{1.287}$$

Magnetic Energy Method

The instantaneous magnetic energy stored in a magnetic device is

$$W_m = \frac{1}{2}LI_m^2 = \frac{1}{2}\int\int\int_V (\mathbf{B} \cdot \mathbf{H}^*)dV, \tag{1.288}$$

yielding the inductance

$$L = \frac{2W_m}{I_m^2} = \frac{1}{I_m^2}\int\int\int_V (\mathbf{B} \cdot \mathbf{H}^*)dV, \tag{1.289}$$

where I_m is the amplitude of the current flowing in the closed path, and W_m is the energy stored in the magnetic field produced by the current flowing through the inductor

$$W_m = \frac{\mu}{2}\int\int\int_V H^2 dV = \frac{1}{2\mu}\int\int\int_V B^2 dV. \tag{1.290}$$

For a linear inductor, $B = \mu H$ and

$$W_m = \frac{1}{2}LI_m^2 = \frac{1}{2}\int\int\int_V \mu H^2 dV = \frac{1}{2\mu}\int\int\int_V B^2 dV, \tag{1.291}$$

resulting in

$$L = \frac{2W_m}{I_m^2} = \frac{1}{I_m^2}\int\int\int_V \mu H^2 dV = \frac{1}{\mu I_m^2}\int\int\int_V B^2 dV. \tag{1.292}$$

The magnetic energy method to determine the inductance is impractical in many situations because of the lack of finite volume over which to integrate the magnetic field.

Example 1.5

Internal Self-Inductance of Round Conductor Determine the internal self-inductance of a round solid conductor of radius r_o and length l at low frequencies.

Solution: The cylindrical coordinates (r, φ, z) will be used to solve this problem. Assume that a sinusoidal current $i = I_m \sin \omega t$ flows through the conductor. From Example 1.1, the magnetic field intensity inside the conductor with the current amplitude I_m at low frequencies is given by

$$H_m(r) = \frac{I_m}{2\pi r_o^2} r \quad \text{for} \quad 0 \le r \le r_o \tag{1.293}$$

and the magnetic flux density at low frequencies is

$$B_m(r) = \mu H_m(r) = \frac{\mu_r \mu_0 I_m}{2\pi r_o^2} r \quad \text{for} \quad 0 \le r \le r_o. \tag{1.294}$$

Note that $dy = r\tan(d\varphi) \approx r d\varphi$ and $dV = (dr)(dy)(dz) = (dr)(rd\varphi)(dz) = rdrd\varphi dz$. The internal self-inductance of round solid wire is

$$L_{int} = \frac{1}{I_m^2}\int\int\int_V \mathbf{H}\cdot\mathbf{B}dV = \frac{1}{I_m^2}\int\int\int_V \mu H_m(r)^2 dV = \frac{1}{I_m^2}\int\int\int_V \frac{\mu I_m^2 r^2}{(2\pi r_o)^2} dV$$

$$= \frac{\mu}{(2\pi r_o^2)^2}\int_0^l\int_0^{2\pi}\int_0^{r_o} r^2 r dr d\varphi dz = \frac{\mu}{(2\pi r_o^2)^2}\int_0^l dz \int_0^{2\pi} d\varphi \int_0^{r_o} r^3 dr = \frac{\mu l}{8\pi}. \tag{1.295}$$

For copper conductors, $L_{int}/l = \mu_0/(8\pi) = 4\pi \times 10^{-7}/(8\pi) = 10^{-7}/2 = 50$ nH/m $= 0.5$ nH/cm. At high frequencies, the current density is not uniform, the magnetic field density does not increase linearly inside the conductor and is altered, the energy stored in the wire decreases, and the internal inductance decreases with frequency due to skin effect, as explained in Chapter 3.

The amplitude of the magnetic field intensity outside the conductor is

$$H_m(r) = \frac{I_m}{2\pi r} \quad \text{for} \quad r \ge r_o. \tag{1.296}$$

The external inductance is

$$L_{ext} = \frac{1}{I_m^2}\int\int\int_V \mu H_m^2 dV = \frac{1}{I_m^2}\int\int\int_V \frac{\mu I_m^2}{4\pi^2 r^2} dV = \frac{\mu}{4\pi^2}\int_0^l dz \int_0^{2\pi}\int_{r_O}^{r_1} \frac{dr}{r}$$

$$= \frac{\mu l}{2\pi}(\ln r_1 - \ln r_o) = \frac{\mu l}{2\pi}\ln\left(\frac{r_1}{r_o}\right). \tag{1.297}$$

As $r_1 \to \infty$, $L_{ext} \to \infty$.

Vector Magnetic Potential Method

The inductance can be determined using the vector magnetic potential **A**

$$L = \frac{1}{I_m^2}\int\int\int_V \mathbf{A}\cdot\mathbf{J}dV. \tag{1.298}$$

The vector magnetic potential is given by

$$\mathbf{A}\,(\mathbf{r}) = \frac{\mu}{4\pi}\int\int\int_V \frac{\mathbf{J}(\mathbf{r})}{R} dV. \tag{1.299}$$

Hence, the inductance is given by

$$L = \frac{1}{I^2}\int\int\int_V \left[\frac{\mu}{4\pi}\int\int\int_V \frac{\mathbf{J}(\mathbf{r})}{R} dV\right]\cdot\mathbf{J}(\mathbf{r})dV. \tag{1.300}$$

Small-Signal Inductance

For nonlinear inductors, $\lambda = f(i)$ is a nonlinear function. The small-signal (or incremental) inductance of a nonlinear inductor is defined as the ratio of the infinitesimal change in the flux linkage $d\lambda$ to the infinitesimal change in the current dl producing it at a given operating point $Q(I_{DC}, \lambda_{DC})$

$$L = \left.\frac{d\lambda}{di}\right|_Q. \tag{1.301}$$

Inductors with ferrous cores are nonlinear because the permeability depends on the applied magnetic field H. Figure 1.20 shows a plot of the magnetic flux linkage λ as a function of current i for nonlinear inductors. At low values of current, the core is not saturated and the relative permeability is high,

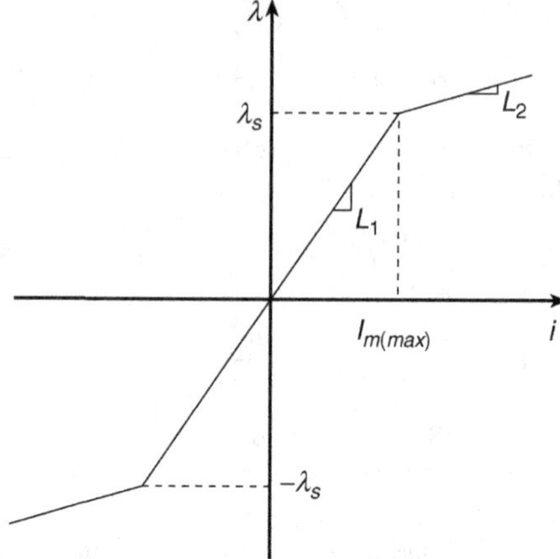

Figure 1.20 Magnetic flux linkage λ as a function of current i producing the flux linkage for nonlinear inductors

resulting in a high slope of the λ-i curve and a large inductance L_1. When the core saturates, the relative permeability μ_{rc} becomes equal to 1, the slope of the λ-i curve decreases, and the inductance decreases to a lower value L_2.

Example 1.6

An inductor is wound on a CC core (see Fig. 2.9) whose cross-sectional area is 2 cm × 2 cm, $l_c =$ 16 cm, core window is 3 cm × 3 cm, and $\mu_{rc} = 100$. The inductor has 10 turns. The core has no air gap. There is a magnetic flux in the core ϕ_c and a leakage flux ϕ_l in the air around the winding. Estimate the inductance using the reluctance method.

Solution: The total magnetic flux consists of the magnetic flux inside the core ϕ_c and the leakage magnetic flux ϕ_l

$$\phi = \phi_c + \phi_l. \tag{1.302}$$

The total reluctance $\mathcal{R}$ is equal to the parallel combination of the core reluctance $\mathcal{R}_c$ and the leakage reluctance $\mathcal{R}_l$. Hence, the inductance is

$$L = \frac{\lambda}{i} = \frac{N\phi}{i} = \frac{N(\phi_c + \phi_l)}{i} = N^2\left(\frac{1}{R_c} + \frac{1}{R_l}\right) = \frac{N^2}{\mathcal{R}} = N^2\left(\frac{\mu_{rc}\mu_0 A_c}{l_c} + \frac{\mu_0 A_l}{l_l}\right)$$

$$= \frac{\mu_{rc}\mu_0 A_c N^2}{l_c}\left[1 + \frac{1}{\mu_{rc}}\left(\frac{A_l}{A_c}\right)\left(\frac{l_c}{l_l}\right)\right] = L_c\left[1 + \frac{1}{\mu_{rc}}\left(\frac{A_l}{A_c}\right)\left(\frac{l_c}{l_l}\right)\right]. \tag{1.303}$$

Let $l_l = l_c/2$ and $A_l = 4A_c$. In this case, the inductance is given by

$$L = \frac{\mu_{rc}\mu_0 A_c N^2}{l_c}\left[1 + \frac{1}{\mu_{rc}}\left(\frac{A_l}{A_c}\right)\left(\frac{l_c}{l_l}\right)\right]$$

$$= \frac{100 \times 4\pi \times 10^{-7} \times 4 \times 10^{-4} \times 10^2}{16 \times 10^{-2}}\left(1 + \frac{4 \times 2}{100}\right) = 3.1416 \times 10^{-6}(1 + 0.08)$$

$$= 33.929\ \mu\text{H}. \tag{1.304}$$

The inductance is increased by 8% due to the leakage magnetic flux. As μ_{rc} increases, the leakage flux effect on the inductance is reduced.

1.8.2 Inductance of Solenoid

A solenoid is made up of insulated wire wound in the shape of a cylinder to obtain N turns. Each turn acts like a magnetic dipole. A series connection of winding turns is used to increase the magnetic flux density B. The magnetic field lines are parallel to the solenoid axis. The magnetic field outside the solenoid is nearly zero. Neglecting the end effects, the amplitude of the magnetic flux density inside a long solenoid is uniform and it is given by

$$B_m = \mu H_m = \frac{\mu N I_m}{l_c}. \tag{1.305}$$

The amplitude of the magnetic flux inside the solenoid is

$$\phi_m = A_c B_m = \frac{\mu N I_m A_c}{l_c} = \frac{\pi \mu N r^2 I_m}{l_c}. \tag{1.306}$$

The amplitude of the flux linkage is

$$\lambda_m = N\phi_m = \frac{\mu N^2 I_m A_c}{l_c} = \frac{\pi \mu_{rc} \mu_0 N^2 r^2 I_m}{l_c} \tag{1.307}$$

where $A_c = \pi r^2$ is the core cross-sectional area of a round core and r is the mean coil radius. A long, tightly wound solenoid can be modeled by an equivalent current sheet that carries total current NI_m. The magnetic field for an infinitely long solenoid is uniform throughout the inside of the solenoid. The inductance of a long solenoid (theoretically, almost infinitely long) with a core and without an air gap in the core at low frequencies is

$$L_\infty = \frac{\lambda_m}{I_m} = \frac{\mu_{rc}\mu_0 A_c N^2}{l_c} = \frac{\pi \mu_{rc}\mu_0 r^2 N^2}{l_c} = \frac{N^2}{l_c/(\mu_{rc}\mu_0 A_c)} = \frac{N^2}{\mathcal{R}}$$
$$= \frac{\mu_{rc}\mu_0 A_c N^2}{Np} = \frac{\mu_{rc}\mu_0 A_c N}{p} \quad \text{for} \quad l_c \gg 2d, \tag{1.308}$$

where $l_c = Np$ is the mean core length and p is the winding pitch, equal to the distance between the centers of two adjacent conductors, μ_{rc} is the relative permeability of the core, and N is the total number of turns. The inductance L is proportional to the core relative permeability μ_{rc}, the square of the number of turns N^2, and the ratio of the core cross-sectional area to the MPL A_c/l_c.

The inductance of intermediate length-to-radius ratio is lower than that of an infinitely long round solenoid. As the inductor length increases, the magnetic field intensity $H = Ni/l_c$ decreases. Therefore, the inductance and the magnetic energy density $w_m = \frac{1}{2}BH = \frac{1}{2}\mu H^2$ also decrease. The inductance and the magnetic energy density stored in the magnetic field are inversely proportional to the length of the magnetic field. As r/l_c increases, L/L_∞ decreases, where L_∞ is the inductance of an infinitely long solenoid. For example, $K = L/L_\infty = 0.85$ for $r/l_c = 0.2$, $K = 0.74$ for $r/l_c = 0.4$, $K = 0.53$ for $r/l_c = 1$, $K = 0.2$ for $r/l_c = 5$, and $K = 0.12$ for $r/l_c = 10$. For r/l_c up to 2 or 3, a first-order approximation is

$$K = \frac{L}{L_\infty} \approx \frac{1}{1 + 0.9\frac{r}{l_c}}, \tag{1.309}$$

resulting in $L \approx KL_\infty = L_\infty/(1 + 0.9r/l_c)$. The inductance of a round single-layer solenoid of a finite length l_c with intermediate length-to-resistance ratio can be approximated by Wheeler's or Nagaoka formula [44], which is correct to within 1% for $r/l_c < 1.25$ or $l_c/(2r) > 0.4$

$$L = \frac{L_\infty}{1 + 0.9\frac{r}{l_c}} = \frac{\mu_{rc}\mu_0 A_c N^2}{l_c\left(1 + 0.9\frac{r}{l_c}\right)} = \frac{\pi\mu_{rc}\mu_0 r^2 N^2}{l_c\left(1 + 0.9\frac{r}{l_c}\right)} = \frac{\pi\mu_{rc}\mu_0 r^2 N^2}{l_c + 0.9r} \ \text{(H)}$$
$$= \frac{0.4\pi^2\mu_{rc} r^2 N^2}{l_c + 0.9r} \ (\mu\text{H}) \quad \text{for} \quad \frac{r}{l_c} < 1.25. \tag{1.310}$$

Figure 1.21 shows a plot of L/L_∞ as a function of r/l_c. As the ratio of the external diameter to the internal diameter decreases, the inductance also decreases.

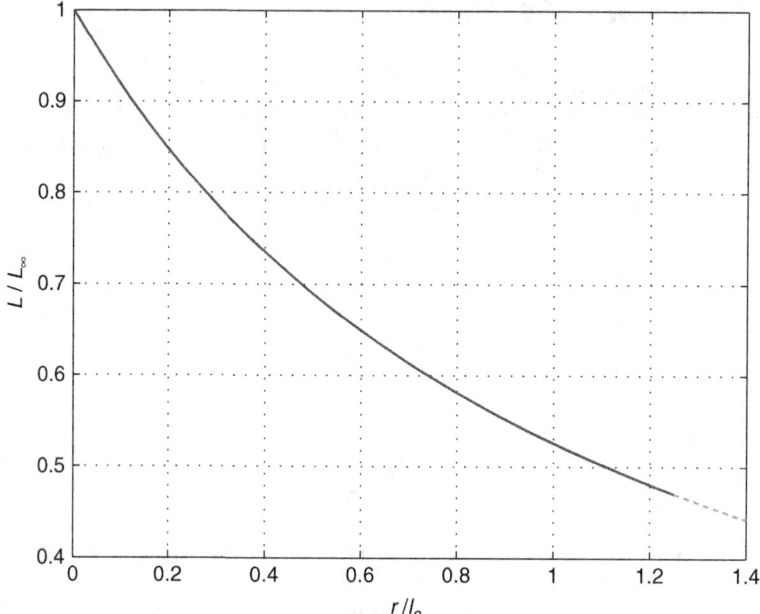

Figure 1.21 Plot of L/L_{∞} as a function of r/l_c

The inductance of a multilayer solenoid is given by

$$L = \frac{0.8\mu\pi r^2 N^2}{l_c + 0.9r + b}, \tag{1.311}$$

where b is the thickness of all layers (or coil build) and r is the average radius of the winding.

A more accurate equation for the inductance of a multilayer inductor is

$$L = \frac{\mu\pi r^2 N^2}{l_c} \frac{1}{1 + 0.9\left(\frac{r}{l_c}\right) + 0.32\left(\frac{b}{r}\right) + 0.84\left(\frac{b}{l_c}\right)}. \tag{1.312}$$

The inductance predicted by this equation is within 2% of the exact value.

The inductance of a short solenoid is given by

$$L = \mu r N^2 \left[\ln\left(\frac{8r}{a}\right) - 2\right] \tag{1.313}$$

where a is the wire radius and r is the outer radius of the solenoid and N is the number of turns. This equation is valid for $N = 1$.

Example 1.7

An air-core solenoid has $N = 20$, $l_c = 15$ cm, and $r = 3$ cm. Find the inductance.

Solution: The inductance of the solenoid is

$$L = \frac{\pi\mu_{rc}\mu_0 r^2 N^2}{l_c\left(1 + 0.9\frac{r}{l_c}\right)} = \frac{\pi \times 1 \times 4\pi \times 10^{-7} \times (3 \times 10^{-2})^2 \times 20^2}{15 \times 10^{-2}\left(1 + 0.9 \times \frac{3\times10^{-2}}{15\times10^{-2}}\right)} = 8.0295\ \mu\text{H}. \tag{1.314}$$

Note that the inductance calculated in this example is about 15% less than the inductance calculated for a very long inductor because $K = 0.8475$.

1.8.3 Inductance of Inductor with Toroidal Core

A toroidal inductor with a rectangular cross section is shown in Fig. 1.22. The dimensions of the magnetic core are: a is the inner radius, b is the outer radius, and h is the toroid height. The toroid is symmetrical about its axis. The RHR shows that the magnetic flux ϕ is mainly circulating in the φ direction. An idealized toroidal inductor can be thought as a finite length solenoid bent around to close on itself to form a doughnut shape. Assume that the inductor is closely wound coil. Consider a general contour C of radius r, $a \leq r \leq b$. Applying Ampère's law,

$$\oint_C \mathbf{H} \cdot d\mathbf{l} = NI. \tag{1.315}$$

It can be observed that

$$dl = rd\varphi. \tag{1.316}$$

Hence,

$$\oint_C \mathbf{H} \cdot d\mathbf{l} = \int_0^{2\pi} Hrd\varphi = Hr\int_0^{2\pi} d\varphi = 2\pi rH. \tag{1.317}$$

Since the path of integration encircles the total current NI, we obtain

$$2\pi rH = NI. \tag{1.318}$$

Hence, the magnetic field intensity inside the toroidal core is

$$H = \frac{NI}{2\pi r} \quad \text{for} \quad a \leq r \leq b \tag{1.319}$$

and the magnetic flux density inside the toroidal core is given by

$$B = \mu H = \frac{\mu NI}{2\pi r} \quad \text{for} \quad a \leq r \leq b. \tag{1.320}$$

Since $dS = (dh)(dr)$, the magnetic flux inside the toroidal core is

$$\phi = \int\int_S BdS = \int_a^b \int_0^h \left(\frac{\mu NI}{2\pi r}\right)(dh)(dr) = \frac{\mu NI}{2\pi}\int_0^h dh \int_a^b \frac{dr}{r} = \frac{\mu NIh}{2\pi}\int_a^b \frac{dr}{r}$$
$$= \frac{\mu NIh}{2\pi}\ln\left(\frac{b}{a}\right). \tag{1.321}$$

The flux linkage of the toroidal inductor is

$$\lambda = N\phi = \frac{\mu hN^2 I}{2\pi}\ln\left(\frac{b}{a}\right), \tag{1.322}$$

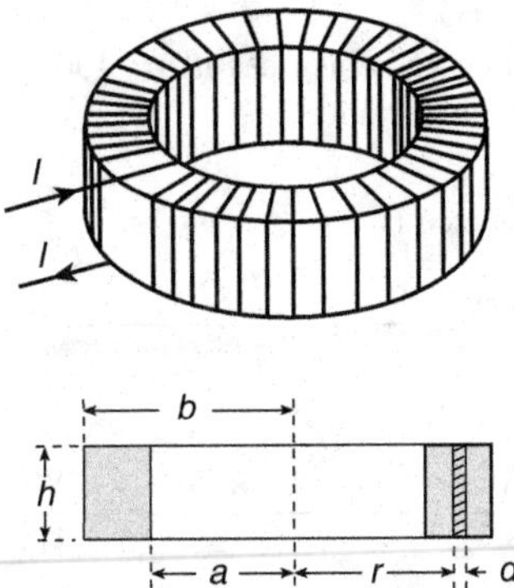

Figure 1.22 Toroidal inductor

resulting in the inductance of a toroidal coil

$$L = \frac{\lambda}{I} = \frac{\mu_{rc}\mu_0 h N^2}{2\pi} \ln\left(\frac{b}{a}\right). \tag{1.323}$$

Alternatively, we may assume that the magnetic flux density within the core is uniform and equal to the average value of the magnetic flux density

$$B = \frac{\phi}{A_c}, \tag{1.324}$$

where $A_c = h(b - a)$ is the cross-sectional area of the core. The magnetic field intensity is

$$H = \frac{B}{\mu} = \frac{\phi}{\mu A_c} = \frac{\phi}{\mu h(b - a)}. \tag{1.325}$$

Using Ampère's law,

$$\oint_C \mathbf{H} \cdot d\mathbf{l} = NI \tag{1.326}$$

yielding

$$\frac{\phi l_c}{\mu A_c} = NI. \tag{1.327}$$

The mean radius of the core is $R = (a + b)/2$ and $l_c = 2\pi R = \pi(a + b)$ is the mean length of the magnetic flux. The magnetic flux inside the core is

$$\phi = \frac{\mu NIA_c}{l_c} = \frac{\mu NIh(b - a)}{\pi(b + a)}, \tag{1.328}$$

which gives the flux linkage

$$\lambda = N\phi = \frac{\mu Ih(b - a)N^2}{\pi(b + a)}, \tag{1.329}$$

Hence, the inductance is

$$L = \frac{\lambda}{I} = \frac{\mu h(b - a)N^2}{\pi(b + a)}. \tag{1.330}$$

An empirical equation for an inductor with a toroidal core is

$$L = 4.6\mu_{rc} N^2 h \log\left(\frac{OD}{ID}\right) \times 10^{-13}\ \text{H}, \tag{1.331}$$

where H is the height of the core in meter, and ID and OD are the inner and outer diameters of the core in meters.

Example 1.8

An inductor is wound on a toroidal core, which has $\mu_{rc} = 150$, $h = 1$ cm, $a = 4$ cm, and $b = 5$ cm. The inductor has 20 turns. Find the inductance.

Solution: Using (1.322), the inductance is

$$L = \frac{\mu_{rc}\mu_0 h N^2}{2\pi} \ln\left(\frac{b}{a}\right) = \frac{150 \times 4\pi \times 10^{-7} \times 10^{-2} \times 20^2}{2\pi} \ln\left(\frac{5}{4}\right) = 26.777\ \mu\text{H}. \tag{1.332}$$

From (1.330),

$$L = \frac{\mu_{rc}\mu_0 h(b - a)N^2}{\pi(b + a)} = \frac{150 \times 2\pi \times 10^{-7} \times 10^{-2} \times (5 - 4) \times 10^{-2} 20^2}{\pi(5 + 4) \times 10^{-2}}$$

$$= 26.667\ \mu\text{H}. \tag{1.333}$$

1.8.4 Inductance of Inductor with Torus Core

The inductance of a torus coil (with a round cross section) can be described by the expression for the inductance of a long solenoid

$$L = \frac{\mu_{rc}\mu_0 A_c N^2}{l_c} = \frac{\mu_{rc}\mu_0 A_c N^2}{2\pi R}, \tag{1.334}$$

where $R = (a+b)/2$ is the mean radius of the core, $l_c = 2\pi R = \pi(a+b)$, $r_o = (b-a)/2$ is the radius of the core cross section, and $A_c = \pi r_o^2 = \pi(b-a)^2/4$ is the cross-sectional area of the core. Hence,

$$L = \frac{\mu_{rc}\mu_0 N^2 (b-a)^2}{4(a+b)}. \tag{1.335}$$

An alternative expression for the inductance of an inductor with a torus core is

$$L = \mu N^2 [R - \sqrt{R^2 - r_o^2}]. \tag{1.336}$$

1.8.5 Inductance of Inductor with Pot Core

The geometry of an inductor with a pot core is very complex and the inductance of these inductors can be determined only approximately. The core cross-sectional area of the pot core is approximately equal to the cross-sectional area of the center post

$$A_c = \frac{\pi d^2}{4}, \tag{1.337}$$

where d is the diameter of the center post. The average diameter of the mean magnetic path is given by

$$D_{av} = \frac{D_i + D_o}{2}, \tag{1.338}$$

where D_i is the inner diameter of the outer core area and D_o is the outer diameter of the outer core area. The mean MPL is given by

$$l_c = D_{av} + 4h = \frac{D_i + D_o}{2} + 4h, \tag{1.339}$$

where h is the height of the core halve. The inductance of an inductor with a pot core can be approximated by

$$L = \frac{\mu_{rc}\mu_0 A_c N^2}{l_c} = \frac{\pi \mu_{rc}\mu_0 d^2 N^2}{2(D_i + D_o) + 16h}. \tag{1.340}$$

1.8.6 Inductance Factor

Equation (1.308) for the inductance can be written as

$$L = \frac{\mu_{rc}\mu_0 A_c N^2}{l_c} = A_L N^2. \tag{1.341}$$

The *specific inductance* of a core, also called the *core inductance factor*, is defined as the inductance per single turn

$$A_L = \frac{L}{N^2} = \frac{\mu_{rc}\mu_0 A_c}{l_c} = \frac{1}{\mathcal{R}} = \mathcal{P} \left(\frac{\mathrm{H}}{\mathrm{turn}^2}\right). \tag{1.342}$$

Each core of different materials, shapes, and sizes will have a unique value of A_L, some of which are not easy to predict analytically, especially for complex core shapes. Core manufacturers give the values of A_L in data sheets.

The specific inductance (or the inductance index) A_L is usually specified in Henry per turn, in millihenry per 1000 turns, or in microhenry per 100 turns for cores without and with air gaps. If the specific inductance A_L is expressed in Henry per turn, the number of turns is given by

$$N = \sqrt{\frac{L(\mathrm{H})}{A_L}}. \tag{1.343}$$

If the specific inductance $A_{L(1000)}$ is expressed in millihenry per 1000 turns, the inductance is given by

$$L = \frac{A_{L(1000)}N^2}{(1000)^2}\ (\mathrm{mH}) \tag{1.344}$$

and the number of turns is

$$N = 1000\sqrt{\frac{L(\mathrm{mH})}{A_{L(1000)}}}. \tag{1.345}$$

For most ferrite cores, the specific inductance $A_{L(100)}$ is expressed in μH per 100 turns. In this case, the inductance is given by

$$L = \frac{A_{L(100)}N^2}{(100)^2}\ (\mu\mathrm{H}). \tag{1.346}$$

To compute the required number of turns N for a desired inductance L in microhenries, the following formula can be used for ferrite cores

$$N = 100\sqrt{\frac{L(\mu\mathrm{H})}{A_{L(100)}}}. \tag{1.347}$$

Common values of $A_{L(100)}$ are 16, 25, 40, 63, 100, 250, 400, and so on.

Air-core inductors are linear devices because the relationship $B = \mu_0 H$ is linear. In general, inductors with magnetic cores are nonlinear devices as the B–H relationship is nonlinear. However, for $B < B_s$, inductors can be modeled as linear devices.

Example 1.9

The relative permeability of the Ferroxcube ferrite magnetic core material is $\mu_{rc} = 1800$. The rectangular toroidal core made up of this material has the inner diameter $d = 13.1$ mm, the external diameter $D = 23.7$ mm, and the height $h = 7.5$ mm. Find the specific inductance of this core. What is the inductance of the inductor with this core if the number of turns is $N = 10$?

Solution: The MPL of a toroidal core is

$$l_c = \pi\frac{d+D}{2} = \pi\frac{13.1+23.7}{2} = 57.805\ \mathrm{mm} \tag{1.348}$$

and the cross-sectional area of the core is

$$A_c = h\frac{(D-d)}{2} = 7.5\times 10^{-3}\times\frac{(23.7-13.1)\times 10^{-3}}{2} = 39.75\times 10^{-6}\ \mathrm{m}^2. \tag{1.349}$$

Hence, the specific inductance of the core is

$$A_L = \frac{\mu_{rc}\mu_0 A_c}{l_c} = \frac{1800\times 4\pi\times 10^{-7}\times 39.75\times 10^{-6}}{57.805\times 10^{-3}} = 1.5554\ \mu\mathrm{H/turn}\ . \tag{1.350}$$

The inductance at $N = 10$ is

$$L = N^2 A_L = 10^2\times 1.5554\times 10^{-6} = 155.54\ \mu\mathrm{H}. \tag{1.351}$$

1.9 Air Gap in Magnetic Core

1.9.1 Inductance

Gapped core inductors and transformers are useful in a variety of applications, particularly those in which core saturation has to be avoided. The introduction of an air gap into the core of an inductor permits much higher levels of magnetic flux density at the expense of considerably reduced inductance. The overall reluctance of the gapped core $\mathcal{R}$ can be controlled by an air gap length l_g. Therefore, magnetic flux ϕ, magnetic flux density B, and inductance L can also be controlled by the length of the air gap l_g. In addition, gapped cores exhibit enhanced thermal stability and more predictable effective permeability, the overall reluctance, and the inductance.

Air gaps can be bulk or distributed. In a gapped core, a small section of the magnetic flux path is replaced by a nonmagnetic medium, such as air or nylon. It is often filled with a spacer. The air-gap length l_g is usually twice the spacer thickness. Some cores have prefabricated air gaps. Standard values of the air-gap length l_g are 0.5, 0.6, 0.7, ..., 5 mm. The same magnetic flux flows in the core and in the gap. Adding an air gap in a core is equivalent to adding a large gap reluctance in series with the core reluctance, that is, a series reluctor. As a result, the magnitude of the magnetic flux ϕ_m at a fixed value of NI_m is reduced. This effect is analogous to adding a series resistor in an electric circuit to reduce the magnitude of the current at a fixed source voltage.

Figure 1.23a illustrates an inductor whose core has an air gap. An equivalent magnetic circuit of an inductor with an air gap is shown in Fig. 1.23b. The reluctance of the air gap is

$$\mathcal{R}_g = \frac{l_g}{\mu_0 A_c}, \tag{1.352}$$

the reluctance of the core is

$$\mathcal{R}_c = \frac{l_c - l_g}{\mu_{rc}\mu_0 A_c} \approx \frac{l_c}{\mu_{rc}\mu_0 A_c}, \tag{1.353}$$

and the overall reluctance of the core with the air gap is

$$\mathcal{R} = \mathcal{R}_c + \mathcal{R}_g = \frac{l_c}{\mu_{rc}\mu_0 A_c} + \frac{l_g}{\mu_0 A_c} = \frac{l_g + l_c/\mu_{rc}}{\mu_0 A_c} = \frac{l_c}{\mu_{rc}\mu_0 A_c}\left(1 + \frac{\mu_{rc} l_g}{l_c}\right) = F_g \mathcal{R}_c, \tag{1.354}$$

where the *air gap factor* is

$$F_g = \frac{\mathcal{R}}{\mathcal{R}_c} = \frac{\mathcal{R}_c + \mathcal{R}_g}{\mathcal{R}_c} = 1 + \frac{\mathcal{R}_g}{\mathcal{R}_c} = 1 + \frac{\mu_{rc} l_g}{l_c}. \tag{1.355}$$

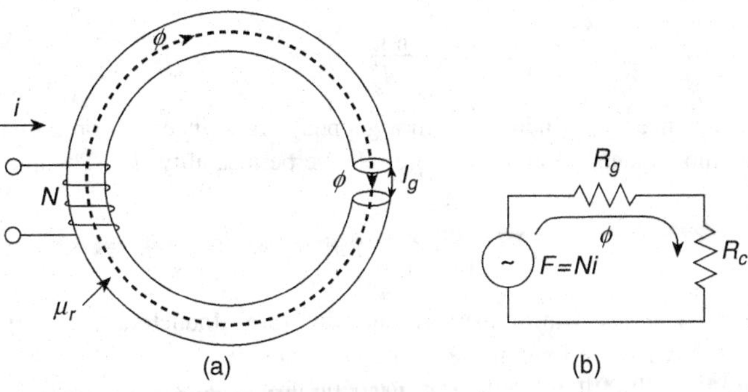

Figure 1.23 Inductor with an air gap. (a) Inductor. (b) Magnetic circuit of an inductor with an air gap

The inductance of a coil with a magnetic core having an air gap at low frequencies is expressed as

$$L=\frac{N^2}{\mathcal{R}}=\frac{N^2}{\mathcal{R}_g+\mathcal{R}_c}=\frac{N^2}{\mathcal{R}_g}\frac{1}{1+\mathcal{R}_c/\mathcal{R}_g}=\frac{N^2}{\frac{l_g}{\mu_0 A_c}+\frac{l_c}{\mu_{rc}\mu_0 A_c}}=\frac{\mu_{rc}\mu_0 A_c N^2}{l_c+\mu_{rc}l_g}=\frac{\mu_0 A_c N^2}{l_g+\frac{l_c}{\mu_{rc}}}$$

$$=\frac{\mu_0 A_c N^2}{l_g\left(1+\frac{l_c}{l_g\mu_{rc}}\right)}=\frac{N^2}{\mathcal{R}_g}\frac{1}{1+\frac{l_c}{l_g\mu_{rc}}}=\frac{\mu_{rc}\mu_0 A_c N^2}{l_c\left(1+\frac{\mu_{rc}l_g}{l_c}\right)}=\frac{\mu_{re}\mu_0 A_c N^2}{l_c}=\frac{\mu_{rc}\mu_0 A_c N^2}{l_c F_g}, \quad (1.356)$$

where the *effective relative permeability* of a core with an air gap is

$$\mu_{re}=\frac{\mu_{rc}}{1+\frac{\mu_{rc}l_g}{l_c}}=\frac{\mu_{rc}}{F_g}. \quad (1.357)$$

For $\mu_{rc}l_g/l_c \gg 1$,

$$\mu_{re}\approx\frac{l_c}{l_g} \quad (1.358)$$

and

$$L\approx\frac{N^2}{\mathcal{R}_g}=\frac{\mu_0 A_c N^2}{l_g}. \quad (1.359)$$

Thus, the inductance with high-permeability cores is dominated by the air gap.

The air gap causes a considerable decrease in the effective relative permeability. However, it produces a more stable effective permeability and reluctance, resulting in a more predictable and stable inductance. The relative permittivity depends on the temperature. As the temperature increases, the relative permeability increases to reach a maximum value and then decreases to 1. In addition, the relative permeability has a large tolerance, typically $\mu_{rc}=\mu_{rc(nom)}\pm 25\%$. For example, inductors used in resonant circuits should be predictable and stable. Usually, at least 95% of the inductance comes from the air gap for high-permeability cores. The length of the air gap is given by

$$l_g=\frac{\mu_0 A_c N^2}{L}-\frac{l_c}{\mu_{rc}}. \quad (1.360)$$

The number of turns of an inductor whose core has an air gap is given by

$$N=\sqrt{\frac{L\left(l_g+\frac{l_c}{\mu_{rc}}\right)}{\mu_0 A_c}}. \quad (1.361)$$

For high-permeability cores, $l_g \gg l_c/\mu_{rc}$, $\mathcal{R}_g \gg \mathcal{R}_c$,

$$\mathcal{R}\approx\mathcal{R}_g, \quad (1.362)$$

and

$$L\approx\frac{\mu_0 A_c N^2}{l_g}=\frac{N^2}{\mathcal{R}_g}. \quad (1.363)$$

Therefore, the inductance of an inductor with a gapped core is inversely proportional to the air-gap length l_g and is almost independent of the core relative permeability μ_{rc}. The number of turns is

$$N\approx\sqrt{\frac{Ll_g}{\mu_0 A_c}} \quad \text{for} \quad l_g \gg \frac{l_c}{\mu_{rc}}. \quad (1.364)$$

The core permeability varies with temperature and flux level. Inductors that carry DC currents and have DC magnetic flux require long air gaps to avoid saturation.

In some applications, the fringing effect of magnetic flux is reduced by multiple air gaps of length $l_{g1}, l_{g2}, l_{g3}, \ldots, l_{gn}$ In this case, the total air-gap length is equal to sum of all air gaps

$$l_g=l_{g1}+l_{g2}+l_{g3}+\ldots+l_{gn}, \quad (1.365)$$

resulting in the effective relative permeability

$$\mu_{re} = \frac{\mu_{rc}}{1 + \frac{\mu_{rc}(l_{g1}+l_{g2}+\ldots+l_{gn})}{l_c}} \tag{1.366}$$

and the inductance of the inductor with the distributed air gap in the core

$$L = \frac{\mu_{rc}\mu_0 A_c N^2}{l_c\left[1 + \frac{\mu_{rc}(l_{g1}+l_{g2}+\ldots+l_{gn})}{l_c}\right]} = \frac{\mu_0 A_c N^2}{l_{g1} + l_{g2} + \ldots + l_{gn} + \frac{l_c}{\mu_{rc}}}. \tag{1.367}$$

1.9.2 Magnetic Field in Air Gap

Using Ampere's law, the MMF of the inductor with an air gap can be written as

$$\mathcal{F} = \int_{l_c} \mathbf{H}\cdot d\mathbf{l} + \int_{l_g} \mathbf{H}\cdot d\mathbf{l} = Ni = H_c l_c + H_g l_g = \frac{B_c l_c}{\mu_{rc}\mu_0} + \frac{B_g l_g}{\mu_0} = \frac{\phi_c l_c}{A_c\mu_{rc}\mu_0} + \frac{\phi_g l_g}{A_a\mu_0}$$
$$= \mathcal{R}_c\phi_c + \mathcal{R}_g\phi_g \approx (\mathcal{R}_c + \mathcal{R}_g)\phi_c. \tag{1.368}$$

The magnetic flux is

$$\phi = \frac{\mathcal{F}}{\mathcal{R}} = \frac{\mathcal{F}}{\mathcal{R}_g + \mathcal{R}_c} = \frac{Ni}{\mathcal{R}_g + \mathcal{R}_c}. \tag{1.369}$$

The air gap reduces the amount of magnetic flux because it increases the overall reluctance. If $\mathcal{R}_g \gg \mathcal{R}_c$,

$$\phi \approx \frac{\mathcal{F}}{\mathcal{R}_g} = \frac{Ni}{\mathcal{R}_g}. \tag{1.370}$$

Neglecting the fringing flux, $A_g = A_c$, $\lambda_g = \lambda_c$, and $B_g = B_c$. Hence,

$$Ni = B_c\left(\frac{l_c}{\mu_{rc}\mu_0} + \frac{l_g}{\mu_0}\right) = \frac{B_c}{\mu_0}\left(\frac{l_c}{\mu_{rc}} + l_g\right). \tag{1.371}$$

The magnetic flux density in the core with an air gap is given by

$$B_c = B_g = \frac{\mu_0 Ni}{l_g + \frac{l_c}{\mu_{rc}}}. \tag{1.372}$$

The ratio of the magnetic flux density in the ungapped core to the magnetic flux density in the gapped core is

$$\frac{B_{c(ungapped)}}{B_{c(gapped)}} = \frac{l_c}{l_g + \frac{l_c}{\mu_{rc}}} \approx \frac{l_c}{l_g}. \tag{1.373}$$

The maximum flux density in the core with an air gap, which is caused by the DC component of the inductor current I_L and the amplitude of the AC component of the inductor current I_m, is expressed by

$$B_{c(pk)} = B_{DC} + B_m = \frac{\mu_0 N(I_L + I_m)}{l_g + \frac{l_c}{\mu_{rc}}} \le B_s \quad \text{for} \quad T \le T_{max}. \tag{1.374}$$

The magnetic flux density and the magnetic field intensity in the core are

$$B_c = \frac{\phi_c}{A_c} \tag{1.375}$$

and

$$H_c = \frac{B_c}{\mu_{rc}\mu_0}. \tag{1.376}$$

Assuming a uniform magnetic flux density in the air gap and neglecting the fringing flux, magnetic flux, magnetic flux density, and magnetic field intensity in the air gap are

$$\phi_g = \phi_c = A_c B_c = A_g B_g, \tag{1.377}$$

$$B_g = \frac{A_c}{A_g} B_c \approx B_c, \tag{1.378}$$

and

$$H_g = \frac{B_g}{\mu_0} = \frac{B_c}{\mu_0} = \frac{\mu_{rc}\mu_0 H_c}{\mu_0} = \mu_{rc} H_c. \tag{1.379}$$

The maximum MMF is

$$\mathcal{F}_{max} = N_{max} I_{Lmax} = \phi(\mathcal{R}_g + \mathcal{R}_c) = B_{pk} A_c (\mathcal{R}_g + \mathcal{R}_c) \approx B_{pk} A_c \mathcal{R}_g = \frac{B_{pk} l_g}{\mu_0}, \tag{1.380}$$

where $\mathcal{R}_g = l_g/(\mu_0 A_c)$. To avoid core saturation, the maximum number of turns is given by

$$N_{max} = \frac{B_{pk} l_g}{\mu_0 I_{Lmax}}. \tag{1.381}$$

As the air-gap length l_g increases, NI_{Lmmax} can be increased and the core losses can be decreased. However, the number of turns N must be increased to achieve a specified inductance L. The increased number of turns increases the winding loss. In addition, the leakage inductance increases and the air gap radiates a larger amount of electromagnetic interference (EMI).

The behavior of an inductor with an air gap is similar to an amplifier with negative feedback

$$A_f = \frac{A}{1 + \beta A} = \frac{\mu_{rc}}{1 + \frac{l_g}{l_c}\mu_{rc}}. \tag{1.382}$$

Thus, μ_{rc} is analogous to A and l_g/l_c is analogous to β.

Power losses associated with the air gap consist of winding loss, core loss, and hardware loss (e.g., power loss in clamps or bolts). The magnetic field around the core gap can be strong and create localized losses in the winding close to the gap.

Example 1.10

A PQ4220 Magnetics core has $\mu_{rc} = 2500$, $l_c = 4.63$ cm, and $A_c = 1.19$ cm^2. The inductor wound on this core has $N = 10$ turns. The required inductance should be $L = 55.6$ μH. Find the length of the air gap l_g.

Solution: The length of the air gap in the core is

$$l_g = \frac{\mu_0 A_c N^2}{L} - \frac{l_c}{\mu_{rc}} = \frac{4\pi \times 10^{-7} \times 1.19 \times 10^{-4} \times 10^2}{55.6 \times 10^{-6}} - \frac{4.63 \times 10^{-2}}{2500}$$
$$= (0.2689564 - 0.01852) \times 10^{-3} = 0.2504 \text{ mm}. \tag{1.383}$$

1.10 Fringing Flux

1.10.1 Fringing Flux Factor

A fringing flux is present around the air gap whenever the core is excited, as shown in Fig. 1.24. It induces currents in the winding conductors and other conductors, which may cause intense local heating in the vicinity of the gap. Figure 1.25 depicts the fringing flux in an inductor with an EE core and an air gap. The magnetic flux lines bulge outward because the magnetic lines repel each other when passing through a nonmagnetic material. As a result, the cross-sectional area of the magnetic field and the effective length are increased and the flux density is decreased. Typically, 10% is added to the air gap cross-sectional area. This effect is called the *fringing flux effect*. The percentage of the fringing flux in the total magnetic flux increases as the air-gap length l_g increases. As the gap length l_g is increased, the radius of the magnetic flux in the gap also increases. A rule of thumb is that foil

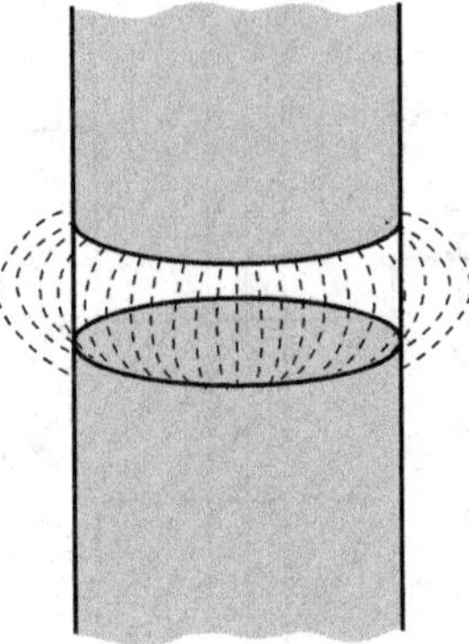

Figure 1.24 Fringing magnetic flux around the periphery of an air gap

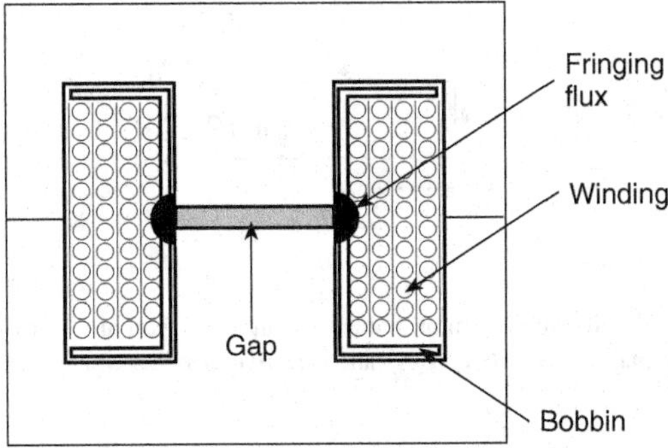

Figure 1.25 Fringing magnetic flux in an inductor with gapped pot core

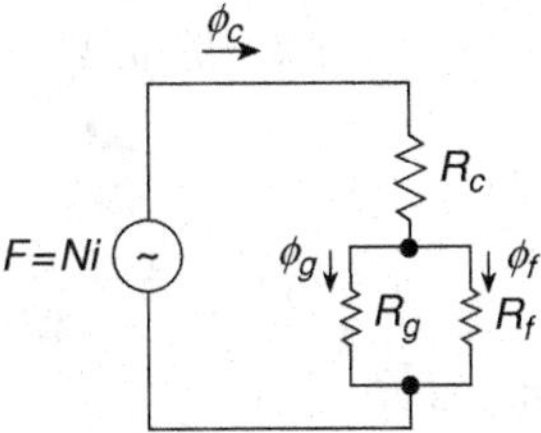

Figure 1.26 Magnetic equivalent circuit of an inductor with an air gap and a fringing magnetic flux

windings should be spaced at least two gap lengths away from the gap to prevent induction of eddy currents in the winding and to prevent overheating. The fringing flux is larger in inductors with a low-permeability core than that in high-permeability core. Figure 1.26 shows a magnetic equivalent circuit for the inductor with an air gap and a fringing flux. The fringing reluctance is shunting the gap reluctance, reducing the equivalent reluctance and slightly increasing the inductance. Thus, the number of turns N required for obtaining a desired inductance should be decreased or the air-gap length l_g should be increased. The effect of the fringing flux on the inductance and the number of turns can be investigated using the concept of the reluctance of the space conducting the fringing magnetic flux. Figure 1.27 shows the magnetic flux distribution in an air-gapped core inductor. The effective cross-sectional area of the gap with fringing flux A_{geff} is obtained by adding the gap length to each of the linear dimensions of the core in the gap area. For a rectangular core, $w_f = l_g$ and $A_{geff} = (a + l_w)(b + l_g)$.

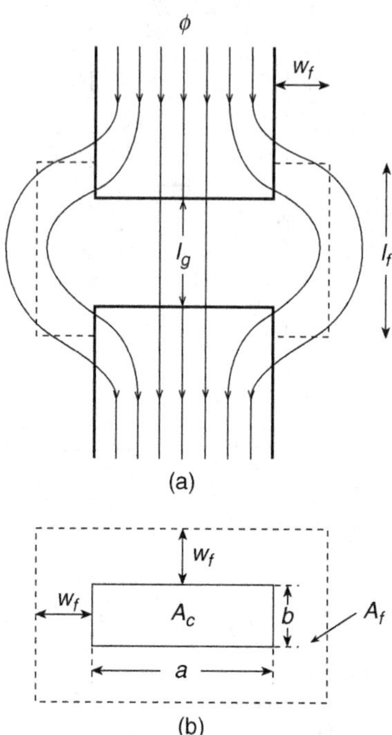

Figure 1.27 Magnetic flux distribution in an inductor with an air gap and a fringing magnetic flux. (a) Flux distribution. (b) Cross-sectional area of the core A_c and effective cross-sectional area of the fringing flux A_f

Due to the continuity of the magnetic flux, the magnetic flux in the core ϕ_c is equal to the sum of the magnetic flux in the air gap ϕ_g and the fringing flux ϕ_f

$$\phi_c = \phi_g + \phi_f. \tag{1.384}$$

The permeance of the core is

$$\mathcal{P}_c = \frac{1}{\mathcal{R}_c} = \frac{\mu_{rc}\mu_0 A_c}{l_c}. \tag{1.385}$$

The permeance of the air gap is

$$\mathcal{P}_g = \frac{1}{\mathcal{R}_g} = \frac{\mu_0 A_c}{l_g}. \tag{1.386}$$

The permeance of the fringing volume is

$$\mathcal{P}_f = \frac{1}{\mathcal{R}_f} = \frac{\mu_0 A_f}{l_f}, \tag{1.387}$$

where A_f is the fringing flux cross-sectional area and l_f is the mean MPL of the fringing flux. Assuming that $A_g = A_c$, the total reluctance is given by

$$\begin{aligned}
\mathcal{R} &= \mathcal{R}_c + \mathcal{R}_g \| \mathcal{R}_f = \mathcal{R}_c + \frac{\mathcal{R}_g \mathcal{R}_f}{\mathcal{R}_g + \mathcal{R}_f} = \frac{l_c}{\mu_{rc}\mu_0 A_c} + \frac{\frac{l_g}{\mu_0 A_g} \times \frac{l_f}{\mu_0 A_f}}{\frac{l_g}{\mu_0 A_g} + \frac{l_f}{\mu_0 A_f}} \\
&= \frac{l_c}{\mu_{rc}\mu_0 A_c} + \frac{1}{\mu_0 (A_g/l_g + A_f/l_f)} = \frac{l_c}{\mu_{rc}\mu_0 A_c} + \frac{l_g}{\mu_0 A_c \left(1 + \frac{A_f/l_f}{Ac/l_g}\right)} = \mathcal{R}_c + \frac{\mathcal{R}_g}{F_f} \\
&= \frac{l_c}{\mu_{rc}\mu_0 A_c}\left(1 + \frac{\mu_{rc} A_c}{l_c} \frac{l_g l_f}{l_f A_g + l_g A_f}\right) = \frac{l_c}{\mu_{rc}\mu_0 A_c}\left(1 + \frac{\mu_{rc} l_g}{l_c} \frac{1}{1 + \frac{l_g A_f}{l_f A_g}}\right),
\end{aligned} \tag{1.388}$$

where the *fringing factor* is defined as

$$F_f = \frac{\mathcal{R}_g}{\mathcal{R}_g \| \mathcal{R}_f} = 1 + \frac{\mathcal{R}_g}{\mathcal{R}_f} = \frac{\mathcal{P}}{\mathcal{P}_g} = \frac{\mathcal{P}_g + \mathcal{P}_f}{\mathcal{P}_g} = 1 + \frac{\mathcal{P}_f}{\mathcal{P}_g} = 1 + \frac{A_f l_g}{A_c l_f} = 1 + \frac{A_f/l_f}{A_c/l_g} > 1. \quad (1.389)$$

Hence, the inductance of an inductor with an air gap and a fringing flux is given by

$$L_f = \frac{N^2}{\mathcal{R}} = \frac{N^2}{\mathcal{R}_c + \frac{\mathcal{R}_g}{F_f}} > \frac{N^2}{\mathcal{R}_c + \mathcal{R}_g}. \quad (1.390)$$

For high-permeability cores, the core reluctance $\mathcal{R}_c$ can be neglected. The total permeance of the air gap and the fringing area is

$$\mathcal{P} = \mathcal{P}_g + \mathcal{P}_f = \frac{\mu_0 A_c}{l_g} + \frac{\mu_0 A_f}{l_f} = \frac{\mu_0 A_c}{l_g}\left(1 + \frac{A_f l_g}{A_c l_f}\right) = \frac{\mu_0 A_c F_f}{l_g} = F_f \mathcal{P}_g. \quad (1.391)$$

Thus, $\mathcal{R} \approx \mathcal{R}_g / F_f$.

Neglecting the permeance of the core, the inductance of the inductor with an air gap and the fringing flux is

$$L_f = \mathcal{P} N^2 = \frac{\mu_0 A_c N^2}{l_g} + \frac{\mu_0 A_f N^2}{l_f} = \frac{\mu_0 A_c N^2}{l_g}\left(1 + \frac{A_f l_g}{A_c l_f}\right) = \frac{\mu_0 A_c N^2 F_f}{l_g} = F_f L. \quad (1.392)$$

Thus, the fringing effect increases the inductance. The number of turns required for obtaining a desired inductance for the inductor with high-permeability core is

$$N = \sqrt{\frac{L}{\mathcal{P}}} = \sqrt{L\mathcal{R}} = \sqrt{\frac{L l_g}{\mu_0 A_c F_f}}. \quad (1.393)$$

The total reluctance of the inductor is

$$\mathcal{R} = \mathcal{R}_c + \mathcal{R}_g \| \mathcal{R}_f = \mathcal{R}_c + \frac{\mathcal{R}_g \mathcal{R}_f}{\mathcal{R}_g + \mathcal{R}_f} = \frac{l_c}{\mu_{rc}\mu_0 A_c} + \frac{l_g}{\mu_0 A_c F_f}$$

$$= \frac{l_g}{\mu_0 A_c}\left(\frac{l_c}{\mu_{rc} l_g} + \frac{1}{F_f}\right) = \mathcal{R}_g\left(\frac{l_c}{\mu_{rc} l_g} + \frac{1}{F_f}\right) = \frac{\mathcal{R}_g}{F_f} \quad \text{for} \quad \frac{l_c}{l_g} \ll \mu_{rc}. \quad (1.394)$$

The number of turns required for obtaining a desired inductance is

$$N = \sqrt{\frac{L}{\mathcal{P}}} = \sqrt{L\mathcal{R}} = \sqrt{\frac{L l_g}{\mu_0 A_c}\left(\frac{l_c}{\mu_{rc} l_g} + \frac{1}{F_f}\right)}. \quad (1.395)$$

If the air gap is enclosed by the winding, the fringing flux is reduced, lowering the value of F_f. However, the inductor losses increase as much as five times. If a winding is in the vicinity of the air gap, the fringing flux penetrates the winding conductor in the transverse direction, causing air-gap loss. To reduce this loss, the copper winding should be moved away from the air gap vicinity by a distance that is equal to two to three times the air-gap length l_g. This prevents the fringing flux of the gap from affecting the current within the winding conductor. The distance of the winding from the core can be increased by increasing the thickness of the bobbin. A reasonable thickness of the bobbin is 2–4 mm. Short distributed air gaps reduce the fringing flux and the power losses significantly. This is because the radial component of the magnetic flux density is reduced. Cores with a large relative permeability require long air gaps, which increase the fringing flux.

1.10.2 Effect of Fringing Flux on Inductance for Round Air Gap

The effect of the fringing flux on the inductance and the number of turns required to obtain a required inductance can be taken into account using the concept of the mean cross-sectional area of the fringing flux A_f and the mean MPL l_f of the fringing magnetic flux ϕ_f. Consider a round core with a single air gap. The ratio of the mean width of the cross-sectional area of the fringing flux to the air-gap length is defined as

$$\alpha = \frac{w_f}{l_g}. \tag{1.396}$$

The cross-sectional area of a round air gap with diameter D_c is

$$A_g = \frac{\pi}{4}D_c^2. \tag{1.397}$$

The cross-sectional area of the fringing flux for a round air gap is

$$A_f = \frac{\pi}{4}(D_c + 2\alpha l_g)^2 - \frac{\pi}{4}D_c^2 = \pi\alpha l_g(D_c + \alpha l_g). \tag{1.398}$$

The ratio of the cross-sectional areas is

$$\frac{A_f}{A_g} = \frac{4\alpha l_g(D_c + \alpha l_g)}{D_c^2}. \tag{1.399}$$

The ratio of the mean MPL of the fringing flux to the air-gap length is defined as

$$\beta = \frac{l_f}{l_g}. \tag{1.400}$$

The fringing flux factor for a round air gap is

$$F_f = 1 + \frac{A_f}{A_g}\frac{1}{\frac{l_f}{l_g}} = 1 + \frac{A_f}{A_g}\frac{1}{\beta} = 1 + \frac{4\alpha l_g(D_c + \alpha l_g)}{\beta D_c^2}. \tag{1.401}$$

In practice, it is difficult to determine the factors α and β. A good choice is $\alpha = 1$ and $\beta = 2$. The inductance with the fringing effect is

$$L_f = LF_f = L\left[1 + \frac{4\alpha l_g(D_c + \alpha l_g)}{\beta D_c^2}\right]. \tag{1.402}$$

Example 1.11

An inductor with an air gap has a round gap with $D_c = 10$ mm and $l_g = 1$ mm. Find F_f and N_f/N.

Solution: Assume $\alpha = 1$ and $\beta = 2$. The fringing factor is

$$F_f = \frac{L_f}{L} = 1 + \frac{4\alpha l_g(D_c + \alpha l_g)}{\beta D_c^2} = 1 + \frac{4 \times 1 \times 1 \times (10 + 1 \times 1)}{2 \times 10^2} = 1.22. \tag{1.403}$$

The ratio of the turns is

$$\frac{N_f}{N} = \frac{1}{\sqrt{F_f}} = \frac{1}{\sqrt{1.22}} = 0.9054. \tag{1.404}$$

1.10.3 Effect of Fringing Flux on Inductance for Rectangular Air Gap

Consider a core with a single rectangular air gap. The ratio of the mean width of the cross-sectional area of the fringing flux to the air-gap length is defined as

$$\alpha = \frac{w_f}{l_g} \tag{1.405}$$

where w_f is the mean width of the fringing flux. The rectangular air gap has dimensions a and b. The cross-sectional area of the air gap is

$$A_g = ab. \tag{1.406}$$

The cross-sectional area of the fringing flux is

$$A_f = (a + 2\alpha l_g)(b + 2\alpha l_g) - ab = 2\alpha l_g(a + b) + 4\alpha^2 l_g^2 = 2\alpha l_g(a + b + 2\alpha l_g), \tag{1.407}$$

resulting in the ratio of the cross-sectional areas

$$\frac{A_f}{A_g} = \frac{2\alpha l_g(a + b + 2\alpha l_g)}{ab} \tag{1.408}$$

The ratio of the mean MPL of the fringing flux to the air-gap length is defined as

$$\beta = \frac{l_f}{l_g} \tag{1.409}$$

where l_f is the mean length of the fringing flux. The fringing flux factor for a rectangular air gap is

$$F_f = 1 + \frac{A_f}{A_g}\frac{1}{\frac{l_f}{l_g}} = 1 + \frac{A_f}{A_g}\frac{1}{\beta} = 1 + \frac{2\alpha l_g(a + b + 2\alpha l_g)}{\beta ab}. \tag{1.410}$$

In practice, it is difficult to know the factors α and β. Reasonable values are $\alpha = 1$ and $\beta = 2$. The inductance with the fringing flux is

$$L_f = LF_f = L\left[1 + \frac{2\alpha l_g(a + b + 2\alpha l_g)}{\beta ab}\right]. \tag{1.411}$$

Assuming that $\alpha = 1$, the effective cross-sectional area of the fringing flux is

$$A_f = (a + 2w_f)(b + 2w_f) - A_c = 2(a + b)w_f + 4w_f^2. \tag{1.412}$$

Hence, the fringing flux factor is

$$F_f = 1 + \left(\frac{l_g}{A_c}\right)\left(\frac{w_f}{l_f}\right)[2(a + b) + 4w_f] \approx 1 + 2(a + b)\left(\frac{l_g}{A_c}\right)\left(\frac{w_f}{l_f}\right). \tag{1.413}$$

Note that F_f is directly proportional to l_g. The inductance is

$$L_f = F_f L = \frac{\mu_0 A_c N^2}{l_g}\left[1 + 2(a + b)\left(\frac{l_g}{A_c}\right)\right] = \frac{\mu_0 A_c N^2}{l_g} + 2(a + b)\mu_0 N^2 \frac{w_f}{l_f}. \tag{1.414}$$

The first term describes the inductance L without fringing flux. The second term is independent of l_g and constitutes a constant excess inductance. The empirical value is $w_f/l_f = 1.1322$.

Example 1.12

An inductor with a single rectangular air gap has $a = 10$ mm, $b = 20$ mm, and $l_g = 1$ mm. Find F_f and N_f/N.

Solution: Assume that $\alpha = 1$ and $\beta = 2$. The fringing flux factor is

$$F_f = \frac{L_f}{L} = 1 + \frac{2\alpha l_g(a + b + 2\alpha l_g)}{\beta ab} = 1 + \frac{2 \times 1 \times (10 + 20 + 2 \times 1 \times 1)}{2 \times 10 \times 20} = 1.16. \tag{1.415}$$

The ratio of the turns is

$$\frac{N_f}{N} = \frac{1}{\sqrt{F_f}} = \frac{1}{\sqrt{1.16}} = 0.9285. \tag{1.416}$$

1.10.4 Method of Effective Air Gap Cross-Sectional Area

Another method of deriving F_f uses an effective area of the air gap and is as follows. The cross-sectional area of the round core is

$$A_c = \frac{\pi D_c^2}{4} \tag{1.417}$$

and the effective cross-sectional area of the air gap is

$$A_e = \frac{\pi (D_c + l_g)^2}{4}. \tag{1.418}$$

Assuming that $l_g/l_f = 1$, we get F_f for a round air gap

$$F_f = \frac{L_f}{L} = \frac{A_e}{A_c} = \left(1 + \frac{l_g}{D_c}\right)^2. \tag{1.419}$$

For the rectangular cross section of the air gap,

$$F_f = \frac{L_f}{L} = \frac{A_e}{A_c} = \frac{(a + l_g)(b + l_g)}{ab}. \tag{1.420}$$

1.10.5 Method of Effective Length of Air Gap

The permeance of the air gap with the physical gap length l_g and the gap cross-sectional area $A_c + A_f$ is

$$\mathcal{P}_g = \frac{\mu_0 (A_c + A_f)}{l_g}, \tag{1.421}$$

where A_f is the effective cross-sectional area of the fringing flux. The permeance of the air gap with the effective gap length l_{eff} and the air gap flux area equal to the core cross-sectional area A_c is

$$\mathcal{P}_g = \frac{\mu_0 A_c}{l_{eff}}. \tag{1.422}$$

The two permeances are equal, which results in

$$\frac{\mu_0 (A_c + A_f)}{l_g} = \frac{\mu_0 A_c}{l_{eff}}. \tag{1.423}$$

The effective air-gap length for a rectangular gap is given by

$$l_{eff} = l_g \frac{A_c}{A_c + A_f} = A_c \frac{l_g}{(a + l_g)(b + l_g)} = A_c \frac{l_g}{ab\left(1 + \frac{l_g}{a}\right)\left(1 + \frac{l_g}{b}\right)}$$

$$= \frac{l_g}{\left(1 + \frac{l_g}{a}\right)\left(1 + \frac{l_g}{b}\right)} = \frac{l_g}{F_f} \tag{1.424}$$

where

$$F_f = \left(1 + \frac{l_g}{a}\right)\left(1 + \frac{l_g}{b}\right). \tag{1.425}$$

1.10.6 Patridge's Fringing Factor

The fringing flux factor given in Refs [6, 7, 13] is described by

$$F_f = 1 + \frac{a l_g}{N_g \sqrt{A_c}} \ln\left(\frac{2w}{l_g}\right) \approx 1 + \frac{l_g}{N_g \sqrt{A_c}} \ln\left(\frac{2w}{l_g}\right) = 1 + \frac{w}{N_g \sqrt{A_c}} \left(\frac{l_g}{w}\right) \ln \frac{2}{\left(\frac{l_g}{w}\right)}, \tag{1.426}$$

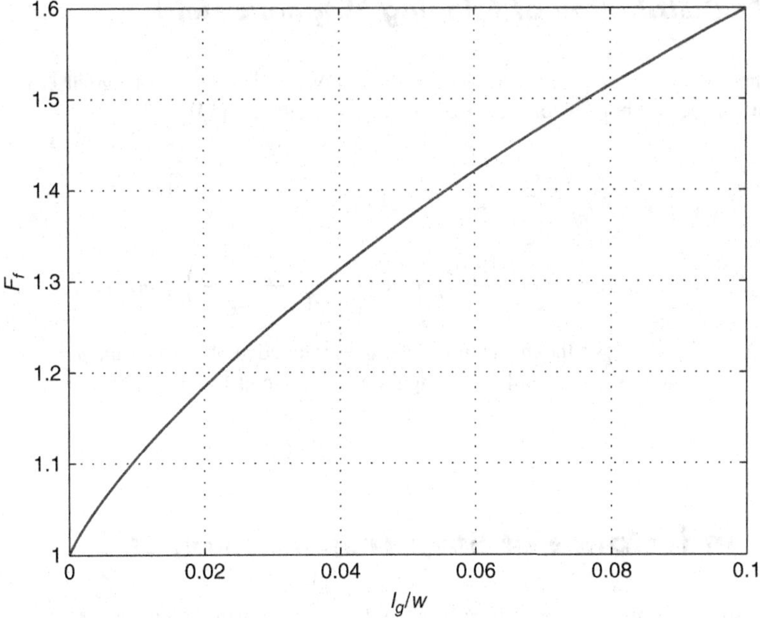

Figure 1.28 Fringing flux factor F_f as a function l_g/w at $A_c = 1$ cm^2 and $w = 2$ cm

where l_g is the total gap length, N_g is the number of air gaps in the magnetic path, w is the width of the core window, $a = 0.85$–0.95 for round cores, and $a = 1$–1.1 for rectangular cores. The typical values of the fringing flux factor F_f are in the range 1.1–1.4. The fringing flux reduces the total reluctance of the magnetic path $\mathcal{R}$, and therefore it increases the inductance L. Figure 1.28 shows a plot of the fringing flux factor F_f as a function of l_g/w at $N_g = 1$, $A_c = 1$ cm^2, and $w = 2$ cm.

The inductance is increased due to the fringing effect and is given by

$$L_f = F_f L = \left[1 + \frac{a l_g}{\sqrt{A_c}} \ln\left(\frac{2w}{l_g}\right)\right] L = \left[1 + \frac{a l_g}{\sqrt{A_c}} \ln\left(\frac{2w}{l_g}\right)\right] \frac{\mu_{rc}\mu_0 A_c N^2}{l_c(1 + \mu_{rc} l_g)}$$

$$= \frac{\mu_0 A_c N^2 F_f}{l_g + \frac{l_c}{\mu_{rc}}}. \tag{1.427}$$

Therefore, the number of turns N to obtain a required inductance L of an inductor with an air gap and the fringing effect should be reduced to

$$N_f = \sqrt{\frac{L_f\left(l_g + \frac{l_c}{\mu_{rc}}\right)}{\mu_0 A_c F_f}} = \frac{N}{\sqrt{F_f}}. \tag{1.428}$$

Fringing flux generates eddy currents, which cause hot spots in both the core and the winding, resulting in power losses. The winding, banding, and clips should be kept away from the fringing flux to reduce power losses. If a long single air gap is used, a high-fringing magnetic field is induced. Many short air gaps (a distributed air gap) along the magnetic path reduce the fringing flux and the winding loss as compared to the winding loss due to a long single air gap. It is important that there is a sufficient distance between air gaps. The distance between adjacent air gaps should be greater than four times the length of one air gap. Otherwise, the fringing fluxes from the adjacent air gaps will overlap and the air gaps will be shunted by the fringing reluctance.

1.10.7 Distribution of Fringing Magnetic Field

Assume that the skin effect in the core is negligible. For a core with a rectangular cross section, the magnitudes of the components of the fringing magnetic field are [54]

$$H_x(x,y) = \frac{H(0,0)}{2\pi} \ln \left[\frac{x^2 + (y - l_g/2)^2}{x^2 + (y + l_g/2)^2} \right] \tag{1.429}$$

and

$$H_y(x,y) = -\frac{H(0,0)}{\pi} \left[\arctan \left(\frac{x l_g}{x^2 + y^2 - l_g^2/4} \right) + m\pi \right], \tag{1.430}$$

where $H(0,0) = 0.9NI/l_g$ is the magnetic field at the center edge of the air gap, $m = 0$ for $x^2 + y^2 \geq l_g^2/4$, and $m = 1$ for $x^2 + y^2 < l_g^2/4$. The magnitude of the total fringing magnetic field is

$$H(x,y) = \sqrt{H_x^2(x,y) + H_y^2(x,y)}. \tag{1.431}$$

1.11 Inductance of Strip Transmission Line

Consider a strip transmission line, where d is the distance between the conductors, l is the length of the strip, and w is the width of the strip. The magnetic field intensity between conducting parallel plates is

$$H = \frac{I}{w} \tag{1.432}$$

resulting in the flux linkage

$$\lambda = \int\int_S \mathbf{B} \cdot d\mathbf{S} = \int_0^l \int_0^d \frac{\mu I}{w} dx dz = \frac{\mu I l d}{w}. \tag{1.433}$$

Hence, the inductance of the strip transmission line is given by

$$L = \frac{\lambda}{I} = \frac{\mu d l}{w}. \tag{1.434}$$

1.12 Inductance of Coaxial Cable

An axial current I flows in the inner conductor of radius a and returns in the outer conductor of radius b, inducing a circumferential magnetic field in the inner conductor, outer conductor, and between the conductors. Assume a uniform current distribution in both the conductors. The magnetic field between the conductors is

$$H_\phi = \frac{I_r}{2\pi r} \quad \text{for} \quad a \leq r \leq b. \tag{1.435}$$

The magnetic flux between radii a and b is

$$\phi_{ext} = \int\int_S \mathbf{B} \cdot d\mathbf{S} = \int_0^{l_w} dz \int_a^b \mu \left(\frac{\mu I}{2\pi r} \right) dr = \frac{\mu l_w I}{2\pi} \ln \left(\frac{b}{a} \right). \tag{1.436}$$

The external inductance of a coaxial cable due to the magnetic energy stored in the magnetic field between the inner and outer conductors is given by

$$L_{ext} = \frac{\phi_{ext}}{I} = \frac{1}{I} \int\int_S \mathbf{B} \cdot d\mathbf{S} = \frac{\mu l_w}{2\pi} \ln \left(\frac{b}{a} \right), \tag{1.437}$$

where a is the radius of the inner conductor, b is the radius of the outer conductor, and l_w is the cable length.

Consider the internal inductance due to the magnetic field inside the inner conductor. The current I_r enclosed by a circle of radius r is

$$I_r = \left(\frac{r}{a}\right)^2 I. \tag{1.438}$$

The magnetic field in the inner conductor is

$$H_\phi(r) = \frac{Ir}{2\pi a^2} \quad \text{for} \quad r < a. \tag{1.439}$$

The instantaneous magnetic energy stored in the magnetic field in the inner conductor is

$$W_m(t) = \frac{1}{2}L_{int(i)}I^2 = \frac{1}{2}\int\int\int_V \mu H^2 dV = \frac{1}{2}\int_0^{l_w} dz \int_0^a \mu\left(\frac{Ir}{2\pi a^2}\right)^2 2\pi r dr = \frac{\mu l_w I^2}{16\pi}. \tag{1.440}$$

Hence, the internal inductance of the inner conductor is

$$L_{int(i)} = \frac{\mu l_w}{8\pi}. \tag{1.441}$$

The magnetic field in the outer conductor is

$$H_\phi(r) = \frac{I}{2\pi(c^2 - b^2)}\left(\frac{c^2}{r} - r\right) \quad \text{for} \quad b \le r \le c \tag{1.442}$$

where c is the outer radius of the outer conductor. The instantaneous magnetic energy stored in the magnetic field in the outer conductor is

$$\frac{1}{2}L_{int(o)}I^2 = \frac{1}{2}\int\int\int_V \mu H^2 dV = \frac{1}{2}\int_0^{l_w} dz \int_0^a \frac{\mu I}{2\pi(c^2 - b^2)}\left(\frac{c^2}{r} - r\right)^2 2\pi r dr \tag{1.443}$$

Hence, the internal inductance of the outer conductor is

$$L_{int(o)} = \frac{\mu l_w}{2\pi}\left[\frac{c^4 \ln(c/b)}{(c^2 - b^2)^2} + \frac{b^2 - 3c^2}{4(c^2 - b^2)}\right]. \tag{1.444}$$

The total inductance of the coaxial transmission line is

$$L = L_{ext} + L_{int} = L_{ext} + L_{int(i)} + L_{int(o)} = \frac{\mu l_w}{2\pi}\ln\left(\frac{b}{a}\right) + \frac{\mu l_w}{8\pi} + \frac{\mu l_w}{2\pi}\left[\frac{c^4 \ln(c/b)}{(c^2 - b^2)^2} + \frac{b^2 - 3c^2}{4(c^2 - b^2)}\right]. \tag{1.445}$$

1.13 Inductance of Two-Wire Transmission Line

The magnetic flux is given by

$$\phi = -\frac{\mu I}{4\pi}\ln\left[\frac{(x - \sqrt{(d/2)^2 - a^2})^2 + y^2}{(x + \sqrt{(d/2)^2 + a^2})^2 + y^2}\right]. \tag{1.446}$$

The magnetic flux difference between its value at $x = d/2 - a$ and $x = -d/2 + a$ is

$$\Delta\phi = -\frac{\mu I}{\pi}\ln\left|\frac{d/2 - a - \sqrt{(d/2)^2 - a^2}}{d/2 - a + \sqrt{(d/2)^2 - a^2}}\right|. \tag{1.447}$$

The internal inductance of a two-wire transmission line for a round conductor of radius a, distance between the conductor centers d, and length l_w is

$$L = \frac{\Delta\phi}{I} = \frac{\mu l_w}{\pi}\cosh^{-1}\left(\frac{d}{2a}\right) = \frac{\mu l_w}{\pi}\ln\left[\frac{d}{2a} + \sqrt{\left(\frac{d}{2a}\right)^2 - 1}\right] \approx \frac{\mu l_w}{\pi}\ln\left(\frac{d}{a}\right) \quad \text{for} \quad \left(\frac{d}{2a}\right)^2 \gg 1, \tag{1.448}$$

where $\cosh^{-1}x \approx \ln(2x)$ for $x \ll 1$.

1.14 Magnetic Energy and Magnetic Energy Density

1.14.1 Magnetic Energy Density

The instantaneous magnetic energy density is

$$w_m(t) = \frac{1}{2}\mathbf{B} \cdot \mathbf{H}. \tag{1.449}$$

For an isotropic medium, $\mathbf{B} = \mu\mathbf{H}$ and the instantaneous magnetic energy density becomes

$$w_m(t) = \frac{1}{2}\mu\mathbf{H} \cdot \mathbf{H} = \frac{1}{2}\mu H^2 = \frac{1}{2}\mathbf{B} \cdot \frac{\mathbf{B}}{\mu} = \frac{1}{2}\frac{B^2}{\mu}. \tag{1.450}$$

For harmonic fields, the instantaneous magnetic energy density is

$$w_m(t) = \frac{1}{2}Re\{\mathbf{B}e^{j\omega t}\} \cdot Re\{\mathbf{H}e^{j\omega t}\}. \tag{1.451}$$

The time-average magnetic energy density for harmonic fields is

$$w_{m(AV)} = \frac{1}{4}Re\{\mathbf{H} \cdot \mathbf{B}^*\}. \tag{1.452}$$

For an isotropic medium,

$$w_{m(AV)} = \frac{1}{4}\mu|H|^2. \tag{1.453}$$

The instantaneous magnetic energy is

$$w_m(t) = \frac{1}{2}\lambda(t)i(t) = \frac{1}{2}N\phi(t)i(t) = \frac{1}{2}NAB(t)i(t) = \frac{1}{2}NA\mu H(t)i(t) = \frac{1}{2}Li^2. \tag{1.454}$$

1.14.2 Magnetic Energy Stored in Inductors with Ungapped Core

Consider first the magnetic energy stored in inductors with ungapped cores. The instantaneous reactive power of an inductor is

$$p(t) = i_L(t)v_L(t) = i_L\left(L\frac{di_L}{dt}\right) = Li_L\frac{di_L}{dt}. \tag{1.455}$$

Power is the time rate of change of energy $P = W/\Delta t$. The instantaneous *magnetic energy* stored in the magnetic field of an inductor without an air gap is given by

$$\begin{aligned} W_m(t) &= \int_0^t p(t)dt = \int_0^t i_L(t)v_L(t)dt = \int_0^t i_L L\frac{di_L}{dt}dt = L\int_0^{i_L} i_L di_L = \frac{1}{2}Li_L^2(t) \\ &= \frac{1}{2}\lambda(t)i_L(t) = \frac{\lambda^2(t)}{2L} = \frac{1}{2}\frac{N^2}{\mathcal{R}}i_L^2(t) = \frac{N^2\phi^2(t)}{2L} = \frac{1}{2}\mathcal{R}\phi^2(t) \\ &= \frac{1}{2}\frac{N^2}{\frac{l_c}{\mu_{rc}\mu_0 A_c}}\left(\frac{l_c H(t)}{N}\right)^2 = \frac{1}{2}\mu_{rc}\mu_0 H^2(t)A_c l_c = \frac{B^2(t)l_c A_c}{2\mu_{rc}\mu_0} = \frac{B^2(t)V_c}{2\mu_{rc}\mu_0}\ \text{(J)}, \end{aligned} \tag{1.456}$$

where $V_c = l_c A_c$ is the core volume, $v_L = Ldi_L/dt$, $i_L = \lambda/L$, $L = N^2/\mathcal{R}$, and $H = B/\mu$. The magnetic energy is proportional to the core volume V_c and the magnetic flux density B, and it is inversely proportional to the core relative permeability μ_{rc}.

The maximum energy stored in an inductor with a core without an air gap is given by

$$W_{c(max)} = \frac{B_{pk}^2 l_c A_c}{2\mu_{rc}\mu_0} = \frac{B_{pk}^2 V_c}{2\mu_{rc}\mu_0}. \tag{1.457}$$

The maximum energy that can be stored in an inductor is limited by the core saturation flux density B_s, the temperature rise caused by core losses, the core volume V_c, and the core relative permeability μ_{rc}. At $B_{pk} = B_s$,

$$W_{c(sat)} = \frac{B_s^2 l_c A_c}{2\mu_{rc}\mu_0} = \frac{B_s^2 V_c}{2\mu_{rc}\mu_0}. \tag{1.458}$$

The *instantaneous magnetic energy density* stored in an inductor with ungapped core is

$$w_m(t) = \frac{W_m(t)}{V_c} = \frac{B^2(t)}{2\mu_{rc}\mu_0} = \frac{1}{2}\mu_{rc}\mu_0 H^2(t) = \frac{1}{2}\mu H^2(t) \quad \left(\frac{\mathrm{J}}{\mathrm{m}^3}\right). \tag{1.459}$$

1.14.3 Magnetic Energy Stored in Inductors with Gapped Core

For an inductor with a gapped core, the magnetic energy stored in the gap is

$$W_g(t) = \frac{B^2(t) l_g A_g}{2\mu_0} = \frac{B^2(t) V_g}{2\mu_0} \approx \frac{B^2(t) l_g A_c}{2\mu_0}, \tag{1.460}$$

where $V_g = l_g A_g \approx l_g A_c$ is the air-gap volume and $A_g \approx A_c$. The instantaneous magnetic energy stored in the core is

$$W_c(t) = \frac{B^2(t) l_c A_c}{2\mu_{rc}\mu_0} = \frac{B^2(t) V_c}{2\mu_{rc}\mu_0} \tag{1.461}$$

where the core volume is $V_c = l_c A_c$. The total energy stored in an inductor with an air gap $W_m(t)$ is equal to the sum of the energy stored in the gap $W_g(t)$ and the energy stored in the core $W_c(t)$

$$W_m(t) = W_g(t) + W_c(t) = \frac{B^2(t) A_c}{2\mu_0}\left(l_g + \frac{l_c}{\mu_{rc}}\right). \tag{1.462}$$

The maximum magnetic energy stored in the core is

$$W_{c(max)} = \frac{B_{pk}^2 l_c A_c}{2\mu_{rc}\mu_0} = \frac{B_{pk}^2 V_c}{2\mu_{rc}\mu_0}. \tag{1.463}$$

For high-permeability cores with $l_g \gg l_c/\mu_{rc}$, almost all the inductor energy is stored in the air gap

$$W_m(t) \approx W_g(t) = \frac{B^2(t) l_g A_g}{2\mu_0} \approx \frac{B^2(t) V_g}{2\mu_0}. \tag{1.464}$$

The maximum magnetic energy that is stored in the air gap is

$$W_{g(max)} = \frac{B_{pk}^2 l_g A_c}{2\mu_0} = \frac{B_{pk}^2 V_g}{2\mu_0}. \tag{1.465}$$

At $B_{pk} = B_s$,

$$W_{g(sat)} = \frac{B_s^2 l_g A_c}{2\mu_0} = \frac{B_s^2 V_g}{2\mu_0}. \tag{1.466}$$

The total maximum energy that can be stored in an inductor with a gapped core is

$$W_{m(max)} = W_{g(max)} + W_{c(max)} = \frac{B_s^2 A_c}{2\mu_0}\left(l_g + \frac{l_c}{\mu_{rc}}\right). \tag{1.467}$$

Hence, the length of the air gap required to obtain a specified maximum magnetic energy $W_{m(max)}$ is

$$l_g = \frac{2\mu_0 W_{m(max)}}{A_c B_s^2} - \frac{l_c}{\mu_{rc}}. \tag{1.468}$$

The ratio of the two energies is

$$\frac{W_{g(max)}}{W_{c(max)}} = \frac{l_g}{l_c}\mu_{rc}. \tag{1.469}$$

The instantaneous *magnetic energy density* stored in the air gap is

$$w_g(t) = \frac{W_g(t)}{V_g} = \frac{B^2(t)}{2\mu_0} = \frac{1}{2}\mu_0 H_g^2(t) \text{ (J/m}^3\text{)} \tag{1.470}$$

and in the core is

$$w_c(t) = \frac{W_c(t)}{V_c} = \frac{B^2(t)}{2\mu_{rc}\mu_0} = \frac{1}{2}\mu_{rc}\mu_0 H_c^2(t) \text{ (J/m}^3\text{)}. \tag{1.471}$$

Example 1.13

An infinitely long round solid straight conductor of radius r_o conducts sinusoidal current $i(t) = I_m \cos \omega t$ in steady state at low frequencies (with no skin effect). Determine the amplitudes of the magnetic energy density and the total magnetic energy stored inside and outside the conductor.

Solution: From Example 1.1, the amplitude of the magnetic field density inside the conductor is given by

$$H_m(r) = \frac{I_m r}{2\pi r_o^2} \quad \text{for} \quad 0 \le r \le r_o. \tag{1.472}$$

The waveform of the magnetic field intensity is

$$H(r,t) = \frac{I_m r}{2\pi r_o^2} \cos \omega t \quad \text{for} \quad 0 \le r \le r_o. \tag{1.473}$$

Thus, the amplitude of the magnetic energy intensity inside the conductor is

$$w_m(r) = \frac{1}{2}\mu H_m^2(r) = \frac{\mu I_m^2 r^2}{8\pi^2 r_o^4} = \frac{\mu I_m^2}{8\pi^2 r_o^2}\left(\frac{r}{r_o}\right)^2 \quad \text{for} \quad 0 \le r \le r_o. \tag{1.474}$$

The maximum magnetic energy density at a given radius r is

$$w_{m(max)}(r) = \frac{\mu I_m^2 r^2}{4\pi^2 r_o^4}. \tag{1.475}$$

Hence, the waveform of the magnetic energy density inside the conductor is

$$w_m(r,t) = \frac{1}{2}\mu H^2(r,t) = \frac{\mu I_m^2 r^2}{8\pi^2 r_o^4}\cos^2 \omega t \quad \text{for} \quad 0 \le r \le r_o. \tag{1.476}$$

The small volume of a cylindrical shell of radius r, thickness dr, and length l_w is

$$dV = (2\pi r)(dr)(l_w) = 2\pi l_w r dr. \tag{1.477}$$

Assuming that μ is uniform for the entire conductor, the amplitude of the magnetic energy stored in the magnetic field inside the conductor is given by

$$W_{m(max)} = \int\int\int_V w_m(r) dV = \frac{1}{2}\int\int\int_V \mu H_m^2(r) dV = \frac{\mu l_w I_m^2}{4\pi r_o^4}\int_0^{r_o} r^3 dr = \frac{\mu l_w I_m^2}{16\pi} \tag{1.478}$$

or using the internal inductance $L_{int} = \mu l_w/(8\pi)$

$$W_{m(max)} = \frac{1}{2}L_{int} I_m^2 = \frac{1}{2}\left(\frac{\mu l_w}{8\pi}\right) I_m^2 = \frac{\mu l_w I_m^2}{16\pi}. \tag{1.479}$$

From Example 1.1, the amplitude of the magnetic field intensity outside the conductor is

$$H_m(r) = \frac{I_m}{2\pi r} \quad \text{for} \quad r \ge r_o. \tag{1.480}$$

The waveform of the magnetic field intensity outside the conductor is

$$H(r,t) = \frac{I_m}{2\pi r}\cos \omega t \quad \text{for} \quad r \ge r_o. \tag{1.481}$$

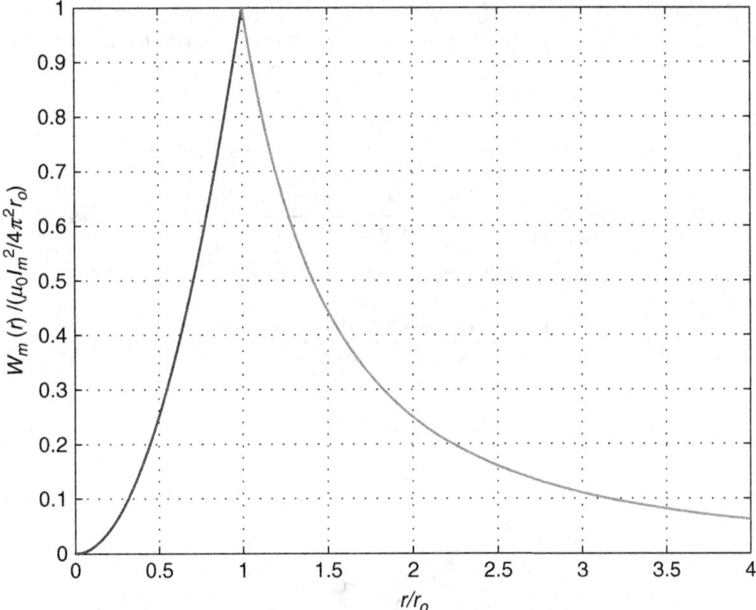

Figure 1.29 Normalized magnetic energy density $w_m(r)/[\mu I_m^2/(4\pi^2 r_o^2)]$ stored inside and outside a long, round, solid conductor conducting a sinusoidal current at low frequencies (with no skin effect)

Therefore, the amplitude of the magnetic energy outside the conductor is

$$w_{m(max)}(r) = \frac{1}{2}\mu H_m^2(r) = \frac{\mu I_m^2}{8\pi^2 r^2} = \frac{\mu I_m^2}{8\pi^2 r_o^2} \frac{1}{\left(\frac{r}{r_o}\right)^2} \quad \text{for} \quad r \geq r_o. \tag{1.482}$$

Hence, the waveform of the magnetic energy density inside the conductor is

$$w_m(r,t) = \frac{1}{2}\mu H_m^2(r,t) = \frac{\mu I_m^2}{8\pi^2 r^2}\cos^2 \omega t \quad \text{for} \quad r \geq r_o. \tag{1.483}$$

Assuming that μ is uniform for the entire area outside the conductor, the magnetic energy stored in the magnetic field outside the conductor is given by

$$W_m = \int\int\int_V w_m(r)dV = \frac{1}{2}\int\int\int_V \mu H_m^2(r)dV = \frac{\mu l_w I_m^2}{8\pi}\int_{r_o}^{\infty}\frac{dr}{r} = \frac{\mu l_w I_m^2}{8\pi}\ln r\Big|_{r_o}^{\infty} = \infty. \tag{1.484}$$

Figure 1.29 shows the normalized magnetic energy density $w_m(r)/[\mu I_m^2/(4\pi^2 r_o^2)]$ inside and outside a long, solid round conductor carrying a sinusoidal current at low frequencies. The reason for the infinite magnetic energy is that the model of the conductor is too ideal, which does not take into account the return path of the current. The presence of this path changes the magnetic field distribution and the stored magnetic energy.

Example 1.14

A Ferroxcube ferrite magnetic core 528T500-4C4 has $A_c = 1.17$ cm^2, $l_c = 8.49$ cm, and $\mu_{rc} = 125$. (a) Determine the maximum magnetic energy that can be stored in the inductor core. (b) Determine the maximum magnetic energy that can be stored in the air gap $l_g = 0.5$ mm. (c) Find the ratio of the maximum magnetic energies.

Solution: The saturation flux density B_s for ferrite cores is $B_s = 0.3\ T$ at room temperature $T = 20\,^{\circ}$C. At $T = 100\,^{\circ}$C, the saturation flux density B_s for ferrite cores decreases by a factor of 2. Thus,

$$B_s = \frac{0.3}{2} = 0.15\ \text{T}. \tag{1.485}$$

The maximum magnetic energy that can be stored in the magnetic core is

$$W_{c(max)} = \frac{B_s^2 l_c A_c}{2\mu_{rc}\mu_0} = \frac{0.15^2 \times 8.49 \times 10^{-2} \times 1.17 \times 10^{-4}}{2 \times 125 \times 4\pi \times 10^{-7}} = 0.711\ \text{mJ}. \tag{1.486}$$

The maximum magnetic energy that can be stored in the air gap is

$$W_{g(max)} = \frac{B_s^2 l_g A_c}{2\mu_0} = \frac{0.15^2 \times 0.5 \times 10^{-3} \times 1.17 \times 10^{-4}}{2 \times 4\pi \times 10^{-7}} = 0.5237\ \text{mJ}. \tag{1.487}$$

Hence,

$$\frac{W_{g(max)}}{W_{c(max)}} = \frac{0.5237}{0.711} = 0.7366. \tag{1.488}$$

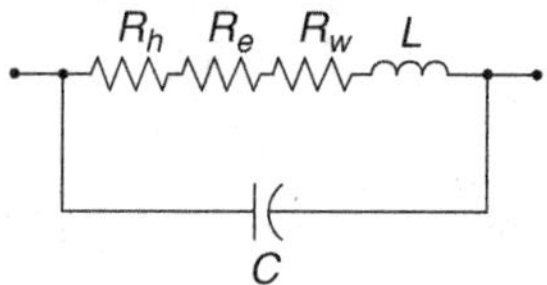

Figure 1.30 Model of an inductor

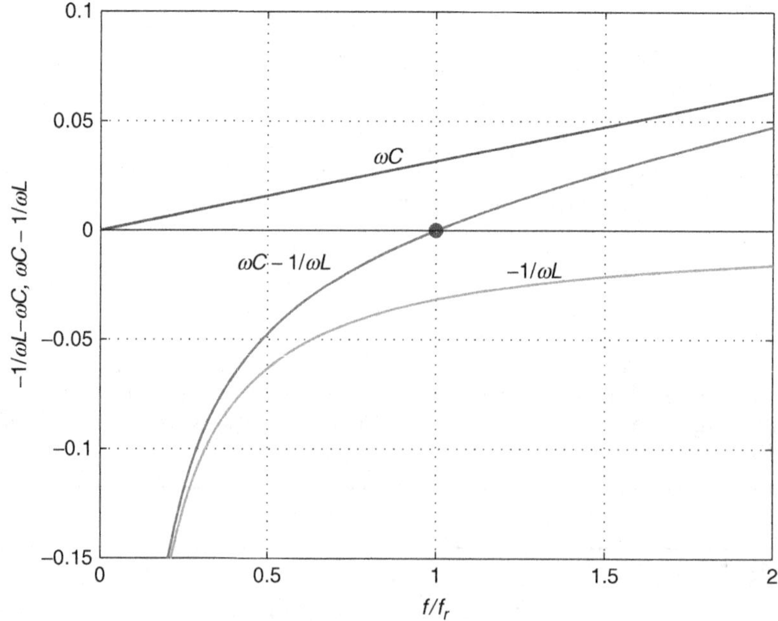

Figure 1.31 Plots of the susceptances $B_C = \omega C$, $B_L = -1/(\omega L)$, and $B = \omega C - 1/(\omega L)$ as functions of frequency for an inductor

1.15 Self-Resonant Frequency

Figure 1.30 shows an equivalent model of an inductor, where L is the inductance, R_w is the winding resistance, $R_{cs} = R_h + R_e$, R_h is the core hysteresis equivalent series resistance (ESR), R_e is the core eddy-current series resistance, and C is the self-capacitance. The distributed capacitance between the winding turns acts like a shunt capacitance, conducting a high-frequency current. This capacitance is called a *stray capacitance* or a *self-capacitance* C [33, 34]. It depends on the winding geometry, the proximity of turns, core, and shield, and the permittivity of the dielectric insulator, in which the winding wire is coated. The core should be insulated to increase the distance between the turns and the core, and therefore reduce the capacitance between the winding and the core. The inductance and the self-capacitance form a parallel resonant circuit, which has the first (fundamental or parallel) SRF

$$f_r = \frac{1}{2\pi\sqrt{LC}}. \tag{1.489}$$

Figure 1.31 shows the plots of the susceptances $B_C = \omega C$, $B_L = -1/(\omega L)$, and $B = B_C + B_L = \omega C - 1/(\omega L)$ as functions of frequency for inductance $L = 1$ μH and $C = 1$ nF. At the SRF f_r, the total susceptance of an inductor is zero. Below the SRF f_r, the inductor reactance is inductive. Above the SRF f_r, the inductor reactance is capacitive. Therefore, the operating frequency range of an inductor is usually from DC to $0.9f_r$.

1.16 Quality Factor of Inductors

A winding represents a series combination of an inductance and a frequency-dependent resistance. The *quality factor* of an inductor with a magnetic core at a given frequency f is defined as

$$Q_{Lo} = \frac{\text{Reactance at } f}{\text{Total resistance at } f} = \frac{X_L}{r_L} = \frac{\omega L}{r_L} = \frac{\omega L}{R_w + R_{cs}} = \frac{1}{\frac{R_w}{\omega L} + \frac{R_{cs}}{\omega L}} = \frac{1}{\frac{1}{\omega L/R_w} + \frac{1}{\omega L/R_{cs}}} = \frac{1}{\frac{1}{Q_{LRw}} + \frac{1}{Q_{LRcs}}} = \frac{Q_{LRw}Q_{LRcs}}{Q_{LRw} + Q_{LRcs}}, \tag{1.490}$$

where $r_L = R_w + R_{cs}$ is the ESR of an inductor at frequency f, R_w is the winding resistance, and R_{cs} is the core series resistance, the quality factor of an inductor due to the winding resistance is

$$Q_{LRw} = \frac{\omega L}{R_w}, \tag{1.491}$$

and the quality factor of an inductor due to the core series resistance is

$$Q_{LRcs} = \frac{\omega L}{R_{cs}}. \tag{1.492}$$

The quality factor of an air-core inductor is defined as

$$Q_{Lo} = Q_{LRw} = \frac{\omega L}{R_w}. \tag{1.493}$$

1.17 Classification of Power Losses in Magnetic Components

Figure 1.32 shows a classification of power losses in magnetic components. These losses can be categorized into winding (or copper) losses P_{Rw} and core losses P_C. The winding losses can be divided into the DC loss and the AC loss.

$$P_{Rw} = P_{wDC} + P_{wAC}. \tag{1.494}$$

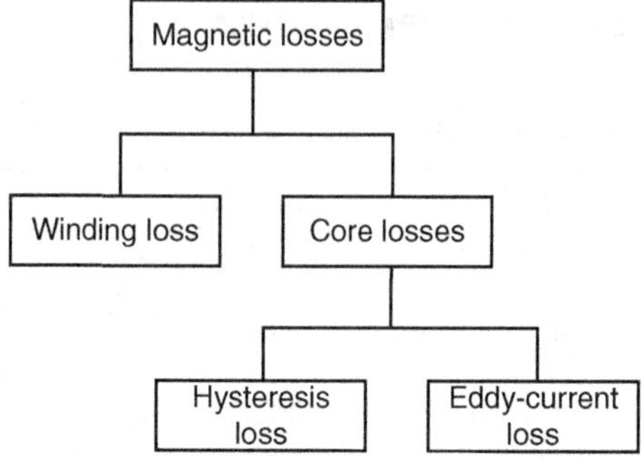

Figure 1.32 Classification of power losses in magnetic components

In turn, core losses can be divided into hysteresis loss P_H and eddy-current loss P_E

$$P_C = P_H + P_E. \tag{1.495}$$

Hence, the total inductor power loss P_L is given by

$$P_L = P_{Rw} + P_C = P_{Rw} + P_H + P_E = P_{wDC} + P_{wAC} + P_H + P_E. \tag{1.496}$$

There are two kinds of eddy-current winding losses: the skin-effect loss and the proximity-effect loss. Both these effects cause current crowding. Eddy-current losses are magnetically induced losses. The winding power losses are increased at high frequencies by eddy currents induced in conductors by magnetic fields. Consider a case where a sinusoidal current is applied to an inductor

$$i_L(t) = I_{Lm} \sin \omega t. \tag{1.497}$$

At a high frequency, the inductor winding carries the applied current $i_L(t)$ and the eddy current $i_{ec}(t) = I_{ecm} \sin \omega t$. The applied current density is uniform and therefore the winding resistance presented to the applied current is equal to the DC winding resistance R_{wDC}. Hence, the power loss due to the applied current is

$$P_{wDC} = \frac{1}{2} R_{wDC} I_{Lm}^2. \tag{1.498}$$

The eddy current density is not uniform and therefore the winding resistance presented to the eddy current R_{wec} is higher than the DC winding resistance presented to the applied sinusoidal current R_{wDC}. The total AC winding resistance R_w is given by

$$R_w = R_{wDC} + R_{wec} = R_{wDC}\left(1 + \frac{R_{wec}}{R_{wDC}}\right) = R_{wDC} + \alpha R_{wDC} = R_{wDC}(1+\alpha) = F_R R_{wDC}, \tag{1.499}$$

where the ratio of the eddy current winding resistance to the DC winding resistance is

$$\alpha = \frac{R_{wec}}{R_{wDC}} \tag{1.500}$$

and the ratio of the AC-to-DC winding resistance is

$$F_R = \frac{R_w}{R_{wDC}} = \frac{R_{wDC} + R_{wec}}{R_{wDC}} = 1 + \frac{R_{wec}}{R_{wDC}} = 1 + \alpha. \tag{1.501}$$

The total winding power loss is equal to the power loss due to the conduction of the applied current P_{wDC} of uniform density and the power losses due to the conduction of the eddy current P_{wec} of nonuniform density

$$P_{wAC} = P_w = P_{wDC} + P_{wec} = \frac{1}{2} R_w I_{Lm}^2 = \frac{I_{Lm}^2}{2}(R_{wDC} + R_{wec}) = \frac{I_{Lm}^2}{2} R_{wDC} + \frac{I_{Lm}^2}{2} \alpha R_{wDC}$$
$$= \frac{I_{Lm}^2}{2} R_{wDC} + \frac{I_{ecm}^2}{2} R_{wDC}. \tag{1.502}$$

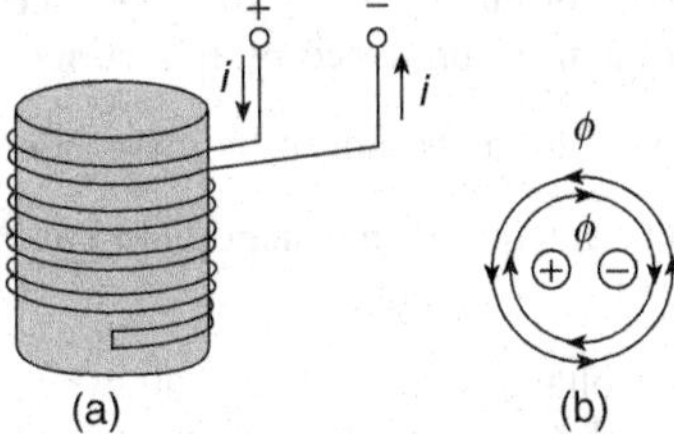

Figure 1.33 Noninductive coil. (a) Bifilar winding. (b) Magnetic flux cancellation

Hence, the amplitude of the eddy current is

$$I_{ecm} = \sqrt{\alpha} I_{Lm} = \sqrt{F_R - 1} \quad I_{Lm} = \sqrt{\alpha} \quad I_{Lm}. \tag{1.503}$$

1.18 Noninductive Coils

In some applications, it is desired to have a noninductive coil. Precision resistors are usually noninductive. For example, current probes require noninductive resistors. A noninductive coil is usually made using closely spaced, parallel windings, called the *bifilar winding*, as illustrated in Fig. 1.33a. Therefore, every coil turn has an adjacent turn, which carries current in the opposite direction. The magnetic fields generated by the adjacent turns cancel each other, as shown in Fig. 1.33b. As a result, the coil does not store magnetic flux and presents no self-inductance.

1.19 Summary

Magnetic Laws

- A field is a spatial distribution of a scalar or vector quantity.
- Field lines may be used for visualization of the behavior of a field.
- A field for which the line integral around the closed path is zero is conservative. A field is conservative or irrotational if $\nabla \times \mathbf{B} = 0$.
- The sources of magnetic fields **H** and **B** are moving charges, that is, the electric current i.
- The source of electric fields **E** and **D** is the electric charge Q.
- The divergence represents the rate of change of a flow.
- The curl represents the rotation of a flow.
- The MMF $\mathcal{F} = Ni$ is a source magnetic flux in a magnetic circuit.
- Magnetic fields can be categorized as self, proximity, mutual, and fringing magnetic fields.
- The instantaneous magnetic field vector is a function of position and time.
- The phasor of magnetic field vector is a function of position only.
- A time-varying current in an inductor produces a changing magnetic flux, which induces a voltage between the terminals of an inductor.
- The RHR states that if the fingers of the right hand are placed around the coil in the direction of the current, the magnetic flux produced by the current is in the direction of the thumb.

- Ampère's law describes the relationship between the (conduction, convection, or displacement) current and the magnetic field intensity produced by this current.
- Both conductive and displacement currents induce the magnetic field.
- Ampère's law states that the line integral of magnetic field intensity **H** around a closed contour C is equal to the current enclosed by the contour.
- Faraday's law states that an AC voltage is induced in a coil, which contains a time-varying magnetic flux, regardless of the source of the magnetic flux.
- According to Faraday's law, the voltage (EMF) induced in closed circuit is equal to the time rate of change of the magnetic flux linkage $v = -d\lambda/dt = -Nd\phi/dt$.
- In the inductor, the induced voltage is proportional to the number of turns N and the time rate of change of the magnetic flux $d\phi/dt$.
- A magnetostatic magnetic field produces no current flow; however, a time-varying magnetic field produces an induced voltage (EMF) in a closed circuit.
- Ohm's law describes the relationship between the conduction (or drift) current density J_{cond} and the electric field intensity E, that is, $\mathbf{J}_{cond} = \sigma\mathbf{E}$.
- Convection current and displacement current do not obey Ohm's law.
- A curl-free vector field is called a irrotational or a conservative field.
- The Biot–Savart's law allows us to calculate the magnetic field intensity produced by a small current element at some point in space. This law states that the differential field intensity $d\mathbf{H}$ produced by the differential current element $Id\mathbf{l}$ at a point P is proportional to the product Idl and sin of the angle θ between the element and the line connecting P and the element and inversely proportional to the square of the distance R between P and the element.
- According to Lenz's law, the direction of the EMF is such that the current forced by the EMF induces a magnetic field that opposes the change in the applied magnetic field. The induced currents never support and always oppose the changes by which they are induced.
- Power is defined as the time rate of change of energy.
- The Poynting vector represents the direction and the magnitude of the surface power flow density of electromagnetic fields at any point in space, that is, the rate of energy transfer per unit area. This vector is equal to the cross product of the electric and magnetic fields $\mathbf{S} = \mathbf{E} \times \mathbf{H}$.
- The Poynting theorem states that the rate of decrease in the energy stored in electric and magnetic fields in volume V, less the energy dissipated as heat, is equal to the power leaving the closed surface S bounding the volume V.
- The magnetic flux density **B** outside a very long current-carrying wire is inversely proportional to the distance from the axis of the wire.
- Joule's law states that the total power loss in a volume V is

$$P_D = \int\int\int_V \mathbf{J} \cdot \mathbf{E} dV. \tag{1.504}$$

Reluctance

- The reluctance is directly proportional to the length of the core mean magnetic path l_c and inversely proportional to the relative permeability μ_{rc} and the core cross-sectional area A_c through which the magnetic flux ϕ flows.

- The magnetic flux always takes the path with the highest permeability μ because the lowest reluctance occurs at the highest permeability.
- The magnetic flux always flows through the lowest reluctance.

Inductance

- The inductor sinusoidal current legs the inductor sinusoidal voltage by 90°.
- The inductance (or self-inductance) of a wire-wound inductor depends on its geometry and is proportional to the square of the number of turns N.
- The inductance is proportional to the ratio of the core cross-sectional area to the MPL A_c/l_c.
- The inductance of an inductor with a ferromagnetic core is μ_{rc} times higher than that of an air-core inductor.
- An inductor has a maximum value of the ampere-turn product $(\mathcal{F} = NI_m)_{max}$ limited by the core saturation flux density B_s.
- The self-inductance can be categorized as an internal inductance and the external inductance.
- The internal inductance is related to the magnetic field inside a conductor itself.
- The external inductance is related to the magnetic field outside a conductor.
- The winding turns should be evenly spaced to achieve consistent inductance and reduce leakage inductance.

Core Saturation

- At the core saturation, the magnetic flux density B approaches its maximum value known as the saturation flux density B_s. For $B > B_s$, $\mu_r = 1$.
- Core saturation can be avoided by reducing the peak value of the magnetic flux ϕ in the core or by increasing core cross-sectional area A_c so that $\phi/A_c < B_s$.
- It is difficult to avoid core saturation during transient circuit operation when the transient inductor current is large.

Air Gap

- An air gap is used to prevent core saturation and to make the inductance almost independent of μ_{rc}, yielding good inductance repeatability.
- Most of the MMF $\mathcal{F} = Ni$ is dropped across the air gap.
- The air gap contains nearly all of the magnetic field energy for high-permeability cores.
- Typically, 95% of inductance comes from the gap for inductors with high-permeability cores.
- An air gap in the core increases the energy storage capability of an inductor or a transformer.
- The core relative permeability μ_{rc} varies considerably with temperature and current. In contrast, the effective relative permeability is less dependent on the temperature and current. Therefore, it is desirable to maintain $\mathcal{R}_c \ll \mathcal{R}_g$ to achieve a predictable and stable inductance.
- The effective relative permeability of the core is proportional to the ratio l_c/l_g.
- The inductance of an inductor with an air gap is lower than the inductance of an inductor without an air gap.

Fringing Effect

- Whenever the core is excited, the fringing flux is present around the air gap, reducing the reluctance, increasing the inductance, and causing power losses.
- Fringing flux represents a larger percentage of the total flux for larger gaps.
- Fringing flux and inductor losses can be reduced by dividing a large air gap into several shorter air gaps.
- The fringing flux reduces the total reluctance $\mathcal{R}$ and increases the inductance L. Therefore, the number of turns should be reduced if the exact value of the inductance is required.
- The fringing field decreases substantially within one gap length distance l_g from the edge of the core.

Power Losses

- Power losses in inductors and transformers consist of winding and core losses.
- Eddy currents are induced in conductors by time-varying magnetic fields.
- Core losses consist of hysteresis loss and eddy-current loss.
- A distributed air gap along the magnetic path reduces the winding loss as compared to the winding loss due to a single gap. This is because the radial component of the magnetic flux is reduced.
- The impact of the radial component of the magnetic flux can be reduced by increasing the distance between the winding and the core. This distance can be increased by increasing the thickness of the bobbin.
- The winding should be moved away from the air gap by a distance, which is equal to twice the air-gap length $2l_g$.

Shielding

- A shield can be used to reduce EM emission by inductors and transformers.
- The thickness of the shield foils should be low compared to the skin depth.
- As the distance between the shield and the inductor decreases, the inductance also decreases.

Self-Resonant Frequency

- The SRF of an inductor is the resonant frequency of the resonant circuit formed by the inductance and the stray capacitance.
- The inductor impedance is capacitive above the SRF $f > f_r$.

1.20 References

[1] P. J. Dowell, "Effects of eddy currents in transformer winding," *Proceedings of the IEE*, vol. 113, no. 8, pp. 1387–1394, August 1966.

[2] J. Jongsma, "High-frequency ferrite power transformer and choke design, Part 3: Transformer winding design, Part 4: Improved method of power choke design" Philips Electronic Components and Materials, Technical Publication, no. 27, Philips, The Netherlands, 1986.

[3] A. Kennelly, F. Laws, and P. Pierce, "Experimental research on skin effect in conductors," *Transactions AIEE*, vol. 34, pp. 1953–2021, 1915.

[4] MIT EE Staff, *Magnetic Components and Circuits*, New York: John Wiley & Sons, 1943.
[5] H. C. Roberts, *Electromagnetic Devices*, New York: John Wiley & Sons, 1941.
[6] G. F. Partridge, *Philosophical Magazine*, vol. 22, 7th Series, pp. 664–678, October 1936.
[7] R. Lee, L. Wilson, and C. E. Carter, *Electronic Transformers and Circuits*, 3rd Ed., New York, John Wiley & Sons, 1988.
[8] J. Lammeraner and M. Stafl, *Eddy Currents*, Cleveland: CRS Press, 1966.
[9] J. Ebert, "Four terminal parameters of HF inductors," *Bulletin de l'Acad'emie Polonaise des Sciences, Séries des Sciences Techniques*, no. 5, 1968.
[10] E. C. Snelling, *Soft Ferrites: Properties and Applications*, London: Iliffe Books Ltd, 1969.
[11] R. L. Stall, *The Analysis of Eddy Currents*, Oxford: Clarendon Press, 1974, pp. 21–27.
[12] J. D. Kraus and D. A. Fleisch, *Electromagnetics with Applications*, 5th Ed., New York: McGraw-Hill, 1992.
[13] W. T. McLyman, *Transformer and Inductor Design Handbook*, 3rd Ed., New York: Marcel Dekker, 2004.
[14] J. K. Watson, *Applications of Magnetism*, Gainesville, FL: John Wiley & Sons, 1985; reprinted by Wiley, 1985.
[15] E. Bennet and S. C. Larsen, "Effective resistance of alternating currents of multilayer windings," *Transactions of the American Institute of Electrical Engineers*, vol. 59, pp. 1010–1017, 1940.
[16] R. L. Perry, "Multiple layer series connected winding design for minimum losses," *IEEE Transactions on Power Applications Systems*, vol. PAS-98, pp. 116–123, January/February 1979.
[17] B. Carsten, "High frequency conductor losses in switch mode magnetics," Proceedings of PCI, Munich, Germany, May 1986, pp. 161–182.
[18] J. P. Vandalec and P. D. Ziogos, "A novel approach for minimizing high-frequency transformer copper loss," *IEEE Transactions on Power Electronics*, vol. 3, pp. 266–276, July 1988.
[19] A. M. Urling, V. A. Niemela, G. R. Skutt, and T. G. Wilson, "Characterizing high frequency effects in transformer windings: a guide to several significant papers," 4th Annual IEEE Transactions on Power Electronics Specialists Conference, March 13–17, 1989, pp. 373–385.
[20] J. A. Ferreira, *Electromagnetic Modelling of Power Electronic Converters*, Boston, MA: Kluwer Academic Publisher, 1989.
[21] J. G. Kassakian, M. F. Schlecht, and G. C. Verghese, *Principles of Power Electronics*, Reading, MA: Addison-Wesley, 1991.
[22] J. D. Jackson, *Classical Electrodynamics*, 3rd Ed., New York: John Wiley & Sons, 2004.
[23] N. Nysveen and M. Hernes, "Minimum loss design of a 100 kHz inductor with foil windings," EPE Conference Proceedings, 1993, pp. 106–111.
[24] S. Ramo, J. R. Whinnery, and T. Van Duzer, *Fields and Waves in Communication Electronics*, 3rd Ed., John Wiley & Sons, New York, 1993.
[25] M. Bartoli, A. Reatti, and M. K. Kazimierczuk, "High-frequency models of ferrite inductors," Proceedings of the IEEE International Conference on Industrial Electronics, Controls, Instrumentation, and Automation (IECON'94), Bologna, Italy, September 5-9, 1994, pp. 1670–1675.
[26] M. Bartoli, A. Reatti, and M. K. Kazimierczuk, "Predicting the high-frequency ferrite-core inductor performance," Proceedings of the Conference of Electrical Manufacturing and Coil Winding, Chicago (Rosemont), IL, September 27-29, 1994, pp. 409–413.
[27] M. Bartoli, A. Reatti, and M. K. Kazimierczuk, "Modeling iron-powder inductors at high-frequencies," Proceedings of the IEEE Industry Applications Society Annual Meeting, Denver, CO, October 2-7, 1994, pp. 1125–1232.
[28] J. A. Ferreira, "Improved analytical modeling of conductive losses in magnetic components," *IEEE Transactions on Power Electronics*, vol. 9, pp. 127–131, January 1994.
[29] M. Bartoli, N. Noferi, A. Reatti, and M. K. Kazimierczuk, "Modeling winding losses in high-frequency power inductors," *Journal of Circuits Systems and Computers*, vol. 5, no. 3, pp. 65–80, March 1995.
[30] N. Mohan, T. M. Underland, and W. P. Robbins, *Power Electronics*, 3rd Ed., New York: John Wiley & Sons, 2003.
[31] M. Bartoli, N. Nefari, A. Reatti, and M. K. Kazimierczuk, "Modeling litz-wire winding losses in high-frequencies power inductors," Proceedings of the IEEE Power Electronics Specialists Conference, Baveno, Italy, June 24-27, 1996, pp. 1690–1696.
[32] R. Petkov, "Optimum design of a high-power high-frequency transformer," *IEEE Transactions on Power Electronics*, vol. 11, no. 1, pp. 33–42, January 1996.
[33] C.-M. Ong, *Dynamic Simulation of Electric Machinery*, Reading, MA: Prentice-Hall, 1998, pp. 38–40, 45–46, and 87–90.
[34] G. Bertotti, *Hysteresis of Magnetism*, San Diego, CA: Academic Press, 1998.
[35] W. G. Hurley, W. H. Wolfe, and J. G. Breslin, "Optimized transformer design: inclusive of high-frequency effects," *IEEE Transactions on Power Electronics*, vol. 13, no. 4, pp. 651–659, July 1998.

[36] W. G. Hurley, E. Gath, and J. G. Breslin, "Optimizing the ac resistance of multilayer transformer windings with arbitrary current waveforms," *IEEE Transactions on Power Electronics*, vol. 15, no. 2, pp. 369–376, March 2000.

[37] M. K. Kazimierczuk, G. Sancineto, U. Reggiani, and A. Massarini, "Small-signal high-frequency model of ferrite inductors," *IEEE Transactions on Magnetics*, vol. 35, pp. 4185–4191, September 1999.

[38] U. Reggiani, G. Sancineto, and M. K. Kazimierczuk, "High-frequency behavior of laminated iron-core inductors for filter applications," Proceedings of the IEEE Applied Power Electronics Conference, New Orleans, LA, February 6-10, 2000, pp. 654–660.

[39] G. Grandi, M. K. Kazimierczuk, A. Massarini, U. Reggiani, and G. Sancineto, "Model of laminated iron-core inductors," *IEEE Transactions on Magnetics*, vol. 40, no. 4, pp. 1839–1845, July 2004.

[40] K. Howard and M. K. Kazimierczuk, "Eddy-current power loss in laminated power cores," Proceedings of the IEEE International Symposium on Circuits and Systems, Sydney, Australia, May 7-9, 2000, paper III-668, pp. 668–672.

[41] A. Reatti and M. K. Kazimierczuk, "Comparison of various methods for calculating the ac resistance of inductors," *IEEE Transactions on Magnetics*, vol. 37, no. 3, pp. 1512–1518, May 2002.

[42] G. Grandi, M. K. Kazimierczuk, A. Massarini, and U. Reggiani, "Stray capacitance of single-layer solenoid air-core inductors," *IEEE Transactions on Industry Applications*, vol. 35, no. 5, pp. 1162–1168, September 1999.

[43] A. Massarini and M. K. Kazimierczuk, "Self-capacitance of inductors," *IEEE Transactions on Power Electronics*, vol. 12, no. 4, pp. 671–676, July 1997.

[44] H. A. Wheeler, "Simple inductance formulas for radio coils," *Proceedings of the IRE*, vol. 16, no. 10, pp. 1398–1400, October 1928.

[45] R. W. Erickson and D. Maksimović, *Fundamentals of Power Electronics*, Norwell, MA: Kluwer Academic Publishers, 2001.

[46] A. Van den Bossche and V. C. Valchev, *Inductors and Transformers for Power Electronics*, Boca Raton, FL: Taylor & Francis, 2005.

[47] A. Van den Bossche and V. C. Valchev, "Improved calculation of winding losses in gapped inductors," *Journal of Applied Physics*, vol. 97, 10Q703, 2005.

[48] F. E. Terman, *Radio Engineers' Handbook*, New York: McGraw-Hill, 1943.

[49] P. Wallmeir, N. Frohleke, and H. Grotstollen, "Improved analytical modeling of conductive losses in gapped high-frequency inductors," *IEEE Transactions on Industry Applications*, vol. 37, no. 4, pp. 558–567, July 2001.

[50] X. K. Mao, W. Chen, and Y. X. Lee, "Winding loss mechanism analysis and design for new structure high-frequency gapped inductor," *IEEE Transactions on Magnetics*, vol. 41, no. 10, pp. 4036–i4038, October 2005.

[51] N. Das and M. K. Kazimierczuk, "An overview of technical challenges in the design of current transformers," Electrical Manufacturing Conference, Indianapolis, IN, October 24-26, 2005.

[52] T. Suetsugu and M. K. Kazimierczuk, "Integration of Class DE inverter for dc-dc converter on-chip power supplies," IEEE International Symposium on Circuits and Systems, Kos, Greece, May 21-24, 2006, pp. 3133–3136.

[53] T. Suetsugu and M. K. Kazimierczuk, "Integration of Class DE synchronized dc-dc converter on-chip power supplies," IEEE Power Electronics Specialists Conference, Jeju, South Korea, June 21-24, 2006.

[54] W. A. Roshen, "Fringing field formulas and winding loss due to an air gap," *IEEE Transactions on Magnetics*, vol 43, no. 8, pp. 3387–3994, August 2007.

[55] M. K. Kazimierczuk, *Pulse-Width Modulated DC-DC Power Converters*, Chichester: John Wiley & Sons, 2008.

[56] M. K. Kazimierczuk, *RF Power Amplifiers*, Chichester: John Wiley & Sons, 2008.

[57] M. K. Kazimierczuk and D. Czarkowski, *Resonant Power Converters*, 2nd Ed., New York: John Wiley & Sons, 2011.

[58] W. G. Hurley and W. H. Wölfe, *Transformers and Inductors for Power Electronics: Theory, Design and Applications*, Chichester: John Wiley & Sons, 2013.

[59] D. Meeker, Finite Element Magnets, Version 4.2 User's Manual, www.femm.foster-miller.net (accessed 2 July 2013).

[60] www.mag-inc.com, www.ferroxcube.com, www.ferrite.de, www.micrometals.com, and www.metglas.com.

1.21 Review Questions

1.1. What is the MMF?

1.2. What is the magnetic flux?

1.3. What is the magnetic field intensity?

1.4. What is the magnetic flux density?

1.5. What is the magnetic linkage?

1.6. Define relative permeability.

1.7. What is the reluctance of a core?

1.8. What is the magnetic circuit? Give an example.

1.9. Can magnetic field exist in a good conductor?

1.10. State Ampère's circuital law.

1.11. State Faraday's law.

1.12. State Lenz's law.

1.13. What is Joule's law?

1.14. What is the point (microscopic) form of Ohm's law?

1.15. Write Maxwell's equations in differential and integral forms.

1.16. Write Maxwell's equations for good conductors.

1.17. State Poynting's law.

1.18. Write Biot–Savart's law.

1.19. Derive Joule's law.

1.20. Define power.

1.21. What is core saturation?

1.22. Define an inductance of a linear inductor.

1.23. Define an inductance of a nonlinear inductor.

1.24. What is the core inductance factor?

1.25. How is the inductance of a coil related to its number of turns?

1.26. What is the effect of an air gap on the inductance?

1.27. What is the fringing factor?

1.28. What is the effect of an air gap on core saturation?

1.29. Where is the magnetic energy stored in an inductor with an air gap?

1.30. Is the magnetic field intensity in the air gap higher or lower than that in the core?

1.31. Is the magnetic flux density in the air gap higher or lower than that in the core?

1.32. What is the volt–second balance?

1.33. Give expressions for magnetic energy in terms of H and B.

1.34. What are the mechanisms of power losses in magnetic components?

1.35. What are winding losses?

1.36. What is hysteresis loss?

1.37. What is eddy-current loss?

1.38. What are the effects of eddy currents on winding conductors and magnetic cores?

1.39. What is the SRF?

1.40. What is the difference between fringing flux and leakage flux?

1.41. The line integral of the magnetic field intensity **H** over a closed contour is zero. What is the net current flowing through the surface enclosed by the contour?

1.22 Problems

1.1. A current flows in the inner conductor of a long coaxial cable and returns through the outer conductor. What is the magnetic field intensity in the region outside the coaxial cable? Explain why.

1.2. Sketch the shape of the magnetic field around a current-carrying conductor and show how the direction of the field is related to the direction of the current in the conductor.

1.3. A toroidal inductor has the number of turns $N = 20$, the inner radius $a = 1$ cm, the outer radius $b = 2$ cm, and the height $h = 1$ cm. The core relative permeability is $\mu_{rc} = 100$. Find the inductance.

1.4. An inductor has $N = 300$ turns, $B = 0.5\ T$, and carries a current I of 0.1 A. The length $l_c = 15$ cm and cross-sectional area $A_c = 4\ \text{cm}^2$. Find the magnetic flux intensity, magnetic flux, and flux linkage.

1.5. An inductor has $\mu_{rc} = 800$, $N = 700$, $\phi = 4 \times 10^{-4}$ Wb, $l_w = 22$ cm, and $A_c = 4 \times 10^{-4}\ \text{m}^2$. Find the current I.

1.6. An inductor has $L = 100\ \mu\text{H}$, $l_c = 2.5$ cm, and $A_c = 2\ \text{cm}^2$. Find the number of turns N.

(a) For $\mu_{rc} = 1$.

(b) For $\mu_{rc} = 25$.

(c) For $\mu_{rc} = 25$ and $l_g = 3$ mm.

(d) For $\mu_{rc} = 2500$ and $l_g = 3$ mm.

1.7. A core has $A_L = 30\ \mu\text{H}/100$ turns. Find N to make an inductor of 1 μH.

1.8. A toroidal core has $N = 500$, $\mu_{rc} = 200$, $A_c = 4\ \text{cm}^2$, $r = 2$ cm, $I_m = 1c\text{A}$, $f = 10$ MHz, $\rho_w = \rho_{Cu} = 1.724 \times 10^{-6}\ \Omega\cdot\text{cm}$, and $\rho_c = 10^5\ \Omega\cdot\text{m}$. Find L, A_L, $\mathcal{R}$, H_m, B_m, ϕ_m, and λ_m.

1.9. A toroidal core of $\mu_{rc} = 3000$ has a mean radius $R = 80$ mm and a circular cross section with radius $b = 25$ mm. The core has an air gap $l_g = 3$ mm and a current I flows in a 500-turn winding to produce a magnetic flux of 10^{-4} Wb. Neglect the leakage flux.

(a) Determine the reluctance of the air gap, the reluctance of the core, and the total reluctance of the core with air gap.

(b) Find B_g and H_g in the air gap and B_c and H_c in the core.

(c) Find the required current I.

1.10. An inductor has $N = 100$, $A_c = 1\ \text{cm}^2$, $B_s = 0.3\ T$, $v_L = 10 \cos \omega t$ (V). Find $\lambda(t)$ and f_{min}.

1.11. Derive an expression for the internal and external inductances of a two-wire transmission line consisting of two parallel conducting wires of radius a that carry currents I in opposite directions. The axis-to-axis distance between the two wires is $d \gg a$.

1.12. The number of turns of a 100-μH inductor is doubled, while maintaining its cross-sectional area, length, and core material. What is the new inductance?

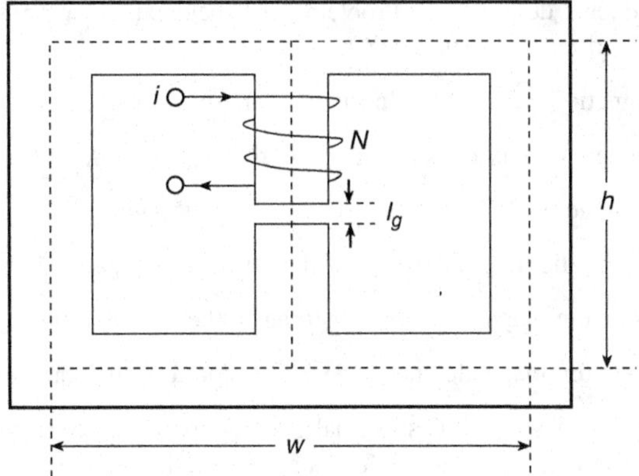

Figure 1.34 An inductor with an EE core and an air gap

1.13. An inductor with an EE magnetic core has an air gap in the center leg of length $l_g = 1$ mm and $\mu_{rc} = 3000$, as shown in Fig. 1.34. The height of the mean magnetic path is $h = 10$ cm. The width of the mean magnetic path at the core base is $w = 16$ cm. The cross-sectional area of all the legs is $A_c = 4$ cm^2. The winding is placed on the center leg. The number of turns is $N = 100$. The inductor current is $i_L = I_m \sin \omega t = 2 \sin 2\pi 60t$ (A).

(a) Draw the magnetic circuit and its single-loop equivalent circuit.

(b) Find the reluctance of each leg, the reluctance of the air gap, and the total reluctance of the core with the air gap.

(c) Find the amplitude of the magnetic flux in each leg and the air gap.

(d) Find the amplitude of the magnetic flux density in each leg and the air gap.

(e) Find the amplitude of the magnetic field intensity in each leg and the air gap.

(f) Determine the inductance.

1.14. An inductor with an air-gapped CC cut core shown in Fig. 1.35 has $\mu_{rc} = 10^5$, $N = 66$, $l_c = 17$ cm, and the length of the air gap on each side of the CC core is 0.5 mm. The cross section of all legs is a rectangular with dimensions $a = 1.28$ cm and $b = 0.98$ cm. Neglect the fringing effect of the magnetic flux.

(a) Determine the reluctance of the core, the reluctance of the gap, and the total reluctance.

(b) Find the ratio of the gap reluctance to the core reluctance.

(c) Determine the inductance.

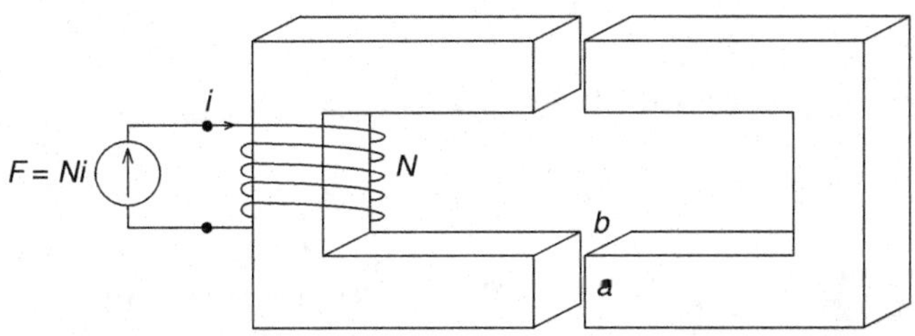

Figure 1.35 Inductor with a CC cut core

1.15. Consider the inductor described in Problem 1.14, neglecting the fringing flux. The inductor current is $i = I_m \sin \omega t = 0.5 \sin \omega t$ (A).

(a) Find the magnetic flux densities in the core and the gaps.

(b) Sketch the magnetic flux densities in the core and the gaps.

(c) Find the magnetic field intensities in the core and the gaps.

(d) Sketch the magnetic field intensities in the core and the gaps.

(e) Find the maximum magnetic energies stored in the core and both gaps.

(f) Find the maximum magnetic energy densities stored in the core and both gaps.

1.16. An air-gapped inductor with a CC supermalloy cut core shown in Fig. 1.35 has $a = 12.8$ mm, $b = 9.8$ mm, $\mu_{rc} = 10^5$, and $N = 66$. The air-gap length on each side is 0.5 mm. Find F_f, L, and L_f.

1.17. An inductor with an air-gapped CC cut core shown in Fig. 1.35 has $\mu_{rc} = 10^5$, $B_s = 1.5$ T, $N = 66$, $l_c = 17$ cm, and the length of the air gap on each side of the CC core is 0.5 mm. The cross section of all legs is rectangular with dimensions $a = 1.28$ cm and $b = 0.98$ cm. The inductor current is sinusoidal. Neglect the fringing effect of the magnetic flux.

(a) Determine the maximum amplitude of magnetic flux density in the core and in the gaps for operation just below core saturation.

(b) Determine the maximum amplitude of magnetic field intensity in the core and in the gaps for operation just below the core saturation.

(c) Determine the amplitudes of magnetic energy densities stored in the core and the air gaps.

(d) Determine the amplitudes of magnetic energies stored in the core and the air gaps.

(e) Determine the maximum amplitude of the inductor current just below the core saturation.

1.18. An inductor has the inductance $L = 100$ μH, ESR $r_L = 1.5$ Ω, and self-capacitance $C = 10$ pF.

(a) Find the quality factor of the inductor at $f = 1$ MHz.

(b) Find the SRF of the inductor.

Magnetic Cores

2.1 Introduction

In this chapter, magnetic properties of materials are reviewed. Magnetic cores used in inductors and transformers are studied [1–81]. The concepts of magnetic dipoles and magnetic domains are presented. Magnetization, relative permeability, and magnetic hysteresis are discussed. Core saturation is explained. Basic properties of various magnetic cores are covered. Hysteresis loss and eddy-current loss in magnetic cores are analyzed. The complex permeability as a function of frequency is also studied.

2.2 Properties of Magnetic Materials

Ferromagnetic chemical elements are

- iron (Fe)
- cobalt (Co)
- nickel (Ni)
- gadolinium (Gd)

Soft magnetic materials can be broadly classified into four categories:

- steels (SiFe)
- FeNi and FeCo alloys
- MnZn and NiZn ferrites
- amorphous (noncrystalline) metallic alloys (metallic glasses)

Almost all materials are poor conductors of magnetic flux ϕ because their relative permeability μ_{rc} is low nearly equal to 1. The purpose of magnetic cores is to reduce the reluctance $\mathcal{R}$, to increase the

High-Frequency Magnetic Components, Second Edition. Marian K. Kazimierczuk.

Companion Website: www.wiley.com/go/kazimierczuk_High2e

inductance L, to increase the magnetic flux ϕ, to obtain a well-defined magnetic path length (MPL), to link the energy stored in the air gap to the winding by a low reluctance flux path, and to contain the magnetic flux in the core, reducing the leakage inductance L_l and electromagnetic interference (EMI) level. Magnetic cores increase the inductance density and magnetic energy density of magnetic devices, reducing the size and weight of power electronic circuits. There are two types of magnetic materials: *soft and hard magnetic materials.* Soft magnetic materials can be easily magnetized and demagnetized so that they can transfer or store magnetic energy in circuits with AC waveforms (sine waves, square waves, rectangular waves, etc.). Therefore, soft magnetic materials can be used in magnetic cores for inductors and transformers. Hard magnetic materials are difficult to magnetize and demagnetize and thus are used as permanent magnets in applications such as brushless and synchronous electric motors.

The major parameters of magnetic materials should be as follows:

- high relative permeability μ_{rc},
- high saturation magnetic flux density B_s,
- low coercivity H_c,
- high electrical resistivity ρ_c,
- high Curie temperature T_c,
- low hysteresis and eddy-current losses per unit volume P_v,
- high upper operating frequency f_H (or wide bandwidth BW).

The figure-of-merit of a magnetic material is defined here as

$$KFOM = \frac{\mu_{rc}\rho_c B_s f_H}{P_v}. \tag{2.1}$$

The major magnetic materials are silicon steel, nickel iron (permalloy), cobalt iron (permendur), ferrites, amorphous metallic alloys (also called metallic glasses), iron powders, moly-permalloy powders (MPPs), and sendust powders. Magnetic cores are conductors of electric current. The range of the electrical resistivity ρ_c of magnetic cores is from 10^{-7} to $10^7\,\Omega\cdot$ m. Table 2.1 lists the values of the electrical resistivity ρ_c of various magnetic materials.

Table 2.1 Resistivities of ferromagnetic materials at $T = 20\,^\circ$C

Material	Resistivity ρ_c ($\Omega\cdot$ cm)
Co ferrite	10^7
Mg ferrite	10^7
NiZn ferrite	$10^3 - 10^7$
Cu ferrite	10^5
Mn ferrite	10^4
MnZn ferrite	$0.5 - 1000$
Zn ferrite	10^2
Fe ferrite	4×10^{-3}
Gd (gadolinium)	131×10^{-6}
Metallic glass	125×10^{-6}
NiFe (Nickel iron)	45×10^{-6}
SiFe (Silicon steel, 2.5% of Si)	45×10^{-6}
50% Co alloy	35×10^{-6}
SiFe (Silicon iron, 1% of Si)	25×10^{-6}
SiFe (Silicon iron, 0.25% of Si)	10×10^{-6}
Fe (iron)	9.61×10^{-6}
Ni (nickel)	6.93×10^{-6}
Co (cobalt)	6.24×10^{-6}
Nanocrystalline	1.2×10^{-6}

The main purpose of the magnetic core is to conduct magnetic flux, that is, to provide an easy path for magnetic flux and to concentrate magnetic field in the core. Magnetic cores conduct magnetic flux in magnetic circuits in a similar fashion as conductors conduct electric currents in electric circuits. The relative permeability μ_{rc} is a measure of how much better a given material is than air for conducting magnetic flux. The relative permeability μ_{rc} of ferromagnetic materials is as much as 10^5 times higher than that of free space. The electrical resistivity of copper ρ_{Cu} is 10^{20} times lower than that of free space. The result of the core presence in a magnetic circuit is an increase in the inductance over that of the identical air-core inductor, and therefore, to achieve better capability to store magnetic energy, while reducing radiation of magnetic field. Useful materials for this purpose have "soft" magnetic properties, which means that they do not become permanently magnetized like "hard" magnetic materials. When a magnetic material is placed in the magnetic field, it becomes magnetized, that is, it becomes a magnet itself. Soft magnetic materials removed from the magnetic field tend to lose their magnetization. The degree of enhancement of magnetic field concentration due to the presence of the core is a function of the relative permeability μ_{rc} of the core material.

Magnetic cores are made of soft ferromagnetic materials, whose relative permeability μ_{rc} is from 5 to 10^5. Ferromagnetic materials are much better conductors of magnetic flux ϕ than any other material. The improvement is by a factor of μ_{rc}. Several metallic elements, such as iron (Fe), cobalt (Co), nickel (Ni), and gadolinium (Gd), act as soft ferromagnetic materials. Many alloys containing these elements are also ferromagnetics. Common alloy elements are silicon, aluminum, manganese, zinc, and chromium. The name iron originates from the Latin word ferrum.

2.3 Magnetic Dipoles

Materials consists of many atoms. An atom consists of a nucleus in the center and electrons rotating on orbits. A nucleus contains positively charged protons and neutral neutrons. Electrons are negatively charged. An atom nucleus spins around itself, the surrounding electrons are orbiting around the nucleus, and each electron spins about its own axis. The motion of electrons around the nucleus is equivalent to the flow of electric current in a small resistanceless loop, called *magnetic dipoles* with south and north poles. Microscopic current loops produce a magnetic field directed from the south poles to the north poles. They behaved like tiny magnets. There are many electrons, and their orbits are randomly oriented, causing their magnetic field cancellations. When an external magnetic field is applied, the orientation of orbits changes, altering the directions of the magnetic fields. By Lenz's law, the orbiting electron magnetic fields tend to oppose the external magnetic field.

A *magnetic domain* is a region with uniform magnetization within a magnetic material, where individual magnetic moments are aligned in parallel in one direction, typically from 10^{12} to 10^{18}. The directions of alignment of various domains are random when no external magnetic field is applied. The size of a magnetic domain is typically from 10^{-6} to 10^{-4} m. Magnetic domains determine the magnetic properties and behavior of ferromagnetic materials. The alignment of the dipoles with the magnetic field **H** is opposed by random thermal motion of electrons caused by ever-present thermal agitation. Magnetic domains exist in the transition metals and the rare earth metals.

An external magnetic field progressively aligns the microscopic current loops in the direction normal to the magnetic field. The magnetic moments are reoriented either by rotation of domains or by growth and contraction of domains. The current loops are displaced from their original positions. This action results in storing energy. Magnetic materials exhibit magnetic polarization when they are subjected to an applied magnetic field. The magnetization of a material occurs when magnetic dipoles are aligned with the applied magnetic field. Magnetization depends on the structure of an atom, the placement of electrons in different energy orbits, crystal environment, and aligning the electron spins parallel to one another. In an atom, there is a central positive nucleus surrounded by electrons in various circular orbits. A fundamental property of atoms is that both electrons and protons spin on their axes. In addition, electrons revolve around the nucleus. The spinning charge appears as current flowing around the loop. A spinning electron is similar to a spinning gyroscope suspended at a point other than its center of gravity. It moves due to a magnetic force, which is coincident with an applied magnetic field. Atoms having more electrons spinning in one direction than another act as small magnets. The spin is a magnetic moment vector, causing an electron or proton to behave like a tiny magnet with a north and south pole. Magnetization of a magnetic material is caused by atomic current loops generated by three mechanisms:

1. electron spin about its own axis;
2. orbital motion of the electrons around the nucleus; and
3. motion of the protons around each other in the nucleus.

An electron moving in circular motion about the nucleus resembles a tiny current loop because it is a moving charge, as shown in Fig. 2.1a. An electron also spins about its own axis, as depicted in Fig. 2.1b. The direction of the loop current is opposite to the direction of the electron movement. Each current loop produces a magnetic field H. The pattern of the magnetic field generated by a loop current is similar to that exhibited by a permanent magnet. Therefore, the miniature current loop can be regarded as a *magnetic dipole* with a north pole and a south pole, as shown in Fig. 2.1c. The magnetic field lines pass through the inside of the circular loop and return externally in the closed paths. The magnetic dipoles may be regarded as tiny magnets. Magnetic dipoles cause magnetization (or magnetic polarization). The tiny current loop has an orbital magnetic moment equal to the product of the loop current I and the loop area $A = \pi r^2$. The average orbit radius is $r \approx 10^{-10}$ m. The *orbital moment* of the electron magnetic dipole is given by

$$\mathbf{m}_o = A\mathbf{I} = \pi r^2 \mathbf{I}. \tag{2.2}$$

The vector of the orbital moment $\mathbf{m}_o$ is perpendicular to the surface of the current loop, and its direction is normal to the loop and is determined by the right-hand rule or the right-hand screw rule. When a magnetic flux density $\mathbf{B}$ is applied to the atom, the dipole of atom experiences a torque (a turning force)

$$\mathbf{T}_q = \mathbf{m}_o \times \mathbf{B} = A\mathbf{I} \times \mathbf{B} = \pi r^2 \mathbf{I} \times \mathbf{B}. \tag{2.3}$$

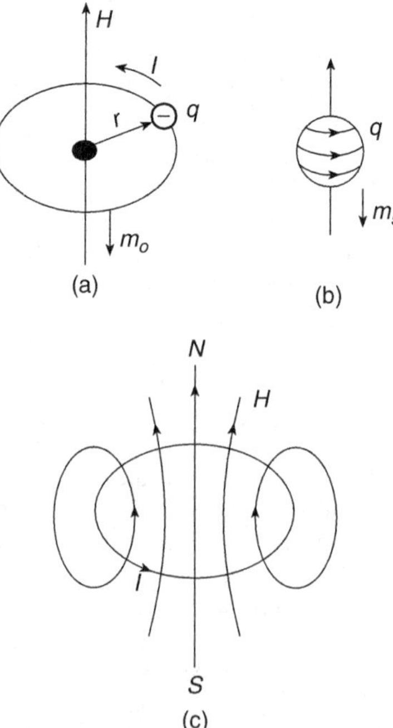

Figure 2.1 Electron orbiting the nucleus and spins about its own axis. (a) Electron orbiting the nucleus and exhibits the orbiting magnetic moment m_o. (b) Electron spins about its own axis and exhibits a spin magnetic moment m_s. (c) Magnetic dipole

This torque tends to align the direction of the orbital magnetic moment with the direction of the applied magnetic field **B**, whose magnitude is given by

$$|\mathbf{T}_q| = |\mathbf{m}_o||\mathbf{B}|\sin\phi = AI|\mathbf{B}|\sin\phi = \pi r^2 I|\mathbf{B}|\sin\phi, \tag{2.4}$$

where ϕ is the angle between the vectors $\mathbf{m}_o$ and $\mathbf{B}$. For full alignment, $\phi = 0$. As a result, the total magnetic field becomes stronger.

A spinning electron behaves like an infinitesimal current loop. For most atoms, the magnetic fields produced by both orbit and spin motions of the electrons cancel out so that the atom exhibits very little magnetic effect. In other atoms, the orbit and spin fields do not cancel each other and therefore the material shows magnetic effects. Although the effect of each atomic current loop is very small, the combined effect of billions of atoms results in a strong magnetic field.

An electron generates the following magnetic moments:

1. an orbital magnetic moment m_o because it rotates around the nucleus and
2. a spin magnetic moment m_s because it spins about its own axis while it orbits the nucleus.

An electron with a $-q$ charge moving with a constant velocity v in a circular orbit of radius r completes one full revolution (the circumference of the orbit) in time

$$T = \frac{2\pi r}{v} = \frac{2\pi}{\omega}, \tag{2.5}$$

where v is the electron orbital speed and $\omega = v/r$. The tiny loop current I flows in the direction opposite to the electron orbiting motion and is given by

$$I = -\frac{q}{T} = -\frac{qv}{2\pi r} = -\frac{q\omega}{2\pi}. \tag{2.6}$$

If the applied magnetic field is zero, the Coulomb's force acting on an electron due to the proton charge q is

$$\mathbf{F} = -\mathbf{a}_r \frac{q^2}{4\pi \epsilon_0 r^2}, \tag{2.7}$$

where $\mathbf{a}_r$ is a unit vector pointing from the proton toward the electron. The centrifugal force acting on the electron due to its circulating motion is

$$\mathbf{F} = m_e \omega^2 r = \frac{m_e v^2}{r}. \tag{2.8}$$

When the electron is located on the orbit, the two forces will counterbalance each other

$$\frac{q^2}{4\pi \epsilon_0 r^2} = \frac{m_e v^2}{r}, \tag{2.9}$$

yielding the orbital electron velocity

$$v = \frac{q}{\sqrt{4\pi \epsilon_0 m_e r}}. \tag{2.10}$$

The time of one cycle of an electron is

$$T = \frac{2\pi r}{v} = \frac{2\pi r}{q}\sqrt{4\pi \epsilon_0 m_e r}. \tag{2.11}$$

Thus, the current induced by one orbiting electron is

$$I = \frac{q}{T} = \frac{q^2}{2\pi r\sqrt{4\pi \epsilon_0 m_e r}} = \frac{q^2}{4\pi r\sqrt{\pi \epsilon_0 m_e r}}. \tag{2.12}$$

The orbital magnetic moment of an electron dipole is

$$m_o = AI = \pi r^2 I = \frac{q^2 r}{4\sqrt{\pi \epsilon_0 m_e r}} = \frac{q^2}{4}\sqrt{\frac{r}{\pi \epsilon_0 m_e}}. \tag{2.13}$$

If the magnetic field **H** normal to the electron orbit plane is applied, the Coulomb's force and the Lorentz force acting on an electron due to the proton charge q and the applied magnetic field H are

$$\mathbf{F} = -\mathbf{a}_r \frac{q^2}{4\pi \epsilon_0 r^2} + qv\mu_0 \mathbf{H} = -\mathbf{a}_r \frac{q^2}{4\pi \epsilon_0 r^2} + q\omega r \mu_0 \mathbf{H}, \tag{2.14}$$

The Coulomb's force and the centrifugal force will counterbalance each other

$$\frac{q^2}{4\pi \epsilon_0 r^2} + qvr\mu_0 H = \frac{m_e v^2}{r}. \tag{2.15}$$

yielding the quadratic equation of v

$$\frac{m_e}{r} v^2 - qr\mu_0 Hv - \frac{q^2}{4\pi \epsilon_0 r^2} = 0. \tag{2.16}$$

Solution of this equation is

$$v = \frac{qr^2\mu_0 H}{2m_e} + \frac{r}{2m_e}\sqrt{(qr\mu_0 H)^2 + \frac{q^2 m_e}{\pi \epsilon_0 r^3}} = \frac{qr}{2m_e}\left[r\mu_0 H + \sqrt{(r\mu_0 H)^2 + \frac{m_e}{\pi \epsilon_0 r^3}}\right]. \tag{2.17}$$

The solution with the minus sign is discarded as nonphysical. The time of one cycle of an electron on its orbit is

$$T = \frac{2\pi r}{v} = \frac{4\pi m_e}{q\left[r\mu_0 H + \sqrt{(r\mu_0 H)^2 + \frac{m_e}{\pi \epsilon_0 r^3}}\right]}. \tag{2.18}$$

The current induced by one orbiting electron is

$$I = \frac{q}{T} = \frac{q^2\left[r\mu_0 H + \sqrt{(r\mu_0 H)^2 + \frac{m_e}{\pi \epsilon_0 r^3}}\right]}{4\pi m_e}. \tag{2.19}$$

The orbital magnetic moment of an electron dipole is

$$m_o = AI = \pi r^2 I = \frac{q^2 r^2\left[r\mu_0 H + \sqrt{(r\mu_0 H)^2 + \frac{m_e}{\pi \epsilon_0 r^3}}\right]}{4m_e}. \tag{2.20}$$

Thus, the electron orbital velocity v increases, T decreases, the current produced by the orbiting electron increases, magnetic field induced by the electron loop increases, and orbital magnetic moment increases. However, the values of the above variables are almost independent of H.

The orbital magnetic moment of the current loop is related to the electron's angular momentum $\mathbf{m}_a$

$$\mathbf{m}_o = A\mathbf{I} = -\frac{q\mathbf{v}}{2\pi r}(\pi r^2) = -\frac{q\mathbf{v}r}{2} = -\frac{q}{2m_e} m_e \mathbf{v} r = -\frac{q}{2m_e}\mathbf{m}_a, \tag{2.21}$$

where m_e is the electron mass and the angular momentum of the electron is

$$\mathbf{m}_a = m_e \mathbf{v} r. \tag{2.22}$$

The mass of an electron is $m_e = 9.11 \times 10^{-31}$ kg.

The orbital angular momentum m_a is quantized. Its minimum step change is

$$\Delta m_{a(min)} = \frac{h}{2\pi} = 1.06 \times 10^{34} \text{ J} \cdot \text{s}, \tag{2.23}$$

where $h = 6.626 \times 10^{-34}$ J · s is the Planck's constant. The orbital angular momentum m_a is always an integer multiple of $h/(2\pi)$

$$m_a = n \times m_{a(min)} = 0, \frac{h}{2\pi}, \frac{2h}{2\pi}, \ldots . \tag{2.24}$$

The minimum step change in the magnitude of the orbital magnetic moment is

$$\Delta m_{o(min)} = \frac{q}{2m_e}\left(\frac{h}{2\pi}\right) = \frac{qh}{4\pi m_e}. \tag{2.25}$$

The orbital magnetic moment is

$$m_o = n \times m_{o(min)} = 0, \frac{q}{2m_e}\left(\frac{h}{2\pi}\right), \frac{q}{2m_e}\left(\frac{2h}{2\pi}\right), \ldots . \tag{2.26}$$

If the magnetic dipoles are aligned in one direction, they form a structure, similar to many parallel solenoids, and therefore induce a magnetic field. A set of magnetic dipoles aligned in the same direction is called a *magnetic domain*.

An electron has an intrinsic angular momentum (spin) and an intrinsic magnetic moment. The spin electron moments tend to align with the external field. Therefore, the total magnetic field becomes stronger. From quantum theory, the spin magnetic moment of an electron is given by

$$m_s = \frac{qh}{4\pi m_e} = m_{o(min)}. \tag{2.27}$$

The spin magnetic moment of an iron electron is $m_s = 9.3 \times 10^{-24}$ A · m^2. If the number of electrons of an atom is odd, the atom has a nonzero spin magnetic moment and external magnetic fields exert torques on the electron spins. If the number of electrons of an atom is even, the spins in every pair of electrons have opposite directions, thereby cancelling each other spin magnetic moment.

The total magnetic moment of an atom is equal to the vector sum of the atom orbital and spin magnetic moments

$$\mathbf{m}_A = \sum_{i=1}^{N_A} \mathbf{m}_i, \tag{2.28}$$

where N_A is the number of magnetic moments in the atom. The orbital magnetic moment m_o is of the same order of magnitude as the spin magnetic moment m_s. The magnetic moment of an electron is of the order 1000 times stronger than that of the nucleus. Therefore, the total magnetic moment of an atom is dominated by the sum of the magnetic moments of its electrons. The magnetic properties of a material are determined by interaction of the magnetic dipoles of its atoms with an external magnetic field. These properties depend on the crystalline structure of a material. Ferrites consist of small crystals (grains) separated by very thin grain boundaries. In most materials, the magnetic moment of one electron in an atom is cancelled by that of another electron orbiting in the opposite direction. Therefore, the net magnetic moment produced by the orbital motion of electrons is either zero or very small in most materials.

In atoms with even number of electrons, pairs of electrons have opposite spins, and thus the spin magnetic moments cancel. In contrast, atoms containing odd number of electrons have an unpaired electron, producing the spin magnetic moment. Magnetic materials can be classified as follows: ferromagnetics, diamagnetics, paramagnetics, antiferromagnetics, ferrimagnetics, and superparamagnetics. In iron, nickel, and cobalt, the orbiting and spin magnetic moments do not cancel each other. These materials are called ferromagnetics.

Magnetic dipoles are a source of magnetization (also called magnetic polarization). An orbiting electron exhibits a magnetic moment in the same direction as that of the applied field. In the presence of an external magnetic field, a magnetic dipole will experience a torque, that is, a turning force. The effect of the torque will tend to align the magnetic dipole (i.e., the magnetic field produced by the orbiting electron) in the direction of the applied magnetic field.

The *magnetization* is the magnetic dipole moment per unit volume. There are n_D magnetic dipoles per cubic meter of a magnetic material. The magnetization (or magnetic polarization) vector **M** of a material is the vector sum of the magnetic dipole moments $\mathbf{m_i}$ of all N_D atoms contained in a unit volume

$$\mathbf{M} = \frac{\sum_{i=1}^{n\Delta V} \mathbf{m}_i}{V_c} = \frac{N_D \mathbf{m}_i}{V_c} = n_D \mathbf{m}_i = n_D A\mathbf{I}. \tag{2.29}$$

The unit of the magnetization vector **M** is A/m. For example, an iron contains $n = 8.5 \times 10^{28}$ atoms/m^3. Each atom of an iron contributes one electron to align its spin magnetic moment m_s along the direction of the applied magnetic field **H**. The maximum magnitude of M for iron is 7.9×10^5 A/m.

Consider a bar of a long cylindrical magnetic domain of length l and cross-sectional area A_D, as shown in Fig. 2.2. The magnetic domain contains N_D magnetic dipoles. The density of dipoles

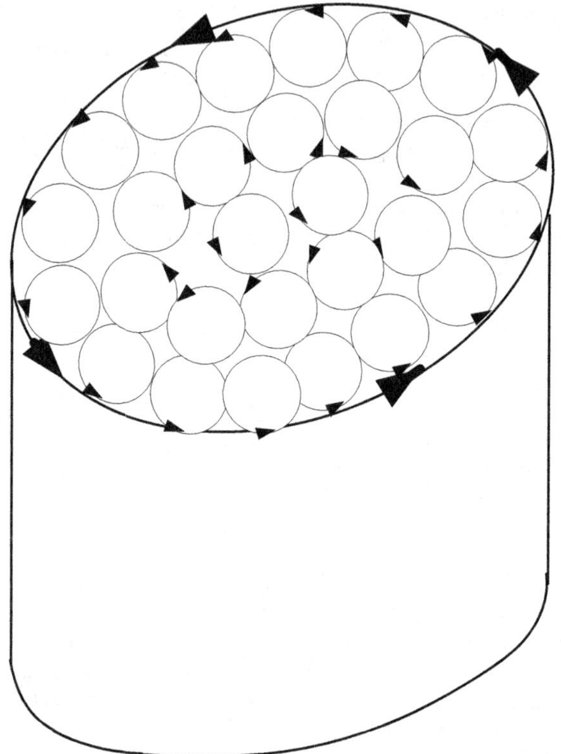

Figure 2.2 Magnetic domain

is $n_D = N_D/lA = N_D/V_D$, where $V_D = lA$ is the domain volume. All dipole moments are aligned along the cylinder axis. In the magnetic domain, the dipoles form many parallel solenoids. In the interior of the domain, the currents of one solenoid cancel the currents flowing in the adjacent solenoids. Therefore, the resultant current in the interior is zero. Only the currents at the domain wall do not cancel each other and produce a net surface current I_{surf}, yielding a large magnetic dipole. The magnetic moment of this dipole is equal to the magnetic moment of the domain composed of individual atom dipoles in the form of large solenoid of cross-sectional area A_D

$$A_D I_{surf} = N_A m_i, \tag{2.30}$$

resulting in the magnetization

$$M = \frac{N_A m_i}{V_D} = \frac{I_{surf}}{L} = J_{surf}. \tag{2.31}$$

Thus, M is equal to the surface current density.

Example 2.1

Find the current and the magnetic moment of the iron electron on the outermost orbit. The outermost orbit radius of iron electron is $r = 140$ pm $= 14 \times 10^{-11}$ m.

Solution: When the applied magnetic field is zero, the Coulomb's force existing between the electron and the proton is counterbalanced by the centrifugal force due to the electron circulating motion

$$F = \frac{q^2}{4\pi \epsilon_0 r^2} = \frac{m_e v^2}{r}, \tag{2.32}$$

yielding the electron velocity on the orbit

$$v = \frac{q}{\sqrt{4\pi\epsilon_0 m_e r}} = \frac{1.60218 \times 10^{-19}}{\sqrt{4\pi \times 8.85418 \times 10^{-12} \times 9.1095 \times 10^{-31} \times 14 \times 10^{-11}}}$$

$$= 1.345 \times 10^6 \text{ m/s}. \tag{2.33}$$

The time of one cycle of an electron is

$$T = \frac{2\pi r}{v} = \frac{2\pi \times 14 \times 10^{-11}}{1.345 \times 10^6} = 6.54 \times 10^{-16} \text{ s}. \tag{2.34}$$

Hence, the current produced by the orbiting electron is

$$I = \frac{q}{T} = \frac{1.60218 \times 10^{-19}}{6.54 \times 10^{-16}} = 0.245 \text{ mA}. \tag{2.35}$$

The amount of current produced by a very small charge of an electron is so large because its speed is very high.

The area of an electron orbit is

$$A = \pi r^2 = \pi \times (14 \times 10^{-11})^2 = 6.1575 \times 10^{-20} \text{m}^2. \tag{2.36}$$

The orbital magnetic moment of an electron is

$$m_o = AI = 0.245 \times 10^3 \times 6.1575 \times 10^{-20} \times 10^{-3} = 1.5086 \times 10^{-23} \text{ A} \cdot \text{m}^2. \tag{2.37}$$

Assuming that the magnetic domain has a shape of a cylinder of length $l_c = 0.25$ mm, the number of "solenoid" turns is

$$N = \frac{l_c}{2r} = \frac{0.25 \times 10^{-3}}{2 \times 14 \times 10^{-11}} = 0.8929 \times 10^6. \tag{2.38}$$

The magnetic field intensity in the domain is

$$H = \frac{NI}{l_c} = \frac{0.8929 \times 10^6 \times 0.245 \times 10^{-3}}{0.25 \times 10^{-3}} = 0.875 \times 10^6 \text{A/m}. \tag{2.39}$$

Hence, the magnetic flux density is

$$B_i = \mu_0 H = 4\pi \times 10^{-7} \times 0.875 \times 10^6 = 1.01 \text{ T}. \tag{2.40}$$

The magnetization due to orbital moments is

$$M = n_D m_o = 8.5 \times 10^{28} \times 1.5086 \times 10^{-23} = 1.2823 \times 10^6 \text{ A/m}. \tag{2.41}$$

2.4 Magnetic Domains

In *ferromagnetic materials*, each atom has a large dipole moment primarily due to the electron spin moment because the orbiting magnetic moment is almost zero. The magnetization is mainly the result of spin interactions among the atoms and within the crystal as a whole due to spin alignment between neighboring atoms. Ferromagnetic materials are made up of microscopic regions consisting of large groups of atoms with parallel spins called ferromagnetic domains, within which all magnetic moments are aligned. They have a variety of shapes and sizes. A magnetic domain acts like a small bar magnet. Large dipole moments are produced by domains as the vector sum of atomic moments. The domain moments, however, vary in direction from domain to domain, and the material as a whole has no magnetic moments. Upon application of an external magnetic field, these domains that have moments of the applied field increase at the expense of their neighbors and the internal field increases greatly over the external field. Spontaneous magnetization of ferromagnetics occurs due to the spin alignment between the neighboring atoms in parallel to each other over small regions, called domains. Domains are individually magnetized regions, arranged in a random manner so that the net magnetic flux in any direction is zero. Domains are created by orbital motion and spin of the electrons. They have

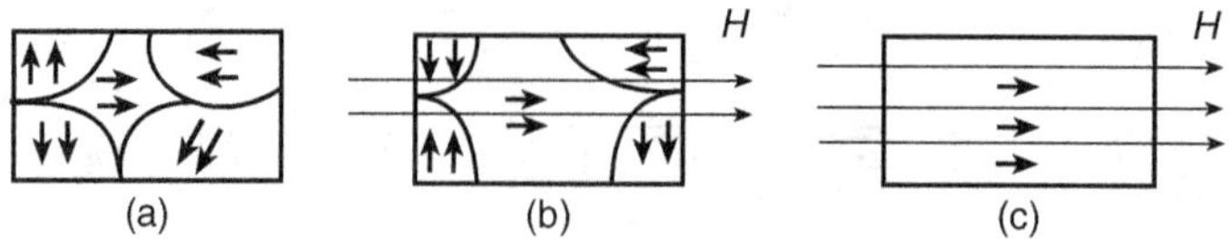

Figure 2.3 Progressive magnetization (polarization) of a ferromagnetic material. (a) An external field $H = 0$ and ferromagnetic domains are randomly oriented. (b) An external field $H > 0$ and domains parallel to the magnetic field expand, while others shrink. (c) At $B = B_s$, all domains are aligned with external field

a special arrangement of electrons in the energy orbits, which results in the parallel alignment of electron spins. Magnetization is due to electron spins and their mutual interactions at the atomic level and interatomic level within the ferromagnetic domains. The domains are approximately 1–10 μm in linear dimension (diameter). The volume of the domains is from 0.001 mm^3 to 1 mm^3. The typical volume of the domains is 1 mm^3. The area of the domain is typically 1000 times smaller than the area of the grain. Each domain contains many magnetic dipoles produced by spinning electrons, which are aligned in parallel by a strong force between neighboring dipoles. This produces a strong local magnetic field. Typically, the magnetic domain contains from 10^{16} to 10^{21} atoms. The domains are divided from each other by thin transition regions termed the *domain walls* or *Bloch walls*, which are of the order of 100 to 150 atoms thick. In these domain boundaries, there is a gradual transition of spin orientation. The atomic magnetic moments reverse the direction across the domain walls. Magnetization process proceeds by motion of the domain walls. This process is discrete in the time domain and the space domain.

When the external magnetic field is zero, the internal structure consists of small, fully magnetized domains oriented at random so that the net magnetic flux measured outside is zero, as illustrated in Fig. 2.3a. As an external magnetic field H is applied to a ferromagnetic material, it interacts with the domain magnetic fields and the domains tend to realign partially with the imposed field. The domains in which the dipoles are parallel to the applied field H grow in size, and the sizes of other domains diminish, as pictured in Fig. 2.3b. The magnetic field produced by the spinning electrons and the original external field combine to result in a stronger total magnetic flux density B. This realignment of large groups of atoms produces a high relative permeability μ_r of the ferromagnetic material. As the magnetic field intensity H is slowly raised by increasing the winding current, domains begin to reorient. The degree of magnetization depends on the degree of alignment of magnetic dipoles. At a high magnetic field intensity H, most domains are aligned with the external field. As the external field H is increased further, the entire body of the magnetic material is magnetized to become a single domain, and all the spinning electrons are aligned, as illustrated in Fig. 2.3c. The magnetic flux density B reaches the *saturation* level B_s and cannot be increased further. Beyond this point, an increase in H causes the same increase in B as in a nonmagnetic material for which $\mu_r = 1$, like in an air core. The range of the saturation flux density B_s is from 0.3 to 2.3 T. When the core enters saturation, the inductance of a coil decreases because the relative permeability μ_{rc} decreases to unity. In addition, the core losses increase. A nonuniform magnetic flux distribution of ferrite cores creates localized magnetic saturation and hot spots in the core material. The saturation flux density B_s decreases as the temperature increases.

2.5 Curie Temperature

Magnetic properties of materials change with temperature. Ferromagnetism is caused by the spontaneous magnetization due to spin alignment between the adjacent atoms in the magnetic domains. Ferromagnetic domains disintegrate abruptly when the temperature exceeds a critical value known as the *Curie temperature*[1] T_c, and therefore the ferromagnetic loses its spontaneous ferromagnetism and

[1]Pierre Curie (1859–1906) was a French physicist who received the Nobel Prize in Physics with his wife Maria Sklodowska-Curie in 1903. Maria Sklodowska-Curie (1867–1934) was a Polish physicist and chemist who also received the Nobel Prize in Chemistry in 1911.

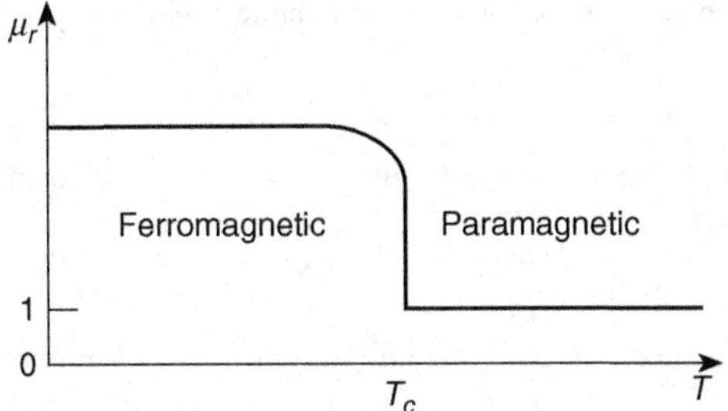

Figure 2.4 Relative permeability μ_r as function of temperature for ferromagnetics. At Curie temperature T_c, ferromagnetics become paramagnetics

Table 2.2 Curie temperatures of ferromagnetic materials

Material	Curie temperature T_c(°C)
Cobalt (Co)	1131
$Fe_{60}Co_{40}$	1087
50% cobalt 50% Fe alloy, Supermendur	950
Iron (Fe)	770
Fe_2O_3	620
Nanocrystalline	600
50%Ni–50%Fe, Orthonol	500
80% Ni, 15% Fe, 5% Mo alloy, Permalloy	460
48% Ni alloy	450
Nickel (Ni)	358
Metallic glass	300
NiZn ferrite	300
MnZn ferrite	150 – 220
Gadolinium (Gd)	19

becomes a paramagnetic. Below T_c, magnetic spins are aligned in the absence of an applied magnetic field. Above T_c, magnetic moments change directions and are randomly aligned in the absence of an applied magnetic field. As a result, the relative permeability drops suddenly at T_c from a high value to $\mu_r \approx 1$, as shown in Fig. 2.4. Spin alignment due to exchange in interactions is sensitive to temperature. Thermal agitation tends to inhibit the process of alignment of magnetic dipoles and domains. When thermal energy $E_T = kT$ exceeds the exchange energy, interactions cease to exist. When a ferromagnetic material is heated above T_c, thermal oscillations of the atomic magnets increase and overcome the coupling magnetic forces that maintain the alignment of the atomic magnets in the domains. There is a very abrupt transition from ferromagnetic to paramagnetic state and vice versa. Below the Curie temperature, a ferromagnetic material exhibits spontaneous magnetization and can be used as a magnetic core. The range of the Curie temperature T_c is usually from 120 to 1131 °C. Curie temperatures for several ferromagnetic materials are given in Table 2.2. The operating temperature of magnetic cores is usually below 100 °C.

2.6 Magnetic Susceptibility and Permeability

The value of the permeability is determined by the orbital and spin angular moments of electrons in a material. The effect of atomic dipole magnets can be described by a quantity called the *magnetization* M defined as the dipole moment per unit volume. Suppose that a magnetic field B_{air} is applied to a magnetic core. The free-space component of the magnetic flux density in the core is given by

$$\mathbf{B}_{air} = \mu_0 \mathbf{H}. \quad (2.42)$$

The internal magnetic flux density induced by the magnetization vector **M**, called the *intrinsic magnetic flux density*, is given by

$$\mathbf{B}_i = \mu_0 \mathbf{M}. \quad (2.43)$$

For a linear, homogeneous, and isotropic medium, the magnetization is proportional to the magnetic field intensity

$$\mathbf{M} = \chi_m \mathbf{H}, \quad (2.44)$$

where χ_m is the *magnetic susceptibility*. It is a dimensionless quantity. The induced (intrinsic) magnetic flux density is

$$\mathbf{B}_i = \mu_0 \mathbf{M} = \mu_0 \chi_m \mathbf{H}. \quad (2.45)$$

In the presence of an external applied magnetic flux, the total magnetic flux density is

$$\mathbf{B} = \mathbf{B}_{air} + B_i = \mu_0 \mathbf{H} + \mu_0 \chi_m \mathbf{H} = \mu_0 (1 + \chi_m) \mathbf{H} = \mu_r \mu_0 \mathbf{H} = \mu \mathbf{H}, \quad (2.46)$$

where the permeability of a magnetic material is

$$\mu = \mu_0 (1 + \chi_m) = \mu_0 \mu_r, \quad (2.47)$$

the relative permeability of a magnetic material is given by

$$\mu_r = \frac{\mu}{\mu_0} = 1 + \chi_m = 1 + \frac{M}{H}, \quad (2.48)$$

$\chi_m = \mu_r - 1$, and $\mu_0 = 4\pi \times 10^{-7}$ A/m. It can be seen that the magnetization may significantly increase the magnetic flux density. For free space, $\chi_m = 0$ and $\mu_r = 1$. The manufacturing tolerance of the relative permeability μ_r of ungapped ferrite cores is usually ±20% or ±25%. The reciprocal of permeability is *reluctivity*

$$\kappa = \frac{1}{\mu} = \frac{1}{\mu_0 \mu_r} = \frac{1}{\mu_0 (1 + \chi_m)}. \quad (2.49)$$

The series complex magnetic susceptibility is

$$\chi_s = \chi_s' - j\chi_s''. \quad (2.50)$$

The parallel complex magnetic susceptibility is

$$\frac{1}{\chi_p} = \frac{1}{\chi_p'} - \frac{1}{j\chi_p''}. \quad (2.51)$$

The inductance of an inductor with a magnetic core can be expressed as

$$L = \frac{\mu_r \mu_0 A_c N^2}{l_c} = \frac{(1 + \chi_m) \mu_0 A_c N^2}{l_c} = \frac{\mu_0 A_c N^2}{l_c} + \frac{\chi_m \mu_0 A_c N^2}{l_c} = L_o + L_{\chi m}, \quad (2.52)$$

where the inductance of air-core inductor is

$$L_o = \frac{\mu_0 A_c N^2}{l_c} \quad (2.53)$$

and the inductance due to the magnetic susceptibility of the core material

$$L_{\chi m} = \frac{\chi_m \mu_0 A_c N^2}{l_c} = \frac{(\mu_r - 1) \mu_0 A_c N^2}{l_c}. \quad (2.54)$$

2.7 Linear, Isotropic, and Homogeneous Magnetic Materials

A magnetic material is *linear* if χ_m is independent of the magnetic field intensity H, that is, if **B** varies linearly with **H**. Otherwise, it is *nonlinear*, i.e., $\chi_m = f(H)$.

A magnetic material is *homogeneous* if χ_m is independent of space coordinates, that is, it does not change from point to point so that $\mu \neq f(x, y, z)$. Otherwise, it is *inhomogeneous*.

A magnetic material is *isotropic* if χ_m is independent of the direction of the magnetic field vector **H**. Otherwise, it is *anisotropic*. In anisotropic materials, magnetic dipoles have a preferential alignment with respect to the crystal structure. If a magnetic medium is *anisotropic*, tensors are used to describe the magnetic susceptibility χ_m and the magnetic permeability μ_r. In this case, $\mathbf{M} = \|\chi\|\mathbf{H}$ and $\mathbf{B} = \mu_0\|\mu_r\|\mathbf{H}$. Otherwise, χ_m and μ_r are scalars.

Ferromagnetic materials are nonlinear because the permeability $\mu(H)$ does not have a fixed value, but it depends on the magnetic field intensity H, that is, $\mu = f(H)$. As a result, $B = \mu(H)H$ is a nonlinear function of H. Almost all ferromagnetic materials are anisotropic. Magnetic anisotropy is the dependence of magnetic properties on a preferred direction. It is the tendency of the magnetization vector in magnetic materials to be aligned in a particular direction. When a magnetic material is magnetized along an *easy axis*, the permeability is higher and saturation is easier than those in other directions. In a *hard axis* or *hard direction*, an extra magnetic field energy is required to rotate the magnetization direction. For anisotropic materials, $\mathbf{B} = \|\mu\|\mathbf{H} = \mu_0\|\mu_r\|\mathbf{H}$. The magnetic susceptibility is a function of frequency and is complex at high frequencies.

2.8 Magnetic Materials

2.8.1 Ferromagnetic Materials

The ferromagnetic chemical, metallic elements are iron (Fe), nickel (Ni), and cobalt (Co). Ferromagnetism is mainly caused by the existence of electron spins. A ferromagnetic in the demagnetized state is divided into many randomly oriented magnetic domains. The magnetic moments are aligned in one direction in each domain. Ferromagnetic materials have large magnetic moments primarily due to spin moments m_s because the orbital magnetic moments m_o are almost zero. Therefore, $m_s \gg m_o$. There is a special quantum phenomenon known as the *strong exchange coupling forces* between the adjacent magnetic dipole moments in a domain in the crystal lattice, holding the dipole moments in parallel. This greatly facilitates the atomic dipole alignment process and locks the magnetic moments into rigid parallel configurations, called *magnetic domains*. Groups of the magnetic dipoles are spontaneously aligned in the same direction and form magnetic domains. The domains are similar to a group parallel solenoids conducting currents and producing magnetic field. When the external magnetic field is zero, the magnetic domains are oriented in random directions and produce internal magnetic field in random directions. Since various domains have random directions, the net magnetic field is zero. In the presence of an external magnetic field, the domains tend to align in the direction of the external magnetic field and produce a very strong internal magnetic field, which adds to the external field. The degree of alignment increases when the magnitude of the external field increases and the temperature decreases. As a result, ferromagnetics have a large and positive magnetic susceptibility $\chi_m = M/H$ in the range from 1 to 1 000 000, resulting in $\mu_r \gg 1$. The values of the relative permeability μ_r for various ferromagnetic materials are given in Table 2.3. For example, $\mu_r = 250$ for cobalt (Co), $\mu_r = 600$ for nickel (Ni), $\mu_r = 5000$ for iron (99.8% pure), $\mu_r = 5000$ for $Co_{50}Fe_{50}$ permendur, $\mu_r = 280\,000$ for iron (99.96% pure), and $\mu_r = 1\,000\,000$ for $Ni_{79}Fe_{15}Mo_5$ Supermalloy. When all the domains are aligned in the direction of the external field, the core saturates and no additional internal field is induced. As the magnetic flux density reaches its saturation value B_s, the core relative permeability μ_r decreases and eventually reaches free-space value $\mu_r = 1$. The values of the saturation flux density B_s for ferromagnetic materials are given in Table 2.4. The relative permeability of ferromagnetic materials is a function of both the applied field and the previous magnetic history.

Table 2.3 Relative permeability of ferromagnetic materials

Material	Relative permeability μ_r
Powder	10 – 60
NiZn ferrite	150
Cobalt	250
Nickel	600
50% Ni, 50% Fe, Orthonol	2 000
MnZn ferrite	1000 – 4000
0.25% Si iron	2 700
48% Ni alloy	4 000
2.5% FeSi steel	5 000
4% FeSi steel	7 000
50% Co alloy	10 000
Metallic glass	10 000
Nanocrystalline	15 000 – 150 000
80% Ni, 4% Mo Alloy	50 000
Mumetal 75% Ni, 5% Cu, 2% Cr	100 000
99.8% Pure iron	5 000
99.96% Pure iron	280 000
79% Ni, 17% Fe, 4% Mo, Permalloy	12 000 – 100 000
79% Ni, 16% Fe, 5% Mo, Supermalloy	1 000 000

Table 2.4 Saturation flux densities of ferromagnetic materials

Material	Saturation flux Density B_s (T) at $T = 20\,^{\circ}$C
50% Co alloy	2.3
0.25% Si iron	2.2
2.5% Si steel	2
Magnesil grain-oriented Si iron	1.5
Metallic glass	1.6
78% Ni–Fe alloy "Permalloy"	1.5
80% Ni, 20%Fe alloy "Supermalloy"	0.6
48% Ni–Fe alloy	1.5
50% Ni, 50% Fe "Orthonol"	1.4 – 1.6
Nanocrystalline	1.2 – 1.5
80% Ni, 4% Mo alloy	0.8
50% Ni–50% Fe alloy "Hipernik"	0.75
MnZn ferrite	0.4 – 0.8
NiZn ferrite	0.3

All spin magnetic moments in the domains are aligned in the direction of the magnetic flux produced by the coil and are equal in magnitude, as shown in Fig. 2.5a. If a ferromagnetic specimen is brought near a strong bar magnet, it will be *attracted* to it. When the external field is reduced to zero, some domains remain aligned in the same direction as before in the presence of the external field.

2.8.2 Antiferromagnetic Materials

In *antiferromagnetic* materials, the magnetic moments of adjacent atoms are aligned in opposite directions and are about equal in magnitude in the absence of an external magnetic field. The net

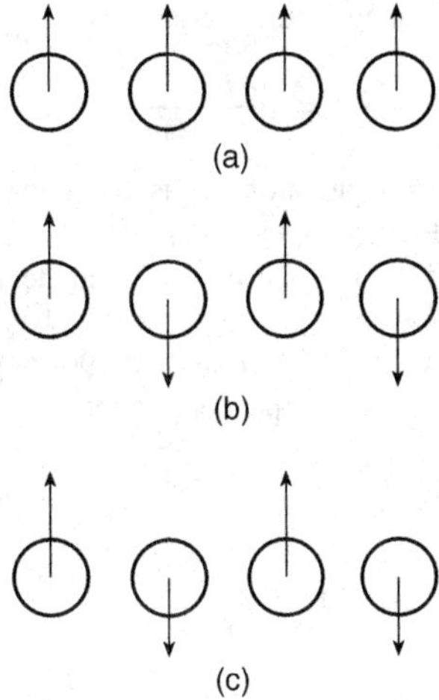

Figure 2.5 Structures of spin magnetic moments under an applied magnetic field. (a) In ferromagnetics. (b) In antiferromagnetics. (c) In ferrimagnetics

magnetic moment is zero in the presence of an applied magnetic field. The presence of an applied magnetic field has a small effect on the material magnetic properties, resulting in relative magnetic permeability μ_r slightly greater than unity. Figure 2.5b shows the structure of spin magnetic moments in antiferromagnetic materials.

2.8.3 Ferrimagnetic Materials

Ferrimagnetic elements, such as chromium and manganese, are neighbors of iron in the periodic table. They are close in the atomic number and have strong coupling forces between atomic dipole moments, but their coupling forces produce antiparallel alignments of electron spins. The spin magnetic moments are very large in magnitude, are greatly unequal, and alternate in direction from atom to atom in the absence of an applied magnetic field, resulting in no net magnetic moment. When a magnetic field is applied, the magnitudes of the magnetic moments are unequal and the directions alternate, resulting in a net nonzero magnetic moment, as shown in Fig. 2.5c. Because of the partial cancellation, the magnetic flux density in ferrimagnetic materials is lower than that in ferromagnetic materials. The relative permeability is much higher than unity ($\mu_r \gg 1$). When a ferrimagnetic material is heated above its Curie temperature, the spin directions suddenly become random and ferrimagnetic becomes paramagnetic. Ferrites are ferrimagnetic materials. They have high resistivity ρ_c and small eddy currents, yielding low eddy-current (ohmic) losses at high frequencies.

2.8.4 Diamagnetic Materials

The diamagnetic elements are bismuth (Bi), copper (Cu), diamond (C), gold (Au), lead (Pb), mercury (Hg), silver (Ag), and silicon (Si). In diamagnetic materials, the magnetic effects are weak. In the absence of an external magnetic field, the orbital and spin magnetic moments cancel each other and the net magnetic moment is zero. An applied magnetic field causes the spin moment to slightly exceed the orbital moment ($m_s > m_o$), resulting in a small net magnetic moment. The small magnetic moment causes small dipole alignment, which produces a small magnetic field. This field *opposes* the applied magnetic field, causing a repulsive effect and slightly reducing the total magnetic field. If a sample of a diamagnetic material is brought near either pole of the strong bar magnet, it will be *repelled* from it. Diamagnetism occurs only in the presence of an external magnetic field. When the external field is reduced to zero, the alignment of dipoles disappears.

The magnetic susceptibility of diamagnetic materials is

$$\chi_m = -\frac{\pi \mu_0 q^2 n N_A r_{av}^2}{6m_e}, \tag{2.55}$$

where N_A is the number of electrons per atom, n is the number of atoms per unit volume, and r_{av} is the average electron orbit. For example, for $n = 5 \times 10^{22}$ m^3, $N_A = 8$, and $r_{av} = 10^{-20}$, we get $\chi_m \approx -10^{-5}$. Diamagnetic materials have a small and negative magnetic susceptibility $\chi_m = M/H \approx -10^{-5} < 0$, resulting in $\mu_r < 1$, where $\mu_r \approx 1 - 10^{-5}$. For example, $\mu_r = 0.99999$ for copper (Cu), $\mu_r = 0.99998$ for silver (Ag), and $\mu_r = 0.99996$ for gold (Au). The relative permeability of diamagnetic materials is constant and independent of the magnitude of the applied magnetic field and temperature.

2.8.5 Paramagnetic Materials

The paramagnetic elements are aluminum (Al), calcium (Ca), chromium (Cr), magnesium (Mg), niobium (Nb), platinum (Pt), titanium (Ti), and tungsten (W). In paramagnetics, the magnetic susceptibility is primarily due to spin magnetic moments ($m_s > m_o$). Energy orbits are only partially occupied by electrons. Electron spins are not totally aligned because of a small magnetic field. When the external field is zero, the individual dipoles are randomly oriented and the net internal magnetic field is zero. When an external field is applied, the magnetic dipoles are aligned in the direction of the external field. The degree of alignment of the dipoles increases as the magnitude of the external field increases and as the temperature decreases. Spin moments in paramagnetics vanish at high temperature because the thermal motion causes the spins to become randomly oriented. The paramagnetic substance is *attracted* to a strong bar magnet. Paramagnetism occurs only in the presence of an external magnetic field. When the external field is reduced to zero, the alignment of dipoles disappears and the dipoles are randomly oriented.

Paramagnetic materials have a small and positive magnetic susceptibility $\chi_m = M/H \approx 10^{-5} > 0$, resulting in $\mu_r > 1$, where $\mu_r \approx 1 + 10^{-5}$. For example, $\mu_r = 1.00002$ for aluminum (Al), $\mu_r = 1.0002$ for titanium (Ti), and $\mu_r = 1.0003$ for platinum (Pt). Paramagnetic materials are linear. Their relative permeabilities are independent of the applied magnetic field.

2.9 Hysteresis

The magnetization curve is called the hysteresis. It is caused by magnetic friction in the core. The word "hysteresis" means "lagging behind." The hysteresis is the relationship between the magnetic flux density B and the magnetic field intensity H. It is determined by the core material characteristics. A ferromagnetic material has a "memory" because it remains magnetized after the external field is removed. In ferromagnetics, the process of repeated magnetization and demagnetization in an AC magnetic field is only partially reversible. The induced magnetization does not vanish even after the removal of the magnetic field. Demagnetization involves either reversing the spins in magnetic domains or moving the domain walls. The B–H curve is a nonlinear and multivalued function, which exhibits saturation, as illustrated in Fig. 2.6. When the magnetic field intensity H is increased, the magnetic flux density B also increases. At high values of H, the core magnetization approaches its saturation value, corresponding to a complete alignment of the magnetic moments, and the magnetic flux density B reaches the saturation value B_s. When the magnetic field intensity H is reduced, many domains tend to remain aligned due to frictional forces resisting the domain movement. Hence, the magnetic field must be reduced significantly before the domains begin to return to their initial unaligned arrangement. When the magnetic field intensity H is reduced to zero, the magnetic flux density B does not become zero because domains remain partially aligned and the core retains some residual magnetic flux density referred to as the *residual flux density*, *remanence*, or the *remnant magnetization* B_r. To reduce B to zero, a negative or reverse magnetic field intensity, called *coercive*

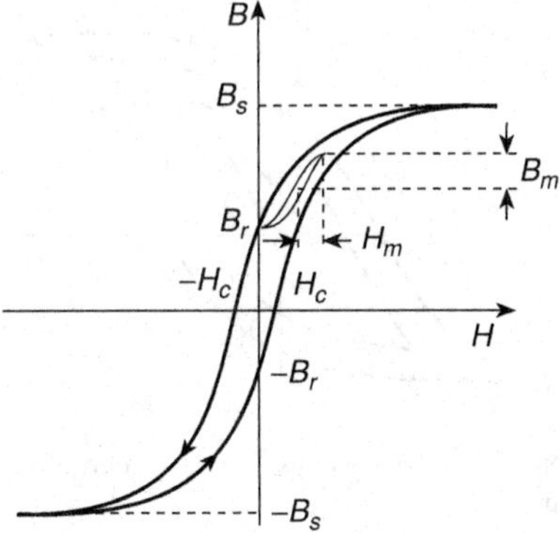

Figure 2.6 Core B–H hysteresis

force $H = -H_c$, must be applied. When the magnetic field intensity is further decreased, the magnetic flux density also decreases and reaches the saturation value $-B_s$. When the magnetic field intensity is increased, the magnetization curve follows a different path. At $H = 0$, $B = \pm B_r$ and at $B = 0$, $H = \pm H_c$. Therefore, the magnetization curve is not fully reversible. The reluctance is low at low values of H and is high in the saturation region. For this reason, magnetic circuits with ferromagnetic cores are nonlinear. When the core is excited from a demagnetized state, the magnetic flux starts to increase from the shortest MPL. For example, the magnetic flux in an inductor with a toroidal core (TC) begins to increase from the inner side of the core. Figure 2.6 also shows a minor hysteresis loop determined by the swing of AC components of $H(t)$ and $B(t)$ for steady-state operation.

The shape and size of the hysteresis loop depend on the properties of the ferromagnetic material and the magnitude of the applied field H. The hysteresis loops for hard and soft magnetic materials are shown in Fig. 2.7. For "hard" ferromagnetic materials such as permanent magnets, the hysteresis loop is wide because it is difficult to move the magnetic walls. For "soft" ferromagnetic materials, the hysteresis loop is narrow and has a small area enclosed because the magnetic walls move easily. Magnetic cores use soft ferromagnetic materials, which have narrow hysteresis loops. The area enclosed by the hysteresis loop represents the energy required to take the material through the hysteresis cycle due to realignment of the domains, producing heat. Therefore, the area of the B–H loop is equal to the hysteresis energy loss per volume for one cycle. When the magnitude B_m is increased, so does the area of the B–H loop. Therefore, the hysteresis energy loss increases. The B–H loop widens as the frequency increases due to eddy currents. A perfect core for inductors and transformers should have very narrow and very high B–H hysteresis. The magnetization process of a ferromagnetic material by carrying it through successive hysteresis loops for growing oscillations is shown in Fig. 2.8.

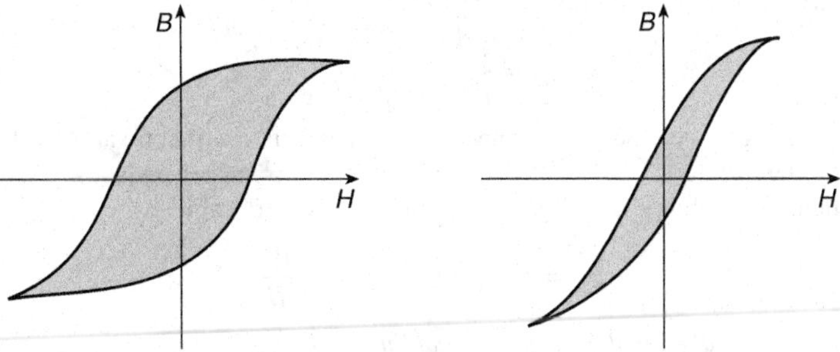

Figure 2.7 Hysteresis loops for hard and soft ferromagnetic materials. (a) Hysteresis loop for hard magnetic material. (b) Hysteresis loop for soft magnetic material

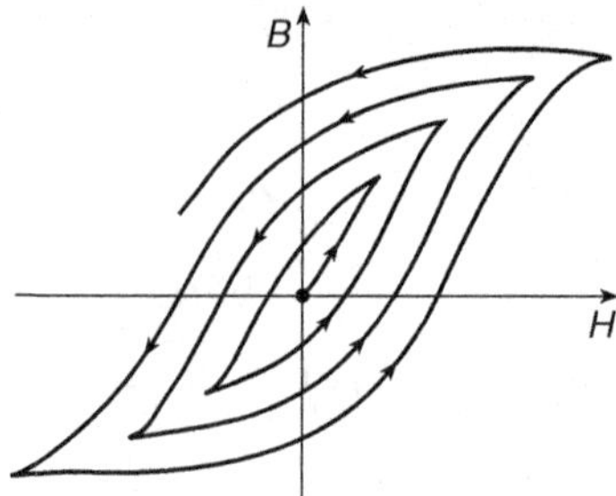

Figure 2.8 Hysteresis loop for growing oscillations. Magnetization of ferromagnetic material by carrying it through successive hysteresis loops

2.10 Low-Frequency Core Permeability

The permeability describes the ability of a magnetic material to conduct magnetic flux. There are several definitions of the permeability for low frequencies, where **B** is in phase with **H**. The *large-signal permeability* is defined as

$$\mu = \mu_r \mu_0 = \frac{B}{H} (\mathrm{H/m}), \tag{2.56}$$

where the *relative permeability* is

$$\mu_r = \frac{\mu}{\mu_0} = \frac{1}{\mu_0} \frac{B}{H}. \tag{2.57}$$

The relative permeability decreases with temperature T.

An inductor with a magnetic core is nonlinear because B is a nonlinear function of H, and therefore μ depends on H. The *small-signal permeability* or the *incremental permeability* at a given operating point of the hysteresis Q is defined as the slope of the B–H curve at the operating point Q

$$\mu_{ss} = \mu_{rss} \mu_0 = \left. \frac{dB}{dH} \right|_{Q(H_{DC},\, B_{DC})}, \tag{2.58}$$

where the *small-signal relative permeability* is

$$\mu_{rss} = \frac{\mu_{ss}}{\mu_0} = \frac{1}{\mu_0} \left. \frac{dB}{dH} \right|_{Q(H_{DC},\, B_{DC})}. \tag{2.59}$$

As the magnetic field intensity H is increased from zero to a maximum value, the relative small-signal permeability increases, reaches a maximum value, then decreases to 1 when the core saturates.

The *initial permeability* is equal to the slope of the magnetization curve at the origin

$$\mu_i = \mu_{ri} \mu_0 = \left. \frac{dB}{dH} \right|_{H \to 0}, \tag{2.60}$$

where the *initial relative permeability* is

$$\mu_{ri} = \frac{\mu_i}{\mu_0} = \frac{1}{\mu_0} \left. \frac{dB}{dH} \right|_{H \to 0}. \tag{2.61}$$

The initial permeability increases with temperature, reaches its maximum value, and then decreases rapidly at Curie temperature.

The maximum permeability is the maximum small-signal permeability

$$\mu_{max} = \mu_{ss(max)}|_{H \to H_P} = \left(\frac{dB}{dH} \right)_{max}, \tag{2.62}$$

where the *maximum small-signal relative permeability* is

$$\mu_{rmax} = \frac{\mu_{ss(max)}}{\mu_0} = \frac{1}{\mu_0} \left(\frac{dB}{dH} \right)_{max} \tag{2.63}$$

and H_p is the magnetic field intensity at which the hysteresis exhibits the largest slope. The permeability of ferromagnetic cores depends on the core material, magnetic flux density B, DC bias, temperature, and frequency. The magnetic flux takes the path of the highest permeability. For high frequencies, **B** is not in phase with **H**. Therefore, the permeability is complex, as described in Section 2.17.

2.11 Core Geometries

The choice of a magnetic core involves the following aspects:

- Core material
- Core geometry
- Core size
- Air gap length.

Cores are made in a wide variety of sizes and shapes, such as TC, pot cores (PCs) with and without air gap, power quality (PQ) cores, rectangular modular (RM) cores, economic transformer design (ETD) cores, CC cores, UU cores, EE cores, EI cores, EC cores, ER cores, UI cores, and planar cores. The windings of E, C, and PCs are wound on a plastic coil former called a *bobbin*, which is placed on the legs of E or C cores or the center post of the pot cores. The bobbin contains and protects the magnet wire and insulation. The bobbin material is a dielectric. Its main characteristics are the dielectric strength and dielectric constant. The dielectric strength is the voltage required to break through the bobbin plastic material, causing an electrical failure. As the frequency increases, the dielectric strength decreases.

2.11.1 Toroidal Cores

A TC is shown in Fig. 2.9. Toroidal inductors and transformers offer a small size (both small volume and weight) and the lowest EMI. The geometry of the core forms the magnetic field lines into a closed circular loops and keeps the majority of the magnetic flux constrained within the core material, reducing EMI radiated by the coil. There is almost no stray magnetic field. Therefore, transformers wound on TCs have low leakage inductance. They are the best cores to build coupled inductors for EMI filters. Single-layer inductor have a low self-capacitance, yielding a high self-resonant frequency. Toroidal inductors with round cross-sectional area have better performance than those with the rectangular cross-sectional area. A shorter turn length per unit area is achieved for round

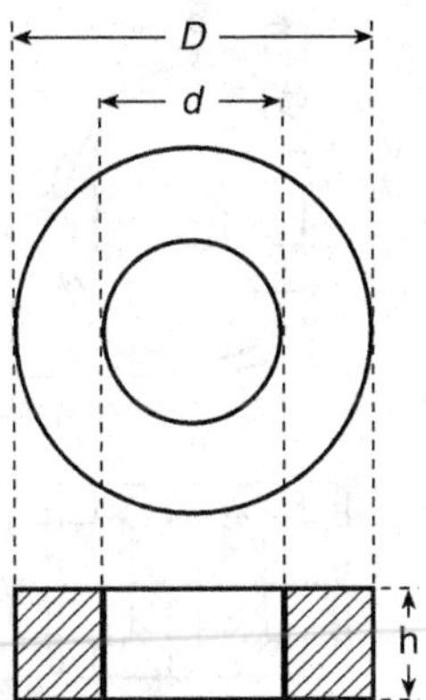

Figure 2.9 Toroidal core

cores, yielding lower winding resistance and power loss. Large surface area of toroidal inductors and transformers results in good heat removal via convection and radiation. To achieve good turn-to-turn coupling, the turns should be evenly spaced. Air gaps are difficult to make in TCs. Toroidal inductors are more expensive than bobbin wound inductors. The magnet wire is wound through the hole of the core. Therefore, windings of TCs are more expensive to make, but this process can be automated. In a TC, an area with one half of the inside diameter must be reserved for the shuttle of winding machine. TCs can be made from a large variety of materials: ferrite, iron powder, silicon steel, nickel iron, and MPP. Ferrite TC often have a dielectric coating made of parylene or epoxy to insulate the windings from the core.

2.11.2 CC and UU Cores

Figure 2.10 shows a CC core or a UU core, where W_a is the window area and A_c is the cross-sectional area of the magnetic path. CC cores have rounded corners and UU cores have sharp corners. The shapes of these cores form a very simple magnetic path. The coil can be wound using conventional wounding machines. The window area W_a can be completely filled with copper wound on a bobbin, resulting in a high window utilization factor. CC and UU cores are usually cut cores. A cut core consists of two matched halves.

2.11.3 Pot Cores

A cross section of a PC is shown in Fig. 2.11. PCs are made of two round halves with internal hollows, which enclose winding(s). Figure 2.12 depicts a construction of an inductor using a PC. A coil is

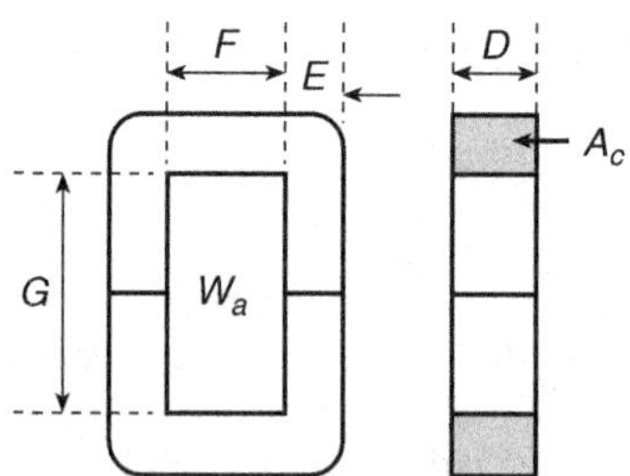

Figure 2.10 CC core

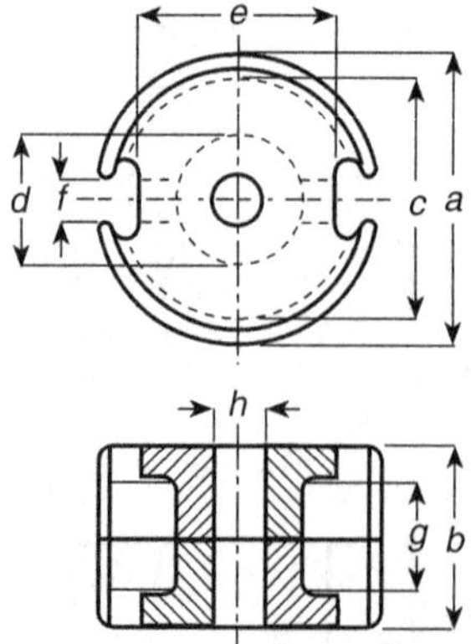

Figure 2.11 Pot core

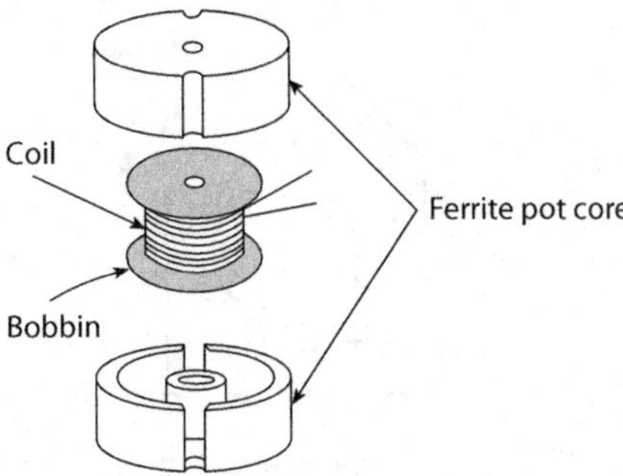

Figure 2.12 Construction of an inductor using a pot core

wound on a bobbin and placed inside a ferrite PC. Almost all the magnetic field is confined to the interior of the core. If the core is grounded, the moderately conductive ferrite acts as an electrostatic shield for the windings as well. PCs are therefore the best shielded of all core types. These cores are used for making high-frequency low-current screened inductors and transformers for applications in DC–DC converters at low-power levels up to 125 W. They are available with an air gap in the center leg and therefore can carry DC and AC currents without saturation. The coil is wound on a plastic bobbin around the center post. Their main advantage is that the coil is almost entirely enclosed by two half core pieces of ferrite, held by a nonconductive screw or clamp. Therefore, the radiated and leakage magnetic fields are reduced, hence minimizing the EMI and radiofrequency interference (RFI) effects. The main disadvantages of PCs are the restricted heat removal from the winding and the core to the outer surface and narrow slots f in the ferrite for winding leads, which can be a problem in high current or multioutput transformers. PCs are available with deliberately introduced air gaps by grinding the center post. When a PC is gapped, heating effects from fringing fields in the windings cause even greater temperature rise. These cores are used for high-Q inductors and low-power transformers and are the most expensive to manufacture. However, they suffer from heat buildup because their windings are surrounded by core material that does not conduct heat well and prevent air circulation.

2.11.4 PQ and RM Cores

The PQ and RM cores are modified versions PCs with a wider notch cut of the ferrite. A cross section of a PQ core is shown in Fig. 2.13. PQ cores are specifically designed for switching-mode power supplies (SMPS). Their geometry is optimized to transfer power with minimal size and weight.

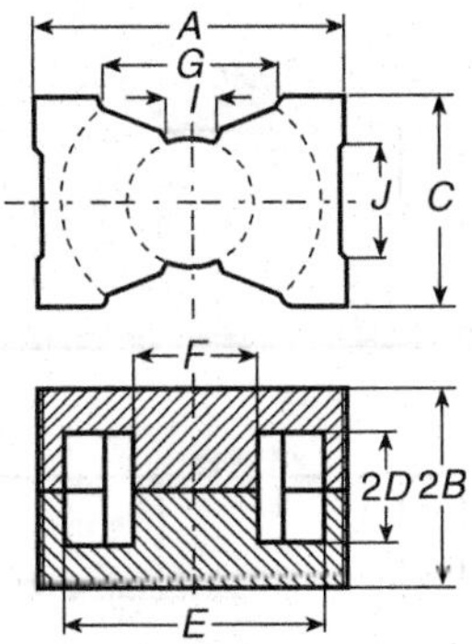

Figure 2.13 Half of PQ core

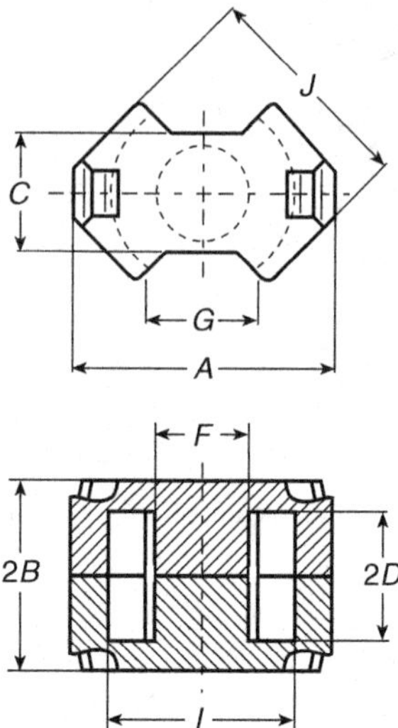

Figure 2.14 Half of RM core

Figure 2.14 depicts a cross section of an RM core. The center post can be ungapped of gapped. They may have a standard height or a low profile suitable for planar inductors and transformers on printed circuit boards (PCBs). These cores allow for a better power dissipation than PCs and a better shielding than E cores. They are also suitable for automated processing. Applications of these cores include energy storage chokes for DC–DC converters, matching transformers, very low loss, highly stable filters and resonant inductors, and low-distortion transmission of modulated signals.

2.11.5 EE and EDT Cores

EE cores are among the most widely used magnetic cores. A cross section of an EE core is shown in Fig. 2.15. The power range of electronic circuits with these cores is from 5 W to 10 kW. The leads can easily enter or leave the bobbin. An EE core consists of a pair of E cores. The center leg can be

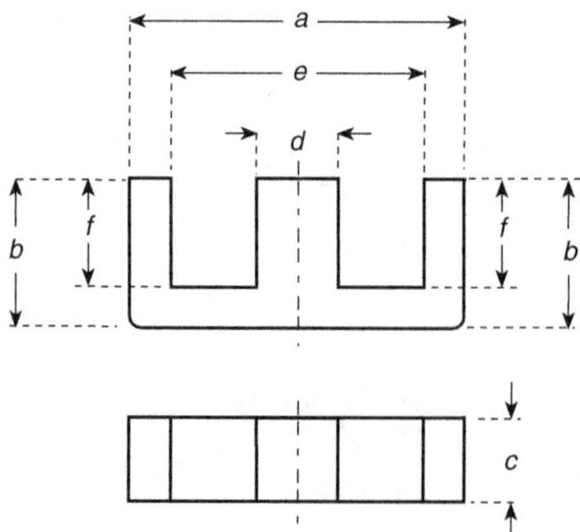

Figure 2.15 Half of EE core

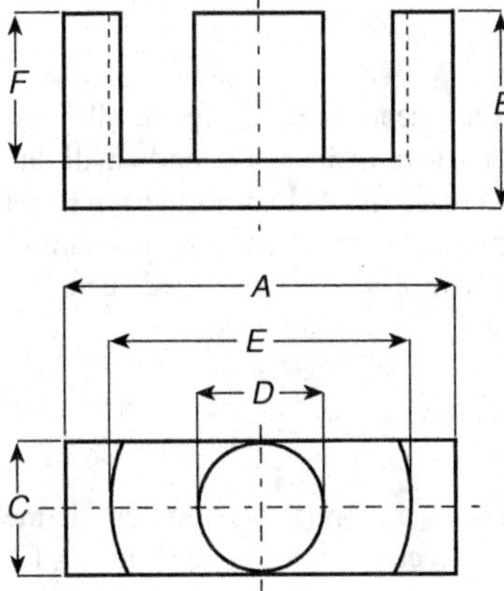

Figure 2.16 Half of EDT core

either round or square. The magnet wire is wound on a plastic bobbin, which is placed on the center leg. The cross-sectional area of the center leg is twice that of each individual outer leg. If a gap is required, the center leg can be shortened so that the air gap is located at the middle of the coil to minimize fringing and reducing EMI. The advantage of a round leg is that the winding mean turn length is 11% shorter than that of a square leg, reducing the magnet wire cost and weight, winding resistance, and winding loss. The EE cores allow for easy airflow, reducing the temperature. However, they produce more EMI and RFI field radiations because the windings are not fully surrounded by magnetic material. Figure 2.16 shows a cross section of an ETD core.

2.11.6 Planar Cores

Planar cores are low-profile cores. They are composed of two flat pieces of magnetic material, one above the coil, and one below the coil that is part of a PCB. The core material is almost exclusively a ferrite. The windings are etched on a PCB. Etching of the windings imposes a limit on the number of turns. Therefore, PCB inductors and transformers are suitable for high-frequency applications only. Typically, planar cores are used to build flat, low-profile, low-volume inductors and transformers on PCBs for applications in high-power density power supplies. Multiple-layer printed boards are often used. Planar cores are excellent for low-cost mass production. They can handle a relatively high power. The power loss density is up to 600 mW/cm^3 due to better power dissipation and lower copper loss.

2.12 Ferromagnetic Core Materials

There are five major types of cores made up of soft magnetic materials:

1. Iron alloy cores.
2. Nickel–iron and cobalt–iron cores.
3. Ferrite cores.
4. Powder cores.
5. Nanocrystalline cores.

2.12.1 Iron Cores

Magnetic properties of materials are dependent on precise alloy composition as well size and shape of particles. Iron cores consist of alloys of iron (Fe) and small amounts of silicon (Si), nickel (Ni), cobalt (Co), and chrome (Cr). These alloys belong to the family of *ferromagnetics*. Iron cores have low resistivity ρ_c, high relative permeability μ_r, and high saturation flux density B_s. The resistivity of iron cores is comparable with that of good conductors, such as aluminum (Al) and copper (Cu).

2.12.2 Ferrosilicon Cores

The silicon magnetic steel is an iron–silicon (FeSi) alloy and is the most widely used soft magnetic material at low frequencies. which affects the crystalline structure. First, silicon increases the resistivity ρ_c and thus reduces eddy currents and eddy-current losses. The addition of 3% of silicon increases the resistivity nearly four times. Second, silicon decreases the magneto-anisotropy energy of iron and increases the relative permeability μ_r and thus decreases hysteresis core losses. Third, silicon decreases magnetostriction and thus reduces the transformer noise (hum). However, silicon decreases the saturation flux density B_s and the Curie temperature T_c. Iron cores with Si $\leq 1\%$ are called low-silicon steels. The typical relative permeability μ_r of iron cores ranges from 2500 to 5000. The typical values of the FeSi core resistivity ρ_c ranges from 10^{-7} to $7 \times 10^{-7}\ \Omega \cdot$ m. The typical values of the saturation flux density B_s of iron lamination cores are high, from 1.5 to 2.2 T and 0.5 to 2 T for silicon steels. The Curie temperature T_c of iron cores is high and ranges from 760 °C to 810 °C. Magnesil is a grain-oriented silicon iron material. Its flux density is up to 1.5 T. It offers moderate performance at the lowest overall core cost. Both hysteresis and eddy-current losses are present in iron cores. However, the eddy-current loss P_{ec} is much higher than the hysteresis loss P_h. The specific power loss is from 0.3 to 3 W/kg at $f = 50$ Hz and $B = 1.5$ T. Iron cores are used at low frequencies up to 500 Hz because of high eddy-current loss due to low resistivity. To reduce the eddy currents, iron cores are made of stacks of many thin laminations insulated from each other by a thin insulating coating. Typical applications of iron cores are power transformers made of low-silicon steels for utility line frequencies of 50 or 60 Hz, aircraft power transformers operated at 400 Hz, electric motors, generators, relays, and filter inductors operated below 5 kHz. Table 2.5 shows the frequency ranges for various ferromagnetic materials.

2.12.3 Amorphous Alloy Cores

Atoms in the amorphous structures do not form a crystal (periodic) structure and are in complete disorder. Amorphous magnetic materials are alloys of iron and other magnetic materials such as cobalt and nickel or transition metals such as boron, silicon, niobium, and manganese. Amorphous metals are called metallic glasses because they have a similar noncrystal structure as glass. Hot molten

Table 2.5 Frequency range of ferromagnetic materials

Material	Frequency range
Iron alloys	50 – 3000 Hz
Ni–iron alloys	50 – 20 000 Hz
Co–iron alloys	1 – 100 kHz
Nanocrystalline	0.4 – 150 kHz
Amorphous alloys	0.4 – 250 kHz
MnZn ferrites	10 – 2000 kHz
Iron powders	0.1 – 100 MHz
NiZn ferrites	0.2 – 100 MHz

metals are compressed on a cooling roller at 1300 °C. Amorphous cores are produced as thin ribbons. METGLAS tape-wound metal magnetic alloys are used for manufacturing magnetic cores. TCs are the most common shapes. The relative permeability μ_r is very high, from 150 000 to 1 000 000. The saturation flux density is in the range from 0.7 to 1.8 T. The resistivity is in the range from 1.2×10^{-6} to 2×10^{-6} $\Omega \cdot$ m. The maximum operating frequency is up to 250 kHz.

2.12.4 Nickel–Iron and Cobalt–Iron Cores

Iron is often alloyed with nickel to achieve a higher permeability or with cobalt to attain a higher saturation magnetic flux density. There are two categories of nickel–iron alloys, which differ by nickel percentage: low-nickel alloys (40–50% Ni) and high-nickel alloys (78–82% Ni). Nickel–iron (NiFe) alloys (40% Ni or 80% Ni and 4% Mo) have the highest permeabilities of any soft magnetic material with $\mu_r = 40\,000$ to 50 000. The saturation flux density is $B_s = 0.8$–1.5 T. The resistivity ρ_c is in the range 4.8×10^{-7} to 5.8×10^{-7} $\Omega \cdot$ m. The Curie temperature T_c is 450 °C. Higher resistivities make these core suitable for *audio-frequency applications.* Some of the iron–nickel alloys have the relative permeability μ_r up to 300 000. Magnetic properties depend on the temperature and material thickness. Cores made of NiFe alloys are sensitive to mechanical stresses. Therefore, these cores are inserted in protective plastic or aluminum cases and space between core and case is filled with damping medium. Nickel–iron alloy cores are used in current transformers because of their high relative permeability μ_r.

Permalloy is a nickel–iron (NiFe) magnetic alloy. The 80% Ni, 20% Fe core material is called Permalloy or Mumetal, the 50% Ni core material is termed Isoperm, and the 36% Ni material is called Invar. It has a higher permeability than silicon steel. Permalloy was discovered in 1914 by Gustav Elmen in Bell Laboratories. In 1923, he enhanced its permeability by heat treatment. Permalloy is used in laminated cores for high-performance transformers.

Supermalloy is a magnetic alloy that consists of 79% Ni, 16% Fe, and 5% Mo. It has the highest permeability and the lowest core losses, but it also has the highest core cost. Usable flux levels are up to 0.6 T.

The applications of nickel–iron core include current transformers, wide-band transformers, linear transformers, pulse transformers, saturable reactors, power inverters, DC–DC converters, and magnetic amplifiers.

Cobalt–iron (CoFe) metal alloys (50% Co) have the highest saturation flux density $B_s = 2.4$ T. The relative permeability is $\mu_r = 10\,000$ and the resistivity is $\rho_c = 3.5 \times 10^{-7}$ $\Omega \cdot$ m. The specific power core loss is 20–24 W/kg at $f = 5$ kHz and $B = 0.25$ T. The typical thickness values of this material are 0.05 and 0.1 mm. Cobalt–iron cores are used in special transformers with low power losses at very high flux densities, in magnetic amplifiers, and in space applications at 20 kHz.

2.12.5 Ferrite Cores

Ferrites are *ferrimagnetic* materials and they are dense, hard, brittle, chemically inert, and homogeneous polycrystalline ceramics made of iron oxides (Fe_2O_4) mixed with oxides or carbonates of either manganese (Mn) and zinc (Zn) or nickel (Ni) and zinc (Zn). These materials are pressed and then fired at temperatures from 1000 °C to 1500 °C. Oxides of metals (i.e., metallic oxides) are called ceramics. Most ceramics are insulators with high resistivity, which reduces eddy currents. A ferrite consists of small crystals of the order 10–20 μm. Magnetic domains are formed in these crystals. In soft ferrites, the crystal structure is cubic. A general chemical composition formula for ferrites is $MMFe_2O_4$, where MM represents the transition metals, such as manganese (Mn), zinc (Zn), nickel (Ni), cobalt (Co), copper (Cu), iron (Fe), or magnesium (Mg). The most common chemical compositions of ferrites are $NiZnFe_2O_4$ and $MnZnFe_2O_4$.

Ferrites show antiparallel alignment of adjacent atomic moments, but the moments are not equal. They have a very high relative permeability μ_r, a high resistivity ρ_c, and a low saturation flux density B_s. The typical range of μ_r is from 40 to 10 000. The typical resistivity ρ_c of ferrite cores is 10^5 $\Omega \cdot$ m, but it can range from 1 to 10^7 Ωm. The resistivity of the ferrites is much higher than that of iron cores, usually by a factor of at least 10^6. Ferrites are insulators and therefore can be used for high-frequency applications because of low eddy-current power losses. The typical saturation magnetic flux density B_s is 0.3 T. The relative dielectric constant of MnZn and NiZn ferrites ϵ_r is in the range 10–18. These cores are used in the high-frequency range from 10 kHz to 50 MHz. Because of high resistivity, eddy-current power loss in ferrite cores is very low. Ferrite cores have only the hysteresis loss. The typical values of the power loss density is $P_v = 100$ mW/cm^3 at $B_m = 100$ mT and $f = 100$ kHz. and $P_v = 300$ mW/cm^3 at $B_m = 50$ mT at $f = 500$ kHz. The quality factor Q of inductors with ferrite cores is high. A unique limiting feature of ferrites is their low Curie temperature $T_c = 150 - 300$ °C. Ferrites lose ferromagnetic properties when the Curie temperature T_c is exceeded. The specific ferrite power loss is 50 mW/cm^3 at $f = 20$ kHz. Ferrite cores have the lowest power losses at frequencies above 500 kHz. These cores are not susceptible to aging effects. The thermal conductivity of ferrites is 35–43 mW/(cm · °C).

The most common alloys are manganese–zinc (MnZn) and nickel–zinc (NiZn). Manganese–zinc (MnZn) soft ferrites have the following content: Mn + Zn + Fe_2O_3. They have a high relative permeability μ_r at low frequencies $\mu_r = 1000 - 4000$, medium saturation flux density $B_s = 0.4$–0.8 T, and relatively low resistivity $\rho_c = 0.5$–2 Ω · m. The Curie temperature is low, from 150 °C to 225 °C. The operating frequencies are usually from 10 kHz to 2 MHz. Examples of MnZn ferrites are the Magnetics K-material and the TDK H7F material, both with $\mu_r = 1500$.

Nickel–zinc (NiZn) soft ferrites have the following chemical formula: Ni + Zn + Fe_2O_3. They have a low relative permeability μ_r at low frequencies ($\mu_r = 40$–300), a constant relative permeability μ_r over a wide frequency range, a high resistivity $\rho_c = 10^3$ to 10^7 Ω · m, and low losses at high frequencies. The saturation flux density is low, $B_s = 0.8$ T. The Curie temperature is $T_c = 300$ °C. The operating frequencies are high, from 100 kHz to 100 MHz. The magnetic properties of NiZn ferrites can change irreversibly in high magnetic fields.

The importance of ferrites is largely due to their high resistivity, within semiconductor temperature range. Gapped ferrite cores are capable of storing substantial amounts of energy. Ferrite cores are made in various shapes, such as TCs, PCs, EE cores, I cores, and planar cores. The reasons for ferrite core popularity in power electronics applications are low cost and low losses. However, these ceramics may break under high mechanical stresses.

The relative permeability of a ferrite μ_{rc} increases with temperature to a maximum value and then decreases to 1. The resistivity of a ferrite ρ_c decreases with frequency. As the temperature increases, the saturation magnetic flux density B_s decreases for a ferrite core. It may decrease as much as to 40% or 50% of the value of B_s at room temperature. Ferrite cores should not be operated at a total loss density level of more than $P_v = P_C/V_c = 300$ mW/cm^3 to prevent the danger of thermal runaway.

Ferrite cores, like other ceramic materials, are brittle and sensitive to mechanical shocks and vibration conditions. Fast changing mechanical loads and high rate cooling can cause cracks or failure of ferrite cores. As the mechanical core stress increases, the core relative permeability decreases. Therefore, the embedding medium of the core should have good elasticity.

Ferrite cores are used in a wide variety of applications in high-frequency power electronics in high-frequency power inductors, high-frequency power transformers, broad band transformers, pulse transformers, gate driver transformers, current transformers, common-mode chokes, resonant inductors, EMI suppression high-impedance filters, antennas, and sensors.

In summary, soft ferrite cores offer desirable high-frequency performance due to their high resistivity ρ_c, but they are limited by a low saturation flux density $B_s < 0.5$ T. Conversely, advanced metal alloys offer a higher saturation flux density $B_s = 2 - 2.2$ T, but suffer from low resistivity ρ_c and consequently high eddy-current loses. Nanoengineered soft magnetic composite materials using metal alloy (cobalt–iron) and metal oxides (ferrite) processed into nanoscale particles may achieve both high resistivity and high saturation flux density.

2.12.6 Powder Cores

Powder cores include iron powder cores, Kool mu cores, and MPP cores. Powder metal cores are manufactured from powdered iron or powdered iron alloys (iron compounds mainly oxides) by grinding the base material into fine powder particles, and then coating these particles with an inert insulating material to separate the particles from each other. This increases the saturation flux B_s, increases the resistivity, reduces eddy-current loss, and controls the size of a *distributed air gap* in the structure. The powders are then pressed into various core shapes under pressure. The air gap is in the nonmagnetic binder holding the magnetic particles together. Forming the air gaps requires a high degree of precision to tradeoff the current capacity and the frequency response. The equivalent permeability of the overall composite core is specified. Inductors with powder cores can conduct high peak currents. The cost of materials for powder cores is very low. The size of the particles is from 5 to 200 μm and is less than the skin depth at high frequencies. The thickness of the insulating material is from 0.1 to 3 μm. The insulating material provides the equivalent of a distributed air gap that lowers the relative permeability μ_r and allows the core to store substantial amount of energy. The small air gaps distributed evenly throughout the cores reduce the effective permeability but increase the amount of DC current that can be passed through the winding before core saturation occurs. The typical values of μ_r are in the range 3–550, depending on the particle size and the amount of magnetically inert material in the mix. The typical resistivity is $\rho_c = 1\ \Omega \cdot \mathrm{m}$. These cores have high values of the saturation magnetic flux density B_s (0.5–1.3 T) and much higher resistivity ρ_c than laminated iron cores. Therefore, they have much lower eddy-current loss than the laminated cores and can be used at high frequencies. The relative permeability μ_r of powder cores is lower than that of iron ferrite cores. Typically, they range from 10 to 550. The standard values of μ_r are 10, 14, 26, 35, 55, 60, 75, 90, 100, 125, 147, 160, 173, 200, 300, and 550. A large effective air gap makes these cores tolerant of DC operation. The Curie temperature is $T_c = 450\,^{\circ}\mathrm{C}$. The energy is stored in nonmagnetic gaps. As the amplitude of the flux density B_m increases, the relative permeability μ_{rc} decreases as much as 40% of its initial value. The size of the air gaps tends to change as the core ages, especially when it is operated at elevated temperatures. At high temperature approaching 100 °C, some of the particles come into contact with each other. As a result, eddy currents and core losses increase, producing more heat and further accelerating the aging process. In new iron powders such as Permalloy, MPP, High-Flux (HF), and Sendust, silicon or aluminum is used to produce distributed air gaps. These new powders are capable of operating at high temperatures of up to 200 °C without degradation. Powder cores exhibit much higher core loss than ferrite cores. However, inductors with low-permeability ungapped powder cores have lower winding loss because the magnetic field distribution is uniform due to lack of fringing flux caused by the air gap. Conversely, high-permeability gapped ferrite cores have higher winding loss due to higher eddy-current winding loss caused by fringing effect. The eddy-current loss in the winding due to air gap can be reduced by segmenting a single air gap, staggering an air gap, using a distributed air gap, using a combination of low and high-permeability cores, introducing a low-permeability material between the gapped core lag and the winding to form two magnetic paths (resulting in a more uniform magnetic field distribution), and forming a winding shape. The core losses of low-permeability powder materials are an order of magnitude higher compared to the high-permeability ferrite materials at the same amplitude of the magnetic flux density B_m and at the same frequency f.

Powder cores are very economical in applications in SMPS. The most common applications of powder cores are in toroidal inductors and adjustable slug-tuned coils. Small toroids are coated with parylene and large cores with epoxy paint. The lack of localized air gap eliminates a fringing effect. Examples of composite powdered-metal cores are Permalloy powder cores and Kool Mμ© cores. Molybdenum permalloy powder (MPP) cores are made of very fine particles of 81% nickel, 17% iron, and 2% molybdenum alloy. MPP cores offer lower losses and higher operating frequencies than iron powder cores. They can be used in frequency range from 1 kHz to 1 MHz. Applications of MPP cores include pulse transformers, flyback transformers, and differential mode EMI noise filters. Kool Mμ (FeSiAl) is a ferrous alloy material. The typical relative permeability μ_{rc} values are 10, 26, 40, 60, and 90. It is well suited for DC bias applications, such as inductors for SMPS, power-factor correction (PFC) circuits, uninterruptible power supply (UPS) systems, flyback transformers, and EMI filters.

2.12.7 Nanocrystalline Cores

Nanocrystalline cores contain ultrafine crystals with a typical size of 7–20 nm and therefore are called nanocrystallines. These cores combine high saturation magnetic flux density B_s of silicon steels and low high-frequency losses of ferrites. The relative permeability μ_r is in the range 15 000–150 000. The saturation magnetic flux B_s is in the range 1.2–1.5 T. The resistivity is $\rho_c = 1.2 \times 10^{-6}\ \Omega \cdot$ m. The Curie temperature is $T_c = 600\ ^\circ$C. The specific power loss is 5 W/kg at $f = 20$ kHz and $B = 0.2$ T. The nanocrystalline cores are used up to 150 kHz. They have a linear hysteresis loop. A very high relative permeability μ_r makes them suitable for applications in current transformers and common-mode EMI filters.

2.13 Superconductors

Ideal conductors have $\sigma = \infty$, $\rho = 0$, $E = 0$ at any frequency, and time-varying $H(t) = 0$. Metallic conductors, such as copper, aluminum, silver, and gold, have a very large conductivity, which is in the range 10^7–10^8 S/m at room temperature. For some materials, the resistivity becomes very low at temperatures near absolute zero, $T = 0$ K $= -273\ ^\circ$C. These materials are called *superconductors*.

Diamagnetic materials have $\mu_r = 1 - 10^{-5} < 1$ and tend to slightly exclude magnetic field from them. As a result, the magnetic field intensity H is lower in a diamagnetic material than that in free space. In diamagnetics, two electrons orbit the nucleus in the opposite directions, but with the same speed. The atoms of diamagnetic materials do not form permanent magnetic moments. When an external magnetic field is applied, a weak magnetic moment is induced in the direction opposite to the applied field. Copper, lead, silver, and gold are examples of diamagnetics.

Superconductors are a special form of diamagnetic materials. Examples of superconductors are tantalum (Ta), titanium (Ti), vanadium (V), niobium (Nb), tin (Sn), lead (Pb), mercury (Hg), aluminum (Al), and ceramic compounds. In these materials, there are circulating microcurrents that oppose the applied magnetic field and, consequently, the superconductor excludes all the external magnetic field. The resistivity of normal materials such as copper steadily decreases as the temperature decreases and reaches a low residual value $\rho_o \approx 0$ at 0 K. In contrast, the resistivity of a superconductor decreases as the temperature T decreases, and then it drops suddenly to zero at the *critical temperature* T_{cr}, as shown in Fig. 2.17. This phenomenon is called *superconductivity*. The resistance of a superconducting wire has been measured to be less than $10^{-26}\ \Omega$. This value is many orders of magnitude lower than that of the best conductors.

The superconductive material remains in the superconducting state below the critical temperature and in the normal state above the critical temperature T_c. This phenomenon was discovered for mercury by the Dutch physicist Heike Kamerlingh Onnes in 1911. He received the Nobel Prize in Physics in 1913. Superconductivity was observed experimentally by Meissner and Ochsenfeld in 1933. The superconductivity occurs when either the number of electrons that are available for current transport is very high or the electron pairs exhibit very strong attractive forces. Metals have many electrons and metallic oxides have fewer electrons, but exhibit strong attractive electron-pair forces. The critical temperature T_c for superconductive metals is from 0.39 to 9.15 K and for ceramic compounds from 89 to 122 K. Aluminum becomes a superconductor at $T_c = 1.2$ K, niobium at $T_c =$ 9.2 K, and niobium–germanium at $T_c = 23$ K. The critical temperature for the ceramic compound $YBa_2Cu_3O_{7-x}$ is about 89 K. Copper and gold are not superconductors. Liquid hydrogen can be used for cooling above $T = 25$ K. When a material makes a transition from normal to superconducting (zero resistance) state, the magnetic field is excluded from its interior. This phenomenon is called the

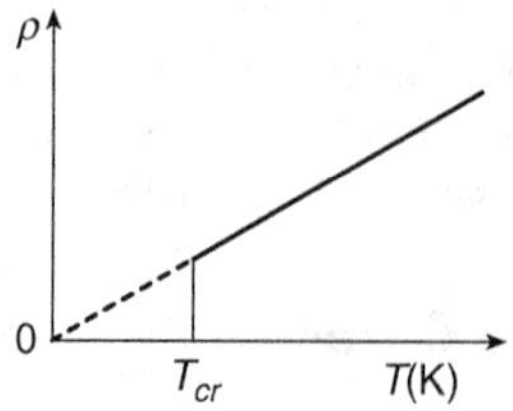

Figure 2.17 Resistivity ρ of a superconductor as a function of temperature T

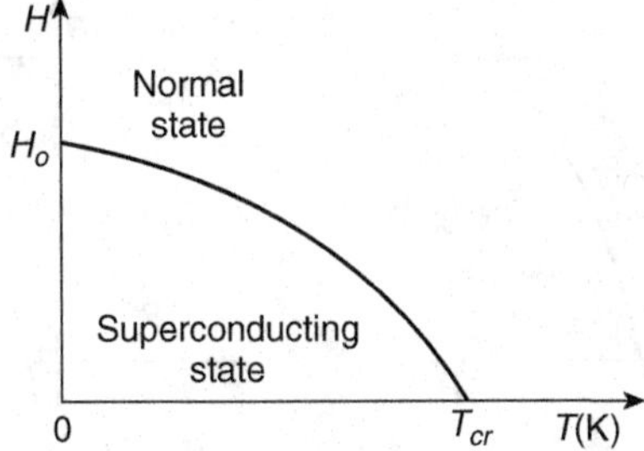

Figure 2.18 Critical magnetic field intensity as a function of temperature T

Meissner effect. The magnetic field decays exponentially inside the superconductor over a distance from 20 to 40 nm. It is described in terms of the London penetration depth. There is some degree of similarity between the Meissner effect in superconductors and skin effect in conductors. The field exclusion in superconductors is equivalent to $\mu_r \to 0$.

In addition to $T < T_{cr}$, the magnetic field intensity and the current density must be below the critical values. The critical magnetic field intensity depends on temperature and is given by

$$H_{cr} = H_o \left[1 - \left(\frac{T}{T_{cr}} \right)^2 \right]. \tag{2.64}$$

Figure 2.18 illustrates (2.64). For example, the metal niobium becomes a superconductor with zero resistance when it is cooled to below 9 K, but its superconductive property ceases when the magnetic flux density B exceeds 0.12 T. Therefore, there is a maximum current the niobium wire can carry and still remain as a superconductor. The superconductor can be divided into two groups: low-temperature superconductors and high-temperature superconductors. The low-temperature superconductors have superconductive properties below 30 K and belong to Type I. These are metals and metal alloys. The high-temperature superconductors have superconductive properties above 30 K and belong to Type II. These are ceramics. In type II superconductors, the magnetic field is not completely excluded from the conductor, but is located in filaments within the material. There is a great need for a high-temperature superconductor that is capable of operating at room temperature.

Karl Alex Muller and Johannes Georg Bednorz from the IBM Zurich Research Laboratory observed that ceramic copper oxide containing barium and lanthanum exhibits a superconductivity at $T_c = 35$ K. They received the Nobel Prize in Physics in 1987 for the discovery of superconductivity in ceramic materials.

Potential applications of superconductors are:

- power plants
- loss-free transmission lines
- rail transportation
- electric machines
- magnetic resonance imaging (MRI)

2.14 Hysteresis Loss

Core loss and increased core temperature are usually the most important limitations in high-frequency applications. Core losses consist of a *hysteresis loss*, an *eddy-current loss*, and *excess loss*. Hysteresis loss is the energy used to align and rotate magnetic moments of the core material. The physics of core losses is very complex. No one has yet developed a theory that allows us to predict core losses from

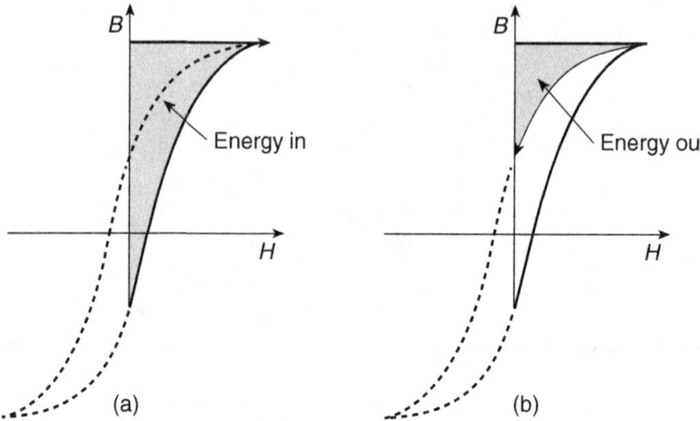

Figure 2.19 Hysteresis power loss. (a) Energy absorbed by the core during core magnetization when H increases from zero to its maximum value. (b) Energy returned from the core during core demagnetization when H decreases from its maximum value to zero

the core material properties, such as chemical content and material structure. The larger the area of the hysteresis, the more loss per cycle and the larger the hysteresis loss. When the winding current is varied through the cycle, during some interval of time, energy flows from the source to the coil, and during some other interval of time, energy is returned to the source. However, the energy flowing in is greater than the energy returned. The energy difference is equal to the core losses, which can be divided into the hysteresis loss and the eddy-current loss.

The area enclosed by the hysteresis curve represents the work required to take the magnetic material through the hysteresis cycle. Figure 2.19 illustrates the hysteresis loss. When the magnetic field intensity H is increased from zero to a maximum value, the energy is supplied to the coil by an external circuit, as shown in Fig. 2.19a. When the magnetic field intensity H is decreased from the maximum value to zero, the energy is returned to the external circuit, as shown in Fig. 2.19b. The difference between the supplied and returned energy is the energy lost in the core.

The hysteresis loss is the energy used to align and rotate magnetic domains in a core. Some energy is absorbed into a magnetic core to overcome the friction forces involved in changing the alignment of the magnetic domains. As an inductor or a transformer is used in an AC current circuit, the core material is going through the hysteresis loop the same number of times in a second as the frequency of the applied current. The energy used to move the domains is dissipated in the core material as heat. Since the hysteresis loss is proportional to the hysteresis area, cores are made of soft ferromagnetics, which have narrow hysteresis loops and correspondingly little energy loss per cycle.

Consider an inductor with a voltage v_L, current i_L, magnetic flux ϕ, magnetic flux density B, magnetic field intensity H, cross-sectional area A_c, and core volume V_c. The flux, voltage, and current of the inductor are

$$\phi = A_c B \tag{2.65}$$

$$v_L(t) = \frac{d\lambda(t)}{dt} = N\frac{d\phi(t)}{dt} = NA_c\frac{dB(t)}{dt} \tag{2.66}$$

and

$$i_L(t) = \frac{l_c H(t)}{N}. \tag{2.67}$$

The instantaneous power delivered to an inductor-carrying current $i_L(t)$ with a voltage between its terminals $v_L(t)$ at any instant of time is given by

$$p_L(t) = i_L(t)v_L(t). \tag{2.68}$$

The energy delivered to an inductor containing a ferromagnetic core during one transverse of the hysteresis loop (i.e., during one cycle) is

$$p_H = \int_0^T p_L(t)dt = \int_0^T i_L v_L dt = \int_0^T i_L d\lambda = N \int_0^T i_L d\phi = N \oint i_L d\phi$$
$$= \mathcal{R} \oint \phi d\phi = NA_c \oint i_L dB = l_c A_c \oint HdB = V_c \oint HdB\,(\mathrm{J}), \tag{2.69}$$

where $V_c = l_c A_c$ is the core volume and $d\lambda = v_L dt = Nd\phi$. This energy is lost as heat because it is used to move magnetic domains and overcome the frictions between the domain walls. It is lost as heat. The hysteresis energy loss per unit volume (or the hysteresis energy loss density) of a magnetic core material is

$$W_h = \frac{W_H}{V_c} = \oint HdB = A_{BH}\,(\mathrm{J/m^3}). \tag{2.70}$$

Hence, the power delivered to the inductor and lost as heat is given by

$$P_H = \frac{W_H}{T} = fW_H = V_c f \oint HdB\,(\mathrm{W}) \tag{2.71}$$

and the hysteresis power loss per unit volume (or the hysteresis power loss density, also called the specific hysteresis power loss) is given by

$$P_h = \frac{P_H}{V_c} = f \oint HdB = fA_{BH}\,(\mathrm{W/m^3}). \tag{2.72}$$

Thus, the hysteresis power loss per unit volume is proportional to the area A_{BH} enclosed by the B–H hysteresis loop and to the frequency f. The higher the frequency f, the more times the B–H loop is swept out, causing a higher hysteresis loss. The larger the flux density amplitude B_m in each cycle, the larger the area A_{BH}, incurring more core loss. In other words, the area of the hysteresis loop corresponds to the hysteresis energy loss per unit volume per cycle. The hysteresis power loss in a core of volume V_c is

$$P_H = P_h V_c\,(\mathrm{W}). \tag{2.73}$$

When the B–H loop is rectangular and the operating point is at the origin, the hysteresis core power loss per unit volume is

$$P_h = f \oint \mathbf{H} \cdot d\mathbf{B} = 4fH_m B_m = 4f\mu_{rc}\mu_0 H_m^2 = \frac{4f\mu_{rc}\mu_0 (NI_{Lm})^2}{l_c^2} = \frac{4fB_m^2}{\mu_{rc}\mu_0}$$
$$= 4fA_{BH}\,(\mathrm{W/m^3}) \tag{2.74}$$

yielding the hysteresis core loss

$$P_H = P_h V_c = P_h l_c A_c = \frac{4A_c \mu_{rc}\mu_0 f N^2 I_{Lm}^2}{l_c}. \tag{2.75}$$

The hysteresis core power loss is

$$P_H = \frac{4V_c \mu_{rc}\mu_0 f N^2 I_{Lm}^2}{l_c^2} = \frac{4A_c \mu_{rc}\mu_0 N^2 f I_{Lm}^2}{l_c} = \frac{R_{hs} I_{Lm}^2}{2} = 8fL. \tag{2.76}$$

Hence, the series equivalent resistance that dissipates the same amount of power as the hysteresis core loss P_H is

$$R_{hs} = \frac{8V_c \mu_{rc}\mu_0 f N^2}{l_c^2} = \frac{8A_c \mu_{rc}\mu_0 f N^2}{l_c} = \frac{8fN^2}{\mathcal{R}_c}. \tag{2.77}$$

When the B–H loop is a parallelogram with an angle α,

$$P_{hs} = f \oint \mathbf{H} \cdot d\mathbf{B} = 4fH_m B_m \sin\alpha = \frac{4fB_m^2}{\mu_{rc}\mu_0} \sin\alpha. \tag{2.78}$$

As the slope of the hysteresis decreases, α also decreases, reducing P_h. The hysteresis power loss density decreases with temperature T because the relative permeability μ_{rc} decreases with temperature.

In general, the hysteresis power loss density is

$$P_h = k_h f B_m^2, \tag{2.79}$$

where k_h is the hysteresis loss coefficient. For example, $k_h = 93.89\ \mathrm{W \cdot s/(T^2 \cdot m^3)}$ for 2.5% silicon steel.

Needless to say that a good magnetic material will have a large permeability, but a very narrow hysteresis loop. Soft iron cores satisfy this condition and are suitable for applications, where the magnetic field is being rapidly reversed. Conversely, hard steel has a wide hysteresis, resulting in large hysteresis losses. Ferrites also have a relatively large hysteresis area. As the operating frequency is increased, the area of the B–H loop also increases, increasing the hysteresis core loss. This fact can be taken into account by replacing f with f^a, where $a > 1$.

Example 2.2

A MnZn ferrite ungapped core has the relative permeability $\mu_{rc} = 2500$, the cross-sectional area $A_c = 80\ \mathrm{mm}^2$, and the mean MPL $l_c = 9.4$ cm. The number of turns of the inductor is $N = 10$. The inductor waveform is $i_L = I_{Lm} \sin \omega t = 0.5 \sin 2\pi 10^5 t$ (A). Determine the hysteresis core loss and the hysteresis core series equivalent resistance.

Solution: The hysteresis core loss is

$$P_H = \frac{4f \mu_{rc} \mu_0 A_c (N I_{Lm})^2}{l_c} = \frac{4 \times 10^5 \times 2500 \times 4\pi \times 10^{-7} \times 80 \times 10^{-6} \times (10 \times 0.5)}{(9.4 \times 10^{-2})^2}$$

$$= 26.737\ \mathrm{W}. \tag{2.80}$$

The hysteresis core loss density is

$$P_h = \frac{4f \mu_{rc} \mu_0 (N I_{Lm})^2}{l_c^2} = \frac{4 \times 10^5 \times 2500 \times 4\pi \times 10^{-7} \times (10 \times 0.5)^2}{(9.4 \times 10^{-2})^2} = 3.555\ \mathrm{MW/m^3}$$

$$= 3.555\ \mathrm{W/cm^3}. \tag{2.81}$$

The core hysteresis series equivalent resistance is

$$R_{hs} = \frac{8f \mu_{rc} \mu_0 A_c N^2}{l_c} = \frac{8 \times 10^5 \times 2500 \times 4\pi \times 10^{-7} \times 80 \times 10^{-6} \times 10^2}{9.4 \times 10^{-2}} = 213.9\ \Omega. \tag{2.82}$$

Example 2.3

A NiZn ferrite ungapped core has the relative permeability $\mu_{rc} = 150$, the cross-sectional area $A_c = 80\ \mathrm{mm}^2$, and the mean MPL $l_c = 9.4$ cm. The number of turns of the inductor is $N = 10$. The inductor waveform is $i_L = I_{Lm} \sin \omega t = 0.5 \sin 2\pi 10^5 t$ (A). Determine the hysteresis core loss and the hysteresis core series equivalent resistance.

Solution: The hysteresis core loss is

$$P_H = \frac{4f \mu_{rc} \mu_0 A_c (N I_{Lm})^2}{l_c} = \frac{4 \times 10^5 \times 150 \times 4\pi \times 10^{-7} \times 80 \times 10^{-6} \times (10 \times 0.5)^2}{(9.4 \times 10^{-2})^2}$$

$$= 1.604\ \mathrm{W}. \tag{2.83}$$

The hysteresis core loss density is

$$P_h = \frac{4f\mu_{rc}\mu_0(NI_{Lm})^2}{l_c^2} = \frac{4 \times 10^5 \times 150 \times 4\pi \times 10^{-7} \times (10 \times 0.5)^2}{(9.4 \times 10^{-2})^2} = 213\ \text{mW/cm}^3. \quad (2.84)$$

The core hysteresis series equivalent resistance is

$$R_{hs} = \frac{8\mu_{rc}\mu_0 A_c f N^2}{l_c} = \frac{8 \times 150 \times 4\pi \times 10^{-7} \times 80 \times 10^{-6} \times 10^5 \times 10^2}{9.4 \times 10^{-2}} = 12.83\ \Omega. \quad (2.85)$$

2.15 Eddy-Current Core Loss

2.15.1 General Expression for Eddy-Current Core Loss

The eddy-current loss is caused by eddy currents flowing in a conductive core. The *eddy-current power loss* (Ohmic loss or resistive loss) is due to the fact that a time-varying magnetic field induces a voltage, in accordance with Faraday's law $v = d\lambda/dt$. This voltage generates circulating currents in a conducting core, called *eddy currents*. Eddy currents follow circular paths normal to the direction of the magnetic flux. These paths resemble eddies in a stream bed. An eddy current is induced in a magnetic core whenever there is a change in a magnetic field. The higher the resistivity of the core material, the lower the eddy-current (resistive) loss. Eddy currents cause heating of the core and may add considerably to the total core losses. They cause a widening of the hysteresis to account for the increased hysteresis core loss. Iron cores are good electrical conductors and an AC magnetic field generates eddy currents within the core. The eddy currents cause conduction loss in the core resistance R_c. Eddy-current loss depends on the size and shape of the core as well as the magnitude and shape of the inductor current.

Figure 2.20 shows an inductor made of a winding and a magnetic core. A time-varying current $i(t)$ flows through the winding, producing an applied magnetic field $H_a(t)$. According to Lenz's law, the applied AC magnetic field $H(t)$ induces eddy currents i_e, which in turn induce a secondary magnetic field $H_i(t)$ that opposes the original applied magnetic field $H_a(t)$.

Consider a cylindrical core of radius r_o and length l_c with a uniform axial time-dependent flux density $B(t)$ and an arbitrary periodic waveform of the inductor voltage [32]. The cylindrical core is shown in Fig. 2.21. The core is made of a uniform magnetic material. The skin effect is assumed to be negligible

$$\delta_c = \sqrt{\frac{\rho_c}{\pi\mu_{rc}\mu_0 f}} \gg r_o. \quad (2.86)$$

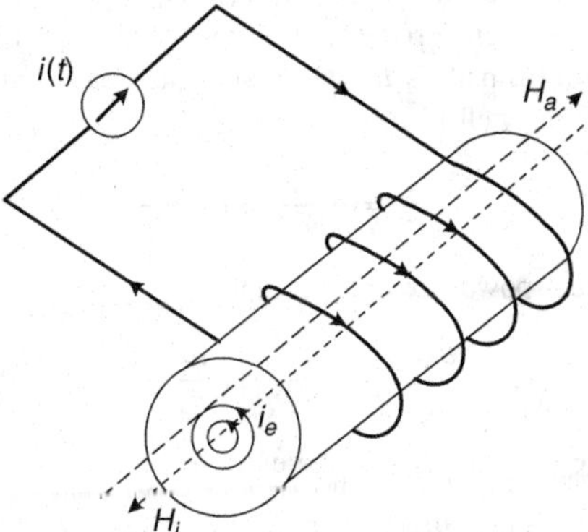

Figure 2.20 Eddy current in cylindrical magnetic core

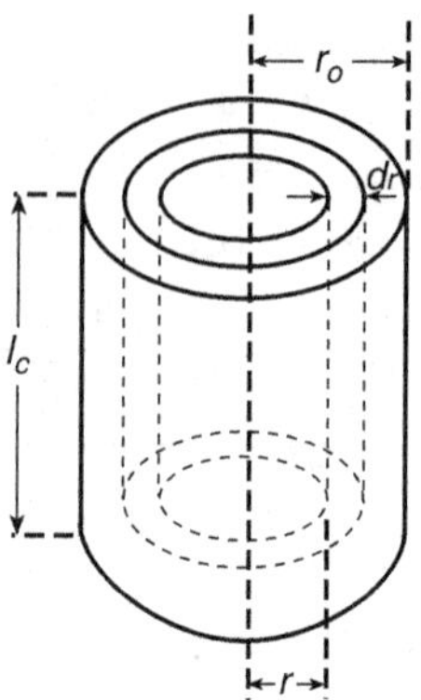

Figure 2.21 Cylindrical magnetic core

This situation corresponds to a low-frequency range in which the magnetic field penetrates the magnetic core completely and uniformly.

Using Faraday's law, we obtain the voltage across the inductor

$$v_L(t) = \frac{d\lambda(t)}{dt} = NA_c\frac{dB(t)}{dt}, \tag{2.87}$$

we get

$$\frac{dB(t)}{dt} = \frac{v_L(t)}{NA_c}, \tag{2.88}$$

where $A_c = \pi r_o^2$ is the core cross-sectional area. The time-dependent magnetic field distribution is uniform inside the cylinder. In the case of a magnetic core, $N = 1$, and therefore $\lambda(t) = \phi(t)$. The instantaneous electric field intensity induced around an elemental cylindrical shell of radius r and thickness dr is given by

$$E(r,t) = \frac{d\phi(t)}{dt} = A_{sh}\frac{dB(t)}{dt} = \pi r^2\frac{dB(t)}{dt} = \frac{A_{sh}v_L(t)}{A_cN} = \frac{\pi r^2 v_L(t)}{A_cN}, \tag{2.89}$$

where $A_{sh} = \pi r^2$ is the cross-sectional area enclosed by the shell. The electric field intensity $E(r,t)$ is proportional to the ratio of surface areas A_{sh}/A_c and the inductor voltage per turn $v_L(t)/N$.

Using Ohm's law, the eddy-current density in the shell is

$$J_e(r,t) = \frac{E(r,t)}{\rho_c} = \frac{A_{sh}}{\rho_c}\frac{dB(t)}{dt} = \frac{\pi r^2}{\rho_c}\frac{dB(t)}{dt} = \frac{A_{sh}v_L(t)}{\rho_cA_cN} = \frac{\pi r^2 v_L(t)}{\rho_cA_cN}. \tag{2.90}$$

Thus, the instantaneous value of the eddy current $J_e(r,t)$ is proportional to the instantaneous inductor voltage per turn $v_L(t)/N$, and the enclosed and it is inversely proportional to the core resistivity ρ_c.

The length of the eddy-current path is $l_i = 2\pi r$ and the area normal to the eddy-current flow is $A_i = l_c dr$. The resistance of the shell is

$$R_{sh} = \rho_c\frac{l_i}{A_i} = \rho_c\frac{2\pi r}{l_c dr}. \tag{2.91}$$

The instantaneous eddy-current power loss in the shell is

$$dp_e(r,t) = \frac{E^2(r,t)}{R_{sh}} = \frac{A_{sh}^2}{R_{sh}}\left(\frac{dB(t)}{dt}\right)^2 = \frac{\pi^2 r^4}{R_{sh}}\left(\frac{dB(t)}{dt}\right)^2 = \frac{\pi l_c r^3 dr}{2\rho_c}\left(\frac{dB(t)}{dt}\right)^2. \tag{2.92}$$

The instantaneous eddy-current power loss in the entire core is

$$p_e(t) = \int_0^{r_o} dp_e(r,t)dr = \frac{\pi l_c}{2\rho_c}\left(\frac{dB(t)}{dt}\right)^2\int_0^{r_o} r^3 dr = \frac{\pi l_c r_o^4}{8\rho_c}\left(\frac{dB(t)}{dt}\right)^2 = \frac{l_c A_c^2}{8\pi\rho_c}\left(\frac{dB(t)}{dt}\right)^2. \tag{2.93}$$

The time-average eddy-current power loss in the core is

$$P_E = \frac{1}{T}\int_0^T p_e(t)dt = \frac{\pi l_c r_o^4}{8\rho_c}\frac{1}{T}\int_0^T \left(\frac{dB(t)}{dt}\right)^2 dt = \frac{V_c r_o^2}{8\rho_c}\frac{1}{T}\int_0^T \left(\frac{dB(t)}{dt}\right)^2 dt$$
$$= \frac{V_c A_c}{8\pi\rho_c}\frac{1}{T}\int_0^T \left(\frac{dB(t)}{dt}\right)^2 dt, \tag{2.94}$$

where $V_c = A_c l_c = \pi r_o^2 l_c$ is the core volume.

The time-average eddy-current power loss density in the magnetic core material is

$$P_e = \frac{P_E}{V_c} = \frac{r_o^2}{8\rho_c}\frac{1}{T}\int_0^T \left(\frac{dB(t)}{dt}\right)^2 dt = \frac{A_c}{8\pi\rho_c}\frac{1}{T}\int_0^T \left(\frac{dB(t)}{dt}\right)^2 dt$$
$$= \frac{A_c}{8\pi\rho_c}\frac{1}{2\pi}\int_0^{2\pi} \left(\frac{dB(\omega t)}{d(\omega t)}\right)^2 d(\omega t). \tag{2.95}$$

Substituting of (2.88) into (2.95), we obtain the time-average eddy-current power loss density in a core material

$$P_e = \frac{1}{8\pi\rho_c A_c N^2}\frac{1}{T}\int_0^T v_L^2(t)dt = \frac{1}{8\pi\rho_c A_c N^2}\frac{1}{2\pi}\int_0^{2\pi} v_L^2(\omega t)d(\omega t). \tag{2.96}$$

2.15.2 Eddy-Current Core Loss for Sinusoidal Inductor Voltage

Consider a sinusoidal inductor voltage given by

$$v_L(t) = V_{Lm}\sin\omega t. \tag{2.97}$$

The inductor current waveform is

$$i_L(t) = \frac{1}{\omega L}\int_0^{\omega t} v_L d(\omega t) + i_L(0) = \frac{V_{Lm}}{\omega L}\int_0^{\omega t}\sin\omega t d(\omega t) + i_L(0) = \frac{V_{Lm}}{\omega L}(1-\cos\omega t) + i_L(0), \tag{2.98}$$

where

$$I_m = \frac{V_{Lm}}{\omega L} \tag{2.99}$$

and

$$i_L(t) = -I_m\cos\omega t. \tag{2.100}$$

Using Faraday's law in (2.87),

$$dB(t) = \frac{1}{NA_c}v_L(t)dt = \frac{V_{Lm}}{NA_c}\sin\omega t dt, \tag{2.101}$$

we obtain

$$B(t) = \int_0^t dB(t) = \frac{1}{NA_c}\int_0^t v_L dt + B(0) = \frac{1}{NA_c}\int_0^t V_{Lm}\sin\omega t dt + B(0)$$
$$= \frac{V_{Lm}}{\omega NA_c}(1-\cos\omega t) + B(0) = B_m(1-\cos\omega t) + B(0), \tag{2.102}$$

where the amplitude of the sinusoidal magnetic flux density is

$$B_m = \frac{V_{Lm}}{\omega NA_c}. \tag{2.103}$$

The peak-to-peak ac magnetic flux density is

$$\Delta B = B(\pi) - B(0) = 2B_m \tag{2.104}$$

producing

$$B(0) = -\frac{\Delta B}{2} = -B_m \tag{2.105}$$

and

$$B(t) = -B_m \cos \omega t = -\frac{V_{Lm}}{\omega N A_c} \cos \omega t. \tag{2.106}$$

Next,

$$\frac{dB(t)}{dt} = \omega B_m \sin \omega t \tag{2.107}$$

and

$$\left(\frac{dB(t)}{dt}\right)^2 = \omega^2 B_m^2 \sin^2 \omega t. \tag{2.108}$$

The eddy-current density in the core is

$$J_e(r,t) = \frac{A_{sh} v_L(t)}{\rho_c A_c N} = \frac{\pi r^2 v_L(t)}{\rho_c A_c N} = \frac{A_{sh} V_{Lm}}{\rho_c A_c N} \sin \omega t = \frac{\pi r^2 V_{Lm}}{\rho_c A_c N} \sin \omega t = J_m(r) \sin \omega t, \tag{2.109}$$

where the amplitude of the eddy-current density is

$$J_m(r) = \frac{A_{sh} V_{Lm}}{\rho_c A_c N} = \frac{\pi r^2 V_{Lm}}{\rho_c A_c N}. \tag{2.110}$$

The core diameter is $d_c = 2r_o$ and the amplitude of the flux density is $B_m = \mu H_m = \mu N I_{Lm}/l_c$. The time-average eddy-current power loss density in the core for uniform magnetic flux density in a core material under sinusoidal drive at $\delta_c \gg r_o$, that is, at low frequencies, is given by

$$\begin{aligned} P_e &= \frac{1}{8\pi \rho_c A_c N^2} \frac{1}{2\pi} \int_0^{2\pi} v_L^2(\omega t) d(\omega t) = \frac{1}{8\pi \rho_c A_c N^2} \frac{1}{2\pi} \int_0^{2\pi} V_{Lm}^2 \sin^2 \omega t d(\omega t) \\ &= \frac{V_{Lm}^2}{16\pi \rho_c A_c N^2} = \frac{A_c \omega^2 B_m^2}{16\pi \rho_c} = \frac{r_o^2 \omega^2 B_m^2}{16\rho_c} = \frac{\pi A_c f^2 B_m^2}{4\rho_c} = \frac{(d_c \omega B_m)^2}{64\rho_c} = \frac{(\pi d_c f B_m)^2}{16\rho_c} \\ &= \frac{\pi A_c (\mu_{rc} \mu_0 f N I_{Lm})^2}{4 l_c^2 \rho_c} = k_{ec} \left(\frac{A_c}{\rho_c}\right) (f B_m)^2 \quad \text{for} \quad \delta_c \gg r_o, \end{aligned} \tag{2.111}$$

where the waveform coefficient for a sinusoidal inductor voltage is

$$k_{ec} = \frac{\pi}{4}. \tag{2.112}$$

The time-average eddy-current power loss in the core for a sinusoidal inductor voltage for $\delta_c \gg r_o$ is

$$\begin{aligned} P_E = P_e V_c &= \frac{l_c V_{Lm}^2}{16\pi \rho_c N^2} = \frac{l_c A_c^2 \omega^2 B_m^2}{16\pi \rho_c} = \frac{\pi l_c A_c^2 f^2 B_m^2}{4\rho_c} = \frac{\pi l_c d_c^4 \omega^2 B_m^2}{256\rho_c} \\ &= \frac{\pi (\mu_{rc} \mu_0 A_c f N I_{Lm})^2}{4 l_c \rho_c} = \frac{\pi l_c f^2 L^2 I_{Lm}^2}{4\rho_c N^2} = \frac{1}{2} R_{cs} I_{Lm}^2. \end{aligned} \tag{2.113}$$

Hence, the low-frequency core eddy-current series equivalent resistance for sinusoidal inductor voltage and current waveforms is

$$R_{es} = \frac{\pi (\mu_{rc} \mu_0 A_c f N)^2}{2\rho_c l_c}. \tag{2.114}$$

The time-average eddy-current power loss density decreases with temperature T. For sinusoidal waveforms of inductor current and voltage, it is given by

$$P_e(T) = \frac{\pi A_c (f B_m)^2}{4\rho_{To}[1 + \alpha_c (T - T_o)]} = \frac{\pi A_c (f B_m)^2}{4\rho_{To}(1 + \alpha_c \Delta T)}, \tag{2.115}$$

where α_c is the temperature coefficient of core material resistivity.

In many applications, the inductor current contains the fundamental component and its harmonics

$$i_L = \sum_{n=1}^{\infty} I_{Lmn} \cos(n\omega t + \phi_n). \tag{2.116}$$

Assuming that μ_{rc} is independent of frequency and neglecting the core nonlinearity, we obtain the core eddy-current power loss density

$$P_e = \frac{\pi A_c (\mu_{rc}\mu_0 N)^2}{4 l_c^2 \rho_c} \sum_{n=1}^{\infty} (nf I_{Lmn})^2$$

$$= \frac{\pi A_c (\mu_{rc}\mu_0 N)^2}{4 l_c^2 \rho_c} [(f I_{Lmn1})^2 + (2f I_{Lmn2})^2 + (3f I_{Lmn3})^2 + \ldots]$$

$$\times \text{ for } \quad \delta_{cn} = \frac{\delta_c}{\sqrt{n}} \gg r_o \tag{2.117}$$

the core eddy-current power loss

$$P_E = \frac{\pi (\mu_{rc}\mu_0 N)^2}{4 l_c \rho_c} \sum_{n=1}^{\infty} (nf I_{Lmn})^2$$

$$= \frac{\pi (\mu_{rc}\mu_0 N)^2}{4 l_c \rho_c} [(f I_{Lmn1})^2 + (2f I_{Lmn2})^2 + (3f I_{Lmn3})^2 + \ldots]$$

$$\text{for } \quad \delta_{cn} = \frac{\delta_c}{\sqrt{n}} \gg r_o. \tag{2.118}$$

2.15.3 Eddy-Current Power Loss in Round Core for Sinusoidal Flux Density

A shorter derivation of eddy-current power loss for a cylindrical core with a sinusoidal flux density is as follows. Assume that the waveform of sinusoidal magnetic flux density in the core is uniform and is given by

$$B(t) = B_m \sin \omega t \quad \text{for} \quad 0 \le r \le r_o = d/2. \tag{2.119}$$

For a core, $N = 1$, and therefore the flux linkage enclosed in the area of radius r is $A_{sh} = \pi r^2$ is

$$\lambda(t) = N\phi(t) = \phi(t) = A_{sh}B(t) = \pi r^2 B_m \sin \omega t \quad \text{for} \quad 0 \le r \le r_o = d/2. \tag{2.120}$$

From Faraday's law, the voltage induced by the enclosed and changing magnetic flux at a radial distance r from the core center at $r = 0$ is

$$v(r,t) = \frac{\lambda(t)}{dt} = \frac{d\phi(t)}{dt} = A_{sh}\frac{dB(t)}{dt} = \pi r^2 \frac{dB(t)}{dt} = A_{sh}\omega B_m \cos \omega t = V_m(r) \cos \omega t, \tag{2.121}$$

where the amplitude of the voltage induced by the magnetic field is

$$V_m(r) = A_{sh}\omega B_m = \pi r^2 \omega B_m. \tag{2.122}$$

The length of the eddy-current path at a radial distance r from the core center is $l_i = 2\pi r$ and the incremental surface area is $dS = drdz$. The incremental resistance is

$$R_{dr} = \rho_c \frac{l_i}{dS} = \rho_c \frac{2\pi r}{drdz}. \tag{2.123}$$

The incremental time-average eddy-current power loss is

$$dP_E = \frac{V_m^2(r)}{2R_{dr}} = (\pi r^2 \omega B_m)^2 \frac{drdz}{2(2\pi r \rho_c)} = \frac{\pi \omega^2 B_m^2}{4\rho_c} r^3 drdz = \frac{\pi^3 (f B_m)^2}{\rho_c} r^3 drdz. \tag{2.124}$$

The cross-sectional area of the core is $A_c = \pi r_o^2 = \pi d^2/4$ and the core volume is $V_c = l_c A_c = \pi d^2 l_c/4$. The time-average eddy-current power loss in the whole core is given by

$$P_E = \int_0^{l_c}\int_0^{r_o} dP_E = \int_0^{l_c} dz \left[\frac{\pi^3 (f B_m)^2}{\rho_c}\int_0^{r_o} r^3 dr\right] = \frac{\pi^3 l_c (f B_m)^2 r_o^4}{4\rho_c} = \frac{\pi^3 l_c (f B_m)^2 d^4}{64\rho_c}$$

$$= \frac{l_c A_c^2 \omega^2 B_m^2}{16\pi \rho_c} = \frac{\pi l_c d_c^4 \omega^2 B_m^2}{256\rho_c} = \frac{\pi l_c A_c^2 (f B_m)^2}{4\rho_c} = \frac{\pi l_c A_c^2 (f B_m)^2}{4\rho_c} = \frac{\pi A_c V_c (f B_m)^2}{4\rho_c}. \tag{2.125}$$

The time-average-specific eddy-current core power loss density is

$$P_e = \frac{P_E}{V_c} = \frac{\pi A_c (f B_m)^2}{4\rho_c} = \frac{A_c \omega^2 B_m^2}{16\pi \rho_c} = \frac{d_c^2 \omega^2 B_m^2}{64\rho_c}. \tag{2.126}$$

2.15.4 Total Core Power Loss for Sinusoidal Inductor Voltage

The total core loss density for a sinusoidal inductor current is

$$P_v = P_h + P_e = f \oint HdB + k_{ec} \frac{A_c f^2 B_m^2}{\rho_c}. \tag{2.127}$$

At low values of B_m, $P_h \gg P_e$. At large values of B_m, $P_e \ll P_h$. The total core loss for a sinusoidal inductor current is

$$P_C = P_H + P_E = (P_h + P_e)V_c = V_c \left(f \oint HdB + k_{ec} \frac{A_c f^2 B_m^2}{\rho_c} \right). \tag{2.128}$$

The core loss depends on the core material, operating frequency, and maximum magnetic flux density.

In general, the total core loss density for sinusoidal voltage is

$$P_v = P_h + P_e = k_h f B_m^2 + k_e f^2 B_m^2 = k_h f B_m^2 + k_e (f B_m)^2. \tag{2.129}$$

The total core loss density for sinusoidal voltage decreases with temperature T to about 100 °C and then increases.

Assuming that the hysteresis is rectangular, the total core power loss density for sinusoidal voltage is equal to the sum of the hysteresis loss density and the eddy-current loss density

$$\begin{aligned} P_v = P_h + P_e &= \frac{4\mu_{rc}\mu_0 f N^2 I_{Lm}^2}{l_c^2} + \frac{\pi \mu_{rc}^2 \mu_0^2 A_c^2 f^2 N^2 I_{Lm}^2}{4\rho_c l_c^2} = \frac{4 f B_m^2}{\mu_{rc}\mu_0} + \frac{\pi A_c f^2 B_m^2}{4\rho_c} \\ &= k_h f B_m^2 + k_e f^2 B_m^2, \end{aligned} \tag{2.130}$$

where

$$k_h = \frac{4}{\mu_{rc}\mu_0} \tag{2.131}$$

and

$$k_e = \frac{\pi A_c}{4\rho_c}. \tag{2.132}$$

Figure 2.22 shows a plot of total core loss density P_v as a function of the amplitude of flux density B_m and frequency f for a ferrite material with $\mu_{rc} = 2500$, $\rho_c = 1\ \Omega \cdot \text{m}$, and $A_c = 1\ \text{cm}^2$.

Equating the hysteresis core loss density and the eddy-current loss density

$$\frac{4 f B_m^2}{\mu_{rc}\mu_0} = \frac{\pi A_c f^2 B_m^2}{4\rho_c}, \tag{2.133}$$

one obtains the critical frequency at which both power loss densities are equal

$$f_{cr} = \frac{16\rho_c}{\pi \mu_{rc}\mu_0 A_c}. \tag{2.134}$$

The total core power loss for sinusoidal voltage is equal to the sum of the hysteresis loss and the eddy-current loss

$$P_C = P_H + P_E = \frac{4\mu_{rc}\mu_0 f A_c N^2 I_{Lm}^2}{l_c} + \frac{\pi A_c^2 \mu_{rc}^2 \mu_0^2 f^2 N^2 I_{Lm}^2}{4\rho_c l_c} = \frac{4 l_c A_c f B_m^2}{\mu_{rc}\mu_0} + \frac{\pi l_c A_c^2 f^2 B_m^2}{4\rho_c}. \tag{2.135}$$

The total series core equivalent resistance is

$$R_{cs} = R_{hs} + R_{es} = \frac{8 f \mu_{rc}\mu_0 A_c N^2}{l_c} + \frac{\pi (\mu_{rc}\mu_0 f A_c N)^2}{2\rho_c l_c} \quad \text{for} \quad \delta_c \gg r_o. \tag{2.136}$$

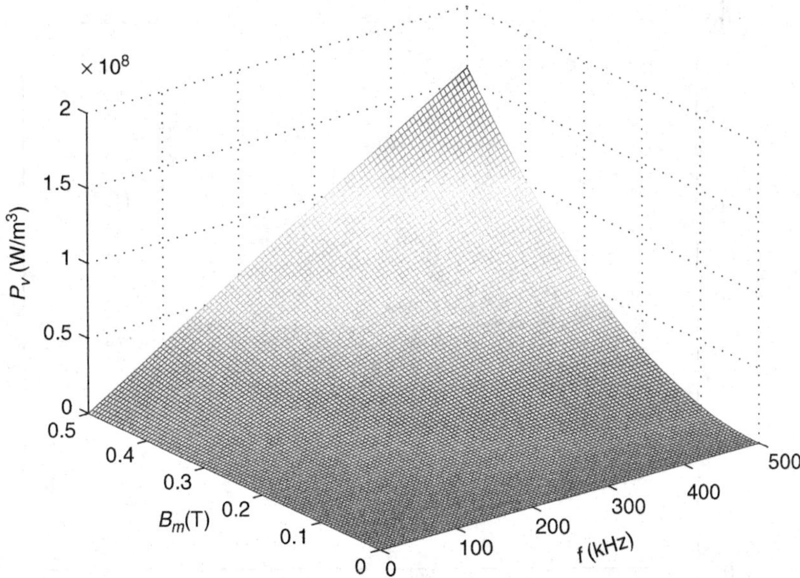

Figure 2.22 Total core loss density P_v as a function of the amplitude of flux density B_m and frequency f for a ferrite material with $\mu_{rc} = 2500$, $\rho_c = 1\Omega \cdot$ m, and $A_c = 1$ cm^2

From the experimental results [76], the total core power loss density for sinusoidal voltage waveforms for grain-oriented silicon steel is

$$P_h = 12.78 \times 10^{-3} f B_m^{0.97} + 37.68 \times 10^{-6} f^2 B_m^{3.13} (\text{W/m}^3), \tag{2.137}$$

for cobalt-based amorphous alloys is

$$P_h = 2.88 \times 10^{-3} f B_m^{1.2} + 1.9 \times 10^{-6} f^2 B_m^{2.75} (\text{W/m}^3), \tag{2.138}$$

and for nanocrystalline alloys is

$$P_h = 0.17 \times 10^{-3} f B_m^{1.11} + 0.71 \times 10^{-6} f^2 B_m^{2.92} (\text{W/m}^3). \tag{2.139}$$

Example 2.4

A MnZn ferrite ungapped core has a cross-sectional area $A_c = 80$ mm^2, a mean MPL $l_c = 94$ mm, the relative permeability $\mu_{rc} = 2500$, the saturation flux density $B_s = 0.32$ T at temperature $T = 20\,°$C, and the resistivity $\rho_c = 1\ \Omega \cdot$ cm. The number of turns is $N = 10$. The inductor voltage waveform is $v_L(t) = V_{Lm} \sin \omega t = 10 \sin 2\pi 10^5 t$ (V). Determine the eddy-current core loss, the eddy-current loss density, the amplitude of the flux density, the core hysteresis series equivalent resistance, and the critical frequency.

Solution: The eddy-current core loss is

$$P_E = \frac{l_c V_{Lm}^2}{16\pi \rho_c N^2} = \frac{94 \times 10^{-3} \times 10^2}{16\pi \times 10^{-2} \times 10^2} = 0.187 \text{ W} = 187 \text{ mW}. \tag{2.140}$$

Figure 2.23 shows the plots of the hysteresis core power loss P_H and the eddy-current core power loss P_E as functions of frequency f for the MnZn ferrite core with $\mu_{rc} = 2500$, $\rho_c = 1\ \Omega \cdot$ cm, $A_c = 80$ mm^2, and $l_c = 9.4$ cm for sinusoidal inductor voltage with $B_m = 0.02$ T. It can be seen that the hysteresis power loss P_H is higher than the eddy-current power loss P_H at low frequencies. Conversely, the hysteresis power loss P_H is lower than the eddy-current power loss P_H at high frequencies.

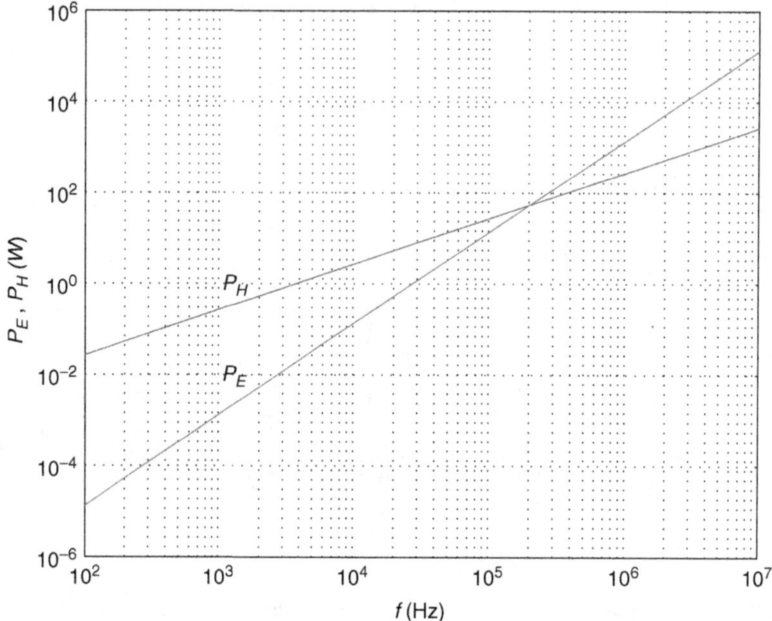

Figure 2.23 Hysteresis core power loss P_H and eddy-current core power loss P_E as functions of frequency for the MnZn ferrite core with $\mu_{rc} = 2500$, $\rho_c = 1\ \Omega \cdot \text{cm}$, $A_c = 80\ \text{mm}^2$, and $l_c = 9.4$ cm for sinusoidal inductor voltage at $B_m = 0.02$ T

The core volume is

$$V_c = l_c A_c = 9.4\ \text{cm} \times 0.8\ \text{cm}^2 = 7.52\ \text{cm}^3. \tag{2.141}$$

The core eddy-current power density is

$$P_e = \frac{P_E}{V_c} = \frac{0.187}{7.52} = 24.87\ \text{mW/cm}^3. \tag{2.142}$$

The magnetic flux density is

$$B_m = \frac{V_{Lm}}{\omega N A_c} = \frac{10}{2\pi \times 10^5 \times 10 \times 80 \times 10^{-6}} = 0.02\ \text{T}. \tag{2.143}$$

The hysteresis loss density is

$$P_h \approx \frac{4fB_m^2}{\mu_{rc}\mu_0} = \frac{4 \times 10^2 \times 0.02^2}{4\pi \times 10^{-7} \times 2500} = 5.093\ \text{mW/cm}^3. \tag{2.144}$$

Hence, the hysteresis power loss is

$$P_H = P_h V_c = 5.093 \times 7.42 = 37.79\ \text{mW}. \tag{2.145}$$

The core eddy-current series equivalent resistance is

$$R_{es} = \frac{\pi(\mu_{rc}\mu_0 f A_c N)^2}{2\rho_c l_c} = \frac{\pi \times (2500 \times 4\pi \times 10^{-7} \times 10^5 \times 80 \times 10^{-6} \times 10)^2}{2 \times 10^{-2} \times 9.4 \times 10^{-2}}$$
$$= 105.553\ \Omega. \tag{2.146}$$

The core hysteresis series equivalent resistance is

$$R_{hs} = \frac{8\mu_{rc}\mu_0 A_c f N^2}{l_c} = \frac{8 \times 4\pi \times 10^{-7} \times 80 \times 10^{-6} \times 10^5 \times 10^2}{9.4 \times 10^{-2}} = 213.896\ \Omega. \tag{2.147}$$

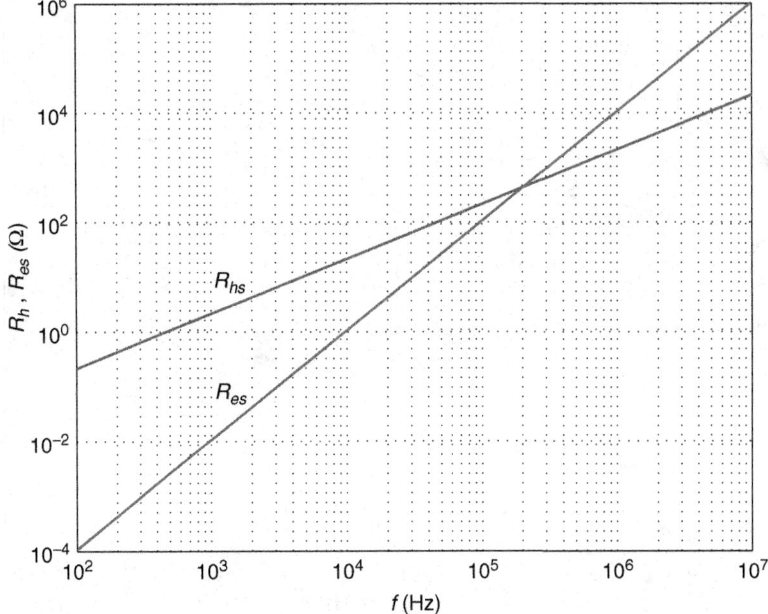

Figure 2.24 Hysteresis core equivalent series resistance R_{hs} and eddy-current core equivalent series resistance R_{es} as functions of frequency for the MnZn ferrite core with $\mu_{rc} = 2500$, $\rho_c = 1\ \Omega \cdot \text{cm}$, $A_c = 80\ \text{mm}^2$, and $l_c = 9.4$ cm

The total series core resistance is

$$R_{cs} = R_{hs} + R_{es} = 213.896 + 105.553 = 419.449\ \Omega. \tag{2.148}$$

The critical frequency at which the hysteresis power loss is equal to the eddy-current power loss is

$$f_{cr} = \frac{16\rho_c}{\pi \mu_{rc} \mu_0 A_c} = \frac{16 \times 10^{-2}}{\pi \times 2500 \times 4\pi \times 10^{-7} \times 80 \times 10^{-6}} = 202.64\ \text{kHz}. \tag{2.149}$$

Figure 2.24 shows the plots of hysteresis core equivalent series resistance R_{hs} and eddy-current core equivalent series resistance R_{es} as functions of frequency for the MnZn ferrite core with $\mu_{rc} = 2500$, $\rho_c = 1\ \Omega \cdot \text{cm}$, $A_c = 80\ \text{mm}^2$, and $l_c = 9.4$ cm.

Example 2.5

A NiZn ferrite core has a cross-sectional area $A_c = 80\ \text{mm}^2$, a mean MPL $l_c = 94$ mm, the relative permeability $\mu_{rc} = 150$, and the resistivity $\rho_c = 10^7\ \Omega \cdot \text{cm}$. The number of turns is $N = 10$. The inductor voltage waveform is $v_L(t) = V_{Lm} \sin \omega t = 10 \sin 2\pi 10^5 t$ (V). Determine the eddy-current core loss, the core hysteresis series equivalent resistance, and the critical frequency.

Solution: The eddy-current core loss is

$$P_E = \frac{l_c V_{Lm}^2}{16\pi \rho_c N^2} = \frac{9.4 \times 10^{-2} \times 10^2}{16\pi \times 10^5 \times 10^2} = 18.7\ \text{nW}. \tag{2.150}$$

Figure 2.25 shows the plots of the hysteresis core power loss P_H and the eddy-current core power loss P_E as functions of frequency f for the NiZn ferrite core with $\mu_{rc} = 150$, $\rho_c = 10^7\ \Omega \cdot \text{cm}$, $A_c = 80\ \text{mm}^2$, and $l_c = 9.4$ cm at $B_m = 0.02$ T.

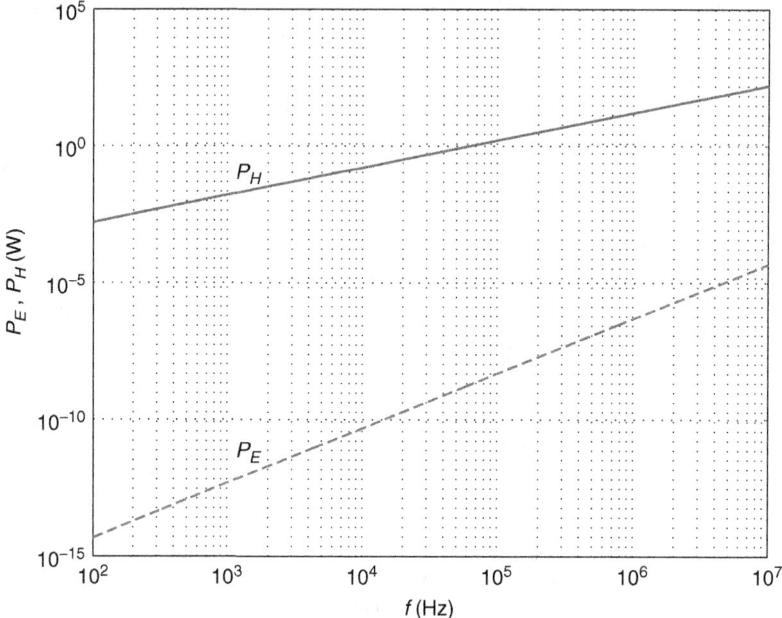

Figure 2.25 Hysteresis core power loss P_H and eddy-current core power loss P_E as functions of frequency for the NiZn ferrite core with $\mu_{rc} = 150$, $\rho_c = 10^7\ \Omega \cdot \text{cm}$, $A_c = 80\ \text{mm}^2$, and $l_c = 9.4$ cm at $B_m = 0.02$ T

The core eddy-current power density is

$$P_e = \frac{P_E}{V_c} = \frac{18.7}{7.52} = 2.487\ \text{nW/cm}^3. \tag{2.151}$$

The core eddy-current series equivalent resistance is

$$R_{es} = \frac{\pi(\mu_{rc}\mu_0 f A_c N)^2}{2\rho_c l_c} = \frac{\pi \times (150 \times 4\pi \times 10^{-7} \times 10^5 \times 80 \times 10^{-6} \times 10)^2}{2 \times 10^5 \times 9.4 \times 10^{-2}}$$
$$= 105.553\ \Omega. \tag{2.152}$$

The critical frequency at which the hysteresis power loss is equal to the eddy-current power loss is given by

$$f_{cr} = \frac{16\rho_c}{\pi\mu_{rc}\mu_0 A_c} = \frac{16 \times 10^5}{\pi \times 150 \times 4\pi \times 10^{-7} \times 80 \times 10^{-6}} = 33.774\ \text{THz}. \tag{2.153}$$

Figure 2.26 shows the plots of hysteresis core power equivalent series resistance R_{hs} and eddy-current core equivalent series resistance R_{es} as functions of frequency for the NiZn ferrite core with $\mu_{rc} = 150$, $\rho_c = 10^7\ \Omega \cdot \text{cm}$, $A_c = 80\ \text{mm}^2$, and $l_c = 9.4$ cm at $B_m = 0.2$ T.

2.15.5 Eddy-Current Core Loss for Square-Wave Inductor Voltage

For a bipolar square-wave inductor voltage

$$v_L(t) = V_{Lm}\,\text{sgn}(t) = \pm V_{Lm}. \tag{2.154}$$

The inductor voltage waveform for the first half of the cycle is

$$v_L(t) = V_{Lm} \quad \text{for} \quad 0 < t \leq \frac{T}{2}, \tag{2.155}$$

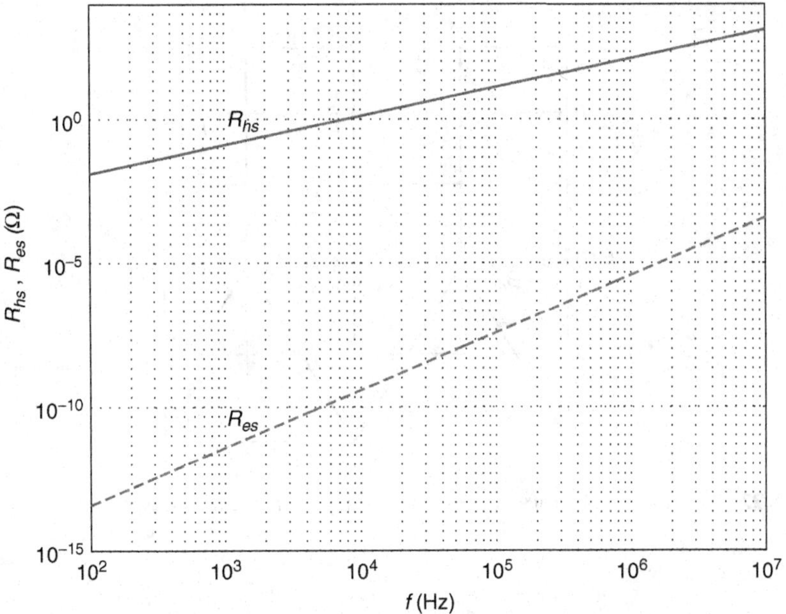

Figure 2.26 Hysteresis core power equivalent series resistance R_{hs} and eddy-current core equivalent series resistance R_{es} as functions of frequency for the NiZn ferrite core with $\mu_{rc} = 150$, $\rho_c = 10^7\ \Omega \cdot \text{cm}$, $A_c = 80\ \text{mm}^2$, and $l_c = 9.4$ cm

yielding the waveform of the magnetic flux density

$$B(t) = \frac{1}{NA_c}\int_0^t v_L dt + B(0) = \frac{V_{Lm}}{NA_c}\int_0^t dt + B(0) = \frac{V_{Lm}}{NA_c}t + B(0) \quad \text{for} \quad 0 < t \le \frac{T}{2}. \tag{2.156}$$

Hence,

$$B\left(\frac{T}{2}\right) = \frac{V_{Lm}T}{2NA_c} + B(0) = \frac{V_{Lm}}{2NA_c f} + B(0). \tag{2.157}$$

The inductor voltage waveform for the second half of the cycle is

$$v_L(t) = -V_{Lm} \quad \text{for} \quad \frac{T}{2} < t \le T, \tag{2.158}$$

producing

$$B(t) = \frac{1}{NA_c}\int_{T/2}^t v_L dt + B\left(\frac{T}{2}\right) = -\frac{V_{Lm}}{NA_c}\int_{T/2}^t dt + B\left(\frac{T}{2}\right)$$
$$= -\frac{V_{Lm}}{NA_c}\left(t - \frac{T}{2}\right) + B\left(\frac{T}{2}\right) \quad \text{for} \quad \frac{T}{2} < t \le T. \tag{2.159}$$

Figure 2.27 shows the waveforms of the square-wave inductor voltage v_L and the resulting triangular magnetic flux density $B(t)$. The peak-to-peak value of the magnetic flux density is

$$\Delta B = B\left(\frac{T}{2}\right) - B(0) = \frac{V_{Lm}}{2fNA_c} + B(0) - B(0) = \frac{V_{Lm}}{2fNA_c} = 2B_m. \tag{2.160}$$

This yields the magnitude of the symmetric triangular waveform of the magnetic flux density

$$B_m = \frac{\Delta B}{2} = \frac{V_{Lm}}{4fNA_c}. \tag{2.161}$$

The minimum and maximum values of the magnetic flux density are

$$B(0) = -B_m = -\frac{V_{Lm}}{4fNA_c} \tag{2.162}$$

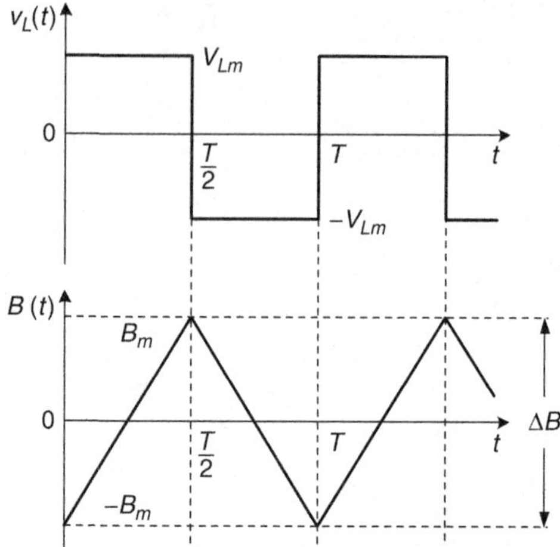

Figure 2.27 Waveforms of the square-wave inductor voltage $v_L(t)$ and the corresponding triangular magnetic flux density $B(t)$

and

$$B\left(\frac{T}{2}\right) = B_m = \frac{V_{Lm}}{4fNA_c}. \tag{2.163}$$

The time-average eddy-current power loss density in a magnetic core material for uniform flux density is obtained from (2.96)

$$P_e = \frac{1}{8\pi\rho_c A_c N^2}\frac{1}{T}\int_0^T v_L^2 dt = \frac{1}{8\pi\rho_c A_c N^2}\frac{1}{T}\int_0^T V_{Lm}^2 dt = \frac{V_{Lm}^2}{8\pi\rho_c A_c N^2}$$
$$= \frac{2A_c f^2 B_m^2}{\pi\rho_c} = \frac{A_c\omega^2 B_m^2}{2\pi^3\rho_c} = k_{ec}\frac{A_c f^2 B_m^2}{\rho_c}, \tag{2.164}$$

where the waveform coefficient for the square-wave inductor voltage is

$$k_{ec} = \frac{2}{\pi}. \tag{2.165}$$

The time-average eddy-current core power loss density is proportional to the inductor voltage per turn squared $(V_{Lm}/N)^2$. The eddy-current loss in the core for a square-wave inductor voltage is

$$P_E = P_e V_c = \frac{2l_c A_c^2 f^2 B_m^2}{\pi\rho_c} = \frac{l_c V_{Lm}^2}{8\pi\rho_c N^2} \quad \text{for} \quad \delta_c \gg d_c. \tag{2.166}$$

The ratio of the core eddy-current power loss density with a sinusoidal voltage waveform to that with a square-wave voltage at the same $(V_{Lm}/N)^2$ is

$$\frac{P_{e(sq)}}{P_{e(\sin)}} = \frac{8}{\pi^2} \approx 0.81 \quad \text{for} \quad \delta_c \gg d_c. \tag{2.167}$$

2.15.6 Eddy-Current Core Loss for Rectangular Inductor Voltage

The bipolar inductor voltage waveform for the first part of the cycle is

$$v_L(t) = V_H \quad \text{for} \quad 0 < t \leq DT, \tag{2.168}$$

producing the waveform of the magnetic flux density

$$B(t) = \frac{1}{NA_c}\int_0^t v_L(t)dt + B(0) = \frac{V_H}{NA_c}\int_0^t dt + B(0) = \frac{V_H}{NA_c}t + B(0) \quad \text{for} \quad 0 < t \leq DT. \tag{2.169}$$

Hence,

$$B(DT) = \frac{V_H DT}{NA_c} + B(0) = \frac{DV_H}{NA_c f} + B(0). \tag{2.170}$$

The bipolar inductor voltage waveform for the second part of the cycle is

$$v_L(t) = -V_L \quad \text{for} \quad DT < t \leq T, \tag{2.171}$$

yielding

$$B(t) = \frac{1}{NA_c}\int_{DT}^t v_L(t)dt + B(DT) = -\frac{V_L}{NA_c}\int_{DT}^t dt + B(DT)$$
$$= -\frac{V_L}{NA_c}(t - DT) + B(DT) \quad \text{for} \quad DT < t \leq T. \tag{2.172}$$

Figure 2.28 shows the waveforms of the rectangular wave inductor voltage $v_L(t)$ and the resulting triangular magnetic flux density $B(t)$. The peak-to-peak value of the magnetic flux density is

$$\Delta B = B(DT) - B(0) = \frac{DV_H}{fNA_c} + B(0) - B(0) = \frac{DV_H}{fNA_c}. \tag{2.173}$$

This yields the magnitude of the unsymmetric triangular waveform of the magnetic flux density

$$B_m = \frac{\Delta B}{2} = \frac{DV_H}{2fNA_c}. \tag{2.174}$$

The minimum and maximum values of the magnetic flux density are given by

$$B(0) = -B_m = -\frac{DV_H}{2fNA_c} \tag{2.175}$$

and

$$B(DT) = B_m = \frac{DV_H}{2fNA_c}. \tag{2.176}$$

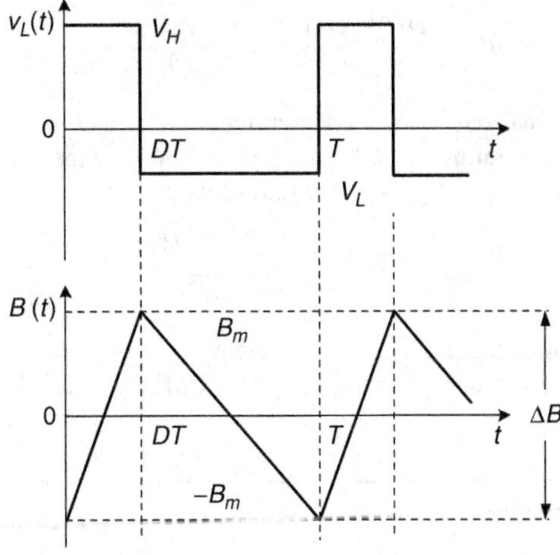

Figure 2.28 Waveforms of the rectangular wave inductor voltage $v_L(t)$ and the corresponding unsymmetric triangular magnetic flux density $B(t)$

The time-average eddy-current power loss density in a magnetic core material for uniform flux density is obtained from (2.96)

$$P_e = \frac{1}{8\pi \rho_c A_c N^2} \frac{1}{T} \int_0^T v_L^2(t)dt = \frac{1}{8\pi \rho_c A_c N^2} \frac{1}{T} \left(\int_0^{DT} V_H^2 dt + \int_{DT}^T V_L^2 dt \right)$$
$$= \frac{DV_H^2 + (1-D)V_L^2}{8\pi \rho_c A_c N^2}. \quad (2.177)$$

The relationship among V_H, V_L, and D is

$$\frac{V_L}{V_H} = \frac{D}{1-D}. \quad (2.178)$$

Hence,

$$P_e = \frac{DV_H^2}{8\pi \rho_c A_c N^2 (1-D)} = \frac{(1-D)V_L^2}{8\pi \rho_c A_c N^2 D}. \quad (2.179)$$

From (2.176),

$$V_H = \frac{2fNA_c B_m}{D}. \quad (2.180)$$

Substitution of this equation into (2.179) produces

$$P_e = \frac{A_c f^2 B_m^2}{2\pi \rho_c D(1-D)} = k_{ec} \frac{A_c f^2 B_m^2}{\rho_c}, \quad (2.181)$$

where the waveform coefficient for the rectangular inductor voltage is

$$k_{ec} = \frac{1}{2\pi D(1-D)}. \quad (2.182)$$

For the buck and buck-boost converters, $V_L = V_O$. From (2.96),

$$P_e = \frac{(1-D)V_O^2}{8\pi \rho_c A_c N^2 D} = \frac{(1-D)A_c f^2 B_m^2}{2\pi \rho_c D^3}. \quad (2.183)$$

Figure 2.29 shows the normalized specific eddy-current core power loss as a function of duty cycle D for the buck and buck-boost converter operating in CCM at a fixed output voltage V_O.

For the boost converter, $V_L = V_I - V_O = -DV_O$. From (2.96),

$$P_e = \frac{D(1-D)V_O^2}{8\pi \rho_c A_c N^2} = \frac{A_c f^2 B_m^2}{2\pi \rho_c D(1-D)}. \quad (2.184)$$

Figure 2.30 shows the normalized specific eddy-current core power loss as a function of duty cycle D for the boost converter operating in CCM at a fixed output voltage V_O.

For the buck, $V_H = V_I - V_O = V_I(1-D)$. From (2.96),

$$P_e = \frac{D(1-D)V_I^2}{8\pi \rho_c A_c N^2}. \quad (2.185)$$

Figure 2.31 shows the normalized specific eddy-current core power loss as a function of duty cycle D for the buck and buck-boost converter operating in CCM at a fixed input voltage V_I.

Example 2.6

A Magnetics ferrite material type K has the resistivity $\rho_c = 20\ \Omega\cdot$m and the relative permeability $\mu_{rc} = 1500$. Find the skin depth at $f = 500$ kHz. Is the skin effect a problem for ferrite cores?

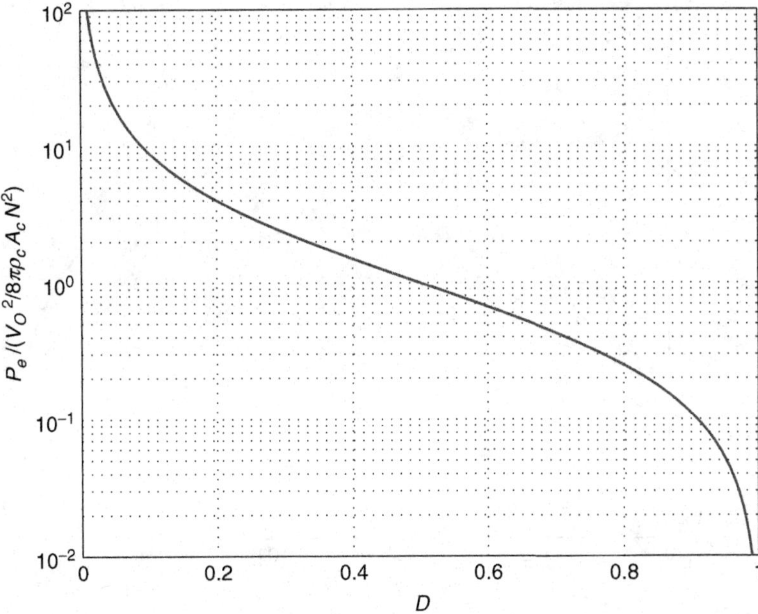

Figure 2.29 Normalized specific eddy-current core power loss as a function of duty cycle D for the buck and buck-boost converter operating in continuous conduction mode (CCM) at a fixed output voltage V_O

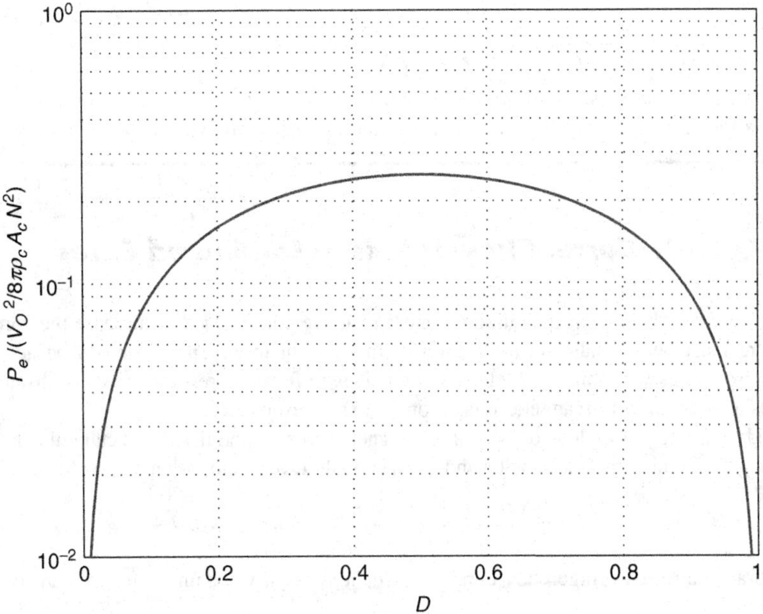

Figure 2.30 Normalized specific eddy-current core power loss as a function of duty cycle D for the boost converter operating in CCM at a fixed output voltage V_O

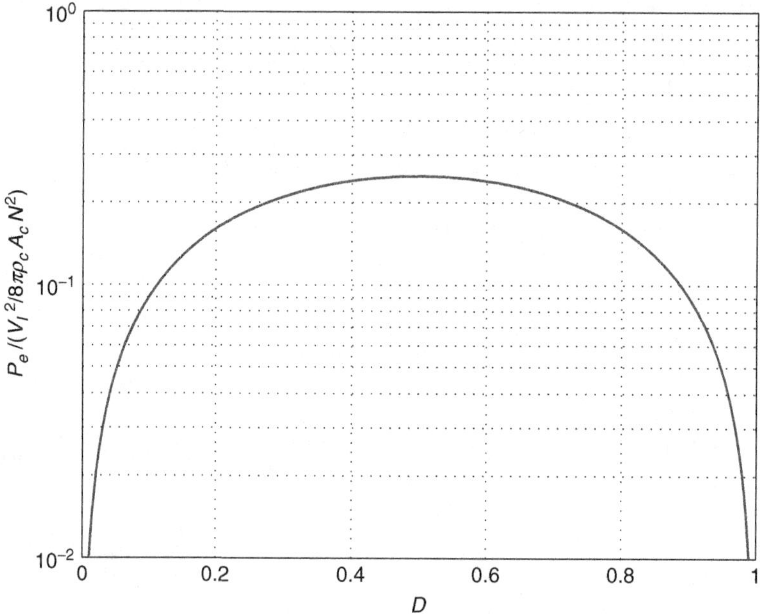

Figure 2.31 Normalized specific eddy-current core power loss as a function of duty cycle D for the buck and buck-boost converter operating in CCM at a fixed input voltage V_I

Solution: The skin depth of the ferrite material at $f = 500$ kHz is

$$\delta_w = \sqrt{\frac{\rho_c}{\pi\mu_0\mu_r f}} = \sqrt{\frac{20}{\pi \times 4\pi \times 10^{-7} \times 1500 \times 0.5 \times 10^6}} = 8.129 \text{ cm}. \tag{2.186}$$

In most practical situations, the skin effect in ferrite cores may be neglected because the ferrite skin depth δ_c is much longer than the core thickness due to large resistivity of ferrites.

2.15.7 Eddy-Current Power Loss in Laminated Cores

To minimize the eddy-current magnitude, a more resistive core is used to increase the core resistance R_c or there must be a smaller cross section of the current loop. The latter can be achieved using laminated iron cores with thin isolated magnetic plates. In ferrite cores, there are also hysteresis losses and losses associated with magnetic relaxation and spin resonances.

The eddy-current power loss density in a magnetic core material for a rectangular lamination of thickness h for a uniform flux density under sinusoidal current and voltage waveforms is given by

$$P_e = \frac{(\pi h f B_m)^2}{6\rho_c} = \frac{(h\omega B_m)^2}{24\rho_c} \quad \text{for} \quad \delta_c \gg h. \tag{2.187}$$

In general, the time-average eddy-current power loss density for uniform flux density in the core is given by

$$P_e = k_{ec}\frac{A_c f^2 B_m^2}{\rho_c}, \tag{2.188}$$

where k_{ec} is the waveform coefficient, which depends on the shape of the inductor voltage waveform. The time-average eddy-current core loss is

$$P_E = P_e V_c = k_{ec}\frac{V_c A_c f^2 B_m^2}{\rho_c} = k_{ec}\frac{l_c A_c^2 f^2 B_m^2}{\rho_c}. \tag{2.189}$$

When the core material resistivity ρ_c is infinity, eddy currents are zero, yielding zero eddy-current power loss. However, almost all magnetic cores are conductors (i.e., $\rho_c < \infty$). Laminated cores are studied in Chapter 6.

2.15.8 Excess Core Loss

The magnetic domains are divided from each other by boundaries, called *domain walls* or Bloch walls. Magnetic materials have defects in the structure. The excess loss is caused by the dynamic losses of the magnetic domains when a magnetic material is subjected to a changing external magnetic field. In iron-based magnetic materials, the excess losses are due to discontinuous (discrete) movements of Bloch walls. When the magnetic field is changing, the magnetization changes in many small discontinuous jumps as the domain overcomes the defects. This phenomenon is called the Barkhausen effect. The jumps are very fast, inducing eddy currents and producing Joule's losses [44]. The core excess power loss density for sinusoidal waveforms is described by

$$P_{ex} = k_{ex}(fB_m)^{\frac{3}{2}} (W/m)^3. \tag{2.190}$$

where k_{ex} is a constant and B_m is the amplitude of the magnetic flux density. The total core power loss density for sinusoidal waveforms is given by

$$P_v = P_h + P_e + P_{ex} = k_h f B_m^2 + k_e(fB_m)^2 + k_{ex}(fB_m)^{\frac{3}{2}} (W/m)^3. \tag{2.191}$$

2.16 Steinmetz Empirical Equation for Total Core Loss

2.16.1 Losses of Ungapped Cores

The total time-averaged core material power loss per unit volume, or the core material power loss density, or the specific core material loss, due to both mechanisms, hysteresis and eddy currents, for a *sinusoidal* magnetic flux density $B(t) = B_m \sin \omega t$ of frequency f can be described by the Steinmetzs[2] empirical equation [20]

$$P_v = \frac{P_C}{V_c} = P_h + P_e = kf^a B_m^b \, (\text{mW/cm}^3), \tag{2.192}$$

where B_m is the amplitude of the AC sinusoidal component of the magnetic flux density usually given in millitorr, f is the frequency of sinusoidal excitation of the core inductor usually expressed in kilohertz, P_C is the total core loss, V_c is the core volume, and k, a, and b are constants for a given core material. Typically, a is somewhat greater than 1 and b is somewhat greater than 2.

For ferrite cores, $a = 1.3$ and $b = 2.4$–2.8. These constants are either given by manufacturers of magnetic materials in tables for several frequency ranges (i.e., Magnetics and Micrometals) or can be determined from the empirical data provided by the manufacturers using curve fitting techniques. In general, the coefficients k, a, and b are functions of frequency. Core material manufacturers provide information about the total core material loss (due to both hysteresis and eddy current) either in watts per unit volume (W/m^3) or in watts per unit weight (W/kg) in the form of tables or graphs of specific power loss as a function of the AC component of the flux density amplitude B_m with frequency f as a parameter. Manufacturers provide specific power loss for specific core materials measured always with sinusoidal excitation using TCs.

Core losses are dependent on core volume V_c. The total power loss in the core of volume V_c is given by

$$P_C = P_v V_c = kf^a B_m^b V_c, \tag{2.193}$$

[2] Charles Proteus Steinmetz (1865–1923) was a mathematician and electrical engineer. He was born in Wroclaw, Poland, and contributed to the expansion of the electrical power industry in the United States.

where the empirical coefficients k, a, and b depend on core material and frequency range. For TCs,

$$P_C = P_v V_c = k f^a B_m^b V_c = k f^a B_m^b A_c l_c. \tag{2.194}$$

The specific core loss in terms of the amplitude of a sinusoidal inductor current I_{Lm} for ungapped core is given by

$$P_v = \frac{P_C}{V_c} = P_h + P_e = k f^a B_m^b = k f^a \left(\frac{\mu N I_{Lm}}{l_c} \right)^b \ (\text{mW/cm}^3), \tag{2.195}$$

where $B_m = \mu N I_{Lm}/l_c$ is the amplitude of the AC component of the magnetic flux density usually in millitorr.

The total power loss density of the Ferroxcube MnZn ferrite material 3F3 is given by an empirical equation

$$P_v = 1.5 \times 10^{-6} f^{1.3} B_m^{2.5} (\text{mW/cm}^3), \tag{2.196}$$

where B_m is the amplitude of the AC component of the magnetic flux density in millitorr and f is the operating frequency in kHz. The 3F3 ferrite has $\mu_{rc} = 2000$, $\rho_c = 2\Omega \cdot \text{m}$, and $T_c = 200\,^\circ\text{C}$. As B_m doubles, P_v increases by a factor of $2^{2.5} = 5.657$. As the frequency f quadruples, P_v increases by a factor $4^{1.3} = 6.063$. Figure 2.32 shows measured plots of P_v for the ferrite material 3F3 as a function of B_m at fixed frequencies f. Figure 2.33 depicts the calculated plots of P_v for the ferrite material 3F3.

The total power loss density of the 3F3 magnetic material can be also expressed by

$$P_v = 47.434 f^{1.3} B_m^{2.5} (\text{mW/cm}^3), \tag{2.197}$$

where B_m is the amplitude of the AC component of the magnetic flux density in T and f is the operating frequency in kHz.

The total power loss density of the Ferroxcube ferrite material 3C80 is given by an empirical equation

$$P_v = 16.7 \times 10^{-3} f^{1.3} B_m^{2.5} (\text{mW/cm}^3), \tag{2.198}$$

where B_m is the amplitude of the sinusoidal AC component of the magnetic flux density in T and f is the operating frequency in Hz.

The total power loss density of the Ferroxcube ferrite material 3C85 is given by an empirical equation

$$P_v = 12 \times 10^{-3} f^{1.3} B_m^{2.55} (\text{mW/cm}^3), \tag{2.199}$$

where B_m is the amplitude of the sinusoidal AC component of the magnetic flux density in T and f is the operating frequency in Hz.

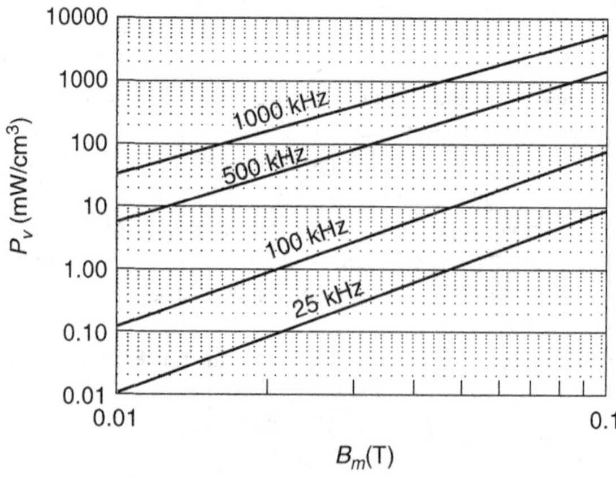

Figure 2.32 Measured total core loss density P_v as a function of the amplitude of the flux density B_m at fixed frequencies f for Ferroxcube ferrite material 3F3

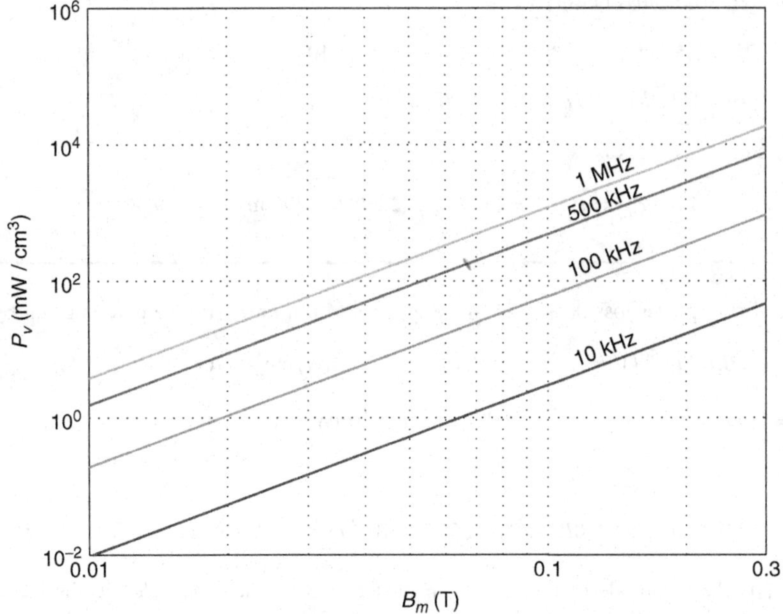

Figure 2.33 Calculated total core loss density P_v as a function of the amplitude of flux density B_m at fixed frequencies f for Ferroxcube ferrite material 3F3

Assume that a sinusoidal current flows through an inductor with a core

$$i_L = I_{Lm} \sin \omega t. \tag{2.200}$$

Using the principle of the conservation of energy, we can express the core power loss as

$$P_C = \frac{R_{cs} I_m^2}{2}, \tag{2.201}$$

where R_{cs} is the series equivalent resistance of an inductor due to the core power loss. Thus,

$$R_{cs} = \frac{2P_C}{I_{Lm}^2} = \frac{2P_v V_c}{I_{Lm}^2} = \frac{2kf^a B_m^b l_c A_c}{I_{Lm}^2} = \frac{2k\mu^2 N^2 A_c f^a B_m^{b-2}}{l_c} = \frac{2k\mu^b N^b A_c f^a I_{Lm}^{b-2}}{l_c^{b-2}}, \tag{2.202}$$

where $B_m = \mu N I_{Lm}/l_c$. The sum of the core series equivalent resistance R_{cs} and the winding resistance R_w form the total series resistance of an inductor.

Example 2.7

A Ferroxcube 3F3 ferrite material is used to make a TC with the following parameters: $d = 13.1$ mm, $D = 23.7$ mm, $h = 7.5$ mm, and $\mu_{rc} = 4000$. A sinusoidal current flows through the inductor at $f = 500$ kHz and $B_m = 0.2$ T. Find the total power loss in the core without an air gap.

Solution: The mean MPL of the TC is

$$l_c = \frac{\pi(d + D)}{2} = \frac{\pi(13.1 + 23.7)}{2} = 57.805 \text{ mm} = 57.805 \times 10^{-3} \text{ m}. \tag{2.203}$$

The cross-sectional area of the core is

$$A_c = \frac{h(D - d)}{2} = \frac{7.5 \times 10^{-3} \times (23.7 - 13.1) \times 10^{-3}}{2} = 39.75 \times 10^{-6} \text{ m}^2. \tag{2.204}$$

The volume of the core is

$$V_c = l_c A_c = 57.805 \times 10^{-3} \times 39.75 \times 10^{-6} = 2.298 \times 10^{-6} \text{m}^3 = 2.298 \text{ cm}^3. \tag{2.205}$$

The total core power loss density is

$$P_v = 1.5 \times 10^{-6} f^{1.3} B_m^{2.5} (\text{mW/cm}^3) = 1.5 \times 10^{-6} \times 500^{1.3} \times 200^{2.5} (\text{mW/cm}^3)$$
$$= 2737.33\ (\text{mW/cm}^3) = 2.73733\ (\text{W/cm}^3). \tag{2.206}$$

The total core power loss is

$$P_C = P_v V_c = 2737.33 \times 2.298 = 6290\ \text{mW} = 6.29\ \text{W}. \tag{2.207}$$

The total core power loss density in the Magnetics MnZn ferrite P-type core is given by

$$P_v = 0.158 f^{1.36} (10B_m)^{2.86} = 114.46 f^{1.36} B_m^{2.86} (\text{mW/cm}^3) \quad \text{for} \quad f < 100\ \text{kHz}, \tag{2.208}$$

$$P_v = 0.0434 f^{1.63} (10B_m)^{2.62} = 18.0924 f^{1.63} B_m^{2.62} (\text{mW/cm}^3) \quad \text{for} \quad 100\ \text{kHz} \le f < 500\ \text{kHz}, \tag{2.209}$$

and

$$P_v = 7.36 \times 10^{-7} f^{3.47} (10B_m)^{2.54} = 255.2 \times 10^{-6} f^{3.47} B_m^{2.54} (\text{mW/cm}^3) \quad \text{for} \quad f \ge 500\ \text{kHz}, \tag{2.210}$$

where B_m is the amplitude of the AC component of the magnetic flux density in T and f is the operating frequency in kHz. Note that the amplitude of the magnetic flux density B_m given in the Magnetics data sheets is expressed in kiloGauss, where 1 T = 10 kGauss. To express B_m in T using Magnetics data sheets, B_m is multiplied by 10 in the above equations. The specific core loss P_v described by (2.209) through (2.210) is illustrated in Fig. 2.34. Figure 2.35 shows a 3D plot of total core loss density P_v as a function of flux density B_m and frequency f at $T = 80\,^\circ$C for Magnetics P-type ferrite material. The resistivity of this ferrite material is $\rho_c = 5\Omega \cdot$ m, the Curie temperature is $T_c > 210\,^\circ$C, and the frequency range is $f \le 1.2$ MHz.

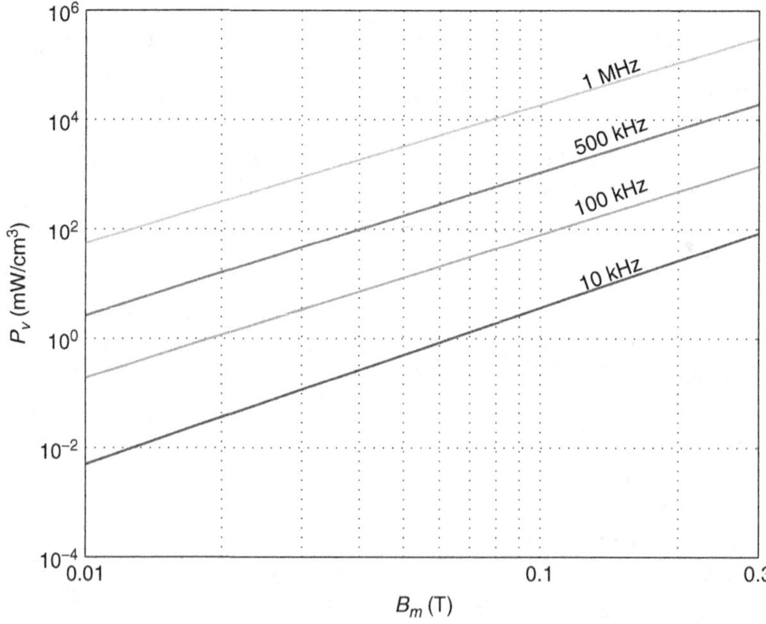

Figure 2.34 Total core loss density P_v as a function of the amplitude of flux density B_m at fixed frequencies f and $T = 80\,^\circ$C for Magnetics P-type MnZn ferrite material

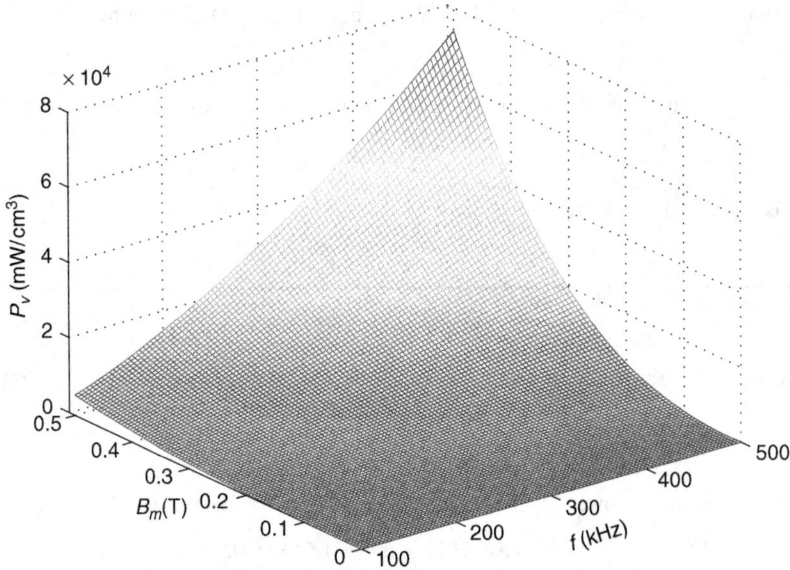

Figure 2.35 Total core loss density P_v as a function of the amplitude of flux density B_m and frequency f at $T = 80\,^{\circ}\mathrm{C}$ for Magnetics P-type MnZn ferrite material

---

Example 2.8

The Magnetics MnZn ferrite P-type PC OP-42616-UG has $A_c = 93.9\ \mathrm{mm}^3$, $l_c = 37.6$ mm, and $V_c = 3530\ \mathrm{mm}^3 = 3.35\ \mathrm{cm}^3$. The inductor conducts a sinusoidal current at $f = 100$ kHz.

(a) Calculate the total core power loss at $B_m = 0.1$ T.

(b) Calculate the total core power loss at $B_m = 0.5$ T.

Solution: (a) The total core power loss density in the Magnetics ferrite P-type core at $f = 100$ kHz and $B_m = 0.1$ T is

$$P_v = 0.0434 f^{1.63}(10B_m)^{2.62}(\mathrm{mW/cm^3})$$
$$= 0.0434 \times 100^{1.63} \times (10 \times 0.1)^{2.62}(\mathrm{mW/cm^3}) = 78.97(\mathrm{mW/cm^3}). \tag{2.211}$$

The total core power loss is

$$P_C = P_v V_c = 78.97 \times 3.53 = 278.78\ \mathrm{mW}. \tag{2.212}$$

(b) The total core power loss density in the Magnetics ferrite P-type core at $f = 100$ kHz and $B_m = 0.5$ T is

$$P_v = 0.0434 f^{1.63}(10B_m)^{2.62}\ (\mathrm{mW/cm^3})$$
$$= 0.0434 \times 100^{1.63} \times (10 \times 0.5)^{2.62}\ (\mathrm{mW/cm^3})$$
$$= 5.355 \times 10^3 (\mathrm{mW/cm^3}) = 5.355\ (\mathrm{W/cm^3}). \tag{2.213}$$

The total core power loss is

$$P_C = P_v V_c = 5.355 \times 10^3 \times 3.53 = 18.9 \times 10^3\ \mathrm{mW} = 18.9\ \mathrm{W}. \tag{2.214}$$

The specific power loss in MATGLAS alloy core material 2705M is given by

$$P_v = 3.2 \times 10^{-6} f^{1.8} B_m^2 \,(\text{mW/cm}^3), \tag{2.215}$$

where f is in Hz and B_m is in millitorr. The total core power loss density per unit volume for an iron core is

$$P_v = 3.5 \times 10^{-6} f^2 B_m^2 \,(\text{mW/cm}^3), \tag{2.216}$$

where f is in Hz and B_m is in Tesla.

Example 2.9

An iron core has the saturation magnetic flux density $B_s = 1.5$ T and the volume $V_c = 200$ cm^3. The core is operated at the US line frequency $f = 60$ Hz and conducts a sinusoidal current. Find the core loss.

Solution: The total core power loss density in the core at $f = 60$ Hz and $B_m = 1.5$ T is

$$P_v = 3.5 \times 10^{-6} f^2 B_m^2 = 3.5 \times 10^{-6} \times 60^2 \times 1.5^2 = 0.02835 (\text{mW/cm}^3), \tag{2.217}$$

where f is in Hz and B_m is in T. The total core power loss is

$$P_C = P_v V_c = 0.02835 \times 200 = 5.67 \text{ W}. \tag{2.218}$$

The power loss density for the Magnetics K-type MnZn ferrite material is

$$P_v = 0.00133 f^{2.19} (10 B_m)^{3.1} (\text{mW/cm}^3), \tag{2.219}$$

where B_m is the amplitude of the AC component of the magnetic flux density in T and f is the operating frequency in kHz.

The power loss density of the Micrometals iron powder Carbonyl E material Mix No. 2 is given by

$$\begin{aligned} P_v &= k \times f^a B_m^b = 8.86 \times 10^{-10} f^{1.14} (10 B_m)^{2.19} (\text{W/cm}^3) \\ &= 1.37225 \times 10^{-7} f^{1.14} (B_m)^{2.19} (\text{W/cm}^3) \quad \text{for} \quad 400 \text{ kHz} \le f \le 10 \text{ MHz}, \end{aligned} \tag{2.220}$$

where f is in Hz and B_m is in millitorr. This material has $\mu_{rc} = 10$ and is recommended for the frequency range from 0.25 to 10 MHz.

2.16.2 Losses of Gapped Cores

The specific gapped core loss in terms of the amplitude of a sinusoidal inductor current I_{Lm} for gapped core is given by

$$P_{v(cg)} = \frac{P_{Cg}}{V_c} = P_h + P_e = k f^a B_m^b = k f^a \left(\frac{\mu_0 N I_{Lm}}{l_g + \frac{l_c}{\mu_{rc}}} \right)^b (\text{mW/cm}^3), \tag{2.221}$$

where $B_m = \mu N I_{Lm}/(l_g + l_c/\mu_{rc})$ is the amplitude of the AC component of the magnetic flux density, usually in millitorr. Figure 2.36 shows gapped core loss P_{Cg} as a function of the gap length l_g at fixed amplitude of sinusoidal inductor current $I_{Lm} = 0.1$ A and at $f = 100$ kHz. The Magnetics 0R421616 core is used in the calculations. The core parameters are: $l_c = 37.6$ mm, $V_c = 3530$ mm^3, $\mu_{rc} = 2300$, and $l_g = 1$ mm. The core is made of the ferrite R-type magnetic material with $k = 0.074$, $a = 1.43$, and $b = 2.85$ for $f = 0$ to 100 kHz. It can be seen that the core loss decreases as the gap length l_g increases.

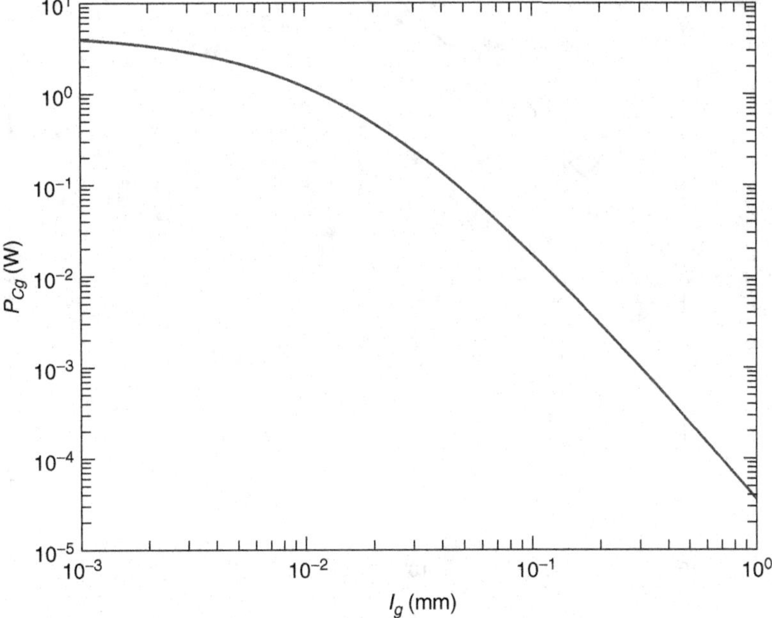

Figure 2.36 Gapped core loss P_{Cg} as a function of the gap length l_g at fixed amplitude of sinusoidal inductor current $I_{Lm} = 0.1$ A and $f = 100$ kHz

Figure 2.37 shows ungapped core loss P_C, gapped core loss P_{Cg}, and winding loss P_w as functions of frequency. The Magnetics 0R421616 core is used in the calculations. The core parameters are: $l_c = 37.6$ mm, $V_c = 3530$ mm^3, $\mu_{rc} = 2300$, and $l_g = 1$ mm. The core is made of the ferrite R-type magnetic material with $k = 0.074$, $a = 1.43$, and $b = 2.85$ for $f = 0$ to 100 kHz, $k = 0.036$, $a = 1.64$, and $b = 2.68$ for $f = 100$ to 500 kHz, and $k = 0.014$, $a = 1.84$, and $b = 2.28$ for $f \geq 500$ kHz. The inductor wound with the AWG26 copper wire with the porosity factor $\eta = 0.85$, $N = 40$, and $N_l = 4$. The amplitude of the sinusoidal inductor current is $I_{Lm} = 1.9$ A. Dowell's equation (discussed in Chapter 5) is used to compute the winding loss P_w.

2.17 Core Losses for Nonsinusoidal Inductor Current

The periodic nonsinusoidal magnetic flux density can be expanded into Fourier series

$$B(t) = B_{DC} + \sum_{n=1}^{\infty} B_{mn} \cos n\omega t. \tag{2.222}$$

Hence, the core power loss density can be estimated by Fourier expansion of the magnetic flux density waveform

$$P_v = k_1 f^{a_1} B_{m1}^{b_1} + k_2 (2f)^{a_2} B_{m2}^{b_2} + k_3 (3f)^{a_3} B_{m3}^{b_3} + \ldots = \sum_{n=1}^{\infty} k_n (nf)^{a_n} B_{mn}^{b_n}, \tag{2.223}$$

where k_n, a_n, and b_n are the coefficients at the corresponding frequencies nf. The amplitudes of the flux density at harmonic frequencies nf for an ungapped core are given by

$$B_{mn} = \frac{\mu_{rc}\mu_0 N I_{mn}}{l_c}, \tag{2.224}$$

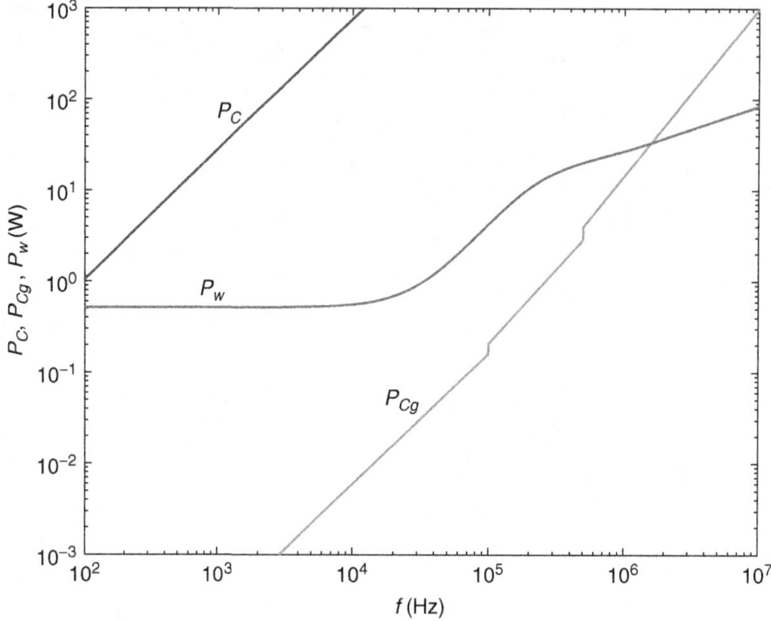

Figure 2.37 Ungapped core loss P_C, gapped core loss P_{Cg}, and winding loss P_w as functions of frequency at fixed amplitude of sinusoidal inductor current. The Magnetics 0R421616 core and AWG26 copper wire are used in the calculations

where I_{mn} are the amplitudes of the inductor current at frequencies nf. Hence, the power loss density in an ungapped core is

$$P_v = k_1 f^{a_1}\left(\frac{\mu_{rc}\mu_0 N I_{m1}}{l_c}\right)^{b_1} + k_2(2f)^{a_2}\left(\frac{\mu_{rc}\mu_0 N I_{m2}}{l_c}\right)^{b_2} + k_3(3f)^{a_3}\left(\frac{\mu_{rc}\mu_0 N I_{m3}}{l_c}\right)^{b_3} + \ldots$$
$$= \sum_{n=1}^{\infty} k_n (nf)^{a_n}\left(\frac{\mu_{rc}\mu_0 N I_{mn}}{l_c}\right)^{b_n}. \quad (2.225)$$

The amplitudes of the flux density at harmonic frequencies nf for a gapped core are given by

$$B_{mn} = \frac{\mu_0 N I_{mn}}{l_g + \frac{l_c}{\mu_{rc}}} \quad (2.226)$$

and the power loss density in a gapped core is

$$P_v = k_1 f^{a_1}\left(\frac{\mu_0 N I_{m1}}{l_g + \frac{l_c}{\mu_{rc}}}\right)^{b_1} + k_2(2f)^{a_2}\left(\frac{\mu_0 N I_{m2}}{l_g + \frac{l_c}{\mu_{rc}}}\right)^{b_2} + k(3f)^{a_3}\left(\frac{\mu_0 N I_{m3}}{l_g + \frac{l_c}{\mu_{rc}}}\right)^{b_3} + \ldots$$
$$= \sum_{n=1}^{\infty} k_n (nf)^{a_n}\left(\frac{\mu_0 N I_{mn}}{l_g + \frac{l_c}{\mu_{rc}}}\right)^{b_n}. \quad (2.227)$$

At high frequencies, μ_{rc} decreases and becomes complex. Core losses are described by nonlinear equations and therefore Fourier analysis approach has disadvantages.

2.18 Complex Permeability of Magnetic Materials

The complex permeability is used to characterize magnetic materials over a wide frequency range. At low frequencies, the permeability μ is a real number and therefore vectors **H** and **B** are parallel

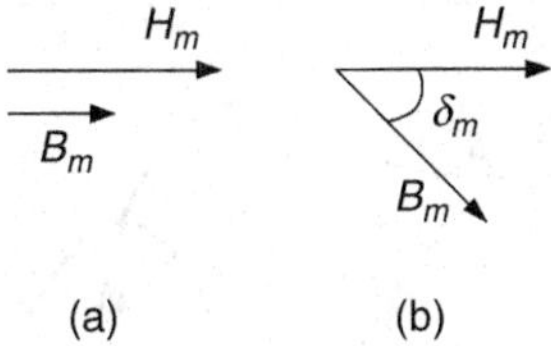

Figure 2.38 Phasor of magnetic field intensity H_m and magnetic flux density B_m. (a) At low frequencies. (b) At high frequencies

to each other, as shown in Fig. 2.38a. At high frequencies, the permeability μ is a complex quantity and therefore vectors **H** and **B** for inductors and transformers with magnetic cores are not parallel to each other, as shown in Fig. 2.38b. Both the real and imaginary parts of complex permeability μ are functions of frequency. The complex permeability may be used to represent all types of magnetic core losses, especially for ferrite cores [28]. The core loss processes are conveniently modeled as the imaginary component of the complex permeability. The real and imaginary parts of complex permeability are either series or parallel terms.

2.18.1 Series Complex Permeability

The phasor of magnetic field intensity is

$$\mathbf{H} = H_m e^{j0} \tag{2.228}$$

and the phasor of magnetic flux density is

$$\mathbf{B} = B_m e^{j(-\delta_m)}. \tag{2.229}$$

Hence, the series complex permeability of a magnetic material is given by

$$\mu_s = \frac{\mathbf{B}}{\mathbf{H}} = \frac{B_m}{H_m} e^{j(-\delta_m)} = \frac{B_m}{H_m}(\cos\delta_m - j\sin\delta_m) = \mu_{so}\cos\delta_m - j\mu_{so}\sin\delta_m, \tag{2.230}$$

where $\mu_{so} = B_m/H_m$. Both μ_{so} and δ_m are functions of frequency. At low frequencies, $\mu_{so} = \mu = \mu_s^{'}$ and $\delta_m = 0$. The *series complex permeability* of a magnetic material is given by

$$\mu_s = \mu_s^{'} - j\mu_s^{''} = \mu_0(\mu_{rs}^{'} - j\mu_{rs}^{''}) = \mu_0\sqrt{\mu_{rs}^{'2} + \mu_{rs}^{''2}}\, e^{j\arctan\left(-\frac{\mu_{rs}^{''}}{\mu_{rs}^{'}}\right)}, \tag{2.231}$$

where $\mu_s^{'} = Re\{\mu_s\} = \mu_{so}\cos\delta_m$ is the inductive (real) series permeability, $\mu_s^{''} = Im\{\mu_s\} = \mu_{so}\sin\delta_m$ is the resistive (imaginary) series permeability, $\mu_{rs}^{'} = Re\{\mu_{rs}\}$ is the relative inductive (real) series permeability, and $\mu_{rs}^{''} = Im\{\mu_{rs}\}$ is the relative resistive (imaginary) series permeability. When the frequency of an applied AC magnetic field matches the natural precession frequency, ferromagnetic resonance occurs. Precession is a change in the orientation of the rotational axis of a rotating electron. At the ferromagnetic resonant (FMR) frequency, the real part of the complex permeability $\mu_s^{'}$ is zero and the imaginary component of the complex permeability $\mu_s^{''}$ reaches its maximum value.

Figure 2.39 depicts typical plots of $\mu_{rs}^{'}$ and $\mu_{rs}^{''}$ as functions of frequency f given in manufacturer's data sheets. It can be seen that $\mu_{rs}^{'}$ is independent of frequency over a wide frequency range and then decreases with frequency like the magnitude of the transfer function of a low-pass filter, and $\mu_{rs}^{''}$ first increases and then decreases with frequency like the magnitude of the transfer function of a band-pass filter. For low frequencies, $\mu_{rs}^{'} \approx \mu_{rso}$ and $\mu_{rs}^{''} \approx 0$. For ferrites, $\mu_r^{''}$ becomes significant for $f \geq 10$ kHz.

Phasor diagrams for series and parallel permeabilities are depicted in Fig. 2.40. Figure 2.41 shows a polar plot of the series complex permeability μ_s. The plots of the complex permeability measured at a low value of the magnetic flux density B_m are given in data sheets of magnetic materials.

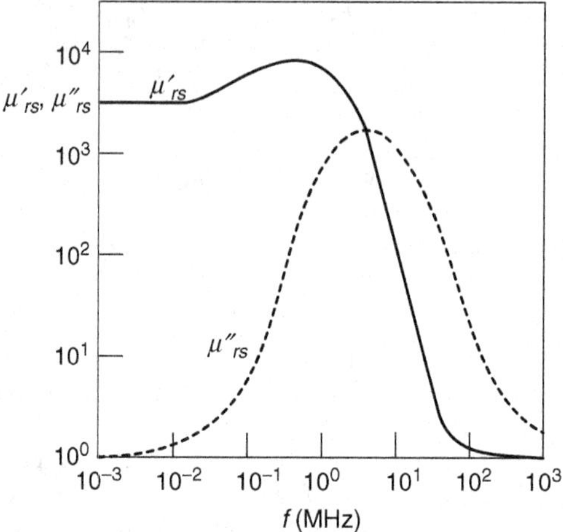

Figure 2.39 Series real relative permeability μ'_{rs} and series imaginary relative permeability μ''_{rs} as functions of frequency f for a magnetic material

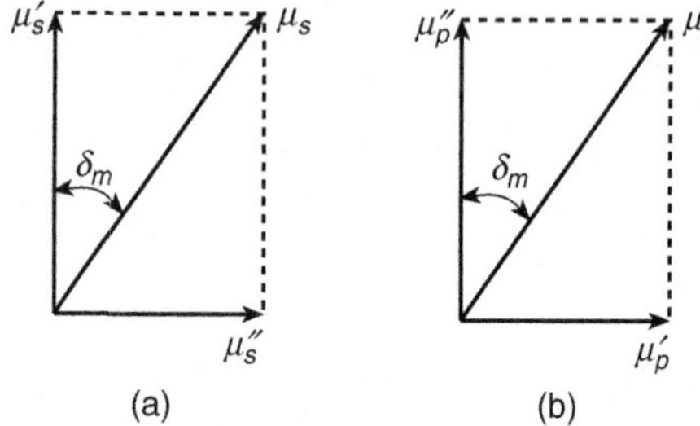

Figure 2.40 Phasor diagrams of series and parallel complex permeabilities. (a) Phasor diagram for series complex permeability. (b) Phasor diagram for parallel complex permeability

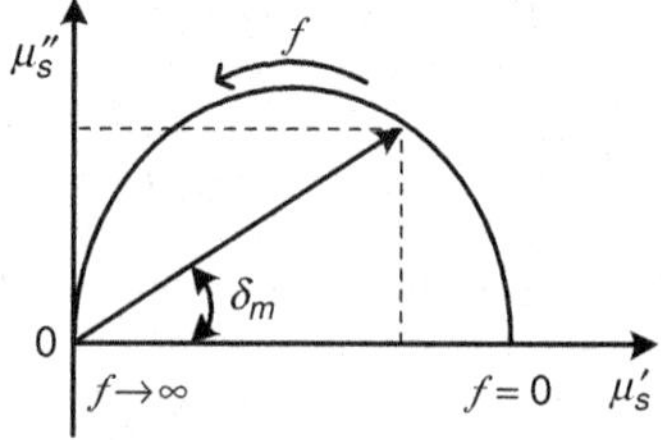

Figure 2.41 Polar plot of series imaginary part of permeability μ''_s as a function of series real part of permeability μ'_s

The complex magnetic flux density is

$$B = \mu_s H = (\mu'_s - j\mu''_s)H. \tag{2.232}$$

The *series complex magnetic susceptibility* is given by

$$\chi_s = \frac{M}{H} = \chi'_s - j\chi''_s = \mu'_{rc} - 1 - j\mu''_{rs}, \tag{2.233}$$

where $\chi'_s = \mu'_{rs} - 1$ and $\chi''_s = \mu''_{rs}$.

For each ferromagnetic material,

$$\mu f_r^2 = Const, \tag{2.234}$$

where f_r is the FMR frequency. Landau and Lifshitz predicted the existence of ferromagnetic resonance in 1935. If the frequency of the AC magnetic field coincides with the precession frequency of electrons, the energy is transferred from the AC magnetic field to the precessing electrons. The precessing electrons dissipate energy in internal frictions as heat. Thus, the magnetic material absorbs the energy from the AC magnetic field. Variations of the power absorbed by the magnetic material in terms of frequency of the magnetic field is similar to the curve of the parallel resonant circuit.

An external magnetic field B exerts a torque on the magnetic moment of an electron. The angular frequency of the precession of the magnetic moment about the external magnetic field B is known as the *Larmor frequency* and it is given by

$$\omega_L = \gamma B = \frac{gq}{2m_e} B, \tag{2.235}$$

where g a constant (usually equal to 1) and $\gamma = gq/2m_e$ is the gyromagnetic ratio.

The real and imaginary components of series relative permeability are given by LLG equations

$$\mu'_{rs} = \frac{4\pi M_s}{H_k} \frac{\omega_r^2(\omega_r^2 - \omega^2)}{(\omega_r^2 - \omega^2)^2 + (4\pi\lambda\omega)^2} + 1 = 4\pi \left(\frac{M_s}{H_k}\right) \frac{1 - \left(\frac{\omega}{\omega_r}\right)^2}{\left[1 - \left(\frac{\omega}{\omega_r}\right)^2\right]^2 + \left(\frac{4\pi\lambda}{\omega_r}\right)^2 \left(\frac{\omega}{\omega_r}\right)^2} + 1 \tag{2.236}$$

and

$$\mu''_{rs} = \frac{4\pi M_s}{H_k} \frac{\omega_r^2(4\pi\lambda\omega)}{(\omega_r^2 - \omega^2)^2 + (4\pi\lambda\omega)^2} = \frac{4\pi}{\omega_r} \left(\frac{M_s}{H_k}\right) \frac{4\pi\lambda\left(\frac{\omega}{\omega_r}\right)}{\left[1 - \left(\frac{\omega}{\omega_r}\right)^2\right]^2 + \left(\frac{4\pi\lambda}{\omega_r}\right)^2 \left(\frac{\omega}{\omega_r}\right)^2}, \tag{2.237}$$

where $M_s \approx B_s/\mu_0$ is the saturation magnetization, H_k is the magnetic field corresponding to M_s, $\gamma = q/(2m_e)$ is the gyromagnetic ratio, λ is given by

$$\lambda = \alpha\gamma M_s, \tag{2.238}$$

the FMR frequency is

$$\omega_r = \sqrt{\frac{\gamma^2 M_s H_k}{\mu_0}}, \tag{2.239}$$

and α is the damping constant. Figure 2.42 shows example plots of the real and imaginary series relative permeabilities μ'_{rs} and μ''_{rs} as functions of frequency f for $M_s = 0.5$ T, $\alpha = 1/137$, $\gamma = 2$, $\mu'_{rc} = 1300$, $f_r = 1$ MHz, $q = 1.60128 \times 10^{-19}$ C, and $m_e = 9.1095 \times 10^{-31}$ kg. For $f \ll f_r$, $\mu'_{rs} \approx \mu'_r$. For $f > f_r$, $\mu'_{rs} > 0$. and $\mu''_{rs} \approx 0$. At $f = f_r$, $\mu'_{rs} = 0$. For $f > f_r$, $\mu'_{rs} < 0$. Figure 2.43 shows a polar plot of the real and imaginary series relative permeabilities μ'_{rs} and μ''_{rs} as functions of frequency f described by LLG model.

An approximation of the series complex relative permeability is given by Smit and Wijn [5]

$$\mu_{rs} = \mu'_{rs} - j\mu''_{rs} \approx \frac{\mu'_{rso}}{1 + \left(\frac{f}{f_H}\right)^2} - j\frac{\mu'_{rso}\left(\frac{f}{f_H}\right)}{1 + \left(\frac{f}{f_H}\right)^2}, \tag{2.240}$$

where

$$\mu'_{rs} \approx \frac{\mu'_{rso}}{1 + \left(\frac{f}{f_H}\right)^2}, \tag{2.241}$$

$$\mu''_{rs} \approx \frac{\mu'_{rso}\left(\frac{f}{f_H}\right)}{1 + \left(\frac{f}{f_H}\right)^2}, \tag{2.242}$$

$$\frac{\mu''_{rs}}{\mu'_{rs}} \approx \frac{f}{f_H}, \tag{2.243}$$

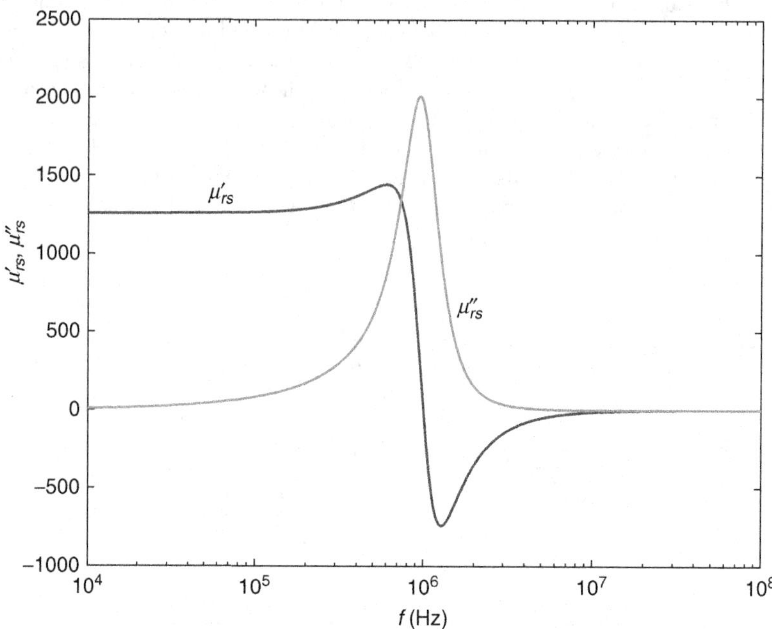

Figure 2.42 Real and imaginary series relative complex permeability μ'_{rs} and μ''_{rs} as functions of frequency f described by Landau–Lifschitz–Gilbert (LLG) model

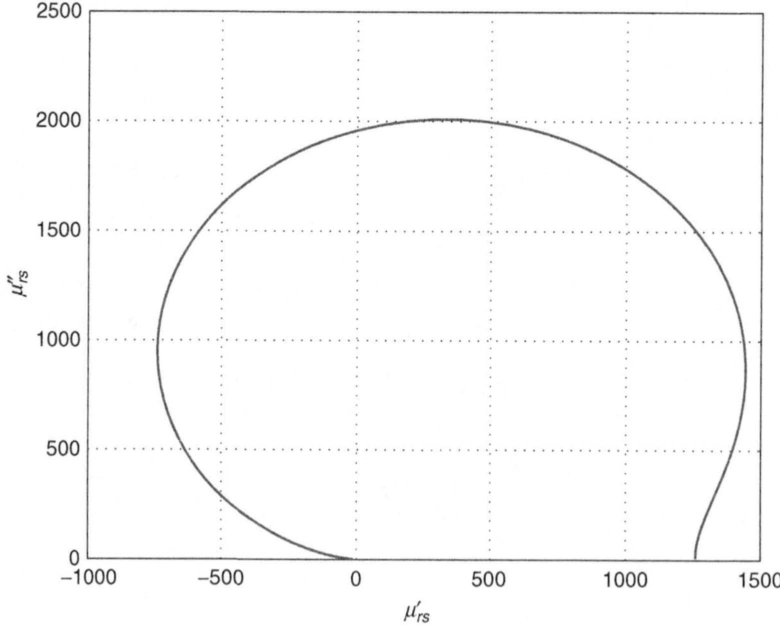

Figure 2.43 Polar plot of real and imaginary series relative complex permeabilities μ'_{rs} and μ''_{rs} as functions of frequency f described by LLG model

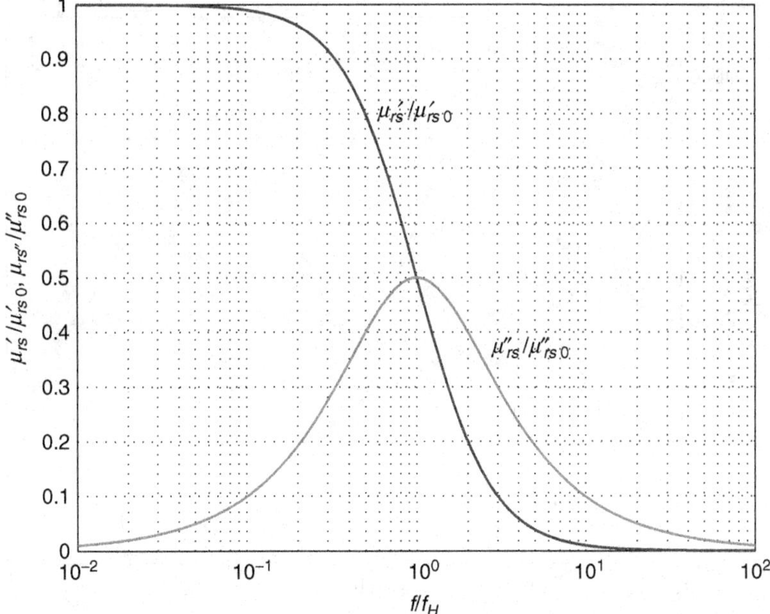

Figure 2.44 Normalized series real relative permeability $\mu_{rs}^{'}/\mu_{rso}^{'}$ and normalized series imaginary relative permeability $\mu_{rs}^{''}/\mu_{rso}^{'}$ as functions of normalized frequency f/f_H

and

$$|\mu_{rs}| = \sqrt{(\mu_{rs}^{'})^2 + (\mu_{rs}^{''})^2} \tag{2.244}$$

where $\mu_{rso}^{'}$ is the low-frequency real series relative permeability and f_H is the 3-dB frequency of $\mu_{rs}^{'}$. Figure 2.44 shows plots of normalized series real relative permeability $\mu_{rs}^{'}/\mu_{rso}^{'}$ and normalized series imaginary relative permeability $\mu_{rs}^{''}/\mu_{rso}^{'}$ as functions of normalized frequency f/f_H calculated from (2.241) and (2.242). Figure 2.45 shows plots of $|\mu_{rs}|$ as a function of frequency f. Figure 2.46 shows a polar plot of series relative permeability.

2.18.2 Loss Angle and Quality Factor

The *magnetic loss tangent* of a magnetic core material is defined as

$$\tan\delta_m = \frac{R_{cs}}{\omega L_s} = \frac{\mu_{rs}^{''}}{\mu_{rs}^{'}} = \frac{\chi_s^{''}}{\chi_s^{'}} = \frac{1}{Q_m} \approx \frac{f}{f_H}, \tag{2.245}$$

where δ_m is the loss angle. Figure 2.47 shows a plot of $\tan\delta_m$ as a function of frequency f. When the core resistance R_{cs} is zero, the loss angle $\delta_m = 0$. As the frequency increases, $\tan\delta_m$ increases.

The quality factor of a magnetic core material is

$$Q_m = \frac{1}{\tan\delta_m} = \frac{\mu_{rs}^{'}}{\mu_{rs}^{''}} = \frac{\chi_s^{'}}{\chi_s^{''}} = \frac{\omega L_s}{R_{cs}} \approx \frac{1}{\frac{f}{f_H}}. \tag{2.246}$$

Figure 2.48 shows a plot of the quality factor Q_m as a function of frequency f. The frequency at which $Q_m = 1$, that is, $\mu_{rs}^{'} = \mu_{rs}^{''}$, is called the *relaxation frequency*. As the frequency increases, Q_m decreases.

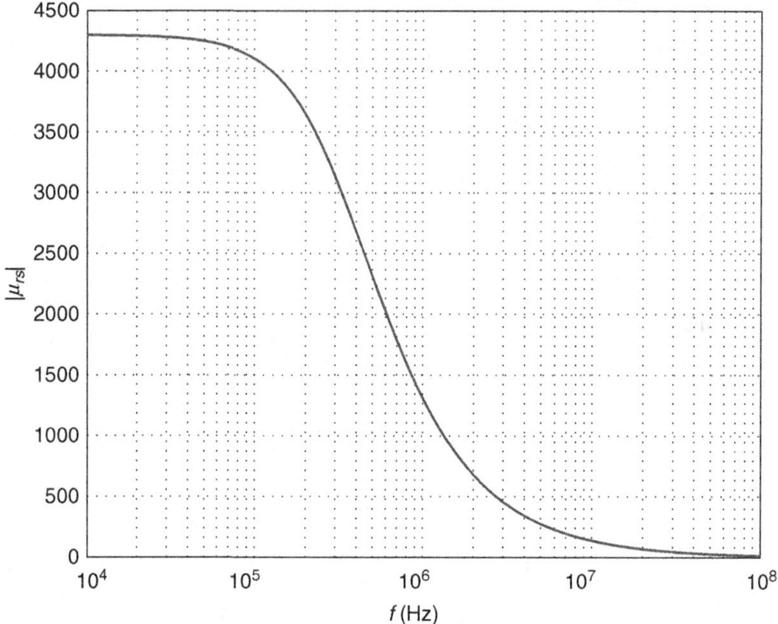

Figure 2.45 Modulus of series relative permeabilities as functions of frequency f at the low-frequency real permeability $\mu'_{rso} = 4300$ and $f_H = 320$ kHz

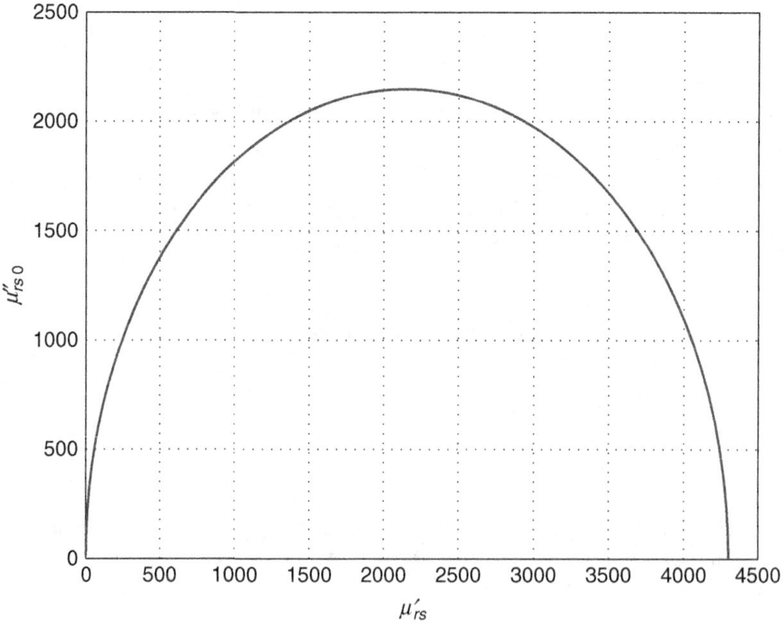

Figure 2.46 Polar plot of series relative permeabilities at the low-frequency real permeability $\mu'_{rso} = 4300$ and $f_H = 320$ kHz

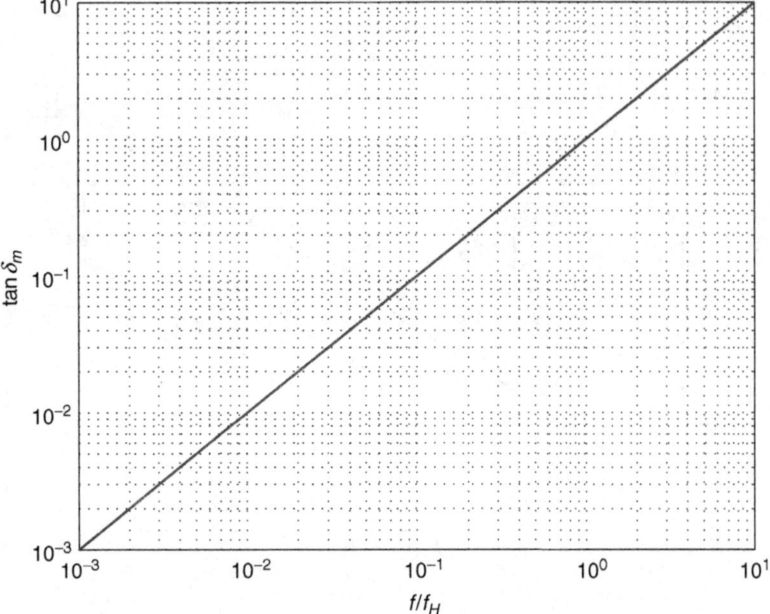

Figure 2.47 Plots of $\tan\delta_m = \mu''_{rs}/\mu'_{rs}$ as a function of frequency f at $\mu'_{rso} = 4300$ and $f_H = 320$ kHz

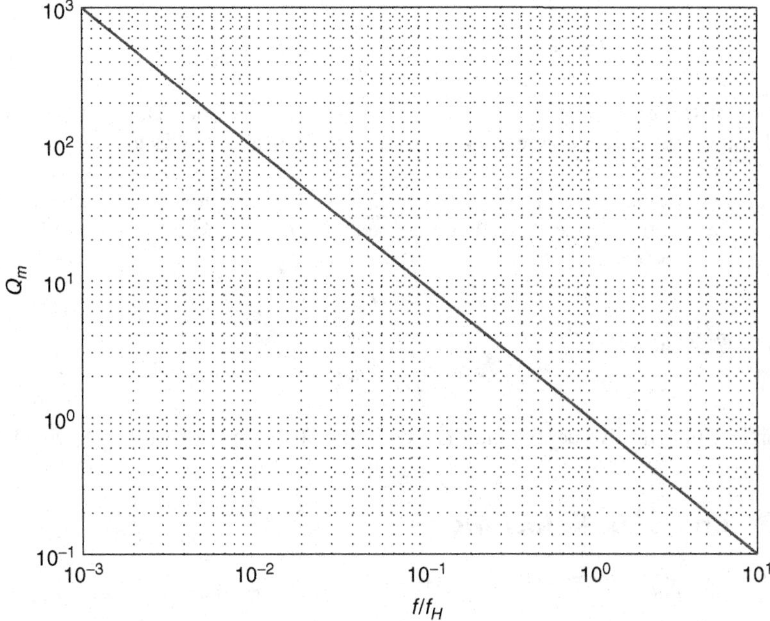

Figure 2.48 Plots of $Q_m = \mu'_{rs}/\mu''_{rs}$ as a function of frequency f at $\mu'_{rso} = 4300$ and $f_H = 320$ kHz

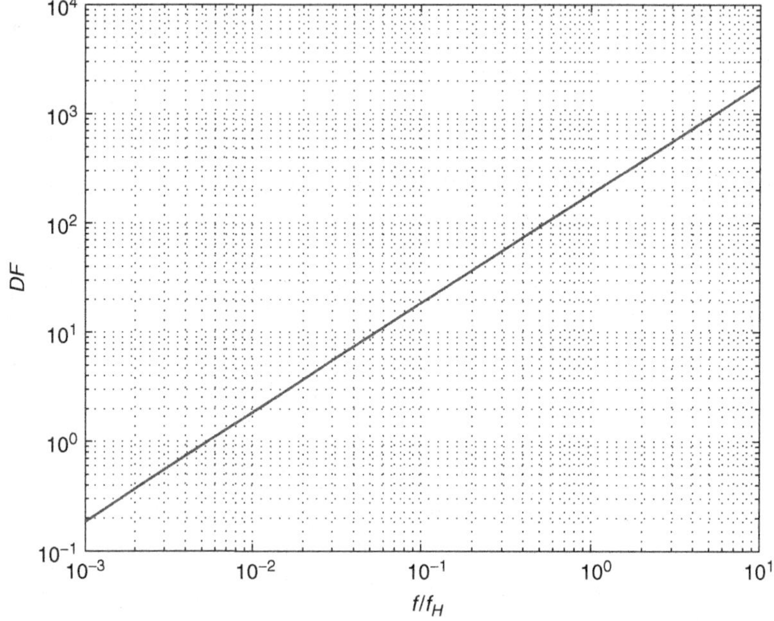

Figure 2.49 Plot of dissipation factor DF as a function of frequency f at $\mu'_{rso} = 4300$ and $f_H = 320$ kHz

The *dissipation factor* or *loss factor* of a magnetic material is defined as

$$DF = \frac{\tan \delta_m}{\mu'_{rso}\mu_0} = \frac{1}{Q_m \mu'_{rso}\mu_0} \approx \frac{1}{\mu'_{rso}\mu_0} \cdot \left(\frac{f}{f_H}\right). \tag{2.247}$$

The dissipation factor DF is a figure-of-merit of a magnetic material. Different materials may be compared using this factor. Figure 2.49 shows a plot of dissipation factor DF as a function of frequency f.

Gapped cores may be used to increase Q_m, enhance the temperature stability, and achieve tighter tolerance of the effective permeability. For a gapped core, the magnetic loss tangent is

$$\tan \delta_{mg} = \frac{\mu_{re}}{\mu_{rc}} \tan \delta_m \tag{2.248}$$

and the quality factor is

$$Q_{mg} = \frac{\mu_{rc}}{\mu_{re}} Q_m. \tag{2.249}$$

For example, if the effective permeability is reduced by 50%, the effective Q_{mg} is doubled.

2.18.3 Complex Reluctance

The *complex reluctance* is

$$\mathcal{R} = \frac{l_c}{\mu_s A_c} = \frac{l_c}{\mu_{rs}\mu_0 A_c} = \frac{l_c}{(\mu'_{rc} - j\mu''_{rc})\mu_0 A_c}. \tag{2.250}$$

The *complex permeance* is

$$\mathcal{P} = \frac{\mu_s A_c}{l_c} = \frac{(\mu'_{rc} - j\mu''_{rc})\mu_0 A_c}{l_c}. \tag{2.251}$$

2.18.4 Complex Inductance

The *complex inductance* of an inductor with an ungapped core is

$$L_c = \frac{\mu_s A_c N^2}{l_c} = \frac{\mu_{rs}\mu_0 A_c N^2}{l_c} = \mu_{rs} L_o = \frac{(\mu'_{rs} - j\mu''_{rs})\mu_0 A_c N^2}{l_c}$$

$$= \frac{\mu'_{rs}\mu_0 A_c N^2}{l_c} - \frac{j\mu''_{rs}\mu_0 A_c N^2}{l_c} = \mu'_{rs} L_o - j\frac{R_{cs}}{\omega}, \tag{2.252}$$

where the inductance of the corresponding air-core coil (at $\mu_r = 1$) is

$$L_o = \frac{\mu_0 A_c N^2}{l_c}. \tag{2.253}$$

2.18.5 Complex Impedance of Inductor

Neglecting the winding resistance, the impedance of an inductor with a magnetic core is equal to the series combination of L_s and R_{cs} is

$$Z = j\omega L_c = j\omega L_o(\mu'_{rs} - j\mu''_{rs}) = j\omega\frac{\mu_0 A_c N^2}{l_c}(\mu'_{rs} - j\mu''_{rs})$$

$$= \frac{\omega\mu_0\mu''_{rs} A_c N^2}{l_c} + \frac{j\omega\mu_0\mu'_{rs} A_c N^2}{l_c} = R_{cs} + j\omega L_s = |Z|e^{j\phi}, \tag{2.254}$$

where the series resistance of the core (which is equal to the sum of the series hysteresis resistance R_{hs} and the series eddy-current resistance) is given by

$$R_{cs} = R_{hs} + R_{es} = \frac{\mu_0\omega\mu''_{rs} A_c N^2}{l_c} = \omega\mu''_{rs} L_o \approx \omega\mu'_{rs} L_o \frac{\left(\frac{f}{f_H}\right)}{1 + \left(\frac{f}{f_H}\right)^2} = \omega_H L_o \frac{\mu'_{rso}}{1 + \left(\frac{f_H}{f}\right)^2}. \tag{2.255}$$

and the series inductance is

$$L_s = \frac{\mu'_{rs}\mu_0 A_c N^2}{l_c} = \mu'_{rs} L_o \approx 1 + \left(\frac{f}{f_H}\right)^2 L_o = \frac{\mu'_{rso} L_o}{1 + \left(\frac{f}{f_H}\right)^2}\frac{\mu_0 A_c N^2}{l_c} \tag{2.256}$$

the modulus of the impedance is

$$|Z| = \sqrt{R_{cs}^2 + (\omega L_s)^2} = \omega L_o\sqrt{(\mu'_{rs})^2 + (\mu''_{rs})^2} \tag{2.257}$$

and

$$\phi = \arctan\left(\frac{\omega L_s}{R_{cs}}\right) = \arctan\left(\frac{\mu'_{rs}}{\mu''_{rs}}\right) = 90^\circ + \delta_m. \tag{2.258}$$

Figure 2.50 shows a series equivalent circuit of magnetic cores and the corresponding phasor diagram of current and voltages. Figures 2.51–2.54 show plots of the series core resistance R_{cs}/L_o, L_s/L_o, $|Z|/L_o$, and ϕ as functions of frequency f, respectively.

The core loss caused by a sinusoidal inductor current $i_L = I_{Lm}\sin\omega t$ is

$$P_{cs} = \frac{R_{cs} I_{Lm}^2}{2} = \frac{\omega\mu_0\mu''_{rs} A_c N^2 I_{Lm}^2}{2l_c}. \tag{2.259}$$

The core loss is proportional to frequency f and μ''_{rs}, where μ''_{rs} is a function of frequency and behaves like the magnitude of the transfer function of a second-order band-pass filter versus frequency.

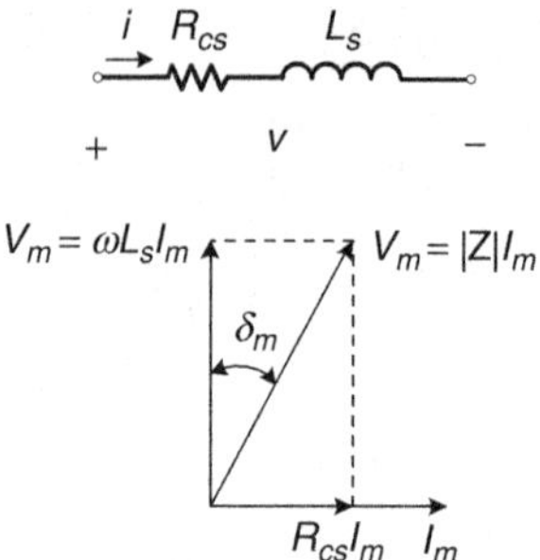

Figure 2.50 Series equivalent circuit of an inductor and the corresponding phasor diagram of voltage and current

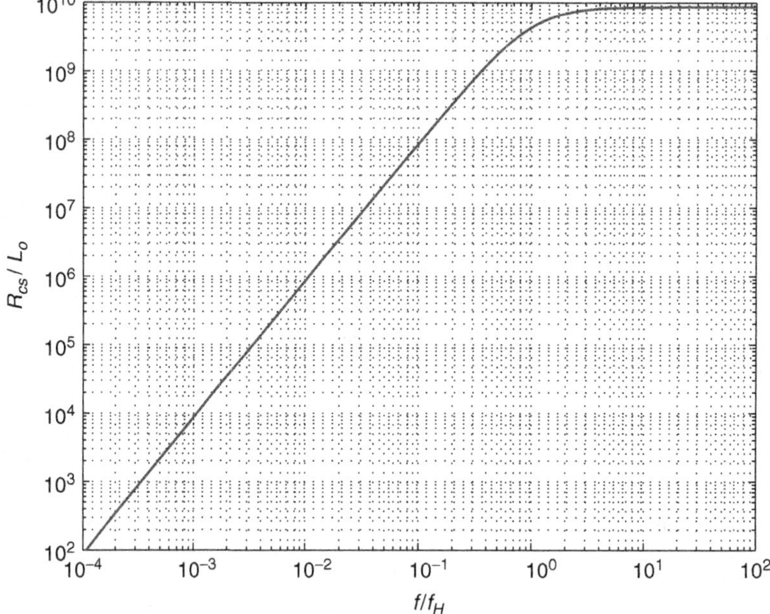

Figure 2.51 Series core resistance R_{cs} as a function of frequency f at the maximum value of the imaginary series relative permeability $\mu'_{rso} = 4300$ and $f_H = 320$ kHz

2.18.6 Approximation of Series Complex Permeability

The inductive (real) series relative permeability can be approximated by the magnitude of the transfer function of a first-order low-pass filter

$$\mu'_{rs} = \frac{\mu'_{rso}}{\sqrt{1 + \left(\frac{f}{f_H}\right)^2}}, \tag{2.260}$$

where $f_H = \omega_H/(2\pi)$ is the 3-dB frequency of μ'_{rs}. Figure 2.55 shows idealized plots of the real relative permeability μ'_{rs} as a function of frequency f. For ferrites, the frequency at which $\mu'_{rs} = 1$ is

$$f_1 = \mu'_{rso} f_H \approx 6.25 \text{ GHz}. \tag{2.261}$$

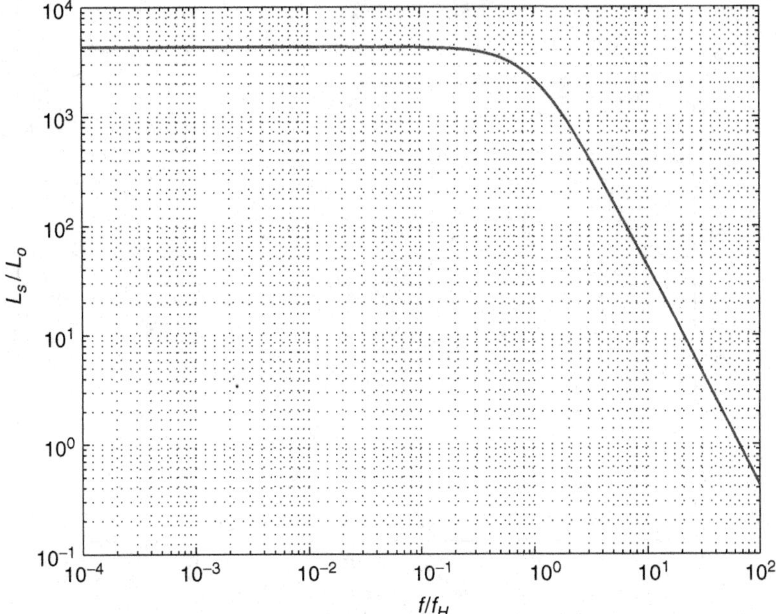

Figure 2.52 Series normalized inductance L_s/L_o as a function of frequency f at $\mu'_{rso} = 4300$ and $f_H = 320$ kHz

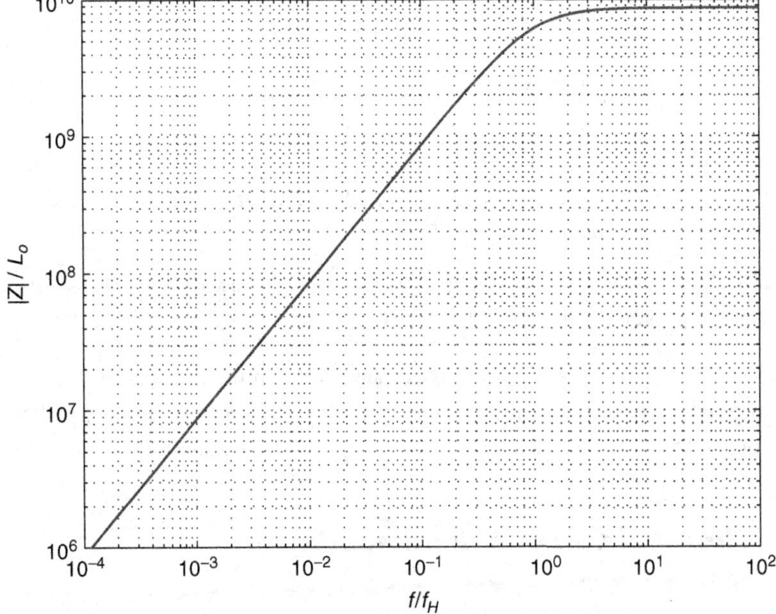

Figure 2.53 Modulus of normalized inductor impedance $|Z|/L_o$ as a function of frequency f at $\mu'_{rso} = 4300$ and $f_H = 320$ kHz

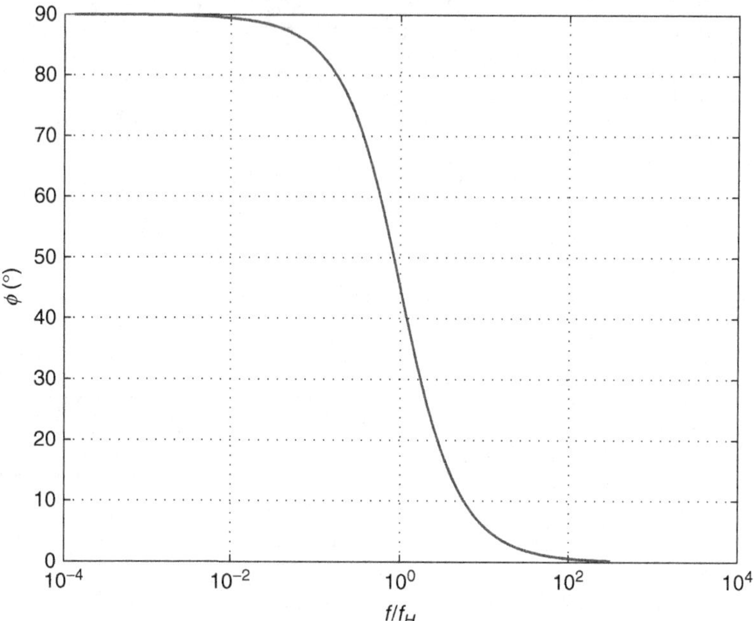

Figure 2.54 Phase of inductor impedance ϕ as a function of frequency f at $\mu'_{rso} = 4300$ and $f_H = 320$ kHz

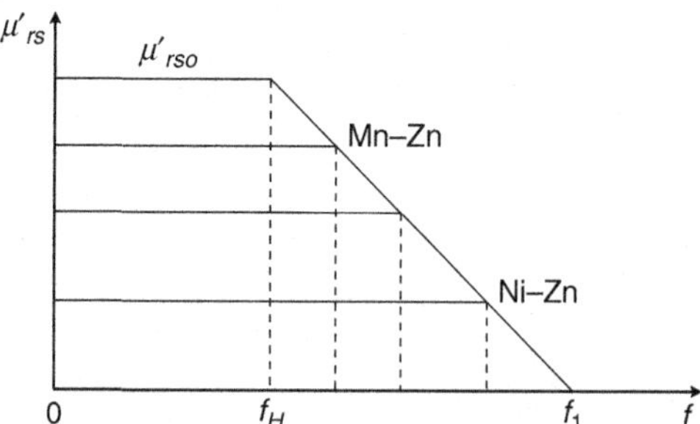

Figure 2.55 Real relative series permeability μ'_{rs} as a function of frequency f for selected values of low-frequency real relative permeability μ'_{rso}

As f_H increases, μ_{rso} decreases. There is an exchange between the low-frequency real relative permeability μ'_{rs} and the bandwidth $BW = f_H$. MnZn ferrites have large values of μ_{ro}, but low values of f_H. In contrast, NiZn ferrites have low values of μ'_{rs}, but high values of f_H.

2.18.7 Parallel Complex Permeability

The *series complex reluctivity* is given by

$$\kappa_s = \frac{\mathbf{H}}{\mathbf{B}} = \kappa'_s + j\kappa''_s = \frac{1}{\mu_s} = \frac{1}{\mu'_s - j\mu''_s} = \frac{\mu'_s + j\mu''_s}{(\mu'_s)^2 + (\mu''_s)^2}. \tag{2.262}$$

The *parallel complex permeability* of a magnetic material is given by

$$\frac{1}{\mu_p} = \frac{1}{\mu'_p} - \frac{1}{j\mu''_p} = \frac{1}{\mu_0\mu'_{rp}} - \frac{1}{j\mu_0\mu''_{rp}} = \frac{1}{\mu'_p} + j\frac{1}{\mu''_p} \tag{2.263}$$

and the relative parallel complex permeability is

$$\frac{1}{\mu_{rp}} = \frac{1}{\mu'_{rp}} - \frac{1}{j\mu''_{rp}}. \tag{2.264}$$

Hence,

$$\kappa'_s = \frac{1}{\mu'_p} = \frac{\mu'_s}{(\mu'_s)^2 + (\mu''_s)^2} \tag{2.265}$$

and

$$\kappa''_s = \frac{1}{\mu''_p} = \frac{\mu''_s}{(\mu''_s)^2 + (\mu''_s)^2}. \tag{2.266}$$

If a parallel equivalent circuit of an inductor is used, the admittance of the parallel combination of L_p and R_p is

$$Y = \frac{1}{j\omega L_o \mu_{rp}} = \frac{1}{j\omega L_o}\left(\frac{1}{\mu'_{rp}} - \frac{1}{j\mu''_{rp}}\right) = \frac{1}{\omega\mu''_{rp}L_o} + \frac{1}{j\omega\mu'_{rp}L_o} = \frac{1}{R_p} + \frac{1}{j\omega L_p}, \tag{2.267}$$

where

$$R_p = R_c = \omega\mu''_{rp}L_o \tag{2.268}$$

and

$$L_p = L = \mu'_{rp}L_o. \tag{2.269}$$

Figure 2.56 shows a parallel equivalent circuit of an inductor along with the corresponding phasor diagram of currents and voltage.

The magnetic loss tangent of a magnetic material is

$$\tan\delta_m = \frac{\omega L_p}{R_p} = \frac{\omega L}{R_c} = \frac{\mu'_{rp}}{\mu''_{rp}} = \frac{1}{Q_m}. \tag{2.270}$$

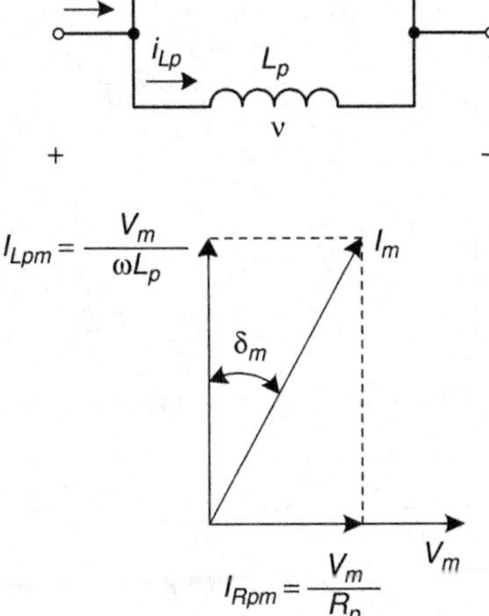

Figure 2.56 Parallel equivalent circuit of magnetic cores and phasor diagram of currents and voltage

2.18.8 Relationships Between Series and Parallel Complex Permeabilities

The relationships between the components of the relative series and parallel complex permeabilities are

$$\mu'_{rp} = \mu'_{rs}(1 + \tan^2\delta_m) = \mu'_{rs}\left[1 + \left(\frac{\mu''_{rs}}{\mu'_{rs}}\right)^2\right] = \mu'_{rso} \tag{2.271}$$

and

$$\mu''_{rp} = \mu''_{rs}\left(1 + \frac{1}{\tan^2\delta_m}\right) = \mu''_{rs}\left[1 + \left(\frac{\mu'_{rs}}{\mu''_{rs}}\right)^2\right] = \frac{\mu'_{rso}}{1 + \left(\frac{f}{f_H}\right)^2}. \tag{2.272}$$

Figure 2.57 shows the plots of the real and imaginary components of the parallel relative permeability μ'_{rp} and μ''_{rp} as functions of frequency f at $\mu'_{rso} = 4300$ and $f_H = 320$ kHz. It can be observed that the real component of the parallel permeability μ'_{rp} is constant, independent of frequency.

The combination of series components L_s and R_{cs} can be converted into the combination of parallel components L_p and R_{cp} as follows:

$$Q_m = \frac{\omega L_s}{R_{cs}} = \frac{\mu'_{rs}}{\mu''_{rs}} = \frac{1}{\tan\delta_m} = \frac{R_{cp}}{\omega L_p} = \frac{\mu''_p}{\mu'_p} \tag{2.273}$$

$$R_{cp} = R_c = R_{cs}(1 + Q_m^2) = R_{cs}\left(1 + \frac{1}{\tan^2\delta_m}\right) = R_{cs}\left[1 + \left(\frac{\mu'_{rs}}{\mu''_{rs}}\right)^2\right]$$

$$= \frac{\omega\mu''_{rs}\mu_0 A_c N^2}{l_c}\left[1 + \left(\frac{\mu'_{rs}}{\mu''_{rs}}\right)^2\right] = \omega_H \mu'_{rso} L_o \tag{2.274}$$

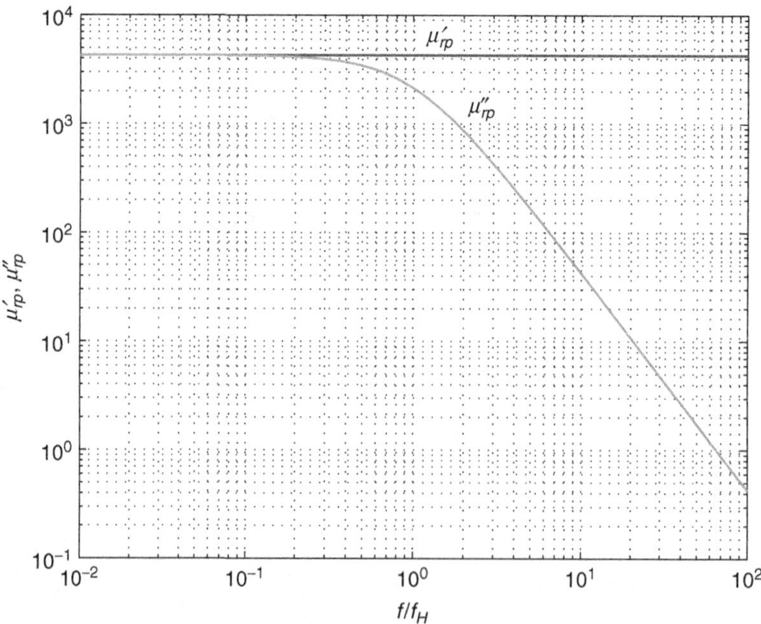

Figure 2.57 Real and imaginary components of parallel relative permeability as functions of frequency f at $\mu'_{rso} = 4300$ and $f_H = 320$ kHz

and

$$L_p = L_s\left(1+\frac{1}{Q_m^2}\right) = L_s(1+\tan^2\delta_m) = L_s\left[1+\left(\frac{\mu''_{rs}}{\mu'_{rs}}\right)^2\right]$$

$$= \frac{\mu'_{rs}\mu_0 A_c N^2}{l_c}\left[1+\left(\frac{\mu''_{rs}}{\mu'_{rs}}\right)^2\right] = \mu'_{rs}L_o\left[1+\left(\frac{\mu''_{rs}}{\mu'_{rs}}\right)^2\right] = \mu'_{rso}L_o. \quad (2.275)$$

Thus, both the parallel resistance R_{cp} and inductance L_p are constant, independent of frequency.

2.19 Cooling of Magnetic Cores

Excessive temperature rise will change the magnetic characteristics of a core or even damage it. For any ferromagnetic material, there is a characteristic temperature, termed the *Curie temperature* T_c, above which the alignment of domains is disrupted and the ferromagnetic material becomes paramagnetic, whose relative permeability μ_r is only slightly higher than 1. Aluminum, platinum, and magnesium are examples of paramagnetics. For most ferromagnetic materials, the Curie temperatures are approximately 500 °C. The maximum operating temperature of magnetic cores is usually below 100 °C. The total core loss per unit volume decreases with temperature.

The temperature rise of a core due to the core power loss is given by

$$\Delta T = R_{thc}P_C, \quad (2.276)$$

where $R_{thc} = l/(kA)$ is the thermal resistance of a core. The typical values of the thermal resistance for ferrite cores is from 10 to 100 °C/W.

Example 2.10

The Magnetics PC 3222 has the thermal resistance $R_{thc} = 19$ °C/W. The total core power loss is $P_C = 3$ W. Find the temperature rise due to the core power loss.

Solution: The temperature rise due to the core power loss is

$$\Delta T = R_{thc}P_C = 19 \times 3 = 57\,^\circ\mathrm{C}. \quad (2.277)$$

The combined losses in an inductor or a transformer must be dissipated through the surface of the wound magnetic component A_t. The dominant heat transfer mechanism is convection. Warm air expands and rises, resulting in natural air circulation. The temperature rise of the core due to the core loss P_C is

$$\Delta T = \left(\frac{0.1P_C}{A_s}\right)^{0.826}, \quad (2.278)$$

where P_C is in watts and the core surface A_s is in square meter or

$$\Delta T = \left(\frac{P_C}{A_s}\right)^{0.826}, \quad (2.279)$$

with P_C in milliwatt and the core surface A_s in square centimeter.

The temperature rise of the core due to the core loss and winding losses P_{cw} can be estimated by the following expression:

$$\Delta T = \left[\frac{\text{Total core and winding loss(mW)}}{\text{Core surface area(cm}^2\text{)}}\right]^{0.826} = \left[\frac{P_{cw}(\mathrm{mW})}{A_t(\mathrm{cm}^2)}\right]^{0.826} = \psi^{0.826}\ (^\circ\mathrm{C}) \quad (2.280)$$

where $\psi = P_{cw}/A_t$ is the surface power loss density. The required core surface area A_t for a given core loss P_C and a given temperature rise ΔT is

$$A_t = \frac{P_{cw}(\text{mW})}{\Delta T^{1.2}}(\text{cm}^2). \tag{2.281}$$

Example 2.11

The total core and winding power loss is $P_{cw} = 3$ W and the temperature rise due to the core power loss is $\Delta T = 40\,^\circ\text{C}$. What is the core surface area?

Solution: The core surface area is

$$A_t = \frac{P_{cw}(\text{mW})}{\Delta T^{1.2}}(\text{cm}^2) = \frac{3000}{40^{1.2}} = 35.86\ (\text{cm}^2). \tag{2.282}$$

The temperature rise of the inductor caused by both the core and winding losses is given by

$$\Delta T = (R_{thc} + R_{thw})(P_{Rw} + P_C) = R_{th}(P_{Rw} + P_C), \tag{2.283}$$

where R_{thw} is the thermal resistance of the winding and $R_{th} = R_{thc} + R_{thw}$ is the thermal resistance of the core and the winding(s). The heat generated in the core and the windings will be first conducted to the inductor surface, then to the air, and then removed by convection. The cooler air will absorb the heat and rise due to natural air circulation in the vertical direction. The external surface of the inductor or the transformer should be large enough to transfer the heat to the air.

2.20 Summary

Magnetization

- In an atom, electrons are orbiting around the nucleus, and each electron spins about its own axis.
- For atoms of most materials, the orbiting and spin magnetic moments of electrons tend to cancel each other, and therefore these materials show no magnetic effects.
- A current loop is a magnetic dipole, producing a magnetic field with south and north poles. The pattern of its magnetic field is similar to a permanent magnet.
- Magnetic field can induce magnetic dipoles, that is, magnetic polarization.
- The relative permeability μ_r of most materials is either slightly more or slightly less than unity.
- As the outer electron spins about its own axis, it causes a spin magnetic moment.
- As the outer electron orbits the nucleus, it causes an orbital magnetic moment, similar to a small loop with current, producing magnetic field and creating a small magnetic dipole.
- Ferromagnetism is mainly caused by electron spins.
- The magnetic domain is a region in a ferromagnetic or ferrimagnetic material in which all magnetic dipole moments are spontaneously aligned in parallel in the same direction.
- The magnetic dipoles form a structure similar to a group of parallel solenoids, and therefore produce a magnetic field in the direction.
- The degree of alignment of magnetic dipoles and magnetic domains increases when the magnitude of the external flux density B increases for $B < B_s$ and the temperature T decreases. The alignment is only independent of temperature for diamagnetics.

- Domain walls or Bloch walls are transition regions that separate the magnetic domains from each other.
- Typically, a magnetic domain contains 10^{17} to 10^{21} atoms.
- A magnetic material is magnetized when the magnetization **M** is not equal to zero everywhere in the material.
- The Curie temperature T_c is the temperature at which a ferromagnetic material loses completely its spontaneous magnetization and becomes paramagnetic. The spin alignment is destroyed by ever-present thermal agitation at T_c.
- The magnetization vector **M** is the vector sum of all the magnetic dipole moments in a unit volume of a material. Its unit is A/m.

Magnetic Elements

- Ferromagnetic elements are iron (Fe), cobalt (Co), and nickel (Ni).
- In ferromagnetics, the orbiting and spin magnetic moments do not cancel each other, and therefore these materials exhibit strong magnetic effects.
- Magnetic materials can be classified as ferromagnetics ($\chi_m \gg 1$ and $\mu_r \gg 1$), paramagnetics ($\chi_m > 0$ and $\mu_r > 1$), and diamagnetics ($\chi_m < 0$ and $\mu_r < 1$).
- Ferromagnetics exhibit spontaneous magnetization and form magnetic domains.
- Magnetic domains are groups of atoms with their magnetic dipoles oriented in the same direction and produce magnetic field.
- The relative permeability of cobalt (Co) is $\mu_r = 250$, nickel (Ni) is $\mu_r = 600$, soft iron (Fe) is $\mu_r = 5000$, and silicon iron is $\mu_r = 7000$.
- Diamagnetics have μ_r slightly less than 1 ($\mu_r = 1 - \epsilon$).
- Paramagnetics have μ_r slightly greater than 1 ($\mu_r = 1 + \epsilon$).
- Ferromagnetics have μ_r in the range from 5 to 10^6 ($\mu_r \gg 1$).
- The reciprocal of permeability is reluctivity.
- Most materials are poor conductors of magnetic flux because their $\mu_r = 1$.
- Ferromagnetic materials are good conductors of magnetic flux because their $\mu_r \gg 1$.
- The range of the relative permeability μ_r for ferromagnetic materials is from 1 to 100 000.
- The magnetization is the magnetic dipole moment per unit volume.
- A magnetic material is linear if μ does not vary with H, homogeneous if μ is the same at all points, and isotropic if μ does not vary with direction.

Hysteresis

- The H–B magnetization curve is called the hysteresis.
- The hysteresis is a double-valued function.
- As the magnetic field intensity H increases, the magnetic flux density B approaches nearly a maximum value called the saturation flux density B_s.
- Saturation of a magnetic core material occurs when all magnetic domains are magnetized and aligned with the external magnetic field. This imposes a limit on the minimum cross-sectional area of the core.

- The saturation flux density is from 1.6 to 2.2 T for iron alloys, from 0.2 to 0.5 T for ferrites, and from 1.2 to 1.3 T for amorphous alloys.
- The magnetization vector **M** is defined as the vector sum of magnetic moments per unit volume.

Magnetic Cores

- Magnetic core may be considered as a conductor of magnetic flux.
- Magnetic cores increase the inductance density and magnetic energy density of magnetic devices such as inductors and transformers, thus reducing the size and weight of power electronics circuits.
- The main purpose of a magnetic core is to provide an easy path for the magnetic flux, to concentrate the magnetic field in it, to guide the magnetic flux, and to obtain a well-defined MPL.
- There are five types of magnetic cores: iron cores, nickel and cobalt cores, ferrite cores, powdered cores, and nanocrystalline cores.
- Ferrite cores are molded ceramics.
- The magnetic core with a high permeability provides a greater degree of coupling between the transformer windings.
- Ferromagnetic elements and their alloys are used to manufacture practical magnetic cores with high permeability ($\mu_r \gg 1$).
- The addition of a few percent of silicon to iron results in higher resistivity ρ_c, lower eddy-current losses, higher relative permeability μ_r, and improved material stability with age.
- Ferrites have low electrical conductivity and are useful in a variety of high-frequency applications.
- Ferrites have low magnetic saturation flux density B_s, typically from 0.3 to 0.5 T.
- Powder cores include iron powder cores, Kool mu cores, and MPP cores.
- Powder cores are manufactured from very fine particles of magnetic material.
- Powder cores have a distributed air gap.
- The relative permeability μ_r of powder cores ranges from 10 to 550. Therefore, powder cores are low-permeability cores.

Core Losses

- Power losses in inductors and transformers consist of winding and core losses.
- Core losses consist of hysteresis loss and eddy-current loss.
- Eddy-current loss depends on (i) the size and shape of the core, especially cross-sectional area A_c, (ii) on the magnitude, shape, and frequency of the inductor or transformer current, and (iii) resistivity of the magnetic core material ρ_c.
- At low frequencies, ferrite core loss is almost entirely hysteresis loss.
- At 200–300 kHz, eddy-current loss overtakes hysteresis loss in ferrite cores.
- The Steinmetz empirical equation describes core loss for sinusoidal magnetic flux density $B(t) = B_m \cos \omega t$. It requires three parameters, which are usually provided by manufacturers.
- Powder cores exhibit much higher core loss (typically an order of magnitude) than ferrite cores at the same amplitude of flux density B_m and operating frequency f.
- Powder cores are usually used in high-frequency applications.
- Power loss of the gapped core is lower than that of the ungapped core at the same magnetic flux density.

- The resistivity of ferrite cores is very high and it is of the order of 10^7 times that of iron cores.
- The core loss depends on the core material, operating frequency, and maximum AC magnetic flux density.

Hysteresis Loss

- Hysteresis core loss is due to energy used to align and rotate elementary magnetic particles of magnetic materials.
- The hysteresis energy loss per cycle per unit volume is equal to the area of the $B-H$ loop.
- The specific hysteresis loss is proportional to the $B-H$ hysteresis area and frequency.
- The width of the hysteresis increases as frequency increases.

Eddy-Current Loss

- Metallic soft materials cannot be used at high frequencies because of the eddy-current losses.
- Soft ferrites are ceramic insulators and are used at high frequencies because of large resistivity and low-eddy-current loss.
- Eddy-current core loss is due to eddy currents circulating in the core.
- Eddy-current power loss density is proportional to the amplitude of the inductor voltage per turn squared $(V_{Lm}/N)^2$.
- Eddy-current power loss density can be reduced using electrically insulated laminations.
- The eddy-current loss increases with the square of frequency, and the hysteresis loss is directly proportional to frequency. As the frequency increases, the exponent a in f^a of the total core loss density P_v should change from 1 to 2 at a crossover frequency.
- Eddy-current power loss density is given by the general equation

$$P_e = k_{ec}\left(\frac{A_c}{\rho_c}\right)(fB_m)^2, \tag{2.284}$$

  where k_{ec} is the waveform coefficient of inductor voltage.
- The eddy-current loss density is directly proportional to the cross-sectional area of the core.
- The amplitude of the magnetic flux density B_m determines the cores' size and core losses.
- Cores with a lower permeability have a larger skin depth, which leads to a lower eddy-current loss.

Ferrites

- The frequency range of ferrite cores is from 20 kHz to several MHz.
- The dominant component of ferrite core losses is the hysteresis loss.
- The eddy-current loss in ferrite cores is low because their resistivity ρ_c is very high.
- The specific ferrite core losses increase roughly as the 1.6-th power of frequency and the 2.7-th power of the AC peak magnetic flux density.
- NiZn ferrites have lower relative permeability, higher resistivity, higher operating frequencies, and lower eddy-current loss than MnZn ferrites.
- The permeability of ferrite increases with temperature to a maximum value and then decreases to 1.

- The resistivity of ferrites decreases with frequency.
- The resistivity of MnZn ferrite is of the order 10^{-3} $\Omega \cdot$ m and the resistivity of NiZn ferrite is of the order 30×10^{-3} $\Omega \cdot$ m.
- Ferrite core loss is usually lower that that of powder core losses.
- Core losses are directly proportional to the core volume.
- Most switching power supply inductors and transformers use ferrite cores.
- The ferrite saturation flux density B_s is about 0.5 T at $T = 25\,^{\circ}$C and 0.2–0.3 T at $T = 100\,^{\circ}$C.
- Ferromagnetic materials have large and positive susceptibility, whereas paramagnetic material have small and positive susceptibility.
- PCs are used for inductors and transformers, especially for applications in DC–DC converters at low power levels.
- Good cores should have high relative permeability μ_{rc}, high saturation flux density B_s, low losses, wide bandwidth, and high Curie temperature T_c.
- High-frequency cores are made of ferrites and powders.
- MnZn ferrite is a high-permeability low-frequency material.
- NiZn ferrite is a low-permeability high-frequency material.
- Core loss in ferrite cores is higher than the winding loss above a few hundred kilohertz.
- The maximum operating frequency of ferrite cores is several megahertz.
- Low-frequency cores are made of iron, silicon, chrome, and alloys.
- DC currents in inductors and transformers cause asymmetrical magnetization of magnetic core. Air gaps are required in these applications.
- Magnetic devices that are expected to store magnetic energy usually contain an air gap.
- Core loss densities of ferrite materials rapidly increase with frequency, reducing the maximum magnitude of the AC component of magnetic flux.
- Core cooling is nearly proportional to its heat-radiating surface area.

Superconductors

- A superconductor below the critical temperature shows no electric resistance.
- The magnetic field exclusion in superconductors (Meissner effect) is equivalent to $\chi_r \rightarrow -1$, $\mu_r \rightarrow$ 0i, and $B = 0$.
- Superconductivity disappears above the critical magnetic field intensity and above the critical current density.
- Superconductor behaves like a perfect diamagnetic and cannot contain magnetic field.

Complex Permeability

- The complex permeability is used to characterize magnetic materials over a wide frequency range.
- Both the real and imaginary series permeabilities of magnetic materials depend on the magnitude of the magnetic flux density and frequency.
- The plot of the real series permeability as a function of frequency is similar to the magnitude of a low-pass filter.

- The real series permeability of magnetic materials describes the inductance as a function of frequency.
- The plot of the imaginary series permeability as a function of frequency is similar to the magnitude of a band-pass filter.
- The imaginary series permeability of magnetic materials describes core resistance and core losses as functions of frequency.

2.21 References

[1] J. Lammeraner and M. Stafl, *Eddy Currents*, Cleveland, OH: CRS Press, 1966.

[2] E. C. Snelling, *Soft Ferrites. Properties and Applications*, London: Iliffe Books Ltd, 1969; 2nd ED, London, UK: Butterworth, 1988; 2nd Ed PSMA reprint, P.O Box 418, Mendham, NJ, 07945, 2010.

[3] R. L. Stall, *The Analysis of Eddy Currents*, Oxford: Clarendon Press, 1974, pp. 21–27.

[4] J. L. Snoek, "Dispertion and absorption in magnetic ferrites at frequencies above 1 Mc/s," *Physica*, vol. 14, no. 4, pp. 207–217, May 1948.

[5] J. Smit and H. P. J. Wijn, *Ferrites*, John Wiley & Sons, p. 290, 1959.

[6] B. Lax and K. J. Button, *Microwave Ferrites and Ferrimagnetics*, New York: McGraw-Hill, 1962.

[7] R. M. Bozorth, *Ferromagnetism*, Princeton, NJ: Van Nostrand, 1968.

[8] J. K. Watson, *Applications of Magnetism*, Gainesville, FL, 1985; also New York: John Wiley & Sons, 1985.

[9] W. T. McLyman, *Transformer and Inductor Design Handbook*, 3rd Ed., New York: Marcel Dekker, 2004.

[10] E. Bennett and S. C. Larsen, "Effective resistance of alternating currents of multilayer windings," *Transactions of the American Institute of Electrical Engineers*, vol. 59, pp. 1010–1017, 1940.

[11] E. C. Cherry, "The duality between inter-linked electric and magnetic circuits," *Proceedings of the Physical Society*, vol. 62, pp. 1001–111, 1949.

[12] C. J. Carpenter, "Magnetic equivalent circuits," *Proceedings of the Institute of Electrical and Electronics Engineers*, vol. 115, pp. 1503–1511, October 1968.

[13] A. Konrad, "Eddy currents and modeling," *IEEE Transactions on Magnetics*, vol. 21, no. 5, pp. 1805–1810, September 1985.

[14] G. E. Fish, "Soft magnetic materials," *Proceedings of the IEEE*, vol. 78, no. 6, pp. 947–972, June 1990.

[15] V. J. Thottuveil, T. G. Wilson, and H. A. Oven, Jr., "High-frequency measurement techniques for magnetic cores," *IEEE Transactions on Power Electronics*, vol. 5, no. 1, pp. 41–53, January 1990.

[16] J. G. Zhu, S. Y. R. Hui, and V. S. Ransden, "Discrete modeling of magnetic cores including hysteresis eddy currents and anomalous losses," *Proceedings of IEEE*, vol. 140, pp. 317–322, July 1993.

[17] J. G. Zhu, S. Y. R. Hui, and V. S. Ransden, "A dynamic equivalent circuit model for solid magnetic core for high switching frequency operations," *IEEE Transactions on Power Electronics*, vol. 10, pp. 791–795, Nov. 1995.

[18] J. G. Zhu, S. Y. R. Hui, and V. S. Ransden, "A general dynamic circuit model of magnetic cores for low- and high-frequency applications – Part I: Theoretical calculation of the equivalent core loss resistance," *IEEE Transactions on Power Electronics*, vol. 11, pp. 246–250, Mar. 1996; Port II: circuit model formulation and implementation,'' *IEEE Transactions on Power Electronics*, vol. 11, pp. 251–259, March 1996.

[19] R. L. Perry, "Multiple layer series connected winding design for minimum losses," *IEEE Transactions on Power Applications Systems*, vol. PAS-98, pp. 116–123, January/February 1979.

[20] C. P. Steinmetz, "On the law of hysteresis," *AIEE*, vol. 9, pp. 3–64, 1892. Reprinted under the title "A Steinmetz contribution to the ac power revolution," Proc. IEEE, vol 72, pp. 196–221, Feb. 1984.

[21] K. H. Carpenter and S. W. Warren, "A wide bandwidth dynamic hysteresis model for magnetization in soft ferrites," *IEEE Transactions on Magnetics*, vol. 28, no. 5, pp. 2037–2041, September 1992.

[22] D. C. Jiles and D. L. Altherton, "Ferromagnetic hysteresis," *IEEE Transactions on Magnetics*, vol. 19, no. 5, pp. 2183–2185, September 1983.

[23] S. Mulder, "Fit formulae for power loss in ferrites and their use in transformer design," Proceedings of PCIM, 1993, pp. 343–353.

[24] D. C. Jiles and D. L. Altherton, "Theory of ferromagnetic hysteresis," *Journal of Applied Physics*, vol. 55, pp. 2115–2120, March 1984.

[25] D. C. Jiles and D. L. Altherton, "Theory of ferromagnetic hysteresis," *Journal of Magnetism and Magnetic Materials*, vol. 61, pp. 48–60, 1986.

[26] M. L. Hodgdon, "Mathematical theory and calculations of magnetic hysteresis in soft ferrites," *IEEE Transactions on Magnetics*, vol. 24, no. 6, pp. 3120–3122, November 1988.

[27] M. L. Hodgdon, "Applications of a theory of ferromagnetic hysteresis," *IEEE Transactions on Magnetics*, vol. 27, pp. 4404–4406, November 1991.

[28] J. P. Vandalec and P. D. Ziogos, "A novel approach for minimizing high-frequency transformer copper loss," *IEEE Transactions on Power Electronics*, vol. 3, pp. 266–276, July 1988.

[29] A. M. Urling, V. A. Niemela, G. R. Skutt, and T. G. Wilson, "Characterizing high frequency effects in transformer windings: a guide to several significant papers," *IEEE Transactions on Power Electronics Specialists Conference*, 1989, pp. 373–385.

[30] J. A. Ferreira, *Electromagnetic Modelling of Power Electronic Converters*, Boston, MA: Kluwer Academic Publisher, 1989.

[31] J. G. Kassakian, M. F. Schlecht, and G. C. Verghese, *Principles of Power Electronics*, Reading, MA: Addison-Wesley, 1991.

[32] W. Roshen, "Ferrite core loss for power magnetic component design," *IEEE Transactions on Magnetics*, vol. 27, no. 6, pp. 4407–4415, November 1991.

[33] M. Bartoli, A. Reatti, and M. K. Kazimierczuk, "High-frequency models of ferrite inductors," Proceedings of the IEEE International Conference on Industrial Electronics, Controls, Instrumentation, and Automation (IECON'94), Bologna, Italy, September 5–9, 1994, pp. 1670–1675.

[34] M. Bartoli, A. Reatti, and M. K. Kazimierczuk, "Predicting the high-frequency ferrite-core inductor performance," Proceedings of the Conference of Electrical Manufacturing and Coil Winding, Chicago (Rosemont), IL, September 27–29, 1994, pp. 409–413.

[35] M. Bartoli, A. Reatti, and M. K. Kazimierczuk, "Modeling iron-powder inductors at high-frequencies," *Proceedings of the IEEE Industry Applications Society Annual Meeting, Denver, CO, October 2–7*, 1994, pp. 1125–1232.

[36] J. A. Ferreira, "Improved analytical modeling of conductive losses in magnetic components," *IEEE Transactions on Power Electronics*, vol. 9, pp. 127–131, January 1994.

[37] M. Bartoli, N. Noferi, A. Reatti, and M. K. Kazimierczuk, "Modeling winding losses in high-frequency power inductors," *Journal of Circuits Systems and Computers*, vol. 5, no. 3, pp. 65–80, March 1995.

[38] N. Mohan, T. M. Underland, and W. P. Robbins, *Power Electronics*, 3rd Ed., New York: John Wiley & Sons, 2003.

[39] R. Petkov, "Optimum design of a high-power high-frequency transformers," *IEEE Transactions on Power Electronics*, vol. 11, no. 1, pp. 33–42, January 1996.

[40] M. Bartoli, N. Nefari, A. Reatti, and M. K. Kazimierczuk, "Modeling litz-wire winding losses in high-frequencies power inductors," Proceedings of the IEEE Power Electronics Specialists Conference, Baveno, Italy, June 24–27, 1996, pp. 1690–1696.

[41] G. Bartotti, *Hysteresis of Magnetism*, San Diego, CA: Academic Press, 1998.

[42] C.-M. Ong, *Dynamic Simulation of Electric Machinery*, Reading: Prentice Hall, 1998, pp. 38–40, 45–46, and 87–90.

[43] M. Mordjaoui, M. Chabane, and M. Boudjema, and B. Daira, "Qualitative ferromagnetic hysteresis modeling," *Journal of Computer Science*, vol. 3, no. 6, pp. 399–405, 2007.

[44] G. Bertotti, *Hysteresis of Magnetism*, San Diego, CA: Academic Press, 1998.

[45] W. G. Hurley, W. H. Wolfe, and J. G. Breslin, "Optimized transformer design: inclusive of high-frequency effects," *IEEE Transactions on Power Electronics*, vol. 13, no. 4, pp. 651–659, July 1998.

[46] W. G. Hurley, E. Gath, and J. G. Breslin, "Optimizing the ac resistance of multilayer transformer windings with arbitrary current waveforms," *IEEE Transactions on Power Electronics*, vol. 15, no. 2, pp. 369–376, March 2000.

[47] M. K. Kazimierczuk, G. Sancineto, U. Reggiani, and A. Massarini, "Small-signal high-frequency model of ferrite inductors," *IEEE Transactions on Magnetics*, vol. 35, pp. 4185–4191, September 1999.

[48] U. Reggiani, G. Sancineto, and M. K. Kazimierczuk, "High-frequency behavior of laminated iron-core inductors for filter applications," Proceedings of the IEEE Applied Power Electronics Conference, New Orleans, LA, February 6–10, 2000, pp. 654–660.

[49] G. Grandi, M. K. Kazimierczuk, A. Massarini, U. Reggiani, and G. Sancineto, "Model of laminated iron-core inductors," *IEEE Transactions on Magnetics*, vol. 40, no. 4, pp. 1839–1845, July 2004.

[50] K. Howard and M. K. Kazimierczuk, "Eddy-current power loss in laminated power cores," Proceedings of the IEEE International Symposium on Circuits and Systems, Sydney, Australia, May 7–9, 2000, paper III-668, pp. 668–672.

[51] A. Reatti and M. K. Kazimierczuk, "Comparison of various methods for calculating the ac resistance of inductors," *IEEE Transactions on Magnetics*, vol. 37, pp. 1512–1518, May 2002.

[52] C. R. Sullivan, "Optimal choice for the number of strands in a litz-wire transformer winding," *IEEE Transactions on Power Electronics*, vol. 14, no. 2, pp. 283–291, March 1999.

[53] J. Reinert, A. Brockmeyer, and R. W. De Doncker, "Calculations of losses in ferro- and ferrimagnetic materials based on modified Steinmetz equation," *IEEE Transactions on Industry Applications*, vol. 37, no. 4, pp. 1055–1061, July/August 2001.

[54] G. Grandi, M. K. Kazimierczuk, A. Massarini, and U. Reggiani, "Stray capacitance of single-layer solenoid air-core inductors," *IEEE Transactions on Industry Applications*, vol. 35, no. 5, pp. 1162–1168, September 1999.

[55] A. Massarini and M. K. Kazimierczuk, "Self-capacitance of inductors," *IEEE Transactions on Power Electronics*, vol. 12, no. 4, pp. 671–676, July 1997.

[56] N. Hamilton, "The small-signal frequency response of ferrites," High-Frequency Electronics, pp. 36–52, June 2011.

[57] R. W. Erickson and D. Maksimović, *Fundamentals of Power Electronics*, Norwell, MA: Kluwer Academic Publishers, 2001.

[58] A. Van den Bossche and V. C. Valchev, *Inductors and Transformers for Power Electronics*, Boca Raton, FL: Taylor & Francis, 2005.

[59] F. E. Terman, *Radio Engineers' Handbook*, New York: McGraw-Hill, 1943.

[60] J. C. Maxwell, *A Treatise of Electricity and Magnetism*, 3rd Ed., New York: Dover Publishing, 1997.

[61] J. D. Adam, L. E. Devis, G. F. Dionne, E. F. Schoemann, and S. N. Stitzer, "Ferrite devices and materials," *IEEE Transactions on Microwave Theory and Techniques*, vol. 50, pp. 721–703, 2002.

[62] N. Das and M. K. Kazimierczuk, "An overview of technical challenges in the design of current transformers," Electrical Manufacturing Conference, Indianapolis, IN, October 24–26, 2005.

[63] D. Zhang and C. F. Foo, "A practical method to determine intrinsic complex permeability and permittivity for Mn-Zn ferrites," *IEEE Transactions on Magnetics*, vol. 41, no. 4, pp. 1226–1232, April 2005.

[64] J. Shenhui and J. Quinxing, "An alternative method to determine the initial permeability of ferrite cores using network analyzer," *IEEE Transaction on Electromagnetic Compatibility*, vol. 47, no. 3, pp. 652–657, August 2005.

[65] R. F. Huang, D. M. Zhang, and K.-J. Tseng, "An efficient finite-difference-based Newton-Raphson method to determine intrinsic complex permeabilities and permittivities for Mn-Zn ferrites," *IEEE Transactions on Magnetics*, vol. 42, no. 6, pp. 1665–1660, June 2006.

[66] R. F. Huang, D. M. Zhang, and K.-J. Tseng, "Determination of dimension-independent magnetic dielectric properties for ferrite cores and its EMI applications," *IEEE Transaction on Electromagnetic Compatibility*, vol. 50, no. 3, pp. 597–602, August 2008.

[67] M. Yamaguchi, K. Suezawa, Y. Takahashi, K. I. Arai, S. Kikuchi, Y. Shimada, S. Tanabe, and K. I. Ito, "Magnetic thin-film inductors for RF-integrated circuits," *Journal of Magnetism and Magnetic Materials*, vol. 215–216, pp. 807–810, June 2, 2000.

[68] M. Yamaguchi, Y. Mayazawa, K. Kaminishi, H. Kikuchi, S. Yabukami, K. I. Arai, and T. Suzuki, "Soft magnetic applications to the RF range," *Journal of Magnetism and Magnetic Materials*, vol. 26, no. 1–2, pp. 170–177, January 2004.

[69] T. Suetsugu and M. K. Kazimierczuk, "Integration of Class DE inverter for dc-dc converter on-chip power supplies," IEEE International Symposium on Circuits and Systems, Kos, Greece, May 21–24, 2006, pp. 3133–3136.

[70] T. Suetsugu and M. K. Kazimierczuk, "Integration of Class DE synchronized dc-dc converter on-chip power supplies," IEEE Power Electronics Specialists Conference, Jeju, South Korea, pp. 433–437, June 21–24, 2006.

[71] W. S. Roshen, "A practical, accurate and very general core loss model for nonsinusoidal waveforms," *IEEE Transactions on Power Electronics*, vol. 22, no. 1, pp. 30–40, January 2007.

[72] Z. Gmyrek, A. Boglietti, and A. Cavagninro, "Iron loss prediction with PWM supply using low- and high-frequency measurements: analysis and results comparison," *IEEE Transactions on Industrial Electronics*, vol. 55, no. 4, pp. 1722–1728, April 2008.

[73] J. Cole, S. D. Sudkoff, and R. R. Chan, "A field-extrema hysteresis loss model for high-frequency ferromagnetic materials," *IEEE Transactions on Industrial Electronics*, vol. 44, no. 7, pp. 1728–1736, July 2008.

[74] C. A. Balanis, *Advanced Engineering Electromagnetics*, 2nd Ed. Hoboken, NJ, John Wiley & Sons, New York, 2012.

[75] W. A. Roshen, "High-frequency tunneling magnetic loss in soft ferrites," *IEEE Transactions on Magnetics*, vol. 43, no. 3, pp. 968–973, March 2007.

[76] K. Sokalski, J. Szczyglowski, and M. Najgebauer, "Losses scaling in soft magnetic materials," *COMPEL*, vol. 26, no. 3, pp. 640–649, 2007.

[77] K. Sokalski and J. Szczyglowski, "Formula for energy loss in soft magnetic materials and scaling," *Acta Physica Polonica A*, vol. 115, no. 5, pp. 920–924, 2009.

[78] K. Sokalski and J. Szczyglowski, "Data collapse of energy loss in soft magnetic materials as a way for testing measurement set," *Acta Physica Polonica A*, vol. 117, no. 3, pp. 497–499, 2010.

[79] T. Miyazaki and H. Jin, *The Physics of Ferromagnetism*, Berlin: Springer, 2012.

[80] P. R. Wilson, J. N. Neil, and A. D. Brown, "Modeling frequency-dependent losses in ferrite cores," *IEEE Transactions on Magnetics*, vol. 40, no. 3, pp. 1537–1541, May 2004.

[81] www.mag-inc.com, www.micrometals.com, www.ferroxcube.com, www.fair-rite.com, www.ferrite.de, www.metglas.com, www.cmi-ferrite.com, www.tokin.com, www.hitachimetals.co.jp, www.jfe-steel.co.jp, www.vacuumschmelze.de, www.coilcraft.com, www.eilor.co.il. www. mkmagnetics.com, and www.magmet.com.

2.22 Review Questions

2.1. What is an orbital magnetic and spin magnetic moment?

2.2. What is a spin magnetic moment?

2.3. What is a magnetic dipole?

2.4. Sketch the magnetic field distribution for the magnetic dipole.

2.5. What is a magnetic domain?

2.6. What are magnetic domain walls?

2.7. What is a ferromagnetic material?

2.8. What is the difference between unmagnetized and magnetized domains in ferromagnetic materials?

2.9. What is the difference between soft and hard ferromagnetic materials?

2.10. What is the magnetic permeability? Derive an expression for it.

2.11. What is the relative permeability?

2.12. List the ferromagnetic elements.

2.13. Classify different types of magnetic materials.

2.14. What is the range of relative permeability μ_r for ferromagnetic materials?

2.15. What are the typical values of the relative permeability of diamagnetics and paramagnetics?

2.16. What causes saturation of magnetic flux density B_s?

2.17. What is the Curie temperature?

2.18. What is the reason for adding silicon to iron in magnetic cores?

2.19. What are the major types of magnetic cores?

2.20. What is the range of resistivity of ferrite cores?

2.21. What is the hysteresis?

2.22. What causes magnetic hysteresis?

2.23. Sketch a typical magnetization curve.

2.24. What is the difference between the hysteresis of soft and hard magnetic materials?

2.25. What is the effect of frequency on hysteresis?

2.26. What is the effect of frequency on hysteresis loss?

2.27. What is the effect of the amplitude of the magnetic flux density B_m on hysteresis loss?

2.28. What are the characteristics of ferrite cores?

2.29. What are the characteristics of powder cores?

2.30. What are the characteristics of iron cores?

2.31. What shape of core is the best for EMI filters?

2.32. What is superconductivity?

2.33. How does the resistivity of a superconductor change with temperature?

2.34. How does the resistivity of a superconductor change with the magnetic flux density?

2.35. What is the complex permeability?

2.36. What is the complex reluctivity?

2.23 Problems

2.1. Find the relative permeability for an iron core with μ_{rc} at $H = 250$ A/m and $B = 0.8$ T.

2.2. The total flux emitted from a magnet pole is 2×10^{-4} Wb and the magnet has a cross-sectional area of 1 cm^2.

(a) Find the flux density within the magnet.

(b) The flux spreads out so that at a certain distance from the pole, it is distributed over an area 2 cm by 2 cm. Find the flux density in this area.

2.3. The Philips magnetic material 3F3 has $\mu_{rc} = 1800$. What is the ratio of the intrinsic flux density to the free-space component of the magnetic flux density?

2.4. The Magnetics F-type core material has the following coefficients $k = 0.0573$, $a = 1.66$, and $b = 2.68$. The volume of a core 0F-44221-UG is $V_c = 2$ cm^3. A sinusoidal current frequency flows through the inductor. Find the total core power loss at $f = 100$ kHz for the following values of the magnetic flux density:

(a) $B_m = 0.1$ T

(b) $B_m = 0.2$ T

(c) $B_m = 0.3$ T

(d) $B_m = 0.4$ T

(e) $B_m = 0.45$ T.

(f) $B_m = 0.5$ T.

2.5. Derive an expression for the eddy-current power loss density for a bipolar square-wave inductor voltage.

2.6. A round silicon steel core has $\rho_c = 10^{-5}$ $\Omega \cdot$ cm, $A_c = 4$ cm^2, $B_m = 0.5$ T, and $f = 60$ Hz. Find the time-average eddy-current power loss density for $v_L = V_{Lm} \sin 2\pi 60t$.

2.7. A round solid MnZn ferrite core has $\rho_c = 10^6$ $\Omega \cdot$ cm, $B_m = 0.3$ T, $\mu_{rc} = 1800$, $A_c = 4$ cm^2, $V_c = 4$ cm^3. The inductor is driven by a sinusoidal voltage at $f = 1$ MHz.

(a) Determine the time-average hysteresis power loss density.

(b) Determine A_{BH}.

(c) Determine the time-average eddy-current power loss density.

(d) Determine the ratio of the hysteresis power loss to the eddy-current power loss.

(e) Determine the total power core loss.

2.8. A cylindrical NiZn ferrite core has $\rho_c = 10^5$ $\Omega \cdot$ cm, $\mu_{rc} = 150$, $A_c = 1.5$ cm^2, $V_c = 5$ cm^2, and $f = 500$ kHz, $B_m = 0.3$ T at $T = 20\,^{\circ}$C.

(a) Determine the hysteresis core power loss.

(b) Determine the eddy-current core power loss.

(c) Determine the total core power loss.

2.9. An inductor is made up of the Magnetics P-type material core and has a single turn. It conducts current $i(t) = 5 \sin 2\pi \times 3 \times 10^5 t$ A. The round core dimensions are $l_c = 6$ cm and $A_c = 4.2$ cm^2.

(a) Determine the specific core power loss using Steinmetz equation.

(b) Determine the total core power loss using Steinmetz equation.

(c) Determine the specific and total hysteresis core power loss using physics equation.

(d) Determine the specific and total eddy-current core power loss using physics equation.

(e) Determine the total specific and total core power losses using physics equation.

3

Skin Effect

3.1 Introduction

A winding conductor and a core conductor, each carrying time-varying currents, experience magnetic fields due to their own currents and also magnetic fields due to all current-carrying conductors in their vicinities. In turn, by Faraday's law, these magnetic fields induce electromotive forces that induce eddy currents in conductors [1–40]. By Lenz's law, they oppose the penetration of the conductor by the magnetic field and convert ohmic losses by converting electromagnetic energy into heat. There are two kinds of eddy-current effects: skin effect and proximity effect. The skin effect in a conductor is caused by the magnetic field induced by its own current. The proximity effect is caused by magnetic fields induced by currents flowing in adjacent conductors. Both of these effects cause nonuniform current density in conductors at high frequencies. These are high-frequency phenomena and limit the ability of conductors to conduct high-frequency currents. The skin effect and the proximity effect are orthogonal to each other and can be considered separately. The proximity effect is predominant in multilayer inductors at high frequencies.

An isolated round and straight conductor carrying time-varying current $i(t)$ induces a time-varying circular magnetic field $H(t)$ both inside and outside the conductor. The magnitude of the magnetic field depends on the rate of change of the current $i(t)$ and the radial distance r from the conductor center. At low frequencies, the current density in the conductor is uniform and the magnitude of the magnetic field increases linearly with the distance r inside the conductor. Outside the conductor, the magnitude of the magnetic field decreases inversely with r. The magnetic field $H(t)$ existing inside the conductor induces eddy currents $i_E(t)$ in the conductor. The eddy current $i_E(t)$ adds to the applied current $i(t)$ in the area close to the surface and subtracts from it in the middle of the conductor. The net current $i(t)$ remains unchanged, but the radial current density is not uniform. It is larger at the surface than in the interior of the conductor. The current density distribution becomes more nonuniform as the frequency of the external current $i(t)$ increases. The magnetic field outside the conductor remains unchanged and depends only on the net current $i(t)$.

The skin effect arises because a conductor carrying a time-varying current is immersed in its own magnetic field (self-field), which causes eddy currents in the conductor itself. In accordance with Lenz's law, the eddy currents produce a secondary magnetic field that opposes the primary magnetic field. When a time-varying (AC) current flows in a conductor, a magnetic field is induced in the conductor by its own current, which causes extra circulating current in the conductor. As a result, the current tends to flow near the surface and the current density tends to decrease from the surface

High-Frequency Magnetic Components, Second Edition. Marian K. Kazimierczuk.

Companion Website: www.wiley.com/go/kazimierczuk_High2e

to the center of the conductor. At very high frequencies (VHFs), the entire current flows in a very narrow skin on the conductor. This phenomenon is called the *skin effect*. The interior of a wire does not contribute to conduction of current. As the frequency increases, the conductor's effective resistance increases and so does the power loss. At low frequencies, the current takes the path of the lowest resistance. At high frequencies, the current takes the path of the lowest inductance.

The *winding power loss*, also called the *copper loss*, is caused by the current flow through the winding resistance. At DC and low frequencies, the current density J is uniformly distributed throughout the conductor's cross-sectional area. When a conducting material is subjected to an alternating magnetic field, *eddy currents* are induced in it. At high frequencies, the current density J becomes nonuniform due to eddy currents caused by two mechanisms: the *skin effect* and the *proximity effect*.

3.2 Resistivity of Conductors

Conductors are used to make windings of inductors and transformers. The conductivity σ or resistivity $\rho = 1/\sigma$ of a material depends on the number of electrons in the valence shell. Very good conductors have from one to three valence electrons. Examples of good conductors are silver (Ag) with the electron configuration $2+8+18+18+1=47$, copper (Cu) with the electron configuration $2+8+18+1=29$, gold (Au) with the electron configuration $2+8+18+32+1=79$, aluminum (Al) with the electron configuration $2+8+3=13$, zinc (Zn) with the electron configuration $2+8+18+2=30$, cobalt (Co) with the electron configuration $2+8+15+2=27$, and iron (Fe) with the electron configuration $2+8+14+2=26$. Valence electrons in conductors are loosely bound to the atoms and can easily become free electrons at room temperature. These free electrons are movable charges and therefore become current carriers. They may conduct current according to Ohm's law $J = \sigma E$. Copper is the most common conductor used to make windings of magnetic components. The concentration of free electrons in copper is $n_{Cu} = 8.45 \times 10^{28}$ free electrons/m^3 and in aluminum is $n_{Al} = 18.1 \times 10^{28}$ free electrons/m^3. The melting temperature of copper is 1084.62 °C or 1985.32 F.

The resistance of a conductor of length l_w, uniform cross-sectional area A_w, and resistivity ρ_w at DC or low frequencies (for uniform current density) is given by

$$R_{wDC} = \rho_w \frac{l_w}{A_w} = \rho_w \frac{N l_T}{A_w}, \tag{3.1}$$

where l_T is the mean turn length (MTL) and N is the number of turns. For a round conductor of diameter d, $A_w = \pi d^2/4$ and the DC resistance is

$$R_{wDC} = \rho_w \frac{l_w}{A_w} = \rho_w \frac{4 l_w}{\pi d^2} = \rho_w \frac{4 N l_T}{\pi d^2}. \tag{3.2}$$

The electrical resistivity is the property of a material and is defined as

$$\rho_w = \frac{1}{\sigma_w} = \frac{E}{J} = R_{wDC} \frac{A_w}{l_w} (\Omega \cdot \text{m}), \tag{3.3}$$

where E is the electric field intensity in a material and J is the uniform current density. The conductivity is the inverse of resistivity $\sigma_w = J/E = 1/\rho_w = l_w/RA_w$ (S/m). Table 3.1 lists resistivities, conductivities, and temperature coefficients of several conductors.

3.2.1 Temperature Dependance of Resistivity

The resistivity of conductors is proportional to temperature T and it is inversely proportional to the mean length of free path of electrons between the collisions l_n and to the square of the amplitude of vibrations A^2. As the temperature T increases, there are more collisions of electrons, the electron

Table 3.1 Resistivities and conductivities of conductors at T = 20 °C

Conductor	ρ_w (Ω·m)	σ_w (S/m)	α (1/°C)
Silver (Ag)	1.59×10^{-8}	6.29×10^{7}	0.0037
Copper (Cu)	1.724×10^{-8}	5.8×10^{7}	0.00393
Gold (Au)	2.439×10^{-8}	4.098×10^{7}	0.0034
Aluminum (Al)	2.65×10^{-8}	3.77×10^{7}	0.0037
Zinc (Zn)	5.92×10^{-8}	1.69×10^{7}	0.0037
Cobalt (Co)	6.24×10^{-8}	1.6×10^{7}	0.0037
Nickel (Ni)	6.99×10^{-8}	1.43×10^{7}	0.00641
Iron (Fe)	9.71×10^{-8}	1.03×10^{7}	0.0061

mobility μ_n decreases, and the resistivity ρ increases. The resistivity of a conductor increases with temperature T

$$\rho_T = \rho_{To} + \Delta\rho_T = \rho_{To}\left(1 + \frac{\Delta\rho_T}{\rho_{To}}\right) = \rho_{T_o}[1 + \alpha(T - T_o)](\Omega \cdot \text{m}), \tag{3.4}$$

where α is the temperature coefficient of the conductor resistivity (in 1/K or in 1/°C) and ρ_{To} is the conductor resistivity at temperature T_o. Typically, $\alpha = 0.003$–0.006 (1/°C) for metals near room temperature. For copper, $\alpha_{Cu} = 0.00393$ (1/°C). The resistivity of copper as a function of temperature T is given by

$$\rho_{Cu(T)} = \rho_{Cu(20\,°\text{C})}[1 + 0.00393(T - 20)](\Omega \cdot \text{m}). \tag{3.5}$$

The resistivity of aluminum as a function of temperature T is given by

$$\rho_{Al(T)} = \rho_{Al(20\,°\text{C})}[1 + 0.0037(T - 20)](\Omega \cdot \text{m}). \tag{3.6}$$

Figure 3.1 shows the copper resistivity ρ_{Cu} as a function of temperature T.

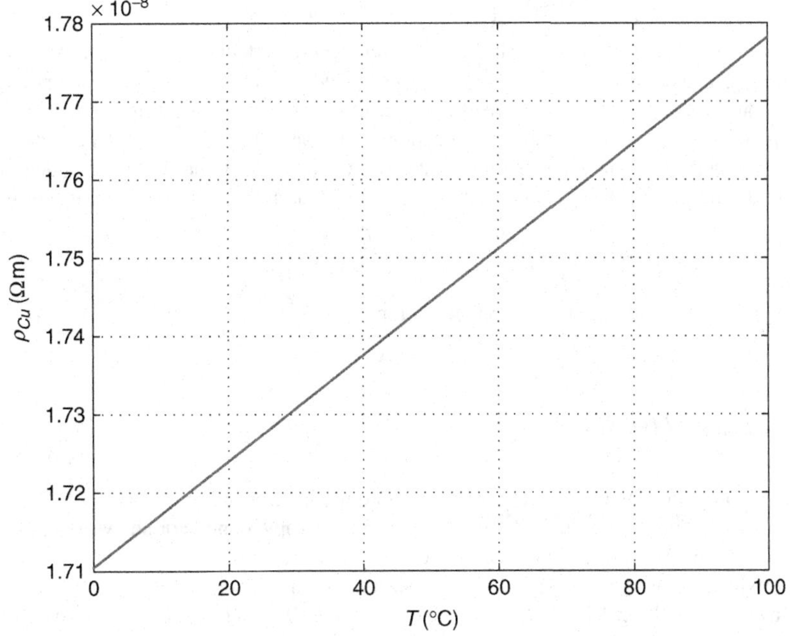

Figure 3.1 Copper resistivity ρ_{Cu} as a function of temperature T

Windings are usually made of copper because of its low resistivity. The resistivity of soft-annealed copper is $\rho_{Cu} = 1.724 \times 10^{-8}\ \Omega\cdot\text{m} = 17.24\ \text{n}\Omega\cdot\text{m}$ (or conductivity $\sigma_{Cu} = 5.8 \times 10^7$ S/m) at $T = 20\,^\circ\text{C}$, $\rho_{Cu} = 2.012 \times 10^{-8}\ \Omega\cdot\text{m} = 20.12\ \text{n}\Omega\cdot\text{m}$ (or $\sigma_{Cu} = 4.97 \times 10^7$ S/m) at $T = 60\,^\circ\text{C}$, and $\rho_{Cu} = 2.266 \times 10^{-8}\ \Omega\cdot\text{m} = 22.66 \times 10^{-8}\ \text{n}\Omega\cdot\text{m}$ (or $\sigma_{Cu} = 4.34 \times 10^7$ S/m) at $T = 100\,^\circ\text{C}$. Note that $\rho_{Cu(100\,^\circ\text{C})}/\rho_{Cu(20\,^\circ\text{C})} = 1.314$.

The DC and low-frequency resistance of a conductor increases with temperature T

$$R_{wDC(T)} = R_{wDC(To)} + \Delta R_{wDC(T)} = R_{wDC(To)}\left(1 + \frac{\Delta R_{wDC(T)}}{R_{To}}\right) = R_{wDC(T_o)}[1 + \alpha(T - T_o)](\Omega). \tag{3.7}$$

The DC and low-frequency winding power loss increases with temperature and is given by

$$P_{wDC(T)} = I_{DC}^2 R_{wDC(T)} = I_{DC}^2 R_{wDC(T_o)}[1 + \alpha(T - T_o)](\text{W}). \tag{3.8}$$

The AC-to-DC resistance ratio at the temperature T is derived in Section 5.28

$$F_{R(T)} = \frac{R_{w(T)}}{R_{wDC(T)}} = \frac{P_{w(T)}}{P_{wDC(T)}} = 1 + \frac{F_{R(To)} - 1}{[1 + \alpha(T - T_o)]^2}, \tag{3.9}$$

resulting in

$$\begin{aligned} R_{w(T)} = R_{wDC(T)} F_{R(T)} &= P_{wDC(T)}\left\{1 + \frac{F_{R(To)} - 1}{[1 + \alpha(T - T_o)]^2}\right\} \\ &= P_{wDC(To)}\left\{1 + \alpha(T - T_o) + \frac{F_{R(To)} - 1}{1 + \alpha(T - T_o)}\right\} \end{aligned} \tag{3.10}$$

and

$$\begin{aligned} P_{w(T)} = P_{wDC(T)} F_{R(T)} &= P_{wDC(T)}\left\{1 + \frac{F_{R(To)} - 1}{[1 + \alpha(T - T_o)]^2}\right\} \\ &= P_{wDC(To)}\left\{1 + \alpha(T - T_o) + \frac{F_{R(To)} - 1}{1 + \alpha(T - T_o)}\right\}. \end{aligned} \tag{3.11}$$

The *ampacity*, *current-carrying capacity*, or *current rating* is defined as the maximum rms electric current $I_{rms(max)}$ that a conductor can continuously carry while remaining below its temperature rating. It depends on the conductor electrical resistance, insulation temperature rating, frequency, and waveform shape of the AC current, ambient temperature, and heat transfer ability. Conducting a current above the ampacity (current-carrying limit) may cause overheating and progressive deterioration or immediate destruction of the conductor or its insulation. The maximum rms current density is

$$J_{rms(max)} = \frac{I_{rms(max)}}{A_w}, \tag{3.12}$$

where $I_{rms(max)}$ is the maximum rms applied current.

3.3 Skin Depth

The magnetic field does not penetrate completely into the interior of a conductor at high frequencies. Let us consider a semiinfinite linear, homogeneous, and isotropic conductor shown in Fig. 3.2. A y–z plane crossing $x = 0$ forms an interface between free space and an infinitely thick plane conductor of conductivity $\sigma = 1/\rho$ with $\sigma \gg \omega\,\epsilon$. Assume that y and z conductor dimensions are so large that the magnetic field intensity H has only one component $H = H_z$ in the z-direction and is only a function of x, f, and t in the conductor $H_z(x, f, t)$. A one-dimensional (1D) model can be used in this case. Also, assume that the magnetic field intensity in free space is sinusoidal and parallel to the conductor in the z-direction

$$\mathbf{H}(t) = \mathbf{a_z} H_m \cos \omega t \quad \text{for} \quad x \le 0, \tag{3.13}$$

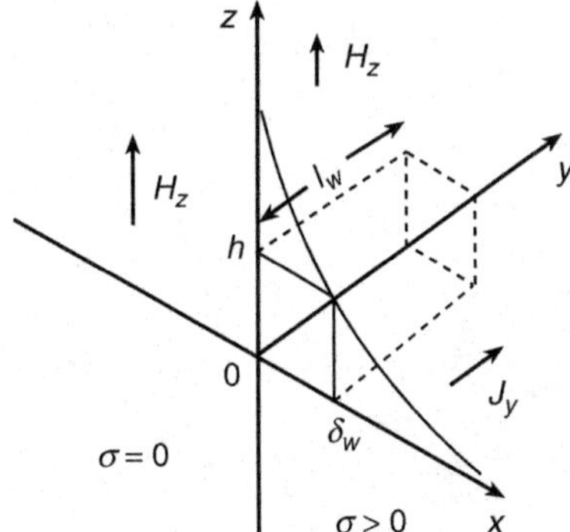

Figure 3.2 Semiinfinite conductor exposed to magnetic field. Exponential decay of magnetic field intensity and current density. The total current flowing through the depth $0 \le x < \infty$ is equivalent to a constant current density $J_m(0)$ flowing through the section $0 \le x \le \delta$

where $H_m = H_z(0)$ is the amplitude of the magnetic field intensity in free space. The magnetic field intensity in the conductor is described by a second-order ordinary differential, linear, and homogeneous equation with a complex coefficient γ for a one-dimensional field distribution in the direction x, which is the scalar *Helmholtz equation*, *one-dimensional diffusion equation*, or the *wave equation* in phasor form

$$\frac{d^2H_z(x)}{dx^2} = j\omega\mu\sigma H_z(x) = \gamma^2 H_z(x) \quad \text{for} \quad x \ge 0 \tag{3.14}$$

where the *complex propagation constant* is given by

$$\gamma^2 = j\omega\mu\sigma \tag{3.15}$$

or

$$\gamma = \sqrt{j\omega\mu\sigma} = (1+j)\sqrt{\frac{\omega\mu\sigma}{2}} = (1+j)\sqrt{\pi\mu\sigma f} = \frac{1+j}{\delta_w} = \frac{\sqrt{2}}{\delta_w}e^{j\frac{\pi}{4}} \tag{3.16}$$

in which

$$\sqrt{j} = \sqrt{e^{j\frac{\pi}{2}}} = e^{j\frac{\pi}{4}} = \cos\left(\frac{\pi}{4}\right) + j\sin\left(\frac{\pi}{4}\right) = \frac{1+j}{\sqrt{2}} \tag{3.17}$$

and δ_w is the *skin depth* or the *depth of penetration* of a conductor given by

$$\delta_w = \sqrt{\frac{2}{\omega\mu\sigma_w}} = \sqrt{\frac{2\rho_w}{\omega\mu}} = \frac{1}{\sqrt{\pi\mu\sigma f}} = \sqrt{\frac{\rho_w}{\pi\mu f}} = \sqrt{\frac{\rho_w}{\pi\mu_r\mu_0 f}}. \tag{3.18}$$

The skin depth decreases when μ_r and f increase and when ρ_w increases. The skin depth describes the degree of penetration of a conductor by the magnetic flux and the eddy current. In other words, the skin depth is the distance at which the amplitude of the electromagnetic wave traveling in a lossy conductor is reduced to $1/e$ of its original value. The skin effect is negligible only if the skin depth δ_w is much greater than the conductor thickness.

The skin depth depends on temperature

$$\delta_{w(T)} = \sqrt{\frac{\rho_{w(T)}}{\pi\mu_{r(T)}\mu_0 f}} = \sqrt{\frac{\rho_{w(T_o)}[1+\alpha(T-T_o)]}{\pi\mu_{r(T)}\mu_0 f}}. \tag{3.19}$$

The skin depth increases with temperature because the resistivity ρ_w increases with temperature and the relative permeability μ_r decreases with temperature T. Figure 3.3 shows the skin depth δ_{Cu} as a function of temperature T for copper.

The skin depth of a copper conductor at $T = 20\,^{\circ}\mathrm{C}$ is

$$\delta_{Cu(20\,^{\circ}\mathrm{C})} = \sqrt{\frac{\rho_{Cu(20\,^{\circ}\mathrm{C})}}{\pi\mu_0 f}} = \sqrt{\frac{1.724\times10^{-8}}{\pi\times4\times\pi\times10^{-7}f}} = \frac{66.083}{\sqrt{f}}\,(\mathrm{mm}) \tag{3.20}$$

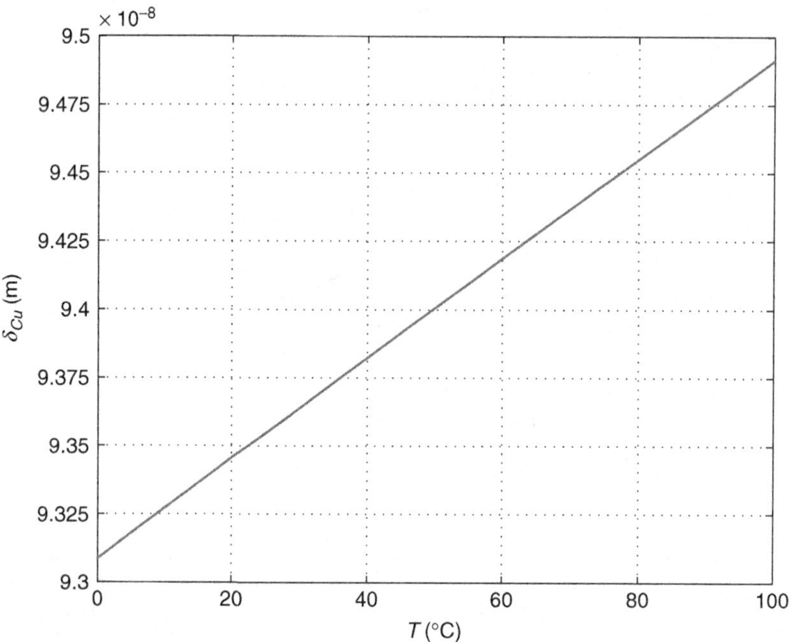

Figure 3.3 Skin depth δ_{Cu} as a function of temperature T for copper

Table 3.2 Skin depth for copper at T = 20 °C

Frequency f	Skin depth δ_{Cu}
60 Hz	8.53 mm
400 Hz	3.3 mm
1 kHz	2.53 mm
10 kHz	0.66 mm
20 kHz	0.467 mm
100 kHz	0.209 mm
1 MHz	0.066 mm
10 MHz	20.9 μm
100 MHz	66 μm
1 GHz	2.09 μm
2 GHz	1.478 μm
10 GHz	0.66 μm

and the skin depth of a copper conductor at $T = 100\,^{\circ}\mathrm{C}$ is

$$\delta_{Cu(100\,^{\circ}\mathrm{C})} = \sqrt{\frac{\rho_{Cu(100\,^{\circ}\mathrm{C})}}{\pi \mu_0 f}} = \sqrt{\frac{2.3 \times 10^{-8}}{\pi \times 4 \times \pi \times 10^{-7} f}} = \frac{76.328}{\sqrt{f}} \,(\mathrm{mm}). \tag{3.21}$$

Figure 3.4 shows the skin depth δ_w as a function of frequency for copper at $T = 20\,^{\circ}\mathrm{C}$. Table 3.2 gives the skin depth for copper at selected frequencies and temperature $T = 20\,^{\circ}\mathrm{C}$.

Solution of (3.14) gives the phasor of the magnetic field intensity in the good conductor

$$H_z(x) = H_m e^{-\gamma x} = H_m e^{-x(1+j)\sqrt{\pi\mu\sigma f}} = H_m e^{-\frac{(1+j)x}{\delta_w}} = H_m e^{-\frac{x}{\delta_w}} e^{-j\frac{x}{\delta_w}}$$
$$= H_m(x) e^{-j\frac{x}{\delta_w}} \quad \text{for} \quad x \geq 0, \tag{3.22}$$

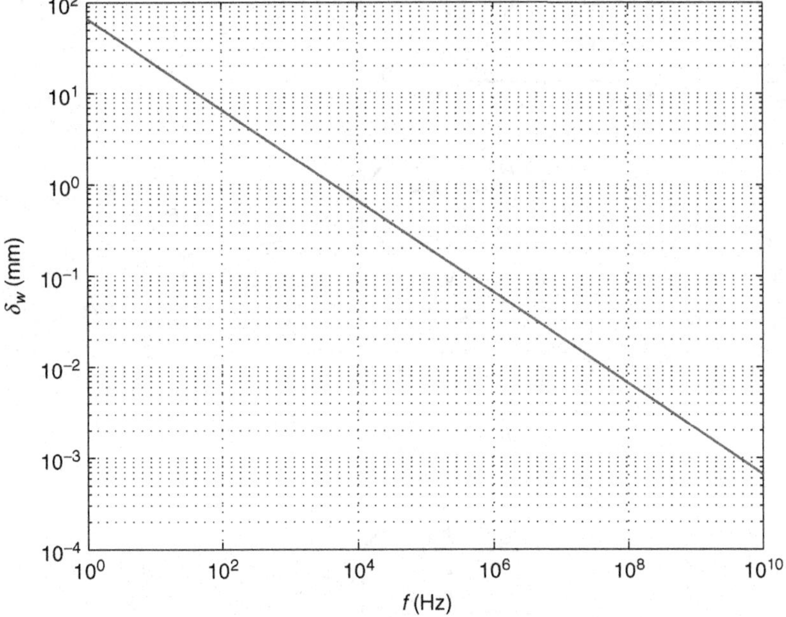

Figure 3.4 Skin depth δ_w as a function of frequency f for copper at $T = 20\,^\circ$C

where the space-dependent amplitude of the magnetic field intensity in the conductor is

$$H_m(x) = H_m e^{-\frac{x}{\delta_w}} \quad \text{for} \quad x \geq 0. \tag{3.23}$$

The wave of the magnetic field is attenuated rapidly with the wavelength much shorter than its value in air. The amplitude of the magnetic field intensity is only a function of x and it does not vary in the transverse direction, that is, in the $y-z$ plane. The magnetic field intensity in the time domain is given by

$$H_z(x,t) = Re\{H_z(x)e^{j\omega t}\} = Re\{H_m e^{-\frac{x}{\delta_w}} e^{j(\omega t - \frac{x}{\delta_w})}\} = H_m e^{-\frac{x}{\delta_w}} \cos\left(\omega t - \frac{x}{\delta_w}\right)$$

$$= H_m(x) \cos\left(\omega t - \frac{x}{\delta_w}\right) \quad \text{for} \quad x \geq 0. \tag{3.24}$$

This equation describes the *traveling wave* of the magnetic field $H(x,t) = H_z(x,t)$ in the conductor in the x-direction in steady state with the velocity $v = \omega\delta_w = \sqrt{4\pi f \rho_w/\mu_0}$. Figure 3.5a shows the amplitude of the magnetic field intensity $H_m(x)$ as a function of x. It can be seen that the amplitude of the magnetic field intensity $H_m(x)$ decreases exponentially with x in the conductor and is reduced to $H_m(0)/e = 0.37\,H_m(0)$ at the distance $x = \delta_w$. The physical explanation of the decrease in H is its partial cancellation by induced current in the positive y-direction. As the frequency increases, the skin depth δ_w decreases and the penetration of the conductor by the magnetic field decreases. The line tangent at $x = 0$ crosses the horizontal axis at $x = \delta_w$. Figure 3.6 shows the waveforms of normalized magnetic field intensity $H(x,\omega t)/H_m$ at $x/\delta_w = 0$, 1, and 2. It can be observed that the phase of the magnetic field wave changes as the wave diffuses into the conductor in addition to the decrease in the amplitude.

The current is induced in the conductor by the magnetic field and flows in the y-direction. Substituting (3.3) into Maxwell's equation, the phasor of the total current density flowing through the

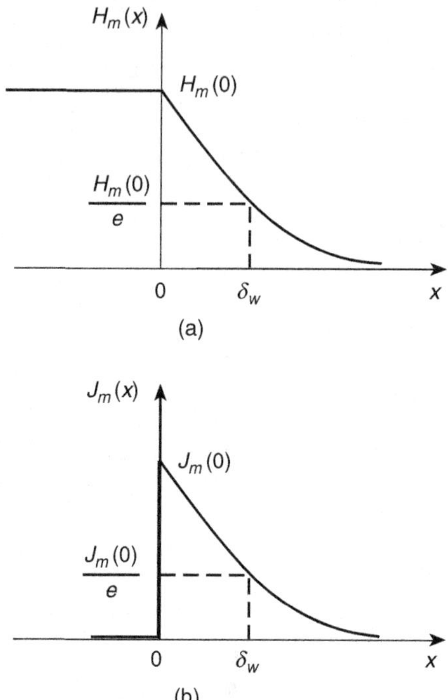

Figure 3.5 Distributions of amplitudes of magnetic field intensity $H_m(x) = H_z(x)$ and current density $J_m(x) = J_y(x)$ at the interface between a free space and a good conductor. (a) $H_m(x)$ as a function of x. (b) $J_m(x)$ as a function of x

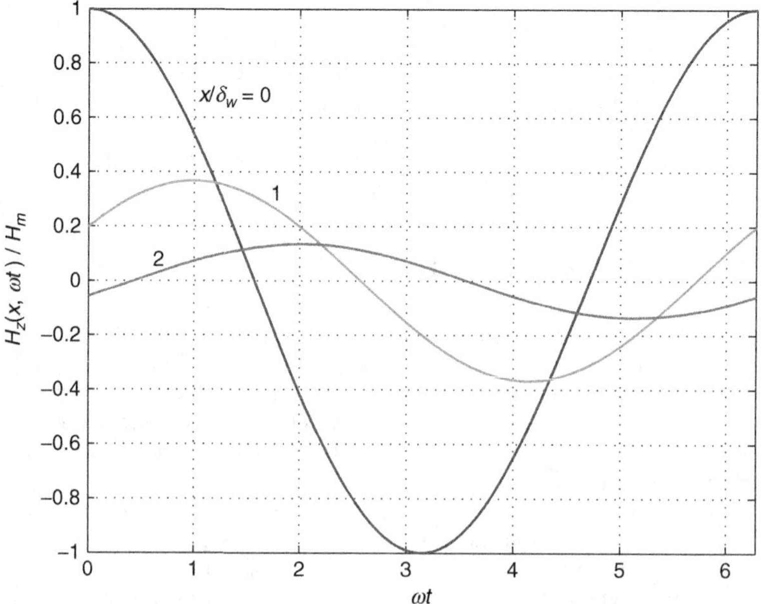

Figure 3.6 Waveforms of normalized magnetic field intensity $H(x, \omega t)/H_m$ at $x/\delta_w = 0$, 1, and 2

resistive and inductive parts of the conductor in the x-direction can be derived as

$$\mathbf{J}(x) = \mathbf{J}_y(x) = -\mathbf{a}_y \frac{\partial H_z(x)}{\partial x} = \mathbf{a}_y H_m \gamma e^{-\gamma x} = \mathbf{a}_y \frac{1+j}{\delta_w} H_m e^{-\frac{x(1+j)}{\delta_w}}$$

$$= \mathbf{a}_y \frac{\sqrt{2}}{\delta_w} H_m e^{-\frac{x}{\delta_w}} e^{j\left(\frac{\pi}{4}-\frac{x}{\delta_w}\right)} = \mathbf{a}_y J_m(x) e^{j\left(\frac{\pi}{4}-\frac{x}{\delta_w}\right)} \quad \text{for} \quad x \geq 0, \tag{3.25}$$

resulting in the current density in the time domain

$$\mathbf{J}(x,t) = Re\{\mathbf{J}(x)e^{j\omega t}\} = \mathbf{a}_y Re\left\{\frac{1+j}{\delta_w} H_m e^{-\frac{x}{\delta_w}} e^{j\left(\omega t-\frac{x}{\delta_w}\right)}\right\}$$

$$= \mathbf{a}_y Re\left\{\frac{\sqrt{2}}{\delta_w} H_m e^{-\frac{x}{\delta_w}} e^{j\left(\omega t+\frac{\pi}{4}-\frac{x}{\delta_w}\right)}\right\} = \mathbf{a}_y \sqrt{2\pi f \mu \sigma} H_m e^{-\frac{x}{\delta_w}} \cos\left(\omega t - \frac{x}{\delta_w} + \frac{\pi}{4}\right)$$

$$= \mathbf{a}_y J_m(0) e^{-\frac{x}{\delta_w}} \cos\left(\omega t - \frac{x}{\delta_w} + \frac{\pi}{4}\right) = \mathbf{a}_y J_m(x) \cos\left(\omega t - \frac{x}{\delta_w} + \frac{\pi}{4}\right) \quad \text{for} \quad x \geq 0, \tag{3.26}$$

where the space-dependent amplitude of the current density is given by

$$J_m(x) = \frac{\sqrt{2} H_m}{\delta_w} e^{-\frac{x}{\delta_w}} = \sqrt{2\pi f \mu \sigma_w} H_m e^{-\frac{x}{\delta_w}} = J_m(0) e^{-\frac{x}{\delta_w}} \quad \text{for} \quad x \geq 0. \tag{3.27}$$

Figure 3.5b depicts the distribution of $J_m(x)$ as a function of x. It can be seen that the amplitude of the current density $J_m(x)$ decreases exponentially with x in the conductor and is reduced to $J_m(0)/e = 0.37\, J_m(0)$ at the distance $x = \delta_w$ and it is reduced to $J_m(0)/e^3 \approx 0.05\, J_m(0)$ for $x = 3\delta_w$. At $x = 5\delta_w$, $J_m = J_m(0)/e^5 \approx 6.7 \times 10^{-3}\, J_m(0)$. As the frequency f increases, the amplitude of the current density on the conductor surface

$$J_m(0) = \frac{\sqrt{2} H_m}{\delta_w} = \sqrt{2\pi f \mu \sigma_w} H_m = \sqrt{\frac{2\pi f \mu}{\rho_w}} H_m \tag{3.28}$$

increases and the skin depth δ_w decreases such that the area under the curve, representing the amplitude of the total current in the conductor, remains constant. In addition, the magnetic field intensity H lags the current density J by $\pi/4$. It can be shown that $0.95I$ flows in the range $0 \leq x \leq 3\delta_w$ and $0.99I$ flows in the range $0 \leq x \leq 5\delta_w$. The phasor of the current density on the conductor surface is

$$J(0) = \frac{(1+j)H_m}{\delta_w} = J_m(0)e^{j\frac{\pi}{4}} = \sqrt{2\pi f \mu \sigma_w} H_m e^{j\frac{\pi}{4}} = \sqrt{\frac{2\pi f \mu}{\rho_w}} H_m e^{j\frac{\pi}{4}}. \tag{3.29}$$

The total time-varying current flowing in the y-direction through a cross-sectional area $0 \leq x < \infty$ and $0 \leq z \leq h$ in the time domain is given by

$$i_y(t) = \int_0^h \int_0^\infty J(x,t)\,dxdz = \int_0^h \int_0^\infty J_m(0) e^{-\frac{x}{\delta_w}} \cos\left(\omega t - \frac{x}{\delta_w} + \frac{\pi}{4}\right) dxdz$$

$$= \frac{J_m(0) h \delta_w}{\sqrt{2}} \cos \omega t = I_m \cos \omega t, \tag{3.30}$$

where

$$I_m = \frac{J_m(0) h \delta_w}{\sqrt{2}} = \frac{J_m(0) A_e}{\sqrt{2}} = H_m h \tag{3.31}$$

in which $A_e = h\delta_w$.

An alternative and useful interpretation of the skin effect is obtained if the total current is assumed to flow uniformly over one skin depth ($0 \leq x \leq \delta_w$) only. The total current I can be found by integrating the actual current density $J_m(x)$ over the infinite depth in the x-direction and h meters into the z-direction. The amplitude of the total current flowing in the y-direction through the conductor cross-sectional area $0 \leq x < \infty$ and $0 \leq z \leq h$ is

$$I_m = \int_{z=0}^{z=h} \int_{x=0}^{x=\infty} J_m(x)\,dxdz = \int_{z=0}^{z=h} \int_{x=0}^{x=\infty} J_m(0) e^{-\frac{x}{\delta_w}} dxdz = J_m(0) h \delta_w = J_m(0) A_e, \tag{3.32}$$

where $A_e = h\delta_w$ is the effective cross-sectional area of a conductor. The area of a rectangle of height $J_m(0)$ and width δ_w is equal to the area under the exponential curve. Thus, the current of uniform density $J_m(0)$ flowing in the rectangular conductor section whose cross-sectional area is $A_e = h\delta_w$ is the same as that with nonuniform density $J_m(x) = J_m(0)e^{-\frac{x}{\delta_w}}$ flowing in the area of thickness h and infinite depth. Therefore, we may compute the effective resistance as that of a conductor of thickness δ_w, in which the current density is uniform, equal to that at the surface $J_m(0)$. In other words, the resistance of a flat slab much thicker than δ_w to AC current is exactly equal to a plate of thickness δ_w to DC current. The equivalent resistance of the strip with a uniform current density (like that current at low frequencies) is given by

$$R_w = R_{LF(equiv)} = \frac{\rho_w l_w}{A_e} = \frac{\rho_w l_w}{h\delta_w} = \frac{l_w}{h}\sqrt{\pi\mu\rho_w f}, \tag{3.33}$$

where the effective cross-sectional area of the strip is $A_e = h\delta_w$ and the length of the strip is l_w. As the frequency f increases, the skin depth δ_w decreases, reducing the effective area A_e. Therefore, the effective resistance R_w increases with frequency f.

Since $\mathbf{J_m}(x)\mathbf{J_m^*}(x) = |J_m(x)|^2$, the total time-average power dissipated in the volume $0 \le x < \infty$, $0 \le y \le l_w$, and $0 \le z \le h$ is given by

$$P_D = \frac{1}{2}\int\int\int_V \rho_w |J_m(x)|^2 dV = \frac{\rho_w}{2}\int_0^{l_w}\int_0^h\int_0^\infty \left|\frac{(1+j)H_m}{\delta_w}e^{-\frac{x}{\delta_w}(1+j)}\right|^2 dxdydz$$

$$= \frac{l_w h \rho_w H_m^2}{2\delta_w} = \frac{l_w h H_m^2}{2}\sqrt{\pi f\mu\rho_w} = \frac{\rho_w J_m^2(0) l_w h\delta_w}{4}. \tag{3.34}$$

The time-average power dissipated in the strip of length l_w, height h, and depth δ_w is

$$P_D = I_{rms}^2 R_w = \left(\frac{H_m h}{\sqrt{2}}\right)^2 R_w = \frac{[J_m(0)A_e]^2}{4}R_w = \frac{J_m^2(0)A_e^2}{4}\frac{\rho_w l_w}{A_e} = \frac{\rho_w J_m^2(0)A_e l_w}{4}$$

$$= \frac{\rho_w J_m^2(0)\delta_w h l_w}{4} = \frac{\rho_w J_m^2(0)}{4}Vol = \frac{\rho_w H_m^2}{2\delta_w^2}Vol = \frac{\pi\mu f H_m^2}{2}Vol = \frac{h l_w H_m^2}{2}\sqrt{\pi\mu\rho_w f}, \tag{3.35}$$

where $Vol = \delta_w h l_w$ is the volume of the strip. Thus, the power dissipated in the strip is equal to the power dissipated in the conductor with infinite depth. The time-average power density decays with distance x as

$$p_D(x) = \frac{\rho_w}{2}|J(x)|^2 = \frac{\rho_w J_m^2(0)}{2}e^{-\frac{2x}{\delta_w}} = \frac{\rho_w H_m^2}{\delta_w^2}e^{-\frac{2x}{\delta_w}}$$

$$= \pi\mu f H_m^2 e^{-\frac{2x}{\delta_w}} = p_D(0)e^{-\frac{2x}{\delta_w}} \quad \text{for} \quad x \ge 0. \tag{3.36}$$

The power density has its maximum value at the interface ($x = 0$) and decays with increasing x. For $x = \delta_w$, the power density decreases by a factor $1/e^2 \approx 0.1369$.

The amplitude of the electric field at the conductor surface is

$$E_m(0) = \rho_w J(0). \tag{3.37}$$

The voltage across the length l_w at the interface of free space and the conductor is

$$V_m = E_m(0)l_w = \rho_w J(0)l_w = \rho_w l_w \frac{(1+j)H_m}{\delta_w}. \tag{3.38}$$

Hence, the impedance of a slab of length l_w and height h is given by

$$Z = \frac{V_m}{I_m} = \frac{\rho_w(1+j)}{\delta_w}\frac{l_w}{h} = \frac{\rho_w}{\delta_w}\frac{l_w}{h} + j\frac{\rho_w}{\delta_w}\frac{l_w}{h} = R_w + jX_L = R_w + j\omega L, \tag{3.39}$$

where the AC resistance R_w and the reactance $X_L = \omega L$ of the slab are

$$R_w = \omega L = \frac{\rho_w}{\delta_w}\frac{l_w}{h} = \frac{\rho_w l_w}{A_e} = \frac{l_w}{h}\sqrt{\pi\mu\rho_w f} \tag{3.40}$$

and the inductance of the slab is

$$L = \frac{X_L}{\omega} = \frac{\rho_w l_w}{\omega \delta_w h} = \frac{l_w}{2h}\sqrt{\frac{\mu \rho_w}{\pi f}}. \quad (3.41)$$

Thus, the expression for the AC resistance R_w is equivalent to that for the DC resistance of a planar conductor of length l_w and cross-sectional area $A_e = h\delta_w$. Both the AC resistance R_w and the reactance X_L of the slab are proportional to $\sqrt{f}$, and the inductance L of the slab is inversely proportional to $\sqrt{f}$.

For all more complex geometrical structures, the analysis of the skin effect is much more difficult and the expressions for the skin depth are much more complicated.

3.4 AC-to-DC Winding Resistance Ratio

The DC winding resistance is given by

$$R_{wDC} = \frac{\rho_w l_w}{A_w}, \quad (3.42)$$

where ρ_w is the winding conductor resistivity, l_w is the winding conductor length, and A_w is the cross-sectional area of the winding conductor. The AC winding resistance can be expressed by

$$R_w = R_{wDC} + \Delta R_w = R_{wDC}\left(1 + \frac{\Delta R_w}{R_{wDC}}\right) = F_R R_{wDC} = F_R \frac{\rho_w l_w}{A_w}. \quad (3.43)$$

The *DC-to-AC resistance ratio* or *AC resistance factor* is defined as the AC-to-DC winding resistance ratio

$$F_R = \frac{R_w}{R_{wDC}} = \frac{R_{wDC} + \Delta R_w}{R_{wDC}} = 1 + \frac{\Delta R_w}{R_{wDC}}. \quad (3.44)$$

When an inductor or a transformer is operated at DC or at a low frequency, that is, a frequency at which the winding conductor skin depth δ_w is larger than the conductor diameter or thickness the current density is uniform, and the winding power loss can be expressed by

$$P_{wDC} = R_{wDC} I_{DC}^2, \quad (3.45)$$

where $I_{rms} = I_{DC}$ is the rms value of the current flowing in the conductor and R_{wDC} is the DC resistance of the winding. When the operating frequency is high, the current density is not uniform, the winding AC resistance R_w is higher than the DC resistance R_{wDC} due to the skin and proximity effects, and the winding power loss is given by

$$P_w = R_w I_{rms}^2 = F_R R_{wDC} I_{rms}^2 = F_R \frac{\rho_w l_w}{A_w} I_{rms}^2. \quad (3.46)$$

From (3.45) and (3.46), the ratio of the winding AC power loss to the winding DC power loss is found to be equal to the ratio of the AC resistance to the DC resistance of the conductor of the same cross-sectional area, carrying currents of the same rms value. For $I_{rms} = I_{DC}$, the AC-to-DC resistance ratio or the AC-to-DC power loss ratio is given by

$$F_R = \frac{P_w}{P_{wDC}} = \frac{R_w I_{rms}^2}{R_{wDC} I_{DC}^2} = \frac{R_w I_{rms}^2}{R_{wDC} I_{rms}^2} = \frac{R_w}{R_{wDC}}. \quad (3.47)$$

Hence, eddy currents increase the AC copper loss over the DC copper loss by a factor of F_R.

3.5 Skin Effect in Long Single Round Conductor

The skin effect increases the AC resistance at high frequencies by restricting the conducting area of the wire to a thin skin on its surface. In a single, long, straight, isolated, solid, round conductor

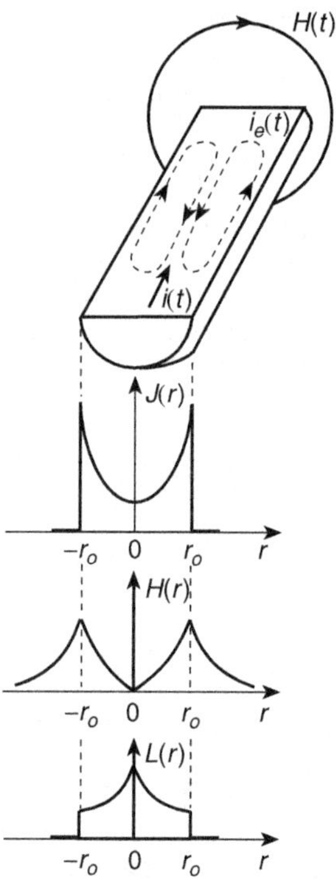

Figure 3.7 Skin effect in a long straight round single conductor at high frequencies

placed in free space and carrying a time-varying current $i(t)$, the net current tends to flow near the surface and less in the middle due to the skin effect, as depicted in Fig. 3.7. The current $i(t)$ generates the circular magnetic field $H(t)$ around the current path, both inside and outside the conductor. This field, in turn, generates opposing eddy currents within the conductor, which flow in opposite direction to the applied current $i(t)$ in the center of the wire and in the same direction near the surface. As a result, there is cancellation of current flow in the center of the conductor and a concentration of current flow near the surface. The largest current density is at the surface of the conductor and it decays with distance toward the interior, as illustrated in Fig. 3.7. The net current is the sum of the eddy current and the applied current. It is cancelled at the center of the wire and crowded near the outer skin. Thus, the current distribution is nonuniform. As a result, the total current-carrying area is less than the area of the full wire and therefore the AC resistance is higher than the DC resistance. As the frequency increases, the skin depth decreases. Therefore, the current density becomes more nonuniform. At VHFs, the current flows in a thin skin at the surface, and the inner region of the conductor plays no role in the current conduction.

An alternate explanation of the current distribution in a conductor at high frequencies can be made using the concept of distributed inductance. Consider two small isolated conductor elements of the same area, one close to the center and the other close to the surface of the bigger conductor. The concentric magnetic flux lines within the conductor link the current with the current toward the center, which is being linked by more flux than the current toward the surface. The inductance of the inner small conductor is larger than that of the outer conductor. The voltage drop per unit length is the same for both small conductors. With a larger inductance in the center and a smaller inductance toward the surface, the current density redistributes itself so that a smaller current flows in the center, where the reactance is higher, whereas a larger current flows at the surface, where the reactance is lower. The skin effect does not change the magnitude of the total current in the conductor. It only changes the current density. The skin effect reduces the inductance because the nonuniform current density alters the internal magnetic flux linkage.

3.6 Current Density in Single Round Conductor

3.6.1 Bessel Differential Equation

Consider a single, straight, solitary, round, and solid conductor of radius $r_o = d/2$ and length l_w that carries a sinusoidal current

$$i = I \cos \omega t \tag{3.48}$$

of amplitude I and frequency $f = \omega/(2\pi)$. The conductor is infinitely long, that is, $l_w \gg r_o$. The current generates a concentric, circular magnetic field $H(r, \omega t)$ inside the wire and in the surrounding space. The current density inside the round conductor is given by

$$J(x,t) = J(r)\cos(\omega t + \Psi), \tag{3.49}$$

where $J(r)$ is the amplitude and Ψ is the phase of the current density.

Bessel functions are often used to solve numerous physical problems in space and time, involving cylindrical and spherical geometries. The derivation of the Bessel equation for round solid conductor is given in the appendix at the end of this chapter. The analytical solution of Maxwell's equations in cylindrical coordinates for the round conductor leads to the 1D current density distribution as a function of the conductor radius r. The current density inside the conductor $J_z = J(r)$ is described by the *modified Bessel second-order ordinary differential equation* in the phasor form with a real variable r (i.e., the conductor radius) and a complex coefficient $k = \sqrt{j\omega\mu\sigma}$ in cylindrical coordinates

$$\frac{d^2J}{dr^2} + \frac{1}{r}\frac{dJ}{dr} - k^2 J = 0, \tag{3.50}$$

where the square of the complex propagation constant is given by

$$k^2 = \gamma^2 = j\omega\mu\sigma = \frac{j\omega\mu}{\rho_w} = j\frac{2}{\delta_w^2}, \tag{3.51}$$

which depends on the current frequency f and the medium properties σ and μ. Since

$$\sqrt{\omega\mu\sigma} = \sqrt{2\pi f\mu\sigma} = \frac{\sqrt{2}}{\delta_w} \tag{3.52}$$

and

$$\sqrt{j} = \sqrt{\frac{2j}{2}} = \sqrt{\frac{1+2j-1}{2}} = \sqrt{\frac{1+2j+j^2}{2}} = \sqrt{\frac{(1+j)^2}{2}} = \frac{1+j}{\sqrt{2}}, \tag{3.53}$$

the complex propagation constant can be expressed by

$$k = \sqrt{j\omega\mu\sigma} = \sqrt{j2\frac{\omega\mu\sigma}{2}} = \sqrt{j}\frac{\sqrt{2}}{\delta_w} = \frac{1+j}{\delta_w}. \tag{3.54}$$

The solution of modified Bessel equation in (3.50) for steady state is given by

$$\begin{aligned}
J(r) &= AI_0(kr) = AI_0(\sqrt{j\omega\mu\sigma}r) = AI_0\left(\sqrt{j}\frac{\sqrt{2}r}{\delta_w}\right) = AI_0\left[(1+j)\left(\frac{r_o}{\delta_w}\right)\left(\frac{r}{r_o}\right)\right] \\
&= A\left(1 + \frac{k^2r^2}{2^2} + \frac{k^4r^4}{2^4(2!)^2} + \frac{k^6r^6}{2^6(3!)^2} + \frac{k^8r^8}{2^8(4!)^2} + \dots\right) \\
&= A\left(1 + j\frac{\omega\mu\sigma r^2}{2^2} - \frac{\omega^2\mu^2\sigma^2r^4}{2^4(2!)^2} - j\frac{\omega^3\mu^3\sigma^3r^6}{2^6(3!)^2} + \dots\right),
\end{aligned} \tag{3.55}$$

where

$$kr = \sqrt{j\omega\mu\sigma}r = \alpha\sqrt{j} = (1+j)\frac{r}{\delta_w} = \sqrt{2j}\left(\frac{r_o}{\delta_w}\right)\left(\frac{r}{r_o}\right) = (1+j)\left(\frac{r_o}{\delta_w}\right)\left(\frac{r}{r_o}\right), \tag{3.56}$$

I_0 is the modified Bessel function of the first kind and order zero, and A is a constant, determined by the initial condition. For $r = 0$, $kr = 0$ and $I_0(0) = 1$. Therefore, the current density in the center of the conductor $r = 0$ is given by

$$J(0) = AI_0(0) = A(1 + 0 + 0 + 0 + \ldots) = A, \tag{3.57}$$

resulting in the solution of the Bessel equation

$$J(r) = J_0(0)I_0(kr). \tag{3.58}$$

The first term in (3.55) represents the current density through the DC-resistive part of the conductor. The terms with powers 4, 8, ... represent the current density through the AC-resistive part of the conductor. The terms with powers 2, 6, ... represent the current density through the inductive part of the conductor.

When m is an integer, the modified Bessel function of the first kind of order m can be written as a power series

$$\begin{aligned} I_m(kr) &= \sum_{n=0}^{\infty} \frac{1}{n!(m+n)!}\left(\frac{kr}{2}\right)^{m+2n} = \left(\frac{kr}{2}\right)^m \sum_{n=0}^{\infty} \frac{1}{n!(m+n)!}\left(\frac{kr}{2}\right)^{2n} \\ &= \left(\frac{kr}{2}\right)^m \frac{1}{m!}\left[1 + \left(\frac{kr}{2}\right)^2 \frac{1}{1!(m+1)} + \left(\frac{kr}{2}\right)^4 \frac{1}{2!(m+1)(m+2)} \right. \\ &\quad \left. + \left(\frac{kr}{2}\right)^6 \frac{1}{3!(m+1)(m+2)(m+3)} + \ldots\right]. \end{aligned} \tag{3.59}$$

For $m = 0$ and $m = 1$, this power series becomes

$$\begin{aligned} I_0(kr) &= \sum_{n=0}^{\infty} \frac{1}{(n!)^2}\left(\frac{kr}{2}\right)^{2n} \\ &= 1 + \left(\frac{kr}{2}\right)^2 \frac{1}{(1!)^2} + \left(\frac{kr}{2}\right)^4 \frac{1}{(2!)^2} + \left(\frac{kr}{2}\right)^6 \frac{1}{(3!)^2} + \left(\frac{kr}{2}\right)^8 \frac{1}{(4!)^2} + \ldots \\ &= 1 + \frac{(kr)^2}{4} + \frac{(kr)^4}{64} + \frac{(kr)^6}{384} + \frac{(kr)^8}{6144} + \ldots \end{aligned} \tag{3.60}$$

and

$$\begin{aligned} I_1(kr) = I_0'(kr) &= \sum_{n=0}^{\infty} \frac{1}{n!(n+1)!}\left(\frac{kr}{2}\right)^{2n+1} \\ &= \frac{kr}{2} + \left(\frac{kr}{2}\right)^3 \frac{1}{1!2!} + \left(\frac{kr}{2}\right)^5 \frac{1}{2!3!} + \left(\frac{kr}{2}\right)^7 \frac{1}{3!4!} + \left(\frac{kr}{2}\right)^9 \frac{1}{4!5!} + \ldots \\ &= \frac{kr}{2} + \frac{(kr)^3}{16} + \frac{(kr)^5}{384} + \frac{(kr)^7}{18\,432} + \ldots = \frac{kr}{2}\left[1 + \frac{(kr)^2}{8} + \frac{(kr)^3}{192} + \frac{(kr)^5}{9216} + \ldots\right]. \end{aligned} \tag{3.61}$$

3.6.2 Kelvin Functions

Let us introduce a new quantity

$$\alpha = r\sqrt{\omega\mu\sigma} = \frac{\sqrt{2}r}{\delta_w} = \sqrt{2}\left(\frac{r_o}{\delta_w}\right)\left(\frac{r}{r_o}\right). \tag{3.62}$$

The modified Bessel function can be expressed by Kelvin functions *ber* and *bei*

$$I_0 = ber(\alpha) + jbei(\alpha), \tag{3.63}$$

where the real part of the modified Bessel function I_0 is

$$ber(\alpha) = Re\{I_0\} = \sum_{n=0}^{\infty} \frac{(-1)^n \alpha^{4n}}{2^{4n}[(2n)!]^2} = \frac{1}{2}\left[I_0\left(\alpha e^{j\frac{\pi}{4}}\right) + I_0\left(\alpha e^{-j\frac{\pi}{4}}\right)\right] \tag{3.64}$$

and the imaginary part of the modified Bessel function I_0 is

$$bei(\alpha) = Im\{I_0\} = \sum_{n=0}^{\infty} \frac{(-1)^n \alpha^{2(2n+1)}}{2^{2(2n+1)}[(2n+1)!]^2} = \frac{1}{2j}\left[I_0\left(\alpha e^{j\frac{\pi}{4}}\right) - I_0\left(\alpha e^{-j\frac{\pi}{4}}\right)\right]. \tag{3.65}$$

The solution of the Bessel equation can be expressed by

$$\begin{aligned} J(r) &= A\left[\left(1 - \frac{\alpha^4}{2^4(2!)^2} + \frac{\alpha^8}{2^8(4!)^2} + \ldots\right) + j\left(\frac{\alpha^2}{2^2} - \frac{\alpha^6}{2^6(3!)^2} + \ldots\right)\right] \\ &= A[ber(\alpha) + jbei(\alpha)] = J(0)[ber(\alpha) + jbei(\alpha)]. \end{aligned} \tag{3.66}$$

3.6.3 Approximations of Bessel's Equation Solution

For low frequencies, $r_o/\delta_w \ll 1$, $kr \ll 1$, and the modified Bessel function can be approximated by taking only first several terms of the series. Taking only first two terms,

$$I_0(kr) \approx 1 - \frac{\alpha^4}{2^4(2!)^2} = 1 - \frac{(\sqrt{\mu\sigma\omega}r)^4}{64} = 1 - \frac{(\mu\sigma\omega)^2 r^4}{64} = 1 - \frac{1}{32}\frac{r^4}{\delta_w^2} \quad \text{for} \quad \frac{r_o}{\delta_w} \leq 2.5. \tag{3.67}$$

For high frequencies, $r_o/\delta_w \gg 1$, $kr \gg 1$, and theoretically an infinite number of terms of the modified Bessel function is required. In this case, a useful approximation of the modified Bessel function is given by

$$I_m(kr) \approx \frac{e^{kr}}{\sqrt{2\pi kr}}\left(1 - \frac{4m^2 - 1}{8kr}\right). \tag{3.68}$$

For the modified Bessel function of the first kind and order zero, $m = 0$, and therefore

$$I_0(kr) \approx \frac{e^{kr}}{\sqrt{2\pi kr}}\left(1 + \frac{1}{8kr}\right). \tag{3.69}$$

Another approximation of the modified Bessel function is

$$I_m(kr) = \sqrt{\frac{2}{\pi kr}} \cos\left(\pi kr - \frac{m\pi}{2} - \frac{\pi}{4}\right). \tag{3.70}$$

For $m = 0$ and $m = 1$,

$$I_0(kr) = \sqrt{\frac{2}{\pi kr}} \cos\left(\pi kr - \frac{\pi}{4}\right) \tag{3.71}$$

and

$$I_1(kr) = \sqrt{\frac{2}{\pi kr}} \cos\left(\pi kr - \frac{3\pi}{4}\right). \tag{3.72}$$

3.6.4 Current Density J(r)/J(0)

The current density normalized with respect to the current density in the conductor center is

$$\begin{aligned} \frac{J(r)}{J(0)} &= I_0(kr) = I_0\left[(1+j)\left(\frac{r_o}{\delta_w}\right)\left(\frac{r}{r_o}\right)\right] = \left|\frac{J(r)}{J(0)}\right| e^{j\theta} = ber(\alpha) + jbei(\alpha) \\ &= ber\left(\frac{\sqrt{2}r}{\delta_w}\right) + jbei\left(\frac{\sqrt{2}r}{\delta_w}\right), \end{aligned} \tag{3.73}$$

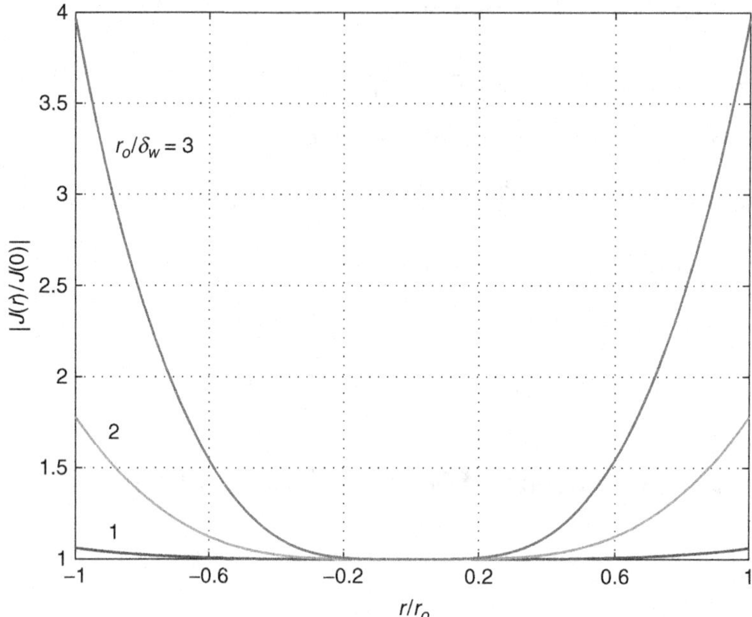

Figure 3.8 Normalized magnitude of the current density $|J(r)/J(0)|$ as a function of r/r_o of a solitary single round conductor at various values of r_o/δ_w

where kr is given by (3.56). Figures 3.8 and 3.9 show the normalized magnitude of the current density $|J(r)/J(0)|$ and the phase θ as functions of r/r_o at various values of r_o/δ_w. The current density $J(r)$ is nearly uniform for $r_o/\delta_w \leq 1$, whereas it is not uniform for $r_o/\delta_w > 1$. The amplitude of the current density at the surface $J(r_o)$ is higher than that at the center $J(0)$. In addition, the current density at the surface lags the current density at the center.

3.6.5 Current Density $J(r)/J(r_o)$

The current density on the conductor surface ($r = r_o$) is expressed as

$$J(r_o) = AI_0(kr_o) \tag{3.74}$$

producing

$$A = \frac{J(r_o)}{I_0(kr_o)} \tag{3.75}$$

and

$$J(r) = AI_0(kr) = \frac{J(r_o)}{I_0(kr_o)} I_0(kr). \tag{3.76}$$

Hence, the current density normalized with respect to the current density at the surface is

$$\frac{J(r)}{J(r_o)} = \frac{I_0(kr)}{I_0(kr_o)} = \frac{I_0\left[(1+j)\left(\frac{r_o}{\delta_w}\right)\left(\frac{r}{r_o}\right)\right]}{I_0\left[(1+j)\left(\frac{r_o}{\delta_w}\right)\right]} = \left|\frac{J(r)}{J(r_o)}\right| e^{j\Psi} = \frac{ber(\alpha) + jbei(\alpha)}{ber(\alpha_o) + jbei(\alpha_o)}, \tag{3.77}$$

where

$$\alpha_o = \frac{d}{\sqrt{2}\delta_w} = \frac{\sqrt{2}r_o}{\delta_w}, \tag{3.78}$$

$$kr_o = \sqrt{j}\alpha_o = \sqrt{2j}\left(\frac{r_o}{\delta_w}\right) = (1+j)\left(\frac{r_o}{\delta_w}\right) \tag{3.79}$$

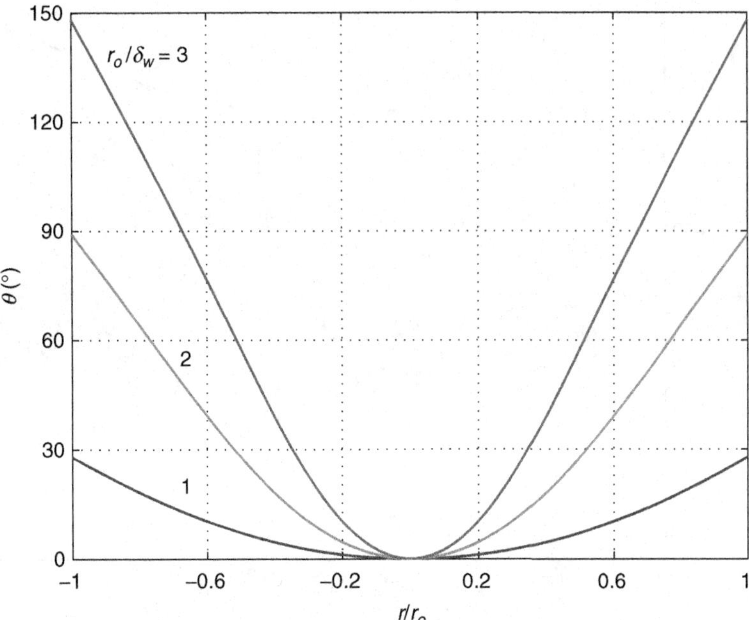

Figure 3.9 Phase θ of the normalized current density $J(r)/J(0)$ as a function of r/r_o of a solitary single round conductor at various values of r_o/δ_w

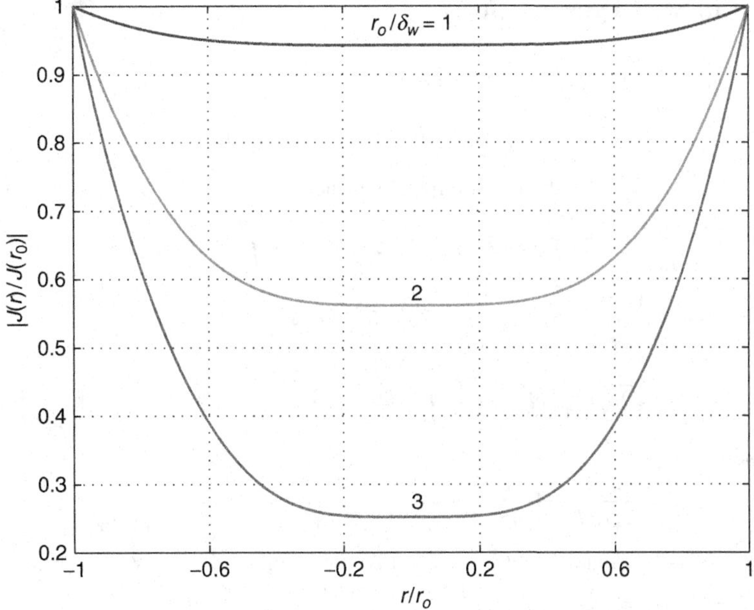

Figure 3.10 Normalized magnitude of the current density $|J(r)/J(r_o)|$ as a function of r/r_o of a solitary single round conductor at various values of r_o/δ_w

and kr is given by (3.56). Figures 3.10 and 3.11 show the normalized magnitude of the current density $|J(r)/J(r_o)|$ and the phase Ψ as functions of r/r_o for selected values of r_o/δ_w. For $r_o/\delta_w \le 1$, the current density is nearly uniform. For $r_o/\delta_w > 1$, the current density in the center is lower than that at the surface and the current density at the center leads the current density at the surface.

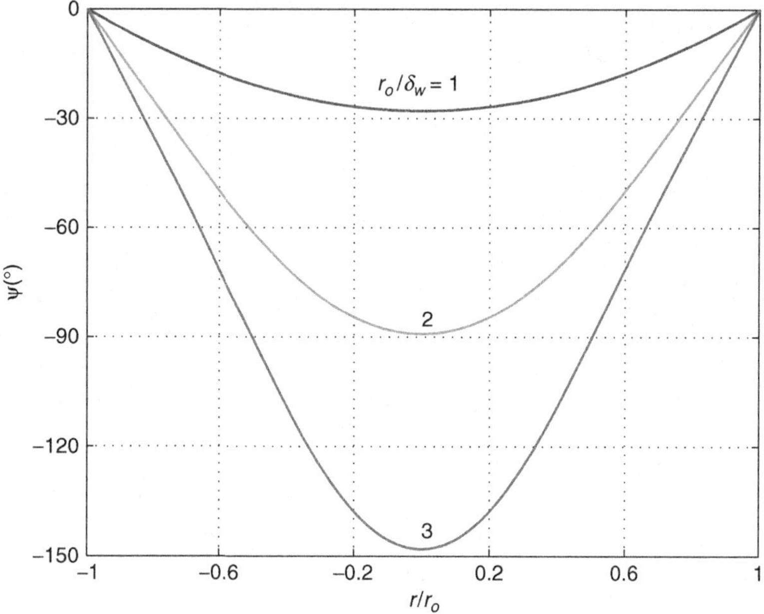

Figure 3.11 Phase Ψ of the normalized current density distribution $J(r)/J(r_o)$ as a function of r/r_o of a solitary single round conductor at various values of r_o/δ_w

3.6.6 Current Density $J(r)/J_{DC}$

Using (3.55) and the relationship

$$\int xI_0(x)dx = xI_0'(x) = xI_1(x) \tag{3.80}$$

we obtain the current flowing through the round conductor

$$\begin{aligned} I &= \int_0^{r_o} J(r)2\pi r dr = \int_0^{r_o} AI_0(kr)2\pi r dr = \frac{1}{k^2}\int_0^{kr_o} AI_0(kr)2\pi(kr)d(kr) \\ &= \frac{2\pi A}{k^2}\int_0^{kr_o} I_0(kr)(kr)d(kr) = A\frac{2\pi r_o}{k}I_1(kr_o) \\ &= A\pi\sqrt{2}\delta_w r_o\left[bei'\left(\frac{\sqrt{2}r_o}{\delta_w}\right) - jber'\left(\frac{\sqrt{2}r_o}{\delta_w}\right)\right] \end{aligned} \tag{3.81}$$

which gives

$$A = \frac{k}{2\pi r_o}\frac{I}{I_1(kr_o)} = \frac{kH(r_o)}{I_1(kr_o)} = \frac{I}{\sqrt{2}\pi\delta_w r_o[bei'(\alpha_o) - jber'(\alpha_o)]}, \tag{3.82}$$

where $H(r_o) = I/(2\pi r_o)$ and I_1 is the modified Bessel function of the first kind of order 1. Hence, the analytical solution of (3.50) for steady state is given by

$$J(r) = \frac{Ik}{2\pi r_o}\frac{I_0(kr)}{I_1(kr_o)} = \frac{I}{\sqrt{2}\pi\delta_w r_o}\frac{ber(\alpha) + jbei(\alpha)}{[bei'(\alpha_o) - jber'(\alpha_o)]} \quad \text{for} \quad 0 \le r \le r_o. \tag{3.83}$$

The DC and low-frequency current density in the round conductor is

$$J_{DC} = \frac{I}{\pi r_o^2}. \tag{3.84}$$

Thus, the normalized current density distribution is given by

$$\frac{J(r)}{J_{DC}} = \frac{kr_o}{2}\frac{I_0(kr)}{I_1(kr_o)} = \frac{(1+j)}{2}\left(\frac{r_o}{\delta_w}\right)\frac{I_0(kr)}{I_1(kr_o)} = \frac{(1+j)}{2}\left(\frac{r_o}{\delta_w}\right)\frac{I_0\left[(1+j)\left(\frac{r_o}{\delta_w}\right)\left(\frac{r}{r_o}\right)\right]}{I_1\left[(1+j)\left(\frac{r_o}{\delta_w}\right)\right]}$$

$$= \left|\frac{J(r)}{J_{DC}}\right| e^{j\phi} = \frac{\alpha_o}{2}\frac{I_0(kr)}{I_1(kr_o)} = \frac{r_o}{\sqrt{2}\delta_w}\frac{ber(\alpha) + jbei(\alpha)}{bei'(\alpha_o) - jber'(\alpha_o)} = \frac{\alpha_o}{2}\frac{ber(\alpha) + jbei(\alpha)}{bei'(\alpha_o) - jber'(\alpha_o)}$$

$$= \frac{\alpha_o}{2}\frac{ber(\alpha)bei'(\alpha_o) - bei(\alpha)ber'(\alpha_o) + j[ber(\alpha)ber'(\alpha_o) + bei(\alpha)bei'(\alpha_o)]}{[ber'(\alpha_o)]^2 + [bei'(\alpha_o)]^2}. \quad (3.85)$$

Figures 3.12 and 3.13 show the normalized amplitude of current density $|J(r)/J_{DC}|$ and phase ϕ as functions of r/r_o at different values of r_o/δ_w. For $r_o/\delta_w \leq 1$, the current density $|J|$ is nearly uniform and the skin effect loss is negligible.

The normalized current density distribution on the surface of the round conductor is given by

$$\frac{J(r_o)}{J_{DC}} = \frac{kr_o}{2}\frac{I_0(kr_o)}{I_1(kr_o)} = \frac{(1+j)}{2}\left(\frac{r_o}{\delta_w}\right)\frac{I_0(kr_o)}{I_1(kr_o)} = \frac{(1+j)}{2}\left(\frac{r_o}{\delta_w}\right)\frac{I_0\left[(1+j)\left(\frac{r_o}{\delta_w}\right)\right]}{I_1\left[(1+j)\left(\frac{r_o}{\delta_w}\right)\right]}$$

$$= \left|\frac{J(r_o)}{J_{DC}}\right| e^{j\phi(r_o)}. \quad (3.86)$$

For $r_o/\delta_w = 1$, $|J(r_o)/J_{DC}| = 1.1$. For $r_o/\delta_w = 2$, $|J(r_o)/J_{DC}| = 1.55$. For $r_o/\delta_w = 3$, $|J(r_o)/J_{DC}| = 2.3$. For $r_o/\delta_w = 5$, $|J(r_o)/J_{DC}| = 3.7$. Figure 3.14 shows the normalized current density on the conductor surface $|J(r_o)/J_{DC}|$ as a function of r_o/δ_w.

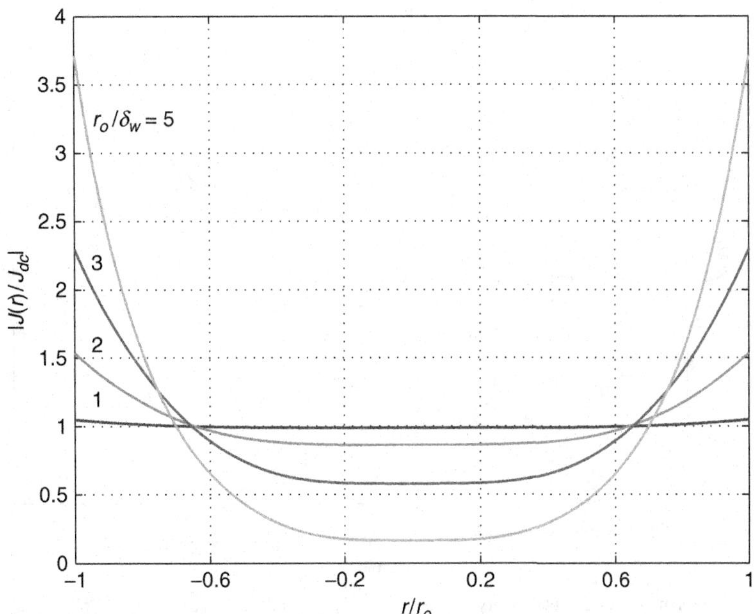

Figure 3.12 Normalized magnitude of the current density $|J(r)/J_{DC}|$ as a function of r/r_o of a solitary single round conductor at various values of r_o/δ_w

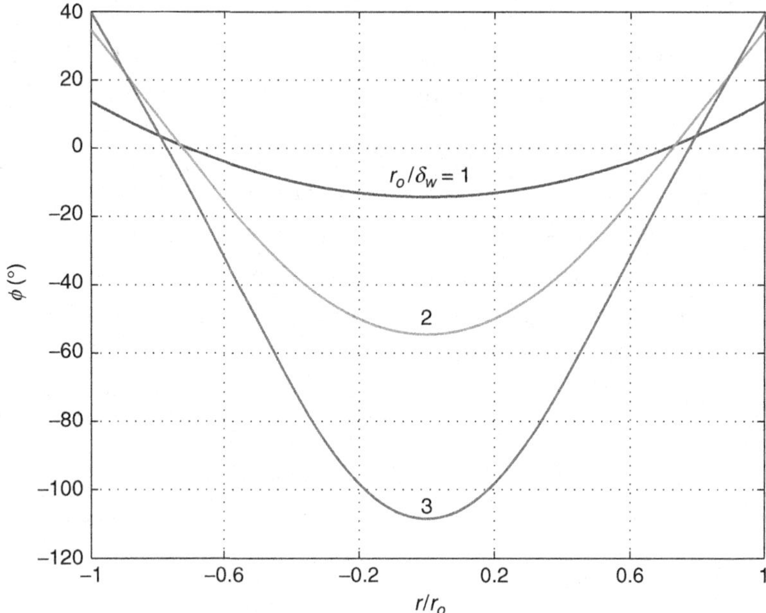

Figure 3.13 Phase ϕ of the normalized current density $J(r)/J_{DC}$ as a function of r/r_o of a solitary single round conductor at various values of r_o/δ_w

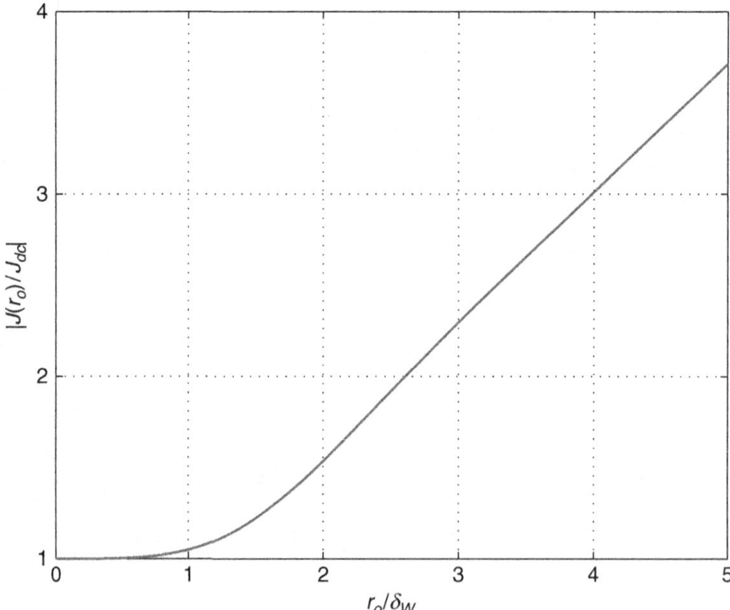

Figure 3.14 Normalized current density on the surface of round conductor $|J(r_o)/J_{DC}|$ of as a function of r_o/δ_w

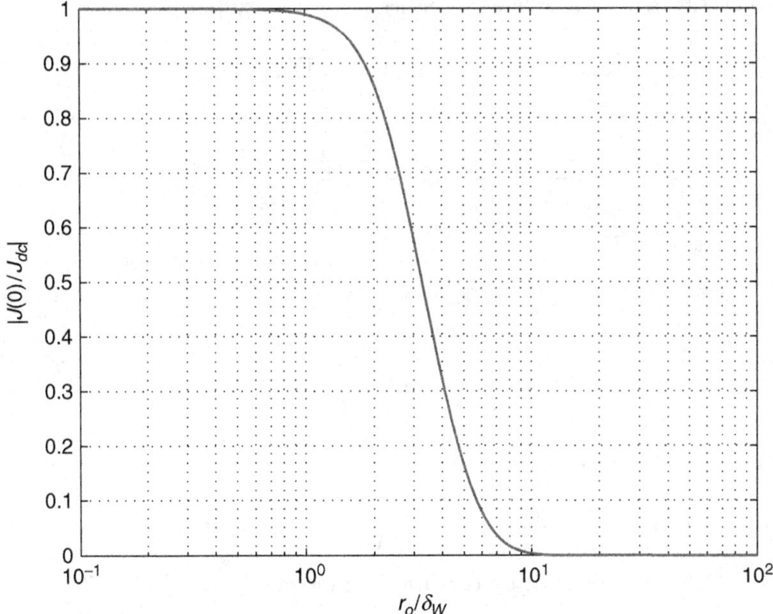

Figure 3.15 Normalized current density at the center of round conductor $|J(0)/J_{DC}|$ of as a function of r_o/δ_w

The normalized current density distribution at the center of the round conductor is given by

$$\frac{J(0)}{J_{DC}} = \frac{kr_o}{2}\frac{I_0(0)}{I_1(kr_o)} = \frac{(1+j)}{2}\left(\frac{r_o}{\delta_w}\right)\frac{I_0(0)}{I_1(kr_o)} = \frac{(1+j)}{2}\left(\frac{r_o}{\delta_w}\right)\frac{I_0(0)}{I_1\left[(1+j)\left(\frac{r_o}{\delta_w}\right)\right]}$$

$$= \left|\frac{J(0)}{J_{DC}}\right| e^{j\phi(0)}. \tag{3.87}$$

Figure 3.15 shows the normalized current density on the conductor surface $|J(0)/J_{DC}|$ as a function of d/δ_w.

The derivatives of the *ber* and *bei* functions are given by

$$ber'(\alpha) = \sum_{n=0}^{\infty} \frac{(-1)^n 4n\alpha^{(4n-1)}}{2^{4n}[(2n)!]^2} = \frac{1}{2}\left[e^{j\frac{\pi}{4}} I_1\left(\alpha e^{j\frac{\pi}{4}}\right) + e^{-j\frac{\pi}{4}} I_1\left(\alpha e^{-j\frac{\pi}{4}}\right)\right] \tag{3.88}$$

and

$$bei'(\alpha) = \sum_{n=0}^{\infty} \frac{(-1)^n 2(2n+1)\alpha^{[2(2n+1)-1]}}{2^{2(2n+1)}[(2n+1)!]^2} = \frac{1}{2j}\left[e^{j\frac{\pi}{4}} I_1\left(\alpha e^{j\frac{\pi}{4}}\right) - e^{-j\frac{\pi}{4}} I_1\left(\alpha e^{-j\frac{\pi}{4}}\right)\right]. \tag{3.89}$$

3.6.7 Approximation of Current Density in Round Conductor

The expression for the current density is cumbersome to use. Therefore, its approximation is useful. For low frequencies ($\delta_w \gg r_o$),

$$I_0(kr) \approx 1 + \frac{(kr)^2}{4} + \frac{(kr)^4}{64} \tag{3.90}$$

and

$$I_1(kr_o) \approx \frac{kr_o}{2} + \frac{(kr_o)^3}{16} + \frac{(kr_o)^5}{384} = \frac{kr_o}{2}\left[1 + \frac{(kr_o)^2}{8} + \frac{(kr_o)^4}{192}\right]. \tag{3.91}$$

The current density in the wire at low frequencies can be approximated by

$$J(r) = \frac{Ik}{2\pi r_o}\frac{I_0(kr)}{I_1(kr_o)} \approx \frac{I}{2\pi r_o}\frac{1+\frac{(kr)^2}{4}+\frac{(kr)^4}{64}}{\frac{kr_o}{2}\left[1+\frac{(kr_o)^2}{8}+\frac{(kr_o)^4}{192}\right]}$$

$$= \frac{I}{\pi r_o^2}\frac{1+\frac{(kr)^2}{4}+\frac{(kr)^4}{64}}{1+\frac{(kr_o)^2}{8}+\frac{(kr_o)^4}{192}} = J_{DC}\frac{1+\frac{(kr)^2}{4}+\frac{(kr)^4}{64}}{1+\frac{(kr_o)^2}{8}+\frac{(kr_o)^4}{192}} \quad \text{for} \quad 0 \le r \le r_o. \tag{3.92}$$

For high frequencies ($\delta_w \ll r_o$),

$$I_0(kr) \approx \frac{e^{kr}}{\sqrt{2\pi kr}}\left(1+\frac{1}{8kr}\right) \tag{3.93}$$

and

$$I_1(kr_o) \approx \frac{e^{kr_o}}{\sqrt{2\pi kr_o}}\left(1-\frac{3}{8kr_o}\right) \tag{3.94}$$

yielding the current density at high frequencies

$$J(r) = I\frac{k}{2\pi r_o}\frac{I_0(kr)}{I_1(kr_o)} \approx I\frac{k}{2\pi r_o}\sqrt{\frac{r_o}{r}}e^{-k(r_o-r)} \quad for \quad 0 \le r \le r_o. \tag{3.95}$$

The electric field intensity inside the conductor is given by

$$E(r) = E_z(r) = E(r_o)\frac{I_0(kr)}{I_0(kr_o)}. \tag{3.96}$$

3.6.8 Impedance of Round Conductor

The resistance of the outer shell of the round conductor of cross-sectional area S and length l_w is given by

$$R_s = \rho_w\frac{l_w}{S}. \tag{3.97}$$

The current on the conductor surface S is

$$I_s = SJ(r_o). \tag{3.98}$$

The voltage across the conductor can be found on the conductor surface as

$$V = R_sI_s = \frac{\rho_w l_w}{S}J(r_o)S = \rho_w l_w J(r_o) = I\rho_w l_w\frac{k}{2\pi r_o}\frac{I_0(kr_o)}{I_1(kr_o)}. \tag{3.99}$$

The impedance of an isolated, straight, and round conductor is given by

$$Z = \frac{V}{I} = \rho_w l_w\frac{k}{2\pi r_o}\frac{I_0(kr_o)}{I_1(kr_o)} = \frac{l_w[ber(\alpha_o)+jbei(\alpha_o)]}{\sqrt{2}\pi\sigma_w\delta_w r_o[bei'(\alpha_o)-jber'(\alpha_o)]}$$

$$= \frac{l_w\{ber(\alpha_o)bei'(\alpha_o)-bei(\alpha_o)ber'(\alpha_o)+j[ber(\alpha_o)ber'(\alpha_o)+bei(\alpha_o)bei'(\alpha_o)]\}}{\sqrt{2}\pi\sigma_w\delta_w r_o\{[ber'(\alpha_o)]^2+[bei'(\alpha_o)]^2\}}$$

$$= R_{wDC}\frac{\alpha_o}{2}\frac{ber(\alpha_o)bei'(\alpha_o)-bei(\alpha_o)ber'(\alpha_o)+j[ber(\alpha_o)ber'(\alpha_o)+bei(\alpha_o)bei'(\alpha_o)]}{[ber'(\alpha_o)]^2+[bei'(\alpha_o)]^2}$$

$$= R_w + jX_L = R_w + j\omega L = |Z|e^{j\phi_Z}, \tag{3.100}$$

where the DC resistance of the round conductor is

$$R_{wDC} = \frac{\rho_w l_w}{\pi r_o^2}, \tag{3.101}$$

the AC resistance of the round conductor is

$$R_w = Re\{Z\}, \tag{3.102}$$

and the reactance of the round conductor is

$$X_L = \omega L = Im\{Z\}. \tag{3.103}$$

Hence, the wire impedance normalized with respect to R_{wDC} is

$$\frac{Z}{R_{wDC}} = \frac{kr_o}{2}\frac{I_0(kr_o)}{I_1(kr_o)} = \frac{(1+j)}{2}\left(\frac{r_o}{\delta_w}\right)\frac{I_0(kr_o)}{I_1(kr_o)} = \frac{J(r_o)}{J_{DC}} = F_R + jF_X, \tag{3.104}$$

where the AC-to-DC resistance ratio, the skin-effect factor, or AC resistance factor is the ratio of the AC-to-DC wire resistance at any frequency

$$F_R = \frac{R_w}{R_{wDC}} = Re\left\{\frac{Z}{R_{wDC}}\right\} = \frac{1}{2}Re\left\{\frac{(kr_o)I_0(kr_o)}{I_1(kr_o)}\right\} = Re\left\{\frac{J(r_o)}{J_{DC}}\right\}$$
$$= \frac{\alpha_o}{2}\frac{ber(\alpha_o)bei'(\alpha_o) - bei(\alpha_o)ber'(\alpha_o)}{[ber'(\alpha_o)]^2 + [bei'(\alpha_o)]^2} \tag{3.105}$$

and the ratio of the wire reactance $X_L = \omega L$ to the DC wire resistance at any frequency is

$$F_X = \frac{X_L}{R_{wDC}} = \frac{\omega L}{R_{wDC}} = Im\left\{\frac{Z}{R_{wDC}}\right\} = \frac{1}{2}Im\left\{\frac{(kr_o)I_0(kr_o)}{I_1(kr_o)}\right\} = Im\left\{\frac{J(r_o)}{J_{DC}}\right\}$$
$$= \frac{\alpha_o}{2}\frac{ber(\alpha_o)ber'(\alpha_o) + bei(\alpha_o)bei'(\alpha_o)}{[ber'(\alpha_o)]^2 + [bei'(\alpha_o)]^2}. \tag{3.106}$$

Figures 3.16 through 3.19 show plots of R_w/R_{wDC}, X_L/R_{wDC}, $|Z|/R_{wDC}$, and ϕ_Z as functions of r_o/δ_w, respectively. It follows from Fig. 3.16 that the AC resistance of the round conductor is equal to the DC resistance in the range

$$0 \le \frac{r_o}{\delta_w} \le 1. \tag{3.107}$$

The condition for the upper frequency at which the skin effect may be neglected is

$$r_o = \delta_w = \sqrt{\frac{\rho_w}{\pi\mu_0 f_{sk}}} \tag{3.108}$$

yielding the upper frequency at which the skin effect is negligible for the round conductor is given by

$$f_{sk} = \frac{\rho_w}{\pi\mu_0 r_o^2}. \tag{3.109}$$

Above this frequency, the conductor resistance increases with frequency.

The low-frequency inductance of the round conductor due to its internal magnetic field inside the conductor is

$$L_{LF} = \frac{\mu l_w}{8\pi}. \tag{3.110}$$

The inductance at any frequency is

$$L = \frac{F_X R_{wDC}}{\omega} = \frac{F_X \rho_w l_w}{\pi r_o^2 \omega} \tag{3.111}$$

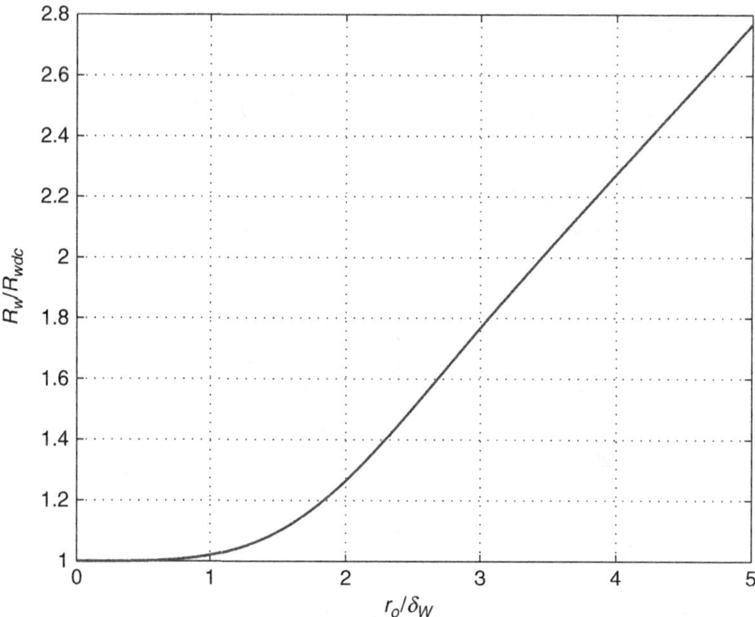

Figure 3.16 Series resistance normalized with respect to the DC resistance R_w/R_{wDC} as a function of r_o/δ_w of a solitary single round conductor

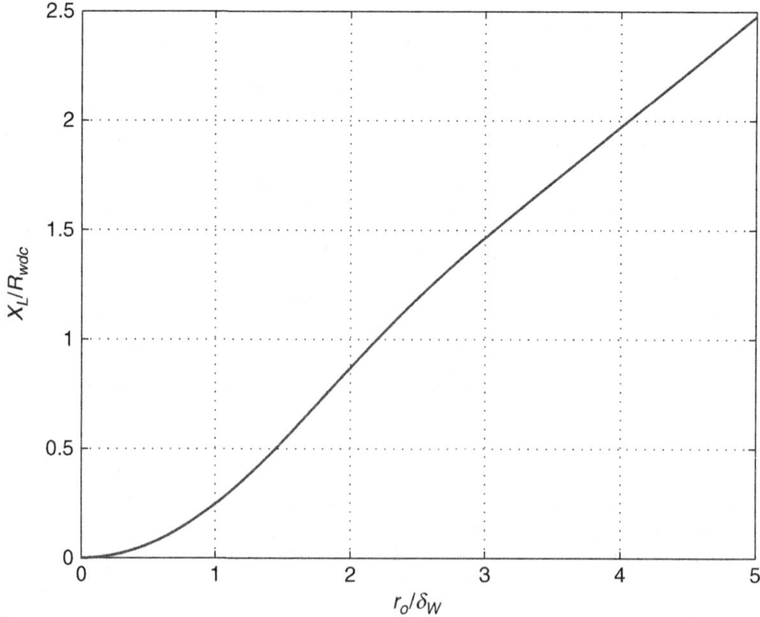

Figure 3.17 Series reactance normalized with respect to the DC resistance X_L/R_{wDC} as a function of r_o/δ_w of a solitary single round conductor

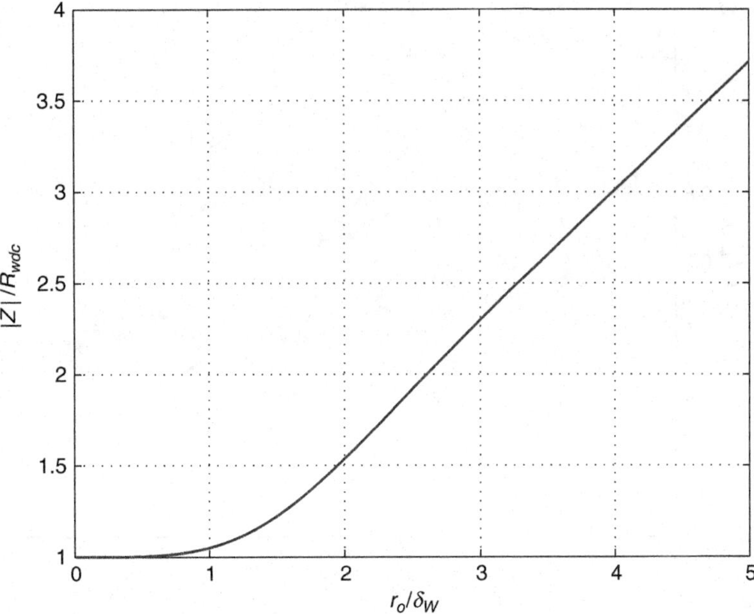

Figure 3.18 Magnitude of the impedance normalized with respect to the DC resistance $|Z|/R_{wDC}$ as a function of r_o/δ_w of a solitary single round conductor

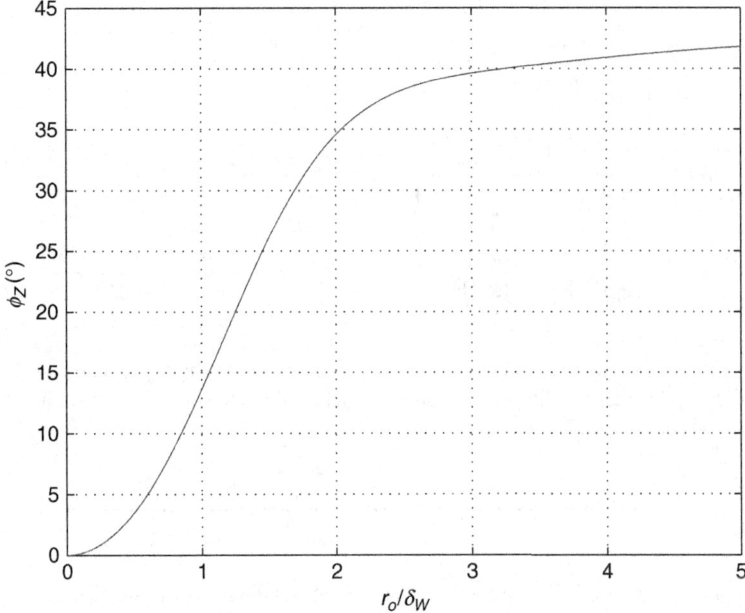

Figure 3.19 Phase of the impedance ϕ_Z as a function of r_o/δ_w of a solitary single round conductor

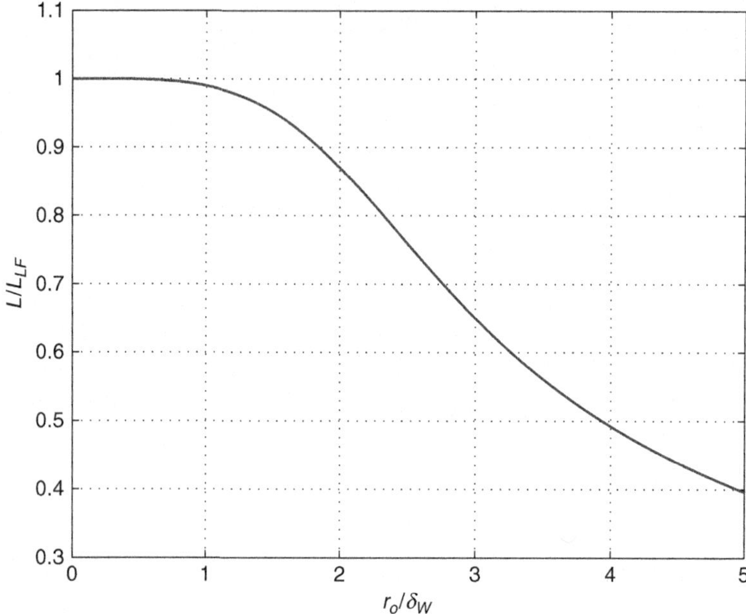

Figure 3.20 Normalized inductance L/L_{LF} as a function of r_o/δ_w of a solitary single round conductor

yielding the normalized wire inductance at any frequency

$$F_L = \frac{L}{L_{LF}} = \frac{4F_X}{\left(\frac{r_o}{\delta_w}\right)^2} = \frac{4}{\left(\frac{r_o}{\delta_w}\right)^2} Im\left\{\frac{Z}{R_{wDC}}\right\} = \frac{2}{\left(\frac{r_o}{\delta_w}\right)^2} Im\left\{\frac{(kr_o)I_0(kr_o)}{I_1(kr_o)}\right\}$$
$$= \frac{2}{\left(\frac{r_o}{\delta_w}\right)} Im\left\{\frac{(1+j)I_0\left[(1+j)\left(\frac{r_o}{\delta_w}\right)\right]}{I_1\left[(1+j)\left(\frac{r_o}{\delta_w}\right)\right]}\right\} = \frac{2\rho_w\alpha_o}{\pi r_o^2 \mu f}\,\frac{ber(\alpha_o)ber'(\alpha_o) + bei(\alpha_o)bei'(\alpha_o)}{[ber'(\alpha_o)]^2 + [bei'(\alpha_o)]^2}$$
$$= \frac{4}{\alpha_o}\,\frac{ber(\alpha_o)ber'(\alpha_o) + bei(\alpha_o)bei'(\alpha_o)}{[ber'(\alpha_o)]^2 + [bei'(\alpha_o)]^2} = 2\sqrt{2}\left(\frac{\delta_w}{r_o}\right)\frac{ber(\alpha_o)ber'(\alpha_o) + bei(\alpha_o)bei'(\alpha_o)}{[ber'(\alpha_o)]^2 + [bei'(\alpha_o)]^2}. \quad (3.112)$$

Figure 3.20 shows the normalized inductance L/L_{LF} as a function of r_o/δ_w. It can be seen that L/L_{LF} decreases as r_o/δ_w increases. The above analysis indicates that the solution in cylindrical coordinates involves a great deal of complexity.

Example 3.1

A round copper conductor has the length $l_w = 1$ m, the radius $r_o = 1$ mm, and temperature $T = 20\,^\circ$C. Calculate low-frequency resistance and inductance of the conductor. Plot the resistance and the inductance of the conductor as functions of frequency.

Solution: The DC and low-frequency resistance of the conductor is

$$R_{wDC} = \frac{\rho_{Cu} l_w}{\pi r_o^2} = \frac{1.724 \times 10^{-8} \times 1}{\pi \times (10^{-3})^2} = 5.4877\ \text{m}\Omega. \quad (3.113)$$

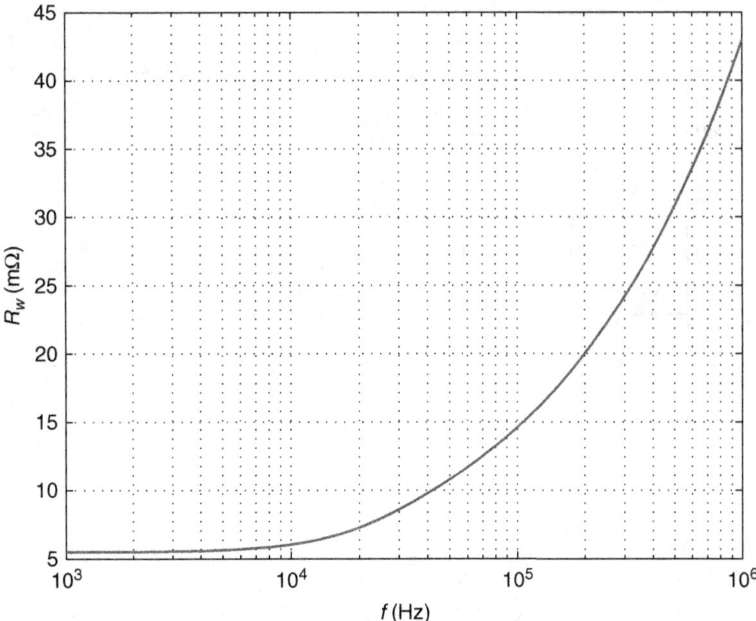

Figure 3.21 Resistance R_w as a function of frequency of a solitary single round conductor for $l_w = 1$ m, $r_o = 1$ mm, and $T = 20\,^\circ$C

The low-frequency inductance is

$$L_{LF} = \frac{\mu_0 l_w}{8\pi} = \frac{4\pi \times 10^{-7} \times 1}{8\pi} = 50\,\text{nH}. \tag{3.114}$$

Plots of the resistance R_w and the inductance L are shown in Figs 3.21 and 3.22, respectively. As the frequency is increased from 10 kHz to 1 MHz, the resistance increases from 5.5 mΩ to 43 mΩ and the inductance decreases from 50 nH to 7 nH. Hence,

$$F_R = \frac{R_w}{R_{wDC}} = \frac{43}{5.5} = 7.818. \tag{3.115}$$

3.6.9 Approximation of Resistance and Inductance of Round Conductor

At low frequencies ($\delta_w > r_o$),

$$F_R = \frac{R_w}{R_{wDC}} \approx 1 \tag{3.116}$$

and

$$F_X = \frac{X_L}{R_{wDC}} = \frac{\omega L}{R_{wDC}} \approx \left(\frac{r_o}{2\delta_w}\right)^2 = \frac{\pi r_o^2 \mu f}{4\rho_w} \tag{3.117}$$

resulting in the low-frequency inductance

$$L_{LF(int)} \approx \frac{\mu r_o^2 R_{wDC}}{8\rho_w} = \frac{\mu l_w}{8\pi}. \tag{3.118}$$

Hence, the inductance per unit length of a straight, round conductor with $\mu = \mu_0$ is

$$\frac{L_{LF}}{l_w} \approx \frac{\mu_0}{8\pi} = \frac{4\pi \times 10^{-7}}{8\pi}\ (\text{H/m}) = 0.05\ (\mu\text{H/m}) = 0.5\ (\text{nH/cm}). \tag{3.119}$$

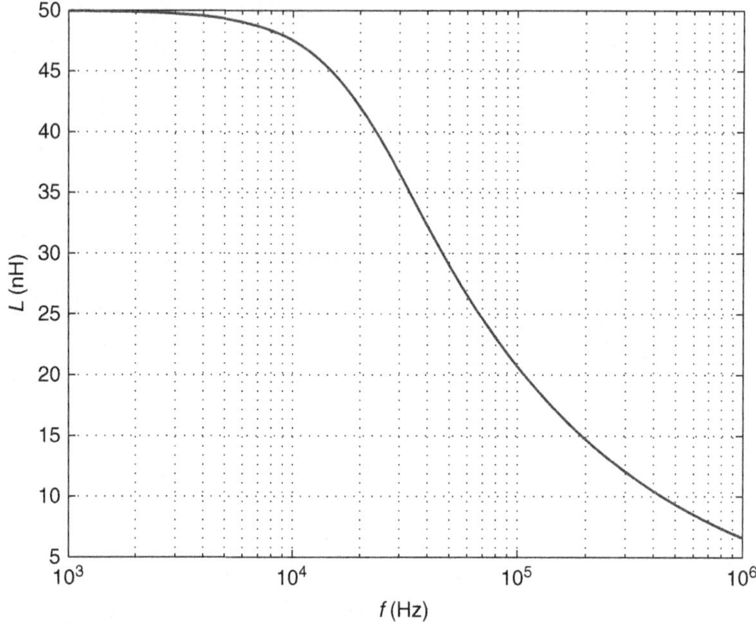

Figure 3.22 Inductance L as a function of frequency of a solitary single round conductor for $l_w = 1$ m, $r_o = 1$ mm, and $T = 20\,^{\circ}$C

At low frequencies,

$$F_L = \frac{L}{L_{LF}} \approx 1. \tag{3.120}$$

Expanding the Bessel functions into series and taking the first terms for $r_o < \delta_w$, we obtain

$$F_R = \frac{R_w}{R_{wDC}} \approx 1 + \frac{1}{48}\left(\frac{r_o}{\delta_w}\right)^4 = 1 + \frac{1}{768}\left(\frac{d_o}{\delta_w}\right)^4 \tag{3.121}$$

and

$$F_X = \frac{X_L}{R_{wDC}} \approx \frac{1}{4}\left(\frac{r_o}{\delta_w}\right)^2\left[1 - \frac{1}{96}\left(\frac{r_o}{\delta_w}\right)^4\right] = \frac{1}{16}\left(\frac{d_o}{\delta_w}\right)^2\left[1 - \frac{1}{1536}\left(\frac{d_o}{\delta_w}\right)^4\right] \tag{3.122}$$

where $d_o = 2r_o$. The internal impedance of the wire at low frequencies is

$$Z_{LF} = R_w + j\omega L \approx R_{wDC}\left[1 + \frac{1}{48}\left(\frac{r_o}{\delta_w}\right)^4\right] + j\frac{w\mu}{8\pi}. \tag{3.123}$$

At high frequencies ($\delta_w \ll r_o$),

$$F_R = \frac{R_w}{R_{wDC}} \approx \frac{r_o}{2\delta_w} \tag{3.124}$$

$$R_w = F_R R_{wDC} \approx \frac{\rho_w l_w}{2\pi r_o \delta_w} = \frac{l_w}{2r_o}\sqrt{\frac{\rho_w \mu f}{\pi}} \tag{3.125}$$

$$F_X = \frac{X_L}{R_{wDC}} = \frac{\omega L}{R_{wDC}} \approx \frac{r_o}{2\delta_w} = \frac{r_o}{2}\sqrt{\frac{\pi \mu f}{\rho_w}} \tag{3.126}$$

$$L_{HF} \approx \frac{r_o}{2\delta_w}\frac{R_{wDC}}{\omega} = \frac{\mu l_w}{4\pi}\left(\frac{\delta_w}{r_o}\right) = \frac{\rho_w l_w}{\pi r_o^2}\frac{r_o}{2\delta_w}\frac{1}{2\pi f} = \frac{l_w}{4\pi r_o}\sqrt{\frac{\mu \rho_w}{\pi f}}. \tag{3.127}$$

and

$$F_L = \frac{L}{L_{LF}} \approx \frac{2}{\left(\frac{r_o}{\delta_w}\right)}. \quad (3.128)$$

At high frequencies, the resistance R_w and reactance X_L are proportional to $\sqrt{f}$ and the inductance L is inversely proportional to $\sqrt{f}$. Since at high frequencies $R_w/R_{wDC} = \omega L/R_{wDC} \approx r_o/(2\delta_w)$, the angle $\phi \approx 45^\circ$.

At high frequencies ($r_o > \delta_w$), the approximation of the Bessel functions using the asymptotic representation gives

$$F_R = \frac{R_w}{R_{wDC}} \approx \frac{1}{4} + \frac{r_o}{2\delta_w} + \frac{3}{32}\frac{1}{\frac{r_o}{\delta_w}} \quad (3.129)$$

and

$$F_X = \frac{X_L}{R_{wDC}} \approx \frac{r_o}{2\delta_w} - \frac{3}{32}\frac{1}{\frac{r_o}{\delta_w}} + \frac{3}{32}\frac{1}{\left(\frac{r_o}{\delta_w}\right)^2}. \quad (3.130)$$

At $r_o = 2\delta_w$, $F_R = 1$, $R_w = R_{wDC}$ and

$$\delta_w = \frac{r_o}{2} = \sqrt{\frac{\rho_w}{\pi\mu f}}. \quad (3.131)$$

Hence, the critical frequency at which the wire resistance begins to increase due to the skin effect is

$$f_c = \frac{4\rho_w}{\pi\mu r_o^2}. \quad (3.132)$$

The internal impedance of the round wire is

$$Z_{HF} = R + j\omega L \approx \frac{(1+j)\rho_w l_w}{2\pi r_o \delta_w}. \quad (3.133)$$

Thus, the resistance and reactance are equal at high frequencies for good conductors. In addition, the resistance of a round wire with radius r_o and length l_w is equal to the resistance of a rectangular conductor of width $2\pi r_o$, length l_w, and thickness δ_w.

The approximate expression for the AC-to-DC resistance ratio of a round wire is

$$F_R \approx \begin{cases} 1 + \frac{(r_o/\delta_w)^4}{48+0.8(r_o/\delta_w)^4}, & \text{for } 0 < r_o/\delta_w \le 1.7 \\ \frac{1}{4} + \frac{1}{2}\left(\frac{r_o}{\delta_w}\right) + \frac{3}{32}\left(\frac{\delta_w}{r_o}\right), & \text{for } 1.7 < r_o/\delta_w \le \infty. \end{cases} \quad (3.134)$$

3.6.10 Simplified Derivation of Round Wire Resistance

At high frequencies, the current is assumed to flow uniformly over the skin depth ($r_o - \delta_w \le r \le r_o$) only near the surface of a straight, solitary, round conductor, and it is zero in the middle of the conductor ($0 \le r \le r_o - \delta_w$). Therefore, $H = 0$ for $0 \le r \le r_o - \delta_w$. The DC resistance of the round conductor is

$$R_{wDC} = \frac{\rho_w l_w}{A_w} = \frac{\rho_w l_w}{\pi r_o^2} = \frac{4\rho_w l_w}{\pi d^2}. \quad (3.135)$$

The effective cross-sectional area of the AC current flow is

$$A_e = \pi r_o^2 - \pi(r_o - \delta_w)^2 = \pi(2r_o\delta_w - \delta_w^2) = \pi\delta_w(d - \delta_w) \approx \pi d\delta_w \quad \text{for} \quad d \gg \delta_w. \quad (3.136)$$

The conductor AC resistance can be approximated by

$$R_w = \frac{\rho_w l_w}{A_e} = \frac{\rho_w l_w}{\pi[r_o^2 - (r_o - \delta_w)^2]} = \frac{\rho_w l_w}{\pi(2r_o\delta_w - \delta_w^2)} = \frac{\rho_w l_w}{\pi(d\delta_w - \delta_w^2)} = \frac{\rho_w l_w}{\pi\delta_w(d - \delta_w)}$$
$$\approx \frac{\rho_w l_w}{2\pi r_o\delta_w} = \frac{\rho_w l_w}{\pi d\delta_w} = \frac{l_w}{2r_o}\sqrt{\frac{\mu_0\rho_w f}{\pi}} = \frac{l_w}{d}\sqrt{\frac{\mu_0\rho_w f}{\pi}}. \quad (3.137)$$

The AC resistance R_w is proportional to $\sqrt{f}$ and inversely proportional to r_o, although the DC resistance is inversely proportional to r_o^2. A silver-plated solid or tubular conductors are used at VHF and microwave frequencies, providing a better conductor than copper. The AC-to-DC resistance ratio for a straight wire at high frequencies is given by

$$F_R = \frac{R_w}{R_{wDC}} = \frac{A_w}{A_e} = \frac{r_o^2}{r_o^2 - (r_o - \delta_w)^2} = \frac{\left(\frac{r_o}{\delta_w}\right)^2}{\left(\frac{r_o}{\delta_w}\right)^2 - \left(\frac{r_o}{\delta_w} - 1\right)^2} \approx \frac{r_o}{2\delta_w} = \frac{r_o}{2}\sqrt{\frac{\pi\mu_0 f}{\rho_w}}. \quad (3.138)$$

A simpler derivation is as follows. The effective cross-sectional area of the AC current flow is

$$A_e \approx 2\pi r_o \delta_w. \quad (3.139)$$

The conductor AC resistance is

$$R_w = \frac{\rho_w l_w}{A_e} \approx \frac{\rho_w l_w}{2\pi r_o \delta_w}. \quad (3.140)$$

Hence, the AC-to-DC resistance ratio is

$$F_R = \frac{R_w}{R_{wDC}} \approx \frac{r_o}{2\delta_w} = \frac{d}{4\delta_w} = \frac{r_o}{2}\sqrt{\frac{\pi\mu_0 f}{\rho_w}} = \frac{d}{4}\sqrt{\frac{\pi\mu_0 f}{\rho_w}}. \quad (3.141)$$

The AC-to-DC resistance ratio is a function of temperature T because the skin depth depends on the temperature

$$F_{R(T)} = \frac{R_{w(T)}}{R_{wDC(T)}} \approx \frac{r_o}{2\delta_{w(T)}} = \frac{r_o}{2}\sqrt{\frac{\pi\mu_0 f}{\rho_{w(T)}}} = \frac{r_o}{2}\sqrt{\frac{\pi\mu_0 f}{\rho_{w(To)}[1+\alpha(T-T_o)]}}$$

$$= \frac{r_o}{2\delta_{w(To)}}\frac{1}{\sqrt{1+\alpha(T-T_o)}} = F_{R(To)}\frac{1}{\sqrt{1+\alpha(T-T_o)}}. \quad (3.142)$$

The winding resistance at temperature T is

$$R_{w(T)} = R_{wDC(T)}F_{R(T)} = R_{wDC(To)}\sqrt{1+\alpha(T-T_o)}\frac{r_o}{2\delta_{w(To)}}\frac{1}{\sqrt{1+\alpha(T-T_o)}}$$

$$= R_{wDC(To)}\frac{r_o}{2}\sqrt{\frac{\pi\mu_0 f}{\rho_{w(To)}}}\sqrt{1+\alpha(T-T_o)}. \quad (3.143)$$

The ratio of the conductor resistance at temperature T to the conductor DC resistance at temperature T_o is

$$\frac{R_{w(T)}}{R_{wDC(To)}} = \frac{r_o}{2}\sqrt{\frac{\pi\mu_o f}{\rho_{w(To)}}}[1+\alpha(T-T_o)] = F_{R(To)}\sqrt{1+\alpha(T-T_o)}. \quad (3.144)$$

Thus, the dependance of the conductor AC resistance on temperature T is different than that for the conductor DC resistance.

Example 3.2

A round copper conductor has the length $l_w = 1$ m, the radius $r_o = 1$ mm, and temperature $T = 20\,°$C. Calculate the approximate resistance of the conductor and F_R at $f = 1$ MHz.

Solution: The DC and low-frequency resistance of the conductor is

$$R_{wDC} = \frac{\rho_{Cu} l_w}{\pi r_o^2} = \frac{1.724\times10^{-8}\times1}{\pi\times(10^{-3})^2} = 5.4877\ \text{m}\Omega. \quad (3.145)$$

The skin depth at $f = 1$ MHz is

$$\delta_w = \sqrt{\frac{\rho_{Cu}}{\pi\mu_0 f}} = \sqrt{\frac{1.724\times10^{-8}}{\pi\times4\times\pi\times10^{-7}\times10^6}} = 0.066\ \text{mm}. \quad (3.146)$$

Hence,

$$\frac{d}{\delta_w} = \frac{2}{0.066} = 30.3. \tag{3.147}$$

The approximate conductor AC resistance is

$$R_w = \frac{\rho_{Cu} l_w}{A_e} \approx \frac{\rho_{Cu} l_w}{\pi \delta_w (d - \delta_w)} = \frac{1.724 \times 10^{-8} \times 1}{\pi \times 0.066 \times 10^{-3} \times (2 - 0.066) \times 10^{-3}} = 42.9919\ \mathrm{m}\Omega. \tag{3.148}$$

The AC-to-DC resistance ratio is

$$F_R = \frac{R_w}{R_{wDC}} = \frac{42.9919}{5.4877} = 7.8342. \tag{3.149}$$

We may use a simpler approximate expression for the conductor AC resistance

$$R_w = \frac{\rho_{Cu} l_w}{A_e} \approx \frac{\rho_{Cu} l_w}{\pi d \delta_w} = \frac{1.724 \times 10^{-8} \times 1}{\pi \times 2 \times 10^{-3} \times 0.066 \times 10^{-3}} = 41.573\ \mathrm{m}\Omega. \tag{3.150}$$

The AC-to-DC resistance ratio is

$$F_R = \frac{R_w}{R_{wDC}} = \frac{41.537}{5.4877} = 7.5757. \tag{3.151}$$

In both the cases, the values of R_w and F_R are very close to those computed from the exact expressions in Example 3.1.

3.7 Magnetic Field Intensity for Round Wire

The magnetic field intensity inside the conductor can be found by solving the equation

$$\frac{dH(r)}{dr} + \frac{H(r)}{r} = J(r) \quad \text{for} \quad r \le r_o. \tag{3.152}$$

Notice that

$$\frac{1}{r}\frac{d}{dr}(rH) = \frac{1}{r}\left(H + r\frac{dH}{dr}\right) = \frac{H}{r} + \frac{dH}{dr}. \tag{3.153}$$

Hence,

$$\frac{1}{r}\frac{d(rH)}{dr} = J(r) = AI_0(kr) = \frac{Ik}{2\pi r_o}\frac{I_0(kr)}{I_1(kr_o)}, \tag{3.154}$$

where

$$A = \frac{Ik}{2\pi r_o I_1(kr_o)} = H(r_o)\frac{k}{I_1(kr_o)}, \tag{3.155}$$

where

$$H(r_o) = \frac{I}{2\pi r_o}. \tag{3.156}$$

Since $d(rH) = J(r)rdr$, we have

$$\int d(rH)dr = \int rJ(r)dr \tag{3.157}$$

yielding

$$rH = A\int rI_0(kr)dr + Const. \tag{3.158}$$

Using

$$\int xI_0(x)x = xI_1(x) \tag{3.159}$$

we obtain

$$rH(r) = \frac{A}{k^2}\int (kr)I_0(kr)d(kr) + Const = \frac{A}{k^2}krI_1(kr) + Const = \frac{A}{k}rI_1(kr) + Const \tag{3.160}$$

resulting in

$$H(r) = \frac{A}{k} I_1(kr) + \frac{Const}{r} = H(r_o)\frac{I_1(kr)}{I_1(kr_o)} + \frac{Const}{r}. \tag{3.161}$$

Because

$$I_1(kr) = \frac{kr}{2}(1 + \ldots) \tag{3.162}$$

we get

$$I_1(0) = 0. \tag{3.163}$$

Therefore,

$$H(0) = \frac{Const}{r}. \tag{3.164}$$

Since

$$\lim_{r\to 0} H(r) = 0 \tag{3.165}$$

we obtain

$$Const = 0. \tag{3.166}$$

The magnetic field intensity inside the conductor at any frequency is given by

$$\frac{H(r)}{H(r_o)} = \frac{I_1(kr)}{I_1(kr_o)} = \frac{I_1\left[(1+j)\left(\frac{r_o}{\delta_w}\right)\left(\frac{r}{r_o}\right)\right]}{I_1\left[(1+j)\left(\frac{r_o}{\delta_w}\right)\right]} = \left|\frac{H(r)}{H(r_o)}\right| e^{\phi_H} \quad \text{for} \quad r \leq r_o. \tag{3.167}$$

Figure 3.23 shows the plots of $|H(r)/H(r_o)|$ as a function of r/r_o for selected values of r_o/δ_w.

The magnetic field intensity outside the conductor is described by

$$\frac{dH(r)}{dr} + \frac{H(r)}{r} = 0 \tag{3.168}$$

which becomes

$$\frac{1}{r}\frac{d}{dr}(rH) = 0 \quad \text{for} \quad r \geq r_o. \tag{3.169}$$

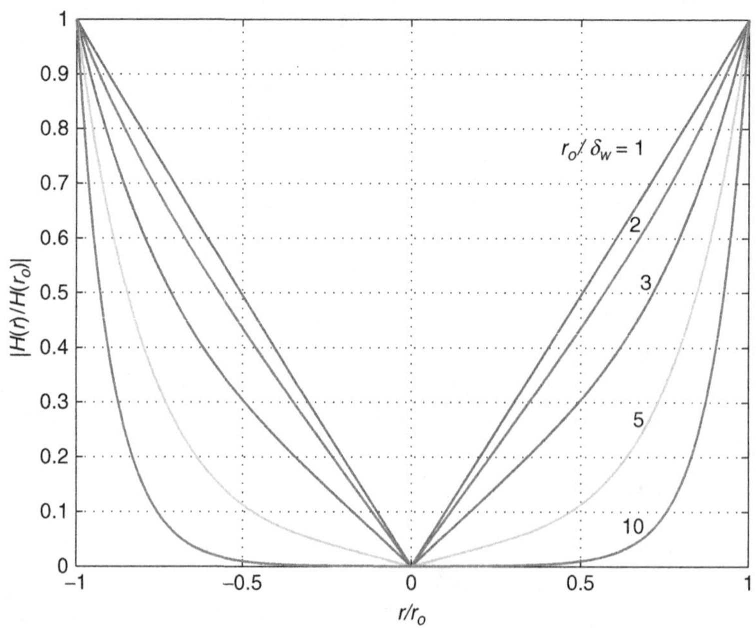

Figure 3.23 Normalized magnitude of the magnetic field intensity $|H(r)/H(r_o)|$ as a function of r/r_o at various values of r_o/δ_w for a solitary single round conductor

Since $d(rH) = 0$, we have

$$\int d(rH)dr = \int 0 \tag{3.170}$$

and

$$rH(r) = Const \tag{3.171}$$

producing

$$H(r) = \frac{Const}{r}. \tag{3.172}$$

From the continuity condition,

$$H(r_o^-) = H(r_o^+) \tag{3.173}$$

yielding

$$\frac{Const}{r_o} = \frac{I}{2\pi r_o} \tag{3.174}$$

which gives

$$Const = \frac{I}{2\pi}. \tag{3.175}$$

Thus, the magnetic field outside the conductor is expressed as

$$H(r) = \frac{I}{2\pi r} = H(r_o)\left(\frac{r_o}{r}\right) \quad \text{for} \quad r \geq r_o. \tag{3.176}$$

The approximation of the magnetic field intensity inside the wire for low frequencies is given by

$$H(r) = \left(\frac{I}{2\pi r_o}\right)\frac{I_1(kr)}{I_1(kr_o)} \approx H(r_o)\frac{\frac{kr}{2} + \frac{(kr)^3}{16}}{\frac{kr_o}{2} + \frac{(kr_o)^3}{16}} = H(r_o)\left(\frac{r}{r_o}\right)\frac{1 + \frac{(kr)^2}{8}}{1 + \frac{(kr_o)^2}{8}}$$
$$\approx H(r_o)\left(\frac{r}{r_o}\right)\left[1 - \frac{1}{8}k^2(r_o^2 - r^2)\right] = H(r_o)\left[1 - \frac{1}{8}j\omega\mu\sigma(r_o^2 - r^2)\right] \quad \text{for} \quad 0 \leq r \leq r_o. \tag{3.177}$$

Hence, the magnitude of the magnetic field inside the wire at low frequencies is

$$|H(r)| = |H(r_o)|\left(\frac{r}{r_o}\right)\sqrt{1 + \frac{(\omega\mu)^2}{64\rho_w^2}(r_o^2 - r^2)} \quad \textit{for} \quad 0 \leq r \leq r_o. \tag{3.178}$$

For high frequencies ($kr \gg 1$ and $kr_o \gg 1$),

$$I_1(kr) \approx \frac{e^{kr}}{\sqrt{2\pi kr}} \tag{3.179}$$

and

$$I_1(kr_o) \approx \frac{e^{kr_o}}{\sqrt{2\pi kr_o}} \tag{3.180}$$

producing the magnetic field intensity inside the wire at high frequencies

$$H(r) = \left(\frac{I}{2\pi r_o}\right)\frac{I_1(kr)}{I_1(kr_o)} \approx H(r_o)\sqrt{\frac{r_o}{r}}e^{-k(r_o - r)} \quad \text{for} \quad r_o - \delta_w \leq r \leq r_o. \tag{3.181}$$

3.8 Other Methods of Determining the Round Wire Inductance

The current enclosed by a cylindrical shell of radius r is

$$I(r) = I_{enc}(r) = \int_0^r J(r)2\pi r dr = \frac{kI}{r_o I_1(kr_o)}\int_0^r I_0(kr)rdr = I\left(\frac{r}{r_o}\right)\frac{I_1(kr)}{I_1(kr_o)}, \tag{3.182}$$

where

$$\int I_0(x)dx = xI_1(x). \tag{3.183}$$

Hence,

$$\frac{dI(r)}{dr} = kI\left(\frac{r}{r_o}\right)\frac{I_0(kr)}{I_1(kr_o)}. \tag{3.184}$$

The ratio of the current flowing through a cylindrical shell of radii r and $r + dr$ is given by

$$\frac{dI(r)}{I} = k\left(\frac{r}{r_o}\right)\frac{I_0(kr)}{I_1(kr_o)}dr. \tag{3.185}$$

The magnetic flux is generated between the cylindrical shell of radius r and the conductor surface of radius r_o by the current enclosed inside the shell

$$\phi(r) = \int\int_S B(r)dS = \int_0^{l_w}\int_r^{r_o} \mu H(r)drdz = \frac{\mu l_w I}{2\pi k r_o I_1(kr_o)}\int_r^{r_o} I_1(kr)d(kr)$$
$$= \frac{\mu l_w I}{2\pi k r_o I_1(kr_o)}[I_0(kr_o) - I_0(kr)], \tag{3.186}$$

where

$$\int I_1(x)dx = I_0(x). \tag{3.187}$$

The distributed magnetic linkage is

$$\lambda(r) = \int_0^r \phi(r)\frac{dI(r)}{I} = \frac{\mu l_w I}{2\pi(kr_o)^2 I_1^2(kr_o)}\int_0^{kr}[I_0(kr_o) - I_0(kr)](kr)I_0(kr)d(kr)$$
$$= \frac{(kr)\mu l_w I}{2\pi(kr_o)^2 I_1^2(kr_o)}\left\{I_0(kr_o)I_1(kr) - \frac{1}{2}(kr)[I_0^2(kr) - I_1^2(kr)]\right\}. \tag{3.188}$$

The distributed inductance is

$$L(r) = \frac{\lambda(r)}{I_{enc}(r)} = \frac{\mu l_w}{2\pi(kr_o)I_1(kr_o)I_1(kr)}\left\{I_0(kr_o)I_1(kr) - \frac{1}{2}(kr)[I_0^2(kr) - I_1^2(kr)]\right\}$$
$$= \frac{\mu l_w}{2\pi}\left\{\frac{1}{kr_o}\frac{I_0(kr_o)}{I_1(kr_o)} - \frac{1}{2}\left(\frac{r}{r_o}\right)\left[\frac{I_0^2(kr)}{I_1(kr_o)I_1(kr)} - \frac{I_1(kr)}{I_1(kr_o)}\right]\right\}. \tag{3.189}$$

The inductance of a single, solid, round conductor can be derived for low frequencies as follows. The current through the entire conductor of radius r_o is I and the current enclosed by the radius $r < r_o$ is $I_{enc}(r)$. At low frequencies, the current density J in the conductor is uniform, and therefore the current density of the entire conductor is

$$J(r_o) = \frac{I}{\pi r_o^2} \tag{3.190}$$

and the current density of a part of the conductor of radius r is

$$J(r) = \frac{I_{enc}(r)}{\pi r^2}. \tag{3.191}$$

Since $J(r) = J(r_o)$,

$$\frac{I_{enc}(r)}{\pi r^2} = \frac{I}{\pi r_o^2} \tag{3.192}$$

resulting in the current enclosed by the radius r

$$I_{enc}(r) = I\left(\frac{r}{r_o}\right)^2 \quad \text{for} \quad 0 \le r \le r_o. \tag{3.193}$$

The magnetic field intensity inside the conductor at radius r is

$$H(r) = \frac{I_{enc}(r)}{2\pi r} = \frac{Ir}{2\pi r_o^2} \quad \text{for} \quad 0 \le r \le r_o. \tag{3.194}$$

The magnetic field intensity $H(r)$ increases linearly with increasing radius r. It is zero at the center and a maximum at the surface. The magnetic flux density is

$$B(r) = \mu H(r) = \frac{\mu I r}{2\pi r_o^2} \quad \text{for} \quad 0 \le r \le r_o. \tag{3.195}$$

All of the flux links are of the current I. The differential magnetic flux linkage is equal to the differential flux times the fraction of the total current that is actually enclosed by the differential flux. The total flux linkage can be determined by integrating (summing) over the interior of the conductor. Consider an annular shell of radii r and $r + dr$. The incremental magnetic flux passing through the strip of width dr, length l_w, and surface $dS = l_w dr$ is given by

$$d\phi = B(r)dS = B(r)drdz = \frac{\mu I r}{2\pi r_o^2} drdz. \tag{3.196}$$

At low frequencies, the current density in the conductor is uniform. Therefore,

$$\frac{I_{enc}}{I} = \left(\frac{r}{r_o}\right)^2 \quad \text{for} \quad 0 \le r \le r_o. \tag{3.197}$$

The incremental flux linkage of radius r for the shell is expressed as

$$d\lambda_1 = \left(\frac{I_{enc}}{I}\right)^2 d\phi = \left(\frac{r}{r_o}\right)^2 d\phi = \left(\frac{r}{r_o}\right)^2 \frac{\mu I r}{2\pi r_o^2} drdz = \frac{\mu I r^3}{2\pi r_o^4} drdz \tag{3.198}$$

resulting in the total flux linkage

$$\lambda_{int} = \int\int_S d\lambda_1 = \int_0^{lw} \int_0^{ro} \frac{\mu I r^3}{2\pi r_o^4} drdz = \frac{\mu l_w I}{8\pi}. \tag{3.199}$$

The low-frequency internal self-inductance is

$$L_{LF(int)} = \frac{\lambda_{int}}{I} = \frac{\mu l_w}{8\pi}. \tag{3.200}$$

The amplitude of the magnetic field intensity outside the conductor is

$$H_m(r) = \frac{I_m}{2\pi r} \quad \text{for} \quad r \ge r_o \tag{3.201}$$

and the amplitude of the magnetic field density outside the conductor is

$$B_m(r) = \mu H_m(r) = \frac{\mu I_m}{2\pi r} \quad \text{for} \quad r \ge r_o. \tag{3.202}$$

The external self-inductance is

$$L_{ext} = \frac{\lambda_{ext}}{I} = \frac{1}{I}\int_0^{lw}\int_{ro}^{\infty} \frac{\mu I}{2\pi r} drdz = \frac{\mu}{2\pi}\int_0^{lw}\int_{ro}^{\infty}\frac{dr}{r}dz = \frac{\mu l_w}{2\pi}\int_{ro}^{\infty}\frac{dr}{r}$$
$$= \frac{\mu l_w}{2\pi} \ln\, r|_{ro}^{\infty} = \infty. \tag{3.203}$$

This result is due to a nonphysical situation. The circuit is not closed and there is no return current. The external self-inductance related to the external magnetic field contained in the range $0 \le r \le 5r_o$ can be estimated as

$$L_{ext} = \frac{\lambda_{ext}}{I} = \frac{1}{I}\int_0^{lw}\int_{ro}^{5ro} \frac{\mu I}{2\pi r} drdz = \frac{\mu}{2\pi}\int_0^{lw}\int_{ro}^{5ro}\frac{dr}{r}dz = \frac{\mu l_w}{2\pi}\ln 5 = 0.256\mu l_w. \tag{3.204}$$

The complete current loop is required to determine the inductance. Not knowing this loop, we can still estimate the external inductance for a segment of wire using the concept of "partial inductances"

$$L_{ext} = \frac{\mu_0 l_w}{2\pi}\left[\ln\left(\frac{2l_w}{r_o}\right) - 1\right]. \tag{3.205}$$

The total inductance L consists of the external inductance L_{ext} and the internal inductance L_{int}

$$L = L_{int} + L_{ext} = \frac{\mu l_w}{2\pi} + \frac{\mu_0 l_w}{2\pi}\left[\ln\left(\frac{2l_w}{r_o}\right) - 1\right]. \tag{3.206}$$

The magnetic energy method to determine the external inductance is often impractical because of the lack of a finite volume over which to integrate the magnetic field.

Another method for determining the inductance of a long straight wire is as follows. Consider an annular shell of radii r and $r + dr$ inside the conductor. Assuming a uniform current density, the ratio of the current flowing through the shell cross-sectional area to the total current is

$$\frac{dI}{I} = \frac{2\pi r dr}{\pi r_o^2} = \frac{2rdr}{r_o^2}. \tag{3.207}$$

The current $I_{enc}(r)$ flowing inside the shell causes a magnetic flux between the shell of radius r and the conductor surface of radius r_o. This magnetic flux is described by

$$\phi(r) = \int\int_S B(r)dS = \int_0^{l_w}\int_r^{r_o} B(r)drdz = l_w \int_r^{r_o} B(r)dr = \frac{\mu I l_w}{2\pi r_o^2}\int_r^{r_o} rdr$$
$$= \frac{\mu I l_w}{4\pi r_o^2}(r_o^2 - r^2). \tag{3.208}$$

Hence, the incremental flux linkage for the shell is given by

$$d\lambda_2 = \phi(r)\frac{dI}{I} = \phi(r)\frac{2rdr}{r_o^2} = \frac{\mu I l_w}{4\pi r_o^2}(r_o^2 - r^2)\frac{2rdr}{r_o^2} = \frac{\mu I l_w}{2\pi r_o^4}(r_o^2 r - r^3)dr. \tag{3.209}$$

The total flux linkage of the conductor is given by

$$\lambda = \int_0^{r_o} d\lambda_2 = \frac{\mu I l_w}{2\pi r_o^4}\int_0^{r_o}(r_o^2 r - r^3)dr = \frac{\mu l_w I}{8\pi}, \tag{3.210}$$

yielding the low-frequency internal inductance of a long, straight wire

$$L_{LF} = \frac{\lambda}{I} = \frac{\mu l_w}{8\pi}. \tag{3.211}$$

The distributed flux linkage of the conductor is given by

$$\lambda(r) = \int_0^{r} d\lambda_2 = \frac{\mu I l_w}{2\pi r_o^4}\int_0^{r}(r_o^2 r - r^3)dr = \frac{\mu I l_w r^2}{4\pi r_o^4}\left(r_o^2 - \frac{r^2}{2}\right). \tag{3.212}$$

As r increases from 0 to r_o, the magnetic flux ϕ decreases from $\mu l_w I/(4\pi)$ to 0 and the flux linkage increases from 0 to $\mu l_w I/(8\pi)$.

The distributed low-frequency inductance is

$$L_{LF}(r) = \frac{\lambda(r)}{I_{enc}(r)} = \frac{\mu l_w}{4\pi r_o^2}\left(r_o^2 - \frac{r^2}{2}\right) = \frac{\mu l_w}{4\pi}\left(1 - \frac{r^2}{2r_o^2}\right). \tag{3.213}$$

As r increases from 0 to r_o, the distributed low-frequency inductance decreases from $\mu l_w/(4\pi)$ to $\mu l_w/(8\pi)$.

The amplitude of the magnetic field intensity inside the round conductor is

$$H(r) = \frac{I}{2\pi r_o^2}r \quad \text{for} \quad 0 \le r \le r_o. \tag{3.214}$$

The magnetic energy density stored in the magnetic field outside the round conductor is

$$w_m(r) = \frac{1}{2}\mu H(r)^2 = \frac{\mu I^2}{8\pi^2}r^2 \quad \text{for} \quad 0 \le r \le r_o. \tag{3.215}$$

Since H is a function of r only, we choose a small volume dV to be a cylindrical shell of length l_w, radius r, and thickness dr along the radial direction. Thus, $dV = (2\pi r)(l_w)(dr)$. Assuming that μ is uniform for the entire conductor, the magnetic energy stored in the magnetic field inside the conductor is

$$W_m = \frac{1}{2}\int\int\int_V \mu H^2(r)dV = \frac{\mu l_w I^2}{4\pi r_o^4}\int_0^{r_o} r^3 dr = \frac{\mu l_w I^2}{16\pi}. \tag{3.216}$$

On the other hand, the magnetic energy stored in the magnetic field inside the conductor is given by

$$W_m = \frac{1}{2}LI^2. \tag{3.217}$$

Equating the two expressions for the magnetic energy W_m stored inside the conductor, we have

$$\frac{1}{2}LI^2 = \frac{\mu l_w I^2}{16\pi}. \quad (3.218)$$

Hence, we obtain the inductance of the conductor due to the magnetic field inside the conductor for low frequencies (with no skin effect present)

$$L_{LF} = \frac{\mu l_w}{8\pi}. \quad (3.219)$$

This expression is different than that obtained from the solution of the Bessel equation and approximated for low frequencies.

The amplitude of the magnetic field intensity outside the round conductor is

$$H(r) = \frac{I}{2\pi r} \quad \text{for} \quad r \geq r_o. \quad (3.220)$$

The magnetic energy stored in the magnetic field outside the round conductor is

$$w_m(r) = \frac{1}{2}\mu H^2(r) = \frac{\mu I^2}{8\pi^2 r^2} \quad \text{for} \quad r \geq r_o. \quad (3.221)$$

Using a small volume of a cylindrical shell outside the conductor $dV = 2\pi r l_w dr$, we obtain the total energy stored in the magnetic field outside the round conductor

$$W_m = \int\int\int_V w_m(r)dV = \frac{\mu l_w I^2}{4\pi}\int_{r_o}^{\infty}\frac{dr}{r} = \frac{\mu l_w I^2}{4\pi}\ln r|_{r_o}^{\infty} = \infty. \quad (3.222)$$

Let us consider the inductance of a round, long wire at high frequencies. Assume that the current density is uniform over the skin depth $r_o - \delta_w \leq r \leq r_o$ and it is zero in the middle of the conductor $0 \leq r < r_o - \delta_w$. The current density at the surface is

$$J(r_o) = \frac{I}{\pi[r_o^2 - (r_o - \delta_w)^2]} = \frac{I}{\pi(2r_o\delta_w - \delta_w^2)} \approx \frac{I}{2\pi r_o \delta_w}. \quad (3.223)$$

The current density at radius r in the skin depth range is

$$J(r) = \frac{I_{enc}(r)}{\pi[r^2 - (r_o - \delta_w)^2]} = \frac{I_{enc}(r)}{\pi[r^2 - r_o^2 + 2r_o\delta_w - \delta_w^2]}. \quad (3.224)$$

Setting $J(r) = J(r_o)$, we obtain

$$I_{enc}(r) = I\frac{r^2 - (r_o - \delta_w)^2}{r_o^2 - (r_o - \delta_w)^2} = I\frac{r^2 - r_o^2 + 2r_o\delta_w - \delta_w^2}{2r_o\delta_w - \delta_w^2}. \quad (3.225)$$

The magnetic field intensity within the skin depth area is given by

$$H(r) = \frac{I_{enc}(r)}{2\pi r} = I\frac{r^2 - (r_o - \delta_w)^2}{2\pi r[r_o^2 - (r_o - \delta_w)^2]} = I\frac{r^2 - r_o^2 + 2r_o\delta_w - \delta_w^2}{2\pi r[2r_o\delta_w - \delta_w^2]}. \quad (3.226)$$

The magnetic field intensity is proportional to r^2 within the skin depth area, where $H(r_o - \delta_w) = 0$ and $H(r_o) = I/(2\pi r_o)$. Using a small volume of the cylindrical shell $dV = 2\pi l_w r dr$, the magnetic energy stored in the magnetic field inside the skin depth conductor area is

$$W_m = \frac{1}{2}\int\int\int_V \mu H^2(r)dV = \frac{\mu l_w I^2}{4\pi[r_o^2 - (r_o - \delta_w)^2]^2}\int_{r_o-\delta_w}^{r_o}\frac{[r^2 - (r_o - \delta_w)^2]^2}{r}dr$$

$$= \frac{\mu l_w I^2}{4\pi[r_o^2 - (r_o - \delta_w)^2]}\left[-\frac{r_o^2}{2} + \frac{3r_o\delta_w}{2} - \frac{3\delta_w^2}{4} - \frac{(r_o - \delta_w)^4\ln\left(1 - \frac{\delta_w}{r_o}\right)}{r_o^2 - (r_o - \delta_w)^2}\right] \approx \frac{\mu l_w I^2}{12\pi}\left(\frac{\delta_w}{r_o}\right), \quad (3.227)$$

where $2r_o \gg \delta_w$ and $\ln(1 + x) \approx x - x^2/2$ for $x \ll 1$, resulting in $\ln(1 - \delta_w/r_o) \approx -\delta_w/r_o - \delta_w^2/(2r_o^2)$. The energy stored in the magnetic field in the skin depth area is

$$\frac{1}{2}LI^2 \approx \frac{\mu l_w I^2}{12\pi}\left(\frac{\delta_w}{r_o}\right). \quad (3.228)$$

Hence, the inductance related to the magnetic field in the skin depth area at high frequencies is

$$L_{HF} \approx \frac{\mu l_w}{6\pi}\left(\frac{\delta_w}{r_o}\right) = \frac{l_w}{6\pi r_o}\sqrt{\frac{\mu\rho_w}{\pi f}}. \tag{3.229}$$

Another way of determining the inductance is given below. The magnetic flux density is given by

$$B(r) = \mu H(r) = \frac{\mu I_{enc}(r)}{2\pi r} = \mu I\frac{r^2 - (r_o - \delta_w)^2}{2\pi r[r_o^2 - (r_o - \delta_w)^2]}. \tag{3.230}$$

The magnetic energy density inside the round conductor is

$$w_m(r) = \frac{1}{2}\mu H(r)^2 = \frac{\mu I^2}{8\pi^2 r_o^4}. \tag{3.231}$$

Since $dS = l_w dr$ and $d\phi = BdS = Bl_w dr$, the magnetic flux linkage is

$$\lambda = \int_S \left[\frac{r^2 - (r_o - \delta_w)^2}{r_o^2 - (r_o - \delta_w)^2}\right] BdS = \frac{\mu l_w I}{2\pi[r_o^2 - (r_o - \delta_w)^2]^2}\int_{r_o-\delta_w}^{r_o}\frac{[r^2 - (r_o - \delta_w)^2]^2}{r}dr$$
$$= \frac{\mu l_w I}{2\pi[r_o^2 - (r_o - \delta_w)^2]}\left[-\frac{r_o^2}{2} + \frac{3r_o\delta_w}{2} - \frac{3\delta_w^2}{4} - \frac{(r_o - \delta_w)^4 \ln\left(1 - \frac{\delta_w}{r_o}\right)}{r_o^2 - (r_o - \delta_w)^2}\right] \approx \frac{\mu I l_w}{6\pi}\left(\frac{\delta_w}{r_o}\right). \tag{3.232}$$

Hence, the high-frequency inductance related to the magnetic field inside the conductor is given by

$$L_{HF} = \frac{\lambda}{I} \approx \frac{\mu l_w}{6\pi}\left(\frac{\delta_w}{r_o}\right) = \frac{l_w}{6\pi r_o}\sqrt{\frac{\mu\rho_w}{\pi f}} = \frac{l_w}{6\pi r_o}\sqrt{\frac{\mu\rho_w}{\pi f}}. \tag{3.233}$$

3.9 Power Loss Density in Round Conductor

Assume that the current excitation of the round conductor is $i = I_m \sin\omega t = I \sin\omega t$. The low-frequency power dissipated in the round conductor is

$$P_{DC} = \frac{1}{2}R_{wDC}I_m^2 = \frac{1}{2}l_w A\rho_w J_m^2 = \frac{1}{2}\pi r_o^2 l_w \rho_w J_m^2 = \frac{1}{2}Vol\,\rho_w J_m^2 = \frac{\rho_w l_w I^2}{2\pi r_o^2}, \tag{3.234}$$

where $A = \pi r_o^2$ and $J_m = I_m/A$. This results in the DC power density per unit volume

$$p_{DC} = \frac{P_{DC}}{Vol} = \frac{1}{2}\rho_w J_m^2 = \frac{1}{2}\rho_w\left(\frac{I}{\pi r_o^2}\right)^2. \tag{3.235}$$

The time-average power density dissipated at any frequency is given by

$$p_D(r) = \frac{1}{2}\rho_w J(r)J^*(r) = \frac{1}{2}\rho_w|J(r)|^2 = \frac{1}{2}\rho_w\left(\frac{I}{2\pi r_o}\right)^2 kk^*\frac{I_0(kr)I_0^*(kr)}{I_1(kr_o)I_1^*(kr_o)}$$
$$= \frac{1}{2}p_{DC}\left(\frac{r_o}{\delta_w}\right)^2\frac{I_0(kr)I_0(k^*r)}{I_1(kr_o)I_1(k^*r_o)}, \tag{3.236}$$

where

$$k^* = (1 - j)\left(\frac{r_o}{\delta_w}\right)\left(\frac{r}{\delta_w}\right). \tag{3.237}$$

Figure 3.24 shows a plot of p_D/p_{DC} as a function of r/r_o.

The time-average power dissipated in the conductor is described by

$$P_D = \int\int\int_V p_D(r)dV = \frac{\rho_w}{2}\int_0^{l_w}\int_0^{2\pi}\int_0^{r}|J(r)|^2 rdrd\phi dz$$
$$= \frac{\rho_w l_w I^2}{4\sqrt{2}\pi r_o\delta_w}\frac{\left[\frac{\sqrt{2}}{(1+j)}I_0(k^*r_o)I_1(kr_o) + \frac{\sqrt{2}}{(1-j)}I_0(kr_o)I_1(k^*r_o)\right]}{I_1(k^*r_o)I_1(kr_o)}. \tag{3.238}$$

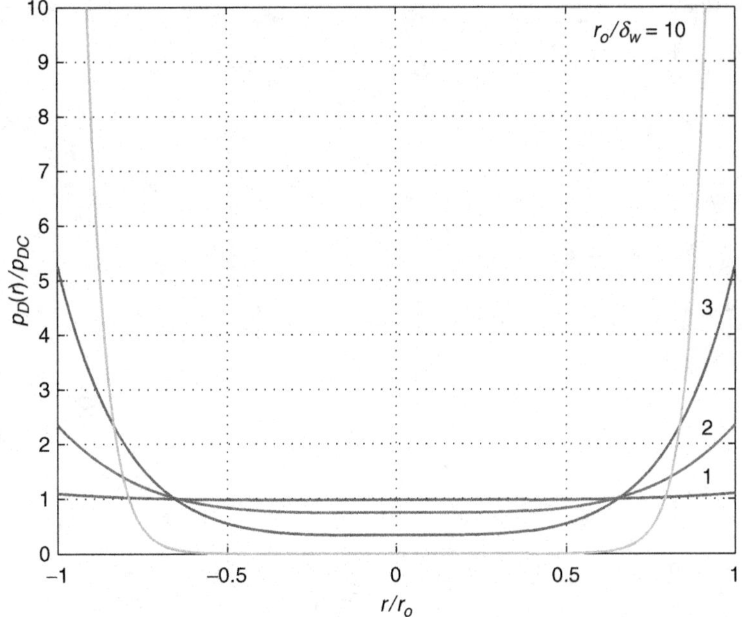

Figure 3.24 Normalized time-average power density $p_D(r)/P_{DC}$ as a function of r/r_o of a solitary single round conductor at various values of r_o/δ_w

The time-average power normalized with respect to the DC power is

$$\frac{P_D}{P_{DC}} = \frac{1}{4}\left(\frac{r_o}{\delta_w}\right)\frac{[(1-j)I_0(k^*r_o)I_1(kr_o) + (1+j)I_0(kr_o)I_1(k^*r_o)]}{I_1(k^*r_o)I_1(kr_o)}. \tag{3.239}$$

Figure 3.25 shows a plot of P_D/P_{DC} as a function of r/r_o. Note that

$$F_R = \frac{R_w}{R_{wDC}} = \frac{P_D}{P_{DC}} = Re\left\{\frac{J(r_o)}{J_{dc}}\right\} = Re\left\{\frac{kr_o}{2}\frac{I_0(kr_o)}{I_1(kr_o)}\right\}. \tag{3.240}$$

For low frequencies ($\delta_w \gg r_o$),

$$I_0(kr) \approx 1 + \frac{(kr)^2}{4} + \frac{(kr)^4}{64} \tag{3.241}$$

and

$$I_1(kr_o) \approx \frac{kr_o}{2} + \frac{(kr_o)^3}{16} + \frac{(kr_o)^5}{384} = \frac{kr_o}{2}\left[1 + \frac{(kr_o)^2}{8} + \frac{(kr_o)^4}{192}\right]. \tag{3.242}$$

The current density in the wire at low frequencies can be approximated by

$$J(r) = \frac{Ik}{2\pi r_o}\frac{I_0(kr)}{I_1(kr_o)} \approx \frac{I}{2\pi r_o}\frac{1 + \frac{(kr)^2}{4} + \frac{(kr)^4}{64}}{\frac{kr_o}{2}\left[1 + \frac{(kr_o)^2}{8} + \frac{(kr_o)^4}{192}\right]}$$

$$= \frac{I}{\pi r_o^2}\frac{1 + \frac{(kr)^2}{4} + \frac{(kr)^4}{64}}{1 + \frac{(kr_o)^2}{8} + \frac{(kr_o)^4}{192}} = J_{DC}\frac{1 + \frac{(kr)^2}{4} + \frac{(kr)^4}{64}}{1 + \frac{(kr_o)^2}{8} + \frac{(kr_o)^4}{192}} \quad \text{for} \quad 0 \le r \le r_o. \tag{3.243}$$

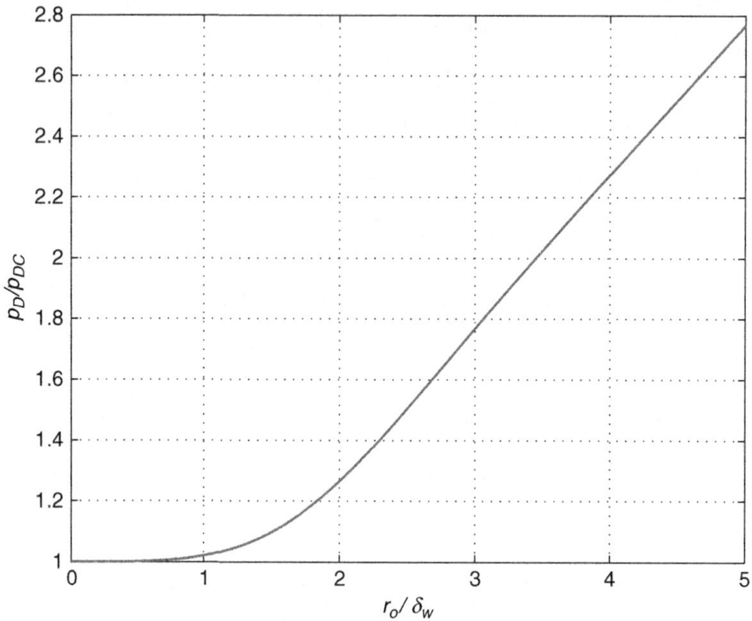

Figure 3.25 Normalized time-average power P_D/P_{DC} as a function of r_o/δ_w

The time-average power density dissipated in the conductor is

$$p_D(r) = \frac{1}{2}\rho_w J(r)J^*(r) = \frac{\rho_w}{2}\left(\frac{I}{\pi r_o^2}\right)^2 \frac{\left[1 + \frac{(kr)^2}{4} + \frac{(kr)^4}{64}\right]\left[1 + \frac{(k^*r)^2}{4} + \frac{(k^*r)^4}{64}\right]}{\left[1 + \frac{(kr_o)^2}{8} + \frac{(kr_o)^4}{192}\right]\left[1 + \frac{(k^*r_o)^2}{8} + \frac{(k^*r_o)^4}{192}\right]}$$

$$\approx p_{DC}\frac{1 + \frac{r^2}{4}(k^2 + k^{*2}) + \frac{r^4}{64}(k^4 + k^{*4} + 4k^2k^{*2})}{1 + \frac{r_o^2}{8}(k^2 + k^{*2}) + \frac{r_o^4}{192}(k^4 + k^{*4} + 3k^2k^{*2})}$$

$$= p_{DC}\frac{1 + \frac{1}{8}\left(\frac{r}{\delta_w}\right)^4}{1 + \frac{1}{48}\left(\frac{r_o}{\delta_w}\right)^4} \quad \text{for} \quad 0 \le r \le r_o, \tag{3.244}$$

where

$$(kk^*)^2 = \left[\frac{(1+j)(1-j)}{\delta_w^2}\right]^2 = \frac{4}{\delta_w^4} \tag{3.245}$$

$$k^2 + k^{*2} = \frac{(1+j)^2 + (1-j)^2}{\delta_w^2} = 0 \tag{3.246}$$

and

$$k^4 + k^{*4} = -\frac{8}{\delta^4}. \tag{3.247}$$

For $r = r_o$,

$$p_D(r_o) \approx p_{DC}\left[1+\frac{1}{8}\left(\frac{r_o}{\delta_w}\right)^4\right]\left[1-\frac{1}{48}\left(\frac{r_o}{\delta_w}\right)^4\right] \approx p_{DC}\left[1+\frac{5}{48}\left(\frac{r_o}{\delta_w}\right)^4\right]. \tag{3.248}$$

The time-average power dissipated in the wire at low frequencies is given by

$$P_D = \frac{1}{2}\int\int\int_V \rho_w J(r)J^*(r)dV = \int\int\int_V p_D(r)dV$$

$$\approx p_{DC}2\pi l_w \frac{1}{1+\frac{1}{48}\left(\frac{r_o}{\delta_w}\right)^4}\int_0^{r_o}\left[1+\frac{1}{8}\left(\frac{r}{\delta_w}\right)^4\right] rdr \approx p_{DC}2\pi l_w\left(\frac{r_o^2}{2}\right)\frac{\left[1+\frac{1}{24}\left(\frac{r_o}{\delta_w}\right)^4\right]}{\left[1+\frac{1}{48}\left(\frac{r_o}{\delta_w}\right)^4\right]}$$

$$\approx P_{DC}\left[1+\frac{1}{48}\left(\frac{r_o}{\delta_w}\right)^4\right] = \frac{\rho_w l_w I^2}{2\pi r_o^2}\left[1+\frac{1}{48}\left(\frac{r_o}{\delta_w}\right)^4\right]. \tag{3.249}$$

For high frequencies,

$$I_0(kr) \approx \frac{e^{kr}}{\sqrt{2\pi kr}}\left(1+\frac{1}{8kr}\right) \tag{3.250}$$

and

$$I_1(kr_o) \approx \frac{e^{kr_o}}{\sqrt{2\pi kr_o}}\left(1-\frac{3}{8kr_o}\right) \tag{3.251}$$

yielding the current density at high frequencies

$$J(r) = I\frac{k}{2\pi r_o}\frac{I_0(kr)}{I_1(kr_o)} \approx I\frac{k}{2\pi r_o}\sqrt{\frac{r_o}{r}}e^{-k(r_o-r)} \quad for \quad 0 \le r \le r_o. \tag{3.252}$$

The time-average power density at high frequencies is

$$p_D(r) = \frac{\rho_w}{2}J(r)J^*(r) = \frac{\rho_w I^2}{2}\frac{kk^*}{(2\pi r_o)^2}\left(\frac{r_o}{r}\right)e^{-(k+k^*)(r_o-r)}$$

$$= \frac{\rho_w I^2}{(2\pi r_o\delta_w)^2}\left(\frac{r_o}{r}\right)e^{-\frac{2(r_o-r)}{\delta_w}} = p_D(r_o)\left(\frac{r_o}{r}\right)e^{-\frac{2(r_o-r)}{\delta_w}} \quad \text{for} \quad 0 \le r \le r_o, \tag{3.253}$$

where

$$p_D(r_o) = \frac{\rho_w I^2}{(2\pi r_o\delta_w)^2} \tag{3.254}$$

$$kk^* = \frac{(1+j)(1-j)}{\delta_w^2} = \frac{2}{\delta_w^2} \tag{3.255}$$

and

$$k+k^* = \frac{1+j+1-j}{\delta_w} = \frac{2}{\delta_w}. \tag{3.256}$$

For $r = r_o - \delta_w$,

$$p_D(r_o-\delta_w) = \frac{\rho_w I^2}{(2\pi r_o\delta_w)^2}\left(\frac{r_o}{r_o-\delta_w}\right) = p_D(r_o)\frac{1}{1-\frac{\delta_w}{r_o}}e^{-2}. \tag{3.257}$$

The total time-average power dissipated in the wire of length l_w at high frequencies is given by

$$P_D = \frac{1}{2}\int\int\int_V \rho_w J(r)J^*(r)dV = \int\int\int_V p_D(r)dV = \frac{\rho_w l_w I^2}{4\pi^2 r_o\delta_w^2}e^{-\frac{2r_o}{\delta_w}}\int_0^{r_o}\frac{1}{r}e^{\frac{2r}{\delta_w}}2\pi rdr$$

$$= \frac{l_w\rho_w I^2}{4\pi r_o\delta_w}\left(1-e^{-\frac{2r_o}{\delta_w}}\right) \approx \frac{l_w\rho_w I^2}{4\pi r_o\delta_w} = \frac{\rho_w l_w}{2\pi r_o\delta_w}\frac{I^2}{2} = R_w\frac{I^2}{2}. \tag{3.258}$$

3.10 Skin Effect in Single Rectangular Plate

3.10.1 Magnetic Field Intensity in Single Rectangular Plate

A single conducting plate placed in free space is shown in Fig. 3.26. Assume that a sinusoidal current

$$i(t) = I \cos \omega t \tag{3.259}$$

flows through the plate in the $-y$-direction, causing a magnetic field inside and outside the plate in the y–z plane in the z-direction for $x > 0$ and in $-z$-direction for $x < 0$. Invoking the odd symmetry of the magnetic field intensity, $H(x) = -H(-x)$ and using Ampère's circuital law

$$\oint_C \mathbf{H} \cdot d\mathbf{l} = I \quad \text{for} \quad a \gg w \tag{3.260}$$

for the magnetic field intensity on the plate surface, we get

$$aH\left(\frac{w}{2}\right) + aH\left(-\frac{w}{2}\right) = I. \tag{3.261}$$

Assuming that $w \ll a$ and $w \ll b$, we obtain the amplitude of the magnetic field intensity at the surface of the plate

$$H\left(\frac{w}{2}\right) = -H\left(-\frac{w}{2}\right) = \frac{I}{2a} = H_m. \tag{3.262}$$

The magnetic field intensity H outside the conductor is assumed to be uniform in the y–z plane and parallel to the current sheet. The amplitude of this field changes only along the x-axis. Therefore, a one-dimensional solution is sufficient. The magnetic field is described by the ordinary second-order differential Helmholtz equation

$$\frac{d^2H(x)}{dx^2} = j\omega\mu_w\sigma_w H(x) = \gamma^2 H(x) \tag{3.263}$$

whose general solution is given by

$$H(x) = H_1 e^{\gamma x} + H_2 e^{-\gamma x}. \tag{3.264}$$

From the boundary conditions,

$$H\left(\frac{w}{2}\right) = H_1 e^{\gamma \frac{w}{2}} + H_2 e^{-\gamma \frac{w}{2}} \tag{3.265}$$

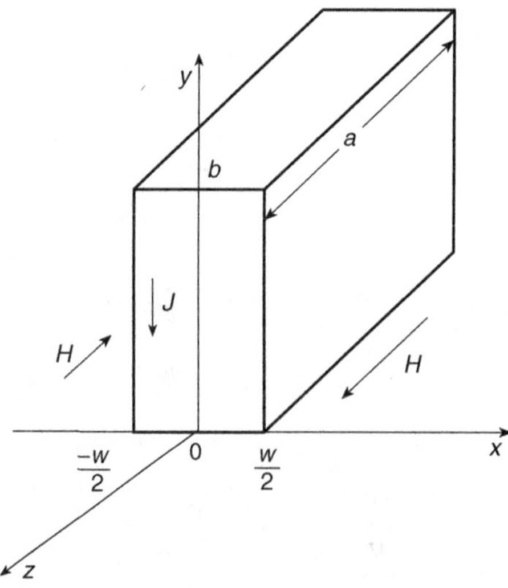

Figure 3.26 Single isolated plate carrying current

and

$$H\left(-\frac{w}{2}\right) = H_1 e^{-\gamma \frac{w}{2}} + H_2 e^{\gamma \frac{w}{2}}. \tag{3.266}$$

Adding these two equations, we obtain

$$H\left(\frac{w}{2}\right) + H\left(-\frac{w}{2}\right) = (H_1 + H_2)(e^{\gamma \frac{w}{2}} + e^{-\gamma \frac{w}{2}}) = 0 \tag{3.267}$$

which produces

$$H_1 = -H_2 = \frac{H_m}{2}. \tag{3.268}$$

Substitution of (3.268) into (3.265) gives

$$H\left(\frac{w}{2}\right) = H_1(e^{\gamma \frac{w}{2}} - e^{-\gamma \frac{w}{2}}) = 2H_1 \frac{(e^{\gamma \frac{w}{2}} - e^{-\gamma \frac{w}{2}})}{2} = 2H_1 \sinh\left(\gamma \frac{w}{2}\right) \tag{3.269}$$

from which

$$H_1 = -H_2 = \frac{H\left(\frac{w}{2}\right)}{2\sinh\left(\gamma \frac{w}{2}\right)} = \frac{I}{2a} \frac{1}{2\sinh\left(\gamma \frac{w}{2}\right)}. \tag{3.270}$$

Hence, the magnetic field intensity distribution inside the plate is

$$H(x) = 2H_1 \frac{e^{\gamma x} - e^{-\gamma x}}{2} = H_m \frac{\sinh(\gamma x)}{\sinh\left(\gamma \frac{w}{2}\right)} = \frac{I}{2a} \frac{\sinh(\gamma x)}{\sinh\left(\gamma \frac{w}{2}\right)} = \frac{I}{2a} \frac{\sinh\left[(1+j)\left(\frac{x}{\delta_w}\right)\right]}{\sinh\left[\frac{(1+j)}{2}\left(\frac{w}{\delta_w}\right)\right]}$$

$$= \frac{I}{2a} \frac{\sinh\left[(1+j)\left(\frac{w}{\delta_w}\right)\left(\frac{x}{w}\right)\right]}{\sinh\left[\frac{(1+j)}{2}\left(\frac{w}{\delta_w}\right)\right]} = |H(x)|e^{j\psi}. \tag{3.271}$$

Figures 3.27 and 3.28 show the plots of $2a|H(x)|/I$ and the phase ψ of the magnetic field intensity $H(x)$ as functions of x/w for selected values of w/δ_w.

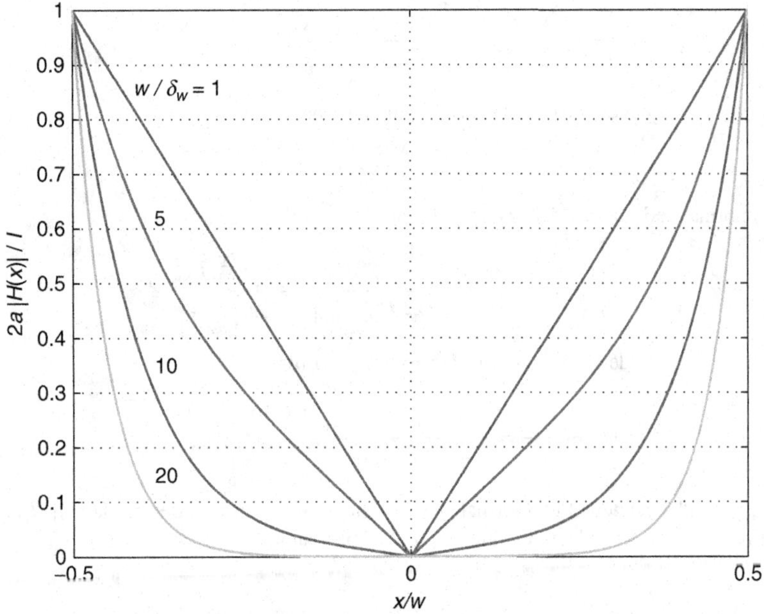

Figure 3.27 Plots of $2a|H(x)|/I$ as a function of x/w

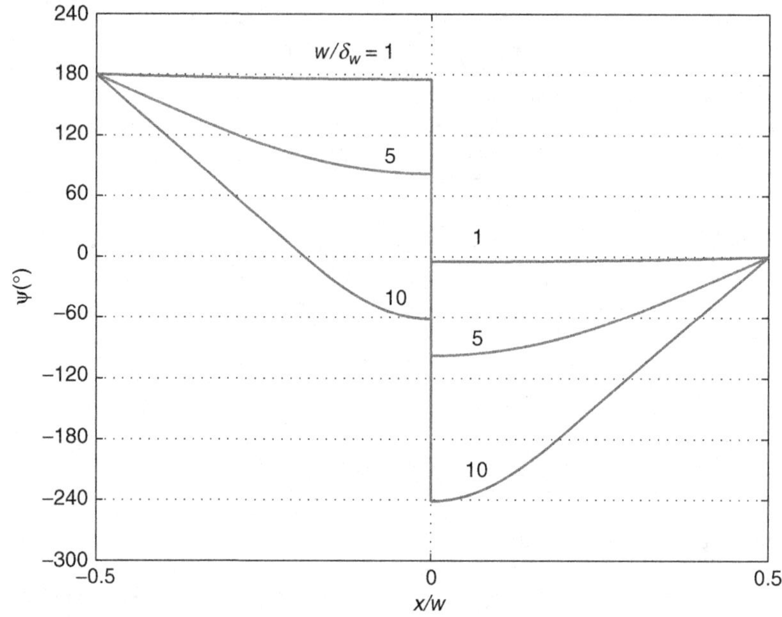

Figure 3.28 Plots of the phase of magnetic field intensity ψ as a function of x/w

3.10.2 Current Density in Single Rectangular Plate

The current density in the plate is expressed as

$$J(x) = J_y(x) = -\frac{dH}{dx} = -\gamma H_m \frac{\cosh(\gamma x)}{\sinh\left(\gamma \frac{w}{2}\right)} = -\frac{\gamma I}{2a} \frac{\cosh(\gamma x)}{\sinh\left(\gamma \frac{w}{2}\right)}$$

$$= -\frac{(1+j)I}{2a\delta_w} \frac{\cosh\left[(1+j)\left(\frac{w}{\delta_w}\right)\left(\frac{x}{w}\right)\right]}{\sinh\left[\frac{(1+j)}{2}\left(\frac{w}{\delta_w}\right)\right]}. \tag{3.272}$$

The DC and low-frequency current density is uniform and given by

$$J_{DC} = \frac{I}{aw}. \tag{3.273}$$

Hence, the normalized current density in the plate is

$$\frac{J(x)}{J_{DC}} = -\frac{(1+j)}{2}\left(\frac{w}{\delta_w}\right) \frac{\cosh\left[(1+j)\left(\frac{w}{\delta_w}\right)\left(\frac{x}{w}\right)\right]}{\sinh\left[\frac{(1+j)}{2}\left(\frac{w}{\delta_w}\right)\right]}. \tag{3.274}$$

The normalized amplitude of the current density is given by

$$\frac{|J(x)|}{J_{DC}} = \frac{1}{2}\left(\frac{w}{\delta_w}\right) \frac{\cosh\left(\frac{x}{\delta_w}\right) + \cos\left(\frac{x}{\delta_w}\right)}{\cosh\left(\frac{x}{\delta_w}\right) - \cos\left(\frac{x}{\delta_w}\right)}. \tag{3.275}$$

Figures 3.29 and 3.30 depict the normalized amplitude of the current density $|J(x)|/J_{DC}$ and phase ϕ as functions of x/w for selected values of w/δ_w for a single plate.

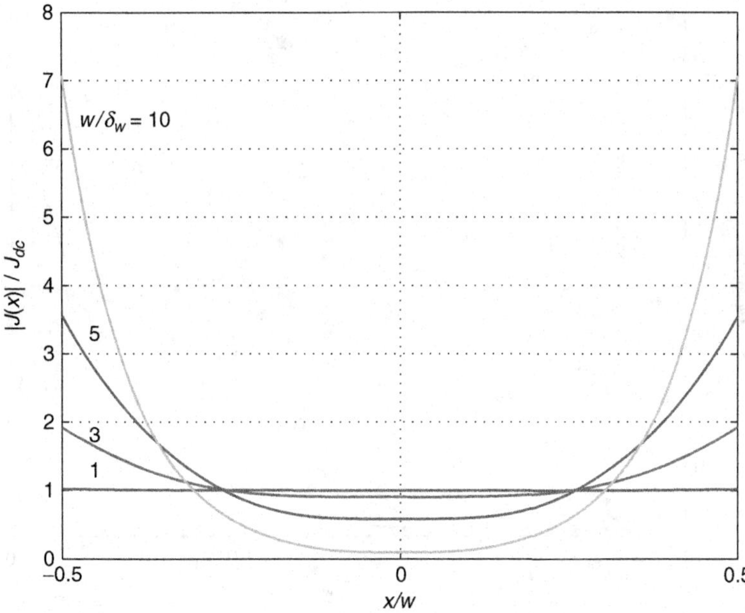

Figure 3.29 Plots of $|J(x)|/J_{DC}$ as a function of x/w at selected values of w/δ_w for a single plate

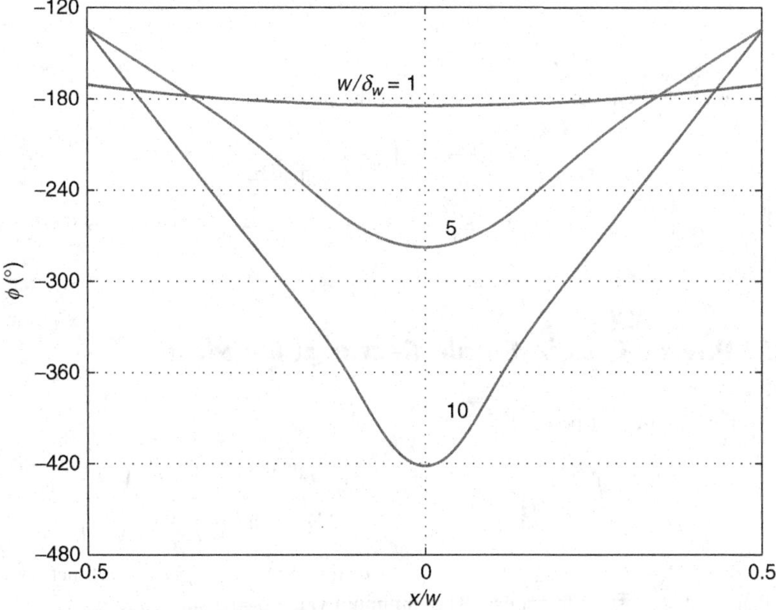

Figure 3.30 Plots of phase ϕ as a function of x/w at selected values of w/δ_w for a single plate

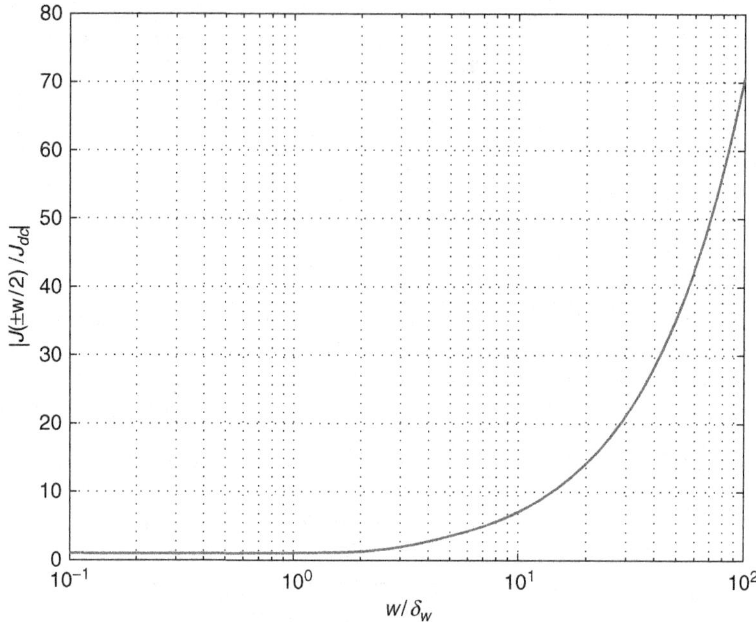

Figure 3.31 Plot of $J(\pm w/2)|/J_{DC}$ as a function of w/δ_w for a single plate

The normalized current density at each surface of the plate is

$$\frac{J\left(\pm\frac{w}{2}\right)}{J_{DC}} = -\frac{(1+j)}{2}\left(\frac{w}{\delta_w}\right)\coth\left[\frac{(1+j)}{2}\frac{w}{2}\right] \tag{3.276}$$

and the normalized current density at the center of the plate is

$$\frac{J(0)}{J_{DC}} = -\frac{(1+j)}{2}\left(\frac{w}{\delta_w}\right)\frac{1}{\sinh\left[\frac{(1+j)}{2}\frac{w}{2}\right]}. \tag{3.277}$$

Figure 3.31 shows a plot of $|J(\pm w/2)|/J_{DC}$ as a function of w/δ_w. A plot of $|J(0)|/J_{DC}$ is depicted in Fig. 3.32.

3.10.3 Power Loss in Single Rectangular Plate

The time-average skin-effect power loss is

$$P_D = \frac{1}{2}\int_0^a\int_0^b\int_{-w/2}^{w/2}\rho_w|J(x)|^2dxdydz = \frac{\rho_w bI^2}{4a\delta_w}\frac{\sinh\left(\frac{w}{\delta_w}\right)+\sin\left(\frac{w}{\delta_w}\right)}{\cosh\left(\frac{w}{\delta_w}\right)-\cos\left(\frac{w}{\delta_w}\right)}. \tag{3.278}$$

Figure 3.33 shows the time-average skin-effect power loss $4a\delta_w P_D/\rho_w bI^2$ as a function of w/δ_w at fixed δ_w, that is, at a constant frequency. The minimum value of the skin-effect power loss occurs at the optimal thickness of the strip w_{opt} in free space

$$\frac{w_{opt}}{\delta_w} = \pi, \tag{3.279}$$

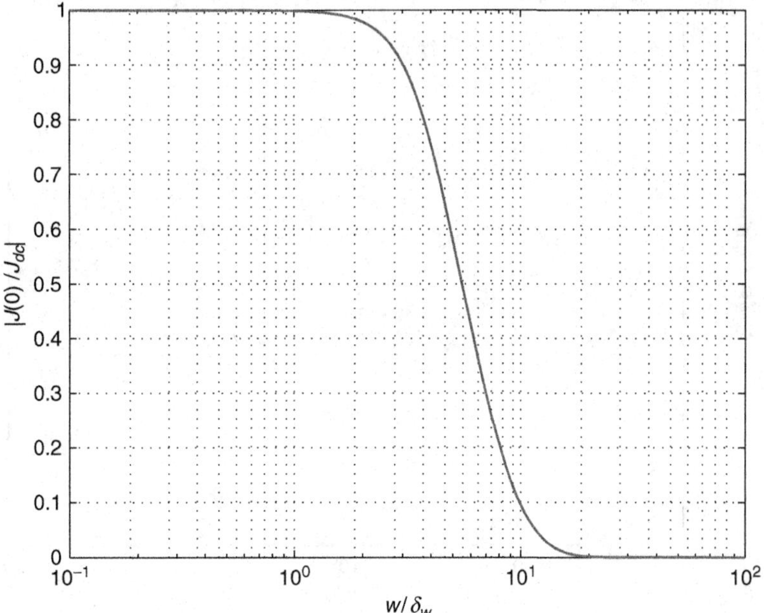

Figure 3.32 Plot of $|J(0)|/J_{DC}$ as a function of w/δ_w for a single plate

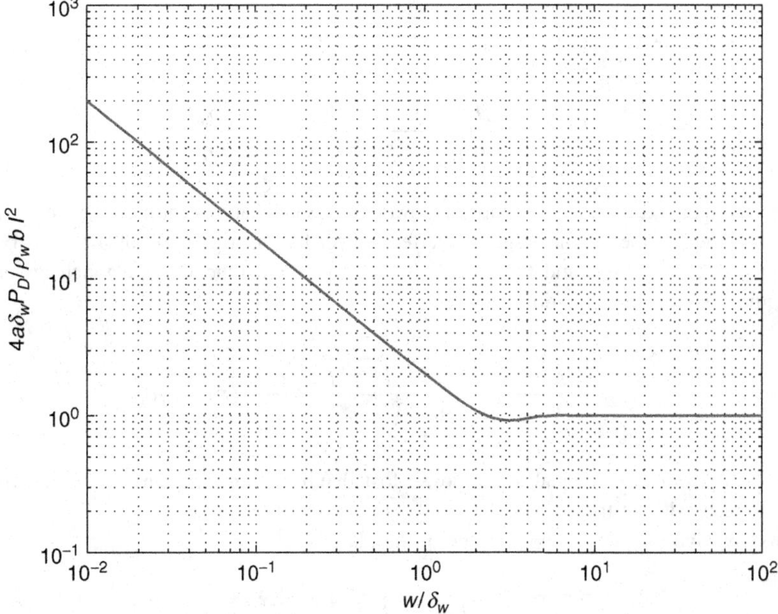

Figure 3.33 Time-average skin-effect power loss $4a\delta_w P_D/b\rho_w I^2$ as a function of w/δ_w at fixed δ_w

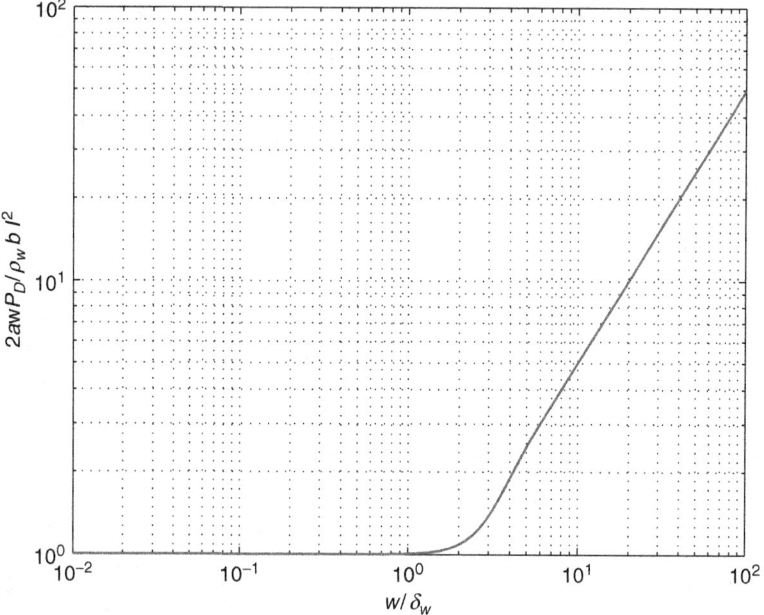

Figure 3.34 Time-average skin-effect power loss $2awP_D/b\rho_w I^2$ as a function of w/δ_w at fixed w

yielding the minimum skin-effect power loss

$$P_{Dmin} = \frac{\rho_w b I^2}{4a\delta_w} \frac{\sinh \pi}{\cosh \pi + 1} = 0.917 \frac{I^2 \rho_w b}{a\delta_w}. \tag{3.280}$$

Thus, the minimum skin-effect power loss is lower by 8.83% than that at high frequencies.

Figure 3.34 depicts the time-average skin-effect power loss as a function of w/δ_w at fixed w. For high frequencies, $w/\delta_w \gg 1$ and the values of the hyperbolic functions are much larger than the values of the trigonometric functions, resulting in

$$P_D \approx \frac{\rho_w b I^2}{4a\delta_w} = \frac{\rho_w b I^2}{4a}\sqrt{\frac{\pi\mu_w f}{\rho_w}} = \frac{bI^2}{4a}\sqrt{\pi\mu_w\rho_w f}. \tag{3.281}$$

Thus, the total power loss is equal to the power loss due to the uniform current flow in $2\delta_w$, that is, in $delta_w$ on each side of the strip.

The time-average magnetic energy stored in the plate is

$$W_m = \frac{1}{2}\int\int\int_V \mu_w |H|^2 dV = \frac{\mu_w}{2}\int_0^a \int_0^b \int_{-w/2}^{w/2} |H|^2 dxdydz$$
$$= \frac{ab\mu_w}{2}\int_{-w/2}^{w/2} \left|\frac{I}{2a}\frac{\sinh(\gamma x)}{\sinh\left(\gamma \frac{w}{2}\right)}\right|^2 dx = \frac{\mu_w b\delta_w I^2}{4a}\frac{\sinh\left(\frac{w}{\delta_w}\right) - \sin\left(\frac{w}{\delta_w}\right)}{\cosh\left(\frac{w}{\delta_w}\right) - \cos\left(\frac{w}{\delta_w}\right)}. \tag{3.282}$$

Figure 3.35 shows the normalized magnetic energy as a function of w/δ_w at fixed δ.

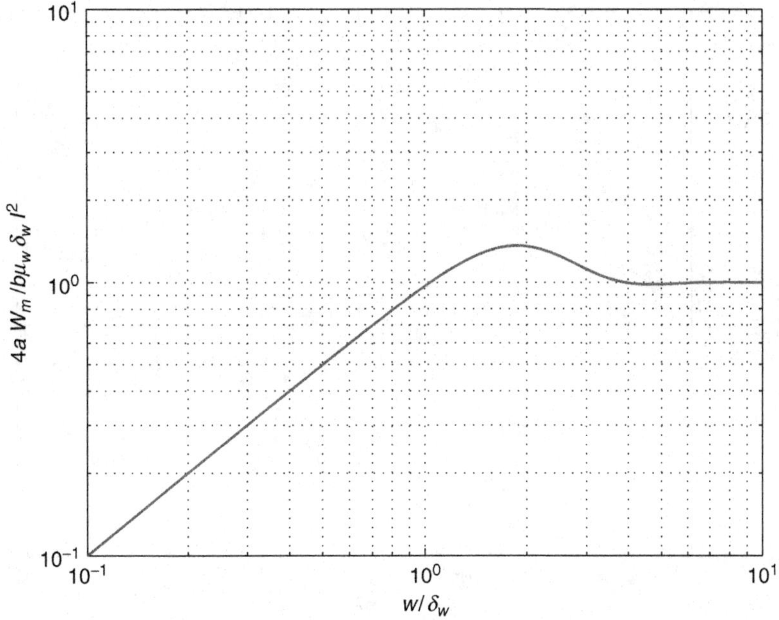

Figure 3.35 Time-average skin-effect energy stored in the plate $4aW_m/b\delta_w\mu_w I^2$ as a function of w/δ_w

3.10.4 Impedance of Single Rectangular Plate

The power loss in the plate at any frequency can be expressed as

$$P_D = \frac{1}{2}I^2R_w = \frac{1}{2}I^2\frac{\rho_w b}{2a\delta_w}\frac{\sinh\left(\frac{w}{\delta_w}\right)+\sin\left(\frac{w}{\delta_w}\right)}{\cosh\left(\frac{w}{\delta_w}\right)-\cos\left(\frac{w}{\delta_w}\right)} = \frac{1}{2}I^2\frac{\rho_w b}{2aw}\left(\frac{w}{\delta_w}\right)\frac{\sinh\left(\frac{w}{\delta_w}\right)+\sin\left(\frac{w}{\delta_w}\right)}{\cosh\left(\frac{w}{\delta_w}\right)-\cos\left(\frac{w}{\delta_w}\right)}$$

$$= \frac{P_{wDC}}{2}\left(\frac{h}{\delta_w}\right)\frac{\sinh\left(\frac{w}{\delta_w}\right)+\sin\left(\frac{w}{\delta_w}\right)}{\cosh\left(\frac{w}{\delta_w}\right)-\cos\left(\frac{w}{\delta_w}\right)} \tag{3.283}$$

yielding the resistance at any frequency

$$R_w = \frac{\rho_w b}{2a\delta_w}\frac{\sinh\left(\frac{w}{\delta_w}\right)+\sin\left(\frac{w}{\delta_w}\right)}{\cosh\left(\frac{w}{\delta_w}\right)-\cos\left(\frac{w}{\delta_w}\right)} = \frac{\rho_w b}{2aw}\left(\frac{w}{\delta_w}\right)\frac{\sinh\left(\frac{w}{\delta_w}\right)+\sin\left(\frac{w}{\delta_w}\right)}{\cosh\left(\frac{w}{\delta_w}\right)-\cos\left(\frac{w}{\delta_w}\right)}. \tag{3.284}$$

The DC resistance of the plate is

$$R_{wDC} = \rho_w\frac{b}{aw}. \tag{3.285}$$

The ratio of the resistance at any frequency to the DC resistance is

$$F_R = \frac{R_w}{R_{wDC}} = \frac{1}{2}\left(\frac{w}{\delta_w}\right)\frac{\sinh\left(\frac{w}{\delta_w}\right)+\sin\left(\frac{w}{\delta_w}\right)}{\cosh\left(\frac{w}{\delta_w}\right)-\cos\left(\frac{w}{\delta_w}\right)}. \tag{3.286}$$

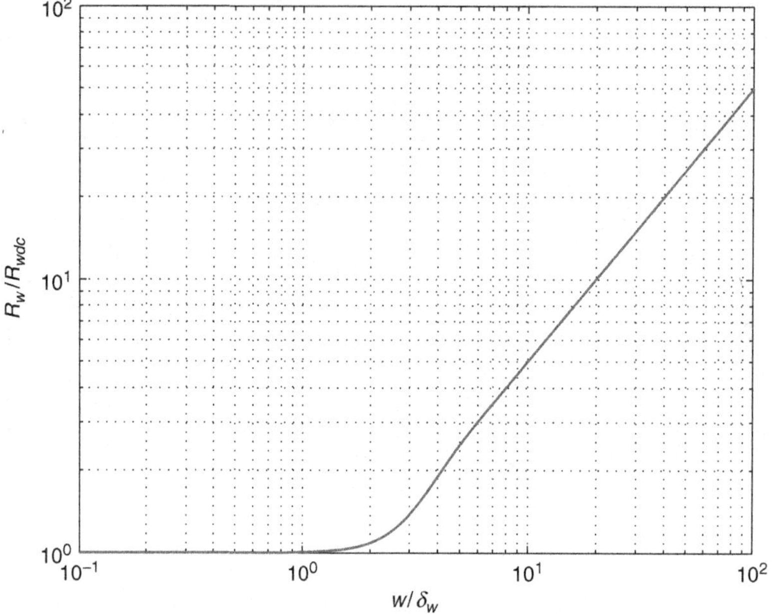

Figure 3.36 Ratio of R_w/R_{wDC} as a function of w/δ_w

Figure 3.36 shows the ratio of R_w/R_{wDC} as a function of w/δ_w. It can be seen that the AC resistance is equal to the DC resistance for the range

$$0 \leq \frac{w}{\delta_w} \leq 1. \tag{3.287}$$

The maximum frequency at which the skin effect can be neglected is determined by the condition

$$w = \delta_w = \sqrt{\frac{\rho_w}{\pi \mu_0 f_{sk}}} \tag{3.288}$$

producing the maximum frequency at which the skin effect is negligible for a single plate

$$f_{sk} = \frac{\rho_w}{\pi \mu_0 w^2}. \tag{3.289}$$

Assuming that the current flows uniformly over the skin depth δ_w only near both the surfaces of effective area $A_e = 2\delta_w a$, the plate AC resistance at high frequencies is

$$R_w \approx \frac{\rho_w b}{2a\delta_w} \approx \frac{b}{2a}\sqrt{\pi \mu_w \rho_w f} \tag{3.290}$$

yielding

$$F_R = \frac{R_w}{R_{wDC}} = \frac{w}{2\delta_w} = \frac{b}{2}\sqrt{\frac{\pi \mu_0 f}{\rho_w}}. \tag{3.291}$$

The magnetic energy stored inside the plate can be expressed as

$$\frac{1}{2}LI^2 = \frac{\mu_w b \delta_w I^2}{4a} \frac{\sinh\left(\frac{w}{\delta_w}\right) - \sin\left(\frac{w}{\delta_w}\right)}{\cosh\left(\frac{w}{\delta_w}\right) - \cos\left(\frac{w}{\delta_w}\right)} \tag{3.292}$$

yielding the internal inductance

$$L = \frac{X_L}{\omega} = \frac{\mu_w b \delta_w}{2a} \frac{\sinh\left(\frac{w}{\delta_w}\right) - \sin\left(\frac{w}{\delta_w}\right)}{\cosh\left(\frac{w}{\delta_w}\right) - \cos\left(\frac{w}{\delta_w}\right)}. \tag{3.293}$$

The reactance X_L due to the internal inductance L normalized with respect to the DC resistance R_{wDC} is given by

$$F_X = \frac{X_L}{R_{wDC}} = \frac{1}{2}\left(\frac{w}{\delta_w}\right)\frac{\sinh\left(\frac{w}{\delta_w}\right) - \sin\left(\frac{w}{\delta_w}\right)}{\cosh\left(\frac{w}{\delta_w}\right) - \cos\left(\frac{w}{\delta_w}\right)}. \tag{3.294}$$

Figure 3.37 depicts the ratio X_L/R_{wDC} as a function of w/δ_w. The total internal impedance is

$$Z = \frac{1}{2}R_{wDC}\left(\frac{w}{\delta_w}\right)\left[\frac{\sinh\left(\frac{w}{\delta_w}\right) + \sin\left(\frac{w}{\delta_w}\right)}{\cosh\left(\frac{w}{\delta_w}\right) - \cos\left(\frac{w}{\delta_w}\right)} + j\frac{\sinh\left(\frac{w}{\delta_w}\right) - \sin\left(\frac{w}{\delta_w}\right)}{\cosh\left(\frac{w}{\delta_w}\right) - \cos\left(\frac{w}{\delta_w}\right)}\right]$$
$$= R_{wDC} + jX_L = |Z|e^{j\phi_Z}. \tag{3.295}$$

Figures 3.38 and 3.39 show $|Z|/R_{wDC}$ and ϕ_Z as functions of w/δ_w. Figure 3.40 depicts the ratio X_L/R_w.

An alternative method of deriving an expression for the plate impedance is given below. The electric field intensity is given by

$$E = \rho_w J = \rho_w \frac{dH}{dx} = -\frac{\rho_w \gamma I}{2a}\frac{\cosh(\gamma x)}{\sinh(\gamma \frac{w}{2})}. \tag{3.296}$$

The plate impedance is

$$Z = \frac{V}{I} = \frac{bE\left(\frac{w}{2}\right)}{I} = \frac{b\rho_w\gamma}{2a}\coth\left(\gamma\frac{w}{2}\right) = \frac{b\rho_w}{2a\delta_w}(1+j)\coth\left(\gamma\frac{w}{2}\right)$$
$$= \frac{1}{2}R_{wDC}\left(\frac{w}{\delta_w}\right)(1+j)\coth\left(\gamma\frac{w}{2}\right) = R + jX_L. \tag{3.297}$$

Using the relationship

$$\coth(1+j)x = \frac{\sinh 2x - j\sin 2x}{\cosh 2x - \cos 2x} \tag{3.298}$$

one obtains (3.295).

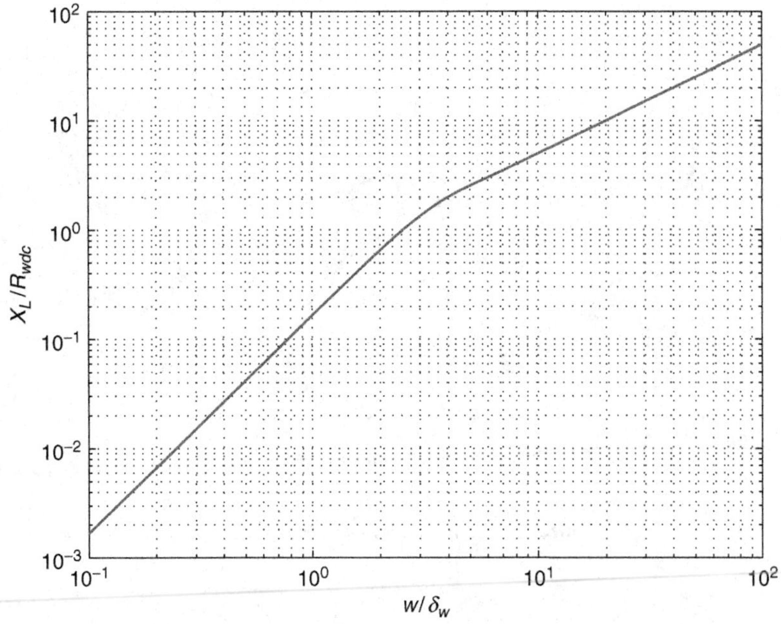

Figure 3.37 Ratio X_L/R_{wDC} as a function of w/δ_w

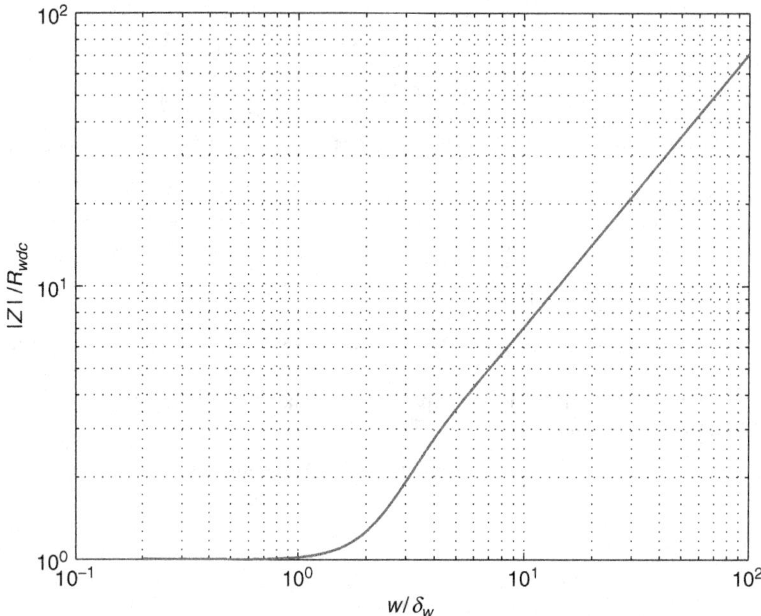

Figure 3.38 Plot of $|Z|/R_{wDC}$ as a function of w/δ_w

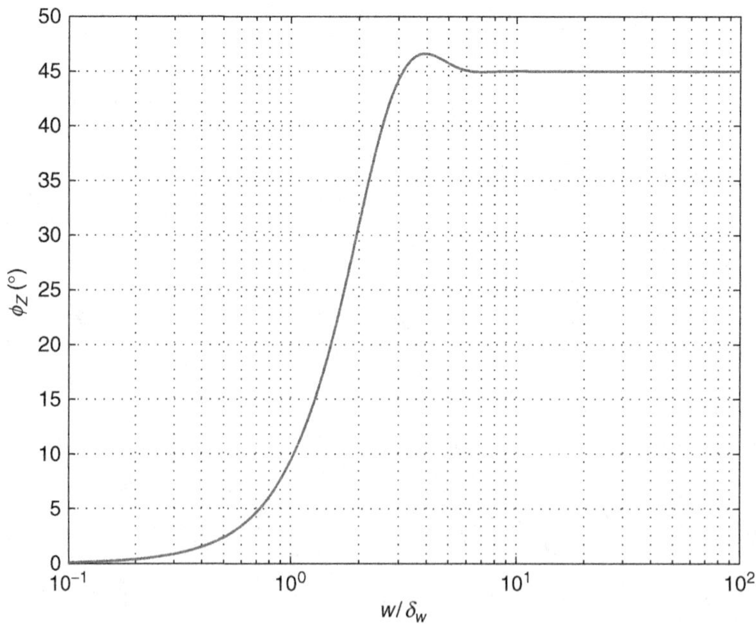

Figure 3.39 Plot of ϕ_Z as a function of w/δ_w

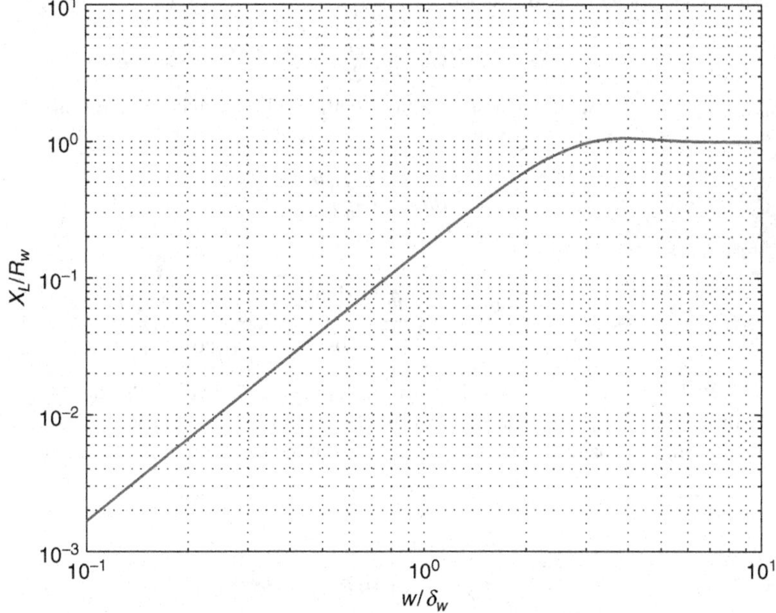

Figure 3.40 Ratio X_L/R_w as a function of w/δ_w

3.11 Skin Effect in Rectangular Foil Conductor Placed Over Ideal Core

Consider a conducting rectangular strip, which has a length b, width a, and thickness w. It is assumed that $w \ll a$ and $w \ll b$. One side of the conductor is placed on an ideal magnetic core with $\mu_{rc} \to \infty$ and the other side is placed in free space. This can be a single-layer winding of an indictor. The magnetic field intensity in the core and at the inner surface of the conductor is zero and there is a sinusoidal magnetic field intensity in free space along the length l of the strip. A sinusoidal current flows along the width w. The problem can be solved using 1D field analysis in Cartesian coordinates. The DC resistance of a rectangular foil conductor is

$$R_{wDC} = \rho_w \frac{w}{hl} \tag{3.299}$$

and the DC power loss is

$$P_{wDC} = R_{wDC} I_{DC}^2. \tag{3.300}$$

The AC power at low frequencies when the current distribution is uniform is

$$P_{w(LF)} = \frac{R_{w(LF)} I_m^2}{2} = R_{w(LF)} I_{rms}^2. \tag{3.301}$$

If $I_{rms} = I_{DC}$, the power loss is the same at the DC and at low frequencies.

The time-average power loss at high frequencies due to the skin effect is given by [9]

$$P_w = \frac{I_m^2 \rho_w b}{2a\delta_w} \frac{\sinh 2\left(\frac{w}{\delta_w}\right) + \sin 2\left(\frac{w}{\delta_w}\right)}{\cosh 2\left(\frac{w}{\delta_w}\right) - \cos 2\left(\frac{w}{\delta_w}\right)} = P_{wDC}\left(\frac{h}{\delta_w}\right) \frac{\sinh 2\left(\frac{w}{\delta_w}\right) + \sin 2\left(\frac{w}{\delta_w}\right)}{\cosh 2\left(\frac{w}{\delta_w}\right) - \cos 2\left(\frac{w}{\delta_w}\right)}, \tag{3.302}$$

where the low-frequency and DC resistance is

$$P_{wDC} = \frac{I_m^2 \rho_w b}{2ah} \tag{3.303}$$

and the high-frequency resistance, which is equal to the DC resistance of the conductor with its thickness equal to one skin depth δ_w

$$P_{w(HF)} = \frac{I_m^2 \rho_w b}{2a\delta_w}. \tag{3.304}$$

The AC-to-DC resistance ratio

$$F_R = \frac{P_w}{P_{wDC}} = \frac{R_w}{R_{wDC}} = \left(\frac{w}{\delta_w}\right) \frac{\sinh 2\left(\frac{w}{\delta_w}\right) + \sin 2\left(\frac{w}{\delta_w}\right)}{\cosh 2\left(\frac{w}{\delta_w}\right) - \cos 2\left(\frac{w}{\delta_w}\right)} \tag{3.305}$$

The power loss P_w is twice the power loss in a strip with magnetic field on both sides of the conductor. In this case, the current flows only on one side of the strip at high frequencies. Figures 3.41–3.44 shows the plots of the normalized power loss. The minimum value of the resistance occurs at

$$\frac{w_{opt}}{\delta_w} = \frac{\pi}{2} \tag{3.306}$$

and is given by

$$P_{w(min)} = 0.92 P_{w(HF)}. \tag{3.307}$$

As the relative permeability μ_r decreases from infinity to 1, the optimum foil conductor thickness w_{opt}/δ_w increases from $\pi/2$ to π and the minimum resistance decreases by a factor of 2. Actually, the conductor resistance depends on μ_r. As the relative permeability μ_r decreases from infinity to 1, the AC resistance decreases for $h/\delta_w > 0.5$.

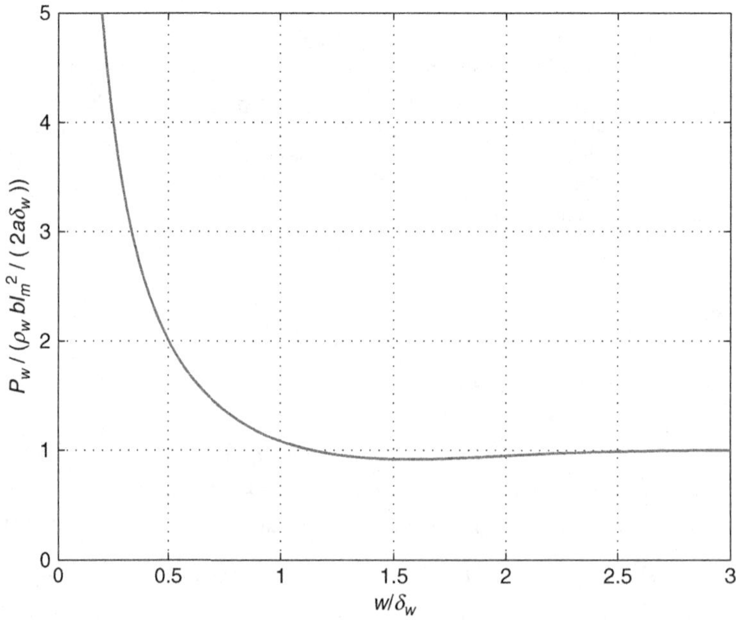

Figure 3.41 Normalized power loss $P_w/(\rho_w b I_m^2/(2a\delta_w))$ as a function of w/δ_w

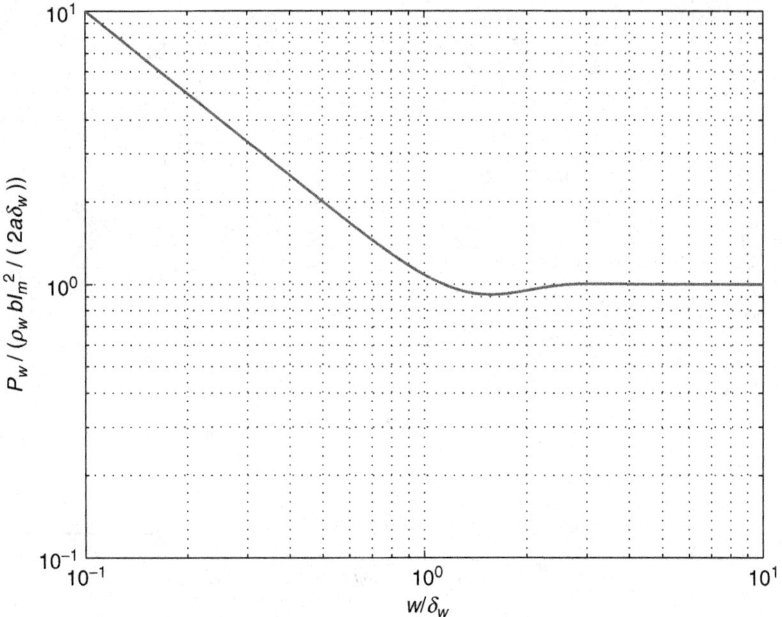

Figure 3.42 Normalized power loss $P_w/(\rho_w b I_m^2/(2a\delta_w))$ as a function of w/δ_w

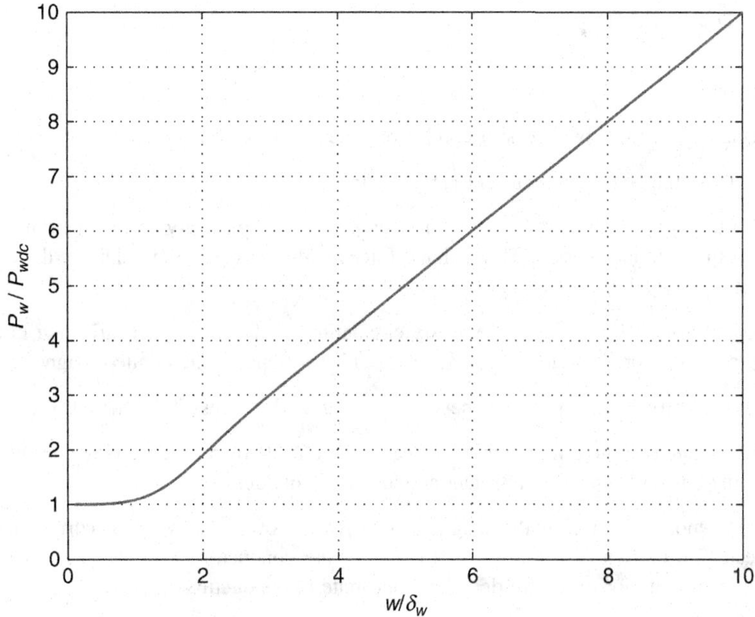

Figure 3.43 Normalized power loss P_w/P_{wDC} as a function of w/δ_w

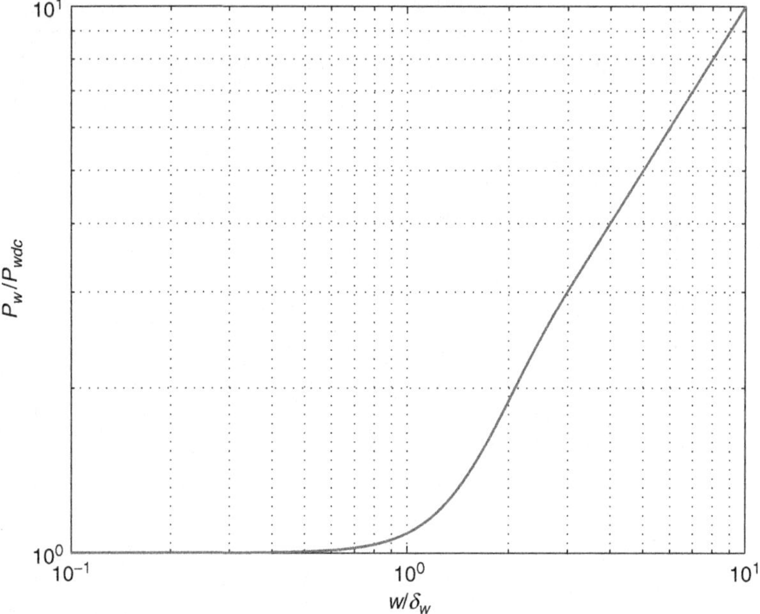

Figure 3.44 Normalized power loss P_w/R_{wDC} as a function of w/δ_w

3.12 Summary

- The resistivity of conductors increases with temperature.
- The resistance and the power loss of conductors increase with temperature.
- The conductor ampacity is the current-carrying limit.
- According to Faraday's law, if a conductor is subjected to a time-varying magnetic field, a voltage is induced in a conductor loop. This voltage forces eddy currents to circulate in the conductor in closed paths.
- According to Lenz's law, eddy currents are circulating in closed loops, nearly like in turns of an inductor, and therefore produce own magnetic field that opposes the applied magnetic field.
- Skin and proximity effects occur whenever the conductor thickness is a fraction of the skin depth.
- The induced magnetic field in a conductor may be due to its own time-varying current or due to a time-varying current of another adjacent conductor or conductors.
- The phenomenon of inducing the magnetic field by its own time-varying current is called the skin effect. The current distribution is affected by the conductor's own magnetic field. At high frequencies, the interior of a wire does not contribute to conduction.
- Skin effect in conductors is caused by self-induced eddy currents.
- The phenomenon of inducing the magnetic field by the time-varying current of another adjacent conductor (or conductors) is called the proximity effect.
- Proximity effect in conductors is caused by externally induced eddy currents.
- The skin and proximity losses are orthogonal to each other and can be determined independently.

- When an AC magnetic field is applied in parallel to the surface of a plate conductor, the magnetic field magnitude is attenuated in the conductor.
- In accordance with Lenz's law, eddy currents produce their own magnetic field to oppose the original field.
- For an infinite frequency, no magnetic field is penetrating into a conductor.
- There are two effects of eddy currents: skin effect and proximity effect. Both of these effects cause current crowding.
- The skin effect takes place when the time-varying magnetic field due to the current flow in a conductor induces eddy currents in the conductor itself.
- Bessel functions are often used as solutions in numerous physical problems, involving cylindrical and spherical geometries. They are also used to describe the current density and the magnetic field intensity in a round conductor at high frequencies.
- When a high-frequency current flows through a single solid round wire, the current flows closer and closer to the outer part of the conductor as the frequency is increased. Therefore, the current and the magnetic flux inside the conductor are reduced.
- Eddy currents cause nonuniform distribution of current density, an increase in conductor AC resistance, an increase in conduction power loss, and a reduction in inductance.
- The skin-effect eddy current and the proximity-eddy current are orthogonal.
- According to Lenz's law, the direction of eddy currents is such that they oppose the change that causes them.
- Both the skin effect and the proximity effect contribute to the nonuniformity of current distribution in conductors.
- The skin effect is the tendency of AC current to flow near the surface of a conductor, thereby restricting the current flow to a small part of the total cross-sectional area and increasing the conductor effective resistance.
- The skin effect and the proximity effect limit the effective capability of a conductor to conduct high-frequency currents.
- The skin depth has a physical meaning. At $x = \delta_w$, the current density in a conductor is reduced to $1/e = 0.37$ of its value on the conductor surface.
- The skin depth is a function of frequency f and conductor properties ρ and μ_r.
- The skin depth δ_w is proportional to $1\sqrt{f}$.
- For $\delta_w > d$, the skin effect can be neglected and the AC winding resistance is approximately equal to the DC winding resistance.
- The low-frequency resistance of a round wire is $R_{LF} \approx \rho_w l/(\pi r_o^2)$ and the low-frequency inductance is $L_{LF} \approx \mu l/(8\pi)$, that is, $L_{LF}/l = 50$ nH/m.
- At high frequencies, the current flows in the area close to the conductor surface due to the skin effect.
- The high-frequency resistance of a round wire is $R_{HF} \approx (l/r_o)\sqrt{f\mu\rho_w/(2\pi)}$, and the high-frequency inductance of a round wire is $L_{HF} \approx (l/r_o)\sqrt{\mu\rho_w/(8\pi^3 f)}$.
- The skin depth δ_w increases as the conductor resistivity ρ_w increases and the conductor relative permeability μ_r decreases.
- The skin depth δ_w increases with temperature T because the conductor resistivity ρ_w increases with temperature T and the relative permeability μ_r decreases with temperature T.

- For $\delta_w < d$, the skin effect increases the AC winding resistance over the DC winding resistance, increasing the copper loss.
- At high frequencies, the resistance is equal to the reactance of good conductors.
- At high frequencies, the resistance of a round wire with radius r_o and length l_w is equal to the resistance of a rectangular conductor of width $2\pi r_o$, length l_w, and thickness δ_w.
- Since the current flows on the surface of the conductors, high-frequency inductors are often silver-plated or are made of litz wire.
- Litz wire increases the effective conduction area at high frequencies and thereby reduces the copper loss.

3.13 Appendix

3.13.1 Derivation of Bessel Equation for Long Round Wire

Consider a infinitely long round solid straight conductor of radius r_o located along the z-axis. The current density and the electric field intensity have only z-components. The current density J in the conductor at high frequencies is not uniform; it depends on the distance from the conductor axis r, that is, $J(r)$. The cross-sectional area of a shell of radius r and thickness dr is $S = 2\pi r dr$. The current flowing through the shell is

$$I_{enc} = \int_0^r J dS = \int_0^r J 2\pi r dr. \tag{3.308}$$

From Ampère's circuital law,

$$\oint_C \mathbf{H} \cdot d\mathbf{l} = H 2\pi r = I_{enc}, \tag{3.309}$$

resulting in the magnetic flux intensity

$$H = \frac{I_{encl}}{2\pi r} = \frac{1}{2\pi r}\int_0^r J 2\pi r dr = \frac{1}{r}\int_0^r J r dr \tag{3.310}$$

and

$$Hr = \int_0^r J r dr. \tag{3.311}$$

Differentiating both sides of this equation,

$$r\frac{dH}{dr} + H\frac{dr}{dr} = Jr, \tag{3.312}$$

producing

$$J = \frac{dH}{dr} + \frac{H}{r}. \tag{3.313}$$

Hence,

$$\frac{dJ}{dt} = \frac{d}{dr}\left(\frac{dH}{dt}\right) + \frac{1}{r}\frac{dH}{dt} = \frac{d^2H}{drdt} + \frac{1}{r}\frac{dH}{dt}. \tag{3.314}$$

The magnetic flux flowing through the cross-sectional area of the shell of length l and thickness dr is

$$\phi = BdS = \mu H(ldr). \tag{3.315}$$

From Faraday's law, the voltage induced by the time-varying magnetic flux in the shell is

$$v = \frac{d\phi}{dt} = \mu dr\frac{dH}{dt}, \tag{3.316}$$

yielding the electric field intensity

$$E = \frac{v}{l} = \mu dr \frac{dH}{dt}. \tag{3.317}$$

The current density on the outer surface of the shell J_o in terms of the current density on its inner surface J is

$$J_o = J + \frac{\partial J}{\partial r} dr, \tag{3.318}$$

resulting in the electric field intensity on the outer surface

$$E_o = \rho_w \left(J + \frac{\partial J}{\partial r} dr \right). \tag{3.319}$$

Therefore, the change in the electric field from the inner to the outer shell surface is

$$E = \rho_w \frac{\partial J}{\partial r} dr. \tag{3.320}$$

Equating the equations for the electric field intensity given by (3.317) and (3.320) in the shell, we obtain

$$\rho_w \frac{\partial J}{\partial r} dr = \mu dr \frac{dH}{dt}. \tag{3.321}$$

Hence,

$$\frac{dH}{dt} = \frac{\rho_w}{\mu} \frac{dJ}{dr} \tag{3.322}$$

Thus, the time rate of change of H is equal to the rate of change of J with respect to r multiplied by a factor ρ_w/μ.

Rearrangement of (3.322) gives

$$\frac{dJ}{dr} = \frac{\rho_w}{\mu} \frac{dJ}{dt}. \tag{3.323}$$

The rate of change of J with respect of r is equal to the time rate of change of H times μ/ρ_w.

Differentiating (3.322),

$$\frac{d}{dr}\left(\frac{dH}{dt} \right) = \frac{\rho_w}{\mu} \frac{d^2 J}{dr^2} \tag{3.324}$$

results in

$$\frac{d^2 H}{drdt} = \frac{\rho_w}{\mu} \frac{d^2 J}{dr^2}. \tag{3.325}$$

Substitution of (3.322) and (3.325) in (3.314) yields

$$\frac{dJ}{dt} = \frac{\rho_w}{\mu} \frac{d^2 J}{dr^2} + \frac{1}{r} \frac{\rho_w}{\mu} \frac{dJ}{dr}. \tag{3.326}$$

Rearrangement of this produces the Bessel's equation in the time domain

$$\frac{d^2 J}{dr^2} + \frac{1}{r} \frac{dJ}{dr} = \frac{\mu}{\rho_w} \frac{dJ}{dt}. \tag{3.327}$$

For sinusoidal current flowing through the conductor, phasor description is useful. Using the transformation from the time domain to the frequency domain $dJ/dt \rightarrow j\omega J$, we get

$$\frac{d^2 J}{dr^2} + \frac{1}{r} \frac{dJ}{dr} = \frac{\mu}{\rho_w} j\omega J = \frac{j\omega\mu}{\rho_w} J = j\omega\mu\sigma J. \tag{3.328}$$

Hence, we obtain the Bessel equation in the phasor form as

$$\frac{d^2 J}{dr^2} + \frac{1}{r} \frac{dJ}{dr} - j\omega\mu\sigma J = 0. \tag{3.329}$$

3.14 References

[1] P. J. Dowell, "Effects of eddy currents in transformer winding," *Proceedings of the IEE*, vol. 113, no. 8, pp. 1387–1394, August 1966.

[2] E. Bennet and S. C. Larsen, "Effective resistance of alternating currents of multilayer windings," *Transactions of the American Institute of Electrical Engineers*, vol. 59, pp. 1010–1017, 1940.

[3] J. Jongsma, "High-frequency ferrite power transformer and choke design, Part 3: transformer winding design, Part 4: improved method of power choke design" Philips Electronic Components and Materials, Technical Publication, no. 27, Philips, The Netherlands, 1986.

[4] A. Kennelly, F. Laws, and P. Pierce, "Experimental research on skin effect in conductors," *Transactions of the AIEE*, vol. 34, p. 1915, 1915.

[5] J. Lammeraner and M. Stafl, *Eddy Currents*, Cleveland: CRS Press, 1966.

[6] E. C. Snelling, *Soft Ferrites: Properties and Applications*, London: Iliffe Books Ltd, 1969.

[7] R. L. Stall, *The Analysis of Eddy Currents*, Oxford: Clarendon Press, 1974, pp. 21–27.

[8] W. T. McLyman, *Transformer and Inductor Design Handbook*, 3rd Ed., New York: Marcel Dekker, 2004.

[9] R. L. Perry, "Multiple layer series connected winding design for minimum losses," *IEEE Transactions on Power Applications Systems*, vol. PAS-98, pp. 116–123, January/February 1979.

[10] B. Carsten, "High frequency conductor losses in switch mode magnetics," Proceedings of PCI, Munich, Germany, May 1986, pp. 161–182.

[11] J. P. Vandalec and P. D. Ziogos, "A novel approach for minimizing high-frequency transformer copper loss," *IEEE Transactions on Power Electronics*, vol. 3, pp. 266–276, July 1988.

[12] A. M. Urling, V. A. Niemela, G. R. Skutt, and T. G. Wilson, "Characterizing high frequency effects in transformer windings: a guide to several significant papers," 4th Annual IEEE Transactions on Power Electronics Specialists Conference, Baltimore, MD, March 13–17, 1989, pp. 373–385.

[13] J. A. Ferreira, *Electromagnetic Modelling of Power Electronic Converters*, Boston, MA: Kluwer Academic Publisher, 1989.

[14] E. E. Kreisis, T. D. Tsiboukis, S. M. Panas, and J. A. Tegopoulos, "Eddy currents: theory and applications," *Proceedings of the IEEE*, vol. 88, no. 10, pp. 1559–1589, October 1992.

[15] M. Bartoli, A. Reatti, and M. K. Kazimierczuk, "High-frequency models of ferrite inductors," Proceedings of the IEEE International Conference on Industrial Electronics, Controls, Instrumentation, and Automation (IECON'94), Bologna, Italy, September 5-9, 1994, pp. 1670–1675.

[16] M. Bartoli, A. Reatti, and M. K. Kazimierczuk, "Predicting the high-frequency ferrite-core inductor performance," Proceedings of the Conference of Electrical Manufacturing and Coil Winding, Chicago (Rosemont), IL, September 27-29, 1994, pp. 409–413.

[17] M. Bartoli, A. Reatti, and M. K. Kazimierczuk, "Modeling iron-powder inductors at high-frequencies," Proceedings of the IEEE Industry Applications Society Annual Meeting, Denver, CO, October 2-7, 1994, pp. 1125–1232.

[18] J. A. Ferreira, "Improved analytical modeling of conductive losses in magnetic components," *IEEE Transactions on Power Electronics*, vol. 9, pp. 127–131, January 1994.

[19] M. Bartoli, N. Noferi, A. Reatti, and M. K. Kazimierczuk, "Modeling winding losses in high-frequency power inductors," *Journal of Circuits Systems and Computers*, vol. 5, no. 3, pp. 65–80, March 1995.

[20] N. Mohan, T. M. Underland, and W. P. Robbins, *Power Electronics*, 3rd Ed., New York: John Wiley & Sons, 2003.

[21] M. Bartoli, N. Nefari, A. Reatti, and M. K. Kazimierczuk, "Modeling litz-wire winding losses in high-frequencies power inductors," Proceedings of the IEEE Power Electronics Specialists Conference, Baveno, Italy, June 24-27, 1996, pp. 1690–1696.

[22] R. Petkov, "Optimum design of a high-power high-frequency transformers," *IEEE Transactions on Power Electronics*, vol. 11, no. 1, pp. 33–42, January 1996.

[23] N. H. Kutkut, "A simple technique to evaluate winding losses including two-dimensional edge effects," *IEEE Transactions on Power Electronics*, vol. 13, no. 4, pp. 950–658, September 1998.

[24] W. G. Hurley, W. H. Wolfe, and J. G. Breslin, "Optimized transformer design: inclusive of high-frequency effects," *IEEE Transactions on Power Electronics*, vol. 13, no. 4, pp. 651–659, July 1998.

[25] W. G. Hurley, E. Gath, and J. G. Breslin, "Optimizing the ac resistance of multilayer transformer windings with arbitrary current waveforms," *IEEE Transactions on Power Electronics*, vol. 15, no. 2, pp. 369–376, March 2000.

[26] M. K. Kazimierczuk, G. Sancineto, U. Reggiani, and A. Massarini, "Small-signal high-frequency model of ferrite inductors," *IEEE Transactions on Magnetics*, vol. 35, pp. 4185–4191, September 1999.

[27] U. Reggiani, G. Sancineto, and M. K. Kazimierczuk, "High-frequency behavior of laminated iron-core inductors for filter applications," Proceedings of the IEEE Applied Power Electronics Conference, New Orleans, LA, February 6-10, 2000, pp. 654–660.

[28] G. Grandi, M. K. Kazimierczuk, A. Massarini, U. Reggiani, and G. Sancineto, "Model of laminated iron-core inductors," *IEEE Transactions on Magnetics*, vol. 40, no. 4, pp. 1839–1845, July 2004.

[29] K. Howard and M. K. Kazimierczuk, "Eddy-current power loss in laminated power cores," Proceedings of the IEEE International Symposium on Circuits and Systems, Sydney, Australia, May 7-9, 2000, paper III-668, pp. 668–672.

[30] A. Reatti and M. K. Kazimierczuk, "Comparison of various methods for calculating the ac resistance of inductors," *IEEE Transactions on Magnetics*, vol. 37, pp. 1512–1518, May 2002.

[31] G. Grandi, M. K. Kazimierczuk, A. Massarini, and U. Reggiani, "Stray capacitance of single-layer solenoid air-core inductors," *IEEE Transactions on Industry Applications*, vol. 35, no. 5, pp. 1162–1168, September 1999.

[32] A. Massarini and M. K. Kazimierczuk, "Self-capacitance of inductors," *IEEE Transactions on Power Electronics*, vol. 12, no. 4, pp. 671–676, July 1997.

[33] T. L. Simpson, "Effect of a conducting shield on the inductance of an air-core solenoid," *IEEE Transactions on Magnetics*, vol. 35, no. 1, pp. 508–515, January 1999.

[34] H. A. Wheeler, "Formulas for the skin effect," *Proceedings of the IRE*, vol. 30, pp. 412–424, September 1942.

[35] R. W. Erickson and D. Maksimović, *Fundamentals of Power Electronics*, Norwell, MA: Kluwer Academic Publishers, 2001.

[36] A. Van den Bossche and V. C. Valchev, *Inductors and Transformers for Power Electronics*, Boca Raton, FL: Taylor & Francis, 2005.

[37] J. C. Maxwell, *A Treatise of Electricity and Magnetism*, 3rd Ed., New York: Dover Publishing, 1997.

[38] R. Wrobel, A. Mlot, and P. H. Meller, "Contribution of end-winding proximity losses to temperature variation in electromagnetic devices," *IEEE Transactions on Industrial Electronics*, vol. 59, no. 2, pp. 848–857, February 2012.

[39] T. Suetsugu and M. K. Kazimierczuk, "Integration of Class DE inverter for dc-dc converter on-chip power supplies," IEEE International Symposium on Circuits and Systems, Kos, Greece, May 21-24, 2006, pp. 3133–3136.

[40] T. Suetsugu and M. K. Kazimierczuk, "Integration of Class DE synchronized dc-dc converter on-chip power supplies," 37th IEEE Power Electronics Specialists Conference, Jeju, South Korea, June 21-24, 2006, pp. 1–5.

3.15 Review Questions

3.1. What is the skin effect?

3.2. What is the skin depth?

3.3. In which part of a single conductor does the eddy current tend to oppose the applied current $i(t)$?

3.4. In which part of a single conductor does the eddy current tend to add to the applied current $i(t)$?

3.5. What is the current density distribution in a single round conductor at high frequencies?

3.6. How is the skin depth related to frequency?

3.7. How is the skin depth related to the conductor resistivity?

3.8. How is the skin depth related to the conductor relative permeability?

3.9. How is the AC-to-DC winding resistance ratio related to the skin effect for a single round conductor?

3.10. How is the conductor AC resistance related to the skin effect?

3.11. Why is the AC resistance higher than the DC resistance of a round conductor at high frequencies?

3.12. Does the resistance of a conductor depend on temperature?

3.13. What is the effective cross-sectional area of a round conductor at high frequencies?

3.16 Problems

3.1. An iron steel core has $\rho_c = 7 \times 10^{-7}$ Ω·m, $l_c = 168$ mm, $l_g = 0.4$ mm, $\mu_{rc} = 300$, and $f = 60$ Hz. Find the skin depth of the core.

3.2. The average resistivity of a human body is $\rho = 10$ Ω·m. Calculate the skin depth δ at $f = 60$ Hz, 1 MHz, and 1 GHz.

3.3. An iron steel core has $\rho_c = 9 \times 10^{-7}$ Ω·m, $\mu_{rc} = 400$, $l_c = 150$ mm, $l_g = 0.5$ mm, and $f = 60$ Hz. Find the skin depth of the core.

3.4. A toroidal core has $N = 500$, $\mu_{rc} = 200$, and $A_c = 4\ \text{cm}^2$. The resistivity of the copper winding conductor is $\rho_{Cu} = 1.724 \times 10^8$ Ω·m and the resistivity of the core material is $\rho_c = 10^5$ Ω·m. Find δ_w, δ_c, and δ_c/δ_w.

3.5. An Mn–Zn ferrite core has $\rho_c = 100$ Ω cm and $\mu_{rc} = 1000$. Find the skin depth at 1 and 10 MHz.

3.6. A standard annealed copper wire has a resistance of 1 Ω at $T = 20\,^\circ$C. What is the copper resistance at $T = 90\,^\circ$C?

3.7. An RF air-core inductor is wound using a copper wire and three turns of a single layer. The bare wire diameter is $d = 0.643$ mm and the wire length $l_w = 14.6$ mm. The operating frequency is $f = 740$ MHz. The inductor current is sinusoidal. Find the winding AC resistance. Calculate the winding AC resistance and the AC resistance factor F_R.

3.8. A copper straight round wire with diameter $d = 1$ mm and length $l = 0.5$ m conducts current $i = 8\sin(2\pi \times 2 \times 10^5 t)$ A.

(a) Find the AC resistance of the wire R_w.

(b) Find the AC-to-DC resistance ratio F_R.

(c) Find the power loss in the wire P_w.

3.9. The current density in a solid round straight conductor of radius r_o is approximated at a high frequency by an expression $\mathbf{J} = J_o\left(\frac{r}{r_o}\right)\mathbf{a_z}$.

(a) Determine the magnetic field intensity inside the conductor.

(b) Determine the magnetic field intensity outside the conductor.

3.10. The current density in a solid round straight conductor of radius r_o is approximated at a high frequency by an expression $\mathbf{J} = J_o\left(\frac{r}{r_o}\right)^2\mathbf{a_z}$.

(a) Determine the magnetic field intensity inside the conductor.

(b) Determine the magnetic field intensity outside the conductor.

3.11. A copper straight round wire AWG18 with a diameter $d = 1.02$ mm and a length $l_c = 0.9$ m and conductor current $i = 5\sin 2\pi \times 10^6 t$ A.

(a) Estimate the effective cross-sectional area of the current flow.

(b) Estimate the wire AC resistance.

(c) Estimate F_R.

(d) Estimate the AC power loss P_w.

3.12. A copper bare round conductor has the diameter $d = 1$ mm, the electron mobility $\mu_n = 0.0032$ m^2/V·s, the resistivity $\rho_{Cu} = 1.724 \times 10^{-8}$ Ω·m, and the conductivity $\sigma_{Cu} = 8.8$ S/m. The wire is subjected to the electric field $E = 0.1$ V/m.

(a) Determine the volume charge density of free electrons.

(b) Determine the volume density of free electrons.

(c) Determine the current density.

(d) Determine the current.

(e) Determine the average drift velocity.

4

Proximity Effect

4.1 Introduction

A time-varying current i_1 in one conductor generates both external and internal time-varying magnetic fields H_1. This field, in turn, induces a time-varying current i_2 known as *eddy current* in a nearby conductor, causing power losses. In a conductor, the induced magnetic field may be due to the own time-varying current of the conductor or the time-varying current in other adjacent conductors. Proximity or closeness of other current-carrying conductors affects the ability of the conductor to carry high-frequency current. The magnetic fields induced by conductors in close proximity will add or subtract depending on their directions. The *proximity effect* in inductors and transformers [1–42] is caused by the time-varying magnetic fields arising from currents flowing in adjacent winding layers in a multiple-layer winding. Currents flow in the opposite directions in the same conductor of a multiple-layer winding except for the first layer. As a result, the amplitudes and rms values of eddy currents caused by magnetic fields in adjacent layers due to the proximity effect increase significantly as the number of layers N_l increases. Therefore, the power loss due to the proximity effect in multiple-layer windings is much higher than the power loss due to the skin effect.

In general, the proximity effect occurs when the current in nearby conductors causes a time-varying magnetic field and induces a circulating current inside the conductor. It is similar to the skin effect, but it is caused by the current carried by nearby conductors. In other words, the proximity effect causes magnetic fields due to high-frequency currents in one conductor to induce voltages in adjacent conductors, which in turn cause eddy currents in the adjacent conductors or in adjacent winding layers in multilayer inductors and transformers. Each conductor is subjected to its own field and the fields generated by other conductors. Eddy currents are induced in a conductor by a time-varying magnetic field whether or not the conductor carries current. If the conductor carries current, the skin-effect eddy current and the proximity-effect eddy current superimpose to form the total eddy current. If the conductor does not carry current, then only the proximity-effect eddy current is induced. The skin-effect current and the proximity-effect current are orthogonal to each other. This is because the current density due to the skin effect exhibits an even symmetry and the current density due to the proximity effect exhibits an odd symmetry. The proximity effect causes nonuniform current density in the cross section of conductors, increasing significantly the winding loss at high frequencies. When two or more conductors are brought into close proximity, their magnetic fields may add or subtract. The high-frequency current will concentrate within a conductor, where the magnetic fields are additive.

High-Frequency Magnetic Components, Second Edition. Marian K. Kazimierczuk.

Companion Website: www.wiley.com/go/kazimierczuk_High2e

The magnitude of the proximity effect depends on (i) frequency, (ii) conductor geometry (shape and size), (iii) arrangement of conductors, and (iv) spacing. Mathematically, the proximity effect is very complex.

4.2 Orthogonality of Skin and Proximity Effects

Assume that a sinusoidal current flows in a conductor. The phasor of total current density $\mathbf{J}$ can be expressed as the sum of the skin current density $\mathbf{J_s}$ and the proximity current density $\mathbf{J_p}$

$$\mathbf{J} = \mathbf{J_s} + \mathbf{J_p}. \tag{4.1}$$

The power loss density is [20]

$$P_v = \frac{1}{2}\rho_w \int\int\int_V |\mathbf{J}| dV = \frac{1}{2}\rho_w \int\int\int_V |\mathbf{J}\cdot\mathbf{J}^*| dV$$
$$= \frac{1}{2}\rho_w \int\int\int_V (J_s + J_p)(J_s^* + J_p^*) dV. \tag{4.2}$$

If the conductor has an axis of symmetry and the applied field is uniform and parallel to symmetry axis, the distribution of the skin current density is an even function and the distribution of the proximity current is an odd function. Therefore,

$$P_v = \frac{1}{2}\rho_w \int\int\int_V (J_s + J_p)(J_s^* + J_p^*) dV = \frac{1}{2}\rho_w \int\int\int_V (J_s J_s^* + J_p J_p^*) dV = P_{vs} + P_{vp} \tag{4.3}$$

where the skin-effect power loss density is

$$P_{vs} = \frac{1}{2}\rho_w \int\int\int_V J_s J_s^* dV \tag{4.4}$$

and the proximity-effect power loss density is

$$P_{vp} = \frac{1}{2}\rho_w \int\int\int_V J_p J_p^* dV. \tag{4.5}$$

4.3 Proximity Effect in Two Parallel Round Conductors

The current density J is high where the magnetic field intensity H is high. If two conductors carry current in opposite directions, their magnetic fields add between the conductors. Therefore, magnetic flux linkages around the adjacent parts of the conductors decrease and in the remote parts increase, causing concentration of current in the adjacent parts of the conductors, as shown in Fig. 4.1a. If the currents in the conductors flow in the same direction, the action is reversed. The magnetic fields oppose each other between the conductors and concentration of current occurs in the more remote parts of the conductors, as shown in Fig. 4.1b. To determine the inductance of the two parallel conductors, the magnetic flux ϕ through the area S between the two conductors should be determined because all the magnetic flux lines cross the surface S.

At low frequencies, the current density is uniform in both conductors and the proximity and skin effects can be neglected. It is assumed that there is no interaction between the two conductors. The low-frequency inductance of two long parallel round conductors conducting currents in opposite directions with the proximity effect neglected is given by

$$L = L_i + L_e = \frac{\lambda}{I} = \frac{2}{I}\left[\int_0^{l_w}\int_0^a \left(\frac{r}{a}\right)^2 \left(\frac{\mu I r}{2\pi a^2}\right) dr dz + \int_0^{l_w}\int_a^{d-a} \frac{\mu I}{2\pi r} dr dz\right]$$
$$= \frac{\mu l_w}{4\pi} + \frac{\mu l_w}{\pi}\ln\left(\frac{d}{a} - 1\right) \approx \frac{\mu l_w}{4\pi} + \frac{\mu l_w}{\pi}\ln\frac{d}{a} \quad \text{for} \quad d \gg a \tag{4.6}$$

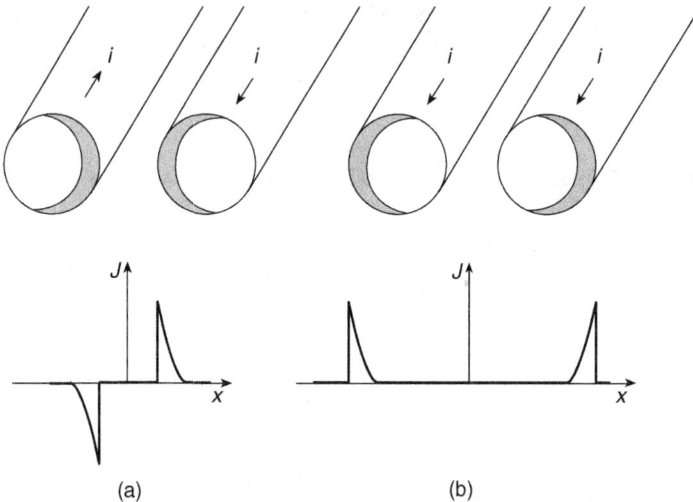

Figure 4.1 Proximity effect in two parallel round conductors at high frequencies. (a) Two conductors with currents flowing in opposite directions. (b) Two conductors with currents flowing in the same direction

where d is the separation between the conductor centers, a is the radius of the conductors, L_i is the internal self-inductance of both conductors due to the flux linkage inside the conductors, and L_e is the external self-inductance due to the flux linkage between the conductors. The inductance of two parallel conductors conducting currents in the same direction is ideally zero because the magnetic fluxes between the conductors cancel out each other. At high frequencies, the current density in the conductors is not uniform due to the proximity effect.

4.4 Proximity Effect in Coaxial Cable

Coaxial cables are often used to connect electrical devices. Consider a coaxial cable shown in Fig. 4.2. The radius of the inner conductor is a, the inner radius of the outer conductor is b, and the outer radius of the outer conductor is c. In a coaxial cable, the outer cylindrical conductor carries the return current of the inner conductor. Both conductors carry the same current I in opposite directions. For example, the inner conductor carries current from a source to a load and the outer conductor carries the same current back to the source. At high frequencies, the magnetic flux linkages in the inner and outer conductors tend toward zero as the current in the inner conductor concentrates at its surface due to skin effect (like in a single conductor) and the current in the outer conductor concentrates at its inner surface due to proximity effect and skin effect, as depicted in Fig. 4.2. The outer conductor acts like a shield.

At low frequencies, the current densities in both conductors are uniform. The magnetic field in the inner conductor is proportional to the radius r and is inversely proportional to the radius r between the conductors and in the outer conductor. The magnetic field is generated in the inner conductor by the current I flowing in the inner conductor. This field is given by

$$H = \frac{Ir}{2\pi a^2} \quad \text{for} \quad 0 \leq r \leq a. \tag{4.7}$$

From Ampère's law, the magnetic field in the annular space between the two conductors is also generated by the current I flowing in the inner conductor. This field is expressed as

$$H = \frac{I}{2\pi r} \quad \text{for} \quad a \leq r \leq b \tag{4.8}$$

where r is measured from the common center of both conductors. The magnetic field in the outer conductor is generated by the current flowing in the inner conductor and the amount of the current

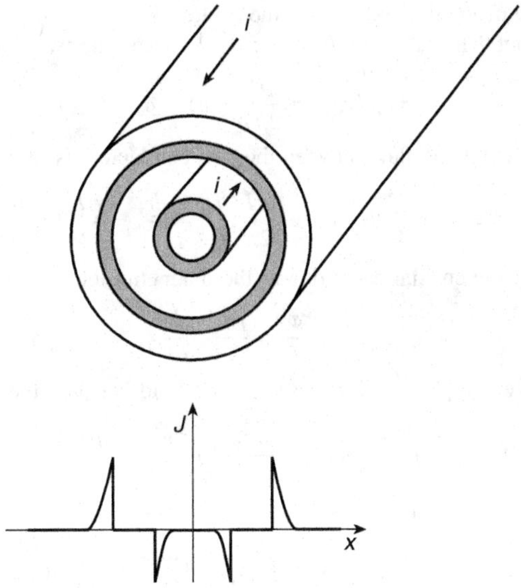

Figure 4.2 Proximity effect in coaxial cable at high frequencies

in the outer conductor that is enclosed by the magnetic flux. At low frequencies, the current in the outer conductor is uniform

$$J_z = \frac{-I}{\pi(c^2 - b^2)}. \tag{4.9}$$

The current enclosed by the magnetic flux is

$$I_{enc} = I - \frac{I}{\pi(c^2 - b^2)} \int_b^r 2\pi r dr = I \frac{c^2 - r^2}{c^2 - b^2}. \tag{4.10}$$

Hence, the magnetic field intensity is

$$H = \frac{I_{enc}}{2\pi r} = \frac{I}{2\pi(c^2 - b^2)} \frac{c^2 - r^2}{r} \quad \text{for} \quad b \le r \le c. \tag{4.11}$$

From Ampère's law, the magnetic field is zero outside the outer conductor because the net current through the area enclosed by a circular path surrounding the cable is zero, $\int_S BdS = 0$. The total inductance per unit length of the coaxial cable at low frequencies is

$$\frac{L}{l_w} = \frac{\mu}{8\pi} + \frac{\mu}{2\pi}\ln\left(\frac{b}{a}\right) + \frac{\mu}{2\pi(c^2 - b^2)^2}\left(c^4 \ln \frac{c}{b} + b^2c^2 - \frac{b^4}{4} - \frac{3}{4}c^4\right). \tag{4.12}$$

The first term represents the inductance due to the magnetic flux linkage in the inner conductor. The second term represents the inductance due to the flux linkage between the two conductors. This term is independent of frequency. It remains the same for high frequencies. The third term represents the inductance due to the flux linkage in the outer conductor.

At high frequencies, the magnetic field is generated between the conductors by the current in the inner conductor and is given by

$$H(r) = \frac{I}{2\pi r} \quad \text{for} \quad a \le r \le b. \tag{4.13}$$

Conversely, the current in the outer conductor does not generate the magnetic field in the inner conductor and between the conductors. The strongest magnetic field intensity is in the space between the inner and outer conductors. Therefore, the largest current densities are at the surface of the inner conductor and at the inner surface of the outer conductor. At high frequencies, the current density distribution in the inner conductor is determined by the skin effect only, whereas the current in the outer conductor is determined by both the proximity effect and the skin effect. The coaxial cable is self-shielded because the magnetic field outside the cable is zero because the contributions by the two conductors will cancel out. The magnetic field density between the two conductors is expressed as

$$B(r) = \mu B(r) = \frac{\mu I}{2\pi r} \quad \text{for} \quad a \leq r \leq b. \tag{4.14}$$

Because $dS = l_w dr$, the magnetic flux between the two conductors is

$$\phi = \lambda = \int_S B(r)dS = \frac{\mu I}{2\pi}\int_a^b \frac{l_w dr}{r} = \frac{\mu I l_w}{2\pi}\ln\frac{b}{a}. \tag{4.15}$$

Hence, the inductance of the coaxial cable due to the magnetic field between the conductors is

$$L = \frac{\phi}{I} = \frac{\mu l_w}{2\pi}\ln\frac{b}{a}. \tag{4.16}$$

Since $dV = 2\pi r l_w dr$, the energy stored in the magnetic field between the two conductors is

$$W_m = \frac{1}{2}\int_V \mu H^2 dV = \frac{\mu I^2 l_w}{4\pi}\int_a^b \frac{dr}{r} = \frac{\mu I^2 l_w}{4\pi}\ln\frac{b}{a}. \tag{4.17}$$

Since the flux linkage as a function of r is expressed as

$$\lambda(r) = \int_S B(r)dS = \frac{\mu I}{2\pi}\int_a^r \frac{l_w dr}{r} = \frac{\mu I l_w}{2\pi}\ln\frac{r}{a}, \tag{4.18}$$

the distributed inductance between the two conductors at low frequencies is given by

$$L(r) = \frac{\lambda(r)}{I} = \frac{\mu l_w}{2\pi}\ln\frac{r}{a}. \tag{4.19}$$

Thus, the lowest inductance is in the area between the conductors close to the inner conductor and the largest inductance is in the area close to the outer conductor.

4.5 Proximity and Skin Effects in Two Parallel Plates

4.5.1 Magnetic Field in Two Parallel Plates

Figure 4.3 shows two rectangular parallel conducting plates. Let us assume that the height b of the plates approaches infinity and the distance between the plates is constant. Consider the case in which

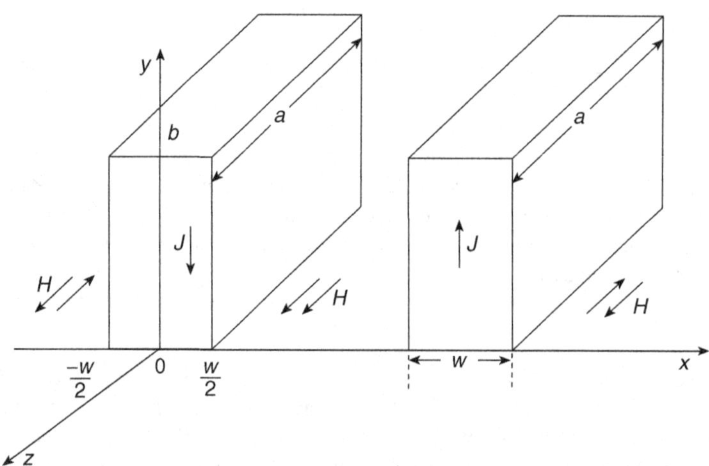

Figure 4.3 Two parallel plates carrying currents in opposite directions

the currents $i(t) = I\cos\omega t$ flow in the plates in opposite directions: the current in the left plate flows in the $-y$-direction and in the right plate in the y-direction. The currents cause the magnetic field inside and outside the plates. The magnetic fields generated by both plates add up between the plates, causing larger current density $J(x)$ in the plate areas, where the plates are close to each other. Using Ampère's circuital law,

$$\oint_C \mathbf{H}\cdot d\mathbf{l} = I, \tag{4.20}$$

we have

$$aH\left(\frac{w}{2}\right) - aH\left(-\frac{w}{2}\right) = aH\left(\frac{w}{2}\right) = I \quad \text{for} \quad a \gg w \tag{4.21}$$

because of the boundary condition

$$H\left(-\frac{w}{2}\right) = 0. \tag{4.22}$$

Thus,

$$H_m = H\left(\frac{w}{2}\right) = \frac{I}{a}. \tag{4.23}$$

The magnetic field inside the plates is described by the Helmholtz equation

$$\frac{d^2H(x)}{dx^2} = j\omega\mu_w\sigma_w H(x) = \gamma^2 H(x). \tag{4.24}$$

A general solution of the Helmholtz equation is

$$H(x) = H_1 e^{\gamma x} + H_2 e^{-\gamma x} \tag{4.25}$$

where H_1 and H_2 are constants, which can be evaluated by applying the boundary conditions at the conductor surfaces. From the boundary conditions,

$$H\left(\frac{w}{2}\right) = H_1 e^{\gamma\frac{w}{2}} + H_2 e^{-\gamma\frac{w}{2}} = \frac{I}{a} \tag{4.26}$$

and

$$H\left(-\frac{w}{2}\right) = H_1 e^{-\gamma\frac{w}{2}} + H_2 e^{\gamma\frac{w}{2}} = 0. \tag{4.27}$$

Hence,

$$H_2 = -H_1 e^{-\gamma w} \tag{4.28}$$

producing

$$H_1 = \frac{I}{a}\frac{e^{\gamma\frac{w}{2}}}{e^{\gamma w} - e^{-\gamma w}} \tag{4.29}$$

and

$$H_2 = -\frac{I}{a}\frac{e^{-\gamma\frac{w}{2}}}{e^{\gamma w} - e^{-\gamma w}}. \tag{4.30}$$

The magnetic field is given by

$$H(x) = \frac{I}{a}\frac{\sinh\left[\gamma\left(x + \frac{w}{2}\right)\right]}{\sinh(\gamma w)} = \frac{I}{a}\frac{\sinh\left[(1+j)\left(\frac{w}{\delta_w}\right)\left(\frac{x}{w} + \frac{1}{2}\right)\right]}{\sinh\left[(1+j)\frac{w}{\delta_w}\right]}. \tag{4.31}$$

Figure 4.4 depicts plots of $a|H(x)|/I$ as a function of x/w for selected values of w/δ_w.

4.5.2 Current Density in Two Parallel Plates

The current density is given by

$$J(x) = -\frac{dH(x)}{dx} = -\frac{\gamma I}{a}\frac{\cosh\left[\gamma\left(x + \frac{w}{2}\right)\right]}{\sinh(\gamma w)} = -\left(\frac{1+j}{\delta_w}\right)\left(\frac{I}{a}\right)\frac{\cosh\left[\gamma\left(x + \frac{w}{2}\right)\right]}{}$$

$$= -\left(\frac{1+j}{\delta_w}\right)\left(\frac{I}{a}\right)\frac{\cosh\left[(1+j)\left(\frac{w}{\delta_w}\right)\left(\frac{x}{w} + \frac{1}{2}\right)\right]}{\sinh\left[(1+j)\frac{w}{\delta_w}\right]}$$

$$= -\left(\frac{I}{wa}\right)\left(\frac{w}{\delta_w}\right)(1+j)\frac{\cosh\left[(1+j)\left(\frac{w}{\delta_w}\right)\left(\frac{x}{w}+\frac{1}{2}\right)\right]}{\sinh\left[(1+j)\frac{w}{\delta_w}\right]}. \quad (4.32)$$

The DC current density in each plate is

$$J_{DC} = \frac{I}{aw}. \quad (4.33)$$

Hence, the normalized current density in each plate is

$$\frac{J(x)}{J_{DC}} = -w\gamma\frac{\cosh\left[\gamma\left(x+\frac{w}{2}\right)\right]}{\sinh(\gamma w)} = -(1+j)\left(\frac{w}{\delta_w}\right)\frac{\cosh\left[\gamma\left(x+\frac{w}{2}\right)\right]}{\sinh(\gamma w)}$$

$$= -(1+j)\left(\frac{w}{\delta_w}\right)\frac{\cosh\left[(1+j)\left(\frac{w}{\delta_w}\right)\left(\frac{x}{w}+\frac{1}{2}\right)\right]}{\sinh\left[(1+j)\frac{w}{\delta_w}\right]}. \quad (4.34)$$

Figure 4.5 shows the normalized current density $|J(x)|/J_{DC}$ as a function of x/w for selected values of w/δ_w.

The current density due to the skin and proximity effects at the surfaces of the conductor, that is, for $x = w/2$ and $x = -w/2$, is given by

$$J_{sp}\left(\frac{w}{2}\right) = -\frac{\gamma I}{a}\frac{\cosh(\gamma w)}{\sinh(\gamma w)} = -\frac{I(1+j)}{aw}\left(\frac{w}{\delta_w}\right)\frac{\cosh\left[(1+j)\frac{w}{\delta_w}\right]}{\sinh\left[(1+j)\left(\frac{w}{\delta_w}\right)\right]} \quad (4.35)$$

$$\frac{J_{sp}\left(\frac{w}{2}\right)}{J_{DC}} = -(1+j)\left(\frac{w}{\delta_w}\right)\frac{\cosh\left[(1+j)\frac{w}{\delta_w}\right]}{\sinh\left[(1+j)\left(\frac{w}{\delta_w}\right)\right]} \quad (4.36)$$

$$J_{sp}\left(-\frac{w}{2}\right) = -\frac{\gamma I}{a}\frac{w}{\sinh(\gamma w)} = -\frac{I(1+j)}{aw}\left(\frac{1}{\delta_w}\right)\frac{1}{\sinh\left[(1+j)\left(\frac{w}{\delta_w}\right)\right]} \quad (4.37)$$

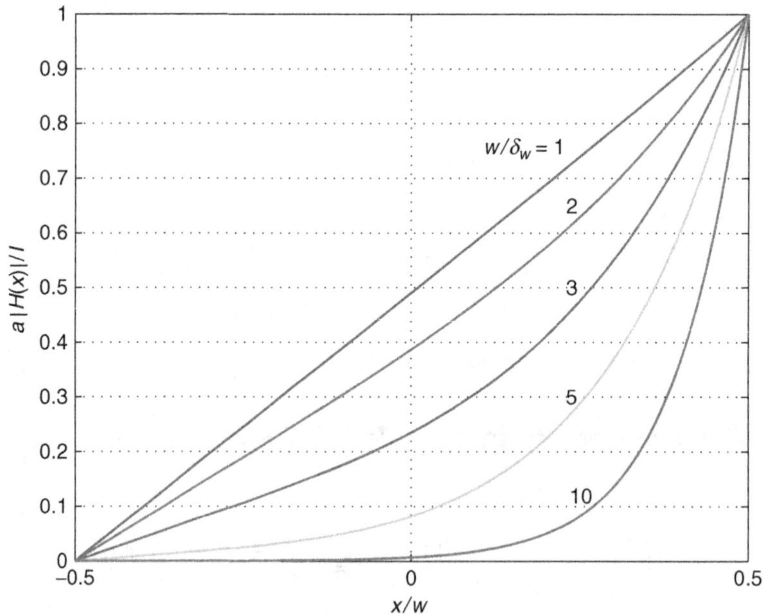

Figure 4.4 Plots of $a|H(x)|/I$ as a function of x/w for selected values of w/δ_w in the left plate due to the proximity effect

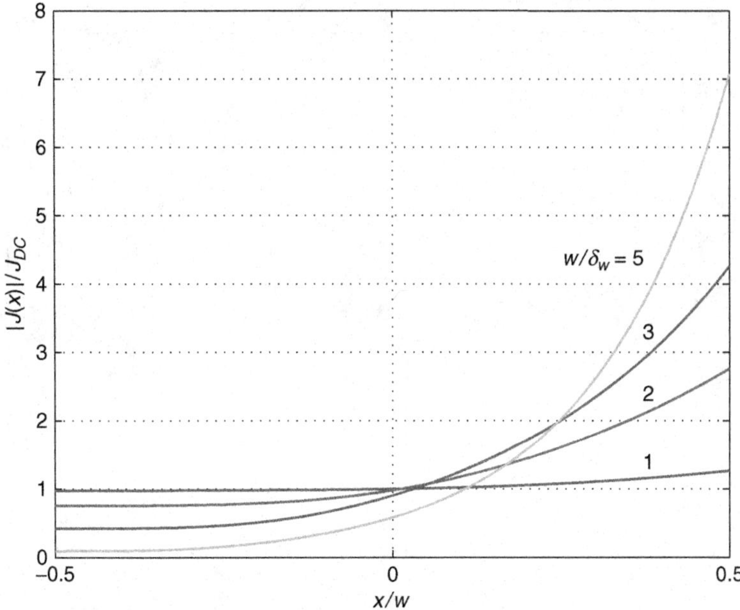

Figure 4.5 Plots of $|J(x)|/J_{DC}$ as a function of x/w for selected values of w/δ_w in the left plate

and

$$\frac{J_{sp}\left(-\frac{w}{2}\right)}{J_{DC}} = -(1+j)\left(\frac{1}{\delta_w}\right)\frac{1}{\sinh\left[(1+j)\left(\frac{w}{\delta_w}\right)\right]}. \tag{4.38}$$

From Chapter 3, the current density distribution due to the skin effect only is given by

$$J_s(x) = -\frac{\gamma}{2a}\frac{\cosh(\gamma x)}{\sinh(\gamma\frac{w}{2})} = -\left(\frac{I}{aw}\right)\left(\frac{w}{\delta_w}\right)\frac{(1+j)}{2}\frac{\cosh\left[(1+j)\left(\frac{x}{\delta_w}\right)\right]}{\sinh\left[\frac{(1+j)}{2}\left(\frac{w}{\delta_w}\right)\right]}. \tag{4.39}$$

The current density due to proximity effect only is

$$\begin{aligned} J_p(x) = J(x) - J_s(x) &= -\frac{\gamma I}{a}\frac{\cosh\left[\gamma(x+\frac{w}{2})\right]}{\sinh(\gamma w)]} + \frac{\gamma I}{2a}\frac{\cosh(\gamma x)}{\sinh(\gamma\frac{w}{2})} \\ &= \left(\frac{I}{aw}\right)\left\{\left(\frac{w}{\delta_w}\right)\frac{(1+j)}{2}\frac{\cosh\left[(1+j)\left(\frac{x}{\delta_w}\right)\right]}{\sinh\left[\frac{(1+j)}{2}\left(\frac{w}{\delta_w}\right)\right]}\right. \\ &\quad \left. -(1+j)\left(\frac{w}{\delta_w}\right)\frac{\cosh\left[(1+j)\left(\frac{w}{\delta_w}\right)\left(\frac{x}{w}+\frac{1}{2}\right)\right]}{\sinh\left[(1+j)\frac{w}{\delta_w}\right]}\right\}. \end{aligned} \tag{4.40}$$

Hence, one obtains the normalized proximity current density

$$\frac{J_p(x)}{J_{DC}} = \left(\frac{w}{\delta_w}\right)\frac{(1+j)}{2}\frac{\cosh\left[(1+j)\left(\frac{x}{\delta_w}\right)\right]}{\sinh\left[\frac{(1+j)}{2}\left(\frac{w}{\delta_w}\right)\right]} - (1+j)\left(\frac{w}{\delta_w}\right)\frac{\cosh\left[(1+j)\left(\frac{w}{\delta_w}\right)\left(\frac{x}{w}+\frac{1}{2}\right)\right]}{\sinh\left[(1+j)\frac{w}{\delta_w}\right]}. \tag{4.41}$$

Figures 4.6 and 4.7 show plots of $|J_{sp}(x)|/J_{DC} = |J(x)|/J_{DC}$, $|J_s(x)|/J_{DC}$, and $|J_p(x)|/J_{DC}$ as functions of x/w at $w/\delta_w = 5$. Note that $J_p(0) = 0$. Figure 4.8 shows a plot of $|J_p(x)|/J_{DC}$ as a function of x/w at $w/\delta_w = 5$.

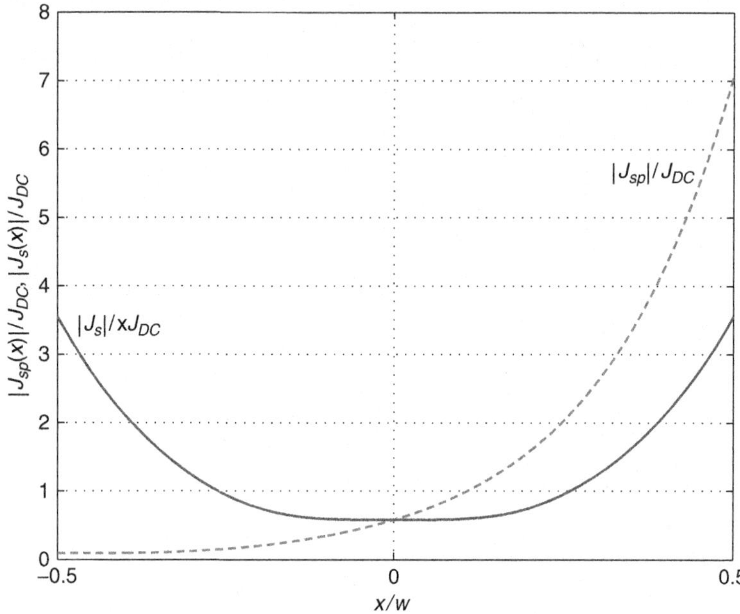

Figure 4.6 Plots of $|J_{sp}(x)|/J_{DC}$ and $|J_s(x)|/J_{DC}$ as functions of x/w at $w/\delta_w = 5$ in the left plate

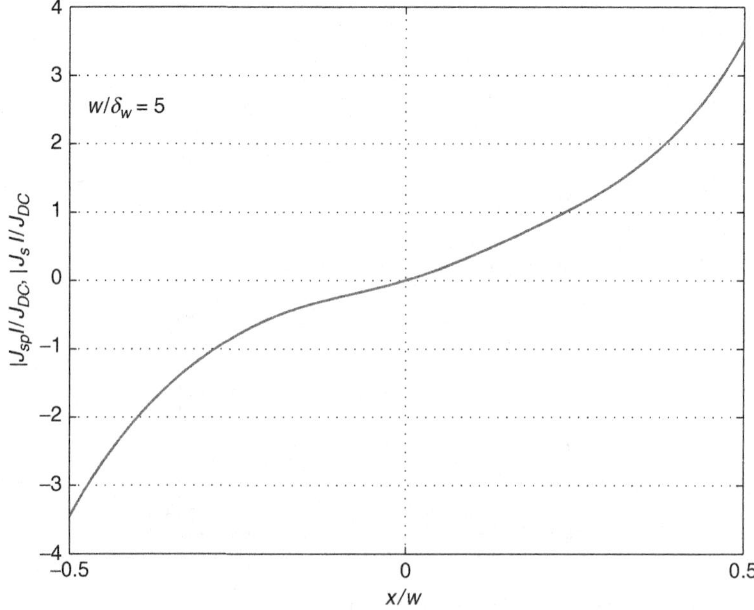

Figure 4.7 Plots of $|J_{sp}(x)|/J_{DC} - |J_s(x)|/J_{DC}$ as a function of x/w at $w/\delta_w = 5$ in the left plate

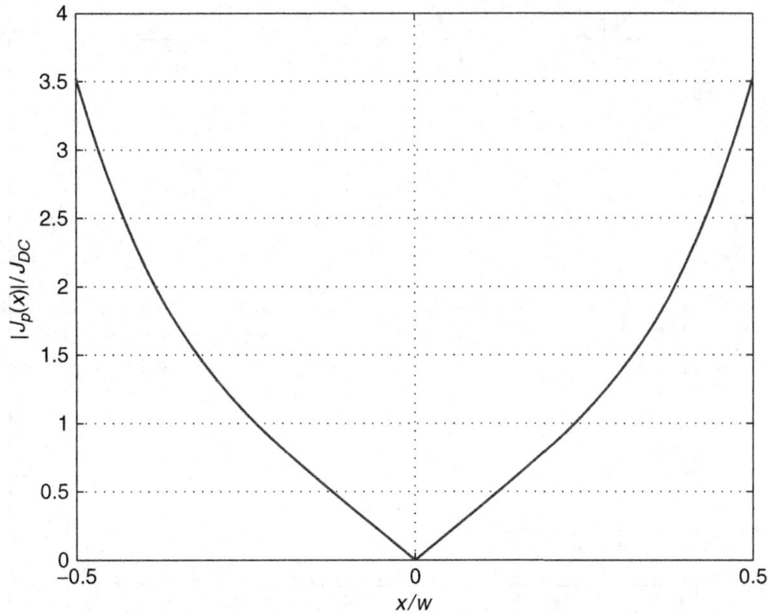

Figure 4.8 Plots of $|J_p(x)|/J_{DC}$ as a function of x/w at $w/\delta_w = 5$ in the left plate

The current density due to the skin effect only for $x = \pm w/2$ and $x = 0$ is

$$J_s\left(\pm\frac{w}{2}\right) = -\frac{\gamma I}{2a}\frac{\cosh(\gamma\frac{w}{2})}{\sinh(\gamma\frac{w}{2})} \tag{4.42}$$

and

$$J_s(0) = J_{sp}(0) = -\frac{\gamma I}{2a}\frac{1}{\sinh[\gamma\frac{w}{2}]}. \tag{4.43}$$

Hence,

$$\frac{J_{sp}\left(-\frac{w}{2}\right)}{J_s\left(-\frac{w}{2}\right)} = \frac{1}{1+\sinh^2\left(\gamma\frac{w}{2}\right)} \tag{4.44}$$

$$\frac{J_{sp}\left(\frac{w}{2}\right)}{J_s\left(\frac{w}{2}\right)} = 1+\tanh^2\left(\gamma\frac{w}{2}\right) \tag{4.45}$$

and

$$\frac{J_{sp}(0)}{J_s(0)} = 1. \tag{4.46}$$

Figures 4.9 and 4.10 show plots of $|J_{sp}(-w/2)/J_s(-w/2)|$ and $|J_{sp}(w/2)/J_s(w/2)|$ as functions of w/δ_w.

4.5.3 Power Loss in Two Parallel Plates

The power density distribution is given by

$$P(x) = \frac{1}{2}\rho_w|J(x)|^2 = \frac{1}{2}\frac{\rho_w I^2}{a^2}\left|\frac{\gamma\cosh[\gamma\left(x+\frac{w}{2}\right)]}{\sinh(\gamma w)}\right|^2. \tag{4.47}$$

A plot of $P(x)/\rho_w$ as a function of x/w at fixed values of w/δ_w in the left plate is shown in Fig. 4.11.

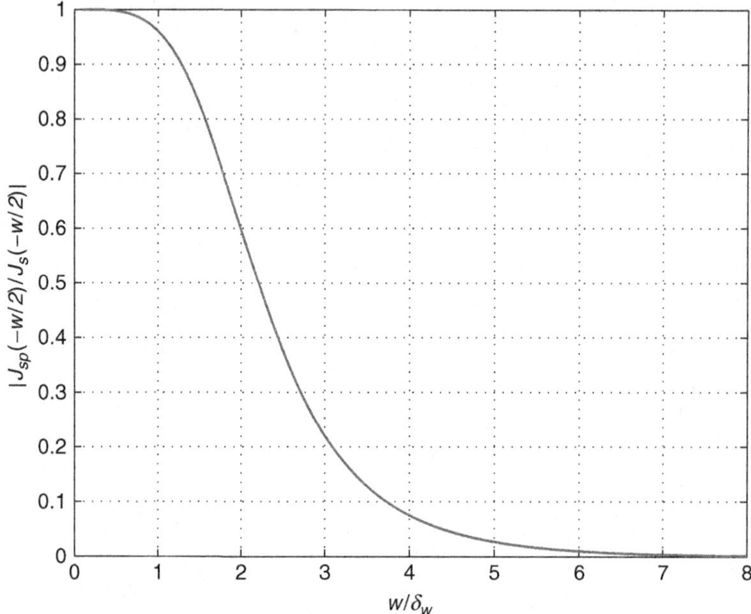

Figure 4.9 Plot of $|J_{sp}(-w/2)/J_s(-w/2)|$ as a function of w/δ_w

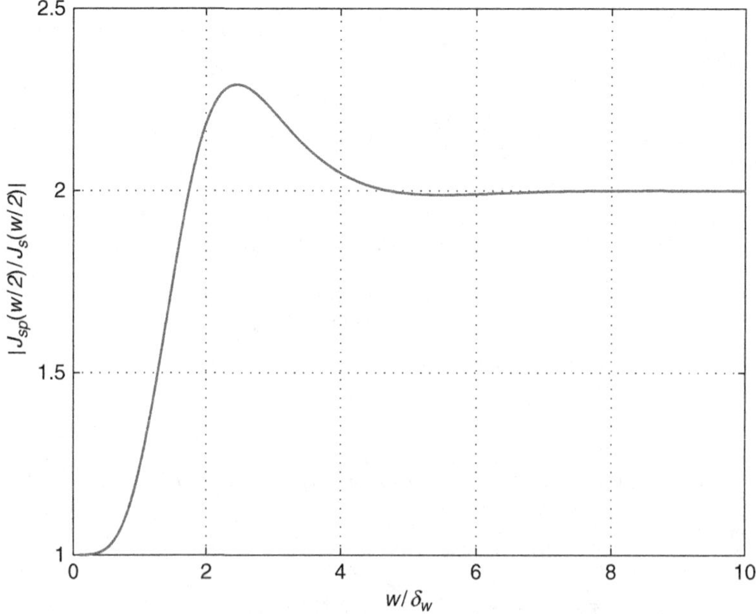

Figure 4.10 Plot of $|J_{sp}(w/2)/J_s(w/2)|$ as a function of w/δ_w

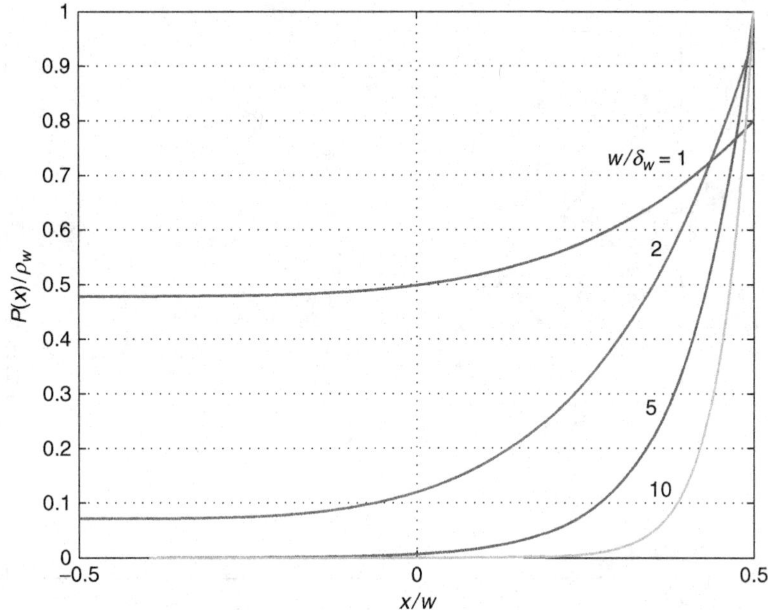

Figure 4.11 Plot of $P(x)/\rho_w$ as a function of x/w at fixed values of w/δ_w in the left plate

The total power loss in each conductor due to the proximity and skin effects is

$$P_{sp} = \frac{1}{2}\int\int\int_V \rho_w |J(x)|^2 dV = \frac{\rho_w}{2}\int_0^a\int_0^b\int_{-w/2}^{w/2}\frac{I^2}{a^2}|\gamma\frac{\cosh[\gamma(x+\frac{w}{2})]}{\sinh(\gamma w)}|^2 dxdydz$$

$$= \frac{\rho_w b I^2}{2a\delta_w}\frac{\sinh\left(\frac{2w}{\delta_w}\right)+\sin\left(\frac{2w}{\delta_w}\right)}{\cosh\left(\frac{2w}{\delta_w}\right)-\cos\left(\frac{2w}{\delta_w}\right)} = \frac{\rho_w b I^2}{2aw}\left(\frac{w}{\delta_w}\right)\frac{\sinh\left(\frac{2w}{\delta_w}\right)+\sin\left(\frac{2w}{\delta_w}\right)}{\cosh\left(\frac{2w}{\delta_w}\right)-\cos\left(\frac{2w}{\delta_w}\right)}. \quad (4.48)$$

Figure 4.12 shows the power loss due to the skin and proximity effects $a\delta_w P_{sp}/b\rho_w I^2$ as a function of w/δ_w at fixed values of the skin depth δ_w. Figure 4.13 depicts the power loss due to the skin and proximity effects as a function of w/δ_w at fixed plate width w.

The power due to the skin effect is the same at that for a single plate and is given by

$$P_s = \frac{\rho_w b I^2}{4a\delta_w}\frac{\sinh\left(\frac{w}{\delta_w}\right)+\sin\left(\frac{w}{\delta_w}\right)}{\cosh\left(\frac{w}{\delta_w}\right)-\cos\left(\frac{w}{\delta_w}\right)} = \frac{\rho_w b I^2}{4aw}\left(\frac{w}{\delta_w}\right)\frac{\sinh\left(\frac{w}{\delta_w}\right)+\sin\left(\frac{w}{\delta_w}\right)}{\cosh\left(\frac{w}{\delta_w}\right)-\cos\left(\frac{w}{\delta_w}\right)}. \quad (4.49)$$

Figure 4.14 depicts a plot of the normalized skin-effect power loss $a\delta_w P_s/\rho_w b I^2$ as a function of w/δ_w at fixed skin depth δ_w. Figure 4.15 shows the skin-effect power loss at fixed plate width w.

The proximity loss is

$$P_p = P_{sp} - P_s = \frac{\rho_w b I^2}{4a\delta_w}\frac{\sinh\left(\frac{w}{\delta_w}\right)-\sin\left(\frac{w}{\delta_w}\right)}{\cosh\left(\frac{w}{\delta_w}\right)+\cos\left(\frac{w}{\delta_w}\right)} = \frac{\rho_w b I^2}{4aw}\left(\frac{w}{\delta_w}\right)\frac{\sinh\left(\frac{w}{\delta_w}\right)-\sin\left(\frac{w}{\delta_w}\right)}{\cosh\left(\frac{w}{\delta_w}\right)+\cos\left(\frac{w}{\delta_w}\right)}. \quad (4.50)$$

Figure 4.16 shows the power loss due to proximity effect $a\delta_w P_p/\rho_w b I^2$ as a function of w/δ_w at fixed δ_w. Figure 4.17 depicts the power loss due to the proximity effect as a function of w/δ_w at fixed values of w.

The ratio of P_p to P_{sp} is given by

$$\frac{P_p}{P_{sp}} = \frac{1}{2}\frac{\sinh\left(\frac{w}{\delta_w}\right)-\sin\left(\frac{w}{\delta_w}\right)}{\cosh\left(\frac{w}{\delta_w}\right)+\cos\left(\frac{w}{\delta_w}\right)}\;\frac{\cosh\left(\frac{2w}{\delta_w}\right)-\cos\left(\frac{2w}{\delta_w}\right)}{\sinh\left(\frac{2w}{\delta_w}\right)+\sin\left(\frac{2w}{\delta_w}\right)}. \quad (4.51)$$

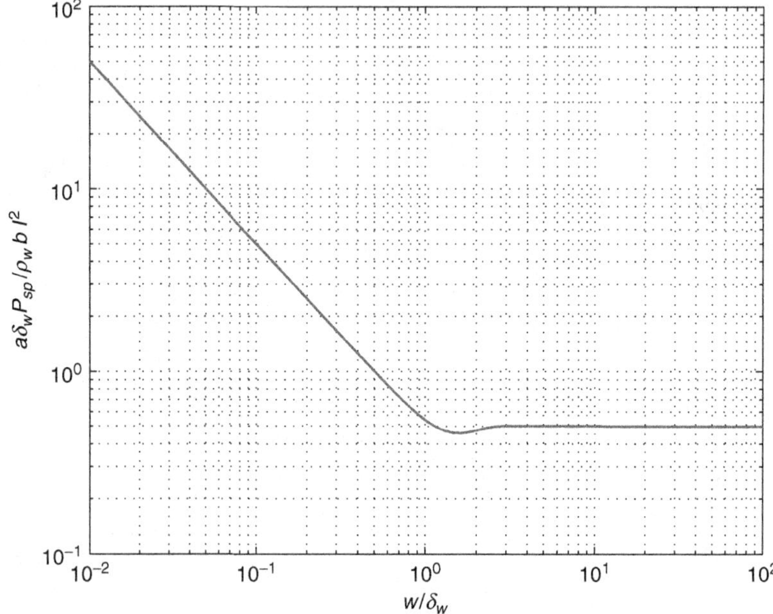

Figure 4.12 Power loss due to skin and proximity effects $a\delta_w P_{sp}/b\rho_w I^2$ as a function of w/δ_w at fixed skin depth δ_w

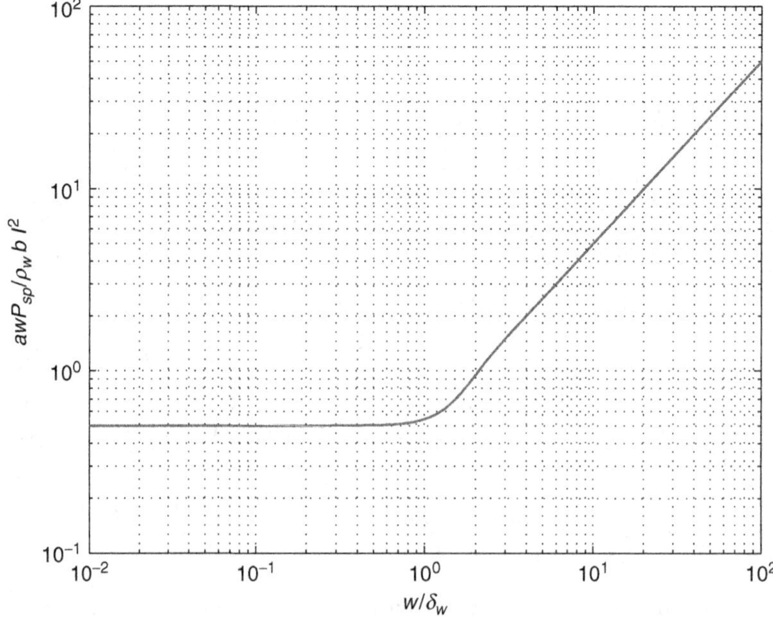

Figure 4.13 Power loss due to skin and proximity effects $awP_{sp}/b\rho_w I^2$ as a function of w/δ_w at fixed plate width w

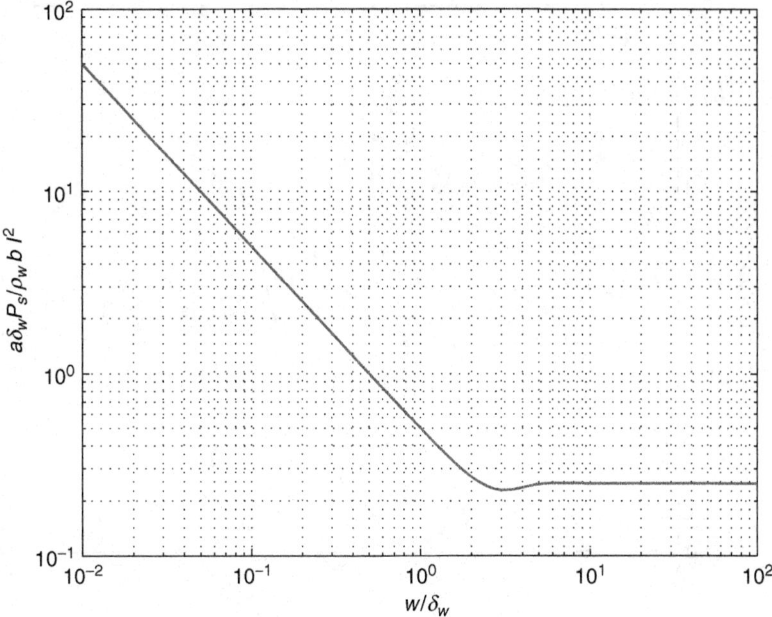

Figure 4.14 Power loss due to skin effect $a\delta_w P_s/b\rho_w I^2$ as a function of w/δ_w at fixed skin depth δ_w

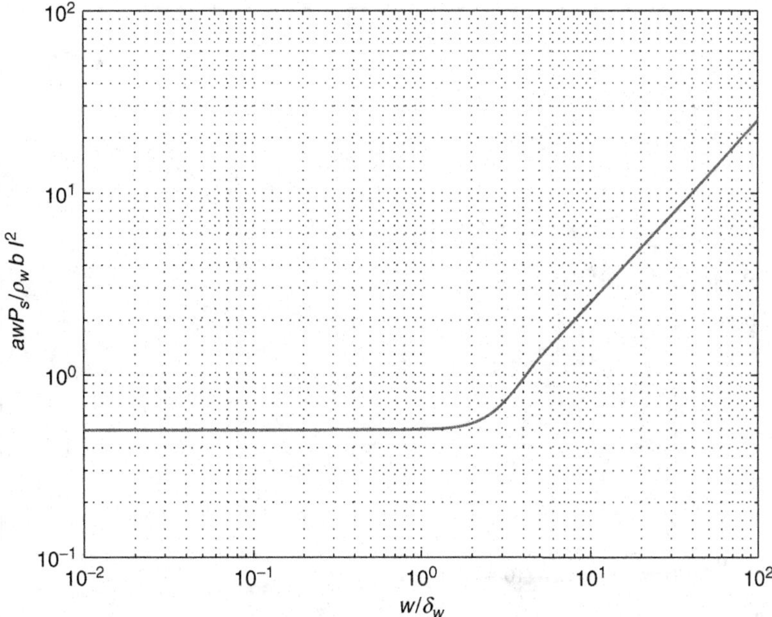

Figure 4.15 Power loss due to skin effect $awP_s/b\rho_w I^2$ as a function of w/δ_w at fixed plate width w

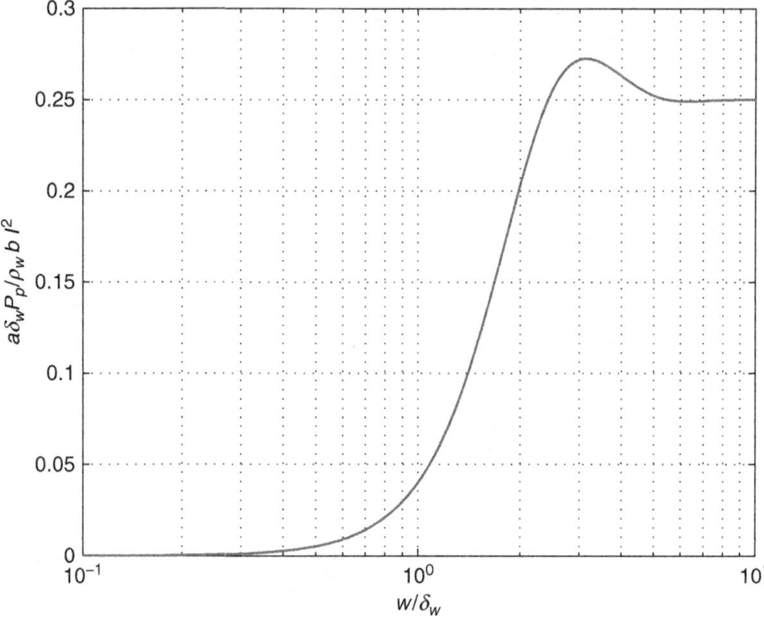

Figure 4.16 Power loss due to proximity effect $a\delta_w P_p/b\rho_w I^2$ as a function of w/δ_w at fixed skin depth δ_w

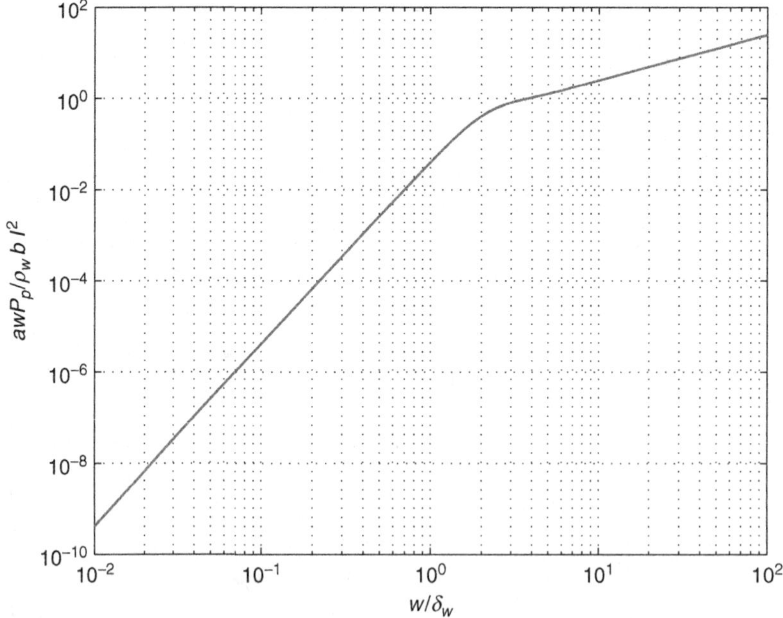

Figure 4.17 Power loss due to proximity effect $awP_p/b\rho_w I^2$ as a function of w/δ_w at fixed plate width w

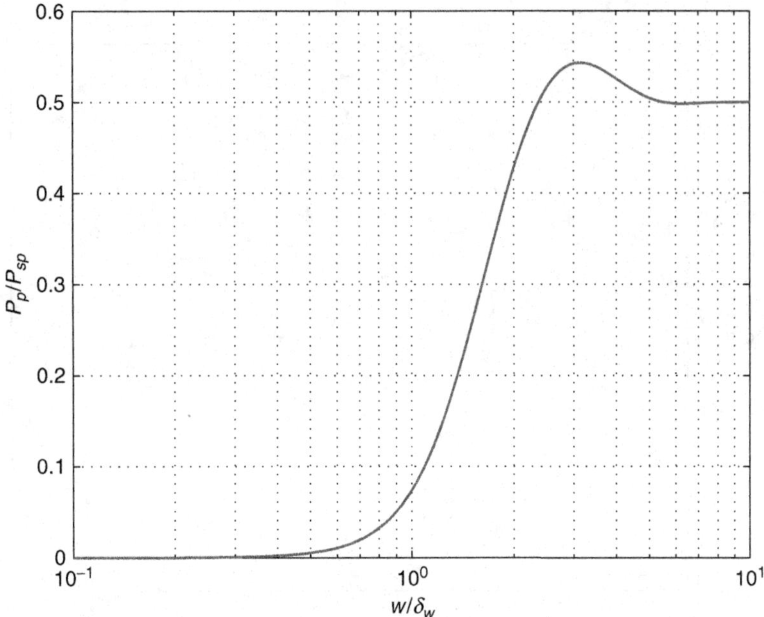

Figure 4.18 Ratio P_p/P_{sp} as a function of w/δ_w

Figure 4.18 shows the ratio P_p/P_{sp} as a function of w/δ_w. The ratio of P_s to P_{sp} is

$$\frac{P_s}{P_{sp}} = \frac{1}{2}\frac{\sinh\left(\frac{w}{\delta_w}\right) + \sin\left(\frac{w}{\delta_w}\right)}{\cosh\left(\frac{w}{\delta_w}\right) - \cos\left(\frac{w}{\delta_w}\right)} \frac{\cosh\left(\frac{2w}{\delta_w}\right) - \cos\left(\frac{2w}{\delta_w}\right)}{\sinh\left(\frac{2w}{\delta_w}\right) + \sin\left(\frac{2w}{\delta_w}\right)}. \tag{4.52}$$

Figure 4.19 depicts the ratio P_s/P_{sp} as a function of w/δ_w. The ratio of P_p to P_{sp} is given by

$$\frac{P_p}{P_s} = \frac{1}{2}\frac{\sinh\left(\frac{w}{\delta_w}\right) - \sin\left(\frac{w}{\delta_w}\right)}{\cosh\left(\frac{w}{\delta_w}\right) + \cos\left(\frac{w}{\delta_w}\right)} \frac{\cosh\left(\frac{w}{\delta_w}\right) - \cos\left(\frac{w}{\delta_w}\right)}{\sinh\left(\frac{w}{\delta_w}\right) + \sin\left(\frac{w}{\delta_w}\right)}. \tag{4.53}$$

Figure 4.20 shows the ratio P_p/P_s as a function of w/δ_w.

The power loss in each plate can be expressed as

$$P_{sp} = \frac{1}{2}I^2 R_w = \frac{\rho_w b I^2}{2a\delta_w} \frac{\sinh\left(\frac{2w}{\delta_w}\right) + \sin\left(\frac{2w}{\delta_w}\right)}{\cosh\left(\frac{2w}{\delta_w}\right) - \cos\left(\frac{2w}{\delta_w}\right)}. \tag{4.54}$$

The DC and low-frequency resistance of each plate is

$$R_{wDC} = \rho_w \frac{b}{wa}. \tag{4.55}$$

The DC power loss in each plate is

$$P_{wDC} = R_{wDC} I_{DC}^2 = \rho_w \frac{b}{wa} I_{DC}^2. \tag{4.56}$$

Assuming that $I_{rms} = I_{DC}$, the normalized resistance of each plate is

$$F_R = \frac{P_{sp}}{P_{wDC}} = \frac{R_w}{R_{wDC}} = \left(\frac{w}{\delta_w}\right) \frac{\sinh\left(\frac{2w}{\delta_w}\right) + \sin\left(\frac{2w}{\delta_w}\right)}{\cosh\left(\frac{2w}{\delta_w}\right) - \cos\left(2\frac{2w}{\delta_w}\right)}. \tag{4.57}$$

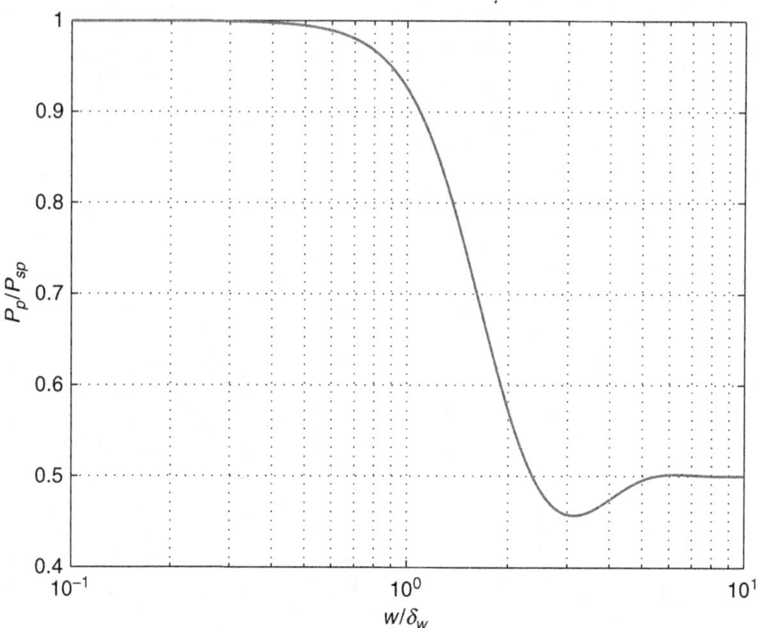

Figure 4.19 Ratio P_s/P_{sp} as a function of w/δ_w at fixed δ_w

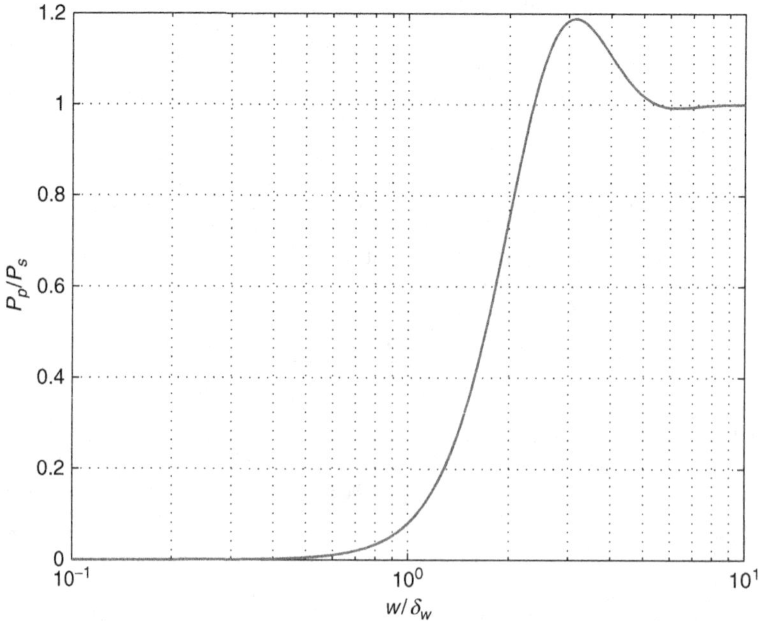

Figure 4.20 Ratio P_p/P_s as a function of w/δ_w

The skin and proximity effects are orthogonal. It can be shown that

$$\int\int_S J_s J_p = 0. \tag{4.58}$$

4.5.4 Impedance of Each Plate

The electric field intensity is

$$E = \rho_w J = -\rho_w \frac{dH}{dx} = -\frac{\rho_w \gamma I}{a} \frac{\cosh[\gamma (x + \frac{w}{2})]}{\sinh(\gamma w)}. \tag{4.59}$$

The impedance of each plate is

$$Z = \frac{V}{I} = \frac{bE\left(\frac{w}{2}\right)}{a} = \frac{b\rho_w \gamma}{a} \coth(\gamma w) = R_{wDC}\left(\frac{w}{\delta_w}\right)(1+j)\coth(\gamma w) = F_R + jF_X \tag{4.60}$$

where the AC resistance factor is

$$F_R = \frac{R_w}{R_{wDC}} = \left(\frac{w}{\delta_w}\right) \frac{\sinh\left(\frac{2w}{\delta_w}\right) + \sin\left(\frac{2w}{\delta_c}\right)}{\cosh\left(\frac{2w}{\delta_w}\right) - \cos\left(\frac{2w}{\delta_w}\right)} \tag{4.61}$$

and the normalized reactance is

$$F_X = \frac{X_L}{R_{wDC}} = \left(\frac{w}{\delta_w}\right) \frac{\sinh\left(\frac{2w}{\delta_w}\right) - \sin\left(\frac{2w}{\delta_c}\right)}{\cosh\left(\frac{2w}{\delta_w}\right) - \cos\left(\frac{2w}{\delta_w}\right)}. \tag{4.62}$$

At high frequencies, the resistance of each plate is

$$F_R = F_X \approx \frac{w}{\delta_w}. \tag{4.63}$$

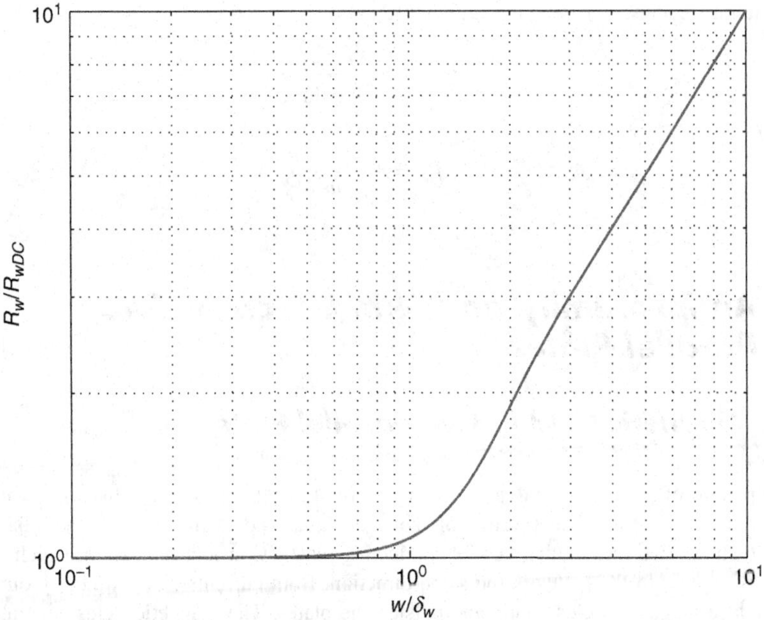

Figure 4.21 R_w/R_{wDC} as a function of w/δ_w

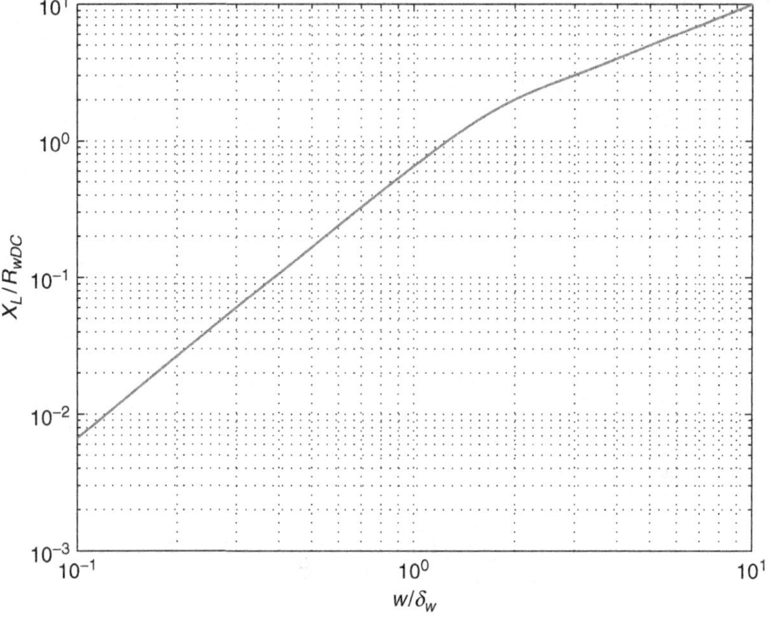

Figure 4.22 X_L/R_{wDC} as a function of w/δ_w

Figures 4.21 and 4.22 show plots of R_w/R_{wDC} and X_L/R_{wDC} as functions of w/δ_w. Plots of $|Z|/R_{wDC}$ and ϕ_Z are shown in Figures 4.23 and 4.24. The AC resistance of the plate is equal to its DC resistance for the range

$$0 \leq \frac{w}{\delta_w} \leq 1. \tag{4.64}$$

The maximum frequency of this range is determined by conduction is

$$w = \delta_w = \sqrt{\frac{\rho_2}{\pi \mu_0 f_{sk}}} \tag{4.65}$$

producing

$$f_{sk} = \frac{\rho_w}{\pi \mu_0 w^2}. \tag{4.66}$$

4.6 Antiproximity and Skin Effects in Two Parallel Plates

4.6.1 Magnetic Field in Two Parallel Plates

A typical pattern of inductor winding is shown in Fig. 4.25. It can be seen that the current flows in the same direction in the adjacent turns of the same layer and in the turns of the adjacent layers. Figure 4.26 shows two rectangular parallel conducting plates. Consider the case in which the currents $i(t) = I \cos \omega t$ flow in the plates in the same direction. Both currents flow in the $-y$-direction. The currents induce magnetic fields inside and outside the plates. The magnetic fields generated by both plates subtract from each other between the plates and add to each other outside the plates, causing larger current density $J(x)$ in the plate areas, where the plates are far from each other. The magnetic

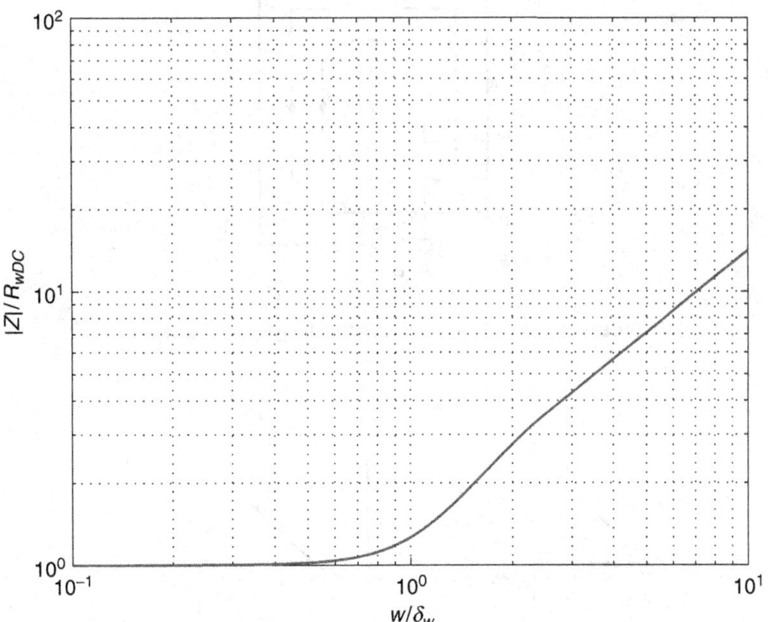

Figure 4.23 $|Z|/R_{wDC}$ as a function of w/δ_w

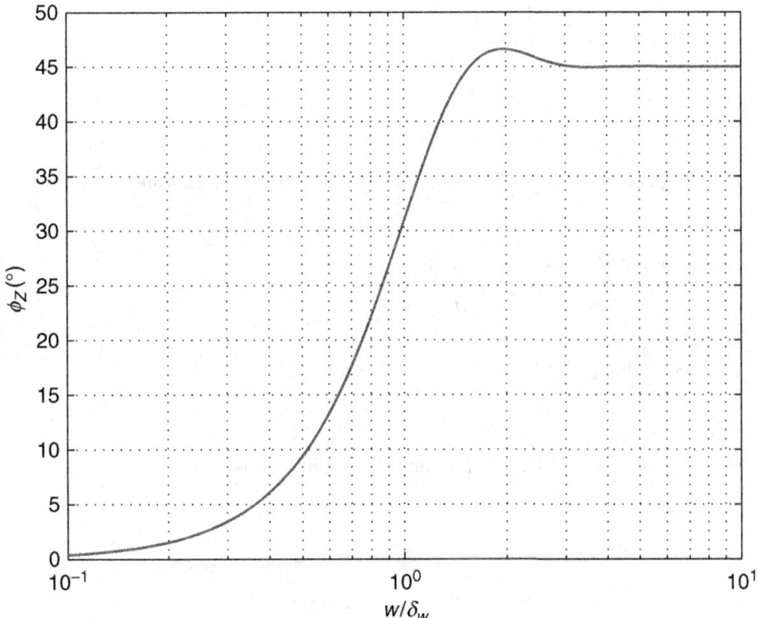

Figure 4.24 ϕ_Z as a function of w/δ_w

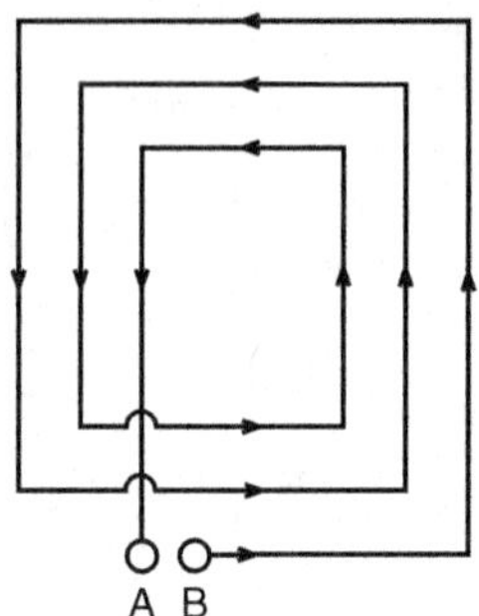

Figure 4.25 Typical pattern of multilayer inductor winding

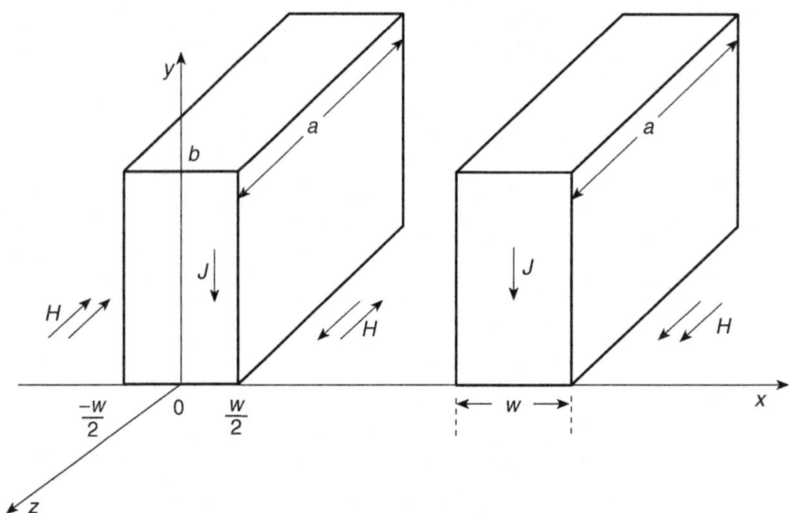

Figure 4.26 Two parallel plates carrying currents in the same direction

field is described by the Helmholtz equation

$$\frac{d^2H(x)}{dx^2} = \gamma^2 H(x). \tag{4.67}$$

A general solution of this equation is

$$H(x) = H_1 e^{\gamma x} + H_2 e^{-\gamma x} \tag{4.68}$$

where constants are determined from the boundary conditions as

$$H_1 = \frac{I}{a} \frac{e^{\gamma \frac{w}{2}}}{e^{\gamma w} - e^{-\gamma w}} \tag{4.69}$$

and

$$H_2 = -\frac{I}{a} \frac{e^{-\gamma \frac{w}{2}}}{e^{\gamma w} - e^{-\gamma w}}. \tag{4.70}$$

The magnetic field is given by

$$H(x) = \frac{I}{a} \frac{\sinh\left[\gamma \left(x - \frac{w}{2}\right)\right]}{\sinh(\gamma w)}. \tag{4.71}$$

Figure 4.27 shows plots of $a|H(x)|/I$ as a function of x/w for selected values of w/δ_w.

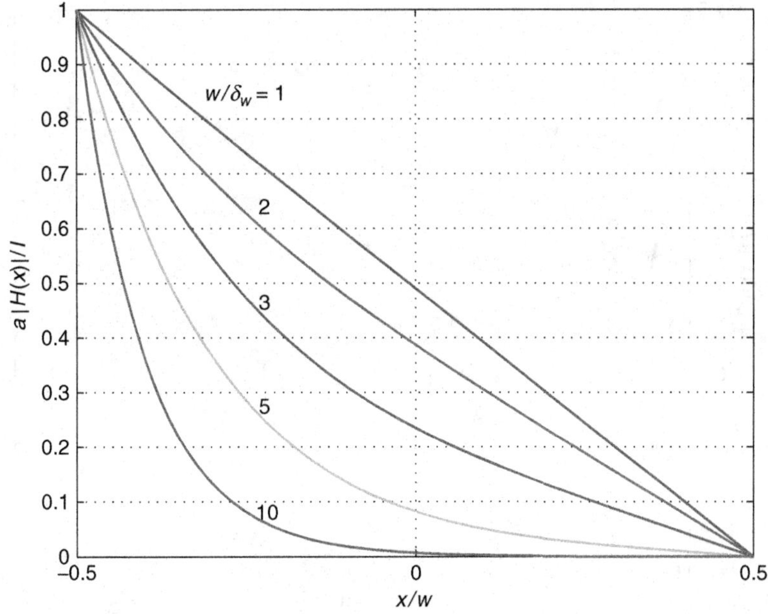

Figure 4.27 Plots of $a|H(x)|/I$ as a function of x/w for selected values of w/δ_w in the left plate due to antiproximity effect

4.6.2 Current Density in Two Parallel Plates

The current density in the left plate is

$$J(x) = -\frac{dH(x)}{dx} = -\frac{\gamma I}{a}\frac{\cosh\left[\gamma\left(x - \frac{w}{2}\right)\right]}{\sinh(\gamma w)}. \tag{4.72}$$

Figure 4.28 depicts plots of $a|J(x)|/I$ as a function of x/w for different values of w/δ_w.

The current density due to the skin and proximity effects on the surfaces of the conductor, that is, for $x = w/2$ and $x = -w/2$, is given by

$$J_{sp}\left(\frac{w}{2}\right) = -\frac{\gamma I}{a}\frac{\cosh(\gamma \frac{w}{2})}{\sinh(\gamma w)} \tag{4.73}$$

and

$$J_{sp}\left(-\frac{w}{2}\right) = -\frac{\gamma I}{a}\frac{1}{\sinh(\gamma w)}. \tag{4.74}$$

Hence,

$$\frac{J_{sp}\left(-\frac{w}{2}\right)}{J_s\left(-\frac{w}{2}\right)} = 1 + \tanh^2\left(\gamma\frac{w}{2}\right) \tag{4.75}$$

$$\frac{J_{sp}\left(\frac{w}{2}\right)}{J_s\left(\frac{w}{2}\right)} = \frac{1}{1 + \sinh^2\left(\gamma\frac{w}{2}\right)} \tag{4.76}$$

and

$$\frac{J_{sp}(0)}{J_s(0)} = 1. \tag{4.77}$$

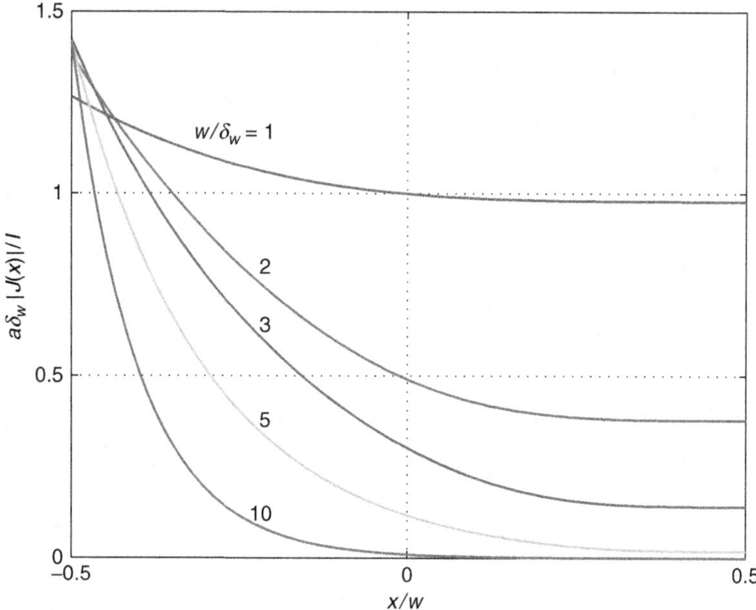

Figure 4.28 Plots of $a\delta_w|J(x)|/I$ as a function of x/w for selected vales of w/δ_w in the left plate due to antiproximity effect

4.6.3 Power Loss in Two Parallel Plates

The local power density is

$$P(x) = \frac{1}{2}\rho_w|J(x)|^2. \tag{4.78}$$

Figure 4.29 shows the plots of $P(x)/\rho_w$ as a function of x/w at various values of w/δ_w in the left plate.

The total power loss in each conductor due to the proximity and skin effects is

$$P_{sp} = \frac{\rho_w I^2 b}{2a\delta_w}\frac{\sinh\left(\frac{2w}{\delta_w}\right) + \sin\left(\frac{2w}{\delta_w}\right)}{\cosh\left(\frac{2w}{\delta_w}\right) - \cos\left(\frac{2w}{\delta_w}\right)}. \tag{4.79}$$

Comparing the power loss for a single conductor and two conductors, we can see that the presence of the second conductor has increased the loss in the first. This increase is simply the proximity loss set up in the first conductor by the magnetic field of the second. The proximity loss is

$$P_p = \frac{\rho_w I^2 b}{4a\delta_w}\frac{\sinh\left(\frac{w}{\delta_w}\right) - \sin\left(\frac{w}{\delta_w}\right)}{\cosh\left(\frac{w}{\delta_w}\right) + \cos\left(\frac{w}{\delta_w}\right)}. \tag{4.80}$$

Hence,

$$P_s = P_{sp} - P_p = \frac{\rho_w b I^2}{4a\delta_w}\frac{\sinh\left(\frac{w}{\delta_w}\right) + \sin\left(\frac{w}{\delta_w}\right)}{\cosh\left(\frac{w}{\delta_w}\right) - \cos\left(\frac{w}{\delta_w}\right)}. \tag{4.81}$$

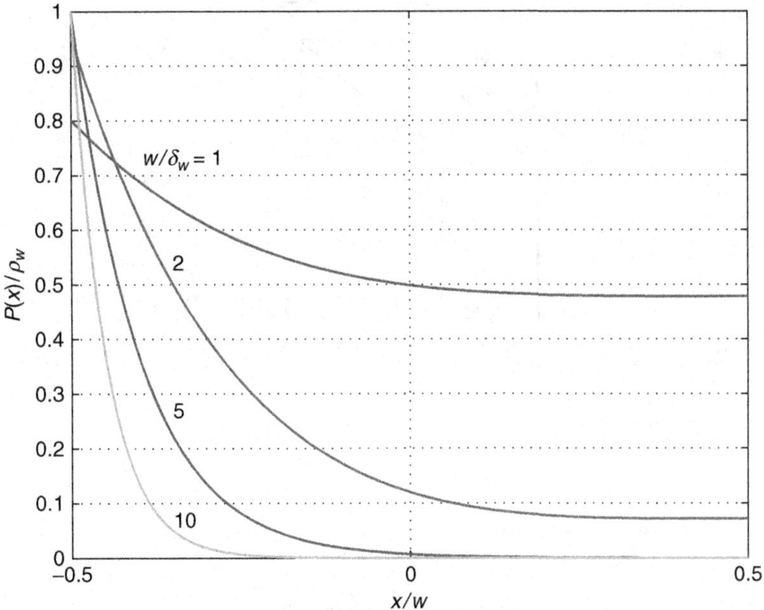

Figure 4.29 Plot of $P(x)/\rho_w$ as a function of x/w at fixed values of w/δ_w in the left plate due to antiproximity effect

4.7 Proximity Effect in Open-Circuit Conductor

Eddy currents are increased by adjacent layers. The amplitude of eddy currents increases due to current crowding as the number of layers is increased. Figure 4.30 illustrates the proximity effect in two adjacent foil conductors at high frequencies. In this case, the skin depth $delta_w$ is much lower than the thickness of the conductors h. The conductor located on the left-hand side carries a high-frequency current $i(t)$ forced by an AC current source. This current induces a magnetic field in the space between the conductors, which in turn induces an eddy current (image current or loop current) in an open-circuit conductor located on the right-hand side. According to Lenz's law, the induced eddy current generates a magnetic field, which opposes the original magnetic field, causing the eddy current. The current in the open-circuit conductor follows a closed loop and is an antimirror image of the forced current equal to $-i(t)$ in the area close to the first conductor and is a mirror image of the applied current $i(t)$ in the area located far from the first conductor. This is a transformer action. The net current in the open-circuit conductor is zero. The eddy current causes a copper loss. The magnetic coupling between the currents reduces the magnetic field, and thus reduces the overall inductance. An equivalent circuit of the two conductors of Fig. 4.30 is a transformer as shown in Fig. 4.31. The impedance of the secondary winding $Z_2 = R_2 + j\omega L_2$ reflected to the primary side in terms of the mutual inductance M is given by

$$Z_R = \frac{(\omega M)^2}{R_2 + j\omega L_2} = \frac{R_2(\omega M)^2}{R_2^2 + (\omega L_2)^2} - j\frac{\omega L_2(\omega M)^2}{R_2^2 + (\omega L_2)^2}. \tag{4.82}$$

Hence, the input impedance of the transformer is

$$Z_i = R_1 + j\omega L_2 + Z_R = R_1 + \frac{(\omega M)^2 R_2}{R_2^2 + (\omega L_2)^2} + j\omega\left[L_1 - \frac{L_2(\omega M)^2}{R_2^2 + (\omega L_2)^2}\right] = R_i + j\omega L_i. \tag{4.83}$$

Thus, the presence of the second conductor causes the input resistance R_i to increase and the input inductance L_i to decrease.

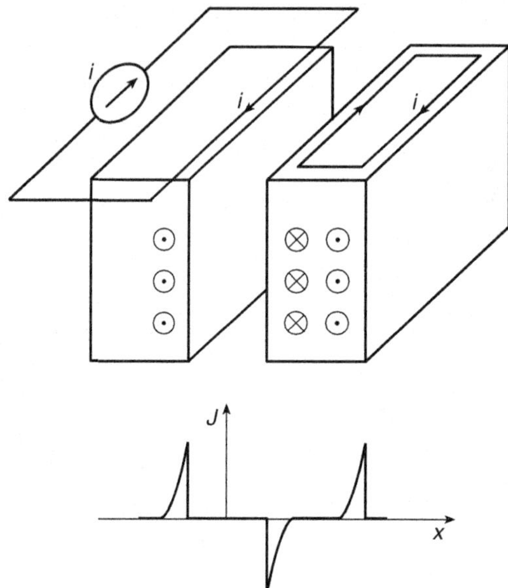

Figure 4.30 Proximity effect in closely spaced conductors at high frequencies, one conductor carries forced current $i(t)$ and the other conductor is open-circuited

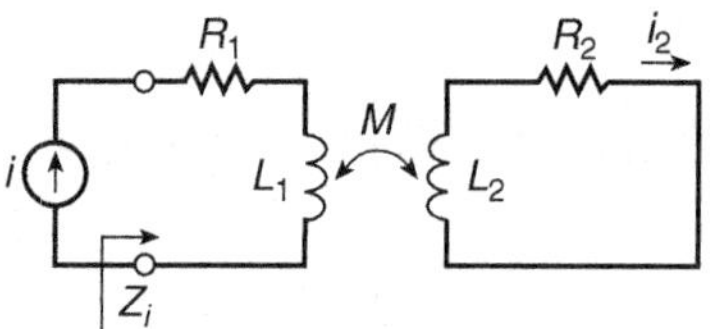

Figure 4.31 Equivalent circuit of the conductor carrying current and the eddy current induced in another conductor due to proximity effect as shown in Fig. 4.30

4.8 Proximity Effect in Multiple-Layer Inductor

To illustrate the proximity effect, a three-layer foil winding inductor with foil thickness $h \gg \delta_w$ is depicted in Fig. 4.32. The top view of the inductor with Ampèrian contours is depicted in Fig. 4.33. A magnetic core is inside the winding. As the layer number m increases from the innermost to the outermost layer, the amount of current enclosed by the Ampèrian contour increases, and therefore the flux density also increases. A high-frequency external sinusoidal current of rms value $I_1 = I_{Lm}/\sqrt{2}$

$$i(t) = I_{Lm} \sin \omega t = \sqrt{2} I_1 \sin \omega t \tag{4.84}$$

is forced to flow through the inductor, that is, through all the layers. The lines of magnetic flux $\phi(t)$ are parallel to the winding layers and normal to the turns and the direction of the current flow. According to Ampère's law, as the number of layer increases from the core to outermost layer, a larger amount of current is encircled and therefore the amount of the magnetic flux increases. Assuming that the core relative permeability is infinity, the magnetic field inside the core is zero ($H = B/\mu = 0$). The current $i(t)$ flows on the outer (right) surface of the innermost layer. Due to the proximity effect, current $-i(t)$ is induced on the innermost (left) surface of the adjacent conductor in the second layer. Therefore, current $i(t)$ is induced on the right surface of the second layer. Because the net current through all layers is $i(t)$, current $2i(t)$ flows on the right surface of the second layer. Hence, current $-2i(t)$ is induced on the left surface of the third layer, and current $3i(t)$ flows on the right surface

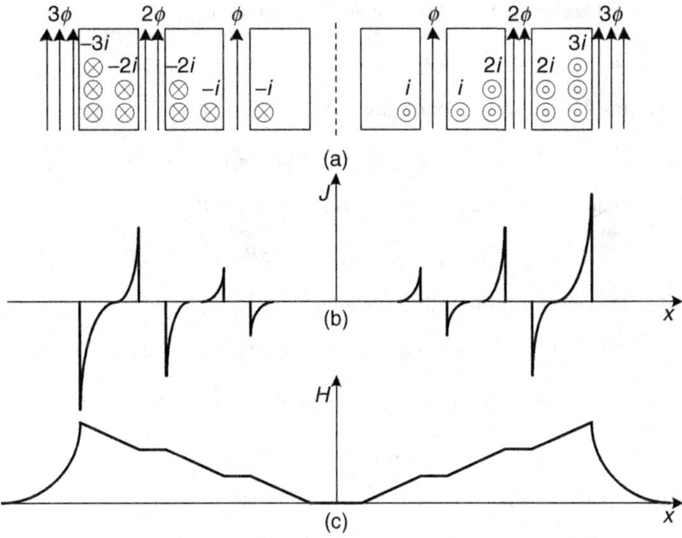

Figure 4.32 Proximity effect in three-layer foil winding inductor at high frequencies. (a) Inductor. (b) Current density J. (c) Magnetic field intensity H

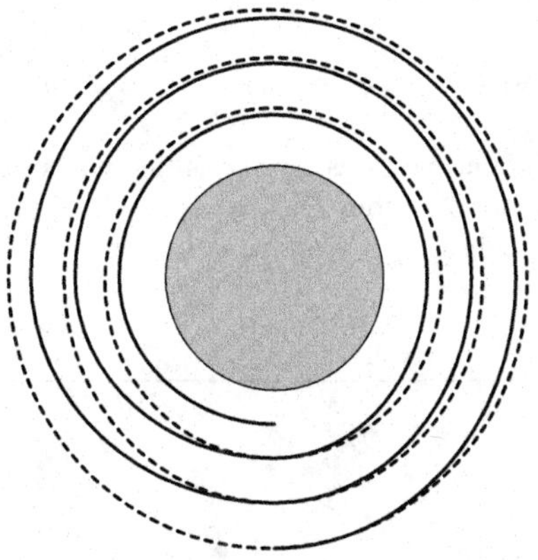

Figure 4.33 Top view of the inductor layers (solid line) with Ampèrian contours (broken lines)

of the third layer so that the net current is $i(t)$. The square of the rms value of the current in the first layer of copper winding is given by

$$I_{1(rms)}^2 = I_1^2 \tag{4.85}$$

producing the copper loss in the first layer with AC resistance R_{w1}

$$P_{L1} = R_{w1} I_{1(rms)}^2 = R_{w1} I_1^2. \tag{4.86}$$

The square of the rms current in the second layer is

$$I_{2(rms)}^2 = I_1^2 + (2I_1)^2 = I_1^2 + 4I_1^2 = 5I_1^2 \tag{4.87}$$

resulting in the copper loss in the second layer with AC resistance R_{w1}

$$P_{L2} = R_{w1} I_{2(rms)}^2 = 5R_{w1} I_1^2. \tag{4.88}$$

The square of the rms current in the third layer is

$$I_{3(rms)}^2 = (2I_1)^2 + (3I_1)^2 = 4I_1^2 + 9I_1^2 = 13I_1^2 \tag{4.89}$$

producing the copper loss in the third layer of AC resistance R_{w1}

$$P_{L3} = R_{w1}I^2_{3(rms)} = 13R_{w1}I_1^2. \tag{4.90}$$

The square of the rms current in the fourth layer is

$$I^2_{4(rms)} = (3I_1)^2 + (4I_1)^2 = 9I_1^2 + 16I_1^2 = 25I_1^2 \tag{4.91}$$

producing the copper loss in the fourth layer of AC resistance R_{w1}

$$P_{L4} = R_{w1}I^2_{4(rms)} = 25R_{w1}I_1^2. \tag{4.92}$$

The square of the rms current in the fifth layer is

$$I^2_{5(rms)} = (4I_1)^2 + (5I_1)^2 = 16I_1^2 + 25I_1^2 = 41I_1^2 \tag{4.93}$$

yielding the copper loss in the fifth layer of AC resistance R_{w1}

$$P_{L5} = R_{w1}I^2_{5(rms)} = 41R_{w1}I_1^2. \tag{4.94}$$

In general, the square of the rms current in the mth layer is given by

$$I^2_{m(rms)} = [(m-1)^2 + m^2]I_1^2 = [2m^2 - 2m + 1]I_1^2. \tag{4.95}$$

Hence, the ratio of the rms value of the current in the mth layer to the rms value of the current in the first layer is equal to the ratio of the amplitude of the current in the mth layer $I_{m(m)}$ to the amplitude of the current in the first layer I_{m1}

$$\frac{I_{m(rms)}}{I_1} = \frac{I_{m(m)}}{I_{m1}} = \sqrt{(m-1)^2 + m^2} = \sqrt{2m^2 - 2m + 1}. \tag{4.96}$$

Thus, the rms value of the current increases greatly as the number of layers is increased. The rms currents in the winding layers form a series $I_n^2/I_1^2 = 1, 5, 13, 25, 41, 61, 85, 113, \ldots$. Figure 4.34 depicts the normalized rms current $I_{m(rms)}/I_1$ in the mth layer.

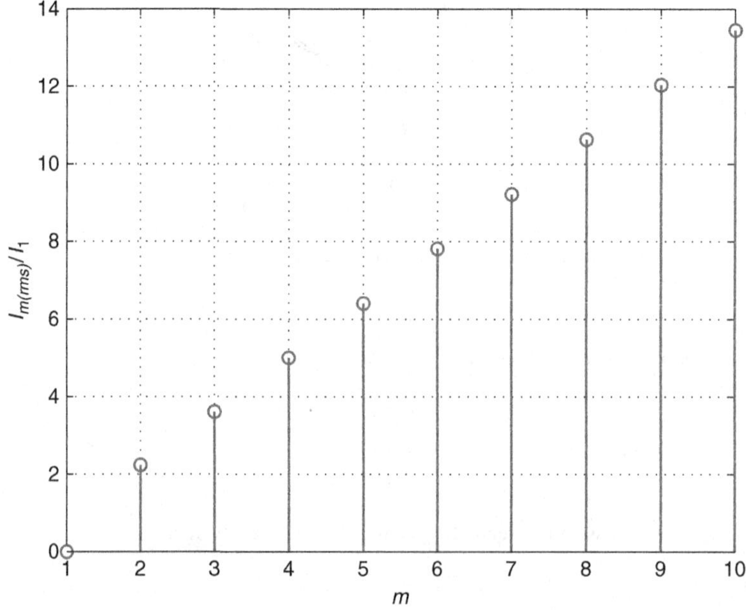

Figure 4.34 Normalized rms current $I_{m(rms)}/I_1$ in the mth layer

The copper loss in the mth layer of AC resistance R_{w1} is given by

$$P_{Lm} = R_{w1}I^2_{m(rms)} = R_{w1}I_1^2[(m-1)^2 + m^2] = R_{w1}I_1^2[2m^2 - 2m + 1]$$
$$= P_{L1}[(m-1)^2 + m^2] = P_{L1}[2m^2 - 2m + 1]. \quad (4.97)$$

Thus, the ratio of the AC power in the mth layer to the AC power in the first layer is

$$\frac{P_{Lm}}{P_{L1}} = (m-1)^2 + m^2 = 2m^2 - 2m + 1. \quad (4.98)$$

Using the equation derived in the appendix of this chapter,

$$\sum_{m=1}^{N_l} m^2 = 1^2 + 2^2 + 3^2 + \ldots + N_l^2 = \frac{N_l(N_l+1)(2N_l+1)}{6} \quad (4.99)$$

and

$$\sum_{m=1}^{N_l} (m-1)^2 = \frac{(N_l-1)(N_l-1+1)[2(N_l-1)+1]}{6} = \frac{(N_l-1)N_l(2N_l-1)}{6}, \quad (4.100)$$

we have

$$\sum_{m=1}^{N_l}[(m-1)^2 + m^2] = \frac{(N_l-1)N_l(2N_l-1)}{6} + \frac{N_l(N_l+1)(2N_l+1)}{6} = \frac{N_l(2N_l^2+1)}{3}. \quad (4.101)$$

Thus, the total copper loss in the winding consisting of N_l layers is given by

$$P_w = \sum_{m=1}^{N_l} P_{Lm} = R_{w1}I_1^2 \sum_{m=1}^{N_l}[(m-1)^2 + m^2] = I_1^2 R_{w1}\frac{N_l(2N_l^2+1)}{3} = P_{L1}\frac{N_l(2N_l^2+1)}{3}. \quad (4.102)$$

Figure 4.35 shows a plot normalized proximity effect power loss P_{Lm}/P_{L1} in the mth layer.

The total DC resistance is

$$R_{wDC} = R_{wDC1}N_l. \quad (4.103)$$

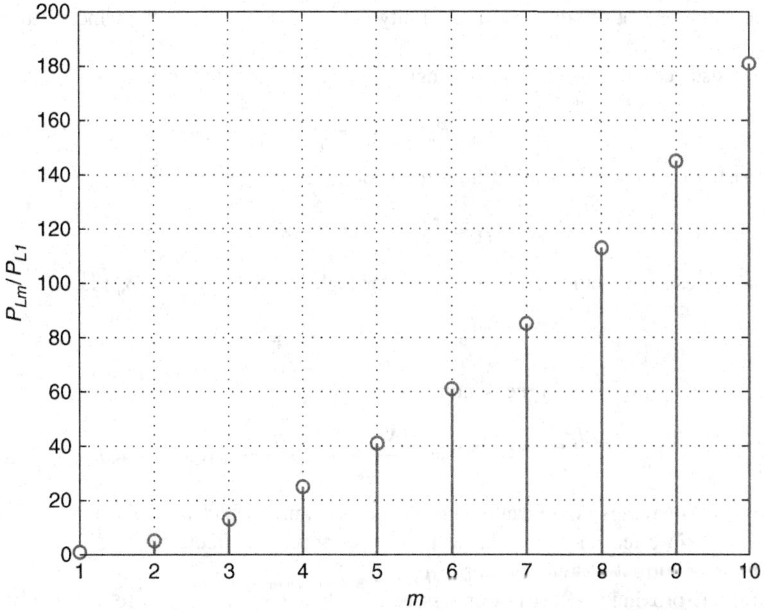

Figure 4.35 Normalized proximity effect power loss P_{Lm}/P_{L1} in the mth layer

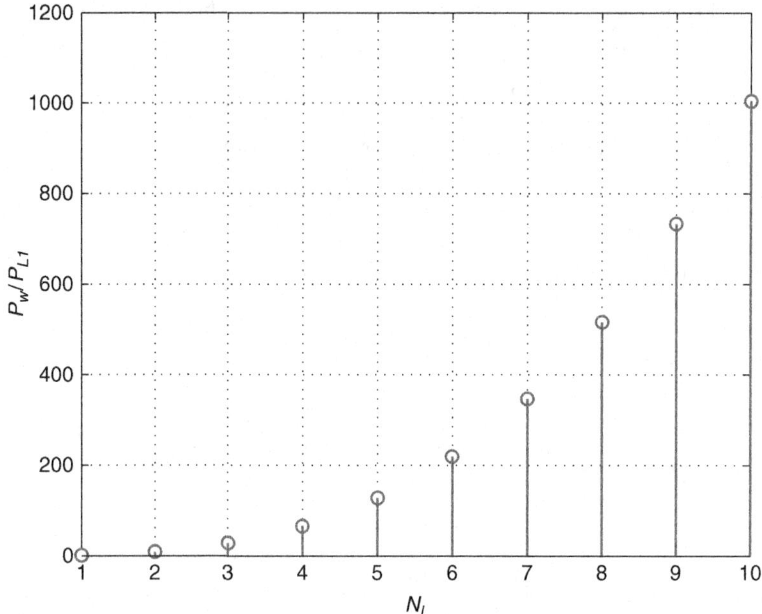

Figure 4.36 Normalized proximity effect power loss P_w/P_{L1} in the winding that consists of N_l layers

Assuming that $I_{DC} = I_1 = I_{1rms}$, the total DC power loss in the winding is

$$P_{wDC} = R_{wDC}I_1^2 = R_{wDC1}N_l I_1^2. \tag{4.104}$$

Hence, the AC-to-DC power loss in the winding is

$$F_R = \frac{P_w}{P_{wDC}} = \frac{2N_l^2+1}{3}\frac{R_{w1}}{R_{wDC1}}. \tag{4.105}$$

Figure 4.36 shows a plot of normalized proximity power loss P_w/P_{L1} in the winding that consists of N_l layers.

The DC resistance of a single layer of length $l_T = l_w/N$ and width b is

$$R_{wDC1} = \frac{\rho_w l_T}{bh}. \tag{4.106}$$

The AC resistance of the first layer at high frequencies is

$$R_{w1} = \frac{\rho_w l_T}{b\delta_w} \quad \text{for} \quad h \gg \delta_w. \tag{4.107}$$

The AC-to-DC resistance ratio of the first layer for high frequencies ($h \gg \delta_w$) is

$$\frac{R_{w1}}{R_{wDC1}} \approx \frac{\rho_w l_T}{b\delta_w} \times \frac{bh}{\rho_w l_T} = \frac{h}{\delta_w} \tag{4.108}$$

and the AC-to-DC power loss in the winding is

$$F_R \approx \frac{2N_l^2+1}{3}\left(\frac{h}{\delta_w}\right) = \frac{(2N_l^2+1)h}{3}\sqrt{\frac{\pi\mu f}{\rho_w}} \quad \text{for} \quad \delta_w \ll h. \tag{4.109}$$

The copper loss increases significantly due to the proximity effect as the number of layers N_l is increased. The skin effect increases the winding resistance with frequency and the proximity effect increases the rms current in higher order layers.

In general, the proximity-effect power loss per unit length of winding wire is given by

$$\frac{P_p}{l_w} = K\rho_w H_m^2 \tag{4.110}$$

where K is a unitless factor, ρ_w is the winding conductor resistivity, and H_m is the amplitude of the external magnetic field intensity.

Example 4.1

An inductor consists of five layers. The rms value of the external sinusoidal current flowing through the inductor is $I_1 = 1$ A. The winding resistance of a single layer is $R_{w1} = 0.1\ \Omega$. The frequency is high so that $h \gg \delta_w$. Calculate the rms value of the current and the copper power loss in each layer. Find the total copper power loss.

Solution The rms value of the sinusoidal current in the first layer is

$$I_1 = 1\text{ A} \tag{4.111}$$

and the copper power loss in the first layer is

$$P_{L1} = R_{w1} I_1^2 = 0.1 \times 1^2 = 0.1\text{ W}. \tag{4.112}$$

For the second layer,

$$I_2 = \sqrt{(2-1)^2 + 2^2} I_1 = \sqrt{1+4} I_1 = \sqrt{5} I_1 = \sqrt{5} \times 1 = 2.236\text{ A} \tag{4.113}$$

and

$$P_{L2} = R_{w1} I_2^2 = 0.1 \times [(2-1)^2 + 2^2] \times 1^2 = 0.1 \times 5 \times 1^2 = 0.5\text{ W}. \tag{4.114}$$

For the third layer,

$$I_3 = \sqrt{(3-1)^2 + 3^2} I_1 = \sqrt{4+9} I_1 = \sqrt{13} I_1 = \sqrt{13} \times 1 = 3.6055\text{ A} \tag{4.115}$$

and

$$P_{L3} = R_{w1} I_3^2 = 0.1 \times [(3-1)^2 + 3^2] \times 1^2 = 0.1 \times 13 \times 1^2 = 1.3\text{ W}. \tag{4.116}$$

For the fourth layer,

$$I_4 = \sqrt{(4-3)^2 + 4^2} I_1 = \sqrt{9+16} I_1 = 5 I_1 = 5 \times 1 = 5\text{ A} \tag{4.117}$$

and

$$P_{L4} = R_{w1} I_4^2 = 0.1 \times [(4-1)^2 + 4^2] \times 1)^2 = 0.1 \times 25 \times 1)^2 = 2.5\text{ W}. \tag{4.118}$$

For the fifth layer,

$$I_5 = \sqrt{(5-4)^2 + 5^2} I_1 = \sqrt{16+25} I_1 = \sqrt{41} I_1 = \sqrt{41} \times 1 = 6.403\text{ A} \tag{4.119}$$

and

$$P_{L5} = R_{w1} I_5^2 = 0.1 \times [(5-1)^2 + 5^2] \times 1^2 = 0.1 \times 41 \times 1^2 = 4.1\text{ W}. \tag{4.120}$$

The total copper loss in all five layers is

$$P_w = P_{L1} \frac{N_l(2N_l^2 + 1)}{3} = 0.1 \times \frac{5 \times (2 \times 5^2 + 1)}{3} = 8.5\text{ W}. \tag{4.121}$$

Denote the number of turns in each layer of an inductor by N_{tl}. In our example, each layer has only one turn, that is, $N_{tl} = 1$. Assume that $\mu_{rc} \gg 1$ for the cores such that the reluctance of the core $\mathcal{R}_c$ is much lower than the reluctance of the spaces between the layers $\mathcal{R}_g$. Therefore, the magnetomotive force (MMF) induced by the leakage flux in the core $\mathcal{F}_c$ is much lower than that induced in the air areas between the layers $\mathcal{F}_g$. According to Ampère's law, the MMF due to the current enclosed by a loop is

$$Ni = \mathcal{F} = \mathcal{F}_g + \mathcal{F}_c = \phi_l (R_g + R_c) = \phi_l \left(\frac{l_g}{\mu_0 A_g} + \frac{l_c}{\mu_{rc} \mu_0 A_c} \right)$$
$$\approx \frac{\phi_l}{\mu_0 A_g} \left(l_g + \frac{l_c}{\mu_{rc}} \right) \approx \frac{\phi_l l_g}{\mu_0 A_g} = \phi_l R_g \approx \mathcal{F}_g \quad \text{for} \quad l_g \gg \frac{l_c}{\mu_{rc}}. \tag{4.122}$$

According to Ampère's law, the closed line integral of the magnetic field intensity along the mean magnetic path within the core and the area between the layers is equal to ampere-turns,

$$Ni = H_g l_g + H_c l_c = \frac{B}{\mu_0} l_g + \frac{B}{\mu_{rc}\mu_0} l_c = \frac{\phi_l}{\mu_0 A_g}\left(l_g + \frac{l_c}{\mu_{rc}}\right) \approx H_g l_g \quad \text{for} \quad l_g \gg \frac{l_c}{\mu_{rc}}. \tag{4.123}$$

For simplicity, we can write

$$\mathcal{F} \approx \mathcal{F}_g = Ni \approx H_g l_g. \tag{4.124}$$

The current enclosed by the leakage flux of the first layer of the winding is given by

$$I_1 = N_{tl} I_1 = I_1 \tag{4.125}$$

resulting in

$$\mathcal{F}_1 = I_1 = N_{tl} I_1 = I_1. \tag{4.126}$$

Similarly, the current enclosed by the leakage flux of the first and second layers of the windings is

$$I_2 = 2N_{tl} I_1 = 2I_1 \tag{4.127}$$

yielding

$$\mathcal{F}_2 = I_2 = 2N_{tl} I_1 = 2I_1. \tag{4.128}$$

The current enclosed by the leakage flux of the first, second, and third layers of the windings is

$$I_3 = 3N_{tl} I_1 = 3I_1 \tag{4.129}$$

producing

$$\mathcal{F}_3 = I_3 = 3N_{tl} I_1 = 3I_1. \tag{4.130}$$

Finally, the current enclosed by all the four layers is

$$I_4 = 4N_{tl} I_1 = 4I_1 \tag{4.131}$$

resulting in

$$\mathcal{F}_4 = I_4 = 4N_{tl} I_1 = 4I_1. \tag{4.132}$$

Hence, the magnetic field intensity between the mth and $(m+1)$th layer is

$$H_{gm} = \frac{Ni}{l_g} = \frac{mM_{tl} i}{l_g}. \tag{4.133}$$

4.9 Self-Proximity Effect in Rectangular Conductors

Consider a rectangular solid conductor of breadth c, thickness $2w$, and length l_w, where $c \gg 2w$. Assume that the skin effect is negligible, that is, $\delta_w \gg w$. The current flowing through the conductor in the direction l_w is sinusoidal

$$i(t) = I_{Lm} \sin \omega t. \tag{4.134}$$

The applied current flows through the entire cross-sectional area of the conductor

$$A_{tot} = 2cw. \tag{4.135}$$

It is assumed that the current density is uniform. Therefore, the current density is uniform in the entire conductor. The amplitude of the current density is

$$J_m = \frac{I_{Lm}}{A_{tot}} = \frac{I_{Lm}}{2cw}. \tag{4.136}$$

Hence, the waveform of the current density is

$$J(t) = J_m \sin \omega t = \frac{I_{Lm}}{A_{tot}} \sin \omega t = \frac{I_{Lm}}{2cw} \sin \omega t \quad \text{for} \quad 0 \le x \le w. \tag{4.137}$$

The enclosed current flows through the cross-sectional area

$$A_{enc} = 2cx \quad \text{for} \quad 0 \le x \le w. \tag{4.138}$$

The amplitude of the current density enclosed by the Ampèrian contour is

$$J_{m(enc)} = \frac{I_{Lm(enc)}}{A_{enc}} = \frac{I_{Lm(enc)}}{2cx} \quad \text{for} \quad 0 \le x \le w. \tag{4.139}$$

Equating the amplitudes of the current densities given in (4.136) and (4.139)

$$\frac{I_{Lm}}{2cw} = \frac{I_{Lm(enc)}}{2cx} \tag{4.140}$$

gives the amplitude of the enclosed current

$$I_{Lm(enc)}(x) = I_{Lm}\left(\frac{x}{w}\right) \quad \text{for} \quad 0 \le x \le w. \tag{4.141}$$

The waveform of the current enclosed by an Ampèrian contour is

$$i_{enc}(x,t) = I_{Lm(enc)}(x) \sin\ \omega t = I_{Lm}\left(\frac{x}{w}\right) \sin\ \omega t \quad \text{for} \quad 0 \le x \le w. \tag{4.142}$$

From Ampère's law, the amplitude of the enclosed current is

$$I_{Lm(enc)} = \oint_C \mathbf{H}_m(x) \cdot d\mathbf{l} = H_m(x) 2c. \tag{4.143}$$

Hence, the amplitude of the magnetic field intensity is

$$H_m(x) = \frac{I_{Lm(enc)}}{2c} = \frac{I_{Lm}}{2c}\left(\frac{x}{w}\right) \quad \text{for} \quad 0 \le x \le w. \tag{4.144}$$

The maximum amplitude of the magnetic field intensity occurs at $x = w$ and is given by

$$H_{m(max)} = H_m(w) = \frac{I_{Lm}}{2c}. \tag{4.145}$$

The waveform of the magnetic field is

$$H(x,t) = H_m(x) \sin\ \omega t = \frac{I_m}{2c}\left(\frac{x}{w}\right) \sin\ \omega t \quad \text{for} \quad 0 \le x \le w. \tag{4.146}$$

The amplitude of the magnetic flux density is

$$B_m(x) = \mu_0 H_m(x) = \frac{\mu_0 I_{Lm}}{2c}\left(\frac{x}{w}\right) \quad \text{for} \quad 0 \le x \le w. \tag{4.147}$$

The maximum amplitude of the magnetic flux density is

$$B_{m(max)} = B_m(w) = \frac{\mu_0 I_{Lm}}{2c}, \tag{4.148}$$

yielding

$$B_m(x) = B_{m(max)}\left(\frac{x}{w}\right) \quad \text{for} \quad 0 \le x \le w. \tag{4.149}$$

The waveform of the magnetic flux density is

$$B(x,t) = B_m(x) \sin\ \omega t = B_{m(max)}\left(\frac{x}{w}\right) \sin\ \omega t \quad \text{for} \quad 0 \le x \le w. \tag{4.150}$$

Now consider the magnetic flux passing through a thin cross-sectional area of length l_w and thickness dx

$$dS = l_w dx. \tag{4.151}$$

The amplitude of the magnetic flux that passes through the incremental surface dS is

$$\phi_m(x) = \int_{-x}^{x} B_m(x) dS = 2\int_0^x B_m(x) dS = 2\int_0^x \frac{\mu_0 I_{Lm}}{2c}\left(\frac{x}{w}\right) l_w dx = \frac{\mu_0 l_w I_{lm}}{2cw} x^2$$
$$= \frac{\mu_0 w l_w I_{Lm}}{2c}\left(\frac{x}{w}\right)^2 = w l_w B_{m(max)}\left(\frac{x}{w}\right)^2 = A_\phi B_{m(max)}\left(\frac{x}{w}\right)^2 \quad \text{for} \quad 0 \le x \le w, \tag{4.152}$$

where $A_\phi = wl_w$. The waveform of the magnetic flux is

$$\phi(x,t) = \phi_m(x)\sin \omega t = A_\phi B_{m(max)}\left(\frac{x}{w}\right)^2 \cos \omega t \quad \text{for} \quad 0 \le x \le w, \tag{4.153}$$

The resistance of the thin layer is

$$R_{dx} = \frac{\rho_w l_w}{cdx}. \tag{4.154}$$

The voltage waveform developed by the magnetic flux in the thin layer is

$$v(t) = \frac{d\phi(x)}{dt} = \omega\phi_m(x)\cos \omega t = V_m(x)\cos \omega t, \tag{4.155}$$

where the amplitude of the voltage developed by the magnetic flux in the thin layer is

$$V(x) = \omega\phi_m(x) = \omega A_\phi B_{m(max)}\left(\frac{x}{w}\right)^2 \quad \text{for} \quad -w \le x \le w. \tag{4.156}$$

The time-average eddy-current power loss in the thin layer is

$$dP_{PE} = \frac{V^2(x)}{2R_{dx}} = \frac{c\omega^2 A_\phi^2 B_{m(max)}^2}{2\rho_w l_w}\left(\frac{x}{w}\right)^4 dx \quad \text{for} \quad -w \le x \le w. \tag{4.157}$$

Hence, the time-average eddy-current power loss in the entire conductor is

$$P_{PE} = \int_{-w}^{w} dP_{PE} = 2\int_0^w dP_{PE} = \frac{c\omega^2 A_\phi^2 B_{m(max)}^2}{\rho_w l_w w^4}\int_0^w x^4 dx = \frac{c\omega^2 A_\phi^2 B_{m(max)}^2}{5\rho_w l_w} w$$
$$= \frac{c\omega^2 l_w B_{m(max)}^2 w^3}{5\rho_w} = \frac{l_w \mu_0^2 \omega^2 I_{Lm}^2 w^3}{20\rho_w c} = \frac{l_w \mu_0^2 \omega^2 I_{Lrms}^2 w^3}{10\rho_w c}. \tag{4.158}$$

Assume that $I_{DC} = I_{Lrms} = I_{Lm}/\sqrt{2}$. The power loss due to the flow of applied current is

$$P_{LF} = P_{wDC} = R_{wDC} I_{DC}^2 = \frac{\rho_w l_w I_{DC}^2}{2wc} = \frac{\rho_w l_w I_{Lrms}^2}{2wc} = \frac{\rho_w l_w I_{Lm}^2}{4wc}. \tag{4.159}$$

The total power loss is equal to the sum of the power loss due to the flow of the applied current and the proximity current

$$P_w = P_{wDC} + P_{PE} = \frac{\rho_w l_w I_{Lrms}^2}{2wc} + \frac{l_w \mu_0^2 \omega^2 I_{Lrms}^2 w^3}{10\rho_w c} = \frac{\rho_w l_w I_{Lrms}^2}{2wc}\left(1 + \frac{\mu_0^2\omega^2 w^4}{5\rho_w^2}\right)$$
$$= P_{wDC}\left[1 + \frac{4}{5}\left(\frac{w}{\delta_w}\right)^4\right] = \frac{l_w \rho_w I_{Lrms}^2}{2c}\left(\frac{1}{w} + \frac{\omega^2\mu_0^2 w^3}{5\rho_w^2}\right) = \frac{l_w \rho_w I_{Lm}^2}{4c}\left(\frac{1}{w} + \frac{4w^3}{5\delta_w^4}\right). \tag{4.160}$$

Taking the derivative of P_w with respect to w and setting the result to zero,

$$\frac{dP_w}{dw} = \frac{\rho_w l_w I_{Lm}^2}{4c}\left(-\frac{1}{w^2} + \frac{12w^2}{5\delta_w^4}\right) = 0, \tag{4.161}$$

we obtain the optimum value of w as

$$\frac{w_{opt}}{\delta_w} = \sqrt[4]{\frac{5}{12}} = 0.803. \tag{4.162}$$

The DC and low-frequency resistance is

$$R_{wDC} = \frac{\rho_w l_w}{2wc} \tag{4.163}$$

The eddy-current resistance is

$$R_e = \frac{2P_{PE}}{I_{Lm}^2} = \frac{\mu_0^2\omega^2 w^3 l_w}{10\rho_w c}. \tag{4.164}$$

The total resistance is equal to the resistance presented to the applied AC current and eddy current

$$R_w = R_{wDC} + R_e = \frac{\rho_w l_w}{2wc} + \frac{\mu_0^2\omega^2 w^3 l_w}{10\rho_w c} = \frac{\rho_w l_w}{2wc}\left(1 + \frac{\mu_0^2\omega^2 w^4}{5\rho_w^2}\right) = R_{wDC}\left[1 + \frac{4}{5}\left(\frac{w}{\delta_w}\right)^4\right]$$

$$= \frac{\rho_w l_w}{2c}\left(\frac{1}{w} + \frac{\mu_0^2\omega^2 w^3}{5\rho_w^2}\right). \tag{4.165}$$

The derivative of the resistance with respect to the width w is

$$\frac{dR_w}{dw} = \frac{\rho_w l_w}{2c}\left(-\frac{1}{w^2} + \frac{3\mu_0^2\omega^2 w^2}{5\rho_w^2}\right) = \frac{\rho_w l_w}{2c}\left(-\frac{1}{w^2} + \frac{12w^2}{5\delta_w^4}\right) = 0, \tag{4.166}$$

yielding the optimum value of w as

$$\frac{w_{opt}}{\delta_w} = \sqrt[4]{\frac{5}{12}} = 0.803 \tag{4.167}$$

and the total optimum conductor thickness

$$\frac{2w_{opt}}{\delta_w} = 2\sqrt[4]{\frac{5}{12}} = 1.606. \tag{4.168}$$

The AC-to-DC resistance ratio is

$$F_R = \frac{R_w}{R_{wDC}} = 1 + \frac{4\mu_0^2\omega^2 w^4}{5\rho_w} = 1 + \frac{4}{5}\left(\frac{w}{\delta_w}\right)^4. \tag{4.169}$$

The AC-to-DC resistance at $w = w_{opt}$ is

$$F_{Rv} = \frac{R_w}{R_{wDC}} = 1 + \frac{1}{3} = \frac{4}{3}. \tag{4.170}$$

The boundary between the low- and high-frequency ranges is defined as

$$F_{RB} = 1 + \frac{4}{5}\left(\frac{w}{\delta_w}\right)^4 = 1.05, \tag{4.171}$$

yielding

$$\left(\frac{w}{\delta_w}\right)_B = \frac{1}{2}. \tag{4.172}$$

Since $\delta_w = \sqrt{2\rho_w/(\mu_0\omega)}$, the boundary frequency is

$$f_B = \frac{\rho_w}{4\pi w^2}. \tag{4.173}$$

4.10 Summary

- In accordance with Lenz's law, eddy currents produce their own magnetic field to oppose the original field.
- There are two effects of eddy currents: skin effect and proximity effect. Both of these effects cause current crowding. However, the skin effect is much stronger than the proximity effect.
- A conductor carrying no net current, but immersed in a magnetic field generated by other sources produces power loss.
- The proximity effect takes place when a second conductor is under the influence of a time-varying magnetic field produced by a nearby first conductor carrying a time-varying current.
- The proximity-effect loss is caused by magnetic fields of adjacent layers and adjacent windings.
- The average current density due to the proximity effect is always zero.

- Eddy currents are induced in the second conductor whether or not the second conductor carries current. If the second conductor does not carry current, then only the proximity effect is present in the second conductor. If the second conductor carries current, then the total eddy current consists of the proximity-eddy current and the skin-effect eddy current.
- The skin-effect eddy current and the proximity-effect eddy current are orthogonal to each other and can be calculated independently.
- Eddy currents cause nonuniform distribution of current density, an increase in the conductor AC resistance, an increase in the conduction power loss, and a reduction in the inductance.
- According to Lenz's law, the direction of eddy currents is such that they oppose the change that causes them.
- The skin effect and the proximity effect both contribute to the nonuniformity of current distribution in conductors.
- The skin effect and the proximity effect limit the effective capability of a conductor to conduct high-frequency currents.
- For $\delta_w > d$, the proximity effects can be neglected, and the AC winding resistance is approximately equal to the DC winding resistance.
- In single-layer inductors, the proximity effect is absent, and the AC winding resistance at high frequencies is higher than the DC resistance due to the skin effect only.
- The proximity-effect loss in multiple-layer windings is much higher than the skin-effect loss.
- In multiple-layer inductors and transformers, the current flows in both directions in higher order winding layers, significantly increasing the rms currents and copper loss.
- The proximity effect dominates over the skin effect in a winding that has many layers.
- Due to the proximity effect, the current density may be higher in the conductor areas, where the conductors are close to each other.
- Increasing the distance between the conductors in the same layer reduces the proximity-effect loss.
- Increasing the distance between the layers reduces the proximity-effect loss.
- Interleaved windings in transformers reduce the copper loss caused by the proximity effect at high frequencies, if the currents through the primary and secondary windings are in phase. Each layer operates as a single-layer winding, and the proximity effect is nearly eliminated.
- In transformers with a single-layer primary and a single-layer secondary, the proximity effect is absent.
- To reduce the copper loss due to the proximity effect, it is highly advantageous to reduce the number of winding layers, increase the winding width, and interleave the windings.
- Litz wire increases the effective conduction area at high frequencies and thereby reduces the copper loss.
- Eddy currents cause power loss in electrostatic shields.

4.11 Appendix

4.11.1 Derivation of Proximity Power Loss

Since

$$(n+1)^2 = n^2 + 2n + 1 \tag{4.174}$$

then

$$(n+1)^2 - n^2 = 2n + 1. \tag{4.175}$$

Thus,

$$2^2 - 1^2 = 2 \times 1 + 1 \tag{4.176}$$

$$3^2 - 2^2 = 2 \times 2 + 1 \tag{4.177}$$

$$4^2 - 3^2 = 2 \times 3 + 1 \tag{4.178}$$

and

$$(n+1)^2 - n^2 = 2n + 1. \tag{4.179}$$

Adding (4.176) through (4.179),

$$(n+1)^2 - 1^2 = 2(1 + 2 + 3 + \ldots + n) + n, \tag{4.180}$$

we obtain

$$1 + 2 + 3 + \ldots + n = \frac{(n+1)^2 - 1^2 - n}{2} = \frac{n(n+1)}{2}. \tag{4.181}$$

Notice that the sum of the left-hand sides in (4.176) through (4.179) is

$$2^2 - 1^2 + 3^3 - 2^2 + 4^2 - 3^2 + 5^2 - 4^2 + \ldots + (n+1)^2 - n^2$$
$$= 4 - 1 + 9 - 4 + 16 - 9 + 25 - 16 + \ldots + (n+1)^2 - n^2 = (n+1)^2 - 1^2. \tag{4.182}$$

Because

$$(n+1)^3 = n^3 + 3n^2 + 3n + 1 \tag{4.183}$$

we have

$$(n+1)^3 - n^3 = 3n^2 + 3n + 1. \tag{4.184}$$

Thus,

$$2^3 - 1^3 = 3 \times 1^2 + 3 \times 1 + 1 \tag{4.185}$$

$$3^3 - 2^3 = 3 \times 2^2 + 3 \times 2 + 1 \tag{4.186}$$

$$4^3 - 3^3 = 3 \times 3^2 + 3 \times 3 + 1 \tag{4.187}$$

and

$$(n+1)^3 - n^3 = 3n^2 + 3n + 1. \tag{4.188}$$

Adding (4.185) through (4.188),

$$(n+1)^3 - 1^3 = 3(1^2 + 2^2 + 3^2 + \ldots + n^2) + 3(1 + 2 + 3 + \ldots + n) + n, \tag{4.189}$$

we get

$$1^2 + 2^2 + 3^2 + \ldots + n^2 = \frac{(n^3 + 3n^2 + 3n) - \frac{3n(n+1)}{2} - n}{3} = \frac{n(n+1)(2n+1)}{6}. \tag{4.190}$$

Using this equation, the proximity power loss can be derived.

4.12 References

[1] P. J. Dowell, "Effects of eddy currents in transformer winding," *Proceedings of the IEE*, vol. 113, no. 8, pp. 1387–1394, August 1966.

[2] J. Jongsma, "High-frequency ferrite power transformer and choke design, Part 3: Transformer winding design, Part 4: Improved method of power choke design," Philips Electronic Components and Materials, Technical Publication, no. 27, Philips, The Netherlands, 1986.

[3] A. Kennelly, F. Laws, and P. Pierce, "Experimental research on skin effect in conductors," *Transactions of AIEE*, vol. 34, p. 1915, 1915.

[4] J. Lammeraner and M. Stafl, *Eddy Currents*, Cleveland: CRS Press, 1966.

[5] J. Ebert, "Four terminal parameters of HF inductors," *Bulletin de l'Acad'emie Polonaise des Sciences, Séries des Sciences Techniques*, no. 5, 1968.

[6] E. C. Snelling, *Soft Ferrites: Properties and Applications*, London: Iliffe Books Ltd, 1969.

[7] R. L. Stall, *The Analysis of Eddy Currents*, Oxford: Clarendon Press, 1974, pp. 21–27.

[8] W. T. McLyman, *Transformer and Inductor Design Handbook*, 3rd Ed., New York: Marcel Dekker, 2004.

[9] J. K. Watson, *Applications of Magnetism*, Gainesville, FL: Continental Media & Beyond, 1985.

[10] E. Bennett and S. C. Larsen, "Effective resistance of alternating currents of multilayer windings," *Transactions of the American Institute of Electrical Engineers*, vol. 59, pp. 1010–1017, 1940.

[11] R. L. Perry, "Multiple layer series connected winding design for minimum losses," *IEEE Transactions on Power Applications Systems*, vol. PAS-98, pp. 116–123, January/February 1979.

[12] B. Carsten, "High frequency conductor losses in switch mode magnetics," Proceedings of PCI, Munich, Germany, 1986, pp. 161–182.

[13] J. P. Vandalec and P. D. Ziogos, "A novel approach for minimizing high-frequency transformer copper loss," *IEEE Transactions on Power Electronics*, vol. 3, pp. 266–276, July 1988.

[14] A. M. Urling, V. A. Niemela, G. R. Skutt, and T. G. Wilson, "Characterizing high frequency effects in transformer windings: a guide to several significant papers," IEEE Transactions on Power Electronics Specialists Conference, 1989, pp. 373–385.

[15] J. A. Ferreira, *Electromagnetic Modelling of Power Electronic Converters*, Boston, MA: Kluwer Academic Publisher, 1989.

[16] E. E. Kreizis, T. D. Tsiboukis, S. M. Panas, and J. A. Tegopoulos, "Eddy currents: theory and applications," *Proceedings of the IEEE*, vol. 80, no. 10, pp. 1559–1589, October 1992.

[17] M. Bartoli, A. Reatti, and M. K. Kazimierczuk, "High-frequency models of ferrite inductors," Proceedings of the IEEE International Conference on Industrial Electronics, Controls, Instrumentation, and Automation (IECON'94), Bologna, Italy, September 5-9, 1994, pp. 1670–1675.

[18] M. Bartoli, A. Reatti, and M. K. Kazimierczuk, "Predicting the high-frequency ferrite-core inductor performance," Proceedings of the Conference of Electrical Manufacturing and Coil Winding, Chicago (Rosemont), IL, September 27-29, 1994, pp. 409–413.

[19] M. Bartoli, A. Reatti, and M. K. Kazimierczuk, "Modeling iron-powder inductors at high-frequencies," Proceedings of the IEEE Industry Applications Society Annual Meeting, Denver, CO, October 2-7, 1994, pp. 1125–1232.

[20] J. A. Ferreira, "Improved analytical modeling of conductive losses in magnetic components," *IEEE Transactions on Power Electronics*, vol. 9, pp. 127–131, January 1994.

[21] M. Bartoli, N. Noferi, A. Reatti, and M. K. Kazimierczuk, "Modeling winding losses in high-frequency power inductors," *Journal of Circuits Systems and Computers*, vol. 5, no. 3, pp. 65–80, March 1995.

[22] N. Mohan, T. M. Underland, and W. P. Robbins, *Power Electronics*, 3rd Ed., New York: John Wiley & Sons, 2003.

[23] M. Bartoli, N. Nefari, A. Reatti, and M. K. Kazimierczuk, "Modeling litz-wire winding losses in high-frequencies power inductors," Proceedings of the IEEE Power Electronics Specialists Conference, Baveno, Italy, June 24-27, 1996, pp. 1690–1696.

[24] K. O'Meara, "Proximity loss in ac magnetic devices," Power Conversion and Intelligent Motion (PCIM) Magazine, pp. 52–57, December 1996.

[25] C.-M. Ong, *Dynamic Simulation of Electric Machinery*, Reading, MA: Prentice Hall, 1998, pp. 38–40, 45–46, and 87–90.

[26] W. G. Hurley, W. H. Wolfe, and J. G. Breslin, "Optimized transformer design: inclusive of high-frequency effects," *IEEE Transactions on Power Electronics*, vol. 13, no. 4, pp. 651–659, July 1998.

[27] W. G. Hurley, E. Gath, and J. G. Breslin, "Optimizing the ac resistance of multilayer transformer windings with arbitrary current waveforms," *IEEE Transactions on Power Electronics*, vol. 15, no. 2, pp. 369–376, March 2000.

[28] M. K. Kazimierczuk, G. Sancineto, U. Reggiani, and A. Massarini, "Small-signal high-frequency model of ferrite inductors," *IEEE Transactions on Magnetics*, vol. 35, pp. 4185–4191, September 1999.

[29] U. Reggiani, G. Sancineto, and M. K. Kazimierczuk, "High-frequency behavior of laminated iron-core inductors for filter applications," Proceedings of the IEEE Applied Power Electronics Conference, New Orleans, LA, February 6-10, 2000, pp. 654–660.

[30] G. Grandi, M. K. Kazimierczuk, A. Massarini, U. Reggiani, and G. Sancineto, "Model of laminated iron-core inductors," *IEEE Transactions on Magnetics*, vol. 40, no. 4, pp. 1839–1845, July 2004.

[31] K. Howard and M. K. Kazimierczuk, "Eddy-current power loss in laminated power cores," Proceedings of the IEEE International Symposium on Circuits and Systems, Sydney, Australia, May 7-9, 2000, paper III-668, pp. 668–672.

[32] A. Reatti and M. K. Kazimierczuk, "Comparison of various methods for calculating the ac resistance of inductors," *IEEE Transactions on Magnetics*, vol. 37, pp. 1512–1518, May 2002.
[33] G. Grandi, M. K. Kazimierczuk, A. Massarini, and U. Reggiani, "Stray capacitance of single-layer solenoid air-core inductors," *IEEE Transactions on Industry Applications*, vol. 35, no. 5, pp. 1162–1168, September 1999.
[34] A. Massarini and M. K. Kazimierczuk, "Self-capacitance of inductors," *IEEE Transactions on Power Electronics*, vol. 12, no. 4, pp. 671–676, July 1997.
[35] R. W. Erickson and D. Maksimović, *Fundamentals of Power Electronics*, Norwell, MA: Kluwer Academic Publishers, 2001.
[36] A. Van den Bossche and V. C. Valchev, *Inductors and Transformers for Power Electronics*, Boca Raton, FL: Taylor & Francis, 2005.
[37] F. E. Terman, *Radio Engineers' Handbook*, New York: McGraw-Hill, 1943.
[38] J. C. Maxwell, *A Treatise of Electricity and Magnetism*, 3rd Ed., New York: Dover Publishing, 1997.
[39] Y.-J. Kim and M. G. Allen, "Integrated solenoid-type inductors for high frequency applications and their characteristics," 48th IEEE 1998 Electronic Components and Technology Conference, May 25–28, 1998, pp. 1247–1252.
[40] N. Das and M. K. Kazimierczuk, "An overview of technical challenges in the design of current transformers," Electrical Manufacturing Conference, Indianapolis, IN, October 24-26, 2005.
[41] T. Suetsugu and M. K. Kazimierczuk, "Integration of Class DE inverter for dc-dc converter on-chip power supplies," 48th IEEE International Symposium on Circuits and Systems, Kos, Greece, May 21-24, 2006, pp. 3133–3136.
[42] T. Suetsugu and M. K. Kazimierczuk, "Integration of Class DE synchronized dc-dc converter on-chip power supplies," IEEE Power Electronics Specialists Conference, Jeju, South Korea, June 21-24, 2006.

4.13 Review Questions

4.1. What is the proximity effect?

4.2. What is the antiproximity effect?

4.3. What is the current density distribution in two parallel round conductors conducting current in the same direction?

4.4. What is the current density distribution in two parallel conductors conducting current in opposite directions?

4.5. What is the current density distribution in a coaxial cable at high frequencies?

4.6. What is the current density distribution in two parallel plates conducting current in the same direction?

4.7. What is the current density distribution in two parallel plates conducting current in the opposite directions?

4.8. Is the skin effect present in two parallel plates?

4.9. What is the expression for the rms value of the current in the mth layer?

4.10. What is the expression for the winding loss in the mth layer?

4.11. What is the expression for the winding loss in the N_l-layer inductor?

4.14 Problems

4.1. Two infinitely long parallel wires carry currents of equal magnitude. What is the resultant magnetic field intensity due to the two wires at a point midway between the wires, compared with the magnetic field intensity due to one of them alone, if the currents are flowing in the same direction and in the opposite directions.

4.2. A 10-layer inductor is driven by a sinusoidal current source of rms value of 1 A and very high frequency so that $d \gg \delta_w$. Its single-layer resistance is 0.1 Ω. Find the total power loss.

4.3. An inductor consists of 22 layers and carries a high-frequency current $i_L = 2\sqrt{2} \sin \omega t$ A. The winding power loss in the first layer due to the proximity effect is $P_1 = 10$ mW.

(a) Find the rms current in twenty-second layer.

(b) Find the proximity-effect winding power loss in the twenty-second layer.

(c) Find the total proximity-effect winding power loss in all 22 layers.

4.4. An inductor conducting a high-frequency sinusoidal current of an amplitude of 1 A consists of 32 layers. The power loss in the first layer due to proximity effect is $P_1 = 1$ mW.

(a) Find the rms value of the current in the thirty-second layer.

(b) Find the proximity-effect power loss in the thirty-second layer.

(c) Find the proximity-effect power loss in all 32 layers.

4.5. An inductor consists of 15 foil layers and conducts an applied sinusoidal current of amplitude 1 A. The AC power loss in the first layer is 20 mW.

(a) Find the rms value of the current in the fifteenth layer.

(b) Find the proximity-effect power loss in the fifteenth layer.

(c) Find the proximity-effect power loss in all 15 layers.

4.6. An inductor conducting a high-frequency sinusoidal current of 2 A consists of 25 layers. The resistance of a single layer is 0.07 Ω. Find the proximity-effect power loss in the winding.

5

Winding Resistance at High Frequencies

5.1 Introduction

In this chapter, winding power losses are analyzed [1–106]. Winding resistance of inductors at high frequencies is studied. Dowell's equation [5] is used to develop the theory of high-frequency winding resistance. Both the skin and proximity effects are taken into account. Winding power losses in individual layers of inductors are explored. Expressions are given for the winding resistance of foil, strip, square, round, and multistrand wires, including litz wire, as functions of size and frequency. The optimization of the winding conductor thickness or diameter is performed. An expression for inductance is given. The application of litz wire at high frequencies is explored. Winding losses for inductor currents with harmonics are also studied.

5.2 Eddy Currents

The power loss in the nth layer of an inductor or transformer winding is the result of the superposition of two distributions of eddy-current densities caused by two effects: the *skin effect* and the *proximity effect*. The first distribution of current density in the nth layer is due to the magnetic flux set up within the winding conductor of the nth layer by the time-varying current in the nth layer itself. The current density due to the skin effect is the same in all layers and independent of n. It is the only eddy-current distribution in a single-layer winding. The second distribution of the current density in the nth layer is the distribution of eddy currents attributed to the magnetic fields set up within the winding conductor of the nth layer by the time-varying currents in all other layers of the winding. The current densities caused by the skin effect and the proximity effect are orthogonal. In multilayer windings, the loss caused by the proximity effect is many times greater than the loss caused by the skin effect [1].

High-Frequency Magnetic Components, Second Edition. Marian K. Kazimierczuk.

Companion Website: www.wiley.com/go/kazimierczuk_High2e

5.3 Magnetic Field Intensity in Multilayer Foil Inductors

A winding of an inductor or a transformer is a set of conductor turns that share the same current. Consider a multiple-layer inductor winding depicted in Fig. 5.1. Each layer of the winding consists of a single turn made up of a foil. The magnetic field H and current density J are derived under the following assumptions in the subsequent analysis:

1. The winding consists of a straight and parallel foil conductor of the same thickness h. It forms one or several parallel layers. As a result, there is only the axial component of the magnetic field. The magnetic field is parallel (tangential) to the winding conductor surface and perpendicular to the current flow.
2. The curvature, edge, and end effects of the conductors are neglected. It is assumed that $h \ll b$ and $h \ll l_w$, where l_w is the winding conductor length. The layer thickness h is much smaller than the curvature radius.
3. The magnetic field is everywhere parallel to the conducting layers, inside the conductors, between the layers, and inside the magnetic core.
4. The net charge density of the conductor is zero, that is, the amounts of positive and negative charges are equal.
5. The core is ideal and ungapped, that is, $\mu_{rc} = \infty$ and $\rho_c = \infty$. Therefore, the magnetic field intensity in the core $H = B/(\mu_{rc}\mu_0) = 0$ and at the inner surface of the innermost layer is assumed to be zero. This means that a practical core has a very high relative permeability μ_{rc}.
6. The winding occupies the full bobbin window and it is placed between zero and the maximum value of the magnetomotive force.
7. Self-capacitance of inductors is neglected.
8. The system is magnetoquasistatic, that is, $\omega\epsilon\rho \ll 1$.
9. The sinusoidal current flows through the winding conductor

$$i_L = I_m \cos \omega t. \tag{5.1}$$

10. The alternating magnetic flux density

$$B_z(x,t) = B(x,t) = B_m(x)\cos(\omega t + \phi_B), \tag{5.2}$$

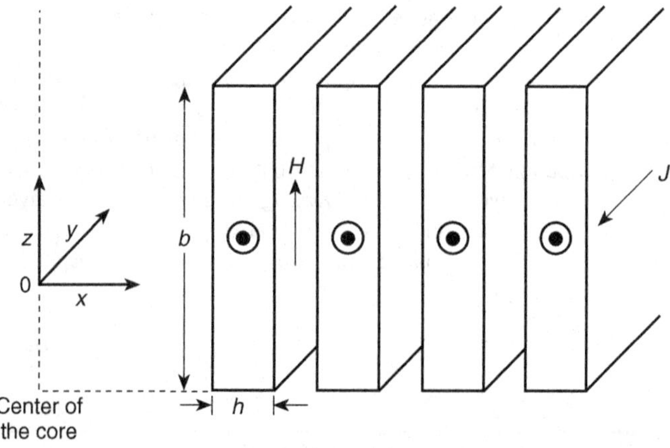

Figure 5.1 Inductor winding made of straight, parallel foil conductor

and the alternating magnetic field intensity

$$H_z(x,t) = H(x,t) = H_m(x)\cos(\omega t + \phi_B), \tag{5.3}$$

are everywhere (inside and outside of a conductor) parallel to the foil winding.

Simulations show that the distributions of magnetic flux lines, winding conductor current density, and winding conductor power loss density depend on frequency. The magnetic flux lines at low frequencies are less parallel to the foil turns than those at high frequencies. At high frequencies, the magnetic flux lines are nearly parallel to the foil winding conductor. At low frequencies, the current density and the power loss density are higher in the middle of the winding conductor than those at the end of turns. As the frequency increases, the current density and the power loss density in the foil winding conductor become more uniform.

Under these assumptions, a one-dimensional (1D) solution of Maxwell's equations can be used to determine magnetic field and current density distribution in the winding. The electromagnetic field in the winding conductor ($\sigma_w < \infty$) can be treated as a uniform attenuated plane wave. Winding conductors are usually linear, homogeneous, and isotropic. Maxwell's equations simplified for good conductors are used to describe electric field, magnetic fields, and current density of an inductor

$$\nabla \times \mathbf{E} = -\mu_0 \frac{\partial \mathbf{H}}{\partial t} \tag{5.4}$$

and

$$\nabla \times \mathbf{H} = \sigma \mathbf{E} = \mathbf{J}, \tag{5.5}$$

where the displacement current is neglected. The Cartesian coordinates can be used and one-dimensional solution can be obtained. Due to symmetry,

$$E_x = 0, \quad E_z = 0, \quad \frac{\partial E_y}{\partial z} = 0 \tag{5.6}$$

and

$$H_x = 0, \quad H_y = 0, \quad \frac{\partial H_z}{\partial y} = 0. \tag{5.7}$$

Also, $J_x = J_z = 0$. The magnetic and electric fields contain only z and y components, respectively, and

$$E = E(x) = E_y = E_y(x) \tag{5.8}$$

and

$$H = H(x) = H_z = H_z(x). \tag{5.9}$$

Also, $J = J(x) = J_y = J_y(x)$. Thus, the magnetic field intensity contains only the z component, and the electric field intensity and the current density contain only the y components.

For steady-state sinusoidal waveforms of the fields, the symbol $\partial/\partial t$ can be replaced by the symbol $j\omega$ and Maxwell's equations can be expressed in phasor forms

$$\frac{dE_y}{dx} = -j\omega\mu_0 H_z \tag{5.10}$$

and

$$-\frac{dH_z}{dx} = \sigma_w E_y = J_y. \tag{5.11}$$

From (5.11),

$$E_y = -\frac{1}{\sigma_w}\frac{dH_z}{dx}. \tag{5.12}$$

Since each field has only one component, the subscripts can be dropped. Substituting this expression in (5.10) and setting $H_z = H$ yield the second-order ordinary differential equation, called the *Helmholtz equation*, or *one-dimensional diffusion equation*,

$$\frac{d^2H}{dx^2} = j\omega\mu_0\sigma_w H = \gamma^2 H, \tag{5.13}$$

where the *complex propagation constant* is

$$\gamma = \sqrt{j\omega\mu_0\sigma_w} = \sqrt{\frac{j\omega\mu_0}{\rho_w}} = (1+j)\sqrt{\frac{\omega\mu\sigma_w}{2}} = \frac{\sqrt{2j}}{\delta_w} = \frac{1+j}{\delta_w} = \alpha + j\beta, \tag{5.14}$$

the *skin depth* or *penetration depth* of the winding conductor, which is a measure of how far the wave penetrates into the conductor, is given by

$$\delta_w = \frac{1}{\alpha} = \sqrt{\frac{2}{\omega\mu_0\sigma_w}} = \frac{1}{\sqrt{\pi f \mu_0 \sigma_w}} = \sqrt{\frac{\rho_w}{\pi\mu_0 f}}, \tag{5.15}$$

and the effective skin depth of a conductor with a porosity factor η is

$$\delta_p = \frac{\delta_w}{\sqrt{\eta}} = \sqrt{\frac{\rho_w}{\pi\mu_0 f\eta}} \tag{5.16}$$

$\alpha = Re\{\gamma\}$ is the attenuation constant, and $\beta = Im\{\gamma\}$ is the phase constant. The skin depth $delta_w$ decreases as frequency f increases and the resistivity ρ_w decreases.

The general solution of the 1D Helmholtz equation is given by

$$H_z(x) = H(x) = H_1 e^{\gamma x} + H_2 e^{-\gamma x}, \tag{5.17}$$

which becomes

$$H(x) = H_1 e^{\frac{1+j}{\delta_w}x} + H_2 e^{-\frac{1+j}{\delta_w}x}, \tag{5.18}$$

where H_1 and H_2 are complex constants, which can be evaluated by applying the initial conditions at the left and right surfaces of each winding conductor layer.

Consider the initial conditions at the left (inner) and right (outer) surfaces of the nth layer. The Ampèrian loops C1 and C2 for two adjacent layers are depicted in Fig. 5.2. The initial conditions at both surfaces of the nth winding layer are illustrated in Fig. 5.3. Ampère's law at the outer surface of the nth layer is

$$\oint_{Cn} \mathbf{H} \cdot d\mathbf{l} = ni_l = nI_m \cos \omega t. \tag{5.19}$$

There is one turn per layer. Each layer constitutes one winding turn. Figure 5.4 shows the distribution of the magnitude of the magnetic field in the first four layers, assuming uniform current density in the winding, which occurs for low frequencies ($\delta_w \ll h$). It is assumed that the relative permeability of the core is $\mu_{rc} = \infty$. Therefore, the magnetic field intensity in the core is zero. Thus, the magnetic field on the inner surface of the first layer is $H(0) = 0$. The magnitude of the magnetic field intensity between the layers is constant, as shown in Fig. 5.4. The magnitude of the magnetic field between the nth layer and the $(n+1)$th layer is equal to that on the outer conductor surface of the nth layer,

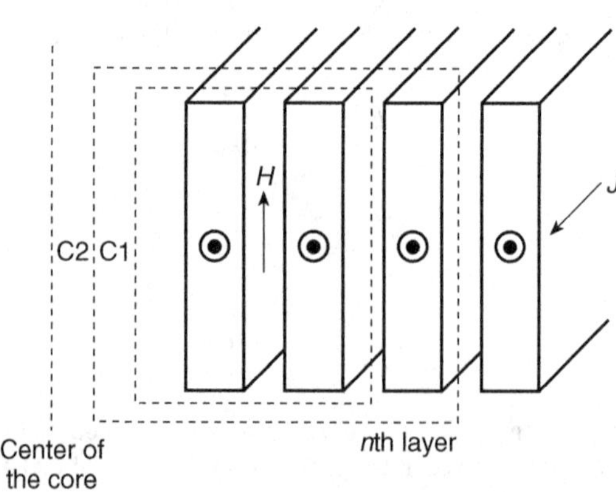

Figure 5.2 Ampèrian loops C1 and C2 for two adjacent layers used to determine the initial conditions at the left and right surfaces of the nth layer of an inductor winding

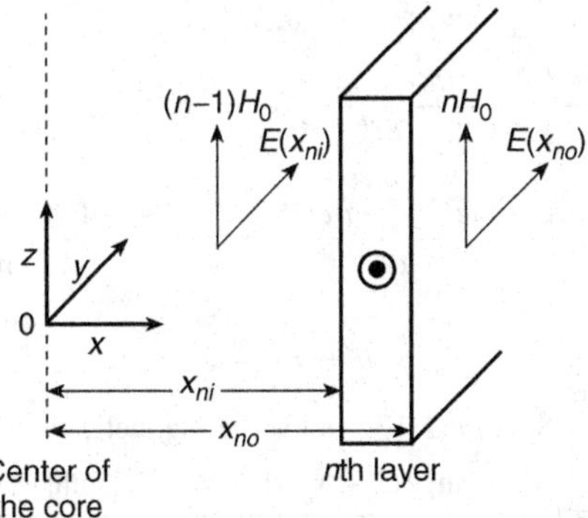

Figure 5.3 Boundary conditions for the nth layer of an inductor winding

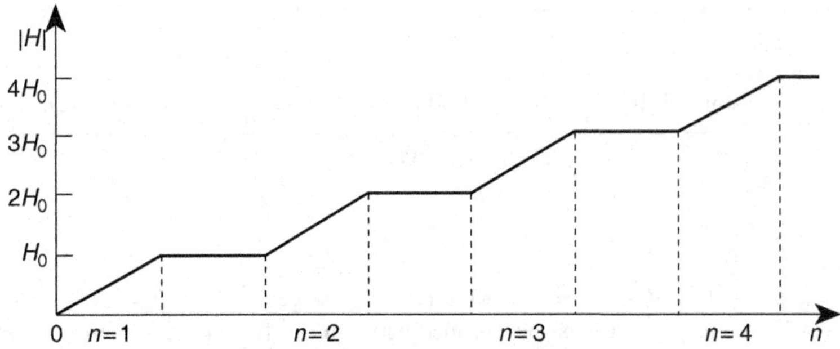

Figure 5.4 Magnitude of the magnetic field $|H|$ in an inductor for the first four layers, assuming uniform current density, that is, at low frequencies ($\delta_w \ll h$) and $\mu_{rc} = \infty$

for example, nH_0, where the amplitude of the magnetic field intensity between the first layer and the second layer is

$$H_0 = \frac{I_m}{b}. \tag{5.20}$$

At low frequencies, the magnitude of the magnetic field intensity in the layers increases linearly. The initial condition for H at the outer surface of the nth winding layer is

$$H(x_{n_o})b = nH_0 b = nI_m, \tag{5.21}$$

yielding

$$H(x_{n_o}) = nH_0 = \frac{nI_m}{b}. \tag{5.22}$$

Similarly, the initial condition for H at the inner surface of the nth winding layer is

$$H(x_{n_i})b = (n-1)H_0 b = (1-n)I_m, \tag{5.23}$$

producing

$$H(x_{n_i}) = (n-1)H_0 = \frac{(1-n)I_m}{b}, \tag{5.24}$$

where b is the winding breadth, x_{n_i} and x_{n_o} are the distances from the core center to the inner surface and the outer surface of the nth layer, respectively. Using (5.17), (5.22), and (5.24), we obtain a set of two equations

$$H(x_{n_i}) = H_1 e^{\gamma x_{n_i}} + H_2 e^{-\gamma x_{n_i}} = (n-1)H_0 \tag{5.25}$$

and

$$H(x_{n_o}) = H_1 e^{\gamma x_{no}} + H_2 e^{-\gamma x_{no}} = nH_0. \tag{5.26}$$

Solution of this set of equations produces constants

$$H_1 = H_0 \frac{ne^{-\gamma x_{n_i}} - (n-1)e^{-\gamma x_{n_o}}}{e^{-\gamma h} - e^{\gamma h}} = H_0 \frac{ne^{-\gamma x_{n_i}} - (n-1)e^{-\gamma x_{n_o}}}{2\sinh \gamma h} \tag{5.27}$$

and

$$H_2 = H_0 \frac{(n-1)e^{\gamma x_{n_o}} - ne^{\gamma x_{n_i}}}{e^{\gamma h} - e^{-\gamma h}} = H_0 \frac{(n-1)e^{\gamma x_{n_o}} - ne^{\gamma x_{n_i}}}{2\sinh \gamma h}, \tag{5.28}$$

where the foil thickness is

$$h = x_{n_o} - x_{n_i}. \tag{5.29}$$

Substitution of (5.27) and (5.28) into (5.17) yields the magnetic field intensity in the nth layer

$$H(x) = H_0 \frac{n \sinh[\gamma(x - x_{n_i})] + (n-1)\sinh[\gamma(x_{n_o} - x)]}{\sinh \gamma h}. \tag{5.30}$$

Hence, the magnetic field intensity in the nth layer normalized with respect to H_0 is

$$\begin{aligned}\frac{H(x)}{H_0} &= \frac{n \sinh\left[(1+j)\frac{h}{\delta_w}\left(\frac{x}{h} - \frac{x_{n_i}}{h}\right)\right] - (n-1)\sinh\left[(1+j)\frac{h}{\delta_w}\left(\frac{x}{h} - \frac{x_{n_o}}{h}\right)\right]}{\sinh\left[(1+j)\frac{h}{\delta_w}\right]} \\ &= \frac{n \sinh\left[(1+j)\frac{h}{\delta_w}\left(\frac{x}{h} - 2n + 2\right)\right] - (n-1)\sinh\left[(1+j)\frac{h}{\delta_w}\left(\frac{x}{h} - 2n + 1\right)\right]}{\sinh\left[(1+j)\frac{h}{\delta_w}\right]} \\ &= \left|\frac{H(x)}{H_0}\right| e^{j\phi_H},\end{aligned} \tag{5.31}$$

where $x_{n_i}/h = 2(n-1)$ and $x_{n_o}/h = 2n - 1$ for $a = h$. In general, $x_{n_i}/h = (n-1)(h+a)/h$ and $x_{n_o}/h = [nh + (n-1)a]/h$ and a is the distance between the foil layers. Figures 5.5 through 5.10 show the plots of $|H(x)/H_0|$ and ϕ_H as functions of x/h at selected values of h/δ_w for the first three layers from the inductor center with $n = 1$, 2, and 3, respectively. Figure 5.11 shows plots of $|H(x)/H_0|$ as a function of x/h at selected values of h/δ_w for all the first three layers, that is,

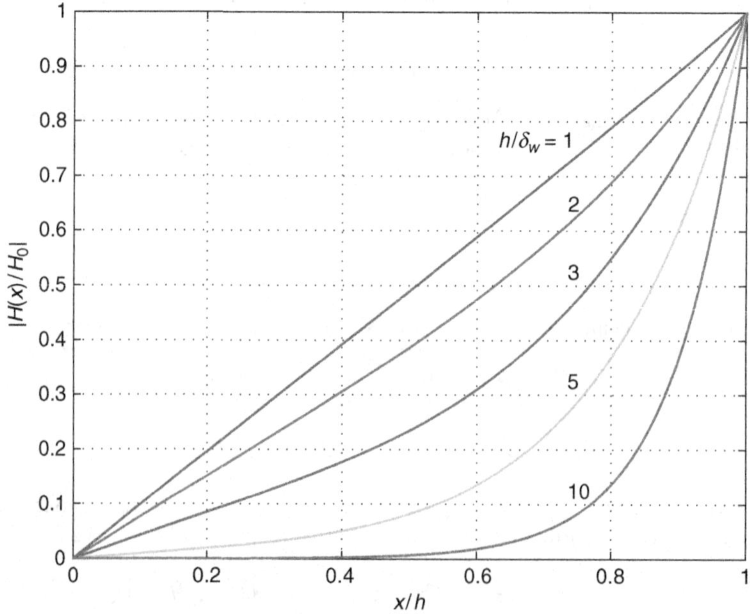

Figure 5.5 Plots of $|H(x)/H_0|$ as a function of x/h at selected values of h/δ_w for the first layer ($n = 1$)

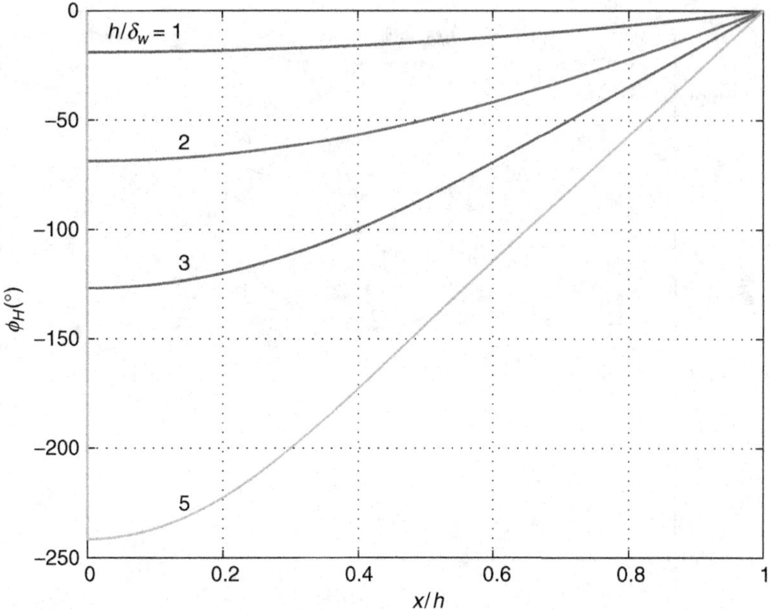

Figure 5.6 Plots of ϕ_H as a function of x/h at selected values of h/δ_w for the first layer ($n = 1$)

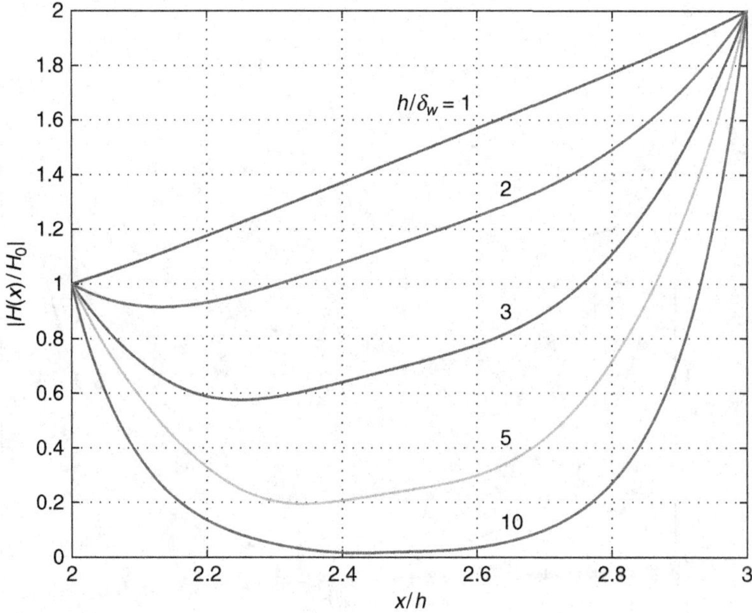

Figure 5.7 Plots of $|H(x)/H_0|$ as a function of x/h at selected values of h/δ_w for the second layer ($n = 2$)

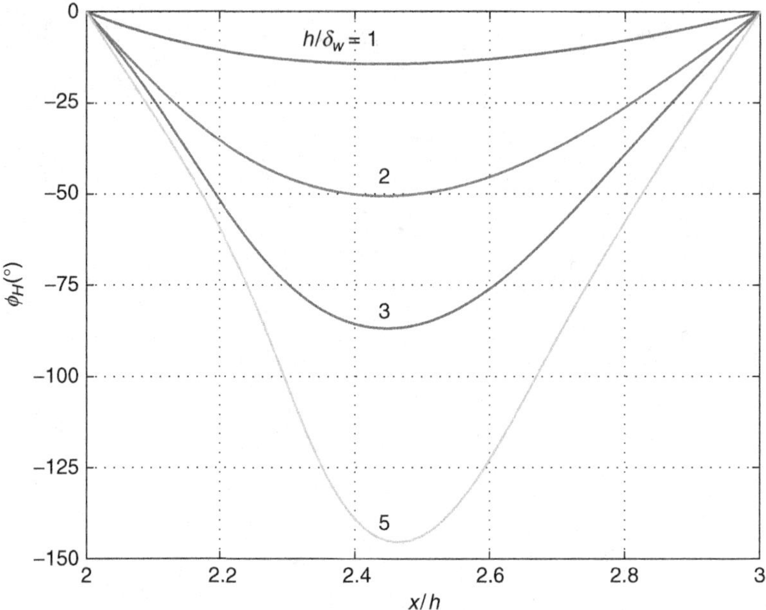

Figure 5.8 Plots of ϕ_H as a function of x/h at selected values of h/δ_w for the second layer ($n = 2$)

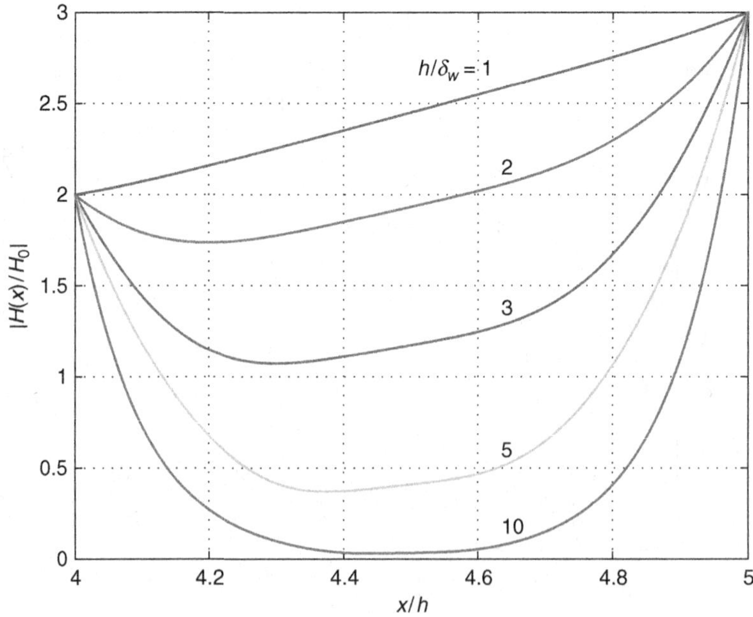

Figure 5.9 Plots of $|H(x)/H_0|$ as a function of x/h at selected values of h/δ_w for the third layer ($n = 3$)

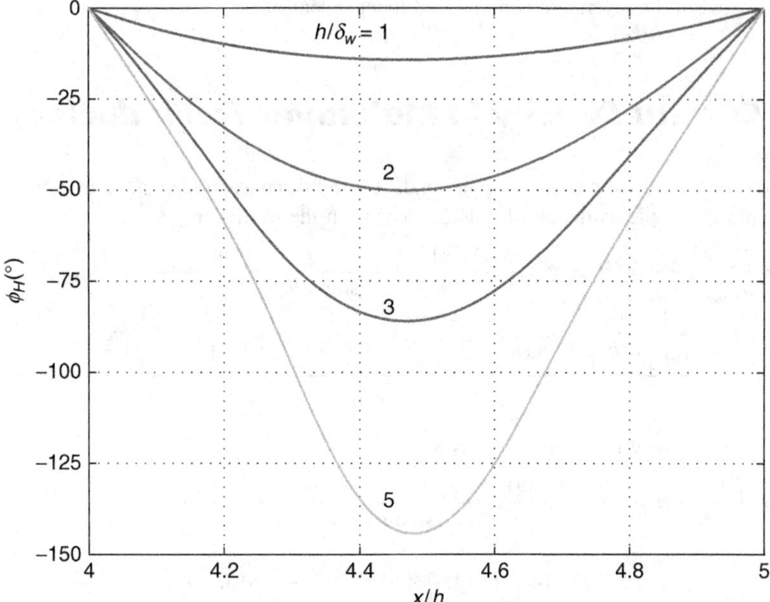

Figure 5.10 Plots of ϕ_H as a function of x/h at selected values of h/δ_w for the third layer ($n = 3$)

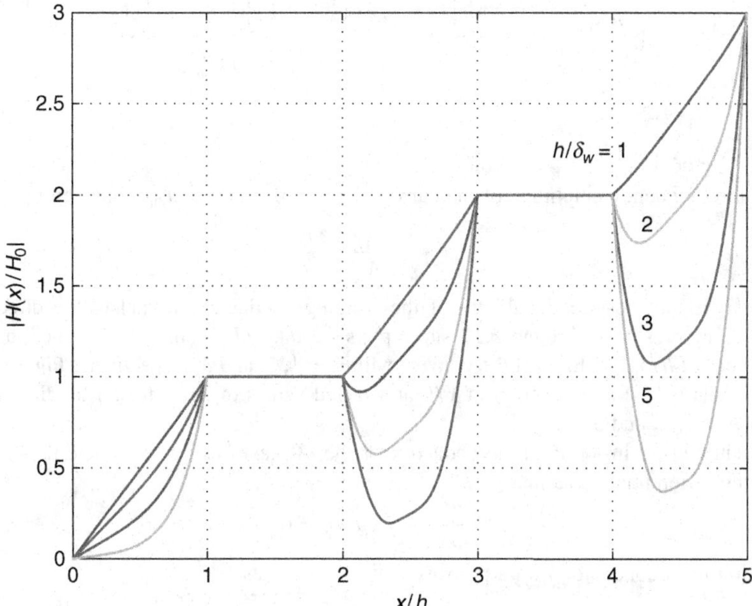

Figure 5.11 Plots of $|H(x)/H_0|$ as a function of x/h at selected values of h/δ_w for the first three layers ($n = 1$, 2, and 3)

for $n = 1$, 2, and 3. Neglecting the end effects, the amplitudes and the phases of the magnetic field intensities between the layers are constant and independent of x.

5.4 Current Density in Multilayer Foil Inductors

Foil inductors are made of a single strip of conductor foil, such as copper and aluminum. Substitution of (5.30) into (5.11) yields the electric field intensity in the nth layer

$$E(x) = -\rho_w \frac{dH(x)}{dx} = H_0 \rho_w \gamma \frac{(n-1)\cosh[\gamma(x_{no} - x)] - n\cosh[\gamma(x - x_{ni})]}{\sinh \gamma h}$$

$$= \frac{I_m \rho_w}{b} \frac{(1+j)}{\delta_w} \frac{(n-1)\cosh\left[(1+j)\frac{h}{\delta_w}\left(\frac{x_{no}}{h} - \frac{x}{h}\right)\right] - n\cosh\left[(1+j)\frac{h}{\delta_w}\left(\frac{x}{h} - \frac{x_{ni}}{\delta_w}\right)\right]}{\sinh\left[(1+j)\frac{h}{\delta_w}\right]}. \quad (5.32)$$

The AC current density in the nth layer is given by

$$J(x) = \frac{E(x)}{\rho_w} = H_0 \gamma \frac{(n-1)\cosh[\gamma(x_{no} - x)] - n\cosh[\gamma(x - x_{ni})]}{\sinh \gamma h}$$

$$= \frac{I_m}{b} \frac{(1+j)}{\delta_w} \frac{(n-1)\cosh\left[(1+j)\frac{h}{\delta_w}\left(\frac{x_{no}}{h} - \frac{x}{h}\right)\right] - n\cosh\left[(1+j)\frac{h}{\delta_w}\left(\frac{x}{h} - \frac{x_{ni}}{\delta_w}\right)\right]}{\sinh\left[(1+j)\frac{h}{\delta_w}\right]}. \quad (5.33)$$

Thus, the current density in the nth layer normalized with respect to the DC current density or low-frequency current density is

$$\frac{J(x)}{J_{DC}} = (1+j)\left(\frac{h}{\delta_w}\right) \frac{(n-1)\cosh\left[(1+j)\frac{h}{\delta_w}\left(\frac{x_{no}}{h} - \frac{x}{h}\right)\right] - n\cosh\left[(1+j)\frac{h}{\delta_w}\left(\frac{x}{h} - \frac{x_{ni}}{h}\right)\right]}{\sinh\left[(1+j)\frac{h}{\delta_w}\right]}$$

$$= \left|\frac{J(x)}{J_{DC}}\right| e^{j\phi_J}, \quad (5.34)$$

where the low-frequency amplitude of the current density, denoted as J_{DC}, is given by

$$J_{DC} = \frac{I_m}{A_w} = \frac{I_m}{hb} \quad (5.35)$$

and $A_w = hb$ is the cross-sectional area of the winding conductor, normal to the direction of the current flow. Figures 5.12 through 5.17 show plots of $|J(x)/J_{DC}|$ and ϕ_J as functions of x/h at selected values of h/δ_w for the first three layers with $n = 1$, 2, and 3, respectively. Figure 5.18 shows the plots of $|J(x)/J_{DC}|$ as a function of x/h at selected values of h/δ_w for all the first three layers, that is, for $n = 1$, 2, and 3.

The current density inside the foil conductor for the nth layer can also be determined by solving the following differential equation:

$$\frac{d^2J}{dx^2} - j\omega\mu\sigma J = 0. \quad (5.36)$$

This equation for a conductor with a porosity factor η becomes

$$\frac{d^2J}{dx^2} - j\omega\mu\sigma\eta J = 0, \quad (5.37)$$

where ηJ is the average current density along the conductor of the porous layer.

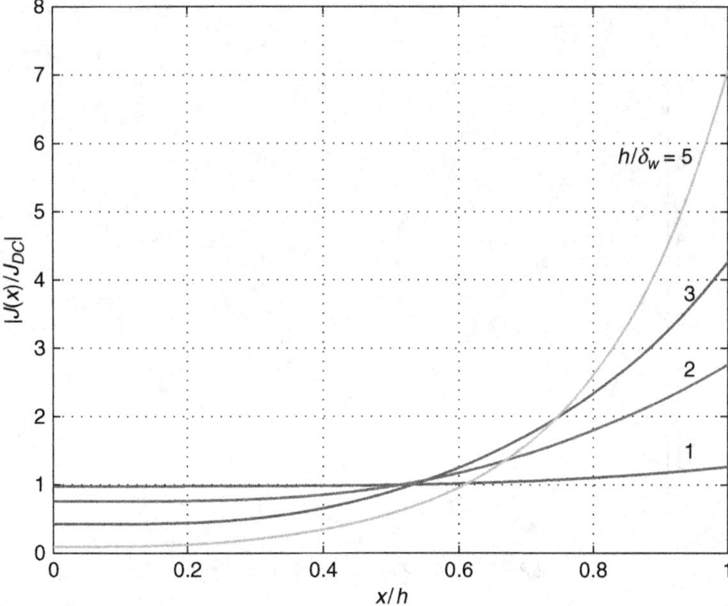

Figure 5.12 Plots of $|J(x)/J_{DC}|$ as a function of x/h at selected values of h/δ_w for the first layer ($n = 1$)

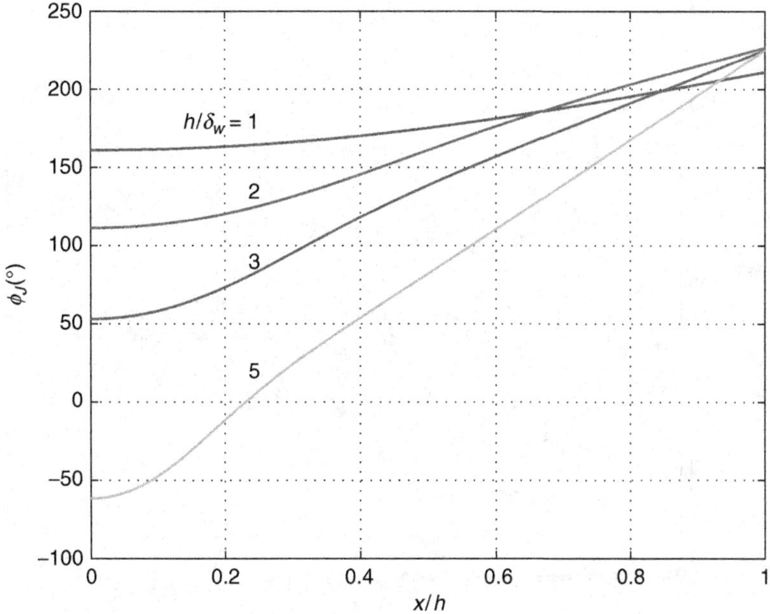

Figure 5.13 Plots of ϕ_J as a function of x/h at selected values of h/δ_w for the first layer ($n = 1$)

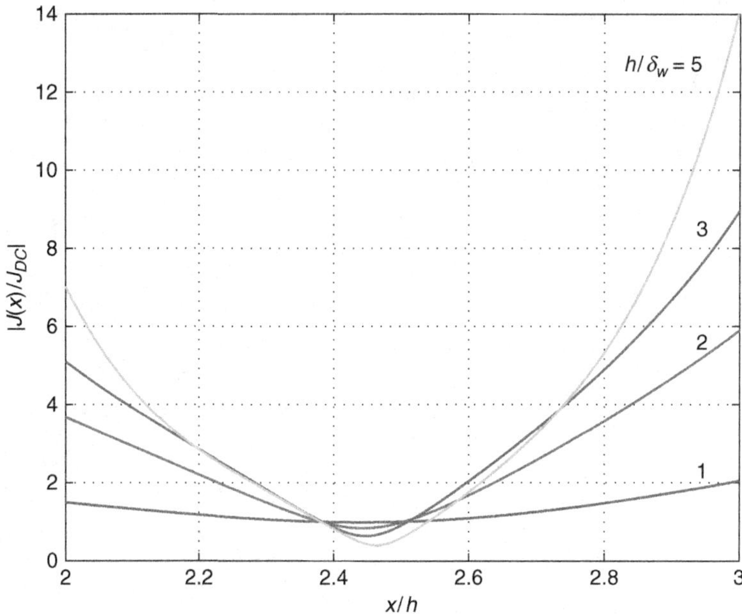

Figure 5.14 Plots of $|J(x)/J_{DC}|$ as a function of x/h at selected values of h/δ_w for the second layer ($n = 2$)

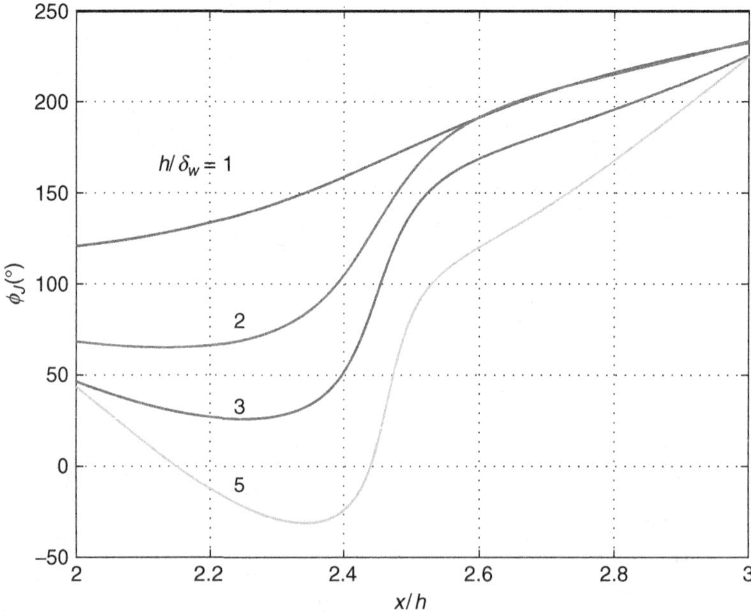

Figure 5.15 Plots of ϕ_J as a function of x/h at selected values of h/δ_w for the second layer ($n = 2$)

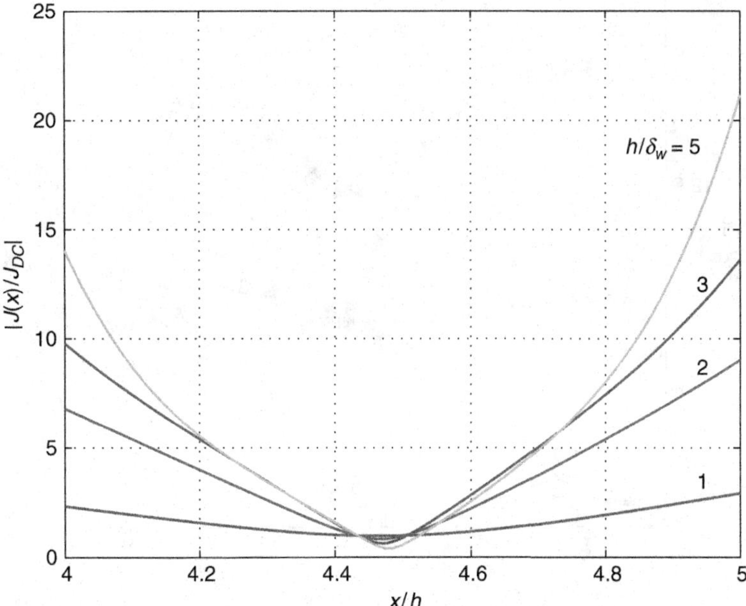

Figure 5.16 Plots of $|J(x)/J_{DC}|$ as a function of x/h at selected values of h/δ_w for the third layer ($n = 3$)

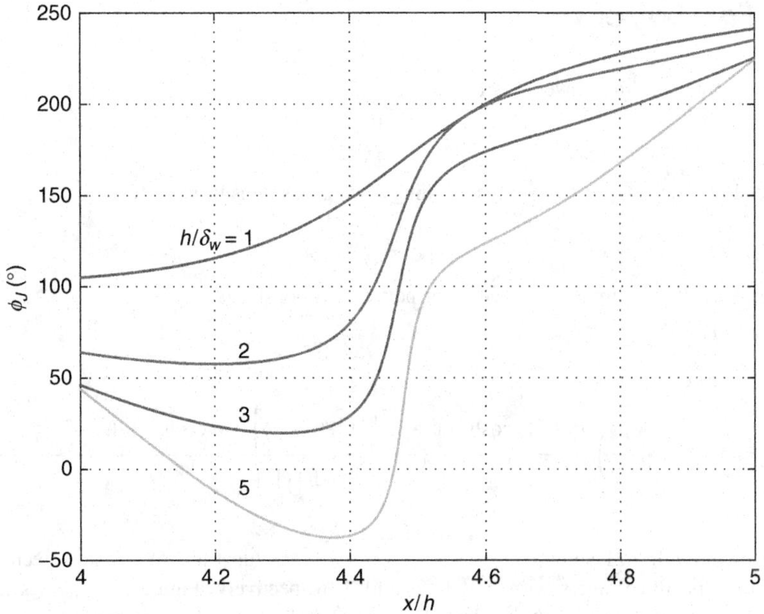

Figure 5.17 Plots of ϕ_J as a function of x/h at selected values of h/δ_w for the third layer ($n = 3$)

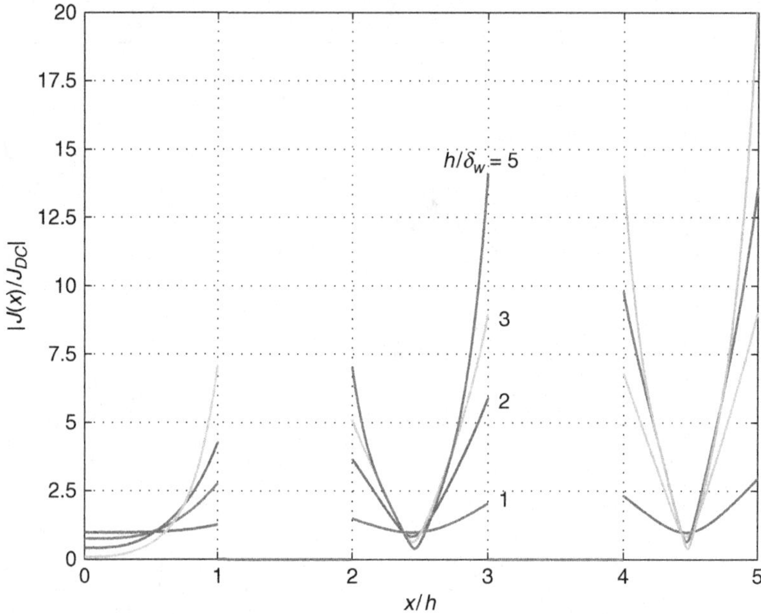

Figure 5.18 Plots of $|J(x)/J_{DC}|$ as a function of x/h at selected values of h/δ_w for $n = 1$, 2, and 3

5.5 Winding Power Loss Density in Individual Foil Layers

The time-average winding power loss density is

$$p_w(x) = \frac{1}{2}\rho_w |J(x)|^2. \tag{5.38}$$

The DC or low-frequency time-average winding power loss density is uniform

$$p_{wDC} = \frac{1}{2}\rho_w J_{DC}^2. \tag{5.39}$$

Hence, the normalized time-average winding power loss density is

$$\frac{p_w(x)}{p_{wDC}} = \frac{|J(x)|^2}{J_{DC}^2}$$
$$= \left| (1+j)\left(\frac{h}{\delta_w}\right) \frac{(n-1)\cosh\left[(1+j)\frac{h}{\delta_w}\left(\frac{x_{no}}{h} - \frac{x}{h}\right)\right] - n\cosh\left[(1+j)\frac{h}{\delta_w}\left(\frac{x}{h} - \frac{x_{ni}}{h}\right)\right]}{\sinh\left[(1+j)\frac{h}{\delta_w}\right]} \right|^2. \tag{5.40}$$

Figures 5.19 through 5.21 show the plots of $p_w(x)/p_{wDC}$ as functions of x/h at selected values of h/δ_w for the first three layers with $n = 1$, 2, and 3, respectively. Figure 5.22 shows the plots of $p_w(x)/p_{wDC}$ as a function of x/h at selected values of h/δ_w for all the first three layers, that is, for $n = 1$, 2, and 3.

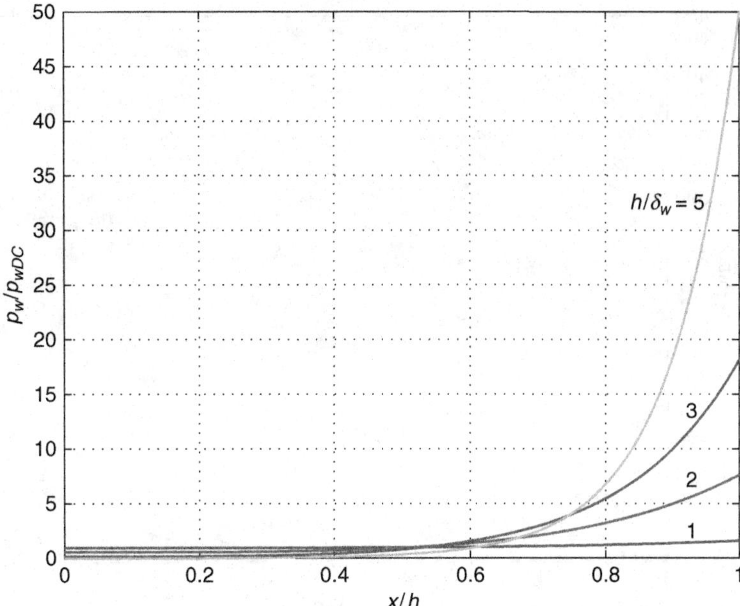

Figure 5.19 Plots of $p_w(x)/p_{wDC}$ as a function of x/h at selected values of h/δ_w for the first layer ($n = 1$)

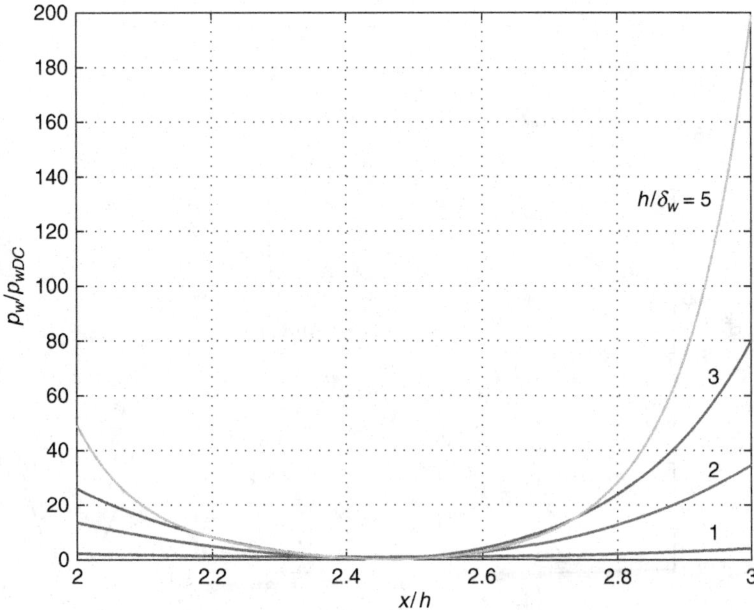

Figure 5.20 Plots of $p_w(x)/p_{wDC}$ as a function of x/h at selected values of h/δ_w for the second layer ($n = 2$)

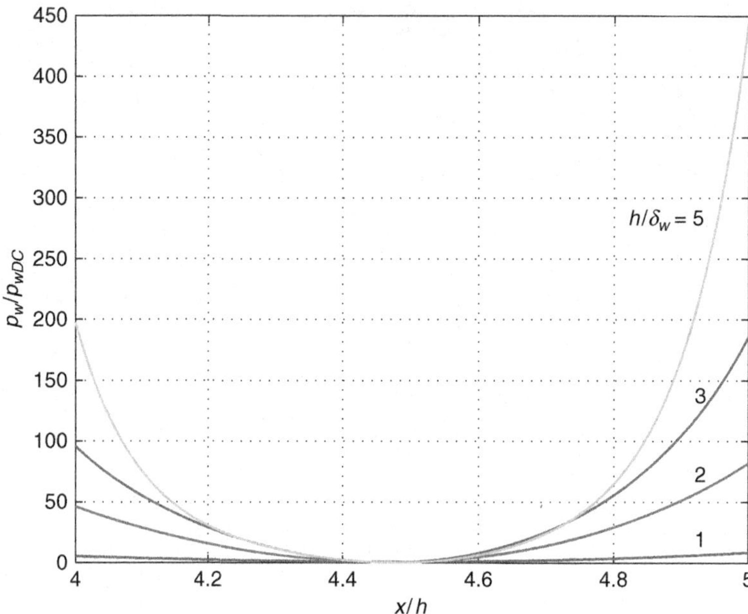

Figure 5.21 Plots of $p_w(x)/p_{wDC}$ as a function of x/h at selected values of h/δ_w for the third layer ($n = 3$)

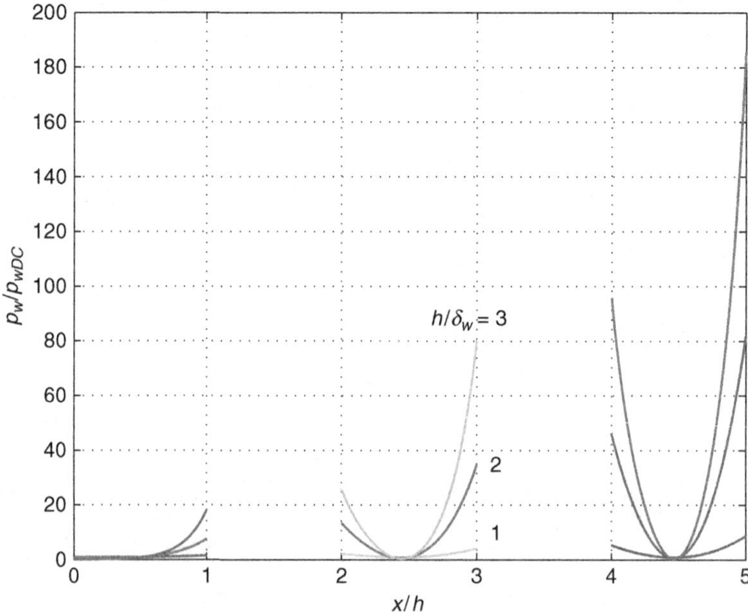

Figure 5.22 Plots of $p_w(x)/p_{wDC}$ as a function of x/h at selected values of h/δ_w for the first third layer ($n = 1$, 2, and 3)

5.6 Complex Winding Power in nth Layer

The Poynting vector theorem for linear, time-invariant media is given by

$$-\oint_S (\mathbf{E} \times \mathbf{H}) \cdot d\mathbf{s} = \int\int\int_V \mathbf{J} \cdot \mathbf{E} dV + \frac{\partial}{\partial t} \int\int\int_V \left(\frac{\mathbf{D} \cdot \mathbf{E}}{2} + \frac{\mathbf{B} \cdot \mathbf{H}}{2} \right) dV$$
$$= \int\int\int_V \mathbf{E} \cdot \mathbf{J} dV + \frac{\partial}{\partial t} \int\int\int_V \left(\frac{1}{2} \epsilon_0 E^2 \right) dV + \frac{\partial}{\partial t} \int\int\int_V \left(\frac{1}{2} \mu_0 H^2 \right) dV, \quad (5.41)$$

where $\mathbf{S} = \mathbf{E} \times \mathbf{H}$ (W/m^2) represents the instantaneous surface power density vector, that is, the instantaneous power per unit area. The power flowing into the closed surface S at any instant is equal to the sum of the ohmic power dissipated as heat and the electric and magnetic energies stored in the electric and magnetic fields within the enclosed volume V. The first term is the instantaneous ohmic power dissipated within the volume V. It is the Joule's law. The second and the third terms are the instantaneous powers, which cause an increase in the electric and magnetic energies stored in the volume V, respectively, or the time rate of increase of the energies stored per unit volume. The integration of the Poynting vector over a closed surface S gives the total power crossing the surface S. The direction of power flow at any point is perpendicular to both the **E** and **H** vectors.

For sinusoidal field waveforms, the field vectors are phasors and the Poynting vector theorem is expressed as

$$\oint_S (\mathbf{E} \times \mathbf{H}^*) \cdot d\mathbf{s} = -\int\int\int_V (\rho_w |J|^2 + j\omega\epsilon_0 |E|^2 + j\omega\mu_0 |H|^2) dV. \quad (5.42)$$

The first term on the right-hand side of the equation is real and describes the ohmic power dissipated as heat in the volume V. The second and the third terms are imaginary and describe the time rate of change in the energy stored in the electric and magnetic fields, respectively.

The complex power crossing the surface S is given by

$$P = \frac{1}{2} \oint_S (\mathbf{E} \times \mathbf{H}^*) \cdot d\mathbf{s} = P_w + j2\omega(W_m + W_e). \quad (5.43)$$

The impedance of the winding is

$$Z = \frac{V}{I} = \frac{VI^*}{II^*} = \frac{VI^*}{|I|^2} = \frac{\frac{1}{2}VI^*}{\frac{1}{2}|I|^2} = \frac{P}{\frac{1}{2}|I|^2} = \frac{2P}{|I|^2}. \quad (5.44)$$

The time-average power loss in a winding conductor is

$$P_w = P_{AV} = \frac{1}{2} Re \left\{ \oint_S (\mathbf{E} \times \mathbf{H}^*) \cdot d\mathbf{s} \right\} = \frac{1}{2} \int\int\int_V \sigma_w |E|^2 dV = \frac{1}{2} \int\int\int_V \rho_w (JJ^*) dV$$
$$= \frac{1}{2} \int\int\int_V \rho_w |J|^2 dV. \quad (5.45)$$

Only the left and right surfaces of the conductor contribute to the integral because the Poynting vector has only normal components to these surfaces. The Poynting vector over the other four surfaces has no component normal to those surfaces.

The complex power of the nth layer is

$$P_{Zn} = -[bl_T E(x_{n_o}) H^*(x_{n_o}) - bl_T E(x_{n_i}) H^*(x_{n_i})], \quad (5.46)$$

where l_T is the mean turn length (MTL). It can be shown that $H^*(x_{n_i}) = H(x_{n_i})$ and $H^*(x_{n_o}) = H(x_{n_o})$. Hence, the complex power in the nth layer given by

$$P_{Zn} = bl_T [E(x_{n_i}) H(x_{n_i}) - E(x_{n_o}) H(x_{n_o})]$$
$$= \frac{\rho_w l_T I_m^2 \gamma}{2b} \left[(2n^2 - 2n + 1) \cosh \gamma h - \frac{2n(n-1)}{\sinh \gamma h} \right]$$
$$= \frac{\rho_w l_T I_m^2 \gamma}{2b} \left[\coth \gamma h + 2(n^2 - n) \tanh \frac{\gamma h}{2} \right]$$

$$= \frac{\rho_w l_T I_m^2}{2bh}\left(\frac{h}{\delta_w}\right)(1+j)\left[\coth\gamma h + 2(n^2-n)\tanh\frac{\gamma h}{2}\right]. \tag{5.47}$$

The DC power loss in the nth layer is

$$P_{wn(LF)} = P_{wDCn} = \frac{\rho_w l_T I_m^2}{2bh} = \frac{\rho_w l_T I_{DC}^2}{bh} = R_{wDCn} I_{DC}^2, \tag{5.48}$$

where the DC resistance of the nth layer is

$$R_{wDCn} = R_{wDC1} = R_{wDC2} = \frac{\rho_w l_T}{bh} \tag{5.49}$$

and $I_m/\sqrt{2} = I_{DC}$. Hence, the normalized complex power in the nth layer is

$$\frac{P_{Zn}}{P_{wDCn}} = \left(\frac{h}{\delta_w}\right)(1+j)\left[\coth\gamma h + 2(n^2-n)\tanh\frac{\gamma h}{2}\right]$$
$$= \left(\frac{h}{\delta_w}\right)(1+j)\left\{\coth\left[(1+j)\frac{h}{\delta_w}\right] + 2(n^2-n)\tanh\left[\frac{(1+j)}{2}\frac{h}{\delta_w}\right]\right\}. \tag{5.50}$$

The following equations are useful to determine the real and imaginary parts of the impedance of the nth layer:

$$\coth\gamma h = \coth\left[(1+j)\frac{h}{\delta_w}\right] = \frac{\sinh\left(\frac{2h}{\delta_w}\right) - j\sinh\left(\frac{2h}{\delta_w}\right)}{\cosh\left(\frac{2h}{\delta_w}\right) - \cos\left(\frac{2h}{\delta_w}\right)} \tag{5.51}$$

and

$$\tanh\frac{\gamma h}{2} = \tanh\left[\frac{(1+j)}{2}\frac{h}{\delta_w}\right] = \frac{\sinh\left(\frac{h}{\delta_w}\right) + j\sinh\left(\frac{h}{\delta_w}\right)}{\cosh\left(\frac{h}{\delta_w}\right) + \cos\left(\frac{h}{\delta_w}\right)}. \tag{5.52}$$

From (5.50), (5.51), and (5.52), one obtains the AC-to-DC impedance ratio of the nth layer

$$F_{Zn} = \frac{P_{Zn}}{P_{wDCn}} = \frac{R_{wn}}{R_{wDCn}} + j\frac{X_{Ln}}{R_{wDCn}} = \frac{R_{wn}}{R_{wDCn}} + j\frac{\omega L_n}{R_{wDCn}}$$
$$= \left(\frac{h}{\delta_w}\right)\left[\frac{\sinh\left(\frac{2h}{\delta_w}\right) + \sin\left(\frac{2h}{\delta_w}\right)}{\cosh\left(\frac{2h}{\delta_w}\right) - \cos\left(\frac{2h}{\delta_w}\right)} + 2(n^2-n)\frac{\sinh\left(\frac{h}{\delta_w}\right) - \sin\left(\frac{h}{\delta_w}\right)}{\cosh\left(\frac{h}{\delta_w}\right) + \cos\left(\frac{h}{\delta_w}\right)}\right]$$
$$+ j\left(\frac{h}{\delta_w}\right)\left[\frac{\sinh\left(\frac{2h}{\delta_w}\right) - \sin\left(\frac{2h}{\delta_w}\right)}{\cosh\left(\frac{2h}{\delta_w}\right) - \cos\left(\frac{2h}{\delta_w}\right)} + 2(n^2-n)\frac{\sinh\left(\frac{h}{\delta_w}\right) + \sin\left(\frac{h}{\delta_w}\right)}{\cosh\left(\frac{h}{\delta_w}\right) + \cos\left(\frac{h}{\delta_w}\right)}\right]. \tag{5.53}$$

5.7 Winding Resistance of Individual Foil Layers

From (5.53), the AC-to-DC resistance of the nth layer of the winding is equal to the real part of the normalized impedance of the nth layer

$$F_{Rn} = \frac{Re\{Z_{wn}\}}{R_{wDCn}} = \frac{R_{wn}}{R_{wDCn}} = \frac{P_{wn}}{P_{wDCn}}$$
$$= \left(\frac{h}{\delta_w}\right)\left[\frac{\sinh\left(\frac{2h}{\delta_w}\right) + \sin\left(\frac{2h}{\delta_w}\right)}{\cosh\left(\frac{2h}{\delta_w}\right) - \cos\left(\frac{2h}{\delta_w}\right)} + 2(n^2-n)\frac{\sinh\left(\frac{h}{\delta_w}\right) - \sin\left(\frac{h}{\delta_w}\right)}{\cosh\left(\frac{h}{\delta_w}\right) + \cos\left(\frac{h}{\delta_w}\right)}\right]. \tag{5.54}$$

The normalized AC-to-DC resistance of the nth layer of the winding is equal to time-average real power loss in the nth layer. The ratio F_{Rn} is a function of the normalized layer thickness h/δ_w and the layer number n. Figure 5.23 depicts the 3D plots of F_{Rn} as a function of h/δ_w and n. Plots of

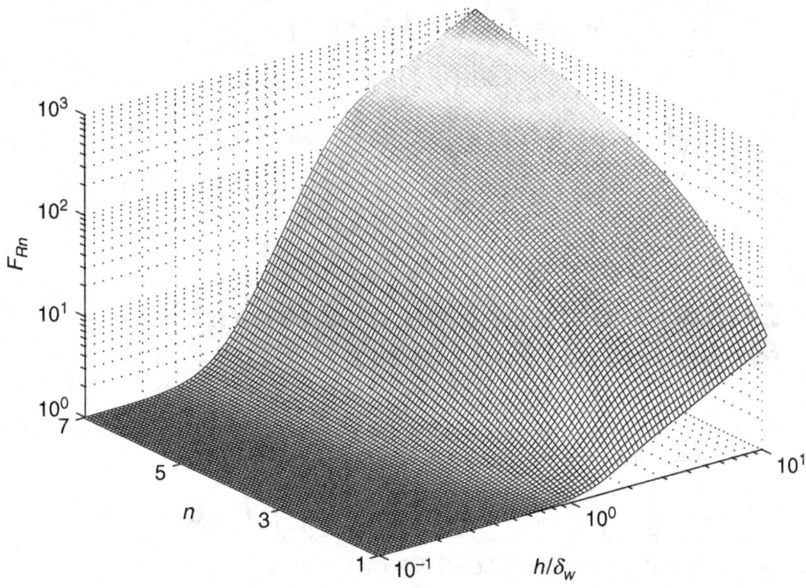

Figure 5.23 3D plots of F_{Rn} as a function of h/δ_w and n

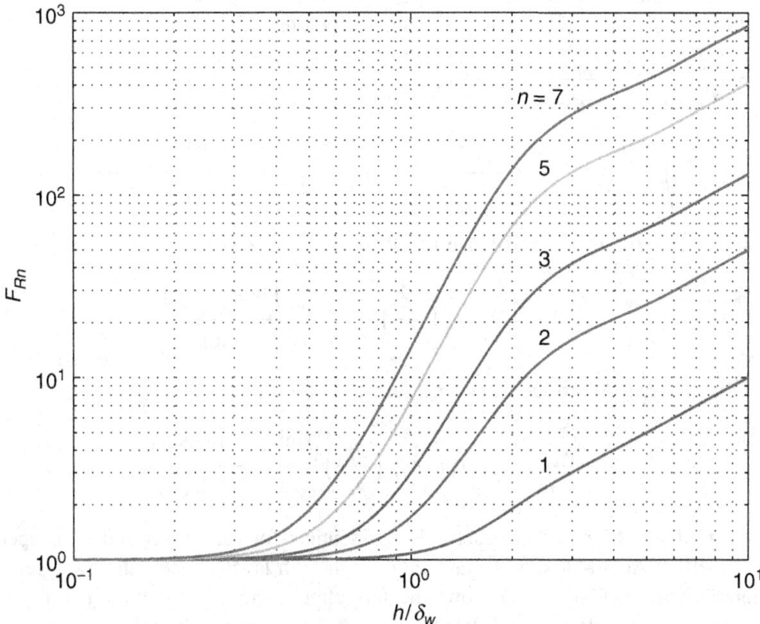

Figure 5.24 Plots of F_{Rn} as a function of h/δ_w for the first, second, third, fifth, and seventh layers

F_{Rn} as a function of h/δ_w for several individual layers are shown in Fig. 5.24. It can be seen that F_{Rn} significantly increases as the layer number n increases from the innermost layer to the outermost layer.

There are other forms of F_{Rn} in the literature [1, 80]

$$F_{Rn} = \frac{P_{wn}}{P_{wDCn}} = \frac{R_{wn}}{R_{wDCn}} = \left(\frac{h}{\delta_w}\right)\left[(2n^2 - 2n + 1)\frac{\sinh\left(\frac{2h}{\delta_w}\right) + \sin\left(\frac{2h}{\delta_w}\right)}{\cosh\left(\frac{2h}{\delta_w}\right) - \cos\left(\frac{2h}{\delta_w}\right)}\right.$$

$$-4(n^2-n)\frac{\sinh\left(\frac{h}{\delta_w}\right)\cos\left(\frac{h}{\delta_w}\right)+\cosh\left(\frac{h}{\delta_w}\right)\sin\left(\frac{h}{\delta_w}\right)}{\cosh\left(\frac{2h}{\delta_w}\right)-\cos\left(\frac{2h}{\delta_w}\right)}\Bigg]$$

$$=\frac{1}{2}\left(\frac{h}{\delta_w}\right)\left[\frac{\sinh\left(\frac{h}{\delta_w}\right)+\sin\left(\frac{h}{\delta_w}\right)}{\cosh\left(\frac{h}{\delta_w}\right)-\cos\left(\frac{h}{\delta_w}\right)}+(2n-1)^2\frac{\sinh\left(\frac{h}{\delta_w}\right)-\sin\left(\frac{h}{\delta_w}\right)}{\cosh\left(\frac{h}{\delta_w}\right)+\cos\left(\frac{h}{\delta_w}\right)}\right]. \tag{5.55}$$

5.8 Orthogonality of Skin and Proximity for Individual Foil Layers

The skin-effect current density J_s is orthogonal to the proximity-effect current density J_p. Therefore, the time-average power loss in a winding conductor is

$$P_w=P_{w(AV)}=\frac{1}{2}\int\int\int_V\rho_w(J_sJ_s^*+J_pJ_p^*)dV=\frac{1}{2}\int\int\int_V\rho_w(|J_s|^2+|J_p|^2)dV. \tag{5.56}$$

The time-average real power loss in the nth layer is

$$P_{wn}=R_{wn}I_{rms}^2=[R_{skin}+R_{prox(n)}]I_{rms}^2, \tag{5.57}$$

where R_{skin} is the resistance of each layer due to the skin effect and is the same for each layer and $R_{prox(n)}$ is the resistance of each layer and increases from the innermost layer to the outermost layer.

The AC-to-DC resistance of the nth layer can also be presented as

$$F_{Rn}=\frac{P_{wn}}{P_{wDCn}}=\frac{R_{wn}}{R_{wDCn}}=F_S+F_{Pn}$$
$$=\left(\frac{h}{\delta_w}\right)\frac{\sinh\left(\frac{2h}{\delta_w}\right)+\sin\left(\frac{2h}{\delta_w}\right)}{\cosh\left(\frac{2h}{\delta_w}\right)-\cos\left(\frac{2h}{\delta_w}\right)}+2n(n-1)\left(\frac{h}{\delta_w}\right)\frac{\sinh\left(\frac{h}{\delta_w}\right)-\sin\left(\frac{h}{\delta_w}\right)}{\cosh\left(\frac{h}{\delta_w}\right)+\cos\left(\frac{h}{\delta_w}\right)}, \tag{5.58}$$

where the skin-effect AC-to-DC resistance factor for each layer is

$$F_S=\frac{R_{skin}}{R_{wDCn}}=\left(\frac{h}{\delta_w}\right)\frac{\sinh\left(\frac{2h}{\delta_w}\right)+\sin\left(\frac{2h}{\delta_w}\right)}{\cosh\left(\frac{2h}{\delta_w}\right)-\cos\left(\frac{2h}{\delta_w}\right)} \tag{5.59}$$

and the proximity effect AC-to-DC resistance factor for the nth layer is

$$F_{Pn}=\frac{R_{prox(n)}}{R_{wDCn}}=2n(n-1)\left(\frac{h}{\delta_w}\right)\frac{\sinh\left(\frac{h}{\delta_w}\right)-\sin\left(\frac{h}{\delta_w}\right)}{\cosh\left(\frac{h}{\delta_w}\right)+\cos\left(\frac{h}{\delta_w}\right)}. \tag{5.60}$$

The skin-effect AC-to-DC resistance factor F_S is identical for each layer and is independent of n. The proximity-effect AC-to-DC resistance factor for the nth layer of the winding depends on n and increases rapidly with n. Figure 5.25 shows the skin effect factor F_S and the proximity effect factor F_{Pn} as functions of h/δ_w for second, third, and fifth layers. It can be seen that the skin effect is negligible for $h/\delta_w < 1$ and F_S increases rapidly for $h/\delta_w \geq 1$. Moreover, the proximity effect is negligible for $h/\delta_w < 1$ and F_{Pn} increases rapidly for $h/\delta_w \geq 1$. The proximity effect does not exist for $n = 1$. Figure 5.26 shows the proximity effect factor F_{Pn} as a function of h/δ_w for second, third, fourth, and fifth layers in log–log scale. It can be seen that F_{Pn} increases rapidly with h/δ_w for $1 < h/\delta_w < 2$ and increases with h/δ_w at a lower rate for $h/\delta_w > 2$.

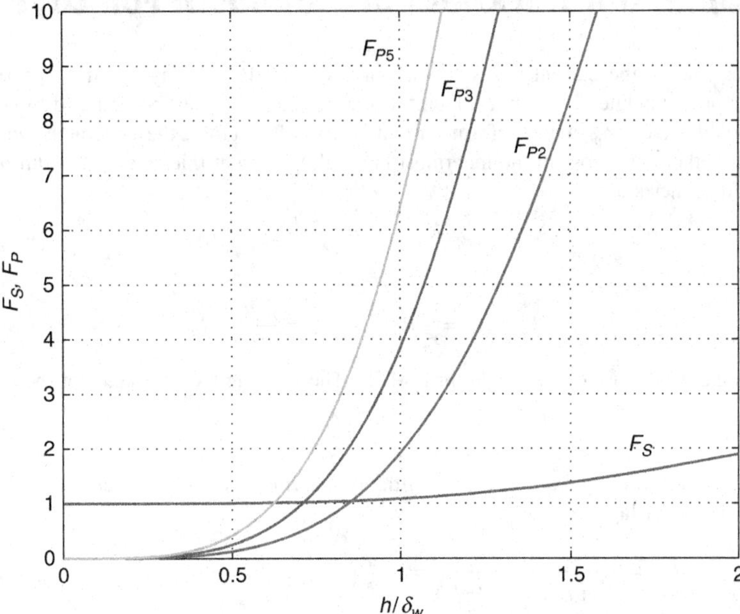

Figure 5.25 Skin-effect factor F_S and proximity-effect factor F_{Pn} as a function of h/δ_w for second, third, and fifth layers

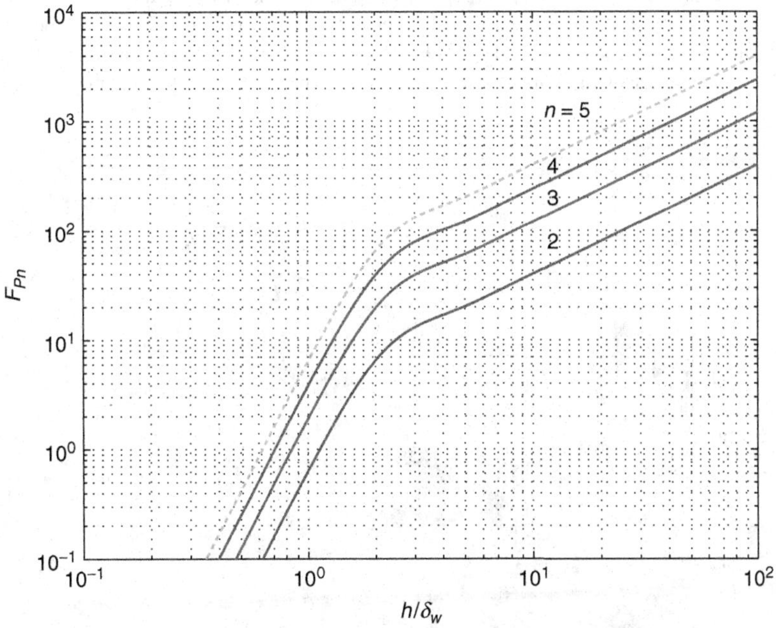

Figure 5.26 Proximity effect factor F_{Pn} as a function of h/δ_w for second, third, fourth, and fifth layers in log–log scale

5.9 Optimum Thickness of Individual Foil Layers

At high frequencies, the current flows on the outer side of the first layer with respect to the core. The conductor of thickness h with a nonuniform current density may be replaced by an equivalent conductor of thickness δ_w with uniform current density. Therefore, the equivalent winding ohmic resistance and the power loss in the innermost layer of equivalent thickness δ_w, width b, and length l_T at high frequencies are

$$R_{w1(HF)} = \frac{\rho_w l_T}{\delta_w b} \tag{5.61}$$

and

$$P_{w1(HF)} = \frac{1}{2} R_{w1(HF)} I_{Lm}^2 = \frac{\rho_w l_w I_{Lm}^2}{2\delta_w b}, \tag{5.62}$$

where l_T is the mean length of a single turn (MTL) The winding DC resistance of the single turn is

$$R_{wDC1} = \frac{\rho_w l_T}{bh}. \tag{5.63}$$

From (5.54), the winding resistance of the nth layer normalized with respect to the high-frequency resistance of the first layer $R_{w1(HF)}$ is given by

$$F_{rn} = \frac{R_{wn}}{R_{w1(HF)}} = \frac{R_{wn}}{\left(\frac{\rho_w l_T}{\delta_w b}\right)} = \frac{F_{Rn}}{\left(\frac{h}{\delta_w}\right)}$$
$$= \frac{\sinh\left(\frac{2h}{\delta_w}\right) + \sin\left(\frac{2h}{\delta_w}\right)}{\cosh\left(\frac{2h}{\delta_w}\right) - \cos\left(\frac{2h}{\delta_w}\right)} + 2(n^2 - n)\frac{\sinh\left(\frac{h}{\delta_w}\right) - \sin\left(\frac{h}{\delta_w}\right)}{\cosh\left(\frac{h}{\delta_w}\right) + \cos\left(\frac{h}{\delta_w}\right)}. \tag{5.64}$$

Figure 5.27 depicts a 3D plot of $R_{wn}/[\rho_w l_T/(\delta_w b)]$ as functions of h/δ_w and n. Plots of $F_{Rn}/(h/\delta_w)$ as a function of h/δ_w for several individual layers are shown in Fig. 5.28. It can be seen that the resistance of each layer $F_{Rn}/(h/\delta_w)$ decreases with h/δ_w, reaches a minimum value, then increases, and finally is independent of h/δ_w. Figure 5.29 shows these plots in the vicinity of the minimum

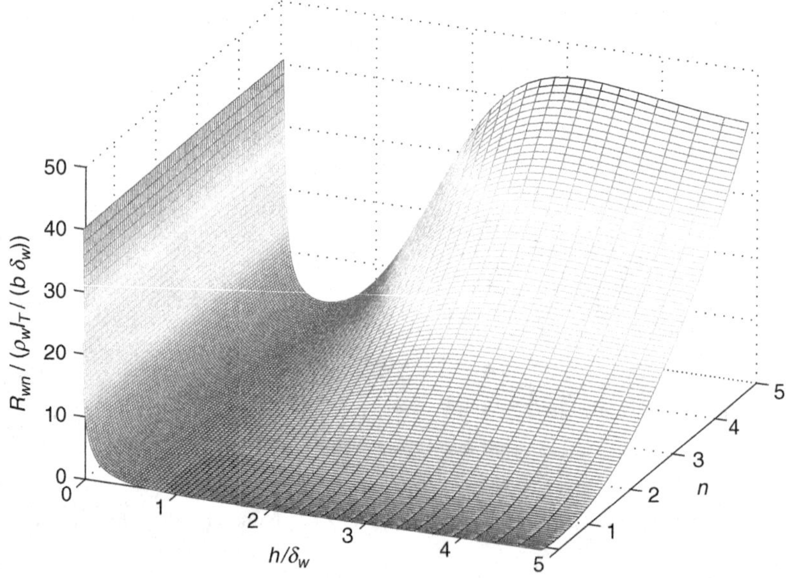

Figure 5.27 Three-dimensional plots of $R_{wn}/(\rho_w l_T/\delta_w b)$ as functions of h/δ_w and n

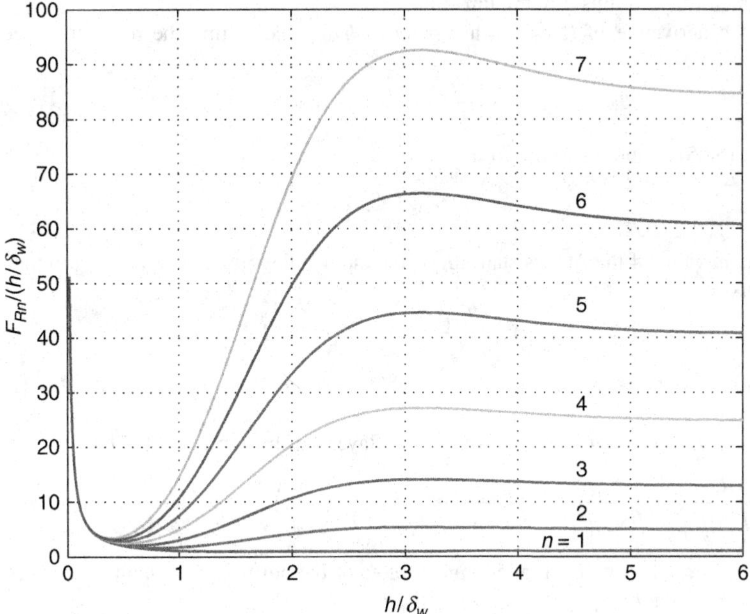

Figure 5.28 Plots of $F_{Rn}/(h/\delta_w)$ as functions of h/δ_w for the first seven layers

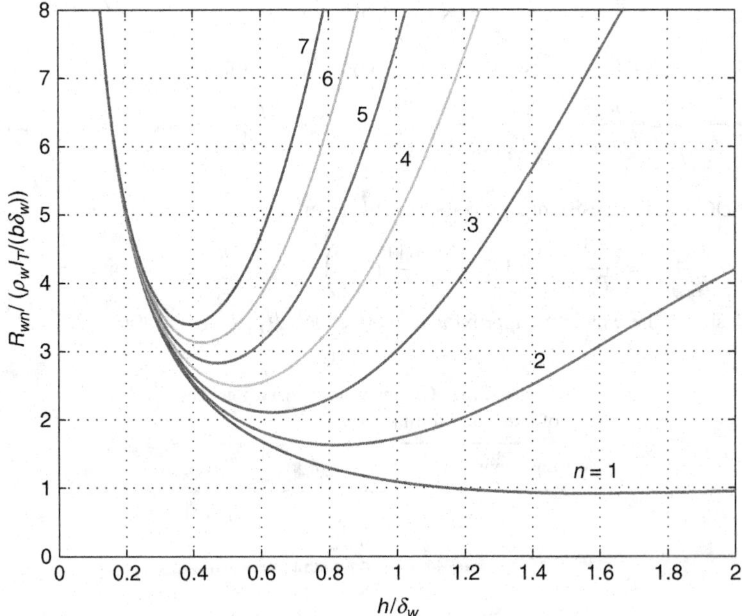

Figure 5.29 Enlarged plots of $R_{wn}/(\rho_w l_T/(\delta_w b))$ as functions of h/δ_w for the first seven layers

values. For a fixed skin depth δ_w (i.e., for fixed frequency f), there is an optimum thickness of each layer at which R_{wn} exhibits a minimum value.

Taking the derivative of (5.64) with respect to h/δ_w and setting the result to be equal to zero, we get

$$\cos\left(\frac{h}{\delta_w}\right) = \frac{n-1}{n}\cosh\left(\frac{h}{\delta_w}\right). \quad (5.65)$$

For $n = 1$, (5.65) simplifies to the form

$$\cos\left(\frac{h}{\delta_w}\right) = 0. \quad (5.66)$$

The minimum value of the AC resistance R_{w1} and the winding power loss $P_{w1(min)}$ in the first winding layer occurs at

$$\frac{h_{opt1}}{\delta_w} = \frac{\pi}{2} \quad \text{for} \quad n = 1. \quad (5.67)$$

Equation (5.65) is transcendental for $n \geq 2$, which can only be solved numerically. The results are given in Table 5.1.

The normalized winding resistance of the first layer can be approximated by

$$F_{r1} = \frac{R_{w1}}{\left(\frac{\rho_w l_T}{\delta_w b}\right)} = \frac{R_{w1}}{R_{w1(HF)}} = \frac{P_{w1}}{P_{w1(HF)}} \approx \frac{1}{\frac{h}{\delta_w}} + \frac{1}{18}\left(\frac{h}{\delta_w}\right)^3 \quad \text{for} \quad \frac{h}{\delta_w} \leq 2 \quad \text{and} \quad n = 1. \quad (5.68)$$

Figure 5.30 shows the exact and approximate plots of the normalized winding resistance for the first layer. The derivative of F_{r1} with respect to the h/δ_w is

$$\frac{dF_{r1}}{d\left(\frac{h}{\delta_w}\right)} = -\frac{1}{\left(\frac{h}{\delta_w}\right)} + \frac{1}{6}\left(\frac{h}{\delta_w}\right)^2 = 0 \quad (5.69)$$

yielding the optimum foil thickness normalized with respect to the skin depth

$$\frac{h_{opt}}{\delta_w} = \sqrt[4]{6} \approx 1.565 \quad \text{for} \quad n = 1. \quad (5.70)$$

The normalized winding resistance of the nth layer can be approximated by

$$F_{rn} = \frac{R_{wn}}{\left(\frac{\rho_w l_T}{\delta_w b}\right)} = \frac{R_{wn}}{R_{w1(HF)}} = \frac{P_{wn}}{P_{w1(HF)}} \approx \frac{1}{\frac{h}{\delta_w}} + \frac{n(n-1)}{3}\left(\frac{h}{\delta_w}\right)^3 \quad \text{for} \quad \frac{h}{\delta_w} < 1.5 \quad \text{and} \quad n \geq 2. \quad (5.71)$$

The AC-to-DC winding resistance ratio for the nth layer is

$$F_{Rn} = \frac{R_{wn}}{R_{wDC1}} = \frac{P_{wn}}{P_{wDC1}} \approx 1 + \frac{n(n-1)}{3}\left(\frac{h}{\delta_w}\right)^4 \quad \text{for} \quad \frac{h}{\delta_w} < 1.5 \quad \text{and} \quad n \geq 2. \quad (5.72)$$

Figure 5.31 shows the exact and approximate plots of $R_{w3}/(\rho_w l_T/b\delta_w)$ at $n = 3$.

Table 5.1 Optimum normalized thickness of individual foil layers n

Layer number n	h_{optn}/δ_w	$F_{rn(min)}$
1	$\pi/2$	0.91715
2	0.823767	1.6286
3	0.634444	2.1062
4	0.535375	2.4932
5	0.471858	2.8276
6	0.426676	3.1339
7	0.392413	3.3989
8	0.365274	3.6511
9	0.343089	3.8870
10	0.324512	4.1093

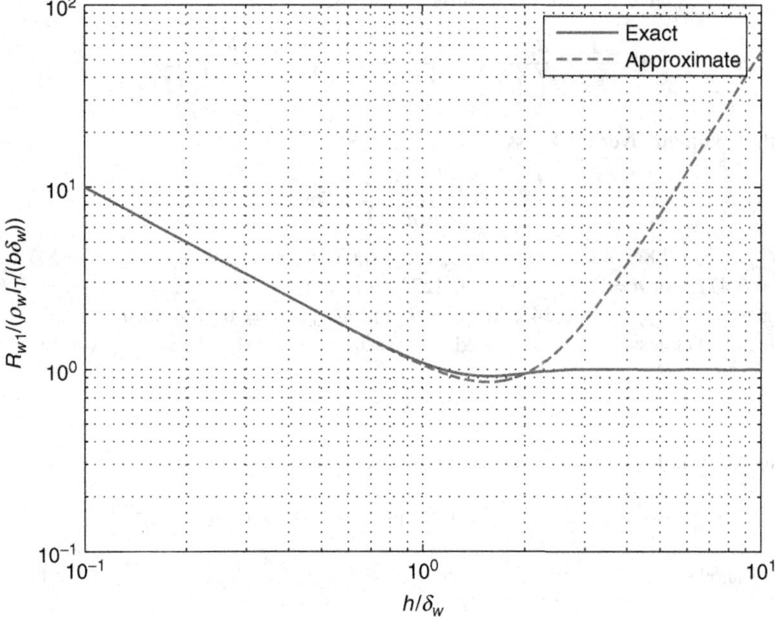

Figure 5.30 Exact and approximate plots of $R_{w1}/(\rho_w l_T/b\delta_w)$ as a function of h/δ_w for the first layer

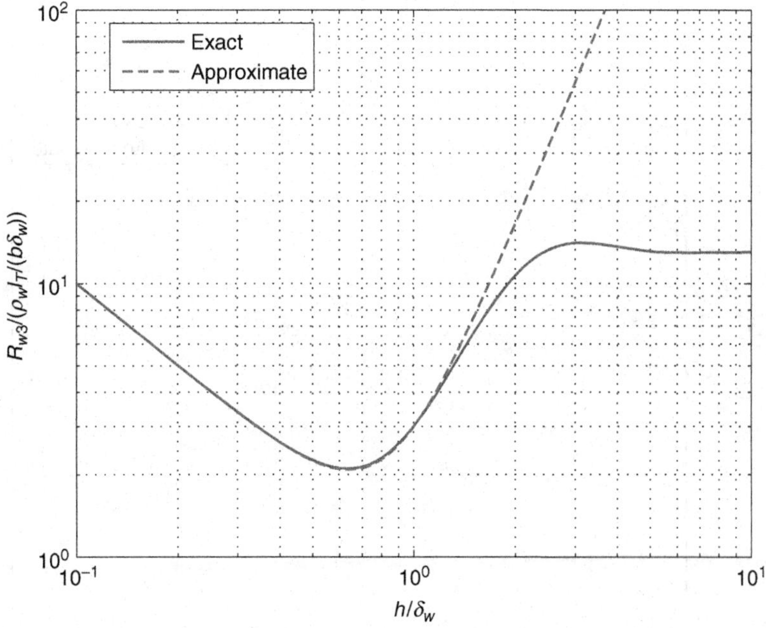

Figure 5.31 Exact and approximate plots of $R_{w3}/(\rho_w l_T/b\delta_w)$ as a function of h/δ_w at $n = 3$

An analytical expression for h_{optn}/δ_w can be derived as follows. The derivative of (5.71) with respect to h/δ_w, equated to zero, is given by

$$\frac{1}{\frac{\rho_w l_T}{\delta_w b}}\frac{R_{wn}}{d\left(\frac{h}{\delta_w}\right)} = -\frac{1}{\left(\frac{h}{\delta_w}\right)^2} + n(n-1)\left(\frac{h}{\delta_w}\right)^2 = 0, \tag{5.73}$$

yielding the optimum layer thickness

$$\frac{h_{optn}}{\delta_w} = \frac{1}{\sqrt[4]{n(n-1)}} \quad \text{for} \quad n \geq 2. \tag{5.74}$$

Hence, $h_{opt2}/\delta_w = 0.8409$ for $n = 2$, $h_{opt3}/\delta_w = 0.6389$ for $n = 3$, $h_{opt4}/\delta_w = 0.5373$ for $n = 4$, $h_{opt5}/\delta_w = 0.4729$ for $n = 5$, $h_{opt6}/\delta_w = 0.4273$ for $n = 6$, $h_{opt7}/\delta_w = 0.3928$ for $n = 7$, $h_{opt8}/\delta_w = 0.3656$ for $n = 8$, $h_{opt9}/\delta_w = 0.3433$ for $n = 9$, and $h_{opt10}/\delta_w = 0.3247$ for $n = 10$.

At high foil thicknesses, the normalized AC resistance at a fixed $delta_w$ is given by

$$F_{rn} = \frac{R_{wn}}{R_{w1(HF)}} = \frac{P_{wn}}{P_{w1(HF)}} \approx n^2 + (n-1)^2 \quad \text{for} \quad 5 \leq \frac{h}{\delta_w} < \infty. \tag{5.75}$$

The AC-to-DC winding resistance ratio at large foil thicknesses is

$$F_{Rn} = \frac{R_{wn}}{R_{wDC1}} = \frac{P_{wn}}{P_{wDC1}} \approx \left(\frac{h}{\delta_w}\right)[n^2 + (n-1)^2] \quad \text{for} \quad 5 \leq \frac{h}{\delta_w} < \infty. \tag{5.76}$$

Figure 5.32 shows the exact and approximate plots of $R_{w3}/(\rho_w l_T/b\delta_w)$ at $n = 3$ for high foil thicknesses.

Substitution of (5.70) in (5.64) produces the minimum value of $F_{r1(min)}$ at h_{opt1}/δ_w

$$F_{r1(min)} = 0.91715 \quad \text{for} \quad n = 1. \tag{5.77}$$

Substituting (5.70) in (5.58), we obtain the value of F_{R1} at h_{opt1}/δ_w

$$F_{R1} = 1.4407 \quad \text{for} \quad n = 1. \tag{5.78}$$

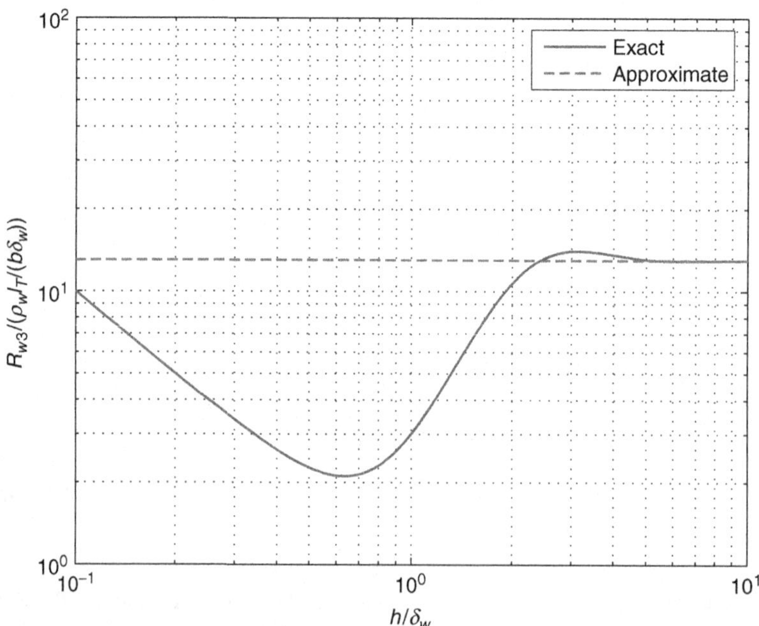

Figure 5.32 Exact and approximate plots of $R_{w3}/(\rho_w l_T/b\delta_w)$ as a function of h/δ_w at $n = 3$ for large foil thicknesses

Substitution of (5.74) into (5.71) and (5.72) produces the minimum normalized winding resistance and the minimum normalized power loss in the nth layer

$$F_{rn(min)} = \frac{R_{wn(min)}}{R_{w1(HF)}} = \frac{P_{wn(min)}}{P_{w1(HF)}} \approx \frac{4}{3}\sqrt[4]{n(n-1)} \quad \text{for} \quad n \geq 2 \tag{5.79}$$

and

$$F_{Rn}\left(\frac{h_{optn}}{\delta_w}\right) = \frac{R_{wn}}{R_{wDC1}} = \frac{P_{wn}}{P_{wDC1}} = \frac{4}{3} \quad \text{for} \quad n \geq 2. \tag{5.80}$$

The ratio of the approximate optimum foil thickness of the nth layer h_{optn} to the exact optimum thickness of the first layer given by (5.67) is

$$\frac{h_{optn}}{h_{opt1}} = \frac{2}{\pi\sqrt[4]{n(n-1)}} \quad \text{for} \quad n \geq 2. \tag{5.81}$$

The relationship between F_{Rn} and F_{rn} for foil and rectangular windings is

$$F_{rn} = \frac{F_{Rn}}{\frac{h}{\delta_w}}. \tag{5.82}$$

5.10 Winding Inductance of Individual Layers

From (5.53), the reactance of the inductance due to magnetic field stored in the conductor of the nth layer is

$$\frac{X_{Ln}}{R_{wDCn}} = \frac{\omega L_n}{R_{wDCn}} = \left(\frac{h}{\delta_w}\right)\left[\frac{\sinh\left(\frac{2h}{\delta_w}\right) - \sin\left(\frac{2h}{\delta_w}\right)}{\cosh\left(\frac{2h}{\delta_w}\right) - \cos\left(\frac{2h}{\delta_w}\right)} + 2(n^2 - n)\frac{\sinh\left(\frac{h}{\delta_w}\right) + \sin\left(\frac{h}{\delta_w}\right)}{\cosh\left(\frac{h}{\delta_w}\right) + \cos\left(\frac{h}{\delta_w}\right)}\right]. \tag{5.83}$$

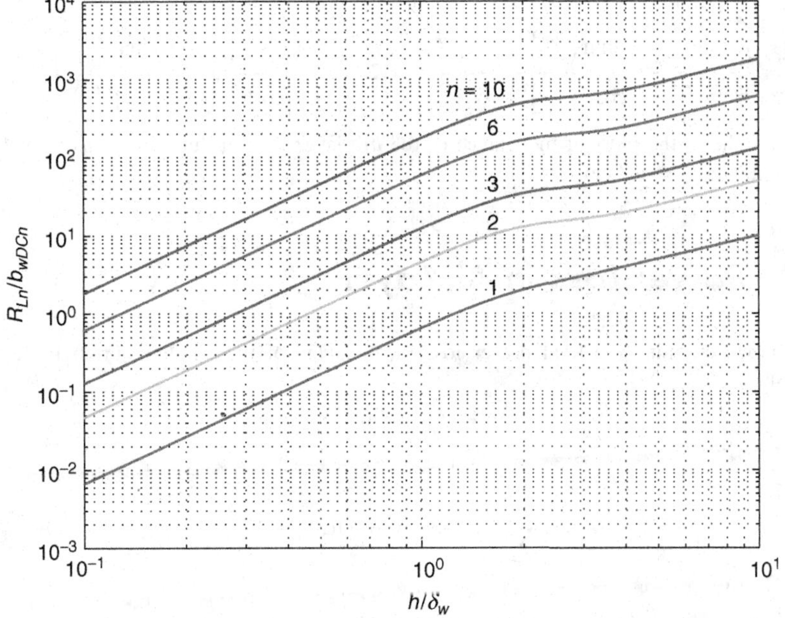

Figure 5.33 Normalized reactance of the nth layer $F_{Xn} = X_{Ln}/R_{wDCn}$ as a function of h/δ_w

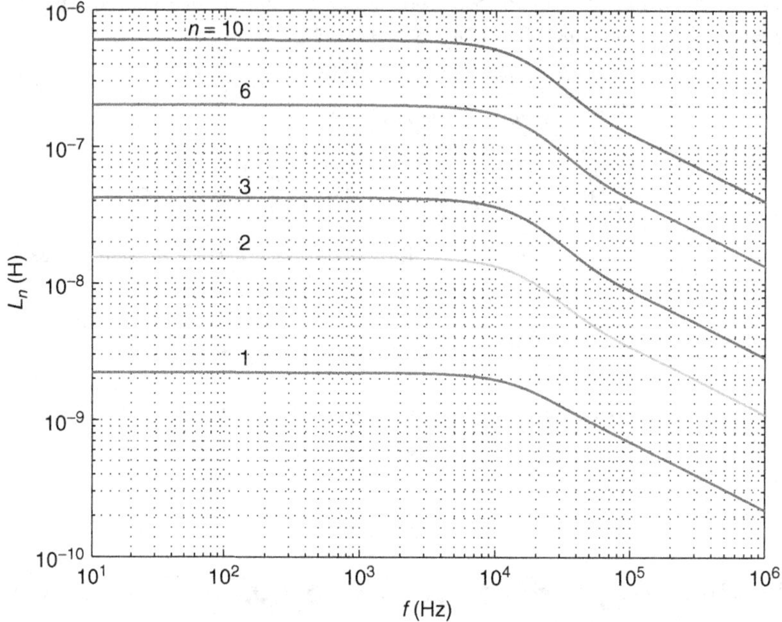

Figure 5.34 Inductance of the nth layer $F_{Xn} = X_{Ln}/R_{wDCn}$ as a function of h/δ_w at $h = 1$ mm, $b = 1$ mmm, and $l_T = 5.3$ mm

Figure 5.33 shows the plots of normalized reactance of the nth layer $F_{Xn} = X_{Ln}/R_{wDCn}$ as a function of h/δ_w. The inductance due to magnetic field stored in the conductor of the nth layer is

$$L_n = \frac{X_{Ln}}{\omega} = \frac{R_{wDCn}}{\omega}\left(\frac{h}{\delta_w}\right)\left[\frac{\sinh\left(\frac{2h}{\delta_w}\right) - \sin\left(\frac{2h}{\delta_w}\right)}{\cosh\left(\frac{2h}{\delta_w}\right) - \cos\left(\frac{2h}{\delta_w}\right)} + 2(n^2 - n)\frac{\sinh\left(\frac{h}{\delta_w}\right) + \sin\left(\frac{h}{\delta_w}\right)}{\cosh\left(\frac{h}{\delta_w}\right) + \cos\left(\frac{h}{\delta_w}\right)}\right]$$

$$= \frac{l_T}{2b}\sqrt{\frac{\mu_0\rho_w}{\pi f}}\left[\frac{\sinh\left(\frac{2h}{\delta_w}\right) - \sin\left(\frac{2h}{\delta_w}\right)}{\cosh\left(\frac{2h}{\delta_w}\right) - \cos\left(\frac{2h}{\delta_w}\right)} + 2(n^2 - n)\frac{\sinh\left(\frac{h}{\delta_w}\right) + \sin\left(\frac{h}{\delta_w}\right)}{\cosh\left(\frac{h}{\delta_w}\right) + \cos\left(\frac{h}{\delta_w}\right)}\right]. \tag{5.84}$$

Figure 5.34 shows the plots of inductance of the nth layer L_n as a function of h/δ_w at $h = 1$ mm, $b = 10$ mm, and $l_T = 5.3$ cm.

5.11 Power Loss in All Layers

We can now determine the total power loss in all layers of the winding. From the appendix in Chapter 4,

$$\sum_{n=1}^{N_l} n = \frac{N_l(N_l + 1)}{2} \tag{5.85}$$

and

$$\sum_{n=1}^{N_l} n^2 = \frac{N_l(N_l + 1)(2N_l + 1)}{6}, \tag{5.86}$$

we get

$$\sum_{n=1}^{N_l}(n^2 - n) = \frac{N_l(N_l + 1)(2N_l + 1)}{6} - \frac{N_l(N_l + 1)}{2} = \frac{N_l(N_l^2 - 1)}{3}. \tag{5.87}$$

Hence, the complex power of the entire winding is

$$P = \sum_{n=1}^{N_l} P_{wn} = \frac{\rho_w l_T I_m^2 \gamma}{2b} \sum_{n=1}^{N_l} \left[\coth \gamma h + 2(n^2 - n) \tanh \frac{\gamma h}{2} \right]$$

$$= \frac{\rho_w l_T N_l I_m^2 \gamma}{2b} \left[\coth \gamma h + \frac{2(N_l^2 - 1)}{3} \tanh \frac{\gamma h}{2} \right]. \tag{5.88}$$

The DC power loss in the entire winding is

$$P_{wDC} = R_{wDC} I_{DC}^2 = \frac{\rho_w l_T N_l I_{DC}^2}{hb} = \frac{\rho_w l_T N_l I_m^2}{2hb} \tag{5.89}$$

where $I_{DC} = I_{rms} = I_m/\sqrt{2}$. The time-average AC power loss in the winding at any frequency is

$$P_w = P_{AV} = Re\{P\}$$

$$= P_{wDC} \left(\frac{h}{\delta_w} \right) \left[\frac{\sinh\left(\frac{2h}{\delta_w}\right) + \sin\left(\frac{2h}{\delta_w}\right)}{\cosh\left(\frac{2h}{\delta_w}\right) - \cos\left(\frac{2h}{\delta_w}\right)} + \frac{2(N_l^2 - 1)}{3} \frac{\sinh\left(\frac{h}{\delta_w}\right) - \sin\left(\frac{h}{\delta_w}\right)}{\cosh\left(\frac{h}{\delta_w}\right) + \cos\left(\frac{h}{\delta_w}\right)} \right]. \tag{5.90}$$

When the winding conducts a DC current I_{DC} and a sinusoidal current I_{rms}, the total winding loss is

$$P_w = R_{wDC} I_{DC}^2 + R_w I_{rms}^2 = R_{wDC} I_{DC}^2 + F_R R_{wDC} I_{rms}^2 = R_{wDC} I_{DC}^2 \left(1 + F_R \frac{I_{rms}^2}{I_{DC}^2} \right)$$

$$= P_{wDC} \left(1 + F_R \frac{I_{rms}^2}{I_{DC}^2} \right). \tag{5.91}$$

If $I_{rms} = I_{DC}$,

$$R_w = R_{wDC}(1 + F_R). \tag{5.92}$$

5.12 Impedance of Foil Winding

The complex power dissipated and stored in the nth foil layer is given by

$$P_{Zn} = V_{wn} I^* = Z_{wn} I I^* = Z_{wn} |I|^2 = Z_{wn} I_{rms}^2 = \frac{1}{2} Z_{wn} I_m^2. \tag{5.93}$$

Therefore, from (5.46), the complex impedance of the nth foil layer is

$$Z_{wn} = \frac{2P_{Zn}}{I_m^2} = \frac{\rho_w l_T \gamma}{b} \left[\coth \gamma h + 2n(n-1) \tanh \frac{\gamma h}{2} \right] = R_{n(AC)} + j\omega L_n. \tag{5.94}$$

The complex impedance of all N_l identical layers of a foil winding is expressed by

$$Z = \sum_{n=1}^{N_l} Z_{wn} = \frac{\rho_w l_T \gamma}{b} \sum_{n=1}^{N_l} \left[\coth \gamma h + 2(n^2 - n) \tanh \frac{\gamma h}{2} \right]$$

$$= \frac{\rho_w l_T N_l \gamma h}{bh} \left[\coth \gamma h + \frac{2(N_l^2 - 1)}{3} \tanh \frac{\gamma h}{2} \right]$$

$$= R_{wDC} \gamma h \left[\coth \gamma h + \frac{2(N_l^2 - 1)}{3} \tanh \frac{\gamma h}{2} \right], \tag{5.95}$$

where the DC and low-frequency winding resistance is

$$R_{wDC} = \frac{\rho_w l_T N_l}{hb} = \frac{\rho_w l_w}{A_w} \tag{5.96}$$

and $l_w = N_l l_T$ is the length of the foil winding conductor and A_w is the cross-sectional area of bare foil conductor.

5.13 Resistance of Foil Winding

The winding power loss in the nth layer is

$$P_{wn} = R_{wn} I_{rms}^2 = \frac{1}{2} R_{wn} I_m^2, \tag{5.97}$$

where R_{wn} is the AC winding resistance of the nth foil layer. The winding power loss in the nth layer at DC and low frequencies is

$$P_{wDCn} = R_{wn} I_{DC}^2. \tag{5.98}$$

The DC and low-frequency winding resistance of the entire winding is equal to the sum of the DC resistances of all layers

$$R_{wDC} = \sum_{n=1}^{N_l} R_{wDCn} = \sum_{n=1}^{N_l} \frac{\rho_w l_T}{bh} = \frac{\rho_w l_T N_l}{bh} = \frac{\rho_w l_w}{bh}. \tag{5.99}$$

The high-frequency winding resistance of the entire winding is equal to the sum of the AC resistances of all layers

$$R_w = \sum_{n=1}^{N_l} R_{wn} = Re\left\{\frac{\rho_w l_T \gamma h}{bh} \sum_{n=1}^{N_l} \left[\coth \gamma h + 2n(n-1)\tanh \frac{\gamma h}{2}\right]\right\}. \tag{5.100}$$

Assuming that $I_{DC} = I_{rms}$, we obtain the ratio of the AC-to-DC resistance of the nth layer

$$\frac{P_w}{P_{wDC}} = \frac{R_w I_{rms}^2}{R_{wDC} I_{DC}^2} = \frac{R_w}{R_{wDC}}. \tag{5.101}$$

The ratio of the AC-to-DC resistance of all uniform layers N_l is

$$F_R = \frac{R_w}{R_{wDC}} = \frac{P_w}{P_{wDC}} = \left(\frac{h}{\delta_w}\right)\left[\frac{\sinh\left(\frac{2h}{\delta_w}\right) + \sin\left(\frac{2h}{\delta_w}\right)}{\cosh\left(\frac{2h}{\delta_w}\right) - \cos\left(\frac{2h}{\delta_w}\right)} + \frac{2(N_l^2-1)}{3} \frac{\sinh\left(\frac{h}{\delta_w}\right) - \sin\left(\frac{h}{\delta_w}\right)}{\cosh\left(\frac{h}{\delta_w}\right) + \cos\left(\frac{h}{\delta_w}\right)}\right]. \tag{5.102}$$

The winding AC power loss is given by

$$P_{w=R_w I_{rms}^2} = F_R R_{wDC} I_{rms}^2. \tag{5.103}$$

Expression (5.102) is the same as Dowell's equation [5], who has derived the impedance from the voltage, which has been determined using the magnetic flux cutting the winding space. Figure 5.35 shows a three-dimensional plot of F_R as a function of h/δ_w and N_l for foil winding. It can be seen that F_R increases when h/δ_w and N_l increase. Figure 5.36 shows the plots of F_R as a function of h/δ_w at fixed numbers of layers N_l for foil winding.

The foil thickness at the boundary between the mid-frequency range and the high-frequency range is determined by

$$h = 3\delta_w = \pi\sqrt{\frac{\rho_w}{\pi \mu_0 f_H}} \frac{h}{\delta_w} = \pi \tag{5.104}$$

yielding the frequency at the foil thickness at the boundary between the mid-frequency range and the high-frequency range

$$f_H = \frac{\pi \rho_w}{\mu_0 h^2}. \tag{5.105}$$

5.14 Dowell's Equation

Using a 1D solution in Cartesian coordinates, the impedance of a multilayer foil conductor of N_l layers carrying a sinusoidal current

$$i_l = I_m \sin \omega t \tag{5.106}$$

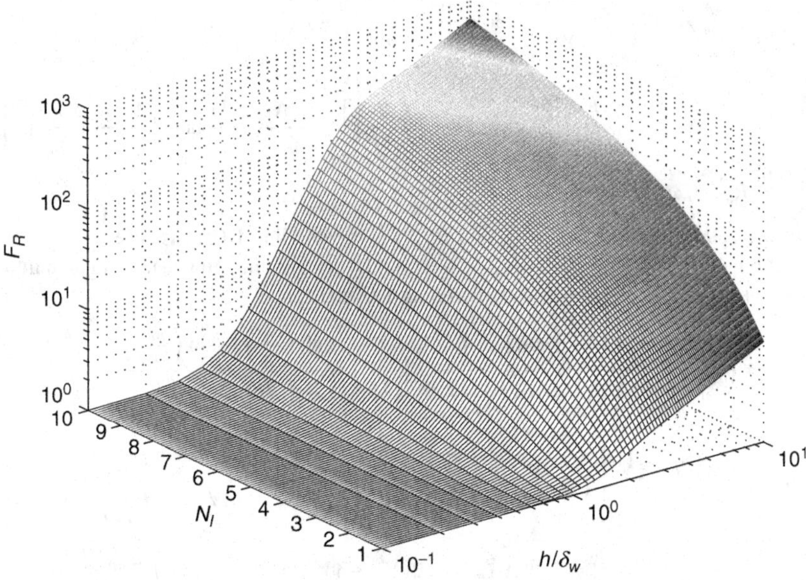

Figure 5.35 3D plot of F_R as a function of h/δ_w and N_l for uniform foil winding thickness

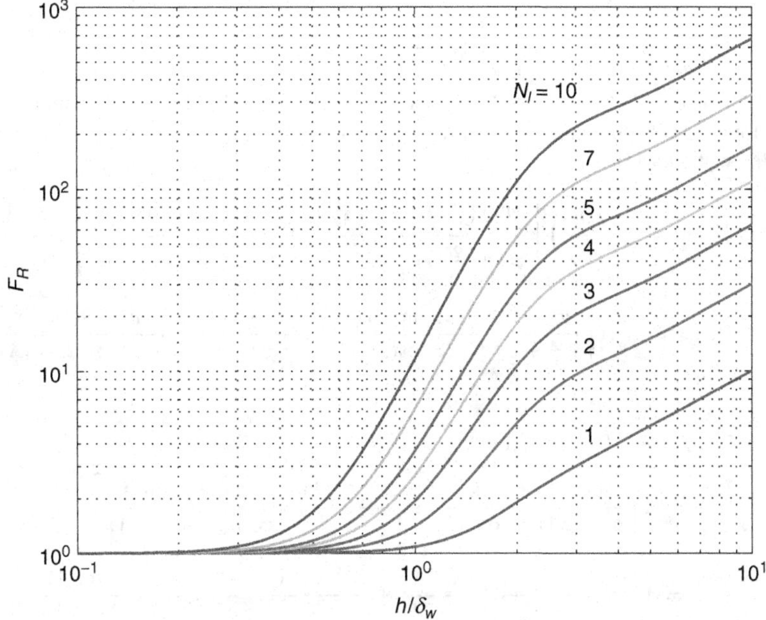

Figure 5.36 Plots of F_R as a function of h/δ_w at fixed numbers of layers N_l for uniform foil winding thickness

is given by Dowell [5]

$$Z = \frac{\rho_w l_w \gamma h}{hb}\left[\coth \quad (\gamma h) + \frac{2(N_l^2 - 1)}{3}\tanh\left(\frac{\gamma h}{2}\right)\right]$$

$$= R_{wDC}(1+j)\left(\frac{h}{\delta_w}\right)\left\{\coth\left[(1+j)\left(\frac{h}{\delta_w}\right)\right] + \frac{2(N_l^2-1)}{3}\tanh\left[\frac{1+j}{2}\left(\frac{h}{\delta_w}\right)\right]\right\}$$

$$= R_w + j\omega L = R_w + jX_L, \tag{5.107}$$

where $X_L = \omega L$, h is the foil thickness, and R_{wDC} is the winding DC resistance.

Substitution of (5.51) and (5.52) in (5.107) yields the winding impedance of a multilayer foil conductor normalized with respect to the winding DC resistance

$$\frac{Z}{R_{wDC}} = \left(\frac{h}{\delta_w}\right)\frac{\sinh\left(\frac{2h}{\delta_w}\right)+\sin\left(\frac{2h}{\delta_w}\right)}{\cosh\left(\frac{2h}{\delta_w}\right)-\cos\left(\frac{2h}{\delta_w}\right)} + j\left(\frac{h}{\delta_w}\right)\frac{\sinh\left(\frac{2h}{\delta_w}\right)-\sin\left(\frac{2h}{\delta_w}\right)}{\cosh\left(\frac{2h}{\delta_w}\right)-\cos\left(\frac{2h}{\delta_w}\right)}$$

$$+\frac{2(N_l^2-1)}{3}\left[\left(\frac{h}{\delta_w}\right)\frac{\sinh\left(\frac{h}{\delta_w}\right)-\sin\left(\frac{h}{\delta_w}\right)}{\cosh\left(\frac{h}{\delta_w}\right)+\cos\left(\frac{h}{\delta_w}\right)} + j\left(\frac{h}{\delta_w}\right)\frac{\sinh\left(\frac{h}{\delta_w}\right)+\sin\left(\frac{h}{\delta_w}\right)}{\cosh\left(\frac{h}{\delta_w}\right)+\cos\left(\frac{h}{\delta_w}\right)}\right]$$

$$= \left(\frac{h}{\delta_w}\right)\frac{\sinh\left(\frac{2h}{\delta_w}\right)+\sin\left(\frac{2h}{\delta_w}\right)}{\cosh\left(\frac{2h}{\delta_w}\right)-\cos\left(\frac{2h}{\delta_w}\right)} + \frac{2(N_l^2-1)}{3}\left(\frac{h}{\delta_w}\right)\frac{\sinh\left(\frac{h}{\delta_w}\right)-\sin\left(\frac{h}{\delta_w}\right)}{\cosh\left(\frac{h}{\delta_w}\right)+\cos\left(\frac{h}{\delta_w}\right)}$$

$$+j\left[\left(\frac{h}{\delta_w}\right)\frac{\sinh\left(\frac{2h}{\delta_w}\right)-\sin\left(\frac{2h}{\delta_w}\right)}{\cosh\left(\frac{2h}{\delta_w}\right)-\cos\left(\frac{2h}{\delta_w}\right)} + \frac{2(N_l^2-1)}{3}\left(\frac{h}{\delta_w}\right)\frac{\sinh\left(\frac{h}{\delta_w}\right)+\sin\left(\frac{h}{\delta_w}\right)}{\cosh\left(\frac{h}{\delta_w}\right)+\cos\left(\frac{h}{\delta_w}\right)}\right]$$

$$= \frac{R_w}{R_{wDC}} + \frac{jX_L}{R_{wDC}} = \frac{R_w}{R_{wDC}} + \frac{j\omega L}{R_{wDC}}. \tag{5.108}$$

Using a 1D model, the winding AC resistance caused by both the skin and proximity effects for a wide and long (ideally, infinitely wide, and long) foil winding with a single turn per layer is given by Dowell's equation [5]

$$R_w = Re\{Z\} = R_{wDC}\left(\frac{h}{\delta_w}\right)\left[\frac{\sinh\left(\frac{2h}{\delta_w}\right)+\sin\left(\frac{2h}{\delta_w}\right)}{\cosh\left(\frac{2h}{\delta_w}\right)-\cos\left(\frac{2h}{\delta_w}\right)} + \frac{2(N_l^2-1)}{3}\frac{\sinh\left(\frac{h}{\delta_w}\right)-\sin\left(\frac{h}{\delta_w}\right)}{\cosh\left(\frac{h}{\delta_w}\right)+\cos\left(\frac{h}{\delta_w}\right)}\right]$$

$$= R_{wDC}A\left[\frac{e^{2A}-e^{-2A}+2\sin(2A)}{e^{2A}+e^{-2A}-2\cos(2A)} + \frac{2(N_l^2-1)}{3}\frac{e^{A}-e^{-A}-2\sin(A)}{e^{A}+e^{-A}+2\cos(A)}\right]$$

$$= F_R R_{wDC}, \tag{5.109}$$

where the *AC-to-DC winding resistance ratio* is

$$F_R = \frac{R_w}{R_{wDC}} = A\left[\frac{\sinh(2A)+\sin(2A)}{\cosh(2A)-\cos(2A)} + \frac{2(N_l^2-1)}{3}\frac{\sinh(A)-\sin(A)}{\cosh(A)+\cos(A)}\right]$$

$$= A\left[\frac{e^{2A}-e^{-2A}+2\sin(2A)}{e^{2A}+e^{-2A}-2\cos(2A)} + \frac{2(N_l^2-1)}{3}\frac{e^{A}-e^{-A}-2\sin(A)}{e^{A}+e^{-A}+2\cos(A)}\right], \tag{5.110}$$

the winding DC resistance of the foil is

$$R_{wDC} = \frac{\rho_w l_w}{A_w} = \frac{\rho_w l_w}{hb} = \frac{\rho_w l_T N_l}{hb}, \tag{5.111}$$

the foil thickness normalized with respect to the foil skin depth is

$$A = \frac{h}{\delta_w} = h\sqrt{\frac{\pi \mu f}{\rho_{wf}}}, \tag{5.112}$$

N_l is the number of winding layers, h is the foil thickness, b is the foil width (or the foil winding breadth), and l_w is the foil winding length. For the foil winding, the factor F_R depends on h/δ_w and N_l.

The winding AC resistance described by (5.109) represents the winding AC resistance due to the skin effect and the winding AC resistance due to the proximity effect

$$R_w = R_{skin} + R_{prox} = F_S R_{wDC} + F_P R_{wDC} = (F_S + F_P) R_{wDC} = F_R R_{wDC}. \tag{5.113}$$

Thus,

$$F_R = \frac{R_w}{R_{wDC}} = F_S + F_P, \tag{5.114}$$

where the *skin-effect factor* is

$$F_S = A\frac{\sinh(2A) + \sin(2A)}{\cosh(2A) - \cos(2A)} = A\frac{e^{2A} - e^{-2A} + 2\sin(2A)}{e^{2A} + e^{-2A} - 2\cos(2A)} \tag{5.115}$$

and the *proximity-effect factor* is

$$F_P = \frac{2(N_l^2 - 1)A}{3}\left[\frac{\sinh(A) - \sin(A)}{\cosh(A) + \cos(A)}\right] = \frac{2(N_l^2 - 1)A}{3}\left[\frac{e^A - e^{-A} - 2\sin(A)}{e^A + e^{-A} + 2\cos(A)}\right]. \tag{5.116}$$

Figure 5.37 shows the plots of F_S, F_P, and F_R versus $A = h/\delta_{wf}$ for $N_l = 3$. It can be seen that $F_P \approx 0$ and $F_S \approx F_R \approx 1$ for $A = h/\delta_{wf} \leq 1$. The proximity effect factor F_P is much higher than the skin effect factor F_S for $N_l \geq 2$ and $A \geq 2$. Therefore, the proximity-effect power loss is a dominant component in the winding power losses for multilayer inductors. For $N_l = 1$, the proximity effect is not present, and therefore the winding power loss in single-layer inductors is only due to skin effect.

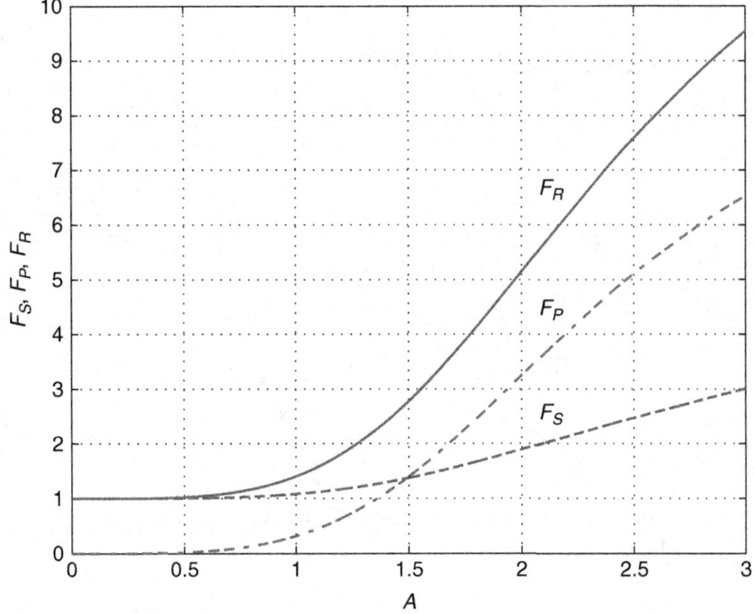

Figure 5.37 Plots of F_S, F_P, and F_R as functions of $A = h/\delta_w$ for $N_l = 3$

5.15 Approximation of Dowell's Equation

5.15.1 Approximation of Dowell's Equation for Low and Medium Frequencies

The trigonometric and hyperbolic functions in Dowell's equation can be expanded into Maclaurin's series, as shown in the appendix. Hence,

$$F_{RS} = \frac{F_S}{A} = \frac{\sinh(2A) + \sin(2A)}{\cosh(2A) - \cos(2A)} \approx \frac{1}{A} + \frac{2A^3}{15} \quad \text{for} \quad A \leq 1.5 \tag{5.117}$$

and

$$F_{RP} = \frac{3F_P}{2(N_l^2 - 1)A} = \frac{\sinh(A) - \sin(A)}{\cosh(A) + \cos(A)} \approx \frac{A^3}{6} \quad \text{for} \quad A \leq 1.5. \tag{5.118}$$

Dowell's equation for foil windings at low and medium frequencies can be approximated by Snelling [8]

$$F_R = \frac{R_w}{R_{wDC}} \approx A\left[\frac{1}{A} + \frac{2A^3}{15} + \frac{2(N_l^2 - 1)}{3}\frac{A^3}{6}\right] \approx A\left(\frac{1}{A} + \frac{5N_l^2 - 1}{45}A^3\right)$$

$$= 1 + \frac{5N_l^2 - 1}{45}A^4 = 1 + \frac{5N_l^2 - 1}{45}\left(\frac{h}{\delta_w}\right)^4 \quad \text{for} \quad A \leq 1.5. \tag{5.119}$$

Figure 5.38 shows the plots of exact Dowell's equation for F_R and its approximation for low and medium frequencies as functions of $A = h/\delta_w$ at $N_l = 3$. It can be seen that the exact and approximate plots of F_R are similar for low and medium values of $A \leq 1.5$.

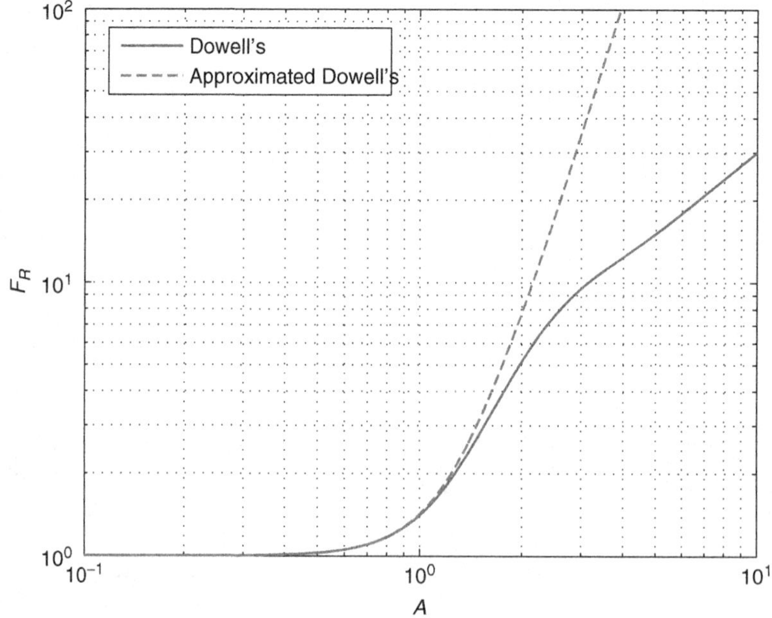

Figure 5.38 Plots of exact Dowell's equation for F_R and its approximation for low and medium frequencies as functions of $A = h/\delta_w$ for $N_l = 3$

5.15.2 Approximation of Dowell's Equation for High Frequencies

The AC-to-DC winding resistance ratio can be expressed as

$$F_R = \frac{R_w}{R_{wDC}} = A\left[F_{RS} + \frac{2(N_l^2 - 1)}{3}F_{RP}\right], \tag{5.120}$$

where

$$F_{RS} = \frac{F_S}{A} = \frac{\sinh(2A) + \sin(2A)}{\cosh(2A) - \cos(2A)} = \frac{e^{2A} - e^{-2A} + 2\sin(2A)}{e^{2A} + e^{-2A} - 2\cos(2A)} \tag{5.121}$$

and

$$F_{RP} = \frac{3F_P}{2A(N_l^2 - 1)} = \frac{\sinh(A) - \sin(A)}{\cosh(A) + \cos(A)} = \frac{e^A - e^{-A} - 2\sin(A)}{e^A + e^{-A} + 2\cos(A)}. \tag{5.122}$$

Figure 5.39 shows the plots of F_{RS} and F_{RP} as functions of $A = h/\delta_{wf}$. It can be seen that $F_{RS} \approx 1$ and $F_{RP} \approx 1$ for $A = h/\delta_{wf} > 3$. Hence, the AC-to-DC winding resistance ratio at high frequencies for foil windings is

$$F_R = \frac{R_w}{R_{wDC}} \approx \frac{(2N_l^2 + 1)}{3}A = \frac{(2N_l^2 + 1)}{3}\frac{h}{\delta_{wf}} \quad \text{for} \quad A = \frac{h}{\delta_{wf}} > 3. \tag{5.123}$$

The winding AC resistance of foil windings at high frequencies may be approximated by

$$R_w \approx R_{wDC}A\left[1 + \frac{2(N_l^2 - 1)}{3}\right] = \frac{(2N_l^2 + 1)}{3}AR_{wDC} = \frac{(2N_l^2 + 1)}{3}A\frac{\rho_{wf}l_w}{b_{wf}h}$$

$$= \frac{(2N_l^2 + 1)}{3}\frac{\rho_{wf}l_w}{b_{wf}\delta_{wf}} = \frac{(2N_l^2 + 1)}{3}\frac{l_w}{b_{wf}}\sqrt{\pi\mu\rho_w f} \quad \text{for} \quad A = \frac{h}{\delta_{wf}} > 3. \tag{5.124}$$

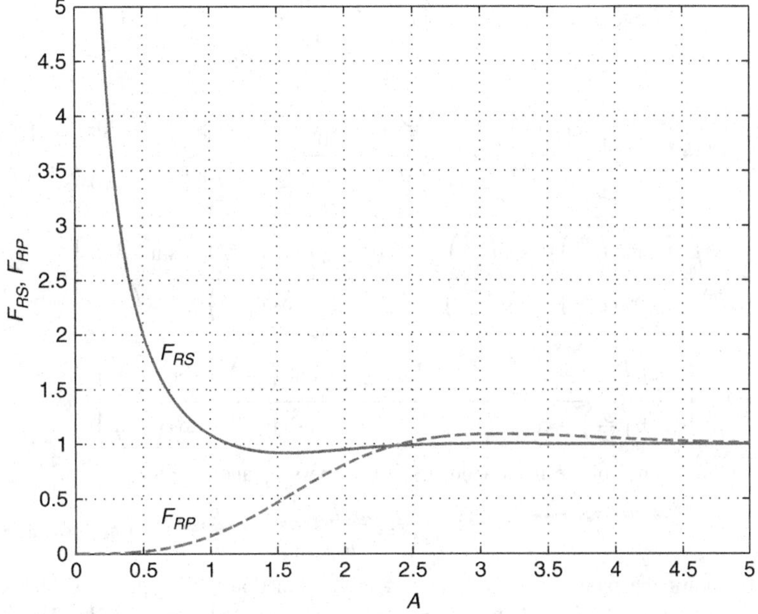

Figure 5.39 Plots of F_{RS} and F_{RP} as functions of $A = h/\delta_w$

For $N_l = 1$,

$$R_w \approx AR_{wDC} = \frac{h}{\delta_{wf}} R_{wDC} = \frac{\rho_{wf} l_w}{b_{wf}\delta_{wf}} \quad \text{for} \quad A = \frac{h}{\delta_{wf}} > 3. \tag{5.125}$$

For $N_l = 2$,

$$R_w \approx 3AR_{wDC} = 3\frac{h}{\delta_{wf}} R_{wDC} = 3\frac{\rho_{wf} l_w}{b_{wf}\delta_{wf}} \quad \text{for} \quad A = \frac{h}{\delta_{wf}} > 3. \tag{5.126}$$

For $N_l = 3$,

$$R_w \approx 6.333AR_{wDC} = 6.333\frac{h}{\delta_{wf}} R_{wDC} = 6.333\frac{\rho_{wf} l_w}{b_{wf}\delta_{wf}} \quad \text{for} \quad A = \frac{h}{\delta_{wf}} > 3. \tag{5.127}$$

For $N_l = 4$,

$$R_w \approx 11AR_{wDC} = 11\frac{h}{\delta_{wf}} R_{wDC} = 11\frac{\rho_{wf} l_w}{b_{wf}\delta_{wf}} \quad \text{for} \quad A = \frac{h}{\delta_{wf}} > 3. \tag{5.128}$$

The multilayer foil winding impedance at high frequencies normalized with respect to the winding DC resistance R_{wDC} can be approximated by

$$F_Z = \frac{Z}{R_{wDC}} \approx (1+j)\left(\frac{h}{\delta_w}\right)\left[1 + \frac{2(N_l^2-1)}{3}\right] = (1+j)\left(\frac{h}{\delta_w}\right)\left(\frac{2N_l^2+1}{3}\right)$$
$$= \left(\frac{h}{\delta_w}\right)\left(\frac{2N_l^2+1}{3}\right) + j\left(\frac{h}{\delta_w}\right)\left(\frac{2N_l^2+1}{3}\right) = \frac{R_w}{R_{wDC}} + \frac{j\omega L}{R_{wDC}} \quad \text{for} \quad \frac{h}{\delta_w} \geq 3. \tag{5.129}$$

5.16 Winding AC Resistance with Uniform Foil Thickness

The foil winding AC resistance for a sinusoidal inductor current waveform is

$$R_w = F_R R_{wDC}$$
$$= \frac{\rho_w l_w}{bh}\left(\frac{h}{\delta_w}\right)\left[\frac{\sinh\left(\frac{2h}{\delta_w}\right) + \sin\left(\frac{2h}{\delta_w}\right)}{\cosh\left(\frac{2h}{\delta_w}\right) - \cos\left(\frac{2h}{\delta_w}\right)} + \frac{2(N_l^2-1)}{3}\frac{\sinh\left(\frac{h}{\delta_w}\right) - \sin\left(\frac{h}{\delta_w}\right)}{\cosh\left(\frac{h}{\delta_w}\right) + \cos\left(\frac{h}{\delta_w}\right)}\right]$$
$$= \frac{\rho_w l_w}{b\delta_w}\left[\frac{\sinh\left(\frac{2h}{\delta_w}\right) + \sin\left(\frac{2h}{\delta_w}\right)}{\cosh\left(\frac{2h}{\delta_w}\right) - \cos\left(\frac{2h}{\delta_w}\right)} + \frac{2(N_l^2-1)}{3}\frac{\sinh\left(\frac{h}{\delta_w}\right) - \sin\left(\frac{h}{\delta_w}\right)}{\cosh\left(\frac{h}{\delta_w}\right) + \cos\left(\frac{h}{\delta_w}\right)}\right]$$
$$= R_{\delta f}\left[\frac{\sinh\left(\frac{2h}{\delta_w}\right) + \sin\left(\frac{2h}{\delta_w}\right)}{\cosh\left(\frac{2h}{\delta_w}\right) - \cos\left(\frac{2h}{\delta_w}\right)} + \frac{2(N_l^2-1)}{3}\frac{\sinh\left(\frac{h}{\delta_w}\right) - \sin\left(\frac{h}{\delta_w}\right)}{\cosh\left(\frac{h}{\delta_w}\right) + \cos\left(\frac{h}{\delta_w}\right)}\right] \tag{5.130}$$

where the DC resistance of the foil conductor of thickness δ_w and length l_w is

$$R_{\delta f} = \frac{\rho_w l_w}{b\delta_w}. \tag{5.131}$$

Figure 5.40 shows the plots of F_R, R_{wDC}, and R_w as functions of copper foil thickness h for a uniform foil winding with $b = 10$ mm, $l_T = 25$ mm, $N_l = 6$, and $f = 100$ kHz. The winding DC resistance $R_{wDC} = \rho_w l_w/(bh)$ decreases as h increases. The AC-to-DC winding resistance ratio F_R is equal to 1 for low values of h and then increases with h. Therefore, the winding AC resistance R_w decreases with increasing h, reaches a minimum value $R_{w\min}$ at an optimum foil thickness h_{opt}, and then increases with h, reaches a local maximum value, and then remains constant.

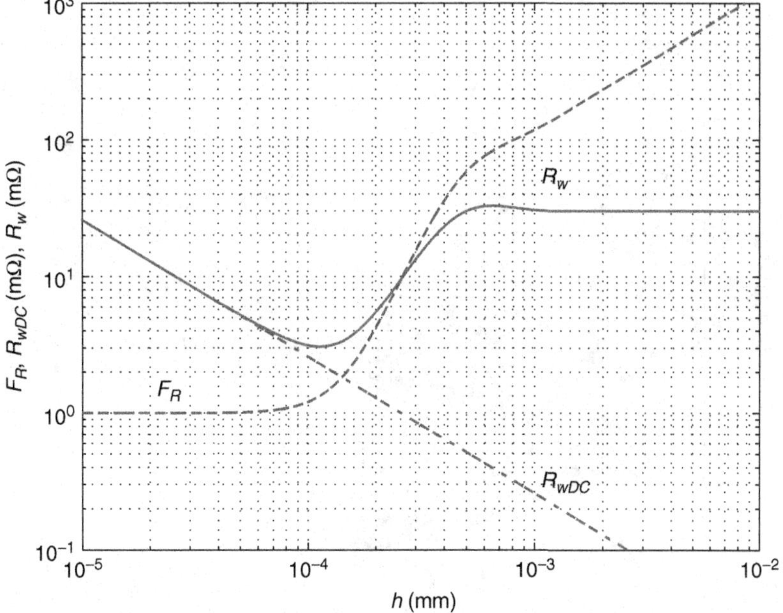

Figure 5.40 Plots of F_R, R_{wDC}, and $R_w = F_R R_{wDC}$ as functions of copper foil thickness h for uniform foil winding with $b_{wf} = 10$ mm, $l_T = 25$ mm, $N_l = 6$, and $f = 100$ kHz

The normalized winding AC resistance for uniform foil winding at a fixed skin depth δ_w, that is, at a fixed frequency, can be expressed as

$$F_r = \frac{R_w}{\rho_w l_w/(b\delta_w)} = \frac{R_w}{R_{\delta_f}} = \frac{F_R}{\frac{h}{\delta_w}}$$
$$= \frac{\sinh\left(\frac{2h}{\delta_w}\right) + \sin\left(\frac{2h}{\delta_w}\right)}{\cosh\left(\frac{2h}{\delta_w}\right) - \cos\left(\frac{2h}{\delta_w}\right)} + \frac{2(N_l^2 - 1)}{3} \frac{\sinh\left(\frac{h}{\delta_w}\right) - \sin\left(\frac{h}{\delta_w}\right)}{\cosh\left(\frac{h}{\delta_w}\right) + \cos\left(\frac{h}{\delta_w}\right)}. \tag{5.132}$$

Figures 5.41 and 5.42 shows 3D and 2D plots of $F_r = R_w/(\rho_w l_w/b\delta_w)$ for uniform foil winding thickness, respectively. It can be seen that there is a minimum value of the normalized winding AC resistance F_r at an optimum foil thickness h_{opt} for a fixed frequency.

5.16.1 Optimum Uniform Foil Thickness of Inductor Winding for Sinusoidal Inductor Current

The optimization of the winding conductor size is an important aspect of inductor and transformer designs. Let us assume that the current through an inductor winding is sinusoidal of constant frequency f (i.e., $i_l = I_m \sin \omega t$). In this case, the skin depth δ_w is also constant. The objective is to find the optimum foil conductor thickness so that the winding loss is minimized for a given frequency and a given number of winding layers.

An analytical optimization of Dowell's equation leads to a transcendental equation and can only be solved numerically for $N_l \geq 2$. Therefore, low- and medium-frequency Dowell's approximate equation (5.119) may be used to determine the optimum uniform foil thickness h_{opt} analytically, which gives the minimum foil winding resistance $R_{w\min}$. The optimum foil thickness occurs in the vicinity of the boundary between the low and medium values of $A = h/\delta_w$. Hence, Dowell's approximate equation (5.119) is suitable for determining h_{opt}. Once h_{opt} is known, the minimum foil

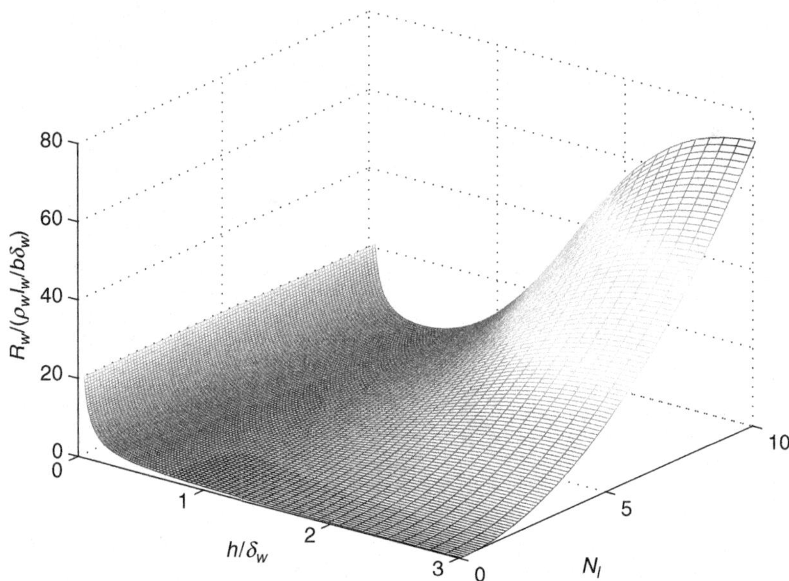

Figure 5.41 Normalized winding AC resistance $F_r = R_w/(\rho_w l_w/b_{wf}\delta_w)$ versus h/δ_w and N_l for a uniform foil winding

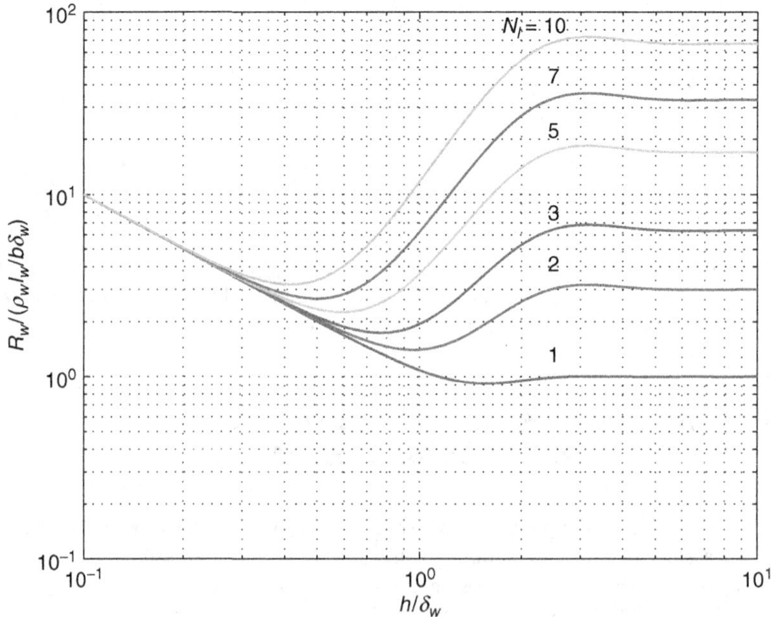

Figure 5.42 Normalized winding AC resistance $F_r = R_w/(\rho_w l_w/b_{wf}\delta_w)$ versus h/δ_w and various values of the number of layers N_l for uniform foil winding

winding resistance may be determined either from the exact Dowell's equation (5.109) or approximate Dowell's equation (5.119).

Using (5.119), the foil winding AC resistance for low and medium values of h/δ_w is

$$R_w = F_R R_{wDC} \approx \frac{\rho_w l_w}{b_{wf} h}\left[1 + \frac{5N_l^2 - 1}{45}\left(\frac{h}{\delta_w}\right)^4\right] = \frac{\rho_w l_w}{b_{wf}}\left[\frac{1}{h} + \frac{(5N_l^2 - 1)h^3}{45\delta_w^4}\right]. \tag{5.133}$$

Assume that the number of layers N_l is constant. Taking the derivative of R_w with respect to h and setting the result to zero, one obtains

$$\frac{dR_w}{dh} = \frac{\rho_w l_w}{b_{wf}}\left[-\frac{1}{h^2} + \frac{(5N_l^2 - 1)h^2}{15\delta_w^4}\right] = 0, \tag{5.134}$$

yielding

$$\frac{(5N_l^2 - 1)}{15}\left(\frac{h}{\delta_w}\right)^4 = 1. \tag{5.135}$$

Hence, the minimum winding AC resistance for sinusoidal inductor current occurs at the normalized optimum foil thickness

$$\frac{h_{opt}}{\delta_w} = \sqrt[4]{\frac{15}{5N_l^2 - 1}} \approx \frac{1.97}{\sqrt[4]{5N_l^2 - 1}}. \tag{5.136}$$

Figure 5.43 shows the normalized optimum uniform foil thickness h_{opt}/δ_w as a function of number of layers N_l. Table 5.2 gives the normalized optimum uniform foil thickness h_{opt}/δ_w calculated from (5.136) for different numbers of layers N_l. It can be seen that h_{opt}/δ_w decreases as N_l increases. The optimum foil thickness is

$$h_{opt} = \delta_w \sqrt[4]{\frac{15}{5N_l^2 - 1}} = \sqrt{\frac{\rho_w}{\pi \mu_0 f}}\sqrt[4]{\frac{15}{5N_l^2 - 1}}. \tag{5.137}$$

The optimum foil thickness is inversely proportional to the square root of frequency f and approximately inversely proportional to the square root of N_l. Figure 5.44 shows a plot of the optimum thickness h_{opt} of the uniform copper foil as functions of frequency f for $N_l = 5$.

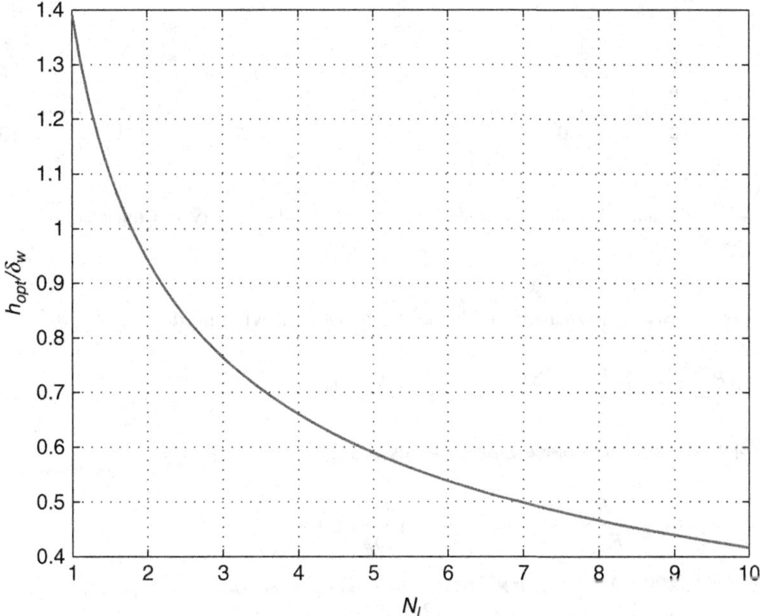

Figure 5.43 Normalized optimum uniform foil thickness h_{opt}/δ_w as a function of number of layers N_l

Table 5.2 Optimum uniform foil thickness h_{opt}/δ_w for different number of layers N_l

Number of layers N_l	h_{opt}/δ_w
1	1.3916
2	0.9426
3	0.7641
4	0.6601
5	0.5898
6	0.5380
7	0.4979
8	0.4657
9	0.4389
10	0.4164

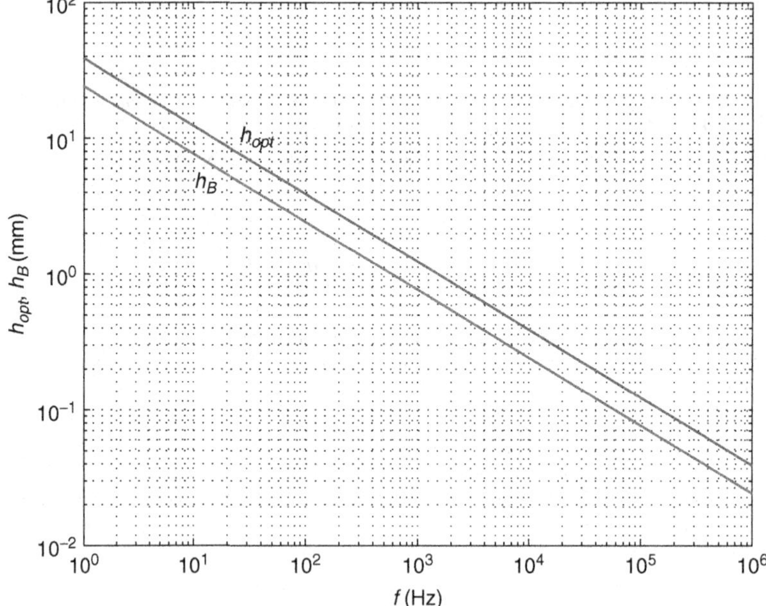

Figure 5.44 Optimum uniform copper foil thickness h_{opt} and boundary foil thickness h_B as functions of frequency f for $N_l = 5$

Substitution of (5.136) in (5.133) yields the AC-to-DC foil winding resistance ratio at the optimum foil thickness

$$F_{Rv} = \frac{R_{w\,min}}{R_{wDC}} = \frac{R_{wDC} + \Delta R_{w\,min}}{R_{wDC}} = 1 + \frac{\Delta R_{w\,min}}{R_{wDC}} = 1 + \frac{1}{3} = \frac{4}{3} \quad \text{for} \quad h = h_{opt}. \tag{5.138}$$

The AC-to-DC winding resistance ratio in terms of h_{opt} is

$$F_R = \frac{R_w}{R_{wDC}} = 1 + \frac{5N_l^2 - 1}{45}\left(\frac{h_{opt}}{\delta_w}\right)^4\left(\frac{h}{h_{opt}}\right)^4 = 1 + \frac{1}{3}\left(\frac{h}{h_{opt}}\right)^4. \tag{5.139}$$

The minimum winding AC resistance is

$$R_{w\,min} = \frac{4}{3}\frac{\rho_w l_w}{b h_{opt}} = \frac{4\rho_w l_w}{3b\delta_w}\sqrt[4]{\frac{5N_l^2 - 1}{15}}. \tag{5.140}$$

The average current density is

$$J = \frac{I}{A_w} = \frac{I}{bh}. \quad (5.141)$$

Hence, the minimum foil width at the optimum foil thickness h_{opt} is

$$b_{min} = \frac{I_{max}}{J_{max}h_{opt}}. \quad (5.142)$$

The typical maximum foil thickness is $J_{max} = 5\text{–}8\ \text{A/mm}^2$.

Example 5.1

Design an inductor with $L = 400\ \mu\text{H}$. A sinusoidal current of amplitude $I_{Lm} = 10$ A flows through an inductor at $f = 180$ kHz. The winding is made of a copper foil. The core relative permeability is $\mu_{rc} = 1800$, the bobbin breadth $b = 2$ cm, the mean magnetic path is $l_c = 2$ cm, and the core cross-sectional area $A_c = 4\ \text{cm}^2$. The MTL is $l_T = 10$ cm. Find:

(a) The number of layers N_l.
(b) The optimum foil thickness h_{opt}.
(c) The winding AC resistance R_w.
(d) The AC power loss P_w.
(e) The inductor quality factor related to the winding resistance.
(f) The average current density.

Solution: (a) The number of layers is

$$N = \sqrt{\frac{l_c L}{\mu_{rc}\mu_0 A_c}} = \sqrt{\frac{2 \times 10^{-2} \times 400 \times 10^{-6}}{1800 \times 4\pi \times 10^{-7} \times 4 \times 10^{-4}}} = 2.9735. \quad (5.143)$$

Pick $N = N_l = 3$. The skin depth of copper at $f = 180$ kHz is

$$\delta_w = \sqrt{\frac{\rho_{Cu}}{\pi\mu_0 f}} = \sqrt{\frac{1.724 \times 10^{-8}}{\pi \times 4\pi \times 10^{-7} \times 180 \times 10^3}} = 0.15575\ \text{mm}. \quad (5.144)$$

(b) The normalized optimum foil thickness is

$$\frac{h_{opt}}{\delta_w} = \sqrt[4]{\frac{15}{5N_l^2 - 1}} = \sqrt[4]{\frac{15}{5 \times 3^2 - 1}} = 0.7641. \quad (5.145)$$

Hence, the optimum foil thickness is

$$h_{opt} = 0.7641\delta_w = 0.7641 \times 0.15575 = 0.119\ \text{mm}. \quad (5.146)$$

The closest standard foil thickness is $h = 0.12$ mm.

(c) The foil winding DC resistance is

$$R_{wDC} = \frac{\rho_{Cu} N l_T}{b h_{opt}} = \frac{1.724 \times 10^{-8} \times 3 \times 10^{-2}}{2 \times 10^{-2} \times 0.12 \times 10^{-6}} = 2.155\ \text{m}\Omega. \quad (5.147)$$

The AC-to-DC winding resistance ratio at the optimum foil thickness is $F_{Rv} = 4/3$. The foil winding AC resistance is

$$R_w = F_{Rv} R_{wDC} = \frac{4}{3} \times 2.155 = 2.8733\ \text{m}\Omega. \quad (5.148)$$

(d) The winding power loss is

$$P_w = \frac{R_w I_{Lm}^2}{2} = \frac{2.8733 \times 10^{-3} \times 10^2}{2} = 143.66\ \text{mW}. \quad (5.149)$$

(e) The inductor quality factor related to the winding resistance is

$$Q_{Lw} = \frac{\omega L}{R_w} = \frac{2\pi \times 180 \times 10^3 \times 400 \times 10^{-6}}{0.028733} = 15\,745. \quad (5.150)$$

(f) The average current density is

$$J_{AV} = \frac{I_{max}}{A_w} = \frac{I_{max}}{bh_{opt}} = \frac{10}{20 \times 0.12} = 4.167\ \text{A/mm}^2. \quad (5.151)$$

5.16.2 Boundary Between Low and Medium Frequencies for Foil Windings

The entire range of F_R as a function of h/δ_w can be divided into three ranges of h/δ_w: low range of h/δ_w, medium range of h/δ_w, and (iii) high range of h/δ_w. For a fixed value of the foil thickness h, there are three frequency ranges: (i) low-frequency range, (ii) medium-frequency range, and (iii) high-frequency range. In the low range of h/δ_w, $F_R \leq 1.05$, the current density distribution in the winding conductor is uniform, both the skin and proximity effects are negligible, and $R_w \approx R_{wDC}$.

In the medium range of h/δ_w, the current density distribution in the winding conductor is nonuniform, and F_R increases when h/δ_w increases. Above the low range of h/δ_w, the current density distribution in the winding conductor is also nonuniform; it is very low in the center of the foil cross-sectional area, and it is high near both foil surfaces. Therefore, F_R increases when h/δ_w increases. However, the rate of increase is lower than that in the medium range of h/δ_w for $N_l \geq 2$. For $N_l = 1$, the proximity effect does not exists and, therefore, the rate of change of F_R in the medium- and high-frequency ranges remains constant.

The definition of the boundary between the low range of h/δ_w and the medium range of h/δ_w is as follows:

$$F_{RB} = 1.05. \quad (5.152)$$

The approximate Dowell's equation is valid for $F_{RB} = 1.05$. Hence,

$$F_{RB} \approx 1 + \frac{5N_l^2 - 1}{45}\left(\frac{h_B}{\delta_w}\right)^4 = 1.05, \quad (5.153)$$

yielding the foil thickness at the boundary between the low and medium values of h/δ_w

$$\frac{h_B}{\delta_w} = \sqrt[4]{\frac{9}{4(5N_l^2 - 1)}}. \quad (5.154)$$

Table 5.3 gives the ratio h_B/δ_w at different numbers of layers N_l for foil windings.

Table 5.3 Boundary between low and medium values of $(h/\delta_w)_B$ versus number of layers N_l at $F_R = 1.05$ for foil winding

Number of layers N_l	$(h/\delta_w)_B$
1	0.8666
2	0.5866
3	0.4755
4	0.4108
5	0.3670
6	0.3348
7	0.3099
8	0.2898
9	0.2732
10	0.2591

The foil boundary thickness is

$$h_B = \delta_w \sqrt[4]{\frac{9}{4(5N_l^2 - 1)}} = \sqrt{\frac{\rho_w}{\pi \mu_0 f}} \sqrt[4]{\frac{9}{4(5N_l^2 - 1)}}. \quad (5.155)$$

Figure 5.44 shows a plot of the boundary thickness h_B of the copper foil as a function of frequency f for $N_l = 5$. The boundary frequency is

$$f_L = \frac{\rho_w}{\pi \mu_0 h_B^2} \sqrt{\frac{9}{4(5N_l^2 - 1)}}. \quad (5.156)$$

The ratio of the optimum foil thickness to the boundary foil thickness is

$$\frac{h_{opt}}{h_B} = \sqrt[4]{\frac{20}{3}} \approx 1.61. \quad (5.157)$$

The ratio of the boundary winding AC resistance R_{wB} to the the minimum winding AC resistance for foil inductors is

$$\frac{R_{wB}}{R_{w\,min}} = \frac{R_{wDCB} F_{RB}}{R_{wDCopt} F_{Rv}} = \frac{h_{opt}}{h_B} \frac{F_{RB}}{F_{Rv}} = \frac{1.05 \times 3}{4} \sqrt[4]{\frac{20}{3}} \approx 1.26539. \quad (5.158)$$

Example 5.2

A foil winding inductor has $L = 400\ \mu\text{H}$ and $N_l = 3$. A sinusoidal current of amplitude $I_{Lm} = 10$ A flows through an inductor at $f = 180$ kHz. The winding is made of a copper foil. The bobbin breadth is $b = 2$ cm. The MTL is $l_T = 10$ cm. Find the foil thickness at the boundary between low and medium frequencies h_B, F_{RB}, and AC power loss.

Solution: The skin depth is

$$\delta_w = \sqrt{\frac{\rho_{Cu}}{\pi \mu_0 f}} = \sqrt{\frac{1.724 \times 10^{-8}}{\pi \times 4\pi \times 10^{-7} \times 180 \times 10^3}} = 0.15575 \text{ mm}. \quad (5.159)$$

The normalized boundary foil thickness is

$$\left(\frac{h}{\delta_w}\right)_B = \sqrt[4]{\frac{9}{4(5N_l^2 - 1)}} = \sqrt[4]{\frac{9}{4 \times (5 \times 3^2 - 1)}} = 0.4755. \quad (5.160)$$

Hence, the boundary foil thickness is

$$h_B = 0.4755\delta_w = 0.4755 \times 0.15575 = 0.074 \text{ mm}. \quad (5.161)$$

The DC foil winding resistance at h_B is

$$R_{wDC} = \frac{\rho_{Cu} N l_T}{b h_B} = \frac{1.724 \times 10^{-8} \times 3 \times 10^{-2}}{2 \times 10^{-1} \times 0.074 \times 10^{-3}} = 3.491 \text{ m}\Omega. \quad (5.162)$$

The AC-to-DC winding resistance ratio at the boundary foil thickness is $F_{RB} = 1.05$. Hence, the winding AC resistance is

$$R_w = F_{RB} R_{wDC} = 1.05 \times 3.491 = 3.6656 \text{ m}\Omega. \quad (5.163)$$

The winding power loss is

$$P_w = \frac{R_w I_{Lm}^2}{2} = \frac{3.6656 \times 10^{-3} \times 10^2}{2} = 183.28 \text{ mW}. \quad (5.164)$$

The average amplitude of the current density is

$$J_{m(AV)} = \frac{I_{Lm}}{A_w} = \frac{I_{Lm}}{bh} = \frac{10}{20 \times 0.12} = 4.167 \text{ A/mm}^2. \quad (5.165)$$

5.17 Transformation of Foil Conductor to Rectangular, Square, and Round Conductors

Dowell's equation [5] was derived for the winding with a single wide foil per layer (i.e., nonporous layer) and can be extended to porous conductor layers (having many turns in each layer). The winding conductors may have different shapes, such as rectangular, square, and round. A porosity factor $\eta = b_w/b_{wf}$ (i.e., a layer fill factor) can be introduced when converting one sheet of a foil conductor into several equivalent conductors in each layer to ensure that the DC resistance of the winding R_{wDC} remains unchanged. The equivalence of a porous layer to nonporous layer can be accomplished by stretching the porous conductor in each layer by a factor of $1/\eta$ and reducing the resistivity of the porous layer conductor by a factor η to obtain approximate analytical descriptions of the winding resistance. A transformation of foil winding to various shapes of a winding conductor is illustrated in Fig. 5.45. A wide sheet of a foil shown in Fig. 5.45a is first replaced by a narrow sheet of a foil, as depicted in Fig. 5.45b. Next the narrow sheet of foil is replaced by several narrower rectangular conductors to obtain multiple turns in each layer, as shown in Fig. 5.45c. Then, the rectangular conductors are replaced with square conductors depicted in Fig. 5.45d. Finally, the square conductors are replaced by round conductors of the same cross-sectional area, as shown in Fig. 5.45e.

The procedure for transforming a *rectangular winding conductor* is as follows. Consider two foil windings with a single turn per layer. Both foil windings have the same length l_w and thickness h. However, the widths and the resistivities of the foils are different. The wider foil has the width (or the winding breadth) b_{wf} and the resistivity ρ_{wf}. The narrower foil conductor has the width $b_w < b_{wf}$ and the resistivity $\rho_w < \rho_{wf}$. Consequently, the narrow conductor skin depth δ_w is lower than the wide conductor skin depth δ_{wf}, that is, $\delta_w < \delta_{wf}$ due to a lower resistivity than that of the wide conductor. The resistivities are such that the DC resistances of both windings remain the same. The DC resistance of the wide foil winding is given by

$$R_{wDCf} = \frac{\rho_{wf} l_w}{h b_{wf}} \tag{5.166}$$

and the DC resistance of the narrow foil winding is

$$R_{wDC} = \frac{\rho_w l_w}{h b_w}. \tag{5.167}$$

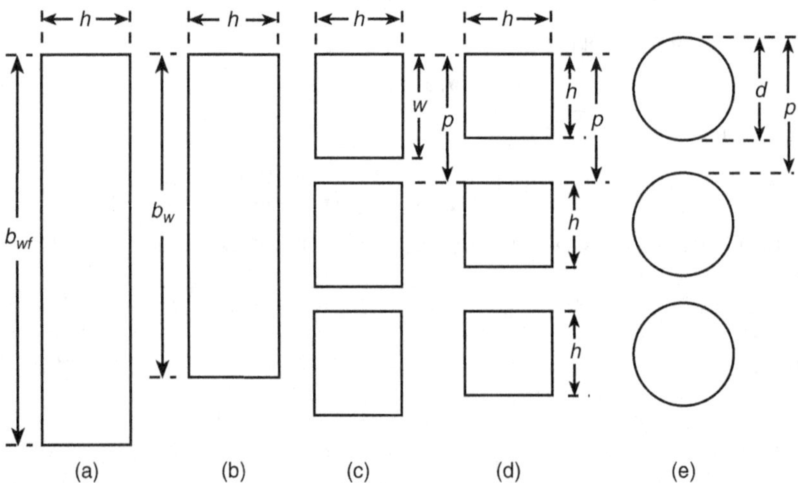

Figure 5.45 Transformations of a foil winding to rectangular, square, and round wires windings. (a) Wide foil winding. (b) Narrow foil winding. (c) Rectangular winding conductor. (d) Square winding conductor. (e) Round winding wire

The narrow foil is cut into narrower rectangular conductors (strips) of width w and stretched so that each layer has N_{tl} turns, the width of each conductor is

$$w = \frac{b_w}{N_{tl}} \tag{5.168}$$

and the distance between the centers of two adjacent conductors, called the winding pitch, is defined by

$$p = \frac{b_{wf}}{N_{tl}}. \tag{5.169}$$

As mentioned earlier, the DC resistances of both foil windings are the same $R_{wDC} = R_{wDCf}$

$$\frac{\rho_w l_w}{hb_w} = \frac{\rho_{wf} l_w}{hb_{wf}} \tag{5.170}$$

which gives the ratio of two resistivities

$$\frac{\rho_w}{\rho_{wf}} = \frac{b_w}{b_{wf}} = \frac{wN_{tl}}{pN_{tl}} = \frac{w}{p} = \eta, \tag{5.171}$$

where the layer *porosity factor*, *spacing factor*, or *fill factor* is defined as

$$\eta = \frac{w}{p}. \tag{5.172}$$

The typical value of the porosity factor is $\eta = 0.8$. The minimum distance between the turns of the same layer is determined by the thickness of the winding conductor insulation

$$g_{min} = p_{min} - w \tag{5.173}$$

Hence, the maximum porosity factor is

$$\eta_{max} = \frac{w}{p_{min}} = \frac{w}{w + g_{min}} = \frac{1}{1 + \frac{g_{min}}{w}}. \tag{5.174}$$

The ratio of the skin depths is given by

$$\frac{\delta_w}{\delta_{wf}} = \sqrt{\frac{\rho_w}{\rho_{wf}}} = \sqrt{\frac{w}{p}} = \sqrt{\eta}. \tag{5.175}$$

Thus, the effective skin depth of a porous layer is

$$\delta_p = \frac{\delta_{wf}}{\sqrt{\eta}} = \sqrt{\frac{\rho_{wf}}{\pi \mu_0 \eta f}}. \tag{5.176}$$

The complex propagation constant with the porosity factor is

$$\gamma = \sqrt{\frac{j\omega\mu_0}{\rho_w}} = \sqrt{\frac{j\omega\mu_0\eta}{\rho_{wf}}} = \frac{1+j}{\delta_{wf}}\sqrt{\eta} = \frac{1+j}{\delta_w}. \tag{5.177}$$

In summary, the layer porosity factor $\eta = N_{tl}w/b_w = b_w/b_{wf}$ is modeled by a decrease in the resistivity of the multiple-turn layer $\rho_w = \eta\rho_{wf} = (w/p)\rho_{wf}$. The reduced resistivity causes a decrease in the skin depth of the multiturn layer $\delta_w = \sqrt{\eta}\delta_{wf}$, reducing the value of $A = \sqrt{\eta}h/\delta_w$. At high frequencies, the layer porosity factor η gives a good approximation only for high values of η when conductors are closely packed [8, 61].

5.18 Winding AC Resistance of Rectangular Conductor

For a *rectangular winding conductor* or *strip winding conductor* shown in Fig. 5.46, the variable A is given by

$$A = A_r = \frac{h}{\delta_{wf}} = h\sqrt{\frac{\pi\mu f}{\rho_w}\frac{w}{p}} = \frac{h}{\delta_w}\sqrt{\frac{w}{p}} = \frac{h}{\delta_w}\sqrt{\eta}, \tag{5.178}$$

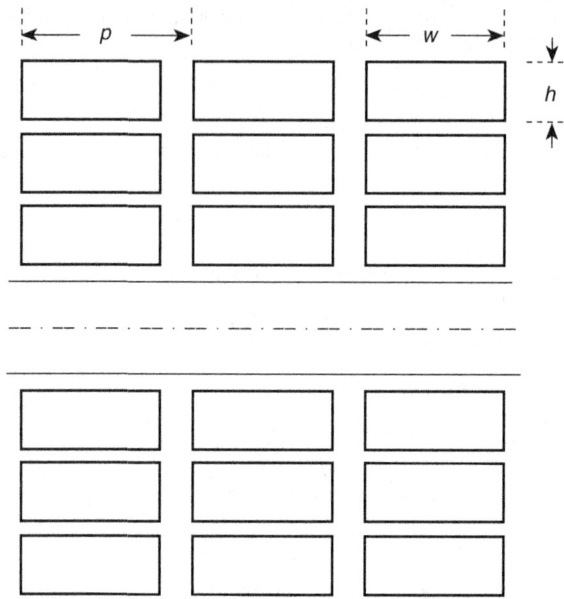

Figure 5.46 Cross section of an inductor with a rectangular wire winding

where h is the conductor height, w is the conductor width, p is the distance between the centers of two adjacent conductors in the same layer called the winding pitch. The minimum value of the distance is $p_{min} = d + 2t$, where t is the conductor insulation. The DC resistance of the rectangular winding conductor is

$$R_{wDC} = \frac{\rho_w l_w}{wh} = \frac{\rho_w l_T N}{wh}. \tag{5.179}$$

For the *rectangular winding conductor* of a fixed width w, the AC-to-DC winding resistance ratio can be written as

$$F_R = \frac{R_w}{R_{wDC}} = A\left[\frac{\sinh(2A)+\sin(2A)}{\cosh(2A)-\cos(2A)} + \frac{2(N_l^2-1)}{3}\frac{\sinh(A)-\sin(A)}{\cosh(A)+\cos(A)}\right]$$
$$= \left(\frac{h}{\delta_w}\right)\sqrt{\eta}\left[\frac{\sinh(2A)+\sin(2A)}{\cosh(2A)-\cos(2A)} + \frac{2(N_l^2-1)}{3}\frac{\sinh(A)-\sin(A)}{\cosh(A)+\cos(A)}\right]. \tag{5.180}$$

Figure 5.47 shows the plots of the normalized winding resistance R_w/R_{wDC} as a function of h/δ_w at $\eta = w/p = 0.8$ for the rectangular winding consisting of different number of layers N_l. Figure 5.48 shows the plots of R_w/R_{wDC} versus h/δ_w at fixed values of the porosity factor $\eta = 0.5$, 0.75, and 1. It can be seen that R_w/R_{wDC} increases as $\eta = w/p$ approaches 1.

The frequency range of the AC-to-DC winding resistance ratio can be divided into (i) low-frequency range, (ii) mid-frequency range, and (iii) high-frequency range.

Low-Frequency Range

In the *low-frequency range*, the current density distribution is nearly uniform and the winding AC resistance is equal to the winding DC resistance and is independent of the number of layers N_l. Thus,

$$R_w \approx R_{wDC} \quad \text{for} \quad 0 \le \frac{h}{\delta_w} \le \frac{h_{lfmax}}{\delta_w}. \tag{5.181}$$

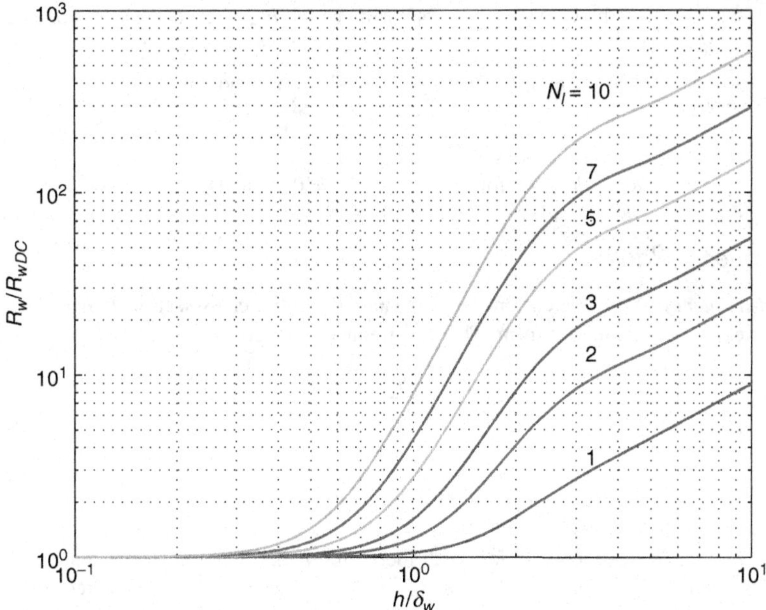

Figure 5.47 Normalized AC-to-DC winding resistance ratio $F_R = R_w/R_{wDC}$ versus h/δ_w at $\eta = w/p = 0.8$ and various values of the number of layers N_l for a rectangular conductor

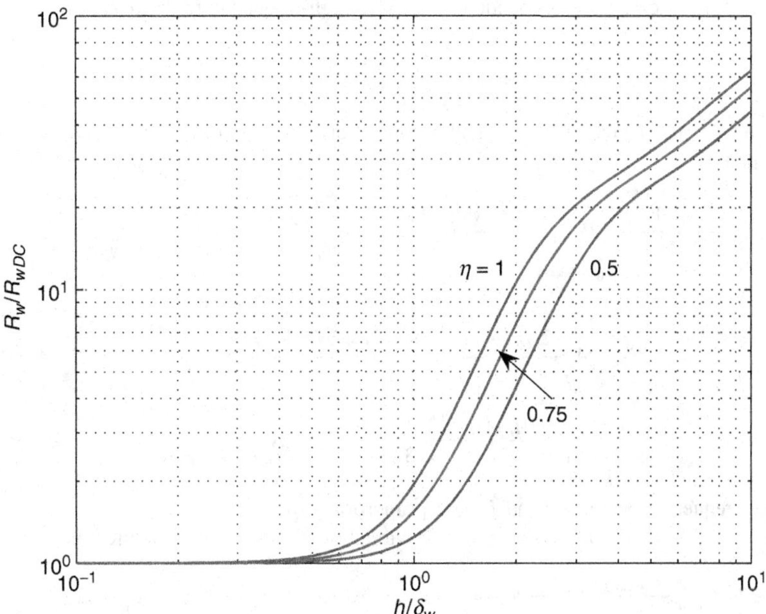

Figure 5.48 Normalized winding AC resistance $F_R = R_w/R_{wDC}$ versus h/δ_w at $\eta = w/p = 0.5$, 0.75, and 1 for $N_l = 3$ for a rectangular conductor

The boundary between the low-frequency range and the mid-frequency range depends on the number of layers N_l. As the number of layers N_l increases, h_{lfmax}/δ_w decreases. For example,

$$R_w \approx R_{wDC} \quad \text{for} \quad \frac{h}{\delta_w} \leq \frac{h_{lfmax}}{\delta_w} = 1 \quad \text{and} \quad N_l = 1 \tag{5.182}$$

and

$$R_w \approx R_{wDC} \quad \text{for} \quad \frac{h}{\delta_w} \leq \frac{h_{lfmax}}{\delta_w} = 0.5 \quad \text{and} \quad N_l = 10. \tag{5.183}$$

Mid-Frequency Range

In the *mid-frequency range*, the current flows through the entire cross-sectional area of the conductor, but the current density is not uniform. The mid-frequency range occurs for

$$\frac{h_{lfmax}}{\delta_w} < \frac{h}{\delta_w} < \frac{h_{hfmin}}{\delta_w} = 3. \tag{5.184}$$

Thus,

$$0.5 < \frac{h}{\delta_w} < 3 \quad \text{for} \quad N_l = 3. \tag{5.185}$$

The ratio $F_R = R_w/R_{wDC}$ increases fast with increasing h/δ_w. The higher the number of layers N_l, the higher is the rate of increase in F_R as h/δ_w increases. For $N_l = 1$, the mid-frequency range does not exist.

High-Frequency Range

In the *high-frequency range*, the current flows in the areas close to the surface, and therefore the winding AC resistance R_w is independent of h. The high-frequency range occurs for

$$\frac{h}{\delta_w} > \frac{h_{hfmin}}{\delta_w} = 3. \tag{5.186}$$

The winding AC resistance of a rectangular conductor winding in the high-frequency range can be approximated by

$$R_w \approx \frac{(2N_l^2+1)}{3}A_r R_{wDC} = \frac{(2N_l^2+1)l_w}{3}\sqrt{\frac{\pi\rho_w\mu_0 f}{wp}} \quad \text{for} \quad A_r = \frac{h}{\delta_w}\sqrt{\frac{w}{p}} > 3 \tag{5.187}$$

and

$$F_R \approx A_r\frac{(2N_l^2+1)}{3} = \frac{h}{\delta_w}\sqrt{\frac{w}{p}}\frac{(2N_l^2+1)}{3}$$
$$= h\sqrt{\frac{\pi\mu f}{\rho_w}}\sqrt{\frac{w}{p}}\frac{(2N_l^2+1)}{3} \quad \text{for} \quad A_r = \frac{h}{\delta_w}\sqrt{\frac{w}{p}} > 3. \tag{5.188}$$

In the high-frequency range, R_w and F_R are proportional to $\sqrt{f}$.

For the *rectangular winding conductor*, the winding resistance can be written as

$$R_w = F_R R_{wDC} = F_R\frac{\rho_w l_w}{wh} = F_R\frac{\rho_w l_w}{w\delta_w\left(\frac{h}{\delta_w}\right)}$$
$$= \frac{\rho_w l_w}{wh}\left(\frac{h}{\delta_w}\right)\left(\sqrt{\frac{w}{p}}\right)\left[\frac{\sinh(2A)+\sin(2A)}{\cosh(2A)-\cos(2A)} + \frac{2(N_l^2-1)}{3}\frac{\sinh(A)-\sin(A)}{\cosh(A)+\cos(A)}\right]$$
$$= \frac{\rho_w l_w}{w\delta_w\left(\frac{h}{\delta_w}\right)}\left(\frac{h}{\delta_w}\right)\left(\sqrt{\frac{w}{p}}\right)\left[\frac{\sinh(2A)+\sin(2A)}{\cosh(2A)-\cos(2A)} + \frac{2(N_l^2-1)}{3}\frac{\sinh(A)-\sin(A)}{\cosh(A)+\cos(A)}\right]$$

$$= R_{\delta_r}\sqrt{\eta}\left[\frac{\sinh(2A)+\sin(2A)}{\cosh(2A)-\cos(2A)} + \frac{2(N_l^2-1)}{3}\frac{\sinh(A)-\sin(A)}{\cosh(A)+\cos(A)}\right], \tag{5.189}$$

where the DC resistance of a strip conductor of thickness δ_w and length l_w is

$$R_{\delta_r} = \frac{\rho_w l_w}{w\delta_w}. \tag{5.190}$$

The normalized rectangular conductor winding AC resistance is

$$F_r = \frac{R_w}{(\rho_w l_w/w\delta_w)} = \frac{F_R}{\frac{h}{\delta_w}} = \frac{R_w}{R_{\delta_r}}$$

$$= \sqrt{\eta}\left[\frac{\sinh(2A)+\sin(2A)}{\cosh(2A)-\cos(2A)} + \frac{2(N_l^2-1)}{3}\frac{\sinh(A)-\sin(A)}{\cosh(A)+\cos(A)}\right], \tag{5.191}$$

where ρ_w, w, and l_w are constant for a given winding and δ_w is also constant for a given operating frequency f and ρ_w. Thus, R_w depends only on h at a constant value of δ_w, that is, a constant frequency f. Figure 5.49 shows a 3D representation of (5.191) for $\eta = w/p = 0.8$. The normalized AC resistance as a function of the normalized rectangular conductor thickness h/δ_w is illustrated in Fig. 5.50 for $\eta = w/p = 0.8$. As the ratio h/δ_w is increased at its low values, $F_R = 1$ and R_w decreases because the DC resistance R_{wDC} decreases when h increases.

The conclusions from Fig. 5.50 are as follows:

1. At low values of h/δ_w ($h/\delta_w < 0.3$ for $N_l \leq 10$), the current density is uniform, $F_R \approx 1$, $R_w \approx R_{wDC}$, and the winding resistance is independent of frequency and the number of winding layers N_l.
2. For $N_l = 1$, the lowest winding AC resistance R_w occurs at $h_{opt}/\delta_w = \pi/2 = 1.57$ as given in (5.67).
   As h/δ_w increases, the winding AC resistance first decreases due to decreasing DC resistance with increasing h, reaches a minimum value $R_{w\,min}$ at h_{opt}/δ_w, next increases, and then remains constant. Thus, the optimum value of the normalized rectangular conductor thickness h_{opt}/δ_w

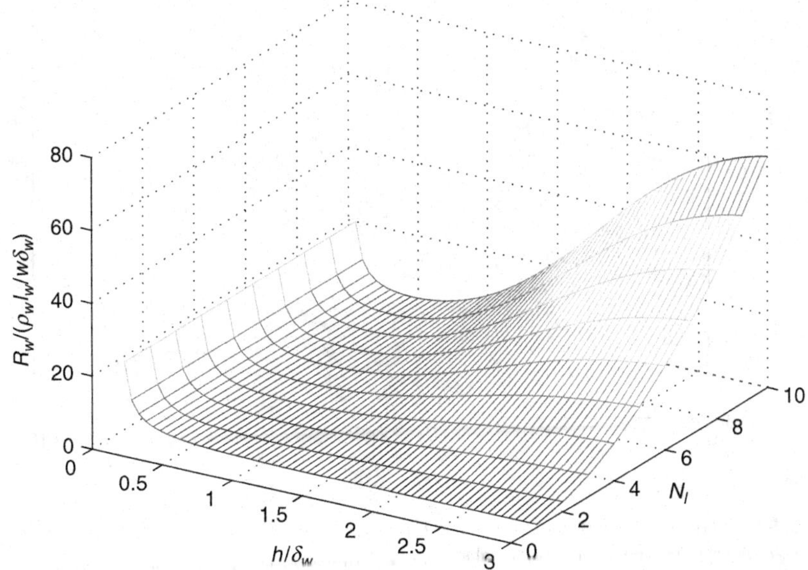

Figure 5.49 Normalized winding AC resistance $F_r = R_w/(\rho_w l_w/w\delta_w) = F_R/(h/\delta_w)$ versus h/δ_w and N_l at $\eta = w/p = 0.8$ for a rectangular winding conductor

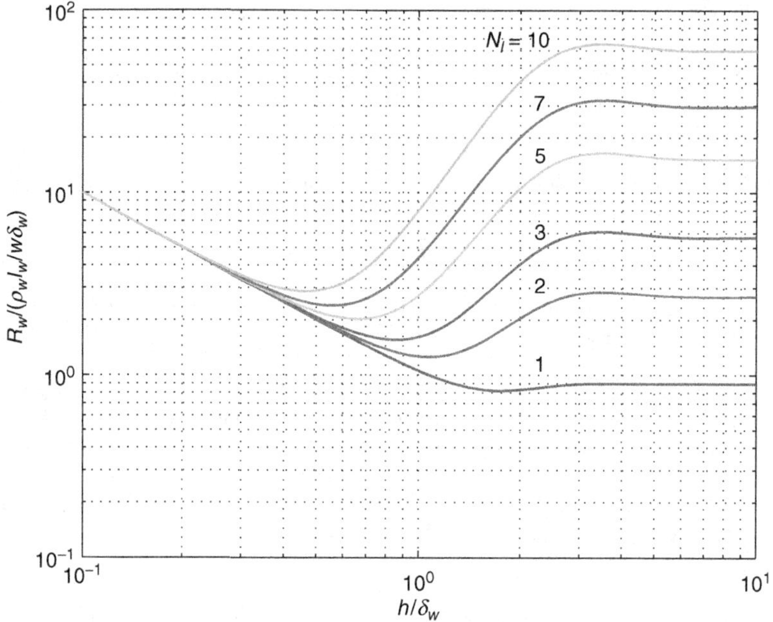

Figure 5.50 Normalized winding AC resistance $F_r = R_w/(\rho_w l_w/w\delta_w) = F_R/(h/\delta_w)$ versus h/δ_w at $\eta = w/p = 0.8$ and various values of the number of layers N_l for a rectangular winding conductor

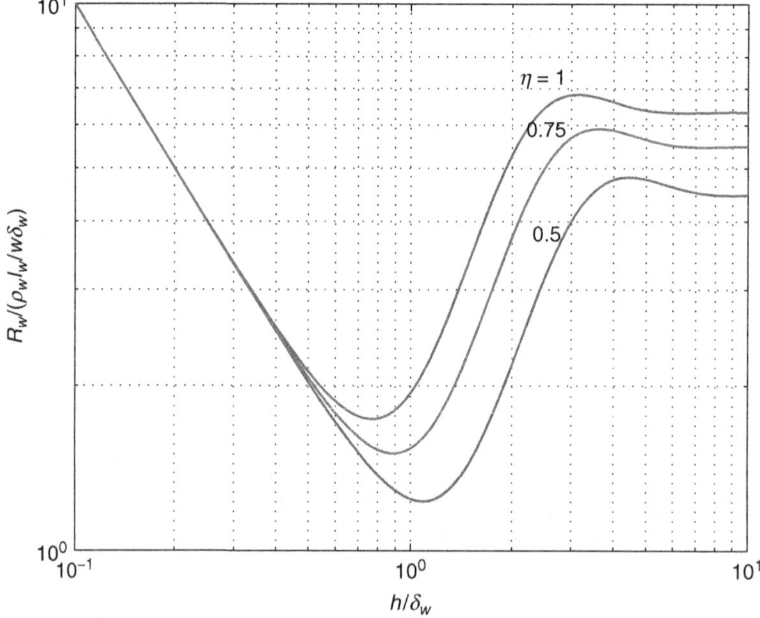

Figure 5.51 Normalized winding AC resistance $F_r = R_w/(\rho_w l_w/w\delta_w) = F_R/(h/\delta_w)$ versus h/δ_w at $N_l = 3$ and $\eta = w/p = 0.5, 0.75$, and 1 for a rectangular winding conductor

exists, resulting in the minimum winding AC resistance $R_{w\,min}$, and the minimum winding AC power loss $P_{w\,min}$. (i) As the ratio h/δ_w increases at its low values ($h/\delta_w < 0.3$ for $N_l \le 10$), the current density is uniform and the AC resistance $R_w \approx R_{wDC}$ decreases because the DC resistance $R_{wDC} = \rho_w l_w/(hw)$ decreases. (ii) As the ratio h/δ_w increases at its high values ($h/\delta_w > 3$), the current flows only at both surfaces of the foil and the AC resistance R_w is independent of thickness constant h for a given number of layer N_l. As the number of layers N_l increases, the AC resistance R_w increases because the proximity effect is stronger. (iii) As the ratio h/δ_w increases for its medium values ($h/\delta_w > h_{opt}/\delta_w$), the AC resistance increases because the current density becomes less and less uniform, but the current flows through the entire cross-sectional area of the winding conductor.

3. As N_l is increased, h_{opt}/δ_w decreases. As $\eta = w/p$ is decreased, h_{opt}/δ_w increases.

Figure 5.51 shows the normalized thickness h/δ_w of the rectangular winding at $N_l = 3$ and $\eta = w/p = 0.5$, 0.75, and 1. It can be seen that h_{opt}/δ_w decreases as $\eta = w/p$ increases and that the minimum winding resistance decreases as η decreases.

For high frequencies, (5.191) can be approximated by

$$F_r = \frac{R_w}{(\rho_w l_w/w\delta_w)} = \frac{F_R}{\frac{h}{\delta_w}} \approx \sqrt{\frac{w}{p}}\left[1 + \frac{2(N_l^2-1)}{3}\right] = \sqrt{\frac{w}{p}}\left(\frac{2N_l^2+1}{3}\right) \quad \text{for} \quad \frac{h}{\delta_w}\sqrt{\frac{w}{p}} \ge 3. \tag{5.192}$$

5.18.1 Optimum Thickness of Rectangular Conductor for Sinusoidal Inductor Current

The AC resistance of a rectangular conductor winding is

$$R_w = F_R R_{wDC} \approx R_{wDC}\left[1 + \frac{5N_l^2-1}{45}A^4\right] = \frac{\rho_w l_w}{wh}\left[1 + \frac{(5N_l^2-1)\eta^2 h^4}{45\delta_w^4}\right]$$
$$= \frac{\rho_w l_w}{w}\left[\frac{1}{h} + \frac{(5N_l^2-1)\eta^2 h^3}{45\delta_w^4}\right]. \tag{5.193}$$

Assume that the number of layers N_l and the porosity factor $\eta = w/p$ are constant. Setting the derivative of R_w with respect to the conductor thickness h to zero, we have

$$\frac{dR_w}{dh} = \frac{\rho_w l_w}{w}\left[\frac{-1}{h^2} + \frac{(5N_l^2-1)\eta^2 h^2}{15\delta_w^4}\right] = 0, \tag{5.194}$$

producing

$$\frac{(5N_l^2-1)\eta^2}{15}\left(\frac{h}{\delta_w}\right)^4 = 1. \tag{5.195}$$

Hence, the optimum rectangular winding conductor thickness for sinusoidal current is

$$\frac{h_{opt}}{\delta_w} = \sqrt[4]{\frac{15}{\eta^2(5N_l^2-1)}} = \frac{1.97}{\sqrt[4]{\eta^2(5N_l^2-1)}}. \tag{5.196}$$

Substituting (5.196) into (5.193) gives

$$F_{Rv} = \frac{R_{w\,min}}{R_{wDC}} = 1 + \frac{1}{3} = \frac{4}{3} = 1.333 \quad \text{for} \quad h = h_{opt}. \tag{5.197}$$

The AC-to-DC winding resistance ratio in terms of h_{opt} is

$$F_R = \frac{R_w}{R_{wDC}} = 1 + \frac{\eta^2(5N_l^2-1)}{45}\left(\frac{h_{opt}}{\delta_w}\right)^4\left(\frac{h}{h_{opt}}\right)^4 = 1 + \frac{1}{3}\left(\frac{h}{h_{opt}}\right)^4. \tag{5.198}$$

The minimum winding AC resistance is given by

$$R_{w\,\min} = \frac{4\rho_w l_w}{wh_{opt}} = \frac{4\rho_w l_w}{3w\delta_w}\sqrt[4]{\frac{\eta^2(5N_l^2-1)}{15}}. \tag{5.199}$$

The average current density is

$$J = \frac{I}{A_w} = \frac{I}{wh}. \tag{5.200}$$

Hence, the minimum foil width at the optimum foil thickness h_{opt} is

$$w_{min} = \frac{I_{max}}{J_{max}h_{opt}}. \tag{5.201}$$

Example 5.3

A sinusoidal current flows through an inductor at $f = 100$ kHz. The winding is made of a copper tape with $\eta = 0.85$. Determine the optimum tape thickness h_{opt}. (a) For $N_l = 2$. (b) For $N_l = 10$.

Solution: The skin depth of copper at $f = 100$ kHz for a sinusoidal current at $T = 20\,^\circ$C is

$$\delta_w = \sqrt{\frac{\rho_w}{\pi\mu_0 f}} = \sqrt{\frac{1.724\times10^{-8}}{\pi\times4\times\pi\times10^{-7}\times10^5}} = 0.2089 \text{ mm}. \tag{5.202}$$

(a) For $N_l = 2$,

$$\frac{h_{opt}}{\delta_w} = \sqrt[4]{\frac{15}{\eta^2(5N_l^2-1)}} = \sqrt[4]{\frac{15}{0.85^2(5\times2^2-1)}} = 1.0224. \tag{5.203}$$

Hence,

$$h_{opt} = 1.0224\delta_w = 1.0224\times0.2089 = 0.2136 \text{ mm}. \tag{5.204}$$

(b) For $N_l = 10$,

$$\frac{h_{opt}}{\delta_w} = \sqrt[4]{\frac{15}{\eta^2(5N_l^2-1)}} = \sqrt[4]{\frac{15}{0.85^2(5\times10^2-1)}} = 0.4516, \tag{5.205}$$

yielding

$$h_{opt} = 0.4516\delta_w = 0.4516\times0.2089 = 0.09434 \text{ mm}. \tag{5.206}$$

Example 5.4

An inductor has $L = 1.4$ mH, $\mu_{rc} = 2000$, a Siemens ferrite EC ungapped core BB66339 with MnZn ferrite material N27, $A_c = 121$ mm^2, $l_c = 89.3$ mm, $l_T = 6.66$ cm, $b_b = 2.5$ cm, width of the bare strip $w = 4.9$ mm, and insulation thickness $t = 50$ μm. The inductor current is $i_L = 5\sin 2\pi 10^5 t$ (A). Find:

(a) The optimum strip thickness.
(b) The winding AC resistance.
(c) The winding power loss.

(d) The average current density.
(e) The inductor quality factor related to the winding AC resistance.

Solution: (a) The number of turns per layer is

$$N_{tl} = \frac{b_b}{w + 2t} = \frac{25}{4.9 + 2 \times 0.05} = \frac{25}{5} = 5. \tag{5.207}$$

The number of layers is

$$N_l = \sqrt{\frac{l_c L}{\mu_{rc}\mu_0 A_c N_{tl}^2}} = \sqrt{\frac{89.3 \times 10^{-3} \times 1.4 \times 10^{-3}}{2000 \times 4\pi \times 10^{-7} \times 89.3 \times 10^{-3} \times 5^2}} = 4.05. \tag{5.208}$$

Select $N_l = 4$. The skin depth of copper at $f = 100$ kHz is

$$\delta_w = \sqrt{\frac{\rho_w}{\pi \mu_0 f}} = \sqrt{\frac{1.724 \times 10^{-8}}{\pi \times 4 \times \pi \times 10^{-7} \times 10^5}} = 0.2089 \text{ mm}. \tag{5.209}$$

The porosity factor is

$$\eta = \frac{w}{p} = \frac{w}{w + 2t} = \frac{4.9}{4.9 + 2 \times 0.05} = 0.98. \tag{5.210}$$

The normalized optimum strip thickness is

$$\frac{h_{opt}}{\delta_w} = \sqrt[4]{\frac{15}{\eta^2 (5N_l^2 - 1)}} = \sqrt[4]{\frac{15}{0.98^2 \times (5 \times 4^2 - 1)}} = 0.8775. \tag{5.211}$$

Hence, the optimum strip thickness is

$$h_{opt} = 0.8775\delta_w = 0.8775 \times 0.2089 = 0.18338 \text{ mm} = 183.38 \text{ µm}. \tag{5.212}$$

Select $h = 180$ µm.

(b) The number of turns is

$$N = N_l N_{tl} = 4 \times 5 = 20. \tag{5.213}$$

The total strip length is

$$l_w = l_T N = 6.66 \times 10^{-2} \times 20 = 1.332 \text{ m}. \tag{5.214}$$

The winding DC resistance is

$$R_{wDC} = \frac{\rho_{Cu} l_w}{wh} = \frac{1.724 \times 10^{-8} \times 1.332}{4.9 \times 10^{-3} \times 0.18 \times 10^{-3}} = 0.026 \ \Omega. \tag{5.215}$$

The winding AC resistance is

$$R_w = F_{Rv} R_{wDC} = \frac{4}{3} \times 0.026 = 0.0347 \ \Omega. \tag{5.216}$$

(c) The winding AC power loss is

$$P_w = \frac{R_w I_{Lm}^2}{2} = \frac{0.0347 \times 5^2}{2} = 0.43375 \text{ W}. \tag{5.217}$$

(d) The average amplitude of the current density in the winding is

$$J_{m(AV)} = \frac{I_{Lm}}{A_w} = \frac{I_{Lm}}{wh} = \frac{5}{4.9 \times 0.18} = 5.69 \text{ A/mm}^2. \tag{5.218}$$

(e) The inductor quality factor related to the winding resistance is

$$Q_{Lw} = \frac{\omega L}{R_w} = \frac{2\pi \times 100 \times 10^3 \times 1.4 \times 10^{-3}}{0.0347} = 25\,350. \tag{5.219}$$

5.18.2 Boundary Between Low and Medium Frequencies for Rectangular Wire Winding

Using the approximate Dowell's equation for low and medium frequencies, the boundary between the low and medium frequencies can be defined as

$$F_{RB} \approx 1 + \frac{\eta^2(5N_l^2 - 1)}{45}\left(\frac{h_B}{\delta_w}\right)^4 = 1.05, \tag{5.220}$$

yielding the rectangular conductor thickness at the boundary between the low and medium values of h/δ_w

$$\frac{h_B}{\delta_w} = \sqrt[4]{\frac{9}{4\eta^2(5N_l^2 - 1)}}. \tag{5.221}$$

The boundary frequency is

$$f_L = \frac{3\rho_w}{2\pi\eta\mu_0 h_B^2\sqrt{(5N_l^2 - 1)}}. \tag{5.222}$$

The ratio of the optimum strip thickness to the boundary strip thickness is

$$\frac{h_{opt}}{h_B} = \sqrt[4]{\frac{20}{3}} \approx 1.61. \tag{5.223}$$

The ratio of the boundary winding AC resistance R_{wB} to the the minimum winding AC resistance $R_{w\,\min}$ for strip inductor windings is

$$\frac{R_{wB}}{R_{w\,\min}} = \frac{R_{wDCB}F_{RB}}{R_{wDCopt}F_{Rv}} = \frac{h_{opt}}{h_B}\frac{F_{RB}}{F_{Rv}} = \frac{1.05 \times 3}{4}\sqrt[4]{\frac{20}{3}} \approx 1.26539. \tag{5.224}$$

5.19 Winding Resistance of Square Wire

For a *square wire winding* shown in Fig. 5.52, $w = h$. Hence, (5.178) becomes

$$A = A_s = \frac{h}{\delta_w}\sqrt{\frac{h}{p}} = \frac{h}{\delta_w}\sqrt{\frac{h}{p}} = \frac{h}{\delta_w}\sqrt{\eta}, \tag{5.225}$$

where p is the distance between the centers of the adjacent square conductors in the same layer and $\eta = h/p$ is the porosity factor. The DC resistance of the square wire winding is given by

$$R_{wDC} = \frac{\rho_w l_w}{h^2} = \frac{\rho_w l_T N}{h^2}. \tag{5.226}$$

The AC-to-DC winding resistance ratio for the square winding conductor is

$$F_R = \frac{R_w}{R_{wDC}} = A\left[\frac{\sinh(2A) + \sin(2A)}{\cosh(2A) - \cos(2A)} + \frac{2(N_l^2 - 1)}{3}\frac{\sinh(A) - \sin(A)}{\cosh(A) + \cos(A)}\right]$$
$$= \left(\frac{h}{\delta_w}\right)\sqrt{\eta}\left[\frac{\sinh(2A) + \sin(2A)}{\cosh(2A) - \cos(2A)} + \frac{2(N_l^2 - 1)}{3}\frac{\sinh(A) - \sin(A)}{\cosh(A) + \cos(A)}\right]. \tag{5.227}$$

Figure 5.53 shows the plots of the normalized winding resistance R_w/R_{wDC} as a function of h/δ_w at $h/p = 0.8$ and various numbers of layers N_l for a square wire. The AC-to-DC square wire winding resistance ratio $F_R = R_w/R_{wDC}$ is constant for low values of h/δ_w and then increases as h/δ_w increases.

At low and medium frequencies,

$$R_w \approx \frac{\rho_w l_w}{h^2}\left[1 + \frac{\eta^2(5N_l^2 - 1)}{45}\left(\frac{h}{\delta_w}\right)^4\right]. \tag{5.228}$$

Figure 5.52 Cross section of an inductor with a square wire winding

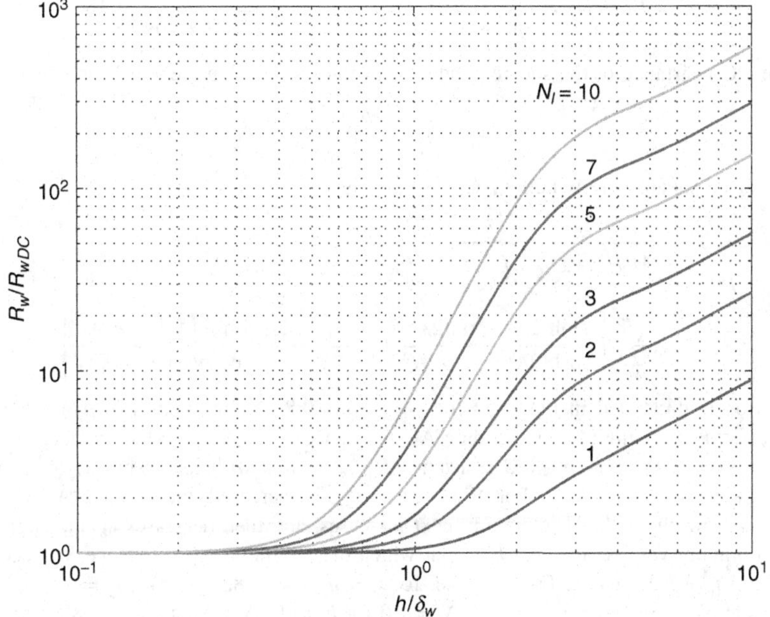

Figure 5.53 Normalized winding AC resistance R_w/R_{wDC} versus h/δ_w at $\eta = h/p = 0.8$ and various numbers of winding layers N_l for a square wire

The winding AC-to-DC resistance ratio is

$$F_R \approx 1 + \frac{\eta^2(5N_l^2 - 1)}{45}\left(\frac{h}{\delta_w}\right)^4. \tag{5.229}$$

At high frequencies, the winding AC resistance can be approximated by

$$R_w \approx \frac{(2N_l^2+1)l_w}{3}\sqrt{\frac{\pi\rho_w\mu_0 f}{ph}} \quad \text{for} \quad A_s > 3 \tag{5.230}$$

and the winding AC-to-DC resistance ratio is

$$F_R \approx A_s\frac{(2N_l^2+1)}{3} = \frac{h}{\delta_w}\sqrt{\frac{h}{p}}\frac{(2N_l^2+1)}{3} \quad \text{for} \quad A_s > 3. \tag{5.231}$$

5.19.1 Winding AC Resistance of Square Conductor

For a square wire winding,

$$\begin{aligned}
R_w &= F_R R_{wDC} = F_R\frac{\rho_w l_w}{h^2} = F_R\frac{\rho_w l_w}{\delta_w^2\left(\frac{h}{\delta_w}\right)^2} \\
&= \frac{\rho_w l_w}{h^2}\left(\frac{h}{\delta_w}\right)\sqrt{\eta}\left[\frac{\sinh(2A)+\sin(2A)}{\cosh(2A)-\cos(2A)} + \frac{2(N_l^2-1)}{3}\frac{\sinh(A)-\sin(A)}{\cosh(A)+\cos(A)}\right] \\
&= \frac{\rho_w l_w}{\delta_w^2\left(\frac{h}{\delta_w}\right)^2}\left(\frac{h}{\delta_w}\right)\sqrt{\eta}\left[\frac{\sinh(2A)+\sin(2A)}{\cosh(2A)-\cos(2A)} + \frac{2(N_l^2-1)}{3}\frac{\sinh(A)-\sin(A)}{\cosh(A)+\cos(A)}\right] \\
&= R_{\delta_s}\frac{\sqrt{\eta}}{\frac{h}{\delta_w}}\left[\frac{\sinh(2A)+\sin(2A)}{\cosh(2A)-\cos(2A)} + \frac{2(N_l^2-1)}{3}\frac{\sinh(A)-\sin(A)}{\cosh(A)+\cos(A)}\right],
\end{aligned} \tag{5.232}$$

where the DC resistance of the square conductor of thickness δ_w and length l_w is

$$R_{\delta_s} = \frac{\rho_w l_w}{\delta_w^2}. \tag{5.233}$$

The normalized square wire winding AC resistance is

$$\begin{aligned}
F_r &= \frac{R_w}{(\rho_w l_w/\delta_w^2)} = \frac{F_R}{\left(\frac{h}{\delta_w}\right)^2} = \frac{R_w}{R_{\delta_s}} \\
&= \frac{\sqrt{\eta}}{\frac{h}{\delta_w}}\left[\frac{\sinh(2A)+\sin(2A)}{\cosh(2A)-\cos(2A)} + \frac{2(N_l^2-1)}{3}\frac{\sinh(A)-\sin(A)}{\cosh(A)+\cos(A)}\right],
\end{aligned} \tag{5.234}$$

This equation is illustrated in Fig. 5.54 for $\eta = h/p = 0.8$. At low values of h/δ_w, the proximity and skin effects are negligible, the winding AC resistance R_w is equal to the winding DC resistance R_{wDC}. For $N_l = 1$, the winding AC resistance R_w decreases with increasing h/δ_w. However, the decrease is slower for $h/\delta_w > 2$. For $N_l > 1$, as h/δ_w is increased, the winding AC resistance R_w first decreases, reaches a local minimum, next increases, and then decreases again. Therefore, there is a local optimum value (valley) of h_v/δ_w at which the minimum AC resistance R_{wv} occurs. As N_l is increased, h_v/δ_w decreases. For example, for $\eta = h_v/p = 0.8$, we have $h_v = 1.06\delta_w$ at $N_l = 2$, $h_v = 0.8\delta_w$ at $N_l = 5$, and $h_v = 0.7\delta_w$ at $N_l = 10$ for $h_v = 0.6\delta_w$. As $\eta = h_v/p$ is decreased, h_v/δ_w increases. For instance, for $N_l = 10$, $h_v = 0.6\delta_w$ at $\eta = h_v/p = 0.9$ and $h_v = 0.8\delta_w$ at $\eta = h_v/p = 0.5$. The normalized thickness of the square conductor is

$$\frac{h_v}{\delta_w} \approx 0.5 \quad \text{to} \quad 1.06 \quad \text{for} \quad 2 \le N_l \le 10. \tag{5.235}$$

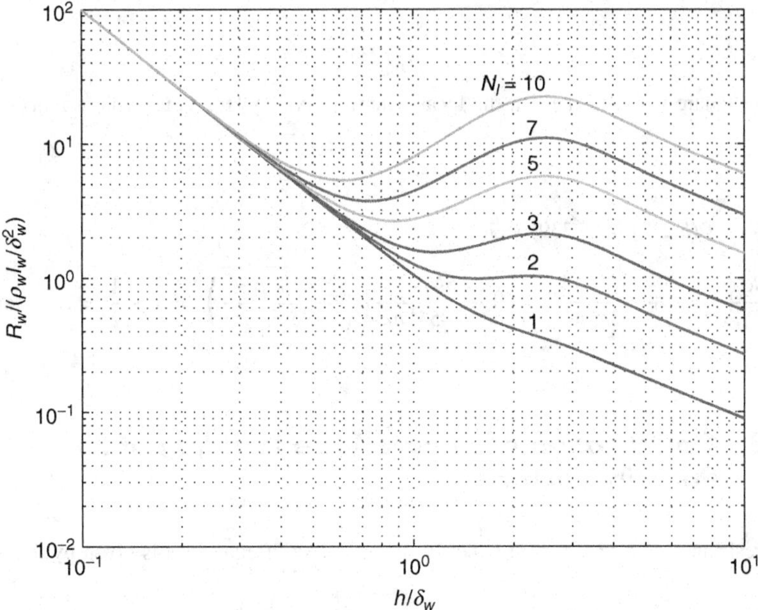

Figure 5.54 Normalized winding AC resistance $F_r = R_w/(\rho_w l_w/\delta_w^2) = F_R/(h/\delta_w)^2$ versus h/δ_w at $\eta = h/p = 0.8$ and different number of layers N_l for a square wire

Figure 5.54 shows that

$$F_{Rv} = \frac{R_{wv}}{R_{wDC}} \approx 1 \quad \text{to} \quad 1.6 \quad \text{for} \quad 2 \le N_l \le 10. \tag{5.236}$$

For high frequencies, (5.234) can be expressed as

$$F_r = \frac{R_w}{(\rho_w l_w/\delta_w^2)} = \frac{F_R}{\left(\frac{h}{\delta_w}\right)^2} \approx \frac{\sqrt{\frac{h}{p}}}{\frac{h}{\delta_w}}\left[1 + \frac{2(N_l^2-1)}{3}\right] = \frac{\sqrt{\frac{h}{p}}}{\frac{h}{\delta_w}}\left(\frac{2N_l^2+1}{3}\right) \quad \text{for} \quad \frac{h}{\delta_w}\sqrt{\frac{h}{p}} \ge 3. \tag{5.237}$$

5.19.2 Optimization of Square Wire Winding at Fixed Pitch

The approximate Dowell's equation for the AC-to-DC resistance ratio of a square wire winding at low and midfrequencies is

$$F_R \approx 1 + \frac{5N_l^2-1}{45}A_s^4 = 1 + \frac{5N_l^2-1}{45\delta_w^4 p^2}h^6. \tag{5.238}$$

The approximate equation for the AC resistance of a square wire winding at low and midfrequencies is given by

$$R_w = F_R R_{wDC} \approx \frac{\rho_w l_w}{h^2}\left[1 + \frac{5N_l^2-1}{45\delta_w^4 p^2}h^6\right] = \rho_w l_w\left[\frac{1}{h^2} + \frac{(5N_l^2-1)h^4}{45\delta_w^4 p^2}\right]. \tag{5.239}$$

Assume that the pitch p and the number of layers N_l are constant. The derivative of the AC resistance with respect to h is

$$\frac{dR_w}{dh} = \rho_w l_w\left[-\frac{2}{h^3} + \frac{4(5N_l^2-1)h^3}{45\delta_w^4 p^2}\right] = 0, \tag{5.240}$$

resulting in

$$\frac{4(5N_l^2 - 1)h^6}{45\delta_w^4 p^2} = 2. \tag{5.241}$$

Hence, the square wire thickness at the local minimum (valley) for sinusoidal current is

$$h_v = \sqrt[6]{\frac{45\delta_w^4 p^2}{2(5N_l^2 - 1)}}. \tag{5.242}$$

Substitution of (5.242) in (5.239) yields

$$F_{Rv} = \frac{R_{wv}}{R_{wDC}} = 1 + \frac{5N_l^2 - 1}{45\delta_w^4 p^2}h^6 = 1 + \frac{5N_l^2 - 1}{45\delta_w^4 p^2}\left(\sqrt[6]{\frac{45\delta_w^4 p^2}{2(5N_l^2 - 1)}}\right)^6 = 1 + \frac{1}{2} = 1.5 \quad \text{for} \quad h = h_v. \tag{5.243}$$

5.19.3 Optimization of Square Wire Winding at Fixed Porosity Factor

The AC resistance of a square wire winding at low and midfrequencies is given by the approximate equation

$$R_w = F_R R_{wDC} \approx \frac{\rho_w l_w}{h^2}\left[1 + \frac{5N_l^2 - 1}{45}\eta^2\left(\frac{h}{\delta_w}\right)^4\right] = \rho_w l_w \left[\frac{1}{h^2} + \frac{(5N_l^2 - 1)\eta^2 h^2}{45\delta_w^4}\right]. \tag{5.244}$$

Let us assume that the number of layers N_l and the porosity factor η are constant. Setting the derivative of winding resistance R_w with respect to the conductor thickness h to zero, we get

$$\frac{dR_w}{dh} = \rho_w l_w \left[-\frac{2}{h^3} + \frac{2(5N_l^2 - 1)\eta^2 h}{45\delta_w^4}\right] = 0, \tag{5.245}$$

resulting in

$$\frac{2(5N_l^2 - 1)\eta^2}{45}\left(\frac{h}{\delta_w}\right)^4 = 2. \tag{5.246}$$

Hence, the square wire thickness at the local minimum (valley) for sinusoidal current is

$$\frac{h_v}{\delta_w} = \sqrt[4]{\frac{45}{\eta^2(5N_l^2 - 1)}} = \frac{2.59}{\sqrt[4]{\eta^2(5N_l^2 - 1)}}. \tag{5.247}$$

Substitution of (5.247) in (5.244) yields

$$F_{Rv} = \frac{R_{wv}}{R_{wDC}} = 1 + 1 = 2 \quad \text{for} \quad h = h_v. \tag{5.248}$$

The local minimum winding resistance occurs at h_v and is given by

$$R_{wv} = F_{Rv} R_{wDC} = \frac{2\rho_w l_w}{h_v^2} = \frac{2\rho_w l_w}{\delta_w^2 \sqrt{\frac{45}{\eta^2(5N_l^2-1)}}}. \tag{5.249}$$

The AC-to-DC winding resistance ratio in terms of h_v is

$$F_R = \frac{R_w}{R_{wDC}} = 1 + \frac{\eta^2(5N_l^2 - 1)}{45}\left(\frac{h_v}{\delta_w}\right)^4\left(\frac{h}{h_v}\right)^4 = 1 + \left(\frac{h}{h_v}\right)^4. \tag{5.250}$$

The maximum average current density is

$$J_{Max(AV)} = \frac{I_{max}}{A_w} = \frac{I_{max}}{h_{opt}^2}. \tag{5.251}$$

Example 5.5

A sinusoidal current flows through an inductor at $f = 100$ kHz. The winding is made of a copper square wire with $\eta = 0.9$. Determine the optimum conductor thickness h_{opt}. (a) For $N_l = 3$. (b) For $N_l = 7$.

Solution: The skin depth of copper at $f = 100$ kHz for a sinusoidal current at $T = 20\,^{\circ}$C is

$$\delta_w = \sqrt{\frac{\rho_w}{\pi \mu_0 f}} = \sqrt{\frac{1.724 \times 10^{-8}}{\pi \times 4 \times \pi \times 10^{-7} \times 10^5}} = 0.2089 \text{ mm}. \tag{5.252}$$

(a) For $N_l = 1$,

$$\frac{h_v}{\delta_w} = \sqrt[4]{\frac{45}{\eta^2(5N_l^2 - 1)}} = \sqrt[4]{\frac{45}{0.9^2(5 \times 3^2 - 1)}} = 0.8054, \tag{5.253}$$

resulting in

$$h_{opt} = 0.8054\delta_w = 0.8054 \times 0.2089 = 0.1683 \text{ mm}. \tag{5.254}$$

(b) For $N_l = 7$,

$$\frac{h_v}{\delta_w} = \sqrt[4]{\frac{45}{\eta^2(5N_l^2 - 1)}} = \sqrt[4]{\frac{45}{0.9^2(5 \times 7^2 - 1)}} = 0.5249, \tag{5.255}$$

producing

$$h_{opt} = 0.5249\delta_w = 0.5249 \times 0.2089 = 0.1096 \text{ mm}. \tag{5.256}$$

Example 5.6

Design an inductor with an optimum square wire winding thickness. The inductance is $L = 5$ mH. The inductor consists of four layers. The winding is made of copper magnet wire and the porosity factor is $\eta = 0.95$. The relative permeability of the core material is $\mu_{rc} = 50$, the mean turn ratio is $l_T = 5.3$ cm, the core length is $l_c = 3.76$ cm, and the core cross-sectional area is $A_c = 93.9$ mm^2. The inductor conducts a current $i_L = 0.1 \sin(2\pi 10^5 t)$ (A). Find:

(a) The optimum wire thickness.
(b) The winding AC resistance.
(c) The winding AC power loss.
(d) The average current density.
(e) The quality factor related to the winding resistance.

Solution: (a) The number of turns is

$$N = \sqrt{\frac{L l_c}{\mu_{rc}\mu_0 A_c}} = \sqrt{\frac{5 \times 10^{-3} \times 37.6 \times 10^{-3}}{50 \times 4\pi \times 10^{-3} \times 93.9 \times 10^{-6}}} = 179. \tag{5.257}$$

The skin depth of copper at $f = 100$ kHz is

$$\delta_w = \sqrt{\frac{\rho_w}{\pi \mu_0 f}} = \sqrt{\frac{1.724 \times 10^{-8}}{\pi \times 4\pi \times 10^{-7} \times 100 \times 10^3}} = 0.20897 \text{ mm}. \tag{5.258}$$

The normalized square wire thickness at the local optimum is

$$\frac{h_v}{\delta_w} = \sqrt[4]{\frac{45}{\eta^2(5N_l^2 - 1)}} = \sqrt[4]{\frac{45}{0.95^2 \times (5 \times 4^2 - 1)}} = 0.8913. \tag{5.259}$$

Hence, the square wire thickness at the local optimum is

$$h_v = 0.8913\delta_w = 0.8913 \times 0.20897 = 0.18626 \text{ mm} = 186.26 \text{ µm}. \quad (5.260)$$

Pick a wire with a standard thickness of bare conductor $h = 0.2019$ mm.

(b) The winding wire length is

$$l_w = N l_T = 179 \times 5.3 \times 10^{-2} = 9.487 \text{ m}. \quad (5.261)$$

Pick $l_w = 9.5$ cm. The winding DC resistance is

$$R_{wDC} = \frac{\rho_{Cu} l_w}{h_v^2} = \frac{1.724 \times 10^{-8} \times 9.5}{(0.2019 \times 10^{-3})^2} = 4.7208 \quad \text{m}\Omega. \quad (5.262)$$

The winding AC resistance is

$$R_w = 2R_{wDC} = 2 \times 4.7208 = 9.4416 \ \Omega. \quad (5.263)$$

(c) The winding AC power loss is

$$P_w = \frac{R_w I_{Lm}^2}{2} = \frac{9.4416 \times 0.1^2}{2} = 47.2 \text{ mW}. \quad (5.264)$$

(d) The average amplitude of the current density is

$$J_{m(AV)} = \frac{I_{Lm}}{h_v^2} = \frac{0.1}{(0.2019 \times 10^{-3})^2} = 2.8824 \text{ A/mm}^2. \quad (5.265)$$

(e) The quality factor of the inductor due to the winding resistance is

$$Q_{Lw} = \frac{\omega L}{R_w} = \frac{2\pi \times 10^5 \times 5 \times 10^{-3}}{9.4416} = 332. \quad (5.266)$$

5.19.4 Critical Thickness of Square Winding Resistance

The square wire winding has a local optimum thickness h_v, resulting a local minimum winding resistance R_{wv}. The winding resistance increases for $h > h_v$, reaches a maximum value, and then decreases with increasing h. At a critical conductor thickness h_{cr}, the winding resistance is equal to the local minimum winding resistance R_{wv}. For $h > h_{cr}$, the winding resistance is lower than the local minimum winding resistance R_{wv}.

From (5.123), the approximate high-frequency AC resistance for a square wire winding is given by

$$R_w \approx \frac{2N_l^2 + 1}{3} A_s R_{wDC} = \frac{(2N_l^2 + 1)\sqrt{\eta}\rho_w l_w}{3\delta_w h}. \quad (5.267)$$

Equating two resistances given by (5.249) and (5.267), we obtain

$$\frac{(2N_l^2 + 1)\sqrt{\eta}\rho_w l_w}{3\delta_w h} = \frac{2\rho_w l_w}{\delta_w^2 \sqrt{\frac{45}{\eta^2 (5N_l^2 - 1)}}}. \quad (5.268)$$

Hence, the normalized critical thickness of the square wire at which $R_w = R_{wv}$ is

$$\frac{h_{cr}}{\delta_w} = \frac{(2N_l^2 + 1)}{6} \sqrt{\frac{45}{\eta(5N_l^2 - 2)}}. \quad (5.269)$$

Example 5.7

An inductor made of a copper square wire carries a sinusoidal current at $f = 100$ kHz and $T = 20$ °C. The inductor has $N_l = 3$, $b_b = 12$ mm, and $l_T = 2$ cm. Find:

(a) The local optimum wire thickness.

(b) The winding AC resistance.

Solution: (a) The skin depth of copper at $f = 100$ kHz is

$$\delta_w = \sqrt{\frac{\rho_w}{\pi \mu_0 f}} = \sqrt{\frac{1.724 \times 10^{-8}}{\pi \times 4 \times \pi \times 10^{-7} \times 100 \times 10^3}} = 0.209 \text{ mm}. \tag{5.270}$$

Assume the porosity factor $\eta = 0.8$. The normalized optimum square wire thickness is

$$\frac{h_v}{\delta_w} = \sqrt[4]{\frac{45}{\eta^2(5N_l^2 - 1)}} = \sqrt[4]{\frac{45}{0.8^2 \times (5 \times 3^2 - 1)}} = 1.2626. \tag{5.271}$$

Hence, the optimum square wire thickness is

$$h_v = 1.2626\delta_w = 1.2626 \times 0.209 = 0.2639 \text{ mm} = 263.9 \text{ μm}. \tag{5.272}$$

Pick $h = 0.27$ mm. The insulation thickness is $t = 0.03$ mm. The porosity factor is

$$\eta = \frac{h}{h + 2t} = \frac{0.27}{0.27 + 2 \times 0.03} = 0.8182. \tag{5.273}$$

The number of turns per layer is

$$N_{tl} = \frac{b_b}{h + 2t} = \frac{10}{0.27 + 2 \times 0.03} = 30. \tag{5.274}$$

(b) The number of turns is

$$N = N_{tl}N_l = 30 \times 3 = 90. \tag{5.275}$$

The required height of the bobbin window is

$$h_b = N_l(h + 2t) = 3 \times (0.27 + 2 \times 0.03) = 0.99 \text{ mm}. \tag{5.276}$$

The length of the square wire is

$$l_w = N l_T = 90 \times 0.02 = 1.8 \text{ m}. \tag{5.277}$$

The winding DC resistance is

$$R_{wDC} = \frac{\rho_{Cu} l_w}{h^2} = \frac{1.724 \times 10^{-8} \times 1.8}{(0.27 \times 10^{-3})^2} = 0.42567 \ \Omega. \tag{5.278}$$

The winding AC resistance is

$$R_w = F_{Rv} R_{wDC} = 2 \times 0.42567 = 0.85135 \ \Omega. \tag{5.279}$$

5.19.5 Boundary Between Low and Medium Frequencies for Square Wire Winding

Using the approximate Dowell's equation for low and mid-frequencies, the boundary is

$$F_{RB} \approx 1 + \frac{\eta^2(5N_l^2 - 1)}{45}\left(\frac{h_B}{\delta_w}\right)^4 = 1.05, \tag{5.280}$$

yielding the square conductor thickness between the low and medium values of h/δ_w

$$\frac{h_B}{\delta_w} = \sqrt[4]{\frac{9}{4\eta^2(5N_l^2 - 1)}}. \tag{5.281}$$

The boundary frequency is

$$f_L = \frac{3\rho_w}{2\pi \mu_0 \eta h_B^2 \sqrt{(5N_l^2 - 1)}}. \tag{5.282}$$

The ratio of the local optimum square wire thickness to the boundary square wire thickness is

$$\frac{h_v}{h_B} = \sqrt[4]{20} \approx 2.115. \quad (5.283)$$

The ratio of the boundary winding AC resistance R_{wB} to the the minimum winding AC resistance for square wire inductors is

$$\frac{R_{wB}}{R_{wv}} = \frac{R_{wDCB}F_{RB}}{R_{wDCv}F_{Rv}} = \left(\frac{h_v}{h_B}\right)^2 \frac{F_{RB}}{F_{Rv}} = \frac{1.05}{2}\sqrt{20} \approx 2.34787. \quad (5.284)$$

5.20 Winding Resistance of Round Wire

To find the exact current distribution in a round conductor, it would inevitably lead to a solution involving Bessel functions [50]. In a wire of circular cross section, the radial distribution of the current density is a Bessel function of argument proportional to the square root of frequency. At DC, the current density is uniform, whereas at high frequencies the current is concentrated close to the surface. Another approach for finding the winding AC resistance is a 2D or 3D numerical simulation using a FEA. Dowell derived a 1D analytical equation (5.109) for the AC resistance of foil conductor windings. This solution was transformed to square wire windings. To adopt his equation for a round wire winding shown in Fig. 5.55, a round conductor can be approximated by a square conductor of the same cross-sectional area [5]. This approach guarantees that the DC resistance of the round and square conductors remain unchanged. Figure 5.56 shows cross section of round and square wires. The cross-sectional area for the square wire is

$$S_s = h^2 \quad (5.285)$$

and the cross-sectional area of the round wire is

$$S_o = \pi\left(\frac{d}{2}\right)^2. \quad (5.286)$$

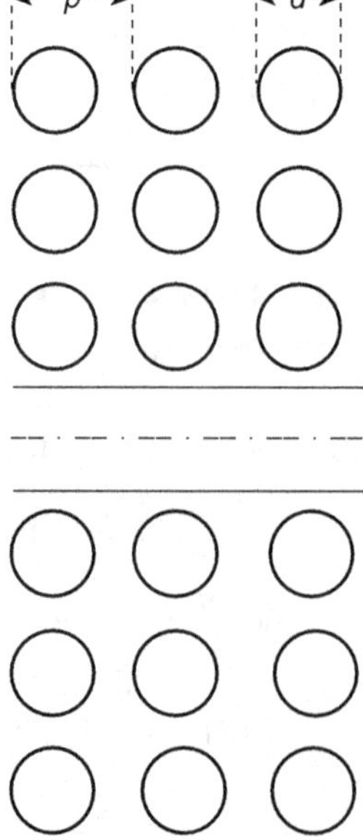

Figure 5.55 Cross section of an inductor with a round wire winding

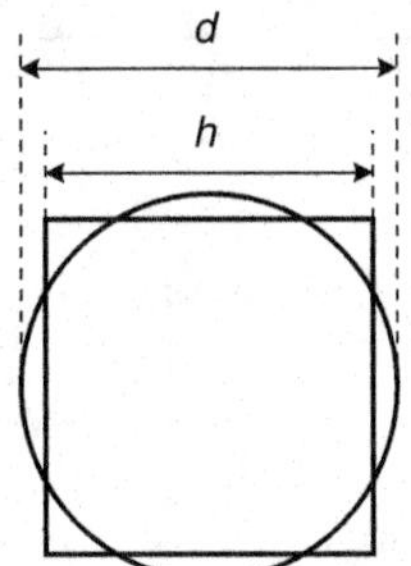

Figure 5.56 Cross sections of round and square wires

Hence, equating the areas S_s and S_o, one obtains

$$h^2 = \pi \left(\frac{d}{2}\right)^2, \quad (5.287)$$

yielding the relationship between h and the effective round wire diameter

$$d = h\sqrt{\frac{4}{\pi}} = \frac{2h}{\sqrt{\pi}} \approx 1.1284h \quad (5.288)$$

from which the effective square conductor height is

$$h = d\sqrt{\frac{\pi}{4}} = \frac{d\sqrt{\pi}}{2} \approx 0.8862d. \quad (5.289)$$

Note that if the cross-sectional areas of the square and round wires are the same, the DC resistances of both windings are the same. Substitution of (5.289) into (5.225) gives

$$A = A_{o1} = \frac{h}{\delta_w}\sqrt{\frac{h}{p}} = \frac{d}{\delta_w}\sqrt{\frac{\pi}{4}}\sqrt{\frac{d}{p}\sqrt{\frac{\pi}{4}}} = \left(\frac{\pi}{4}\right)^{\frac{3}{4}}\frac{d}{\delta_w}\sqrt{\frac{d}{p}} = \left(\frac{\pi}{4}\right)^{\frac{3}{4}}\frac{d}{\delta_w}\sqrt{\eta} \approx 0.8343\frac{d}{\delta_w}\sqrt{\frac{d}{p}}, \quad (5.290)$$

where p is the distance between the centers of the adjacent winding round conductors in the same layer and $\eta = d/p$ is the porosity factor. The maximum value of the porosity factor is $\eta_{max} = d/d_o$. Another possible definition of A is

$$A = A_{o2} = \frac{h}{\delta_w}\sqrt{\frac{d}{p}} = \left(\frac{h}{\delta_w}\right)\left(\frac{\sqrt{\pi}}{2}\right)\sqrt{\frac{d}{p}} \approx 0.8862\frac{d}{\delta_w}\sqrt{\frac{d}{p}}. \quad (5.291)$$

The DC resistance of a round wire winding is given by

$$R_{wDC} = \frac{4\rho_w l_w}{\pi d^2} = \frac{4\rho_w l_T N}{\pi d^2}. \quad (5.292)$$

The AC-to-DC winding resistance ratio for the round wire winding conductor is

$$F_R = \frac{R_w}{R_{wDC}} = A\left[\frac{\sinh(2A)+\sin(2A)}{\cosh(2A)-\cos(2A)} + \frac{2(N_l^2-1)}{3}\frac{\sinh(A)-\sin(A)}{\cosh(A)+\cos(A)}\right]$$
$$= \left(\frac{\pi}{4}\right)^{\frac{3}{4}}\left(\frac{d}{\delta_w}\right)\sqrt{\eta}\left[\frac{\sinh(2A)+\sin(2A)}{\cosh(2A)-\cos(2A)} + \frac{2(N_l^2-1)}{3}\frac{\sinh(A)-\sin(A)}{\cosh(A)+\cos(A)}\right]. \quad (5.293)$$

Figure 5.57 shows plots of the normalized winding resistance R_w/R_{wDC} as a function of d/δ_w at $\eta = d/p = 0.8$ and different numbers of winding layers N_l for a round wire. It is obvious that large diameter wires have a large AC-to-DC resistance ratio F_R, which increases greatly with frequency. At high frequencies, the winding AC resistance can be approximated by

$$R_w \approx \frac{4l_w(2N_l^2+1)}{3\pi}\left(\frac{\pi}{4}\right)^{\frac{3}{4}}\sqrt{\frac{\pi\rho_w\mu_0 f}{pd}} \quad \text{for} \quad A_o > 3 \quad (5.294)$$

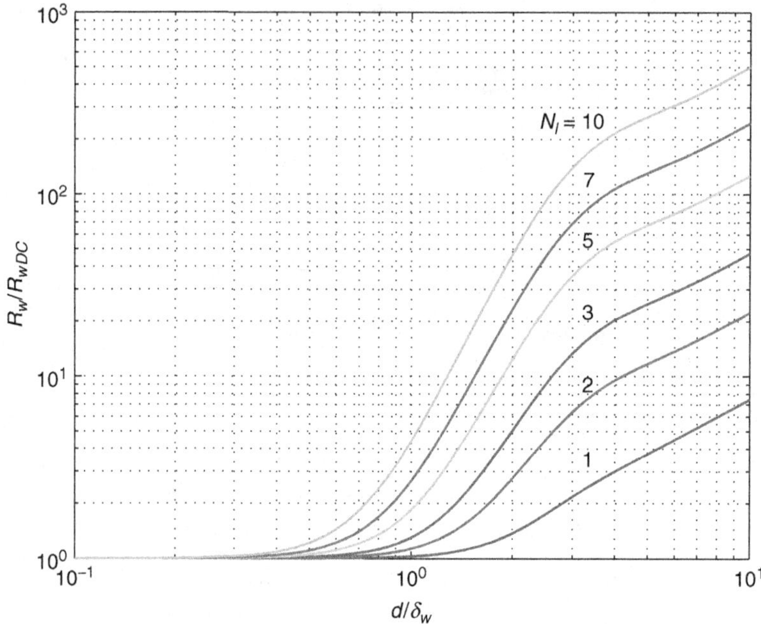

Figure 5.57 Normalized winding AC resistance $F_R = R_w/R_{wDC}$ versus d/δ_w at $\eta = d/p = 0.8$ and different number of layers N_l for a round wire

and the AC-to-DC winding resistance ratio by

$$F_R \approx A_o \frac{(2N_l^2+1)}{3} = \left(\frac{\pi}{4}\right)^{\frac{3}{4}} \frac{d}{\delta_w} \sqrt{\frac{d}{p}} \frac{(2N_l^2+1)}{3} \approx 0.8343 \frac{d}{\delta_w} \sqrt{\frac{d}{p}} \frac{(2N_l^2+1)}{3} \quad \text{for} \quad A_o > 3. \tag{5.295}$$

It is worth to note that F_R for a winding is much higher than that of a single, straight, and solitary round conductor at high frequencies (analyzed in Chapter 3), even for a single-layer winding. For example, F_R for a single-layer winding with $\eta = 0.8$ is 2.5 higher than that of a straight wire (Chapter 3) at the same high frequency. This is because the current flows only on the top and the bottom of a winding conductor. The increase of F_R is 7.5 times for two layers and 45 times for five layers.

For an inductor with a single layer ($N_l = 1$),

$$F_R \approx A_o = \left(\frac{\pi}{4}\right)^{\frac{3}{4}} \frac{d}{\delta_w} \sqrt{\frac{d}{p}} \quad \text{for} \quad A_o > 3, \tag{5.296}$$

resulting in the winding AC resistance

$$R_w = F_R R_{wDC} = \left(\frac{\pi}{4}\right)^{\frac{3}{4}} \frac{d}{\delta_w} \sqrt{\frac{d}{p}} \times \frac{4\rho_w l_w}{\pi d^2} = 4\left(\frac{\pi}{4}\right)^{\frac{3}{4}} \sqrt{\frac{d}{p}} \frac{\rho_w l_w}{\pi d \delta_w} \quad \text{for} \quad A_o > 3. \tag{5.297}$$

The winding AC resistance can be estimated using the skin depth

$$R_w \approx \frac{\rho_w l_w}{\pi d \delta_w}. \tag{5.298}$$

The ratio of (5.297) to (5.298) is

$$S = 4\left(\frac{\pi}{4}\right)^{\frac{3}{4}} \sqrt{\frac{d}{p}} = 3.33 \sqrt{\frac{d}{p}}. \tag{5.299}$$

Thus, the two equations give significantly different results.

Example 5.8

An inductor is made of a copper round conductor with $d = 0.511$ mm, $N_l = 5$, and $d/p = 0.8$. The winding wire length is $l_w = 1$ m. Find the winding AC resistance for a sinusoidal current at frequency $f = 10$ MHz.

Solution: The skin depth of the winding wire is

$$\delta_w = \sqrt{\frac{\rho_w}{\pi \mu_0 f}} = \sqrt{\frac{1.724 \times 10^{-8}}{\pi \times 4\pi \times 10^{-7} \times 10 \times 10^6}} = 20.897\ \mu\text{m}. \tag{5.300}$$

The winding DC resistance is

$$R_{wDC} = \frac{4\rho_w l_w}{\pi d^2} = \frac{4 \times 1.724 \times 10^{-8} \times 1}{\pi \times (0.511 \times 10^{-3})^2} = 84.1\ \text{m}\Omega. \tag{5.301}$$

The ratio of the wire diameter to the skin depth is

$$\frac{d}{\delta_w} = \frac{0.511 \times 10^{-3}}{20.897 \times 10^{-6}} = 24.45. \tag{5.302}$$

Since $d/\delta_w \gg 1$, the ratio of the AC-to-DC winding resistance is

$$F_R = \frac{R_w}{R_{wDC}} \approx \left(\frac{\pi}{4}\right)^{\frac{3}{4}} \frac{d}{\delta_w} \sqrt{\frac{d}{p}} \frac{(2N_l^2 + 1)}{3}$$
$$= \left(\frac{\pi}{4}\right)^{\frac{3}{4}} \frac{0.511 \times 10^{-3}}{20.897 \times 10^{-6}} \times \sqrt{0.8} \times \frac{(2 \times 5^2 + 1)}{3} = 310. \tag{5.303}$$

The winding AC resistance for the sinusoidal current is

$$R_w = F_R R_{wDC} = 310 \times 0.0841 = 26.088\ \Omega. \tag{5.304}$$

5.20.1 AC Resistance of Round Wire Winding

For a *round wire winding*,

$$R_w = F_R R_{wDC} = F_R \frac{4\rho_w l_w}{\pi d^2} = F_R \frac{4\rho_w l_w}{\pi \delta_w^2 \left(\frac{d}{\delta_w}\right)^2}$$
$$= \frac{4\rho_w l_w}{\pi d^2} \left(\frac{\pi}{4}\right)^{\frac{3}{4}} \sqrt{\frac{d}{p}} \frac{d}{\delta_w} \left[\frac{\sinh(2A) + \sin(2A)}{\cosh(2A) - \cos(2A)} + \frac{2(N_l^2 - 1)}{3} \frac{\sinh(A) - \sin(A)}{\cosh(A) + \cos(A)}\right]$$
$$= \frac{4\rho_w l_w}{\pi \delta_w^2 \left(\frac{d}{\delta_w}\right)^2} \left(\frac{\pi}{4}\right)^{\frac{3}{4}} \sqrt{\frac{d}{p}} \frac{d}{\delta_w} \left[\frac{\sinh(2A) + \sin(2A)}{\cosh(2A) - \cos(2A)} + \frac{2(N_l^2 - 1)}{3} \frac{\sinh(A) - \sin(A)}{\cosh(A) + \cos(A)}\right]$$
$$= \frac{R_{\delta_o}}{\frac{d}{\delta_w}} \left(\frac{\pi}{4}\right)^{\frac{3}{4}} \sqrt{\eta} \left[\frac{\sinh(2A) + \sin(2A)}{\cosh(2A) - \cos(2A)} + \frac{2(N_l^2 - 1)}{3} \frac{\sinh(A) - \sin(A)}{\cosh(A) + \cos(A)}\right], \tag{5.305}$$

where the DC resistance of diameter δ_w and length l_w is

$$R_{\delta_o} = \frac{4\rho_w l_w}{\pi \delta_w^2}. \tag{5.306}$$

The normalized round wire winding AC resistance is

$$F_r = \frac{R_w}{(4\rho_w l_w/\pi \delta_w^2)} = \frac{F_R}{\left(\frac{d}{\delta_w}\right)^2} = \frac{R_w}{R_{\delta_o}}$$

$$= \frac{\left(\frac{\pi}{4}\right)^{\frac{3}{4}}\sqrt{\frac{d}{p}}}{\frac{d}{\delta_w}}\left[\frac{\sinh(2A)+\sin(2A)}{\cosh(2A)-\cos(2A)} + \frac{2(N_l^2-1)}{3}\frac{\sinh(A)-\sin(A)}{\cosh(A)+\cos(A)}\right]. \quad (5.307)$$

This equation is illustrated in Fig. 5.58 for $\eta = d/p = 0.8$. For $d/\delta_w \ll 1$, the winding AC resistance is independent of N_l because the proximity effect can be neglected. For $N_l = 1$, the winding AC resistance decreases, when d/δ_w is increased. For $N_l > 1$, the winding AC resistance first decreases, next increases, and then decreases again as d/δ_w is increased. Hence, there is a local optimum of d_{opt}/δ_w, which decreases with increasing N_l and increasing d/p. Figure 5.59 shows plots of F_R, R_{wDC}, and R_w as functions of diameter d at $N_l = 4$, $l_T = 6.6$ cm, $N = 100$, $\eta = d/p = 0.9$, and $f = 100$ kHz for round wire winding.

For high frequencies, (5.307) can be approximated by

$$F_r = \frac{R_w}{(4\rho_w l_w/\pi \delta_w^2)} = \frac{F_R}{\left(\frac{d}{\delta_w}\right)^2} \approx \frac{\left(\frac{\pi}{4}\right)^{\frac{3}{4}}\sqrt{\frac{d}{p}}}{\frac{d}{\delta_w}}\left[1 + \frac{2(N_l^2-1)}{3}\right]$$

$$= \frac{\left(\frac{\pi}{4}\right)^{\frac{3}{4}}\sqrt{\frac{d}{p}}}{\frac{d}{\delta_w}}\left(\frac{2N_l^2+1}{3}\right) \quad \text{for} \quad \left(\frac{\pi}{4}\right)^{\frac{3}{4}}\frac{d}{\delta_w}\sqrt{\frac{d}{p}} \geq 3. \quad (5.308)$$

The accuracy of Dowell's equation decreases as the porosity factor $\eta = w/p$, $\eta = h/p$, or $\eta = d/p$ decreases, and as the number of layers N_l decreases. Dowell's equation underestimates the winding AC resistance at low values of the porosity factor $\eta < 0.4$ [28].

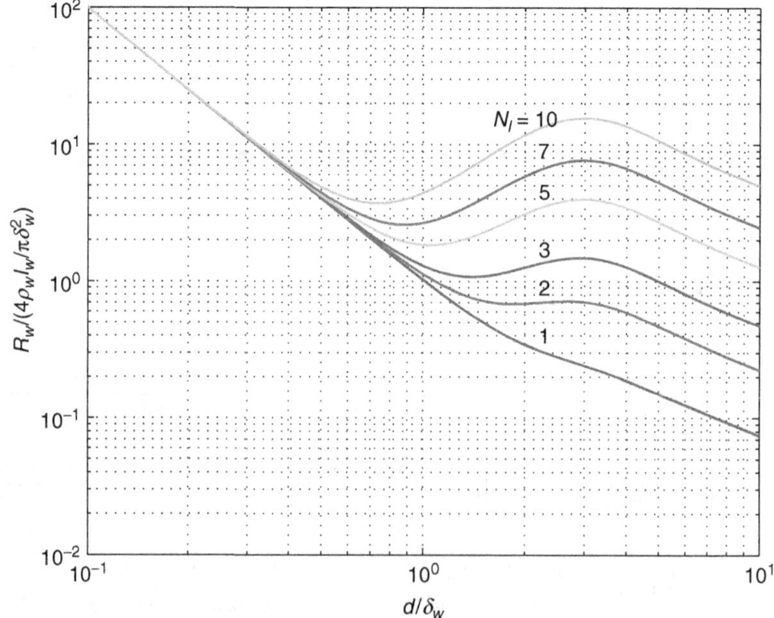

Figure 5.58 Normalized winding AC resistance $F_r = R_w/(4\rho_w l_w/\pi \delta_w^2) = F_R/(d/\delta_w)^2$ as a function of d/δ_w for $\eta = d/p = 0.8$ and different numbers of layers N_l for a round wire winding

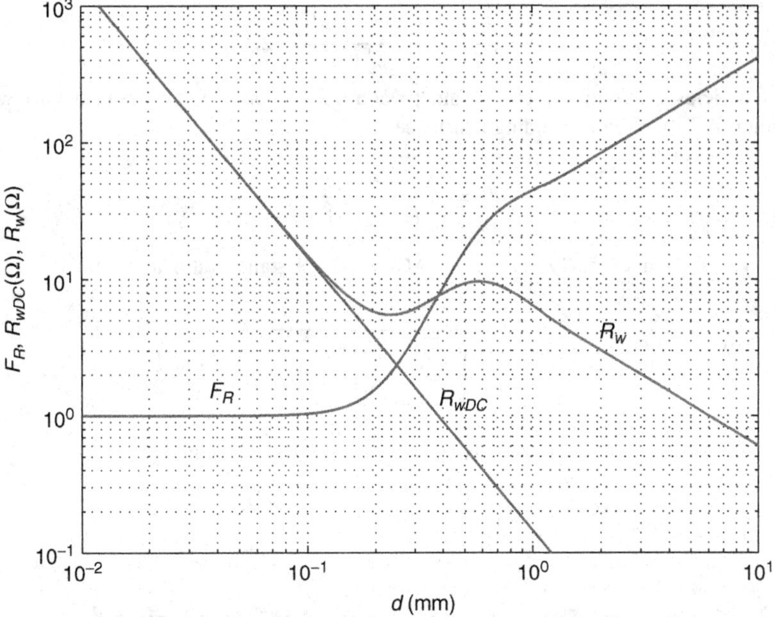

Figure 5.59 Plots of F_R, R_{wDC}, and R_w as functions of diameter d at $N_l = 4$, $l_T = 6.6$ cm, $N = 100$, $\eta = d/p = 0.9$, and $f = 100$ kHz, for a round wire winding

The normalized diameter of the bare round wire at which the local maximum value of the winding AC resistance occurs can be determined by the least-square function fit. The result is

$$\frac{d_h}{\delta_w} = \frac{2.274 - \frac{0.2576}{N_l}}{\left(\frac{\pi}{4}\right)^{0.75}\sqrt{\eta}}. \tag{5.309}$$

5.20.2 Optimum Diameter of Round Wire at Fixed Pitch

Dowell's equation for F_R for a round wire winding for low and medium values of d/δ_w is given by Snelling [8]

$$F_R \approx 1 + \frac{5N_l^2 - 1}{45}A_{o1}^4 = 1 + \left(\frac{\pi}{4}\right)^3 \frac{(5N_l^2 - 1)}{45\delta_w^4 p^2} d^6 \quad \text{for} \quad \frac{d}{\delta_w} \le 2. \tag{5.310}$$

Hence, the AC resistance of the round wire winding for low and medium frequencies is

$$R_w = F_R R_{wDC} \approx \frac{4\rho_w l_w}{\pi d^2}\left[1 + \left(\frac{\pi}{4}\right)^3 \frac{(5N_l^2 - 1)}{45\delta_w^4 p^2} d^6\right]$$
$$= \frac{4\rho_w l_w}{\pi}\left[\frac{1}{d^2} + \left(\frac{\pi}{4}\right)^3 \frac{(5N_l^2 - 1)d^4}{45\delta_w^4 p^2}\right]. \tag{5.311}$$

Assume that the number of layers N_l and the pitch p are constant. Taking the derivative of this equation with respect to d and equating the result to zero, one obtains

$$\frac{dR_w}{dd} = \frac{4\rho_w l_w}{\pi}\left[\frac{-2}{d^3} + \left(\frac{\pi}{4}\right)^3 \frac{4(5N_l^2 - 1)d^3}{45\delta_w^4 p^2}\right] = 0, \tag{5.312}$$

producing

$$\left(\frac{\pi}{4}\right)^3 \frac{4(5N_l^2 - 1)d^6}{45\delta_w^4 p^2} = 2. \quad (5.313)$$

The solution yields the optimum strand diameter at a fixed value of δ_w (or frequency f), which gives the minimum AC round wire winding resistance

$$d_v = \sqrt[6]{\frac{45\delta_w^4 p^2}{\left(\frac{\pi}{4}\right)^3 2(5N_l^2 - 1)}}. \quad (5.314)$$

Substitution (5.314) into (5.317) gives the AC-to-DC resistance ratio at d_v for solid round wire windings

$$F_{Rv} = \frac{R_{wv}}{R_{wDC}} \approx 1 + \left(\frac{\pi}{4}\right)^3 \frac{5N_l^2 - 1}{45\delta_w^4 p^2}\left(\sqrt[6]{\frac{45\delta_w^4 p^2}{\left(\frac{\pi}{4}\right)^3 2(5N_l^2 - 1)}}\right)^6 = 1 + \frac{1}{2} = 1.5 \quad \text{for} \quad d = d_v. \quad (5.315)$$

The minimum winding AC resistance occurs at $d = d_v$ and is given by

$$R_{w\,min} = F_{Rv} R_{wDCv} = \frac{8\rho_w l_w}{\pi d_{opt}^2} \quad (5.316)$$

5.20.3 Optimum Diameter of Round Wire at Fixed Porosity Factor

An approximate expression of Dowell's equation for F_R for a round wire winding for low and medium values of d/δ_w is given by Snelling [8]

$$F_R \approx 1 + \frac{5N_l^2 - 1}{45} A_{o1}^4 = 1 + \left(\frac{\pi}{4}\right)^3 \frac{\eta^2(5N_l^2 - 1)}{45}\left(\frac{d}{\delta_w}\right)^4 \quad \text{for} \quad \frac{d}{\delta_w} \leq 2. \quad (5.317)$$

The AC resistance of the round wire winding for low and medium frequencies is

$$R_w = F_R R_{wDC} \approx \frac{4\rho_w l_w}{\pi d^2}\left[1 + \left(\frac{\pi}{4}\right)^3 \frac{\eta^2(5N_l^2 - 1)}{45}\left(\frac{d}{\delta_w}\right)^4\right]$$
$$= \frac{4\rho_w l_w}{\pi}\left[\frac{1}{d^2} + \left(\frac{\pi}{4}\right)^3 \frac{\eta^2(5N_l^2 - 1)d^2}{45\delta_w^4}\right]. \quad (5.318)$$

Assume that the number of layers N_l and the porosity factor η are constant. Taking the derivative of this equation with respect to d and equating the result to zero, we get

$$\frac{dR_w}{dd} = \frac{4\rho_w l_w}{\pi}\left[\frac{-2}{d^3} + \left(\frac{\pi}{4}\right)^3 \frac{2\eta^2(5N_l^2 - 1)d}{45\delta_w^4}\right] = 0, \quad (5.319)$$

yielding

$$\left(\frac{\pi}{4}\right)^3 \frac{2\eta^2(5N_l^2 - 1)d}{45}\left(\frac{d}{\delta_w}\right)^4 = 2. \quad (5.320)$$

The solution produces the optimum strand diameter at a fixed value of δ_w (or frequency f), which gives the minimum AC round wire winding resistance

$$\frac{d_v}{\delta_w} = \sqrt[4]{\frac{45}{\left(\frac{\pi}{4}\right)^3 \eta^2(5N_l^2 - 1)}} = \frac{3.10445}{\sqrt[4]{\eta^2(5N_l^2 - 1)}}. \quad (5.321)$$

Substitution (5.321) into (5.317) gives the AC-to-DC resistance ratio at d_v/δ_w for solid round wire windings

$$F_{Rv} = \frac{R_{wv}}{R_{wDC}} = 1 + 1 = 2 \quad \text{for} \quad d = d_v. \quad (5.322)$$

The minimum winding AC resistance occurs at $d = d_v$ and is given by

$$R_{w\min} = F_{Rv} R_{wDCv} = \frac{8\rho_w l_w}{\pi d_{opt}^2} = \frac{8\rho_w l_w}{\pi \delta_w^2 \sqrt{\frac{45}{\left(\frac{\pi}{4}\right)^3 \eta^2 (5N_l^2-1)}}}. \quad (5.323)$$

The winding AC-to-DC resistance ratio in terms of d_v and d is

$$F_R \approx 1 + \left(\frac{d}{d_v}\right)^4. \quad (5.324)$$

The maximum average current density is

$$J_{max(AV)} = \frac{I_{max}}{A_w} = \frac{I_{max}}{4\pi d_{opt}^2}. \quad (5.325)$$

Example 5.9

A sinusoidal current flows through an inductor at $f = 100$ kHz. The winding is made of a round copper wire with $\eta = 0.8$. Find the optimum wire diameter. (a) For $N_l = 4$. (b) For $N_l = 10$.

Solution: The skin depth of copper at $f = 100$ kHz for a sinusoidal current at $T = 20$ °C is

$$\delta_w = \sqrt{\frac{\rho_w}{\pi \mu_0 f}} = \sqrt{\frac{1.724 \times 10^{-8}}{\pi \times 4 \times \pi \times 10^{-7} \times 10^5}} = 0.2089 \text{ mm}. \quad (5.326)$$

(a) For $N_l = 4$,

$$\frac{d_v}{\delta_w} = \sqrt[4]{\frac{45}{\left(\frac{\pi}{4}\right)^3 \eta^2 (5N_l^2 - 1)}} = \sqrt[4]{\frac{45}{\left(\frac{\pi}{4}\right)^3 \times 0.8^2 (5 \times 4^2 - 1)}} = 1.1642, \quad (5.327)$$

yielding

$$d_{opt} = 1.1642\delta_w = 1.1642 \times 0.2089 = 0.2432 \text{ mm}. \quad (5.328)$$

A wire AWG30 (AWG, American Wire Gauge) can be selected with the inner bare copper diameter $d_i = 0.2546$ mm and the outer insulated wire diameter $d_o = 0.294$ mm.

(b) For $N_l = 10$,

$$\frac{d_v}{\delta_w} = \sqrt[4]{\frac{45}{\left(\frac{\pi}{4}\right)^3 \eta^2 (5N_l^2 - 1)}} = \sqrt[4]{\frac{45}{\left(\frac{\pi}{4}\right)^3 \times 0.8^2 (5 \text{ times}^6 - 1)}} = 0.7344, \quad (5.329)$$

yielding

$$d_{opt} = 0.7344\delta_w = 0.7344 \times 0.2089 = 0.1534 \text{ mm}. \quad (5.330)$$

A wire AWG34 can be selected with the inner bare copper diameter $d_i = 0.1601$ mm and the outer insulated wire diameter $d_o = 0.191$ mm.

Pick a wire of AWG28 whose $d = 0.321$ mm.

Example 5.10

Design an inductor with a local optimum round wire winding diameter of an inductance $L = 21.27$ mH for a resonant circuit operating at $f = 200$ kHz. The inductor consists of three layers ($N_l = 3$). The inductor winding is made of copper magnet wire with a porosity factor. The 0R42616 Magnetics ferrite gapped pot core is used with the 00B261601 bobbin. The relative permeability of the core ferrite is $\mu_{rc} = 2500$, the air-gap length is $l_g = 0.1$ mm, the MTL is $l_T = 5.3$ cm, the core length is $l_c = 37.6$ mm, the core cross-sectional area is $A_c = 93.9$ mm^2, and the bobbin width is $b_b = 9.93$ mm. The inductor current is $i_L = 0.1\cos(2\pi ft)$ (A). Find:

(a) The optimum wire diameter
(b) The winding DC resistance
(c) The winding AC resistance
(d) The winding power loss
(e) The average current density
(f) The inductor quality factor related to the winding resistance

Solution: (a) The number of turns is

$$N = \sqrt{\frac{L\left(l_g + \frac{l_c}{\mu_{rc}}\right)}{\mu_0 A_c}} = \sqrt{\frac{21.27 \times 10^{-3}\left(0.1 \times 10^{-3} + \frac{37.6\times10^{-3}}{2500}\right)}{4\pi \times 10^{-7} \times 93.9 \times 10^{-6}}} = 144. \tag{5.331}$$

The skin depth of copper at $f = 200$ kHz is

$$\delta_w = \sqrt{\frac{\rho_w}{\pi \mu_0 f}} = \sqrt{\frac{1.724 \times 10^{-8}}{\pi \times 4 \times \pi \times 10^{-7} \times 200 \times 10^3}} = 0.14776 \text{ mm}. \tag{5.332}$$

$$\frac{d_v}{\delta_w} = \sqrt[4]{\frac{45}{\left(\frac{\pi}{4}\right)^3 \eta^2 (5N_l^2 - 1)}} = \sqrt[4]{\frac{45}{\left(\frac{\pi}{4}\right)^3 0.85^2 \times (5 \times 3^2 - 1)}} = 1.129456. \tag{5.333}$$

Hence, the optimum round wire thickness is

$$d_v = 1.129456\delta_w = 1.129456 \times 0.14776 = 0.166894 \text{ mm} = 167 \text{ µm}. \tag{5.334}$$

Pick AWG34 wire with bare wire diameter $d_i = 0.1601$ mm and the insulated outer wire diameter $d_o = 0.191$ mm.

(b) The winding DC resistance is

$$R_{wDC} = \frac{4\rho_{Cu} l_w}{\pi d_i^2} = \frac{4 \times 1.724 \times 10^{-8} \times 0.96}{\pi \times (0.1601 \times 10^{-3})^2} = 6.5513\ \Omega. \tag{5.335}$$

(c) The winding AC resistance is

$$R_w = F_{Rv} R_{wDC} = 2 \times 6.5513 = 13.1026\ \Omega. \tag{5.336}$$

(d) The winding power loss is

$$P_w = \frac{R_w I_{Lm}^2}{2} = \frac{13.1026 \times 0.1^2}{2} = 65.513 \text{ mW}. \tag{5.337}$$

The average amplitude of the current density is

$$J_{m(AV)} = \frac{I_{Lm}}{A_w} = \frac{4I_{Lm}}{\pi d_i^2} = \frac{4 \times 0.1}{\pi d_i^2} = \frac{4 \times 0.1}{\pi \times 0.1601^2} = 4.967 \text{ A/mm}^2. \tag{5.338}$$

(f) The inductor quality factor related to the winding resistance is

$$Q_{Lw} = \frac{\omega L}{R_w} = \frac{2\pi \times 200 \times 10^3 \times 21.27 \times 10^{-3}}{13.1026} = 2040. \tag{5.339}$$

5.20.4 Critical Round Wire Diameter

The winding resistance R_w for high frequencies may be approximated by

$$R_w \approx \frac{2N_l^2 + 1}{3} A R_{wDC} = \frac{4\rho_w l_w}{\pi d^2} \frac{2N_l^2 + 1}{3} \left(\frac{\pi}{4}\right)^{\frac{3}{4}} \sqrt{\eta} \frac{d}{\delta_w}. \tag{5.340}$$

Let

$$\frac{8\rho_w l_w}{\pi\delta_w^2\sqrt{\frac{45}{\left(\frac{\pi}{4}\right)^3\eta^2(5N_l^2-1)}}} = \frac{4\rho_w l_w}{\pi d^2}\frac{2N_l^2+1}{3}\left(\frac{\pi}{4}\right)^{\frac{3}{4}}\sqrt{\eta}\frac{d}{\delta_w}. \tag{5.341}$$

Hence, the critical value of the wire diameter at which $R_w = R_{w\,\min}$ is

$$\frac{d_{cr}}{\delta_w} = \frac{2N_l^2+1}{6\left(\frac{\pi}{4}\right)^{\frac{3}{4}}}\sqrt{\frac{45}{\eta(5N_l^2-1)}}. \tag{5.342}$$

5.20.5 Boundary Between Low and Medium Frequencies for Round Wire Winding

The approximate Dowell's equation is valid for $F_{RB} = 1.05$. Hence,

$$F_{RB} \approx 1 + \left(\frac{\pi}{4}\right)^3\frac{\eta^2(5N_l^2-1)}{45}\left(\frac{d_B}{\delta_w}\right)^4 = 1.05, \tag{5.343}$$

yielding the round wire diameter at the boundary between the low and medium values of d/δ_w

$$\frac{d_B}{\delta_w} = \sqrt[4]{\frac{9}{4\left(\frac{\pi}{4}\right)^3\eta^2(5N_l^2-1)}}. \tag{5.344}$$

The boundary frequency is

$$f_L = \frac{\rho_w}{\pi\mu_0 d_B^2}\sqrt{\frac{9}{4\left(\frac{\pi}{4}\right)^3\eta^2(5N_l^2-1)}}. \tag{5.345}$$

The ratio of the optimum round solid wire diameter to the boundary round solid wire diameter is

$$\frac{d_v}{d_B} = \sqrt[4]{20} \approx 2.115. \tag{5.346}$$

The ratio of the boundary winding AC resistance R_{wB} to the the minimum winding AC resistance for round wire inductors is

$$\frac{R_{wB}}{R_{w\,\min}} = \frac{R_{wDCB}F_{RB}}{R_{wDCopt}F_{Rv}} = \left(\frac{d_{opt}}{d_B}\right)^2\frac{F_{RB}}{F_{Rmin}} = \frac{1.05}{2}\sqrt{20} \approx 2.34787. \tag{5.347}$$

5.21 Inductance

The magnetic energy is stored in the magnetic field between the layers and inside the winding conductor, and it is proportional to H^2. The reactance of the winding inductance due to the magnetic field stored inside the conductor for the foil conductor winding with a single turn per layer is given by

$$\begin{aligned} X_{L_l} &= Im\{Z\} = X_{L1} + X_{L2} + X_{L3} + \ldots + X_{LN_l} = \sum_{n=1}^{N_l} X_{Ln} = F_X R_{wDC} \\ &= R_{wDC}\left(\frac{h}{\delta_w}\right)\left[\frac{\sinh\left(\frac{2h}{\delta_w}\right) - \sin\left(\frac{2h}{\delta_w}\right)}{\cosh\left(\frac{2h}{\delta_w}\right) - \cos\left(\frac{2h}{\delta_w}\right)} + \frac{2(N_l^2-1)}{3}\frac{\sinh\left(\frac{h}{\delta_w}\right) + \sin\left(\frac{h}{\delta_w}\right)}{\cosh\left(\frac{h}{\delta_w}\right) + \cos\left(\frac{h}{\delta_w}\right)}\right] \end{aligned} \tag{5.348}$$

where X_{L1}, X_{L2}, X_{L3}, ... are the inductive reactances of the layers.

The internal winding inductance is given by

$$L_l = \frac{R_{wDC}A}{\omega}\left[\frac{\sinh\left(\frac{2h}{\delta_w}\right) - \sin\left(\frac{2h}{\delta_w}\right)}{\cosh\left(\frac{2h}{\delta_w}\right) - \cos\left(\frac{2h}{\delta_w}\right)} + \frac{2(N_l^2-1)}{3}\frac{\sinh\left(\frac{h}{\delta_w}\right) + \sin\left(\frac{h}{\delta_w}\right)}{\cosh\left(\frac{h}{\delta_w}\right) + \cos\left(\frac{h}{\delta_w}\right)}\right]$$

$$= \frac{R_{wDC}A}{\omega}\left[\frac{e^{2A}-e^{-2A}-2\sin(2A)}{e^{2A}+e^{-2A}-2\cos(2A)}+\frac{2(N_l^2-1)}{3}\frac{e^{A}-e^{-A}+2\sin(A)}{e^{A}+e^{-A}+2\cos(A)}\right]$$

$$= F_L R_{wDC} = \frac{R_{wDC}A}{\omega}\left[F_{LS}+\frac{2(N_l^2-1)}{3}F_{LP}\right], \tag{5.349}$$

where the inductance normalized with respect to the winding DC resistance is

$$F_L = \frac{L_l}{R_{wDC}} = \frac{A}{\omega}\left[\frac{\sinh\left(\frac{2h}{\delta_w}\right)-\sin\left(\frac{2h}{\delta_w}\right)}{\cosh\left(\frac{2h}{\delta_w}\right)-\cos\left(\frac{2h}{\delta_w}\right)}+\frac{2(N_l^2-1)}{3}\frac{\sinh\left(\frac{h}{\delta_w}\right)+\sin\left(\frac{h}{\delta_w}\right)}{\cosh\left(\frac{h}{\delta_w}\right)+\cos\left(\frac{h}{\delta_w}\right)}\right]$$

$$= \frac{A}{\omega}\left[\frac{e^{2A}-e^{-2A}-2\sin(2A)}{e^{2A}+e^{-2A}-2\cos(2A)}+\frac{2(N_l^2-1)}{3}\frac{e^{A}-e^{-A}+2\sin(A)}{e^{A}+e^{-A}+2\cos(A)}\right], \tag{5.350}$$

the inductance due to the skin effect normalized with respect to the winding DC resistance is

$$F_{LS} = \frac{L_{ls}}{R_{wDC}} = \frac{\sinh\left(\frac{2h}{\delta_w}\right)-\sin\left(\frac{2h}{\delta_w}\right)}{\cosh\left(\frac{2h}{\delta_w}\right)-\cos\left(\frac{2h}{\delta_w}\right)} = \frac{e^{2A}-e^{-2A}-2\sin(2A)}{e^{2A}+e^{-2A}-2\cos(2A)}, \tag{5.351}$$

and the inductance due to the proximity effect normalized with respect to the winding DC resistance is

$$F_{LP} = \frac{L_{lp}}{R_{wDC}} = \frac{\sinh\left(\frac{h}{\delta_w}\right)+\sin\left(\frac{h}{\delta_w}\right)}{\cosh\left(\frac{h}{\delta_w}\right)+\cos\left(\frac{h}{\delta_w}\right)} = \frac{e^{A}-e^{-A}+2\sin(A)}{e^{A}+e^{-A}+2\cos(A)}. \tag{5.352}$$

Figure 5.60 shows the plots of F_{LS} and F_{LP} versus A. It can be seen that $F_{LS} \approx 1$ and $F_{LP} \approx 1$ for $A > 4$. Hence, the winding inductance can be approximated by

$$L_l \approx \frac{R_{wDC}A}{\omega}\left[1+\frac{2(N_l^2-1)}{3}\right] = \frac{2N_l^2+1}{3}\frac{R_{wDC}A}{\omega} \quad \text{for} \quad A > 4. \tag{5.353}$$

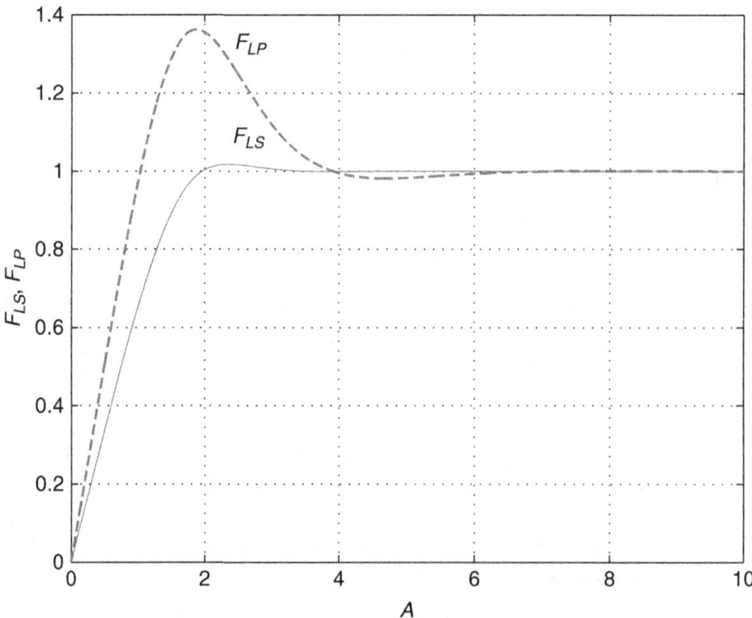

Figure 5.60 Plots of F_{LS} and F_{LP} as functions of A

For the foil winding conductors,

$$L_l \approx \frac{2N_l^2+1}{3}\frac{R_{wDC}A}{\omega} = \frac{2N_l^2+1}{3}\frac{\rho_w l_w}{\omega bh}\frac{h}{\delta_w} = \frac{2N_l^2+1}{3}\frac{\rho_w l_w}{\omega b\delta_w} = \frac{2N_l^2+1}{3}\frac{R_\delta}{\omega}$$

$$= \frac{2N_l^2+1}{6}\frac{l_w}{b}\sqrt{\frac{\mu_0\rho_w}{\pi f}} \quad \text{for} \quad A > 4. \tag{5.354}$$

For rectangular winding conductors,

$$L_l \approx \frac{2N_l^2+1}{3}\frac{R_{wDC}A_r}{\omega} = \frac{2N_l^2+1}{3}\frac{\rho_w l_w}{\omega wh}\frac{h}{\delta_w}\sqrt{\frac{w}{p}} = \frac{2N_l^2+1}{6} l_w\sqrt{\frac{\mu_0\rho_w}{\pi pwf}} \quad \text{for} \quad A > 4. \tag{5.355}$$

For square winding conductors,

$$L_l \approx \frac{2N_l^2+1}{3}\frac{R_{wDC}A_s}{\omega} = \frac{2N_l^2+1}{3}\frac{\rho_w l_w}{\omega h^2}\frac{h}{\delta_w} = \frac{2N_l^2+1}{6} l_w\sqrt{\frac{\mu_0\rho_w}{\pi phf}} \quad \text{for} \quad A > 4. \tag{5.356}$$

For a round winding conductors,

$$L_l \approx \frac{2N_l^2+1}{3}\frac{R_{wDC}A_o}{\omega} = \frac{2N_l^2+1}{3}\frac{4\rho_w l_w}{\omega\pi d^2}\left(\frac{\pi}{4}\right)^{\frac{3}{4}}\frac{d}{\delta_w}\sqrt{\frac{d}{p}}$$

$$= \frac{2N_l^2+1}{3}\left(\frac{\pi}{4}\right)^{\frac{3}{4}}\frac{2l_w}{\pi}\sqrt{\frac{\mu_0\rho_w}{\pi pdf}} \quad \text{for} \quad A > 4. \tag{5.357}$$

Figure 5.61 shows the normalized winding inductance L_l/R_{wDC} versus d/δ_w at $d = 1$ mm and $d/p = 0.8$ and different number of layers N_l for a round conductor. It can be seen that the inductance is independent of d/δ_w for $d/\delta_w < 2$ and decreases with d/δ_w for $d/\delta_w > 2$. It also increases with the number of layers N_l.

The reduction of the inductance with frequency occurs only in the current-carrying conductors. There is no reduction in the inductance in air gaps, interlayer gaps, and primary–secondary interface gap.

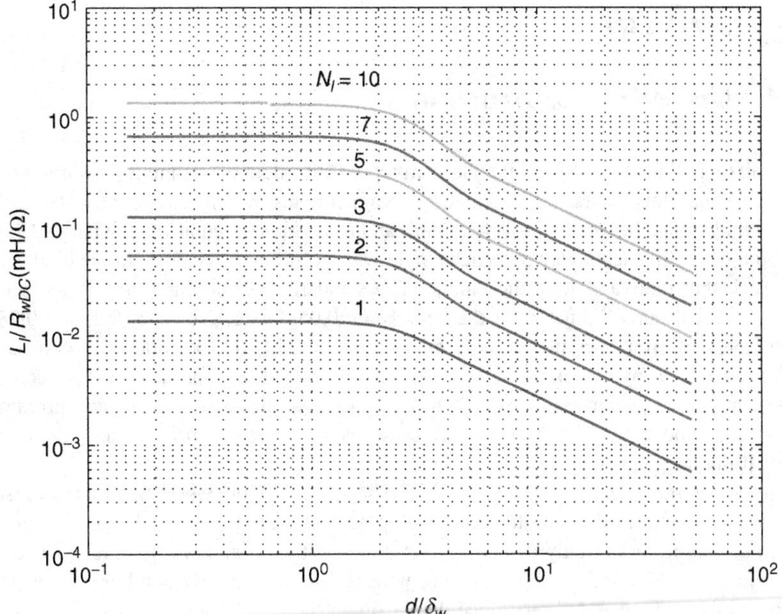

Figure 5.61 Normalized winding inductance L_l/R_{wDC} versus d/δ_w at $d = 1$ mm, $d/p = 0.8$, and different number of layers N_l for a round conductor

5.22 Solution for Round Conductor Winding in Cylindrical Coordinates

The magnetic field distribution for a round conductor winding can be described by modified Bessel's ordinary differential equation in the cylindrical coordinates

$$\frac{d^2H}{dr^2} + \frac{1}{r}\frac{dH}{dr} = j\omega\mu\sigma H = k^2 H. \tag{5.358}$$

The solution for this equation is given by Hurley *et al.* [56]

$$H(r) = AI_0(kr) + BK_0(kr), \tag{5.359}$$

where

$$A = H_o \frac{mK_0(r_i) - (m-1)K_0(kr_o)}{I_0(kr_o)K_0(kr_i) - K_0(kr_o)I_0(kr_i)} \tag{5.360}$$

$$B = H_o \frac{(m-1)I_0(kr_o) - nI_0(kr_i)}{I_0(kr_o)K_0(kr_i) - K_0(kr_o)I_0(kr_i)} \tag{5.361}$$

$$H_o = \frac{N_{tl}I}{w} \tag{5.362}$$

$k = \sqrt{j\omega\mu\sigma}$, N_{tl} is the number of turns in a layer, m is the layer number, r_i is the inner radius of the layer, r_o is the outer radius of the layer, and I_0 and K_0 are the modified Bessel functions of the first and second kinds of order zero, respectively. The electric field intensity is

$$E(r) = -\rho_w \frac{dH(r)}{dr} = -\rho_w m[AI_1(kr) - BK_1(kr)], \tag{5.363}$$

where I_1 and K_1 are the modified Bessel function of the first kind and the first order, respectively. Hence, one can find the Poynting vector $\mathbf{E} \times \mathbf{H}$.

5.23 Litz Wire

5.23.1 Litz-Wire Construction

The term "litz wire" is derived from the German word "Litzendraht" meaning woven wire. It consists of a bundle of fine films insulated and enameled, magnet wires, which are bunched, stranded together, twisted or braided together into a uniform pattern and are externally connected in parallel. These wires are braided together in such a way that each single wire occupies every place in the total cross section for periodic length intervals along the bundle. The standard twist configuration is 18–36 twists per meter. The nominal outer diameters of the strands are 0.05, 0.071, 0.1, 0.15, 0.2, and 0.28 mm. The typical numbers of strands n in a bundle are 7, 19, 25, 37, 60, 100, 200, and 400. Litz wire is often wrapped with a nylon textile or yarn for added strength and protection. Litz construction is designed to minimize power losses exhibited in solid wires due to skin effect and proximity effect. If a multistrand conductor is used, the overall cross-sectional area is spread among many conductors with a small diameter. For this reason, a stranded and twisted wire in the form of a rope results in a more uniform current density distribution than a solid wire. Moreover, litz wires are assembled so that each single strand, in the longitudinal development of wire, occupies all positions in the wire's cross section. Therefore, not only the skin effect, but also the proximity effect, is drastically reduced at high frequencies. Figure 5.62 shows a solid wire and litz wire; the shaded area indicates the skin depth. Litz wire has a lower AC resistance than a single round solid wire of the same cross-sectional area. It is important to position each individual strand in a uniform pattern moving from the center to the outside and back in a given length so that each strand takes up all possible positions in the cross section in order to reduce the proximity effect. Litz wire intended for higher frequencies requires more

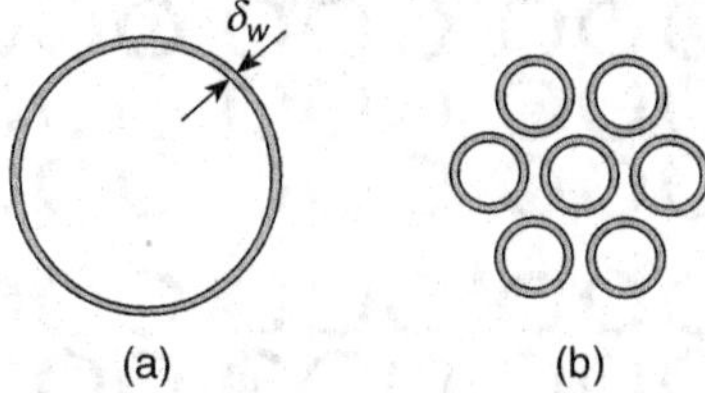

Figure 5.62 Solid and litz wires. (a) Solid wire. (b) Litz wire. Shaded areas indicate the skin depth

strands of less diameter. Polyurethane is the film most often used for insulating individual strands because of its solderability and low dielectric losses. Litz wire is expensive. The DC resistance of a litz wire is higher than that of a solid wire with the same length and equivalent cross-sectional area because each strand path is longer than the average wire length. In addition, the utilization of the winding space inside a bobbin width is reduced with respect to a solid wire. Typically, the winding made of litz wire takes 40% more space than that of sold wire. The fill factor η is usually less than 80%. Litz wire is fairly easy to wind and can be used for various window geometries.

5.23.2 Model of Litz-Wire and Multistrand Wire Windings

Figure 5.63 shows a transformation of the foil winding into a litz-wire winding. First, the foil conductor winding is converted into a square wire winding. Next, the square wire winding is converted into the round solid wire. The round solid wire is converted into round multistrand bundle. Finally, the bundle of round wire winding is converted into the round strands, which forms a square. The total number of strand turns is kN, where N is the number of bundle turns. Figure 5.64 shows a model of a litz-wire winding. Each bundle has k strands. The strands in each bundle are arranged in a square with $\sqrt{k}$ strands on each side of the bundle. It is assumed that the individual strands are parallel to the axis of the bundle and are uniformly spaced. The effective number of layers of the litz-wire winding with a square arrangement of the strands in a bundle is given by

$$N_{ll} = N_l\sqrt{k}, \tag{5.364}$$

where N_l is the number of bundle layers.

The winding window area of a bobbin is

$$A_w = b_b h_b, \tag{5.365}$$

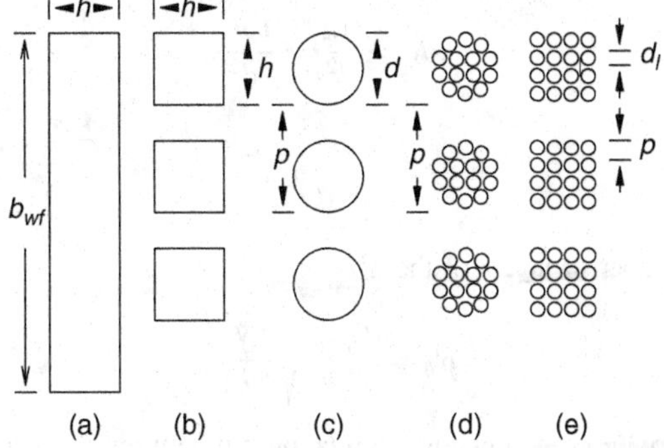

Figure 5.63 Transformation of foil winding into litz-wire winding. (a) Foil winding. (b) Square wire winding. (c) Solid wire winding. (d) Round bundle of multistrand wire winding. (e) Square bundle of multistrand wire winding

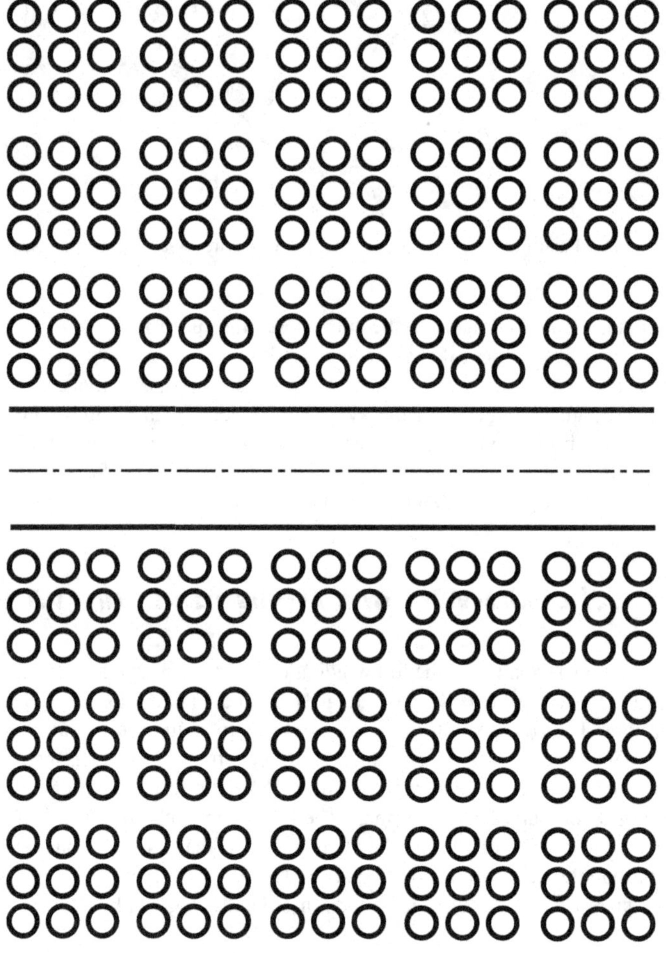

Figure 5.64 Model of litz-wire winding

where b_b is the bobbin window width and h_b is the bobbin window height. The effective area of each strand of a square arrangement of the strands in a bundle, including the air area, is d_o^2, where d_o is the strand diameter with insulation. The total number of litz-wire strand turns is

$$kN = \frac{A_w}{d_o^2} = \frac{b_b h_b}{d_o^2}, \tag{5.366}$$

yielding

$$d_o = \sqrt{\frac{b_b h_b}{kN}}. \tag{5.367}$$

Hence, the effective number of layers of the litz-wire winding is

$$N_{ll} = \frac{h_b}{d_o} = \sqrt{\frac{kN\, h_b}{b_b}}. \tag{5.368}$$

Equating (5.364) and (5.368), one obtains the relationship between the number of bundle layers N_l and the number of bundle turns N

$$N_l = \sqrt{\frac{N\, h_b}{b_b}}. \tag{5.369}$$

For $h_b = b_b$,

$$N_l = \sqrt{N}. \tag{5.370}$$

5.23.3 Litz-Wire Winding Resistance

The DC resistance of a single strand is

$$R_{sDC} = \frac{4\rho_w l_{wl}}{\pi d_l^2} = \frac{4\rho_w N l_T}{\pi d_l^2}, \tag{5.371}$$

where d_l is the diameter of the strand, N is the total number of turns, l_T is the MTL, and $l_{wl} = N l_T$ is the total strand length. The DC resistance of the bundle made up of k strands connected in parallel is given by

$$R_{wDC} = \frac{R_{sDC}}{k} = \frac{4\rho_w l_{wl}}{k\pi d_l^2} = \frac{4\rho_w N l_T}{k\pi d_l^2}. \tag{5.372}$$

The power loss in each strand is

$$P_s = \frac{1}{2} R_s \left(\frac{I_m}{k}\right)^2. \tag{5.373}$$

The power loss in all k strands is

$$P_k = kP_s = \frac{1}{2} R_s \left(\frac{I_m^2}{k}\right). \tag{5.374}$$

The cross-sectional area of the bare solid wire is $A_w = \pi d^2/4$ and the cross-sectional area of the bare litz-wire strand is $A_{wl} = \pi d_l^2/4$. Equating the DC resistances of the litz-wire winding and the solid wire winding, we get

$$\frac{4\rho_w l_w}{\pi d^2} = \frac{4\rho_w l_{wl}}{k\pi d_l^2}, \tag{5.375}$$

one obtains $\pi d^2 = k\pi d_l^2$ and the number of strands

$$k = \left(\frac{l_{wl}}{l_w}\right)\left(\frac{d}{d_l}\right)^2, \tag{5.376}$$

where d is the diameter of the conductor of the solid wire and d_l is the diameter of the conductor of the litz-wire strand. For $l_{wl} = l_w$,

$$k = \left(\frac{d}{d_l}\right)^2. \tag{5.377}$$

For the litz wire,

$$A = A_l = \left(\frac{\pi}{4}\right)^{0.75} \frac{d_l}{\delta_w}\sqrt{\eta} = \left(\frac{\pi}{4}\right)^{0.75} \frac{d}{\delta_w \sqrt{k}}\sqrt{\eta}. \tag{5.378}$$

Dowell's equation for the litz-wire winding is

$$F_{R(litz)} = \frac{R_w}{R_{wDC}} = \frac{P_w}{P_{wDC}} = A\left[\frac{\sinh(2A) + \sin(2A)}{\cosh(2A) - \cos(2A)} + \frac{2(N_l^2 k - 1)}{3}\,\frac{\sinh(A) - \sin(A)}{\cosh(A) + \cos(A)}\right]$$

$$= A\left[\frac{\sinh(2A) + \sin(2A)}{\cosh(2A) - \cos(2A)} + \frac{2(N_{ll}^2 - 1)}{3}\,\frac{\sinh(A) - \sin(A)}{\cosh(A) + \cos(A)}\right]. \tag{5.379}$$

Figure 5.65 shows a 3D plot of AC-to-DC winding resistance ratio $F_{R(litz)}$ as a function of d/δ_w and N_l at $k = 225$ and $\eta = 0.9$ for litz-wire winding. Figure 5.66 shows the plots of skin effect AC-to-DC resistance ratio F_S, proximity effect AC-to-DC resistance ratio F_P, and total AC-to-DC resistance ratio F_R for litz-wire winding at $N_l = 2$, $k = 30$, $N_{ll} = 12$, and $\eta = 0.8$. It can be seen that the skin effect is negligible for $d/\delta_w > 2$.

Figure 5.67 depicts the plots of the AC-to-DC winding resistance ratio F_R as a function of d/δ_w at $N_l = 3$ at $k = 9$, and $\eta = 0.9$ for the solid wire winding and the litz-wire winding. It can be seen

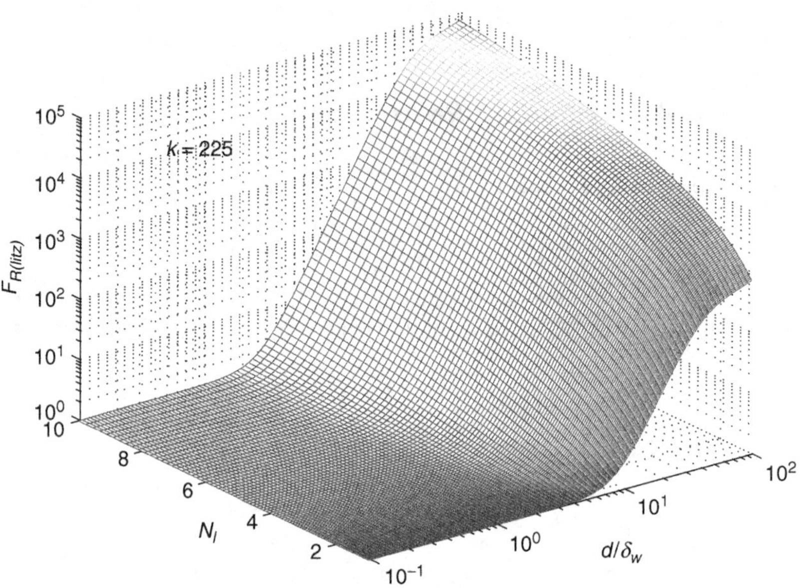

Figure 5.65 3D plot of AC-to-DC winding resistance ratio $F_{R(litz)}$ as a function of d/δ_w and N_l at $k = 225$ and $\eta = 0.9$ for litz-wire winding

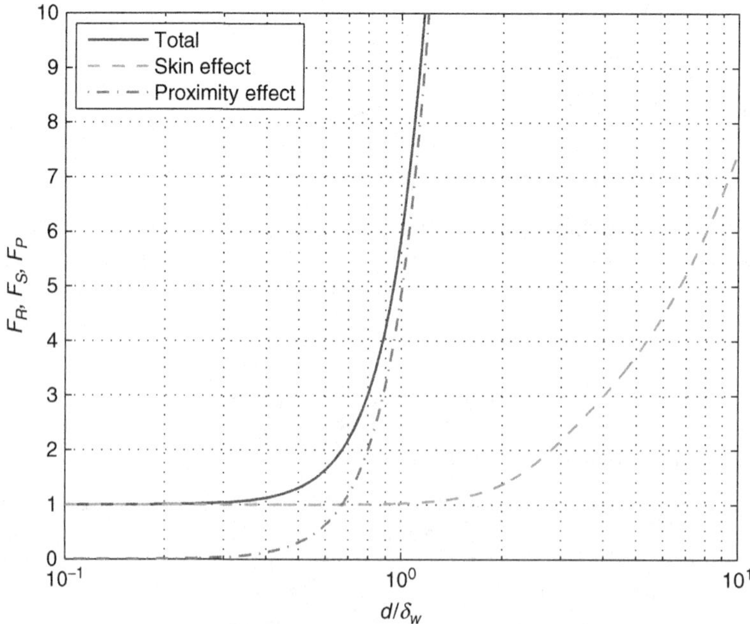

Figure 5.66 Plot of skin effect AC-to-DC winding resistance ratio F_S, proximity effect AC-to-DC winding resistance ratio F_P, and total AC-to-DC winding resistance ratio F_R as functions of d/δ_w at $N_l = 2$, $k = 30$, $N_{ll} = 12$, and $\eta = 0.8$ for litz-wire winding

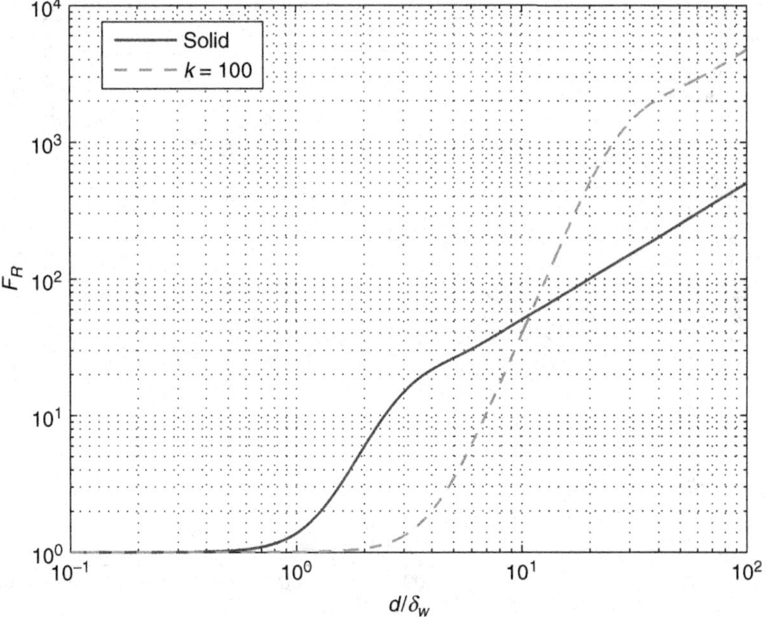

Figure 5.67 Plots of AC-to-DC winding resistance ratio F_R as a function of d/δ_w at $N_l = 3$, $k = 100$, and $\eta = 0.9$ for solid wire winding and litz-wire winding

that the flat part of the plots, where the AC resistance of the litz-wire winding is equal to the DC resistance, is larger than that of the solid wire winding. It can also be observed that there is a range of d/δ_w in which the resistance of litz-wire winding is lower than that of the solid wire winding. However, above a certain frequency, the resistance of the solid wire winding is lower than that of the litz-wire winding. This is because the litz-wire winding has a larger number of the effective layers than the solid wire winding with the cross-sectional conductor area. The frequency corresponding to crossing point will be called the maximum frequency of the litz-wire winding f_{max}. In the example shown in Fig. 5.67, the resistance of the litz-wire winding is lower than that of the solid wire winding in the range $0.5 \leq d/\delta_w \leq 10$.

The ratio of the maximum low frequency of the litz-wire winding to that of the solid round wire winding is

$$\frac{f_{L(litz)}}{f_{L(solid)}} = \sqrt{k}. \tag{5.380}$$

This ratio is illustrated in Fig. 5.68. It increases when k increases.

The ratio of the AC-to-DC resistance factor of the litz-wire winding $F_{R(litz)}$ to the AC-to-DC resistance factor of the solid wire winding $F_{R(solid)}$ is defined as

$$\xi = \frac{F_{R(litz)}}{F_{R(solid)}}. \tag{5.381}$$

Figure 5.69 shows the plots of the ratio of the AC litz-wire winding resistance to the AC solid wire winding resistance as a function of d/δ_w at $N_l = 4$, $\eta = 0.9$, and $k = 9$, 16, and 25 for the litz-wire winding. It can be seen that as d/δ_w is increased, the ratio ξ is first constant at $\xi = 1$, then decreases, reaches a minimum value, next increases, and finally remains constant at $\xi > 1$. When $\xi < 1$, the litz-wire winding resistance is lower than that of the solid wire winding. When $\xi > 1$, the litz-wire winding resistance is several times higher than that of the solid wire winding. The value of d/δ_w at which ξ has the minimum value is the optimum value $(d/\delta_w)_{opt}$ of the litz-wire winding. This value corresponds to the optimum frequency f_{opt} at a fixed value of d. In addition, as the number of strands k is increased, the minimum value of ξ decreases and f_{opt} increases.

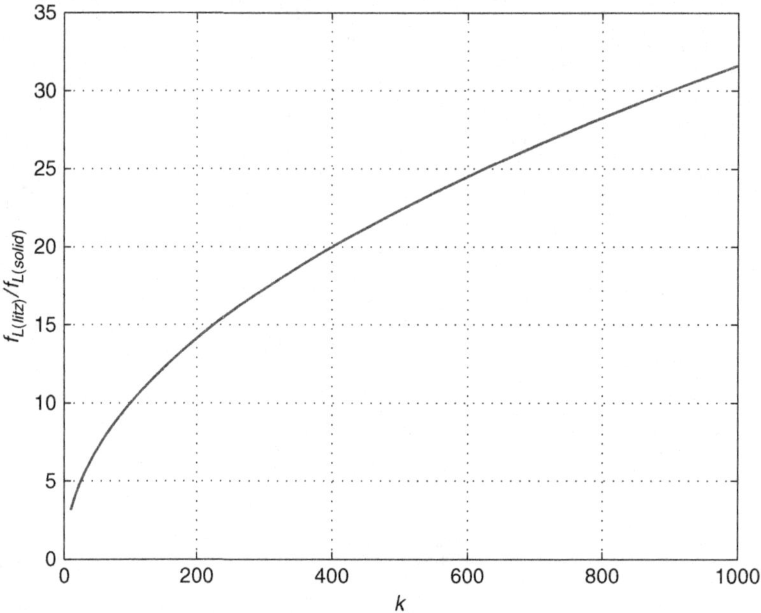

Figure 5.68 Ratio of litz-wire boundary frequency to the solid wire boundary frequency as a function of the number of strands k at selected numbers of layers N_l at the same cross-sectional conductor area

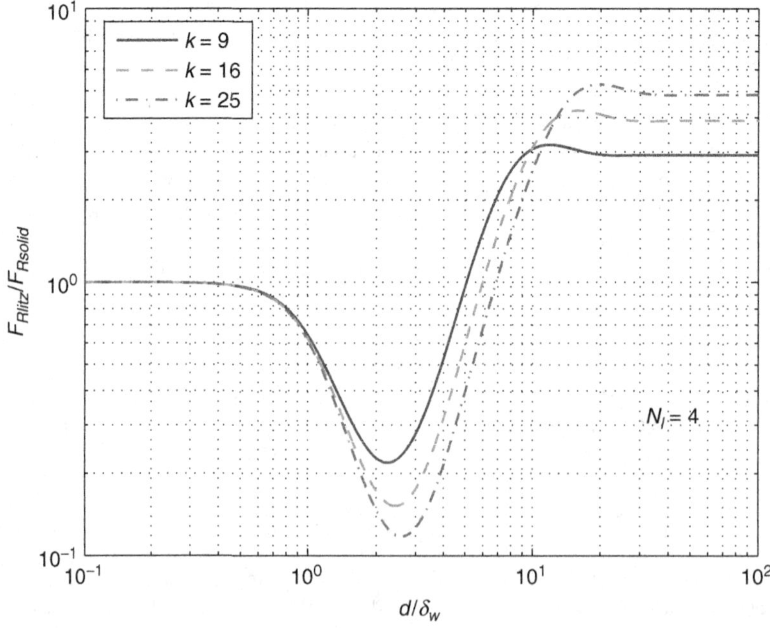

Figure 5.69 Plots of AC-to-DC winding resistance ratio of the litz-wire winding to that of the solid wire winding F_{Rlitz}/F_{Rsolid} as a function of d/δ_w at $N_l = 4$, $\eta = 0.9$, and $k = 9$, 16, and 25 for litz-wire winding

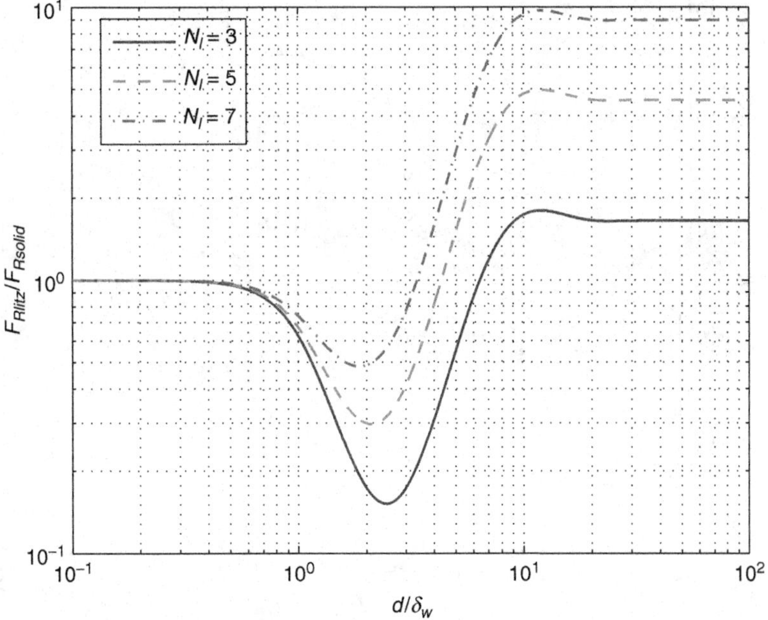

Figure 5.70 Plots of AC-to-DC winding resistance ratio of the litz-wire winding to that of the solid wire winding F_{Rlitz}/F_{Rsolid} as a function of d/δ_w at $k = 9$, $\eta = 0.9$, and $N_l = 3$, 5, and 7 for litz-wire winding

Figure 5.70 depicts the plots of the AC-to-DC resistance ratio of the litz-wire winding to that of the solid wire winding as a function of d/δ_w at $k = 9$, $\eta = 0.9$, and $N_l = 3$, 5, and 7 for the litz-wire winding. As the number of bundle layers N_l is increased, the minimum value of ξ increases and the value of ξ at high values of d/δ_w increases.

5.23.4 Optimum Strand Diameter at Fixed Porosity Factor

The litz-wire winding resistance is

$$R_w = R_{wDC} R_{R(litz)}$$
$$= \frac{4\rho_w l_w}{\pi k d_l^2} \left(\frac{\pi}{4}\right)^{\frac{3}{4}} \frac{d_l\sqrt{\eta}}{\delta_w} \left[\frac{\sinh(2A) + \sin(2A)}{\cosh(2A) - \cos(2A)} + \frac{2(N_{ll}^2 - 1)}{3} \frac{\sinh(A) - \sin(A)}{\cosh(A) + \cos(A)}\right]$$
$$= \left(\frac{\pi}{4}\right)^{\frac{3}{4}} \frac{4\rho_w l_w \sqrt{\eta}}{\pi k d_l \delta_w} \left[\frac{\sinh(2A) + \sin(2A)}{\cosh(2A) - \cos(2A)} + \frac{2(N_{ll}^2 - 1)}{3} \frac{\sinh(A) - \sin(A)}{\cosh(A) + \cos(A)}\right]. \quad (5.382)$$

The normalized litz-wire winding resistance is given by

$$F_r = \frac{R_w}{4\rho_w l_w/(\pi \delta_w^2 k)}$$
$$= \left(\frac{\pi}{4}\right)^{\frac{3}{4}} \frac{\sqrt{\eta}}{\left(\frac{d_l}{\delta_w}\right)} \left[\frac{\sinh(2A) + \sin(2A)}{\cosh(2A) - \cos(2A)} + \frac{2(N_{ll}^2 - 1)}{3} \frac{\sinh(A) - \sin(A)}{\cosh(A) + \cos(A)}\right]. \quad (5.383)$$

Figure 5.71 illustrates F_r for the litz wire winding as a function of d_l/δ_w at $k = 100$ and $\eta = 0.7$.

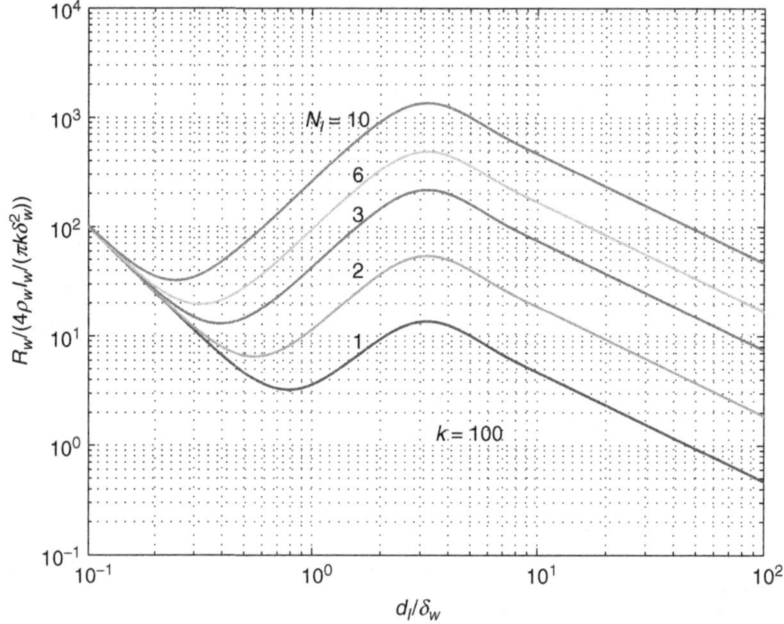

Figure 5.71 Plots of F_r for the litz-wire winding as a function of d_l/δ_w at $k = 100$ and $\eta = 0.7$

5.23.5 Approximated Optimum Strand Diameter

An approximate expression of Dowell's equation for F_R for a litz-wire winding for low and medium values of d_l/δ_w is given by

$$F_R \approx 1 + \frac{5N_{ll}^2 - 1}{45} A_{o1}^4 = 1 + \frac{5N_l^2 k - 1}{45} \eta^2 \left(\frac{\pi}{4}\right)^3 \left(\frac{d_l}{\delta_w}\right)^4 \quad \text{for} \quad \frac{d_l}{\delta_w} \le 2. \tag{5.384}$$

Figure 5.72 shows the exact and approximate plots of AC-to-DC winding resistance ratio F_R as a function of d/δ_w at $N_l = 2$, $k = 30$, and $\eta = 0.8$ for litz-wire winding.

The DC resistance of the litz-wire winding is

$$R_{wDC(litz)} = \frac{4\rho_w l_w}{\pi d_l^2 k}. \tag{5.385}$$

The AC resistance of the litz-wire winding at low and medium frequencies is

$$R_{w(litz)} = F_R R_{wDC(litz)} = \frac{4\rho_w l_w}{\pi d_l^2 k} \left[1 + \left(\frac{\pi}{4}\right)^3 \frac{\eta^2 (5N_l^2 k - 1)}{45} \left(\frac{d_l}{\delta_w}\right)^4\right]$$
$$= \frac{4\rho_w l_w}{\pi k} \left[\frac{1}{d_l^2} + \left(\frac{\pi}{4}\right)^3 \frac{\eta^2 (5N_l^2 k - 1)}{45\delta_w^4} d_l^2\right]. \tag{5.386}$$

Taking the derivative of this equation with respect to d_l and equating the result to zero, we get

$$\frac{dR_w}{dd_l} = \frac{4\rho_w l_w}{\pi k} \left[\frac{-2}{d_l^3} + \left(\frac{\pi}{4}\right)^3 \frac{2\eta^2 (5N_l^2 k - 1)}{45\delta_w^4} d_l\right] = 0. \tag{5.387}$$

The solution of this equation produces the optimum strand diameter at fixed values of k and δ_w (or frequency f), which gives the minimum AC litz-wire winding resistance

$$\frac{d_{lopt}}{\delta_w} = \sqrt[4]{\frac{45}{\eta^2 (5N_{ll}^2 - 1)}} = \sqrt[4]{\frac{45}{\eta^2 (5N_l^2 k - 1)\left(\frac{\pi}{4}\right)^3}} = \frac{3.10445}{\sqrt[4]{\eta^2 (5N_l^2 k - 1)}}. \tag{5.388}$$

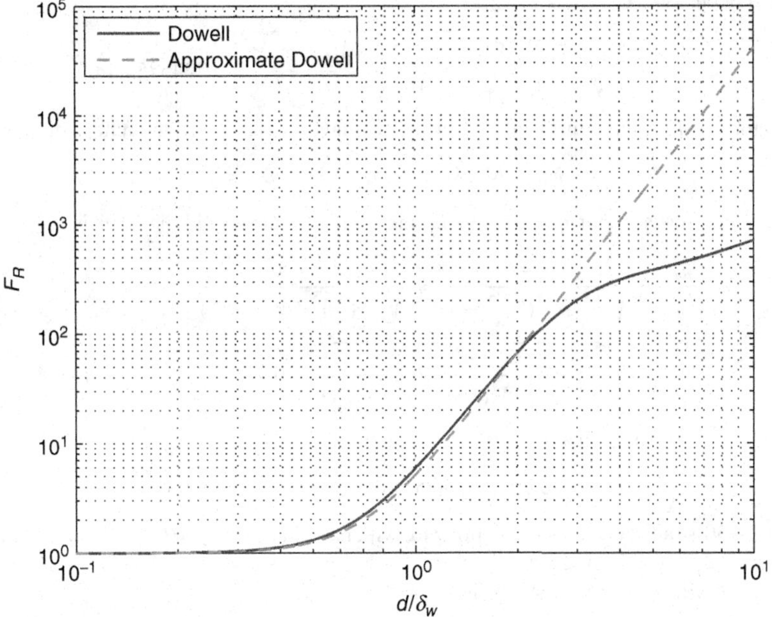

Figure 5.72 Exact and approximate AC-to-DC winding resistance ratio F_R as a function of d/δ_w at $N_l = 2$, $k = 30$, and $\eta = 0.8$ for litz-wire winding

The AC-to-DC resistance ratio at the d_{lopt} is

$$F_{Rv} = 2 \quad \text{for} \quad d_l = d_{lopt}. \tag{5.389}$$

The critical value of the strand diameter at which $F_r = F_r[d_{str(opt)}]$ is given by

$$\frac{d_{str(cr)}}{\delta_w} = \frac{2N_{ll}^2 + 1}{6\left(\frac{\pi}{4}\right)^{\frac{3}{4}}}\sqrt{\frac{45}{\eta(5N_{ll}^2 - 1)}}. \tag{5.390}$$

Substitution of $d_l = d/\sqrt{k}$ and (5.389) in (5.384) gives

$$F_{Rv} \approx 1 + \frac{5N_l^2 k - 1}{45}\eta^2\left(\frac{\pi}{4}\right)^3\left(\frac{d}{\delta_w\sqrt{k}}\right)^4 = 2. \tag{5.391}$$

Hence, the number of strands at the optimum strand diameter is

$$k_{cr} = \frac{5N_l^2 + \sqrt{(5N_l^2)^2 - \frac{180}{\eta^2\left(\frac{\pi}{4}\right)^3}\left(\frac{\delta_w}{d_i}\right)^4}}{\frac{90}{\eta^2\left(\frac{\pi}{4}\right)^3}\left(\frac{\delta_w}{d_i}\right)^4}. \tag{5.392}$$

Example 5.11

An inductor made of a copper litz wire carries a sinusoidal current at $f = 73$ kHz and $T = 20\,^\circ$C. The inductance is $L = 1.48$ mH. The total number of bundle turns is $N = 24$. The number of bundle layers is $N_l = 2$. The MTL is $l_T = 6.66$ cm. The amplitude of the inductor current is $I_{Lm} = 0.9169$ A. Find the optimum strand diameter d_{lopt} and the winding resistance $R_{w(litz)}$.

The skin depth is

$$\delta_w = \sqrt{\frac{\rho_{Cu}}{\pi \mu_0 f}} = \sqrt{\frac{1.724 \times 10^{-8}}{\pi \times 4 \times \pi \times 73 \times 10^3}} = 0.244585 \text{ mm}. \tag{5.393}$$

The inductor has 12 bundle turns in each of the two layers. The effective number of layers is

$$N_{ll} = N_l\sqrt{k} = 2 \times \sqrt{30} = 10.95 \approx 11. \tag{5.394}$$

Assuming the porosity factor $\eta = 0.7$, the normalized optimum strand diameter is

$$\frac{d_{lopt}}{\delta_w} = \sqrt[4]{\frac{45}{\eta^2\left(\frac{\pi}{4}\right)^3(5N_{ll}^2 - 1)}} = \sqrt[4]{\frac{45}{\eta^2\left(\frac{\pi}{4}\right)^3(5N_l^2 k - 1)}}$$

$$= \sqrt[4]{\frac{45}{0.7^2 \times \left(\frac{\pi}{4}\right)^3(5 \times 2^2 \times 30 - 1)}} = 0.75. \tag{5.395}$$

Hence, the optimum strand diameter is

$$d_{lopt} = 0.75\delta_w = 0.75 \times 0.224584 = 0.183439 \text{ mm}. \tag{5.396}$$

We select the 30-strand AWG32 copper litz-wire with bare diameter $d_l = 0.2019$ mm and the insulated wire diameter $d_o = 0.241$ mm.

The DC litz-wire winding resistance is

$$R_{wDC(litz)} = \frac{\rho_{Cu}}{\pi k d_l^2} = \frac{1.724 \times 10^{-8}}{\pi \times 30 \times (0.2019 \times 10^{-3})^2} = 0.02869\ \Omega = 0.02869\ \text{m}\Omega. \tag{5.397}$$

The maximum porosity factor is

$$\eta_{max} = \left(\frac{d_l}{d_o}\right)^2 = \left(\frac{0.2019}{0.241}\right)^2 = 0.70184 \approx 0.702. \tag{5.398}$$

The actual normalized strand diameter is

$$\frac{d_l}{\delta_w} = \frac{0.2019}{0.244584} = 0.82548. \tag{5.399}$$

The AC-to-DC resistance ratio is

$$F_{R(litz)} \approx 1 + \frac{5N_l^2 k - 1}{45}\left(\frac{\pi}{4}\right)^3 \eta_{max}^2 \left(\frac{d_l}{\delta_w}\right)^4.$$

$$= 1 + \frac{5 \times 2^2 \times 30 - 1}{45}\left(\frac{\pi}{4}\right)^3 0.70184^2 \times (0.82548)^4 = 2.475. \tag{5.400}$$

Thus, the AC litz-wire winding resistance is

$$R_{w(litz)} = F_{R(litz)} R_{wDC} = 2.475 \times 0.02869 = 0.071\ \Omega = 71\ \text{m}\Omega. \tag{5.401}$$

The power loss in the inductor winding is

$$P_{w(litz)} = \frac{R_{w(litz)} I_{Lm}^2}{2} = \frac{0.071 \times 0.9169^2}{2} = 0.02985 \text{ W} = 29.85 \text{ mW}. \tag{5.402}$$

5.23.6 Optimum Strand Diameter at Variable Porosity Factor

The effective diameter of the strand normalized with respect to the skin depth is

$$A = A_{str} = \left(\frac{\pi}{4}\right)^{\frac{3}{4}} \frac{d_l}{\delta_w}\sqrt{\frac{\eta}{k}} = \left(\frac{\pi}{4}\right)^{\frac{3}{4}} \frac{d_l}{\delta_w}\sqrt{\frac{d_l}{pk}} \tag{5.403}$$

The approximate winding AC resistance is

$$R_w = R_{wDC}F_R \approx R_{wDC}\left(1 + \frac{5N_{ll}^2 - 1}{45}\right)A_{str}^4 = \frac{4\rho_w l_w}{\pi k}\left[\frac{1}{d_l^2} + \frac{\pi^3(5N_{ll}^2 - 1)d_l^4}{2880\delta_w^4 p^2 k^2}\right]. \quad (5.404)$$

Taking the derivative of this equation with respect to d_l and equating the result to zero,

$$\frac{dR_w}{dd_l} = \frac{4\rho_w l_w}{\pi k}\left[\frac{-2}{d_l^3} + \frac{4\pi^3(5N_{ll}^2 - 1)d_l^3}{2880\delta_w^4 p^2 k^2}\right] = 0, \quad (5.405)$$

the optimum strand diameter is obtained

$$d_{l(opt)} = \sqrt[6]{\frac{1440\delta_w^4 p^2 k^2}{\pi^3(5N_{ll}^2 - 1)}}. \quad (5.406)$$

5.23.7 Boundary Between Low and Medium Frequencies for Litz-Wire Windings

The approximate Dowell's equation is valid for $F_{RB} = 1.05$. Hence,

$$F_{RB} \approx 1 + \left(\frac{\pi}{4}\right)^3 \frac{\eta^2(5N_l^2 k - 1)}{45}\left(\frac{d_{lB}}{\delta_w}\right)^4 = 1.05, \quad (5.407)$$

yielding the strand diameter at the boundary between the low and medium values of d/δ_w for litz-wire windings

$$\frac{d_{lB}}{\delta_w} = \sqrt[4]{\frac{9}{4\left(\frac{\pi}{4}\right)^3\eta^2(5N_l^2 k - 1)}}. \quad (5.408)$$

The boundary frequency is

$$f_L = \frac{\rho_w}{\pi\mu_0 d_B^2}\sqrt{\frac{9}{4\left(\frac{\pi}{4}\right)^3\eta^2(5N_l^2 k - 1)}}. \quad (5.409)$$

The ratio of the optimum strand diameter to the boundary strand diameter is

$$\frac{d_{lopt}}{d_{lB}} = \sqrt[4]{20} \approx 2.115. \quad (5.410)$$

The ratio of the boundary winding AC resistance R_{wB} to the minimum winding AC resistance for litz-wire inductors is

$$\frac{R_{wB}}{R_{w\min}} = \frac{R_{wDCB}F_{RB}}{R_{wDCopt}F_{Rv}} = \left(\frac{d_{lopt}}{d_{lB}}\right)^2 \frac{F_{RB}}{F_{Rv}} = \frac{1.05}{2}\sqrt{20} \approx 2.34871. \quad (5.411)$$

5.23.8 Approximation of Litz-Wire Winding Resistance for Low and Medium Frequencies

For litz-wire winding, the total number of strand turns is Nk. Therefore, the AC-to-DC winding resistance ratio for litz wire is given by Sullivan [48]

$$F_R = \frac{R_w}{R_{wDC}} = 1 + \frac{\pi^2\mu_0^2\omega^2(Nk)^2 d^6}{768\rho_w^2 b_c^2} = 1 + \frac{\pi^4\mu_0^2 f^2(Nk)^2 d^6}{192\rho_w^2 b_c^2}. \quad (5.412)$$

The porosity factor related to the strand insulation is

$$\eta_s = \frac{d_l}{d_t} = \frac{d_l}{d_l + 2t} = \frac{1}{1 + \frac{2t}{d_l}}. \quad (5.413)$$

The diameter of the strand with the insulation thickness t is

$$d_t = d_l + 2t = d_l\left(1 + \frac{2t}{d_l}\right) = \frac{d_l}{\eta_s}. \tag{5.414}$$

Consider a rectangular bobbin window of width b_b and height h_c. The bobbin width factor is defined as

$$\eta_b = \frac{b_b}{b_c} \tag{5.415}$$

The number of turns of strands per strand layer is

$$N_{tl} = \frac{b_b}{d_t} = \frac{\eta_b b_c}{d_t} = \frac{\eta_s \eta_b b_c}{d_t} \tag{5.416}$$

and the number of strand layers is

$$N_{ll} = \frac{h_b}{d_t} = \frac{h_b}{d_t} = \frac{h_b}{d_t}. \tag{5.417}$$

The total number of strand turns is

$$Nk = N_{ll}N_{tl} = N_{ll}\frac{b_c}{d_t} = N_{ll}\frac{\eta_s\eta_b b_c}{d_l} = N_{ll}\frac{\eta b_c}{d_l}, \tag{5.418}$$

where $\eta = \eta_s \eta_b$.

The AC-to-DC resistance ratio for litz-wire windings becomes

$$F_R = \frac{R_w}{R_{wDC}} = 1 + \frac{\pi^4 \mu_0^2 f^2 N^2 k^2 d_l^6}{192 \rho_w^2 b_c^2} = 1 + \frac{\pi^2 \omega^2 \mu_0^2 \eta^2 N_{ll}^2 d_l^4}{768 \rho_w^2} = 1 + \frac{\pi^2 \eta^2 N_{ll}^2}{192}\left(\frac{d_l}{\delta_w}\right)^4. \tag{5.419}$$

Figure 5.73 shows the exact and approximate AC-to-DC winding resistance ratio F_R as a function of d/δ_w at $N_l = 2$, $k = 30$, and $\eta = 0.8$ for litz-wire winding. Plots shown in Figs 5.72 and 5.73 are nearly identical.

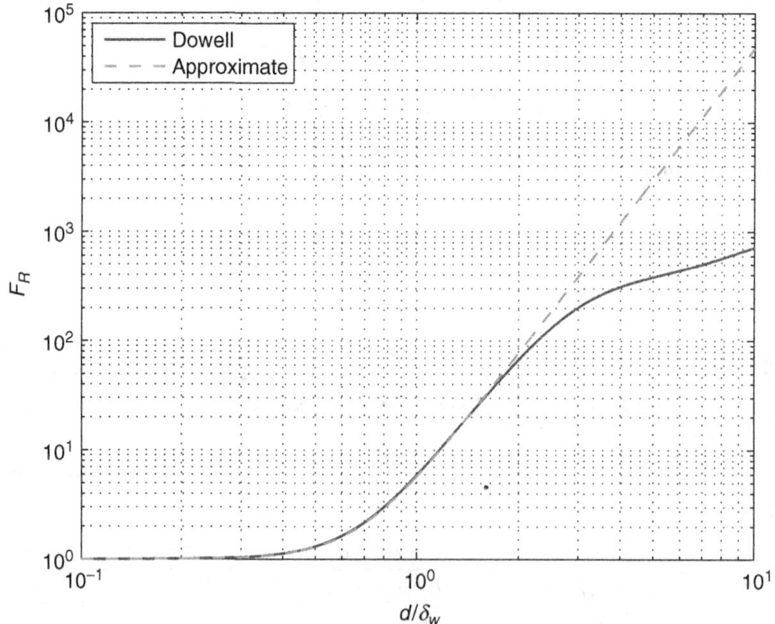

Figure 5.73 Exact and approximate AC-to-DC winding resistance ratio F_R as a function of d/δ_w at $N_l = 2$, $k = 30$, and $\eta = 0.8$ for litz-wire winding

The winding AC resistance of the litz-wire winding is

$$R_w = F_R R_{wDC} = \frac{4\rho_w l_w}{\pi k d_l^2}\left[1 + \frac{\pi^2\eta^2 N_{ll}^2}{192}\left(\frac{d_l}{\delta_w}\right)^4\right] = \frac{4\rho_w l_w}{\pi k}\left(\frac{1}{d_l^2} + \frac{\pi^2\eta^2 N_{ll}^2 d_l^2}{192\delta_w^4}\right). \tag{5.420}$$

Taking the derivative of R_w with respect to d_l and setting it to zero, we get

$$\frac{dR_w}{dd_l} = \frac{4\rho_w l_w}{\pi k}\left(\frac{-2}{d_l^3} + \frac{\pi^2\eta^2 N_{ll}^2 d_l}{96\delta_w^4}\right) = 0. \tag{5.421}$$

Hence, the optimum strand diameter is

$$\frac{d_{lopt}}{\delta_w} = \sqrt[4]{\frac{192}{(\pi\eta N_{ll})^2}} \approx \frac{2.1}{\sqrt{\eta N_{ll}}}. \tag{5.422}$$

The AC-to-DC resistance litz-wire winding ratio at d_{lopt}/δ_w is

$$F_{Rv} = 2. \tag{5.423}$$

5.24 Winding Power Loss for Inductor Current with Harmonics

In many applications such as pulse-width modulated (PWM) DC–DC converters, the inductor current i_L contains a DC component, a fundamental component, and many harmonics. If we know the DC component and the amplitudes or rms values of harmonics of the inductor current and the winding DC and AC resistances at the frequencies of harmonics nf_s, the winding power loss for each harmonic and the total winding power loss can be determined. The DC power loss in a winding is

$$P_{wDC} = R_{wDC} I_L^2. \tag{5.424}$$

Figure 5.74 shows the spectra of inductor current, inductor winding resistance, and winding power loss. The inductor current waveform can be expanded into a Fourier series

$$i_L = I_L + \sum_{n=1}^{\infty} I_{mn}\cos(n\omega t + \phi_n) = I_L + \sqrt{2}\sum_{n=1}^{\infty} I_n \cos(n\omega t + \phi_n), \tag{5.425}$$

where I_L is the DC component of the inductor current, I_{mn} is the amplitude, and $I_n = I_{mn}/\sqrt{2}$ is the rms value of the nth harmonic of the inductor current. The winding power loss due to the DC current and all harmonics of the inductor current is given by

$$\begin{aligned}
P_w &= R_{wDC} I_L^2 + \sum_{n=1}^{\infty} R_{wn} I_n^2 = R_{wDC} I_L^2 + R_{w1} I_1^2 + R_{w2} I_2^2 + R_{w3} I_3^2 + \ldots \\
&= R_{wDC} I_L^2\left[1 + \frac{R_{w1}}{R_{wDC}}\left(\frac{I_1}{I_L}\right)^2 + \frac{R_{w2}}{R_{wDC}}\left(\frac{I_2}{I_L}\right)^2 + \frac{R_{w3}}{R_{wDC}}\left(\frac{I_3}{I_L}\right)^2 + \ldots\right] \\
&= R_{wDC}\left[I_L^2 + \sum_{n=1}^{\infty} F_{Rn} I_n^2\right] = R_{wDC} I_L^2\left[1 + \sum_{n=1}^{\infty} F_{Rn}\left(\frac{I_n}{I_L}\right)^2\right] = P_{wDC}\left[1 + \sum_{n=1}^{\infty} F_{Rn}\left(\frac{I_n}{I_L}\right)^2\right] \\
&= P_{wDC}\left(1 + \sum_{n=1}^{\infty}\gamma_n^2 F_{Rn}\right) = P_{wDC}\left[1 + \frac{1}{2}\sum_{n=1}^{\infty} F_{Rn}\left(\frac{I_{mn}}{I_L}\right)^2\right] = F_{Rh} P_{wDC},
\end{aligned} \tag{5.426}$$

where $R_{wn} = R_{w1}, R_{w2}, R_{w3}, \ldots$ are the winding resistances at frequencies $nf_1 = f_1, f_2 = 2f_1, f_3 = 3f_1, \ldots F_{Rn} = R_{wn}/R_{wDC}$, the harmonic winding loss factor is defined as the ratio of the winding DC and AC power losses to the winding DC power loss

$$F_{Rh} = \frac{P_w}{P_{wDC}} = 1 + \sum_{n=1}^{\infty} F_{Rn}\left(\frac{I_n}{I_L}\right)^2 = 1 + \sum_{n=1}^{\infty}\gamma_n^2 F_{Rn}, \tag{5.427}$$

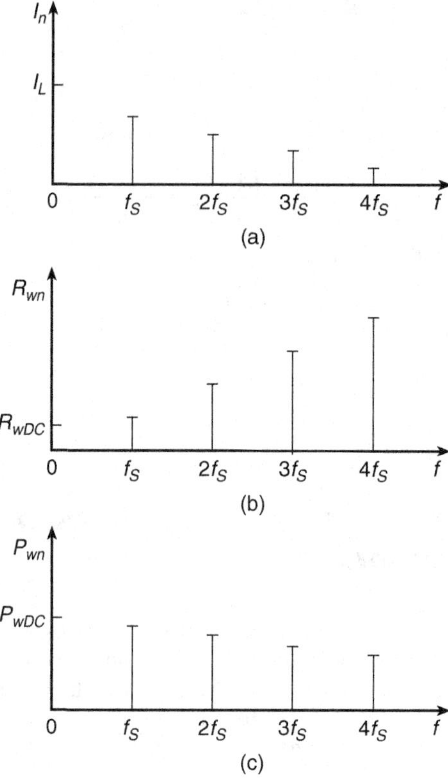

Figure 5.74 Spectra of the inductor current, winding resistance, and winding power loss. (a) Spectrum of the inductor current. (b) Spectrum of the inductor winding resistance. (c) Spectrum of the inductor winding power loss

and

$$\gamma_n = \frac{I_n}{I_L}. \tag{5.428}$$

The skin depth of the winding conductor at the nth harmonic is

$$\delta_{wn} = \sqrt{\frac{\rho_{Cu}}{\pi \mu_0 n f_s}} = \frac{\delta_{w1}}{\sqrt{n}}, \tag{5.429}$$

yielding

$$A_n = \frac{h}{\delta_{wn}} = \frac{h\sqrt{n}}{\delta_{w1}} = \sqrt{n}A, \tag{5.430}$$

where for a rectangular winding conductor

$$A = A_r = \frac{h}{\delta_w}\sqrt{\frac{w}{p}} \tag{5.431}$$

for a square winding conductor

$$A = A_s = \frac{h}{\delta_w}\sqrt{\frac{h}{p}} \tag{5.432}$$

and for a round winding conductor

$$A = A_o = \left(\frac{\pi}{4}\right)^{\frac{3}{4}} \frac{d}{\delta_{w1}}\sqrt{\frac{d}{p}}. \tag{5.433}$$

The normalized winding resistance at the fundamental frequency and harmonic frequencies is

$$F_{Rn} = \frac{R_{wn}}{R_{wDC}} = F_{sn} + F_{pn}$$
$$= \sqrt{n}A\left[\frac{\sinh(2A\sqrt{n}) + \sin(2A\sqrt{n})}{\cosh(2A\sqrt{n}) - \cos(2A\sqrt{n})} + \frac{2(N_l^2 - 1)}{3}\frac{\sinh(A\sqrt{n}) - \sin(A\sqrt{n})}{\cosh(A\sqrt{n}) + \cos(A\sqrt{n})}\right]$$
$$= \sqrt{n}A\left[\frac{e^{2A\sqrt{n}} - e^{-2A\sqrt{n}} + 2\sin(2A\sqrt{n})}{e^{2A\sqrt{n}} + e^{-2A\sqrt{n}} - 2\cos(2A\sqrt{n})} + \frac{2(N_l^2 - 1)}{3}\frac{e^{A\sqrt{n}} - e^{-A\sqrt{n}} - 2\sin(A\sqrt{n})}{e^{A\sqrt{n}} + e^{-A\sqrt{n}} + 2\cos(A\sqrt{n})}\right]. \quad (5.434)$$

This equation can be approximated by Hurley *et al.* [56]

$$F_{Rn} = \frac{R_w}{R_{wDC}} \approx 1 + \left(\frac{\pi}{4}\right)^3 \frac{\eta^2(5N_l^2 - 1)n^2}{45}\left(\frac{d}{\delta_{w1}}\right)^4 \quad \text{for} \quad \frac{d}{\delta_{w1}} \leq 2. \quad (5.435)$$

5.24.1 Copper Power Loss in PWM DC–DC Converters for Continuous Conduction Mode

Figure 5.75 shows a typical inductor current waveform i_L in nonisolated PWM DC–DC power converters for the continuous conduction mode (CCM). The inductor current waveform i_L consists of the DC component I_L and the AC component i_l. The two components are independent. The AC component of the inductor current waveform $i_l = i_L - I_L$ is given by

$$i_l = \begin{cases} \frac{\Delta i_L}{DT}t - \frac{\Delta i_L}{2}, & \text{for} \quad 0 < t \leq DT \\ \frac{-\Delta i_L}{(1-D)T}(t - DT) + \frac{\Delta i_L}{2}, & \text{for} \quad DT < t \leq T. \end{cases} \quad (5.436)$$

The trigonometric Fourier series of the AC component of the inductor current is

$$i_l(t) = \sum_{n=1}^{\infty}(a_n \cos n\omega t + b_n \sin n\omega t) = \sum_{n=1}^{\infty} I_{mn} \sin(n\omega t + \phi_n) \quad (5.437)$$

where the Fourier coefficients are

$$a_n = \frac{2}{T}\int_0^T i_l \cos n\omega t dt = \frac{\Delta i_L}{2\pi^2 D(1-D)}\frac{\cos(2\pi nD) - 1}{n^2}, \quad (5.438)$$

$$b_n = \frac{2}{T}\int_0^T i_l \sin n\omega t dt = \frac{\Delta i_L}{2\pi^2 D(1-D)}\frac{\sin(2\pi nD)}{n^2}, \quad (5.439)$$

the amplitudes of the fundamental component and the harmonics are

$$I_{mn} = \sqrt{a_n^2 + b_n^2} = \frac{\Delta i_L}{\pi^2 D(1-D)}\frac{\sin(n\pi D)}{n^2} = \frac{\Delta i_L}{n\pi(1-D)}\frac{\sin(n\pi D)}{n\pi D}$$
$$= \frac{\Delta i_L}{n\pi(1-D)}\text{sinc}(n\pi D), \quad (5.440)$$

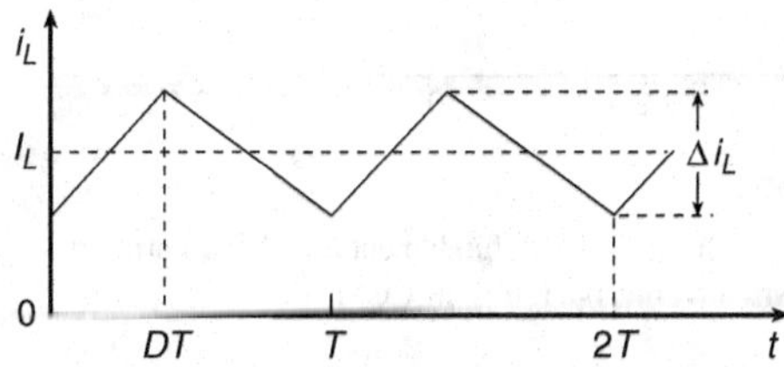

Figure 5.75 Inductor current waveform i_L in nonisolated PWM DC–DC power converters for CCM

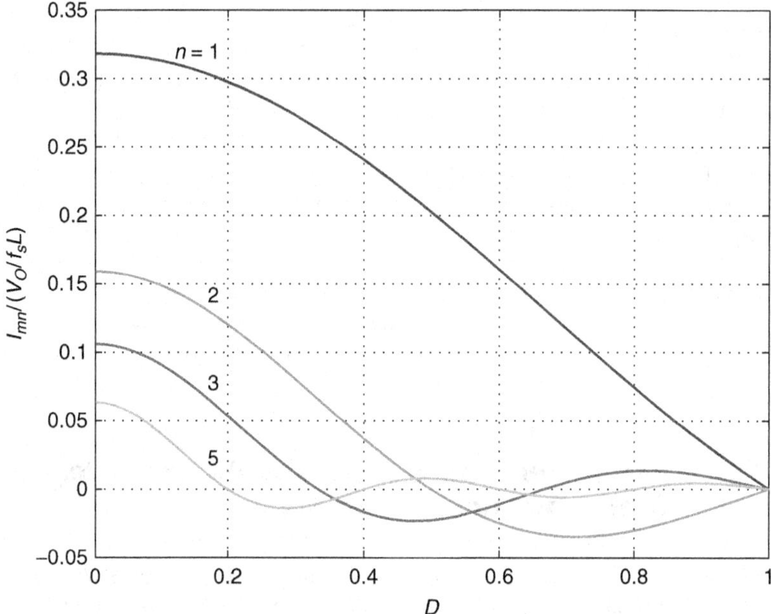

Figure 5.76 Normalized amplitudes of the fundamental and the harmonics $I_{mn}/[V_O/(f_sL)]$ as a function of the duty cycle D for the buck and buck–boost converters operating in CCM

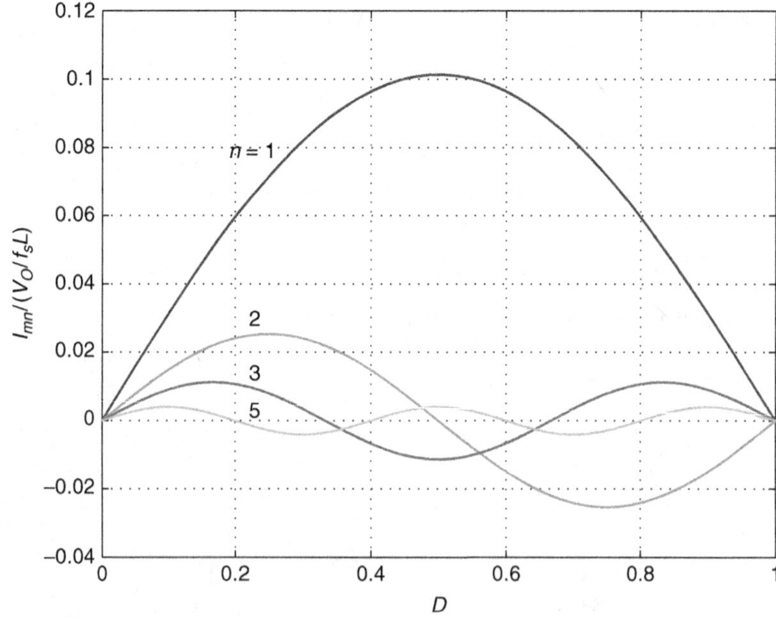

Figure 5.77 Normalized amplitudes of the fundamental and the harmonics $I_{mn}/[V_O/(f_sL)]$ as a function of the duty cycle D for the boost converter operating in CCM

and the phases of the fundamental component and the harmonics are

$$\phi_n = -\tan^{-1}\left(\frac{b_n}{a_n}\right) = -\tan^{-1}\left[\frac{\sin(2\pi nD)}{\cos(2\pi nD) - 1}\right] = \frac{\pi}{2} - \pi nD. \tag{5.441}$$

The trigonometric Fourier series of the AC component of the inductor current becomes

$$i_l = \frac{\Delta i_L}{\pi^2 D(1-D)} \sum_{n=1}^{\infty} \frac{1}{n^2}\{[\cos(2\pi nD) - 1]\cos n\omega t + \sin(2\pi nD)\sin n\omega t\}$$

$$= \frac{\Delta i_L}{\pi^2 D(1-D)} \sum_{n=1}^{\infty} \frac{\sin(n\pi D)}{n^2} \sin(n\omega t - n\pi D). \tag{5.442}$$

The Fourier series of the inductor current in nonisolated PWM DC–DC power converters for CCM at any duty cycle $D = t_{on}/T$ is given by

$$i_L = I_L + i_l = I_L + \frac{\Delta i_L}{\pi^2 D(1-D)} \sum_{n=1}^{\infty} \frac{\sin(n\pi D)}{n^2} \sin(n\omega t + \phi_n)$$

$$= I_L\left[1 + \frac{1}{\pi^2 D(1-D)}\left(\frac{\Delta i_L}{I_L}\right) \sum_{n=1}^{\infty} \frac{\sin(n\pi D)}{n^2} \sin(n\omega t - n\pi D)\right]. \tag{5.443}$$

For buck and buck–boost converters, the peak-to-peak value of the inductor current is given by Kazimierczuk [83]

$$\Delta i_L = \frac{V_O(1-D)}{f_s L}, \tag{5.444}$$

yielding the amplitudes of the fundamental and the harmonics of the inductor current

$$I_{mn} = \frac{V_O}{f_s L}\frac{\sin(n\pi D)}{n^2\pi^2 D} = \frac{V_O}{f_s L}\frac{\text{sinc}(n\pi D)}{n\pi}. \tag{5.445}$$

Figure 5.76 shows the plots of the normalized amplitudes of the fundamental and the harmonics $I_{mn}/[V_O/(f_s L)]$ as a function of the duty cycle D for the buck and buck–boost converters operating in CCM.

For boost converter, the peak-to-peak value of the inductor current is [83]

$$\Delta i_L = \frac{V_O D(1-D)}{f_s L}, \tag{5.446}$$

resulting in the amplitudes of the fundamental and the harmonics of the inductor

$$I_{mn} = \frac{V_O}{f_s L}\frac{\sin(n\pi D)}{n^2\pi^2}. \tag{5.447}$$

Figure 5.77 shows the plots of the normalized amplitudes of the fundamental and the harmonics $I_{mn}/[V_O/(f_s L)]$ as a function of the duty cycle D for the boost converter operating in CCM.

The normalized winding power loss at DC and harmonics P_{wn} is

$$\frac{P_{wn}}{P_{wDC}} = 1 + \frac{1}{2\pi^2 D^2(1-D)^2}\left(\frac{\Delta i_L}{I_L}\right)^2 F_{Rn}\left[\frac{\sin(n\pi D)}{n^2}\right]^2. \tag{5.448}$$

The copper power loss in the inductor winding for PWM DC–DC converters operating in CCM is given by

$$P_w = P_{wDC} + P_{wAC} = R_{wDC}I_L^2 + \frac{1}{2}\sum_{n=1}^{\infty} R_{wn}I_{mn}^2 = R_{wDC}I_L^2\left[1 + \frac{1}{2}\sum_{n=1}^{\infty}\frac{R_{wn}}{R_{wDC}}\left(\frac{I_{mn}}{I_L}\right)^2\right]$$

$$= P_{wDC}\left[1 + \frac{1}{2}\sum_{n=1}^{\infty} F_{Rn}\left(\frac{I_{mn}}{I_L}\right)^2\right] = R_{wDC}I_L^2 + \frac{(\Delta i_L)^2 R_{wDC}}{2\pi^4 D^2(1-D)^2}\sum_{n=1}^{\infty} F_{Rn}\left[\frac{\sin(n\pi D)}{n^2}\right]^2$$

$$= R_{wDC}I_L^2\left\{1 + \frac{1}{2\pi^4 D^2(1-D)^2}\left(\frac{\Delta i_L}{I_L}\right)^2 \sum_{n=1}^{\infty} F_{Rn}\left[\frac{\sin(n\pi D)}{n^2}\right]^2\right\}$$

$$= P_{wDC} \left\{ 1 + \frac{1}{2\pi^4 D^2 (1-D)^2} \left(\frac{\Delta i_L}{I_L} \right)^2 \sum_{n=1}^{\infty} F_{Rn} \left[\frac{\sin(n\pi D)}{n^2} \right]^2 \right\} = F_{Rh} P_{wDC}, \tag{5.449}$$

where

$$F_{Rh} = \frac{P_w}{P_{wDC}} = 1 + \frac{1}{2} \sum_{n=1}^{\infty} F_{Rn} \left(\frac{I_{mn}}{I_L} \right)^2 = 1 + \frac{1}{2\pi^4 D^2 (1-D)^2} \left(\frac{\Delta i_L}{I_L} \right)^2 \sum_{n=1}^{\infty} F_{Rn} \left[\frac{\sin(n\pi D)}{n^2} \right]^2. \tag{5.450}$$

If both the DC and AC currents flow through the winding,

$$F_{Rh} = \frac{P_w}{P_{wDC}} = \frac{P_{wDC} + P_{wAC}}{P_{wDC}} = 1 + \frac{P_{wAC}}{P_{wDC}} = 1 + F_{RhAC}, \tag{5.451}$$

where

$$F_{RhAC} = \frac{P_{wAC}}{P_{wDC}} = \frac{1}{2\pi^4 D^2 (1-D)^2} \left(\frac{\Delta i_L}{I_L} \right)^2 \sum_{n=1}^{\infty} F_{Rn} \left[\frac{\sin(n\pi D)}{n^2} \right]^2. \tag{5.452}$$

Figure 5.78 shows the plots of P_w/P_{wDC} as a function of d/δ_w for CCM at $D = 0.5$, $\Delta i_L/I_L = 0.2$, and $n = 100$ for a round solid wire. Figure 5.79 shows the plots of the normalized winding power loss P_w/P_{wDC} as a function of duty cycle D at the number of layers $N_l = 6$, normalized peak-to-peak inductor current $\Delta i_L/I_L = 0.2$, the number of harmonics $n = 100$, and selected values of d/δ_w for round solid wire. It can be seen that the lowest winding loss occurs at $D = 0.5$. For $d/\delta_w \le 1$, $P_w \approx P_{wDC}$ for $\Delta i_L/I_L \le 0.2$.

Figure 5.80 depicts the plots of normalized winding power loss P_w/P_{wDC} as a function of duty cycle D at selected values of the layers N_l and at $d/\delta_w = 5$, $\Delta i_L/I_L = 0.2$, and the number of harmonics $n = 100$ for solid round wire. As the number of layers N_l increases, the normalized winding loss P_w/P_{wDC} also increases.

Figure 5.81 shows the plots of the normalized winding power loss P_w/P_{wDC} as a function of duty cycle D at selected values of normalized peak-to-peak inductor current $\Delta i_L/I_L = 0.2$ and fixed values of $d/\delta_w = 5$, the number of layers $N_l = 6$, and the number of harmonics $n = 100$ for round

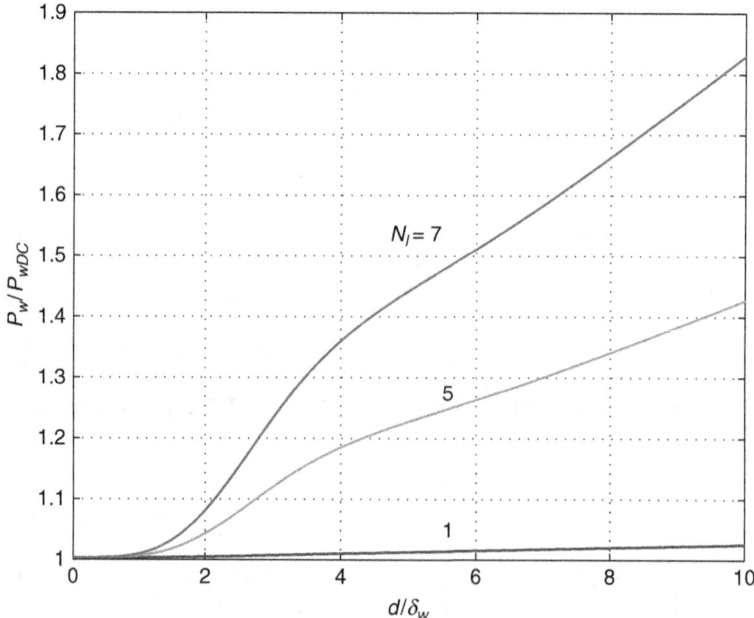

Figure 5.78 Normalized winding power loss P_w/P_{wDC} as a function of d/δ_w for CCM at selected numbers of layers of N_l at $D = 0.5$, $\Delta i_L/I_L = 0.2$, and $n = 100$ for solid round wire

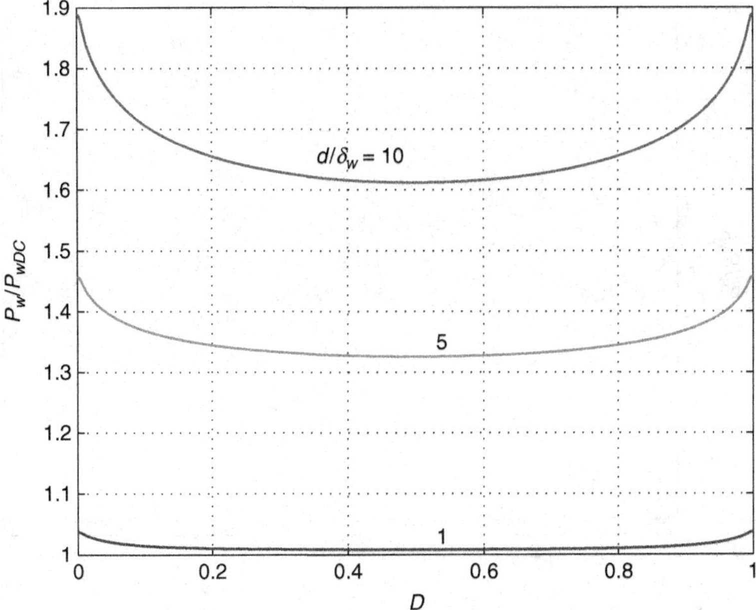

Figure 5.79 Normalized winding power loss P_w/P_{wDC} as a function of duty cycle D for CCM at selected values of d/δ_w for $N_L = 6$ and $\Delta i_L/I_L = 0.2$ for solid round wire

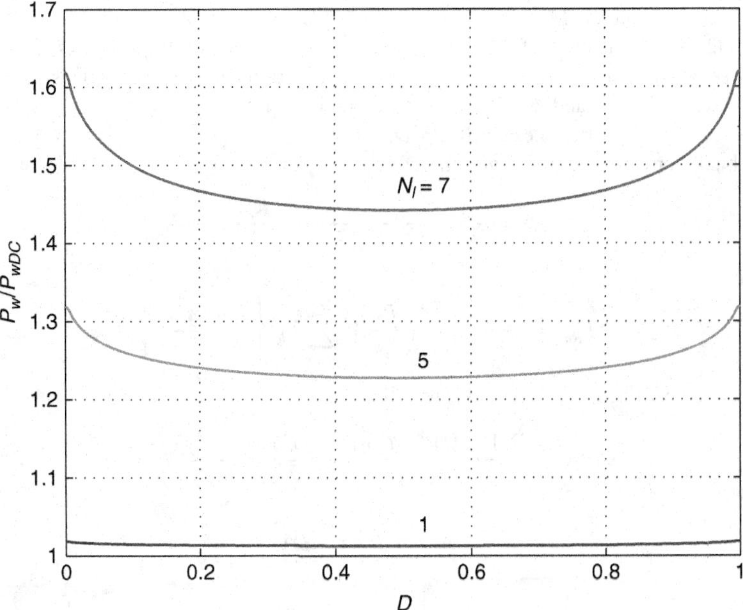

Figure 5.80 Normalized winding power loss P_w/P_{wDC} as a function of duty cycle D for CCM at selected values of N_l and fixed values of $d/\delta_w = 5$ and $\Delta i_L/I_L = 0.2$ for solid round wire

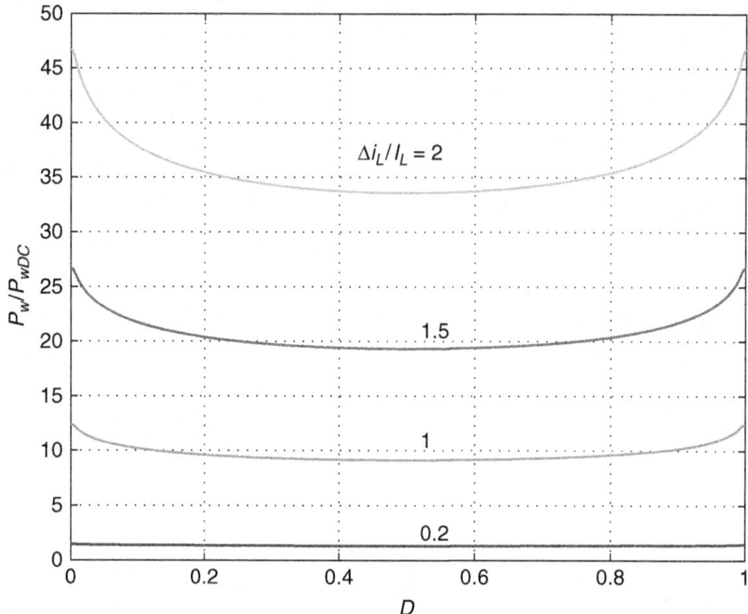

Figure 5.81 Normalized winding power loss as a function of duty cycle D for CCM at selected values of $\Delta i_L/I_L$ and fixed values of $d/\delta_w = 5$ and $N_l = 6$ for solid round wire

solid wire. It can be seen that P_w/P_{wDC} increases as $\Delta i_L/I_L$ increases. At $\Delta i_L/I_L = 2$, the mode of operation is close to the DCM.

Figure 5.82 depicts the plots of the normalized winding power loss P_w/P_{wDC} as a function of duty cycle D at selected values of harmonics n used in calculations and at fixed values of $d/\delta_w = 5$, the number of layers $N_l = 5$, and $\Delta i_L/I_L = 0.2$ for round solid wire. It can be seen that it is sufficient to take into account 10 harmonics for $0.1 \leq D \leq 0.9$.

For the buck converter, $I_L = I_O$, and therefore

$$\frac{\Delta i_L}{I_L} = \frac{\Delta i_L}{I_O} = \frac{V_O(1-D)}{I_O f_s L} = \frac{R_L(1-D)}{f_s L}. \tag{5.453}$$

This gives

$$F_{Rh} = 1 + \frac{1}{2\pi^4 D^2}\left(\frac{R_L}{f_s L}\right)^2 \sum_{n=1}^{\infty} F_{Rn}\left[\frac{\sin(n\pi D)}{n^2}\right]^2. \tag{5.454}$$

For the boost converter, $I_O = I_L(1-D)$, resulting in

$$\Delta i_L = \frac{V_O D(1-D)}{f_s L} = \frac{I_O R_L(1-D)}{f_s L} = \frac{I_L R_L (1-D)^2}{f_s L}. \tag{5.455}$$

Therefore,

$$\frac{\Delta i_L}{I_L} = \frac{R_L(1-D)^2}{f_s L}. \tag{5.456}$$

Hence,

$$F_{Rh} = 1 + \frac{(1-D)^2}{2\pi^4}\left(\frac{R_L}{f_s L}\right)^2 \sum_{n=1}^{\infty} F_{Rn}\left[\frac{\sin(n\pi D)}{n^2}\right]^2. \tag{5.457}$$

For the buck–boost converter, $I_L = DI_O/(1-D)$. Thus,

$$\Delta i_L = \frac{I_L R_L (1-D)^2}{D f_s L}, \tag{5.458}$$

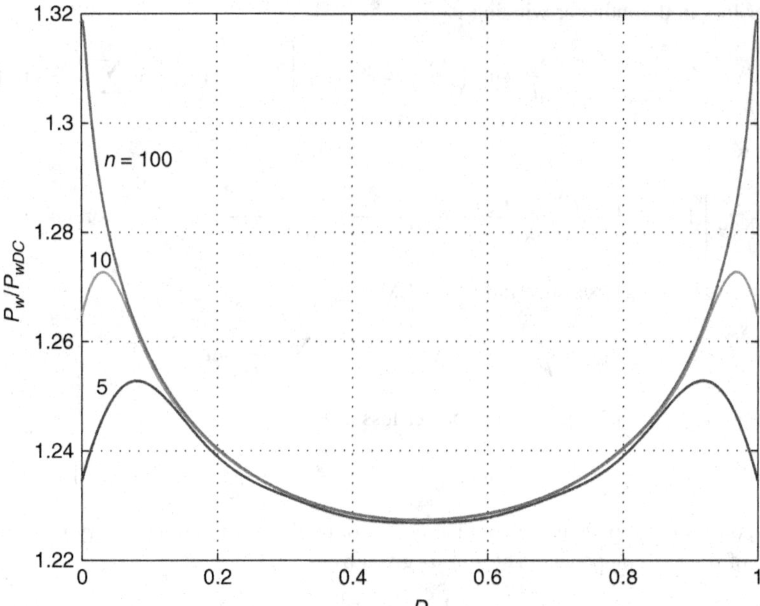

Figure 5.82 Normalized winding power loss P_w/P_{wDC} as a function of duty cycle D for CCM at selected values of harmonics n and fixed values of $d/\delta_w = 5$, $\Delta i_L/I_L = 0.2$, and $N_l = 5$ for solid round wire

producing

$$\frac{\Delta i_L}{I_L} = \frac{R_L(1-D)^2}{Df_s L}. \tag{5.459}$$

Hence,

$$F_{Rh} = 1 + \frac{(1-D)^2}{2\pi^4 D^4}\left(\frac{R_L}{f_s L}\right)^2 \sum_{n=1}^{\infty} F_{Rn}\left[\frac{\sin(n\pi D)}{n^2}\right]^2. \tag{5.460}$$

Let us consider a buck, a boost, or a buck–boost converter operated in CCM at duty cycle $D = 0.5$. For the odd symmetry of i_L, the inductor current waveform can be represented by a Fourier series as

$$\begin{aligned} i_L &= I_L + \frac{4\Delta i_L}{\pi^2}\sum_{n=1}^{\infty}\frac{1}{n^2}\sin\left(\frac{n\pi}{2}\right)\sin(n\omega t) \\ &= I_L + \frac{4\Delta i_L}{\pi^2}\left(\sin\omega t - \frac{1}{9}\sin 3\omega t + \frac{1}{25}\sin 5\omega t - \frac{1}{49}\sin 7\omega t + \ldots\right) \\ &= I_L\left[1 + \frac{4\Delta i_L}{\pi^2 I_L}\left(\sin\omega t - \frac{1}{9}\sin 3\omega t + \frac{1}{25}\sin 5\omega t - \frac{1}{49}\sin 7\omega t + \ldots\right)\right]. \end{aligned} \tag{5.461}$$

For the even symmetry of i_L,

$$\begin{aligned} i_L &= I_L + \frac{4\Delta i_L}{\pi^2}\sum_{n=1}^{\infty}\frac{1}{n^2}\sin\left(\frac{n\pi}{2}\right)\sin\left(n\omega t - \frac{n\pi}{2}\right) \\ &= I_L - \frac{4\Delta i_L}{\pi^2}\left(\cos\omega t + \frac{1}{9}\cos 3\omega t + \frac{1}{25}\cos 5\omega t + \frac{1}{49}\cos 7\omega t + \ldots\right) \\ &= I_L\left[1 - \frac{4\Delta i_L}{\pi^2 I_L}\left(\cos\omega t + \frac{1}{9}\cos 3\omega t + \frac{1}{25}\cos 5\omega t + \frac{1}{49}\cos 7\omega t + \ldots\right)\right]. \end{aligned} \tag{5.462}$$

The power loss in the inductor winding is given by

$$P_w = R_{wDC} I_L^2 + \frac{8\Delta i_L^2}{\pi^4} \sum_{n=1}^{\infty} \frac{R_{wn}}{n^4} \sin^2\left(\frac{n\pi}{2}\right) = R_{wDC} I_L^2 \left[1 + \frac{8}{\pi^4}\left(\frac{\Delta i_L}{I_L}\right)^2 \sum_{n=1}^{\infty} \frac{F_{Rn}}{n^4} \sin^2\left(\frac{n\pi}{2}\right)\right] \tag{5.463}$$

which gives

$$P_w = R_{wDC} I_L^2 \left[1 + \frac{8}{\pi^4}\left(\frac{\Delta i_L}{I_L}\right)^2 \left(\frac{R_{w1}}{R_{wDC}} + \frac{R_{w3}}{81R_{wDC}} + \frac{R_{w5}}{625R_{wDC}} + \ldots\right)\right] \quad \text{for} \quad n = 1, 3, 5, \ldots \ . \tag{5.464}$$

The AC-to-DC winding resistance ratio for CCM is

$$F_{Rh} = \frac{P_w}{P_{wDC}} = 1 + \frac{8}{\pi^4}\left(\frac{\Delta i_L}{I_L}\right) \sum_{n=1}^{\infty} \frac{F_{Rn}}{n^4} \sin^2\left(\frac{n\pi}{2}\right). \tag{5.465}$$

Assuming $\Delta i_L / I_L = 0.2$, the winding power loss at $D = 0.5$ is

$$P_w = R_{wDC} I_L^2 \left[1 + 0.003285\left(\frac{R_{w1}}{R_{wDC}} + \frac{R_{w3}}{81R_{wDC}} + \frac{R_{w5}}{625R_{wDC}} + \ldots\right)\right]. \tag{5.466}$$

Thus, the power of the fundamental and harmonics of the inductor current is very small compared to the power of the DC component. The sum is given by

$$\sum_{k=1}^{\infty} \frac{1}{(2k-1)^4} = \frac{\pi^4}{96} = 1.014678, \tag{5.467}$$

where the order of odd harmonics is given by $n = 2k - 1$ and $k = 1, 2, 3, \ldots$. If $R_{wn} = R_{wDC}$, $F_{Rn} = 1$, and the winding power loss becomes

$$\begin{aligned} P_w &= R_{wDC} I_L^2 + \frac{8\Delta i_L^2 R_{wn}}{\pi^4} = R_{wDC} I_L^2 \left[1 + \frac{8}{\pi^4}\left(\frac{\Delta i_L}{I_L}\right)^2 \frac{R_{wn}}{R_{wDC}} \sum_{n=1}^{\infty} \frac{1}{n^4}\right] \\ &= P_{wDC}\left[1 + \frac{1}{12}\left(\frac{\Delta i_L}{I_L}\right)^2\right] \quad \text{for} \quad n = 1, 3, 5, \ldots \end{aligned} \tag{5.468}$$

For $D = 0.5$, $\Delta i_L / I_L = 0.2$, and $R_{wn} = R_{wDC}$,

$$P_w = P_{wDC}\left[1 + \frac{1}{12}\left(\frac{1}{5}\right)^2\right] = P_{wDC}\left(1 + \frac{1}{300}\right) = 1.003333 P_{wDC} \tag{5.469}$$

and

$$F_{Rh} = \frac{P_w}{P_{wDC}} = 1.003333. \tag{5.470}$$

5.24.2 Copper Power Loss in PWM DC–DC Converters for DCM

Figure 5.83 depicts a typical inductor waveform i_L in PWM DC–DC power converters for the discontinuous conduction mode (DCM). In general, this is a periodic nonsymmetrical triangle waveform with time interval between pulses when the current is zero. The inductor current waveform is given by

$$i_L = \begin{cases} \frac{\Delta i_L}{DT} t, & \text{for} \quad 0 < t \le DT \\ -\frac{\Delta i_L}{D_1 T}(t - DT) + \Delta i_L, & \text{for} \quad DT < t \le (D + D_1)T \\ 0, & \text{for} \quad (D + D_1)T < t \le T. \end{cases} \tag{5.471}$$

The Fourier series of the inductor current is given by

$$i_L = I_L + \sum_{n=1}^{\infty} (a_n \cos n\omega t + b_n \sin n\omega t) = I_L + \sum_{n=1}^{\infty} I_{mn} \cos(n\omega t + \phi_n) \tag{5.472}$$

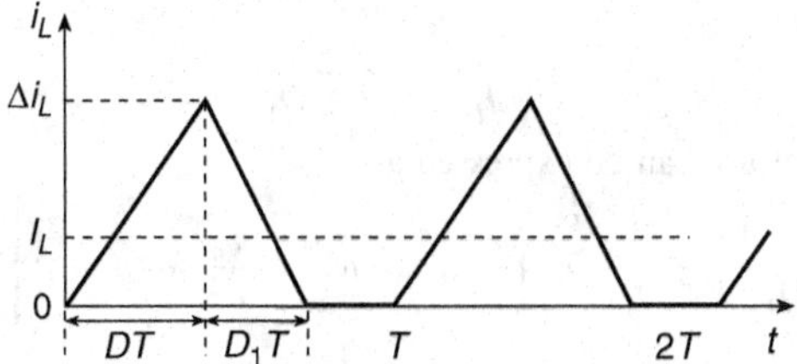

Figure 5.83 Inductor current waveform i_L in PWM DC–DC power converters for DCM

where

$$I_{mn} = \sqrt{a_n^2 + b_n^2} \tag{5.473}$$

and

$$\phi_n = -\arctan\left(\frac{b_n}{a_n}\right). \tag{5.474}$$

The DC component of the inductor current is

$$I_L = \frac{1}{T}\int_0^{(D+D_1)T} i_L dt = \frac{\Delta i_L (D + D_1)}{2}. \tag{5.475}$$

The Fourier series coefficients are

$$a_n = \frac{2}{T}\int_0^{(D+D_1)T} i_L \cos(n\omega t)dt = \frac{\Delta i_L}{2\pi^2 n^2 D D_1} X_n = \frac{I_L}{\pi^2 n^2 D D_1 (D + D_1)} X_n \tag{5.476}$$

and

$$b_n = \frac{2}{T}\int_0^{(D+D_1)T} i_L \sin(n\omega t)dt = \frac{\Delta i_L}{2\pi^2 n^2 D D_1} Y_n = \frac{I_L}{\pi^2 n^2 D D_1 (D + D_1)} Y_n, \tag{5.477}$$

where

$$X_n = D_1(\cos 2\pi nD - 1) - D[\cos 2\pi n(D + D_1) - \cos 2\pi nD] \tag{5.478}$$

and

$$Y_n = D_1 \sin 2\pi nD - D[\sin 2\pi n(D + D_1) - \sin 2\pi nD]. \tag{5.479}$$

The amplitudes of the fundamental component and the harmonics of the inductor current of the PWM DC–DC converters operating in DCM are given by

$$\begin{aligned} I_{mn} &= \sqrt{a_n^2 + b_n^2} = \frac{\Delta i_L}{2\pi^2 n^2 D D_1}\sqrt{X_n^2 + Y_n^2} = \frac{I_L}{\pi^2 n^2 D D_1 (D + D_1)}\sqrt{X_n^2 + Y_n^2} \\ &= \frac{I_L\sqrt{4D_1^2 \sin^2 \pi nD + 4D^2 \sin^2 \pi nD_1 - 2DD_1[4 \sin \pi nD \sin \pi nD_1 \cos \pi n(D + D_1)]}}{\pi^2 n^2 D D_1 (D + D_1)}. \end{aligned} \tag{5.480}$$

The phases of the harmonics are

$$\phi_n = -\arctan\left(\frac{b_n}{a_n}\right) = -\arctan\left(\frac{Y_n}{X_n}\right). \tag{5.481}$$

The trigonometric Fourier series of the inductor current of PWM DC–DC converters for DCM is expressed as

$$\begin{aligned} i_L &= \frac{\Delta i_L (D + D_1)}{2} + \frac{\Delta i_L}{2\pi^2 D D_1}\sum_{n=1}^{\infty}\left[\frac{X_n}{n^2}\cos(n\omega t) + \frac{Y_n}{n^2}\sin(n\omega t)\right] \\ &= I_L + \frac{\Delta i_L}{2\pi^2 D D_1}\sum_{n=1}^{\infty}\left[\frac{X_n}{n^2}\cos(n\omega t) + \frac{Y_n}{n^2}\sin(n\omega t)\right] \\ &= I_L\left\{1 + \frac{1}{2\pi^2 D D_1}\left(\frac{\Delta i_L}{I_L}\right)\sum_{n=1}^{\infty}\left[\frac{X_n}{n^2}\cos(n\omega t) + \frac{Y_n}{n^2}\sin(n\omega t)\right]\right\}. \end{aligned} \tag{5.482}$$

Since

$$\frac{\Delta i_L}{I_L} = \frac{2}{D + D_1}, \tag{5.483}$$

the inductor current waveform can be expressed as

$$i_L = I_L \left\{ 1 + \frac{1}{\pi^2 D D_1 (D + D_1)} \sum_{n=1}^{\infty} \left[\frac{X_n}{n^2} \cos(n\omega t) + \frac{Y_n}{n^2} \sin(n\omega t) \right] \right\}$$
$$= I_L \left[1 + \sum_{n=1}^{\infty} \left(\frac{I_{mn}}{I_L} \right) \cos(n\omega t + \phi_n) \right] = I_L \left[1 + \sum_{n=1}^{\infty} \gamma_n \cos(n\omega t + \phi_n) \right], \tag{5.484}$$

where

$$\gamma_n = \frac{I_{mn}}{I_L}$$
$$= \frac{\sqrt{4D_1^2 \sin^2 \pi n D + 4D^2 \sin^2 \pi n D_1 - 2DD_1 [4 \sin \pi n D \sin \pi n D_1 \cos \pi n (D + D_1)]}}{2\pi^2 n^2 D D_1}. \tag{5.485}$$

The relationship between D_1 and D is converter dependent [83]. For the buck converter,

$$D_1 = D \left(\frac{V_I}{V_O} - 1 \right) = \frac{D}{2} \left(\sqrt{1 + \frac{8 f_s L}{D R_L}} - 1 \right). \tag{5.486}$$

For the boost converter,

$$D_1 = \frac{D}{\frac{V_O}{V_I} - 1} = \frac{2D}{\sqrt{1 + \frac{2D^2 R_L}{f_s L}} - 1}. \tag{5.487}$$

For the buck–boost converter,

$$D_1 = \sqrt{\frac{2 f_s L}{2 R_L}}. \tag{5.488}$$

The copper power loss in inductor winding of PWM DC–DC converters in DCM is given by

$$P_w = R_{wDC} \frac{[\Delta i_L (D + D_1)]^2}{4} + \frac{(\Delta i_L)^2 R_{wDC}}{8\pi^4 D^2 D_1^2} \sum_{n=1}^{\infty} \frac{F_{Rn} (X_n^2 + Y_n^2)}{n^4}$$
$$= R_{wDC} I_L^2 \left[1 + \frac{1}{8\pi^4 D^2 D_1^2} \left(\frac{\Delta i_L}{I_L} \right)^2 \sum_{n=1}^{\infty} \frac{F_{Rn} (X_n^2 + Y_n^2)}{n^4} \right]$$
$$= P_{wDC} \left[1 + \frac{1}{2\pi^4 D^2 D_1^2 (D + D_1)^2} \sum_{n=1}^{\infty} \frac{F_{Rn} (X_n^2 + Y_n^2)}{n^4} \right] = F_{Rh} P_{wDC}, \tag{5.489}$$

where

$$F_{Rh} = \frac{P_w}{P_{wDC}} = \frac{1}{2} \sum_{n=1}^{\infty} \frac{R_{wn}}{R_{wDC}} \left(\frac{I_{mn}}{I_L} \right)^2 = \frac{1}{2} \sum_{n=1}^{\infty} F_{Rn} \left(\frac{I_{mn}}{I_L} \right)^2 = \frac{1}{2} \sum_{n=1}^{\infty} F_{Rn} \gamma_n^2$$
$$= 1 + \frac{1}{2\pi^4 D^2 D_1^2 (D + D_1)^2} \sum_{n=1}^{\infty} \frac{F_{Rn} (X_n^2 + Y_n^2)}{n^4}. \tag{5.490}$$

The relationship between D_1 and D is converter dependent [83]. Figure 5.84 shows the plots of P_w/P_{wDC} as a function of d/δ_w for DCM at $D = 0.4$, $D_1 = 0.3$, and $n = 100$ for round solid wire. Figure 5.85 shows the plots of P_w/P_{wDC} as a function of d/δ_w for DCM at $D = 0.4$, $D_1 = 0.3$, and $n = 100$ for round solid wire in the enlarged scale.

The optimum thickness of the winding foil conductor of an inductor used in PWM DC–DC power converters operating in DCM at $D_1 = D$ is [56]

$$\frac{h_{opt}}{\delta_w} = \sqrt{\frac{5\pi^2 (D + D_1)}{5N_l^2 - 1}} = \sqrt{\frac{10\pi^2 D}{5N_l^2 - 1}} \approx \sqrt{\frac{5\pi^2}{5N_l^2 - 1}} \approx \frac{\pi}{\sqrt{N_l^2 - \frac{1}{5}}} \approx \frac{\pi}{N_l}. \tag{5.491}$$

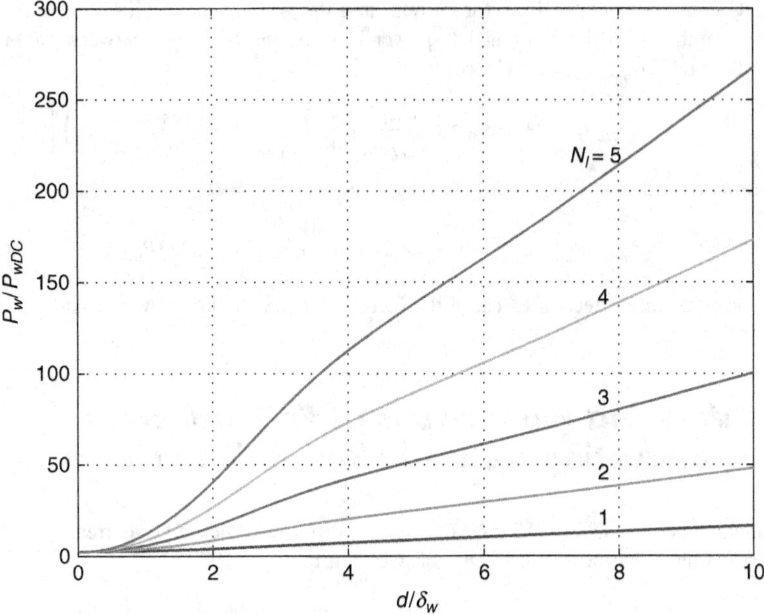

Figure 5.84 Normalized winding power loss P_w/P_{wDC} as a function of d/δ_w for DCM at selected numbers of layers of N_l at $d/p = 0.8$, $D = 0.4$, $D_1 = 0.3$, and $n = 100$ for solid round wire

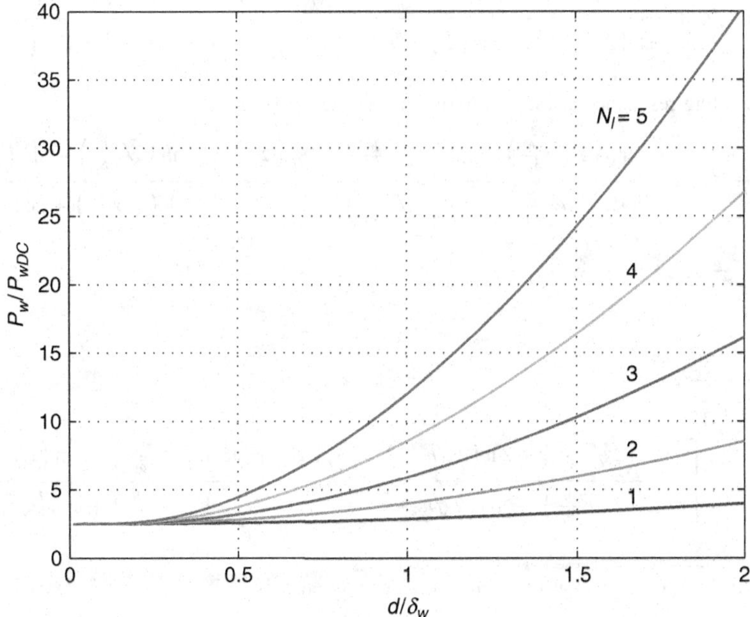

Figure 5.85 Normalized winding power loss P_w/P_{wDC} as a function of d/δ_w for DCM at selected numbers of layers of N_l at $d/p = 0.8$, $D = 0.4$, $D_1 = 0.3$, and $n = 100$ for solid round wire in enlarged scale

For DCM operation, the amplitudes of the fundamental and harmonics may become comparable with the DC component of the inductor current, and the winding power loss of the AC harmonics may be comparable with the winding DC power loss. At the boundary between DCM and CCM, $D = D' = 0.5$ and $\Delta i_{Lmax}/I_L = 2$. Hence,

$$P_w = R_{wDC} I_L^2 \left[1 + 0.3285 \left(\frac{R_{w1}}{R_{wDC}} + \frac{R_{w3}}{81 R_{wDC}} + \frac{R_{w5}}{625 R_{wDC}} + \ldots \right) \right]. \tag{5.492}$$

For uniform current density, $R_{wn} = R_{wDC}$. Hence, for $D = D' = 0.5$,

$$P_w = R_{wDC} I_L^2 \left(1 + \frac{1}{12} \times 2^2 \right) = \frac{4}{3} P_{wDC} = 1.3333 P_{wDC}. \tag{5.493}$$

The skin and proximity effects may cause the increase of the winding power loss.

5.25 Winding Power Loss of Foil Inductors Conducting DC and Harmonic Currents

In general, an inductor in DC-to-DC PWM power converters conducts a DC current I_L and AC current that consists of many harmonics. The DC foil winding resistance is

$$R_{wDC} = \frac{\rho_w l_w}{A_w} = \frac{\rho_w l_w}{bh} = \frac{\rho_w l_w}{\delta_w b \frac{h}{\delta_w}}, \tag{5.494}$$

where b is the foil width, l_w is the foil length, and $A_w = bh$ is the cross-sectional area of the foil. Using Dowell's equation, the winding AC-to-DC resistance ratio at the nth harmonic frequency is

$$F_{Rn} = \sqrt{n} \left[\frac{\sinh\left(2\sqrt{n}\frac{h}{\delta_w}\right) + \sin\left(2\sqrt{n}\frac{h}{\delta_w}\right)}{\cosh\left(2\sqrt{n}\frac{h}{\delta_w}\right) - \cos\left(2\sqrt{n}\frac{h}{\delta_w}\right)} + \frac{2(N_l^2 - 1)}{3} \frac{\sinh\left(\sqrt{n}\frac{h}{\delta_s}\right) - \sin\left(\sqrt{n}\frac{h}{\delta_w}\right)}{\cosh\left(\sqrt{n}\frac{h}{\delta_w}\right) + \cos\left(\sqrt{n}\frac{h}{\delta_w}\right)} \right] \tag{5.495}$$

The foil winding resistance at the nth harmonic frequency is

$$R_{wn} = \frac{\rho_w l_w}{b\delta_w} \sqrt{n} \left[\frac{\sinh\left(2\sqrt{n}\frac{h}{\delta_w}\right) + \sin\left(2\sqrt{n}\frac{h}{\delta_w}\right)}{\cosh\left(2\sqrt{n}\frac{h}{\delta_w}\right) - \cos\left(2\sqrt{n}\frac{h}{\delta_w}\right)} + \frac{2(N_l^2 - 1)}{3} \frac{\sinh\left(\sqrt{n}\frac{h}{\delta_s}\right) - \sin\left(\sqrt{n}\frac{h}{\delta_w}\right)}{\cosh\left(\sqrt{n}\frac{h}{\delta_w}\right) + \cos\left(\sqrt{n}\frac{h}{\delta_w}\right)} \right]$$

$$= \frac{\rho_w l_w}{b\delta_w} F_{rn}, \tag{5.496}$$

where

$$F_{rn} = \frac{R_{wn}}{\frac{\rho_w l_w}{b\delta_w}}$$

$$= \sqrt{n} \left[\frac{\sinh\left(2\sqrt{n}\frac{h}{\delta_w}\right) + \sin\left(2\sqrt{n}\frac{h}{\delta_w}\right)}{\cosh\left(2\sqrt{n}\frac{h}{\delta_w}\right) - \cos\left(2\sqrt{n}\frac{h}{\delta_w}\right)} + \frac{2(N_l^2 - 1)}{3} \frac{\sinh\left(\sqrt{n}\frac{h}{\delta_w}\right) - \sin\left(\sqrt{n}\frac{h}{\delta_w}\right)}{\cosh\left(\sqrt{n}\frac{h}{\delta_w}\right) + \cos\left(\sqrt{n}\frac{h}{\delta_w}\right)} \right]. \tag{5.497}$$

The foil winding power loss at the nth harmonic frequency is

$$P_{wn} = R_{wn} I_n^2 = \frac{\rho_w l_w}{b\delta_w} F_{rn} I_n^2. \tag{5.498}$$

The total foil winding power loss is

$$P_w = P_{wDC} + P_{wAC} = R_{wDC} I_L^2 + \sum_{n=1}^{\infty} R_{wn} I_n^2 = \frac{\rho_w l_w I_L^2}{b\delta_w} \left(\frac{1}{\frac{h}{\delta_w}} + \sum_{n=1}^{\infty} F_{rn} \frac{I_n^2}{I_L^2} \right)$$

$$= \frac{\rho_w l_w I_L^2}{b\delta_w}\left(\frac{1}{\frac{h}{\delta_w}} + \sum_{n=1}^{\infty} F_{rn}\gamma_n^2\right), \tag{5.499}$$

where

$$\gamma_n = \frac{I_n}{I_L} = \frac{I_{mn}}{\sqrt{2}I_L} \quad \text{for} \quad n = 1, 2, 3, \ldots \tag{5.500}$$

5.25.1 Optimum Foil Thickness of Inductors Conducting DC and Harmonic Currents

The AC-to-DC winding resistance ratio at nth harmonic frequency in (5.495) can be approximated using Maclaurin's series

$$F_{Rn} = \frac{R_{wn}}{R_{wDC}} \approx 1 + \frac{n^2(5N_l^2 - 1)}{45}\left(\frac{h}{\delta_w}\right)^4. \tag{5.501}$$

The winding AC resistance at the nth frequency harmonic is

$$R_{wn} = F_{Rn}R_{wDC} \approx \frac{\rho_w l_w}{b_{wf} h}\left[1 + \frac{n^2(5N_l^2 - 1)}{45}\left(\frac{h}{\delta_w}\right)^4\right]. \tag{5.502}$$

For low and medium foil thicknesses, the normalized resistance of the inductor foil winding at the nth harmonic frequency can be approximated by

$$F_{rn} = \frac{R_{wn}}{\frac{\rho_w l_w}{b\delta_w}} \approx \frac{1}{\frac{h}{\delta_w}} + \frac{n^2(5N_l^2 - 1)}{45}\left(\frac{h}{\delta_w}\right)^3 \quad \text{for} \quad \frac{h}{\delta_w} \leq 1.5. \tag{5.503}$$

The total power winding loss caused by both the DC current and harmonics for multilayer inductors is

$$P_w = \frac{\rho_w l_w I_L^2}{b\delta_w}\left(\frac{1}{\frac{h}{\delta_w}} + \sum_{n=1}^{\infty}\gamma_n^2 F_{rn}\right) \approx \frac{\rho_w l_w I_L^2}{b\delta_w}\left\{\frac{1}{\frac{h}{\delta_w}} + \sum_{n=1}^{\infty}\gamma_n^2\left[\frac{1}{\frac{h}{\delta_w}} + \frac{n^2(5N_l^2 - 1)}{45}\left(\frac{h}{\delta_w}\right)^3\right]\right\}. \tag{5.504}$$

The optimum foil thickness of a multilayer inductor conducting both the DC current and harmonics may be determined by taking the derivative of P_w with respect to h/δ_w and setting the result to zero

$$\frac{1}{\frac{\rho_w l_w I_L^2}{b\delta_w}}\frac{dP_w}{d\left(\frac{h}{\delta_w}\right)} = -\frac{1}{\left(\frac{h}{\delta_w}\right)^2}\left(1 + \sum_{n=1}^{\infty}\gamma_n^2\right) + \frac{(5N_l^2 - 1)}{15}\left(\frac{h}{\delta_w}\right)^2\sum_{n=1}^{\infty} n^2\gamma_n^2 = 0. \tag{5.505}$$

The normalized optimum foil thickness of a multilayer inductor is

$$\frac{h_{opt}}{\delta_w} = \sqrt[4]{\frac{15}{5N_l^2 - 1}}\sqrt[4]{\frac{1 + \sum_{n=1}^{\infty}\gamma_n^2}{\sum_{n=1}^{\infty} n^2\gamma_n^2}} = k_f k_w, \tag{5.506}$$

where the foil winding factor is

$$k_f = \sqrt[4]{\frac{15}{5N_l^2 - 1}} \tag{5.507}$$

and the waveform factor of the inductor current is

$$k_w = \sqrt[4]{\frac{1 + \sum_{n=1}^{\infty}\gamma_n^2}{\sum_{n=1}^{\infty} n^2\gamma_n^2}}. \tag{5.508}$$

Figure 5.86 shows the plots of winding DC power loss P_{wDC}, winding AC power loss P_{wAC}, and the total winding power loss $P_w = P_{wDC} + P_{wAC}$ as functions of foil thickness h for the 15-layer inductor ($N_l = 15$) used in the PWM converter operated in DCM.

For the rectangular (strip) conductor winding, the normalized optimum foil thickness of a multilayer inductor is

$$\frac{h_{opt}}{\delta_w} = \sqrt[4]{\frac{15}{\eta^2(5N_l^2 - 1)}}\sqrt[4]{\frac{1 + \sum_{n=1}^{\infty}\gamma_n^2}{\sum_{n=1}^{\infty} n^2\gamma_n^2}} = k_r k_w, \tag{5.509}$$

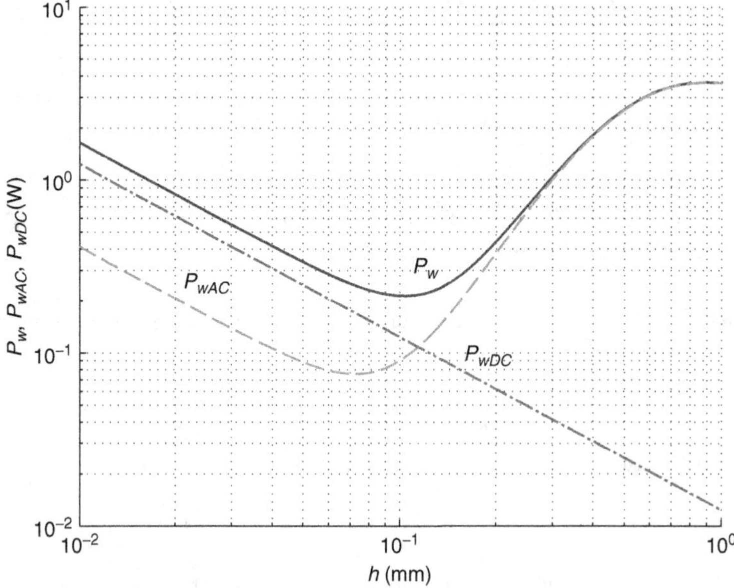

Figure 5.86 Winding DC power loss P_{wDC}, winding AC power loss P_{wAC}, and the total winding power loss $P_w = P_{wDC} + P_{wAC}$ as functions of foil thickness h for the 15-layer inductor ($N_l = 15$) used in the PWM converter operated in DCM

where the strip winding factor is

$$k_r = \sqrt[4]{\frac{15}{\eta^2(5N_l^2 - 1)}}. \tag{5.510}$$

5.26 Winding Power Loss of Round Wire Inductors Conducting DC and Harmonic Currents

For low and medium foil thicknesses, the normalized resistance of the winding DC resistance made of a round wire is

$$R_{wDC} = \frac{\rho_w l_w}{A_w} = \frac{4\rho_w l_w}{\pi d^2} = \frac{4\rho_w l_w}{\pi \delta_w^2 \left(\frac{d}{\delta_w}\right)^2}, \tag{5.511}$$

where d is the bare wire diameter and l_w is the wire length. The winding power loss at the nth harmonic frequency is

$$P_{wn} = R_{wn} I_n^2 = \frac{4\rho_w l_w}{\pi \delta_w^2} F_{rn} I_n^2, \tag{5.512}$$

where

$$F_{rn} = \frac{R_{wn}}{\frac{4\left(\frac{\pi}{4}\right)^{\frac{3}{4}} \sqrt{\eta}\rho_w l_w}{\pi \delta_w^2}}$$

$$= \frac{\sqrt{n}}{\frac{d}{\delta_w}} \left[\frac{\sinh(2\sqrt{n}A) + \sin(2\sqrt{n}A)}{\cosh(2\sqrt{n}A) - \cos(2\sqrt{n}A)} + \frac{2(N_l^2 - 1)}{3} \frac{\sinh(\sqrt{n}A) - \sin(\sqrt{n}A)}{\cosh(\sqrt{n}A) + \cos(\sqrt{n}A)}\right] \tag{5.513}$$

and

$$A = \left(\frac{\pi}{4}\right)^{\frac{3}{4}} \frac{d}{\delta_w} \sqrt{\eta}. \tag{5.514}$$

The winding power loss due to both the DC current I_L and harmonics flowing through the inductor is

$$P_w = P_{wDC} + P_{wAC} = R_{wDC} I_L^2 + \sum_{n=1}^{\infty} R_{wn} I_n^2 = \frac{4\rho_w l_w I_L^2}{\pi \delta_w^2} \left[\frac{1}{\left(\frac{d}{\delta_w}\right)^2} + \sum_{n=1}^{\infty} F_{rn} \gamma_n^2 \right]. \quad (5.515)$$

5.26.1 Optimum Diameter of Inductors Conducting DC and Harmonic Currents

The winding AC-to-DC resistance ratio at the nth harmonic frequency is

$$F_{Rn} = \frac{R_{wn}}{R_{wDC}} \approx 1 + \left(\frac{\pi}{4}\right)^{\frac{3}{4}} \frac{n^2(5N_l^2 - 1)}{45} \left(\frac{h}{\delta_w}\right)^4. \quad (5.516)$$

For low and medium diameters, the normalized resistance of the inductor winding at the nth harmonic frequency is given by

$$F_{rn} = \frac{R_{wn}}{\frac{4\rho_w l_w}{\pi \delta_w^2}} \approx \frac{1}{\left(\frac{d}{\delta_w}\right)^2} + \frac{\left(\frac{\pi}{4}\right)^3 n^2 \eta^2 (5N_l^2 - 1)}{15} \left(\frac{d}{\delta_w}\right)^2 \quad \text{for} \quad \frac{d}{\delta_w} \le 1.5. \quad (5.517)$$

The power loss due to both the DC and AC currents with harmonics is

$$P_w = \frac{4\rho_w l_w I_L^2}{\pi \delta_w^2} \left\{ \frac{1}{\left(\frac{d}{\delta_w}\right)^2} + \sum_{n=1}^{\infty} \gamma_n^2 \left[\frac{1}{\left(\frac{d}{\delta_w}\right)^2} + \frac{\left(\frac{\pi}{4}\right)^3 n^2 \eta^2 (5N_l^2 - 1)}{45} \left(\frac{d}{\delta_w}\right)^2 \right] \right\}. \quad (5.518)$$

The optimum diameter is determined by taking the derivative of P_w with respect to d/δ_w and setting the result to zero

$$\frac{1}{\frac{4\rho_w l_w I_L^2}{\pi \delta_w^2}} \frac{dP_w}{d\left(\frac{d}{\delta_w}\right)} = -\frac{1}{\left(\frac{d}{\delta_w}\right)^3} \left(1 + \sum_{n=1}^{\infty} \gamma_n^2\right) + \frac{\left(\frac{\pi}{4}\right)^3 \eta^2 (5N_l^2 - 1)}{45} \left(\frac{d}{\delta_w}\right) \sum_{n=1}^{\infty} n^2 \gamma_n^2 = 0. \quad (5.519)$$

Hence, one obtains the optimum normalized wire diameter for multilayer inductors conducting both the DC current and harmonics

$$\frac{d_v}{\delta_w} = \frac{1}{\left(\frac{\pi}{4}\right)^{0.75} \sqrt{\eta}} \sqrt[4]{\frac{45}{5N_l^2 - 1}} \sqrt[4]{\frac{1 + \sum_{n=1}^{\infty} \gamma_n^2}{\sum_{n=1}^{\infty} n^2 \gamma_n^2}} = k_o k_w, \quad (5.520)$$

where the round wire winding factor is

$$k_o = \frac{1}{\left(\frac{\pi}{4}\right)^{0.75} \sqrt{\eta}} \sqrt[4]{\frac{45}{5N_l^2 - 1}}. \quad (5.521)$$

Figure 5.87 shows the plots of winding DC power loss P_{wDC}, winding AC power loss P_{wAC}, and the total winding power loss $P_w = P_{wDC} + P_{wAC}$ as functions of wire diameter d for the four-layer inductor ($N_l = 4$) used in the PWM converter operated in DCM.

5.27 Effective Winding Resistance for Nonsinusoidal Inductor Current

The rms value of the nonsinusoidal periodic inductor current i_L is given by

$$I_{Lrms} = \sqrt{I_L^2 + I_1^2 + I_2^2 + \ldots} = \sqrt{I_L^2 + \sum_{n=1}^{\infty} I_n^2} = \sqrt{I_L^2 \left[1 + \sum_{n=1}^{\infty} \left(\frac{I_n}{I_L}\right)^2\right]}. \quad (5.522)$$

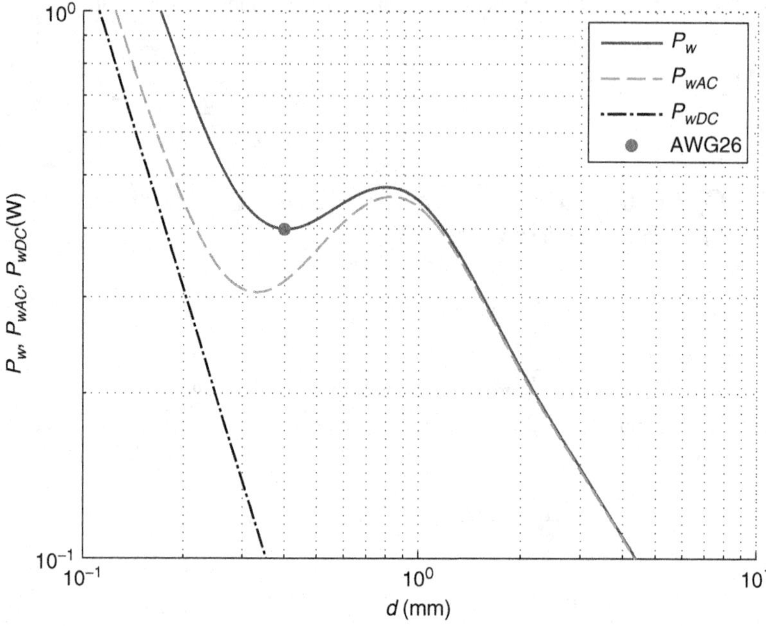

Figure 5.87 Winding DC power loss P_{wDC}, winding AC power loss P_{wAC}, and the total winding power loss $P_w = P_{wDC} + P_{wAC}$ as functions of wire diameter d for the four-layer inductor ($N_l = 4$) used in the PWM converter operated in DCM

The total winding power loss can be expressed as

$$P_w = R_{eff} I_{Lrms}^2 = R_{eff}\left(I_L^2 + \sum_{n=1}^{\infty} I_n^2\right) = R_{eff} I_L^2 \left[1 + \sum_{n=1}^{\infty}\left(\frac{I_n}{I_L}\right)^2\right], \tag{5.523}$$

where R_{eff} is the winding effective resistance. From (5.426), the total winding power loss is also given by

$$P_w = R_{wDC}\left(I_L^2 + \sum_{n=1}^{\infty} F_{Rn} I_n^2\right) = R_{wDC} I_L^2 \left[1 + \sum_{n=1}^{\infty} F_{Rn}\left(\frac{I_n}{I_L}\right)^2\right]. \tag{5.524}$$

Using the principle of conservation of energy, we can equate the right-hand sides of (5.523) and (5.524) to obtain

$$R_{eff} I_L^2 \left[1 + \sum_{n=1}^{\infty}\left(\frac{I_n}{I_L}\right)^2\right] = R_{wDC} I_L^2 \left[1 + \sum_{n=1}^{\infty} F_{Rn}\left(\frac{I_n}{I_L}\right)^2\right]. \tag{5.525}$$

The effective-to-DC winding resistance ratio for a nonsinusoidal periodic inductor current is given by Hurley *et al.* [56]

$$K_{Rh} = \frac{R_{eff}}{R_{wDC}} = \frac{I_L^2 + \sum_{n=1}^{\infty} F_{Rn} I_n^2}{I_L^2 + \sum_{n=1}^{\infty} I_n^2} = \frac{I_L^2 + \sum_{n=1}^{\infty} F_{Rn} I_n^2}{I_{Lrms}^2} = \frac{1 + \sum_{n=1}^{\infty} F_{Rn}\left(\frac{I_n}{I_L}\right)^2}{1 + \sum_{n=1}^{\infty}\left(\frac{I_n}{I_L}\right)^2}$$

$$= \frac{F_{Rh}}{1 + \sum_{n=1}^{\infty}\left(\frac{I_n}{I_L}\right)^2}, \tag{5.526}$$

where F_{Rh} is given by (5.427). The effective winding resistance is

$$R_{eff} = K_{Rh} R_{wDC}, \tag{5.527}$$

yielding

$$P_w = R_{eff} I_{Lrms}^2 = K_{Rh} R_{wDC} I_{Lrms}^2 = K_{Rh} R_{wDC} I_L^2 \left(\frac{I_{Lrms}}{I_L}\right)^2 = K_{Rh} \left(\frac{I_{Lrms}}{I_L}\right)^2 P_{wDC}. \tag{5.528}$$

Hence,

$$K_{Rh} = \frac{P_w}{P_{wDC}} \left(\frac{I_L}{I_{Lrms}}\right)^2. \tag{5.529}$$

The normalized effective winding resistance can be written as

$$K_{Rh} = \frac{R_{eff}}{R_{wDC}} = \frac{I_L^2 + \sum_{n=1}^{\infty} F_{Rn} I_n^2}{I_L^2 + \sum_{n=1}^{\infty} I_n^2} = \frac{I_L^2 + \sum_{n=1}^{\infty} I_n^2 + \frac{(5N_l^2-1)}{45}\left(\frac{h}{\delta_w}\right)^4 \sum_{n=1}^{\infty} n^2 I_n^2}{I_L^2 + \sum_{n=1}^{\infty} I_n^2}$$

$$= \frac{I_{Lrms} + \frac{(5N_l^2-1)}{45}\left(\frac{h}{\delta_w}\right)^4 \sum_{n=1}^{\infty} n^2 I_n^2}{I_{Lrms}} = 1 + \frac{(5N_l^2-1)}{45}\left(\frac{h}{\delta_w}\right)^4 \frac{\sum_{n=1}^{\infty} n^2 I_n^2}{I_{Lrms}}. \tag{5.530}$$

The derivative of the inductor current waveform given by (5.425) is

$$\frac{di_L}{dt} = -\omega \sum_{n=1}^{\infty} n I_{mn} \sin(n\omega t + \phi_n). \tag{5.531}$$

Hence, the rms value of the inductor current derivative is

$$I'_{Lrms} = \sqrt{\frac{\omega^2}{2} \sum_{n=1}^{\infty} n^2 I_{mn}^2} = \sqrt{\omega^2 \sum_{n=1}^{\infty} n^2 I_n^2}, \tag{5.532}$$

yielding

$$\sum_{n=1}^{\infty} n^2 I_n^2 = \frac{I'^2_{Lrms}}{\omega^2}. \tag{5.533}$$

This produces

$$K_{Rh} = \frac{R_{eff}}{R_{wDC}} = 1 + \frac{n^2(5N_l^2-1)}{45}\left(\frac{h}{\delta_w}\right)^4 \left(\frac{I'_{Lrms}}{\omega I_{Lrms}}\right)^2. \tag{5.534}$$

The effective winding resistance is

$$R_{eff} = K_{Rh} R_{wDC} = \frac{\rho_w l_w}{bh}\left[1 + \frac{n^2(5N_l^2-1)}{45}\left(\frac{h}{\delta_w}\right)^4 \left(\frac{I'_{Lrms}}{\omega I_{Lrms}}\right)^2\right]. \tag{5.535}$$

The DC resistance of a foil of thickness δ_w

$$R_{\delta f} = \frac{\rho_w l_w}{b\delta_w}. \tag{5.536}$$

Thus, the normalized effective winding resistance is

$$\frac{R_{eff}}{R_{\delta f}} = \frac{1}{\frac{h}{\delta_w}}\left[1 + \frac{n^2(5N_l^2-1)}{45}\left(\frac{h}{\delta_w}\right)^4 \left(\frac{I'_{Lrms}}{\omega I_{Lrms}}\right)^2\right]$$

$$= \frac{1}{\frac{h}{\delta_w}} + \frac{n^2(5N_l^2-1)}{45}\left(\frac{h}{\delta_w}\right)^3 \left(\frac{I'_{Lrms}}{\omega I_{Lrms}}\right)^2. \tag{5.537}$$

The derivative of the normalized effective winding resistance at a fixed frequency f is

$$\frac{d}{d\left(\frac{h}{\delta_w}\right)}\left(\frac{R_{eff}}{R_{d\delta f}}\right) = -\frac{1}{\left(\frac{h}{\delta_w}\right)^2} + \frac{n^2(5N_l^2-1)}{45}\left(\frac{h}{\delta_w}\right)^2 \left(\frac{I'_{Lrms}}{\omega I_{Lrms}}\right)^2. \tag{5.538}$$

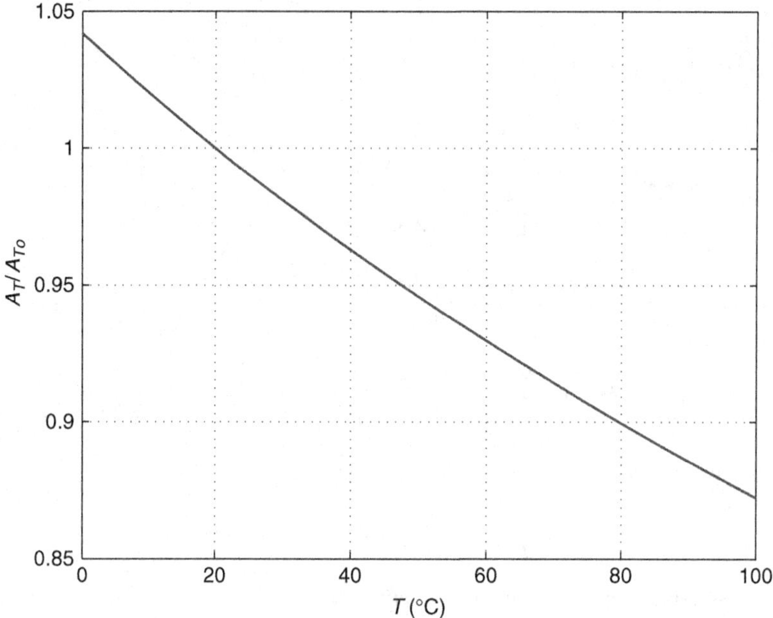

Figure 5.88 Ratio A_T/A_{To} as a function of temperature T for copper at $T_o = 20\ ^\circ\mathrm{C}$ for copper

Hence, the optimum foil thickness at which the minimum value of the effective winding resistance occurs is

$$\frac{h_{opt}}{\delta_w} = \sqrt[4]{\frac{15}{5N_l^2 - 1}}\sqrt[4]{\frac{\omega I_{Lrms}}{I'_{Lrms}}} = \sqrt[4]{\frac{15}{5N_l^2 - 1}}\sqrt[4]{\frac{I_L^2 + \sum_{n=1}^{\infty} I_n^2}{\sum_{n=1}^{\infty} n^2 I_n^2}} = \sqrt[4]{\frac{15}{5N_l^2 - 1}}\sqrt[4]{\frac{1 + \sum_{n=1}^{\infty}\left(\frac{I_n}{I_L}\right)^2}{\sum_{n=1}^{\infty} n^2\left(\frac{I_n}{I_L}\right)^2}}$$

$$= \sqrt[4]{\frac{15}{5N_l^2 - 1}}\sqrt[4]{\frac{1 + \sum_{n=1}^{\infty} \gamma_n^2}{\sum_{n=1}^{\infty} n^2\gamma_n^2}}. \qquad (5.539)$$

This equation is the same as that derived from minimization of winding power loss.

5.28 Thermal Effects on Winding Resistance

Consider an inductor with a foil winding. The ratio of the foil thickness h to the skin depth of the foil conductor at temperature T is

$$A_T = \frac{h}{\delta_{w(T)}} = h\sqrt{\frac{\pi\mu_0 f}{\rho_{w(To)}[1 + \alpha(T - T_o)]}} = \frac{h}{\delta_{w(To)}}\frac{1}{\sqrt{1 + \alpha(T - T_o)}}$$

$$= A_{To}\frac{1}{\sqrt{1 + \alpha(T - T_o)}}. \qquad (5.540)$$

Figure 5.88 shows the ratio A_T/A_{To} as a function of temperature T for copper at $T_o = 20\ ^\circ\mathrm{C}$ for copper. The AC-to-DC resistance ratio at temperature T is

$$F_{R(T)} = \frac{R_{w(T)}}{R_{wDC(T)}} = A_T F(A_T) = h\sqrt{\frac{\pi\mu_0 f}{\rho_{w(To)}[1 + \alpha(T - T_o)]}}F(A_T)$$

$$= \frac{h}{\delta_{w(To)}} \frac{1}{\sqrt{1+\alpha(T-T_o)}} F(A_T) = \frac{A_{To}}{\sqrt{1+\alpha(T-T_o)}} F(A_T) \quad (5.541)$$

where the exact Dowell's equation is

$$F(A_T) = F_{RS}(A_T) + \frac{2(N_l^2-1)}{3} F_{RP}(A_T). \quad (5.542)$$

Figure 5.89 shows the plots of ratios $F_{R(To)} = F_{R(T)}$ as functions of A_{To} for copper $T = 20\ ^\circ$C and $T = 100\ ^\circ$C. The winding AC resistance at temperature T is

$$R_{w(T)} = F_{R(T)} R_{wDC(T)} = R_{wDC(To)}[1+\alpha(T-T_o)]h\sqrt{\frac{\pi\mu_0 f}{\rho_{w(To)}[1+\alpha(T-T_o)]}} F(A_T)$$

$$= R_{wDC(To)} \frac{h}{\delta_{w(To)}} \sqrt{1+\alpha(T-T_o)} F(A_T) = R_{wDC(To)} A_{To} \sqrt{1+\alpha(T-T_o)} F(A_T). \quad (5.543)$$

Figure 5.90 shows the plots of ratios $R_{w(T)}/R_{w(To)}$ at $f = 300$ kHz and $R_{wDC(T)}/R_{wDC(To)}$ as functions of temperature T at $T = 20\ ^\circ$ and $N_l = 15$. Figure 5.91 shows the plots of winding AC resistance R_w as functions of temperature A_{To} for $b = 2.95$ mm, $h = 1.2$ mm, $l_T = 0.05$ m, and $N_l = 15$ for copper at $T = 20\ ^\circ$C, and $T = 100\ ^\circ$C.

The approximate AC-to-DC resistance ratio at temperature T is given by

$$F_{R(T)} \approx 1 + \frac{5N_l^2-1}{45} A_T^4 = 1 + \frac{5N_l^2-1}{45}\left(\frac{h}{\delta_w(T_o)}\right)^4 \frac{1}{[1+\alpha(T-T_o)]^2}$$

$$= 1 + \frac{F_{R(To)}-1}{[1+\alpha(T-T_o)]^2} \quad \text{for} \quad A_T \le 2.5. \quad (5.544)$$

The winding resistance at temperature T is

$$R_{w(T)} = R_{wDC(T)} F_{R(T)} = R_{wDC(To)}[1+\alpha(T-T_o)]\left\{1 + \frac{5N_l^2-1}{45} \frac{A_T^4}{[1+\alpha(T-T_o)]^2}\right\}$$

$$= R_{wDC(To)}\left[1+\alpha(T-T_o) + \frac{F_{R(To)}-1}{1+\alpha(T-T_o)}\right] \quad \text{for} \quad A_T \le 2.5. \quad (5.545)$$

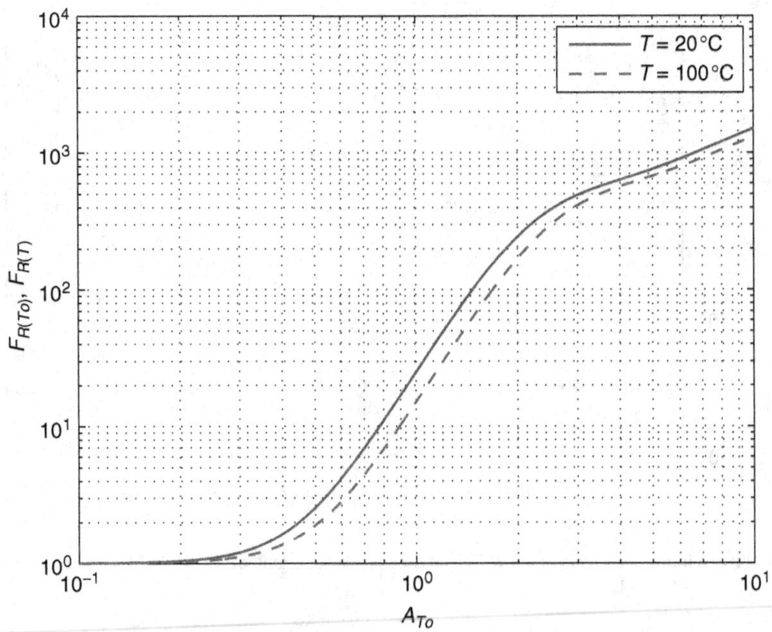

Figure 5.89 Ratios $F_{R(To)} = R_{w(T_o)}/R_{wDC(To)}$ and $F_{R(T)} = R_{w(T)}/R_{wDC(T)}$ as functions of $A_{To} = h/\delta_{w(To)}$ for copper at $N_l = 15$, $T = 20\ ^\circ$C and $T = 100\ ^\circ$C for copper

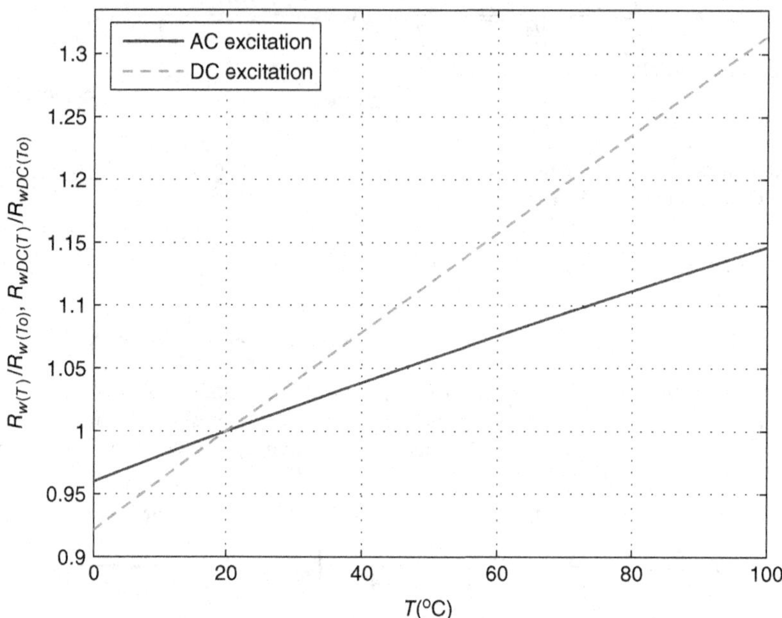

Figure 5.90 Ratios $R_{w(T)}/R_{w(To)}$ at $f = 300$ kHz and $R_{wDC(T)}/R_{wDC(To)}$ as functions of temperature T at $T = 20^\circ$ and $N_l = 15$

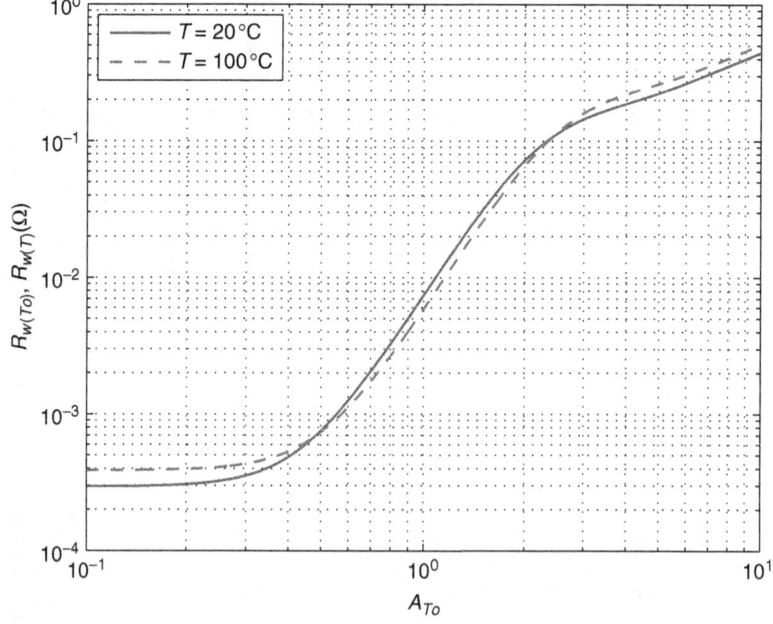

Figure 5.91 Winding AC resistances $R_{w(To)}$ and $R_{w(T)}$ as functions of $A_{To} = h/\delta_{w(T_o)}$ for copper at $N_l = 15$, $T = 20\,^\circ$C and $T = 100\,^\circ$C

5.29 Thermal Model of Inductors

There are three mechanisms of heat transfer: conduction, convection, and radiation. A thermal resistance θ is used to model heat transfer by both conduction and convection. In most applications, heat is transferred by conduction and convection. The thermal resistance of a layer of material describing heat transfer by conduction is expressed as

$$\theta_{cond} = \frac{l}{kA} = \frac{\rho_\theta l}{A} \quad (^\circ\text{C/W}), \tag{5.546}$$

where l is the length, A is the cross-sectional area, k is the thermal conductivity, and ρ_θ is the thermal resistivity of the material. The unit of the thermal conductivity k is W/m·K. The unit of the thermal resistivity ρ_θ is m·K/W or m·C/W. The typical values of the thermal conductivity are 200 W/m·K for aluminum and 380 W/m·K for copper. The difference of temperature between two sides of a piece of material conducting power P_D is

$$\Delta T = T_1 - T_2 = \theta_{cond} P_D. \tag{5.547}$$

The thermal resistance describing heat transfer by convection is

$$\theta_{conv} = \frac{1}{hA}, \tag{5.548}$$

where A is the cross-sectional area of the body-ambient interface and h is a constant. The unit of h is W/K·m^2. The difference of the surface temperature T_s and the ambient temperature T_A is

$$\Delta T = T_s - T_A = \theta_{conv} P_D = \frac{P_D}{hA}. \tag{5.549}$$

Thus, the temperature difference ΔT is inversely proportional to the surface area A.

Figure 5.92 depicts a thermal model of inductors in steady state. There are two sources of heat: the core power loss P_C and the winding power loss P_w. For steady state, all the energy generated by the heat sources is transferred to the ambient through the outer surface of an inductor. A thermal resistance θ can be used to model heat transfer by both the conduction and the convection. The thermal resistance of the core is

$$\theta_c = \frac{r_c}{k_c A_{sc}}, \tag{5.550}$$

where k_c is the thermal conductivity of the core, r_c is the core radius or thickness, and A_{sc} is the area of the outer core surface. The ferrite core thermal resistance θ_c usually ranges from 10 to 100 °C/W.

The winding thermal resistance is

$$\theta_w = \frac{t_w}{k_w A_{sw}}, \tag{5.551}$$

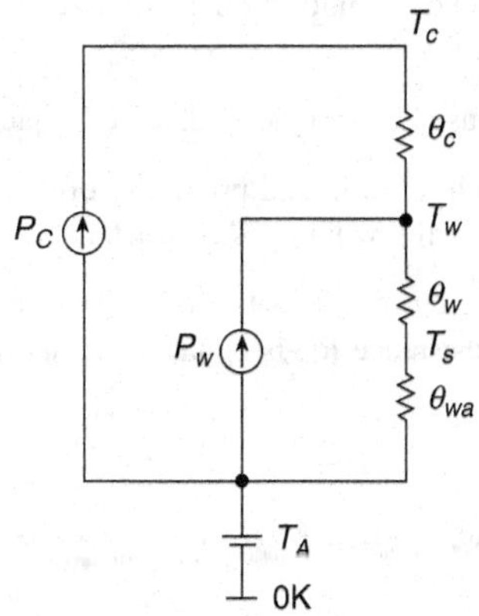

Figure 5.92 Thermal model of inductors for steady state

where k_w is the winding thermal conductivity, t_w is the winding thickness, and A_{sw} is the area of the inductor–air interface. The thermal conductivity of copper is $k_{Cu} = 401$ W/m·K. The thermal resistivity of copper is equal to 0.25 K·cm/W.

The winding-ambient resistance is

$$\theta_{wa} = \frac{1}{hA_t}, \tag{5.552}$$

where h is a constant of convection and A_t is the total outer area of an inductor. For natural convection, $h = 2$–$25\,\text{W/K·m}^2$. The temperature rise is

$$\Delta T = T_c - T_A = \theta_c P_C + (\theta_w + \theta_{wa})(P_C + P_w). \tag{5.553}$$

A simple equation for the temperature rise ΔT of the outer inductor surface A_t above the ambient temperature T_A is expressed by McLyman [63]

$$\Delta T = T_S - T_A = 450\left(\frac{P_C + P_w}{A_t}\right)^{0.826} = 450\left(\frac{P_{cw}}{A_t}\right)^{0.826}, \tag{5.554}$$

where P_{cw} is in watts and A_t is in square centimeter. The surface power loss density is

$$\psi = \frac{P_{cw}}{A_t} = \left(\frac{\Delta T}{450}\right)^{1.21} \quad \left(\frac{\text{W}}{^\circ\text{C}}\right). \tag{5.555}$$

For $\Delta T = 50\,^\circ$C, $\psi = 70$ mW/°C.

In inductors with a toroidal core, all the heat from the core must flow through the winding, which is exposed to the ambient.

5.30 Summary

- Eddy currents are induced in conductors by time-varying magnetic fields.
- The frequency range of the single-layer winding AC resistance R_w can be divided into low-frequency and high-frequency ranges.
- In single-layer winding inductors, the winding AC resistance is equal to the DC resistance in low-frequency range and increases with frequency in high-frequency range due to the skin effect. The proximity effect is not present in single-layer winding inductors.
- The frequency range of the multilayer winding AC resistance R_w can be divided into low-frequency, mid-frequency, and high-frequency ranges.
- In multilayer inductors, the winding AC resistance is equal to the DC resistance in low-frequency range and increases with frequency in mid-frequency range and high-frequency range due to both the skin and proximity effects.
- The proximity effect loss is caused by magnetic fields of adjacent layers and adjacent windings.
- In the low-frequency range, both the skin and proximity effects can be neglected and therefore the winding ac resistance is equal to the winding DC resistance.
- In the low-frequency range, the current density distribution in the winding conductor is uniform, and therefore the winding ac resistance R_w is equal to the winding DC resistance R_{wDC}.
- In the low-frequency range, the winding AC resistance R_w is independent of frequency for a sinusoidal current.
- In the low-frequency range, the winding AC resistance R_w is independent of the number of layers N_l.
- In the mid-frequency range, the current density distribution is not uniform, but the current flows through the entire cross section of a winding conductor.

- In the high-frequency range, the current density distribution is not uniform and the current flows only near the surface of a winding conductor.
- In the low- and high-frequency ranges, the winding AC resistance R_w increases as the number of layers N_l increases at a fixed frequency of a sinusoidal current.
- In the mid-frequency range, the rate of increase in R_w/R_{wDC} with frequency increases as the number of layers N_l is increased.
- In the high-frequency range, the rate of increase in R_w/R_{wDC} with frequency increases as the number of layers N_l increases.
- In the high-frequency range, the rate of increase of R_w/R_{wDC} with frequency is lower than that in the mid-frequency range for multilayer inductors.
- The rate of increase in R_w/R_{wDC} increases as the porosity factor η increases.
- Eddy-current winding loss and inductance decrease when the winding breadth increases and the winding height decreases. The winding breadth should be several times greater than the winding height.
- As h/δ_w increases at a fixed frequency, the winding AC resistance R_w first decreases, reaches a minimum value, then increases, and finally remains constant for inductors with foil and rectangular winding conductors and multiple layers.
- There is an optimum normalized height h_{opt}/δ_w of the foil and rectangular conductors for a fixed frequency of a sinusoidal current at which a minimum conductor AC resistance R_w occurs.
- As the number of layers N_l increases, the optimum normalized height h_{opt}/δ_w decreases for inductors with foil and rectangular winding conductors.
- As the porosity factor η increases at a fixed frequency, the optimum normalized height h_{opt}/δ_w also increases for inductors with foil and rectangular winding conductors.
- As h/δ_w increases at a fixed frequency, the winding AC resistance R_w first decreases, reaches a minimum value, then increases, and finally decreases again for inductors with square winding conductor and multiple layers.
- As the normalized diameter d/δ_w increases at a fixed frequency, the winding AC resistance R_w first decreases, reaches a minimum value, then increases, and finally decreases again for inductors with round winding conductor and multiple layers.
- In order to minimize the winding AC resistance and winding power loss, cores should be long with narrow windows to reduce the number of winding layers N_l.
- Litz wire consists of multiple strands insulated electrically from each other and connected.
- There is a frequency range in which litz-wire winding has a lower resistance than that of a solid wire winding.
- Litz-wire windings are inferior to solid wire windings at high frequencies.

5.31 Appendix

5.31.1 Derivation of Dowell's Equation Approximation

Maclaurin's series of function $f(x)$ is

$$f(x)=\sum_{n=0}^{\infty}\frac{f^{(n)}}{n!}x^n=f(0)+f'(0)x+\frac{f''(0)}{2!}x^2+\frac{f'''(0)}{3!}x^3+\ldots+\frac{f^{(n)}(0)}{n!}x^n+\ldots \tag{5.556}$$

The expansions of trigonometric and hyperbolic functions present in Dowell's equation into Maclaurin's series for $-\infty < x < \infty$ are as follows:

$$\sin x = \sum_{n=0}^{\infty} \frac{(-1)^n}{(2n+1)!} x^{2n+1} = x - \frac{x^3}{3!} + \frac{x^5}{5!} - \frac{x^7}{7!} + \ldots \tag{5.557}$$

$$\cos x = \sum_{n=0}^{\infty} \frac{(-1)^n}{(2n)!} x^{2n} = 1 - \frac{x^2}{2!} + \frac{x^4}{4!} - \frac{x^6}{6!} + \ldots \tag{5.558}$$

$$\sin 2x = \sum_{n=0}^{\infty} \frac{(-1)^n}{(2n+1)!} (2x)^{2n+1} = 2x - \frac{8x^3}{3!} + \frac{32x^5}{5!} + \ldots \tag{5.559}$$

$$\cos 2x = \sum_{n=0}^{\infty} \frac{(-1)^n}{(2n)!} (2x)^{2n} = 1 - \frac{4x^2}{2!} + \frac{16x^4}{4!} - \frac{64x^6}{6!} + \ldots \tag{5.560}$$

$$\sinh x = \sum_{n=0}^{\infty} \frac{1}{(2n+1)!} x^{2n+1} = x + \frac{x^3}{3!} + \frac{x^5}{5!} + \frac{x^7}{7!} + \ldots \tag{5.561}$$

$$\cosh x = \sum_{n=0}^{\infty} \frac{1}{(2n)!} x^{2n} = 1 + \frac{x^2}{2!} + \frac{x^4}{4!} + \ldots \tag{5.562}$$

$$\sinh 2x = \sum_{n=0}^{\infty} \frac{1}{(2n+1)!} (2x)^{2n+1} = 2x + \frac{8x^3}{3!} + \frac{32x^5}{5!} + \ldots \tag{5.563}$$

$$\cosh 2x = \sum_{n=0}^{\infty} \frac{1}{(2n)!} (2x)^{2n} = 1 + \frac{4x^2}{2!} + \frac{16x^4}{4!} + \frac{64x^6}{6!} + \ldots \tag{5.564}$$

Hence,

$$F_{RS} = \frac{\sinh 2x + \sin 2x}{\cosh 2x - \cos 2x} \approx \frac{2x + \frac{8x^3}{3!} + \frac{32x^5}{5!} + 2x - \frac{8x^3}{3!} + \frac{32x^5}{5!}}{1 + \frac{4x^2}{2!} + \frac{16x^4}{4!} - 1 + \frac{4x^2}{2!} - \frac{16x^4}{4!}} = \frac{1}{x} + \frac{2x^3}{15} \quad \text{for} \quad x \leq 1.5 \tag{5.565}$$

and

$$F_{RP} = \frac{\sinh x - \sin x}{\cosh x + \cos x} \approx \frac{x + \frac{x^3}{3!} - x + \frac{x^3}{3!}}{1 + \frac{x^2}{2!} + 1 - \frac{x^2}{2!}} = \frac{x^3}{6} \quad \text{for} \quad x \leq 1.5 \tag{5.566}$$

Thus, Dowell's equation can be approximated by

$$F_R = A\left[\frac{\sinh(2A) + \sin(2A)}{\cosh(2A) - \cos(2A)} + \frac{2(N_l^2 - 1)}{3} \frac{\sinh(A) - \sin(A)}{\cosh(A) + \cos(A)}\right]$$
$$= A\left[F_S + \frac{2(N_l^2 - 1)}{3} F_P\right] \approx A\left[\frac{1}{A} + \frac{2A^3}{15} + \frac{2(N_l^2 - 1)}{3} \frac{A^3}{6}\right] = 1 + \frac{2A^4}{15} + \frac{(N_l^2 - 1)A^4}{9}$$
$$= 1 + \frac{5N_l^2 + 1}{45} A^4 \quad \text{for} \quad A \leq 1.5 \tag{5.567}$$

where

$$F_S \approx \frac{1}{A} + \frac{2A^3}{15} \tag{5.568}$$

and

$$F_P \approx \frac{A^3}{6}. \tag{5.569}$$

However, the expression with a negative sign gives a better approximation of F_R

$$F_R \approx 1 + \frac{5N_l^2 - 1}{45} A^4 \quad \text{for} \quad A \leq 1.5. \tag{5.570}$$

Therefore, this approximation is used is this book.

5.32 References

[1] E. Bennett and S. C. Larson, "Effective resistance to alternating currents of multilayer windings," *Transactions of the American Institute of Electrical Engineers*, vol. 59, no. 12, pp. 1010–1017, 1940.

[2] H. A. Wheeler, "Simple inductance formulas for radio coils," *Proceedings of the IRE*, vol. 16, no. 10, pp. 1398–1400, October 1928.

[3] H. A. Wheeler, "Formulas for the skin effect," *Proceedings of the IRE*, vol. 30, pp. 412–424, September 1942.

[4] S. Ryzko and J. Ebert, "Effect of a screen on a single layer solenoid at high frequencies," *Bulletin de L'Acadmie Polonaise Des Sciences: Série des Sciences Techniques*, no. 8, 1965.

[5] P. J. Dowell, "Effects of eddy currents in transformer winding," *Proceedings of the IEE*, vol. 113, no. 8, pp. 1387–1394, August 1966.

[6] J. Lammeraner and M. Stafl, *Eddy Currents*. Cleveland, OH: CRS Press, 1966.

[7] J. Ebert, "Four terminal parameters of HF inductors," *Bulletin de L'Acadmie Polonaise Des Sciences: Série des Sciences Techniques*, no. 5, 1968.

[8] E. C. Snelling, *Soft Ferrites: Properties and Applications*, 2nd Ed., London: Butterworth, 1988.

[9] R. L. Stall, *The Analysis of Eddy Currents*. Oxford: Clarendon Press, 1974, pp. 21–27.

[10] M. P. Perry and T. B. Jones, "Eddy current induction in a solid conducting cylinder with a transverse magnetic field," *IEEE Transactions on Magnetics*, vol. 14, no. 4, pp. 227–232, 1978.

[11] R. Reeves, "Air-core foil-wound inductors," *Proceedings of the IEE*, vol. 125, no. 5, pp. 460–464, May 1978.

[12] M. P. Perry, "Multiple layer series connected winding design for minimum losses," *IEEE Transactions on Power Apparatus and Systems*, vol. PAS-98, no. 1, pp. 116–123, January/February 1979.

[13] P. S. Venkatraman, "Winding eddy current losses in switch mode power transformers due to rectangular wave currents," *Proceedings Powercon 11, Section A-1*, 1984, pp. 1–11.

[14] J. K. Watson, *Applications of Magnetism*, Gainesville, FL, 1985; also John Wiley & Sons, 1985.

[15] B. Carsten, "High frequency conductor losses in switch mode magnetics," Proceedings of PCI, Munich, Germany, 1986, pp. 161–182.

[16] N. R. Coonrod, "Transformer computer design aid for frequency switching power supplies," *IEEE Transactions on Power Electronics*, vol. 1, no. 4, pp. 248–256, October 1986.

[17] J. Jongsma, "High-frequency ferrite power transformer and choke design, Part 3: Transformer winding design, Part 4: Improved method of power choke design" Philips Electronic Components and Materials, Technical Publication, no. 27, Philips, The Netherlands, 1986, pp. 1–28.

[18] J. P. Vandalec and P. D. Ziogos, "A novel approach for minimizing high-frequency transformer copper loss," *IEEE Transactions on Power Electronics*, vol. 3, no. 3, pp. 266–277, July 1988.

[19] C. W. T. McLyman, *Transformer and Inductor Design Handbook*, 2nd Ed., New York: Marcel Dekker, 1988.

[20] A. M. Urling, V. A. Niemela, G. R. Skutt, and T. G. Wilson, "Characterizing high frequency effects in transformer windings: a guide to several significant papers," *4th Annual IEEE Power Electronics Specialists Conference*, Baltimore, MD, March 13–17, 1989, pp. 373–385.

[21] A. Kennelly, F. Laws, and P. Pierce, "Experimental research on skin effect in conductors," *Transactions of the AIEE*, vol. 34, p. 1915, 1915.

[22] P. N. Murgatroy, "Calculation of proximity loss in multistrand conductor bunches," *IEE Proceedings - Part A: Physical Science, Measurement and Instrumentation, Management and Education*, vol. 136, no. 3, pp. 115–120, May 1989.

[23] P. D. Evens and W. M. Chew, "Reduction of proximity losses in coupled inductors," *IEE Proceedings - Part B: Electric Power Applications*, vol. 138, no. 2, pp. 51–58, March 1991.

[24] J. A. Ferreira, *Electromagnetic Modelling of Power Electronic Converters*, Boston, MA: Kluwer Academic Publisher, 1989.

[25] J. A. Ferreira, "Analytical computation of ac resistance of round and rectangular litz wire windings," *IEE Proceedings - Part B: Electric Power Applications*, vol. 139, no. 1, pp. 21–25, 1992.

[26] A. W. Lotfi, P. M. Gradzki, and F. C. Lee, "Proximity effects on coils for high frequency power amplifier," *IEEE Transactions on Magnetics*, vol. 28, no. 5, pp. 2169–2171, September 1992.

[27] J. A. Ferreira, "Improved analytical modeling of conductive losses in magnetic components," *IEEE Transactions on Power Electronics*, vol. 9, pp. 127–131, January 1994.

[28] A. W. Lotfi and F. C. Lee, "A high frequency model for litz wire for switch-mode magnetics," *Proceedings of the 28th IEEE Industry Applications Society Annual Meeting*, vol. 2, October 1993, pp. 1169–1175.

[29] M. Bartoli, A. Reatti, and M. K. Kazimierczuk, "High-frequency models of ferrite inductors," *Proceedings of the IEEE International Conference on Industrial Electronics, Controls, Instrumentation, and Automation (IECON'94), Bologna, Italy, September 5-9*, 1994, pp. 1670–1675.

[30] M. Bartoli, A. Reatti, and M. K. Kazimierczuk, "Predicting the high-frequency ferrite-core inductor performance," *Proceedings of the Conference of Electrical Manufacturing and Coil Winding*, Chicago (Rosemont), IL, September 27-29, 1994, pp. 409–413.

[31] M. Bartoli, A. Reatti, and M. K. Kazimierczuk, "Modeling iron-powder inductors at high-frequencies," *Proceedings of the IEEE Industry Applications Society Annual Meeting, Denver, CO, October 2-7*, 1994, pp. 1225–1232.

[32] M. Bartoli, N. Noferi, A. Reatti, and M. K. Kazimierczuk, "Modeling winding losses in high-frequency power inductors," *Journal of Circuits Systems and Computers*, vol. 5, no. 4, pp. 607–626, December 1995.

[33] M. Bartoli, N. Noferi, A. Reatti, and M. K. Kazimierczuk, "Modeling litz-wire winding losses in high-frequencies power inductors," *Proceedings of the IEEE Power Electronics Specialists Conference, Baveno, Italy, June 23-27*, 1996, pp. 1690–1696.

[34] K. W. E. Cheng and P. D. Evens, "Calculation of winding losses in high-frequency toroidal inductors using multistrand conductors," *IEE Proceedings, Electric Power Applications, Part B*, vol. 142, no. 5, pp. 313–222, September 1995.

[35] R. Petkov, "Optimum design of a high-power high-frequency transformers," *IEEE Transactions on Power Electronics*, vol. 11, no. 1, pp. 33–42, January 1996.

[36] J. Schultz, J. Roudet, and A. Schellmanns, "Modeling litz-wire windings," *Proceedings of the IEEE Industry Applications Society Annual Meeting, New Orleans, LA, October 5-9*, 1997, vol. 2, pp. 1190–1195.

[37] J. C. Maxwell, *A Treatise of Electricity and Magnetism*, 3rd Ed., New York, NY: Dover Publishing, 1997.

[38] A. Massarini and M. K. Kazimierczuk, "Self-capacitance of inductors," *IEEE Transactions on Power Electronics*, vol. 12, no. 4, pp. 671–676, July 1997.

[39] G. Grandi, M. K. Kazimierczuk, A. Massarini, and U. Reggiani, "Stray capacitance of single-layer solenoid air-core inductors," *IEEE Transactions on Industry Applications*, vol. 35, no. 5, pp. 1162–1168, September 1999.

[40] C.-M. Ong, *Dynamic Simulation of Electric Machinery*, Reading, MA: Prentice Hall, 1998, pp. 38–40, 45–46, and 87–90.

[41] J. C. Stones, N. Williams, and C. Pollack, "Reducing losses in foil wound air-cored inductors," *IEEE Proceedings Power Electronics and Variable Speed Drives Conference*," Publication No. 456, pp. 399–4004, 1998.

[42] W. G. Hurley, W. H. Wolfe, and J. G. Breslin, "Optimized transformer design: inclusive of high-frequency effects," *IEEE Transactions on Power Electronics*, vol. 13, no. 4, pp. 651–659, July 1998.

[43] N. H. Kutkut, N. H. Novotny, D. W. Divan, and E. Yeow, "Analysis of winding losses in high frequency foil wound inductors," *IEEE Proceedings of Industry Applications Society Annual Meeting, Orlando, USA*, 1995, pp. 859–867.

[44] N. H. Kutkut, "A simple technique to evaluate winding losses including two-dimensional edge effects," *IEEE Transactions on Power Electronics*, vol. 13, no. 5, pp. 950–958, September 1998.

[45] N. H. Kutkut and D. M. Divan, "Optimal air-gap design in high-frequency foil windings", *IEEE Transactions on Power Electronics*, vol. 13, no. 5, pp. 942–949, September 1998.

[46] T. L. Simpson, "Effect of a conducting shield on the inductance of an air-core solenoid," *IEEE Transactions on Magnetics*, vol. 35, no. 1, pp. 508–515, January 1999.

[47] J. Hu and C. R. Sullivan, "Optimization of shapes for round-wire high-frequency gapped-inductor windings," *Proceedings of the IEEE Industry Society Annul Meeting*, 1998, pp. 907–912.

[48] C. R. Sullivan, "Optimal choice for the number of strands in a litz-wire transformer winding," *IEEE Transactions on Power Electronics*, vol. 14, no. 2, pp. 283–291, March 1999.

[49] C. R. Sullivan, "Cost-constrained selection of strand diameter and number in a litz-wire transformer winding," *IEEE Transactions on Power Electronics*, vol. 16, no. 2, pp. 281–288, March 1999.

[50] M. K. Kazimierczuk, G. Sancineto, U. Reggiani, and A. Massarini, "High-frequency small-signal model of ferrite inductors," *IEEE Transactions on Magnetics*, vol. 35, no. 5, pp. 4185–4191, September 1999.

[51] U. Reggiani, G. Sancineto, and M. K. Kazimierczuk, "High-frequency behavior of laminated iron-core inductors for filter applications," *Proceedings of the IEEE Applied Power Electronics Conference, New Orleans, LA, February 6-10*, 2000, pp. 654–660.

[52] M. Albach, "Two-dimensional calculation of winding losses in transformer," *31st IEEE Power Electronics Specialists Conference, Galway, Ireland, June 18-23*, 2000, pp. 1639–1644.

[53] M. Albach, "The influence of air gap and winding position on the proximity losses in high frequency transformers," *32nd Annual IEEE Power Electronics Specialists Conference*, Vancouver, BC, Canada, June 17–21, 2001, vol. 3, pp. 1485–1490.

[54] W. Chen, J. He, H. L. Luo, and J. Hu, and C. Wen, "Winding loss analysis and new air-gap arrangement for high-frequency inductor," *Proceedings of the IEEE PESC*, 2001, pp. 2084–2089.

[55] P. Wallmeir, N. Frohleke, and H. Grostollen, "Improved analytical modeling of conductive losses in gapped high-frequency inductors," *IEEE Transactions on Industry Applications*, vol. 37, no. 4, pp. 558–567, July 2001.

[56] W. G. Hurley, E. Gath, and J. G. Breslin, "Optimizing the ac resistance of multilayer transformer windings with arbitrary current waveforms," *IEEE Transactions on Power Electronics*, vol. 15, no. 2, pp. 369–376, March 2000.

[57] K. Howard and M. K. Kazimierczuk, "Eddy-current power loss in laminated power cores," *Proceedings of the IEEE International Symposium on Circuits and Systems, Sydney, Australia, May 7-9*, 2000, Paper III-668, pp. 668–672.

[58] F. Tourkhani and P. Viarouge, "Accurate analytical model of winding losses in round litz wire winding," *IEEE Transactions on Magnetics*, vol. 37, no. 1, pp. 438–443, January 2001.

[59] S. Gross and T. Weller, "Determining the RF resistance and Q-factor of air-core inductors," *Microwave and Optical Technology Letters*, vol. 29, no. 2, pp. 89–93, April 20, 2001.

[60] A. Reatti and M. K. Kazimierczuk, "Comparison of various methods for calculating the ac resistance of inductors," *IEEE Transactions on Magnetics*, vol. 37, no. 3, pp. 1512–1518, May 2002.

[61] F. Robert, "A theoretical discussion about the layer copper factor used in winding loss calculation," *IEEE Transactions on Magnetics*, vol. 38, no. 5, pp. 3177–3179, September 2002.

[62] N. Mohan, T. M. Undeland, and W. P. Robbins, *Power Electronics*, 3rd Ed., New York: John Wiley & Sons, 2003.

[63] W. T. McLyman, *Transformer and Inductor Design Handbook*, 3rd Ed., New York: Marcel Dekker, 2004.

[64] X. Nan and C. R. Sulivan, "An improved calculation of proximity-effect loss in high-frequency windings of round conductors," *IEEE Power Electronics Specialists Conference*, 2003, vol. 2, pp. 853–860.

[65] G. Grandi, M. K. Kazimierczuk, A. Massarini, U. Reggiani, and G. Sancineto, "Model of laminated iron-core inductors," *IEEE Transactions on Magnetics*, vol. 40, no. 4, pp. 1839–1845, July 2004.

[66] R. G. Medhurst, "HF resistance and self-capacitance of single-layer solenoids," Wireless Engineers, pp. 35–43, February 1947, and pp. 80–92, March 1947.

[67] F. Robert, P. Mathys, and J. P. Schauwavers, "Ohmic losses calculation in SMPS transformers: numerical study of Dowell's approach accuracy," *IEEE Transactions on Magnetics*, vol. 34, no. 4, pp. 1255–1257, July 1998.

[68] J. P. Schauwavers, F. Robert, and P. Mathys, "A closed-form formula for 2-D ohmic losses calculation in SMPS transformer foils," *IEEE Transactions on Power Electronics*, vol. 16, no. 3, pp. 437–444, March 2001.

[69] R. W. Erickson and D. Maksimović, *Fundamentals of Power Electronics*, Norwell, MA: Kluwer Academic Publishers, 2001.

[70] A. Van den Bossche and V. C. Valchev, *Inductors and Transformers for Power Electronics*, Boca Raton, FL: CRC, Taylor & Francis, 2005.

[71] V. Zubeck, "Eddy current losses in transformer low voltage foil coils," *Journal of Electrical Engineering*, vol. 56, no. 3-4, pp. 95–90, 2005.

[72] X. K. Mao, W. Chen, and Y. X. Li, "Winding loss mechanism analysis and design for new structure high-frequency gapped inductor," *IEEE Transactions on Magnetics*, vol. 41, no. 10. pp. 4036–4038, October 2005.

[73] F. E. Terman, *Radio Engineers' Handbook*, New York: McGraw-Hill, 1943.

[74] A. D. Podoltsev, I. N. Kucheryavaya, and B. B. Lebedev, "Analysis of effective resistance and eddy-current losses in multiturn winding of high-frequency magnetic components," *IEEE Transactions of Magnetics*, vol. 39, no. 1, pp. 539–548, January 2003.

[75] N. Das and M. K. Kazimierczuk, "An overview of technical challenges in the design of current transformers," *Electrical Manufacturing Conference, Indianapolis, IN, October 24-26*, 2005.

[76] T. Suetsugu and M. K. Kazimierczuk, "Integration of Class DE inverter for dc-dc converter on-chip power supplies," *IEEE International Symposium on Circuits and Systems, Kos, Greece, May 21-24*, 2006, pp. 3133–3136.

[77] T. Suetsugu and M. K. Kazimierczuk, "Integration of Class DE synchronized dc-dc converter on-chip power supplies," *IEEE Power Electronics Specialists Conference, Jeju, South Korea, June 21-24*, 2006, pp. 433–437.

[78] J. Acero, R. Alonso, J. M. Burdio, L. A. Barragan, and D. Puyal, "Frequency-dependent resistance in litz-wire planar windings for domestic induction heating appliances," *IEEE Transactions on Power Electronics*, vol. 21, no. 4, pp. 856–866, April 2006.

[79] Z. Yang, W. Liu, and E. Basham, "Inductor modeling in wireless links for implantable electronics," *IEEE Transactions on Magnetics*, vol. 43, no. 10, pp. 3851–3860, January 2007.

[80] D. C. Penz and I. W. Hofsajer, "Improved ac-resistance of multiple foil winding by varying of thickness of successive layers," *COMPEL: The International Journal for Computation and Mathematics in Electrical and Electronic Engineering*, vol. 27, no. 1, pp. 181–195, 2008.

[81] Z. Gmyrek, A. Boglietti, and A. Cavagnino, "Iron loss prediction with PWM supply using low- and high-frequency measurements: analysis and results comparison," *IEEE Transactions on Industrial Electronics*, vol. 55, no. 4, pp. 1722–1728, April 2008.

[82] W. A. Roshen, "High-frequency fringing fields loss in thick rectangular and round wire windings," *IEEE Transactions on Magnetics*, vol. 44, no. 10, pp. 2396–2401, October 2008.

[83] M. K. Kazimierczuk, *Pulse-Width Modulated DC-DC Power Converters*, New York: John Wiley & Sons, 2008.

[84] X. Nan and C. R. Sullivan, "An equivalent complex permeability model for litz-wire winding," *IEEE Transactions on Industry Applications*, vol. 45, no. 2, pp. 854–860, March/April 2009.

[85] G. S. Dimitrakakis and E. C. Tatakis, "High-frequency losses copper losses in Magnetic components with layered windings," *IEEE Transactions on Magnetics*, vol. 45, no. 8, pp. 3187–3199, August 2009.

[86] M. K. Kazimierczuk and H. Sekiya, "Design of ac resonant inductors using area product method," *IEEE Energy Conversion Conference and Exhibition, San Jose, CA, September 20-24*, 2009, pp. 994–1001.

[87] M. K. Kazimierczuk and R. P. Wojda, "Foil winding resistance and power loss in individual layers of windings," *International Journal of Electronics and Telecommunications*, vol. 56, no. 3, pp. 237–24, 2010.

[88] D. Murthy-Bellur and M. K. Kazimierczuk, "Harmonic winding loss in buck DC-DC converter for discontinuous conduction mode," *IET Proceedings, Power Electronics*, vol. 3, no. 5, pp. 740–754, May 2010.

[89] C. Cerri, S. A. Kovyralovi, and V. M. Primiani, "Modeling of a litz-wire planar winding science," *IET Measurements and Technology*, vol. 4, no. 4, pp. 214–219, 2010.

[90] D. Murthy-Bellur and M. K. Kazimierczuk, "Winding losses caused by harmonics in high-frequency transformers for pulse-width modulated dc-dc flyback converters for discontinuous conduction mode," *IET Proceedings Power Electronics*, vol. 3, no. 5, pp. 804–817, May 2010.

[91] N. Kondrath and M. K. Kazimierczuk, "Inductor winding loss owing to skin and proximity effects including harmonics in non-isolated dc-dc PWM converters in continuous conduction mode," *IET Proceedings Power Electronics*, vol. 3, no, 6, pp. 986–1000, November 2010.

[92] D. Murthy-Bellur, N. Kondrath, and M. K. Kazimierczuk, "Transformer winding losses caused by skin and proximity effects including harmonics in pulse-width modulated dc-dc flyback converters for continuous conduction mode," *IET Proceedings Power Electronics*, vol. 4, no. 5, pp. 804–817, May 2011.

[93] R. Wrobel, N. McNeill, and P. H. Mellor, "Performance analysis and thermal modeling of a high-density prebiased inductor," *IEEE Transactions on Industrial Electronics*, vol. 57, no. 1, pp. 201–208, January 2010.

[94] R. Wrobel and P. H. Mellor, "Thermal design of a high energy density wound components," *IEEE Transactions on Industrial Electronics*, vol. 58, no. 9, pp. 4096–4104, September 2011.

[95] R. Wrobel, A. Mlot, and P. H. Mellor, "Contribution of end-winding proximity loss to temperature variation in electromagnetic devices," *IEEE Transactions on Industrial Electronics*, vol. 59, no. 2, pp. 848–857, February 2012.

[96] D. A. Nagarajan, D. Murthy-Bellur, and M. K. Kazimierczuk, "Harmonic winding loss in the transformer of forward pulse-width modulated dc-dc converter for continuous conduction mode," *IET Proceedings Power Electronics*, vol. 5, no. 2, pp. 221–236, February 2012.

[97] R. P. Wojda and M. K. Kazimierczuk, "Inductor resistance of litz-wire and multi-strand windings," *IET Proceedings Power Electronics*, vol. 5, no. 2, pp. 257–268, February 2012.

[98] R. P. Wojda, "Winding resistance and winding power loss of high-frequency power inductors," PhD dissertation, Wright State University, 2012.

[99] R. P. Wojda and M. K. Kazimierczuk, "Optimum foil thickness of inductors conducting dc and non-sinusoidal periodic currents," *IET Proceedings Power Electronics*, vol. 5, no. 6, pp. 801–812, June 2012.

[100] R. P. Wojda and M. K. Kazimierczuk, "Proximity effect winding loss in different conductors using magnetic field averaging," *COMPEL: The International Journal for Computation and Mathematics in Electrical and Electronic Engineering* vol. 31, no. 6, pp. 1793–1814, 2012.

[101] R. P. Wojda and M. K. Kazimierczuk, "Analytical optimization of solid-round wire windings," *IEEE Transactions on Industrial Electronics*, vol. 59, no. 3, pp. 1033–1041, March 2013.

[102] R. P. Wojda and M. K. Kazimierczuk, "Analytical winding foil thickness optimization of inductors conducting harmonic currents," *IET Proceedings Power Electronics*, vol. 6, no. 5, pp. 963–973, May 2013.

[103] A. Ayachit and M. K. Kazimierczuk, " Thermal effects on resistance of winding conductors at high frequencies," *IEEE Magnetics Letters*, 2013.

[104] R. P. Wojda and M. K. Kazimierczuk, "Magnetic field distribution and analytical optimization of inductor foil windings," *IEEE Magnetics Letters*, vol. 4, April 2013.

[105] R. P. Wojda and M. K. Kazimierczuk, "Analytical optimization of solid-round-wire windings conducting dc and ac non-sinusoidal periodic currents," *IET Proceedings Power Electronics*, vol. 6, no. 7, pp. 1462–1474, July 2013.

[106] R. P. Wojda and M. K. Kazimierczuk, "Analytical winding size optimization for different conductor shapes using Ampère's law," *IET Proceedings Power Electronics*, vol. 6, no. 6, pp. 963–973, June 2013.

5.33 Review Questions

5.1. What kind of power loss is present in a single-layer inductor winding?

5.2. What kind of power loss is present in a multiple-layer inductor winding?

5.3. How does the normalized winding AC resistance R_w/R_{wDC} depend on frequency?

5.4. How does the number of layers N_l affect the variation of R_w/R_{wDC} with frequency?

5.5. How does the layer porosity factor η affect the variation of R_w/R_{wDC} with frequency?

5.6. How does the winding AC resistance vary with the rectangular conductor height?

5.7. How does the winding AC resistance vary with the square conductor height?

5.8. How does the winding AC resistance vary with the round conductor diameter?

5.9. Is the eddy-current winding loss in inductors carrying nonsinusoidal current different than that in inductors carrying sinusoidal current?

5.10. Does the winding AC loss depend on the duty cycle D of the inductor voltage?

5.11. Is the litz wire winding always superior to solid wire of equal DC resistance?

5.12. What are the mechanisms of heat transfer?

5.34 Problems

5.1. A sinusoidal current flows in an inductor at $f = 40$ kHz. The winding is made of a copper foil and has five layers ($N_l = 5$).

(a) Determine the optimum copper foil thickness h_{opt}.

(b) Determine the copper foil boundary thickness h_B.

5.2. A sinusoidal current flows in an inductor at $f = 120$ kHz. The winding is made of a copper strip and has four layers ($N_l = 4$). The porosity factor is $\eta = 0.95$.

(a) Determine the optimum copper strip thickness h_{opt}.

(b) Determine the copper strip boundary thickness h_B.

5.3. A sinusoidal current flows through an inductor at $f = 75$ kHz. The winding is made of a copper square wire and has three layers ($N_l = 3$). The porosity factor is $\eta = 0.92$.

(a) Determine the optimum copper square wire thickness h_{opt}.

(b) Determine the copper square wire boundary thickness h_B.

5.4. A sinusoidal current flows in an inductor at $f = 200$ kHz. The winding is made of a round copper wire and has five layers ($N_l = 5$).

(a) Determine the optimum copper round wire thickness h_{opt}.

(b) Determine the copper round wire boundary thickness h_B.

5.5. An inductor is made of a copper round wire with inner diameter $d = 0.511$ mm, $N_l = 5$, and $d/p = 0.8$. The winding wire length is $l_w = 1$ m. Find the AC resistance for a sinusoidal current at frequency $f = 100$ MHz.

5.6. An inductor is wound on a toroidal core using a round copper wire and a single layer. The inductor is used in Class E power amplifier at $f = 1$ MHz. The bare wire diameter is $d = 1.15$ mm and the wire length is $l_w = 0.95$ m.

(a) Calculate the winding AC resistance and the AC resistance factor F_R using Dowell's equation.

(b) Calculate the winding AC resistance and the AC resistance factor F_R using skin-depth equation.

(c) Find the ratio of the two resistances.

5.7. An inductor is wound on a toroidal core using a round copper wire and a single layer. The inductor is used in Class E power amplifier at $f = 1$ MHz. The bare wire diameter is $d = 1.02$ mm and the wire length is $l_w = 1$ m.

(a) Calculate the winding AC resistance and the AC resistance factor F_R for using Dowell's equation.

(b) Calculate the winding AC resistance and the AC resistance factor F_R using skin-depth equation.

(c) Find the ratio of the two resistances.

5.8. A foil inductor has $L = 200$ μH, $I_{Lm} = 20$ A, $\mu_{rc} = 2000$, $l_c = 6$ cm, $b = 3$ cm, $A_c = 4.2$ cm^2, and $l_T = 7$ cm.

(a) Find the number of layers.

(b) Find the optimum foil thickness.

(c) Find the winding resistance.

(d) Find the winding power loss.

5.9. Design an optimum foil thickness of an inductor with $L = 14.36$ μH. The inductor winding is made of silver foil of resistivity $\rho_{Au} = 15.87$ nΩ·m at $T = 20$ °C. The core of the inductor is made of nickel–zinc (NiZn) ferrite with relative permeability $\mu_{rc} = 250$, diameter of the cylindrical core is $D = 4$ mm, and the core length is $l_c = 11$ cm. The width of the bobbin is $b_b = 10$ mm and the MTL is $l_T = 6.66$ cm. The inductor current is $i_L = 7.5 \sin(2\pi \times 10^5 t)$ (A).

(a) Find the number of layers.

(b) Find the optimum foil thickness.

(c) Find the winding DC resistance.

(d) Find the winding AC resistance.

(e) Find the AC power loss.

(f) Find the average current density.

5.10. For the inductor given in Problem 5.8.

(a) Find the boundary of $(h/\delta_w)_B$ between the low and medium frequencies.

(b) Find the winding DC resistance at the boundary.

(c) Find the winding AC resistance at the boundary.

(d) Find the winding AC power loss at the boundary.

(e) Find the ratio of the winding AC resistance at the boundary to the winding AC resistance at the optimum foil thickness.

6

Laminated Cores

6.1 Introduction

In this chapter, laminated cores are studied [1–43]. Eddy-current loss can be reduced in two ways. First, a high-resistivity material can be used for cores. This increases the skin depth δ_c and thus reduces w/δ_c, making the distribution of the magnetic flux density B more uniform. If this is done, the condition $w/\delta_c < 1$ is satisfied over a wider frequency range.

To increase the core resistivity ρ_c and therefore to reduce eddy-current amplitude, the iron is alloyed with a small amount of silicon or chrome (1–5%), producing a *magnetic steel*. Ferrite and powder cores are also examples of materials with high resistivity. Second, eddy-current power loss can be greatly reduced by dividing the core into a large number of thin slices that are electrically insulated from each other by an oxide film. These thin insulated sheets are called *laminations*. They should be oriented parallel to the magnetic flux ϕ. Eddy currents flow in a plane parallel to the magnetic flux. Thin laminations have little flux in an individual layer, and therefore, the induced voltage decreases. In addition, the resistance of a single lamination is k times higher than that of the corresponding solid core, where k is the number of sheets. Laminations are insulated from each other by an oxide film or insulating varnish and then stacked together to form the magnetic core. If thin sheets are used, more of the core volume is utilized. The lamination thickness w ranges from 0.1 to 1 mm. Typical thickness is usually 0.5, 0.75, or 1 mm for 50 and 60 Hz, 0.3 or 0.35 mm for frequencies up to 400 Hz, 0.1 mm for frequencies between 400 Hz and 2 kHz, and 0.05 mm for higher frequencies and pulse applications. The lamination thickness can be as low as 12.7 μm. The typical thickness of insulation is 0.015 mm. The thickness of tape-wound metal alloy laminations is 10–100 μm. The most common cores use 97% iron and 3% silicon. They have $\mu_r = 10\,000$ and the lamination thickness equal to 0.3 mm. In powder cores, eddy-current power loss is low because individual particles are insulated from each other. Laminations do not affect the magnetic performance of the core at low frequencies. Line frequency transformers and electric machines use laminated cores.

The skin effect is another reason for using laminations. Otherwise, a solid core would contain flux only in a shell with thickness equal to the core skin depth δ_c. As w/δ_w is increased from 1 to 10, $H(0)/H(w/2)$ and $B(0)/B(w/2)$ decrease from approximately 1 to 0. For example, $B(0)/B(w/2) = 0.98$ at $w/\delta_c = 1$, $B(0)/B(w/2) = 0.78$ at $w/\delta_c = 2$, $B(0)/B(w/2) = 0.46$ at $w/\delta_c = 3$, and $B(0)/B(w/2) = 0.1$ at $w/\delta_c = 8$. When the core resistivity ρ_c is increased, the core skin depth δ_c is also increased. Therefore, there will be a low-frequency range, in which $\delta_c > w$, the distribution of H is nearly uniform, and the skin effect in the core can be neglected.

High-Frequency Magnetic Components, Second Edition. Marian K. Kazimierczuk.

Companion Website: www.wiley.com/go/kazimierczuk_High2e

When the lamination area is cut into half, the flux in each lamination is reduced by a factor of 2. The voltage is reduced twofold. The amplitude of the eddy current is reduced by a factor of 2, assuming that the path of the eddy current remains the same. Therefore, the core loss is reduced fourfold. The core eddy-current losses can be reduced using thinner laminations, using higher core resistivity material ρ_c, reducing the maximum value of the magnetic flux density B_m, and reducing the operating frequency f.

6.2 Low-Frequency Eddy-Current Laminated Core Loss

The derivation of the eddy-current power loss in a laminated iron core for low frequencies (where $\delta_c \gg w$) is as follows. Consider an iron lamination without an air gap, depicted in Fig. 6.1. Assume that $h \gg w$ and $\delta_c > w$, that is, let us assume that the distribution of magnetic field intensity H along the x-axis is uniform. Also, assume that a sinusoidal current flows in the inductor winding

$$i(t) = I_m \sin \omega t \tag{6.1}$$

resulting in the magnetic flux density in the core located inside the winding

$$B(t) = B_m \sin \omega t. \tag{6.2}$$

The distribution of the magnetic flux density $B = B_z$ in the core placed inside the winding is uniform for $w < \delta_c$. The magnetic flux passes through the area

$$A_\phi(x) = hx \quad \text{for} \quad 0 \le x \le \frac{w}{2}. \tag{6.3}$$

and the magnetic flux passing through half of the area encircled by the eddy-current loop is

$$\phi(x,t) = A_\phi B_m \sin \omega t = A_\phi B(t) = hxB_m \sin \omega t \quad \text{for} \quad x > 0. \tag{6.4}$$

By Faraday's law, the voltage induced by the magnetic flux between the bottom and top terminals of the lamination is equal to the derivative of the magnetic flux passing through half of the area enclosed by the eddy-current loop

$$v(x,t) = \frac{d\phi(x,t)}{dt} = A_\phi \frac{dB(t)}{dt} = A_\phi \omega B_m \cos \omega t = hx\omega B_m \cos \omega t = V_m(x) \cos \omega t \tag{6.5}$$

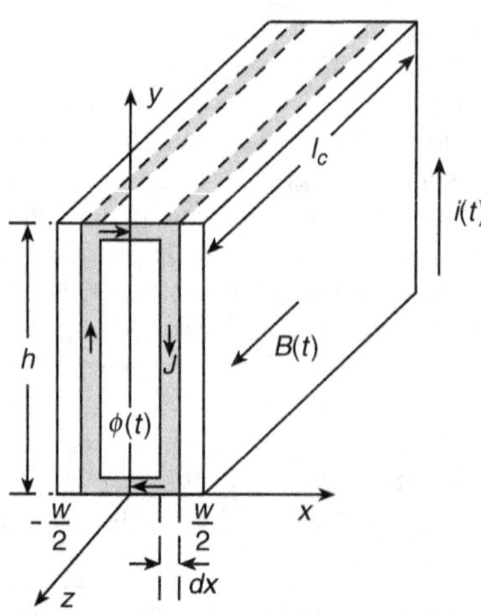

Figure 6.1 Cross section of single lamination used for analyzing eddy-current power loss.

where

$$V_m(x) = \omega B_m hx. \tag{6.6}$$

Thus, $V_m(x)$ is proportional to f, B_m, h, and x. By Lenz's law, the voltage $v(t)$ induces the eddy current of density J, which, in turn, generates the magnetic flux $\phi'(t)$ that tends to oppose the imposed flux $\phi(t)$. Since V_m increases with x, the maximum current density occurs near the surface and $V_m = 0$ at $x = 0$, resulting in zero current density in the middle of the lamination.

The lamination resistance of the incremental strip of thickness dx and area $A_{ex} = l_c(dx)$ is

$$R_{cx} = \rho_c \frac{h}{A_{ex}} = \rho_c \frac{h}{l_c(dx)} \tag{6.7}$$

resulting in the amplitude of the eddy current

$$I_{em}(x) = \frac{V_m}{R_{cx}} = \frac{\omega B_m l_c x(dx)}{\rho_c} \tag{6.8}$$

and in the amplitude of the eddy-current density

$$|J_m(x)| = \frac{I_{em}}{A_{ex}} = \frac{\omega B_m x}{\rho_c}. \tag{6.9}$$

It can be seen that the amplitude of the eddy-current density changes linearly with distance x. It is zero in the middle of the lamination and increases with the distance x from the middle of the lamination. The maximum amplitude of the eddy-current density is at the lamination surface

$$J_{m(max)} = J_m\left(\frac{w}{2}\right) = \frac{\omega B_m w}{2\rho_c}. \tag{6.10}$$

The time-averaged eddy-current power loss in the lamination at low frequencies is

$$\begin{aligned} P_E &= \int\int\int_{V_c} \frac{\rho_c J_m^2(x)}{2} dxdydz = \frac{\rho_c}{2}\int_0^{l_c}\int_0^h\int_{-w/2}^{w/2} \frac{\omega^2 B_m^2 x^2}{\rho_c^2} dxdydz \\ &= \frac{\omega^2 B_m^2}{\rho_c}\int_0^{l_c} dz \int_0^h dy \int_0^{w/2} x^2 dx = \frac{\omega^2 B_m^2 h l_c w^3}{24\rho_c} = \frac{\pi^2 f^2 B_m^2 h l_c w^3}{6\rho_c} = \frac{\pi^2 f^2 B_m^2 w^2 V_c}{6\rho_c} \\ &= \frac{\omega^2 B_m^2 h l_c w^3}{24\rho_c} = \frac{\omega^2 B_m^2 w^2 V_c}{24\rho_c} \quad \text{for} \quad w < \delta_c, \end{aligned} \tag{6.11}$$

yielding the time-averaged eddy-current power loss per unit volume (or the specific eddy-current power loss) for a sinusoidal waveform of $B(t) = B_m \sin\omega t$ at low frequencies

$$P_e = \frac{P_E}{V_c} = \frac{\pi^2 w^2 f^2 B_m^2}{6\rho_c} = \frac{w^2\omega^2 B_m^2}{24\rho_c} \quad \text{for} \quad w < \delta_c = \sqrt{\frac{\rho_c}{\pi\mu_{rc}\mu_0 f}}, \tag{6.12}$$

where $V_c = whl_c$ is the single lamination volume and w is the single lamination thickness. Note that P_E is proportional to w^3 and P_e is proportional to w^2. It is evident from the above expression that the eddy-current power loss density at low frequencies is reduced using thinner laminations and a core material with a high resistivity.

The eddy-current power loss density decreases with temperature T because the core resistivity ρ_c increases with temperature

$$P_{e(W)} = \frac{\pi^2 w^2 f^2 B_m^2}{6\rho_{c(To)}[1 + \alpha_c(T - T_o)]} w < \delta_c. \tag{6.13}$$

For a triangle waveform of $B(t)$, the time-averaged eddy-current power loss per unit volume at low frequencies is given by

$$P_{e(tr)} = \frac{4}{3}\frac{w^2 f^2 B_m^2}{\rho_c}. \tag{6.14}$$

Hence, the ratio of the time-average eddy-current power loss for sinusoidal waveform and triangle waveform of $B(t)$ at low frequencies is

$$\frac{P_e}{P_{e(tr)}} = \frac{8}{\pi^2} = 0.811. \tag{6.15}$$

An alternative method for deriving an expression for P_E is as follows. The resistance of a thin strip with eddy currents is

$$R = \rho_c \frac{h}{l_c dx}. \tag{6.16}$$

Hence, the eddy-current power dissipated in the thin strip is

$$dP_E = \frac{V_m^2}{2R} = \frac{hl_c\omega^2 B_m^2}{2\rho_c} x^2 dx \tag{6.17}$$

resulting in the eddy-current power dissipated in the lamination at low frequencies

$$P_E = \int_{-w/2}^{w/2} dP_E = 2\int_0^{w/2} dP_E = \frac{hl_c\omega^2 B_m^2 hl_c w^3}{\rho_c}\int_0^{w/2} x^2 dx = \frac{\omega^2 B_m^2 hl_c w^3}{24\rho_c}$$
$$= \frac{\omega^2 B_m^2 w^2}{24\rho_c} V_c = \frac{\pi^2 f^2 B_m^2 w^2}{6\rho_c} V_c \quad \text{for} \quad w < \delta_c. \tag{6.18}$$

The loss due to eddy currents is proportional to the square of frequency and the cube of lamination thickness w.

The total power loss density in a laminated core for sinusoidal waveforms at low frequencies is given by

$$P_v = P_h + P_e + P_{ex} = k_h f B_m^2 + \frac{\pi^2 w^2}{6\rho_c}(fB_m)^2 + k_{ex}(fB_m)^{\frac{3}{2}} \,(\mathrm{W/m})^3. \tag{6.19}$$

A laminated iron core M270-35A has resistivity $\rho_c = 5.2083 \times 10^{-7}\ \Omega \cdot \mathrm{m}$, lamination thickness $w = 1$ mm, and total power loss density in a laminated core for sinusoidal waveforms at low frequencies [40]

$$P_v = P_h + P_e + P_{ex} = 118.6 f B_m^2 + \frac{\pi^2 w^2}{6\rho_c}(fB_m)^2 + 6.4158(fB_m)^{\frac{3}{2}} \,(\mathrm{W/m})^3. \tag{6.20}$$

Figure 6.2 shows the plots of hysteresis core loss density P_h, eddy-current core loss density P_e, and excess core loss density P_{ex} as functions of magnetic flux amplitude B_m for $w = 1$ mm at $f = 50$ Hz for laminated iron cores given by (6.20).

A laminated iron core has resistivity $\rho_c = 4.5045 \times 10^{-7}\ \Omega \cdot \mathrm{m}$, lamination thickness $w = 0.35$ mm, and total power loss density in a laminated core for sinusoidal waveforms at low frequencies [42]

$$P_v = P_h + P_e + P_{ex} = 93.89 f B_m^2 + \frac{\pi^2 w^2}{6\rho_c}(fB_m)^2 + 11.5311(fB_m)^{\frac{3}{2}} \,(\mathrm{W/m})^3. \tag{6.21}$$

Figure 6.3 shows the plots of hysteresis core loss density P_h, eddy-current core loss density P_e, and excess core loss density P_{ex} as functions of magnetic flux amplitude B_m for $w = 0.35$ mm at $f = 50$ Hz for laminated iron cores given by (6.21).

An iron-alloy-laminated core has resistivity $\rho_c = 4 \times 10^{-7}\ \Omega \cdot \mathrm{m}$, lamination thickness $w = 0.1$ mm, and total power loss density in a laminated core for sinusoidal waveforms at low frequencies [41]

$$P_v = P_h + P_e + P_{ex} = 111.64 f B_m^2 + \frac{\pi^2 w^2}{6\rho_c}(fB_m)^2 + 3.5547(fB_m)^{\frac{3}{2}} \,(\mathrm{W/m})^3. \tag{6.22}$$

Figure 6.4 shows the plots of hysteresis core loss density P_h, eddy-current core loss density P_e, and excess core loss density P_{ex} as functions of magnetic flux amplitude B_m for $w = 0.1$ mm at $f = 50$ Hz for laminated iron cores given by (6.22). It can be seen that the eddy-current loss density P_h decreases as the lamination thickness w decreases. Figure 6.5 shows the plots of core loss density P_v as a function of the magnetic flux amplitude B_m at $f = 50$ Hz for laminated cores given by (6.21) and (6.22).

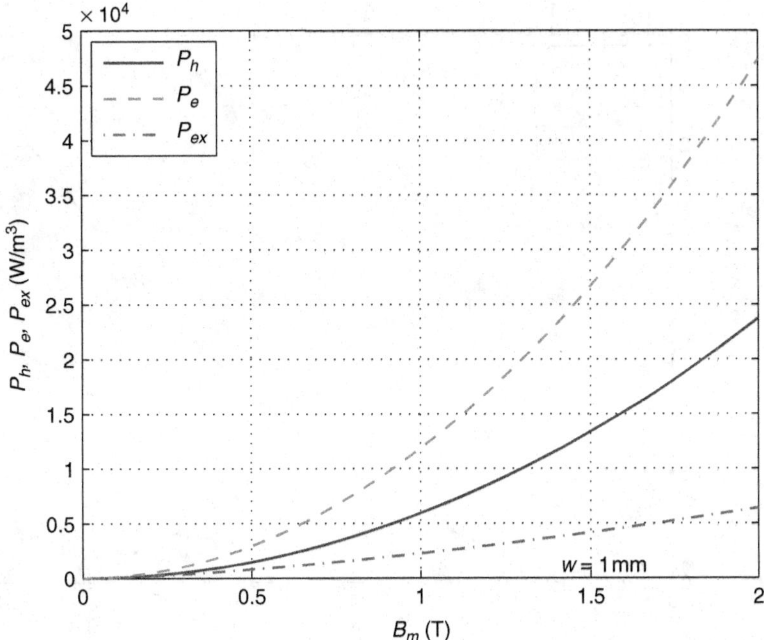

Figure 6.2 Hysteresis core loss density P_h, eddy-current core loss density P_e, and excess core loss density P_{ex} as functions of magnetic flux amplitude B_m for $w = 1$ mm at $f = 50$ Hz for laminated iron cores given by (6.22).

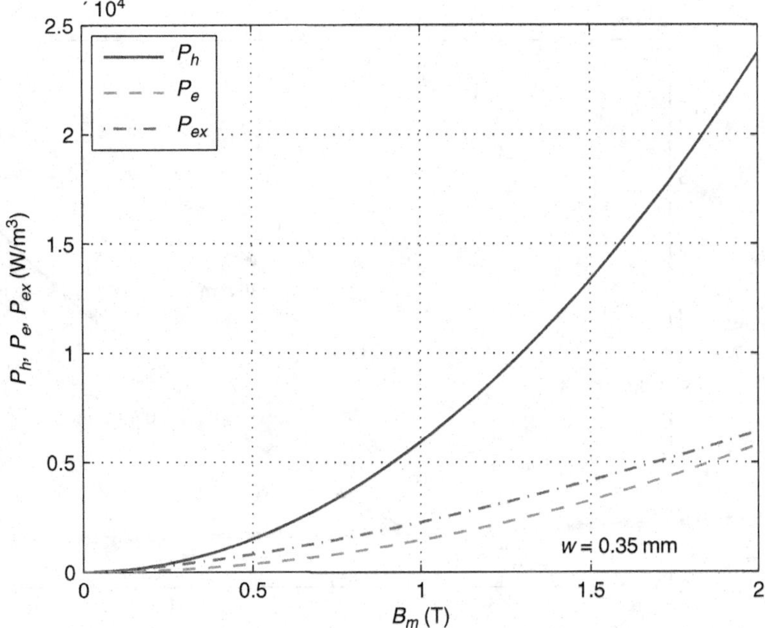

Figure 6.3 Hysteresis core loss density P_h, eddy-current core loss density P_e, and excess core loss density P_{ex} for $w = 1$ mm at $f = 50$ Hz as functions of magnetic flux amplitude B_m for laminated iron cores given by (6.21).

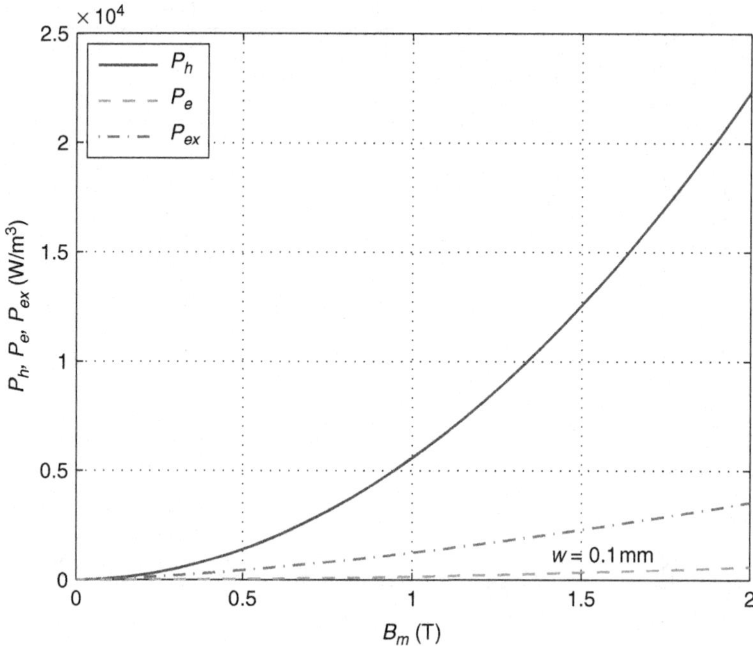

Figure 6.4 Hysteresis core loss density P_h, eddy-current core loss density P_e, and excess core loss density P_{ex} as functions of magnetic flux amplitude B_m for $w = 0.1$ mm at $f = 50$ Hz for laminated iron cores given by (6.22).

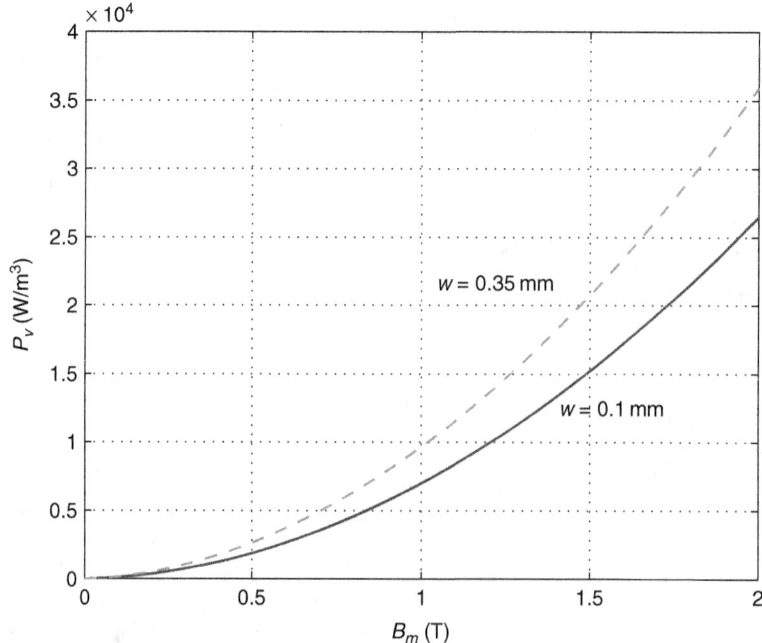

Figure 6.5 Core loss density P_v as a function of magnetic flux amplitude B_m for $w = 0.35$ mm at $f = 50$ Hz for laminated iron cores given by (6.21) and (6.22).

6.3 Comparison of Solid and Laminated Cores

Assuming that a solid core width is $w_s = kw$, the ratio of the eddy-current power loss density in a solid core P_{es} to that in the single lamination P_e at the same f, B_mi, and ρ_c is given by

$$\frac{P_{es}}{P_e} = \left(\frac{w_s}{w}\right)^2 = k^2 \quad \text{for} \quad w < \delta_c = \sqrt{\frac{\rho_c}{\pi \mu_{rc} \mu_0 f}}. \tag{6.23}$$

The ratio of the eddy-current loss in the solid core P_{Es} to the eddy-current loss in a single lamination P_E is

$$\frac{P_{Es}}{P_E} = \left(\frac{w_s}{w}\right)^3 = k^3 \quad \text{for} \quad w < \delta_c = \sqrt{\frac{\rho_c}{\pi \mu_{rc} \mu_0 f}}. \tag{6.24}$$

Hence, the ratio of the eddy-current power loss in a solid core P_{Es} to the eddy-current loss in a laminated core consisting of k laminations P_{El} is given by

$$\frac{P_{Es}}{P_{El}} = \frac{P_{Es}}{kP_E} = \left(\frac{w_s}{w}\right)^2 = k^2 \quad \text{for} \quad w < \delta_c = \sqrt{\frac{\rho_c}{\pi \mu_{rc} \mu_0 f}}. \tag{6.25}$$

Therefore, the eddy-current loss in the solid core is k^2 times higher than that in the laminated core, assuming that both cores have the same conducting volume.

The solid core resistance at low frequencies is

$$R_{csd} = \rho_c \frac{h}{A_{cs}} = \rho_c \frac{h}{w_s l_c}. \tag{6.26}$$

The resistance of a single lamination is

$$R_{csl} = \rho_c \frac{h}{A_{cs}} = \rho_c \frac{h}{w l_c} = kR_{csd}. \tag{6.27}$$

The core resistance of the laminated core with k laminations at low frequencies is

$$R_{cl} = \rho_c \frac{h}{kwl_c} = R_{csd}. \tag{6.28}$$

6.4 Alternative Solution for Low-Frequency Eddy-Current Core Loss

Figure 6.6 shows a rectangular magnetic core. The ratio of the core height h to the core width w is

$$K = \frac{h}{w} = \frac{y}{x} = \frac{dy}{dx}. \tag{6.29}$$

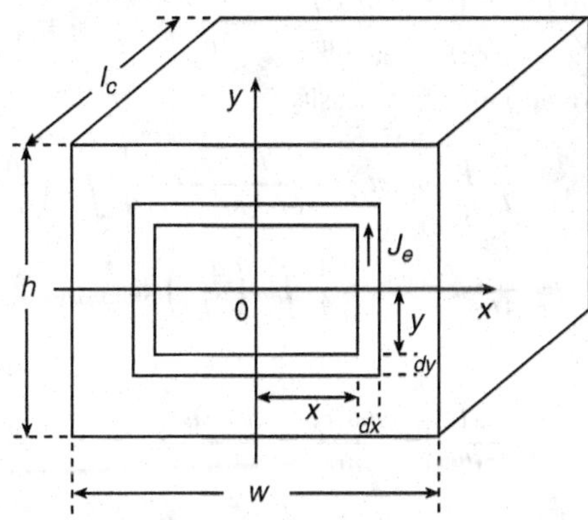

Figure 6.6 Cross section of rectangular magnetic core with eddy-current power loss.

The cross-sectional area of the core perpendicular to the magnetic flux density **B** and parallel to the winding current flow is given by

$$A_c = hw = Kw^2 = \frac{h^2}{K}. \tag{6.30}$$

The core volume is

$$V_c = l_c A_c = l_c hw = l_c K w^2. \tag{6.31}$$

Assume that the magnetic field in the core is uniform, that is, $\delta_c \gg w$. The low-frequency resistance of the shell of volume $dv = l_c dxdy$, which is approximately equal to its DC resistance, is given by

$$R_{sh} = \rho_c \left(\frac{4x}{l_c dy} + \frac{4y}{l_c dx} \right) = \rho_c \left(\frac{4x}{l_c K dx} + \frac{4Kx}{l_c dx} \right) = \frac{4\rho_c}{l_c} \left(K + \frac{1}{K} \right) \frac{x}{dx}$$
$$= \frac{4\rho_c}{K l_c} (K^2 + 1) \frac{x}{dx}. \tag{6.32}$$

The magnetic flux enclosed in the shell is

$$\phi(t) = 4xyB(x) = 4Kx^2 B(t). \tag{6.33}$$

By Faraday's law, the voltage induced by the magnetic flux enclosed in the shell is

$$v(t) = \frac{d\phi(t)}{dt} = 4xy\frac{dB(t)}{dt} = 4Kx^2\frac{dB(t)}{dt}. \tag{6.34}$$

The instantaneous power loss in the shell caused by the circulating eddy currents is

$$dp_e(t) = \frac{v^2(t)}{R_{sh}} = \frac{\left(4Kx^2 \frac{dB}{dt}\right)^2}{4\rho_c \left(K + \frac{1}{K}\right) \frac{x}{l_c dx}} = \frac{4 l_c K^2 \left(\frac{dB}{dt}\right)^2 x^3 dx}{\rho_c \left(K + \frac{1}{K}\right)} = \frac{4 l_c K^3 \left(\frac{dB}{dt}\right)^2 x^3 dx}{\rho_c (K^2 + 1)}. \tag{6.35}$$

The instantaneous eddy-current power loss in the entire core is

$$p_e(t) = \int_0^{w/2} dp_e = \frac{4 l_c K^3 \left(\frac{dB}{dt}\right)^2}{\rho_c (K^2+1)} \int_0^{\frac{w}{2}} x^3 dx = \frac{l_c K^3 w^4 \left(\frac{dB}{dt}\right)^2}{16\rho_c (K^2+1)} = \frac{l_c K A_c^2 \left(\frac{dB}{dt}\right)^2}{16\rho_c (K^2+1)}. \tag{6.36}$$

The time-average eddy-current power loss in the core for steady state is

$$P_{e(av)} = \frac{1}{T} \int_0^T p_e dt = \frac{l_c K^3 w^4}{16\rho_c (K^2+1)} \frac{1}{T} \int_0^T \left(\frac{dB}{dt} \right)^2 dt$$
$$= \frac{l_c K A_c^2}{16\rho_c (K^2+1)} \frac{1}{T} \int_0^T \left(\frac{dB}{dt} \right)^2 dt. \tag{6.37}$$

The instantaneous eddy-current power loss density in the core is the power dissipated per unit volume

$$p_{ev}(t) = \frac{p_e(t)}{V_c} = \frac{p_e(t)}{l_c hw} = \frac{p_e(t)}{l_c K w^2} = \frac{K^2 w^2 \left(\frac{dB}{dt}\right)^2}{16\rho_c (K^2+1)} = \frac{K A_c \left(\frac{dB}{dt}\right)^2}{16\rho_c (K^2+1)}. \tag{6.38}$$

The time-average eddy-current power loss density is

$$p_{ev(av)} = \frac{1}{T} \int_0^T p_{ev} dt = \frac{K^2 w^2}{16\rho_c (K^2+1)} \frac{1}{T} \int_0^T \left(\frac{dB}{dt} \right)^2 dt$$
$$= \frac{K A_c}{16\rho_c (K^2+1)} \frac{1}{T} \int_0^T \left(\frac{dB}{dt} \right)^2 dt. \tag{6.39}$$

From Faraday's law,

$$v(t) = \frac{d\lambda(t)}{dt} = \frac{d\phi(t)}{dt} = \frac{d[A_c B(t)]}{dt} = A_c \frac{dB(t)}{dt}, \tag{6.40}$$

yielding

$$\frac{dB}{dt} = \frac{v(t)}{A_c} \tag{6.41}$$

and

$$\left(\frac{dB}{dt}\right)^2 = \frac{v^2(t)}{A_c^2}. \tag{6.42}$$

Hence, the time-average eddy-current power loss density is

$$P_{ev(av)} = \frac{KA_c}{16\rho_c(K^2+1)}\frac{1}{T}\int_0^T \left(\frac{dB}{dt}\right)^2 dt = \frac{K}{16\rho_c A_c(K^2+1)}\frac{1}{T}\int_0^T v^2(t)dt. \tag{6.43}$$

6.4.1 Sinusoidal Inductor Voltage

Consider a sinusoidal inductor voltage

$$v_L = V_{Lm}\cos\omega t = \frac{d\lambda(t)}{dt} = N\frac{d\phi(t)}{dt} = NA_c\frac{dB(t)}{dt}. \tag{6.44}$$

Hence,

$$dB(t) = \frac{v_L}{NA_c}dt, \tag{6.45}$$

yielding

$$B(t) = \frac{1}{NA_c}\int v_L dt = \frac{V_{Lm}}{NA_c}\int \sin\omega t dt = -\frac{V_{Lm}}{NA_c}\sin\omega t = -B_m\sin\omega t, \tag{6.46}$$

where

$$B_m = \frac{V_{Lm}}{NA_c}. \tag{6.47}$$

Thus,

$$\frac{dB}{dt} = \frac{\omega V_{Lm}}{A_c}\cos\omega t = \omega B_m\cos\omega t, \tag{6.48}$$

producing

$$\left(\frac{dB}{dt}\right)^2 = \omega^2 B_m^2\cos^2\omega t. \tag{6.49}$$

The instantaneous power loss in the core caused by eddy currents is

$$p_e(t) = \frac{l_c K^3 w^4\omega^2 B_m^2}{16\rho_c(K^2+1)}\cos^2\omega t = \frac{l_c KA_c^2 w^4\omega^2 B_m^2}{16\rho_c(K^2+1)}\cos^2\omega t. \tag{6.50}$$

The time-average eddy-current power loss in the core is

$$\begin{aligned} P_{E(av)} &= \frac{l_c K w^4\omega^2 B_m^2}{16\rho_c(K^2+1)}\frac{1}{T}\int_0^T \cos^2\omega t dt = \frac{l_c K^3 w^4\omega^2 B_m^2}{32\rho_c(K^2+1)} = \frac{\pi^2}{8}\frac{l_c K^3 w^4 f^2 B_m^2}{\rho_c(K^2+1)} \\ &= \frac{\pi^2}{8}\frac{l_c K^2 A_c^2 f^2 B_m^2}{\rho_c(K^2+1)} = \frac{\pi^2}{8}\frac{l_c A_c^2 f^2 B_m^2}{\rho_c(1+1/K^2)}. \end{aligned} \tag{6.51}$$

Figure 6.7 shows a plot of normalized eddy-current power loss $8\rho_c P_{ew(av)}/(\pi A_c f^2 B_m^2) = 1/(1 + 1/K^2)$. The time-average eddy-current power loss in the core decreases as A_c decreases and K increases.

The time-average eddy-current power loss density in the core is

$$\begin{aligned} P_{ev(av)} &= \frac{P_{e(av)}}{V_c} = \frac{K^2 w^2\omega^2 B_m^2}{16\rho_c(K^2+1)}\frac{1}{T}\int_0^T \cos^2\omega t dt = \frac{K^2 w^2\omega^2 B_m^2}{32\rho_c(K^2+1)} = \frac{\pi^2}{8}\frac{K^2 w^2 f^2 B_m^2}{\rho_c(K^2+1)} \\ &= \frac{\pi^2}{8}\frac{KA_c f^2 B_m^2}{\rho_c(K^2+1)} = \frac{\pi^2}{8}\frac{A_c f^2 B_m^2}{\rho_c(K+1/K)}. \end{aligned} \tag{6.52}$$

Figure 6.8 shows a plot of normalized eddy-current power loss density $8\rho_c P_{ev(av)}/(\pi^2 A_c f^2 B_m^2) = 1/(K+1/K)$ as a function of K. It can be seen that the maximum time-average eddy-current power density occurs at $K = 1$. The time-average eddy-current power density decreases as A_c decreases and as $K = h/w$ increases for $K > 1$.

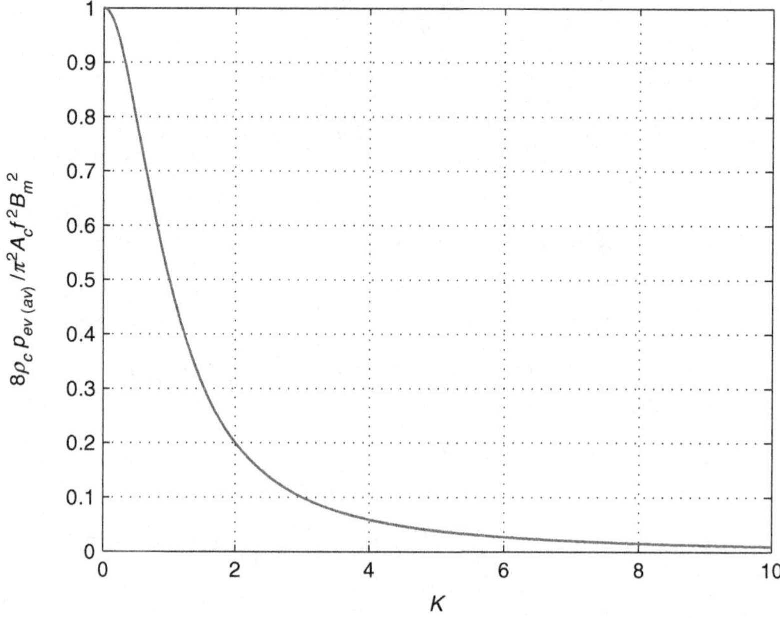

Figure 6.7 Plot of normalized eddy-current core loss $8\rho_c P_{ew(av)}/(\pi A_c f^2 B_m^2) = 1/(1 + 1/K^2) = 1/(1+ 1/K^2)$ as a function of K values of w/δ_c.

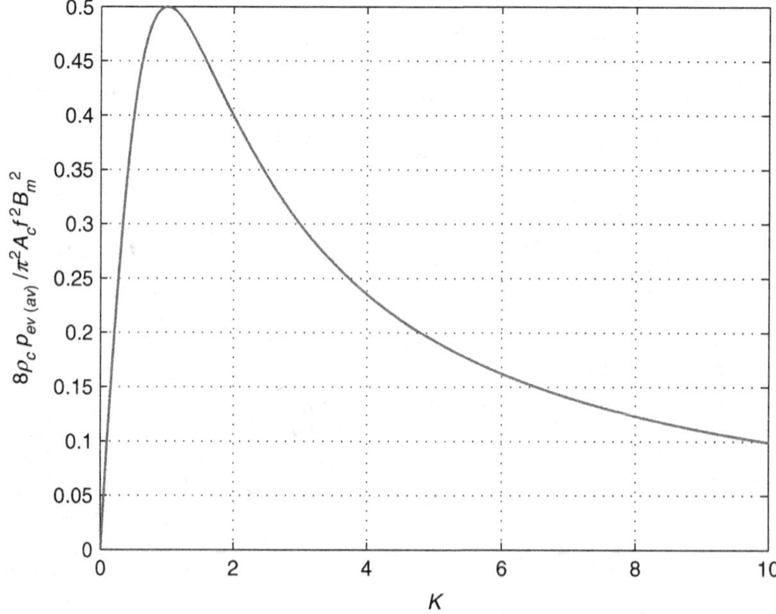

Figure 6.8 Plot of normalized eddy-current power loss density $8\rho_c P_{ev(av)}/(\pi^2 A_c f^2 B_m^2) = 1/(K + 1/K)$ as a function of K values of w/δ_c.

6.4.2 Square-Wave Inductor Voltage

For the bipolar symmetrical square-wave inductor voltage, the slope of the increasing flux density is constant and is given by

$$\frac{dB}{dt} = \frac{4B_m}{T} = 4B_m f, \tag{6.53}$$

yielding

$$\left(\frac{dB}{dt}\right)^2 = \frac{16B_m^2}{T^2} = 16f^2B_m^2. \tag{6.54}$$

Since the inductor voltage waveform is constant in each half of the cycle, the time-average eddy-current core power loss density is equal to the instantaneous core power loss

$$P_{ev(av)} = p_e(t) = \frac{K^2w^2}{16\rho_c(K^2+1)}\left(\frac{dB}{dt}\right)^2 = \frac{K^2w^2f^2B_m^2}{\rho_c(K^2+1)} = \frac{KA_cf^2B_m^2}{\rho_c(K^2+1)}. \tag{6.55}$$

6.4.3 Rectangular Inductor Voltage

For a rectangular inductor voltage waveform with a duty ratio D, the eddy-current core power loss density is

$$P_{ev(av)} = \frac{P_{ev}}{V_c} = \frac{K^2w^2B_m^2}{16\rho_c(K^2+1)D^2T^2}. \tag{6.56}$$

The time-average eddy-current core power loss density is

$$P_{ev(av)} = \frac{p_{ev}}{T} = \frac{1}{T}\int_0^{DT} p_{ev}(t)dt = \frac{K^2w^2DB_m^2}{16\rho_c(K^2+1)D^2T^2} = \frac{K^2w^2f^2B_m^2}{16\rho_c D(K^2+1)}. \tag{6.57}$$

6.5 General Solution for Eddy-Current Laminated Core Loss

6.5.1 Magnetic Field Distribution at High Frequencies

In this section, we will find a general solution valid for both low and high frequencies. At any frequency, the magnetic field intensity is described by the Helmholtz equation

$$\frac{d^2H_z(x)}{dx^2} = \gamma^2H_z(x) = j\mu_c\sigma_c\omega H_z(x) \tag{6.58}$$

where

$$\gamma = \sqrt{j\mu_c\sigma_c\omega} = (1+j)\sqrt{\pi\mu_c\sigma_c f} = \frac{1+j}{\delta_c} \tag{6.59}$$

with the skin depth into the conductor sheets given by

$$\delta_c = \frac{1}{\sqrt{\pi\sigma_c\mu_c f}} = \sqrt{\frac{\rho_c}{\pi\mu_c f}}. \tag{6.60}$$

A general solution of (6.58) is

$$H(x) = H_1e^{\gamma x} + H_2e^{-\gamma x}. \tag{6.61}$$

Hence,

$$H\left(\frac{w}{2}\right) = H_1e^{\gamma\frac{w}{2}} + H_2e^{-\gamma\frac{w}{2}} \tag{6.62}$$

and

$$H\left(-\frac{w}{2}\right) = H_1e^{-\gamma\frac{w}{2}} + H_2e^{\gamma\frac{w}{2}}. \tag{6.63}$$

The condition of even symmetry requires that $H(x) = H(-x)$, from which $H(w/2) = H(-w/2)$. Subtracting (6.63) from (6.62), we get

$$H\left(\frac{w}{2}\right) - H\left(-\frac{w}{2}\right) = H_1 e^{\gamma \frac{w}{2}} + H_2 e^{-\gamma \frac{w}{2}} - H_1 e^{-\gamma \frac{w}{2}} - H_2 e^{\gamma \frac{w}{2}} = (e^{\gamma \frac{w}{2}} - e^{-\gamma \frac{w}{2}})(H_1 - H_2) = 0, \tag{6.64}$$

from which

$$H_1 = H_2. \tag{6.65}$$

Substituting for H_2 in (6.61),

$$H(x) = H_1(e^{\gamma x} + e^{-\gamma x}) = 2H_1 \frac{e^{\gamma x} + e^{-\gamma x}}{2} = 2H_1 \cosh(\gamma x). \tag{6.66}$$

Using the boundary condition,

$$H\left(\frac{w}{2}\right) = 2H_1 \cosh\left(\frac{\gamma w}{2}\right) \tag{6.67}$$

we obtain

$$H_1 = \frac{H\left(\frac{w}{2}\right)}{2\cosh\left(\frac{\gamma w}{2}\right)}. \tag{6.68}$$

The amplitude of the magnetic field intensity at $x = w/2$ is

$$H\left(\frac{w}{2}\right) = H_m = \frac{NI_m}{l_c}. \tag{6.69}$$

Substitution of (6.68) and (6.69) into (6.66) yields the solution of the Helmholtz equation

$$H(x) = H_m \frac{\cosh(\gamma x)}{\cosh(\gamma \frac{w}{2})} = H_m \frac{\cosh(\frac{x}{\delta_c} + j\frac{x}{\delta_c})}{\cosh(\frac{w}{2\delta_c} + j\frac{w}{2\delta_c})} = H_m \frac{\cosh\left[(1+j)\left(\frac{w}{\delta_c}\right)\left(\frac{x}{w}\right)\right]}{\cosh\left[\frac{(1+j)}{2}\left(\frac{w}{\delta_c}\right)\right]}. \tag{6.70}$$

The magnitude of the magnetic field intensity is

$$|H(x)| = H_m \sqrt{\frac{\cosh^2 \frac{x}{\delta_c} - \sin^2 \frac{x}{\delta_c}}{\cosh^2 \frac{w}{2\delta_c} - \sin^2 \frac{w}{2\delta_c}}} = H_m \sqrt{\frac{\cosh\left(\frac{2x}{\delta_c}\right) + \cos\left(\frac{2x}{\delta_c}\right)}{\cosh\left(\frac{w}{\delta_c}\right) + \cos\left(\frac{w}{\delta_c}\right)}}. \tag{6.71}$$

Figure 6.9 shows the plots of $|H(x)|/H_m$ as a function of x/w at different values of w/δ_c.

The current density is

$$J(x) = J_y(x) = -\frac{dH_z(x)}{dx} = -\gamma H_m \frac{\sinh(\gamma x)}{\cosh(\gamma \frac{w}{2})} = -\frac{1+j}{\delta_c} H_m \frac{\sinh(\frac{x}{\delta_c} + j\frac{x}{\delta_c})}{\cosh(\frac{w}{2\delta_c} + j\frac{w}{2\delta_c})}. \tag{6.72}$$

The magnitude of the current density is

$$|J(x)| = \frac{H_m}{\delta_c} \sqrt{\frac{2\left[\cosh\left(\frac{2x}{\delta_c}\right) - \cos\left(\frac{2x}{\delta_c}\right)\right]}{\cosh\left(\frac{w}{\delta_c}\right) + \cos\left(\frac{w}{\delta_c}\right)}}. \tag{6.73}$$

Figure 6.10 shows the plots of $\delta_c|J(x)|/H_m$ as a function of x/w at different values of w/δ_c.

At high frequencies, $\delta_c \ll w$, the distribution of the envelope of magnetic field intensity $H(x)$ is not uniform and the amplitude of the eddy-current density $J(x)$ does not vary linearly with distance x, as illustrated in Fig. 6.11a for a solid core at $w = 8\delta_c$. In this case, $J(w/2)$ is very high, resulting in a high core loss. The distributions of $H(x)/H_o$ and $J(x)$ in the laminated core at $w = 2\delta_c$ are depicted in Fig. 6.11b. Silicon is often added to steel to reduce σ_c, increase δ_c, and reduce w/δ_c, making $H(x)$ more uniform.

Example 6.1

A laminated core is made of iron and 0.25% silicon. Its properties are $\rho_c = 10^{-7}$ Ωm and $\mu_{rc} = 2700$. What is the maximum thickness of the lamination at which the magnetic field intensity is uniform at $f = 10$ kHz?

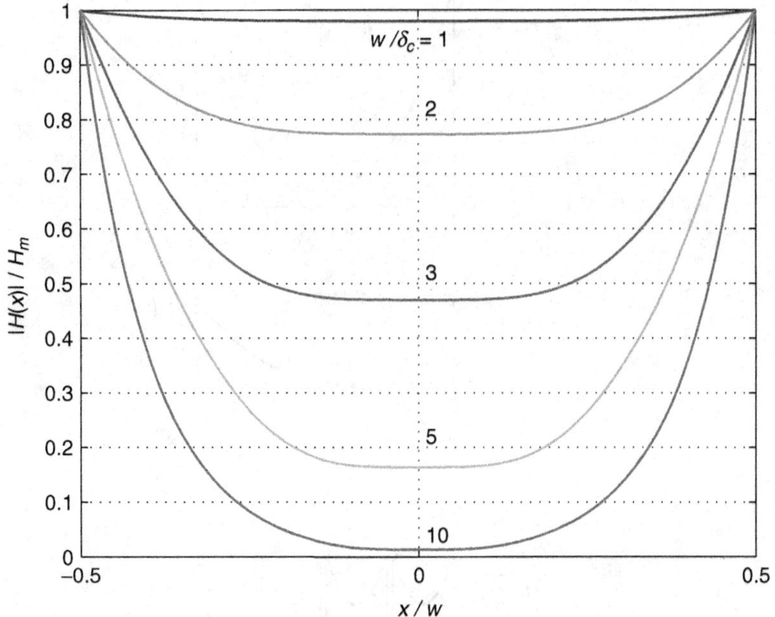

Figure 6.9 Plots of $|H(x)|/H_m$ as a function of x/w at different values of w/δ_c.

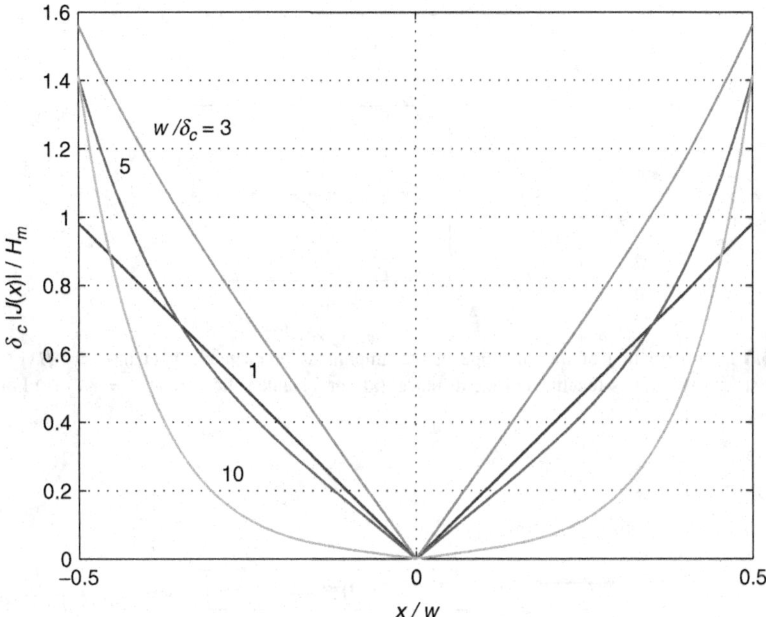

Figure 6.10 Plots of $\delta_c|J(x)|/H_m$ as a function of x/w at different values of w/δ_c.

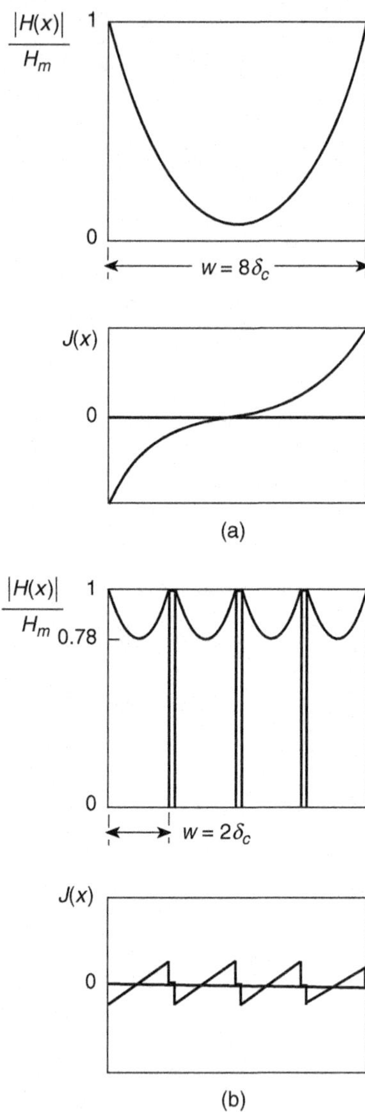

Figure 6.11 Distribution of the envelope of the amplitudes of magnetic field intensity $H(x)/H_m$ and the eddy-current density $J(x)$ as functions of the distance. (a) For a single solid core at $w = 8\delta_c$. (b) For a laminated core at $w = 2\delta_c$.

Solution: The skin depth of the core material is

$$\delta_c = \sqrt{\frac{\rho_c}{\pi \mu_{rc} \mu_0 f}} = \sqrt{\frac{10^{-7}}{\pi \times 2700 \times 4\pi \times 10^{-7} \times 10^4}} = 0.03062 \text{ mm}. \tag{6.74}$$

Hence,

$$w_{max} = \delta_c = 0.03062 \text{ mm}. \tag{6.75}$$

Pick $w = 0.03$ mm.

6.5.2 Power Loss Density Distribution at High Frequencies

Using Ohm's law $\mathbf{E} = \rho_c \mathbf{J}$, we obtain the time-average eddy-current power loss density

$$P_e(x) = \frac{1}{2}\mathbf{E}\cdot\mathbf{J}^* = \frac{\rho_c}{2}\mathbf{J}\cdot\mathbf{J}^* = \frac{\rho_c |J(x)|^2}{2} = \frac{\rho_c H_m^2}{\delta_c^2}\frac{\cosh\left(\frac{2x}{\delta_c}\right) - \cos\left(\frac{2x}{\delta_c}\right)}{\cosh\left(\frac{w}{\delta_c}\right) + \cos\left(\frac{w}{\delta_c}\right)}. \tag{6.76}$$

Figure 6.12 shows the plots of $\delta_c^2 P_e(x)/\rho_c H_m^2$ as a function of x/w for different values of w/δ_w.

Since $\sinh(x/\delta_c + jx/\delta_c) \approx (1+j)x/\delta_c$ and $\cosh(w/\delta_c + jw/\delta_c) \approx 1$ for $w/\delta_c \ll 1$, and $x/\delta_c \ll 1$, the current density at low frequencies is approximated by

$$J(x) = J_y(x) \approx H_m\left(\frac{1+j}{\delta_c}\right)^2 x = \frac{B_m}{\mu_c}\left(\frac{1+j}{\delta_c}\right)^2 x = \frac{2B_m}{\mu_c\delta_c^2}x. \tag{6.77}$$

The complex power for sinusoidal waveforms in phasor form is

$$S_{E(complex)} = \frac{1}{2}\int\int\int_V \mathbf{E}\cdot\mathbf{J}^* dV. \tag{6.78}$$

The time-average eddy-current power loss dissipated in the lamination at low frequencies ($w \ll \delta_c$) is given by

$$\begin{aligned} P_E &= \frac{1}{2}Re\left\{\int\int\int_V \mathbf{E}\cdot\mathbf{J}^* dV\right\} = \frac{1}{2}\int\int\int_V \rho_c |J|^2 dV = \frac{\rho_c}{2}\int\int\int_V |J_y|^2 dV \\ &\approx \frac{\rho_c B_m^2}{2\mu_c^2}\int_0^{l_c}\int_0^h\int_{-w/2}^{w/2} |\left(\frac{1+j}{\delta_c}\right)^2 x|^2 dxdydz = \frac{4\rho_c B_m^2}{2\mu_c^2\delta_c^4}\int_0^{l_c}\int_0^h\int_{-w/2}^{w/2} x^2 dxdydz \\ &= \frac{\rho_c l_c h w^3 B_m^2}{6\mu_c^2\delta_c^4} = \frac{l_c h w^3 \omega^2 B_m^2}{24\rho_c} = \frac{V_c w^2\omega^2 B_m^2}{24\rho_c} = \frac{V_c \pi^2 w^2 f^2 B_m^2}{6\rho_c}\,(\mathrm{W}). \end{aligned} \tag{6.79}$$

This expression is identical to that given by (6.18).

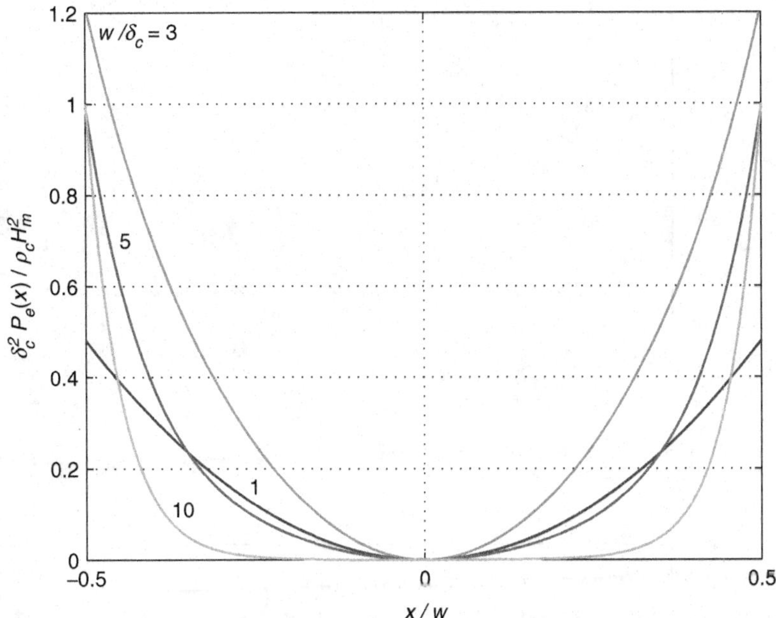

Figure 6.12 Plots of $\delta_c^2 P_e(x)/\rho_c H_m^2$ as a function of x/w at different values of w/δ_c.

Using (6.73), the time-average eddy-current power loss dissipated in the lamination is

$$P_E = \frac{1}{2}\int\int\int_V \rho_c |J(x)|^2 dV = \frac{1}{2}\int_0^{l_c}\int_0^{h}\int_{-w/2}^{w/2} \rho_c |J(x)|^2 dV$$

$$= \frac{\rho_c}{2}\frac{2H_m^2}{\delta_c^2}2\int_0^{l_c}\int_0^{h}\int_0^{w/2} \frac{\cosh\left(\frac{2x}{\delta_c}\right)-\cos\left(\frac{2x}{\delta_c}\right)}{\cosh\left(\frac{w}{\delta_c}\right)+\cos\left(\frac{w}{\delta_c}\right)} dV = \frac{\rho_c H_m^2 l_c h}{\delta_c}\frac{\sinh\left(\frac{w}{\delta_c}\right)-\sin\left(\frac{w}{\delta_c}\right)}{\cosh\left(\frac{w}{\delta_c}\right)+\cos\left(\frac{w}{\delta_c}\right)}. \quad (6.80)$$

Figure 6.13 shows $\delta_c P_E/\rho_c H_m^2 h l_c$ as a function of w/δ_c. As w/δ_c increases from 0, P_E increases from 0, reaches the maximum value, decreases to a minimum value, increases again to a maximum value, and so on.

Taking the derivative of P_E with respect to (w/δ_c) and setting it to zero

$$\frac{dP_E}{d(w/\delta_c)} = \frac{2\rho_c H_m^2 h l_c}{\delta_c}\frac{\sinh\left(\frac{w}{\delta_c}\right)\sin\left(\frac{w}{\delta_c}\right)}{\left[\cosh\left(\frac{w}{\delta_c}\right)+\cos\left(\frac{w}{\delta_c}\right)\right]^2} = 0 \quad (6.81)$$

we obtain

$$\sin\left(\frac{w}{\delta_c}\right) = 0. \quad (6.82)$$

Hence, the maximum and minimum values of P_E occur at

$$\frac{w}{\delta_c} = n\pi \quad (6.83)$$

where n is an integer. The largest maximum value of P_E occurs for $n = 1$, resulting in the most lossy lamination thickness

$$w_{P_E(\max)} = \pi\delta_c = \sqrt{\frac{\pi\rho_c}{\mu_c f}}. \quad (6.84)$$

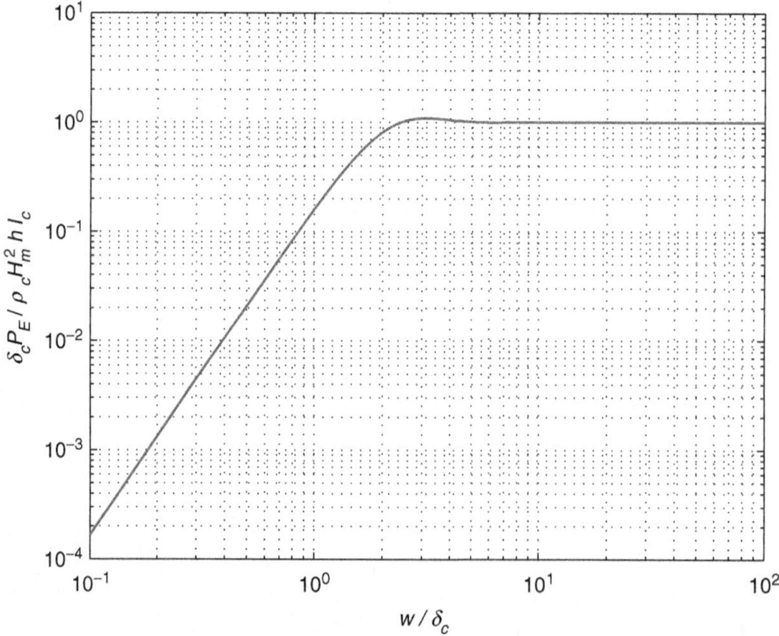

Figure 6.13 Plot of $\delta_c P_E/\rho_c H_m^2 h l_c$ as a function of w/δ_c.

Substitution of (6.84) in (6.80) gives the maximum time-average eddy-current total power loss

$$P_{E(\max)} = \frac{\rho_c H_m^2 h l_c}{\delta_c} \frac{\sinh(\pi)}{\cosh(\pi) - 1} = \frac{1.09 \rho_c H_m^2 h l_c}{\delta_c} = \frac{1.09 \rho_c B_m^2 h l_c}{\mu_c^2 \delta_c} = \frac{1.09 B_m^2 h l_c}{\mu_c} \sqrt{\frac{\pi \rho_c f}{\mu_c}}. \quad (6.85)$$

Using (6.80),

$$\frac{d^2 P_E}{d(w/\delta_c)^2} = \frac{2\rho_c H_m^2 h l_c}{\delta_c \left[\cosh\left(\frac{w}{\delta_c}\right) + \cos\left(\frac{w}{\delta_c}\right)\right]^3} \left\{ \sinh\left(\frac{w}{\delta_c}\right) \sin^2\left(\frac{w}{\delta_c}\right) - \sinh^2\left(\frac{w}{\delta_c}\right) \sin\left(\frac{w}{\delta_c}\right) \right.$$

$$\left. + \left[1 + \cosh\left(\frac{w}{\delta_c}\right) \cos\left(\frac{w}{\delta_c}\right)\right] \left[\sinh\left(\frac{w}{\delta_c}\right) + \sin\left(\frac{w}{\delta_c}\right)\right] \right\}. \quad (6.86)$$

Hence,

$$\frac{d^2 P_E}{d(w/\delta_c)^2}\bigg|_{(w/\delta_c)=n\pi} = (-1)^n \frac{2\rho_c H_m^2 h l_c}{\delta_c} \frac{\sinh(n\pi)}{[\cosh(n\pi) + (-1)^n]^2}. \quad (6.87)$$

For odd values of n ($n = 1, 3, 5, \ldots$),

$$\frac{d^2 P_E}{d(w/\delta_c)^2} < 0 \quad (6.88)$$

resulting in the maximum values of P_E. For even values of n ($n = 2, 4, 6, \ldots$),

$$\frac{d^2 P_E}{d(w/\delta_c)^2} > 0 \quad (6.89)$$

yielding the minimum values of P_E. For $n = 2$,

$$w_{P_E(\min)} = 2\pi\delta_c = 2\sqrt{\frac{\pi \rho_c}{\mu_c f}} \quad (6.90)$$

and

$$P_{E(\min)} = \frac{\rho_c H_m^2 h l_c}{\delta_c} \frac{\sinh(2\pi)}{\cosh(2\pi) + 1} = 0.996 \frac{\rho_c H_m^2 h l_c}{\delta_c}. \quad (6.91)$$

The time-average eddy-current power loss per unit volume at any frequency is

$$P_e = \frac{P_E}{V_c} = \frac{\rho_c}{2V_c} \int_{-w/2}^{w/2} |J|^2 dx = \frac{\rho_c}{l_c h w} \int_0^{w/2} |J|^2 dx$$

$$= \frac{\rho_c H_m^2}{w\delta_c} \frac{\sinh\left(\frac{w}{\delta_c}\right) - \sin\left(\frac{w}{\delta_c}\right)}{\cosh\left(\frac{w}{\delta_c}\right) + \cos\left(\frac{w}{\delta_c}\right)} = \frac{\rho_c H_m^2}{\delta_c^2 \left(\frac{w}{\delta_c}\right)} \frac{\sinh\left(\frac{w}{\delta_c}\right) - \sin\left(\frac{w}{\delta_c}\right)}{\cosh\left(\frac{w}{\delta_c}\right) + \cos\left(\frac{w}{\delta_c}\right)} \ (\mathrm{W/m^3}). \quad (6.92)$$

Figure 6.14 shows $\delta_c^2 P_e / \rho_c H_m^2$ as a function of w/δ_c.

For low frequencies or a very thin plate ($w \ll \delta_c$), we obtain

$$\frac{\sinh\left(\frac{w}{\delta_c}\right) - \sin\left(\frac{w}{\delta_c}\right)}{\cosh\left(\frac{w}{\delta_c}\right) + \cos\left(\frac{w}{\delta_c}\right)} \approx \frac{\frac{w}{\delta_c} + \frac{1}{6}\left(\frac{w}{\delta_c}\right)^3 - \frac{w}{\delta_c} + \frac{1}{6}\left(\frac{w}{\delta_c}\right)^3}{1 + 1} \approx \frac{1}{6}\left(\frac{w}{\delta_c}\right)^3. \quad (6.93)$$

Hence, the time-average eddy-current power loss per unit volume of the core at low frequencies is

$$P_e \approx \frac{\rho_c H_m^2}{w\delta_c} \frac{\left(\frac{w}{\delta_c}\right)^3}{6} = \frac{\rho_c H_m^2 w^2}{6\delta_c^4} = \frac{\pi^2 w^2 f^2 B_m^2}{6\rho_c} \quad (6.94)$$

and P_E is approximated by (6.79). The eddy-current loss is low when the plate is very thin compared to δ_c. In this case, the magnetic field produced by the eddy currents is very low. The eddy currents are restricted due to lack of space or high core resistivity. Hence, they are resistance-limited.

At high frequencies or a very thick plate ($w \gg \delta_c$), $e^{-w/\delta_c} \approx 0$, $\sinh(w/\delta_c) = \cosh(w/\delta_c) \approx e^{w/\delta_c}/2$, resulting in

$$\frac{\sinh\left(\frac{w}{\delta_c}\right) - \sin\left(\frac{w}{\delta_c}\right)}{\cosh\left(\frac{w}{\delta_c}\right) + \cos\left(\frac{w}{\delta_c}\right)} \approx \frac{\frac{e^{w/\delta_c}}{2}}{\frac{e^{w/\delta_c}}{2}} = 1. \quad (6.95)$$

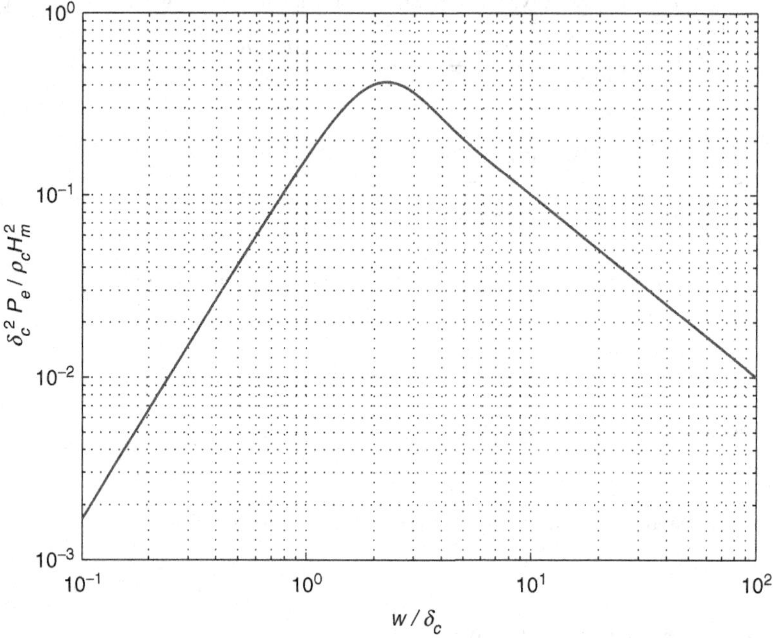

Figure 6.14 Plot of $\delta_c^2 P_e / \rho_c H_m^2$ as a function of w/δ_c.

Therefore, the time-average eddy-current power loss per unit volume of the core at high frequencies is given by

$$P_e \approx \frac{\rho_c H_m^2}{w\delta_c} = \frac{\rho_c B_m^2}{w\mu_c^2 \delta_c} = \frac{H_m^2}{w}\sqrt{\pi \rho_c \mu_c f} = \frac{B_m^2}{w\mu_c}\sqrt{\frac{\pi \rho_c f}{\mu_c}} \left(\frac{\text{W}}{\text{m}^3}\right) \tag{6.96}$$

and the total time-average eddy-current power loss at high frequencies is

$$P_E \approx \frac{\rho_c H_m^2 h l_c}{\delta_c} = H_m^2 h l_c \sqrt{\pi \rho_c \mu_c f} = \frac{B_m^2 h l_c}{\mu_c}\sqrt{\frac{\pi \rho_c \mu_c f}{\mu_c}}. \tag{6.97}$$

At high frequencies, the eddy currents are limited by their own magnetic field and are said to be inductance-limited.

At low frequencies, P_e is proportional to w^2. In contrast, at high frequencies, P_e is inversely proportional to w. Thus, there is the worst case of w at which P_e reaches a maximum value [31]. The lamination thickness for the maximum power loss per unit volume occurs at

$$w_{Pe(\max)} \approx 2.252\delta_c \tag{6.98}$$

resulting in the maximum time-average eddy-current power loss per unit volume

$$P_{e(\max)} \approx 0.4172\frac{\rho_c H_m^2}{\delta_c^2} = 0.4172\frac{\rho_c B_m^2}{\mu_c^2 \delta_c^2}. \tag{6.99}$$

6.5.3 Lamination Impedance at High Frequencies

The DC resistance of a single lamination is

$$R_{cDC} = \frac{\rho_c h}{w l_c}. \tag{6.100}$$

The impedance of a lamination is given by

$$Z = \frac{h\rho_c\gamma}{2l_c}\coth\left(\gamma\frac{w}{2}\right) = \frac{h\rho_c(1+j)}{2l_c\delta_c}\coth\left(\gamma\frac{w}{2}\right)$$
$$= \frac{1}{2}R_{cDC}\left(\frac{w}{\delta_c}\right)(1+j)\coth\left(\gamma\frac{w}{2}\right) = R_c + jX_L. \quad (6.101)$$

The normalized resistance of the lamination is

$$F_R = \frac{R_c}{R_{cDC}} = \frac{1}{2}\left(\frac{w}{\delta_c}\right)\frac{\sinh\left(\frac{w}{\delta_c}\right) - \sin\left(\frac{w}{\delta_c}\right)}{\cosh\left(\frac{w}{\delta_c}\right) + \cos\left(\frac{w}{\delta_c}\right)}. \quad (6.102)$$

Figure 6.15 shows a plot of R_c/R_{cDC} as a function of w/δ_c.

The lamination reactance X_L normalized with respect to the core DC resistance R_{wDC} is given by

$$F_X = \frac{X_L}{R_{cDC}} = \frac{1}{2}\left(\frac{w}{\delta_c}\right)\frac{\sinh\left(\frac{w}{\delta_c}\right) + \sin\left(\frac{w}{\delta_c}\right)}{\cosh\left(\frac{w}{\delta_c}\right) + \cos\left(\frac{w}{\delta_c}\right)} \quad (6.103)$$

yielding the core inductance

$$L = \frac{\mu_c h\delta_c}{4l_c}\frac{\sinh\left(\frac{w}{\delta_c}\right) + \sin\left(\frac{w}{\delta_c}\right)}{\cosh\left(\frac{w}{\delta_c}\right) + \cos\left(\frac{w}{\delta_c}\right)}. \quad (6.104)$$

Figure 6.16 shows a plot of X_L/R_{wDC} as a function of w/δ_c. The core impedance is

$$Z = \frac{1}{2}R_{cDC}\left(\frac{w}{\delta_c}\right)\left[\frac{\sinh\left(\frac{w}{\delta_c}\right) - \sin\left(\frac{w}{\delta_c}\right)}{\cosh\left(\frac{w}{\delta_c}\right) + \cos\left(\frac{w}{\delta_c}\right)} + j\frac{\sinh\left(\frac{w}{\delta_c}\right) + \sin\left(\frac{w}{\delta_c}\right)}{\cosh\left(\frac{w}{\delta_c}\right) + \cos\left(\frac{w}{\delta_c}\right)}\right]$$
$$= R_w + jX_L = |Z|e^{j\phi_Z}. \quad (6.105)$$

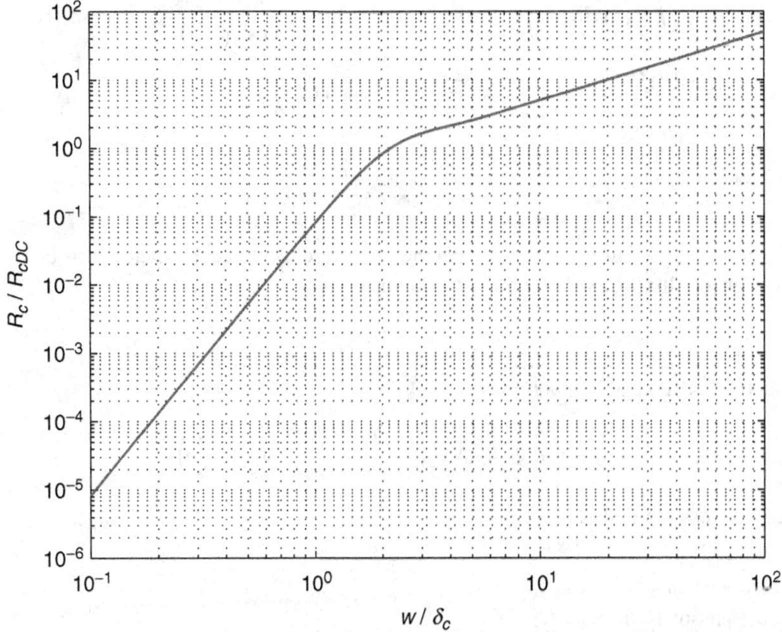

Figure 6.15 Plot of R_c/R_{cDC} as a function of w/δ_c.

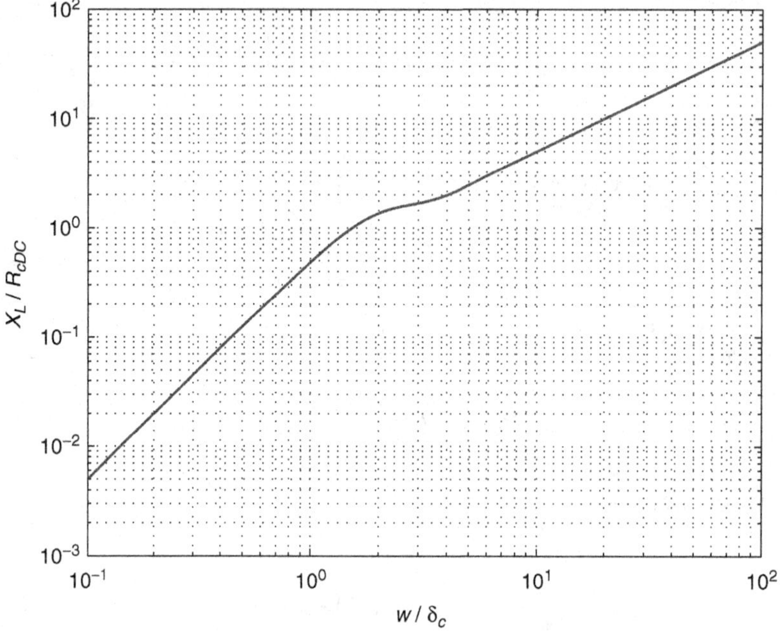

Figure 6.16 Plot of X_L/R_{cDC} as a function of w/δ_c.

Figures 6.17 and 6.18 show plots of $|Z|/R_{cDC}$ and ϕ_Z as functions of w/δ_c. Figure 6.19 depicts the ratio X_L/R_c as a function of w/δ_c.

The magnetic flux flowing through the cross section of a single lamination is given by

$$\phi_{c1} = \int\int_S \mathbf{B}(x) \cdot d\mathbf{S} = \int_0^h dy \int_{-w/2}^{w/2} \mu_c H(x) dx = \frac{h\mu_c H_m}{\cosh\left(\gamma \frac{w}{2}\right)} \int_{-w/2}^{w/2} \cosh(\gamma x) dx$$
$$= \frac{2h\mu_c H_m}{\gamma} \tanh\left(\gamma \frac{w}{2}\right) = \frac{2h\mu_c N I_m}{\gamma l_c} \tanh\left(\gamma \frac{w}{2}\right) \tag{6.106}$$

where

$$H_m = \frac{N I_m}{l_c} \tag{6.107}$$

and N is the number of turns. Since the laminations are close to each other, it can be assumed that the total magnetic flux flows through only n laminations

$$\phi_c = n\phi_{c1}. \tag{6.108}$$

The magnetic flux linkage is given by

$$\lambda_c = N\phi = nN\phi_{c1} = \frac{2nN^2 h\mu_c}{\gamma l_c} \tanh\left(\gamma \frac{w}{2}\right). \tag{6.109}$$

The voltage across the coil is

$$v = \frac{d\lambda_c}{dt} \tag{6.110}$$

which in the phasor form becomes

$$V = j\omega\lambda_c. \tag{6.111}$$

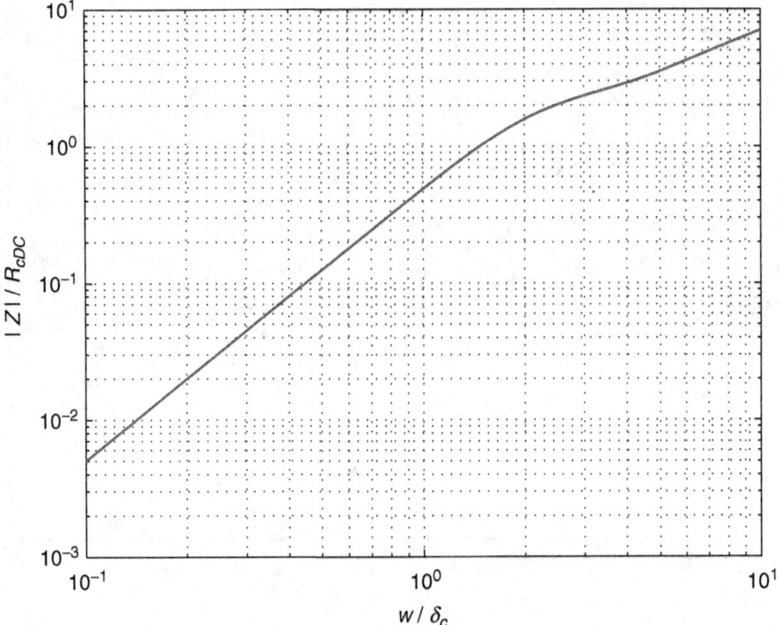

Figure 6.17 Plot of $|Z|/R_{cDC}$ as a function of w/δ_c.

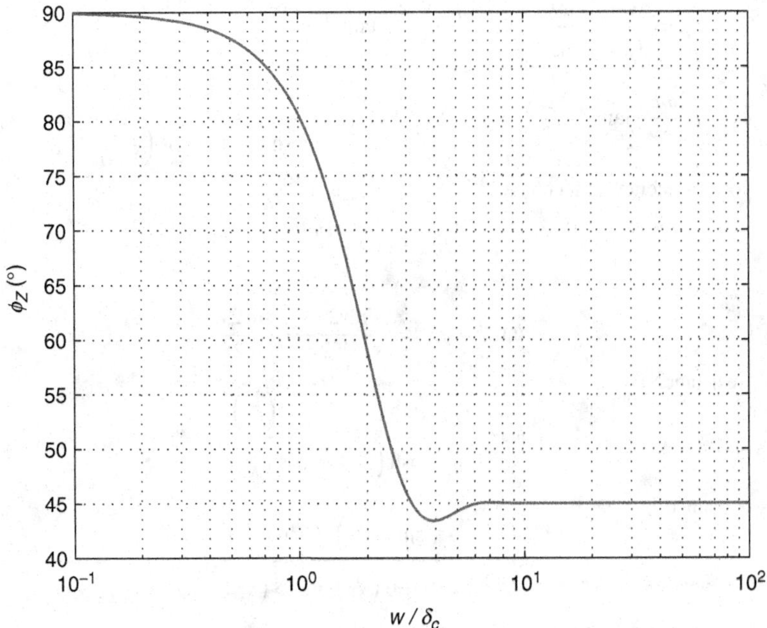

Figure 6.18 Plot of ϕ_Z as a function of w/δ_c.

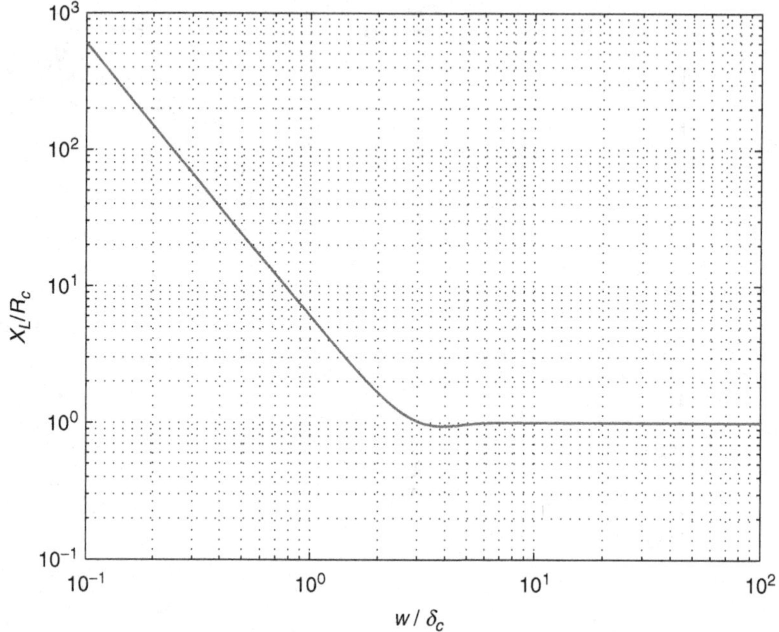

Figure 6.19 Ratio X_L/R_c as a function of w/δ_c.

Hence, the coil impedance is given by

$$Z = \frac{\mathbf{V}}{\mathbf{I}} = \frac{\mathbf{V}}{I_m} = \frac{j\omega\lambda_c}{I_m} = j\omega\frac{2\mu_c N^2 nh}{\gamma l_c}\tanh\left(\gamma\frac{w}{2}\right)$$

$$= j\frac{2\omega L_o}{w\gamma}\tanh\left(\gamma\frac{w}{2}\right) = \omega L_o\left(\frac{\delta_c}{w}\right)(1+j)\frac{\sinh\left(\frac{w}{\delta_c}\right) + j\sin\left(\frac{w}{\delta_c}\right)}{\cosh\left(\frac{w}{\delta_c}\right) + \cos\left(\frac{w}{\delta_c}\right)} = R + jX_L \quad (6.112)$$

where the low-frequency inductance is

$$L_o = \frac{\mu_c N^2 nwh}{l_c} \quad (6.113)$$

and

$$\tanh(1+j)x = \frac{\sinh 2x - j\sin 2x}{\cosh 2x + \cos 2x}. \quad (6.114)$$

The series resistance is

$$R = \frac{\omega L_o}{\frac{w}{\delta_c}}\frac{\sinh\left(\frac{w}{\delta_c}\right) - \sin\left(\frac{w}{\delta_c}\right)}{\cosh\left(\frac{w}{\delta_c}\right) + \cos\left(\frac{w}{\delta_c}\right)}, \quad (6.115)$$

the series reactance is

$$X_L = \frac{\omega L_o}{\frac{w}{\delta_c}}\frac{\sinh\left(\frac{w}{\delta_c}\right) + \sin\left(\frac{w}{\delta_c}\right)}{\cosh\left(\frac{w}{\delta_c}\right) + \cos\left(\frac{w}{\delta_c}\right)}, \quad (6.116)$$

and the inductance at any frequency is

$$L = \frac{X_L}{\omega} = \frac{L_o}{\frac{w}{\delta_c}}\frac{\sinh\left(\frac{w}{\delta_c}\right) + \sin\left(\frac{w}{\delta_c}\right)}{\cosh\left(\frac{w}{\delta_c}\right) + \cos\left(\frac{w}{\delta_c}\right)}. \quad (6.117)$$

Figures 6.20 through 6.24 show the plots R/R_{wDC}, X_L/R_{wDC}, L/L_o, $|Z|/R_{wDC}$, and ϕ_Z as functions of w/δ_c.

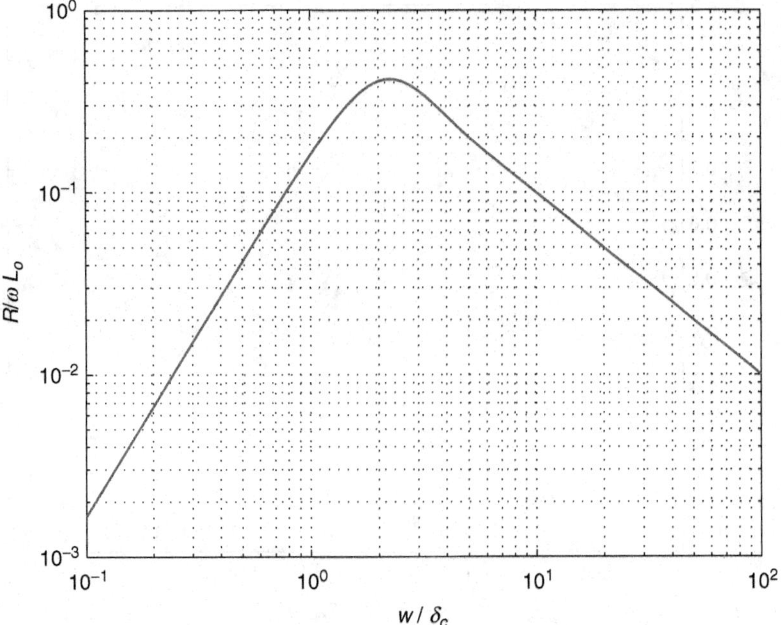

Figure 6.20 Plot of $R/\omega L_o$ as a function of w/δ_c.

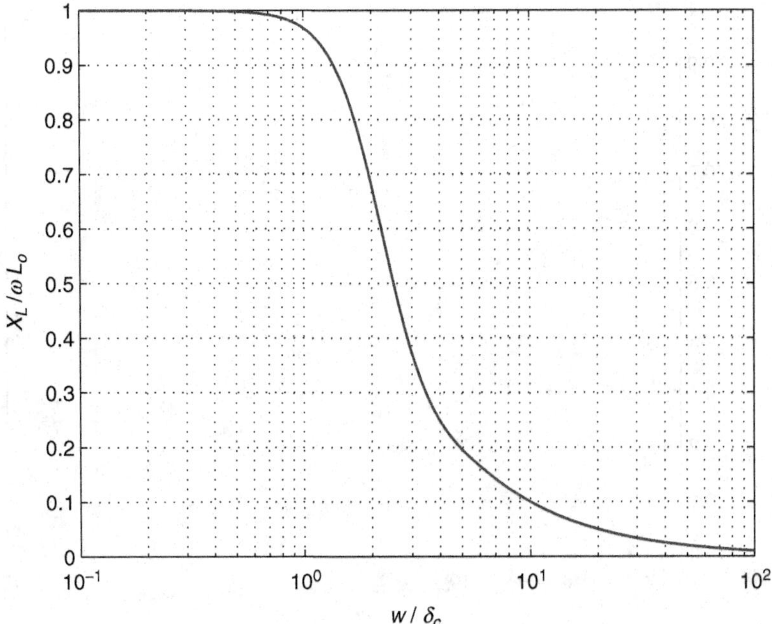

Figure 6.21 Plot of $X_L/\omega L_o$ as a function of w/δ_c.

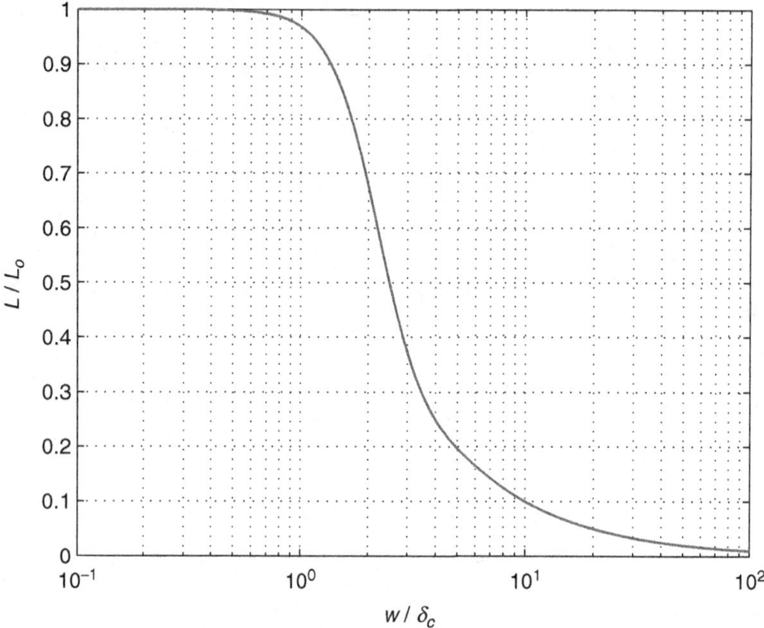

Figure 6.22 Plot of L/L_o as a function of w/δ_c.

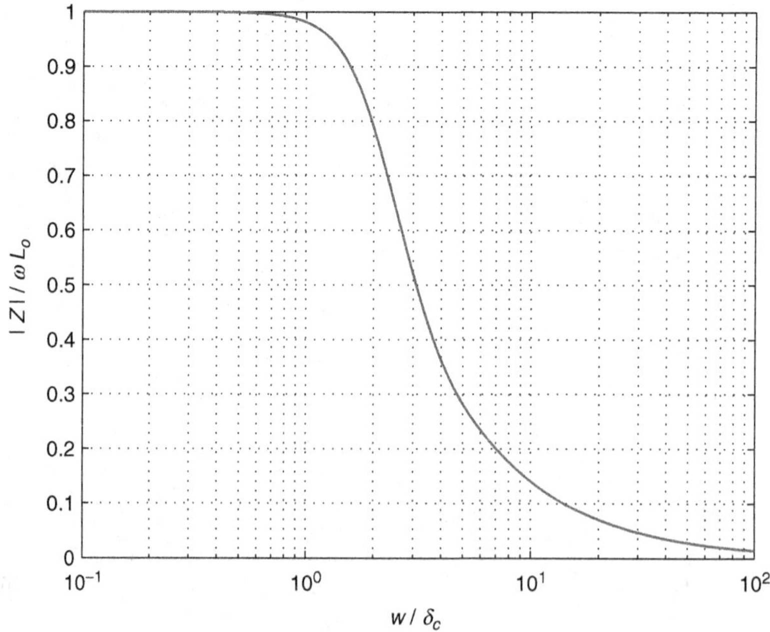

Figure 6.23 Plot of $|Z|/\omega L_o$ as a function of w/δ_c.

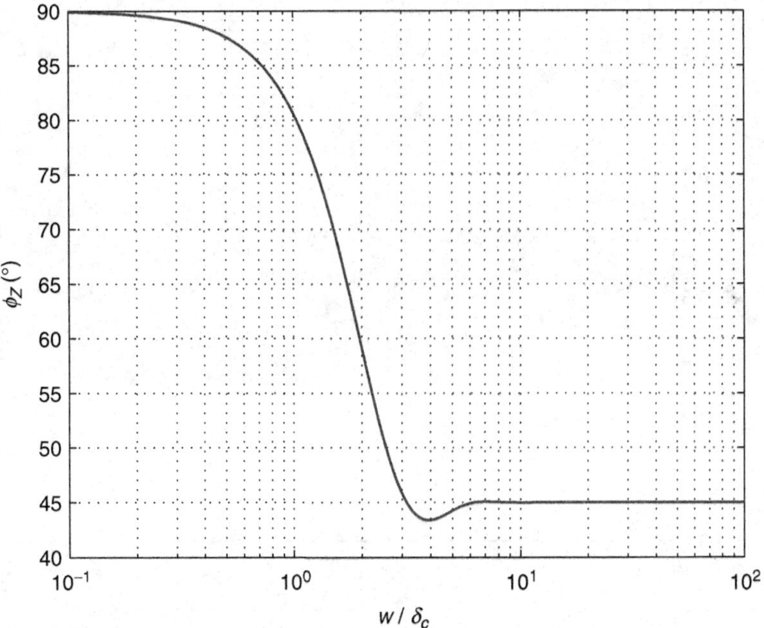

Figure 6.24 Plot of ϕ_Z as a function of w/δ_c.

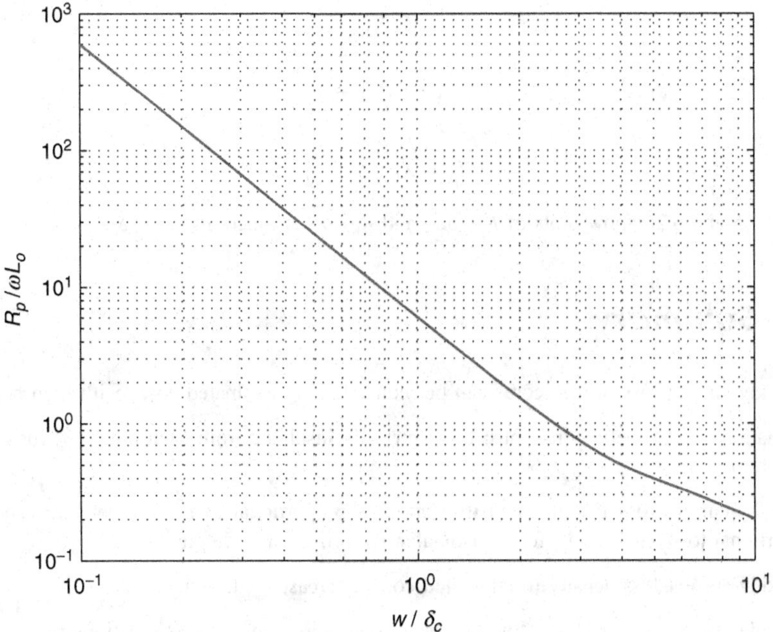

Figure 6.25 Plot of $R_p/\omega L_o$ as a function of w/δ_c.

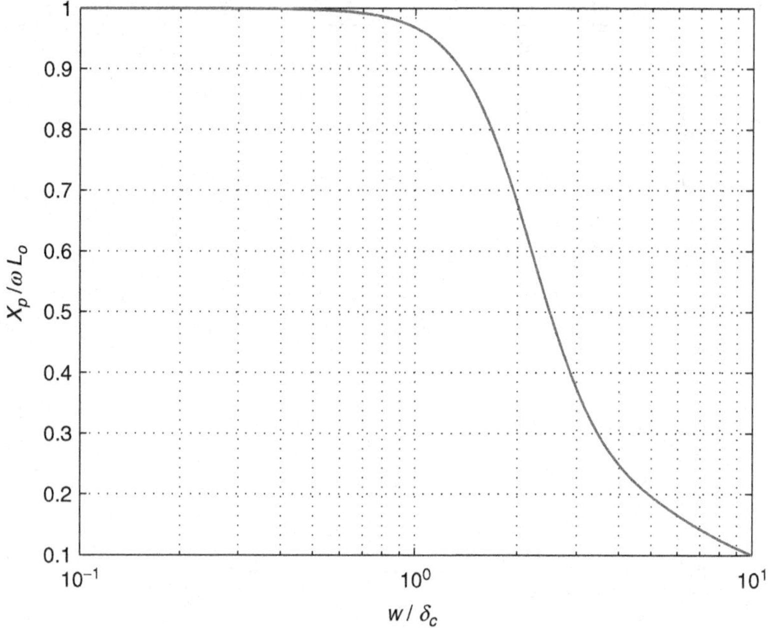

Figure 6.26 Plot of $X_p/\omega L_o$ as a function of w/δ_c.

The series equivalent circuit of the coil can be converted into a parallel equivalent circuit

$$q = \frac{X_L}{R} \tag{6.118}$$

$$R_p = R(1 + q^2) \tag{6.119}$$

and

$$X_p = X_L\left(1 + \frac{1}{q^2}\right). \tag{6.120}$$

Figures 6.25 and 6.26 show plots of $R_p/\omega L_o$ and R_p/ω_o as functions of w/δ_c.

6.6 Summary

- The eddy-current loss in iron cores can be reduced using laminated cores with high resistivity.
- Laminations of magnetic cores must be electrically insulated from each other by, for example, an oxide film.
- The eddy-current loss in laminated iron cores is proportional to f^2, B_m^2, and w^2, and inversely proportional to ρ_c at $w < \delta_w$ for sinusoidal waveforms at low frequencies.
- The eddy-current loss density in laminated cores decreases with temperature.
- The eddy-current power loss in the laminated core consisting of k laminations is k^2 times lower than that of the corresponding solid core.
- The normalized inductance L/L_o of inductors with laminated cores decreases as w/δ_w increases for $w/\delta_c > 1$.
- The magnitude of the impedance of inductors with laminated cores decreases as w/δ_w increases for $w/\delta_c > 1$.

6.7 References

[1] P. J. Dowell, "Effects of eddy currents in transformer winding," *Proceedings of the IEE*, vol. 113, no. 8, pp. 1387–1394, August 1966.

[2] J. Jongsma, "High-frequency ferrite power transformer and choke design, Part 3: transformer winding design, Part 4: improved method of power choke design" Philips Electronic Components and Materials, Technical Publication, no. 27, Philips, The Netherlands, 1986.

[3] A. Kennelly, F. Laws, and P. Pierce, "Experimental research on skin effect in conductors," *Transactions AIEE*, vol. 34, p. 1915.

[4] P. D. Agrawal, "Eddy-current losses in solid and laminated iron," *Transactions AIEE*, vol. 78, pp. 169–171, 1959.

[5] J. Lammeraner and M. Stafl, *Eddy Currents*, Cleveland: CRS Press, 1966.

[6] E. C. Snelling, *Soft Ferrites: Properties and Applications*, London: Iliffe Books Ltd, 1969.

[7] R. L. Stall, *The Analysis of Eddy Currents*, Oxford: Clarendon Press, 1974, pp. 21–27.

[8] W. T. McLyman, *Transformer and Inductor Design Handbook*, 3rd Ed., New York: Marcel Dekker, 2004.

[9] J. K. Watson, *Applications of Magnetism*, University of Florida: Gainesville, FL, 1985.

[10] E. Bennet and S. C. Larsen, "Effective resistance of alternating currents of multilayer windings," *Transactions of the American Institute of Electrical Engineers*, vol. 59, pp. 1010–1017, 1940.

[11] R. L. Perry, "Multiple layer series connected winding design for minimum losses," *IEEE Transactions on Power Applications Systems*, vol. PAS-98, pp. 116–123, January/February 1979.

[12] B. Carsten, "High frequency conductor losses in switch mode magnetics," Proceedings of PCI, Munich, Germany, 1986, pp. 161–182.

[13] J. P. Vandalec and P. D. Ziogos, "A novel approach for minimizing high-frequency transformer copper loss," *IEEE Transactions on Power Electronics*, vol. 3, pp. 266–276, July 1988.

[14] A. M. Urling, V. A. Niemela, G. R. Skutt, and T. G. Wilson, "Characterizing high frequency effects in transformer windings: a guide to several significant papers," IEEE Transactions on Power Electronics Specialists Conference, 1989, pp. 373–385.

[15] J. A. Ferreira, *Electromagnetic Modelling of Power Electronic Converters*, Boston, MA: Kluwer Academic Publisher, 1989.

[16] M. Bartoli, A. Reatti, and M. K. Kazimierczuk, "High-frequency models of ferrite inductors," Proceedings of the IEEE International Conference on Industrial Electronics, Controls, Instrumentation, and Automation (IECON'94), Bologna, Italy, September 5–9, 1994, pp. 1670–1675.

[17] M. Bartoli, A. Reatti, and M. K. Kazimierczuk, "Predicting the high-frequency ferrite-core inductor performance," Proceedings of the Conference of Electrical Manufacturing and Coil Winding, Chicago (Rosemont), IL, September 27–29, 1994, pp. 409–413.

[18] M. Bartoli, A. Reatti, and M. K. Kazimierczuk, "Modeling iron-powder inductors at high-frequencies," Proceedings of the IEEE Industry Applications Society Annual Meeting, Denver, CO, October 2–7, 1994, pp. 1125–1232.

[19] J. A. Ferreira, "Improved analytical modeling of conductive losses in magnetic components," *IEEE Transactions on Power Electronics*, vol. 9, pp. 127–131, January 1994.

[20] M. Bartoli, N. Noferi, A. Reatti, and M. K. Kazimierczuk, "Modeling winding losses in high-frequency power inductors," *Journal of Circuits Systems and Computers*, vol. 5, no. 3, pp. 65–80, March 1995.

[21] N. Mohan, T. M. Underland, and W. P. Robbins, *Power Electronics*, 3rd Ed., New York: John Wiley & Sons, 2003.

[22] M. Bartoli, N. Nefari, A. Reatti, and M. K. Kazimierczuk, "Modeling litz-wire winding losses in high-frequencies power inductors," Proceedings of the IEEE Power Electronics Specialists Conference, Baveno, Italy, June 24–27, 1996, pp. 1690–1696.

[23] R. Petkov, "Optimum design of a high-power high-frequency transformers," *IEEE Transactions on Power Electronics*, vol. 11, no. 1, pp. 33–42, January 1996.

[24] C.-M. Ong, *Dynamic Simulation of Electric Machinery*, Reading: Prentice Hall, 1998, pp. 38–40, 45–46, and 87–90.

[25] W. G. Hurley, W. H. Wolfe, and J. G. Breslin, "Optimized transformer design: inclusive of high-frequency effects," *IEEE Transactions on Power Electronics*, vol. 13, no. 4, pp. 651–659, July 1998.

[26] W. G. Hurley, E. Gath, and J. G. Breslin, "Optimizing the ac resistance of multilayer transformer windings with arbitrary current waveforms," *IEEE Transactions on Power Electronics*, vol. 15, no. 2, pp. 369–376, March 2000.

[27] M. K. Kazimierczuk, G. Sancineto, U. Reggiani, and A. Massarini, "Small-signal high-frequency model of ferrite inductors," *IEEE Transactions on Magnetics*, vol. 35, pp. 4185–4191, September 1999.

[28] U. Reggiani, G. Sancineto, and M. K. Kazimierczuk, "High-frequency behavior of laminated iron-core inductors for filter applications," Proceedings of the IEEE Applied Power Electronics Conference, New Orleans, LA, February 6–10, 2000, pp. 654–660.

[29] G. Grandi, M. K. Kazimierczuk, A. Massarini, U. Reggiani, and G. Sancineto, "Model of laminated iron-core inductors," *IEEE Transactions on Magnetics*, vol. 40, no. 4, pp. 1839–1845, July 2004.

[30] K. Howard and M. K. Kazimierczuk, "Eddy-current power loss in laminated power cores," Proceedings of the IEEE International Symposium on Circuits and Systems, Sydney, Australia, May 7–9, 2000, paper III-668, pp. 668–672.

[31] A. Reatti and M. K. Kazimierczuk, "Comparison of various methods for calculating the ac resistance of inductors," *IEEE Transactions on Magnetics*, vol. 37, pp. 1512–1518, May 2002.

[32] G. Grandi, M. K. Kazimierczuk, A. Massarini, and U. Reggiani, "Stray capacitance of single-layer solenoid air-core inductors," *IEEE Transactions on Industry Applications*, vol. 35, no. 5, pp. 1162–1168, September 1999.

[33] A. Massarini and M. K. Kazimierczuk, "Self-capacitance of inductors," *IEEE Transactions on Power Electronics*, vol. 12, no. 4, pp. 671–676, July 1997.

[34] T. L. Simpson, "Effect of a conducting shield on the inductance of an air-core solenoid," *IEEE Transactions on Magnetics*, vol. 35, no. 1, pp. 508–515, January 1999.

[35] R. W. Erickson and D. Maksimović, *Fundamentals of Power Electronics*, Norwell, MA: Kluwer Academic Publishers, 2001.

[36] A. Van den Bossche and V. C. Valchev, *Inductors and Transformers for Power Electronics*, Boca Raton, FL: Taylor & Francis, 2005.

[37] N. Das and M. K. Kazimierczuk, "An overview of technical challenges in the design of current transformers," Electrical Manufacturing Conference, Indianapolis, IN, October 24–26, 2005.

[38] T. Suetsugu and M. K. Kazimierczuk, "Integration of Class DE inverter for dc-dc converter on-chip power supplies," IEEE International Symposium on Circuits and Systems, Kos, Greece, May 21–24, 2006, pp. 3133–3136.

[39] T. Suetsugu and M. K. Kazimierczuk, "Integration of Class DE synchronized dc-dc converter on-chip power supplies," IEEE Power Electronics Specialists Conference, Jeju, South Korea, June 21–24, 2006.

[40] R. Wrobel, P. H. Mellor, and D. Holliday, "Thermal modeling of a segmented stator winding design," *IEEE Transactions on Industry Applications*, vol. 47, no. 5, pp. 2023–2030, September/October 2011.

[41] R. Wrobel and P. H. Mellor, "Thermal design of high-density wound components," *IEEE Transactions on Industrial Electronics*, vol. 58, no. 9, pp. 4096–4104, September 2011.

[42] R. Wrobel, A. Mlot, and P. H. Mellor, "Contribution to end-winding proximity loss to temperature variation in electromagnetic devices," *IEEE Transactions on Industrial Electronics*, vol. 59, no. 2, pp. 848–857, February 2012.

[43] www.mag-inc.com, www.ferroxcube.com, www.ferrite.de, www.micrometals.com, and www.metglas.com.

6.8 Review Questions

6.1. Explain the origin of eddy currents in magnetic cores.

6.2. Write the equation for the magnitude of eddy currents at low frequencies.

6.3. Write the equation for power loss due to eddy currents per unit volume at low frequencies.

6.4. Why are eddy currents reduced in laminated cores?

6.5. How does core resistivity affect eddy-current loss in laminated cores?

6.6. How does frequency affect eddy-current loss in laminated cores?

6.7. How does eddy-current loss depend on the amplitude of the magnetic flux density?

6.8. How does eddy-current loss depend on the thickness of laminated cores?

6.9. What is the effect of temperature on the eddy-current loss density in laminated cores?

6.10. How many times is the eddy-current power loss reduced in a laminated core compared with the solid core?

6.9 Problems

6.1. A laminated core is made of iron and 1% silicon. Its properties are $\rho = 2.5 \times 10^{-7}\ \Omega \cdot \text{m}$ and $\mu_{rc} = 4500$. What is the maximum thickness of the lamination at $f = 20\,\text{kHz}$.

6.2. A laminated core is made of a silicon steel (an iron and 1% silicon). Its properties are $\rho = 4 \times 10^{-7}\ \Omega \cdot \text{m}$ and $\mu_{rc} = 5000$. What is the maximum thickness of the lamination at $f = 20\,\text{kHz}$.

6.3. Derive an expression for the eddy-current power loss density in a rectangular core. (a) For a sinusoidal inductor voltage. (b) For a bipolar symmetrical square-wave inductor voltage. (c) For a unipolar rectangular wave inductor voltage with a duty ratio D.

6.4. Calculate hysteresis power loss density P_h, eddy-current power loss density P_e, excess power loss density P_{ex}, and the total power loss density P_v for the laminated iron core M270-35A at $B_m = 1\,\text{T}$ and $f = 60\,\text{Hz}$.

Transformers

7.1 Introduction

This chapter covers high-frequency transformers [1–73]. It consists of two or more mutually coupled windings, with or without a magnetic core. A transformer is a magnetic device that transfers energy from one circuit to another through magnetic field. Transformers are widely used in consumer and industrial electronic products to step-down or step-up the AC voltage to a level suitable for low-voltage or high-voltage circuits, respectively. For example, they are used in isolated switch-mode power supplies (SMPS). Transformers are usually the largest, heaviest, and most expensive of all components in a system.

A winding is a set of conductor turns that conduct the same current. Neighboring coils, which share a common magnetic flux, are said to be mutually coupled. A changing current in one coil winding produces a changing magnetic flux in the remaining mutually coupled windings, which induces voltages between the terminals of all windings. A transformer is a system of two or more mutually coupled coils that are wound on the same core. The main functions of transformers are as follows:

- To change the levels of AC voltages and AC currents, yielding step-down and step-up transformers
- To invert voltage or current waveform
- To transform impedance (yielding impedance matching)
- To provide DC electric isolation between parts of a system that operate at different potentials.
- To store magnetic energy
- To transfer magnetic energy
- To provide multiple outputs.

The main purpose of a transformer is to transfer the energy from the input to the output through a magnetic field. The amount of energy transferred by a transformer is determined by the magnetic flux density, frequency, and operating temperature.

High-Frequency Magnetic Components, Second Edition. Marian K. Kazimierczuk.

Companion Website: www.wiley.com/go/kazimierczuk_High2e

7.2 Transformer Construction

A two-winding transformer is shown in Fig. 7.1. It consists of two coils wound around a common magnetic core, usually made of iron or ferrite. Coil 1 is referred to as the *primary winding* and coil 2 as the *secondary winding*. The coil of the primary winding has N_1 turns and the coil of the secondary winding has N_2 turns. A symbol and a magnetic equivalent circuit of a two-winding transformer are shown in Fig. 7.2. A time-varying current i_1 flowing through the primary winding creates a time-varying magnetic flux in both windings. In turn, the time-varying magnetic flux induces a voltage v_2 across the secondary winding. Part of the magnetic flux produced by the primary winding flows through the secondary winding. The magnetic flux common to both windings is called the mutual flux. The two windings sharing the same magnetic flux are said to be *magnetically coupled* or *inductively coupled*. The interaction between the two windings through magnetic field induced by one of them is characterized by a *mutual inductance*. Transformers may be noninverting and inverting. They may be step-down or step-up. The polarity of the induced voltage v_2 is determined by the direction of the two coils and Lenz's law. The dot convention is used to assign the polarity reference of voltage v_2.

7.3 Ideal Transformer

An *ideal transformer* is lossless, it has a perfect coupling ($k = 0$), its self-inductances and mutual inductance are infinite, its self-capacitances are zero, and it stores no magnetic energy. For an

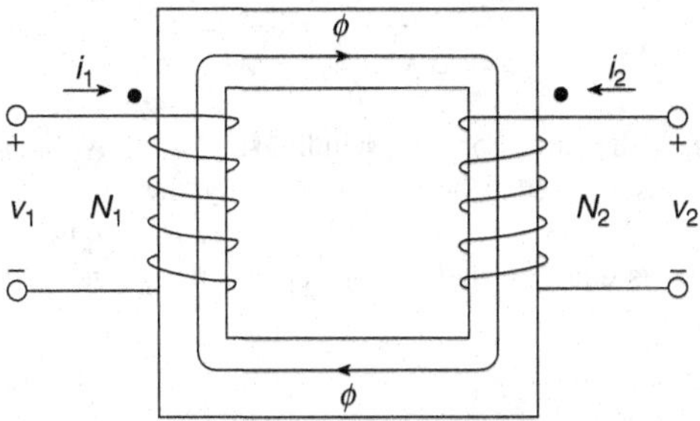

Figure 7.1 Two-winding transformer.

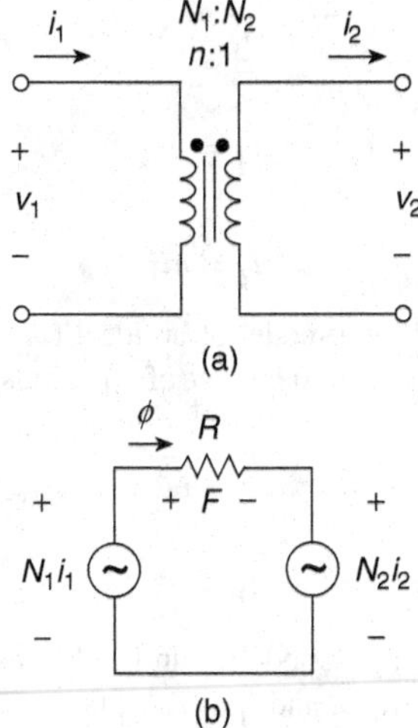

Figure 7.2 Two-winding transformer. (a) Transformer symbol. (b) Transformer magnetic equivalent circuit.

ideal transformer, the core resistivity is infinity ($\rho_c = \infty$), the core resistance is infinity ($R_c = \infty$), the hysteresis loss is zero, the core loss is zero ($P_C = 0$), the winding conductor resistivity is zero ($\rho_w = 0$), the winding resistances are zero ($R_{w1} = R_{w2} = 0$), the total winding loss is zero ($P_w = 0$), the core relative permeability is infinite ($\mu_{rc} = \infty$), the core reluctance is zero ($\mathcal{R} = l_c/\mu_{rc}\mu_0 A_c = 0$), the magnetic field is confined within the core, all of the magnetic flux ϕ is linked to both the windings, the leakage inductances are zero ($L_{l1} = L_{l2} = 0$), the coupling coefficient is 1 ($k = 1$), the magnetizing inductance is infinite ($L_m = \infty$), the magnetic energy stored in the ideal transformer is zero, ($W_m = 0$) the stray capacitances are zero, as the saturation flux density is infinity ($B_s = \infty$), and the bandwidth is infinity ($BW = \infty$).

The primary coil is usually connected to an AC voltage source $v_1(t)$ and the secondary coil is connected to a load resistor R_L. The AC input voltage $v_1(t)$ generates the current $i_1(t)$ in the primary winding, which establishes a magnetic flux $\phi(t)$. The magnetic flux ϕ is related to the AC input voltage v_1 by Faraday's law

$$v_1 = N_1 \frac{d\phi}{dt}. \tag{7.1}$$

The changes in the magnetic flux $\phi(t)$ induce the AC output voltage $v_2(t)$ across the secondary winding according to Faraday's law

$$v_2 = N_2 \frac{d\phi}{dt}, \tag{7.2}$$

which in turn causes the output current $i_2 = v_2/R_L$ to flow. The voltage transfer function of the transformer is equal to the primary-to-secondary turns ratio

$$\frac{v_1}{v_2} = \frac{N_1 \frac{d\phi}{dt}}{N_2 \frac{d\phi}{dt}} = \frac{N_1}{N_2} = n. \tag{7.3}$$

The voltage drop per turn is the same for both windings: $v_t = v_1/N_1 = v_2/N_2$. An ideal transformer is lossless and all of the instantaneous power supplied by the input source $p_1 = i_1 v_1$ is delivered to the load resistance R_L as the output instantaneous power $p_2 = i_2 v_2$. For lossless transformer, the instantaneous output power is equal to the instantaneous input power

$$p_2 = p_1, \tag{7.4}$$

which can be written as

$$i_2 v_2 = i_1 v_1, \tag{7.5}$$

resulting in

$$\frac{v_1}{v_2} = \frac{i_2}{i_1} = \frac{N_1}{N_2} = n. \tag{7.6}$$

Hence, we obtain a set of equations:

$$v_1 = n v_2 \tag{7.7}$$

and

$$i_2 = n i_1. \tag{7.8}$$

This set of equations is represented by a model of an ideal transformer using dependent voltage and current sources, as shown in Fig. 7.3a. Another set of equations is as follows:

$$v_2 = \frac{v_1}{n} \tag{7.9}$$

and

$$i_1 = \frac{i_2}{n}. \tag{7.10}$$

This set of equations is represented by a model of an ideal transformer, shown in Fig. 7.3b.

Because $v_2 = R_L i_2$, $v_1 = n v_2 = n R_L i_2$, and $i_1 = i_2/n$, the transformer input resistance is given by

$$R_i = \frac{v_1}{i_1} = \frac{n R_L i_2}{i_2/n} = n^2 R_L. \tag{7.11}$$

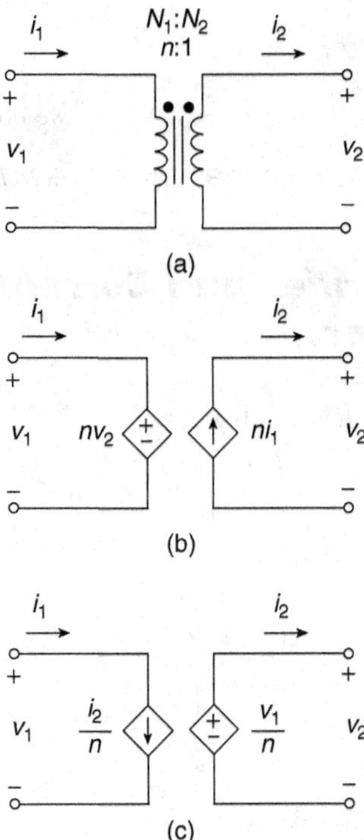

Figure 7.3 Models of an ideal transformer. (a) Symbol. (b) Model consisting of voltage-dependent voltage source and current-dependent current source. (c) Model consisting of current-dependent current source and voltage-dependent voltage source.

When the load is an impedance Z_L and the input voltage is sinusoidal, the transformer input impedance is given by

$$Z_i = \frac{v_1}{i_1} = \left(\frac{N_1}{N_2}\right)^2 Z_L = n^2 Z_L. \tag{7.12}$$

Thus, a transformer may serve as an impedance-matching device. A voltage transformer should never be operated with a short circuit at the output.

From Fig. 7.2b,

$$\mathcal{R}\phi = N_1 i_1 - N_2 i_2, \tag{7.13}$$

where the core reluctance is

$$\mathcal{R} = \frac{l_c}{\mu_{rc}\mu_0 A_c}. \tag{7.14}$$

For $\mu_{rc} = \infty$, $\mathcal{R} = 0$. Hence,

$$N_1 i_1 - N_2 i_2 = 0, \tag{7.15}$$

yielding the transformer turns ratio $i_2/i_1 = N_1/N_2 = n$.

For multiple output ideal transformers with k windings, the turns ratio is

$$V_1 : V_2 : V_3 : \ldots : V_k = N_1 : N_2 : N_3 : \ldots : N_k. \tag{7.16}$$

The voltage drop across each turn is the same

$$v_t = \frac{d\phi}{dt} = \frac{d\left(\frac{\lambda_1}{N_1}\right)}{dt} = \frac{1}{N_1}\frac{d\lambda_1}{dt} = \frac{v_1}{N_1}, \tag{7.17}$$

resulting in

$$V_t = \frac{V_1}{N_1} = \frac{V_2}{N_2} = \frac{V_3}{N_3} = \ldots = \frac{V_k}{N_k}. \tag{7.18}$$

Neglecting the losses,

$$V_1 I_1 = V_2 I_2 + V_3 I_3 + \ldots + V_k I_k. \tag{7.19}$$

Since $V_2 = V_1 N_2/N_1$, $V_3 = V_1 N_3/N_1$, and $V_k = V_1 N_k/N_1$, we obtain for $\mathcal{R} = \infty$

$$N_1 I_1 = N_2 I_2 + N_3 I_3 + \ldots + N_k I_k. \tag{7.20}$$

7.4 Voltage Polarities and Current Directions in Transformers

For the noninverting transformer shown in Fig. 7.4a,

$$\frac{v_1}{v_2} = \frac{i_2}{i_1} = \frac{N_1}{N_2} = n. \tag{7.21}$$

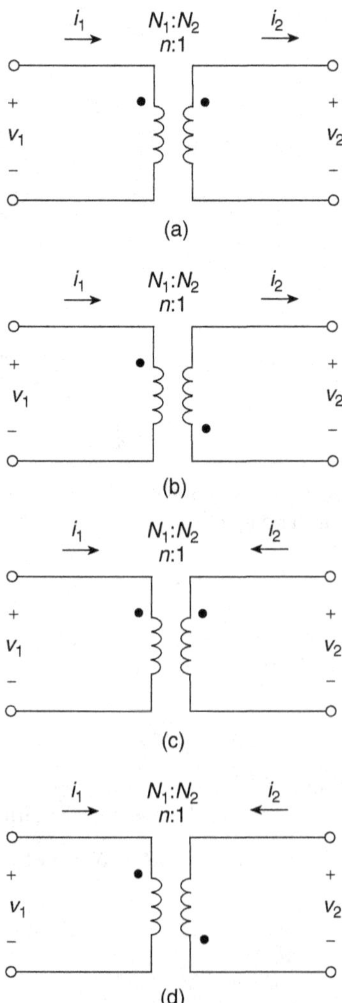

Figure 7.4 Transformers with various voltage polarities and current directions. (a) Noninverting transformer. (b) Inverting transformer. (c) Noninverting transformer. (d) Inverting transformer.

For the inverting transformer shown in Fig. 7.4b,

$$\frac{v_1}{v_2} = \frac{i_2}{i_1} = -\frac{N_1}{N_2} = -n. \tag{7.22}$$

For the noninverting transformer shown in Fig. 7.4c,

$$\frac{v_1}{v_2} = -\frac{i_2}{i_1} = \frac{N_1}{N_2} = n. \tag{7.23}$$

For the inverting transformer of Fig. 7.4d,

$$\frac{v_1}{v_2} = -\frac{i_2}{i_1} = -\frac{N_1}{N_2} = -n. \tag{7.24}$$

7.5 Nonideal Transformers

The main parameters of real transformers are turns ratio n, magnetizing inductance L_m, primary and secondary leakage inductances L_{lp} and L_{ls}, primary and secondary winding resistances R_p and R_s, primary and secondary stray capacitances C_p and C_s, power losses, current, voltage and power levels, insulation breakdown voltage, and operating frequency range. The total magnetic flux ϕ of a transformer consists of a mutual flux ϕ_m and leakage flux ϕ_l

$$\phi = \phi_m + \phi_l = \phi_m + \phi_{l1} + \phi_{l2}, \tag{7.25}$$

where ϕ_{l1} is the leakage flux on the primary side and ϕ_{l2} is the leakage flux on the secondary side. The *mutual flux* ϕ_m of a two-winding transformer is the portion of the total flux that is common to both windings. The *leakage flux* ϕ_l of a two-winding transformer is the portion of the total flux that does not link both windings.

Consider a two-winding linear transformer driven by an AC current source i_1 with an open circuit at the output, as shown in Fig. 7.5a. The self-inductances of the windings are L_1 and L_2. The current through the secondary winding is zero ($i_2 = 0$) and does not induce any magnetic flux. The AC current i_1 flowing in the primary winding induces a magnetic flux ϕ_{11} in the primary winding, which in turn induces an AC voltage v_1 across the primary winding. The magnetic flux induced by current i_1 consists of a *mutual flux* ϕ_{21} and a *primary leakage flux* ϕ_{l1}

$$\phi_{11} = \phi_{l1} + \phi_{21}. \tag{7.26}$$

The mutual flux ϕ_{21} links both the primary and secondary windings. The primary leakage flux ϕ_{l1} links only the turns of the primary winding and not the turns of the secondary winding. The flux linkage of the primary winding is given by

$$\lambda_1 = N_1\phi_{11} = N_1(\phi_{l1} + \phi_{21}). \tag{7.27}$$

The flux linkage in the primary winding is proportional to the current i_1

$$\lambda_1 = N_1\phi_{11} = L_1 i_1 = \frac{N_1^2}{\mathcal{R}_1} i_1, \tag{7.28}$$

$$\mathcal{R}_1 = \frac{l_1}{\mu A_1}. \tag{7.29}$$

and L_1 is the *self-inductance* of the primary winding

$$L_1 = \lambda_1/i_1 = \frac{N_1^2}{\mathcal{R}_1} = \frac{\mu A_1 N_1^2}{l_1}. \tag{7.30}$$

The mean path length l_1 of the magnetic flux linking the primary winding and the cross-sectional area of the flux A_1 are the average or effective values. Figure 7.6a shows a plot of λ_1 as a function of i_1 for a linear transformer. The self-inductance of the primary winding is

$$L_1 = \frac{\lambda_1}{i_1} = \frac{N_1\phi_{l1}}{i_1} + \frac{N_1\phi_{21}}{i_1} = L_{l1} + M_{21}. \tag{7.31}$$

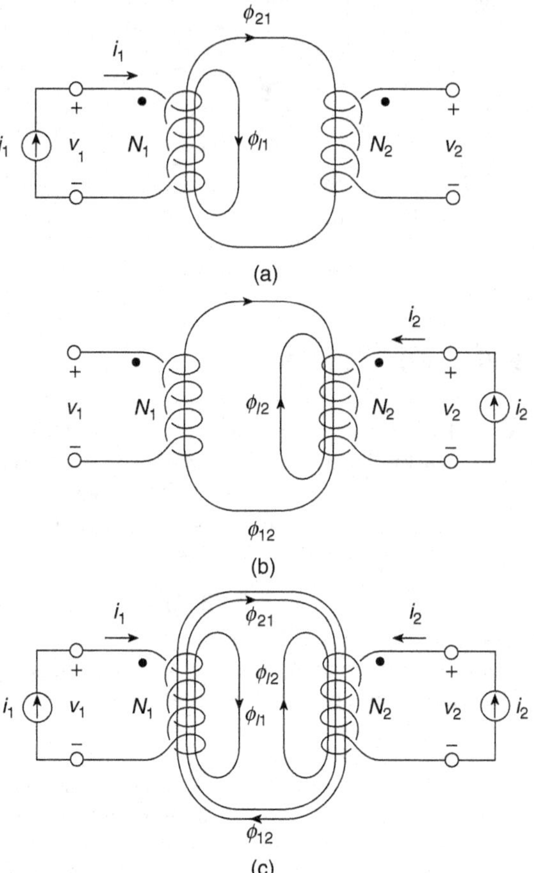

Figure 7.5 Transformer. (a) Transformer driven on the primary winding side by an AC current source i_1 and open-circuited on the secondary side. (b) Transformer driven on the secondary side by an AC current source i_2 and open-circuited on the primary winding side. (c) Two magnetically coupled coils driven on both sides by AC current sources i_1 and i_2.

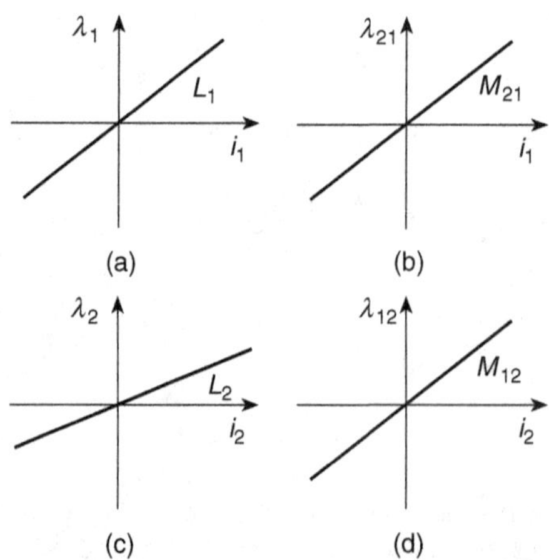

Figure 7.6 Characteristics of self-inductances and mutual inductances of a two-winding linear transformer. (a) $\lambda_1 = L_1 i_1$. (b) $\lambda_{21} = M_{21} i_1$. (c) $\lambda_2 = L_1 i_2$. (d) $\lambda_{12} = M_{12} i_2$.

By Faraday's law, the *self-induced voltage* across the primary winding by the primary current i_1 is

$$v_1 = \frac{d\lambda_1}{dt} = \frac{d(L_1 i_1)}{dt} = L_1 \frac{di_1}{dt}. \quad (7.32)$$

The flux linkage of the secondary winding is equal to the product of the mutual magnetic flux ϕ_{21} and the secondary number of turns N_2

$$\lambda_{21} = N_2 \phi_{21}. \quad (7.33)$$

By Faraday's law, the *mutually induced voltage* across the secondary winding by the primary current i_1 is

$$v_2 = \frac{d\lambda_{21}}{dt} = \frac{d(N_2\phi_{21})}{dt} = N_2 \frac{d\phi_{21}}{dt} = N_2 \frac{d(k_1\phi_{11})}{dt} = k_1 N_2 \frac{d(\phi_{11})}{dt}, \quad (7.34)$$

where $k_1 = \phi_{21}/\phi_{11}$ is the coupling coefficient of the primary winding to the secondary winding. In linear transformers, the mutual flux ϕ_{21} is proportional to the AC input current i_1

$$\lambda_{21} = N_2 \phi_{21} = M_{21} i_1, \quad (7.35)$$

where M_{21} is the *mutual inductance* of the coupled coils. Figure 7.6b shows a plot of λ_{21} as a function of i_1 for a linear transformer. By Faraday's law, the *mutually induced voltage* across the secondary winding by the primary current i_1 can be expressed as

$$v_2 = \frac{d\lambda_{21}}{dt} = \frac{d(M_{21} i_1)}{dt} = \frac{N_1 N_2}{\mathcal{R}_{21}} \frac{di_1}{dt} = M_{21} \frac{di_1}{dt}, \quad (7.36)$$

where the reluctance of the common magnetic path is

$$\mathcal{R}_{21} = \frac{l_{21}}{\mu A_{21}} \quad (7.37)$$

and the mutual inductance is

$$M_{21} = \lambda_{21}/i_1 = \frac{N_1 N_2}{\mathcal{R}_{21}} = \frac{\mu A_{21} N_1 N_2}{l_{21}}, \quad (7.38)$$

l_{21} is the mean length of the mutual flux path, and A_{21} is the cross-sectional area of the mutual flux. The voltage v_2 is proportional to the mutual inductance M_{21} and the rate of change in i_1. Figure 7.7a shows a circuit model, which represents (7.32) and (7.36). The voltage across the self-inductance of the secondary winding is zero because the current through the secondary winding $i_2 = 0$.

Let us now connect an AC current source i_2 in series with the secondary winding and make an open-circuit across the primary winding, as shown in Fig. 7.5b. The AC current i_2 induces a magnetic flux in the secondary winding ϕ_{22}, which in turn induces a voltage v_2. The magnetic flux ϕ_{22} is equal to the mutual magnetic flux ϕ_{12} linking both the secondary and primary windings and the secondary leakage flux ϕ_{l2} linking only the secondary winding

$$\phi_{22} = \phi_{l2} + \phi_{12}. \quad (7.39)$$

The flux linkage of the secondary winding is

$$\lambda_2 = N_2 \phi_{22} = L_2 i_2 = N_2(\phi_{l2} + \phi_{12}) = \frac{N_2^2}{\mathcal{R}_2} i_2, \quad (7.40)$$

where L_2 is the *self-inductance* of the secondary winding and the reluctance of the secondary winding is

$$\mathcal{R}_2 = \frac{l_2}{\mu A_2} \quad (7.41)$$

and the self-inductance of the secondary winding is

$$L_2 = \lambda_2/i_2 = \frac{N_2^2}{\mathcal{R}_2} = \frac{\mu A_2 N_2^2}{l_2}, \quad (7.42)$$

where l_2 is the mean path length of the magnetic flux linking the secondary winding and A_2 is the cross-sectional area of this flux. Figure 7.6c shows a plot of λ_2 as a function of i_2 for a linear transformer. The self-inductance of the secondary winding is

$$L_2 = \frac{\lambda_2}{i_2} = \frac{N_2\phi_{22}}{i_2} = \frac{N_2\phi_{l2}}{i_2} + \frac{N_2\phi_{12}}{i_2} = L_{l2} + M_{12}. \quad (7.43)$$

By Faraday's law, the *self-induced voltage* across the secondary winding by the secondary current i_2 is

$$v_2 = \frac{d\lambda_2}{dt} = \frac{d(L_2 i_2)}{dt} = L_2 \frac{di_2}{dt}. \tag{7.44}$$

The flux linkage in the primary winding is equal to the product of the mutual magnetic flux ϕ_{12} and the primary number of turns N_1

$$\lambda_{12} = N_1 \phi_{12}. \tag{7.45}$$

By Faraday's law, this flux linkage produces a *mutually induced voltage* drop across the secondary winding current i_2

$$v_1 = \frac{d\lambda_{12}}{dt} = \frac{d(N_1\phi_{12})}{dt} = N_1 \frac{d\phi_{12}}{dt} = N_1 \frac{d(k_2\phi_{22})}{dt} = k_2 N_1 \frac{\phi_{22}}{dt}, \tag{7.46}$$

where $k_2 = \phi_{12}/\phi_{22}$ is the coupling coefficient of the secondary winding to the primary winding. The flux linkage in the primary winding λ_1 is proportional to the current i_2 through the secondary winding

$$\lambda_{12} = N_1\phi_{12} = M_{12} i_2, \tag{7.47}$$

where M_{12} is a mutual inductance of the coils. Figure 7.6d shows a plot of λ_{12} as a function of i_1 for a linear transformer. Hence, by Faraday's law, the *mutually induced voltage* drop across the primary winding by the secondary current i_2 is

$$v_1 = \frac{d\lambda_{12}}{dt} = \frac{d(M_{12} i_2)}{dt} = \frac{N_1 N_2}{\mathcal{R}_{12}} \frac{di_2}{dt} = M_{12} \frac{di_2}{dt}. \tag{7.48}$$

Figure 7.7b shows a circuit model, which represents (7.44) and (7.48). The path of the mutual magnetic flux between the two windings is physically the same. Hence, from the principle of reciprocity,

$$\phi_{21} = \phi_{12} = \phi, \tag{7.49}$$

$$\mathcal{R}_{21} = \mathcal{R}_{12} = \mathcal{R}, \tag{7.50}$$

and

$$M_{21} = M_{12} = M = \frac{\lambda_{21}}{i_1} = \frac{\lambda_{12}}{i_2}. \tag{7.51}$$

The mutual inductance of winding 1 with respect to winding 2 is equal to that of winding 2 with respect to winding 1.

Finally, consider a two-winding transformer with an AC current source i_1 connected to the primary winding and an AC current source i_2 connected to the secondary winding, as shown in Figure 7.5c. Using the principle of superposition for a linear transformer, we obtain the magnetic flux in the primary winding induced by currents i_1 and i_2

$$\phi_1 = \phi_{l1} + \phi_{21} + \phi_{12} = \phi_{11} + \phi_{12}, \tag{7.52}$$

resulting in the flux linkage in the primary winding

$$\lambda_1 = N_1\phi_1 = N_1\phi_{11} + N_1\phi_{12} = L_1 i_1 + M_{12} i_2 \tag{7.53}$$

and the voltage across the primary winding

$$v_1 = \frac{d\lambda_1}{dt} = N_1 \frac{d\phi_{11}}{dt} + N_1 \frac{d\phi_{12}}{dt} = L_1 \frac{di_1}{dt} + M \frac{di_2}{dt} = v_{L1} + v_{M1}. \tag{7.54}$$

The voltage across the primary inductance is the sum of the voltage due to the primary winding self-inductance and the voltage due to the mutual inductance.

Similarly, the magnetic flux in the secondary winding induced by currents i_1 and i_2 is given by

$$\phi_2 = \phi_{l2} + \phi_{12} + \phi_{21} = \phi_{22} + \phi_{21}, \tag{7.55}$$

yielding the flux linkage in the secondary winding

$$\lambda_2 = N_2\phi_2 = N_2\phi_{21} + N_2\phi_{22} = M_{21} i_1 + L_2 i_2. \tag{7.56}$$

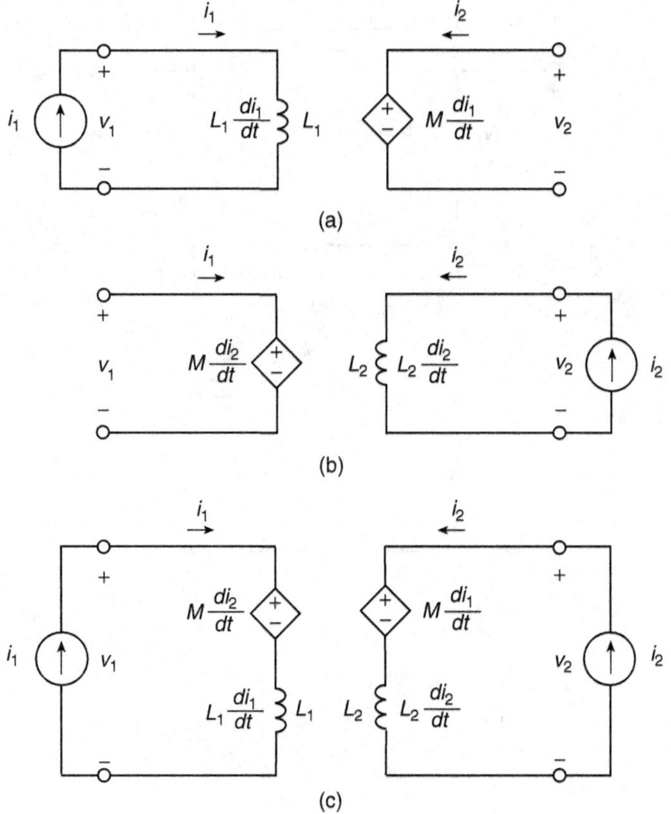

Figure 7.7 Model of a noninverting transformer. (a) For $i_2 = 0$. (b) For $i_1 = 0$. (c) For nonzero currents i_1 and i_2.

and the voltage across the secondary winding

$$v_2 = \frac{d\lambda_2}{dt} = N_2\frac{d\phi_{22}}{dt} + N_2\frac{d\phi_{21}}{dt} = L_2\frac{di_2}{dt} + M\frac{di_1}{dt} = v_{L2} + v_{M2}. \tag{7.57}$$

The voltage across the secondary inductance is the sum of the voltage due to the secondary winding self-inductance and the voltage due to the mutual inductance. Figure 7.7c shows a transformer model that represents (7.54) and (7.57). The input part of the model consists of an input self-inductance L_1 and a dependent voltage source controlled by the derivative of the secondary current di_2/dt. Likewise, the output part of the model consists of an output self-inductance L_2 and a dependent voltage source controlled by the derivative of the primary current di_1/dt.

The Laplace transform equations for a transformer of Fig. 7.1 at $i_1(0^-) = i_2(0^-) = 0$ are obtained by replacing the operator d/dt by s in (7.54) and (7.57), respectively

$$\mathbf{V}_1(s) = sL_1\mathbf{I}_1(s) + sM\mathbf{I}_2(s) \tag{7.58}$$

and

$$\mathbf{V}_2(s) = sM\mathbf{I}_1(s) + sL_2\mathbf{I}_2(s). \tag{7.59}$$

For steady state, $s = j\omega$, yielding

$$V_1(j\omega) = j\omega L_1 I_1(j\omega) + j\omega M I_2(j\omega) \tag{7.60}$$

and

$$V_2(j\omega) = j\omega M I_1(j\omega) + j\omega L_2 I_2(j\omega). \tag{7.61}$$

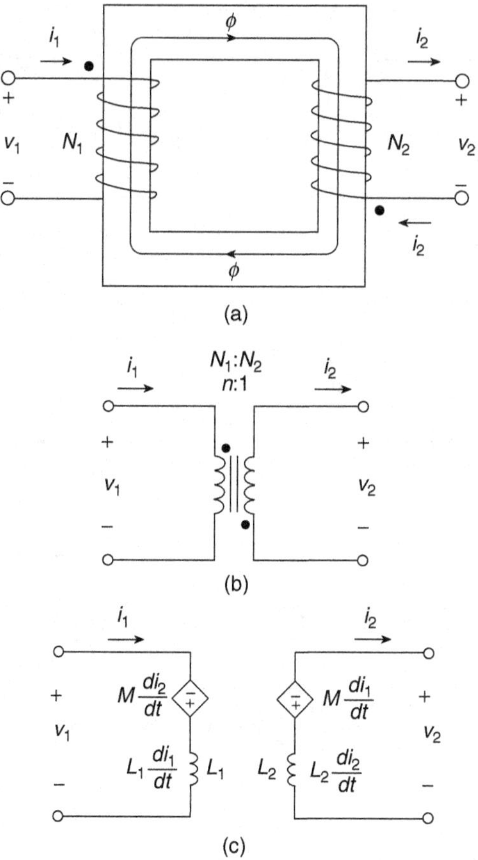

Figure 7.8 Inverting transformer. (a) Physical windings. (b) Symbol. (c) Model of an inverting transformer.

Figure 7.8 shows an inverting transformer, where

$$\frac{v_1}{v_2} = \frac{i_2}{i_1} = -\frac{N_1}{N_2} = -n. \tag{7.62}$$

Also,

$$v_1 = \frac{d\lambda_1}{dt} = N_1\frac{d\phi_{11}}{dt} - N_1\frac{d\phi_{12}}{dt} = L_1\frac{di_1}{dt} - M\frac{di_2}{dt} = v_{L1} - v_{M1} \tag{7.63}$$

and

$$v_2 = \frac{d\lambda_2}{dt} = N_2\frac{d\phi_{22}}{dt} - N_2\frac{d\phi_{21}}{dt} = L_2\frac{di_2}{dt} - M\frac{di_1}{dt} = v_{L2} - v_{M2}. \tag{7.64}$$

7.6 Neumann's Formula for Mutual Inductance

Figure 7.9 shows two magnetically coupled closed loops to illustrate the concept of mutual inductance. Suppose that an inductor L_1 has N_1 turns, conducts current I_1, and produces a flux density $\mathbf{B}_1$. Some of the magnetic flux $\mathbf{B}_1$ passes through inductor L_2 bounded by a closed loop C_2. According to the right-hand screw rule, the direction of the current I_2 in the second inductor is as shown in Fig. 7.9. The mutual magnetic flux is given by

$$\phi_{12} = \int_{S_2} \mathbf{B}_1 \cdot d\mathbf{S}_2. \tag{7.65}$$

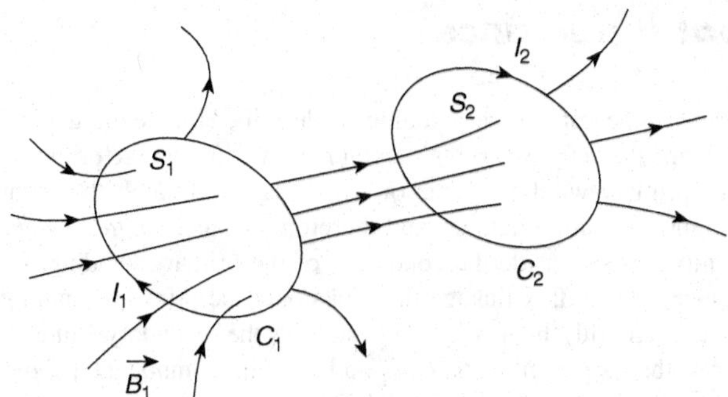

Figure 7.9 Two magnetically coupled closed loops explaining the concept of mutual inductance.

The mutual inductance caused by the current I_1 is

$$M_{12} = \frac{\lambda_{12}}{I_1} = \frac{N_1\phi_{12}}{I_1} = \frac{N_1}{I_1}\int_{S_2} \mathbf{B}_1 \cdot d\mathbf{S}_2. \tag{7.66}$$

The magnetic flux density $\mathbf{B}_1$ can be expressed in terms of the vector magnetic potential $\mathbf{A}_1$

$$\mathbf{B}_1 = \nabla \times \mathbf{A}_1. \tag{7.67}$$

Hence, the mutual inductance induced by inductor L_1 becomes

$$M_{12} = \frac{N_2}{I_1}\int_{S_2} (\nabla \times \mathbf{A}_1) \cdot d\mathbf{S}_2 = \frac{N_2}{I_1}\oint_{C_2} \mathbf{A}_1 \cdot d\mathbf{l}_2. \tag{7.68}$$

The magnetic potential is given by

$$\mathbf{A}_1 = \frac{\mu N_1}{4\pi}\int_V \frac{\mathbf{J}_1}{R} dV = \frac{\mu N_1 I_1}{4\pi}\oint_{C_1} \frac{d\mathbf{l}_1}{R}. \tag{7.69}$$

Hence, *Neumann's formula* for mutual inductance between the two circuits is given by

$$M_{12} = M_{21} = M = \frac{N_2}{I_1}\oint_{C_2}\left(\frac{\mu I_1}{4\pi}\oint_{C_1}\frac{d\mathbf{l}_1}{R}\right)\cdot d\mathbf{l}_2 = \frac{\mu N_1 N_2}{4\pi}\oint_{C_1}\oint_{C_2}\frac{d\mathbf{l}_1 \cdot d\mathbf{l}_2}{R}, \tag{7.70}$$

where R is the distance between the differential lengths $d\mathbf{l}_1$ and $d\mathbf{l}_2$. It can be seen that Neumann's formula is symmetrical with respect to the subscripts 12 and 21. The mutual inductance depends solely on the geometry of the two circuits and the permeability of the medium. If the mutual inductance is multiplied by the current in one circuit, it gives the flux linkage in the other. The mutual inductance between two circuits is 1 H when a current of 1 A in one of the circuits produces a flux linkage of 1 Wb-turn in the other circuit.

By Faraday's induction law, the time-varying magnetic flux ϕ_{12} produces the induced electromotance, electromotive force (EMF), or voltage in the loop C_2, which is equal to the line integral of $\mathbf{E} \cdot d\mathbf{l}$ around the closed path. The electromotance induced in the second circuit by a change in the current I_1 in the first circuit is given by

$$\oint_{C_1} \mathbf{E} \cdot d\mathbf{l} = \frac{d\lambda_{12}}{dt} = N_1\frac{d\phi_{12}}{dt} = M_{12}\frac{dI_1}{dt}. \tag{7.71}$$

Likewise, the electromotance induced in the first circuit by a change in the current I_2 in the second circuit is

$$\oint_{C_2} \mathbf{E} \cdot d\mathbf{l} = \frac{d\lambda_{21}}{dt} = N_2\frac{d\phi_{21}}{dt} = M_{21}\frac{dI_2}{dt}. \tag{7.72}$$

The mutual inductance between two circuits is equal to 1 H if the current changing at the rate of 1 A/s in one circuit induces an electromotance of 1 V in the other circuit.

7.7 Mutual Inductance

The magnetic coupling of the transformer windings is described by the mutual inductance and the coupling coefficient. Assume that a time-varying current i_1 flows through the primary winding. It produces magnetic flux in the primary winding, some of which passes through the secondary winding. This magnetic flux is common to both windings, and therefore is called *mutual magnetic flux*. The current i_1 produces (i) the flux linkage inside the conductor of the primary winding, which gives rise to an internal self-inductance, (ii) the flux linkage that links only the primary winding and gives rise to an external self-inductance, and (iii) the flux linkage that links the secondary winding. The sum of the first two components forms the *leakage inductance* L_l and the third component is the *mutual inductance* M.

For a two-winding transformer, the mutual inductance is

$$M = \frac{\lambda_{21}}{i_1} = \frac{N_2\phi_{21}}{i_1} = \frac{\lambda_{12}}{i_2} = \frac{N_1\phi_{12}}{i_2}. \tag{7.73}$$

The mutual magnetic flux created by the current i_1 is expressed as

$$\phi_{21} = \frac{\mu A_c N_2 i_1}{l_c} = \frac{N_2 i_1}{\mathcal{R}}, \tag{7.74}$$

the flux linkage in the secondary winding is

$$\lambda_{21} = N_1\phi_{21} = \frac{\mu A_c N_1 N_2 i_1}{l_c} = \frac{N_1 N_2}{\mathcal{R}} i_1, \tag{7.75}$$

and the mutual inductance between the two windings is

$$M_{21} = \frac{\lambda_{21}}{i_1} = \frac{\mu A_c N_1 N_2}{l_c} = \frac{N_1 N_2}{\mathcal{R}}, \tag{7.76}$$

where N_1 is the number of turns of the primary winding, A_c is the core cross-sectional area, and l_c is the core magnetic path length. If $N_1 = N_2$, $M_{21} = N_1^2/\mathcal{R} = N_2^2/\mathcal{R}$.

Similarly, the mutual magnetic flux generated by current i_2 is given by

$$\phi_{12} = \frac{\mu A_c N_1 i_2}{l_c} = \frac{N_1 i_2}{\mathcal{R}}. \tag{7.77}$$

The flux linkage of the primary winding is

$$\lambda_{12} = N_2\phi_{12} = \frac{\mu A_c N_1 N_2 i_2}{l_c} = \frac{N_1 N_2}{\mathcal{R}} i_2, \tag{7.78}$$

where N_2 is the number of turns of the secondary winding. The mutual inductance between two windings is

$$M_{12} = \frac{\lambda_{12}}{i_2} = \frac{\mu A_c N_1 N_2}{l_c} = \frac{N_1 N_2}{\mathcal{R}}. \tag{7.79}$$

It follows from (7.79) and (7.76) that $M_{12} = M_{21} = M = N_1 N_2/\mathcal{R}$.

When the coupling coefficient is $k = 1$, $L_{l1} = 0$ and the self-inductance of the primary winding is

$$L_1 = \frac{\mu A_c N_1^2}{l_c} = \frac{N_1^2}{\mathcal{R}}. \tag{7.80}$$

When $k = 1$, $L_{l2} = 0$ and the self-inductance of the secondary winding is

$$L_2 = \frac{\mu A_c N_2^2}{l_c} = \frac{N_2^2}{\mathcal{R}}. \tag{7.81}$$

Assuming $k = 1$, finding $N_1 = \sqrt{\mathcal{R}L_1}$ and $N_2 = \sqrt{\mathcal{R}L_2}$ from the above equations and substituting them into the equation for M, we get

$$M = \sqrt{L_1 L_2} = \frac{N_1 N_2}{\mathcal{R}} = \frac{\mu A_c N_1 N_2}{l_c}. \tag{7.82}$$

Thus, the value of the mutual inductance is a function of self-inductances of the primary and secondary windings. The mutual inductance is proportional to the geometric mean of the self-inductances.

The unit of the mutual inductance of coupled inductors is the henry (H), the same as that for the self-inductance. For two windings of turns N_1 and N_2 sharing a common flux path, the ratio of the self-inductances is related to the turns ratio n by

$$n = \frac{N_1}{N_2} = \sqrt{\frac{L_1}{L_2}}. \tag{7.83}$$

The ratio of the self-inductances and the mutual inductance is

$$L_1 : L_2 : M = N_1^2 : N_2^2 : N_1 N_2. \tag{7.84}$$

If the permeability of the core is nonlinear, the small-signal (incremental) mutual inductance at a given operating point can be determined as

$$M = M_{12} = \frac{d\lambda_{12}}{di_1}. \tag{7.85}$$

7.8 Magnetizing Inductance

Consider a two-winding transformer shown in Fig. 7.2. Assume that the coupling between the windings is perfect ($k = 1$) and therefore the leakage flux is zero ($\phi_l = 0$) so that the entire flux ϕ links both the windings ($\phi = \phi_m$), but core relative permeability is finite and therefore the core reluctance is nonzero, that is, $\mu_{rc} < \infty$ and $\mathcal{R} > 0$. Consequently, the magnetizing inductance L_m is finite. When the secondary winding is open-circuited, the current through the secondary winding is zero ($i_2 = 0$), but there is a nonzero current flowing through the primary winding ($i_1 \neq 0$). This current is called the *magnetizing current* i_{Lm} and flows through the *magnetizing inductance* L_m. The model of the transformer with a core finite permeability, perfect coupling coefficient ($k = 1$), zero core losses, and zero winding loss consists of a magnetizing inductance and an ideal transformer connected in parallel, as shown in Fig. 7.17. The magnetizing inductance L_m can be placed on the primary side as shown in Fig. 7.17a or L_{ms} on the secondary side of the ideal transformer as shown in Fig. 7.17b. When the transformer output is open-circuited, the input current i_1 flows through the magnetizing inductance L_m in the model shown in Fig. 7.17a or through the ideal transformer and the magnetizing inductance L_{ms} as in the model shown in Fig. 7.17b.

Referring to Fig. 7.2b and using Ampère's law, the magnetic flux is given by

$$\phi = \frac{\mathcal{F}_1 - \mathcal{F}_2}{\mathcal{R}} = \frac{N_1 i_1 - N_2 i_2}{\mathcal{R}} = \frac{N_1}{\mathcal{R}}\left(i_1 - \frac{N_2 i_2}{N_1}\right) = \frac{N_1}{\mathcal{R}}\left(i_1 - \frac{i_2}{n}\right), \tag{7.86}$$

where the core reluctance is

$$\mathcal{R} = \frac{l_c}{\mu_{rc}\mu_0 A_c} \tag{7.87}$$

and the turns ratio is

$$n = \frac{N_1}{N_2}. \tag{7.88}$$

Since $\lambda_1 = N_1 \phi$, the transformer input voltage is

$$v_1 = \frac{d\lambda_1}{dt} = N_1 \frac{d\phi}{dt} = N_1 \frac{d}{dt}\left[\frac{N_1}{\mathcal{R}}\left(i_1 - \frac{i_2}{n}\right)\right] = \frac{N_1^2}{\mathcal{R}}\frac{d}{dt}\left(i_1 - \frac{i_2}{n}\right) = L_m \frac{di_{Lm}}{dt}, \tag{7.89}$$

where the *magnetizing inductance* on the primary side is

$$L_m = \frac{N_1^2}{\mathcal{R}} = \frac{\mu_{rc}\mu_0 A_c N_1^2}{l_c} \tag{7.90}$$

and the *magnetizing current* on the primary side is

$$i_{Lm} = i_1 - \frac{i_2}{n}. \tag{7.91}$$

Figure 7.17a shows an equivalent circuit of the transformer with magnetizing inductance L_m on the primary side. When the reluctance $\mathcal{R}$ is zero, that is, as μ_{rc} approaches infinity, the magnetizing

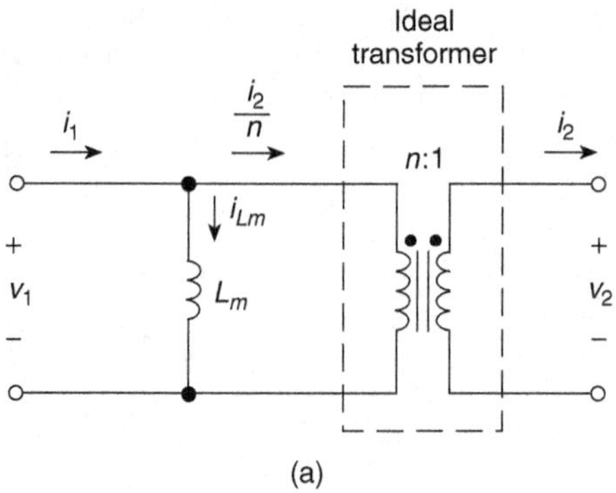

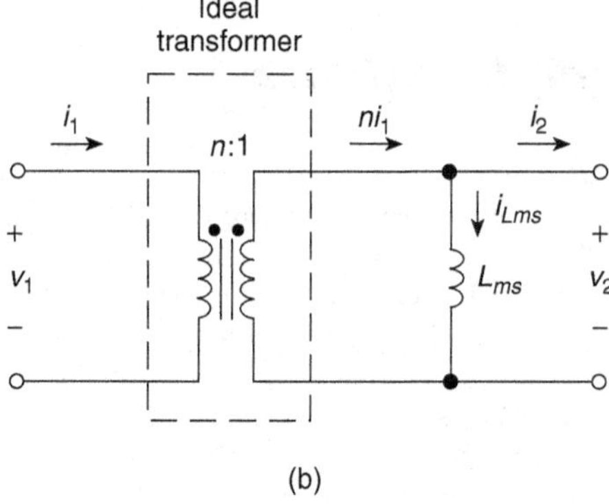

Figure 7.10 Transformer models with finite permeability ($\mu_{rc} < 0$ and therefore finite magnetizing inductance), perfect coupling ($k = 1$ and therefore zero leakage inductances), zero core losses, and zero winding loss. (a) Transformer model with magnetizing inductance L_m on the primary side. (b) Transformer model with magnetizing inductance L_{ms} on the secondary side.

inductance L_m also approaches infinity and the transformer becomes an ideal transformer. If $i_2 = 0$, $L = L_m = N_1^2/\mathcal{R}$.

The magnetizing inductance can be placed on the secondary side. The magnetic flux can be expressed by

$$\phi = \frac{N_1 i_1 - N_2 i_2}{\mathcal{R}} = \frac{N_2}{\mathcal{R}}\left(\frac{N_1 i_1}{N_2} - i_2\right) = \frac{N_2}{\mathcal{R}}(ni_1 - i_2). \tag{7.92}$$

Taking into account that $\lambda_2 = N_2\phi$,

$$v_2 = \frac{d\lambda_2}{dt} = N_2\frac{d\phi}{dt} = N_2\frac{d}{dt}\left[\frac{N_2}{\mathcal{R}}(ni_1 - i_2)\right] = \frac{N_2^2}{\mathcal{R}}\frac{d}{dt}(ni_1 - i_2) = L_{ms}\frac{di_{Lms}}{dt}, \tag{7.93}$$

where the magnetizing inductance on the secondary side is

$$L_{ms} = \frac{N_2^2}{\mathcal{R}} = \frac{\mu_{rc}\mu_0 A_c N_2^2}{l_c} = \frac{\mu_{rc}\mu_0 A_c N_1^2}{l_c}\left(\frac{N_2}{N_1}\right)^2 = \frac{L_m}{n^2} \tag{7.94}$$

and the magnetizing current on the secondary side is

$$i_{Lms} = ni_1 - i_2. \tag{7.95}$$

Figure 7.17b shows an equivalent circuit of the transformer with magnetizing inductance L_{ms} on the secondary side. If $i_1 = 0$, $L_{ms} = N_2^2/\mathcal{R}$.

The mutual inductance is

$$M = \frac{N_1 N_2}{\mathcal{R}} = \frac{N_1 N_2 \mu_{rc}\mu_0 A_c}{l_c} = \frac{L_m N_2}{N_1} = \frac{L_{ms} N_1}{N_2}. \tag{7.96}$$

Hence, we obtain the magnetizing inductance of the primary side

$$L_m = M\frac{N_1}{N_2} = nM \tag{7.97}$$

and the magnetizing inductance of the secondary side

$$L_{ms} = M\frac{N_2}{N_1} = \frac{M}{n}. \tag{7.98}$$

Example 7.1

A transformer has the following parameters: $k = 1$, $\mu_{rc} = 125$, $A_c = 1\ \text{cm}^2$, $l_c = 12\ \text{cm}$, $N_1 = 100$, and $N_2 = 10$. Calculate the magnetizing inductances L_m and L_{ms}.

Solution: The magnetizing inductance on the primary side of the transformer is

$$L_m = \frac{\mu_{rc}\mu_0 A_c N_1^2}{l_c} = \frac{125 \times 4\pi \times 10^{-7} \times 10^{-4} \times 100^2}{12 \times 10^{-2}} = 1.308\ \text{mH}. \tag{7.99}$$

The magnetizing inductance on the secondary side of the transformer is

$$L_{ms} = \frac{\mu_{rc}\mu_0 A_c N_2^2}{l_c} = \frac{125 \times 4\pi \times 10^{-7} \times 10^{-4} \times 10^2}{12 \times 10^{-2}} = 13.089\ \mu\text{H}. \tag{7.100}$$

7.9 Coupling Coefficient

In general, not all of the magnetic flux produced by the primary winding is coupled to the secondary winding. The magnetic flux "leaks," and therefore it is called the *leakage flux*. The mutual inductance can be expressed as

$$M = k\sqrt{L_1 L_2}, \tag{7.101}$$

where k is the *coupling coefficient* of two magnetically coupled inductors. It is a measure of the closeness with which the magnetic fields of two inductors are interlinked. Its range is $0 \leq k \leq 1$. When the two windings are *perfectly coupled* so that there is no leakage of the magnetic flux and all of the flux that links the primary winding also links the secondary winding, $k = 1$ and $M = \sqrt{L_1 L_2}$. It is an ideal situation because it is impossible to wind two coils so that they share the same magnetic flux precisely. When the two windings are *tightly coupled*, nearly all the magnetic flux links both windings, and therefore M is close to its maximum value and k is close to 1. When $k \ll 1$, the windings are *loosely coupled*, and therefore the mutual flux is low and the leakage flux is large. When the two windings have no common flux, that is, $\phi_{21} = \phi_{12} = 0$, then $M = 0$ and $k = 0$. Equation (7.101) is valid only if both inductances are independent of their currents.

The *leakage coefficient* is defined as

$$h = 1 - k = 1 - \frac{M}{\sqrt{L_1 L_2}}. \tag{7.102}$$

For an ideal transformer, $k = 1$ and $h = 0$.

The coupling coefficient of the primary winding is defined as the ratio of the mutual flux ϕ_{21} produced by the current i_1 to the total flux ϕ_{11} produced by i_1

$$k_1 = \frac{\phi_{21}}{\phi_{11}} = \frac{\phi_{21}}{\phi_{l1} + \phi_{21}} = \frac{\dfrac{N_1\phi_{21}}{i_1}}{\dfrac{N_1\phi_{11}}{i_1}} = \frac{\dfrac{\lambda_{21}}{i_1}}{\dfrac{\lambda_1}{i_1}} = \frac{M_{21}}{L_1} = \frac{L_1 - L_{l1}}{L_1} = 1 - \frac{L_{l1}}{L_1}. \tag{7.103}$$

Hence, the leakage inductance of the primary winding is

$$L_{l1} = (1 - k_1)L_1. \tag{7.104}$$

The coupling coefficient of the secondary winding is defined as the ratio of the mutual flux ϕ_{12} produced by the current i_2 to the total flux ϕ_{22} produced by i_2

$$k_2 = \frac{\phi_{12}}{\phi_{22}} = \frac{\phi_{12}}{\phi_{l2} + \phi_{12}} = \frac{\frac{N_2\phi_{12}}{i_2}}{\frac{N_2\phi_{22}}{i_2}} = \frac{\frac{\lambda_{12}}{i_2}}{\frac{\lambda_2}{i_2}} = \frac{M_{12}}{L_2} = \frac{L_2 - L_{l2}}{L_2} = 1 - \frac{L_{l2}}{L_2}. \tag{7.105}$$

Thus, the leakage inductance of the secondary winding is

$$L_{l2} = (1 - k_2)L_2. \tag{7.106}$$

The mutual magnetic flux is $\phi_{21} = \phi_{12} = \phi$. The coupling coefficient of a transformer is defined as the geometric mean of the coefficients k_1 and k_2, that is, the geometric mean of the ratios of mutual to total fluxes

$$k = \sqrt{k_1 k_2} = \sqrt{\left(\frac{\phi_{21}}{\phi_{l1} + \phi_{21}}\right)\left(\frac{\phi_{12}}{\phi_{l2} + \phi_{12}}\right)} = \sqrt{\left(\frac{\phi}{\phi_{l1} + \phi}\right)\left(\frac{\phi}{\phi_{l2} + \phi}\right)}$$

$$= \sqrt{\left(\frac{\phi_{21}}{\phi_{11}}\right)\left(\frac{\phi_{12}}{\phi_{22}}\right)} = \frac{\phi}{\sqrt{\phi_{11}\phi_{22}}}. \tag{7.107}$$

If $\phi_{l1} = 0$ and $\phi_{l2} = 0$, then $k = 1$.

Since

$$\phi_{11} = \frac{\lambda_1}{N_1}, \tag{7.108}$$

$$\phi_{22} = \frac{\lambda_2}{N_2}, \tag{7.109}$$

$$\phi_{21} = \frac{\lambda_{21}}{N_1} = \frac{M_{21} i_1}{N_1}, \tag{7.110}$$

and

$$\phi_{12} = \frac{\lambda_{12}}{N_2} = \frac{M_{12} i_2}{N_2}, \tag{7.111}$$

the coupling coefficient becomes

$$k = \sqrt{\left(\frac{\phi_{21}}{\phi_{11}}\right)\left(\frac{\phi_{12}}{\phi_{22}}\right)} = \sqrt{\frac{\left(\frac{M_{21} i_1}{N_1}\right)\left(\frac{M_{12} i_2}{N_2}\right)}{\left(\frac{\lambda_1}{N_1}\right)\left(\frac{\lambda_2}{N_2}\right)}} = \sqrt{\frac{M_{21} M_{12}}{\left(\frac{\lambda_1}{i_1}\right)\left(\frac{\lambda_2}{i_2}\right)}} = \frac{\sqrt{M_{21} M_{12}}}{\sqrt{L_1 L_2}} = \frac{M}{\sqrt{L_1 L_2}}. \tag{7.112}$$

If $k_1 = k_2$, then $k_1 = k_2 = k$. When the windings have equal number of turns, $k_1 \approx k_2$. Otherwise, k_1 is usually not equal to k_2.

Since $P_{12} = P_{21}$, the product of the self-inductances can be expressed in terms of the permeances

$$L_1 L_2 = (N_1^2 P_1)(N_2^2 P_2) = N_1^2 (P_{l1} + P_{21}) N_2^2 (P_{l2} + P_{12})$$

$$= N_1^2 N_2^2 P_{21}^2 \left(1 + \frac{P_{l1}}{P_{21}}\right)\left(1 + \frac{P_{l2}}{P_{12}}\right) = M^2 \left(1 + \frac{P_{l1}}{P_{21}}\right)\left(1 + \frac{P_{l2}}{P_{12}}\right) = \frac{M^2}{k^2}, \tag{7.113}$$

where

$$P_{21} = P_{12} = \frac{\mu_{rc}\mu_0 A_c}{l_c}, \tag{7.114}$$

$$P_{l1} = \frac{\mu_0 A_{l1}}{l_{l1}}, \tag{7.115}$$

$$P_{l2} = \frac{\mu_0 A_{l2}}{l_{l2}}, \tag{7.116}$$

$$\frac{P_{l1}}{P_{21}} = \frac{1}{\mu_{rc}} \left(\frac{l_c}{l_{l1}}\right)\left(\frac{A_{l1}}{A_c}\right), \tag{7.117}$$

and

$$\frac{P_{l2}}{P_{12}} = \frac{1}{\mu_{rc}} \left(\frac{l_c}{l_{l2}}\right)\left(\frac{A_{l2}}{A_c}\right). \tag{7.118}$$

Thus, the coupling coefficient is

$$k = \frac{M}{\sqrt{L_1 L_2}} = \frac{1}{\sqrt{\left(1 + \frac{P_{l1}}{P_{21}}\right)\left(1 + \frac{P_{l2}}{P_{12}}\right)}}$$

$$= \frac{1}{\sqrt{\left[1 + \frac{1}{\mu_{rc}}\left(\frac{l_c}{l_{l1}}\right)\left(\frac{A_{l1}}{A_c}\right)\right]\left[1 + \frac{1}{\mu_{rc}}\left(\frac{l_c}{l_{l2}}\right)\left(\frac{A_{l2}}{A_c}\right)\right]}} \tag{7.119}$$

where A_{l1} and A_{l2} are the cross-sectional areas of the leakage fluxes and l_{l1} and l_{l2} are the path lengths of the leakage fluxes. The coupling coefficient k increases as the core relative permeability μ_{rc} increases. For $l_{l1} = l_{l2} = l_l$ and $A_{l1} = A_{l2} = A_l$, we get

$$k = \frac{M}{\sqrt{L_1 L_2}} = \frac{1}{1 + \frac{1}{\mu_{rc}}\left(\frac{l_c}{l_l}\right)\left(\frac{A_l}{A_c}\right)}. \tag{7.120}$$

For example, when $l_l/l_c = 3/4$ and $A_l/A_c = 4$,

$$k = \frac{1}{1 + \frac{16}{3\mu_{rc}}}. \tag{7.121}$$

As the relative permeability μ_{rc} increases from 1 to infinity under the above-mentioned assumptions, the coupling coefficient k increases from 0.16 to 1 in the above example.

The ratio of the self-inductance of the primary winding L_1 to the mutual inductance M is

$$\frac{L_1}{M} = \frac{1}{k_1}\frac{N_2}{N_1} = \frac{n}{k_1}. \tag{7.122}$$

The ratio of the mutual inductance M to the self-inductance of the secondary winding L_2 is

$$\frac{M}{L_2} = k_2 \frac{N_1}{N_2} = k_2 n. \tag{7.123}$$

The geometric mean of the two inductance ratios is

$$\sqrt{\frac{L_1}{M} \times \frac{M}{L_2}} = \sqrt{\frac{L_1}{L_2}} = \frac{N_1}{N_2}\sqrt{\frac{k_2}{k_1}} = n\sqrt{\frac{k_2}{k_1}}. \tag{7.124}$$

Hence, the turns ratio is

$$n = \sqrt{\frac{k_1}{k_2}}\sqrt{\frac{L_1}{L_2}}. \tag{7.125}$$

7.10 Leakage Inductance

In a two-winding transformer, not all the magnetic flux induced by one winding links the other winding and vice versa. There is a magnetic flux induced by one winding in the space between the core and the winding ϕ_{cw}, in the space between the layers ϕ_{ll}, within the conductors ϕ_{cond}, and within the insulation between the windings ϕ_{ins}. These flux components are not linked with the other winding. Therefore, the coupling coefficient is $k < 1$. The leakage flux stores magnetic energy and behaves like an inductor. This effect is modeled by leakage inductances L_{l1} and L_{l2} connected in series

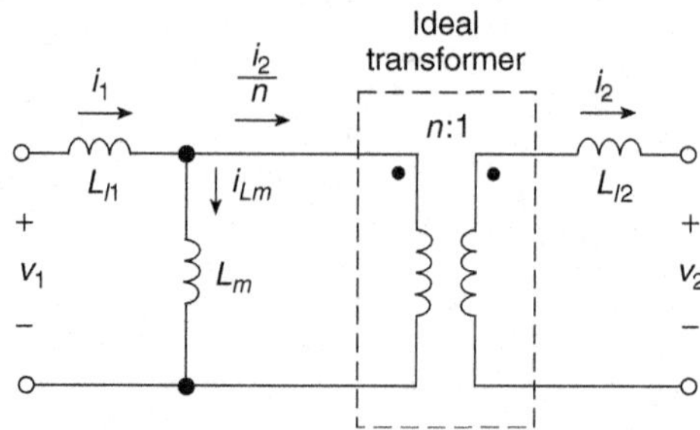

Figure 7.11 Transformer model with magnetizing inductance L_m and leakage inductances L_{l1} and L_{l2}.

with the windings, as shown in Fig. 7.18. The leakage inductance depends on the winding layout, core geometry, and the core relative permeability. To reduce the leakage inductances, the windings should be wide and have a low thickness, the insulation between windings should be reduced, one winding should be placed over the other with good overlapping, bifilar windings should be used, and the number of turns should be reduced. Wide and flat windings with a minimum isolation reduce the leakage inductance. The windings of inductors with toroidal cores should cover the entire or nearly entire magnetic path. In addition, the windings should be made of litz wire or a twisted bundle of insulated wires. The use of wide and thin foil gives the lowest leakage inductances. The coupling coefficient is given by

$$k = \frac{M}{\sqrt{L_1 L_2}} = \frac{M}{\sqrt{(L_m + L_{l1})(L_{ms} + L_{l2})}}, \tag{7.126}$$

where

$$L_1 = L_{l1} + L_m \tag{7.127}$$

and

$$L_2 = L_{l2} + L_{ms} = L_{l2} + \frac{L_m}{n^2}. \tag{7.128}$$

Hence, the leakage inductances are

$$L_{l1} = L_1 - L_m = L_1 - kL_1 = (1-k)L_1 \tag{7.129}$$

and

$$L_{l2} = L_2 - L_{\{ms\}} = L_2 - kL_2 = (1-k)L_2. \tag{7.130}$$

As k decreases from 1 to 0, L_{l1} increases from 0 to L_1 and L_{l2} increases from 0 to L_2. The magnetizing inductances are

$$L_m = kL_1 \tag{7.131}$$

and

$$L_{ms} = kL_2. \tag{7.132}$$

As k decreases from 1 to 0, L_m decreases from L_1 to 0 and L_{ms} increases from L_2 to 0. Figure 7.18 shows a transformer model with leakage inductances. The ratio of the transformer terminal voltages is no longer described by the turns ratio n of the windings. We need to subtract the voltage drops across the leakage inductances from the terminal voltages to obtain the voltages across the windings of the ideal transformer. The turns ratio of a transformer with nonperfect coupling is given by

$$n = k\sqrt{\frac{L_2}{L_1}}. \tag{7.133}$$

The leakage inductance of a transformer is proportional to the height of the windings and inversely proportional to the breadth of the windings. It is also inversely proportional to the square of the number of layers of the windings. One way to ensure tight coupling is using the bifilar winding, where each turn of the secondary winding is side by side with the one on the primary winding. The primary and secondary conductors can be twisted together before winding them. Interleaving the windings reduces the leakage inductance. Fractional layers should be avoided because they increase the leakage inductance and thus reduce the coupling coefficient. In transformers with multiple secondary windings, the lower power secondary winding should be wound on the top of the higher power secondary winding. This improves cross regulation. The typical ratio of the leakage inductance to the magnetizing inductance is $L_l/L_m = 0.5\%$. If a transformer has a large turns ratio n and is made of a foil winding on one side and a wire winding on the other side, it is useful to place the foil winding between the layers of the wire winding to reduce the leakage inductance.

The leakage inductance reduces the output voltage of the transformer and forms a resonant circuit with transformer parasitic capacitances, diode junction capacitances, and transistor capacitances. This leads to ringing.

Example 7.2

A transformer has the following parameters: $\mu_{rc} = 125$, $A_c = 1\ \text{cm}^2$, $l_c = 12$ cm, $N_1 = 100$, $N_2 = 10$, $k_1 = 0.999$, and $k_2 = 0.9995$. Calculate n, k, $\mathcal{R}$, L_1, L_2, M, L_m, L_{ms}, L_{l1}, L_{l2}, L_{lp}, and L_{ls}.

Solution: The transformer turns ratio is

$$n = \frac{N_1}{N_2} = \frac{100}{10} = 10. \tag{7.134}$$

The reluctance of the transformer core is

$$\mathcal{R} = \frac{l_c}{\mu_{rc}\mu_0 A_c} = \frac{12 \times 10^{-2}}{125 \times 4\pi \times 10^{-7} \times 10^{-4}} = 7.6394 \times 10^6\ \text{(A·turns/Wb)}. \tag{7.135}$$

The self-inductance of the primary winding is

$$L_1 = \frac{N_1^2}{\mathcal{R}} = \frac{100^2}{7.639 \times 10^6} = 1.309\ \text{mH}. \tag{7.136}$$

The self-inductance of the secondary winding is

$$L_2 = \frac{N_2^2}{\mathcal{R}} = \frac{10^2}{7.639 \times 10^6} = 13.09\ \mu\text{H}. \tag{7.137}$$

The mutual inductance of the windings is

$$M = \frac{N_1 N_2}{\mathcal{R}} = \frac{100 \times 10}{7.639 \times 10^6} = 130.9\ \mu\text{H}. \tag{7.138}$$

The magnetizing inductances are

$$L_m = k_1 L_1 = 0.999 \times 1.0473 = 1.0463\ \text{mH} \tag{7.139}$$

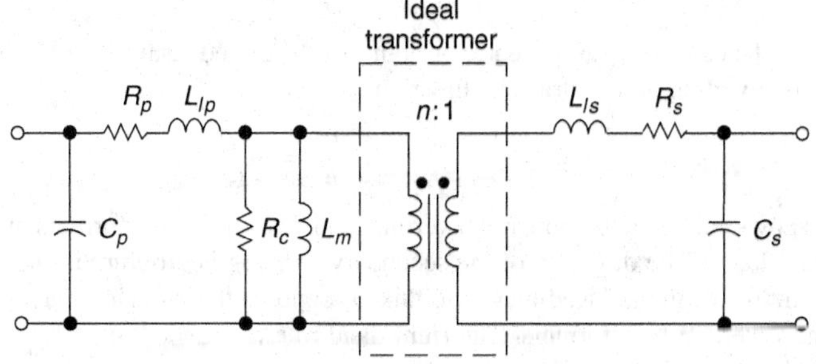

Figure 7.12 Model of a transformer with core resistance, leakage inductances, winding resistances, and stray capacitances.

and

$$L_{ms} = k_2 L_2 = 0.9995 \times 13.09 = 13.0835\ \mu\text{H} \tag{7.140}$$

The primary leakage inductance is

$$L_{l1} = (1 - k_1)L_1 = (1 - 0.999) \times 1.309 \times 10^{-3} = 1.309\ \mu\text{H}. \tag{7.141}$$

The secondary leakage inductance is

$$L_{l2} = (1 - k_2)L_2 = (1 - 0.9995) \times 13.09 \times 10^{-6} = 6.545\ \text{nH}. \tag{7.142}$$

The total leakage inductance on the primary side of the transformer is

$$L_{lpT} = L_{l1} + n^2 L_{l2} = 1.309 \times 10^{-6} + 10^2 \times 6.545 \times 10^{-9} = 1.9635\ \mu\text{H}. \tag{7.143}$$

The total leakage inductance on the secondary side of the transformer is

$$L_{lsT} = L_{l2} + \frac{L_{l1}}{n^2} = 6.545 \times 10^{-9} + \frac{1.309 \times 10^{-6}}{10^2} = 19.635\ \text{nH}. \tag{7.144}$$

7.11 Dot Convention

The *dot convention* is used to indicate the polarity of voltages across the windings of coupled coils, that is, the phase relationship between the primary and secondary voltages. The polarity of the mutually induced voltages depends on the way the coils are wound in relation to the reference direction of coil currents. The polarities of the induced voltages are indicated by a method known as the *dot convention*, in which a dot is placed on one terminal of each winding. If the current enters the dotted terminal of one coil, the polarity of the voltage of the second coil is positive at the dotted terminal of the second coil.

If both the voltages v_1 and v_2 have the same polarity at the dotted terminals, that is, both are positive or both are negative, the transformer is noninverting and its transfer function is given by

$$A_v = \frac{1}{n} = \frac{N_2}{N_1} = \frac{v_2}{v_1} = \frac{i_1}{i_2}. \tag{7.145}$$

Figure 7.10a shows a symbol of a noninverting transformer. Conversely, if the voltages v_1 and v_2 have the opposite polarities at the dotted terminals, for example, one voltage is positive and the other is negative, the transformer is inverting and its transfer function is

$$A_v = \frac{1}{n} = \frac{N_2}{N_1} = -\frac{v_2}{v_1} = -\frac{i_1}{i_2}. \tag{7.146}$$

Figure 7.10b shows a symbol of an inverting transformer.

If one of the currents enters the dotted terminal and the other current exits the dotted terminal as shown in Fig. 7.10a, the transformer is noninverting and its transfer function is expressed as

$$A_v = \frac{1}{n} = \frac{N_2}{N_1} = \frac{v_2}{v_1} = \frac{i_1}{i_2}. \tag{7.147}$$

In contrast, if both the currents i_1 and i_2 enter or both exit the dotted terminals, as shown in Fig. 7.10b, the transformer is inverting and its transfer function is

$$A_v = \frac{1}{n} = \frac{N_2}{N_1} = -\frac{v_2}{v_1} = -\frac{i_1}{i_2}. \tag{7.148}$$

Figure 7.11 shows an inductor and noninverting and inverting transformers with inducing and induced magnetic fluxes. The direction of the secondary winding is different in the two transformers. According to Lenz's law, the induced magnetic flux ϕ_i opposes the inducing magnetic flux ϕ in both noninverting and inverting transformers. The right-hand rule is satisfied in both the cases.

The mutual inductance can be used to relate the voltage v_2 induced in the secondary winding due to a time-varying current i_1 in the primary winding depending on the transformer polarity. If the current enters the dotted terminal of one inductor, the reference polarity of the mutual voltage $v_2 = M di_1/dt$

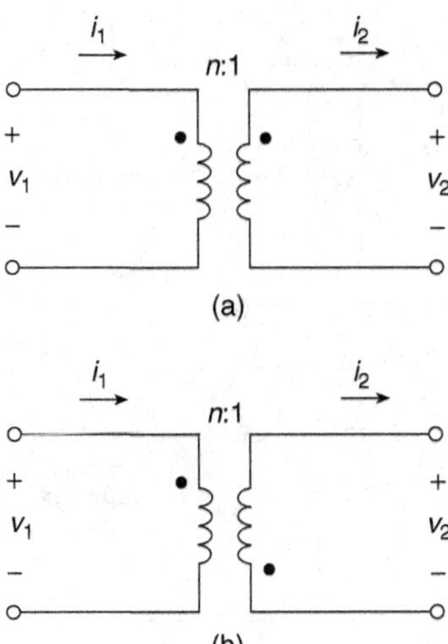

Figure 7.13 Noninverting and inverting transformers. (a) Noninverting transformer ($v_2 = M(di_1/dt)$ and $i_2 > 0$). (b) Inverting transformer ($v_2 = -M(di_1/dt)$ and $i_2 < 0$).

across the second inductor is positive at the dotted terminal of the second inductor. This convention is illustrated in Fig. 7.12.

When the current through the primary winding is i_1 and the secondary winding is an open-circuit coil, then the voltage across the secondary winding is given by

$$v_2 = \pm\frac{d\lambda_{12}}{dt} = \pm\frac{N_2 d\phi_{12}}{dt} = \pm\frac{\mu A_c N_1 N_2}{l_c}\frac{di_1}{dt} = \pm k\sqrt{L_1 L_2}\frac{di_1}{dt} = \pm M\frac{di_1}{dt}. \quad (7.149)$$

Similarly, a current i_2 in L_2 produces an open-circuit voltage v_1 across L_1

$$v_1 = \pm\frac{d\lambda_{21}}{dt} = \pm\frac{N_1 d\phi_{21}}{dt} = \pm\frac{\mu A_c N_1 N_2}{l_c}\frac{di_2}{dt} = \pm k\sqrt{L_1 L_2}\frac{di_2}{dt} = \pm M\frac{di_2}{dt}. \quad (7.150)$$

The "+" sign applies to noninverting transformers and the "−" sign applies to inverting transformers.

If nonzero currents flow in both windings into a noninverting transformer, the transformer is described by the following set of equations:

$$v_1 = L_1\frac{di_1}{dt} + M\frac{di_2}{dt} \quad (7.151)$$

and

$$v_2 = L_2\frac{di_2}{dt} + M\frac{di_1}{dt}. \quad (7.152)$$

In the transformer shown in Fig. 7.1, the dot convention rule indicates that the reference polarity of the voltage induced in the primary winding by the current i_2 is positive at the dotted terminal of the secondary winding. For instance, the voltage induced in the secondary winding by the current i_1 is $M(di_1/dt)$ and its reference polarity is positive at the dotted terminal of the secondary winding.

If nonzero currents flow in both windings of an inverting transformer, the input current flows into the transformer, and the output current flows out of the transformer; the following set of equations describes the transformer:

$$v_1 = L_1\frac{di_1}{dt} - M\frac{di_2}{dt} \quad (7.153)$$

and

$$v_2 = -L_2\frac{di_2}{dt} + M\frac{di_1}{dt}. \quad (7.154)$$

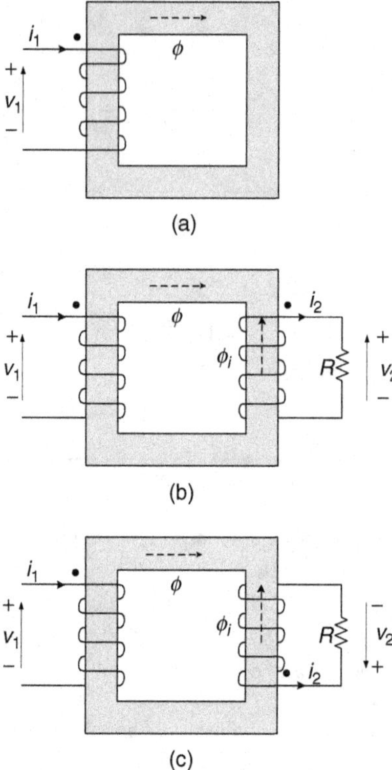

Figure 7.14 Non-inverting and inverting transformers. (a) Inductor with inducing magnetic flux. (b) Non-inverting transformer with inducing magnetic flux ϕ and induced magnetic flux ϕ_i. (c) Inverting transformer with inducing magnetic flux ϕ and induced magnetic flux ϕ_i.

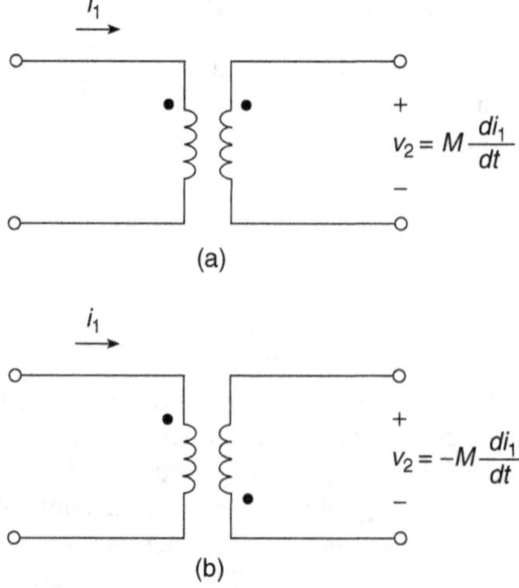

Figure 7.15 Dot convention in transformers. (a) Reference polarity of mutual voltage is positive. (b) Reference polarity of mutual voltage is negative.

7.12 Series-Aiding and Series-Opposing Connections

Two magnetically coupled inductors can be connected in series with their magnetic flux aiding or opposing. Figure 7.13 depicts magnetically coupled inductors connected in series for the two situations. The series-aiding coupled inductors shown in Fig. 7.13a have magnetic fluxes of both windings flowing in the same direction. The voltage across both the inductors is

$$v = \left(L_1\frac{di}{dt} + M\frac{di}{dt}\right) + \left(M\frac{di}{dt} + L_2\frac{di}{dt}\right) = (L_1 + L_2 + 2M)\frac{di}{dt} = L\frac{di}{dt}, \tag{7.155}$$

where the total inductance is

$$L = L_1 + L_2 + 2M = L_1 + L_2 + 2k\sqrt{L_1L_2}. \tag{7.156}$$

The concept of series-aiding coupled inductors is used in integrated inductors to reduce the inductor area [65].

The series-opposing coupled inductors shown in Fig. 7.13b have magnetic fluxes of both the windings flowing in opposite direction.

$$v = \left(L_1\frac{di}{dt} - M\frac{di}{dt}\right) + \left(-M\frac{di}{dt} + L_2\frac{di}{dt}\right) = (L_1 + L_2 - 2M)\frac{di}{dt} = L\frac{di}{dt}, \tag{7.157}$$

where the total inductance is

$$L = L_1 + L_2 - 2M = L_1 + L_2 - 2k\sqrt{L_1L_2}. \tag{7.158}$$

7.13 Equivalent T Network

Figure 7.14a shows a noninverting transformer with common terminals. The windings of this transformer form a series-aiding connection. Equations (7.151) and (7.152) can be written as

$$v_1 = (L_1 - M)\frac{di_1}{dt} + M\left(\frac{di_1}{dt} + \frac{di_2}{dt}\right) \tag{7.159}$$

and

$$v_2 = M\left(\frac{di_1}{dt} + \frac{di_2}{dt}\right) + (L_2 - M)\frac{di_2}{dt}. \tag{7.160}$$

These equations can be represented by the transformer T equivalent circuit as shown Fig. 7.14b.

Figure 7.15a shows an inverting transformer with common terminals. The windings of this transformer form a series-opposing connection. Equations (7.153) and (7.154) can be written as

$$v_1 = (L_1 + M)\frac{di_1}{dt} - M\left(\frac{di_1}{dt} + \frac{di_2}{dt}\right) \tag{7.161}$$

and

$$v_2 = -M\left(\frac{di_1}{dt} + \frac{di_2}{dt}\right) + (L_2 + M)\frac{di_2}{dt}. \tag{7.162}$$

These equations can be represented by the transformer T equivalent circuit as shown Fig. 7.15b.

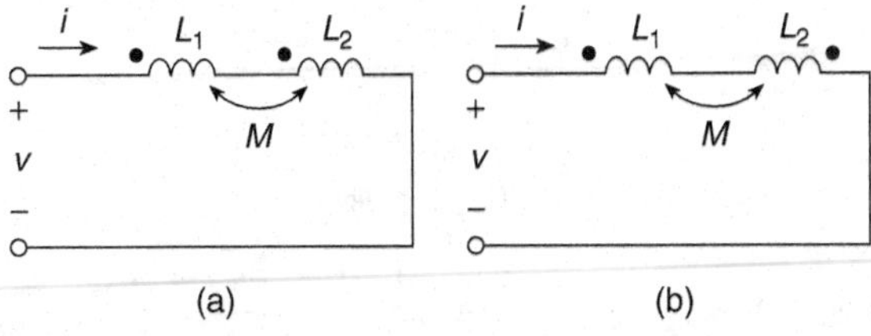

Figure 7.16 Magnetically coupled inductors. (a) Series-aiding connection. (b) Series-opposing connection.

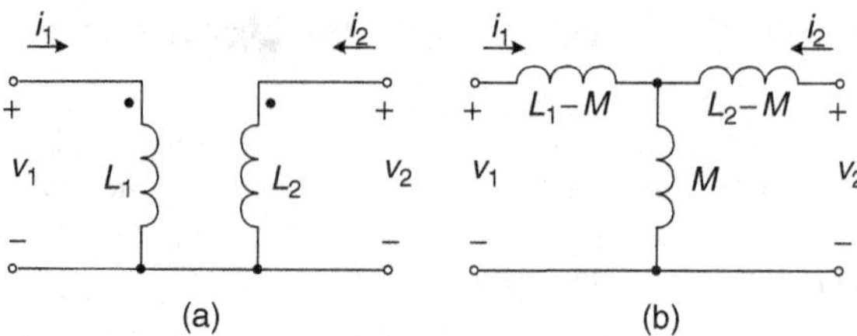

Figure 7.17 Noninverting transformer T network. (a) Transformer with common terminals. (b) Equivalent T network.

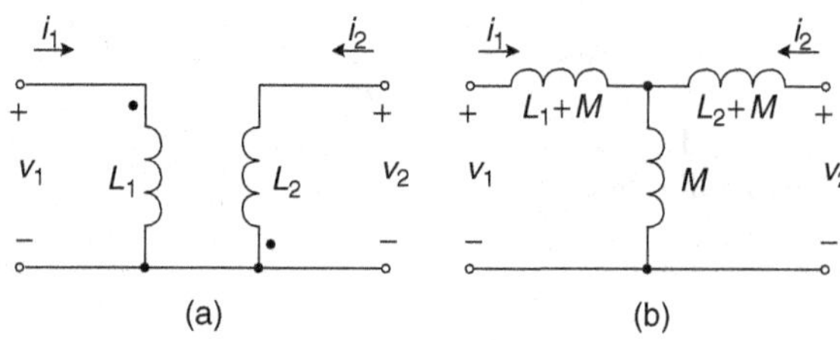

Figure 7.18 Inverting transformer T network. (a) Transformer with common terminals. (b) Transformer T equivalent circuit.

7.14 Energy Stored in Coupled Inductors

Figure 7.16 shows two magnetically coupled inductors. Assume that the energy stored in the magnetic field of the coupled inductors at time $t = 0$ is zero. Suppose that the current in the primary winding at time $t = 0$ is increased from 0 to I_1 at time $t = t_1$, while the secondary winding is open-circuited so that $i_2 = 0$. The instantaneous power delivered by the primary winding terminals is

$$p_1(t) = i_1 v_1 = i_1 L_1 \frac{di_1}{dt} \quad \text{for} \quad 0 \le t \le t_1. \tag{7.163}$$

The magnetic energy entering the primary winding terminals and stored in the magnetic field at time t_1 when $i_1(t) = I_1$ is given by

$$w_1 = \int_0^{t_1} p_1 dt = L_1 \int_0^{I_1} i_1 di_1 = \frac{1}{2} L_1 I_1^2. \tag{7.164}$$

Suppose that the current in the secondary winding is increased from 0 at time t_1 to I_2 at $t = t_2$, while maintaining a constant current $i_1 = I_1$. The instantaneous power entering the secondary winding terminals is

$$p_2(t) = i_2 v_2 = i_2 L_2 \frac{di_2}{dt} \quad \text{for} \quad t_1 \le t \le t_2. \tag{7.165}$$

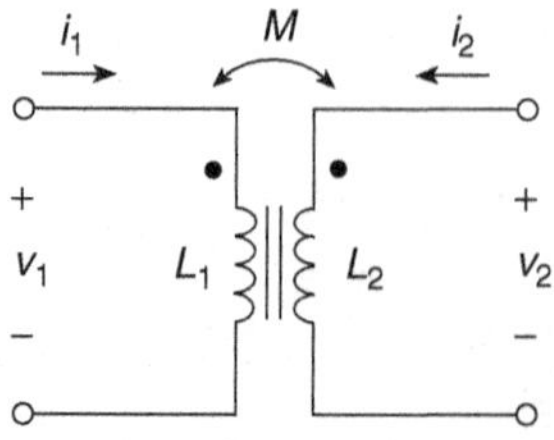

Figure 7.19 Magnetically coupled inductors.

The energy delivered by the secondary winding terminals and stored in the magnetic field at time t_2 is

$$w_2 = \int_{t_1}^{t_2} p_2 dt = L_2 \int_0^{I_2} i_2 di_2 = \frac{1}{2} L_2 I_2^2. \quad (7.166)$$

For time interval $t_1 \le t \le t_2$, the current through the primary winding is

$$i_1 = I_1, \quad (7.167)$$

the voltage across the primary winding is

$$v_1 = L_1 \frac{di_1}{dt} + M \frac{di_2}{dt} = M \frac{di_2}{dt} \quad \text{for} \quad t_1 \le t \le t_2, \quad (7.168)$$

the instantaneous power entering the primary terminals is

$$p_{12}(t) = i_1 v_1 = I_1 v_1 = I_1 M \frac{di_2}{dt} \quad \text{for} \quad t_1 \le t \le t_2, \quad (7.169)$$

and the energy delivered by the primary winding terminals is

$$w_{12} = \int_{t_1}^{t_2} p_{12} dt = I_1 M \int_{t_1}^{t_2} \frac{di_2}{dt} dt = M I_1 \int_0^{I_2} di_2 = M I_1 I_2. \quad (7.170)$$

Hence, the total energy stored in the magnetic field of coupled inductors at time t_2 is

$$w_m(t_2) = w_1 + w_2 + w_{12} = \frac{1}{2} L_1 I_1^2 + \frac{1}{2} L_2 I_2^2 + M I_1 I_2. \quad (7.171)$$

This equation is valid only if each winding current enters the dotted terminal. Otherwise,

$$w_m(t_2) = w_1 + w_2 - w_{12} = \frac{1}{2} L_1 I_1^2 + \frac{1}{2} L_2 I_2^2 - M I_1 I_2. \quad (7.172)$$

In general, the instantaneous magnetic energy stored in a two-winding transformer with either polarity at any time t is given by

$$w_m(t) = w_1 + w_2 \pm w_{12} = \frac{1}{2} L_1 i_1^2 + \frac{1}{2} L_2 i_2^2 \pm M i_1 i_2. \quad (7.173)$$

There are many applications of coupled inductors. For example, electric toothbrushes use the mutual inductance as part of its contactless wireless battery charger.

7.15 High-Frequency Transformer Model

A high-frequency model of a transformer is depicted in Fig. 7.19. It consists of ideal transformer, magnetizing inductance L_m, core loss shunt resistance R_c, leakage inductances L_{lp} and L_{ls}, winding resistances R_p and R_s, and stray capacitances C_p and C_s. A winding-to-winding capacitance can be added between the primary and secondary windings.

The secondary leakage inductance L_{ls} and the secondary winding resistance R_s can be reflected on the primary side of the ideal transformer using the following expressions:

$$R_{s(p)} = n^2 R_s, \quad (7.174)$$

$$L_{ls(p)} = n^2 L_{ls}, \quad (7.175)$$

and

$$C_{s(p)} = \frac{C_s}{n^2}. \quad (7.176)$$

The *load regulation* of the transformer describes the degree to which the output voltage changes between low-load and full-load conditions due to changes of the voltage drops across winding resistances and leakage inductances. It is defined as

$$LR = \frac{v_{oLL} - v_{oFL}}{v_{oFL}} \times 100\%|_f, \tag{7.177}$$

where v_{oLL} is the transformer output voltage at low load and v_{oFF} is the transformer output voltage at full load. When low load becomes no load, $v_{oLL} = v_{oNL}$ and the load regulation is defined as

$$LR = \frac{v_{oNL} - v_{oFL}}{v_{oFL}} \times 100\%|_f. \tag{7.178}$$

7.16 Stray Capacitances

Stray capacitances are normally unintended and unwanted capacitances. Any system of separated conductors represents a capacitance. A voltage applied between two conductors generates an electric field, which stores electric energy. In inductors and transformers, they include turn-to-turn capacitance, layer-to-layer capacitance, winding-to-winding capacitance, winding-to-core capacitance, winding-to-shield, core-to-shield, and the capacitance between the outer winding and surrounding circuitry. The stray capacitances can be reduced by reducing the number of turns in the winding, increasing the number of layers, increasing the dielectric insulation thickness, reducing the winding width, not using bifilar winding, and using a Faraday electrostatic shield. The effects of the stray capacitance include resonant frequency, reduced transformer bandwidth, large current spikes at sudden voltage changes, and electrostatic coupling with surrounding circuits. The stray capacitances greatly reduce the impedance of coupled inductors at high frequencies, allowing the noise generated by the converter switching network to pass freely to the load. It is worth to note that a decrease in the stray capacitances usually results in an increase in the leakage inductances and vice versa. Stray capacitances are analyzed in Chapter 9.

7.17 Transformer Efficiency

The transformer efficiency is given by

$$\eta = \frac{P_o}{P_i} = \frac{P_o}{P_o + P_{Loss}} = \frac{P_o}{P_o + P_w + P_c} = \frac{P_o}{P_o + P_w + P_H + P_E} = \frac{1}{1 + \frac{P_{Loss}}{P_o}}, \tag{7.179}$$

where $P_{Loss} = P_w + P_c = P_w + P_H + P_E$. For sinusoidal current and voltage waveforms, the transformer input power is

$$P_i = \frac{1}{2}Re\{\mathbf{V}_p \cdot \mathbf{I}_p^*\} = \frac{1}{2}V_p I_p \cos\theta_p \tag{7.180}$$

and the transformer output power is

$$P_o = \frac{1}{2}Re\{\mathbf{V}_s \cdot \mathbf{I}_s^*\} = \frac{1}{2}V_s I_s \cos\theta_{Load}. \tag{7.181}$$

7.18 Transformers with Gapped Cores

The reluctance of the air gap is given by

$$\mathcal{R}_g = \frac{l_g}{\mu_0 A_c}. \tag{7.182}$$

Assuming that $\mathcal{R}_c \ll \mathcal{R}_g$ and neglecting leakage inductances, the magnetizing inductance on the primary side is given by

$$L_m = L_1 = \frac{N_1^2}{\mathcal{R}_g} = \frac{\mu_0 A_c N_1^2}{l_g}. \tag{7.183}$$

Ignoring the leakage inductance, the self-inductance of the primary winding is $L_1 \approx L_m$. The self-inductance of the secondary winding is

$$L_2 = L_{ms} = \frac{N_2^2}{\mathcal{R}_g} = \frac{\mu_0 A_c N_2^2}{l_g}. \tag{7.184}$$

The magnetomotive force (MMF) is

$$\mathcal{F}_{max} = N_{1max} I_{Lmmax} = \phi(\mathcal{R}_g + \mathcal{R}_c) = B_{pk} A_c (\mathcal{R}_g + \mathcal{R}_c) \approx B_{pk} A_c \mathcal{R}_g = \frac{B_{ph} l_g}{\mu_0}. \tag{7.185}$$

To avoid core saturation, the maximum number of turns of the primary winding is

$$N_{1max} = \frac{B_{pk} l_g}{\mu_0 I_{Lm(max)}}. \tag{7.186}$$

7.19 Multiple-Winding Transformers

In some applications, transformers with multiple windings are used. Examples of such transformers include multiple-output switching-mode pulse-width modulated (PWM) and resonant DC–DC converters. Also, the flyback converter contains a three-winding transformer. Transformers with n windings are described by the following matrix:

$$\begin{vmatrix} \lambda_1 \\ \lambda_2 \\ \cdot \\ \cdot \\ \cdot \\ \lambda_n \end{vmatrix} = \begin{vmatrix} L_{11} & M_{12} & \cdots & M_{1n} \\ M_{21} & L_{22} & \cdots & M_{2n} \\ \cdot & \cdot & \cdots & \cdot \\ \cdot & \cdot & \cdots & \cdot \\ \cdot & \cdot & \cdots & \cdot \\ M_{n1} & M_{n2} & \cdots & L_{nn} \end{vmatrix} \begin{vmatrix} I_1 \\ I_2 \\ \cdot \\ \cdot \\ \cdot \\ I_n \end{vmatrix}. \tag{7.187}$$

7.20 Autotransformers

Figure 7.20 shows the topologies of step-down and step-up autotransformers. If both the primary and the secondary windings are made of a single winding, the transformer becomes an autotransformer. The connection point between the primary and the secondary is called the *top*. Autotransformers are an integral part of tapped-inductor PWM DC–DC power converter topologies [64]. They can be used when electric isolation is not required.

For an ideal step-down autotransformer shown in Fig. 7.20a, the ratio of the voltages is given by

$$\frac{v_1}{v_2} = \frac{i_2}{i_1} = \frac{N_1 + N_2}{N_2} = 1 + \frac{N_1}{N_2}. \tag{7.188}$$

For an ideal step-up autotransformer shown in Fig. 7.20b, the relationship is given by

$$\frac{v_2}{v_1} = \frac{i_1}{i_2} = \frac{N_1 + N_2}{N_1} = 1 + \frac{N_2}{N_1}. \tag{7.189}$$

Figure 7.21 shows an equivalent circuit of the step-down autotransformer, where L_m is the magnetizing inductance connected in parallel with N_1 turns. The reactance of L_m is given by

$$\omega L_m = \frac{v_{Lm}}{i_{Lm}}. \tag{7.190}$$

Using the relationship for an inverting transformer, we have

$$\frac{v_{Lm}}{v_2} = \frac{i_L}{i_1} = -\frac{N_1}{N_2} = -n. \tag{7.191}$$

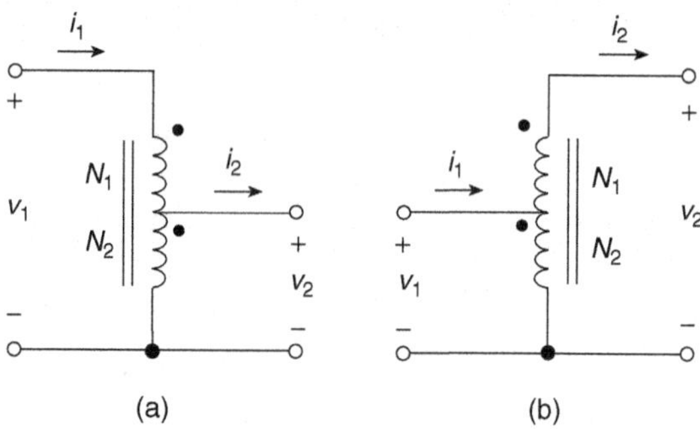

Figure 7.20 Autotransformers. (a) Step-down autotransformer. (b) Step-up autotransformer.

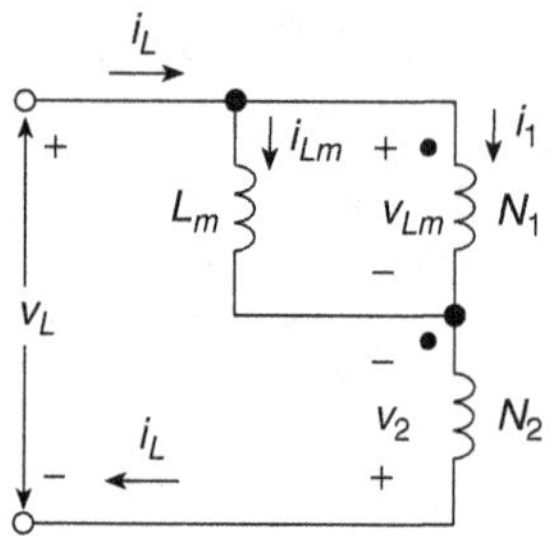

Figure 7.21 Equivalent circuit of a step-down autotransformer.

Using Kirchhoff's current law (KCL),

$$i_L = i_{Lm} + i_1 = i_{Lm} - \frac{i_L}{n}. \tag{7.192}$$

Hence,

$$i_L = \frac{i_{Lm}}{1 + \frac{1}{n}}. \tag{7.193}$$

Using Kirchhoff's voltage law (KVL),

$$v_L = v_{Lm} - v_2 = v_{Lm} + \frac{v_{Lm}}{n} = v_{Lm}\left(1 + \frac{1}{n}\right). \tag{7.194}$$

The reactance of the inductance is given by

$$\omega L = \frac{v_L}{i_L} = \frac{v_{Lm}}{i_{Lm}}\left(1 + \frac{1}{n}\right)^2 = \omega L_m\left(1 + \frac{1}{n}\right)^2. \tag{7.195}$$

Thus, the magnetizing inductance is expressed by

$$L_m = \frac{L}{\left(1 + \frac{1}{n}\right)^2} = \frac{L}{\left(1 + \frac{N_2}{N_1}\right)^2} = \left(\frac{N_1}{N_1 + N_2}\right)^2 L. \tag{7.196}$$

7.21 Measurements of Transformer Inductances

The reactances of a transformer can be determined from open-circuit and short-circuit measurements. In an open-circuit test, the secondary winding is open so that $i_2 = 0$, and an AC voltage is applied to the primary winding. Neglecting the stray capacitances C_p and C_s, the input impedance is

$$Z_i = R_p + j\omega L_{lp} + R_c || j\omega L_m. \tag{7.197}$$

Neglecting R_p, L_{lp}, C_p, and C_s,

$$Y_i = \frac{1}{R_c} + \frac{1}{j\omega L_m}. \tag{7.198}$$

Neglecting R_p, R_c, C_p, and C_s, the measurements at the primary terminals give the primary self-inductance

$$L_1 = L_{l1} + L_m. \tag{7.199}$$

Similarly, when the primary winding is open and AC voltage is applied to the secondary winding, the self-inductance can be measured

$$L_2 = L_{l2} + L_{ms}. \tag{7.200}$$

In a short-circuit test, the secondary winding terminals are short-circuited and a low AC voltage is applied to the primary winding. Under these conditions, the current flowing in the magnetizing inductance L_m is usually negligible. Neglecting R_p and R_c, the measurements at the primary winding terminals give the total leakage inductance on the primary side

$$L_l = L_{l1} + n^2 L_{l2}. \tag{7.201}$$

Figure 7.22 shows two connections of the transformer windings: in series and in inverse series. The mutual inductance can be measured as follows. When the two windings are connected in the series-aiding mode so that the total magnetic flux is equal to the sum of the fluxes in both windings, the inductance is

$$L_a = L_1 + L_2 + 2M. \tag{7.202}$$

and when the two windings are connected in the series-opposing mode so that the fluxes in the two winding oppose to each other, the inductance is

$$L_b = L_1 + L_2 - 2M. \tag{7.203}$$

Hence, the mutual inductance can be calculated as

$$M = \frac{L_a - L_b}{4}. \tag{7.204}$$

The leakage inductances can be determined as follows. The magnetizing inductance on the primary side is

$$L_m = \frac{N_1}{N_2} M \tag{7.205}$$

and the magnetizing inductance on the secondary side is

$$L_{ms} = \frac{N_2}{N_1} M. \tag{7.206}$$

Hence, the leakage inductances are

$$L_{lp} = L_1 - L_m \tag{7.207}$$

and

$$L_{ls} = L_2 - L_{ms}. \tag{7.208}$$

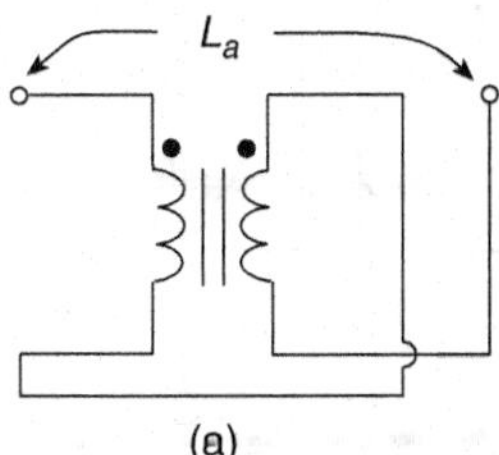

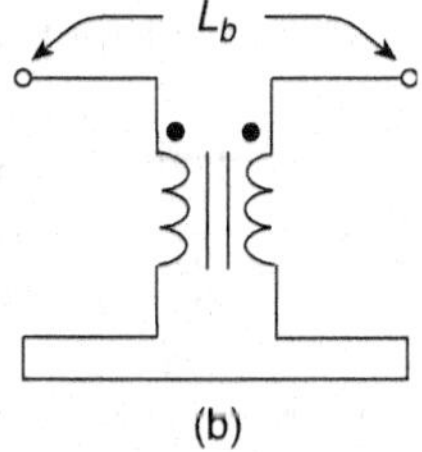

Figure 7.22 Measurement of transformer parameters. (a) Windings connected in series-aiding mode. (b) Windings connected in series-opposing mode.

7.22 Noninterleaved Windings

Figure 7.23 depicts a pattern of a noninterleaved (conventional) transformer windings in which the layers of the primary winding form one group and the layers of the secondary winding form another group. The current in all layers of the primary winding flows in the same direction and the current in the secondary winding flows in the same direction. However, the currents in both windings may flow in the opposite direction (as shown in the figure) or in the same direction. The arrangement of the transformer windings has a significant impact on the winding losses caused by the proximity effect.

Figure 7.24 shows a two-winding transformer with noninterleaved windings. Assume that the number of turns in the primary N_p and the number of turns in the secondary N_s are the same so that the rms current through the primary I_p is the same as the rms current through the secondary I_s. In general, the number of turns in each layer of the primary n_{lp} and the number of turns in each layer of the secondary n_{ls} are the same. In our case, each layer has one turn so that $n_{lp} = n_{ls} = 1$. Assume that the currents through the primary and secondary windings are in phase so that the directions of the primary and secondary currents are as shown in Fig. 7.24. Also, assume that $\mu_{rc} \gg 1$ for the cores so that the reluctance of the core $\mathcal{R}_c$ is much lower than the reluctance of the areas between the layers $\mathcal{R}_g$. Therefore, the MMF induced by the leakage flux in the core $\mathcal{F}_c$ is much lower than that induced in the air areas between the layers $\mathcal{F}_g$. According to Ampère's law, the MMF due to the current enclosed by a loop is

$$\mathcal{F} = \mathcal{F}_g + \mathcal{F}_c \approx \mathcal{F}_g = Ni = H_c l_c + H_g l_g \approx H_g l_g, \tag{7.209}$$

where l_g is the winding width. In other words,

$$\mathcal{F} \approx \mathcal{F}_g = Ni \approx H_g l_g, \tag{7.210}$$

producing

$$H_g = \frac{Ni}{l_g}. \tag{7.211}$$

The current enclosed by the leakage flux in the first layer of the primary winding is given by

$$I_1 = n_{lp} I_p = I_p, \tag{7.212}$$

resulting in

$$\mathcal{F}_1 = I_1 = n_{lp} I_p = I_p. \tag{7.213}$$

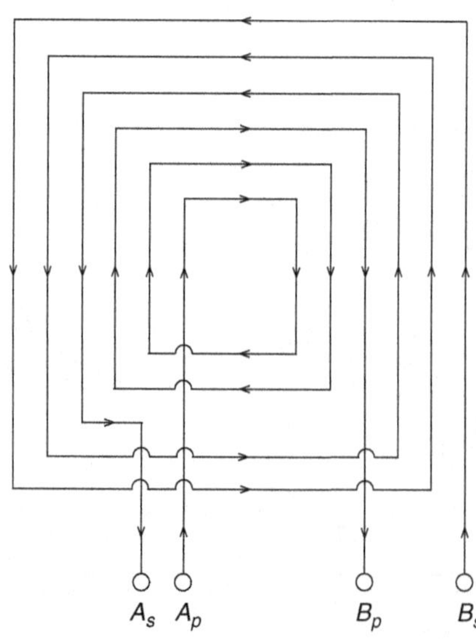

Figure 7.23 Pattern of noninterleaved winding of a transformer.

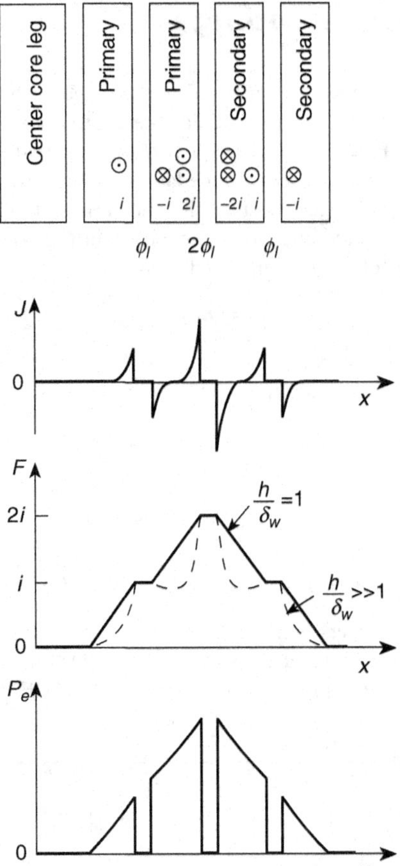

Figure 7.24 Current density J, MMF $\mathcal{F}$, and eddy-current loss density P_e distribution for a transformer with noninterleaved windings.

Likewise, the current enclosed by the leakage flux in the first and in second layers of the primary winding is

$$I_2 = 2n_{lp}I_p = 2I_p, \tag{7.214}$$

yielding

$$\mathcal{F}_2 = I_2 = 2n_{lp}I_p = 2I_p. \tag{7.215}$$

The current enclosed by the leakage flux in the first and second layers of the primary winding and the second layer of the secondary is

$$I_3 = 2n_{lp}I_p - n_{ls}I_s = 2I_p - I_s = I_p, \tag{7.216}$$

producing

$$\mathcal{F}_3 = I_3 = 2n_{lp}I_p - n_{ls}I_s = I_p. \tag{7.217}$$

Finally, the current enclosed by both the layers of the primary and both the layers of the secondary is

$$I_4 = 2n_{lp}I_p - 2n_{ls}I_s = 2I_p - 2I_s = 0, \tag{7.218}$$

resulting in

$$\mathcal{F}_4 = I_4 = 2n_{lp}I_p - 2n_{ls}I_s = 2I_p - 2I_s = 0. \tag{7.219}$$

Figure 7.24 shows the distribution of current density J, MMF $\mathcal{F}$, and eddy-current loss density P_e (W/cm^2) for a transformer with noninterleaved foil windings.

7.23 Interleaved Windings

Figure 7.25 shows a transformer with interleaved primary and secondary windings. An interleaved transformer windings are depicted in Fig. 7.26 for the case, where currents in the adjacent layers flow in opposite directions. Figure 7.27 shows a two-winding transformer with interleaved layers. Interleaving reduces the field gradient between the layers and causes reduction in leakage inductance. All the assumptions remain unchanged. Assume that the currents through the primary and secondary windings are in phase and have the directions as depicted in Fig. 7.27. The current enclosed by the leakage flux of the first layer of the primary winding is given by

$$I_1 = n_{lp} I_p = I_p, \tag{7.220}$$

producing

$$\mathcal{F}_1 = I_1 = n_{lp} I_p = I_p. \tag{7.221}$$

The current enclosed by the leakage flux of the first layer of the primary windings and the second layer of the secondary winding is

$$I_2 = n_{lp} I_p - n_{ls} I_s = I_p - I_s = 0, \tag{7.222}$$

producing

$$\mathcal{F}_2 = I_2 = n_{lp} I_p - n_{ls} I_s = I_p - I_s = 0. \tag{7.223}$$

The behavior of the next two layers is the same as the first two. Figure 7.27 shows the distribution of current density J and MMF $\mathcal{F}$ for a transformer with interleaved windings. It can be seen that the leakage flux ϕ_l, the magnitudes of the current density J, MMF $\mathcal{F}$, and eddy-current power loss

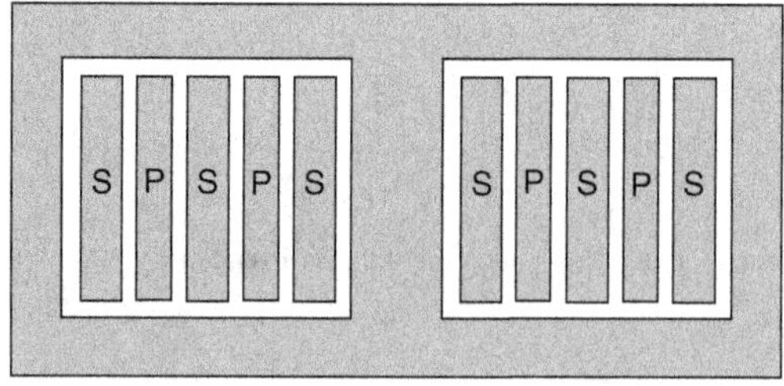

Figure 7.25 Transformer with interleaved primary and secondary windings.

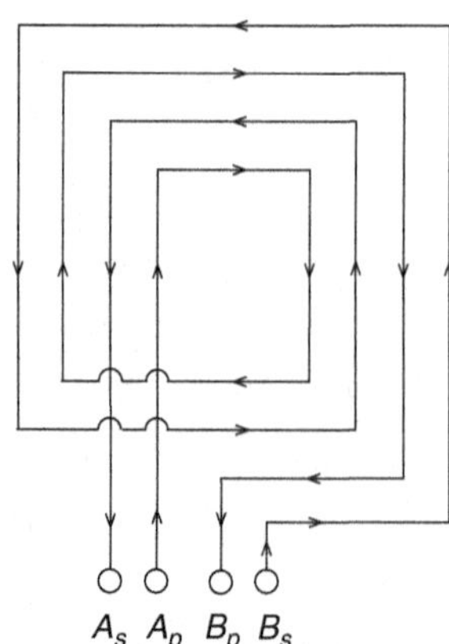

Figure 7.26 Currents in adjacent layers of transformer with interleaved windings flow in opposite directions causing proximity effect.

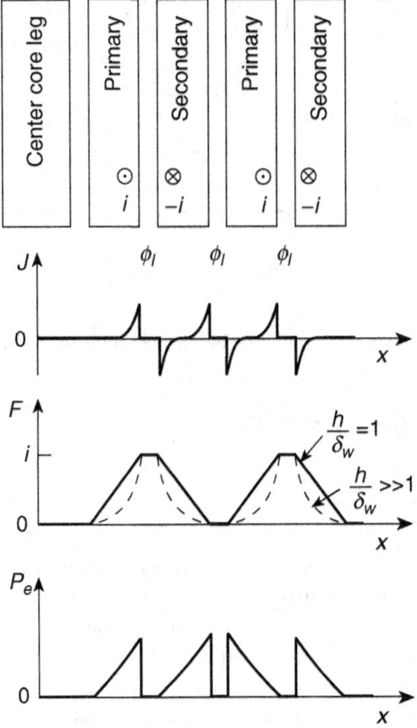

Figure 7.27 Current density J, MMF $\mathcal{F}$, and the eddy-current loss density P_e distribution for a transformer with interleaved windings.

density P_e are the same for all layers. Therefore, each layer operates like a single-layer winding and the proximity effect is virtually eliminated. A disadvantage of interleaving the transformer windings is increased primary-to-secondary capacitance.

Figure 7.28 shows a SPPS configuration of the interleaved transformer winding with turns ratio 2 : 1, in which two secondary turns can be connected in parallel or in series. Figure 7.29 depicts a PSSPPSSP configuration of the interleaved transformer winding with turns ratio 4 : 1, in which two secondary turns can be connected in parallel or in series. In both these transformers, the proximity effect is eliminated. The only high-frequency effect is the skin effect.

If the magnetizing inductance is large enough so that the magnetizing current is negligibly small, the optimum thickness of the foil is $h_{opt1} = \pi/2$ for the first and last layer and $h_{optn} = \pi$ for all other layers. If the magnetizing inductance is small so that the magnetizing is large, interleaving does not help to reduce the wining resistance. The optimum thickness of the foil is the same as that for inductors.

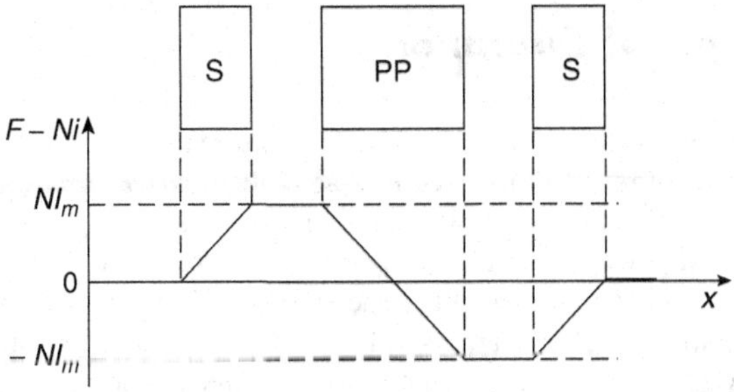

Figure 7.28 SPPS configuration of interleaved transformer winding and its MMF $\mathcal{F}$ distribution.

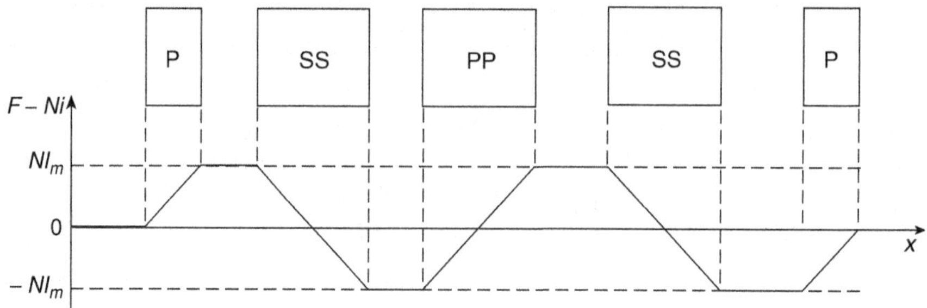

Figure 7.29 PSSPPSSP configuration of interleaved transformer and its MMF $\mathcal{F}$ distribution.

7.24 Wireless Energy Transfer

Wireless energy transfer using magnetically coupled inductors has a wide range of applications, such as contactless battery chargers, implementable biomedical devices, cordless electric toothbrushes, and mobile telephones [57–70]. This method of energy transfer is also preferred in hazardous applications, such as mining due to elimination of sparks and underwater due to elimination of electrical shocks. A wireless energy transfer system consists of a primary transmitting inductor L_1 and a secondary receiving inductor L_2 to form a special transformer with a large separation between the primary and secondary windings. The inductors L_1 and L_2 are magnetically coupled via mutual inductance M. Therefore, one of the most important parameters of a wireless energy transfer system is the mutual inductance $M = k\sqrt{L_1 L_2}$ between both inductors because it determines the power link performance. The energy is transferred inductively through air. The mutual inductance M depends on the magnetic field shared by the two inductors. The shared magnetic field depends on the geometry of the inductors and is sensitive to changes in axial and lateral displacements. The transformer with a large separation of windings usually has a low magnetizing inductance and large leakage inductances. The inductors are usually discrete in high-power applications such as electric vehicle battery chargers or integrated in low-power applications such as biomedical devices and portable consumer electronics. The source of the energy transfer system can be either DC or AC voltage or current. The electric current at the transformer secondary is converted to either DC or AC voltage. A resonant inverter is usually used on the primary winding site and a regulated rectifier is used on the secondary winding site. This permits bidirectional power flow through the transformer. Therefore, the magnetic energy stored in the leakage inductances can be transferred either to the load or to the input source depending on the load requirements. The CLL resonant inverter is often used because the transformer-magnetizing inductance and the transformer leakage inductances are absorbed into the inverter topology [28–66].

7.25 AC Current Transformers

7.25.1 Principle of Operation

Magnetic devices used in instrumentation for observing current waveforms, measuring current peak values, or detecting overcurrent and undercurrent are called *current transformers* (CTs) or *current probes* (CPs). CTs are also used in control circuits to sense a current, for example, in current-mode control of DC–DC PWM power converters. They are used in testing semiconductor power devices, in control and protection of electric motors and generators, and in monitoring the distribution network of electric power. Figure 7.30 shows an AC CP. AC CPs are based on the principle of magnetic coupling. A typical AC CT consists of toroidal magnetic core, wounding, noninductive sense resistor R, and voltmeter. A copper wire of N turns is wound on a toroidal core to form a secondary winding. This winding is terminated with a low-inductance sense resistor R, called a *burden resistor*.

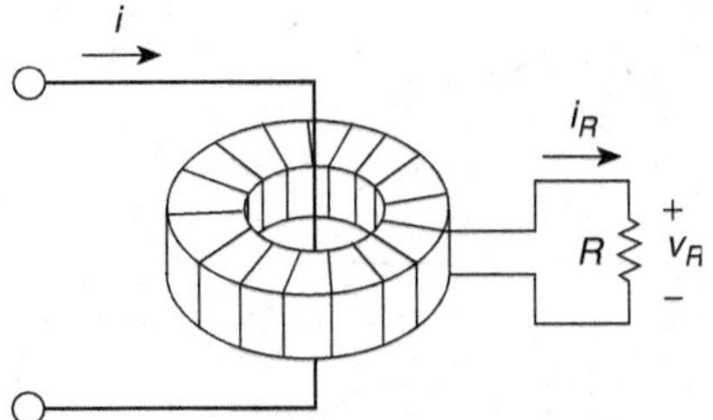

Figure 7.30 Current transformer.

Thick-film chip resistors are made by depositing a thick resistive film in such a way to minimize the parasitic inductance. Coaxial structures are used to make sense resistors. However, the skin effect at high frequencies increases the resistance. The conductor carrying the measured current i acts as a primary for the CT. The winding wound on the toroidal core acts as a secondary for the current transformer. A toroidal core is used to minimize the leakage inductance, and thus increases the accuracy and the upper corner frequency. The toroidal inductor can be clamped around the conductor, which carries the measured current i. The probe is normally connected through a 50-Ω coaxial cable to a high-impedance oscilloscope to observe the measured current waveform. The measured AC current i flowing through the primary produces a time-varying magnetic field. The magnetic toroidal core contains the majority of the magnetic field within it. The magnetic field passes through the secondary winding and, according to the Faraday's law, induces a voltage v_R across the terminals of the secondary winding, which the same as that across the sense resistor v_R. The measured current waveform is given by

$$i = N i_R = \frac{N}{R} v_R. \tag{7.224}$$

Thus, the waveforms of the measured current i and the voltage across the sense resistor v_R have the same shape. The CP can be calibrated by passing a known current through the primary and measuring the voltage across the sense resistor R. The sensitivity of the CP is given by

$$S_p = \frac{v_R}{i} = \frac{R}{N} \quad \left(\frac{\text{V}}{\text{A}}\right). \tag{7.225}$$

For example, if $N = 100$ and $R = 50\ \Omega$, then $S_p = R/N = 50/100 = 0.5$ V/A. Thus, the current flowing through the sense resistor R is much lower than the measured current i.

7.25.2 Model of Current Transformer

Figure 7.31a shows a circuit of a physical current transformer. A model of the CP is shown in Fig. 7.31b, where L_m is the magnetizing inductance reflected to the secondary side of the transformer, R_c is the core parallel equivalent resistance reflected to the secondary side, L_l is the leakage inductance of the secondary winding, R_s is the series resistance of the secondary winding, C is the stray capacitance, and L_R is the inductance of the sense resistor. The turns ratio of the CT is defined as

$$n = \frac{N_s}{N_p}. \tag{7.226}$$

In most cases, $N_p = 1$ and $n = N_s = N$. The magnetizing inductance on the secondary side is given by

$$L_m = \frac{\mu_{rc}\mu_0 N^2 A_c}{l_c}, \tag{7.227}$$

where A_c is the core cross-sectional area and l_c is the core mean length. The model is valid for a wide frequency range. Figure 7.31c shows a model of the CT in which the current source is reflected to secondary side.

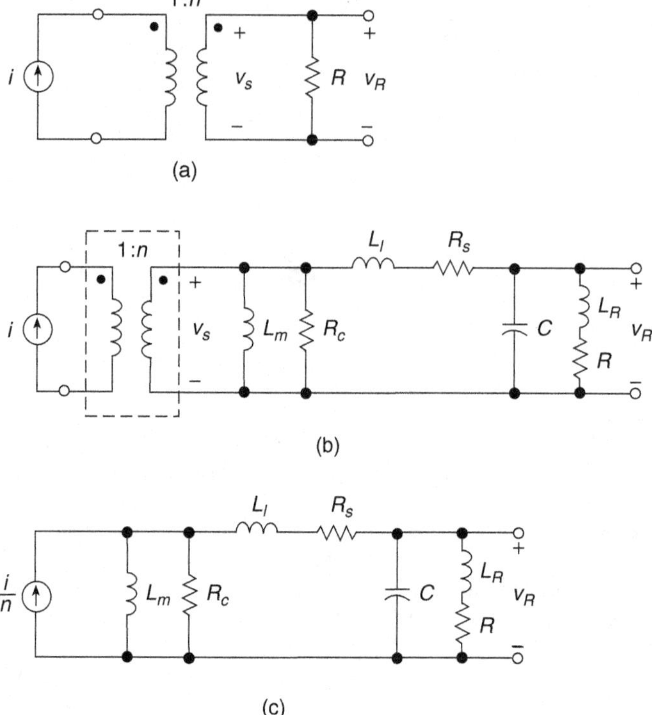

Figure 7.31 Model of a current transformer. (a) Circuit of physical current transformer. (b) Current transformer with parasitic components. (c) Transformer model with current source reflected to the secondary side.

7.25.3 Low-Frequency Response

In order to determine the bandwidth of the CT, low-frequency and high-frequency equivalent circuits are analyzed separately. The low-frequency equivalent circuit is shown in Fig. 7.32a. It is derived from the general CP model shown in Fig. 7.31c by neglecting the leakage inductance L_l and the sense resistor inductance L_R as they present a very low reactances at low frequencies. The current flowing through the sense resistor R is given by

$$i_R = \frac{i}{n} - i_{Lm} - i_{Rc}. \tag{7.228}$$

The magnetizing impedance $j\omega L_m \| R_c$ determines the accuracy of the CT because it shunts part of the current i/n. The values of L_m and R_c are determined by the core material and the construction of the secondary winding. Most CTs are designed using toroidal cores to achieve the highest permeability. In general, the reactance of the magnetizing inductance L_m and the core parallel resistance R_c should be very large so that the current through the sense resistor R is nearly equal to i/n. At low frequencies, the reactance of the magnetizing inductance is comparable to the resistance $R_{Lm} = R_c \| (R + R_s) \approx R$. As a result, more and more current will flow through the magnetizing inductance L_m as the frequency decreases.

The amplitude of the current through the secondary winding and the sense resistor R is

$$I_{sm} = \frac{I_m}{N_s}. \tag{7.229}$$

The amplitude of the voltage across the secondary winding is

$$V_{sm} = I_{sm}(R + R_s). \tag{7.230}$$

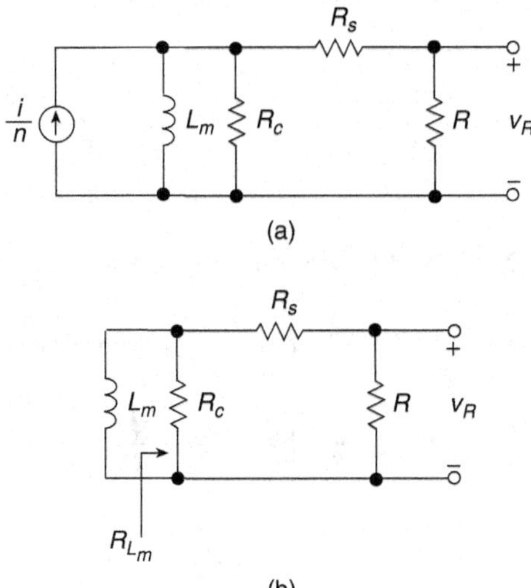

Figure 7.32 Model of a current transformer for low frequencies. (a) Low-frequency model. (b) Low-frequency dead circuit.

To avoid core saturation, the minimum cross-sectional area of the core for sinusoidal current and voltage waveforms is given by

$$A_{c(min)} = \frac{V_{sm(max)}}{\omega_{min} N_s B_s} = \frac{I_{sm(max)}(R + R_s)}{\omega_{min} N_s B_s} = \frac{I_{m(max)}(R + R_s)}{\omega_{min} N_s^2 B_s}, \tag{7.231}$$

where $I_{m(max)}$ is the maximum amplitude of the input current i. The maximum amplitude of the input sinusoidal current is

$$I_{m(max)} = \frac{\omega_{min} A_c B_s N_s^2}{R + R_s}. \tag{7.232}$$

Transimpedance is defined as the input current-to-output voltage transfer function of the probe for low frequencies is given by

$$Z_m(s) = \frac{v_R(s)}{i(s)} = \frac{R}{N} \frac{R_c}{R_c + R + R_s} \frac{s}{s + \omega_L}, \tag{7.233}$$

where

$$\omega_L = \frac{R_c \| (R + R_s)}{L_m}. \tag{7.234}$$

Therefore, the lower 3-dB cutoff frequency is

$$f_L = \frac{R_c \| (R + R_s)}{2\pi L_m} \approx \frac{R}{2\pi L_m}. \tag{7.235}$$

The cutoff frequency f_L can be determined form the dead equivalent circuit shown in Fig. 7.32b. To achieve a low value of the cutoff frequency f_L, the core should be made of a material with high relative permeability μ_{rc} of the order 10^5, such as Permalloy, FeCo amorphous ribbon, nanocrystalline cores, or molybdenum (Mo) alloys. The number of turns N should be large. The resistance R should be low.

7.25.4 High-Frequency Response

A high-frequency model for the CT is shown in Fig. 7.33. The leakage inductance L_l and the stray capacitance are included in the model, whereas L_m and R_s can be neglected at high frequencies. The series inductance of the sense resistor L_R is also neglected because noninductive resistors should be

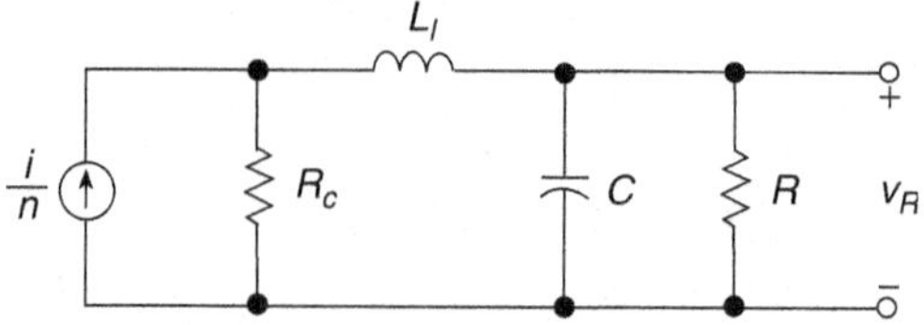

Figure 7.33 Model of a current transformer for high frequencies.

used in CTs. The transimpedance of the CT for high frequencies is given by

$$Z_m(s) = \frac{v_R(s)}{i(s)} = \frac{R_c}{NCL_l} \frac{1}{s^2 + \left(\frac{R_c}{L_l} + \frac{1}{RC}\right)s + \frac{1}{CL_l}\left(\frac{R_c}{R} + 1\right)} = \frac{R_c}{NCL_l} \frac{1}{s^2 + 2\zeta\omega_0 s + \omega_0^2}, \tag{7.236}$$

where

$$\omega_0 = \sqrt{\frac{1}{CL_l}\left(\frac{R_c}{R} + 1\right)} \tag{7.237}$$

and

$$\zeta = \frac{\frac{R_c}{L_l} + \frac{1}{RC}}{2\sqrt{\frac{1}{CL_l}\left(\frac{R_c}{R} + 1\right)}}. \tag{7.238}$$

For $\zeta < 1$, the poles of $Z_m(s)$ are complex imaginary conjugate and the upper 3-dB frequency is $f_H = kf_0$, where k depends on ζ. For $\zeta \geq 1$, the poles of $Z_m(s)$ are real and the denominator of $Z_m(s)$ can be represented as

$$s^2 + \left(\frac{R_c}{L_l} + \frac{1}{RC}\right)s + \frac{1}{CL_l}\left(\frac{R_c}{R} + 1\right) = (s + \omega_{p1})(s + \omega_{p2}) = s^2 + (\omega_{p1} + \omega_{p2})s + \omega_{p1}\omega_{p2}. \tag{7.239}$$

Hence,

$$\omega_{p1} + \omega_{p2} = \frac{R_c}{L_l} + \frac{1}{RC} \tag{7.240}$$

and

$$\omega_{p1}\omega_{p2} = \frac{R_c}{RCL_l} + \frac{1}{CL_l} \approx \frac{R_c}{RCL_l} \quad \text{for} \quad \frac{R_c}{R} \gg 1. \tag{7.241}$$

This inequality is well satisfied in practice. The frequencies of the real poles are

$$f_{p1} = \frac{\omega_{p1}}{2\pi} = \frac{1}{2\pi RC} \tag{7.242}$$

and

$$f_{p2} = \frac{\omega_{p2}}{2\pi} = \frac{R_c}{2\pi L_l}. \tag{7.243}$$

If $f_{p2} \geq 4f_{p1}$, the pole p_1 is dominant and

$$f_H \approx f_{p1} = \frac{1}{2\pi RC}. \tag{7.244}$$

Conversely, if $f_{p1} \geq 4f_{p2}$, the pole p_2 is dominant and

$$f_H \approx f_{p2} = \frac{R_c}{2\pi L_l}. \tag{7.245}$$

If the pole frequencies f_{p1} and f_{p2} are comparable, the upper 3-dB frequency of the CT is given by

$$f_H \approx \frac{1}{\sqrt{\frac{1}{f_{p1}^2} + \frac{1}{f_{p2}^2}}} = \frac{1}{2\pi\sqrt{(RC)^2 + \left(\frac{L_l}{R_c}\right)^2}}. \tag{7.246}$$

To achieve a high value of f_H, the stray capacitance C and the leakage inductance L_l should be reduced. The bandwidth of the CT is given by

$$BW = f_H - f_L = \frac{1}{2\pi\sqrt{(RC)^2 + \left(\frac{L_l}{R_c}\right)^2}} - \frac{R_{Lm}}{2\pi L_m}. \tag{7.247}$$

Example 7.3

Design a high-frequency CT to meet the following specifications: $f_L < 5\,\text{kHz}$ and $f_H > 150\,\text{MHz}$.

Solution: Let us select a ferrite toroidal core 528T500-4C4 manufactured by Ferroxcube. The parameters of the core are: external diameter $D = 39.9\,\text{mm}$, internal diameter $d = 18.6\,\text{mm}$, height $h = 14.7\,\text{mm}$, toroidal mean length $l_c = 8.49\,\text{cm}$, the cross-sectional area $A_c = 1.17\,\text{cm}^2$, $B_s = 0.3\,\text{T}$, $\mu_{rc} = 125$, and inductance per 1000 turns $A_L = 218\,\text{mH/1000 turns}$. We will use a noninductive resistor $R = 10.5\,\Omega$ and the secondary winding with the number of turns $N = 40$ made up of a solid copper wire AWG22 whose diameter without insulation is $d = 0.644\,\text{mm}$.

The sensitivity of the probe is

$$S_p = \frac{v_R}{i} = \frac{R}{N} = \frac{10.5}{40} = 0.2625\,\text{V/A}. \tag{7.248}$$

The magnetizing inductance is

$$L_m = \frac{\mu_{rc}\mu_0 N^2 A_c}{l_c} = \frac{N^2 A_L}{10^6}\,(\text{mH}) = \frac{40^2 \times 218}{10^6} = 0.349\,\text{mH}. \tag{7.249}$$

The resistivity of the ferrite material is $\rho_c = 1000\,\Omega\text{cm}$. Hence, the DC resistance of the core is

$$R_c = \rho_c \frac{l_c}{A_c} = 1000 \times \frac{8.49}{1.17} = 7.256\,\text{k}\Omega. \tag{7.250}$$

The length of a single turn is

$$l_T = 2\left(h + \frac{D-d}{2}\right) = 2\left(14.7 + \frac{39.9 - 18.6}{2}\right) = 5.07\,\text{cm}. \tag{7.251}$$

The length of the secondary winding wire is

$$l_w = N l_T = 40 \times 5.07 = 2.028\,\text{m}. \tag{7.252}$$

The DC and low-frequency series resistance of the secondary winding at $T = 20\,^\circ\text{C}$ is

$$R_s = \rho_{Cu}\frac{l_w}{A_w} = \rho_{Cu}\frac{l_w}{\pi r^2} = 1.724 \times 10^{-8} \times \frac{2.028}{\pi \times (0.322 \times 10^{-3})^2} = 0.1073\,\Omega. \tag{7.253}$$

The total resistance seen by the magnetizing inductance L_m at low frequencies is

$$R_{Lm} = \frac{R_c(R + R_s)}{R_c + R + R_s} = \frac{7.256 \times 10^3 \times (10.5 + 0.1073)}{7.256 \times 10^3 + 10.5 + 0.1073} = 10.59\,\Omega. \tag{7.254}$$

The lower 3-dB cutoff frequency is

$$f_L = \frac{R_{Lm}}{2\pi L_m} = \frac{10.59}{2\pi \times 0.349 \times 10^{-3}} = 4.829\,\text{kHz}. \tag{7.255}$$

The maximum amplitude of the input current i at which the core at the boundary between saturation and nonsaturation regions given by is

$$I_{m(max)} = \frac{\omega_{min} N_s^2 A_c B_s}{R + R_s} = \frac{2\pi \times 4.829 \times 10^3 \times 40^2 \times 1.17 \times 10^{-4} \times 0.3}{10.5 + 0.1073}$$
$$= 160.64\,\text{A}. \tag{7.256}$$

The cross-sectional area of the bare wire of the secondary winding is

$$A_{ws} = \pi r^2 = \pi (0.322 \times 10^{-3})^2 = 0.3257 \text{ mm}^2. \tag{7.257}$$

Assuming the maximum DC current density in the secondary winding $J_{m(max)} = 5$ A/mm^2, the maximum amplitude of the current through the secondary is

$$I_{sm(max)} = A_{ws} J_{m(max)} = 0.3257 \times 5 = 1.6285 \text{ A}. \tag{7.258}$$

Hence, the maximum amplitude of the input current is

$$I_{m(max)} = N_s I_{ms(max)} = 40 \times 1.6285 = 65.14 \text{ A}. \tag{7.259}$$

Thus, the range of the amplitude of the input current is from 0 to 65.14 A. The limitation of the maximum amplitude of the input current is stronger by the maximum current density of the secondary winding than that by the core saturation.

The measured leakage inductance is $L_l = 126$ nH and the stray capacitance is $C = 81$ pF. The frequencies of the high-frequency poles are

$$f_{p1} = \frac{1}{2\pi RC} = \frac{1}{2\pi \times 10.59 \times 81 \times 10^{-12}} = 185.5 \text{ MHz} \tag{7.260}$$

and

$$f_{p2} = \frac{R_c}{2\pi L_l} = \frac{7.256 \times 10^3}{2\pi \times 126 \times 10^{-9}} = 9.165 \text{ GHz}. \tag{7.261}$$

Hence, the upper 3-dB frequency is

$$f_H = \frac{1}{\sqrt{\frac{1}{f_{p1}^2} + \frac{1}{f_{p2}^2}}} = \frac{1}{\sqrt{\frac{1}{(185.5\times 10^6)^2} + \frac{1}{(9.165\times 10^9)^2}}} = 185.465 \text{ MHz}. \tag{7.262}$$

The pole p_1 is dominant, and therefore $f_H \approx f_{p1}$. The bandwidth of the CT is

$$BW = f_H - f_L = 185.465 \times 10^6 - 4.829 \times 10^3 = 185.46 \text{ MHz}. \tag{7.263}$$

7.25.5 Maximum Power Transfer by Current Transformer

Wireless and contactless energy transfer is a very attractive technology. Figure 7.34 shows a CT loaded by a resistor R. A simple low-frequency model of the CT is depicted in Fig. 7.35. It consists of an ideal transformer and the secondary magnetizing inductance L_{ms}. The rms value of the current in the secondary of the ideal transformer is

$$I_s = \frac{I_p}{N}, \tag{7.264}$$

where N is the number of turns of the secondary winding. The rms value of the current through the load resistance R is

$$I_R(s) = \frac{sL_{ms}}{R + sL_{ms}} I_s = \frac{I_p}{N\left(1 + \frac{R}{sL_{ms}}\right)} = \frac{I_p}{N\left(1 + \frac{\omega_L}{s}\right)}, \tag{7.265}$$

where the lower 3-dB frequency is

$$\omega_L = \frac{R}{L_{ms}}. \tag{7.266}$$

Setting $s = j\omega$, we get

$$I_R(j\omega) = \frac{I_p}{N\left(1 + \frac{\omega_L}{j\omega}\right)} = \frac{I_p}{N} \frac{1}{\left(1 - j\frac{\omega_L}{\omega}\right)}. \tag{7.267}$$

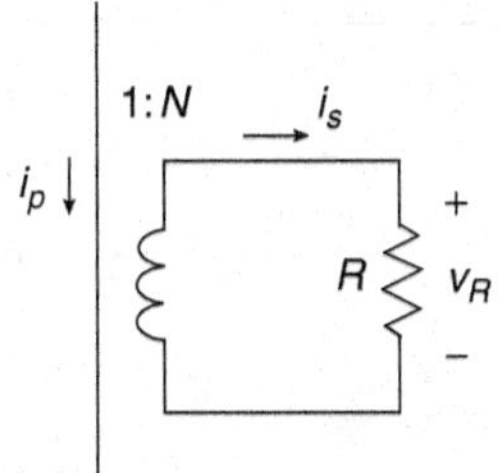

Figure 7.34 Current transformer.

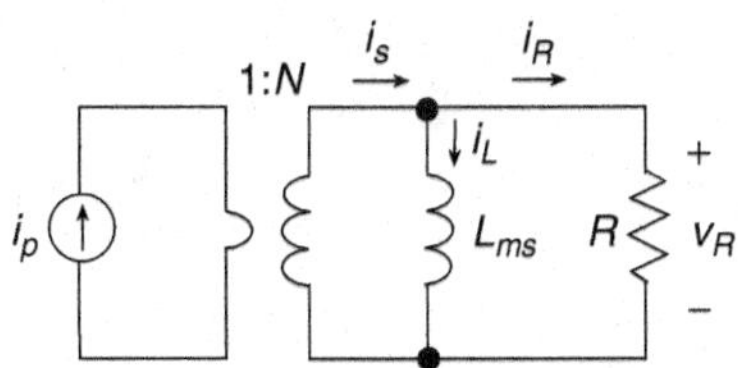

Figure 7.35 Simple low-frequency model of current transformer.

Hence, the magnitude of the current through the load resistance R is

$$|I_R| = \frac{I_p}{N}\frac{1}{\sqrt{1+\left(\frac{\omega_L}{\omega}\right)^2}}. \tag{7.268}$$

The output power is

$$P_O = R|I_R|^2 = \frac{RI_p^2}{N^2}\frac{1}{1+\left(\frac{\omega_L}{\omega}\right)^2}. \tag{7.269}$$

The magnetizing inductance of the CT on the secondary side, with a high-permeability core ($\mu_r \gg 1$) and an air gap of length l_g, is given by

$$L_{ms} = \frac{\mu_0 A_c N^2}{l_g}. \tag{7.270}$$

The number of turns is

$$N = \sqrt{\frac{l_g L_{ms}}{\mu_0 A_c}}. \tag{7.271}$$

Hence, the output power is

$$P_O = \frac{\mu_0 A_c R I_p^2}{l_g L_{ms}}\frac{1}{1+\left(\frac{\omega_L}{\omega}\right)^2} = \frac{\mu_0 A_c I_p^2}{l_g}\frac{\omega_L}{1+\left(\frac{\omega_L}{\omega}\right)^2} = \frac{\mu_0 A_c I_p^2 \omega}{l_g}\frac{\omega_L/\omega}{1+\left(\frac{\omega_L}{\omega}\right)^2}. \tag{7.272}$$

To find the maximum output power at a fixed frequency ω, we take the derivative of P_O with respect to ω_L/ω and set the result to zero

$$\frac{dP_O}{d(\omega_L/\omega)} = 0 \tag{7.273}$$

to obtain

$$\frac{\omega_L}{\omega} = 1. \tag{7.274}$$

Hence, the maximum output power is

$$P_{Omax} = \frac{\mu_0 A_c I_p^2 \omega}{2l_g}. \tag{7.275}$$

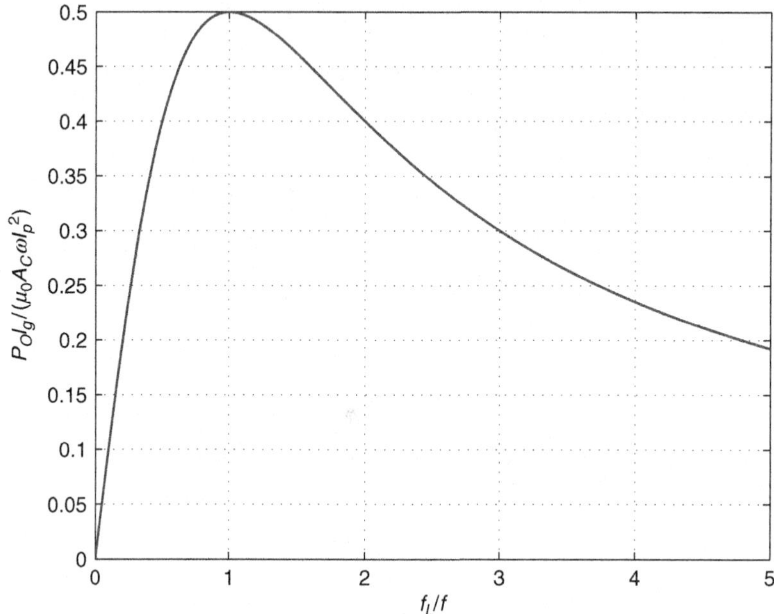

Figure 7.36 Normalized output $P_O l_g/(\mu_0 A_c \omega I_p^2)$ power as a function of the normalized corner frequency f_L/f.

Figure 7.36 shows a plot of normalized output $P_O l_g/(\mu_0 A_c \omega I_p^2)$ power as a function of the normalized corner frequency f_L/f.

For $f_L \ll f$, the magnitude of the current through the load resistance R is

$$I_R \approx \frac{I_p}{N} \tag{7.276}$$

and the output power is

$$P_O = RI_R^2 \approx \frac{RI_p^2}{N^2}. \tag{7.277}$$

The rms value of the sinusoidal voltage across the load resistance R is

$$V_R = RI_R = \frac{R}{N} I_p. \tag{7.278}$$

7.26 Saturable Reactors

A saturable reactor is a variable inductor, in which the magnetic core is deliberately saturated by a DC current. When the magnetic core saturates, the relative permeability μ_{rc} decreases, the inductance decreases, and the inductor reactance decreases. A saturable reactor is usually made of two coupled inductors. One inductor conducts only a DC current and the other inductor conducts only the AC current. Saturable reactors may be used for electronic tuning of resonant circuits by varying their inductance or to control the amplitude of the AC current.

The flux linkage for nonlinear inductors is

$$\lambda(i) = L(i)i. \tag{7.279}$$

Hence, the voltage across the inductor is

$$v(t) = \frac{d\lambda(i)}{dt} = L(i)\frac{di}{dt} + i\frac{dL(i)}{dt} = L(i)\frac{di}{dt} + i\frac{dL(i)}{di}\frac{di}{dt} = \left[L(i) + i\frac{dL(i)}{di}\right]\frac{di}{dt} = L_{eq}\frac{di}{dt}, \tag{7.280}$$

where

$$L_{eq} = L(i) + i\frac{dL(i)}{di} = \frac{\mu_0 A_c N^2}{l_c}\mu_{rc}(i) + \frac{\mu_0 A_c N^2}{l_c}\frac{d\mu_{rc}}{dt} = \frac{\mu_0 A_c N^2}{l_c}\left[\mu_{rc}(i) + \frac{d\mu_{rc}}{dt}\right]. \quad (7.281)$$

7.27 Transformer Winding Power Losses with Harmonics

7.27.1 Winding Power Losses with Harmonics for CCM

In transformers used in isolated PWM DC–DC power converters in continuous conduction mode (CCM), the current waveforms of the primary and secondary windings are pulsating. Therefore, the amplitudes of the harmonics are high, causing high losses. This situation is different in nonisolated PWM converters in CCM. This is because the current waveform of the inductor is nearly constant and amplitudes of harmonics are low.

Primary Winding

Assume that the waveform of the primary winding current for CCM is a rectangular wave with a duty ratio D

$$i_p = \begin{cases} I_{pmax}, & \text{for} \quad -\frac{DT}{2} < t \le \frac{DT}{2} \\ 0, & \text{for} \quad \frac{DT}{2} < t \le \frac{3DT}{2}, \end{cases} \quad (7.282)$$

where I_{pmax} is the peak value of the primary winding current. The DC component is

$$I_{pDC} = \frac{1}{T}\int_{-DT/2}^{DT/2} i_p dt = \frac{2}{T}\int_0^{DT/2} I_{pmax} dt = DI_{pmax} = \alpha_0 I_{pmax} \quad (7.283)$$

where

$$\alpha_0 = \frac{I_{pDC}}{I_{pmax}} = D. \quad (7.284)$$

The amplitudes of the fundamental and the harmonics of the primary current waveform are

$$I_{pmn} = \frac{4}{T}\int_0^{DT/2} i_p \cos \quad n\omega t dt = \frac{4}{T}\int_0^{DT/2} I_{pmax} \cos \quad n\omega t dt$$
$$= I_{pmax}\frac{2\sin(n\pi D)}{n\pi} = I_{pDC}\frac{2\sin(n\pi D)}{n\pi D} = \alpha_n I_{pmax} \quad (7.285)$$

where

$$\alpha_n = \frac{I_{pmn}}{I_{pmax}} = \frac{2\sin(n\pi D)}{n\pi}. \quad (7.286)$$

Figure 7.37 shows the Fourier coefficients α_n as functions of D for the primary winding current waveform in CCM. The Fourier series of the current waveform through the primary winding for CCM is

$$i_p = I_{pDC} + \sum_{n=1}^{n=\infty} I_{pmn} \cos n\omega t = I_{pmax}\left(\alpha_0 + \sum_{n=1}^{n=\infty} \alpha_n \cos n\omega t\right)$$
$$= DI_{pmax}\left[1 + 2\sum_{n=1}^{\infty}\frac{\sin(n\pi D)}{n\pi D}\cos n\omega_s t\right] = I_{pDC}\left[1 + 2\sum_{n=1}^{\infty}\frac{\sin(n\pi D)}{n\pi D}\cos n\omega_s t\right]. \quad (7.287)$$

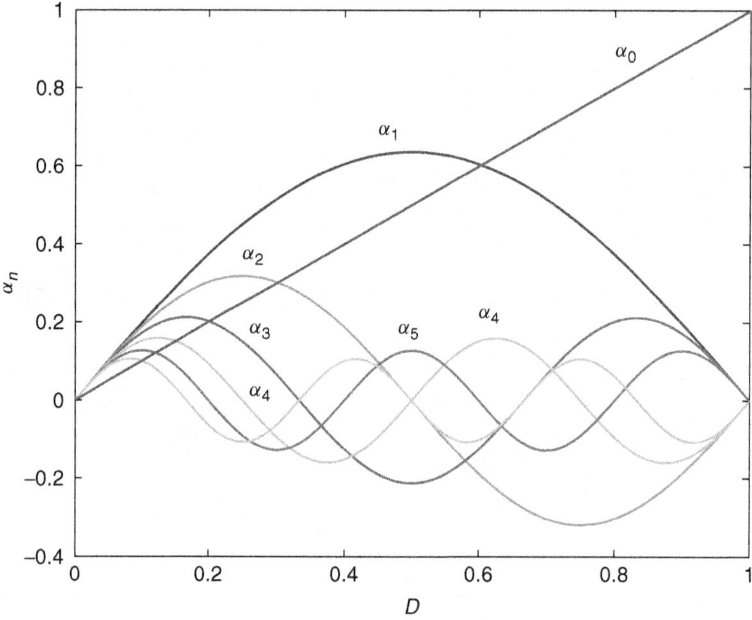

Figure 7.37 Fourier coefficients α_n as functions of D for the primary winding current in CCM.

The power loss in the primary winding for CCM is

$$\begin{aligned} P_{wp} &= R_{wpDC}I_{pDC}^2 + \frac{1}{2}R_{wp1}I_{pm1}^2 + \frac{1}{2}R_{wp2}I_{pm2}^2 + \ldots \\ &= R_{wpDC}I_{pDC}^2 + \ldots + 2R_{wpn}I_{pDC}^2\left[\frac{\sin(n\pi D)}{n\pi D}\right]^2 + \ldots \\ &= R_{wpDC}I_{pDC}^2\left[1 + \frac{1}{2}\sum_{n=1}^{\infty}\frac{R_{wpn}}{R_{wpDC}}\left(\frac{I_{pmn}}{I_{wpDC}}\right)^2\right] \\ &= R_{wpDC}I_{pDC}^2\left\{1 + 2\sum_{n=1}^{\infty}\frac{R_{wpn}}{R_{wpDC}}\left[\frac{\sin(n\pi D)}{n\pi D}\right]^2\right\} \\ &= P_{wpDC}\left\{1 + 2\sum_{n=1}^{\infty}F_{Rpn}\left[\frac{\sin(n\pi D)}{n\pi D}\right]^2\right\} = P_{wpDC}F_{Rph}. \end{aligned} \tag{7.288}$$

The AC -to-DC resistance ratio for the primary winding

$$F_{Rph} = \frac{P_{wp}}{P_{wpDC}} = 1 + \frac{1}{2}\sum_{n=1}^{\infty}\frac{R_{wpn}}{R_{wpDC}}\left(\frac{I_{pmn}}{I_{wpDC}}\right)^2 = 1 + 2\sum_{n=1}^{\infty}F_{Rpn}\left[\frac{\sin(n\pi D)}{n\pi D}\right]^2. \tag{7.289}$$

Figure 7.38 shows the plots of F_{Rph} as a function of D for $d/\delta_w = 1$ and $d/p = 0.8$.

Secondary Winding

Similarly, assume that the waveform of the secondary winding current for CCM is a rectangular wave. In our case, only the amplitudes of the fundamental component and the harmonics is important. Therefore, we may consider the current waveform described by

$$i_s = \begin{cases} 0, & \text{for} \quad -\frac{DT}{2} < t \le \frac{DT}{2} \\ I_{smax}, & \text{for} \quad \frac{DT}{2} < t \le \frac{3DT}{2}, \end{cases} \tag{7.290}$$

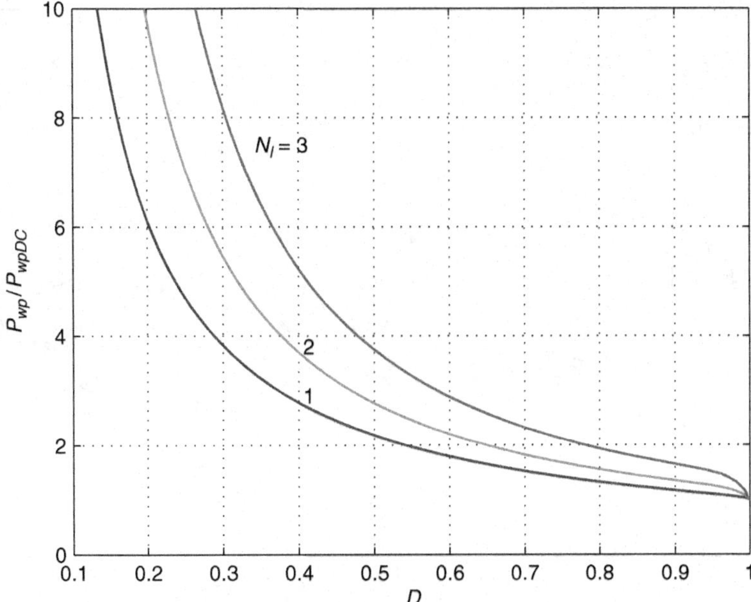

Figure 7.38 Harmonic AC-to-DC resistance ratio of the primary winding $F_{Rph} = P_{wp}/P_{wpDC}$ as a function of D for CCM at $d/\delta_w = 1$ and $d/p = 0.8$.

where I_{smax} is the peak value of the secondary winding current. The DC component of the secondary winding is

$$I_{sDC} = \frac{1}{T}\int_{DT/2}^{3DT/2} i_s dt = \frac{2}{T}\int_{DT/2}^{3DT/2} I_{smax} dt = (1-D)I_{smax} = \beta_0 I_{smax} \quad (7.291)$$

where

$$\beta_0 = \frac{I_{sDC}}{I_{smax}} = 1 - D. \quad (7.292)$$

The amplitudes of the fundamental and the harmonics are

$$I_{smn} = \frac{4}{T}\int_{DT/2}^{3DT/2} i_s \cos n\omega t dt = \frac{4}{T}\int_{DT/2}^{3DT/2} I_{smax} \cos n\omega t dt$$
$$= I_{smax}\frac{2\sin[n\pi(1-D)]}{n\pi} = I_{sDC}\frac{2\sin[n\pi(1-D)]}{n\pi} = \beta_n I_{smax} \quad (7.293)$$

where

$$\beta_n = \frac{I_{pmn}}{I_{pmax}} = -\frac{2\sin(n\pi D)}{n\pi}. \quad (7.294)$$

Figure 7.39 shows the Fourier coefficients β_n as functions of D for the secondary winding current in CCM. The Fourier series of the current waveform of the secondary winding for CCM is

$$i_s = I_{sDC} + \sum_{n=1}^{n=\infty} I_{smn} \cos n\omega t = I_{smax}\left(\beta_0 + \sum_{n=1}^{n=\infty} \beta_n \cos n\omega t\right)$$
$$= (1-D)I_{smax}\left\{1 + 2\sum_{n=1}^{\infty}\frac{\sin[n\pi(1-D)]}{n\pi(1-D)}\cos n\omega_s t\right\}$$
$$= I_{sDC}\left\{1 + 2\sum_{n=1}^{\infty}\frac{\sin[n\pi(1-D)]}{n\pi(1-D)}\cos n\omega_s t\right\}, \quad (7.295)$$

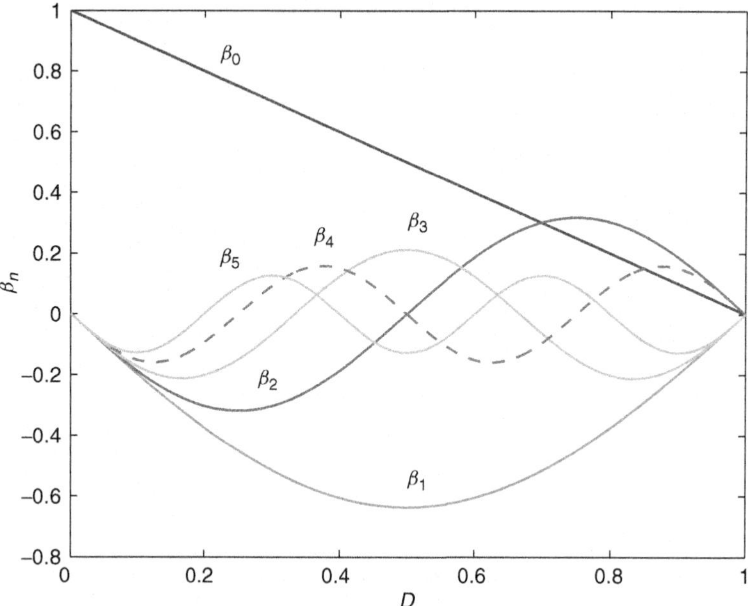

Figure 7.39 Fourier coefficients β_n as functions of D for the secondary winding current in CCM.

where I_{smax} is the peak value of the secondary winding current. The power loss in the secondary winding for CCM is

$$P_{ws} = R_{wsDC} I_{pDC}^2 \left[1 + \frac{1}{2} \sum_{n=1}^{\infty} \frac{R_{wsn}}{R_{wsDC}} \left(\frac{I_{smn}}{I_{wsDC}} \right)^2 \right]$$
$$= R_{wsDC} I_{sDC}^2 \left\{ 1 + 2 \sum_{n=1}^{\infty} F_{Rsn} \left[\frac{\sin(n\pi(1-D))}{n\pi(1-D)} \right]^2 \right\}, \quad (7.296)$$

where I_{sDC} is the DC component of the secondary winding current. The AC-to-DC resistance ratio for the secondary winding is

$$F_{Rsh} = \frac{P_{ws}}{P_{wsDC}} = 1 + \frac{1}{2} \sum_{n=1}^{\infty} \frac{R_{wsn}}{R_{wsDC}} \left(\frac{I_{smn}}{I_{wsDC}} \right)^2 = 1 + 2 \sum_{n=1}^{\infty} F_{Rpn} \left\{ \frac{\sin[n\pi(1-D)]}{n\pi(1-D)} \right\}^2. \quad (7.297)$$

Figure 7.40 shows the plots of F_{Rsh} as a function of D for $d/\delta_w = 1$ and $d/p = 0.8$.

The total power loss in a transformer consists of core loss P_C and winding losses P_w

$$P_{tloss} = P_C + P_w = P_C + P_{wp} + P_{ws}. \quad (7.298)$$

The efficiency of a two-winding transformer is given by

$$\eta_t = \frac{P_{Ot}}{P_{It}} = \frac{P_{Ot}}{P_{Ot} + P_{tloss}} = \frac{P_{Ot}}{P_{Ot} + P_C + P_w} = \frac{P_{Ot}}{P_{Ot} + P_C + P_{wp} + P_{ws}}, \quad (7.299)$$

where P_{Ot} and P_{It} are the input and output powers of the transformer, respectively.

For the flyback PWM converter, the peak value of the primary winding current I_{pmax} is approximately equal to the DC component of the magnetizing current I_{Lm} and the DC component of the primary winding current $I_{pDC} = DI_{pmax} = DI_{Lm}$ is equal to the DC component of the converter input current I_I and it is given by Kazimierczuk [64]

$$I_{pDC} = I_I = \frac{D}{n_t(1-D)} I_O, \quad (7.300)$$

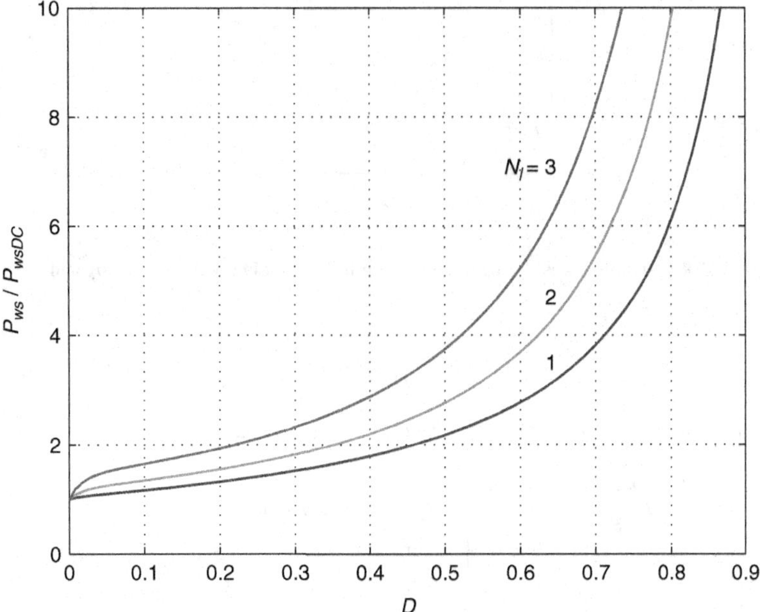

Figure 7.40 Harmonic AC-to-DC resistance ratio of the secondary winding $F_{Rsh} = P_{ws}/P_{wsDC}$ as a function of D for CCM at $d/\delta_w = 1$ and $d/p = 0.8$.

where $n_t = N_p/N_s$ is the turns ratio. Thus, the DC power loss in the primary winding is

$$P_{wpDC} = R_{wpDC} I_{pDC}^2 = R_{wpDC} \left[\frac{DI_O}{n_t(1-D)} \right]^2. \quad (7.301)$$

The maximum DC power loss in the primary winding occurs at the maximum load current I_{Omax}, the minimum DC input voltage V_{Imin}, and the maximum duty cycle D_{max}

$$P_{wpDC(max)} = R_{wpDC} I_{pDC(max)}^2 = R_{wpDC} \left[\frac{D_{max} I_{Omax}}{n(1-D_{max})} \right]^2. \quad (7.302)$$

The DC winding power loss in the primary winding is

$$P_{wp} = R_{wpDC} \frac{D^2 I_O^2}{n_t^2 (1-D)^2} \left\{ 1 + 2 \sum_{n=1}^{\infty} \frac{R_{wpn}}{R_{wpDC}} \left[\frac{\sin(n\pi D)}{n\pi D} \right]^2 \right\} = R_{wpDC} \frac{D^2 I_O^2}{n^2(1-D)^2} F_{Rph}. \quad (7.303)$$

The DC component of the secondary winding current for the flyback converter is $I_{sDC} = I_O$, yielding the DC power loss in the secondary winding

$$P_{sDC} = R_{wsDC} I_O^2 \quad (7.304)$$

and the power loss in the secondary winding for CCM is

$$P_{ws} = R_{wsDC} I_O^2 \left\langle 1 + 2 \sum_{n=1}^{\infty} F_{Rsn} \left\{ \frac{\sin[n\pi(1-D)]}{n\pi(1-D)} \right\}^2 \right\rangle = R_{wsDC} I_O^2 F_{Rsh}. \quad (7.305)$$

In transformer converters such as forward converter, the rise time t_r and fall time t_f of the waveforms of the primary and secondary windings are nonzero for CCM. It can be approximated by a trapezoidal waveform. Figure 7.41 shows a current waveform through the primary winding with rise time t_r and fall time t_f. The following definitions can be used

$$D = \frac{t_o}{T} \quad (7.306)$$

$$D_r = \frac{t_r}{T} \quad (7.307)$$

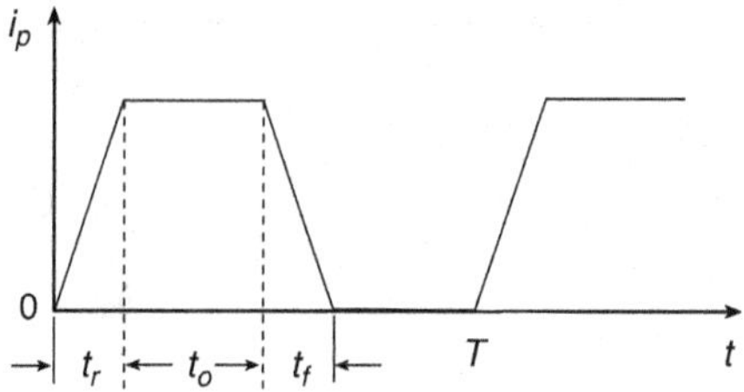

Figure 7.41 Current waveforms through the primary winding with rise time t_r and fall time t_f.

and

$$D_f = \frac{t_f}{T}. \tag{7.308}$$

The current waveform of the primary winding for CCM is given by

$$i_p = \begin{cases} \frac{I_{pmax}}{D_r T} t, & \text{for } 0 < t \le D_r T \\ I_{pmax}, & \text{for } D_r T < t \le (D_r + D)T \\ I_{pmax}\left[1 - \frac{t-(D+D_r)T}{D_f T}\right], & \text{for } (D_r + D)T < t \le (D + D_r + D_f)T \\ 0, & \text{for } (D + D_r + D_f)T < t \le T. \end{cases} \tag{7.309}$$

For $D_f = D_r$, the Fourier series technique leads to

$$I_{pDC} = \left(D + \frac{D_r + D_f}{2}\right) I_{pmax} = D\left(1 + \frac{D_r + D_f}{2D}\right) I_{pmax} = D\left(1 + \frac{D_r}{D}\right) I_{pmax} \tag{7.310}$$

and the amplitude of the nth harmonic of the primary current is

$$I_{pn} = 2(D + D_r) I_{pmax} \frac{\sin[\pi n(D + D_r)]}{\pi n(D + D_r)} \frac{\sin(n\pi D_r)}{n\pi D_r}. \tag{7.311}$$

The Fourier series of the current waveform in the primary winding is

$$i_p = I_{pmax} D \left\{\left(1 + \frac{D_r}{D}\right) + 2\left(1 + \frac{D_r}{D}\right) \sum_{n=1}^{\infty} \frac{\sin[\pi n(D + D_r)]}{\pi n(D + D_r)} \frac{\sin(n\pi D_r)}{n\pi D_r} \cos \quad n\omega t\right\}. \tag{7.312}$$

Figure 7.42 shows a spectrum of the trapezoidal waveform at $D = 0.4$ and $D_r = 0.1$.

7.27.2 Winding Power Losses with Harmonics for DCM

Primary Winding

The current waveform of the primary winding for discontinuous conduction mode (DCM) is given by

$$i_p = \begin{cases} \frac{I_{pmax} t}{DT}, & \text{for } 0 < t \le DT \\ 0, & \text{for } DT < t \le T. \end{cases} \tag{7.313}$$

The DC component of the primary winding current for DCM is

$$I_{pDC} = \frac{1}{T}\int_0^T i_p dt = \frac{I_{pmax}}{DT^2}\int_0^{DT} t dt = \frac{DI_{pmax}}{2}. \tag{7.314}$$

Using the Fourier formula, we have

$$a_n = \frac{2}{T}\int_0^T i_p \cos \quad n\omega t dt = \frac{2I_{pmax}}{DT^2}\int_0^{DT} t \cos \quad n\omega t dt$$
$$= \frac{I_{pmax}}{2\pi^2 n^2 D}(\cos \quad 2\pi nD - 1 + 2\pi nD \sin \quad 2\pi nD) \tag{7.315}$$

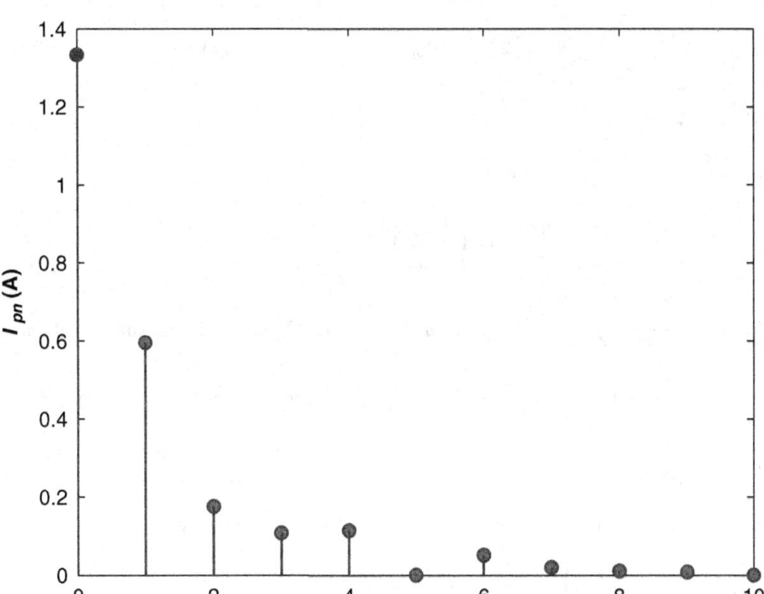

Figure 7.42 Spectrum of the trapezoidal waveform at $D = 0.3$ and $D_r = 0.1$.

and

$$b_n = \frac{2}{T}\int_0^T i_p \sin n\omega t dt = \frac{2I_{pmax}}{DT^2}\int_0^{DT} t \sin n\omega t dt$$
$$= \frac{I_{pmax}}{2\pi^2 n^2 D}(\sin 2\pi nD - 2\pi nD \cos 2\pi nD), \quad (7.316)$$

which gives the amplitudes of the fundamental component and the harmonics of the primary winding current

$$I_{pmn} = \sqrt{a_n^2 + b_n^2} = \frac{I_{pmax}}{2\pi^2 n^2 D}\sqrt{2(1 + 2\pi^2 n^2 D^2 - 2\pi nD \sin \quad 2\pi nD - \cos \quad 2\pi nD)}$$
$$= \frac{I_{pDC}}{\pi^2 n^2 D^2}\sqrt{2(1 + 2\pi^2 n^2 D^2 - 2\pi nD \sin \quad 2\pi nD - \cos \quad 2\pi nD)}. \quad (7.317)$$

For $D = 0.5$,

$$I_{pDC} = \frac{I_{pmax}}{4} \quad (7.318)$$

$$I_{pmn} = \frac{1}{n\pi} \quad \text{for} \quad n = 2, 4, 6, \ldots \quad (7.319)$$

and

$$I_{pmn} = \frac{1}{n^2\pi^2}\sqrt{4 + n^2\pi^2} \qquad \text{for} \qquad n = 1, 3, 5, \ldots \quad . \quad (7.320)$$

The power loss in the primary winding for DCM is

$$P_{wp} = R_{wpDC} I_{pDC}^2 + \frac{1}{2} R_{wp1} I_{pm1}^2 + \ldots = R_{wpDC} I_{pDC}^2 \left(1 + \frac{1}{2}\frac{R_{wp1}}{R_{wpDC}}\frac{I_{pm1}^2}{I_{pDC}^2} + \ldots\right)$$
$$= R_{wpDC} I_{pDC}^2 \left[1 + \frac{1}{2}\sum_{n=1}^{\infty}\left(\frac{R_{wpn}}{R_{wpDC}}\right)\left(\frac{I_{pmn}}{I_{pDC}}\right)^2\right] = P_{wpDC}\left[1 + \frac{1}{2}\sum_{n=1}^{\infty} F_{Rpn}\left(\frac{I_{pmn}}{I_{pDC}}\right)^2\right]$$

$$= P_{wpDC}\left[1 + \frac{1}{\pi^4 D^4}\sum_{n=1}^{\infty}\frac{F_{Rpn}}{n^4}\left(1 + 2\pi^2 n^2 D^2 - \cos \quad 2\pi nD - 2\pi nD \sin \quad 2\pi nD\right)\right]$$

$$= P_{wpDC} F_{Rph}, \tag{7.321}$$

where the AC-to-DC resistance ratio is

$$F_{Rph} = \frac{P_{wp}}{P_{wpDC}} = 1 + \frac{1}{2}\sum_{n=1}^{\infty} F_{Rpn}\left(\frac{I_{pmn}}{I_{pDC}}\right)^2$$

$$= 1 + \frac{1}{\pi^4 D^4}\sum_{n=1}^{\infty}\frac{F_{Rpn}}{n^4}(1 + 2\pi^2 n^2 D^2 - \cos \quad 2\pi nD - 2\pi nD \sin \quad 2\pi nD). \tag{7.322}$$

Figure 7.43 shows the plots of F_{Rph} as a function of D for DCM at $d/\delta_w = 1$ and $d/p = 0.8$.

Secondary Winding

The current waveform of the secondary winding for DCM is given by

$$i_s = \begin{cases} 0, & \text{for} \quad 0 < t \le DT \\ -\frac{I_{smax} t}{D_1 T} + I_{smax}, & \text{for} \quad DT < t \le (D + D_1)T \\ 0, & \text{for} \quad (D + D_1)T < t \le T. \end{cases} \tag{7.323}$$

Using the time-shift theorem, we can write

$$i_s = \begin{cases} -\frac{I_{smax} t}{D_1 T} + I_{smax}, & \text{for} \quad 0 < t \le D_1 T \\ 0, & \text{for} \quad D_1 T < t \le T, \end{cases} \tag{7.324}$$

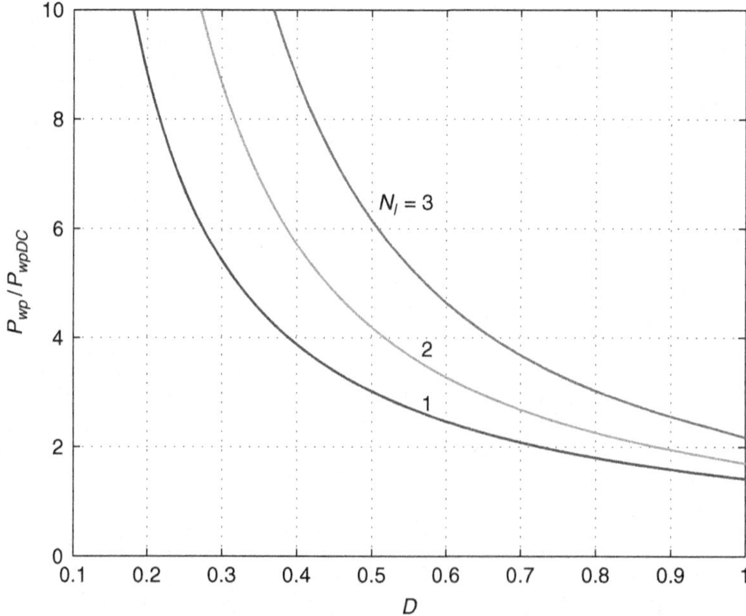

Figure 7.43 Harmonic AC-to-DC resistance ratio of the primary winding $F_{Rph} = P_{wp}/P_{wpDC}$ as a function of D for DCM at $d/\delta_w = 1$ and $d/p = 0.8$.

where D_1 is the duty cycle of the fall time of secondary winding current. The DC component of the secondary winding current for DCM is

$$I_{sDC} = \frac{1}{T}\int_0^T i_s dt = \frac{I_{smax}}{T}\int_0^{D_1 T}\left(1 - \frac{t}{D_1 T}\right) dt = \frac{D_1 I_{smax}}{2}. \quad (7.325)$$

The Fourier coefficients are

$$a_n = \frac{2}{T}\int_0^T i_s \cos \quad n\omega t dt = \frac{2I_{smax}}{T}\int_0^{D_1 T}\left(1 - \frac{t}{D_1 T}\right)\cos \quad n\omega t dt$$
$$= \frac{I_{smax}}{2\pi^2 n^2 D_1}(1 - \cos \quad 2\pi n D_1) \quad (7.326)$$

and

$$b_n = \frac{2}{T}\int_0^T i_s \sin \quad n\omega t dt = \frac{2I_{smax}}{T}\int_0^{D_1 T}\left(1 - \frac{t}{D_1 T}\right)\sin \quad n\omega t dt$$
$$= \frac{I_{smax}}{2\pi^2 n^2 D_1}(2\pi n D_1 - \sin \quad 2\pi n D_1), \quad (7.327)$$

yielding the amplitudes of the fundamental and the harmonics of the secondary winding current

$$I_{smn} = \sqrt{a_n^2 + b_n^2} = \frac{I_{smax}}{2\pi^2 n^2 D_1}\sqrt{2[1 - \cos \quad 2\pi n D_1 + 2\pi n D_1(\pi n D_1 - \sin \quad 2\pi n D_1)]}$$
$$= \frac{I_{sDC}}{\pi^2 n^2 D_1^2}\sqrt{2[1 - \cos \quad 2\pi n D_1 + 2\pi n D_1(\pi n D_1 - \sin \quad 2\pi n D_1)]}. \quad (7.328)$$

The power loss in the secondary winding for DCM is

$$P_{ws} = R_{wsDC} I_{sDC} + \frac{1}{2} R_{ws1} I_{pm1}^2 + \ldots = R_{wsDC} I_{sDC}^2 \left(1 + \frac{1}{2}\frac{R_{ws1}}{R_{wsDC}}\frac{I_{s1}^2}{I_{sDC}^2} + \ldots\right)$$
$$= R_{wsDC} I_{sDC}^2 \left[1 + \frac{1}{2}\sum_{n=1}^{\infty}\left(\frac{R_{wsn}}{R_{wsDC}}\right)\left(\frac{I_{smn}}{I_{wsDC}}\right)^2\right] = P_{wsDC}\left[1 + \frac{1}{2}\sum_{n=1}^{\infty} R_{Rsn}\left(\frac{I_{sn}}{I_{wsDC}}\right)^2\right]$$
$$= P_{wsDC}\left\{1 + \frac{1}{\pi^4 D_1^4}\sum_{n=1}^{\infty}\frac{F_{Rsn}}{n^4}[1 - \cos \quad 2\pi n D_1 + 2\pi n D_1(\pi n D_1 - \sin \quad 2\pi n D_1)]\right\}$$
$$= P_{wsDC} F_{Rsh}, \quad (7.329)$$

where

$$F_{Rsh} = \frac{P_{ws}}{P_{wsDC}} = 1 + \frac{1}{2}\sum_{n=1}^{\infty} F_{Rsn}\left(\frac{I_{smn}}{I_{pDC}}\right)^2$$
$$= 1 + \frac{1}{\pi^4 D_1^4}\sum_{n=1}^{\infty}\frac{F_{Rsn}}{n^4}[1 - \cos \quad 2\pi n D_1 + 2\pi n D_1(\pi n D_1 - \sin \quad 2\pi n D_1)]. \quad (7.330)$$

The relationship between D_1 and D is converter dependent. Figure 7.44 shows a plot of $F_{Rsh} = P_{ws}/P_{wsDC}$ as a function of $D = 1 - D_1$ at $d/\delta_w = 1$ and $d/p = 0.8$.

For the flyback converter, we have [64]

$$I_{sDC} = I_O, \quad (7.331)$$

$$P_{wsDC} = R_{wsDC} I_O^2, \quad (7.332)$$

$$I_{pDC} = I_I = \frac{D^2 V_I}{2 f_s L_m}, \quad (7.333)$$

$$P_{wpDC} = R_{wsDC} I_I^2 = R_{wsDC}\frac{D^4 V_I^2}{4 f_s^2 L_m^2}, \quad (7.334)$$

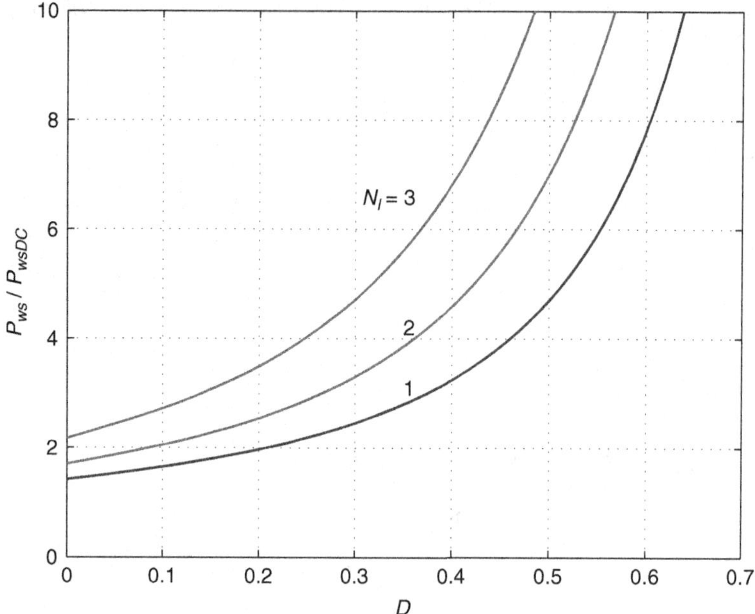

Figure 7.44 Harmonic AC-to-DC resistance ratio of the secondary winding $F_{Rsh} = P_{ws}/P_{wsDC}$ as a function of $D = 1 - D_1$ for DCM at $d/\delta_w = 1$ and $d/p = 0.8$.

and

$$D_1 = \frac{DV_I}{nV_O} = \sqrt{\frac{2f_s L_m I_O}{n_t^2 V_O}}, \tag{7.335}$$

where $n_t = N_p/N_s$ is the turns ratio.

7.28 Thermal Model of Transformers

Figure 7.45 shows a thermal model of a two-winding transformers. It is assumed that the primary winding is close to the core and the secondary winding is close to the outer surface of the transformer. There are three sources of heat in a two-winding transformer: core power loss P_C, primary winding power loss P_{wp}, and the secondary winding power loss P_{ws}. The definitions of thermal resistances are the same as those of an inductor. The thermal resistance of the core is

$$\theta_c = \frac{r_c}{k_c A_{sc}}, \tag{7.336}$$

where k_c is the thermal conductivity of the core, r_c is the radius of the core, and A_{sc} is the area of the outer core surface. The thermal resistance of the primary winding is

$$\theta_{wp} = \frac{t_{wp}}{k_w A_{swp}}, \tag{7.337}$$

where k_w is the winding conductivity, t_{wp} is the primary winding thickness, and A_{swp} is the area of the primary winding outer surface. The thermal resistance of the secondary winding is

$$\theta_{ws} = \frac{t_{ws}}{k_w A_{sws}}. \tag{7.338}$$

where t_{ws} is the secondary winding thickness and A_{sws} is the area of the secondary winding outer surface. The secondary winding-ambient resistance is

$$\theta_{wa} = \frac{1}{hA_t}, \tag{7.339}$$

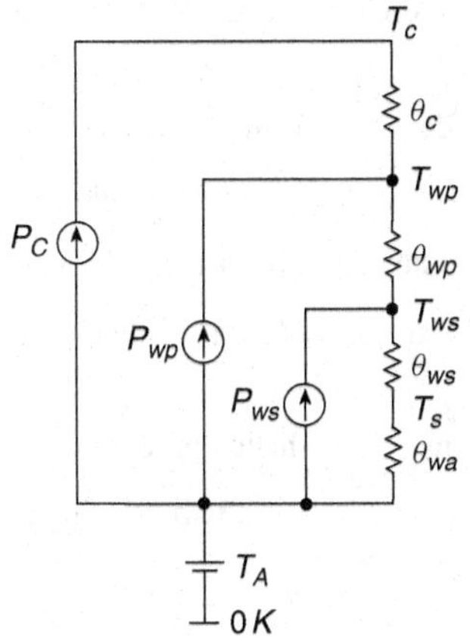

Figure 7.45 Thermal model of transformers for steady state.

where h is a constant and A_t is the area of the outer surface of the transformer. The temperature rise is

$$\Delta T = T_C - T_A = \theta_c P_C + \theta_{wp}(P_C + P_{wp}) + (\theta_{ws} + \theta_{wa})(P_C + P_{wp} + P_{ws}). \quad (7.340)$$

In a transformer, the contribution of the natural heat removal by conduction and convection is nearly the same. In many core geometries, the heat from the inner winding is removed by the core and the heat from the outer winding is removed through its outer surface.

A simple expression for the temperature rise is given by McLyman [58]

$$\Delta T = 450\left(\frac{P_C + P_{wp} + P_{ws}}{A_t}\right)^{0.826} = 450\left(\frac{P_{cw}}{A_t}\right)^{0.826}, \quad (7.341)$$

where P_{cw} is in watts and A_t is in square centimeter.

7.29 Summary

- A transformer is a system of mutually coupled two or more coils usually wound on a common magnetic core.
- A winding is a set of turns that conduct the same current.
- In an ideal transformer, the ratios of the secondary to primary AC voltages and AC currents are governed by the turns ratio $n = N_1/N_2 = V_1/v_2 = I_2/I_1$.
- An ideal transformer stores no magnetic energy.
- The self-inductance of a winding describes the winding ability to induce a voltage across itself due its own time-varying current.
- The mutual inductance of a transformer describes the ability of one winding to induce a voltage across the other winding through a magnetic flux generated by the time-varying current flowing in the first winding.
- The self-induced voltage across the primary winding by the primary current is $v_1 = L_1(di_1/dt)$.
- The mutually induced voltage across the secondary winding by the primary current is $v_2 = M(di_1/dt)$.
- The coupling coefficient $k = M/\sqrt{L_1L_2}$ is a measure of the magnetic coupling between two windings; $0 \le k \le 1$.

- Two windings may be perfectly coupled ($k = 1$), tightly coupled ($k \approx 1$), and loosely coupled ($k \ll 1$).
- The coupling coefficient in an ideal transformer is $k = 1$.
- In mutually coupled inductors, the voltage is induced and not the current.
- The dot convention is used to indicate the polarity of the voltages due to the mutual inductance.
- The current flowing into the dotted terminal of a winding produces a positive voltage at the dotted terminal of the other voltage.
- A transformer can serve as an impedance-matching device.
- The instantaneous magnetic energy stored in a two-winding transformer is given by

$$w(t) = \frac{1}{2}L_1 i_1^2 + \frac{1}{2}L_2 i_2^2 \pm M i_1 i_2. \tag{7.342}$$

- Buck-derived PWM converters, such as forward, half-bridge, full-bridge and push–pull converters, require an ideal transformer.
- The transformer in the flyback converter is required to store energy like an inductor and to magnetically couple the output to the input.
- Faraday's law states that an AC voltage is induced in a coil, which contains a time-varying magnetic flux, regardless of the source of flux.
- Mutual inductance occurs when two coils are placed in a close proximity to one another and share a common magnetic flux.
- The unit of the mutual inductance of coupled inductors is the henry (H), the same as that for self-inductance.
- The self-induced voltage is the product of the self-inductance of the coil L and the first derivative of the current in that coil di/dt. For example, the self-induced voltage across the primary winding by the current in the primary winding i_1 is $v_1 = L_1(di_1/dt)$ with open-circuited on the secondary side.
- The voltage induced in the secondary winding can be related to the time-varying current in the primary winding (and vice versa) by a mutual inductance M.
- The mutually induced voltage is the product of the mutual inductance of the coils M and the first derivative of the current in the other coil. For example, the mutual voltage induced in the secondary winding by the current through the primary i_1 is $M(di_1/dt)$.
- The self-inductance of a transformer winding is proportional to the square of the number of turns of that winding (N^2).
- The value of mutual inductance is a function of self-inductances of the primary and secondary windings.
- The mutual inductance is proportional to the product of the numbers of turns of the primary and secondary windings ($N_1 N_2$).
- For the coupling coefficient $k = 1$, all the magnetic flux that links the primary winding also links the secondary winding.
- In multiple-layer transformers, the winding resistance at high frequencies is higher than the DC resistance due to the skin effect and proximity effects.
- The proximity effect loss in multiple-layer windings is much higher than the skin-effect loss.
- In multiple-layer inductors and transformers, the current flows in both directions in higher order winding layers, significantly increasing the rms currents and copper loss.

- The proximity effect dominates over the skin effect in a winding that has many layers.
- Due to the proximity effect, the current density may be higher in the conductor areas, where the conductors are close to each other.
- Increasing the distance between the conductors in the same layer reduces the proximity-effect loss.
- Increasing the distance between the layers reduces the proximity-effect loss.
- Interleaved windings in transformers reduce the copper loss caused by proximity effect at high frequencies, if the currents through the primary and secondary windings are in phase. Each layer operates as a single-layer winding and the proximity effect is nearly eliminated.
- In transformers with a single-layer primary winding and a single-layer secondary winding, the proximity effect is absent.
- There is an optimum conductor thickness that leads to the minimum copper loss.
- To reduce the copper loss due to the proximity effect, it is highly advantageous to reduce the number of winding layers, increase the winding width, and interleave the windings.
- Litz wire increases the effective conduction area at high frequencies and thereby reduces the copper loss.
- Eddy currents cause power loss in electrostatic shields.
- The current waveforms through primary and secondary windings in PWM DC–DC converters are pulsating, resulting in high amplitudes of current harmonics.
- Power losses due to current harmonics in transformers of PWM DC–DC converters are significant for both CCM and DCM.
- The leakage inductance of transformers is reduced if the primary winding is interleaved with the secondary winding.
- In transformers with multiple secondary windings, the lower power secondary winding should be wound on the top of the higher power secondary winding.
- The leakage inductance can be reduced by twisting the primary and secondary winding wires before winding them on the core.

7.30 References

[1] P. L. Dowell, "Effects of eddy currents in transformer winding," *Proceedings IEE*, vol. 113, no. 8, pp. 1387–1394, August 1966.

[2] E. Bennett and S. C. Larson, "Effective resistance of alternating currents of multilayer windings," *Transactions of the American Institute of Electrical Engineers*, vol. 59, pp. 1010–1017, 1940.

[3] J. Jongsma, "High-frequency ferrite power transformer and choke design, Part 3: transformer winding design, Part 4: improved method of power choke design," Philips Electronic Components and Materials, Technical Publication, no. 27, Philips, The Netherlands, 1986.

[4] A. Kennelly, F. Laws, and P. Pierce, "Experimental research on skin effect in conductors," *Transactions AIEE*, vol. 34, p. 1915, 1915.

[5] J. Lammeraner and M. Stafl, *Eddy Currents*, Cleveland, OH: CRS Press, 1966.

[6] A. Wright, *Current Transformers: Their Transient and Steady State Performance*, London: Chapman and Hall, 1968.

[7] E. C. Snelling, *Soft Ferrites: Properties and Applications*, London: Iliffe Books Ltd, 1969.

[8] R. L. Stall, *The Analysis of Eddy Currents*, Oxford: Clarendon Press, 1974, pp. 21–27.

[9] R. L. Perry, "Multiple layer series connected winding design for minimum losses," *IEEE Transactions on Power Applications Systems*, vol. PAS-98, pp. 116–123, January/February 1979.

[10] N. R. Grossner, *Transformers for Electronic Circuits*, 2nd Ed., New York: McGraw-Hill, 1983.

[11] P. S. Venkatraman, "Winding eddy current losses in switch mode power transformers due to rectangular wave currents," Proceedings Powercon 11, Section A-1, 1984, pp. 1–11.
[12] L. F. Blume, *Transforemer Engineering*, New York: John Wiley & Sons, 1982.
[13] C. P. Steinmetz, "On the law of hysteresis," *AIEE*, vol. 9, pp. 3–64, 1992; Also, "A Steinmetz contribution to the ac power revolution," *Proceedings of the IEEE*, vol. 72, pp. 196–221, 1984.
[14] J. K. Watson, *Applications of Magnetism*, Gainesville, FL: University of Florida, 1985.
[15] W. M. Flanagan, *Handbook of Transformer Applications*, New York: McGraw-Hill, 1986.
[16] B. Carsten, "High frequency conductor losses in switch mode magnetics," Proceedings of PCI, Munich, Germany, May 1986, pp. 161–182.
[17] R. Lee, L. Wilson, and C. H. Carter, *Electronics Transformers and Circuits*, 3rd Ed., New York: John Wiley & Sons, 1988.
[18] A. F. Goldberg, J. G. Kassakian, and M. F. Schlecht, "Issues related to 1-10 MHz transformer design," *IEEE Transactions on Power Electronics*, vol. 4, no. 1, pp. 1130123, January 1989.
[19] J. P. Vandalec and P. D. Ziogos, "A novel approach for minimizing high-frequency transformer copper loss," *IEEE Transactions on Power Electronics*, vol. 3, no. 3, pp. 266–276, July 1988.
[20] J. G. Kassakian and M. F. Schlecht, "High-frequency high-density converters for distributed power-supply systems," *Proceedings of the IEEE*, vol. 76, no. 4, pp. 362–376, April 1988.
[21] A. M. Urling, V. A. Niemela, G. R. Skutt, and T. G. Wilson, "Characterizing high frequency effects in transformer windings: a guide to several significant papers," 4th Annual IEEE Transactions on Power Electronics Specialists Conference, Baltimore, MD, March 13–17, 1989, pp. 373–385.
[22] J. A. Ferreira, *Electromagnetic Modelling of Power Electronic Converters*, Boston, MA: Kluwer Academic Publisher, 1989.
[23] K. D. T. Ngo, R. P. Alley, A. J. Yerman, R. J. Charles, and M. H. Kuo, "Evaluation of trade-offs in transformer design for very-low-voltage power supply with very high efficiency and power density," 5th IEEE Applied Power Electronics Conference, Los Angeles, CA, March 11–16, 1990, pp. 344–353.
[24] J. H. Spreen, "Electrical terminal representation of conductor loss in transformers," *IEEE Transactions on Power Electronics*, vol. 5, no. 4, pp. 424–429, October 1990.
[25] K. D. T. Ngo and R. S. Lai, "Effect of height on power density in high-frequency transformers," 22nd Annual IEEE Power Electronics Specialists Conference, Cambridge, MA, June 24–27, 1991, pp. 667–672.
[26] A. F. Goldberg and M. F. Schlecht, "The relationship between size and power distribution in a 1-10 MHz transformer," *IEEE Transactions on Power Electronics*, vol. 7, no. 1, pp. 63–74, January 1992.
[27] M. K. Kazimierczuk and N. Thirunarayan, "Class D voltage-switching inverter with tapped resonant inductor," *Proceedings IEE Part B: Electric Power Applications*, vol. 140, pp. 177–185, May 1993.
[28] M. K. Kazimierczuk and D. Czarkowski, "Phase-controlled CLL resonant converter," 5th IEEE Applied Power Electronics Conference, San Diego, CA, March 7–11, 1993, pp. 432–438.
[29] W. J. Gu and R. Liu, "A study of volume and weight vs. frequency for high-frequency transformers," 23rd IEEE Power Electronics Specialists Conference, Seattle, WA, June 20–24, 1993, pp. 1123–1129.
[30] B. Cogirore, J. P. Keradec, and J. Barbaroux, "The two-winding transformer: an experimental method to obtain a wide frequency range equivalent circuit," *IEEE Transactions on Instrumentation and Measurement*, vol. 43, no. 2, pp. 364–371, April 1994.
[31] M. Bartoli, A. Reatti, and M. K. Kazimierczuk, "High-frequency models of ferrite inductors," Proceedings of the IEEE International Conference on Industrial Electronics, Controls, Instrumentation, and Automation (IECON'94), Bologna, Italy, September 5–9, 1994, pp. 1670–1675.
[32] M. Bartoli, A. Reatti, and M. K. Kazimierczuk, "Predicting the high-frequency ferrite-core inductor performance," Proceedings of the Conference of Electrical Manufacturing and Coil Winding, Chicago (Rosemont), IL, September 27–29, 1994, pp. 409–413.
[33] M. Bartoli, A. Reatti, and M. K. Kazimierczuk, "Modeling iron-powder inductors at high-frequencies," Proceedings of the IEEE Industry Applications Society Annual Meeting, Denver, CO, October 2-7, 1994, pp. 1125–1232.
[34] J. A. Ferreira, "Improved analytical modeling of conductive losses in magnetic components," *IEEE Transactions on Power Electronics*, vol. 9, pp. 127–131, January 1994.
[35] M. Bartoli, N. Noferi, A. Reatti, and M. K. Kazimierczuk, "Modeling winding losses in high-frequency power inductors," *Journal of Circuits Systems and Computers*, vol. 5, no. 3, pp. 65–80, March 1995.
[36] W. A. Roshen, R. L. Steigerwald, R. J. Charles, W. G. Earls, G. S. Claydon, and C. F. Saj, "High-frequency, high-density MHz magnetic components for low profile converters," *IEEE Transactions on Industry Applications*, vol. 31, no. 4, pp. 869–878, July/August 1995.
[37] M. Bartoli, N. Nefari, A. Reatti, and M. K. Kazimierczuk, "Modeling litz-wire winding losses in high-frequencies power inductors," Proceedings of the IEEE Power Electronics Specialists Conference, Baveno, Italy, June 24-27, 1996, pp. 1690–1696.
[38] C. R. Sullivan and S. R. Sanders, "Design of microfabricated transformers and inductors for high-frequency power conversion," *IEEE Transactions on Power Electronics*, vol. 11, no. 2, pp. 228–238, March 1996.

[39] R. Petkov, "Optimum design of a high-power high-frequency transformers," *IEEE Transactions on Power Electronics*, vol. 11, no. 1, pp. 33–42, January 1996.

[40] W. G. Odendaal and J. Ferreira, "A thermal model for high-frequency magnetic components," *IEEE Transactions on Industry Applications*, vol. 35, no. 4, pp. 932–940, July/August 1999.

[41] M. K. Kazimierczuk, G. Sancineto, U. Reggiani, and A. Massarini, "Small-signal high-frequency model of ferrite inductors," *IEEE Transactions on Magnetics*, vol. 35, pp. 4185–4191, September 1999.

[42] U. Reggiani, G. Sancineto, and M. K. Kazimierczuk, "High-frequency behavior of laminated iron-core inductors for filter applications," Proceedings of the IEEE Applied Power Electronics Conference, New Orleans, LA, February 6–10, 2000, pp. 654–660.

[43] K. Howard and M. K. Kazimierczuk, "Eddy-current power loss in laminated power cores," Proceedings of the IEEE International Symposium on Circuits and Systems, Sydney, Australia, May 7-9, 2000, paper III-668, pp. 668–672.

[44] A. Reatti and M. K. Kazimierczuk, "Comparison of various methods for calculating the ac resistance of inductors," *IEEE Transactions on Magnetics*, vol. 37, pp. 1512–1518, May 2002.

[45] G. Grandi, M. K. Kazimierczuk, A. Massarini, U. Reggiani, and G. Sancineto, "Model of laminated iron-core inductors," *IEEE Transactions on Magnetics*, vol. 40, no. 4, pp. 1839–1845, July 2004.

[46] J. D. Lavers and V. Bolborici, "Loss comparison in the design of high frequency inductors and transformers," *IEEE Transactions on Magnetics*, vol. 35, no. 5, pp. 3541–3543, September 1999.

[47] W. G. Hurley, W. H. Wolfe, and J. G. Breslin, "Optimized transformer design: inclusive of high-frequency effects," *IEEE Transactions on Power Electronics*, vol. 13, no. 4, pp. 651–659, July 1998.

[48] W. G. Hurley, E. Gath, and J. G. Breslin, "Optimizing the ac resistance of multilayer transformer with arbitrary current waveforms," *IEEE Transactions on Power Electronics*, vol. 15, no. 2, pp. 369–376, March 2000.

[49] C. R. Sullivan, "Optimal choice for the number of strands in a litz-wire transformer winding," *IEEE Transactions on Power Electronics*, vol. 14, no. 2, pp. 283–291, March 1999.

[50] G. Grandi, M. K. Kazimierczuk, A. Massarini, and U. Reggiani, "Stray capacitance of single-layer solenoid air-core inductors," *IEEE Transactions on Industry Applications*, vol. 35, no. 5, pp. 1162–1168, September 1999.

[51] A. Massarini and M. K. Kazimierczuk, "Self-capacitance of inductors," *IEEE Transactions on Power Electronics*, vol. 12, no. 4, pp. 671–676, July 1997.

[52] H. Y. Lu, J. G. Zhu, and S. Y. R. Hui, "Experimental determination of stray capacitances in high frequency transformers," *IEEE Transactions on Power Electronics*, vol. 18, no. 5, pp. 1105–1112, September 2003.

[53] L. Dalessandro, F. S. Cavalcante, and J. K. Kolar, "Self-capacitance of high-voltage transformers" *IEEE Transactions on Power Electronics*, vol. 22, no. 5, pp. 2081–2092, September 2007.

[54] R. W. Erickson and D. Maksimović, *Fundamentals of Power Electronics*, 2nd Ed., Norwell, MA: Kluwer Academic Publishers, 2001.

[55] N. Locci and C. Muscas, "Hysteresis and eddy currents in current transformers," *IEEE Transactions on Power Delivery*, vol. 16. no. 2, pp. 154–159, April 2001.

[56] N. Mohan, T. M. Underland, and W. P. Robbins, *Power Electronics*, 3rd Ed., New York: John Wiley & Sons, 2003.

[57] Y. Jang and M. M. Jovanović, "A contactless electrical energy transmission system for portable-telephone battery chargers," *IEEE Transactions on Industrial Electronics*, vol. 50, no. 3, pp. 520–527, June 2003.

[58] W. T. McLyman, *Transformer and Inductor Design Handbook*, 3rd Ed., New York: Marcel Dekker, 2004.

[59] A. Van den Bossche and V. C. Valchev, *Inductors and Transformers for Power Electronics*, Boca Raton, FL: Taylor & Francis, 2005.

[60] N. Das and M. K. Kazimierczuk, "An overview of technical challenges in the design of current transformers," Electrical Manufacturing Conference, Indianapolis, IN, October 24–26, 2005.

[61] J. Lastowiecki and P. Staszewski, "Sliding transformer with long magnetic circuit for contactless electrical energy delivery to mobile receivers," *IEEE Transactions on Industrial Electronics*, vol. 53, no. 6, pp. 1943–1948, December 2006.

[62] E. Labouré, F. Costa, and F. Forest, "Current measurement in static converters and realization of a high-frequency current probe (50 A-300 MHz)," *Proceedings of the EPE'93*, vol. 4, no. 377, 1993, pp. 478–483.

[63] F. Costa, P. Poulichet, F. Mazalerat, and E. Labouré, "The current sensors in power electronics, a review," *EPE Journal*, vol. 11, no. 1, pp. 7–18, 2001.

[64] M. K. Kazimierczuk, *Pulse-Width Modulated DC-DC Power Converters*, Chichester: John Wiley & Sons, 2008.

[65] M. K. Kazimierczuk, *RF Power Amplifiers*, Chichester: John Wiley & Sons, 2008.

[66] M. K. Kazimierczuk and D. Czarkowski, *Resonant Power Converters*, 2nd Ed., New York: John Wiley & Sons, 2011.

[67] G. Cerri, R. De Leo, V. M. Primiani, S. Pennesi, and P. Russo, "Wide-band characterization of current probes," *IEEE Transactions on Electromagnetic Compatibility*, vol. 45, no. 4, pp. 616–625, November 2003.

[68] J. Sallán, J. L Villa, A. Llomabart, and J. F. Sanz, "Optimal design of ICPT system applied to electric vehicle battery charge," *IEEE Transactions on Industrial Electronics*, vol. 56, no. 6, pp. 2140–2149, June 2009.

[69] N. Kondrath and M. K. Kazimierczuk, "Bandwidth of current transformers," *IEEE Transactions on Instrumentation and Measurement*, vol. 58, no. 4, pp. 2008–2016, June 2009.

[70] S.-H. Lee and R. D. Lorenz, "Development of model for 95%-efficiency 220-W wireless power transfer over a 30-cm air gap," *IEEE Transactions on Industry Applications*, vol. 47, no. 6, pp. 2495–2504, November 2011.

[71] D. Murthy-Bellur, N. Kondrath, and M. K. Kazimierczuk, "Transformer winding loss caused by skin and proximity effects including harmonics in PWM dc-dc flyback converter for continuous conduction mode," *Proceedings of the IET Power Electronics*, vol. 3, no. 4, pp. 288–295, 2011.

[72] D. A. Nagarajan, D. Murthy-Bellur, and M. K. Kazimierczuk, "Harmonic winding losses in the transformer of a forward pulse-width modulated dc-dc converter for continuous conduction mode," *Proceedings of the IET Power Electronics*, vol. 5, no. 2, pp. 221–236, 2012.

[73] www.mag-inc.com, www.ferroxcube.com, www.ferrite.de, www.micrometals.com, and www.metglas.com.

7.31 Review Questions

7.1. What is a transformer?

7.2. What is the mutual magnetic flux?

7.3. What is the leakage magnetic flux?

7.4. What are the self-inductances of a transformer?

7.5. What is the mutual inductance of a transformer?

7.6. What are the leakage inductances of a transformer?

7.7. What is the difference between a self-inductance and mutual inductance?

7.8. What is the magnetic coupling?

7.9. What is the coupling coefficient?

7.10. What are perfect coupling, tight coupling, and loose coupling?

7.11. What are the characteristics of an ideal transformer?

7.12. What is the dot convention?

7.13. What is the series-aiding connection?

7.14. What is the series-opposing connection?

7.15. What are the characteristics of a nonideal transformer?

7.16. List stray capacitances of a transformer.

7.17. Draw a high-frequency model of a transformer?

7.18. What is a CT?

7.19. Give an expression for the lower 3-dB corner frequency of the current transformer.

7.20. Are the amplitudes of harmonics in PWM DC–DC converters high?

7.21. What is the effect of harmonics on winding and core power losses in transformers used in PWM DC–DC converters?

7.22. Compare the effect of harmonics on power losses in inductors and transformers of nonisolated and isolated PWM DC–DC converters for CCM.

7.23. Compare the effect of harmonics on power losses in inductors and transformers of nonisolated and isolated PWM DC–DC converters for DCM.

7.32 Problems

7.1. A transformer has $N_1 = 100$, $N_2 = 10$, $\mu_{rc} = 100$, $A_c = 1$ cm^2, and $l_c = 10$ cm. Find L_m and L_{ms}.

7.2. A transformer has $\mu_{rc} = 1800$, $A_c = 1.2$ cm^2, $l_c = 14$ cm, $N_1 = 80$, $N_2 = 8$, $k_1 = 0.99$, and $k_2 = 0.995$. Find n, $\mathcal{R}$, $\mathcal{P}$, M, L_1, L_2, L_m, L_{ms}, L_{l1}, L_{l2}, and k.

7.3. A noninductive burden resistor in CP has a resistance $R = 50$ Ω. The sensitivity of the probe should be $S_p = 0.5$ V/A. What is the required number of turns of the secondary winding?

7.4. A toroidal nanocrystalline core T-46-6 manufactured by MECAGIS Group Arcelor has the following parameters: relative permeability $\mu_{rc} = 70\,000$, external diameter $D = 25$ mm, internal diameter $d = 16$ mm, height $h = 10$ mm, $l_c = 6.4$ cm, and $A_c = 0.36$ cm^2. This core is used to construct a CT. The number of the secondary winding turns is $N = 100$. Determine the magnetizing inductance located on the secondary side of the transformer.

7.5. The secondary winding of a CT is made of copper round wire with radius $r = 0.3$ mm. The number of turns is $N = 100$. A toroidal finished core has the following dimensions: $D = 26$ mm, $d = 15$ mm, and $h = 12.5$ mm. What is the low-frequency resistance of the secondary winding?

7.6. The magnetizing inductance of the secondary winding of a current transformer is $L_m = 494.8$ mH. The resistance of the burden resistor is $R = 50$ Ω. What is the lower 3-dB cutoff frequency of the CT?

8

Integrated Inductors

8.1 Introduction

One of the largest challenges in making completely monolithic integrated circuits is to make high-performance integrated inductors [1–78]. Lack of a good integrated inductor is the most important disadvantage of standard IC processes for many applications. It is a major gap in radio frequency (RF) technology. The wireless communication revolution has increased the interest immensely in the design of radio transceivers. Front-end RF integrated circuits (ICs) require monolithic integrated inductors to construct resonant circuits, power amplifiers, impedance-matching networks, bandpass filters, low-pass filters, couplers, dividers, isolators, circulators, and high-impedance chokes. RF IC inductors are essential elements of wireless communication circuit blocks, such as low-noise amplifiers (LNAs), mixers, intermediate-frequency filters (IFFs), and voltage-controlled oscillators (VCOs) that drive cellular phone transmitters. Monolithic inductors are also used to bias RF amplifiers, RF oscillators, tuning varactors, PIN diodes, transistors, and monolithic circuits. Applications of RF IC inductors include cellular phones, wireless local networks (WLNs), TV tuners, and radars. On-chip and off-chip inductors are used in RF circuits. The structure of planar spiral inductors is based on the standard complementary metal-oxide-semiconductor (CMOS) and BiCMOS technologies as the cost is one of the major factors. Fully integrated radio transceiver is a very strong motivation to develop high-performance IC inductors with a high quality factor Q and a high self-resonant frequency (SRF) f_r. There are four major types of integrated RF inductors: planar spiral inductors, meander inductors, bondwire inductors, and microelectromechanical system (MEMS) inductors. In this chapter, integrated inductors are studied.

8.2 Skin Effect

At low frequencies, the current density J in a conductor is uniform. In this case, the AC resistance of the conductor R_{AC} is equal to the DC resistance R_{DC}, and the conduction power loss is identical for AC and DC currents if $I_{rms} = I_{DC}$. In contrast, the current density J at high frequencies is not uniform due to eddy currents. Eddy currents are present in any conducting material, which is subjected to a time-varying magnetic field. The skin effect is the tendency of high-frequency current to flow mostly near the surface of a conductor. This causes an increase in the effective AC resistance of a conductor R_{AC} above its DC resistance R_{DC}, resulting in an increase in conduction power loss and heat.

High-Frequency Magnetic Components, Second Edition. Marian K. Kazimierczuk.

Companion Website: www.wiley.com/go/kazimierczuk_High2e

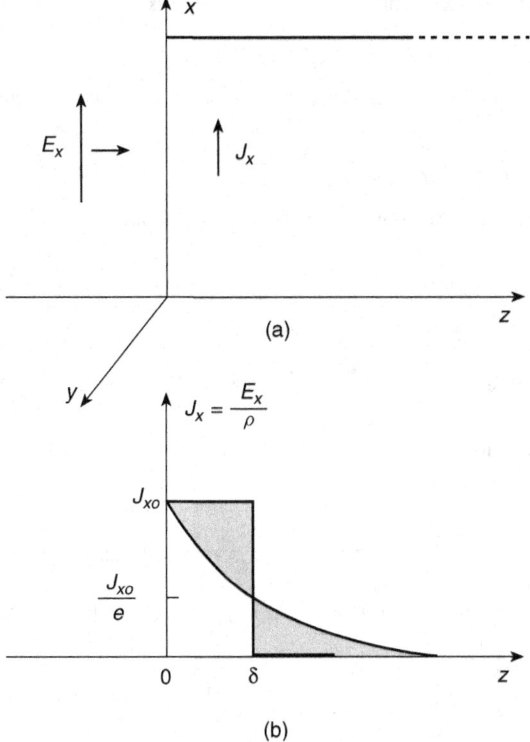

Figure 8.1 Semiinfinite slab. (a) Conductor. (b) Current density distribution

Consider a semiinfinite slab of a conductor with conductivity $\sigma = 1/\rho$ that occupies $z \geq 0$, where ρ is the resistivity of the conductor. The semiinfinite conductor is shown in Fig. 8.1a. The harmonic electric field incident on the conductor is given by

$$E_x(z) = E_{xo} e^{-\frac{z}{\delta}} \tag{8.1}$$

where E_{xo} is the amplitude of the electric field just on the surface of the conductor and the skin depth is

$$\delta = \frac{1}{\sqrt{\pi \mu \sigma f}} = \sqrt{\frac{\rho}{\pi \mu f}}. \tag{8.2}$$

The current density in the conductor is

$$J_x(z) = \sigma E_x(z) = \frac{E_x(z)}{\rho} = J_{xo} e^{-\frac{z}{\delta}} \tag{8.3}$$

where $J_{xo} = \sigma E_{xo} = E_{xo}/\rho$ is the amplitude of the current density at the surface. The current density distribution in the conductor is shown in Fig. 8.1b. The current density J_x exponentially decays as a function of depth z into the conductor.

The current flows in the x-direction through the surface extending into the z-direction from $z = 0$ to $z = \infty$. The surface has a width w in the y-direction. The total current is obtained as

$$I = \int_{z=0}^{z=\infty} \int_{y=0}^{w} J_{xo} e^{-\frac{z}{\delta}} dydz = J_{xo} w\delta. \tag{8.4}$$

Thus, the equivalent current is obtained if we assume that the current density J_x is constant and equal to J_{xo} from the surface down to skin depth δ and is zero for $z > \delta$. The area of a rectangle of sides J_{xo} and δ is equivalent to the area of the exponential curve. It can be shown that 95% of the total current flows in thickness of 3δ and 99.3% of the total current flows in the thickness of 5δ.

The voltage drop across the distance l in the x-direction is

$$V = E_{xo}l = \rho J_{xo}l. \tag{8.5}$$

Hence, the AC resistance of the semiinfinite conductor of width w and length l that extends from $z = 0$ to $z = \infty$ is

$$R_{AC} = \frac{V}{I} = \frac{\rho}{\delta}\left(\frac{l}{w}\right) = \frac{1}{\sigma\delta}\left(\frac{l}{w}\right) = \left(\frac{l}{w}\right)\sqrt{\pi\rho\mu f}. \tag{8.6}$$

As the skin depth δ decreases with increasing frequency, the AC resistance R_{AC} increases.

Example 8.1

Calculate the skin depth at $f = 10$ GHz for (a) copper, (b) aluminum, (c) silver, and (d) gold.

Solution: (a) The resistivity of copper at $T = 20\,^{\circ}\mathrm{C}$ is $\rho_{Cu} = 1.724 \times 10^{-8}$ Ω·m. The skin depth of the copper at $f = 10$ GHz is

$$\delta_{Cu} = \sqrt{\frac{\rho_{Cu}}{\pi\mu_0 f}} = \sqrt{\frac{1.724 \times 10^{-8}}{\pi \times 4\pi \times 10^{-7} \times 10 \times 10^{9}}} = 0.6608\ \mu\mathrm{m}. \tag{8.7}$$

(b) The resistivity of aluminum at $T = 20\,^{\circ}\mathrm{C}$ is $\rho_{Al} = 2.65 \times 10^{-8}$ Ω·m. The skin depth of the aluminum at $f = 10$ GHz is

$$\delta_{Al} = \sqrt{\frac{\rho_{Al}}{\pi\mu_0 f}} = \sqrt{\frac{2.65 \times 10^{-8}}{\pi \times 4\pi \times 10^{-7} \times 10 \times 10^{9}}} = 0.819\ \mu\mathrm{m}. \tag{8.8}$$

(c) The resistivity of silver at $T = 20\,^{\circ}\mathrm{C}$ is $\rho_{Ag} = 1.59 \times 10^{-8}$ Ω·m. The skin depth of the silver at $f = 10$ GHz is

$$\delta_{Ag} = \sqrt{\frac{\rho_{Ag}}{\pi\mu_0 f}} = \sqrt{\frac{1.59 \times 10^{-8}}{\pi \times 4\pi \times 10^{-7} \times 10 \times 10^{9}}} = 0.6346\ \mu\mathrm{m}. \tag{8.9}$$

(d) The resistivity of gold at $T = 20\,^{\circ}\mathrm{C}$ is $\rho_{Au} = 2.44 \times 10^{-8}$ Ω·m. The skin depth of the gold at $f = 10$ GHz is

$$\delta_{Au} = \sqrt{\frac{\rho_{Au}}{\pi\mu_0 f}} = \sqrt{\frac{2.44 \times 10^{-8}}{\pi \times 4\pi \times 10^{-7} \times 10 \times 10^{9}}} = 0.786\ \mu\mathrm{m}. \tag{8.10}$$

8.3 Resistance of Rectangular Trace with Skin Effect

Consider a trace (a slab) of a conductor of thickness h, length l, and width w, as shown in Fig. 8.2a. The DC resistance of the trace is

$$R_{DC} = \rho\frac{l}{wh}. \tag{8.11}$$

The harmonic electric field incident on the conductor is given by

$$E_x(z) = E_{xo}e^{-\frac{z}{\delta}}. \tag{8.12}$$

The current density at high frequencies is given by

$$J_x = J_{xo}e^{-\frac{z}{\delta}} \quad \text{for} \quad 0 \le z \le h \tag{8.13}$$

and

$$J_x = 0 \quad \text{for} \quad z > h. \tag{8.14}$$

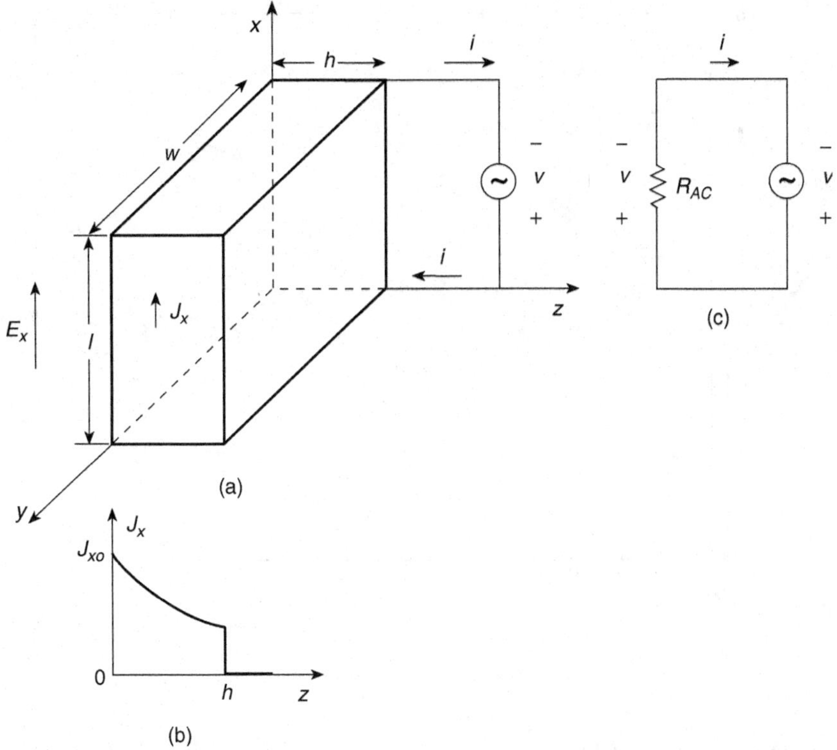

Figure 8.2 Slab conductor. (a) Slab. (b) Current density distribution. (c) Equivalent circuit

The current density distribution in the trace conductor is shown in Fig. 8.2b. Hence, the total current flowing in the conductor is

$$I = \int_{z=0}^{h} \int_{y=0}^{w} J_{xo} e^{-\frac{z}{\delta}} dydz = w\delta J_{xo} \left(1 - e^{-\frac{h}{\delta}}\right) \tag{8.15}$$

Since $E_{xo} = \rho J_{xo}$, the voltage drop across the conductor in the x-direction of the current flow at $z = 0$ is

$$V = E_{xo} l = \frac{J_{xo}}{\sigma} l = \rho J_{xo} l. \tag{8.16}$$

The AC trace resistance is given by

$$R_{AC} = \frac{V}{I} = \frac{\rho l}{w\delta(1 - e^{-h/\delta})} = \frac{l}{w(1 - e^{-h/\delta})}\sqrt{\pi\rho\mu f} \quad \text{for} \quad \delta < h. \tag{8.17}$$

At high frequencies, the AC resistance of the metal trace R_{AC} is increased by the skin effect. The current density in the metal trace increases on the side of the substrate. The equivalent circuit of the trace conductor for high frequencies is shown in Fig. 8.2c.

The ratio of the AC-to-DC resistances is

$$F_R = \frac{R_{AC}}{R_{DC}} = \frac{h}{\delta(1 - e^{-\frac{h}{\delta}})}. \tag{8.18}$$

The ratio R_{AC}/R_{DC} as a function of h/δ is shown in Fig. 8.3. As h/δ increases with increasing frequency, R_{AC}/R_{DC} increases.

For very high frequencies, this expression can be approximated by

$$R_{AC} \approx \frac{h}{\delta} = h\sqrt{\frac{\pi\mu f}{\rho}} \quad \text{for} \quad \delta < \frac{h}{5}. \tag{8.19}$$

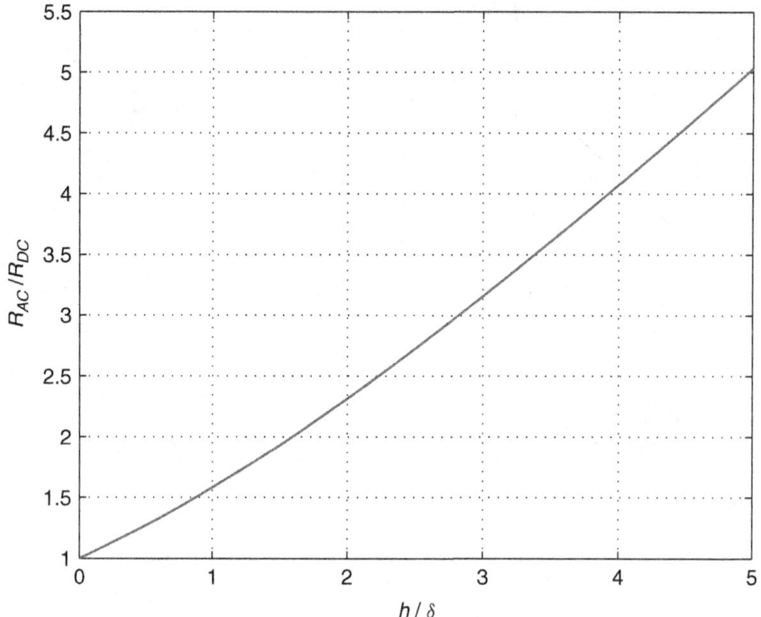

Figure 8.3 R_{AC}/R_{DC} as a function of h/δ

The power loss in the metal trace resistance due to the flow of a sinusoidal current of amplitude I_m is given by

$$P_{Rac} = \frac{1}{2} R_{AC} I_m^2. \tag{8.20}$$

Example 8.2

Calculate the DC resistance of a copper trace with $l = 50$ μm, $w = 1$ μm, and $h = 1$ μm. Find the AC resistance of the trace at 10 GHz. Calculate F_R.

Solution: The resistivity of copper at $T = 20$ °C is $\rho_{Cu} = 1.724 \times 10^{-8}$ Ω·m. The DC resistance of the trace is

$$R_{DC} = \rho_{Cu} \frac{l}{wh} = \frac{1.724 \times 10^{-8} \times 50 \times 10^{-6}}{1 \times 10^{-6} \times 1 \times 10^{-6}} = 0.862\ \Omega. \tag{8.21}$$

The skin depth of the copper at $f = 10$ GHz is

$$\delta_{Cu} = \sqrt{\frac{\rho_{Cu}}{\pi \mu_0 f}} = \sqrt{\frac{1.724 \times 10^{-8}}{\pi \times 4\pi \times 10^{-7} \times 10 \times 10^9}} = 0.6608\ \mu\text{m}. \tag{8.22}$$

The AC resistance of the trace is

$$R_{AC} = \frac{\rho l}{w \delta_{Cu} (1 - e^{-h/\delta_{Cu}})} = \frac{1.724 \times 10^{-8} \times 50 \times 10^{-6}}{1 \times 10^{-6} \times 0.6608 \times 10^{-6} \left(1 - e^{-\frac{1}{0.6608}}\right)} = 1.672\ \Omega. \tag{8.23}$$

The ratio of the AC-to-DC resistances is

$$F_R = \frac{R_{AC}}{R_{DC}} = \frac{1.672}{0.862} = 1.93. \tag{8.24}$$

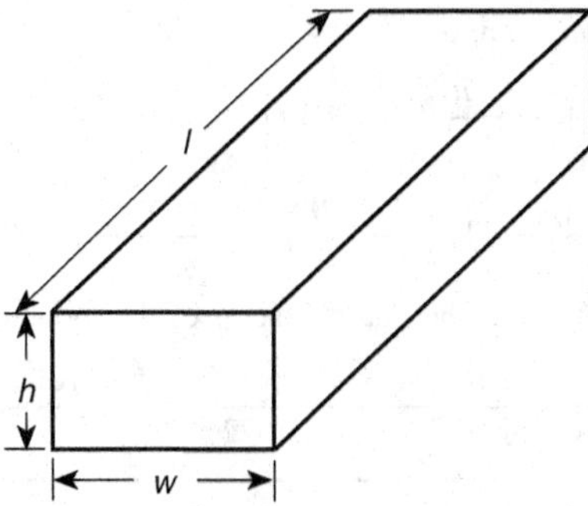

Figure 8.4 Straight rectangular trace inductor

8.4 Inductance of Straight Rectangular Trace

A straight rectangular trace inductor is shown in Fig. 8.4. The self-inductance of a solitary rectangular straight conductor at low frequencies, where the current density is uniform and skin effect can be neglected, is given by Grover's formula [8]

$$L = \frac{\mu_0 l}{2\pi}\left[\ln\left(\frac{2l}{w+h}\right) + \frac{w+h}{3l} + 0.50049\right] \text{(H)} \quad \text{for} \quad w \leq 2l \quad \text{and} \quad h \leq 2l \tag{8.25}$$

where l, w, and h are the conductor length, width, and thickness in meters, respectively. Since $\mu_0 = 2\pi \times 10^{-7}$, the trace inductance is

$$L = 2 \times 10^{-7} l\left[\ln\left(\frac{2l}{w+h}\right) + \frac{w+h}{3l} + 0.50049\right] \text{(H)} \quad \text{for} \quad w \leq 2l \quad \text{and} \quad h \leq 2l \tag{8.26}$$

where all dimensions are in meters. Grover's formula can also be expressed as

$$L = 0.0002l\left[\ln\left(\frac{2l}{w+h}\right) + \frac{w+h}{3l} + 0.50049\right] \text{(nH)} \quad \text{for} \quad w \leq 2l \quad \text{and} \quad h \leq 2l \tag{8.27}$$

where l, w, and h are the conductor length, width, and thickness in micrometers, respectively.

Figures 8.5 through 8.7 show the inductance L versus w, h, and l, respectively. The inductance L of a conducting trace increases as l increases and as w and h decrease. Straight traces are segments of inductors of more complex structures of integrated planar inductors, such as square integrated planar inductors. They are also used as interconnections in ICs.

The quality factor of the straight rectangular trace at frequency $f = \omega/(2\pi)$ is defined as

$$Q = 2\pi\frac{\text{Peak magnetic energy stored in } L}{\text{Energy dissipated per cycle}} = \frac{\omega L}{R_{AC}}. \tag{8.28}$$

Another definition of the quality factor is given by

$$Q = 2\pi\frac{\text{Peak magnetic energy} - \text{peak electric energy}}{\text{Energy dissipated per cycle}}. \tag{8.29}$$

Example 8.3

Calculate the inductance of a straight rectangular trace with $l = 50\ \mu\text{m}$, $w = 1\ \mu\text{m}$, $h = 1\ \mu\text{m}$, and $\mu_r = 1$. Find its quality factor Q at a frequency of 10 GHz. Neglect the substrate loss. Take into account only metal loss.

Solution: The inductance of the straight trace is

$$L = \frac{\mu_0 l}{2\pi}\left[\ln\left(\frac{2l}{w+h}\right) + \frac{w+h}{3l} + 0.50049\right]$$

$$= \frac{4\pi \times 10^{-7} \times 50 \times 10^{-6}}{2\pi}\left[\ln\left(\frac{2\times 50}{1+1}\right) + \frac{1+1}{3\times 50} + 0.50049\right] = 0.04426 \text{ nH}. \quad (8.30)$$

From Example 8.2, $R_{AC} = 1.672\ \Omega$. The quality factor of the inductor is

$$Q = \frac{\omega L}{R_{AC}} = \frac{2\pi \times 10 \times 10^9 \times 0.04426 \times 10^{-9}}{1.672} = 1.7092. \quad (8.31)$$

8.5 Inductance of Rectangular Trace with Skin Effect

The magnetic field distribution for a semiinfinite conductor with sinusoidal current is

$$H_y = -\frac{\sigma\delta E_0}{1+j}e^{-(1+j)x/\delta}. \quad (8.32)$$

The current per unit width is

$$J_{sz} = -H_y(0) = \frac{\sigma\delta E_0}{1+j} \quad (8.33)$$

The time-average stored magnetic energy is

$$\frac{1}{4}L|I|^2 = \frac{1}{4}\int_V \mu|H|^2 dV. \quad (8.34)$$

Thus,

$$\frac{L}{4}\frac{\sigma^2\delta^2 w^2 E_0^2}{2} = \frac{1}{4}wl\mu\int_0^\infty \frac{\sigma^2\delta^2 E_0^2}{2}e^{-2x/\delta}dx \quad (8.35)$$

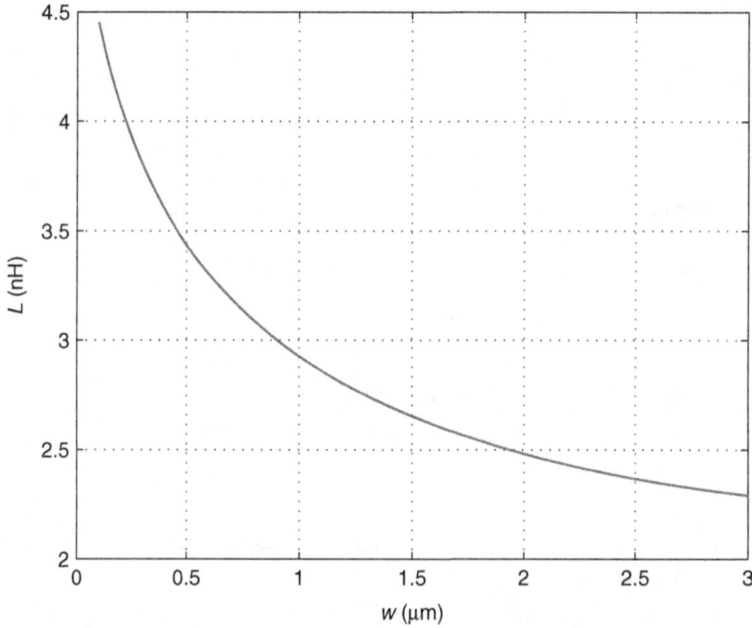

Figure 8.5 Inductance L as a function of trace width w for straight rectangular trace at $h = 1$ μm and $l = 50$ μm

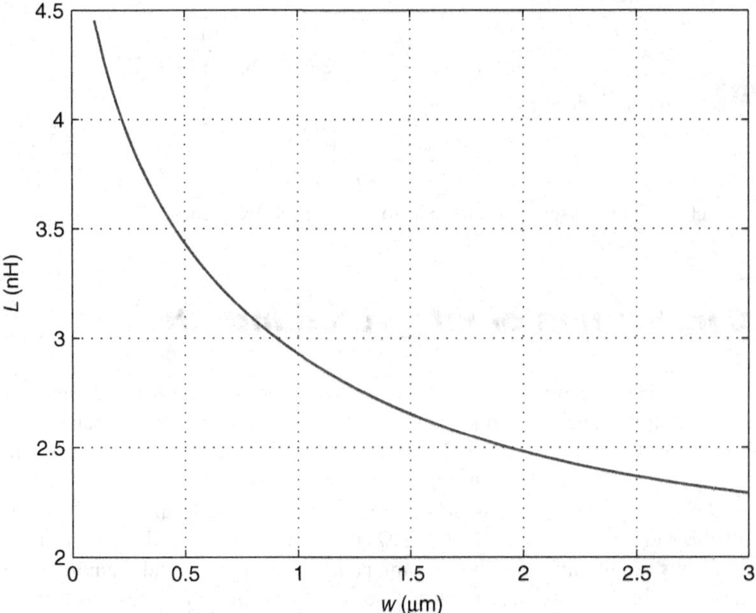

Figure 8.6 Inductance L as a function of trace height h for straight rectangular trace at $w = 1$ μm and $l = 50$ μm

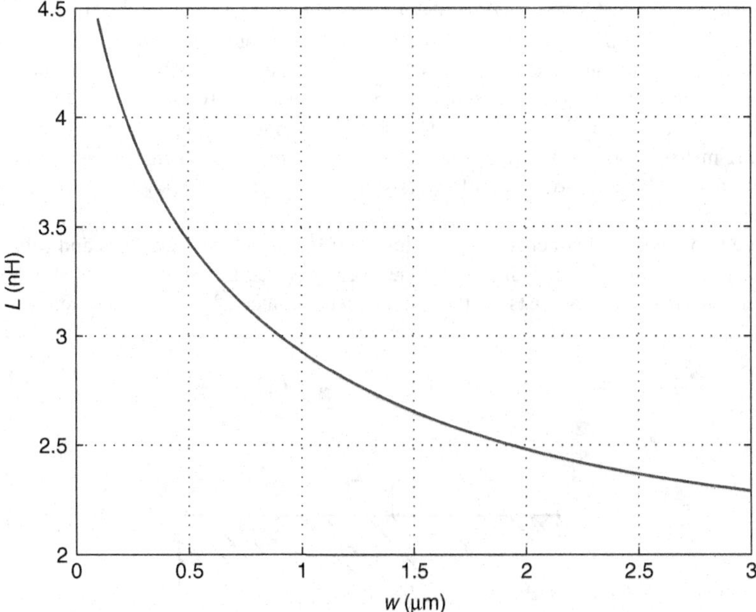

Figure 8.7 Inductance L as a function of trace length l for straight rectangular trace at $w = 1$ μm and $h = 1$ μm

yielding the internal inductance of a straight rectangular trace

$$L = \frac{\mu l}{w} \int_0^\infty e^{-2x/\delta} dx = \frac{\mu l \delta}{2w} = \frac{\mu l}{2w\sqrt{\pi \mu \sigma f}} \tag{8.36}$$

and the reactance of the inductor

$$X_L = \omega L = \frac{l}{w \sigma \delta} = \frac{l}{w}\sqrt{\frac{\pi \mu \sigma f}{\sigma}} = \frac{l}{w}\sqrt{\pi \mu \rho f}. \tag{8.37}$$

Thus, the inductance is inversely proportional to $\sqrt{f}$ at high frequencies.

8.6 Construction of Integrated Inductors

Figure 8.8 shows a cross section of a simple integrated inductor. It consists of a metal trace, silicon dioxide, and a substrate. The metal trace is made of a low-resistivity metal such as aluminum (Al), silver (Ag), copper (Cu), or gold (Au). The conductivities of these materials at room temperature are, respectively, $\sigma_{Al} = 3.65 \times 10^7$ S/m, $\sigma_{Ag} = 6.21 \times 10^7$ S/m, $\sigma_{Cu} = 5.88 \times 10^7$ S/m, and $\sigma_{Au} = 4.55 \times 10^7$ S/m. The typical thickness of the metal trace h ranges from 0.5 to 4 μm. The sheet DC resistance of the metal trace is from 7 to 55 mΩ per square. Most metal layers in ICs are made of aluminum, which is usually mixed with platinum, palladium, tungsten, and titanium. Electromigration in aluminum determines the maximum current density. It is the self-diffusion of metallic atoms in response to continuous current flow. The removal of metal causes an increase in metal resistance, reducing the inductor quality factor. Eventually, the resistance may increase enough to affect the RF circuit performance and cause the product to fail. Therefore, the inductor reliability becomes an important problem. Most IC processes have three to eight metal layers for interconnections. Some IC processes dedicated to wireless communication circuits have a thick top metal layer for making inductors. The metal trace is mounted on a silicon dioxide (SiO_2) layer whose thickness t_{ox} typically ranges from 0.4 to 3 μm. The silicon dioxide acts as an insulation between the metal trace and the substrate. The commonly used materials for substrate layers are silicon (Si), silicon germanium (SiGe), and gallium arsenide (GeAs). Silicon is the most extensively used material for manufacturing the substrates. Its resistivity is $\rho_{Si} = 10^{-4}$ Ω·m for heavy doping concentration $N = 10^{21}$ cm^{-3} and $\rho_{Si} = 100$ Ω·m for light doping concentration $N = 10^{13}$ cm^{-3}. In recent years, gallium arsenide is becoming a favorite material due to its low resistivity. The typical thickness of the substrate is from 500 to 700 μm.

There are two major components of power losses in IC inductors: metal loss and substrate loss. At high frequencies, the metal loss is increased due to eddy currents, resulting in the skin and proximity effects. The substrate loss consists of the magnetically induced loss and electrically induced loss.

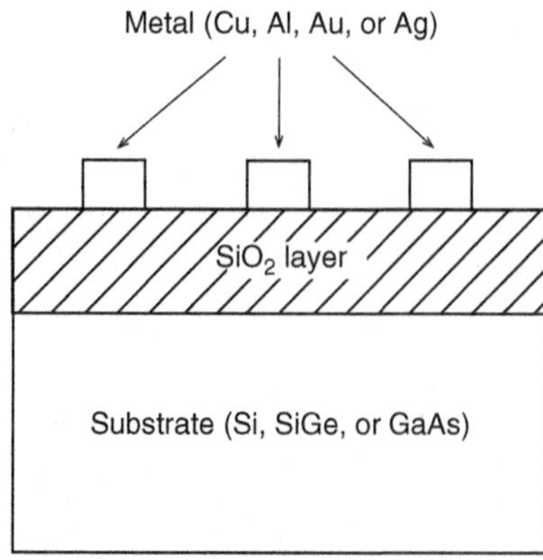

Figure 8.8 Cross section of an integrated inductor

In highly conductive substrates (such as silicon), the magnetically induced loss caused by eddy currents is dominant. In moderately conductive substrates, electrically induced losses caused by conduction and displacement currents are dominant. In inductors with a silicon substrate, the dominant component of the losses is the substrate loss. In inductors with a GsAs substrate, the metal loss is dominant.

8.7 Meander Inductors

Meander inductors have a simple layout as shown in Fig. 8.9. It is sufficient to use only one metal layer because both metal contacts are on the same metal level. This simplifies the technological process by avoiding two levels of photolithography. They usually have a low inductance per unit area and a large AC resistance because the length of the trace is large, causing a large DC resistance. This results in a low quality factor. The meander inductor can be divided into straight segments. The self-inductance of each straight segment is given by (8.27). The total self-inductance is equal to the sum of the self-inductances of all straight segments and is given by Stojanovic *et al.* [55]

$$L_{self} = 2L_a + 2L_b + NL_c + (N+1)L_s \quad (8.38)$$

where

$$L_a = \frac{\mu_0 a}{2\pi}\left[\ln\left(\frac{2a}{w+c}\right) + \frac{w+c}{3a} + 0.50049\right] \text{ (H)} \quad (8.39)$$

$$L_b = \frac{\mu_0 b}{2\pi}\left[\ln\left(\frac{2b}{w+c}\right) + \frac{w+c}{3b} + 0.50049\right] \text{ (H)} \quad (8.40)$$

$$L_c = \frac{\mu_0 c}{2\pi}\left[\ln\left(\frac{2c}{w+c}\right) + \frac{w+c}{3c} + 0.50049\right] \text{ (H)} \quad (8.41)$$

$$L_s = \frac{\mu_0 s}{2\pi}\left[\ln\left(\frac{2s}{w+c}\right) + \frac{w+c}{3s} + 0.50049\right] \text{ (H)} \quad (8.42)$$

N is the number of segments of length c and L_a, L_b, L_c, and L_s are self-inductances of the segments. All dimensions are in meters.

The mutual inductance between two equal parallel conductors (segments) shown in Fig. 8.10 is given by Grover [8]

$$M = \pm\frac{\mu_0 l}{2\pi}\left\{\ln\left[\frac{l}{s} + \sqrt{1 + \left(\frac{l}{s}\right)^2}\right] - \sqrt{1 + \left(\frac{s}{l}\right)^2} + \frac{s}{l}\right\} \text{ (H)} \quad (8.43)$$

where l is the length of the parallel conductors and s is the center-to-center separation of the conductors. All dimensions are in meters. If the currents flow in both parallel segments in the same direction, the mutual inductance is positive. If the currents flow in the parallel segments in opposite directions, the mutual inductance is negative. The mutual inductance of perpendicular conductors is zero. In meander inductors, the currents in longer adjacent segments flow in opposite directions, resulting in negative mutual inductance between two adjacent segments. Therefore, meander inductors suffer from a low inductance.

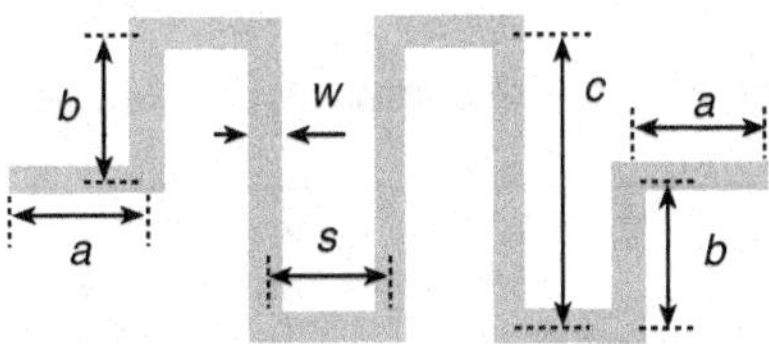

Figure 8.9 Meander inductor

Figure 8.10 Two equal parallel straight rectangular conductors

It can be shown that the input inductance of a transformer with the inductance of each winding equal to L and both currents flowing in the same direction is

$$L_i = L + M. \tag{8.44}$$

The input inductance of a transformer with the inductance of each winding equal to L and the currents flowing in opposite directions is

$$L_i = L - M. \tag{8.45}$$

The total inductance of the meander inductor is the sum of the self-inductances of all segments and the positive and negative mutual inductances.

The monomial equation for the total inductance of the meander inductor is given by Stojanovic *et al.* [55]

$$L = 0.00266a^{0.0603}c^{0.4429}N^{0.954}s^{0.606}w^{-0.173} \quad \text{(nH)} \tag{8.46}$$

where $b = c/2$. All dimensions are in micrometers. The accuracy of this equation is better than 12%. Figures 8.11 through 8.15 show plots of the inductance L as a function of w, s, c, a, and N for the meander inductor.

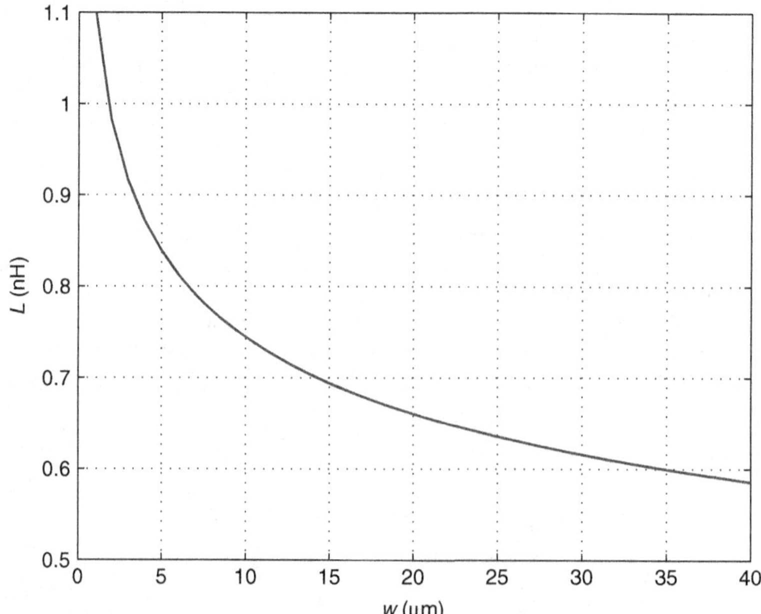

Figure 8.11 Inductance L as a function of trace width w for meander inductor at $N = 5$, $a = 40$ μm, $s = 20$ μm, and $c = 100$ μm

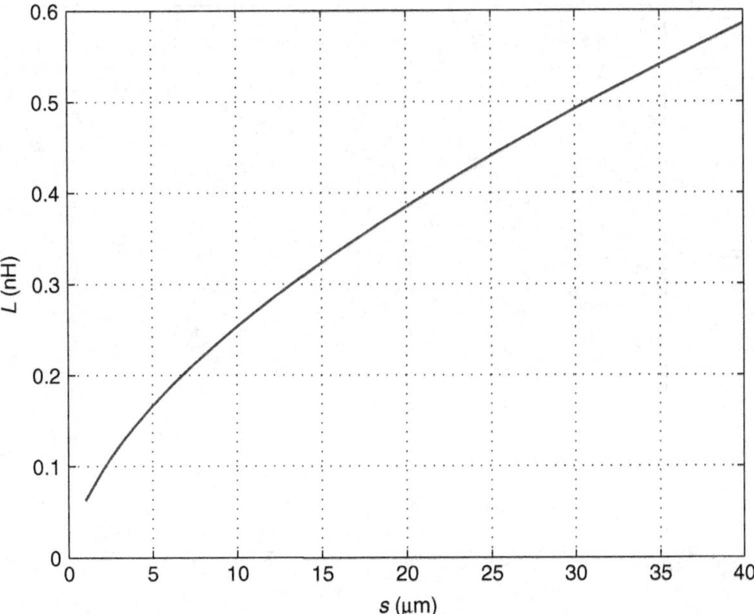

Figure 8.12 Inductance L as a function of separation s for meander inductor at $N = 5$, $a = 40$ μm, $w = 40$ μm, and $c = 100$ μm

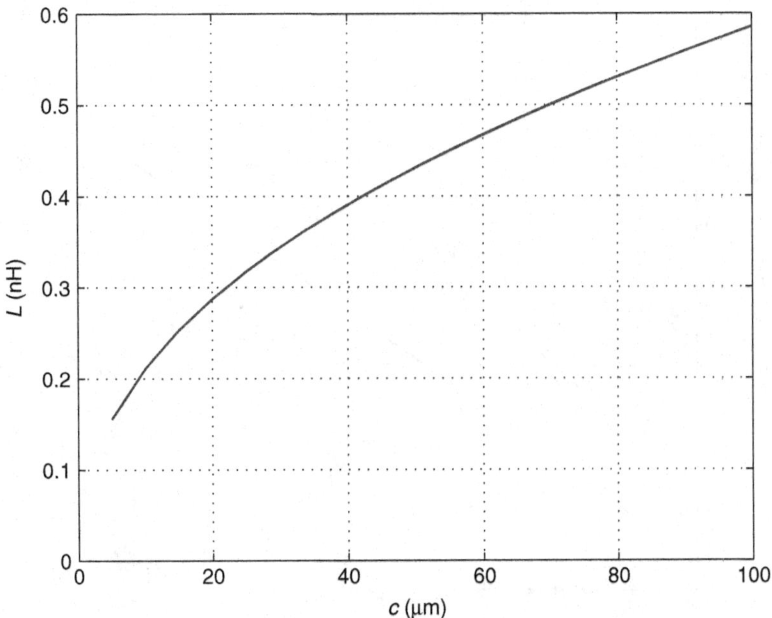

Figure 8.13 Inductance L as a function of segment length c for meander inductor at $N = 5$, $s = 40$ μm, $w = 40$ μm, and $a = 40$ μm

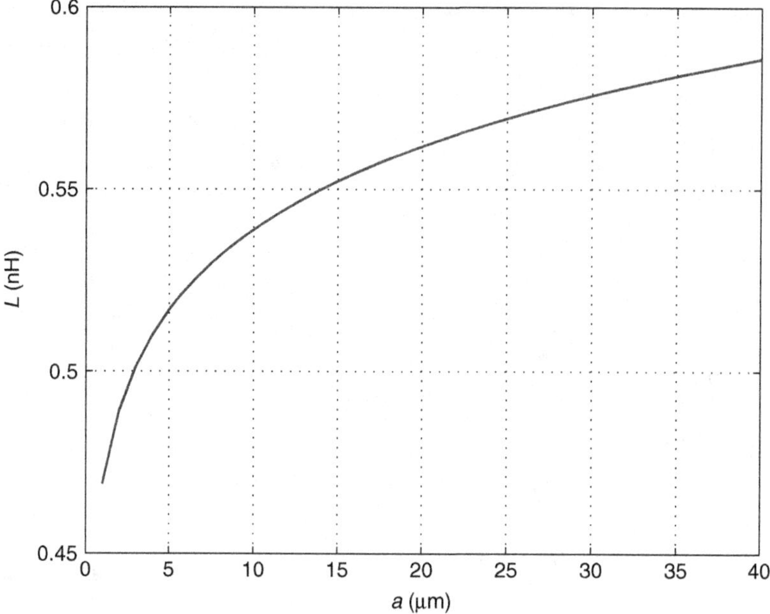

Figure 8.14 Inductance L as a function of a for meander inductor at $N = 5$, $s = 40$ μm, $w = 40$ μm, and $c = 100$ μm

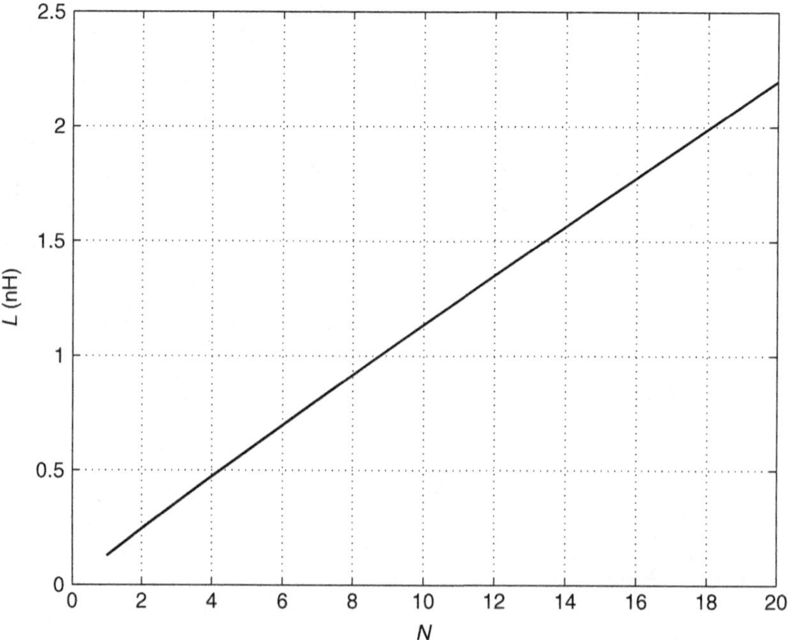

Figure 8.15 Inductance L as a function of N for meander inductor at $s = 40$ μm, $w = 40$ μm, $c = 100$ μm, and $a = 40$ μm

Example 8.4

Calculate the inductance of a meander inductor with $N = 5$, $w = 40$ μm, $s = 40$ μm, $a = 40$ μm, $c = 100$ μm, and $\mu_r = 1$.

Solution: The inductance of the trace is

$$L = 0.00266a^{0.0603}c^{0.4429}N^{0.954}s^{0.606}w^{-0.173}$$
$$= 0.00266 \times 40^{0.0603}100^{0.4429}5^{0.954}40^{0.606}40^{-0.173} = 0.585 \text{ nH}. \tag{8.47}$$

8.8 Inductance of Straight Round Conductor

The inductance of a round straight inductor of length l and conductor radius a, when the effects of nearby conductors (including the return current) and the skin effect can be neglected, is given by Grover [8]

$$L = \frac{\mu_0 l}{2\pi}\left[\ln\left(\frac{2l}{a}\right) + \frac{a}{l} - \frac{3}{4}\right] \text{ (H)} \tag{8.48}$$

where the dimensions are in meters. This equation can be simplified to the form

$$L = 0.0002l\left[\ln\left(\frac{2l}{a}\right) + \frac{a}{l} - 0.75\right] \text{ (nH)} \tag{8.49}$$

where the dimensions are in micrometers.

Another expression for a round straight conductor is

$$L = \frac{\mu l}{8\pi} \quad \text{for} \quad \delta > a. \tag{8.50}$$

The inductance at high frequencies decreases with frequency. The high-frequency inductance of a round straight inductor when the skin effect cannot be neglected is given by

$$L_{HF} = \frac{l}{4\pi a}\sqrt{\frac{\mu\rho}{\pi f}} \quad \text{for} \quad \delta < a. \tag{8.51}$$

Example 8.5

Calculate the inductance of a copper round straight inductor with $l = 2$ mm, $a = 0.1$ mm, and $\mu_r = 1$ for $\delta > h$. Also, calculate the inductance of a copper round straight inductor at $f = 1$ GHz.

Solution: The inductance of the round straight inductor for $\delta > a$ (low-frequency operation) is given by

$$L = \frac{\mu_0 l}{2\pi}\left[\ln\left(\frac{2l}{a}\right) + \frac{a}{l} - 0.75\right] \text{ (H)}$$
$$= \frac{4\pi \times 10^{-7} \times 2 \times 10^{-3}}{2\pi}\left[\ln\left(\frac{2 \times 2}{0.1}\right) + \frac{0.1}{2} - 0.75\right] = 1.1956 \text{ nH}. \tag{8.52}$$

The skin depth in copper at $f = 1$ GHz is

$$\delta = \sqrt{\frac{\rho}{\pi\mu_0 f}} = \sqrt{\frac{1.724 \times 10^{-8}}{\pi \times 4\pi \times 10^{-7} \times 1 \times 10^9}} = 2.089 \text{ μm}. \tag{8.53}$$

Since $a = 0.1$ mm $= 100$ μm, $\delta \ll a$. Therefore, the operation at $f = 1$ GHz is the high-frequency operation. The inductance of the round straight inductor at $f = 1$ GHz is

$$L_{HF} = \frac{l}{4\pi a}\sqrt{\frac{\mu_0 \rho}{\pi f}} = \frac{2 \times 10^{-3}}{4\pi \times 0.1 \times 10^{-3}}\sqrt{\frac{4\pi \times 10^{-7} \times 1.724 \times 10^{-8}}{\pi \times 1 \times 10^{9}}} = 4.179 \text{ pH}. \tag{8.54}$$

Hence, the ratio of the high-frequency inductance to the low-frequency inductance is

$$F_L = \frac{L_{HF}}{L_{LF}} = \frac{4.179 \times 10^{-12}}{1.1956 \times 10^{-9}} = 3.493 \times 10^{-3} = \frac{1}{286}. \tag{8.55}$$

8.9 Inductance of Circular Round Wire Loop

The inductance of a circular loop of round wire with loop radius r and wire radius a is [8]

$$L = \mu_0 a \left[\ln\left(\frac{8r}{a}\right) - 1.75\right] \text{ (H)}. \tag{8.56}$$

All dimensions are in meters. The dependence of the inductance on wire radius is weak.

The inductance of a round wire square loop is

$$L = \frac{\mu_0 l}{2\pi}\left[\ln\left(\frac{l}{4a}\right) - 0.52401\right] \text{ (H)} \tag{8.57}$$

where l is the loop length and a is the wire radius. All dimensions are in meters.

8.10 Inductance of Two-Parallel Wire Loop

Consider a loop formed by two parallel conductors whose length is l, radius is r, and the distance between the conductors is s, where $l \gg s$. The conductors carry equal currents in opposite directions. The inductance of the two-parallel wire loop is [8]

$$L = \frac{\mu_0 l}{\pi}\left[\ln\left(\frac{s}{r}\right) - \frac{s}{l} + 0.25\right] \text{ (H)}. \tag{8.58}$$

8.11 Inductance of Rectangle of Round Wire

The inductance of a rectangle of round wire of radius r with rectangle side lengths x and y is given by Grover [8]

$$L = \frac{\mu_0}{\pi}[x \ln\left(\frac{2x}{a}\right) + y \ln\left(\frac{2y}{a}\right) + 2\sqrt{x^2 + y^2} - x \text{ arcsinh}\left(\frac{x}{y}\right) - y \text{ arcsinh}\left(\frac{y}{x}\right) - 1.75(x + y). \tag{8.59}$$

8.12 Inductance of Polygon Round Wire Loop

The inductance of a single-loop polygon can be approximated by Grover [8]

$$L \approx \frac{\mu p}{2\pi}\left[\ln\left(\frac{2p}{a}\right) - \ln\left(\frac{p^2}{A}\right) + 0.25\right] \text{ (H)} \tag{8.60}$$

where p is the perimeter of the coil, A is the area enclosed by the coil, and a is the wire radius. All dimensions are in meters. The inductance is strongly dependent on the perimeter and weakly dependent on the loop area and wire radius. Inductors enclosing the same perimeter with similar shape have approximately the same inductance.

The inductance of a triangle is [8]

$$L = \frac{\mu_0 l}{2\pi}\left[\ln\left(\frac{l}{3a}\right) - 1.15546\right] \text{ (H)}. \tag{8.61}$$

where a is the wire radius. The inductance of a square is

$$L = \frac{\mu_0 l}{2\pi}\left[\ln\left(\frac{l}{4a}\right) - 0.52401\right] \text{ (H)}. \tag{8.62}$$

The inductance of a pentagon is

$$L = \frac{\mu_0 l}{2\pi}\left[\ln\left(\frac{l}{5a}\right) - 0.15914\right] \text{ (H)}. \tag{8.63}$$

The inductance of a hexagon is

$$L = \frac{\mu_0 l}{2\pi}\left[\ln\left(\frac{l}{6a}\right) + 0.09848\right] \text{ (H)}. \tag{8.64}$$

The inductance of a octagon is

$$L = \frac{\mu_0 l}{2\pi}\left[\ln\left(\frac{l}{8a}\right) + 0.46198\right] \text{ (H)}. \tag{8.65}$$

All dimensions are in meters.

8.13 Bondwire Inductors

Bondwire inductors are frequently used in RF ICs. Figure 8.16 shows a bondwire inductor. It can be regarded as a fraction of a round turn. Bondwires are often used for chip connections. The main advantage of bondwire inductors is a very small series resistance. Standard bondwires have a relatively large diameter of about 25 μm and can handle substantial currents with low loss. They may be placed well above any conductive planes to reduce parasitic capacitances and hence increase the SRF f_r. Typical inductances range from 2 to 5 nH. Since bondwires have much larger cross-sectional area than the traces of planar spiral inductors, these inductors have lower resistances and hence a higher quality factor Q_{Lo}, typically $Q_{Lo} = 20–50$ at 1 GHz. These inductors may be placed well above any conductive planes to reduce parasitic capacitances, yielding a higher SRF f_r. The low-frequency inductance of bondwire inductors is given by

$$L \approx \frac{\mu_0 l}{2\pi}\left[\ln\left(\frac{2l}{a}\right) - 0.75\right] \text{ (H)} \tag{8.66}$$

where l is the length of the bondwire and a is the radius of the bondwire. A standard bondwire inductance with $l = 1$ mm gives $L = 1$ nH or 1 nH/mm. The resistance is

$$R = \frac{l}{2\pi a\delta\sigma} \tag{8.67}$$

where σ is the conductivity and δ is the skin depth. For aluminum, $\sigma = 4 \times 10^7$ S/m. The skin depth for aluminum is $\delta_{Al} = 2.5$ μm at 1 GHz. The resistance per unit length is $R_{Al}/l \approx 0.2$ Ω·/mm at 2 GHz.

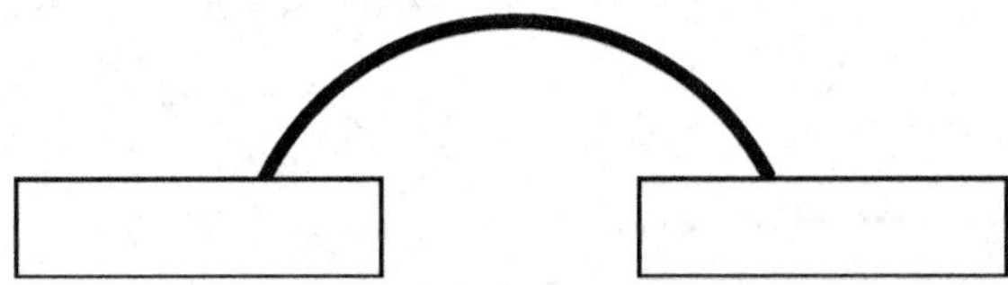

Figure 8.16 Bondwire inductor

Bondwires are often used as interconnects and package leads in ICs. A large number of closely spaced bondwires are often connected in parallel. The mutual inductance of two parallel bond wires is

$$M = \frac{\mu_0 l}{2\pi}\left(\ln\frac{2l}{s} + \frac{s}{l} - 1\right) \tag{8.68}$$

where s is the distance (separation) between the two bondwires. Typically, $M = 0.3$ nH/mm at $s = 0.2$ mm. A typical value of the coupling coefficient k is 0.3.

The major disadvantage of bondwire inductors is low predictability because of possible variations in length and spacing. Bondwire inductance depends on bonding geometry and the existence of neighboring bondwires, making accurate prediction of the inductance value difficult. However, once the configuration for a particular inductance is known, it is rather repeatable in subsequent bondings. The sensitivity of the circuit performance to the variations in the bondwire inductances and mutual inductances should be examined. Common bondwire metals include gold and aluminum. Gold is generally preferred because of its higher conductivity and flexibility. This allows higher quality factor Q_{Lo} with shorter physical lengths to be bonded for a given die height.

Example 8.6

Calculate the inductance of a round bondwire inductor with $l = 2$ mm, $a = 0.2$ mm, and $\mu_r = 1$ for $\delta > a$.

Solution: The inductance of the bondwire inductor for $\delta > a$ is

$$L \approx \frac{\mu_0 l}{2\pi}\left[\ln\left(\frac{2l}{a}\right) - 0.75\right] = \frac{4\pi \times 10^{-7} \times 2 \times 10^{-3}}{2\pi}\left[\ln\left(\frac{2\times 2}{0.2}\right) - 0.75\right] = 0.8982 \text{ nH}. \tag{8.69}$$

8.14 Single-Turn Planar Inductor

The self-inductance is a special case of mutual inductance. Therefore, an expression for the self-inductance of an inductor can be derived using the concept of the mutual inductance. This is demonstrated by deriving an expression for the inductance of a single-turn planar inductor.

Figure 8.17 shows a single-turn strip inductor of width w, inner radius b, and outer radius a. The concept of mutual inductance can be used to determine an external self-inductance. Assume that the

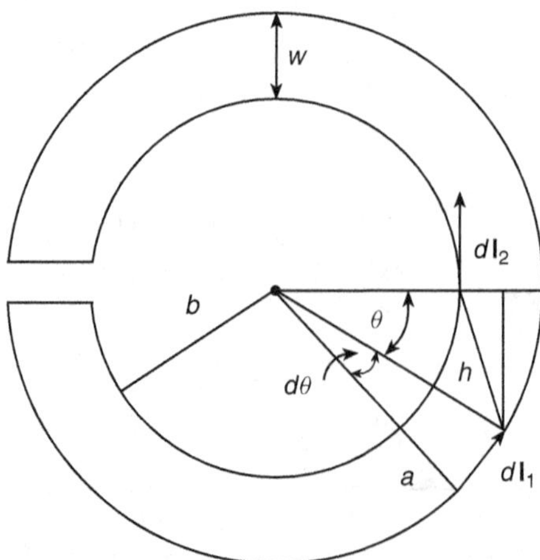

Figure 8.17 Single-turn planar inductor

conductor width w is small compared with loop radius a. In this case, the magnetic field is nearly the same as if the current were concentrated along the center of the conductor. Therefore, the external inductance of the loop can be well approximated by the mutual inductance between two filaments: one through the center of the conductor and the other along inside edge. Consider two circuits: a line circuit along the inner surface of the strip and along the axis (middle) of the strip. The vector magnetic potential caused by circuit 2 is

$$\mathbf{A}_2 = I_2 \oint_{C_2} \frac{\mu d\mathbf{l}_2}{4\pi h}. \tag{8.70}$$

The integration of the vector magnetic potential $\mathbf{A}_2$ along circuit 1 resulting from the current I_2 in circuit 2 gives the mutual inductance between the two circuits

$$M = \frac{1}{I_2} \oint_{C_1} \mathbf{A}_2 \cdot d\mathbf{l}_1. \tag{8.71}$$

Hence, the mutual inductance of two conductors is given by

$$M = \frac{\mu}{4\pi} \oint_{C_1} \oint_{C_2} \frac{d\mathbf{l}_1 \cdot d\mathbf{l}_2}{h}. \tag{8.72}$$

From geometrical consideration and the following substitution

$$\theta = \pi - 2\phi \tag{8.73}$$

we have

$$dl_1 = |d\mathbf{l}_1| = a d\theta = -2a d\phi. \tag{8.74}$$

yielding

$$d\mathbf{l}_1 \cdot d\mathbf{l}_2 = dl_1 dl_2 \cos\theta = 2a \cos 2\phi dl_2 d\phi \tag{8.75}$$

and

$$\begin{aligned} h &= \sqrt{(a\cos\theta - b)^2 + (a\sin\theta)^2} = \sqrt{a^2 + b^2 - 2ab\cos\theta} = \sqrt{a^2 + b^2 + 2ab\cos 2\phi} \\ &= \sqrt{a^2 + b^2 + 2ab(1 - 2\sin^2\phi)} = \sqrt{(a+b)^2 - 4ab\sin^2\phi} = (a+b)\sqrt{1 - k^2\sin^2\phi} \end{aligned} \tag{8.76}$$

where

$$k^2 = \frac{4ab}{(a+b)^2} = 1 - \frac{(a-b)^2}{(a+b)^2}. \tag{8.77}$$

Also,

$$\oint dl_2 = 2\pi b. \tag{8.78}$$

The mutual inductance is

$$\begin{aligned} M &= \frac{\mu}{4\pi} \oint dl_2 \int_0^{2\pi} \frac{a\cos\theta d\theta}{\sqrt{a^2 + b^2 - 2ab\cos\theta}} = \frac{2\mu ab}{a+b} \int_0^{\pi/2} \frac{(2\sin^2\phi - 1)d\phi}{\sqrt{1 - k^2\sin^2\phi}} \\ &= \mu\sqrt{ab}k \int_0^{\pi/2} \frac{(2\sin^2\phi - 1)d\phi}{\sqrt{1 - k^2\sin^2\phi}} = \mu\sqrt{ab}\left[\left(\frac{2}{k} - k\right)K(k) - \frac{2}{k}E(k)\right] \end{aligned} \tag{8.79}$$

where the complete elliptic integrals are

$$E(k) = \int_0^{\pi/2} \sqrt{1 - k^2\sin^2\phi} d\phi \tag{8.80}$$

and

$$K(k) = \int_0^{\pi/2} \frac{d\phi}{\sqrt{1 - k^2\sin^2\phi}}. \tag{8.81}$$

The width of the strip is

$$w = a - b \tag{8.82}$$

resulting in

$$a + b = 2a - w \tag{8.83}$$

$$ab = a(a - w) = a^2\left(1 - \frac{w}{a}\right) \tag{8.84}$$

and

$$k^2 = 1 - \frac{w^2}{(2a - w)^2}. \tag{8.85}$$

If $w/a \ll 1$, $k \approx 1$, $E(k) \approx 1$,

$$k^2 = 1 - \frac{w^2}{4a^2\left(1 - \frac{w}{2a}\right)^2} \approx 1 - \frac{w^2}{4a^2} \tag{8.86}$$

and

$$K(k) \approx \ln\frac{4}{\sqrt{1-k^2}} \approx \ln\left[\frac{4}{\sqrt{1 - \left(1 - \frac{w^2}{4a^2}\right)}}\right] = \ln\left(\frac{8a}{w}\right). \tag{8.87}$$

Hence, the self-inductance of a single-turn inductor is given by

$$L \approx \mu a\left[\ln\left(\frac{8a}{w}\right) - 2\right] = \frac{\mu l}{2\pi}\left[\ln\left(\frac{8a}{w}\right) - 2\right] \quad \text{for} \quad a \gg w \tag{8.88}$$

where $l = 2\pi a$. The self-inductance of an inductor with N turns is

$$L \approx \mu a N^2\left[\ln\left(\frac{8a}{w}\right) - 2\right] = \frac{\mu l N^2}{2\pi}\left[\ln\left(\frac{8a}{w}\right) - 2\right] \quad \text{for} \quad a \gg w. \tag{8.89}$$

Example 8.7

Calculate the inductance of a single-turn round planar inductor with $a = 100$ μm, $w = 1$ μm, and $\mu_r = 1$ for $\delta > a$.

Solution: The inductance of the single-turn round planar inductor for $\delta < a$ is

$$L \approx \mu a\left[\ln\left(\frac{8a}{w}\right) - 2\right] = 4\pi \times 10^{-7} \times 100 \times 10^{-6}\left[\ln\left(\frac{8 \times 100}{1}\right) - 2\right] = 0.58868 \text{ nH}. \tag{8.90}$$

8.15 Inductance of Planar Square Loop

The self-inductance of a square coil made of rectangular wire of length $l \gg w$ and width w is

$$L \approx \frac{\mu_0 l}{\pi}\left[\text{arcsinh}\left(\frac{l}{2w}\right) - 1\right] \text{ (H)}. \tag{8.91}$$

Both l and w are in meters.

8.16 Planar Spiral Inductors

8.16.1 Geometries of Planar Spiral Inductors

Planar spiral inductors are the most widely used RF IC inductors. Figure 8.18 shows square, octagonal, hexagonal, and circular planar spiral inductors. The circular spiral inductor has a shorter conductor

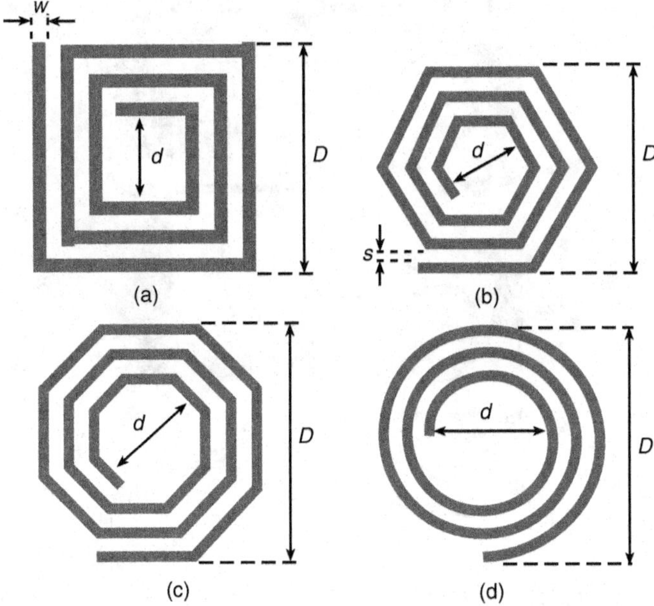

Figure 8.18 Planar spiral integrated inductors. (a) Square inductor. (b) Hexagonal inductor. (c) Octagonal inductor. (d) Circular inductor

than the square conductor and the quality factor is about 10% higher than that of a square spiral inductor with the same outer diameter D. A planar spiral inductor consists of a low-resistivity metal trace (aluminum, copper, gold, or silver), silicon dioxide SiO_2 of thickness t_{ox}, and a silicon substrate, as shown in Fig. 8.19. A metal layer embedded in silicon dioxide is used to form the metal spiral. The topmost metal layer is usually the thickest and thus most conductive. In addition, a larger distance of the topmost metal layer to the substrate reduces the parasitic capacitance $C = \epsilon A_m / t_{ox}$, increasing the SRF $f_r = 1/(2\pi\sqrt{LC})$, where A_m is the metal trace area. The substrate is a thick silicon (Si), gallium arsenide (GaAs), or silicon germanium (SiGe) layer of thickness of the order 500–700 μm. Thin silicon dioxide SiO_2 of thickness 0.4–3 μm is used to isolate the metal strips of the inductor from the silicon substrate. The outer end of the spiral is connected directly to a port. The connection of the innermost turn is made through an *underpass* by another metal layer or via an *air bridge*. The whole spiral structure is connected to pads and surrounded by a ground plane.

Commonly used shapes of spiral inductors are square, rectangular, hexagonal, octagonal, and circular. Hexagonal and octagonal spiral inductors generally have less inductance and less series resistance per turn compared to square spirals. Since the hexagonal and octagonal structures occupy larger chip area, they are rearly used.

A planar spiral integrated inductor can only be fabricated with technologies having two or more metal layers because the inner connection requires a metal layer different than that used for the spiral to connect the inside turn of the coil to outside. Square inductor shapes are the most compatible with IC layout tools. They are easily designed with Manhattan-style physical layout tools, such as MAGIC. Hexagonal, octagonal, and higher order polygon spiral inductors have higher quality factor Q_{Lo} than square spirals.

The parameters of interest for integrated inductors are the inductance L_s, the quality factor Q_{Lo}, and the SRF f_r. The typical inductances are in the range 1–30 nH, the typical quality factors are from 5 to 20, and the typical SRFs are from 2 to 20 GHz. Low values of the quality factor Q is the major disadvantage of planar spiral inductors. The geometrical parameters of an inductor are the number of turns N, the metal strip width w, the metal strip height h, the turn-to-turn space s, the

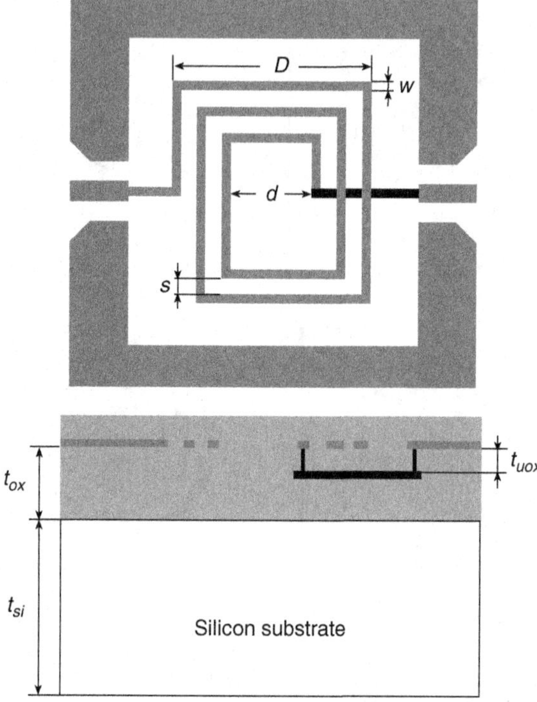

Figure 8.19 Cross section of a planar RF IC inductor

inner diameter d, the outer diameter D, the thickness of the silicon substrate t_{Si}, the thickness of the oxide layer t_{ox}, and the thickness of oxide between the metal strip and the underpass t_{uox}. The typical metal strip width w is 30 μm, the typical turn-to-turn spacing s is 20 μm, and the typical metal strip sheet DC resistance is $R_{sheet} = \rho/h = 0.03\text{–}0.1\ \Omega$. IC inductors require a lot of chip area. Therefore, practical IC inductors have small inductances, typically $L \leq 10$ nH, but they can be as high as 30 nH. The typical inductor size is from 130 × 130 μm to 1000 × 1000 μm. The inductor area is $A = D^2$.

The topmost metal layer is usually used for construction of IC inductors because it is the thickest metal layer and thus has the lowest resistance. In addition, the distance from the topmost layer to the substrate is the largest, reducing the parasitic capacitances.

The preferred metallization for integrated inductors is a low-resistivity, inert metal, such as gold. Other low-resistivity metals such as silver and copper do not offer the same level of resistance to atmospheric sulfur and moisture. Platinum, another noble metal, is two times more expensive than gold and has higher resistivity. The thickness of the metal h should be higher than 2δ, one skin depth δ on the top of the trace and one skin depth on the bottom. The direction of the magnetic flux is perpendicular to the substrate. The magnetic field penetrates the substrate and induces eddy currents. A high substrate conductivity tends to lower the inductor quality factor Q_{Lo}. The optimization of the spiral geometry and the metal trace width leads to minimization of the trace ohmic resistance and the substrate capacitance. To reduce power losses in the substrate, the substrate can be made of high-resistive silicon oxide such as silicon-on-insulator (SOI), thick dielectric layers, or thick and multilayer conductor lines.

Figure 8.20 shows a ground shield inserted between the metal trace and the substrate to reduce the paths for eddy currents. A shield is made of a lower metal layer, a lower layer, or a polysilicon layer. This is patterned to reduce the lengths of eddy current paths, thus reducing power loss. Eddy currents flow in the direction perpendicular to the magnetic field, which is parallel to the metal trace and the substrate.

Figure 8.20 Ground shield in a planar RF IC inductor

8.16.2 Inductance of Square Planar Inductors

The total inductance of spiral inductors is equal to the sum of the self-inductances of the straight segments and the sum of the mutual inductances between the segments. If currents in two parallel segments flow in the same direction, the mutual inductance is positive. If currents in two parallel segments flow in opposite directions, the mutual inductance is negative. The mutual inductance of perpendicular conductors is zero. In square spiral inductors, the currents in adjacent parallel segments flow in the same direction, yielding a high inductance.

Several expressions have been developed to estimate RF spiral planar inductances. In general, the inductance of planar inductors L increases when the number of turns N and the area of an inductor A increase. The inductance decreases when the distance between the ground plane and the metal trace decreases.

The inductance of a single loop of area A and any shape (e.g., square, rectangular, hexagonal, octagonal, or circular) is given by

$$L \approx \mu_0 \sqrt{\pi A}. \tag{8.92}$$

Hence, the inductance of a single-turn round inductor of radius r is

$$L \approx \pi \mu_0 r = 4\pi^2 \times 10^{-7} r = 4 \times 10^{-6} r \text{ (H)}. \tag{8.93}$$

For example, $L = 4$ nH for $r = 1$ mm.

The inductance of an arbitrary planar spiral inductor with N turns, often used in ICs, is given by

$$L \approx \pi \mu_0 r N^2 = 4\pi^2 \times 10^{-7} r N^2 = 4 \times 10^{-6} r N^2 \text{ (H)} \tag{8.94}$$

where N is the number of turns and r is the spiral radius.

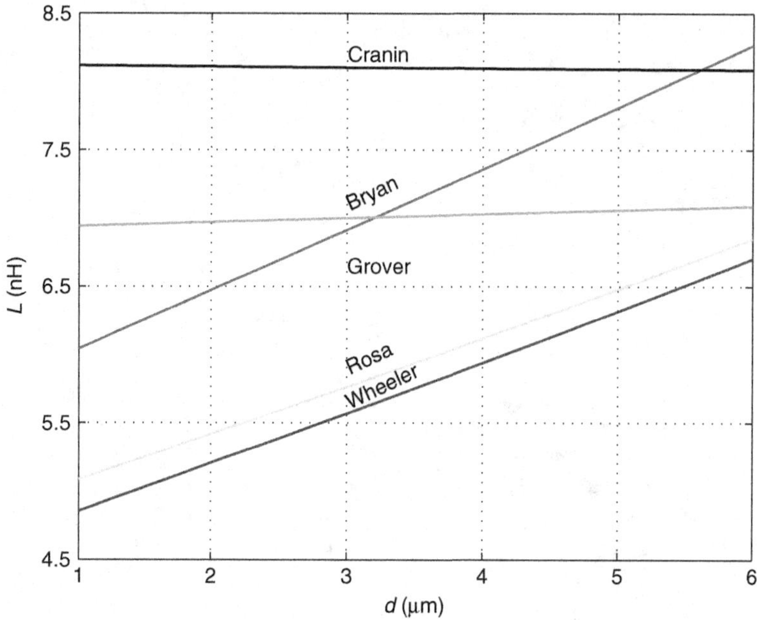

Figure 8.21 Inductance L as a function of inner diameter d for square spiral inductor at $N = 5$, $w = 30$ μm, $s = 20$ μm, and $h = 20$ μm

Bryan's Formula

The inductance of the square planar spiral inductor is given by empirical Bryan's equation [8]

$$L = 6.025 \times 10^{-7}(D + d)N^{\frac{5}{3}} \ln\left[4\left(\frac{D + d}{D - d}\right)\right] \quad \text{(H)} \tag{8.95}$$

where the outermost diameter is

$$D = d + 2N\omega + 2(N - 1)s. \tag{8.96}$$

N is the number of turns, and D and d are the outermost and innermost diameters of the inductor in meters, respectively.

Figures 8.21 through 8.24 illustrate equations developed by various authors for the inductance L as a function of d, w, s, a, and N. The inductance L of spiral planar inductors increases when the trace width w, spacing between the traces s, the trace length l, and the number of turns N increase.

Example 8.8

Calculate the inductance of a square planar spiral inductor with $N = 5$, $d = 60$ μm, $w = 30$ μm, $s = 20$ μm, and $\mu_r = 1$ for $\delta > h$. Use Bryan's formula.

Solution: The outermost diameter is

$$D = d + 2N\omega + 2(N - 1)s = 60 + 2 \times 5 \times 30 + 2 \times (5 - 1) \times 20 = 520\ \mu\text{m}. \tag{8.97}$$

The inductance of the round planar inductor for $\delta > h$ is

$$L = 6.025 \times 10^{-7}(D + d)N^{\frac{5}{3}} \ln\left[4\left(\frac{D + d}{D - d}\right)\right] \quad \text{(H)}$$

$$= 6.025 \times 10^{-7} \times (520 + 60) \times 10^{-6} \times 5^{\frac{5}{3}} \ln\left[4\left(\frac{520 + 60}{520 - 60}\right)\right] = 8.266\ \text{nH}. \tag{8.98}$$

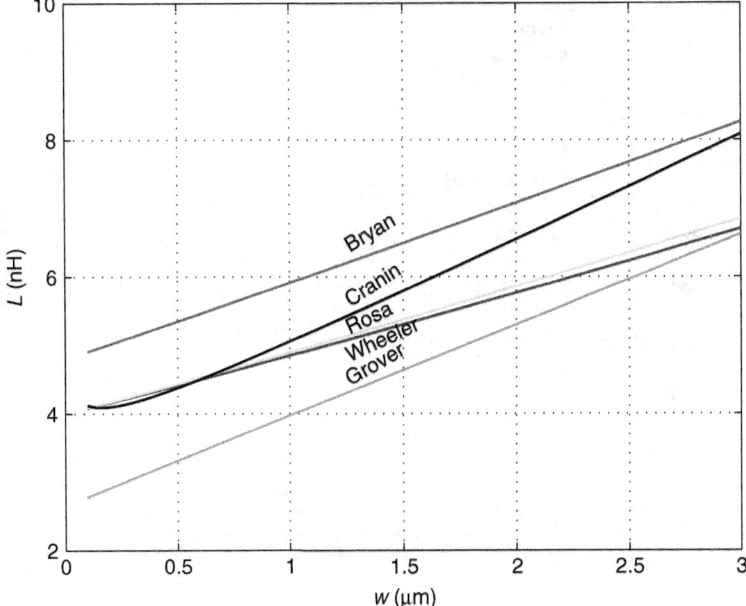

Figure 8.22 Inductance L as a function of trace width w at $N = 5$, $d = 60$ μm, $s = 20$ μm, and $h = 3$ μm

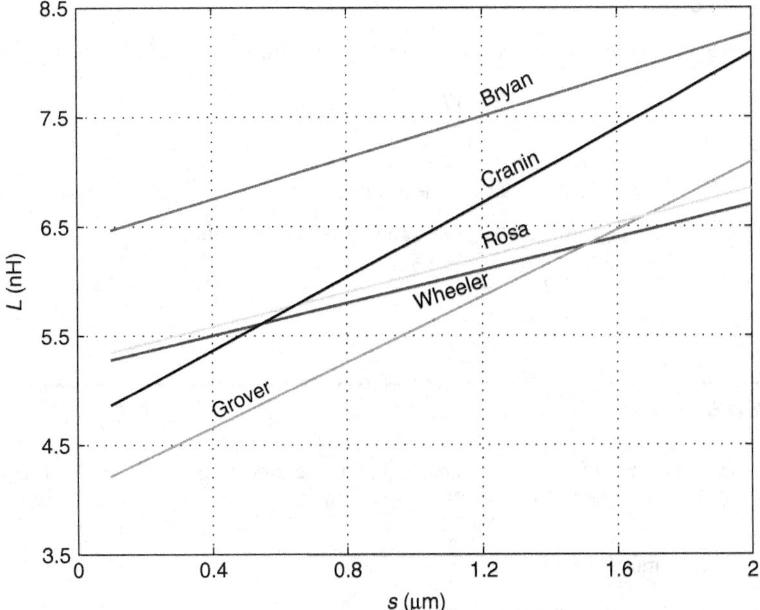

Figure 8.23 Inductance L as a function of trace separation s for square spiral inductor at $N = 5$, $d = 60$ μm, $w = 30$ μm, and $h = 3$ μm

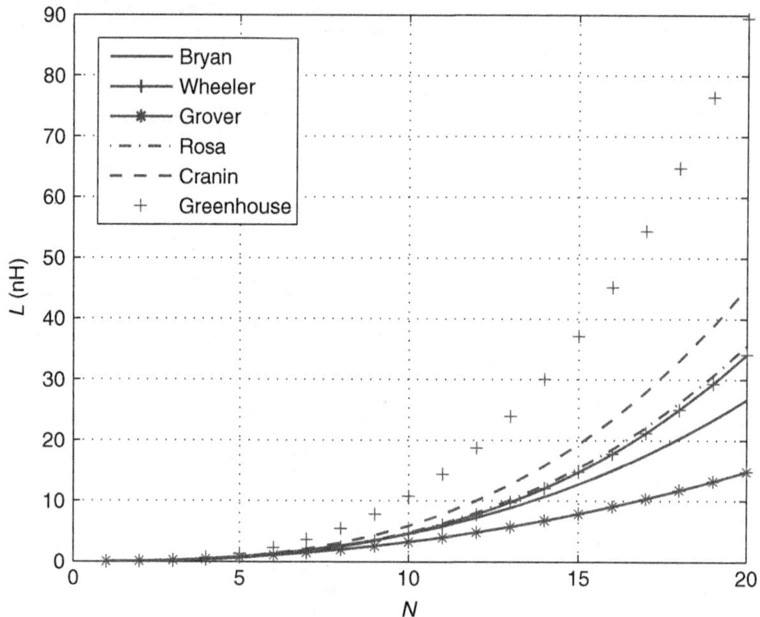

Figure 8.24 Inductance L as a function of number of turns N for square spiral inductor at $d = 60$ μm, $w = 30$ μm, $s = 20$ μm, and $h = 3$ μm

Wheeler's Formula

The inductance of a square spiral inductor is given by modified Wheeler's formula [4, 36]

$$L = 1.17\mu_0 N^2 \frac{D+d}{1 + 2.75\frac{D-d}{D+d}} \text{ (H)} \tag{8.99}$$

where

$$D = d + 2N\omega + 2(N-1)s. \tag{8.100}$$

All dimensions are in m. The inductance in terms of N, w, and s is

$$L = \frac{2.34\mu_0 N^2}{3.75N(w+s) - 3.75s + d} \text{ (H)}. \tag{8.101}$$

Example 8.9

Calculate the inductance of a square planar spiral inductor with $N = 5$, $d = 60$ μm, $w = 30$ μm, $s = 20$ μm, and $\mu_r = 1$ for $\delta > h$. Use Wheeler's formula.

Solution: The outermost diameter is

$$D = d + 2N\omega + 2(N-1)s = 60 + 2 \times 5 \times 30 + 2 \times (5-1) \times 20 = 520 \text{ μm}. \tag{8.102}$$

The inductance of the square planar inductor for $\delta > h$ is

$$L = 1.17\mu_0 N^2 \frac{D+d}{1 + 2.75\frac{D-d}{D+d}} \text{ (H)}$$
$$= 1.17 \times 4\pi \times 10^{-7} \times 5^2 \times \frac{(520+60) \times 10^{-6}}{1 + 2.75 \times \frac{520-60}{520+60}} = 6.7 \text{ nH}. \tag{8.103}$$

Greenhouse Formula

The inductance of spiral inductors is given by Greenhouse's equation [14]

$$L = 10^{-9} DN^2 \left\{ 180 \left[\log \left(\frac{D}{4.75(w+h)} \right) + 5 \right] + \frac{1.76(w+h)}{D} + 7.312 \right\} \quad \text{(nH)}. \tag{8.104}$$

All dimensions are in m.

Example 8.10

Calculate the inductance of a square planar spiral inductor with $N = 5$, $d = 60$ μm, $w = 30$ μm, $s = 20$ μm, $h = 20$ μm, and $\mu_r = 1$ for $\delta > h$. Use Greenhouse's formula.

Solution: The outermost diameter is

$$D = d + 2N\omega + 2(N-1)s = 60 + 2 \times 5 \times 30 + 2 \times (5-1) \times 20 = 520 \ \mu\text{m}. \tag{8.105}$$

The inductance of the single-turn round planar inductor for $\delta > h$ is

$$\begin{aligned} L &= 10^{-9} DN^2 \left\{ 180 \left[\log \left(\frac{D}{4.75(w+h)} \right) + 5 \right] + \frac{1.76(w+h)}{D} + 7.312 \right\} \quad \text{(nH)} \\ &= 10^{-9} \times 520 \times 10^{-6} \times 5^2 \left\{ 180 \left[\log \left(\frac{520}{4.75(30+20)} \right) + 5 \right] + \frac{1.76(30+20)}{520} + 7.312 \right\} \\ &= 12.59 \text{ nH}. \end{aligned} \tag{8.106}$$

Grover's Formula

The inductance of a square or rectangular spiral planar inductor with a rectangular cross section of the trace is expressed as [8, 34]

$$L = \frac{\mu_0 l}{2\pi} \left[\ln \left(\frac{2l}{w+h} \right) + \frac{w+h}{3l} + 0.50049 \right] \quad \text{(H)} \tag{8.107}$$

where the length of the metal trace is

$$l = 2(D + w) + 2N(2N-1)(w+s), \tag{8.108}$$

the outer diameter is

$$D = d + 2N\omega + 2(N-1)s, \tag{8.109}$$

and h is the metal trace height. All dimensions l, w, s, and h are in meters. This equation is identical as that for a straight trace inductor. It accounts for the self-inductance only and neglects mutual inductances between the traces.

Example 8.11

Calculate the inductance of a square planar spiral inductor with $N = 5$, $d = 60$ μm, $w = 30$ μm, $s = 20$ μm, $h = 20$ μm, and $\mu_r = 1$ for $\delta > h$. Use Grover's formula.

Solution: The outer diameter is

$$D = d + 2N\omega + 2(N-1)s = 60 + 2 \times 5 \times 30 + 2 \times (5-1) \times 20 = 520 \ \mu\text{m}. \tag{8.110}$$

The length of the metal trace is

$$l = 2(D + w) + 2N(2N - 1)(w + s) = 2(520 + 30) + 2 \times 5(2 \times 5 - 1)(30 + 20) = 5600\ \mu\text{m}. \tag{8.111}$$

The inductance of the square spiral planar inductor for $\delta > h$ is

$$\begin{aligned} L &= \frac{\mu_0 l}{2\pi}\left[\ln\left(\frac{2l}{w+h}\right) + \frac{w+h}{3l} + 0.50049\right] \text{ (H)} \\ &= \frac{4\pi \times 10^{-7} \times 5600 \times 10^{-6}}{2\pi}\left[\ln\left(\frac{2 \times 5600}{30+20}\right) + \frac{30+20}{3 \times 5600} + 0.50049\right] \\ &= 6.608 \text{ nH}. \end{aligned} \tag{8.112}$$

Rosa's Formula

The equation for the inductance of square planar spiral inductors is expressed as [1, 9, 36]

$$L = 0.3175\mu N^2(D + d)\left\{\ln\left[\frac{2.07(D+d)}{D-d}\right] + \frac{0.18(D-d)}{D+d} + 0.13\left(\frac{D-d}{D+d}\right)^2\right\} \text{ (H)}. \tag{8.113}$$

All dimensions are in meters.

Example 8.12

Calculate the inductance of a square planar spiral inductor with $N = 5$, $d = 60\ \mu\text{m}$, $w = 30\ \mu\text{m}$, $s = 20\ \mu\text{m}$, $h = 20\ \mu\text{m}$, and $\mu_r = 1$ for $\delta > h$. Use Rosa's formula.

Solution: The outer diameter is

$$D = d + 2N\omega + 2(N - 1)s = 60 + 2 \times 5 \times 30 + 2 \times (5 - 1) \times 20 = 520\ \mu\text{m}. \tag{8.114}$$

The inductance of the square spiral planar inductor for $\delta > h$ is

$$\begin{aligned} L &= 0.3175\mu_0 N^2(D + d)\left\{\ln\left[\frac{2.07(D+d)}{D-d}\right] + \frac{0.18(D-d)}{D+d} + 0.13\left(\frac{D-d}{D+d}\right)^2\right\} \\ &= 0.3175 \times 4\pi \times 10^{-7} \times 5^2 \times (520 + 60) \times 10^{-6} \\ &\quad \times \left\{\ln\left[\frac{2.07(520+60)}{520-60}\right] + \frac{0.18 \times (520-60)}{520+60} + 0.13\left(\frac{520-60}{520+60}\right)^2\right\} \\ &= 6.849 \text{ nH}. \end{aligned} \tag{8.115}$$

Cranin's Formula

An empirical expression for the inductance of spiral planar inductors that has less than 10% error for inductors in the range 5–50 nH is given by Craninckx and Steyeart [22]

$$L \approx 1.3 \times 10^{-7}\frac{A_m^{5/3}}{A_{tot}^{1/6} w^{1.75}(w + s)^{0.25}} \text{ (nH)} \tag{8.116}$$

where A_m is the metal area, A_{tot} is the total inductor area, w is the metal trace width, and s is the spacing between metal traces. All dimensions are in meters.

Example 8.13

Calculate the inductance of a square planar spiral inductor with $N = 5$, $d = 60$ μm, $w = 30$ μm, $s = 20$ μm, $h = 20$ μm, and $\mu_r = 1$ for $\delta > h$. Use Cranin's formula.

Solution: The outer diameter is

$$D = d + 2N\omega + 2(N - 1)s = 60 + 2 \times 5 \times 30 + 2 \times (5 - 1) \times 20 = 520 \ \mu\text{m}. \tag{8.117}$$

The length of the metal trace is

$$l = 2(D + w) + 2N(2N - 1)(w + s) = 2(520 + 30) + 2 \times 5(2 \times 5 - 1)(30 + 20) = 5600 \ \mu\text{m}. \tag{8.118}$$

Hence, the total area of the inductor is

$$A_{tot} = D^2 = (520 \times 10^{-6})^2 = 0.2704 \times 10^{-6} \ \text{m}^2. \tag{8.119}$$

The metal area is

$$A_m = \omega l = 30 \times 10^{-6} \times 5600 \times 10^{-6} = 0.168 \times 10^{-6} \ \text{m}^2. \tag{8.120}$$

The inductance of the square spiral planar inductor for $\delta > h$ is

$$L \approx 1.3 \times 10^{-7} \frac{A_m^{5/3}}{A_{tot}^{1/6} w^{1.75} (w + s)^{0.25}} \ \text{(H)}$$

$$= 1.3 \times 10^{-7} \frac{(0.168 \times 10^{-6})^{5/3}}{(0.2704 \times 10^{-6})^{1/6} (30 \times 10^{-6})^{1.75} [(30 + 20) \times 10^{-6}]^{0.25}} = 8.0864 \ \text{nH}. \tag{8.121}$$

Monomial Formula

The data-fitted monomial empirical expression for the inductance of the square spiral inductor is [36]

$$L = 0.00162 D^{-1.21} w^{-0.147} \left(\frac{D + d}{2}\right)^{2.4} N^{1.78} s^{-0.03} \ \text{(nH)}. \tag{8.122}$$

All dimensions are in micrometers.

Example 8.14

Calculate the inductance of a square planar spiral inductor with $N = 5$, $d = 60$ μm, $w = 30$ μm, $s = 20$ μm, $h = 20$ μm, and $\mu_r = 1$ for $\delta > h$. Use monomial formula.

Solution: The outer diameter is

$$D = d + 2N\omega + 2(N - 1)s = 60 + 2 \times 5 \times 30 + 2 \times (5 - 1) \times 20 = 520 \ \mu\text{m}. \tag{8.123}$$

The inductance of the square spiral planar inductor for $\delta < a$ is

$$L = 0.00162 D^{-1.21} w^{-0.147} \left(\frac{D + d}{2}\right)^{2.4} N^{1.78} s^{-0.03} \ \text{(nH)}$$

$$= 0.00162 \times (520)^{-1.21} 30^{-0.147} \left(\frac{520 + 60}{2}\right)^{2.4} 5^{1.78} 20^{-0.03} = 6.6207 \ \text{nH}. \tag{8.124}$$

Jenei's Formula

The total inductance consists of the self-inductance L_{self} and the sum of the positive mutual inductances M^+ and the sum of the negative mutual inductances M^-. The derivation of an expression for the total inductance is given in [40]. The self-inductance of one straight segment is [8]

$$L_{self1} = \frac{\mu_0 l_{seg}}{2\pi}\left[\ln\left(\frac{2l_{seg}}{w+h}\right) + 0.5\right] \tag{8.125}$$

where l_{seg} is the segment length, w is the metal trace width, h is the metal trace thickness. The total length of the conductor is

$$l = (4N+1)d + (4N_i+1)N_i(w+s) \tag{8.126}$$

where N is the number of turns, N_i is the integer part of N, and s is the spacing between the segments. The total self-inductance of a square planar inductor is equal to the sum of $4N$ self-inductances of all the segments.

$$L_{self} = 4N L_{self1} = \frac{\mu_0 l}{2\pi}\left[\ln\left(\frac{2l}{N(w+h)}\right) - 0.2\right]. \tag{8.127}$$

The antiparallel segments contribute to negative mutual inductance M^-. The sum of all interactions is approximately equal to $4N^2$ average interactions between segments of an average length and an average distance. The negative mutual inductance is

$$M^- = \frac{0.47\mu_0 N l}{2\pi}. \tag{8.128}$$

The positive mutual inductance is caused by interactions between parallel segments on the same side of a square. The average distance is

$$b = (w+s)\frac{(3N-2N_i-1)(N_i+1)}{3(2N-N_i-1)}. \tag{8.129}$$

For $N_i = N$,

$$b = \frac{(w+s)(N+1)}{3}. \tag{8.130}$$

The total positive mutual inductance is

$$M^+ = \frac{\mu_0 l(N-1)}{2\pi}\left\{\ln\left[\sqrt{1+\left(\frac{l}{4Nb}\right)^2} + \frac{l}{4Nb}\right] - \sqrt{1+\left(\frac{4Nb}{l}\right)^2} + \frac{4Nb}{l}\right\}. \tag{8.131}$$

The inductance of a square planar inductors is

$$L = L_{self} + M^- + M^+ = \frac{\mu_0 l}{2\pi}(\ln\left[\frac{l}{N(w+h)}\right] - 0.2 - 0.47N$$

$$+ (N-1)\left\{\ln\left[\sqrt{1+\left(\frac{l}{4Nb}\right)^2} + \frac{l}{4Nb}\right] - \sqrt{1+\left(\frac{4Nb}{l}\right)^2} + \frac{4Nb}{l}\right\}. \tag{8.132}$$

All dimensions are in meters.

Example 8.15

Calculate the inductance of a square planar spiral inductor with $N = N_i = 5$, $d = 60$ μm, $w = 30$ μm, $s = 20$ μm, $h = 20$ μm, and $\mu_r = 1$ for $\delta > h$. Use Jenei's formula.

Solution: The total length of the conductor is

$$l = (4N+1)d + (4N_i+1)N_i(w+s) = (4\times 5+1)\times 60\times 10^{-6} + (4\times 5+1)5(30+20)\times 10^{-6}$$

$$= 6500\ \mu\text{m}. \tag{8.133}$$

The average distance is

$$b = (w+s)\frac{(3N-2N_i-1)(N_i+1)}{3(2N-N_i-1)}$$
$$= (30+20)\times 10^{-6}\frac{(3\times 5-2\times 5-1)(5+1)}{3(2\times 5-5-1)} = 100\ \mu\text{m}. \tag{8.134}$$

The inductance of the square spiral planar inductor for $\delta > h$ is

$$L = \frac{\mu_0 l}{2\pi}\left\langle \ln\left[\frac{l}{N(w+h)}\right] - 0.2 - 0.47N \right.$$
$$\left. + (N-1)\left\{\ln\left[\sqrt{1+\left(\frac{l}{4Nb}\right)^2}+\frac{l}{4Nb}\right]-\sqrt{1+\left(\frac{4Nb}{l}\right)^2}+\frac{4Nb}{l}\right\}\right\rangle$$
$$= \frac{4\pi\times 10^{-7}\times 6500\times 10^{-6}}{2\pi}\left\langle \ln\left[\frac{6500}{5(30+20)}\right] - 0.2 - 0.47\times 5 + (5-1)\right.$$
$$\left.\times\left\{\ln\left[\sqrt{1+\left(\frac{6500}{4\times 5\times 100}\right)^2}+\frac{6500}{4\times 5\times 100}\right]-\sqrt{1+\left(\frac{4\times 5\times 100}{6500}\right)^2}+\frac{4\times 5\times 100}{6500}\right\}\right\rangle$$
$$= 6.95\ \text{nH}. \tag{8.135}$$

Dill's Formula

The inductance of a square planar inductor is [10]

$$L = 8.5\times 10^{-10}N^{5/3}\ (\text{H}). \tag{8.136}$$

For $N = 5$, $L = 12.427$ nH.

Terman's Formula

The inductance of a square planar spiral single-turn inductor is

$$L = 18.4173\times 10^{-4}D\left[\log\left(\frac{0.7874\times 10^{-4}D^2}{w+h}\right) - \log(0.95048\times 10^{-4}D)\right]$$
$$+ 10^{-4}\times[7.3137D + 1.788(w+h)]\quad (\text{nH}). \tag{8.137}$$

All dimensions are in micrometers.

The inductance of a square planar spiral multiple-turn inductor is

$$L = 18.4173\times 10^{-4}DN^2\left[\log\left(\frac{0.7874\times 10^{-4}D^2}{w+h}\right) - \log(0.95048\times 10^{-4}D)\right]$$
$$+ 8\times 10^{-4}N^2[0.914D + 0.2235(w+h)]\quad (\text{nH}). \tag{8.138}$$

where the outer diameter is

$$D = d + 2N\omega + 2(N-1)s. \tag{8.139}$$

All dimensions are in meters.

Example 8.16

Calculate the inductance of a square planar spiral inductor with $N = 5$, $D = 520$ μm, $w = 30$ μm, $h = 20$ μm, and $\mu_r = 1$ for $\delta > h$. Use Terman's formula.

Solution: The inductance of a square planar spiral multiple-turn inductor is

$$\begin{aligned} L &= 18.4173 \times 10^{-4} DN^2 \left[\log\left(\frac{0.7874 \times 10^{-4} D^2}{w+h}\right) - \log(0.95048 \times 10^{-4} D)\right] \\ &\quad + 8 \times 10^{-4} N^2 [0.914D + 0.2235(w+h)] \quad \text{(nH)} \\ &= 18.4173 \times 10^{-4} \times 520 \times 5^2 \left[\log\left(\frac{0.7874 \times 10^{-4} \times 520^2}{30+20}\right) - \log(0.95048 \times 10^{-4} \times 520)\right] \\ &\quad + 8 \times 10^{-4} 5^2 \, [0.914 \times 520 + 0.2235(30+20)] = 32.123 \text{ nH}. \end{aligned} \tag{8.140}$$

Rosa's, Wheeler's, Grover's, Bryan's, Jenei's, and monomial formulas give similar values of inductance. Greenhouse's, Cranin's, and Terman's formulas give higher values of inductance than those calculated from the former group of formulas.

8.16.3 Inductance of Hexagonal Spiral Inductors

Wheeler's Formula

The inductance of a hexagonal spiral inductor is given by modified Wheeler's formula [4, 36]

$$L = 1.165 \mu_0 N^2 \frac{D+d}{1 + 3.82\frac{D-d}{D+d}} \quad \text{(H)} \tag{8.141}$$

where

$$D = d + 2N\omega + 2(N-1)s. \tag{8.142}$$

All dimensions are in meters.

Rosa's Formula

The equation for the inductance of hexagonal planar spiral inductors is expressed as [1, 9, 36]

$$L = 0.2725 \mu N^2 (D+d) \left\{ \ln\left[\frac{2.23(D+d)}{D-d}\right] + 0.17\left(\frac{D-d}{D+d}\right)^2 \right\} \quad \text{(H)}. \tag{8.143}$$

All dimensions are in meters.

Example 8.17

Calculate the inductance of a hexagonal planar spiral inductor with $N = 5$, $d = 60$ μm, $w = 30$ μm, $s = 20$ μm, $h = 20$ μm, and $\mu_r = 1$ for $\delta > h$. Use Rosa's formula.

Solution: The outer diameter is

$$D = d + 2N\omega + 2(N-1)s = 60 + 2 \times 5 \times 30 + 2 \times (5-1) \times 20 = 520 \text{ μm}. \tag{8.144}$$

The inductance of the square spiral planar inductor for $\delta > h$ is

$$L = 0.2725\mu N^2(D+d)\left\{\ln\left[\frac{2.23(D+d)}{D-d}\right] + 0.17\left(\frac{D-d}{D+d}\right)^2\right\} \text{ (H)}$$
$$= 0.2725 \times 4\pi \times 10^{-7} \times 5^2(520+60) \times 10^{-6}$$
$$\times \left\{\ln\left[\frac{2.23(520+60)}{520-60}\right] + 0.17\left(\frac{520-60}{520+60}\right)^2\right\}$$
$$= 5.6641 \text{ nH}. \tag{8.145}$$

Grover's Formula

The inductance of a hexagonal planar inductor is [8]

$$L = \frac{2\mu l}{\pi}\left[\left(\frac{l}{6r}\right) + 0.09848\right] \text{ (H)}. \tag{8.146}$$

All dimensions are in micrometers.

Monomial Formula

The data-fitted monomial empirical expression for the inductance of the hexagonal spiral inductor is [36]

$$L = 0.00128 D^{-1.24} w^{-0.174} \left(\frac{D+d}{2}\right)^{2.47} N^{1.77} s^{-0.049} \text{ (nH)}. \tag{8.147}$$

All dimensions are in micrometers.

8.16.4 Inductance of Octagonal Spiral Inductors

Wheeler's Formula

The inductance of an octagonal spiral inductor is given by modified Wheeler's formula [4, 36]

$$L = 1.125\mu_0 N^2 \frac{D+d}{1 + 3.55\frac{D-d}{D+d}} \text{ (H)} \tag{8.148}$$

where

$$D = d + 2N\omega + 2(N-1)s. \tag{8.149}$$

All dimensions are in meters.

Rosa's Formula

The inductance of an octagonal planar spiral inductors is expressed as [1, 9, 36]

$$L = 0.2675\mu N^2(D+d)\left\{\ln\left[\frac{2.29(D+d)}{D-d}\right] + 0.19\left(\frac{D-d}{D+d}\right)^2\right\} \text{ (H)}. \tag{8.150}$$

All dimensions are in meters.

Grover's Formula

The inductance of an octagonal planar inductor is [8]

$$L = \frac{2\mu l}{\pi}\left[\left(\frac{l}{8r}\right) - 0.03802\right] \text{ (H)}. \tag{8.151}$$

All dimensions are in meters.

Monomial Formula

The data-fitted monomial empirical expression for the inductance of an octagonal spiral inductor is [36]

$$L = 0.00132D^{-1.21}w^{-0.163}\left(\frac{D+d}{2}\right)^{2.43}N^{1.75}s^{-0.049} \quad \text{(nH)}. \tag{8.152}$$

All dimensions are in micrometers.

The inductance of octagonal inductors is almost the same as that of hexagonal inductors.

8.16.5 Inductance of Circular Spiral Inductors

Rosa's Formula

The inductance of circular planar spiral inductor is expressed as [1, 9, 36]

$$L = 0.25\mu N^2(D+d)\left\{\ln\left[\frac{2.46(D+d)}{D-d}\right] + 0.19\left(\frac{D-d}{D+d}\right)^2\right\} \text{ (H)}. \tag{8.153}$$

All dimensions are in meters.

Example 8.18

Calculate the inductance of a circular planar spiral inductor with $N = 5$, $d = 60$ μm, $w = 30$ μm, $s = 20$ μm, and $\mu_r = 1$ for $\delta > h$. Use Rosa's formula.

Solution: The outer diameter is

$$D = d + 2N\omega + 2(N-1)s = 60 + 2 \times 5 \times 30 + 2 \times (5-1) \times 20 = 520 \ \mu\text{m}. \tag{8.154}$$

The inductance of the circular spiral planar inductor for $\delta < a$ is

$$\begin{aligned} L &= 0.25\mu N^2(D+d)\left\{\ln\left[\frac{2.46(D+d)}{D-d}\right] + 0.19\left(\frac{D-d}{D+d}\right)^2\right\} \text{ (H)} \\ &= 0.25 \times 4\pi \times 10^{-7} \times 5^2(520+60) \times 10^{-6} \\ &\times \left\{\ln\left[\frac{2.46(520+60)}{520-60}\right] + 0.19\left(\frac{520-60}{520+60}\right)^2\right\} \\ &= 5.7 \text{ nH}. \end{aligned} \tag{8.155}$$

Wheeler's Formula

The inductance of a circular planar inductor is [3]

$$L = 31.33\mu N^2 \frac{a^2}{8a + 11h} \text{ (H)}. \tag{8.156}$$

All dimensions are in meters.

Schieber's Formula

The inductance of a circular planar inductor is [16]

$$L = 0.874\pi \times 10^{-5} DN^2 \text{ (H)}. \tag{8.157}$$

All dimensions are in meters.

Spiral IC inductors have high parasitic resistances and high shunt capacitances, resulting in a low Q_{Lo} and low SRFs. It is difficult to achieve $Q_{Lo} > 10$ and f_r greater than a few gigahertz for planar inductors due to the loss in the substrate and metallization. The RF MEMS technology has a potential to improve the performance of RF IC inductors by removing the substrate under the planar spiral via top-side etching, which decouples the RF IC inductor performance from substrate characteristics.

8.17 Multimetal Spiral Inductors

Multimetal planar spiral inductors (called stacked inductors) are also used to achieve compact high-inductance magnetic devices. A double-layer spiral inductor can be implemented using metal 1 and metal 2 layers, as shown in Fig. 8.25 [49, 50]. An equivalent circuit of a two-layer inductor is shown in Fig. 8.26.

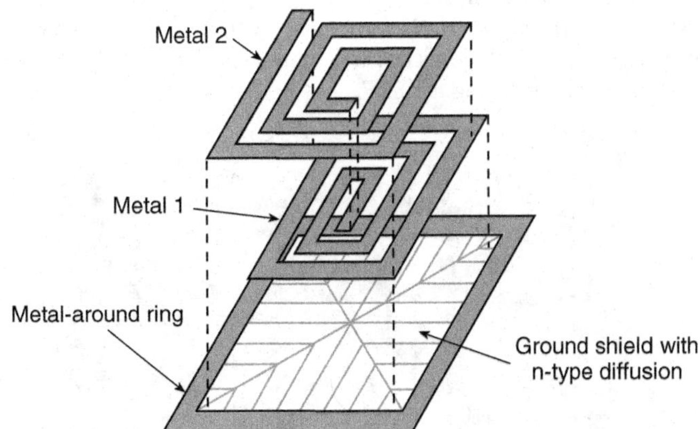

Figure 8.25 Multimetal spiral inductor that uses metal 1 and metal 2

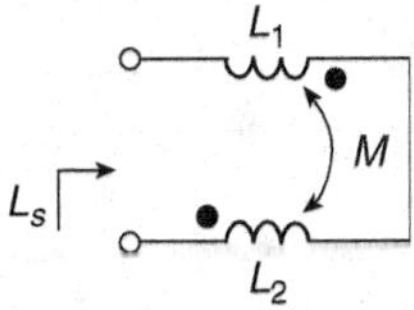

Figure 8.26 Equivalent circuit of two-layer inductor

The impedance of a two-layer inductor is given by

$$Z = j\omega(L_1 + L_2 + 2M) = j\omega(L_1 + L_2 + 2k\sqrt{L_1L_2}) \approx j\omega(L_1 + L_2 + 2\sqrt{L_1L_2}). \tag{8.158}$$

where the coupling coefficient $k \approx 1$. If both inductors are equal,

$$Z \approx j\omega(L + L + 2L) = j\omega(4L) = j\omega L_s. \tag{8.159}$$

Thus, the total inductance L_s for a two-layer inductor increases nearly four times due to mutual inductance. For an m-layer inductor, the total inductance is increased by a factor of m^2 compared to a self-inductance of a single-layer spiral inductor. Modern CMOS technologies provide more than five metal layers and stacking inductors or transformers give large inductances in a small chip area.

A patterned ground shield reduces eddy currents. Many RF IC designs incorporate several inductors on the same die. Since these structures are physically large, substrate coupling can cause significant problems, such as feedback and oscillations.

The spiral inductors have several drawbacks. First, the size is large compared with other inductors for the same number of turns N. Second, the spiral inductor requires a lead wire to connect the inside end of the coil to the outside, which introduces a capacitance between the conductor and the lead wire, and this capacitance is the dominant component of the overall stray capacitance. Third, the direction of the magnetic flux is perpendicular to the substrate, which can interfere with the underlying circuit. Fourth, the quality factor Q_{Lo} is very low. Fifth, the SRF f_r is low.

8.18 Planar Transformers

Monolithic transformers are required in many RF designs. The principle of planar transformers relies on lateral magnetic coupling. They can be used as narrowband or wideband transformers. An interleaved planar spiral transformer is shown in Fig. 8.27. In these transformers, the coupling coefficient k is higher and the number of turns in the primary and secondary windings is the same. The coupling coefficient of these transformers is in the range $0.6 \leq k \leq 0.8$. Figure 8.28 shows a planar spiral transformer, in which the turns ratio $n = N_p : N_s$ is not equal to 1. The coupling coefficient k is low in this transformer, typically $k = 0.4$. Stacked transformers use multiple metal layers. In this case, the self-capacitance increases, reducing the SRF.

Figure 8.27 Interleaved planar spiral transformer

Figure 8.28 Planar spiral transformer in which turns ratio $n = N_p : N_s$ is not equal to 1

8.19 MEMS Inductors

MEMS technology can improve the performance of integrated inductors. MEMS inductors usually are solenoids fabricated using surface micromachining techniques and polymer/metal multilayer process techniques. Figure 8.29 shows an integrated solenoid inductor [32, 33]. These inductors may have air cores or electroplated Ni–Fe Permalloy cores. The winding is made of electroplated copper layers. There is an air gap between the metal winding and the substrate. This geometry gives small inductor size, low stray capacitance C_S, high SRF f_r, low power loss, and high quality factor Q. However, it requires a 2D design approach. MEMS inductors find applications in magnetic actuators.

The magnetic field intensity inside the solenoid is uniform and given by

$$B = \frac{\mu NI}{l_c} \tag{8.160}$$

where l_c is the length of the core and N is the number of turns. The magnetic flux is

$$\phi = A_c B = \frac{\mu NIA_c}{l_c} \tag{8.161}$$

where A_c is the cross-sectional area of the core. The magnetic linkage is

$$\lambda = N\phi = \frac{\mu N^2 IA_c}{l_c}. \tag{8.162}$$

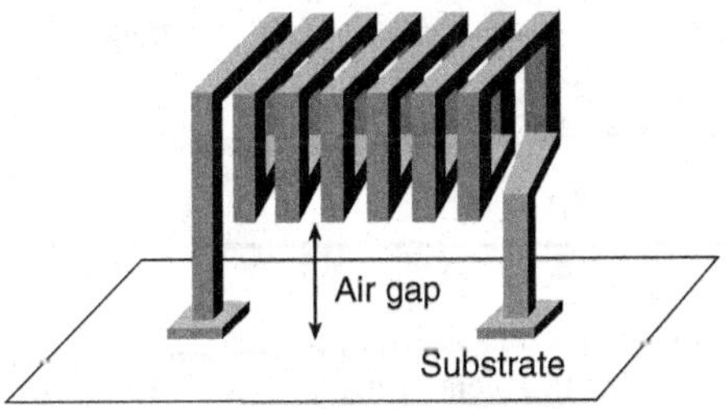

Figure 8.29 Integrated MEMS solenoid inductor [32, 33]

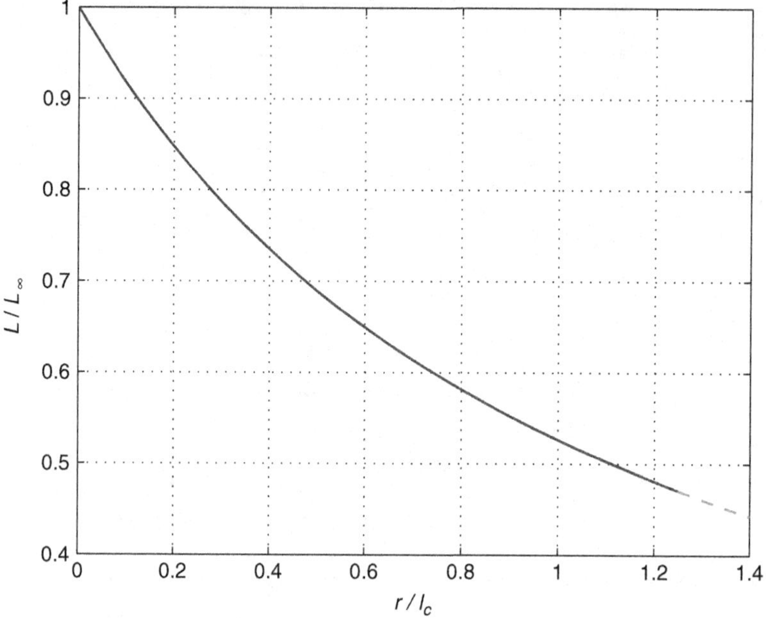

Figure 8.30 Ratio L/L_∞ versus r/l_c

The inductance of the infinitely long MEMS solenoid inductor is given by

$$L = \frac{\lambda}{I} = \frac{\mu_{rc}\mu_0 A_c N^2}{l_c} = \frac{\mu_{rc}\mu_0 A_c N^2}{Np} = \frac{\mu_{rc}\mu_0 A_c N}{p} = \frac{\mu_{rc}\mu_0 A_c N}{w+s} \tag{8.163}$$

where $l_c = Np$, $p = w + s$ is the winding pitch, which is equal to the distance between the centers of two adjacent turn conductors, w is the width of the winding wire, and s is the separation between the turns.

The inductance of a short solenoid is lower than that of an infinitely long one. The inductance of a round solenoid of radius r and length l_c can be determined from Wheeler's equation [3]

$$L = \frac{L_\infty}{1 + 0.9\frac{r}{l_c}} = \frac{\mu A_c N^2}{l_c\left(1 + 0.9\frac{r}{l_c}\right)}. \tag{8.164}$$

For a square solenoid, we can use an equivalent cross-sectional area of the solenoid

$$A = h^2 = \pi r^2 \tag{8.165}$$

yielding

$$r = \frac{h}{\sqrt{\pi}}. \tag{8.166}$$

Figure 8.30 show a plot of L/L_∞ as a function of r/l_c.

Example 8.19

Calculate the inductance of a MEMS solenoid made of $N = 10$ turns of square wire with thickness $h = 8$ μm and space between the turns $s = 8$ μm. The solenoid has a square shape with width $w =$ 50 μm.

Solution: The cross-sectional area of the MEMS solenoid is

$$A_c = w^2 = (50 \times 10^{-6})^2 = 25 \times 10^{-10}\ \text{m}^2. \tag{8.167}$$

The length of the solenoid is

$$l_c = (N-1)(h+s) = (10-1)(8+8) \times 10^{-6} = 144 \times 10^{-6}\ \ \text{m} = 144\ \mu\text{m}. \tag{8.168}$$

The inductance of the solenoid is

$$L = \frac{\mu N^2 A_c}{l_c} = \frac{4\pi \times 10^{-7} \times 10^2 \times 25 \times 10^{-10}}{144 \times 10^{-6}} = 2.182\ \text{nH}. \tag{8.169}$$

Brooks Inductor

The maximum inductance with a given length of wire is obtained for $a/c = 3/2$ and $b = c$, where a is the coil mean radius, b is the coil axial thickness, and c is the coil radial thickness. The inductance in this case is [17, 18]

$$L = 1.353\mu a N^2. \tag{8.170}$$

8.20 Inductance of Coaxial Cable

The inductance of a coaxial cable is

$$L = \frac{\mu l}{2\pi} \ln\left(\frac{b}{a}\right) \tag{8.171}$$

where a is the inner radius of the dielectric, b is the outer radius of the dielectric, and l is the length of the coaxial cable.

8.21 Inductance of Two-Wire Transmission Line

The inductance of a two-parallel wire transmission line is

$$L = \frac{\mu l}{\pi}\left[\ln\left(\frac{d}{2a}\right) + \sqrt{\left(\frac{d}{2a}\right)^2 - 1}\right] \tag{8.172}$$

where a is the radius of the wires, d is the distance between the centers of the wires, and l is the length of the transmission line.

8.22 Eddy Currents in Integrated Inductors

The performance degradation of planar spiral inductors is caused by the following effects:

1. trace resistance due to its finite conductivity ($\sigma < \infty$);
2. eddy-current losses due to magnetic field penetration into the substrate and adjacent traces;
3. substrate ohmic losses; and
4. capacitance formed by metal trace, silicon oxide, and substrate. This capacitance conducts displacement currents, which flow through the substrate and the metal trace.

When a time-varying voltage is applied between the terminals of the inductor, a time-varying current flows along the inductor conductor and induces a magnetic field, which penetrates the metal trace in the direction normal to its surface. As a result, eddy currents are produced within the metal trace. These currents are concentrated near the edges of the trace due to the skin and proximity effects. Eddy currents subtract from the inductor external current on the outside trace edge and add to it on the inside trace edge near the center of the spiral. Therefore, the current density is not uniform. It is high on the inside edge and it is low on the outside edge, increasing the effective resistance of the metal trace. The magnetic field also induces eddy currents circulating below the spiral metal trace in the semiconductor substrate. In CMOS technologies, the substrate has a low resistivity, typically $\rho_{sub} = 0.015\ \Omega\cdot$cm. Therefore, eddy current losses are dominant component of the overall loss. The inductor quality factor Q is usually in the range 3–4. In BiCMOS technologies, the substrate resistivity is high usually in the range 10–30 $\Omega\cdot$cm. Therefore, eddy current losses are reduced to negligible values. The quality factor Q is typically from 5 to 10. The turn-to-substrate capacitance conducts displacement currents, causing power losses. These losses can be reduced by placing patterned ground shields, etching away the semiconductor material below the spiral trace, separating the spiral trace from the substrate by placing a thick oxide layer, using a high-resistivity substrate bulk, or using an insulating substrate such as sapphire. At high frequencies, the effective resistance of the metal traces increases due to skin and proximity effects, resulting in current crowding.

A broken guard ring made up of n^+ or p^+ diffusion regions surrounding the coil and connected to ground is often used. The purpose of the guard ring is to provide a low-impedance path to ground for currents induced in the substrate, reducing the substrate loss and increasing the inductor quality factor.

An effective method for reducing the loss in the substrate is to insert a high-conductivity ground shield between the spiral and the substrate. The ground shield is patterned with slots to reduce the paths for eddy currents.

If n-well is fabricated beneath the oxide, the p-substrate and the n-well form a pn junction. If this junction is reverse biased, it represent a capacitance C_J. The oxide capacitance C_{ox} and the junction capacitance C_J are connected in series, reducing the total capacitance. Therefore, the current flowing from the metal trace to the substrate is reduced, reducing substrate loss and improving the quality factor.

Various structures may be used to block eddy currents. Eddy currents flow in paths around the axis of the spiral. This can be accomplished by inserting narrow strips of n^+ regions perpendicular to the eddy current flow to create blocking pn junctions.

The magnetic field intensity is the strongest in the center of the spiral inductor. The magnetic losses are predominate in the inner turns due to eddy currents, whereas the ohmic losses are dominant in the outer turns. Therefore, the series resistance and the metal power loss can be reduced if the trace width varies from smaller for the inner turns to larger for the outer turns, as illustrated in Fig. 8.31. A planar inductor with a variable trace width is called a *tapered spiral inductor*. This construction leads to a higher quality factor Q of spiral inductors. Also, the spacing between the turns can be increased from the center of the spiral to the outer turns. Tapering is an effective technique in inductors, where the metal loss is dominant and the substrate loss is negligible, for example, when high-resistivity insulating substrates are used. In addition, the inner turns can be removed to produce a hollow spiral inductor. High power losses in integrated inductors lead to a very low quality factor Q, which significantly increases the noise figure and reduces the power gain and selectivity of LNAs. A low inductor Q increases the phase noise of VCOs. The phase noise of an oscillator is proportional to $1/Q_L^2$, where Q_L is the loaded quality factor of the resonant circuit. A low inductor Q also reduces the efficiency of RF power amplifiers.

8.23 Model of RF-Integrated Inductors

A two-port π-model of RF IC planar spiral inductors is shown in Figure 8.32 [27]. The components L_s, R_s, and C_s represent the inductance, resistance, and capacitance of the metal trace. The capacitance C_{ox} represents the metal–oxide–substrate capacitance. The components R_{Si} and C_{Si} represent the resistance and capacitance of the substrate. The inductance L_s is given by one of the equations given in Section 8.15. The AC resistance is given by (8.17).

Figure 8.31 Tapered RF IC spiral inductor with a variable trace width

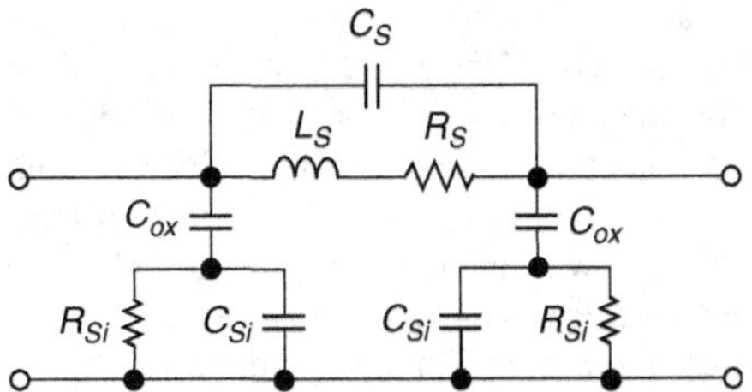

Figure 8.32 Lumped-parameter physical model of an RF IC spiral inductor

The shunt stray capacitance is formed by the overlap of the underpass and the spiral inductor turns. It is given by

$$C_S = \epsilon_{ox} \frac{N w^2}{t_{ox}}. \tag{8.173}$$

The turn-to-turn capacitances C_{tt} are small even for very small spacing s between the turns because the thickness of the metal trace h is small, resulting in a small vertical area. The lateral turn-to-turn capacitances are connected in series and therefore the equivalent interturn capacitance C_{tt}/N is small. The adjacent turns sustain a very small voltage difference. The electric energy stored in turn-to-turn capacitances is proportional to the square of voltage. Also, the displacement current through the turn-to-turn capacitance $i_C = C_{tt} dv/dt \approx C_{tt} \Delta v/\delta t$ is very low. Therefore, the effect of the turn-to-turn capacitances is negligible. The effect of the overlap capacitance is more significant because the potential difference between the underpass and the spiral turns is higher. The stray capacitance allows the current to flow directly from the input port to the output port without passing through the inductance. The current can also flow between one terminal and the substrate through capacitances C_{ox} and C_{Si}.

The capacitance between the metal trace and the silicon substrate is

$$C_{ox} = \epsilon_{ox} \frac{A}{2t_{ox}} = wl \frac{\epsilon_{ox}}{2t_{ox}} \tag{8.174}$$

where w is the trace width, l is the trace length, $A = wl$ is the area of the metal trace, t_{ox} is the thickness of the silicon dioxide SiO_2, and $\epsilon_{ox} = 3.9\epsilon_0$ is the permittivity of the silicon dioxide. The typical value of t_{ox} is 1.8×10^{-8} m.

The capacitance of the substrate is

$$C_{Si} \approx \frac{wlC_{sub}}{2} \tag{8.175}$$

where $C_{sub} = 10^{-3}$ to 10^{-2} fF/μm^2.

The resistance representing the substrate dielectric loss is given by

$$R_{Si} = \frac{2}{\omega l G_{sub}} \tag{8.176}$$

where G_{sub} is the substrate conductance per unit area, typically $G_{sub} = 10^{-7}$ S/μm^2.

8.24 PCB Inductors

A printed circuit board (PCB) transformer is shown in Fig. 8.33. Planar inductors and transformers consist of a flat copper trace (foil) etched on a PCB and two pieces of flat ferrite core, one below and one above the coil. The windings are etched usually on both sides of a PCB. Typically, E cores, PQ cores, and RM cores with short legs are used in PCB inductor and transformers. The cores are often wide. Planar magnetic component technology is suitable for a low number of turns. As the operating frequency increases, the number of turns decreases. At high frequencies, only several turns are often required. Eddy-current losses are lower for thin copper foils than those for circular copper wires because it is easier to satisfy the condition $h < 2\delta_w$. Most planar inductors and transformers are gapped ferrite devices. The minimum total power losses are achieved by reducing the winding loss by lowering the current density and increasing the core loss by allowing the flux density to be higher than that in traditional wire-wound magnetic components. The typical leakage inductance of planar transformers ranges from 0.1 to 1% of the primary winding inductance. Planar cores have a low winding window area and cover a large square area. Planar inductors have a higher level of electromagnetic interference (EMI)/radiofrequency interference (RFI) than inductors with pot cores. Figure 8.34 shows a multilayer PCB inductor, where windings are etched in each layer of a PCB. Planar inductors and transformers are excellent devices for mass production.

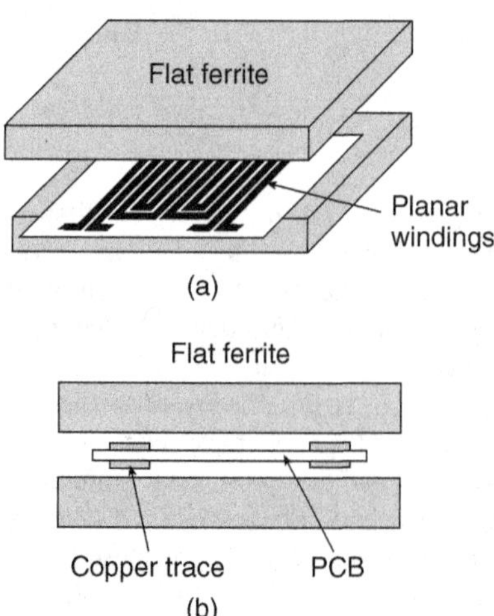

Figure 8.33 PCB transformer. (a) Top view of PCB transformer. (b) Side view of PCB transformer

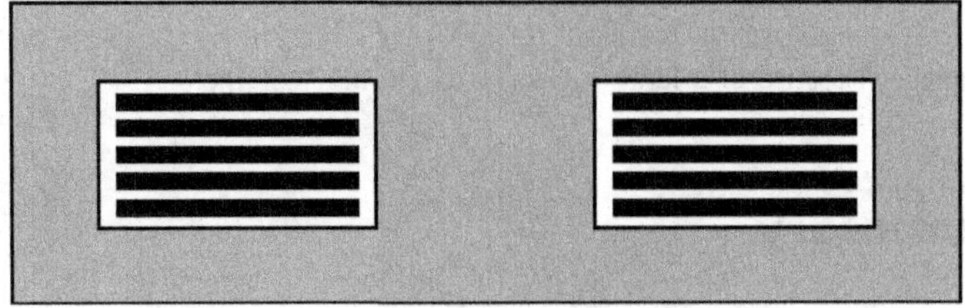

Figure 8.34 Multilayer PCB inductor

The types of inductors that can be used for PCB technology are

- Coreless inductors
- Inductors with planar windings and magnetic plates
- Closed-core structures.

The PCB inductors can be single-sided, double-sided, and multilayer board inductors. Photolithographic manufacturing processes are used, which have been developed for making PCBs. Photomasks are required to produce PCB inductors.

The simplest structure of a PCB inductor is a coreless planar winding. The maximum inductance is achieved by fitting the largest number of winding turns in each of the PCB layers. Typically, there are six layers in a PCB. The maximum power capability of the coreless PCB inductor is limited by the maximum temperature rise.

The planar inductors also use magnetic cores mounted on the top and bottom of the board. The magnetic plates on either side of the planar winding structure increase the inductance. The inductance also increases when the plate thickness increases. PCB inductors with magnetic plates offer better EMI performance than coreless inductors. PCB inductors have immense power density because they avoid the space-consuming bobbins.

Another category of inductors are thin-film inductors, which use NiFe magnetic films. A typical thickness of the film is 2.5 μm. Typical inductance achievable using this technology is in the range 0.5–10 μm. They can be used in the frequency range from 1 to 10 MHz. Their quality factor ranges from 4 to 6. Using this type of inductors, pulse-width modulated (PWM) DC–DC converter can be integrated.

The closed-core structure of a PCB inductor consists of a core area provided by magnetic holes. The inductance increases with the inner core radius and the number of winding turns.

The advantages of planar inductors and transformers are as follows:

- Low profile
- Low volume
- High power density
- Small packages
- Excellent repeatability and consistency of performance
- High magnetic coupling
- Low leakage inductance
- Excellent mechanical integrity
- Low skin and proximity effect power loss due to a low thickness of metal traces
- Reduced winding AC resistance and power losses

- Low cost
- Good heat transfer because of a large surface area per unit volume.

8.25 Summary

- Inductors are key components of radio transmitters and other wireless communication circuits.
- Lack of high-performance IC RF inductors is a major gap in RF technology.
- The most important figure-of-merit of an inductor is the quality factor Q.
- High power losses and a low quality factor Q of planar integrated inductors is the major problem.
- The family of integrated inductors includes planar spiral inductors, planar meander inductors, bondwire inductors, and MEMS inductors.
- Planar inductors are spiral inductors and meander inductors.
- Planar inductors are compatible with IC technology.
- The range of integrated inductances is from 0.1 to 20 nH.
- Integrated inductors require a large amount of chip area.
- The inductance of a spiral inductor is equal to the sum of the self-inductances of the straight segments plus the sum of the mutual inductances between the segments.
- If the currents in two parallel conductors flow in the same direction, the mutual inductance is positive.
- If the currents in two parallel conductors flow in opposite directions, the mutual inductance is negative.
- If two conductors are perpendicular, the mutual inductance is zero.
- For the square spiral inductors, the total mutual inductance is equal to the sum of mutual inductances of parallel segments.
- The mutual inductance of spiral inductors is positive, which increases the overall inductance.
- The turn-to-turn mutual inductance of meander inductors is negative. Therefore, meander inductors suffer from a very low overall inductance.
- Silicon substrate is lossy.
- Integrated inductors have a very low quality factor.
- Planar spiral RF IC inductors are the most widely used inductors.
- Meander inductors require only one metal layer.
- Meander inductors have a low inductance-to-surface area ratio.
- Planar spiral inductors usually require two metal layers.
- Planar spiral inductors have a high inductance per unit area.
- The direction of magnetic flux is perpendicular to the substrate and therefore can interfere with the circuit.
- In square planar inductors, the total self-inductance is equal to the sum of self-inductances of all straight segments.
- In square planar inductors, the mutual inductance is only present between parallel segments.

- The mutual inductance between parallel segments is positive if the currents in parallel conductors flow in the same direction.
- The mutual inductance between parallel segments is negative if the currents in parallel conductors flow in opposite directions.
- The inductance of planar inductors is approximately proportional to the trace length l.
- The inductance of planar inductors increases when the trace width w increases.
- The inductance of planar spiral inductors increases when the spacing s between the traces increases.
- The conductor thickness does not affect the inductance value and significantly reduces the resistance if $h < \delta$.
- The magnetic flux of RF IC inductors penetrates the substrate, where it induces high-loss eddy currents.
- Patterned ground shield can be used to reduce eddy-current loss.
- MEMS solenoid inductors are more complex to fabricate than the planar spiral inductors.
- MEMS inductors have a higher quality factor than the planar integrated inductors.
- The predictability of bondwire inductors is low.
- The parasitic capacitance of integrated planar inductors can be minimized by implementing the spiral with the topmost metal layer, increasing the SRF.

8.26 References

[1] E. B. Rosa, "Calculation of the self-inductance of single-layer coils," *Bulletin of the National Bureau of Standards*, vol. 2, no. 2, pp. 161–187, 1906.

[2] E. B. Rosa, "The self and mutual inductances of linear conductors," *Bulletin of the National Bureau of Standards*, vol. 4, no. 2, pp. 302–344, 1907.

[3] H. A. Wheeler, "Simple inductance formulas for radio coils," *Proceedings of the IRE*, vol. 16, no. 10, pp. 1398–1400, October 1928.

[4] H. A. Wheeler, "Formulas for the skin effect," *Proceedings of the IRE*, vol. 30, pp. 412–424, September 1942.

[5] F. E. Terman, *Radio Engineers' Handbook*, New York: McGraw-Hill, 1943.

[6] R. G. Medhurst, "HF resistance and self-capacitance of single-layer solenoids," *Wireless Engineers*, pp. 35–43, February 1947, and pp. 80–92, March 1947.

[7] H. E. Bryan, "Printed inductors and capacitors," *Tele-Tech and Electronic Industries*, vol. 14, no. 12, p. 68, December 1955.

[8] F. W. Grover, *Inductance Calculations: Working Formulas and Tables*, Princeton, NJ: Van Nostrand, 1946; reprinted by Dover Publications, New York, 1962.

[9] J. C. Maxwell, *A Treatise of Electricity and Magnetism*, Parts III and IV, 1st Ed., 1873, 3rd Ed., 1891; reprinted by New York: Dover Publishing, 1954 and 1997.

[10] H. Dill, "Designing inductors for thin-film applications," *Electronic Design*, pp. 52–59, February 1964.

[11] D. Daly, S. Knight, M. Caulton, and R. Ekholdt, "Lumped elements in microwave integrated circuits," *IEEE Transactions on Microwave Theory and Techniques*, vol. 15, no. 12, pp. 713–721, December 1967.

[12] J. Ebert, "Four terminal parameters of HF inductors," *Bulletin de l Academie Polonaise des Sciences-serie des Sciences Techniques*, no. 5, 1968.

[13] R. A. Pucel, D. J. Massé, and C. P. Hartwig, "Losses in microstrips," *IEEE Transactions on Microwave Theory and Techniques*, vol. 16, no. 6, pp. 342–250, June 1968.

[14] H. M. Greenhouse, "Design of planar rectangular microelectronic inductors," *IEEE Transactions on Parts Hybrids and Packaging*, vol. PHP-10, no. 2, pp. 101–109, June 1974.

[15] N. Saleh, "Variable microelectronic inductors," *IEEE Transactions on Components Hybrids and Manufacturing Technology*, vol. 1, no. 1, pp. 118–124, March 1978.

[16] D. Schieber, "On the inductance of printed spiral coils," *Archiv fur Elektrotechnik*, vol. 68, pp. 155–159, 1985.

[17] B. Brooks, "Design of standards on inductance, and the proposed use of model reactors in the design of air-core and iron-core reactors," *Bureau of Standards Journal of Research*, pp. 289–328, vol. 7, 1931.

[18] P. Murgatroyd, "The Brooks inductor: A study of optimal solenoid cross-sections," *IEE Proceedings: Part B, Electric Power Applications*, vol. 133, no. 5, pp. 309–314, September 1986.

[19] L. Weimer and R. H. Jansen, "Determination of coupling capacitance of underpasses, air bridges and crossings in MICs and MMICS," *Electronic Letters*, vol. 23, no. 7, pp. 344–346, March 1987.

[20] N. M. Nguyen and R. G. Mayer, "Si IC-compatible inductors and LC passive filter," *IEEE Journal of Solid-State Circuits*, vol. 27, no. 10, pp. 1028–1031, August 1990.

[21] P. R. Gray and R. G. Mayer, "Future directions in silicon IC's for RF personal communications," Proceedings of the IEEE 1995 Custom Integrated Circuits Conference, May 1995, pp. 83–90.

[22] J. Craninckx and M. S. J. Steyeart, "A 1.8 GHz CMOS low noise voltage-controlled oscillator with prescalar," *IEEE Journal of Solid-State Circuits*, vol. 30, pp. 1474–1482, December 1995.

[23] C. R, Sullivan and S. R. Sanders, "Design of microfabricated transformers and inductors foe high-frequency conversion," *IEEE Transactions on Power Electronics*, vol. 11, no. 2, pp. 228–238, March 1996.

[24] J. R. Long and M. A. Copeland, "The modeling, characterization, and design of monolithic inductors for silicon RF IC's," *IEEE Journal of Solid-State Circuits*, vol. 32, no. 3, pp. 357–369, May 1997.

[25] J. N. Burghartz, M. Soyuer, and K. Jenkins, "Microwave inductors and capacitors in standard multilevel interconnect silicon technology," *IEEE Transactions on Microwave Theory and Technique*, vol. 44, no. 1, pp. 100–103, January 1996.

[26] K. B. Ashby, I. A. Koullias, W. C. Finley, J. J. Bastek, and S. Moinian, "High Q inductors for wireless applications in a complementary silicon bipolar process," *IEEE Journal of Solid-State Circuits*, vol. 31, no. 1, pp. 4–9, January 1996.

[27] C. P. Yue, C. Ryu, J. Lau, T. H. Lee, and S. S. Wong, "A physical model for planar spiral inductors in silicon," *International Electron Devices Meeting Technical Digest*, December 1996, pp. 155–158.

[28] J. Cronickx and M. S. Steyaert, "A 1.8 GHz low-phase-noise CMOS VCO using optimized hollow spiral inductor," *IEEE Journal of Solid-State Circuits*, vol. 32, no. 5, pp. 736–744, May 1997.

[29] C. P. Yue and S. S. Wang, "On-chip spiral inductors with patterned ground shields for Si-bases RF ICs," *IEEE Journal of Solid-State Circuits*, vol. 33, no. 5, pp. 743–752, May 1998.

[30] F. Mernyei, F. Darrer, M. Pardeon, and A. Sibrai, "Reducing the substrate losses of RF integrated inductors," *IEEE Microwave and Guided Wave Letters*, vol. 8. no. 9, pp. 300–3001, September 1998.

[31] A. M. Niknejad and R. G. Mayer, "Analysis, design, and optimization of spiral inductors and transformers for Si RF IC's," *IEEE Journal of Solid-State Circuits*, vol. 33. no. 10, pp. 1470–1481, October 1998.

[32] Y.-J. Kim and M. G. Allen, "Integrated solenoid-type inductors for high frequency applications and their characteristics," 1998 Electronic Components and Technology Conference, 1998, pp. 1249–1252.

[33] Y.-J. Kim and M. G. Allen, "Surface micromachined solenoid inductors for high frequency applications," *IEEE Transactions on Components Packaging and Manufacturing, Part C*, vol. 21, no. 1, pp. 26–33, January 1998.

[34] C. P. Yue and S. S. Wong, "Design strategy of on-chip inductors highly integrated RF systems," Proceedings of the 36th Design Automation Conference, 1999, pp. 982–987.

[35] M. T. Thomson, "Inductance calculation techniques – Part II: approximations and handbook methods," *Power Control and Intelligent Motion*, pp. 1–11, December 1999.

[36] S. S. Mohan, M. Hershenson. S. P. Boyd, and T. H. Lee, "Simple accurate expressions for planar spiral inductors," *IEEE Journal of Solid-State Circuits*, vol. 34, no. 10, pp. 1419–1424, October 1999.

[37] Y. K. Koutsoyannopoulos and Y. Papananos, "Systematic analysis and modeling of integrated inductors and transformers in RF IC design" *IEEE Transactions on Circuits and Systems-II, Analog and Digital Signal Processing*, vol. 47, no. 8, pp. 699–713, August 2000.

[38] W. B. Kuhn and N. M. Ibrahim, "Analysis of current crowding effects in multiturn spiral inductors," *IEEE Transactions on Microwave Theory and Techniques*, vol. 49, no. 1, pp. 31–38, January 2001.

[39] A. Zolfaghari, A. Chan, and B. Razavi, "Stacked inductors and transformers in CMOS technology," *IEEE Journal of Solid-State Circuits*, vol. 36, no. 4, pp. 620–628, April 2001.

[40] S. Jenei, B. K. J. Nauwelaers, and S. Decoutere, "Physics-based closed-form inductance expressions for compact modeling of integrated spiral inductors," *IEEE Journal of Solid-State Circuits*, vol. 37, no. 1, pp. 77–80, January 2002.

[41] T.-S. Horng, K.-C. Peng, J.-K. Jau, and Y.-S. Tsai, "S-parameters formulation of quality factor for a spiral inductor in generalized tow-port configuration," *IEEE Transactions on Microwave Theory and Technique*, vol. 51, no. 11, pp. 2197–2202, November 2002.

[42] Yu. Cao, R. A. Groves, X. Huang, N. D. Zamder, J.-O. Plouchart, R. A. Wachnik, T.-J. King, and C. Hu, "Frequency-independent equivalent-circuit model for on-chip spiral inductors," *IEEE Journal of Solid-State Circuits*, vol. 38, no. 3, pp. 419–426, March 2003.

[43] J. N. Burghartz and B. Rejaei, "On the design of RF spiral inductors on silicon," *IEEE Transactions on Electron Devices*, vol. 50, no. 3, pp. 718–729, March 2003.

[44] J. Aguilera and R. Berenguer, *Design and Test of Integrated Inductors for RF Applications*, Boston, MA: Kluwer Academic Publishers, 2003.

[45] W. Y. Lin, J. Suryanarayan, J. Nath, S. Mohamed, L. P. B. Katehi, and M. B. Steer, "Toroidal inductors for radio-frequency integrated circuits," *IEEE Transactions on Microwave Circuits and Techniques*, vol. 52, no. 2, pp. 646–651, February 2004.

[46] N. Wong, H. Hauser, T. O'Donnel, M. Brunet, P. McCloskey, and S. C. O'Mathuna, "Modeling of high-frequency micro-transformers," *IEEE Transactions on Magnetics*, vol. 40, pp. 2014–2016, July 2004.

[47] H.-M. Hsu, "Analytical formula for inductance of metal of various widths in planar inductors," *IEEE Transactions on Electron Devices*, vol. 51, no. 8, pp. 1343–1346, August 2004.

[48] H.-M. Hsu, "Investigation of layout parameters of on-chip inductor," *Microelectronics Journal*, vol. 37, pp. 800–803, 2006.

[49] T. Suetsugu and M. K. Kazimierczuk, "Integration of Class DE inverter for dc-dc converter on-chip power supplies," IEEE International Symposium on Circuits and Systems, Kos, Greece, May 21–24, 2006, pp. 3133–3136.

[50] T. Suetsugu and M. K. Kazimierczuk, "Integration of Class DE synchronized dc-dc converter on-chip power supplies," IEEE Power Electronics Specialists Conference, Jeju, South Korea, June 21–24, 2006.

[51] W.-Z. Chen, W.-H. Chen, and K.-C. Hsu, "Three-dimensional fully symmetrical inductors, transformers, and balun in CMOS technology," *IEEE Transactions on Circuits and Systems-I*, vol. 54, no. 7, pp. 1413–1423, July 2007.

[52] Y. Zhuang, M. Vroubel, B. Rejaei, and J. N. Burghartz, "Integrated RF inductors with micro-patterned NiFe core," *Solid-State Electronics*, vol. 51, pp. 405–413, 2007.

[53] A. Massarini and M. K. Kazimierczuk, "Self-capacitance of inductors," *IEEE Transactions on Power Electronics*, vol. 12, no. 4, pp. 671–676, July 1997.

[54] G. Grandi, M. K. Kazimierczuk, A. Massarini, and U. Reggiani, "Stray capacitance of single-layer solenoid air-core inductors," *IEEE Transactions on Industry Applications*, vol. 35, no. 5, pp. 1162–1168, September 1999.

[55] G. Stojanovic, L. Zivanov, and M. Damjanovic, "Compact form of expressions for inductance calculation of meander inductors," *Serbian Journal of Electrical Engineering*, vol. 1, no. 3, pp. 57–68, November 2004.

[56] J.-T. Kuo, K.-Y. Su, T.-Y. Liu, H.-H. Chen and S.-J. Chung, "Analytical calculations for dc inductances of rectangular spiral inductors with finite metal thickness in the PEEC Formulation," *IEEE Microwave and Wireless Components Letters*, pp. 1–3, 2006.

[57] A. Estrov, "Planar magnetics for power converters," *Transactions on Power Electronics*, vol. 4, pp. 46–53, February 1989.

[58] D. van der Linde, C. A. M. Boon, and J. B. Klaasens, "Design of a high-frequency planar power transformer in the multilayer technology," *IEEE Transactions on Power Electronics*, vol. 38, no. 2, pp. 135–141, April 1991.

[59] M. T. Quire, J. J. Barrett, and M. Hayes, "Planar magnetic component Technology – A review," *IEEE Transactions on Components Hybrids and Manufacturing Technology*, vol. 15, nr. 5, pp. 884–892, October 1992.

[60] Y. Fukuda, T. Inoue, T. Mizoguchi, S. Yatabe, and Y. Tachi, "Planar inductor with ferrite layers for dc-dc converter," *IEEE Transactions on Magnetics*, vol. 39, no. 4, pp. 217–233, July 2003.

[61] S.C. O. Mathuna, T. O'Donnell, N. Wang, and K. Rinne, "Magnetics on silicon: an enabling technology for power supply on chip," *IEEE Transactions on Power Electronics*, vol. 20, no. 3, pp. 558–592, May 2005.

[62] I. Kowasa, G. Schrom, F. Paillet, B. Jamieson, T. Kamik, and S. Borkar, "A planar inductor using MnZn ferrite/polyimide composite film for low voltage and large current dc-dc converter," *IEEE Transactions on Magnetics*, vol. 41, no. 10, pp. 4760–4766, October 2005.

[63] D. S. Gardner, H. Tomita, A. Sawabe, T. Inoue, T. Mizoguchi, and M. Sahashi, "Review of on-chip inductor structures with magnetic fields," *IEEE Transactions on Magnetics*, vol. 45, no. 10, pp. 4760–4766, October 2009.

[64] T. Sato, H. Tomita, A. Sawabe, T. Inoue, T. Mizoguchi, and M. Sahashi, "A magnetic thin film inductor and its application to a MHz switching dc-dc converter," *IEEE Transactions on Magnetics*, vol. 30, no. 2, pp. 217–233, March 1994.

[65] C. D. Meyer, S. S. Bedair, B. C. Morgan, and D. P. Arnold, "High inductance density air-core power inductors and transformers designed for operation at 100-500 MHz," *IEEE Transactions on Magnetics*, vol. 46, no. 6, pp. 2236–2239, June 2010.

[66] M. Wang, J. Li, K. D. T. Ngo, and H. Xie, "A surface mountable micro-fabricated power inductor in silicon for ultra-compact power supplies," *IEEE Transactions on Power Electronics*, vol. 26, no. 5, pp. 1310–1315, May 2011.

[67] I. K. Nashadham, "Extrinsic equivalent circuit modeling of SMD inductors for printed circuit applications," *Transactions on Electromagnetic Compatibility*, vol. 43, no. 4, pp. 557–565, November 2001.

[68] I. K. Nashadham and T. Durak, "Measurement-based closed-form modeling of surface-mounted RF components," *Transactions on Microwave Theory and Technique*, vol. 50, no. 10, pp. 2276–2286, October 2002.

[69] C. H. Ahn and Mi. G. Allen, "A comparison of two micromechanical inductors (bar- and meander type) for fully integrated boost dc/dc power converters," *Transactions on Power Electronics*, vol. 11, no. 2, pp. 239–245, February 1996.

[70] J. Wibben and R. Harjani, "A high-efficiency DC-DC converter using 2 nH integrated inductors," *IEEE Journal of Solid-State Circuits*, vol. 43, no. 4, pp. 844–854, April 2008.

[71] T. Komma and H. Gueldner, "The effect of different air-gap positions on the winding losses on modern planar ferrite cores in switch mode power supplies," IEEE SPEEDAM, 2008, pp. 632–637.

[72] D.-H. Shin, C.-S. Kim, J. Jeong, and S. Nam, "Fabrication of double rectangular type FeTaNI film inductors," *IEEE Transactions on Magnetics*, vol. 35, no. 5, pp. 3511–3513, May 1999.

[73] W. M. Chew, P. D. Evens, and W. J. Heffernan, "High frequency inductor design concepts," IEEE Power Electronics Specialists Conference Record, 1991, pp. 673–678.

[74] M. Soma, D. C. Golbraith, and R. L. White, "Radio-frequency coils in implantable devices: Misalignment analysis and design procedure," *IEEE Transactions on Biomedical Engineering*, vol. 34, no. 4, pp. 276–282, April 1987.

[75] W. G. Hurley and M. C. Duffy, "Calculation of self and mutual impedance in planar magnetic structure," *IEEE Transactions on Magnetics*, vol. 31, no. 4, pp. 2416–2522, July 1995.

[76] U.-M. Jow and M. Ghovankoo, "Design and optimization of printed spiral coils for efficient transcutaneous inductive power transmission," *IEEE Transactions on Biomedical Circuits and Systems*, vol. 1, no. 3, pp. 193–202, September 2007.

[77] A. Ramrakhyani, S. Mirabbasi, and M. Chiao, "Design and optimization of resonance-biased efficient wireless power delivery system for biomedical implants," *IEEE Transactions on Biomedical Circuits and Systems*, vol. 5, no. 1, pp. 48–63, February 2011.

[78] R. C. Fitch, Jr., M. K. Kazimierczuk, J. K. Gillespe, A. G. Mattamana, P. L. Orlando, K. S. Groves, and T. K. Quach, "Hybrid integration of microwave circuit solenoid inductors and AlGaN/GaN HEMTs using an SU-8 photosensitive epoxy interposer layer," *The Electrochemical (ECS) Society Transactions*, vol. 33, no. 13, pp. 23–45, Oct. 2010.

8.27 Review Questions

8.1. List the types of integrated inductors.

8.2. What is the range of integrated inductances?

8.3. Is it easy to achieve good performance of integrated inductors?

8.4. What kind of integrated inductors is most widely used?

8.5. What is the main disadvantage of planar integrated inductors?

8.6. How many metal layers are usually used in RF IC planar inductors?

8.7. What is an underpass in spiral planar inductors?

8.8. What is an air bridge in integrated inductors?

8.9. When the mutual inductance between two conductors is zero?

8.10. When the mutual inductance between two conductors is positive and when it is negative?

8.11. Is the quality factor high for integrated inductors?

8.12. What is the SRF of an inductor?

8.13. What is the main disadvantage of bondwire inductors?

8.14. What are the advantages and disadvantages of planar inductors?

8.15. What are the advantages and disadvantages of MEMS inductors?

8.16. How eddy-current losses can be reduced?

8.17. What are the components of the model of planar spiral inductors?

8.28 Problems

8.1. Calculate the skin depth at $f = 1$ GHz for (a) copper, (b) aluminum, (c) silver, and (d) gold.

8.2. Calculate the DC resistance of an aluminum trace with $l = 50$ μm, $w = 1$ μm, and $h = 1$ μm. Find the AC resistance of the trace at 10 GHz. Calculate F_R.

8.3. Calculate the inductance of a straight trace with $l = 100$ μm, $w = 1$ μm, $h = 1$ μm, and $\mu_r = 1$.

8.4. Calculate the inductance of a meander inductor with $N = 10$, $w = 40$ μm, $s = 40$ μm, $a =$ 40 μm, $h = 100$ μm, and $\mu_r = 1$.

8.5. Calculate the inductance of a round straight inductor with $l = 2$ mm and $\mu_r = 1$ for $\delta \gg h$. Find the inductance of the above copper round straight conductor for $a = 20$ μm at $f = 2.4$ GHz.

8.6. Calculate the inductance of a bondwire with $l = 1$ mm and $a = 1$ μm.

8.7. Find the inductance of the spiral planar inductor with $N = 10$ and $r = 100$ μm.

8.8. Calculate an inductance of a square planar spiral inductor with $N = 10$, $s = 20$ μm, $w = 30$ μm, and $d = 40$ μm. Use Bryan's formula.

8.9. Calculate a square planar spiral inductance with $N = 10$, $s = 20$ μm, $w = 30$ μm, and $d =$ 40 μm. Use Wheeler's formula.

8.10. Calculate a square planar spiral inductance with $N = 15$, $s = 20$ μm, $w = 30$ μm, $h = 20$ μm, and $d = 40$ μm. Use Greenhouse's formula.

8.11. Calculate a square planar spiral inductance with $N = 10$, $s = 20$ μm, $w = 30$ μm, $h = 20$ μm, and $d = 40$ μm. Use Rosa's formula.

8.12. Calculate a square planar spiral inductance with $N = 10$, $s = 20$ μm, $w = 30$ μm, $h = 20$ μm, and $d = 40$ μm. Use Cranin's formula.

8.13. Calculate a square planar spiral inductance with $N = 10$, $s = 20$ μm, $w = 30$ μm, $h = 20$ μm, and $d = 40$ μm. Use monomial formula.

8.14. Calculate a square planar spiral inductance with $N = N_i = 10$, $s = 20$ μm, $w = 30$ μm, $h =$ 20 μm, and $d = 40$ μm. Use Jenei's formula.

8.15. Calculate a square planar spiral inductance with $N = 10$, $w = 30$ μm, $h = 20$ μm, and $D =$ 1000 μm. Use Terman's formula.

8.16. Calculate a hexagonal planar spiral inductance with $N = 10$, $s = 20$ μm, $w = 30$ μm, $h =$ 20 μm, and $d = 40$ μm. Use Rosa's formula.

8.17. Calculate a octagonal planar spiral inductance with $N = 5$, $s = 20$ μm, $w = 30$ μm, $h = 20$ μm, and $d = 60$ μm. Use Rosa's formula.

8.18. Calculate a circular planar spiral inductance with $N = 10$, $d = 40$ μm, $w = 30$ μm, and $s =$ 20 μm. Use Rosa's formula.

8.19. Calculate the inductance of a MEMS solenoid with $N = 20$. The thickness of a square wire is $h = 10$ μm and the separation between the turns is $s = 10$ μm. The solenoid has a square shape with $w = 100$ μm.

9

Self-Capacitance

9.1 Introduction

Every system of conductors separated by dielectrics represents capacitance. Whenever a charged conductor is placed in the proximity of other conductors, the lines of electric field leave one conductor and terminate on the surrounding conductors. In general, capacitance of a parallel-plate capacitor is equal to the ratio of the charge stored on one plate to the voltage difference between the two plates. The geometrical structures of inductors and transformers are much more complex than those of the parallel-plate capacitor. Therefore, it is much more difficult to derive expressions for self-capacitance of real magnetic components. The self-capacitance is also called stray capacitance or parasitic capacitance. In this chapter, a high-frequency model of inductors is developed [1–33]. The impedance of inductors is studied as a function of frequency. Physics-based analytical expressions for self-capacitances of foil winding inductors, single-layer, and multilayer round wire inductors are derived using geometrical methods. Finally, an expression for the first self-resonant frequency is given.

9.2 High-Frequency Inductor Model

The distributed inductance, resistance, and capacitance of an inductor are illustrated in Fig. 9.1. The conductor of an inductor has a nonzero resistivity ρ_w and therefore presents a series resistance R_s. Insulated conductors of turns form turn-to-turn capacitances C_{tt}. Therefore, an electric field exists between the turns with different potentials and stores electric energy. At high frequencies, the displacement current flows through the capacitors and bypasses the inductive and resistive conductors. Figure 9.2a shows a high-frequency lumped-circuit inductor model, where L_s is the inductance, R_c is the parallel core resistance, R_w is the winding resistance, and C_s is the self-capacitance. The self-capacitance C_s does not change significantly with frequency. Figure 9.2b shows a high-frequency lumped-circuit inductor, where the parallel circuit L–R_c is converted to a series circuit L_s–R_{cs}. The total inductor AC series resistance at frequencies much lower than the first self-resonant frequency is

$$R_s = R_w + R_{cs}. \tag{9.1}$$

High-Frequency Magnetic Components, Second Edition. Marian K. Kazimierczuk.

Companion Website: www.wiley.com/go/kazimierczuk_High2e

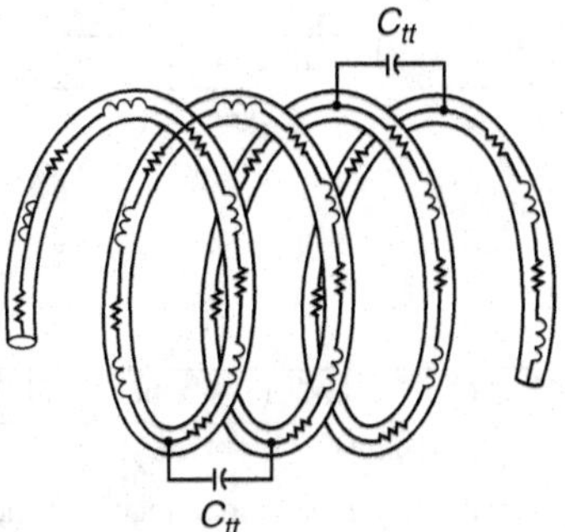

Figure 9.1 Distributed inductance, resistance, and capacitance of an inductor

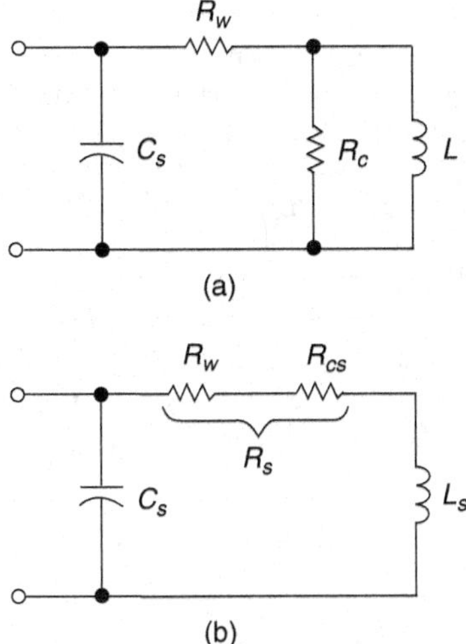

Figure 9.2 High-frequency lumped-circuit models of an inductor. (a) Inductor model with parallel inductance L and parallel core resistance R_c. (b) Inductor model with series inductance L_s and series core resistance R_{cs}

The impedance of the model shown in Fig. 9.2b is given by Bartoli *et al.* [8, 9]

$$Z = \frac{(R_s + j\omega L_s)\frac{1}{j\omega C_s}}{R_s + j\omega L_s + \frac{1}{j\omega C_s}} = \frac{R_s + j\omega K_s}{1 - \omega^2 L_s C_s + j\omega R_s C_s}$$

$$= \frac{R_s + j\omega L_s(1 - \omega^2 L_s C_s - C_s R_s^2/L_s)}{(1 - \omega^2 L_s C_s)^2 + (\omega C_s R_s)^2} = r + jx = \sqrt{r^2 + x^2} e^{j\arctan(x/r)} = |Z| e^{j\phi}, \tag{9.2}$$

where

$$r = \frac{R_s}{(1 - \omega^2 L_s C_s)^2 + (\omega C_s R_s)^2}, \tag{9.3}$$

$$x = \omega L_s \frac{1 - \omega^2 L_s C_s - C_s R_s^2/L_s}{(1 - \omega^2 L_s C_s)^2 + (\omega C_s R_s)^2}, \tag{9.4}$$

$$|Z| = \sqrt{r^2 + x^2}, \tag{9.5}$$

and

$$\phi = \arctan\left(\frac{x}{r}\right). \tag{9.6}$$

Hence,

$$R_s = \frac{1 - \sqrt{1 - 4r^2\omega^2 C_s^2 (1 - \omega^2 L_s C_s)^2}}{2r\omega^2 C_s^2} \tag{9.7}$$

and

$$L_s = \frac{1 + \sqrt{\frac{R_s}{r} - (\omega C_s R_s)^2}}{\omega^2 C_s}$$
$$= \frac{1 + 2x\omega C_s - \sqrt{(1 + 2x\omega C_s)^2 - 4\omega C_s (1 + x\omega C_s)\{x[1 + (\omega C_s R_s)^2] + \omega C_s R_s^2\}}}{2\omega^2 C_s (1 + x\omega C_s)}. \tag{9.8}$$

Figure 9.3 depicts the magnitude and phase of impedance for the inductor high-frequency model. The series resistance r and the series reactance x can be measured with most impedance meters or network analyzers.

The quality factor of an inductor at a frequency $f = \omega/(2\pi)$ is defined as

$$Q_{Lo} = \frac{\omega L_s}{R_s}. \tag{9.9}$$

where L_s and R_s are, in general, frequency dependent.

At the first self-resonant frequency f_r, the inductor impedance is purely resistive, $x(f_r) = 0$, and the phase of the inductor impedance ϕ is zero. Hence,

$$1 - \omega^2 L_s C_s - \frac{C_s R_s^2}{L_s} = 0, \tag{9.10}$$

yielding the self-resonant radial frequency

$$\omega_r = \sqrt{\frac{1}{L_s C_s} - \left(\frac{R_s}{L_s}\right)^2} = \frac{1}{\sqrt{L_s C_s}}\sqrt{1 - \frac{R_s^2}{L_s/C_s}} = \frac{1}{\sqrt{L_s C_s}}\sqrt{1 - \frac{R_s^2}{Z_o^2}} = \frac{1}{\sqrt{L_s C_s}}\sqrt{1 - \frac{1}{Q_o^2}}, \tag{9.11}$$

where the quality factor of the inductor at the self-resonant frequency f_r is

$$Q_o = \frac{\sqrt{\frac{L_s}{C_s}}}{R_s} = \frac{Z_o}{R_s} \tag{9.12}$$

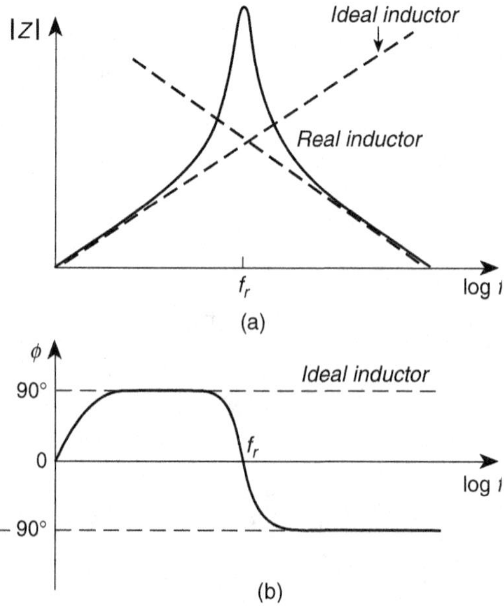

Figure 9.3 Impedance of the high-frequency inductor model. (a) $|Z|$ versus frequency. (b) ϕ versus frequency

and the characteristic impedance of the inductor is

$$Z_o = \sqrt{\frac{L_s}{C_s}}. \tag{9.13}$$

At self-resonant frequency f_r, the impedance of the inductor impedance is

$$Z(f_r) = \frac{R_s Q_o^4}{2Q_o^2 - 1}. \tag{9.14}$$

Hence, the series resistance at the self-resonant frequency is

$$R_s = \frac{Z(f_r)(2Q_o^2 - 1)}{Q_o^4} \tag{9.15}$$

and the series inductance at the self-resonant frequency is

$$L_s = C_s R_s Q_o^2 \approx \frac{Q_o R_s}{\omega_r}. \tag{9.16}$$

Since $C_s R_s^2 / L_s \ll 1$, the first parallel self-resonant frequency can be approximated by

$$f_r \approx \frac{1}{2\pi\sqrt{L_s C_s}}. \tag{9.17}$$

For $\omega = 1/\sqrt{L_s C_s}$, the inductor impedance is capacitive

$$Z = \frac{1}{(\omega C_s)^2 R_s} - j\frac{1}{\omega C_s} = \frac{L_s}{C_s R_s} - j\frac{1}{\omega C_s}, \tag{9.18}$$

$$r = \frac{1}{R_s(\omega C_s)^2} = \frac{L_s}{R_s C_s}, \tag{9.19}$$

and

$$x = -\frac{1}{\omega C_s}. \tag{9.20}$$

Figures 9.4 through 9.7 illustrate $|Z|$, ϕ, r, and x as functions of frequency for an inductor at $L_s = 168$ μH, $C_s = 21.964$ pF, $R_{cs} = 0$, and $R_s = R_w = 1\ \Omega$. At $f = 0$ and $f = f_r$, the inductor impedance is resistive. For $f < f_r$, the inductor impedance is inductive ($x > 0$). For $f > f_r$, the inductor impedance is capacitive ($x < 0$). The useful frequency range of inductors in most applications is from DC to about $0.9 f_r$. The self-resonant frequency f_r can be measured with an impedance analyzer or a Q-meter. The stray capacitance can be measured at a frequency sufficiently higher than the self-resonant frequency f_r. If the inductance L_s is known, it can be calculated from (9.17). Higher order self-resonant frequencies can be found by taking into account transmission line effects. Such effects are outside the scope of this work. At frequencies much lower than f_r, $R_s \approx R_w + R_{cs}$. In reality, L_s, R_w, and R_{cs} are frequency-dependent. For inductors with a magnetic core, the inductance L_s is frequency dependent because the permeability of the core μ depends on the frequency. It usually decreases with frequency. In all inductors, the inductance due to magnetic field inside the winding conductor decreases with frequency. The permittivity of most dielectric materials ϵ used in the construction of inductors is frequency-independent up to 30 MHz. Therefore, the capacitance C_s is frequency-independent up to 30 MHz.

At low frequencies, $2\delta_w \gg d$, skin and proximity effects, core losses, and stray self-capacitance C_s can be neglected. Therefore, $R_c \to \infty$, $R_{cs} = 0$, $R_w \approx R_{wDC}$, and $R_s = R_w \approx R_{wDC}$. The low-frequency inductor model consists of inductance L_s in series with $R_s = R_{wDC}$, and the self-capacitance C_s can be neglected. Figure 9.8 shows the plot of the series reactance x as a function of frequency for an inductor with $L_s = 168$ μH, $C_s = 21.964$ pF, $f_r = 2.62$ MHz, $R_{cs} = 0$, $R_s = R_w$, $N_l = 4$, $n = 20$, and $d_l = 80$ μm. The reactance x is negative above the self-resonant frequency.

The quality factor of an inductor at a frequency f is defined as

$$Q_{Lo} = 2\pi \frac{\text{Peak magnetic energy}}{\text{Energy loss in one cycle}} = 2\pi \frac{\text{Peak electric energy}}{\text{Energy loss in one cycle}}. \tag{9.21}$$

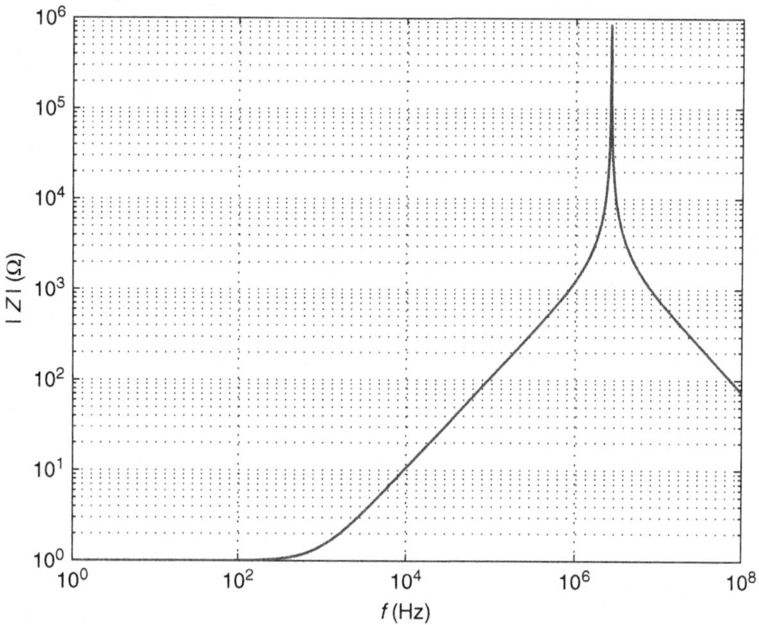

Figure 9.4 Magnitude of the inductor impedance as a function of frequency with $L_s = 168$ μH, $C_s = 21.964$ pF, $R_{cs} = 0$, and $R_s = R_w = 1\ \Omega$

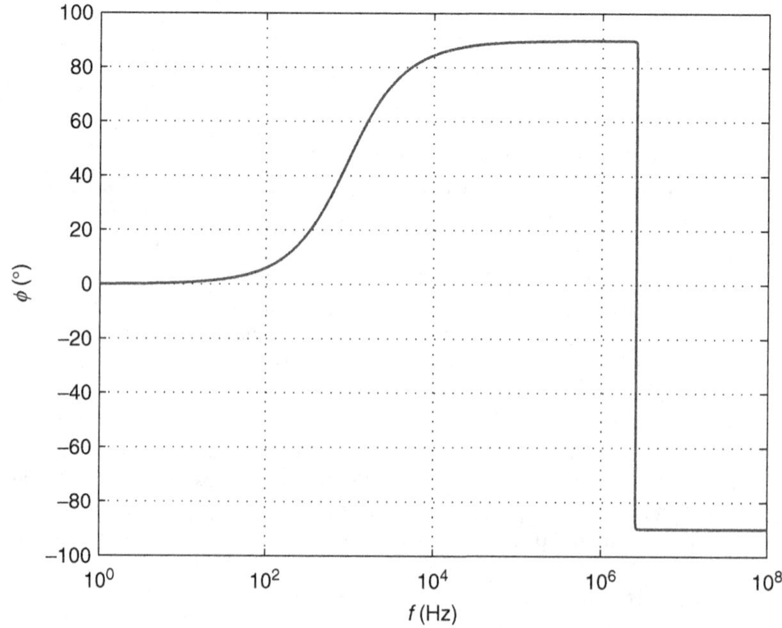

Figure 9.5 Phase of the inductor impedance as a function of frequency with $L_s = 168$ μH, $C_s = 21.964$ pF, $R_{cs} = 0$, and $R_s = R_w = 1\ \Omega$

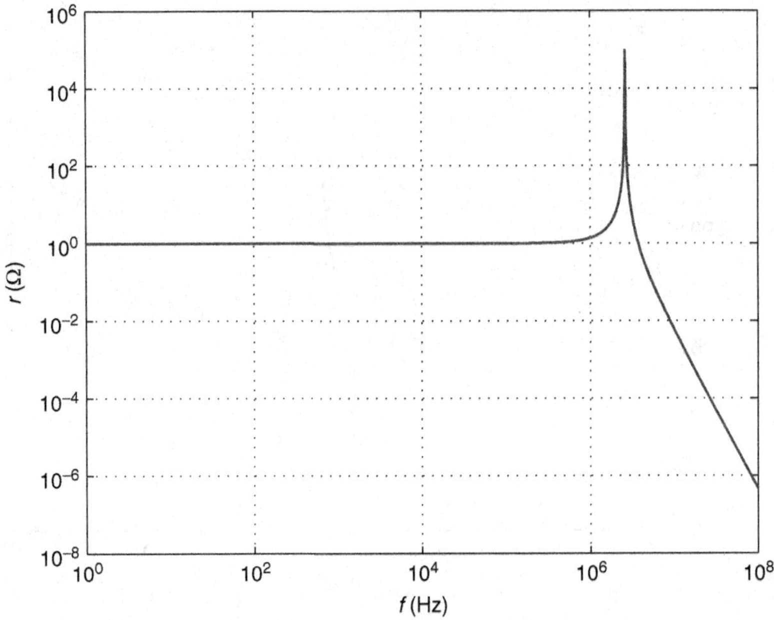

Figure 9.6 Series equivalent resistance of the inductor impedance as a function of frequency with $L_s = 168\ \mu\text{H}$, $C_s = 21.964$ pF, $R_{cs} = 0$, and $R_s = R_w = 1\ \Omega$

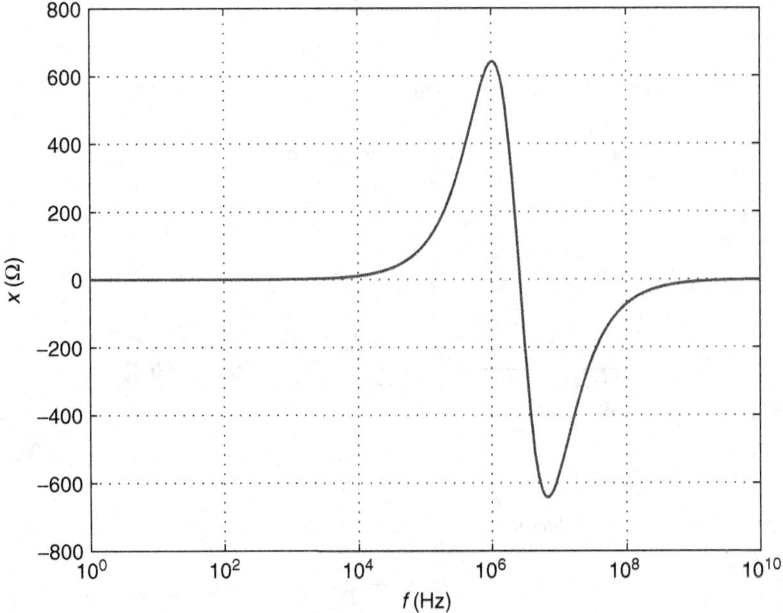

Figure 9.7 Series equivalent reactance of the inductor impedance as a function of frequency with $L_s = 168\ \mu\text{H}$, $C_s = 21.964$ pF, $R_{cs} = 0$, and $R_s = R_w = 1\ \Omega$

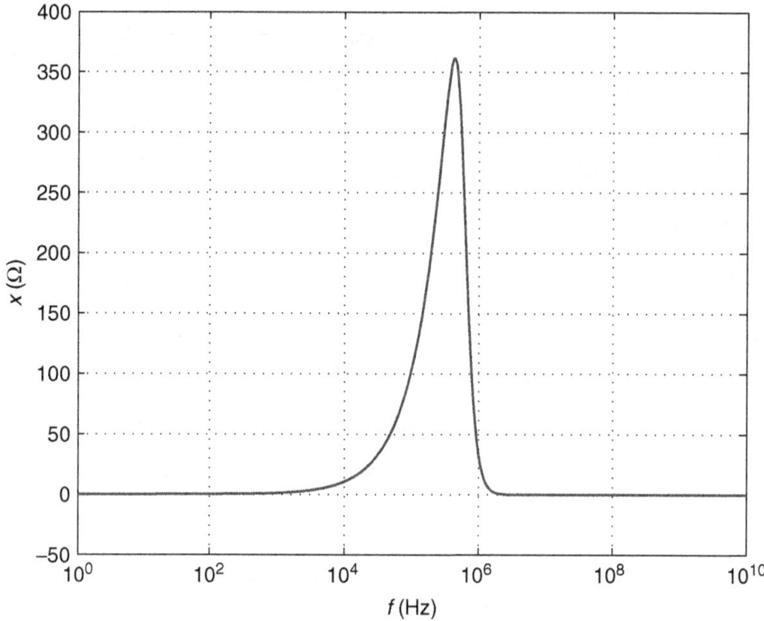

Figure 9.8 Series reactance x as functions of frequency for an inductor with $L_s = 168$ μH, $C_s = 21.964$ pF, $f_r = 2.62$ MHz, $R_{cs} = 0$, $R_s = R_w$, $N_l = 4$, $n = 20$, and $d_l = 80$ μm

The peak magnetic energy stored in the inductance L_s is given by

$$W_m = \frac{1}{2} L_s I_m^2 \tag{9.22}$$

where I_m is the amplitude of the current through the inductance L_s. The energy loss in the resistance R_s in one cycle $T = 1/f$ is

$$W_{Rs} = \frac{1}{2} R_s I_m^2 T = \frac{1}{2f} R_s I_m^2. \tag{9.23}$$

Hence, the quality factor of an inductor is

$$\begin{aligned} Q_{Lo} &= 2\pi \frac{W_m}{W_{Rs}} = \frac{\omega L_s}{R_s} = \frac{\omega L_s}{R_w + R_{cs}} = \frac{1}{\frac{R_w}{\omega L_s} + \frac{R_{cs}}{\omega L_s}} \\ &= \frac{1}{\frac{1}{\omega L_s / R_w} + \frac{1}{\omega L_s / R_{cs}}} = \frac{1}{\frac{1}{Q_{LRw}} + \frac{1}{Q_{LRcs}}} = \frac{Q_{LRw} Q_{LRcs}}{Q_{LRw} + Q_{LRcs}}, \end{aligned} \tag{9.24}$$

where the quality factor of an inductor due to the winding resistance is

$$Q_{LRw} = \frac{\omega L_s}{R_w} \tag{9.25}$$

and the quality factor of an inductor due to the core series resistance is

$$Q_{LRcs} = \frac{\omega L_s}{R_{cs}}. \tag{9.26}$$

The reactance factor of the inductor model at a frequency f is defined as

$$q = \tan \phi = \frac{x}{r} = \frac{\omega L_s \left(1 - \omega^2 L_s C_s - \frac{C_s R_s^2}{L_s}\right)}{R_s} = Q_{Lo} \left(1 - \omega^2 L_s C_s - \frac{C_s R_s^2}{L_s}\right). \tag{9.27}$$

Figure 9.9 depicts the plots of winding resistance R_w and series resistance r as functions of frequency for an inductor with $L_s = 168$ μH, $C_s = 21.964$ pF, $R_{cs} = 0$, $N_l = 4$, $n = 20$, and $d_l = 80$ μm,

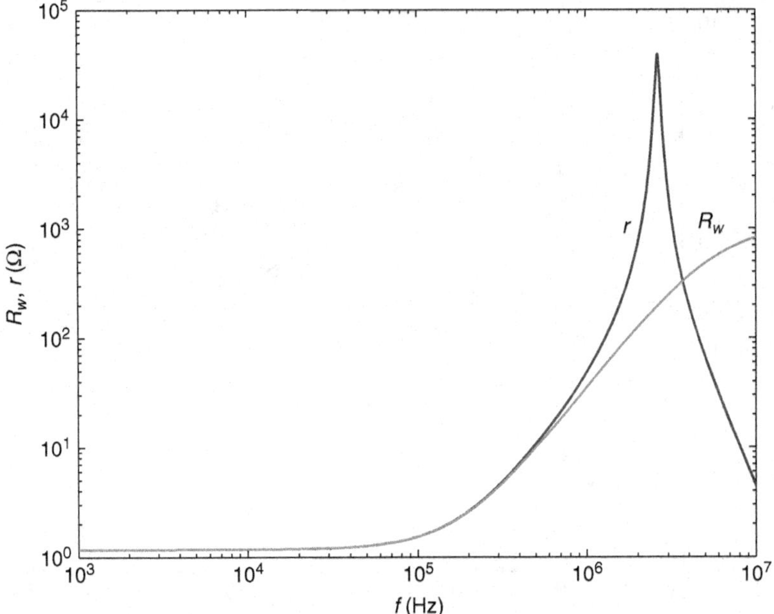

Figure 9.9 Winding resistance R_w and series resistance r as functions of frequency for an inductor with $L_s = 168\ \mu\text{H}$, $C_s = 21.964$ pF, $f_r = 2.62$ MHz, $R_{cs} = 0$, $R_s = R_w$, $N_l = 4$, $n = 20$, and $d_l = 80\ \mu\text{m}$

where R_w is computed from Dowell's equation and the equivalent series resistance r is calculated from (9.3). Both R_w and r vary with frequency. It can be seen that $r \approx R_w$ for $0 \le f/f_r \le 0.4$. For higher frequencies, r and R_w are not equal to each other.

Figure 9.10 shows the inductor quality factor due to the winding resistance $Q_{LRw} = \omega L_s/R_w$ along with ωL_s and R_w, when $R_{cs} = 0$ and R_w varies with frequency. The inductor quality factor Q_{LRw} first increases with frequency, reaches a local maximum value, then decreases, and finally increases again. At low frequencies, R_w remains constant and ωL_s increases with frequency. Therefore, Q_{LRw} also increases with frequency. At about 200 kHz, Q_{LRw} reaches a local maximum value $Q_{LRw(\max)} = 90$. Next, both R_w and ωL_s increase with frequency. However, R_w increases faster than ωL_s, causing Q_{LRw} to decrease with frequency. Finally, Q_{LRw} increases with frequency again because ωL_s increases faster with frequency than R_w.

Figures 9.11 and 9.12 show the magnitude and the phase of the inductor impedance computed from (9.2) for $L_s = 168\ \mu\text{H}$, $C_s = 21.964$ pF, $R_{cs} = 0$, and $R_s = R_w$, where R_w varies with frequency according to Dowell's equation. The self-resonant frequency is $f_r = 1/(2\pi\sqrt{L_s C_s}) = 2.62$ MHz.

Figure 9.13 shows the plots of R_{cs}, L_s, $Q_{LRcs} = \omega L_s/R_{cs}$, where both R_{cs} and ωL_s are functions of frequency because both μ'_{rs} and μ''_{rs} are frequency dependent. Figure 9.14 shows the plots of $R_s = R_w + R_{cs}$, ωL_s, and $Q_{Lo} = \omega L_s/(R_w + R_{cs})$, where R_{cs}, R_w, and L_s are functions of frequency.

The equivalent circuit of an inductor can be converted to a basic parallel resonant circuit R_p–L_p–C_s, as shown in Fig. 9.15. The stray capacitance C_s remains the same in both circuits. The relations among R_p, R_s, L_p, and L_s are

$$R_p = \frac{R_s^2 + \omega^2 L_s^2}{R_s} \tag{9.28}$$

$$L_p = \frac{R_s^2 + \omega^2 L_s^2}{\omega^2 L_s} \tag{9.29}$$

$$R_s = \frac{\omega^2 L_p^2}{R_p^2 + \omega^2 L_p^2} \tag{9.30}$$

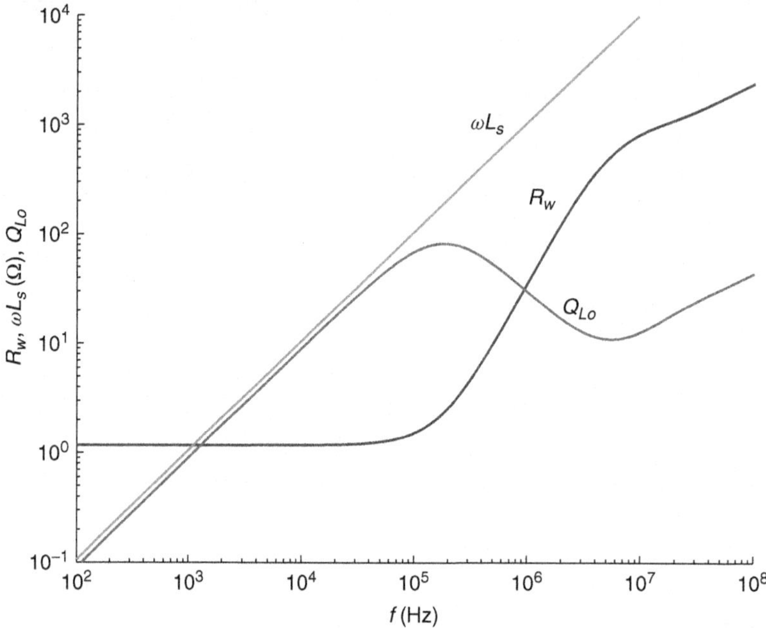

Figure 9.10 Plots of R_w, ωL_s, and $Q_{LRw} = \omega L_s / R_w$ as functions of frequency for an inductor with $L_s = 168$ μH, $C_s = 21.96$ pF, $N_l = 4$, $n = 20$, and $d_l = 80$ μm

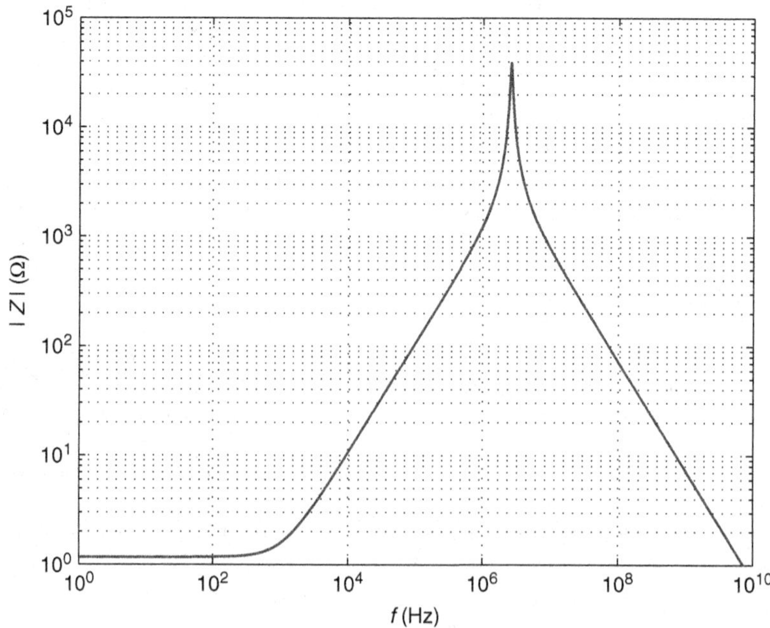

Figure 9.11 Magnitude of an inductor impedance for $L_s = 168$ μH, $C_s = 21.964$ pF, $R_{cs} = 0$, and $R_s = R_w$, where R_w varies with frequency according do Dowell's equation

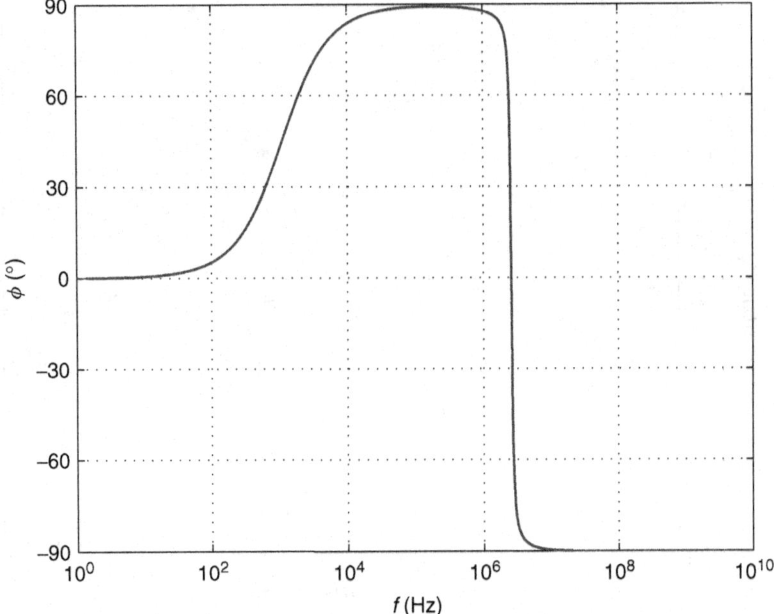

Figure 9.12 Phase of an inductor impedance for $L_s = 168$ μH, $C_s = 21.964$ pF, $R_{cs} = 0$, and $R_s = R_w$, where R_s varies with frequency according to Dowell's equation

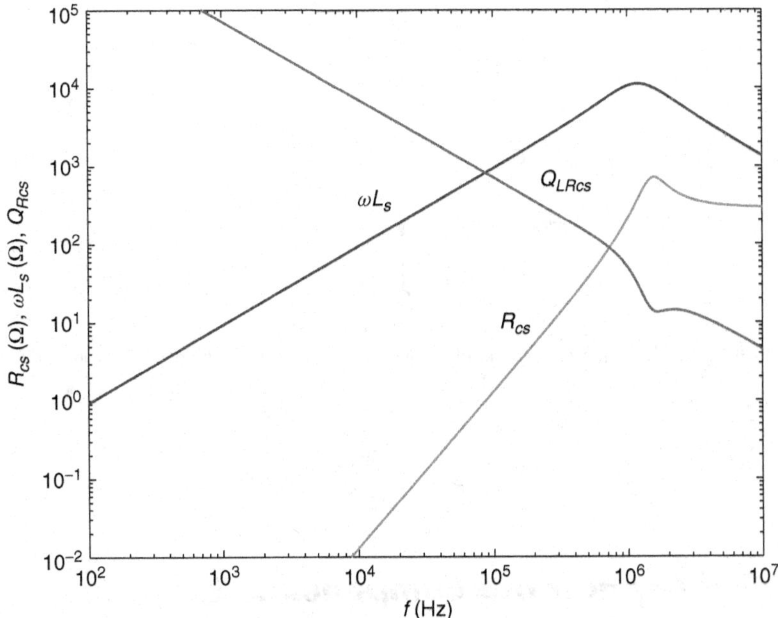

Figure 9.13 Plots of R_{cs}, ωL_s, and $Q_{LRcs} = \omega L_s / R_{cs}$ as functions of frequency for an inductor at $R_w = 0$, where both R_{cs} and L_s are functions of frequency

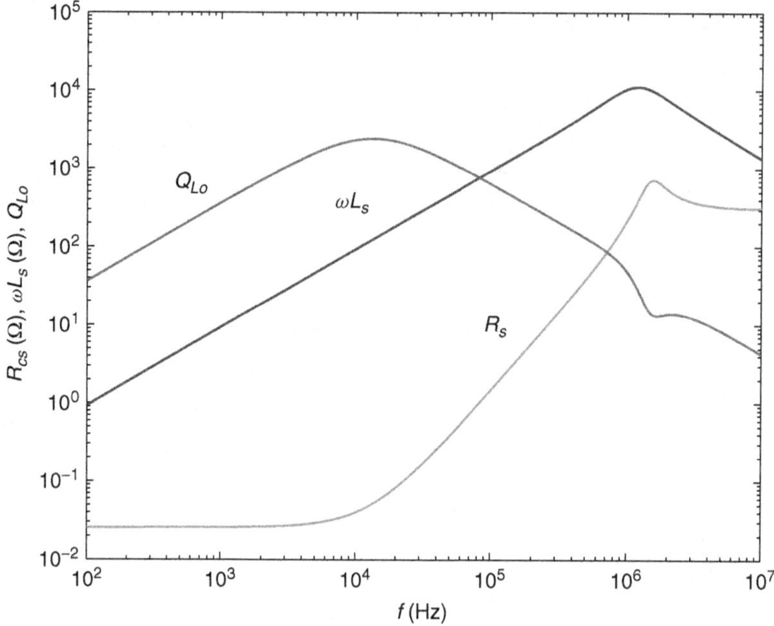

Figure 9.14 Plots of $R_s = R_w + R_{cs}$, ωL_s, and $Q_{Lo} = \omega L_s / R_s$ as functions of frequency for an inductor, where R_{cs}, R_w, and L_s are functions of frequency

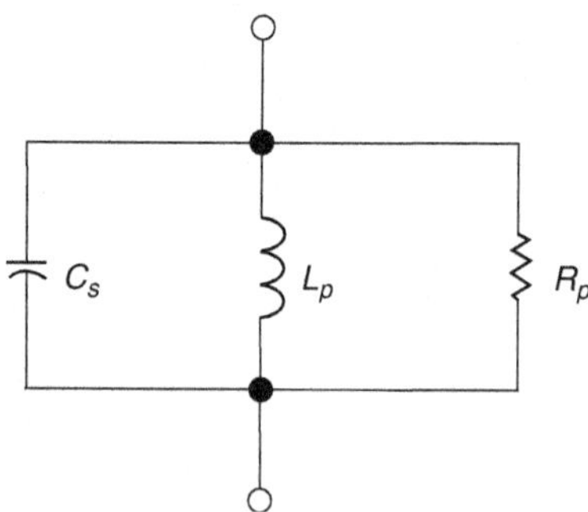

Figure 9.15 Equivalent circuit of an inductor in the form of the parallel resonant circuit

and

$$L_s = \frac{R_p^2 L_p}{R_p^2 + \omega^2 L_p^2}. \tag{9.31}$$

9.3 Self-Capacitance Components

The *stray capacitance* C_s, or *self-capacitance*, models the parasitic capacitive coupling between the terminals of the inductor. This capacitance allows the current to flow directly from one terminal to the other without passing through the inductance. The stray capacitance C_s consists of the following components [18, 22]:

- Turn-to-turn capacitance within the same layer C_{tt}
- Turn-to-core capacitance C_{tc}
- Layer-to-layer capacitance C_{ll}
- Turn-to-screen capacitance C_{tsc}
- Turn-to-shield capacitance C_{ts}
- Outer winding-to-board and surrounding circuitry capacitance
- Winding-to-winding capacitance in transformers.

The stray capacitance is strongly dependent on the following factors:

- The winding geometry and the proximity of conducting surfaces, such as spacing between the wire turns, distance between the turns and the core, the distance between the turns to the shield, and the distance between the turns to the screen.
- The relative permittivity of insulating materials ϵ_r, such as wire-insulating coating, insulation between the layers, and insulation between the windings.

The distributed turn-to-turn capacitance behaves like a shunt capacitance of an inductor, conducting a high-frequency current and bypassing the inductive reactance. In other words, the current "leaks" through the self-capacitance at high frequencies. Magnetic cores are made of conductive material and behave like a lossy capacitance. Therefore, the turn-to-core capacitance depends on the core resistivity ρ_c and core relative permittivity ϵ_r. In order to reduce the effect of stray capacitance and losses in the dielectric insulation between conductors, turns with a high potential difference should be avoided to reduce dv/dt. The current through the capacitance is $i_C = Cdv/dt$. Adjacent layers should be wound in the same direction rather then in the opposite direction.

Consider two conductors with a surface charge ρ_s and a potential difference between the conductors V. Using Gauss's law, the charge stored in one of the conductors is

$$Q = \oint_S \rho_s dS = \oint_S \epsilon \mathbf{E} \cdot d\mathbf{S}. \tag{9.32}$$

The voltage between the conductors is

$$V = -\int_{P_1}^{P_2} \mathbf{E} \cdot d\mathbf{l}, \tag{9.33}$$

where P_1 is a point located on the conductor with charge $-Q$ and P_2 is a point located on the conductor with charge $+Q$. The capacitance of any two conductors is defined as the ratio of the magnitude of the total charge Q on either conductor to the potential difference V between the conductors

$$C = \frac{Q}{V} = \frac{\oint_S \epsilon \mathbf{E} \cdot d\mathbf{S}}{-\int_{P_1}^{P_2} \mathbf{E} \cdot d\mathbf{l}}. \tag{9.34}$$

9.4 Capacitance of Parallel-Plate Capacitor

The geometrical structures of inductors and transformers are complex. Therefore, it is difficult to derive analytical equations for the self-capacitance of these devices. To gain some level of understanding, capacitances of simple geometrical structures are considered first.

The simplest structure of a capacitor is a parallel-plate capacitor. It consists of two conducting plates of area A separated by a distance d. The space between the plates is filled with a dielectric of permittivity ϵ. The lower plate is placed in the $x-z$ plane at $y = 0$, and the upper plate is placed at $y = d$. When a voltage potential V is applied between the two plates from the lower plate to the upper plate, charges Q and $-Q$ are induced on the upper and lower plates, respectively. These charges are uniformly distributed over the conducting plates with surface densities ρ_s and $-\rho_s$. The surface charge density is given by

$$\rho_s = \frac{Q}{A}. \tag{9.35}$$

A uniform electric field intensity is induced between the plates in the dielectric medium in the $-y$ direction

$$\mathbf{E} = -\mathbf{a_y}\frac{\rho_s}{\epsilon} = -\mathbf{a_y}\frac{Q}{\epsilon A}. \tag{9.36}$$

The fringing electric field near the edges can be neglected if the dimensions of the plates are much larger than the separation d between them. In this case, the majority of the electric field lines will exist in the medium between the plates. The potential difference between the plates is given by

$$V = -\int_0^d \mathbf{E}\cdot d\mathbf{l} = -\int_0^d \left(-\mathbf{a_y}\frac{Q}{\epsilon A}\right)\cdot(a_y dy) = \frac{Q}{\epsilon A}d. \tag{9.37}$$

Hence, the capacitance of a parallel-plate capacitor is

$$C = \frac{Q}{V} = \frac{\epsilon_r \epsilon_0 A}{d}, \tag{9.38}$$

where the permittivity of the dielectric placed between the plates is

$$\epsilon = \epsilon_r \epsilon_0, \tag{9.39}$$

the permittivity of free space is

$$\epsilon_0 = \frac{1}{36\pi} \times 10^{-9} = 8.85 \times 10^{-12}\ \text{F/m}, \tag{9.40}$$

and ϵ_r is the relative permittivity (or the relative dielectric constant) of the capacitor dielectric. The capacitance depends on the geometry (i.e., the shape and physical dimensions) of conductors and the relative permittivity of the dielectric ϵ_r.

9.5 Self-Capacitance of Foil Winding Inductors

The turn-to-turn capacitance C_{tt} in foil windings is higher than that in round-wire inductors. The total capacitance between the N turns is $C = C_{tt}/N$. The self-capacitance of foil winding inductors with a magnetic core can be described by the expression for the capacitance of a parallel-plate capacitor

$$C_s = \frac{\epsilon A}{d} = \frac{\epsilon_r \epsilon_0 N w l_T}{d}, \tag{9.41}$$

where the surface area of the foil is

$$A = wNl_T, \tag{9.42}$$

w is the foil width, $d = 2t$ is the thickness of the insulation between the bare foil surfaces, t is the insulating coating thickness, l_T is the mean length of a turn, and N is the number of turns. As the thickness of the insulation between the bare foil surfaces d increases, the self-capacitance C_s decreases and the first self-resonant frequency f_r increases. Foil winding inductors have a relatively high self-capacitance C_s and a low self-resonant frequency f_r. Therefore, extra insulation can be used in addition to the insulating coating to increase the distance d and thereby to reduce the self-capacitance C_s. The following insulation materials are used: kapton ($\epsilon_r = 3.3$–3.5), mylar ($\epsilon_r = 3$–3.5), kraft paper ($\epsilon_r = 1.5$–3), fish paper ($\epsilon_r = 1.5$–3), and nomex ($\epsilon_r = 1.6$–2.9). The self-capacitance of foil winding inductors without a magnetic core is described by

$$C_s = \frac{\epsilon A}{d} = \frac{\epsilon_r \epsilon_0 (N-1) w l_T}{d}. \tag{9.43}$$

Example 9.1

Calculate the stray capacitance C_s of a foil winding inductor with $w = 3$ cm, $l_T = 2$ cm, $N = 11$, and $d = 0.2$ mm. Kapton is used for insulation.

Solution: The stray capacitance is

$$C_s = \frac{\epsilon_r \epsilon_0 (N-1) w l_T}{d} = \frac{3.5 \times 8.85 \times 10^{-12} \times (11-1) \times 3 \times 10^{-2} \times 2 \times 10^{-2}}{0.2 \times 10^{-3}}$$
$$= 92.925 \text{ pF}. \tag{9.44}$$

9.6 Capacitance of Two Parallel Round Conductors

9.6.1 Potential of Infinite Single Straight Round Conductor with Charge

Consider a thin single straight round bare conductor with a positive electric charge. The conductor length is l_w, the total charge on the conductor is Q, and the charge per unit length is $\rho_l = Q/l_w$. Assume a uniform charge distribution in the conductor cross-sectional area. The electric field intensity at a radial distance r from the conductor center is given by Krauss [7]

$$E = E_r = \frac{\rho_l}{2\pi\epsilon r} \frac{l_w/2}{\sqrt{r^2 + \frac{l_w^2}{4}}} = \frac{\rho_l}{2\pi\epsilon r} \frac{1}{\sqrt{\left(\frac{r}{l_w/2}\right)^2 + 1}}. \tag{9.45}$$

Now consider an infinite conductor with a positive charge. The charge per unit length is ρ_l. As l_w approaches infinity in (9.45), the electric field intensity induced by the conductor charge is [7]

$$E = E_r = \frac{\rho_l}{2\pi\epsilon r}. \tag{9.46}$$

The potential difference V_{21} induced by the electric charge between any two points at radial distances r_1 and r_2 from the infinite conductor with charge Q is the work per unit charge W_e required to transport a positive charge from location r_2 to location r_1, that is, $V_{21} = W_e/Q$. For $\rho_l > 0$ and $r_2 > r_1$, the potential is higher at r_1 that that at r_2. The potential difference between the two points is obtained as

$$V_{21} = -\int_{r_2}^{r_1} E_r dr = \frac{\rho_l}{2\pi\epsilon} \int_{r_1}^{r_2} \frac{dr}{r} = \frac{\rho_l}{2\pi\epsilon} \ln \frac{r_2}{r_1}. \tag{9.47}$$

9.6.2 Potential Between Two Infinite Parallel Straight Round Conductors with Nonuniform Charge Density

Figure 9.16a shows two long parallel round bare conductors carrying opposite line charges ρ_l and $-\rho_l$. A two-wire transmission line has this configuration. Each bare conductor has a radius a and a length l_w. The surface of each conductor is equipotential. As the conductors are moved closer to each other, the charge density distribution in the cross-sectional area of the conductors is not uniform. According to Coulomb's law, the opposite charges are attracted to each other, and therefore the charge density is larger in the regions between both the conductors. This is *charge proximity effect.* The two conductors can be modeled as two infinite parallel infinitesimally thin lines of charges ρ_l and $-\rho_l$ located in the zx plane, as illustrated in Fig. 9.16b. A positive line charge ρ_l is located at and $z = 0$,

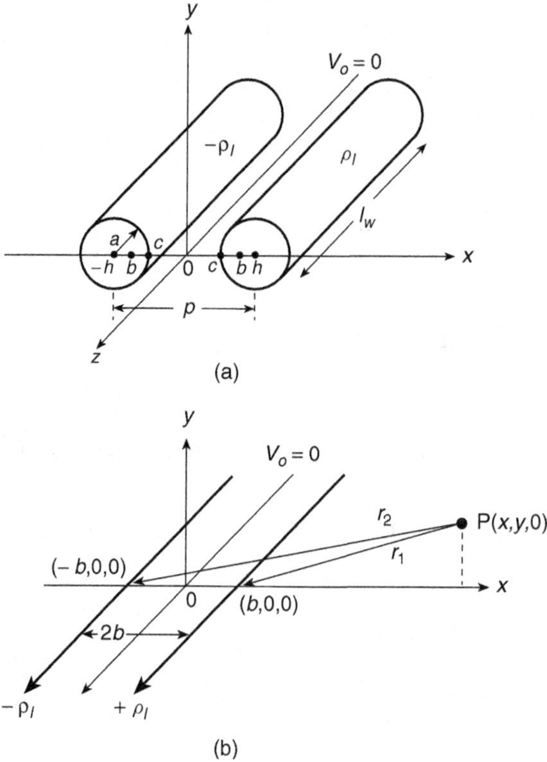

Figure 9.16 Two parallel straight round bare conductors. (a) Parallel round conductors. (b) Model of two parallel wires in the form of two parallel infinitesimally thin lines of charges ρ_l and $-\rho_l$

and a negative line charge $-\rho_l$ is located at $x = -b$, $y = 0$, and $z = 0$. The charge of one conductor is $Q = l_w \rho_l$ and the charge of the other conductor is $Q = -l_w \rho_l$, where ρ_l is the charge per unit length, called the line charge density. Each line charge produces a family of equipotential cylindrical surfaces enveloping each line charge. One of these surfaces is located exactly on the conductor surface, which is identical to the conductor surface. The potential V_0 of the surface at $x = 0$ is zero. The potential field and the electric field intensity are independent of z. Therefore, it is sufficient to determine the potential distribution on the yz plane at $x = 0$.

The distances r_1 and r_2 from the two thin lines and point $P(x, y, 0)$ are, respectively,

$$r_1 = \sqrt{(x - b)^2 + y^2} \tag{9.48}$$

and

$$r_2 = \sqrt{(x + b)^2 + y^2}. \tag{9.49}$$

Let us use the origin as a reference for potential in Fig. 9.16b. If only the line with a positive charge is present in Fig. 9.16b, the potential difference between point $P(x, y, 0)$ and the origin obtained from (9.47) is given by Krauss [7]

$$V_+ = \frac{\rho_l}{2\pi\epsilon} \ln\left(\frac{b}{r_1}\right). \tag{9.50}$$

If only the line with a negative charge is present in Fig. 9.16b, the potential difference between point $P(x, y, 0)$ and the origin is the radial distance r_2 at point $P(x, y, 0)$ is

$$V_- = -\frac{\rho_l}{2\pi\epsilon} \ln\left(\frac{b}{r_2}\right). \tag{9.51}$$

When both lines are present, the combined potential difference between point $P(x, y, 0)$ and the origin is

$$V = V_+ + V_- = \frac{\rho_l}{2\pi\epsilon}\left[\ln\left(\frac{b}{r_1}\right) - \ln\left(\frac{b}{r_2}\right)\right] = \frac{\rho_l}{2\pi\epsilon}\ln\left(\frac{r_2}{r_1}\right) = \frac{\rho_l}{2\pi\epsilon}\ln\sqrt{\frac{(x+b)^2+y^2}{(x-b)^2+y^2}}. \quad (9.52)$$

For a constant value of $V = V_e$, an equipotential surface crossing point $P(x, y, 0)$ is given by

$$V_e = \frac{\rho_l}{2\pi\epsilon_r\epsilon_0}\ln\sqrt{\frac{(x+b)^2+y^2}{(x-b)^2+y^2}} = \frac{\rho_l}{2\pi\epsilon}\ln K. \quad (9.53)$$

Hence,

$$K = \frac{r_2}{r_1} = e^{2\pi\epsilon V_e/\rho_l} = \sqrt{\frac{(x+b)^2+y^2}{(x-b)^2+y^2}}. \quad (9.54)$$

Rearrangement of this equation yields

$$x^2 - 2bx\frac{K^2+1}{K^2-1} + y^2 + b^2 = 0 \quad (9.55)$$

resulting in an equipotential circle

$$\left(x - b\frac{K^2+1}{K^2-1}\right)^2 + y^2 = \left(\frac{2bK}{K^2-1}\right)^2 \quad (9.56)$$

which can be expressed by

$$(x - x_o)^2 + y^2 = r_o^2 \quad (9.57)$$

where the radius of the circle is

$$r_o = \frac{2bK}{K^2-1} \quad (9.58)$$

and the center of the circle is

$$x_o = b\frac{K^2+1}{K^2-1}. \quad (9.59)$$

Figure 9.17 shows the potential distribution induced by two parallel round wires at $b = 5$ and $K = 2$, 3, 5, and 10.

One of the equipotential circles coincides with the positively charged bare conductor surface of radius a if

$$r_o = a = \frac{2bK_p}{K_p^2-1} \quad (9.60)$$

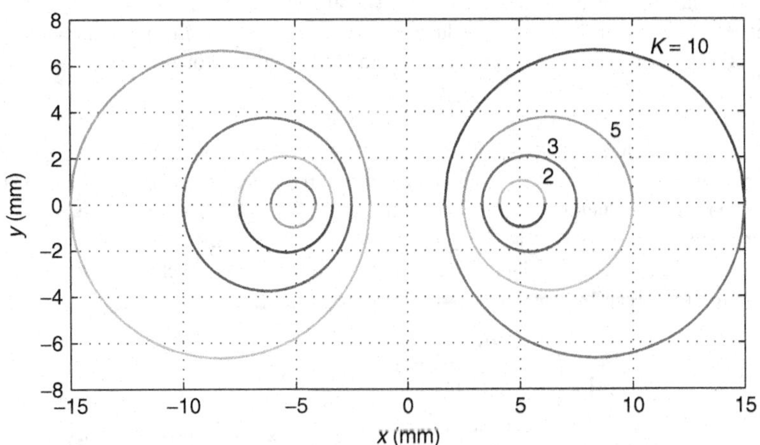

Figure 9.17 Equipotential curves around two parallel conductors carrying the opposite charges for $b = 5$ and $K = 2$, 3, 5, and 10

and the center of the circle is at

$$x_o = h = b\frac{K_p^2 + 1}{K_p^2 - 1}. \tag{9.61}$$

The equipotential conductor surface with a positive charge is described by

$$(x - h)^2 + y^2 = a^2. \tag{9.62}$$

Similarly, the equipotential conductor surface with a negative charge is

$$(x - h)^2 + y^2 = b^2. \tag{9.63}$$

From geometrical consideration, one half of the separation of the charge centers is

$$b = \sqrt{h^2 - a^2} = \sqrt{h^2 - \frac{4b^2 K_p}{(K_p - 1)^2}}, \tag{9.64}$$

yielding the separation of the charge centers

$$2b = \sqrt{p^2 - (2a)^2} = \sqrt{(2h)^2 - (2a)^2}. \tag{9.65}$$

Hence,

$$K_p = \frac{h + \sqrt{h^2 - a^2}}{a} = \frac{h}{a} + \sqrt{\left(\frac{h}{a}\right)^2 - 1}. \tag{9.66}$$

As K_p increases, the conductor radius a approaches zero and h approaches b. Thus, the actual conductors become infinitesimally thin conductors. The potential difference between the surface of the conductor with a positive line charge ρ_l and the origin is given by

$$V_p = \frac{\rho_l}{2\pi\epsilon}\ln K_p = \frac{\rho_l}{2\pi\epsilon}\ln\left[\frac{h}{a} + \sqrt{\left(\frac{h}{a}\right)^2 - 1}\right]. \tag{9.67}$$

9.6.3 Capacitance of Two Parallel Wires with Nonuniform Charge Density

The charge is not uniformly distributed on the wire surface, but it has higher density on the adjacent sides of the conductors. This is the charge proximity effect for electrical charge. As the cylindrical conductors move more closer to each other, the charge distribution is less uniform. The charge density is larger on the inner surfaces of the conductors. Therefore, the effective change centers displace from the geometrical centers of the round conductors toward the inner surfaces, as shown in Fig. 9.16a.

By Gauss's law, the positive charge stored on the round conductor surface is

$$Q = \rho_l l_w. \tag{9.68}$$

The potential is zero along the y axis, that is, $V_0 = 0$ at $x = 0$. The potential difference between any point on the round conductor with a positive charge and the plane at $x = 0$ is V_p. The potential on the surface of the round conductor with a negative charge is $V_n = -V_p$. Therefore, the potential difference between the two conductors is $\Delta V = V_p - V_n = V_p - (-V_p) = 2V_p$. The capacitance between two long parallel straight round conductors placed in a homogeneous dielectric (including the charge proximity effect) is given by

$$C_s = \frac{Q}{2V_p} = \frac{\rho_l l_w}{2V_p} = \frac{\pi\epsilon l_w}{\ln\left[\frac{h}{a} + \sqrt{\left(\frac{h}{a}\right)^2 - 1}\right]} = \frac{\pi\epsilon l_w}{\cosh^{-1}\left(\frac{h}{a}\right)}$$

$$= \frac{\pi\epsilon_r\epsilon_0 l_w}{\ln\left[\frac{p}{2a} + \sqrt{\left(\frac{p}{2a}\right)^2 - 1}\right]} = \frac{\pi\epsilon_r\epsilon_0 l_w}{\cosh^{-1}\left(\frac{p}{2a}\right)} = \frac{\pi\epsilon_r\epsilon_0 l_w}{\ln\left[\frac{p}{d} + \sqrt{\left(\frac{p}{d}\right)^2 - 1}\right]} = \frac{\pi\epsilon_r\epsilon_0 l_w}{\cosh^{-1}\left(\frac{p}{d}\right)}, \tag{9.69}$$

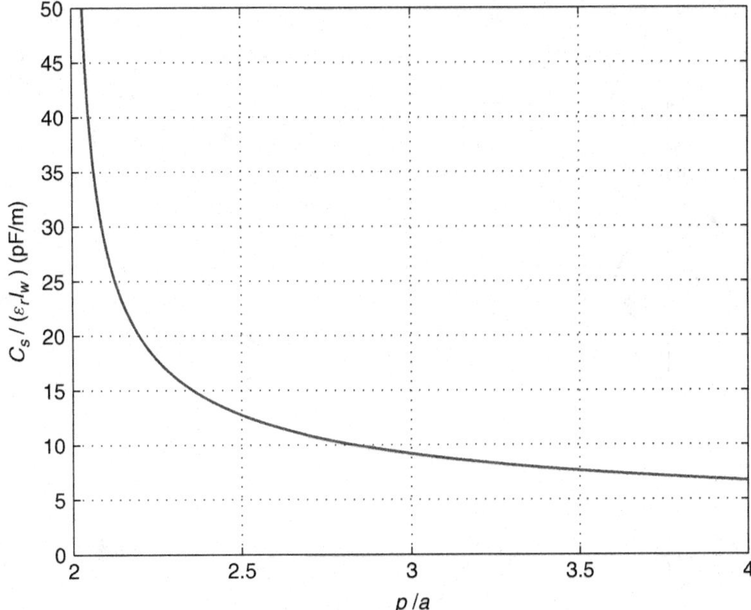

Figure 9.18 Normalized capacitance $C_s/(\epsilon_r l_w)$ of two parallel straight round conductors as a function of p/a

where $\epsilon = \epsilon_r \epsilon_0$ is the permittivity of the dielectric between the two bare conductors, ϵ_r is the relative permittivity, a is the radius of the conductors, l_w is the length of each conductor, h is one-half of the distance between the centers of the conductors, $d = 2a$, $p = 2h \geq d$, and $\cosh^{-1}x = \ln(x + \sqrt{x^2 - 1})$. The effect of the insulating coating on the capacitance is neglected in (9.69). Figure 9.18 shows a plot of $C_s/(\epsilon_r l_w)$ as a function of p/a for two parallel round conductors. As the normalized distance between conductors p/a increases, the normalized capacitance $C_s/(\epsilon_r l_w)$ decreases.

If the two conductors are sufficiently far apart ($p \geq 5a = 2.5d$), the charge proximity effect can be neglected and the charge density is nearly uniform. The following approximation holds true: $\ln(x + \sqrt{x^2 - 1}) \approx \ln(2x)$ for $x \gg 1$, expression (9.69) for two long straight parallel round conductors placed in a homogeneous dielectric medium of permittivity ϵ may be approximated by

$$C_s \approx \frac{\pi \epsilon_r \epsilon_0 l_w}{\ln\left(\frac{p}{a}\right)} = \frac{\pi \epsilon_r \epsilon_0 l_w}{\ln\left(\frac{2p}{d}\right)} \quad \text{for} \quad \left(\frac{p}{2a}\right)^2 = \left(\frac{p}{d}\right)^2 \gg 1 \quad \text{or} \quad p > 3d \quad \text{or} \quad p > 6a. \tag{9.70}$$

Figure 9.19 shows the exact and approximate plots of normalized capacitance $C_s/(\epsilon_r l_w)$ of two parallel round conductors as a function of p/d. It can be observed that the approximate expression (9.69) underestimates the exact stray capacitance C_s for low values of p/d because it neglects the charge proximity effect. For accuracy of 99%, the ratio p/d must be greater than 3.5. For accuracy of 95%, the ratio p/d must be greater than 2.

The capacitance between fractions of surfaces of two long straight parallel round conductors corresponding to angle $2\theta_1$ and placed in a homogeneous dielectric medium of permittivity ϵ is given by

$$C_{2\theta} \approx \frac{2\theta}{2\pi} C_s = \frac{\theta}{\pi} C_s = \frac{\epsilon_r \epsilon_0 l_w \theta}{\ln\left(\frac{p}{a}\right)} = \frac{\epsilon_r \epsilon_0 l_w \theta}{\ln\left(\frac{2p}{d}\right)}. \tag{9.71}$$

The separation of the two surfaces of the two bare parallel straight conductors is given by

$$s = p - 2a = p - d = 2h - 2a = 2h - d. \tag{9.72}$$

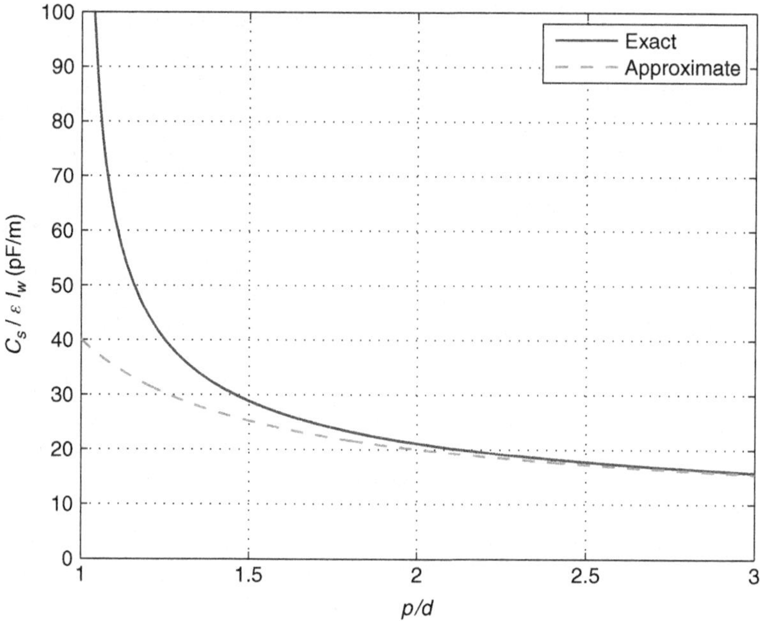

Figure 9.19 Exact and approximate plots of normalized capacitance $C_s/(\epsilon_r l_w)$ of two parallel round conductors as a function of p/d

From (9.70), the capacitance between the two bare parallel straight wires is

$$C_s = \frac{\pi \epsilon_r \epsilon_0 l_w}{\cosh^{-1}\left(\frac{s}{d}+1\right)} = \frac{\pi \epsilon_r \epsilon_0 l_w}{\ln\left[\frac{s}{d}+1+\sqrt{\left(\frac{s}{d}+1\right)^2-1}\right]}. \tag{9.73}$$

Figure 9.20 shows a plot of the normalized capacitance $C_s/(\epsilon_r l_w)$ as a function of the normalized separation of the conductors s/d. As the separation s increases, the charge distribution becomes more uniform and therefore (9.73) can be approximated by

$$C_s \approx \frac{\pi \epsilon_r \epsilon_0 l_w}{\ln\left(\frac{s}{a}\right)} = \frac{\pi \epsilon_r \epsilon_0 l_w}{\ln\left(\frac{2s}{d}\right)} \quad \text{for} \quad s \gg a. \tag{9.74}$$

This expression underestimates the capacitance C_s for low values of s/a.

Example 9.2

Calculate the stray capacitance C_s of two parallel round bare conductors of length $l_w = 1$ m and radius $a = 1$ mm placed in air. The distance between the centers of the conductors is $p = 2.4$ mm. Use exact and approximate equations for the capacitance.

Solution: The stray capacitance is

$$C_s = \frac{\pi \epsilon_0 l_w}{\ln\left[\frac{p}{2a}+\sqrt{\left(\frac{p}{2a}\right)^2-1}\right]} = \frac{\pi \times 8.85 \times 10^{-12} \times 1}{\ln\left[\frac{2.4}{2\times 1}+\sqrt{\left(\frac{2.4}{2\times 1}\right)^2-1}\right]} = 44.673 \text{ pF}. \tag{9.75}$$

The approximate capacitance is

$$C_s = \frac{\pi \epsilon_0 l_w}{\ln\left(\frac{p}{a}\right)} = \frac{\pi \times 8.85 \times 10^{-12} \times 1}{\ln\left(\frac{2.4}{1}\right)} = 31.776 \text{ pF}. \tag{9.76}$$

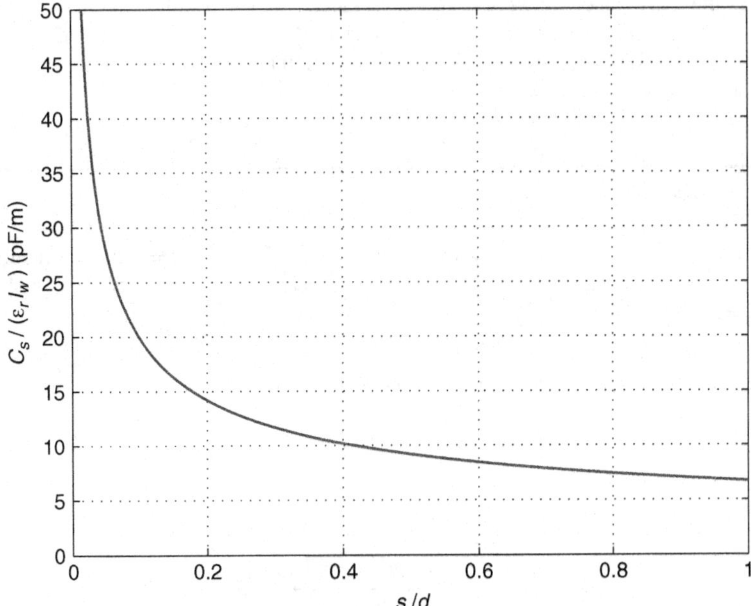

Figure 9.20 Normalized capacitance $C_s/(\epsilon_r l_w)$ of two parallel round conductors as a function of s/d

The approximate equation underestimates the capacitance because $p/a = 2.4 < 6$. The error is 40.59%. Thus, the approximate equation does not give an accurate result when p/a is small.

9.7 Capacitance of Round Conductor and Parallel Conducting Plane

The potential distribution can be obtained using the technique of images. If the conducting plane is replaced by a plane with a zero potential and a conductor with a charge per unit length $-\rho_l$ is placed symmetrically at $x = -h$, we obtain the two parallel conductors.

Using (9.67), the capacitance between a round conductor and a parallel conducting plane is given by

$$C_{cp} = \frac{Q}{V_p} = \frac{\rho_l l_w}{V_p} = \frac{2\pi\epsilon l_w}{\ln\left[\frac{h}{a} + \sqrt{\left(\frac{h}{a}\right)^2 - 1}\right]} = \frac{2\pi\epsilon l_w}{\cosh^{-1}\left(\frac{h}{a}\right)}, \tag{9.77}$$

where a is the radius of the bare conductor and h is the distance between the center of the conductor and the conducting plane. The approximate expression for C_{cp} is

$$C_s \approx \frac{2\pi\epsilon l_w}{\ln\left(\frac{2h}{a}\right)} = \frac{2\pi\epsilon l_w}{\ln\left(\frac{p}{a}\right)} = \frac{2\pi\epsilon l_w}{\ln\left(\frac{2p}{d}\right)} \quad \text{or} \quad p > 3d \quad \text{or} \quad p > 6a. \tag{9.78}$$

The separation of the conductor from the conducting plane is $s = h - a$. Hence, the capacitance between round bare conductor and a conducting plane is

$$C_{cp} = \frac{2\pi\epsilon l_w}{\ln\left[\frac{s}{a} + 1 + \sqrt{\left(\frac{s}{a} + 1\right)^2 - 1}\right]} = \frac{2\pi\epsilon l_w}{\cosh^{-1}\left(\frac{h}{a} + 1\right)}. \tag{9.79}$$

If $h = p/2$, the capacitance between a round conductor and a parallel conducting plane is twice the capacitance between two round conductors, that is, $C_{cp} = 2C_s$. In other words, the capacitance between two round conductors is equal to the series combination of a capacitance between a round conductor and a conducting plane.

Example 9.3

Calculate the stray capacitance C_{cp} of a round conductors of length $l_w = 1$ cm and radius $a = 1$ mm placed in air above a conducting plane. The distance between the center of the conductor and the plane is $h = 2$ mm.

Solution: The conductor-to-plane stray capacitance is

$$C_{cp} = \frac{2\pi\epsilon_0 l_w}{\ln\left[\frac{h}{a} + \sqrt{\left(\frac{h}{a}\right)^2 - 1}\right]} = \frac{2\pi \times 8.85 \times 10^{-12} \times 10^{-2}}{\ln\left[\frac{2}{1} + \sqrt{\left(\frac{2}{1}\right)^2 - 1}\right]} = 0.422 \text{ pF}. \tag{9.80}$$

The approximate stray capacitance is

$$C_s \approx \frac{2\pi\epsilon l_w}{\ln\left(\frac{2h}{a}\right)} = \frac{2\pi \times 8.85 \times 10^{-12} \times 10^{-2}}{\ln\left(\frac{2\times 2}{1}\right)} = 0.401 \text{ pF}. \tag{9.81}$$

9.8 Capacitance of Straight Parallel Wire Pair Over Ground

The capacitance of straight parallel wire pair over ground is given by

$$C_s = \frac{\epsilon l_w \ln\left[\left(\frac{2H}{p}\right)^2 + 1\right]}{\ln^2\left(\frac{2H}{a}\right)}, \tag{9.82}$$

where a is the conductor radius, p is the distance between the centers of the conductors, and H is the distance between the conductor center and the ground plane.

9.9 Capacitance Between Two Parallel Straight Round Conductors with Uniform Charge Density

Let us consider two parallel straight round conductors of radius a placed in a uniform dielectric, as shown in Fig. 9.16a. The distance between the centers of the conductors is $p = 2h$. One conductor has a charge Q and the other $-Q$. The charge density per unit length is $q_l = Q/l_w$. Assume that the charge density is uniform and equal to q_l. Therefore, the charge proximity effect is neglected. We can use the method of charge images. A grounded infinite conducting surface is placed between the two conductors to form a symmetrical structure. The electric field intensities on the straight line connecting the centers of both conductors are

$$E_1 = \frac{q_l}{2\pi\epsilon_r\epsilon_0 r} \tag{9.83}$$

and

$$E_2 = \frac{q_l}{2\pi\epsilon_r\epsilon_0(2h - r)}. \tag{9.84}$$

Thus, the total electric field intensity on the straight line connecting the centers of both conductors is

$$E = E_1 + E_2 = \frac{q_l}{2\pi\epsilon_r\epsilon_0}\left(\frac{1}{r} + \frac{1}{2h-r}\right). \tag{9.85}$$

The voltage between the conductor and the equipotential surface is

$$V_1 = \int_a^h Edl = \int_a^h Edr = \frac{q_l}{2\pi\epsilon_r\epsilon_0}\int_a^h\left(\frac{1}{r} + \frac{1}{2h-r}\right)dr = \frac{q_l}{2\pi\epsilon_r\epsilon_0}\ln\left(\frac{2h}{a} - 1\right). \tag{9.86}$$

Hence, the capacitance between the conductor and the equipotential surface per unit length is

$$\frac{C_{cs}}{l_w} = \frac{q_l}{V_1} = \frac{2\pi\epsilon_r\epsilon_0}{\ln\left(\frac{2h}{a} - 1\right)} = \frac{2\pi\epsilon_r\epsilon_0}{\ln\left(\frac{4h}{d} - 1\right)}, \tag{9.87}$$

yielding the capacitance formed by a round conductor and the equipotential surface

$$C_{cs} = \frac{Q}{V_1} = \frac{q_l l_w}{V_1} = \frac{2\pi\epsilon_r\epsilon_0 l_w}{\ln\left(\frac{2h}{a} - 1\right)} = \frac{2\pi\epsilon_r\epsilon_0 l_w}{\ln\left(\frac{4h}{d} - 1\right)}. \tag{9.88}$$

The capacitance between the two straight parallel round conductors per unit length is

$$\frac{C_{cc}}{l_w} = \frac{q_l}{2V_1} = \frac{\pi\epsilon_r\epsilon_0}{\ln\left(\frac{2h}{a} - 1\right)} = \frac{\pi\epsilon_r\epsilon_0}{\ln\left(\frac{4h}{d_o} - 1\right)}, \tag{9.89}$$

resulting in the capacitance between the two straight parallel round conductors

$$C_{cc} = \frac{Q}{2V_1} = \frac{q_l l_w}{2V_1} = \frac{\pi\epsilon_r\epsilon_0 l_w}{\ln\left(\frac{2h}{a} - 1\right)} = \frac{\pi\epsilon_r\epsilon_0 l_w}{\ln\left(\frac{4h}{d} - 1\right)} = \frac{\pi\epsilon_r\epsilon_0 l_w}{\ln\left(\frac{p}{a} - 1\right)} = \frac{\pi\epsilon_r\epsilon_0 l_w}{\ln\left(\frac{2p}{d} - 1\right)}, \tag{9.90}$$

where $d = 2a$.

Another method of the derivation of a capacitance between two round conductors placed in a uniform dielectric is as follows. Consider a single straight round conductor placed in uniform dielectric. The conductor has a uniform line charge density $\rho_l = Q/l_w$. From Gauss's law, the electric field E induced by the conductor is given by

$$E = \frac{q_l}{2\pi\epsilon_r\epsilon_0 r}, \tag{9.91}$$

where r is the radial distance from the center of the conductor to a point in space located outside the conductor. In general, the potential difference V_{ab} between two points with radial distances r_a and r_b from the center of the conductor is

$$V_{ab} = -\int_b^a Edr = \int_a^b \frac{q_l}{2\pi\epsilon_r\epsilon_0 r}dr = \frac{q_l}{2\pi\epsilon_0\epsilon_r}\ln\left(\frac{r_a}{r_b}\right). \tag{9.92}$$

Next, consider that two parallel straight round conductors of radii a are placed in uniform dielectric with distance p between their centers. The conductors are uniformly charged with line charge densities q_l and $-q_l$. Assume that the charge proximity effect is neglected. The potential difference between the two round conductors is given by

$$V_{12} = -\int_{p-a}^{a} Edr = \int_a^{p-a}\frac{q_l}{2\pi\epsilon_r\epsilon_0 r}dr = \frac{q_l}{2\pi\epsilon_0\epsilon_r}\ln\left(\frac{p-a}{a}\right) = \frac{q_l}{2\pi\epsilon_0\epsilon_r}\ln\left(\frac{p}{a} - 1\right)$$
$$= \frac{q_l}{2\pi\epsilon_0\epsilon_r}\ln\left(\frac{2p}{d} - 1\right). \tag{9.93}$$

The capacitance between the two round conductors per unit length is

$$\frac{C_{cc}}{l_w} = \frac{q_l}{V_{12}} = \frac{\pi\epsilon_0\epsilon_r}{\ln\left(\frac{p}{a} - 1\right)} = \frac{\pi\epsilon_0\epsilon_r}{\ln\left(\frac{2p}{d} - 1\right)}. \tag{9.94}$$

Hence, the capacitance between two round conductors is

$$C_{cc} = \frac{Q}{V_{12}} = \frac{q_l l_w}{V_{12}} = \frac{\pi\epsilon_0\epsilon_r l_w}{\ln\left(\frac{p}{a} - 1\right)} = \frac{\pi\epsilon_0\epsilon_r l_w}{\ln\left(\frac{2p}{d} - 1\right)}. \tag{9.95}$$

9.10 Capacitance of Cylindrical Capacitor

A cross-sectional view of a cylindrical capacitor is shown in Fig. 9.21. A cylindrical capacitor consists of inner round conductor, outer round conductor, and dielectric placed between the two conductors, where a is the outer radius of the inner conductor and b is the inner radius of the outer conductor. The lengths of both the conductors l are identical The capacitor is long so that $l \ll b$. The relative permittivity of the dielectric between the conductors is ϵ_r. When a voltage V is applied between the two conductors, charges Q and $-Q$ accumulate on on the inner surface of the outer conductor and the outer surface of the inner conductor, respectively. The line charge densities are ρ_l and $-\rho_l$ on the outer and inner conductors, respectively. This density is given by

$$\rho_l = \frac{Q}{l}. \tag{9.96}$$

The opposite charges induce a radial electric field $\mathbf{E}$ in the dielectric. From Gauss's law, the electric field intensity in the dielectric between the conductors is

$$\mathbf{E} = -\mathbf{a}_r \frac{\rho_l}{2\pi \epsilon r} = -\mathbf{a}_r \frac{Q}{2\pi \epsilon r l} \quad \text{for} \quad a \le r \le b. \tag{9.97}$$

The voltage between the two conductors is

$$V = -\int_a^b \mathbf{E} \cdot d\mathbf{l} = -\int_a^b \left(-\mathbf{a}_r \frac{Q}{2\pi \epsilon r l}\right) \cdot (\mathbf{a}_r dr) = \frac{Q}{2\pi \epsilon l} \ln\left(\frac{b}{a}\right). \tag{9.98}$$

The capacitance of a cylindrical capacitor is given by

$$C_{cyl} = \frac{Q}{V} = \frac{2\pi \epsilon l}{\ln\left(\frac{b}{a}\right)}. \tag{9.99}$$

The capacitance of a cylindrical capacitor is useful to model the capacitance related to the insulating coating of winding round wire.

9.11 Self-Capacitance of Single-Layer Inductors

An analytical method is used to determine the self-capacitance of a single-layer inductor. A cross-sectional view of a single-layer inductor is shown in Fig. 9.22. Magnet wire uses one to four coating layers of polymer film insulation of thickness f. Magnet wire-insulating coating is made of polyurethane, polyamide, polyester, or polyimide. It is assumed that the capacitances between nonadjacent turns are much lower than those between adjacent turns and can be neglected. Expression (9.69) for the capacitance between two parallel round bare conductors placed in a homogeneous medium can be adopted for the turn-to-turn capacitance of a single-layer inductor without a core or a shield by neglecting the insulating coating and the turn curvature [18]

$$C_{tt} = \frac{\pi \epsilon l_T}{\cosh^{-1}\left(\frac{p}{2a}\right)} = \frac{\pi \epsilon l_T}{\ln\left[\frac{p}{2a} + \sqrt{\left(\frac{p}{2a}\right)^2 - 1}\right]} = \frac{\pi^2 \epsilon D_T}{\ln\left[\frac{p}{2a} + \sqrt{\left(\frac{p}{2a}\right)^2 - 1}\right]} \quad \text{for} \quad t \ll p - 2a, \tag{9.100}$$

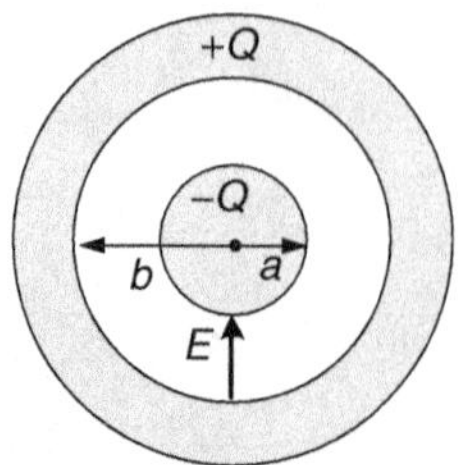

Figure 9.21 Cross-sectional view of a cylindrical capacitor

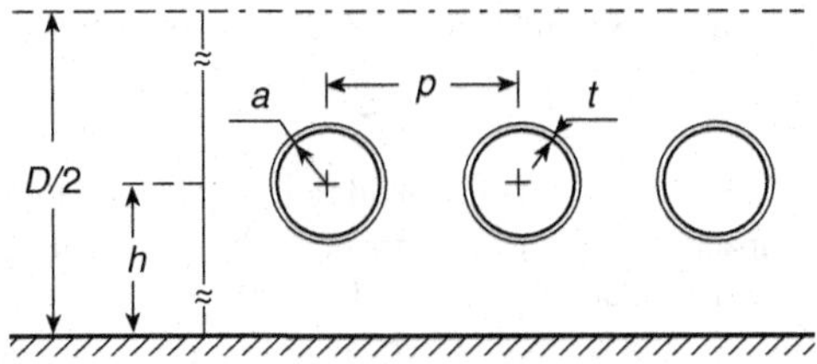

Figure 9.22 Cross-sectional view of a single-layer inductor with a core or a shield

where $l_T = \pi D_T$ is the length of a single turn, D_T is the coil diameter, a is the bare wire radius, $t = p - 2a$ is the thickness of the insulating coating, and p is the winding pitch, that is, the distance between the centerlines of two adjacent turns.

Let us take into account the presence of the wire-insulating coating. A radial electric field is assumed in the insulating coating. When the insulating coating thickness t of relative permittivity ϵ_r is comparable with the air gap $p - 2a$, the capacitance of a cylindrical capacitor with two insulating layers can be used to determine the turn-to-turn capacitance [18]. The equivalent turn-to-turn capacitance is equal to the series combination of the capacitance related to the insulating coating of both adjacent turns and the capacitance related to the air gap between the adjacent turns. Using expression (9.99) for the capacitance of the cylindrical capacitor, the capacitance related to the insulating coating of one turn is given by

$$C_c = \frac{2\pi \epsilon l_T}{\ln\left(1 + \frac{t}{a}\right)}. \tag{9.101}$$

The capacitance of insulating coating of two adjacent turns is a series combination of two capacitances C_c

$$C_{ttc} = \frac{C_c}{2} = \frac{\pi \epsilon l_T}{\ln\left(1 + \frac{t}{a}\right)}. \tag{9.102}$$

The air-gap capacitance gap between two adjacent turns can be determined from (9.100) by replacing the wire radius a with the external coating radius $(a + t)$

$$C_g = \frac{\pi \epsilon_0 l_T}{\ln\left[\frac{p/2a}{1+\frac{t}{a}} + \sqrt{\left(\frac{p/2a}{1+\frac{t}{a}}\right)^2 - 1}\right]}. \tag{9.103}$$

Hence, the turn-to-turn capacitance is equal to the series combination of capacitances C_{ttc} and C_g [18]

$$\begin{aligned} C_{tt} &= \frac{C_{ttc} C_g}{C_{ttc} + C_g} = \frac{C_c C_g}{C_c + 2C_g} = \frac{\pi \epsilon_0 \epsilon l_T}{\ln\left\{\left(1 + \frac{t}{a}\right)^{\epsilon_0}\left[\frac{p/2a}{1+\frac{t}{a}} + \sqrt{\left(\frac{p/2a}{1+\frac{t}{a}}\right)^2 - 1}\right]^{\epsilon}\right\}} \\ &= \frac{\pi \epsilon_0 l_T}{\ln\left\{\left(1 + \frac{t}{a}\right)^{1/\epsilon_r}\left[\frac{p/2a}{1+\frac{t}{a}} + \sqrt{\left(\frac{p/2a}{1+\frac{t}{a}}\right)^2 - 1}\right]\right\}} \\ &= \frac{\pi^2 \epsilon_0 D_T}{\ln\left\{\left[\frac{p/2a}{(1+t/a)^{1-1/\epsilon_r}}\right] + \sqrt{\left[\frac{p/2a}{(1+t/a)^{1-1/\epsilon_r}}\right]^2 - (1 + \frac{t}{a})^{2/\epsilon_r}}\right\}}. \end{aligned} \tag{9.104}$$

The air-gap capacitance C_g is not uniform because the distance of the air-gap changes. Therefore, the series combination of the nonuniform air-gap capacitance C_g with a uniform capacitance C_{tts} does not give a very accurate value of the total turn-to-turn capacitance.

For a single-layer inductor without a shield or a core, the turn-to-turn capacitances are connected in series. Therefore, the total turn-to-turn capacitance of a single-layer inductor of N turns and without a shield or a core is equal to a series combination of $(N - 1)$ turn-to-turn capacitances [18]

$$C_s = \frac{C_{tt}}{N - 1}. \tag{9.105}$$

Neglecting the contribution of the insulating coating and the curvature of both the turns and the shield, the turn-to-shield capacitance C_{ts} can be modeled by a capacitance between a long straight round conductor and a parallel conducting plane as [18]

$$C_{ts} = \frac{2\pi \epsilon l_w}{\cosh^{-1}\left(\frac{h}{a}\right)} = \frac{2\pi^2 \epsilon D_T}{\ln\left[\frac{h}{a} + \sqrt{\left(\frac{h}{a}\right)^2 - 1}\right]} \approx \frac{2\pi^2 \epsilon D_T}{\ln\left(\frac{2h}{a}\right)} \quad \text{for} \quad \left(\frac{h}{a}\right)^2 \gg 1, \tag{9.106}$$

where h is the distance between the shield and the center of the coil winding wire and l_w is the length of the winding wire.

Example 9.4

Calculate the stray capacitance C_s and the first self-resonant frequency of a single-layer air inductor without insulating coating, core, and shield, which has $L = 82.4\ \mu\text{H}$, $N = 16$, $a = 5$ mm, $p = 10.2$ mm, and $D_T = 326$ mm.

Solution: From (9.100), the turn-to-turn capacitance is

$$C_{tt} = \frac{\pi^2 \epsilon_0 D_T}{\ln\left[\frac{p}{2a} + \sqrt{\left(\frac{p}{2a}\right)^2 - 1}\right]} = \frac{\pi^2 \times 8.85 \times 10^{-12} \times 326 \times 10^{-3}}{\ln\left[\frac{10.2}{2\times 5} + \sqrt{\left(\frac{10.2}{2\times 5}\right)^2 - 1}\right]} = 142.6\ \text{pF}. \tag{9.107}$$

The stray capacitance is

$$C_s = \frac{C_{tt}}{N - 1} = \frac{142.6}{16 - 1} = 9.5066\ \text{pF}. \tag{9.108}$$

The first self-resonant frequency is

$$f_r = \frac{1}{2\pi\sqrt{LC_s}} = \frac{1}{2\pi\sqrt{82.4 \times 10^{-6} \times 9.507 \times 10^{-12}}} = 5.686\ \text{MHz}. \tag{9.109}$$

The maximum operating frequency of the inductor is estimated to be $f_{max} = 0.9 f_r = 0.9 \times 5.679 \approx 5.1$ MHz.

An expression for the equivalent shunt capacitance of a single-layer air-core coil is also given by Medhurst [1]

$$C_s = 11.25(D_T + l) + D\sqrt{\frac{D_T}{l}} \tag{9.110}$$

where D_T is the coil diameter and l is the length of an inductor.

Another method for estimating the winding capacitance of a single-layer inductor is as follows. Assume that an inductor has length l, total number of turns N, turn radius r, and wire radius a. The distance between the turns is

$$d = \frac{l}{N}, \tag{9.111}$$

the winding wire length is

$$l_w = 2\pi r N, \tag{9.112}$$

and the total external area of the wire is

$$A = (2\pi a)(2\pi r N) = 4\pi^2 a r N. \tag{9.113}$$

Hence, the stray capacitance of the winding is

$$C_s = \epsilon_0 \frac{A}{d} = \epsilon_0 \frac{4\pi^2 a r N}{l/N} = 4\pi^2 \epsilon_0 \frac{a r N^2}{l}. \tag{9.114}$$

9.12 Self-Capacitance of Multilayer Inductors

9.12.1 Exact Equation for Self-Capacitance of Multilayer Inductors

A cross-sectional view of a multilayer inductor with hexagonal winding pattern is depicted in Fig. 9.23. Enameled copper wire is covered with an insulating coating usually made of polyurethane. There are symmetries in the winding geometry. The lines of the electric field that begin at a turn surrounded by other conductors end in these conductors. No line can go to infinity if the conductors of the coil are close to each other. The conductor surfaces are equipotential, and therefore the lines of the electric field are orthogonal to the conductor surfaces. If we consider two adjacent conductors, the elementary capacitance dC between two opposite corresponding elementary surfaces of the two adjacent conductors dS is

$$dC = \epsilon \frac{dS}{x} \tag{9.115}$$

where x is the length of a line of electric field connecting the two opposite elementary surfaces. The length x is not constant, but it is a function of the elementary surface. The basic cell of the winding with a hexagonal pattern is shown in Fig. 9.24. The cell is the same for adjacent turns of the same layer as well as for two adjacent turns of different layers. Only cells adjacent to the core and the shield differ from the turn-to-turn cells. The basic cells include the portion of the perimeter of the turn that corresponds to an angle of $\pi/3$. In order to derive an expression for dC, (9.115) must be integrated over the angle $\pi/3$.

For the basic cell of Fig. 9.24, three different regions are crossed by the lines of the electric field: two insulating coatings and the air gap between them. Therefore, the elementary capacitance dC between the adjacent turns is equivalent to the capacitance of a series combination of three elementary capacitors, each with a uniform dielectric material. The first capacitance is related to the insulating coating of the first turn, the second capacitance is related to the air gap, and the third capacitance is related to the insulating coating of the second turn. The first and the third capacitances are identical.

We will derive now an expression for the capacitance of the insulating coatings. Figure 9.25 shows an elementary cylindrical surface located between the bare conductor surface and the outer surface of the coating. For a round conductor, the location of each elementary surface can be described in

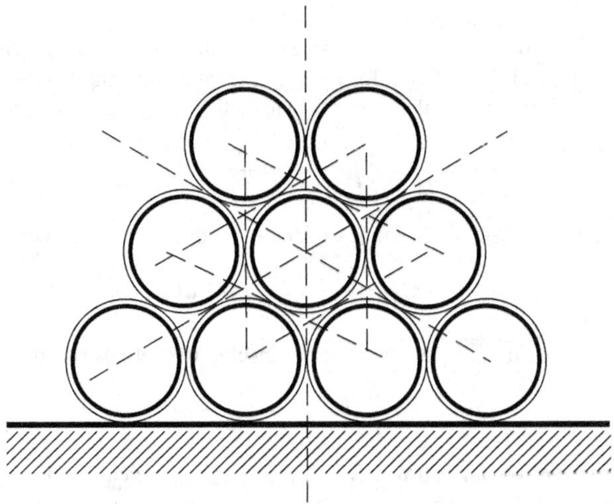

Figure 9.23 Cross-sectional view of a multilayer inductor with hexagonal winding pattern

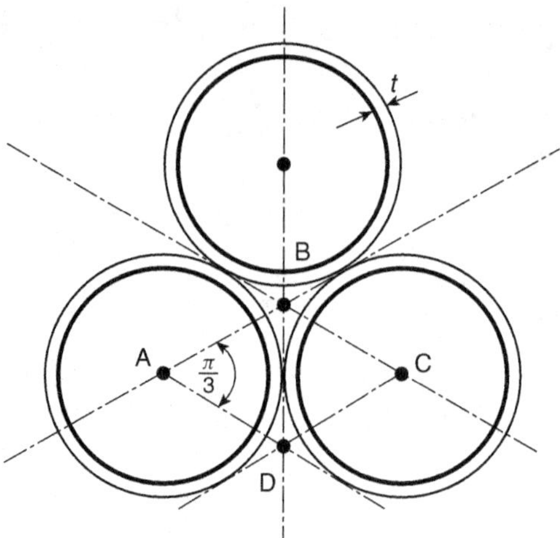

Figure 9.24 Basic cell ABCD that forms the turn-to-turn capacitance for hexagonal winding pattern

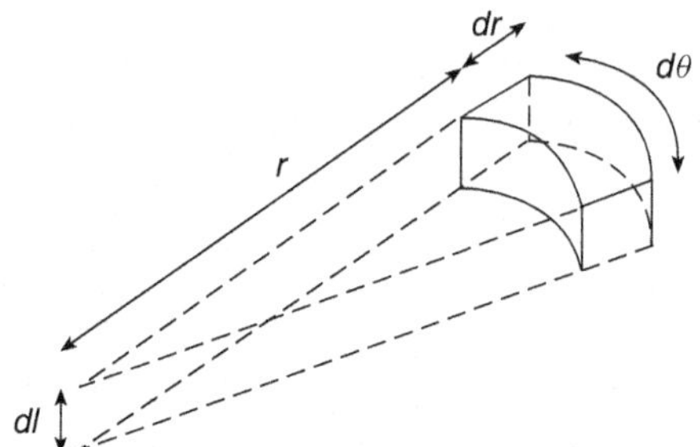

Figure 9.25 Elementary cylindrical surface located inside the insulating coating of a winding wire

the cylindrical coordinate system. The elementary capacitance related to the insulating coating shell is given by

$$dC = \frac{\epsilon r}{dr} d\theta dl. \tag{9.116}$$

Integrating this equation for r ranging from the inner radius of the bare conductor r_i to the outer radius of the wire with insulation r_o, and for l ranging from zero to the mean turn length (MTL) l_T, we obtain the capacitance of the insulating coating related to an elementary angle $d\theta$ for one turn

$$dC_c = \epsilon d\theta \int_0^{l_T} dl \int_{r_i}^{r_o} \frac{r}{dr} = \frac{\epsilon l_T}{\int_{r_i}^{r_o} \frac{dr}{r}} d\theta = \frac{\epsilon l_T}{\ln \frac{r_o}{r_i}} d\theta. \tag{9.117}$$

The capacitance due to the insulating coatings of the two adjacent turns is connected in series

$$dC_{ttc} = \frac{dC_c}{2} = \frac{\epsilon l_T}{2 \ln \frac{r_o}{r_i}} d\theta. \tag{9.118}$$

Hence, the capacitance per unit angle related to the insulating coatings of both adjacent turns is

$$\frac{dC_{ttc}}{d\theta} = \frac{\epsilon l_T}{2 \ln \frac{r_o}{r_i}}. \tag{9.119}$$

Next, the capacitance of the air gap will be determined. It is difficult to determine the locations of the electric field lines in the air gap between the insulating coatings of the adjacent turns. The shortest possible path can be used as a conservative approximation. Figure 9.26 shows one of the paths $x(\theta)$ of an electric field between two adjacent turns for a given angular coordinate θ. The approximation

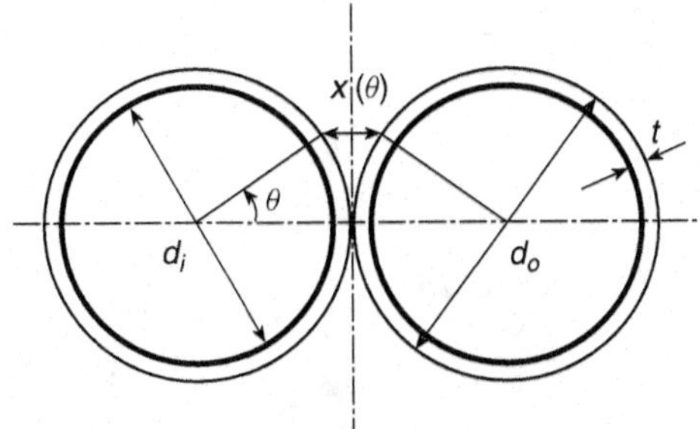

Figure 9.26 Approximated path $x(\theta)$ of an electric field between two adjacent turns for the basic cell with hexagonal winding pattern

is more valid for small values of θ, which produce the largest contributions to the turn-to-turn capacitance. As the angle θ increases, the error caused by the approximation also increases, which leads to somewhat larger than actual capacitances. However, the contributions of the capacitances for larger values of θ to the turn-to-turn capacitance is reduced, reducing the error caused by the approximation. From geometrical considerations in Fig. 9.26, the length of the approximated path of the electric field in terms of the angle θ is given by

$$x(\theta) = d_o(1 - \cos\theta). \tag{9.120}$$

The elementary surface of the wire, including the insulating coating, has a form of an elementary ring of length l_T and it is given by

$$dS = \frac{l_T d_o}{2} d\theta. \tag{9.121}$$

The elementary air-gap capacitance is

$$dC_g(\theta) = \frac{\epsilon_0 dS}{x(\theta)} = \frac{\epsilon_0 l_T d_o}{2x(\theta)} d\theta = \frac{\epsilon_0 l_T d_o}{2d_o(1 - \cos\theta)} d\theta = \frac{\epsilon_0 l_T}{2(1 - \cos\theta)} d\theta. \tag{9.122}$$

As θ increases, dC_g decreases. The elementary air-gap capacitance per unit angle is

$$\frac{dC_g}{d\theta} = \frac{\epsilon_0 dS}{2x(\theta)} = \frac{\epsilon_0 l_T d_o}{2x(\theta)} = \frac{\epsilon_0 l_T}{2(1 - \cos\theta)}. \tag{9.123}$$

The series combination of the elementary capacitances related to the insulating coatings of two turns and the air gap between the two adjacent turns is

$$dC_{tt}(\theta) = \frac{dC_{ttc} dC_g}{dC_{ttc} + dC_g} = \frac{\epsilon_0 l_T}{2\left(1 + \frac{1}{\epsilon_r} \ln \frac{d_o}{d_i} - \cos\theta\right)} d\theta. \tag{9.124}$$

Figure 9.27 shows plots of $C_{ttc}/d\theta$, $C_g/d\theta$, and $C_{tt}/d\theta$ as functions of angle θ for multilayer inductors at $l_T = 44.925$ mm, $\epsilon_r = 3.5$, $d_o = 0.495$ mm, and $d_i = 0.45$ mm. The thickness of the insulating coating t is constant, and the length of the path of the electric field intensity **E** is also constant. Therefore, the elementary capacitance dC_c related to the insulating coatings of two adjacent turns remains constant, independent of the angle θ over the entire basic cell. For $\theta = 0$, the path of the electric field intensity **E** in the air gap is zero, and the corresponding elementary capacitance dC_g approaches infinity. As angle θ increases, the paths of the electric field intensity **E** become longer in the air gap, and therefore the elementary air gap capacitance decreases. Thus, the elementary air-gap capacitance is not uniform. For small values of angle θ, the elementary air-gap capacitance is larger than the elementary insulating coating capacitance. Conversely, for large values of angle θ, the insulating coating capacitance is larger then the air-gap capacitance. The elementary turn-to-turn capacitance dC_{tt} is the series combination of the elementary air-gap capacitance dC_g and the elementary capacitance of the insulating coatings of two turns dC_c. Therefore, for small values of angle θ, the elementary insulating coating capacitances dC_c are dominant, and the elementary turn-to-turn capacitances dC_{tt} are approximately equal to the elementary insulating coating capacitances

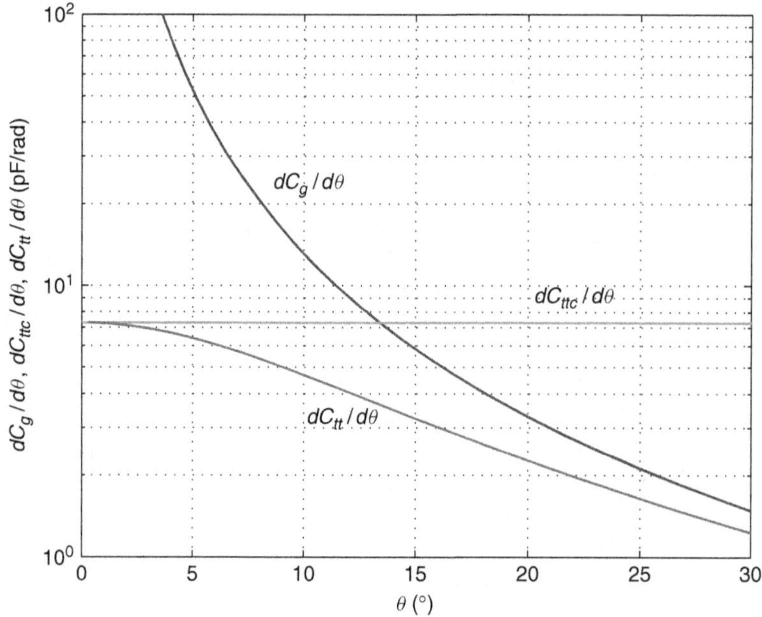

Figure 9.27 Plots of $C_{ttc}/d\theta$, $C_g/d\theta$, and $C_{tt}/d\theta$ as functions of angle θ for the basic cell of multilayer inductors at $l_T = 44.925$ mm, $\epsilon_r = 3.5$, $d_o = 0.495$ mm, and $d_i = 0.45$ mm

dC_c. On the other hand, for large values of angle θ, the elementary air gap capacitances dC_g are dominant, and the elementary turn-to-turn capacitances are equal to the air-gap capacitances dC_g. Half of the turn-to-turn capacitance is equal to the elementary area under the curve of $dC_{tt}/d\theta$.

The following general equation is useful for the derivation of an expression for C_{tt}

$$\int \frac{dx}{a - \cos x} = \frac{2}{\sqrt{a^2 - 1}} \arctan\left(\sqrt{\frac{a+1}{a-1}} \tan \frac{x}{2}\right). \tag{9.125}$$

Let

$$a = 1 + \frac{1}{\epsilon_r \ln \frac{d_o}{d_i}}. \tag{9.126}$$

Integrating dC_{tt}, we obtain the turn-to-turn capacitance of multiple-layer inductors with the hexagonal winding pattern [22]

$$C_{tt} = \int_{-\pi/6}^{\pi/6} dC_{tt} = 2\int_0^{\pi/6} dC_{tt} = \epsilon_0 l_T \int_0^{\pi/6} \frac{1}{1 + \frac{1}{\epsilon_r} \ln \frac{d_o}{d_i} - \cos\theta} d\theta$$

$$= \frac{2\epsilon_0 l_T}{\sqrt{\left(1 + \frac{1}{\epsilon_r} \ln \frac{d_o}{d_i}\right)^2 - 1}} \arctan\left[\frac{\sqrt{3}-1}{\sqrt{3}+1}\sqrt{1 + \frac{2\epsilon_r}{\ln \frac{d_o}{d_i}}}\right]$$

$$= \frac{2\epsilon_r \epsilon_0 l_T}{\sqrt{\left(2\epsilon_r + \ln \frac{d_o}{d_i}\right) \ln \frac{d_o}{d_i}}} \arctan\left[\frac{\sqrt{3}-1}{\sqrt{3}+1}\sqrt{\frac{2\epsilon_r + \ln \frac{d_o}{d_i}}{\ln \frac{d_o}{d_i}}}\right]$$

$$= \frac{2\epsilon_r \epsilon_0 l_T}{\sqrt{\left(2\epsilon_r + \ln \frac{d_i+2t}{d_i}\right) \ln \frac{d_i+2t}{d_i}}} \arctan\left[\frac{\sqrt{3}-1}{\sqrt{3}+1}\sqrt{\frac{2\epsilon_r + \ln \frac{d_i+2t}{d_i}}{\ln \frac{d_i+2t}{d_i}}}\right]$$

$$= \frac{2\pi \epsilon_r \epsilon_0 D_T}{\sqrt{\left(2\epsilon_r + \ln \frac{r_i+t}{r_i}\right) \ln \frac{r_i+t}{r_i}}} \arctan \left[\frac{\sqrt{3}-1}{\sqrt{3}+1} \sqrt{\frac{2\epsilon_r + \ln \frac{r_i+t}{r_i}}{\ln \frac{r_i+t}{r_i}}} \right], \tag{9.127}$$

where $l_T = \pi D_T$, $d_i = 2r_i$, $d_o = 2(r_i + t)$, ϵ_r is the relative permittivity of the insulating coating, D_T is the diameter of the turn, t is the thickness of wire insulation, and r_i is the bare wire radius.

The stray capacitance for a two-layer coreless inductor is [22]

$$C_s = 1.618 C_{tt} \quad \text{for} \quad N \geq 10. \tag{9.128}$$

The stray capacitance for a two-layer inductor with a conductive core is

$$C_s = 1.83 C_{tt} \quad \text{for} \quad N \geq 10. \tag{9.129}$$

The stray capacitance for a three-layer inductor with a conductive core is

$$C_s = 0.5733 C_{tt} \quad \text{for} \quad N \geq 10. \tag{9.130}$$

There are also multilayer inductors with square winding pattern, in which case the above equations should be modified.

Example 9.5

A multilayer inductor has the diameter $D_T = 1.22$ cm, the bare wire radius $r_i = 0.5$ mm, and the insulating coating thickness $t = 0.0175$ mm. Polyurethane is used for an insulating film with relative permittivity $\epsilon_r = 6.2$.

(a) Find the self-capacitance of a two-layer coreless inductor with the number of turns $N = 16$.
(b) Find the self-capacitance of a two-layer inductor with a conductive core and the number of turns $N = 16$.
(c) Find the self-capacitance of a three-layer inductor with a conductive core and the number of turns $N = 16$.

Solution: The turn-to-turn capacitance is

$$C_{tt} = \frac{2\pi \epsilon_r \epsilon_0 D_T}{\sqrt{\left(2\epsilon_r + \ln \frac{r_i+t}{a}\right) \ln \frac{r_i+t}{r_i}}} \arctan \left[\frac{\sqrt{3}-1}{\sqrt{3}+1} \sqrt{\frac{2\epsilon_r + \ln \frac{r_i+t}{r_i}}{\ln \frac{r_i+t}{r_i}}} \right]$$

$$= \frac{2\pi \times 6.2 \times 8.85 \times 10^{-12} \times 1.22 \times 10^{-2}}{\sqrt{\left(2 \times 6.2 + \ln \frac{0.5+0.0175}{0.5}\right) \ln \frac{0.5+0.0175}{0.5}}} \arctan \left[\frac{\sqrt{3}-1}{\sqrt{3}+1} \sqrt{\frac{2 \times 6.2 + \ln \frac{0.5+0.0175}{0.5}}{\ln \frac{0.5+0.0175}{0.5}}} \right]$$

$$= 507.37 \text{ pF}. \tag{9.131}$$

(a) The stray capacitance of the two-layer coreless inductor is

$$C_s = 1.618 C_{tt} = 1.61 \times 507.37 \times 10^{-12} = 820.92 \text{ pF} \quad \text{for} \quad N \geq 10. \tag{9.132}$$

(b) The stray capacitance of the two-layer inductor with a conductive core is

$$C_s = 1.83 C_{tt} = 1.83 \times 507.37 \times 10^{-12} = 928.49 \text{ pF} \quad \text{for} \quad N \geq 10. \tag{9.133}$$

(c) The stray capacitance of the three-layer inductor with a conductive core is

$$C_s = 0.5733 C_{tt} = 0.5733 \times 507.37 \times 10^{-12} = 290.877 \text{ pF} \quad \text{for} \quad N \geq 10. \tag{9.134}$$

9.12.2 Approximate Equation for Turn-to-Turn Self-Capacitance of Multilayer Inductors

Consider the capacitance due to wire insulation. The capacitance of a cylindrical capacitor of the dielectric inner diameter d_i, the dielectric outer diameter d_o, and the length l_T is given by

$$C_{cyl} = \frac{2\pi\epsilon_r\epsilon_0 l_T}{\ln\frac{d_o}{d_i}}. \tag{9.135}$$

The ratio of the capacitance due to the wire insulation corresponding to the angle $2\theta_1$ to the total cylindrical capacitance of wire insulation is

$$\frac{C_c}{C_{cyl}} = \frac{2\theta_1}{2\pi} = \frac{\theta_1}{\pi}. \tag{9.136}$$

Hence, the capacitance due to the coating insulation of a single turn in the middle of the basic cell corresponding to the angle $2\theta_1$, that is, for $-\theta_1 \leq \theta \leq \theta_1$, is expressed as

$$C_c = \frac{\theta_1}{\pi} \times C_{cyl} = \frac{\theta_1}{\pi} \times \frac{2\pi\epsilon_r\epsilon_0 l_T}{\ln\frac{d_o}{d_i}} = \frac{2\epsilon_r\epsilon_0 l_T\theta_1}{\ln\frac{d_o}{d_i}}. \tag{9.137}$$

The equivalent capacitance of the insulating coating of the two turns in the middle of the basic cell for the angle $2\theta_1$, that is, for $-\theta_1 \leq \theta \leq \theta_1$, is a series connection of two capacitances C_c and is given by

$$C_{ttc} = \frac{C_c}{2} = \frac{\epsilon_r\epsilon_0 l_T\theta_1}{\ln\frac{d_o}{d_i}}. \tag{9.138}$$

From (9.122), the elementary capacitance of the air gap is

$$dC_g = \frac{\epsilon_0 l_T}{2(1-\cos\theta)}d\theta. \tag{9.139}$$

Using a general equation for integration

$$\int \frac{dx}{1-\cos x} = -\cot\left(\frac{x}{2}\right), \tag{9.140}$$

the air-gap capacitance of both end parts of the basic cell for the angle $-\pi/6 \leq \theta \leq theta_1$, and the angle $\theta_1 \leq \theta \leq \pi/6$ is given by

$$C_g = \int_{-\frac{\pi}{6}}^{\theta_1} dC_g + \int_{\theta_1}^{\frac{\pi}{6}} dC_g = 2\int_{\theta_1}^{\frac{\pi}{6}} dC_g = 2\int_{\theta_1}^{\frac{\pi}{6}} \frac{\epsilon_0 l_T}{2(1-\cos\theta)}d\theta$$

$$= \epsilon_0 l_T\left(\cot\frac{\theta_1}{2} - \cot\frac{\pi}{12}\right). \tag{9.141}$$

The angle corresponding to the crossing point of the elementary air-gap capacitance $dC_g/d\theta$ and the elementary insulating coating capacitance $dC_{ttc}/d\theta$ occurs when

$$\frac{dC_{ttc}}{d\theta} = \frac{dC_g}{d\theta}. \tag{9.142}$$

Hence,

$$\frac{\epsilon_r\epsilon_0 l_T}{2\ln\frac{d_o}{d_i}} = \frac{\epsilon_0 l_T}{2(1-\cos\theta_1)}, \tag{9.143}$$

yielding

$$\theta_1 = \arccos\left(1 - \frac{\ln\frac{d_o}{d_i}}{\epsilon_r}\right). \tag{9.144}$$

The following approximation can be made:

$$dC_{tt} = \frac{dC_{ttc}dC_g}{dC_{ttc}+dC_g} \approx \frac{dC_{ttc}}{d\theta} \quad \text{for} \quad -\theta_1 \leq \theta \leq \theta_1 \tag{9.145}$$

and

$$dC_{tt} = \frac{dC_{ttc}dC_g}{dC_{ttc} + dC_g} \approx \frac{dC_g}{d\theta} \quad \text{for} \quad -\frac{\pi}{6} \le \theta \le \theta_1 \quad \text{and} \quad \theta_1 \le \theta \le \frac{\pi}{6}. \tag{9.146}$$

The turn-to-turn capacitance is

$$C_{tt} \approx C_{ttc} + C_g = \epsilon_0 l_T \left(\frac{\epsilon_r \theta_1}{\ln \frac{d_o}{d_i}} + \cot \frac{\theta_1}{2} - \cot \frac{\pi}{12} \right) = \epsilon_0 l_T \left(\frac{\epsilon_r \theta_1}{\ln \frac{d_o}{d_i}} + \cot \frac{\theta_1}{2} - 3.732 \right). \tag{9.147}$$

Another derivation of the capacitance related to the wire-insulating coating is as follows:

$$dl \approx b, \tag{9.148}$$

$$\tan \theta \approx d\theta = \frac{b}{r_o} \approx \frac{dl}{r_o}, \tag{9.149}$$

$$b = r_o d\theta, \tag{9.150}$$

$$dS = l_T b = l_T r_o d\theta = \frac{1}{2} l_T d_o d\theta, \tag{9.151}$$

$$dC_c = \frac{\epsilon dS}{t} = \frac{\epsilon l_T d_o d\theta}{2t}, \tag{9.152}$$

$$C_c = \int_{-\pi/2}^{\pi/2} dC_c = 2\int_0^{\pi/2} dC_c = \frac{\epsilon l_T d_o}{t} \int_0^{\pi/2} d\theta = \frac{\pi \epsilon l_T d_o}{2t}, \tag{9.153}$$

$$C_{ttc} = \frac{C_c}{2} = \frac{\pi \epsilon l_T d_o}{4t}. \tag{9.154}$$

The turn-to-core capacitance is

$$C_{tc} = 2C_{ttc} = \frac{\pi \epsilon l_T d_o}{2t}. \tag{9.155}$$

Example 9.6

A three-layer inductor has the inductance $L = 75.1$ μH, the number of circular turns $N = 95$, the turn diameter $D_T = 14.3$ cm, the inner bare wire diameter $d_i = 0.45$ mm, the outer insulated wire diameter $d_o = 0.495$ mm, and the dielectric relative permittivity $\epsilon_r = 3.5$.

(a) Find the self-capacitance of the inductor using approximate equation.
(b) Determine the self-resonant frequency of the inductor using the approximate equation for the self-capacitance.
(c) Find the self-capacitance of the inductor using exact equation.
(d) Determine the self-resonant frequency of the inductor using the exact equation for the self-capacitance.

Solution: The MTL is

$$l_T = \pi D_T = \pi \times 14.3 \text{ mm} = 44.92 \text{ mm}. \tag{9.156}$$

The coating thickness is

$$t = \frac{d_o - d_i}{2} = \frac{0.495 - 0.45}{2} = 0.00225 \text{ mm}. \tag{9.157}$$

The critical angle is

$$\theta_1 = \arccos\left(1 - \frac{\ln \frac{d_o}{d_i}}{\epsilon_r}\right) = \arccos\left(1 - \frac{\ln \frac{0.495}{0.45}}{3.5}\right) = 0.2339 \text{ rad} = 13.4°. \tag{9.158}$$

(a) The capacitance due to wire insulation is

$$C_{ttc} = \epsilon_0 l_T \left(\frac{\epsilon_r \theta_1}{\ln \frac{d_O}{d_i}} \right) = \frac{1}{36\pi} \times 10^{-9} \times 44.92 \times 10^{-3} \left(\frac{3.5 \times 0.2339}{\ln \frac{0.495}{0.45}} \right) = 3.42 \text{ pF}. \quad (9.159)$$

The air-gap capacitance is

$$C_g = \epsilon_0 l_T \left(\cot \frac{\theta_1}{2} - \cot \frac{\pi}{12} \right) = \frac{1}{36\pi} \times 10^{-9} \times 44.92 \times 10^{-3} \left(\cot \frac{0.2339}{2} - \cot \frac{\pi}{12} \right)$$
$$= 1.898 \text{ pF}. \quad (9.160)$$

The ratio of the insulation capacitances to air-gap capacitance is

$$\frac{C_{ttc}}{C_g} = \frac{3.42}{1.898} = 1.8. \quad (9.161)$$

The turn-to-turn capacitance using the approximate equation (9.147) is

$$C_{tt} \approx C_{ttc} + C_g = \epsilon_0 l_T \left(\frac{\epsilon_r \theta_1}{\ln \frac{d_O}{d_i}} + \cot \frac{\theta_1}{2} - \cot \frac{\pi}{12} \right)$$
$$= \frac{1}{36\pi} \times 10^{-9} \times 44.92 \times 10^{-3} \left(\frac{3.5 \times 0.2339}{\ln \frac{0.495}{0.45}} + \cot \frac{0.2339}{2} - \cot \frac{\pi}{12} \right) = 5.318 \text{ pF}. \quad (9.162)$$

The self-capacitance of the three-layer inductor is

$$C_s = 0.5733 C_{tt} = 0.5733 \times 5.318 = 3.0488 \text{ pF} \quad \text{for} \quad N \geq 10. \quad (9.163)$$

(b) The self-resonant frequency is

$$f_r = \frac{1}{2\pi \sqrt{LC_s}} = \frac{1}{2\pi \sqrt{75.1 \times 10^{-6} \times 3.0488 \times 10^{-12}}} = 10.518 \text{ MHz}. \quad (9.164)$$

(c) The turn-to-turn capacitance using the exact equation (9.127) is

$$C_{tt} = \frac{2\pi \epsilon_r \epsilon_0 D_T}{\sqrt{\left(2\epsilon_r + \ln \frac{d_O}{d_i}\right) \ln \frac{d_O}{d_i}}} \arctan \left[\frac{\sqrt{3} - 1}{\sqrt{3} + 1} \sqrt{\frac{2\epsilon_r + \ln \frac{d_O}{d_i}}{\ln \frac{d_O}{d_i}}} \right]$$
$$= \frac{2\pi \times 3.5 \times 8.85 \times 10^{-12} \times 14.3 \times 10^{-3}}{\sqrt{\left(2 \times 3.5 + \ln \frac{0.495}{0.45}\right) \ln \frac{0.495}{0.45}}} \arctan \left[\frac{\sqrt{3} - 1}{\sqrt{3} + 1} \sqrt{\frac{2 \times 3.5 + \ln \frac{0.495}{0.45}}{\ln \frac{0.495}{0.45}}} \right]$$
$$= 3.9363 \text{ pF}. \quad (9.165)$$

(d) The self-capacitance of the inductor is

$$C_s = 0.5733 C_{tt} = 0.5733 \times 3.9363 = 2.257 \text{ pF}. \quad (9.166)$$

The self-resonant frequency is

$$f_r = \frac{1}{2\pi \sqrt{LC_s}} = \frac{1}{2\pi \sqrt{75.1 \times 10^{-6} \times 2.257 \times 10^{-12}}} = 12.225 \text{ MHz}. \quad (9.167)$$

The ratio of the approximate turn-to-turn capacitance to the exact turn-to-turn capacitance is

$$\frac{C_{tt(approx)}}{C_{tt(exact)}} = \frac{5.318}{3.9363} = 1.351. \quad (9.168)$$

9.13 Self-Capacitance of Single-Layer Inductors

9.13.1 Exact Equation for Self-Capacitance of Single-Layer Inductors

A cross section of a single-layer inductor with a magnetic core is depicted in Fig. 9.28. Figure 9.29 shows a basic cell of a single-layer inductor representing the turn-to-turn capacitance C_{tt}. For the basic cell of a single-layer inductor, the angle θ is in the range $-\pi/2 \leq \theta \leq \pi/2$. The following equations hold true

$$dC_{ttc} = \frac{\epsilon l_T}{2\ln\frac{d_o}{d_i}} d\theta \tag{9.169}$$

and

$$dC_g = \frac{\epsilon l_T}{2(1-\cos\theta)} d\theta. \tag{9.170}$$

The series combination of the elementary capacitances of the air gap between the two adjacent turns and the insulation coatings of the two adjacent turns is given by

$$dC_{tt}(\theta) = \frac{dC_{ttc}dC_g}{dC_{ttc}+dC_g} = \frac{\epsilon_0 l_T}{2\left(1+\frac{1}{\epsilon_r}\ln\frac{d_o}{d_i}-\cos\theta\right)} d\theta. \tag{9.171}$$

Figure 9.30 shows the plots of $C_{ttc}/d\theta$, $C_g/d\theta$, and $C_{tt}/d\theta$ as functions of angle θ for single-layer inductors at $l_T = 44.925$ mm, $\epsilon_r = 3.5$, $d_o = 0.495$ mm, and $d_i = 0.45$ mm.

To determine the turn-to-turn capacitance, the following integral is useful

$$\int \frac{dx}{a-\cos x} = \frac{2}{\sqrt{a^2-1}}\arctan\left[\sqrt{\frac{a+1}{a-1}}\tan\left(\frac{x}{2}\right)\right]. \tag{9.172}$$

In our case,

$$a = 1+\frac{1}{\epsilon_r}\ln\left(\frac{d_o}{d_i}\right). \tag{9.173}$$

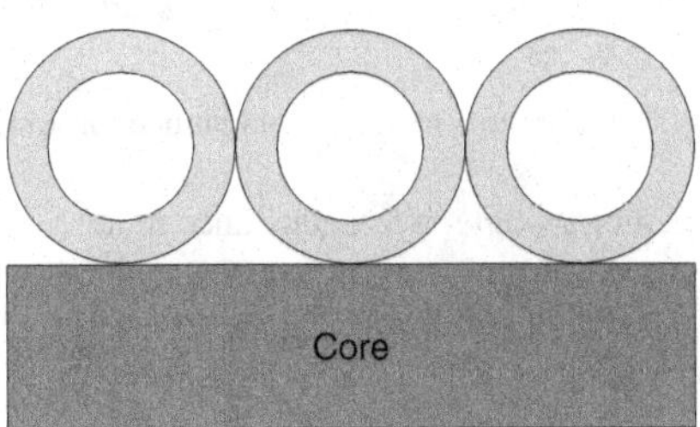

Figure 9.28 Cross-sectional view of a single-layer inductor with a magnetic core

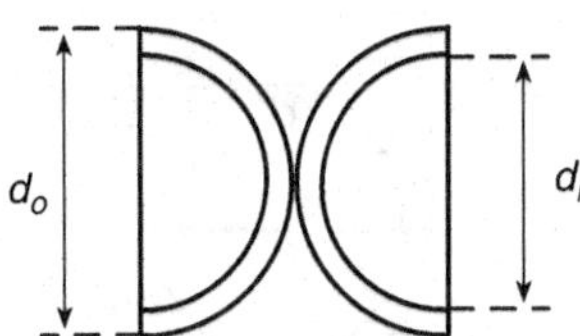

Figure 9.29 Basic cell of single-layer and square-pattern inductors to determine turn-to-turn capacitance C_{tt}

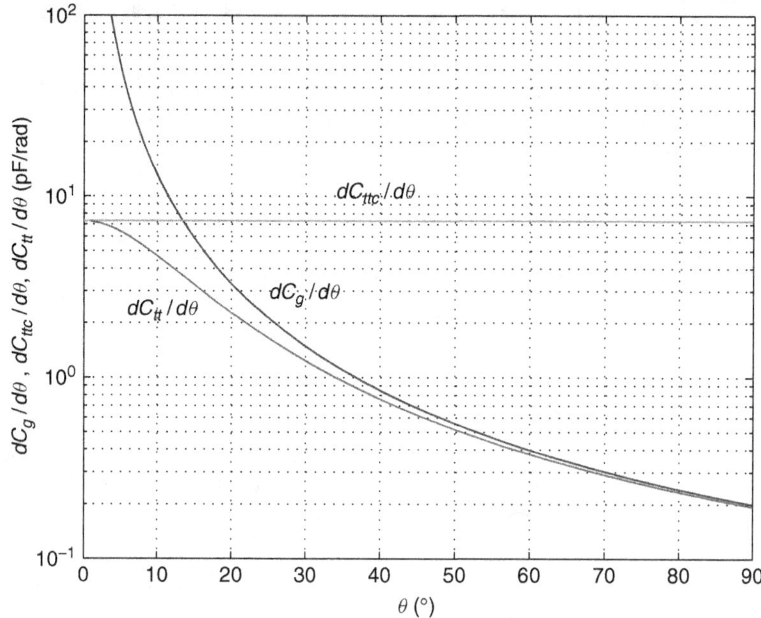

Figure 9.30 Plots of $C_{ttc}/d\theta$, $C_g/d\theta$, and $C_{tt}/d\theta$ as functions of angle θ for single-layer inductors at $l_T = 44.925$ mm, $\epsilon_r = 3.5$, $d_o = 0.495$ mm, and $d_i = 0.45$ mm

Hence, the turn-to-turn capacitance is

$$C_{tt} = \int_{-\frac{\pi}{2}}^{\frac{\pi}{2}} dC_{tt} = 2\int_0^{\frac{\pi}{2}} dC_{tt} = \epsilon_0 l_T \int_0^{\frac{\pi}{2}} \frac{1}{1 + \frac{1}{\epsilon_r}\ln\frac{d_o}{d_i} - \cos\theta} d\theta$$

$$= \frac{2\epsilon_0 l_T}{\sqrt{\left(\frac{1}{\epsilon_r}\ln\frac{d_o}{d_i} + 1\right)^2 - 1}} \arctan\left(\sqrt{1 + \frac{2}{\frac{1}{\epsilon_r}\ln\frac{d_o}{d_i}}}\right). \quad (9.174)$$

This expression is also valid for the turn-to-turn capacitance of inductors with a square winding pattern.

Figure 9.31 shows a basic cell of a single-layer inductance representing the turn-to-core capacitance C_{tc}. This structure of the basic cell of turn-to-core capacitance applies for all structures of inductors, such as square and hexagonal winding patterns.

The turn-to-core capacitance is

$$C_{tc} = 2C_{tt}. \quad (9.175)$$

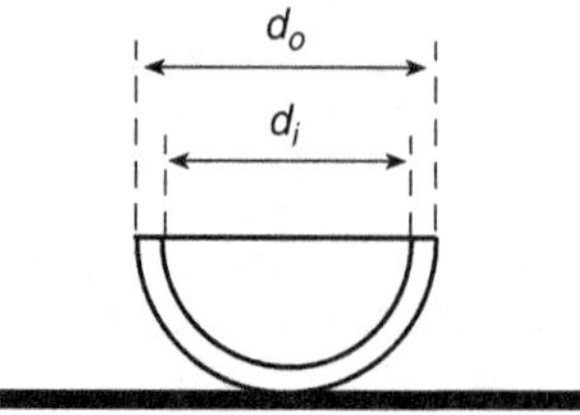

Figure 9.31 Basic cell of a single-layer inductor representing the turn-to-core capacitance C_{tc}

9.13.2 Approximate Equation for Turn-to-Turn Self-Capacitance of Single-Layer Inductors

The capacitance of a cylindrical capacitor of the dielectric inner diameter d_i, the dielectric outer diameter d_o, and the length l_T is given by

$$C_{cyl} = \frac{2\pi \epsilon_r \epsilon_0 l_T}{\ln \frac{d_o}{d_i}}. \tag{9.176}$$

The ratio of the capacitance due to the wire insulation corresponding to the angle $2\theta_1$ to the total cylindrical capacitance is

$$\frac{C_c}{C_{cyl}} = \frac{2\theta_1}{2\pi} = \frac{\theta_1}{\pi}. \tag{9.177}$$

Hence, the capacitance due to the coating insulation corresponding to the angle θ_1 for a single turn is

$$C_c = \frac{\theta_1}{\pi} \times C_{cyl} = \frac{\theta_1}{\pi} \times \frac{\pi \epsilon_r \epsilon_0 l_T}{\ln \frac{d_o}{d_i}} = \frac{2\epsilon_r \epsilon_0 l_T \theta_1}{\ln \frac{d_o}{d_i}}. \tag{9.178}$$

The equivalent capacitance of the insulating coating of the two turns for $0 \le \theta \le \theta_1$ is

$$C_{ttc} = \frac{C_c}{2} = \frac{\epsilon_r \epsilon_0 l_T \theta_1}{\ln \frac{d_o}{d_i}}. \tag{9.179}$$

The elementary capacitance of the air gap is

$$dC_g = \frac{\epsilon_0 l_T}{2(1 - \cos\theta)} d\theta. \tag{9.180}$$

Using the general equation for integration

$$\int \frac{dx}{1 - \cos x} = -\cot\left(\frac{x}{2}\right), \tag{9.181}$$

the air-gap capacitance is

$$C_g = 2\int_{\theta_1}^{\frac{\pi}{2}} dC_g = 2\int_{\theta_1}^{\frac{\pi}{2}} \frac{\epsilon_0 l_T}{2(1 - \cos\theta)} d\theta = \epsilon_0 l_T \left(\cot\frac{\theta_1}{2} - \cot\frac{\pi}{4}\right) = \epsilon_0 l_T \left(\cot\frac{\theta_1}{2} - 1\right). \tag{9.182}$$

The angle corresponding to the crossing point can be obtained as follows:

$$\frac{\epsilon_r \epsilon_0 l_T}{2 \ln \frac{d_o}{d_i}} = \frac{\epsilon_0 l_T}{2(1 - \cos\theta_1)}, \tag{9.183}$$

yielding

$$\theta_1 = \arccos\left(1 - \frac{\ln \frac{d_o}{d_i}}{\epsilon_r}\right). \tag{9.184}$$

If

$$\frac{\epsilon_r \epsilon_0 l_T}{\ln \frac{d_o}{d_i}} = A \frac{\epsilon_0 l_T}{1 - \cos\theta_1}, \tag{9.185}$$

then

$$\theta_1 = \arccos\left(1 - \frac{A \ln \frac{d_o}{d_i}}{\epsilon_r}\right). \tag{9.186}$$

The turn-to-turn capacitance is

$$C_{tt} \approx C_{ttc} + C_g = \epsilon_0 l_T \left(\frac{\epsilon_r \theta_1}{\ln \frac{d_o}{d_i}} + \cot\frac{\theta_1}{2} - 1\right). \tag{9.187}$$

Example 9.7

A single-layer inductor has the inductance $L = 75.1\ \mu$H, the number of round turns $N = 95$, the turn diameter $D_T = 14.3$ cm, the inner bare wire diameter $d_i = 0.45$ mm, the outer insulated wire diameter $d_o = 0.495$ mm, and the dielectric relative permittivity $\epsilon_r = 3.5$.

(a) Find the self-capacitance of the inductor using approximate equation.

(b) Determine the self-resonant frequency of the inductor using the approximate equation for the self-capacitance.

(c) Find the self-capacitance of the inductor using exact equation.

(d) Determine the self-resonant frequency of the inductor using the exact equation for the self-capacitance.

Solution: The MTL is

$$l_T = \pi D_T = \pi \times 14.3 \text{ mm} = 44.92 \text{ mm}. \tag{9.188}$$

The coating thickness is

$$t = \frac{d_o - d_i}{2} = \frac{0.495 - 0.45}{2} = 0.00225 \text{ mm}. \tag{9.189}$$

The critical angle is

$$\theta_1 = \arccos\left(1 - \frac{\ln \frac{d_o}{d_i}}{\epsilon_r}\right) = \arccos\left(1 - \frac{\ln \frac{0.495}{0.45}}{3.5}\right) = 0.2339 \text{ rad} = 13.4^\circ. \tag{9.190}$$

(a) The capacitance due to wire insulation is

$$C_{ttc} = \epsilon_0 l_T \left(\frac{\epsilon_r \theta_1}{\ln \frac{d_o}{d_i}}\right) = \frac{1}{36\pi} \times 10^{-9} \times 44.92 \times 10^{-3} \left(\frac{3.5 \times 0.2339}{\ln \frac{0.495}{0.45}}\right) = 2.42 \text{ pF}. \tag{9.191}$$

The air-gap capacitance is

$$C_g = \epsilon_0 l_T \left(\cot \frac{\theta_1}{2} - 1\right) = \frac{1}{36\pi} \times 10^{-9} \times 44.92 \times 10^{-3} \left(\cot \frac{0.2339}{2} - 1\right) = 2.984 \text{ pF}. \tag{9.192}$$

The ratio of the insulation capacitances to air-gap capacitance is

$$\frac{C_{ttc}}{C_g} = \frac{2.42}{2.984} = 1.146. \tag{9.193}$$

The turn-to-turn capacitance using the approximate equation (9.187) is

$$C_{tt} \approx C_{ttc} + C_g = \epsilon_0 l_T \left(\frac{\epsilon_r \theta_1}{\ln \frac{d_o}{d_i}} + \cot \frac{\theta_1}{2} - 1\right)$$

$$= \frac{1}{36} \times 10^{-9} \times 44.92 \times 10^{-3} \left(\frac{3.5 \times 0.2339}{\ln \frac{0.495}{0.45}} + \cot \frac{0.2339}{2} - 1\right) = 6.396 \text{ pF}. \tag{9.194}$$

The self-capacitance of the inductor is

$$C_s = 1.366 C_{tt} = 1.366 \times 6.396 = 8.737 \text{ pF}. \tag{9.195}$$

(b) The self-resonant frequency is

$$f_r = \frac{1}{2\pi\sqrt{LC_s}} = \frac{1}{2\pi\sqrt{75.1 \times 10^{-6} \times 8.737 \times 10^{-12}}} = 6.814 \text{ MHz}. \tag{9.196}$$

(c) The turn-to-turn capacitance using the exact equation is

$$C_{tt} = \frac{2\pi\epsilon_0 D_T}{\sqrt{\left(\frac{1}{\epsilon_r}\ln \frac{d_o}{d_i} + 1\right)^2 - 1}} \arctan\left(\sqrt{1 + \frac{2}{\frac{1}{\epsilon_r}\ln \frac{d_o}{d_i}}}\right)$$

$$= \frac{2\pi \times 8.85 \times 10^{-12} \times 14.3 \times 10^{-3}}{\sqrt{\left(\frac{1}{3.5} \ln \frac{0.495}{0.45} + 1\right)^2 - 1}} \arctan\left(\sqrt{1 + \frac{2}{\frac{1}{3.5} \ln \frac{0.495}{0.45}}}\right)$$

$$= 4.623 \text{ pF}. \quad (9.197)$$

(d) The self-capacitance of the inductor is

$$C_s = 1.366 C_{tt} = 1.366 \times 4.623 = 6.315 \text{ pF}. \quad (9.198)$$

The self-resonant frequency is

$$f_r = \frac{1}{2\pi\sqrt{LC_s}} = \frac{1}{2\pi\sqrt{75.1 \times 10^{-6} \times 6.315 \times 10^{-12}}} = 7.92 \text{ MHz}. \quad (9.199)$$

The ratio of the approximate turn-to-turn capacitance to the exact turn-to-turn capacitance is

$$\frac{C_{tt(approx)}}{C_{rr(exact)}} = \frac{6.396}{4.623} = 1.3835. \quad (9.200)$$

9.14 Δ-to-Y Transformation of Capacitors

Figure 9.32 shows the Δ-to-Y (or π-to-T) transformation of capacitors, which is useful to determine the overall self-capacitance of inductors. The capacitances in the Y-configuration are related to the capacitances in the Δ-configuration by the following equations:

$$C_1 = C_b + C_c + \frac{C_b C_c}{C_a}, \quad (9.201)$$

$$C_2 = C_a + C_c + \frac{C_a C_c}{C_b}, \quad (9.202)$$

and

$$C_3 = C_a + C_b + \frac{C_a C_b}{C_c}. \quad (9.203)$$

9.15 Overall Self-Capacitance of Single-Layer Inductor with Core

Figure 9.33 shows turn-to-turn and turn-to-core capacitances for a single-layer inductor. Capacitances between nonadjacent turns are neglected. The turn-to-core capacitance is the capacitance formed by a half of the round winding conductor, its insulating coating, and a parallel conducting plane

$$C_{tc} = 2C_{tt}. \quad (9.204)$$

For $N = 1$, the stray capacitance consists of the turn-to-core capacitance

$$C_1 = C_{tc} = 2C_{tt}. \quad (9.205)$$

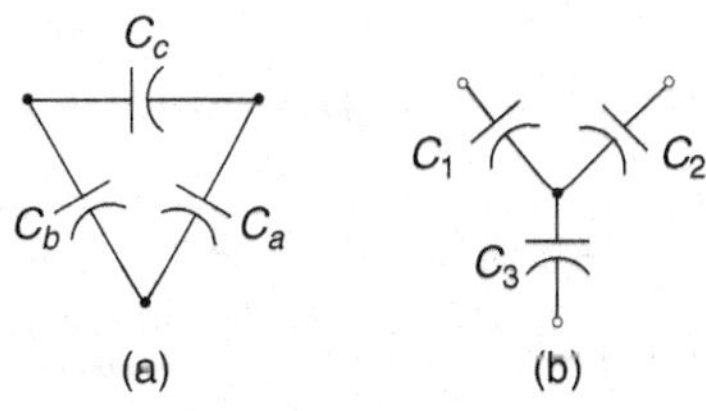

Figure 9.32 Δ-to-Y transformation of capacitors. (a) Δ-configuration of capacitors. (b) Y-configuration of capacitors

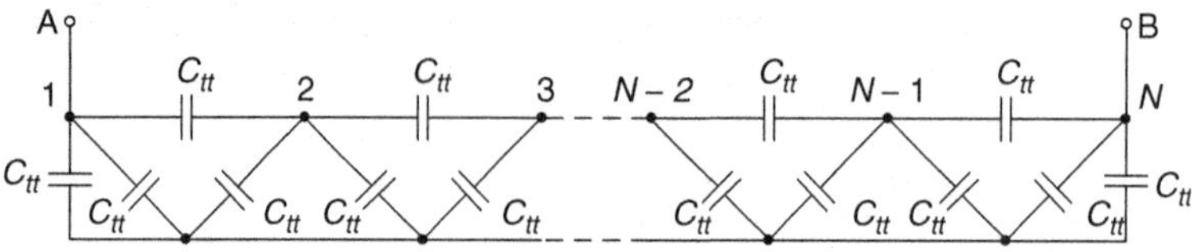

Figure 9.33 Turn-to-turn and turn-to-core capacitances for a single-layer inductor

For $N = 2$, the stray capacitance consists of the turn-to-turn capacitance C_{tt} in parallel with the series combination of two turn-to-core capacitances C_{tc} [32]

$$C_{1,2} = C_{tt} + \frac{C_{ttc}}{2} = C_{tt} + \frac{2C_{tt}2C_{tt}}{2C_{tt} + 2C_{tt}} = C_{tt} + C_{tt} = 2C_{tt}. \tag{9.206}$$

For $N = 3$,

$$C_{1,3} = 1.5C_{tt}. \tag{9.207}$$

The Δ-to-Y transformation can be used to determine the stray capacitance of inductors with many turns. In general, the stray capacitance of a single-layer inductor with a core and N turns is given by Massarini and Kazimierczuk [22]

$$C_{1,N} = C_{tt} + \frac{C_{1,(N-2)}C_{tt}}{2C_{1,(N-2)} + C_{tt}} = C_{tt} + \frac{C_{tt}}{\frac{C_{tt}}{C_{1,(N-2)}} + 2} = C_{tt}\left[\frac{C_{tt} + 3C_{1,(N-1)}}{C_{tt} + 2C_{1,(N-1)}}\right] \quad \text{for} \quad N \geq 4. \tag{9.208}$$

Hence, the stray capacitance can be determined either from the equivalent circuit or from (9.208) for $N = 4$

$$C_{1.4} = C_{tt} + \frac{2C_{tt}C_{tt}}{4C_{tt} + C_{tt}} = C_{tt} + \frac{C_{1,2}C_{tt}}{2C_{1,2} + C_{tt}} = C_{tt} + \frac{2C_{tt}C_{tt}}{4C_{tt} + C_{tt}} = \frac{7}{5}C_{tt} = 1.4C_{tt}. \tag{9.209}$$

For $N = 5$,

$$C_{1,5} = C_{tt} + \frac{C_{1,3}C_{tt}}{2C_{1,3} + C_{tt}} = C_{tt} + \frac{1.5C_{tt}C_{tt}}{2 \times 1.5C_{tt} + C_{tt}} = \frac{11}{8}C_{tt} = 1.375C_{tt}. \tag{9.210}$$

For $N = 6$,

$$C_{1,6} = C_{tt} + \frac{C_{1,4}C_{tt}}{2C_{1,4} + C_{tt}} = C_{tt} + \frac{\frac{7}{5}C_{tt}C_{tt}}{2 \times \frac{7}{5}C_{tt} + C_{tt}} = \frac{26}{19}C_{tt} = 1.36842C_{tt}. \tag{9.211}$$

For $N = 7$,

$$C_{1,7} = C_{tt} + \frac{C_{1,5}C_{tt}}{2C_{1,5} + C_{tt}} = C_{tt} + \frac{\frac{11}{8}C_{tt}C_{tt}}{2 \times \frac{11}{8}C_{tt} + C_{tt}} = \frac{11}{8}C_{tt} = 1.366667C_{tt}. \tag{9.212}$$

For $N = 8$,

$$C_{1,8} = C_{tt} + \frac{C_{1,6}C_{tt}}{2C_{1,6} + C_{tt}} = C_{tt} + \frac{\frac{26}{19}C_{tt}C_{tt}}{2 \times \frac{26}{19}C_{tt} + C_{tt}} = \frac{97}{71}C_{tt} = 1\frac{26}{71}C_{tt} = 1.3661972C_{tt}. \tag{9.213}$$

For $N = 9$,

$$C_{1,9} = C_{tt} + \frac{C_{1,7}C_{tt}}{2C_{1,7} + C_{tt}} = C_{tt} + \frac{\frac{41}{30}C_{tt}C_{tt}}{2 \times \frac{41}{30}C_{tt} + C_{tt}} = \frac{153}{112}C_{tt} = 1\frac{41}{112}C_{tt} = 1.36607C_{tt}. \tag{9.214}$$

For $N = 10$,

$$C_{1,10} = C_{tt} + \frac{C_{1,8}C_{tt}}{2C_{1,8} + C_{tt}} = C_{tt} + \frac{\frac{97}{71}C_{tt}C_{tt}}{2 \times \frac{97}{71}C_{tt} + C_{tt}} = \frac{362}{265}C_{tt} = 1\frac{97}{265}C_{tt} = 1.3660377C_{tt}. \tag{9.215}$$

In general, the stray capacitance of a single-layer inductor with a core or a shield is given by Massarini and Kazimierczuk [22]

$$C_s \approx 1.366C_{tt} \quad \text{for} \quad N \geq 10. \tag{9.216}$$

Figure 9.34 shows a plot of self-capacitance as a function of the number of turns N at $C_{tt} = 5.318$ pF.

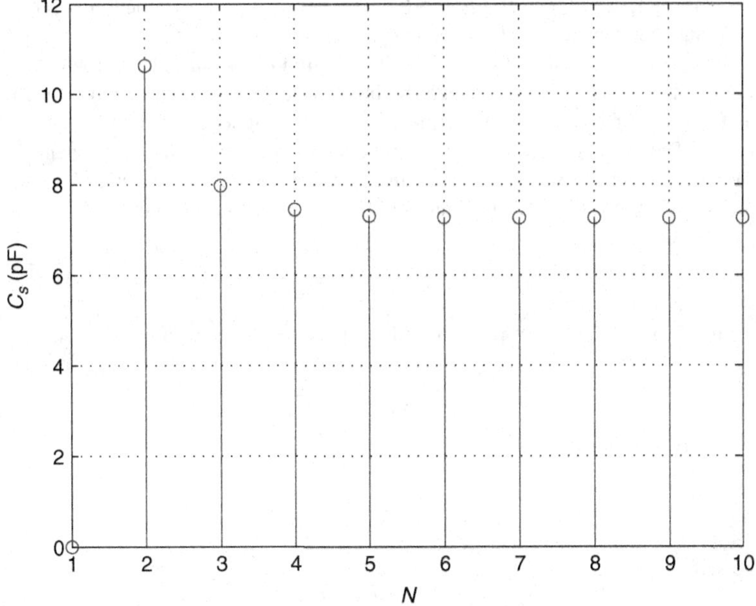

Figure 9.34 Self-capacitance as a function of the number of turns N at $C_{tt} = 5.318$ pF

9.16 Measurement of Self-Capacitance

Method I. The self-capacitance of an inductor can be experimentally determined by measuring the self-resonant frequency f_r with a network analyzer, and the inductance L can be measured with an LCZ impedance meter. The self-resonant frequency f_r can be indicated by the maximum magnitude or the zero phase of the inductor impedance. The self-resonant frequency is

$$f_r = \frac{1}{2\pi\sqrt{L_s C_s}}. \tag{9.217}$$

If the self-resonant frequency f_r and the inductance L_s are measured, the self-capacitance can be calculated from the equation:

$$C_s = \frac{1}{4\pi^2 L f_r^2}. \tag{9.218}$$

Method II. An external and known capacitance C_{ext} can be connected in parallel with the inductor. Then the resonant frequency can be measured

$$f_{r1} = \frac{1}{2\pi\sqrt{L_1(C_s + C_{ext})}}. \tag{9.219}$$

The ratio of the two frequencies is

$$\frac{f_r}{f_{r1}} = \sqrt{1 + \frac{C_{ext}}{C_s}}, \tag{9.220}$$

yielding the self-capacitance

$$C_s = \frac{C_{ext}}{\left(\frac{f_r}{f_{r1}}\right)^2 - 1}. \tag{9.221}$$

Expression (9.219) can be rearranged to the form:

$$C_{ext} = \frac{1}{4\pi^2 L_s} \frac{1}{f_{r1}^2} - C_s = ax + b, \tag{9.222}$$

where $a = 1/(4\pi^2 L_s)$, $x = 1/f_{r1}^2$, and $b = -C_s$. The plot of C_{ext} as a function of $1/f_{r1}^2$ is a straight line at a constant inductance L_s. Measurements of f_{r1} with several external capacitances will give several points located on the straight line. The intersection of the straight line with the vertical axis is equal to $-C_s$. This method is good when the inductance L_s is frequency independent.

Method III. At high frequencies, the permeability of the magnetic core is frequency dependent and can be complex. Therefore, the inductance of an inductor with a magnetic core depends on frequency. A method for measuring the self-capacitance uses two external capacitors with known capacitances C_1 and C_1 [23]. When capacitor C_1 is connected in parallel with an inductor, the measured resonant frequency is

$$f_{r1} = \frac{1}{2\pi\sqrt{L_1(C_s + C_1)}}. \tag{9.223}$$

When capacitor C_2 is connected in parallel with an inductor, the measured resonant frequency is

$$f_{r2} = \frac{1}{2\pi\sqrt{L_2(C_s + C_2)}}. \tag{9.224}$$

The ratio of the two frequencies is

$$\frac{f_{r1}}{f_{r2}} = \sqrt{\frac{L_2(C_s + C_2)}{L_1(C_s + C_1)}}, \tag{9.225}$$

yielding the self-capacitance

$$C_s = \frac{C_2 L_2 f_{r2}^2 - C_1 L_1 f_{r1}^2}{L_1 f_{r1}^2 - L_2 f_{r2}^2} \tag{9.226}$$

If resonant frequencies f_{r1} and f_{r2} are close enough to each other, $L_1 \approx L_2$ and (9.226) simplifies to the form:

$$C_s = \frac{\frac{C_2}{C_1}\left(\frac{f_{r2}}{f_{r1}}\right)^2 - 1}{1 - \left(\frac{f_{r1}}{f_{r2}}\right)^2} C_1. \tag{9.227}$$

Once C_s is known, the inductance can be calculated from (9.223)

$$L_1 = \frac{1}{4\pi^2 f_{r1}^2 (C_s + C_1)} \tag{9.228}$$

and the series resistance can be calculated from the expression:

$$R_s = \frac{L_1}{(C_s + C_1) r(f_{r1})}, \tag{9.229}$$

where $r(f_{r1})$ is the series resistance of the inductor at $f = f_{r1}$.

9.17 Inductor Impedance

The series resistance of the inductor is $R_s = R_w + R_{cs}$. Figure 9.35 shows the plots of winding resistance R_s, series core resistance R_{cs}, and total series resistance R_s as functions of frequency f.

The impedance of the series part of the inductor model is

$$Z_s = R_s + j\omega L_s = |Z_s| e^{j\phi_s}. \tag{9.230}$$

Figures 9.36 and 9.37 show the plots of the magnitude $|Z_s|$ and phase ϕ_s of the series part of the inductor model as functions of frequency f. The total impedance of the inductor model consists of the parallel combination of the impedance Z_s and the reactance of the stray capacitance C_s

Figure 9.38 shows the plots of R_s, X_{Ls}, and X_{Cs} as functions of frequency f. Figures 9.39 and 9.40 show the plots of $|Z|$ and ϕ as functions of frequency f.

The quality factor of the inductor is defined as

$$Q = \frac{X_{Ls}}{R_s} = \frac{\omega L_s}{R_w + R_{cs}}. \tag{9.231}$$

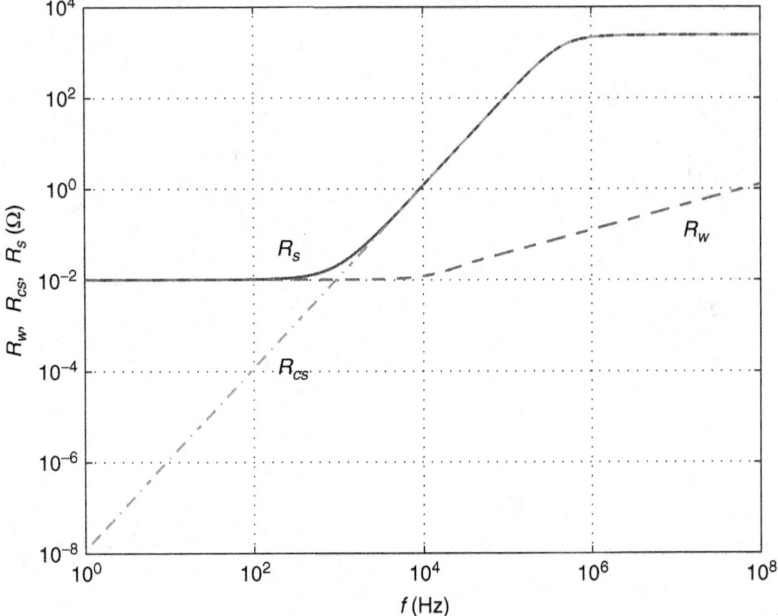

Figure 9.35 Winding resistance R_s, series core resistance R_{cs}, and total series resistance R_s as functions of frequency f

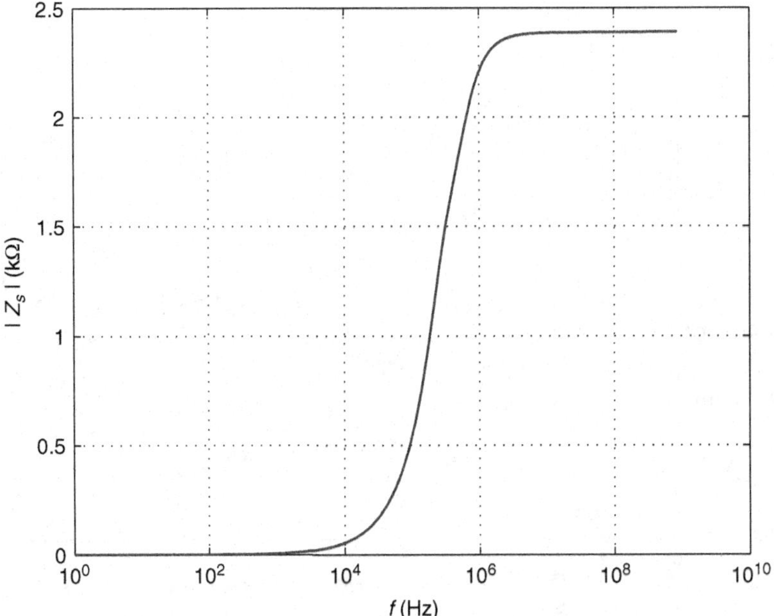

Figure 9.36 Magnitude of the series part of inductor impedance Z_s as a function of frequency f

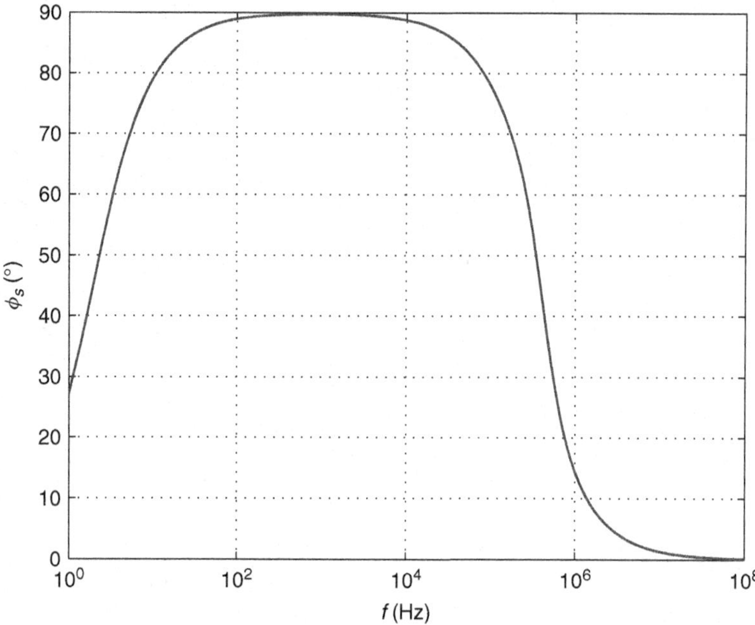

Figure 9.37 Phase of the series part of the inductor impedance Z_s as a function of frequency f

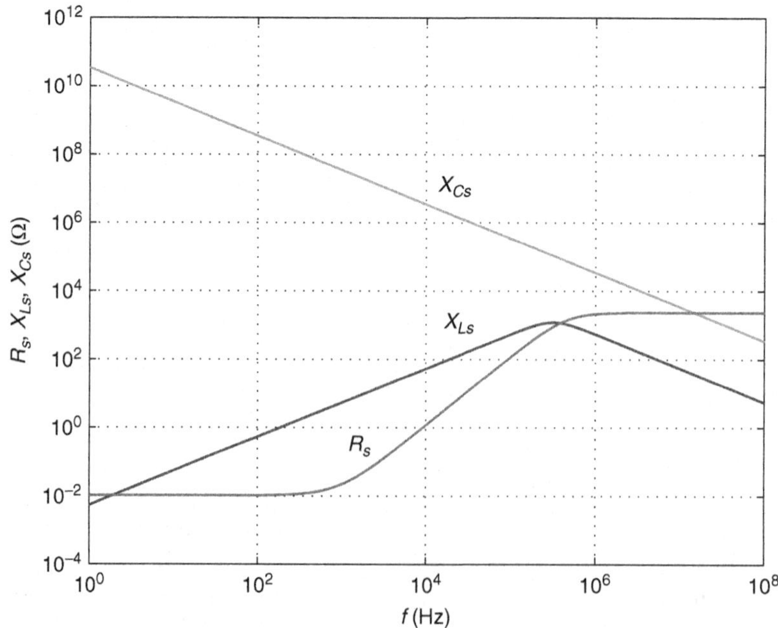

Figure 9.38 Series resistance R_s, reactance of series inductance L_s, and reactance of stray capacitance $C_s = 4.56\,\text{pF}$ as functions of frequency f

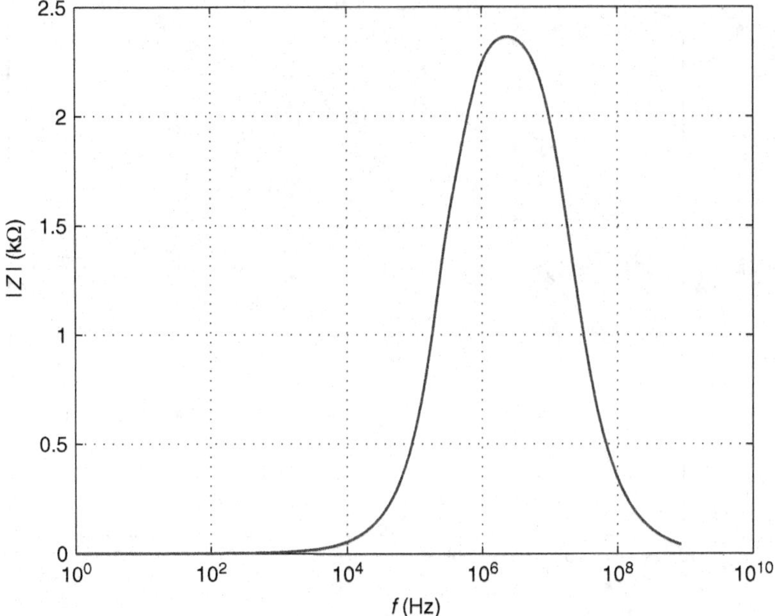

Figure 9.39 Magnitude of the inductor impedance as a function of frequency f

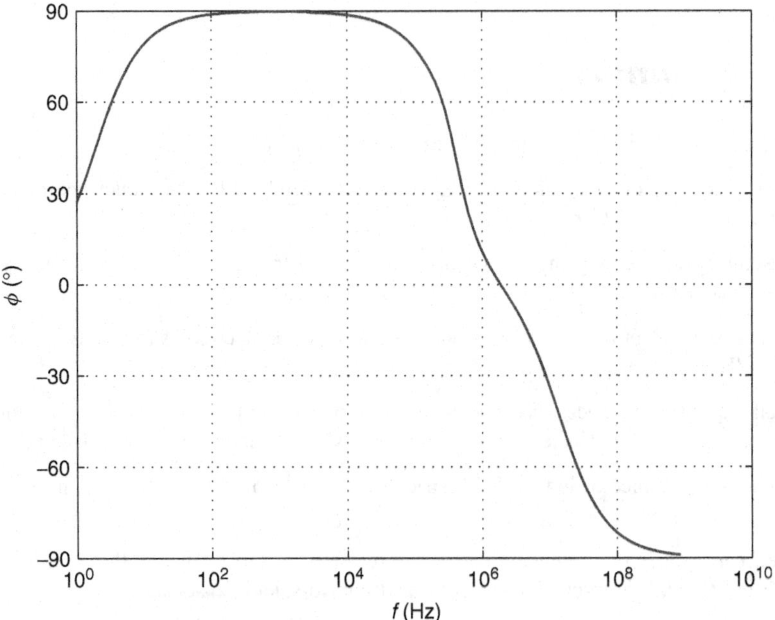

Figure 9.40 Phase of the inductor impedance as a function of frequency f

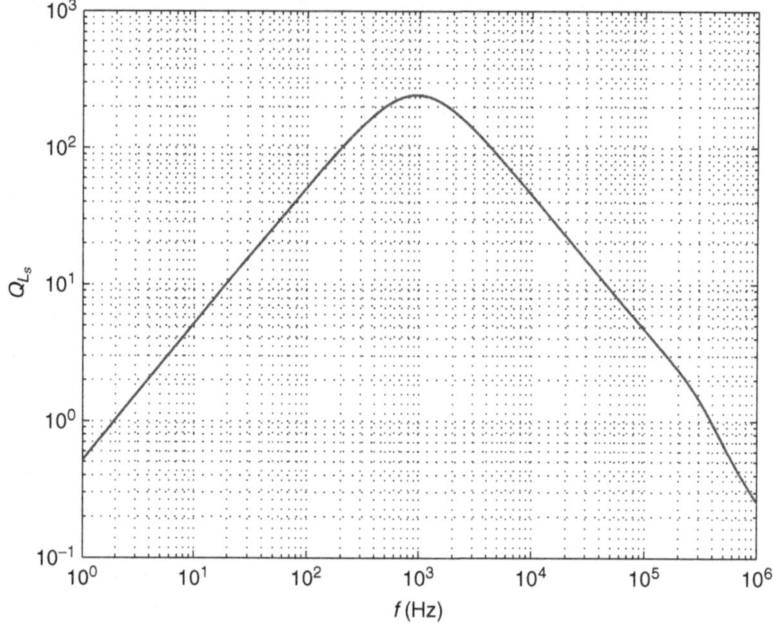

Figure 9.41 Quality factor of an inductor as a function of frequency f

Figure 9.41 shows a plot of the quality factor of an inductor as a function of frequency f.

9.18 Summary

- An inductor has a stray capacitance C_s, called *self-capacitance.*
- The first self-resonant frequency of inductors f_r is determined by the inductance L_s and the self-capacitance C_s ($f_r = 1/\sqrt{L_s C_s}$).
- The useful operating frequency range of inductors and transformers is determined by their self-resonant frequency.
- The first self-resonant frequency of inductors used in power electronics is usually in the range from 1 to 10 MHz.
- The self-capacitance reduces the impedance of an inductor at high frequencies, reducing its ability for filtering at high frequencies and forming the desired inductor current and voltage waveforms.
- A low self-capacitance of inductors is especially important in electromagnetic interference (EMI) filters.
- The self-capacitance of an inductor consists of turn-to-turn capacitance of the same layer, layer-to-layer capacitance, turn-to-core capacitance, and turn-to-shield capacitance.
- Foil inductors have a large self-capacitance C_s and a low self-resonant frequency.
- An extra insulation can be used in foil inductors to reduce the self-capacitance and increase the self-resonant frequency.
- The turn-to-turn capacitance of a single-layer inductor is directly proportional to the turn length l_T.
- The turn-to-turn capacitance of a single-layer inductor decreases as p/a increases.

- The insulation coating causes the turn-to-turn capacitance of a single-layer inductor to increase.
- Single-layer inductors have lower self-capacitance and higher self-resonant frequency than multi-layer inductors.
- The self-capacitance of a transformer consists of the same components as an inductor plus a winding-to-winding capacitance.
- The self-capacitance C_s decreases as the number of turns N increases.

9.19 References

[1] R. G. Medhurst, "HF resistance and self-capacitance of single-layer solenoids," *Wireless Engineers*, pp. 35–43, February 1947, and pp. 80–92, March 1947.

[2] W. B. Boast, *Principles of Electric and Magnetic Fields*, New York: Harper & Brothers, 1956.

[3] J. Ebert, "Four terminal parameters of HF inductors," *Bulletin de l'Acad'emie Polonaise des Sciences, S'erie des Sciences Techniques*, no. 5, 1968.

[4] A. A. Zaky and R. Hawley, *Fundamentals of Electromagnetic Fields Theory*, London: Harrap, 1974.

[5] R. J. Kemp, P. N. Mugatroyd, and N. J. Walkers, "Self resonance in foil inductors," *Electronics Letters*, vol. 11, no. 15, pp. 337–338, July 1975.

[6] W. H. Hayt Jr., *Engineering Electromagnetics*, 4th Ed., New York: McGraw-Hill, 1981.

[7] J. D. Krauss, *Electromagnetics*, 3rd Ed., New York: McGraw-Hill, 1984.

[8] M. Bartoli, A. Reatti, and M. K. Kazimierczuk, "High-frequency models of ferrite inductors," Proceedings of the IEEE International Conference on Industrial Electronics, Controls, and Instrumentation (IECON'94), Bologna, Italy, September 5–9, 1994, pp. 1670–1675.

[9] M. Bartoli, A. Reatti, and M. K. Kazimierczuk, "Predicting the high-frequency ferrite-core inductor performance," Proceedings of the Conference of Electrical Manufacturing and Coil Winding, Chicago (Rosemont), IL, September 27–29, 1994, pp. 409–413.

[10] M. Bartoli, A. Reatti, and M. K. Kazimierczuk, "Modeling iron-powder inductors at high-frequencies," Proceedings of the IEEE Industry Applications Society Annual Meeting, Denver, CO, October 2–7, 1994, pp. 1125–1232.

[11] F. Blache, J. Keradec, and B. Cogitore, "Stray capacitances of two winding transformers: Equivalent circuits, measurements, calculation and lowering," IEEE Industry Applications Society Annual Meeting, Denver, CO, vol. 2, October 2–6, 1994, pp. 1211–1217.

[12] M. Bartoli, N. Noferi, A. Reatti, and M. K. Kazimierczuk, "Modeling winding losses in high-frequency power inductors," *Journal of Circuits Systems and Computers*, vol. 5, no. 3, pp. 65–80, March 1995.

[13] A. Massarini and M. K. Kazimierczuk, "Modeling the parasitic capacitances of inductors," Proceedings of the Capacitors and Resistors Technology Symposium (CARTS'96), New Orleans, LA, March 1996, pp. 78–85.

[14] A. Massarini, M. K. Kazimierczuk, and G. Grandi, "Lumped parameter models for single- and multiple-layer inductors," Proceedings PESC'96, Baveno, Italy, June 23–27, 1996, pp. 295–301.

[15] M. Bartoli, N. Nefari, A. Reatti, and M. K. Kazimierczuk, "Modeling litz-wire winding losses in high-frequencies power inductors," Proceedings of the IEEE Power Electronics Specialists Conference, Baveno, Italy, June 23–27, 1996, pp. 1690–1696.

[16] H. G. Green, "A simplified derivation of the capacitance of a two-wire transmission line," *IEEE Transactions on Microwave Theory and Technique*, vol. 47, no. 3, pp. 365–366, March 1999.

[17] K. Nashadham, "A rigorous experimental characterization of ferrite inductors for RF noise suppression," IEEE Radio and Wireless Conference, Denver, CO, August 1–4, 1999, pp. 271–274.

[18] G. Grandi, M. K. Kazimierczuk, A. Massarini, and U. Reggiani, "Stray capacitance of single-layer solenoid air-core inductors," *IEEE Transactions on Industry Applications*, vol. 35, no. 5, pp. 1162–1168, September 1999.

[19] M. K. Kazimierczuk, G. Sancineto, U. Reggiani, and A. Massarini, "Small-signal high-frequency model of ferrite inductors," *IEEE Transactions on Magnetics*, vol. 35, pp. 4185–4191, September 1999.

[20] U. Reggiani, G. Sancineto, and M. K. Kazimierczuk, "High-frequency behavior of laminated iron-core inductors for filter applications," Proceedings of the IEEE Applied Power Electronics Conference, New Orleans, LA, February 6–10, 2000, pp. 654–660.

[21] G. Grandi, M. K. Kazimierczuk, A. Massarini, U. Reggiani, and G. Sancineto, "Model of laminated iron-core inductors," *IEEE Transactions on Magnetics*, vol. 40, no. 4, pp. 1839–1845, July 2004.

[22] A. Massarini and M. K. Kazimierczuk, "Self-capacitance of inductors," *IEEE Transactions on Power Electronics*, vol. 12, no. 4, pp. 671–676, July 1997.

[23] Q. You and T. W. Holmes, "A study on stray capacitance of inductors by using the finite element method," *IEEE Transactions on Electromagnetic Compatibility*, vol. 43. no. 1, pp. 88–93, February 2001.
[24] Q. You, T. W. Holmes, and K. Nashadham, "RF equivalent modeling of ferrite-core inductors and characterization of core material," *IEEE Transactions on Electromagnetic Compatibility*, vol. 44. no. 1, pp. 258–262, February 2002.
[25] H. Y. Lu, J. G. Zhu, and S. Y. R. Hui, "Experimental determination of stray capacitances in high frequency transformers," *IEEE Transactions on Power Electronics*, vol. 18, no. 5, pp. 1105–1112, September 2003.
[26] A. Van den Bossche and V. C. Valchev, *Inductors and Transformers for Power Electronics*, Boca Raton, FL: Taylor & Francis, 2005.
[27] M. J. Hole and L. C. Appel, "Stray capacitance of a two-layer air-cored inductor," *IEEE Proceedings, Part G - Circuits, Devices and Systems*, vol. 152, no.6, pp. 565–572, December 2005.
[28] T. C. Neugebauer and D. J. Perreault, "Parasitic capacitance cancellation in filter inductors," *IEEE Transactions on Power Electronics*, vol. 21, no. 1, pp. 282–288, January 2006.
[29] D.-M. Fang, Y. Zhou, X.-M. Jing, X.-L. Zhou, and X.-N. Wang, "Modeling, optimization, and performance of high-Q MEMS solenoid inductors," *Microsystem Technologies*, vol. 14, no. 2, pp. 185–191, 2008.
[30] L. Dalessandro, F. S. Cavalcante, and J. W. Kolar, "Self-capacitance of high-voltage transformers," *IEEE Transactions on Power Electronics*, vol. 22, no. 5, pp. 2081–2092, September 2007.
[31] B. A. Mazzeo, "Analytic solutions for capacitance of cylinders near a dielectric interface," *IEEE Transactions on Dielectric and Insulation*, vol. 27, no. 6, pp. 1877–1883, December 2010.
[32] S. Pasko, "Analysis of the effects of EMI filter construction on conducted noise of power converters," PhD Dissertation, Silesian University of Technology, Poland, 2011.
[33] K. Naishadham, "Closed-form design formulas for the equivalent circuit characterization of ferrite inductors," *IEEE Transactions on Electromagnetic Compatibility*, vol. 53, no. 4, pp. 923–932, November 2011.

9.20 Review Questions

9.1. What is the self-capacitance of an inductor?

9.2. What is the self-resonant frequency of an inductor?

9.3. List the components of a self-capacitance of inductors.

9.4. What is the character of the reactance of an inductor above the self-resonant frequency?

9.5. How does the turn length l_w affect the turn-to-turn capacitance in single-layer inductors?

9.6. How does the ratio p/a affect the turn-to-turn capacitance in single-layer inductors?

9.7. How does the number of turns N affect the turn-to-turn capacitance in single-layer inductors?

9.8. How does the insulation coating material affect the turn-to-turn capacitance in single-layer inductors?

9.9. What is the fringing electric field and when may it be neglected?

9.10. Draw the basic cell of multilayer inductors with hexagonal winding pattern.

9.11. Draw the basic cell of the turn-to-turn capacitance of single-layer inductors.

9.12. Draw the basic cell of the turn-to-core capacitance of single-layer inductors.

9.21 Problems

9.1. Calculate the stray capacitance C_s of two parallel round conductors of length $l_w = 0.5$ m and radius $a = 0.5$ mm placed in air. The distance between the centers of the conductors is $p = 1.2$ mm.

9.2. Calculate the stray capacitance C_s of a round conductor of length $l_w = 1$ cm and radius $a = 1$ mm placed in air above a copper plane. The distance between the center of the conductor and the plane is $h = 1.2$ mm.

9.3. Calculate the stray capacitance C_s of an inductor with $N = 10$, $a = 1$ mm, $p = 2.4$ mm, and $D_T = 40$ mm.

9.4. A multilayer inductor has diameter $D_T = 1$ cm, wire radius $a = 0.5$ mm, insulation coating thickness $t = 0.01$ mm, and relative permittivity of the insulation coating $\epsilon_r = 1$.

(a) Find the self-capacitance of a two-layer coreless inductor with the number of turns $N = 16$.

(b) Find the self-capacitance of a two-layer inductor with a conductive core and the number of turns $N = 16$.

(c) Find the self-capacitance of a three-layer inductor with a conductive core and the number of turns $N = 16$.

9.5. A multilayer inductor has diameter $D_T = 1$ cm, the wire radius $a = 0.5$ mm, insulation coating thickness $t = 0.01$ mm, and relative permittivity of the insulation coating $\epsilon_r = 4$.

(a) Find the self-capacitance of a two-layer coreless inductor with number of turns $N = 16$.

(b) Find the self-capacitance of a two-layer inductor with a conductive core and number of turns $N = 16$.

(c) Find the self-capacitance of a three-layer inductor with a conductive core and the number of turns $N = 16$.

Design of Inductors

10.1 Introduction

The subject of this chapter is the design of high-frequency power inductors [1–29]. The core area product method A_p and the core geometrical coefficient K_g method [1] for designing of inductors will be introduced. Expressions for the core area product A_p and the core geometry coefficient K_g are derived. A step-by-step design procedure for inductor designs is given for several typical inductors using the core area product and the core geometrical coefficient methods. The design of an inductor involves the following steps:

- Core material selection
- Core shape selection
- Core size selection
- Winding wire selection.

The first step in designing the magnetic components is the core material selection. The determinants of the core material selection are as follows:

- Operating frequency and its significant harmonics (if any)
- Relative permeability μ_{rc}
- Peak value of the magnetic flux density $B_{pk} < B_s$ to avoid core saturation
- Maximum amplitude of the magnetic flux density B_m to avoid excessive temperature rise ΔT due to excessive core losses
- Selection between ferrite material and iron-powdered material from the point of view of core loss density and saturation magnetic flux density.

High-Frequency Magnetic Components, Second Edition. Marian K. Kazimierczuk.

Companion Website: www.wiley.com/go/kazimierczuk_High2e

The second step is the selection of the core, which involves core shape, core size, core volume, core or bobbin winding window, air gap, and core profile. There are two methods of core size selection: the area product A_p and the core geometry coefficient K_g.

The third step is the winding conductor selection, which involves the selection of conductor shape (round wire or strip), conductor size, conductor insulation, number of turns N, and number of layers N_l. In the case of transformers, the designer must also select winding configuration (interleaved winding or conventional winding) and insulation between the windings.

10.2 Magnet Wire

Magnet wire is enamel-coated wire. Windings are made of magnet round wire, square wire, rectangular (strip) wire, foil, or litz wire. Magnet wire consists of solid copper, circular in cross section, and coated with one or two layers of insulating polymer. The coating can be either a single layer of polyurethane or a layer of polyurethane with a nylon overcoat. Thick coating increases the thermal resistance but reduces the turn-to-turn capacitances. Important characteristics of magnet wire are solderability and dielectric breakdown voltage of the insulating coating. In high-current applications, foil, square, or rectangular conductors may be used. The insulating film on magnet wire is the most vulnerable to thermal overloads. Therefore, the selection of the insulating film is very critical for long life of magnetic devices.

Figure 10.1 shows two winding patterns made of solid round wire: triangular (also called hexagonal) and square. For the triangular winding pattern, a round wire will tend to lie in a groove of the lower layer. The triangular winding pattern gives the tightest packing of wire. For the square winding pattern, the individual wires lie exactly above each other.

In high-frequency applications, where the skin and proximity effects can cause significant losses, litz wire is used. If large wire is made from many small strands that are insulated from each other and are externally connected in parallel, skin proximity effect losses are greatly reduced. This type of wire is called *litz wire*. Strands near the center are gradually brought to the edge of the bundle and vice versa. This prevents current from flowing only in some strands. Litz wire diameter is about 40% greater than that of solid wire. This is a result of its complex twisted structure and the presence of a large amount of internal insulation. It is hard to strip litz wire and it is about three times more expensive than solid wire. The window utilization factor K_u with litz wire is less than 30%. High values of the window utilization factor is obtained with foil winding, up to 80%.

To standardize the diameter of a single, round, solid, nonferrous, electrically conducting wire, the American Wire Gauge (AWG)-standardized system is commonly used in the United States. The range of wire diameters is from AWG0000 to AWG50. The bare wire diameter increases as the gauge number decreases. The higher the number is, the thinner will be the wire. Every six gauge decrease doubles the wire diameter, and every three gauge decrease doubles the wire cross-sectional area. The inner diameter of the AWG50 wire is d_i = 1 mil = 0.001 in. = 0.0254 mm, and the inner diameter of the AWG30 wire is d_i = 0.254 mm. The diameter of the AWG0000 is d_i = 0.46 in. = 18.11 mm and

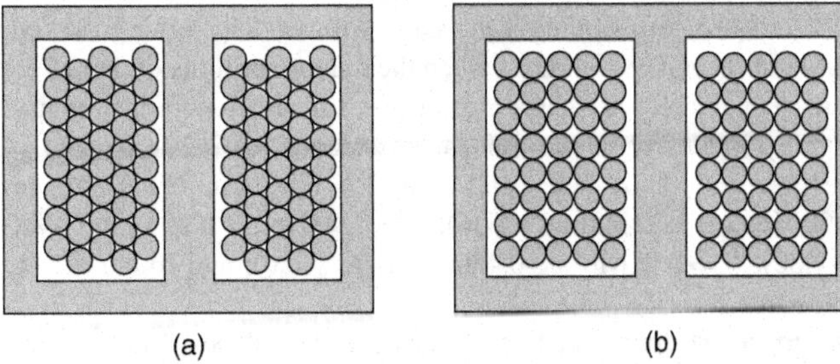

Figure 10.1 Solid round wire winding patterns. (a) Triangular (or hexagonal) pattern. (b) Square pattern

the inner diameter of the AWG36 wire is $d_i = 0.005$ in. $= 0.127$ mm. Hence, the ratio of diameters is 92 and the number of steps is 39. The inner diameter of an AWG bare wire in millimeters is given by

$$d_i = 0.127 \times 92^{\frac{36-AWG}{39}} \text{ (mm)} \tag{10.1}$$

and in inches is

$$d_i = 0.005 \times 92^{\frac{36-AWG}{39}} \text{ (in.)}. \tag{10.2}$$

The diameter of an insulated wire d_o is higher than that of a bare wire d_i. The cross-sectional area of the bare wire determines the current-carrying capability (ampacity). The cross-sectional area of the winding conductor is

$$A_w = \frac{I_{Lm}}{J_m} = \frac{I_{Lrms}}{J_{rms}}, \tag{10.3}$$

where J_m is the amplitude of the DC or low-frequency conductor current density and J_{rms} is the rms value of the DC or low-frequency conductor current density. Typically, the maximum value of the DC and low-frequency conductor current density is from 0.1 to 5 A/mm^2. Household wiring is AWG12 ($d = 2.05$ mm) or AWG14 ($d = 1.63$ mm). Therefore, the maximum uniform current density $J = I_{rms(max)}/A$ at the maximum wall outlet rms current $I_{rms(max)} = 15$ A is 4.5 or 7.2 A/mm^2.

Table 10.1 gives parameters for the standardized AWG round, solid copper wire. The DC resistance of a round magnet wire is

$$R_{wDC} = \frac{\rho_w l_w}{A_w} = \frac{4\rho_w l_w}{\pi d_i^2} = \frac{\rho_w N l_T}{A_w} = \frac{4\rho_w N l_T}{\pi d_i^2}, \tag{10.4}$$

where d_i is the diameter of the bare wire and $l_w = N l_T$ is the length of the winding wire. The DC magnet wire resistance per unit length is given by

$$\frac{R_{wDC}}{l_w} = \frac{\rho_w}{A_w} = \frac{4\rho_w}{\pi d_i^2}. \tag{10.5}$$

The skin and proximity effects on the wire resistance are neglected.

A foil inductor consists of a single strip of conductor foil and a slightly wider strip of a plastic foil. Aluminum foil is often used as a conductor and polypropylene is used as a dielectric. Typically, the thickness of metal foil is from 10 to 100 μm and the thickness of dielectric plastic is from 6 to 20 μm. The width of the metal foil is from 3 to 5 cm.

Copper foil windings offer electrical, thermal, and mechanical advantages. They have high window utilization and very low DC resistance. The core window utilization area with foil conductor winding is better than that with round wire because of the unused air space between round wires and insulation coating. Foil windings are good for high DC currents because the DC resistance is low. Heat dissipation in foil windings is better than that in round wire because the mass of solid conductor can transfer heat from the center of the inductor or transformer more effectively than round magnet wire. The foil extends over the entire inductor width and every layer has two edges in contact with ambient. In multilayer round wire inductors, the innermost turns are isolated from the ambient. The voltage stress between the layers of a foil winding is constant. Foil windings offer good coupling between the transformer primary and secondary windings when the number of turns of one of the windings is low. A foil-wound inductor or transformer has a greater mechanical strength than a wire-wound magnetic component. The typical values of copper foil thickness are from 0.2 to 0.9 mm. Selected values of copper foil thickness are: $h = 0.001$ in. $= 0.0254$ mm, $h = 0.0014$ in. $= 0.03556$ mm, $h = 0.002$ in. $=$ 0.508 mm, $h = 0.005$ in. $= 0.127$ mm (AWG36), $h = 0.008$ in. $= 0.203$ mm (AWG32), $h = 0.01$ in. $= 0.254$ mm (AWG30), $h = 0.016$ in. $= 0.4064$ mm (AWG26), and $h = 0.0216$ in. $= 0.54864$ mm (AWG24).

The hardness of copper is determined by the temper. The standard tempers of copper are hard, half-hard, and soft. Soft copper is the best conductor for most commercial applications because it is easier to solder and to wind in manufacturing magnetic components.

Table 10.1 AWG round copper wire parameters

AWG no	Nominal bare wire diameter d_i (mm)	Nominal outer diameter d_o (mm)	Bare wire area A_w (mm^2)	DC resistance R_{wDC}/l_w (Ω/m)	Turns per cm
55	0.0140	0.0147	0.0001539	112.47	
54	0.0157	0.0165	0.0001936	88.52	
53	0.0178	0.0183	0.0002489	69.45	
52	0.0198	0.0216	0.0003079	55.94	
51	0.0224	0.0241	0.0003941	43.93	
50	0.0254	0.0267	0.0004948	34.71	
49	0.0282	0.0297	0.0006245	27.615	
48	0.0315	0.0328	0.0007793	22.12	
47	0.0356	0.0368	0.0009953	17.36	
46	0.0399	0.04394	0.001246	13.8	
45	0.0447	0.05797	0.00157	11	
44	0.05023	0.0635	0.00202	8.5072	
43	0.05541	0.0685	0.002452	7.0308	
42	0.06334	0.0762	0.003166	5.4429	
41	0.07113	0.086	0.003972	4.3405	
40	0.07937	0.0965	0.004869	3.54	125
39	0.08969	0.109	0.006207	2.775	111
38	0.1007	0.124	0.008107	2.1266	99.3
37	0.1131	0.14	0.01026	1.6801	88.4
36	0.127	0.152	0.01266	1.3608	78.7
35	0.1426	0.17	0.01589	1.0849	70.1
34	0.1601	0.191	0.02011	0.85728	62.4
33	0.1798	0.216	0.02554	0.6748	55.6
32	0.2019	0.241	0.03242	0.53149	49.5
31	0.2268	0.267	0.04013	0.4294	44.1
30	0.2546	0.294	0.05067	0.34022	39.3
29	0.2859	0.33	0.06470	0.26643	35
28	0.321	0.366	0.08046	0.21427	31.1
27	0.3606	0.409	0.1021	0.1687	27.7
26	0.405	0.452	0.1288	0.1345	24.7
25	0.445	0.505	0.1623	0.1062	22
24	0.510	0.566	0.2047	0.08421	19.6
23	0.573	0.632	0.2588	0.0666	17.4
22	0.644	0.701	0.3243	0.05314	15.5
21	0.7229	0.785	0.4116	0.04189	13.8
20	0.812	0.879	0.5188	0.03323	12.3
19	0.912	0.98	0.6531	0.02639	11
18	1.02	1.09	0.8228	0.02095	9.77
17	1.15	1.22	1.039	0.01658	8.7
16	1.29	1.37	1.307	0.01318	7.75
15	1.45	1.53	1.651	0.0104	6.9
14	1.63	1.71	2.082	0.00828	6.14
13	1.83	1.9	2.626	0.00656	5.47
12	2.05	2.13	3.308	0.005209	4.87
11	2.31	2.38	4.168	0.004137	4.34
10	2.59	2.67	5.261	0.003277	3.86
9	2.906	2.9185	6.632	0.002599	3.44
8	3.264	3.2715	8.367	0.002061	3.06
7	3.665	3.6728	10.34555	0.001634	2.73
6	4.115	4.1199	13.3	0.001296	2.43
5	4.621	4.7504	16.77	0.001028	2.16
4	5.189	5.3253	21.15	0.0008152	1.93

(continued overleaf)

Table 10.1 *(continued)*

AWG no	Nominal bare wire diameter d_i (mm)	Nominal outer diameter d_o (mm)	Bare wire area A_w (mm^2)	DC resistance R_{wDC}/l_w (Ω/m)	Turns per cm
3	5.827	5.9705	26.67	0.0006465	1.72
2	6.544	6.69508	33.63	0.0005127	1.53
1	7.348	7.50705	42.41	0.0004066	1.36
0	8.2522	8.42847	53.49	0.0003224	1.21
00	9.266	9.45156	67.43	0.000256	1.08
000	10.404	10.59935	85.3	0.000203	0.891
0000	11.68	11.88965	107.21	0.000161	0.856

10.3 Wire Insulation

An insulation film is placed on the surface of the bare copper wire to avoid short circuits between turns and breakdown between the turns. The following grades of wire insulation are used:

- Single insulation
- Heavy insulation
- Triple insulation
- Quad insulation.

The nominal thickness of insulation is given by

$$t = 25.4 \times 10^{\left(x - \frac{AWG}{44.8}\right)} (\mu\text{m}) \tag{10.6}$$

and

$$t = 10^{\left(x - \frac{AWG}{44.8}\right)} (\text{mils}), \tag{10.7}$$

where $x = 0.518$ is for single insulation and $x = 0.818$ is for heavy insulation. Table 10.2 gives the outer diameter of insulated magnet round wire. The insulation film may be degraded by overheating.

Standard numbers of strands in a bundle of litz wire are 7, 19, 37, and so on. Fine strands are made with copper wire of AWG44-50.

10.4 Restrictions on Inductors

The inductors and transformers should satisfy the following categories of requirements:

- Electrical requirements
- Mechanical requirements
- Thermal requirements

The restrictions encountered in designing inductors are

1. $L = L_{required}$.
2. $B < B_s$.
3. $R_w < R_{w\max}$.
4. $A_w > A_{wmin}$.
5. $J < J_{max}$.

Table 10.2 Maximum diameter of insulated round copper wire d_o (mm)

AWG	Single insulation	Heavy insulation	Triple insulation	Quad insulation
44	0.061	0.069	0.076	0.081
43	0.066	0.074	0.084	0.089
42	0.076	0.081	0.094	0.097
41	0.084	0.091	0.102	0.109
40	0.094	0.102	0.112	0.119
39	0.104	0.114	0.127	0.135
38	0.119	0.130	0.142	0.152
37	0.132	0.145	0.157	0.170
36	0.147	0.160	0.175	0.188
35	0.163	0.178	0.193	0.208
34	0.183	0.198	0.213	0.231
33	0.206	0.224	0.241	0.259
32	0.231	0.249	0.267	0.287
31	0.254	0.274	0.292	0.315
30	0.284	0.305	0.323	0.348
29	0.320	0.340	0.358	0.386
28	0.353	0.376	0.394	0.422
27	0.396	0.419	0.437	0.470
26	0.422	0.470	0.488	0.523
25	0.495	0.523	0.544	0.579
24	0.554	0.582	0.602	0.6401
23	0.620	0.648	0.668	0.709
22	0.693	0.721	0.742	0.782
21	0.777	0.805	0.828	0.869
20	0.866	0.897	0.919	0.963
19	0.970	1.003	1.026	1.072
18	1.082	1.118	1.143	1.189
17	1.214	1.250	1.275	1.321
16	1.356	1.392	1.417	1.466
15	1.521	1.557	1.585	1.636
14	1.702	1.737	1.765	1.816
13	1.905	1.943	1.971	2.022
12	2.134	2.172	2.202	2.256
11	2.391	2.431	2.451	2.517
10	2.677	2.720	2.753	2.809

6. $T < T_{max}$.
7. $P_v < P_{vmax}$.
8. $P_w < P_{w\,max}$.
9. $P_{cw} < P_{cwmax}$.
10. Size, weight, profile, and surface area.
11. Winding, core, and gap power losses.
12. Self-capacitance.
13. Self-resonant frequency.
14. Skin and proximity effects.
15. Electromagnetic interference (EMI) effects.
16. AC inductor current only or both DC and AC inductor currents.
17. Mechanical requirements.

A core should be selected such that it suits the mechanical size requirements. Based on the practical experience, the maximum amplitude of the current density under natural cooling conditions is $J_{m(max)} = 5\text{A/mm}^2$. Prolonged operation above the maximum current density must be avoided. The typical maximum core power loss density is $P_{vmax} = 100\,\text{mW/cm}^3$ for natural convection cooling. The maximum amplitude of the AC magnetic flux for ferrite cores is usually $B_{m(max)} = 0.1\,\text{T}$ (1000 G) at the operating frequency $f = 100\,\text{kHz}$. As the frequency increases, the core loss density P_v increases. Therefore, the amplitude of the AC flux density B_m should be reduced. The maximum amplitude of the AC magnetic flux is usually $B_{m(max)} = 10 - 20\,\text{mT}$ at the operating frequency $f = 1\,\text{MHz}$.

The design process of an inductor includes the following steps:

- Magnetic core material selection
- Core shape and size selection
- Number of turns N
- Wire size selection
- Gap selection (if any).

10.5 Window Utilization Factor

10.5.1 Wire Insulation Factor

The wire insulation factor is defined as

$$K_i = \frac{\text{bare conductor cross-sectional area}}{\text{insulated wire conductor cross-sectional area}} = \frac{A_w}{A_{wo}}, \tag{10.8}$$

where A_w is the cross-sectional area of the bare wire and A_{wo} is the total cross-sectional area of the wire including insulation.

Figure 10.2 shows a cross section of a round magnet wire with insulation. The cross-sectional area of the bare wire is

$$A_w = \frac{\pi d_i^2}{4} \tag{10.9}$$

and the total wire cross-sectional area of magnet wire with insulation is

$$A_{wo} = \frac{\pi d_o^2}{4} = \frac{\pi (d_i + 2t)^2}{4}, \tag{10.10}$$

where d_i is the diameter of bare wire, d_o is the diameter of insulated wire, and t is the wire insulation thickness. Hence, the insulation factor is

$$K_i = \frac{A_w}{A_{wo}} = \left(\frac{d_i}{d_o}\right)^2 = \left(\frac{d_i}{d_i + 2t}\right)^2. \tag{10.11}$$

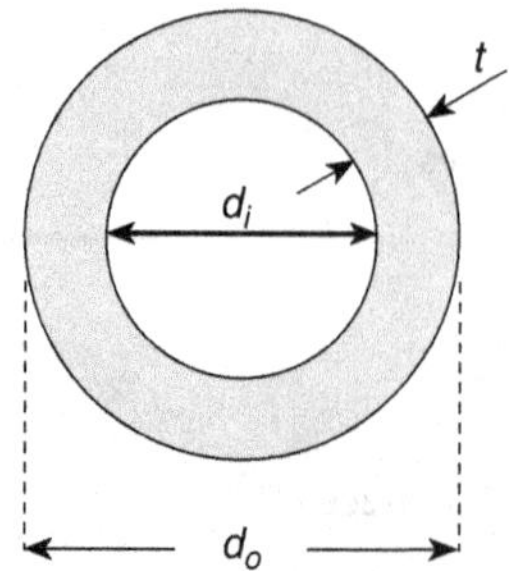

Figure 10.2 Cross section of magnet wire with insulation

Example 10.1

An AWG20 wire has the bare wire diameter $d_i = 0.812$ mm and the insulated wire diameter $d_o = 0.879$ mm with single insulation. Find the wire insulation factor K_i.

Solution: The insulation thickness is

$$t = \frac{d_o - d_i}{2} = \frac{0.879 - 0.812}{2} = 0.0335 \text{ mm}. \tag{10.12}$$

The wire insulation factor is

$$K_i = \left(\frac{d_i}{d_o}\right)^2 = \left(\frac{0.812}{0.879}\right)^2 = 0.8534. \tag{10.13}$$

10.5.2 Air and Wire Insulation Factor

The air and wire insulation factor is defined as

$$K_{ai} = \frac{\text{bare wire cross-sectional area}}{\text{total cell cross-sectional area}} = \frac{A_w}{A_{cell}}, \tag{10.14}$$

where A_{cell} is the cell cross-sectional area. The factor K_{ai} depends on the winding pattern, wire tension, and wire quality.

Triangular Pattern

Figure 10.3 shows a cros section of a basic cell of a winding, which is made of a round wire with a triangular pattern. The height of a single cell is

$$h = \frac{d_o}{2}\tan 60^\circ = \frac{(d_i + 2t)}{2}\tan 60^\circ = \frac{\sqrt{3}}{2}(d_i + 2t) \approx 0.866(d_i + 2t). \tag{10.15}$$

The cell cross-sectional area is

$$A_{cell} = \frac{hd_o}{2} = \frac{h(d_i + 2t)}{2} = \frac{\sqrt{3}}{4}(d_i + 2t)^2. \tag{10.16}$$

The cross-sectional area of the conductor in the single cell is equal to one-half of the cross-sectional area of the bare wire

$$A_w = \frac{1}{2} \times \frac{\pi d_i^2}{4} = \frac{\pi d_i^2}{8}. \tag{10.17}$$

Hence, the air and wire insulation factor for the triangular pattern is

$$K_{ai} = \frac{A_w}{A_{cell}} = \frac{\pi}{2\sqrt{3}}\left(\frac{d_i}{d_o}\right)^2 = \frac{\pi}{2\sqrt{3}}\left(\frac{d_i}{d_i + 2t}\right)^2 = \frac{\pi}{2\sqrt{3}}\left(\frac{d_i}{d_o}\right)^2 \approx 0.9069\left(\frac{d_i}{d_o}\right)^2. \tag{10.18}$$

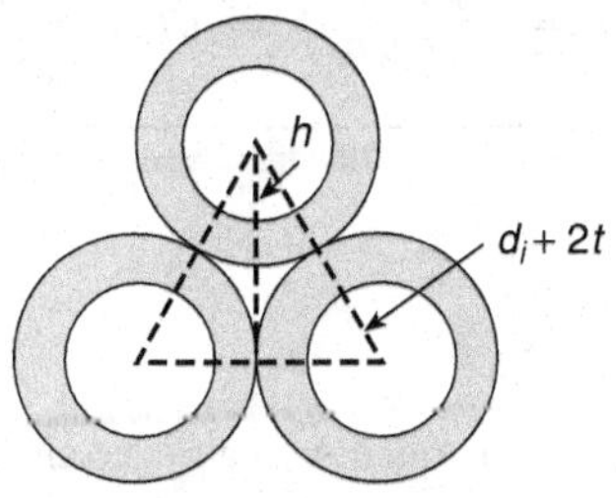

Figure 10.3 Cross section of basic cell of winding with triangular pattern

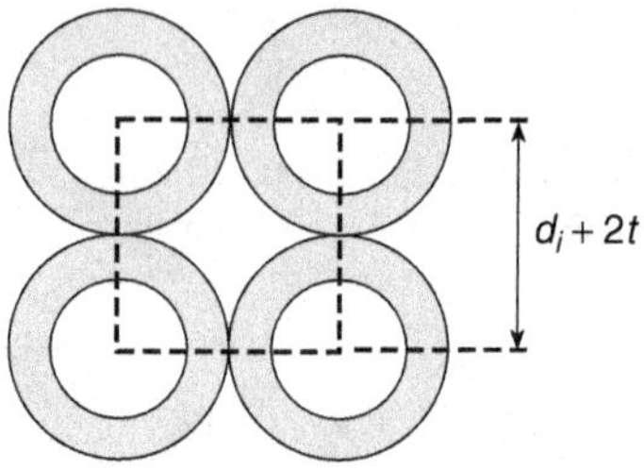

Figure 10.4 Cross section of basic cell of winding with square pattern

Square Pattern

Figure 10.4 shows a cross section of a basic cell of a winding made of a round wire with a square pattern. The cross-sectional area of a single cell is

$$A_{cell} = d_o^2 = (d_i + 2t)^2 \tag{10.19}$$

and the cross-sectional area of the conductor is

$$A_w = \frac{\pi d_i^2}{4}. \tag{10.20}$$

Thus, the air and wire insulation the square pattern is

$$K_{ai} = \frac{A_w}{A_{cell}} = \frac{\pi}{4}\left(\frac{d_i}{d_o}\right)^2 = \frac{\pi}{4}\left(\frac{d_i}{d_i + 2t}\right)^2 = \frac{\pi}{4}\left(\frac{d_i}{d_o}\right)^2 = \frac{\pi}{4}\left(\frac{d_i}{d_o}\right)^2 \approx 0.7854\left(\frac{d_i}{d_o}\right)^2. \tag{10.21}$$

Example 10.2

An AWG20 wire has $d_i = 0.812$ mm and $d_o = 0.879$ mm. Find the air and insulation factor K_{ai} for triangular and a square winding patterns.

Solution: The insulation thickness is

$$t = \frac{d_o - d_i}{2} = \frac{0.879 - 0.812}{2} = 0.0335 \text{ mm}. \tag{10.22}$$

The air and wire insulation factor for a triangle pattern is

$$K_{ai} = \frac{\pi}{2\sqrt{3}}\left(\frac{d_i}{d_o}\right)^2 = \frac{\pi}{2\sqrt{3}}\left(\frac{0.812}{0.879}\right)^2 = 0.9069 \times 0.8534 = 0.7739. \tag{10.23}$$

The air and wire insulation factor for a square pattern is

$$K_{ai} = \frac{\pi}{4}\left(\frac{d_i}{d_o}\right)^2 = \frac{\pi}{4}\left(\frac{0.812}{0.879}\right)^2 = 0.7854 \times 0.8534 = 0.6078. \tag{10.24}$$

10.5.3 Air Factor

The area between insulated round wire is filled with air. The air factor is defined as

$$K_a = \frac{\text{insulated wire cross} - \text{sectional area}}{\text{insulated wire area} + \text{air area}} = \frac{A_{wo}}{A_{cell}} = \frac{A_{wo}}{A_{wo} + A_{air}}. \tag{10.25}$$

For the triangle winding pattern, the air factor is

$$K_a = \frac{A_{wo}}{A_{wo} + A_{air}} = \frac{K_{ai}}{K_i} = \frac{\pi}{2\sqrt{3}} \approx 0.9069. \tag{10.26}$$

For the square winding pattern, the air factor is

$$K_a = \frac{A_{wo}}{A_{wo} + A_{air}} = \frac{K_{ai}}{K_i} = \frac{\pi}{4} \approx 0.7854. \tag{10.27}$$

An alternative derivation of K_a for the square pattern is as follows. Figure 10.5 shows a basic cell of a square winding pattern. The cell cross-sectional area of this cell (including copper, insulation, and air areas) is

$$A_{cell} = d_o^2 \tag{10.28}$$

and the cross-sectional area of the insulated wire is

$$A_{wo} = \frac{\pi d_o^2}{4}, \tag{10.29}$$

resulting in the air factor

$$K_a = \frac{A_{wo}}{A_{cell}} = \frac{\pi}{4}. \tag{10.30}$$

10.5.4 Bobbin Factor

Figure 10.6 shows a cross section of a bobbin (or a coil former). The bobbin factor (if any) is defined as

$$K_b = \frac{\text{bobbin window}}{\text{core window area}} = \frac{W_a - W_b}{W_a} = 1 - \frac{W_b}{W_a}. \tag{10.31}$$

where W_b is the bobbin area.

For a pot core (PC), the core window area is

$$W_a = \frac{\pi(D_o^2 - D_i^2)}{4}, \tag{10.32}$$

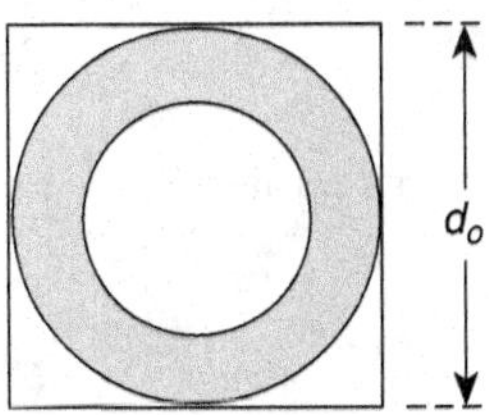

Figure 10.5 Alternative presentation of cross section of basic cell of winding with square pattern

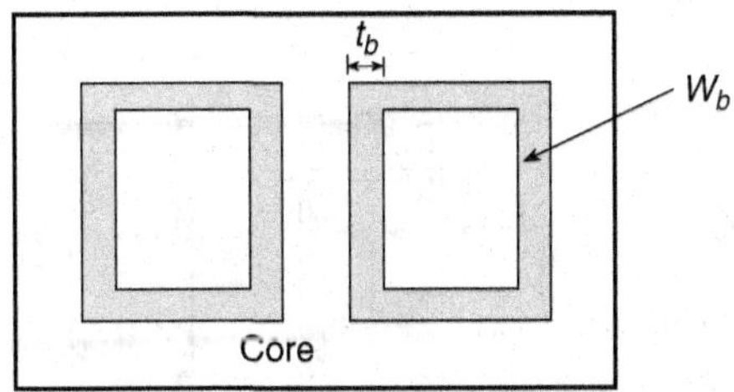

Figure 10.6 Cross section of a bobbin

where D_i is the core inner diameter and D_o is the core outer diameter designed for the bobbin. The bobbin cross-sectional area is

$$W_b = \pi D_i t_b, \tag{10.33}$$

where t_b is the thickness of the bobbin wall. Thus, the bobbin factor is

$$K_b = 1 - \frac{W_b}{W_a} = 1 - \frac{4D_i t_b}{D_o^2 - D_i^2}. \tag{10.34}$$

The typical values of the bobbin factor are from 0.55 to 0.75.

Example 10.3

Calculate the bobbin factor K_b for the Magnetics bobbin 00B362201, which is designed for the Magnetics PC core 43622.

Solution: Figure 10.7 shows a cross section of the bobbin for a PC core. The area of the bobbin 00B362201 is $W_b = 22.677\,\text{mm}^2$. The window area of the PC core 43622 is $W_a = 52.54\,\text{mm}^2$. The bobbin factor is

$$K_b = 1 - \frac{W_b}{W_a} = 1 - \frac{22.677}{52.54} = 0.568. \tag{10.35}$$

10.5.5 Edge Factor

The bobbin is not fully filled with a winding at the edges. The edge factor is

$$K_{ed} = \frac{\text{usable winding area}}{\text{window area}}. \tag{10.36}$$

Typically, $K_{ed} = 0.75$.

10.5.6 Number of Turns

For the square pattern, the number of turns per layer is

$$N_{tl} = \frac{b_b}{d_o} \tag{10.37}$$

and the number of layers is

$$N_l = \frac{h_b}{d_o}, \tag{10.38}$$

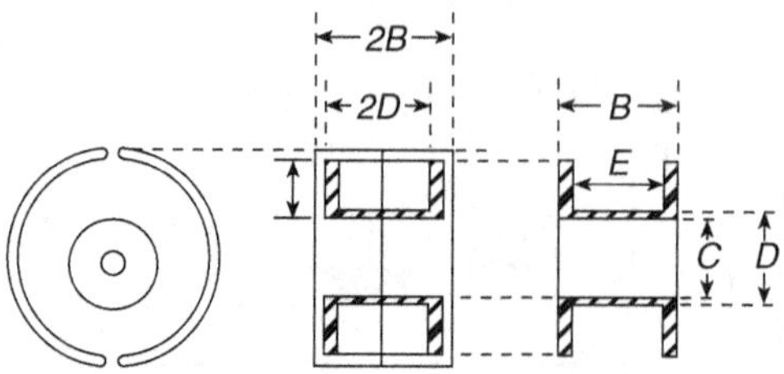

Figure 10.7 Cross section of a bobbin for a PC core

yielding the number of turns

$$N = N_l N_{tl} = \frac{b_b h_b}{d_o^2} = \frac{W_b}{d_o^2}, \tag{10.39}$$

where $W_b = b_b h_b$ is the window cross-sectional area, b_b is the bobbin width, and h_b is the bobbin height. For the square winding pattern of round wire, the cross-sectional area of the copper is

$$A_{Cu} = NA_w = \frac{\pi d_i^2 N}{4}, \tag{10.40}$$

and the core window area is

$$W_b = Nd_o^2, \tag{10.41}$$

yielding the air and wire insulation factor

$$K_{ai} = \frac{A_{Cu}}{W_b} = \frac{\pi}{4}\left(\frac{d_i}{d_o}\right)^2 \approx 0.785\left(\frac{d_i}{d_o}\right)^2. \tag{10.42}$$

For the hexagonal pattern, the number of turns per layer is

$$N_{tl} \approx \frac{b_b}{d_o} \tag{10.43}$$

and the number of layers is

$$N_l = \frac{h_b}{\frac{\sqrt{3}}{2} d_o}, \tag{10.44}$$

producing the number of turns

$$N = N_l N_{tl} = \frac{2}{\sqrt{3}} \frac{b_b h_b}{d_o^2} = \frac{2}{\sqrt{3}} \frac{W_a}{d_o^2}. \tag{10.45}$$

Hence, the air and wire insulation factor is

$$K_{ai} = \frac{A_{Cu}}{W_b} = \frac{2}{\sqrt{3}}\left(\frac{d_i}{d_o}\right)^2 \approx 0.9069\left(\frac{d_i}{d_o}\right)^2. \tag{10.46}$$

10.5.7 Window Utilization Factor

The *core window utilization factor* is defined as

$$K_u = \frac{\text{bare conductor area}}{\text{core window area}} = \frac{A_{Cu}}{W_a} = \frac{NA_w}{W_a} = K_{ai} K_b K_{ed}, \tag{10.47}$$

where $A_{Cu} = NA_w$ is the bare copper cross-sectional area.

If the winding wire consists of S_n strands, the core window utilization factor is given by

$$K_u = \frac{A_{Cu}}{W_a} = \frac{NS_n A_{wso}}{W_a}, \tag{10.48}$$

where A_{wso} is the cross-sectional area of each strand.

Figure 10.8 illustrates the core window area for an EE core and Fig. 10.9 for a toroidal core. The core window area W_a is related to the bare conductor (copper) area, and hence the current capability of an inductor or a transformer. The typical value of the window utilization factor K_u is 0.4 for inductors and it is 0.3 for transformers. For toroidal cores, the typical value of the window utilization is $K_u = 0.15–0.25$. Only about 40% of the winding space is filled with copper for inductors and only 35% of the winding space is filled with copper for transformers.

In transformers, safety requirements impose a minimum distance between windings for creepage and a minimum insulation thickness, which occupy a significant percentage of window area. Triple-insulated wire can be used to satisfy the minimum insulation thickness requirement and eliminate the creepage distance.

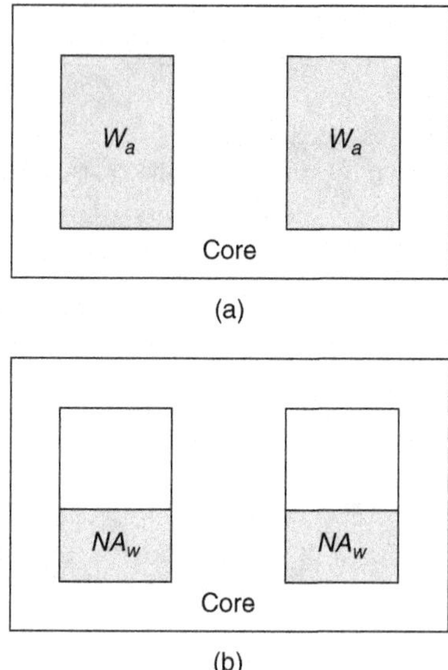

Figure 10.8 Window of EE core. (a) Window cross-sectional area W_a. (b) Bare conductor (copper) area $A_{Cu} = NA_w$ illustrating core window utilization factor $K_u = NA_w/W_a$

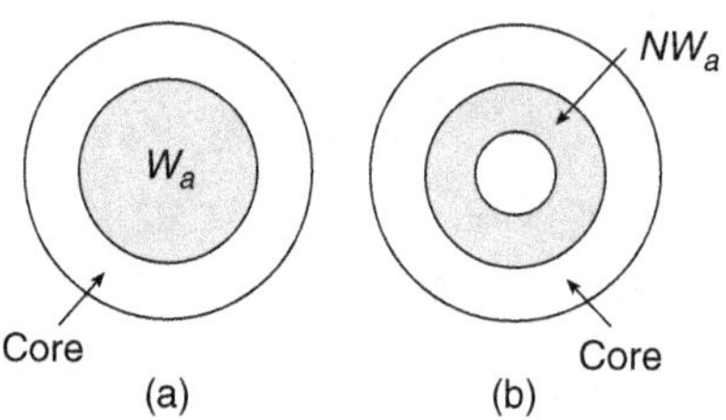

Figure 10.9 Window of toroidal core. (a) Window cross-sectional area W_a. (b) Window utilization $K_u = NA_w/W_a$

Example 10.4

The bobbin factor is $K_b = 0.75$ and the edge factor is $K_{ed} = 0.75$. Find the core window utilization factor K_u. (a) The fill and wire insulation factor for the triangular winding pattern is $K_{ai} = 0.7867$. (b) The fill and wire insulation factor for the square winding pattern is $K_{ai} = 0.6813$.

Solution: The product of the bobbin factor and the edge factor is

$$K_b K_{ed} = 0.75 \times 0.75 = 0.5625. \tag{10.49}$$

(a) The core window utilization factor for the triangular winding pattern is

$$K_u = K_{ai} K_b K_{ed} = 0.7828 \times 0.75 \times 0.75 = 0.4403. \tag{10.50}$$

(b) The core window utilization factor for the square winding pattern is

$$K_u = K_{ai} K_b K_{ed} = 0.6779 \times 0.75 \times 0.75 = 0.3813. \tag{10.51}$$

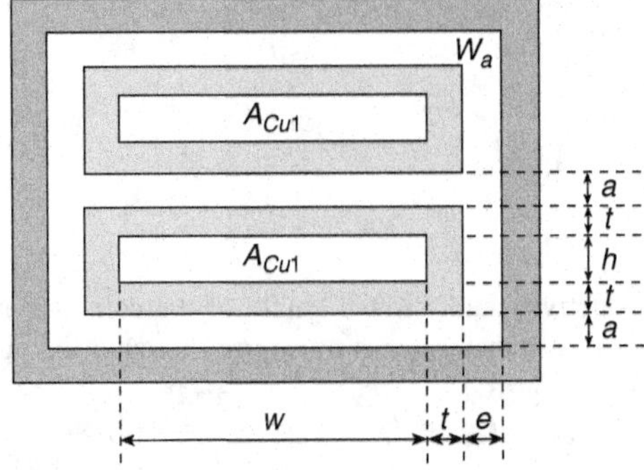

Figure 10.10 Window utilization for inductors with foil winding

10.5.8 Window Utilization Factor for Foil Winding

Consider a foil winding inductor that contains a single foil turn per layer, as shown in Fig. 10.10. The cross-sectional area of a single bare conductor is

$$A_{Cu1} = wh \tag{10.52}$$

where w is the width of the bare foil and h is the height of the bare foil. The cross-sectional area of the single layer, which consists of a bare conductor, its insulation, and the air area between the layers and at the edges, is

$$A_{Layer} = (w + 2t + 2e)(h + a + 2t) \tag{10.53}$$

where a is the height of the air between the layers, t is the insulation thickness, and e is width of the air area at each foil edge. Hence, the window utilization factor is

$$K_u = \frac{A_{Cu}}{W_a} = \frac{N_l A_{Cu1}}{N_l A_{Layer}} = \frac{A_{Cu1}}{A_{Layer}} = \frac{wh}{(w + 2t + 2e)(h + a + 2t)} \tag{10.54}$$

where N_l is the number of layers.

For example, for $w = 2$ cm, $e = 0.3$ mm, $h = 0.2794$ mm, $t = 0.03048$ mm, and $a = 2t = 0.060096$ mm, the window utilization factor is

$$\begin{aligned} K_u &= \frac{wh}{(w + 2t + 2\Delta w)(h + a + 2t)} \\ &= \frac{20 \times 0.2794}{(20 + 2 \times 0.03048 + 2 \times 0.3)(0.2794 + 0.060096 + 2 \times 0.03048)} = 0.675. \end{aligned} \tag{10.55}$$

If a bobbin is used, then the above result should be multiplied by the bobbin factor K_b.

10.6 Temperature Rise of Inductors

10.6.1 Expression for Temperature Rise of Inductors

Thermal management is an important concept of magnetic devices. Cooling of an inductor or a transformer is nearly proportional to its heat radiating surface area A_t. The surface power loss density of an inductor is given by

$$\psi = \frac{P_{cw}}{A_t}, \tag{10.56}$$

where A_t is the outer heat radiating surface area of the inductor and P_{cw} is the total power loss equal to the sum of the core loss P_C and the winding loss P_w

$$P_{cw} = P_C + P_w. \tag{10.57}$$

The thermal resistance is given by

$$R_{Th} = k\frac{l}{A_t}, \tag{10.58}$$

where k is the thermal conductivity and l is the length of the core. For steady state, all the energy generated inside an inductor or a transformer is transferred through the outer surface A_t

$$P_{cw} = P_{out} = hA_t\Delta T = \frac{\Delta T}{R_{th}} = \frac{k}{l}A_t\Delta T, \tag{10.59}$$

where h is a constant. At 45% of conduction and 55% of radiation with the emissivity equal to 95% in free air around, the temperature rise of an inductor for steady-state operation is given by [1]

$$\Delta T = 450\psi^{0.826} = 450\left(\frac{P_{cw}}{A_t}\right)^{0.826} = 450\left(\frac{P_C + P_w}{A_t}\right)^{0.826} \ (^\circ\text{C}), \tag{10.60}$$

where the surface power loss density ψ is in watts per square centimeter. Thus, the temperature rise ΔT is almost inversely proportional to the heat radiating surface area A_t. Core losses are directly proportional to the core volume V_c. As the core becomes larger, its volume V_c increases faster than the heat radiating surface area A_t. Therefore, the temperature of a larger core rises more rapidly than its surface area A_t which cools it.

For example, the volume of a round toroidal (torus) core is

$$V_c = (2\pi R)(\pi r^2) = 2\pi^2 r^2 R \tag{10.61}$$

and its surface area is

$$A_t = (2\pi R)(2\pi r) = 4\pi^2 rR, \tag{10.62}$$

resulting in

$$\frac{V_c}{A_t} = \frac{r}{2}, \tag{10.63}$$

where r is the radius of the ferrite tube and R is the distance from the center of the torus to the center of the tube.

10.6.2 Surface Area of Inductors with Toroidal Core

Figure 10.11 shows a cross section of an inductor with a toroidal core. The height of an inductor with a single-layer winding is

$$H_w = H + 2d_o \tag{10.64}$$

and the outer diameter of the inductor is

$$D_{out} = D_o + 2d_o, \tag{10.65}$$

yielding the area of the side surface

$$A_{outer} = \pi D_{out} H_w = \pi(D_o + 2d_o)(H + 2d_o). \tag{10.66}$$

The inner diameter of the inductor is

$$D_{in} = D_i - 2d_o, \tag{10.67}$$

resulting in the area of the inner surface of the inductor

$$A_{inner} = \pi D_{in} H_w = \pi(D_i - 2d_o)(H + 2d_o). \tag{10.68}$$

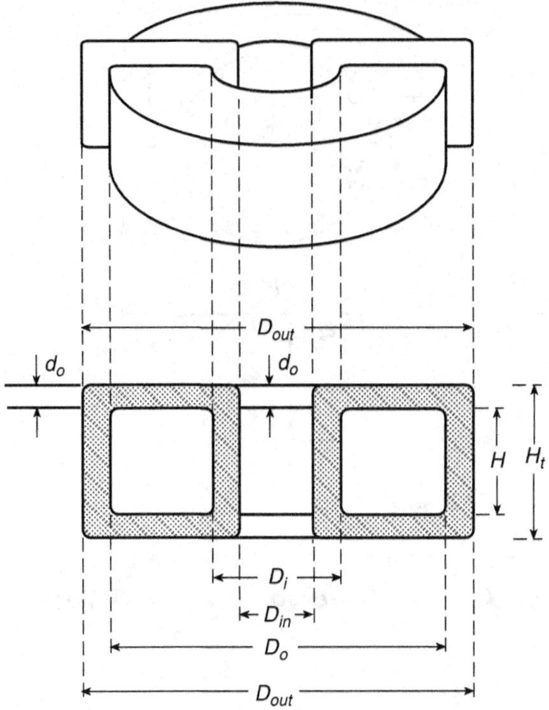

Figure 10.11 Cross section of an inductor with a toroidal core

Hence, the area of the inner and outer sides is

$$A_{io} = A_{inner} + A_{outer} = \pi(H + 2d_o)(D_i + D_o). \tag{10.69}$$

The area of the top and bottom surfaces is

$$A_{tb} = 2\left[\frac{\pi(D_o + 2d_o)^2}{4} - \frac{\pi(D_i - 2d_o)^2}{4}\right] = \frac{\pi}{2}(D_o + D_i)(D_o - D_i + 4d_o). \tag{10.70}$$

Hence, the total surface area of an inductor with toroidal core and single-layer winding is

$$A_t = A_{io} + A_{tb} = \pi(D_o + D_i)\left(\frac{D_o - D_i}{2} + H + 4d_o\right). \tag{10.71}$$

10.6.3 Surface Area of Inductors with Pot Core

Figure 10.12 shows a cross section of a (PC). The area of the top and bottom surfaces is

$$A_{tb} = 2\left(\frac{\pi A^2}{4}\right) = \frac{\pi A^2}{2} \tag{10.72}$$

and the area of the side surface is

$$A_{side} = \pi A(2B) = 2\pi AB. \tag{10.73}$$

Hence, the total area of an inductor with (PC) is

$$A_t = \frac{\pi A^2}{2} + 2\pi AB. \tag{10.74}$$

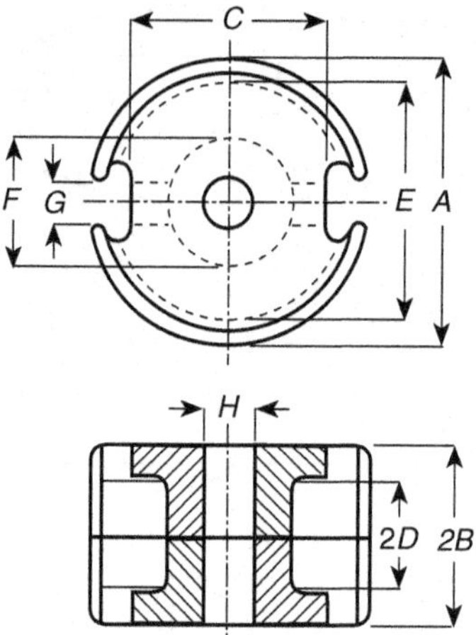

Figure 10.12 Cross section of pot core

10.6.4 Surface Area of Inductors with PQ Core

Figure 10.13 shows a cross section of PQ cores. The area of the left and right side surfaces is

$$A_{lr} = 2(2BC). \tag{10.75}$$

The area of the front and back surfaces is

$$A_{fb} = 2(2BA). \tag{10.76}$$

The area of the top and bottom surfaces is

$$A_{tb} = 2(AC). \tag{10.77}$$

Hence, the total area of an inductor with (PC) is given by

$$A_t = A_{lr} + A_{fb} + A_{tb} = 2(2BC + 2AB + AC). \tag{10.78}$$

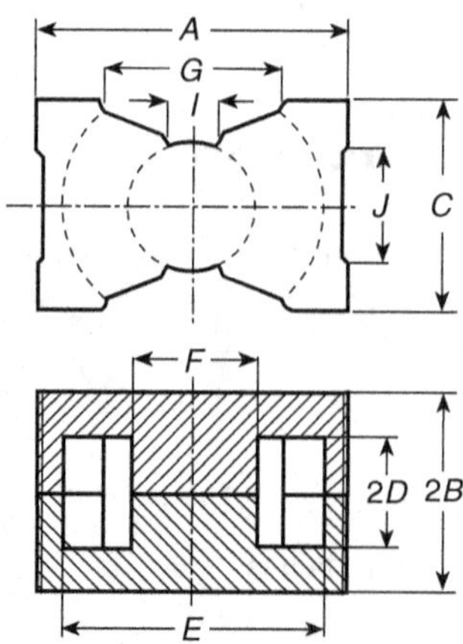

Figure 10.13 Cross section of power quality (PQ) core

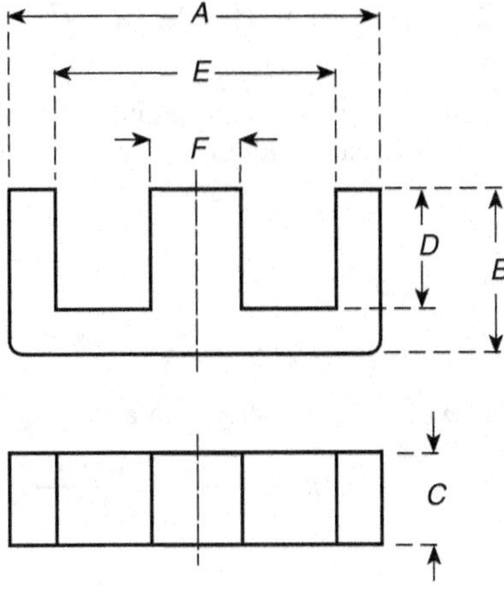

Figure 10.14 Cross section of EE core

10.6.5 Surface Area of Inductors with EE Core

A cross section of an EE core is shown in Fig. 10.14. For an inductor with EE core, we obtain the area of the front and back surfaces of the core

$$A_{fb} = 2(2BA) = 4AB, \tag{10.79}$$

the area of the top and bottom surfaces

$$A_{tb} = 2AC, \tag{10.80}$$

the area of the left and right surfaces of the core,

$$A_{lr} = 2(2BC) = 4BC \tag{10.81}$$

and the left and right surfaces as well as the top and bottom surfaces of extended part of the windings

$$A_{wlr} = 8D\left(\frac{E-F}{2}\right) + 4E\left(\frac{E-F}{2}\right) = 2(E-F)(2D+E). \tag{10.82}$$

Hence, the total surface area of an inductor with EE core is

$$A_t = A_{fb} + A_{tb} + A_{lr} + A_{wlr} = 4AB + 2AC + 4BC + 2(E-F)(2D+E). \tag{10.83}$$

10.7 Mean Turn Length of Inductors

10.7.1 Mean Turn Length of Inductors with Toroidal Cores

Figure 10.11 shows a cross-sectional view of an inductor with a toroidal core. The mean length per turn (MLT) or the mean turn length (MTL) of an inductor with this core is

$$l_T = 2(D_o - D_i) + 2H + 2d. \tag{10.84}$$

Assuming that the turns are tightly wound, the length of the winding wire of an inductor with a toroidal core is

$$l_w = Nl_T = 2N(D_o - D_i + H + d). \tag{10.85}$$

10.7.2 Mean Turn Length of Inductors with PC and PQ Cores

Cross-sectional views of the PC and PQ core are depicted in Figs 10.12 and 10.13, respectively. The mean diameter of the single turn of inductors with these cores is given by

$$D_m = \frac{E+F}{2}. \tag{10.86}$$

The MTL of an inductor with the two cores is

$$l_T = \pi D_m = \frac{\pi(E+F)}{2}. \tag{10.87}$$

Hence, the length of the winding wire of an inductor with a PC, PQ core, or an EE core is

$$l_w = N l_T = \pi D_m N = \frac{\pi N(E+F)}{2}. \tag{10.88}$$

10.7.3 Mean Turn Length of Inductors with EE Cores

Figure 10.14 shows a cross-sectional view of an EE core. The minimum turn length is

$$l_{Tmin} = 2F + 2C \tag{10.89}$$

and the maximum turn length is

$$l_{Tmax} = 2E + 2(C + E - F) = 4E + 2C - 2F, \tag{10.90}$$

resulting the MTL

$$l_T = \frac{l_{Tmin} + l_{Tmax}}{2} = 2(C+E). \tag{10.91}$$

Hence, the length of the winding wire is

$$l_w = N l_T = 2N(C+E). \tag{10.92}$$

10.8 Area Product Method

10.8.1 General Expression for Area Product

Figure 10.15 illustrates the definition of the core area product A_p. It shows the core winding window area W_a and the core cross-sectional area A_c. The area product of the core is defined as

$$A_p = W_a A_c. \tag{10.93}$$

The flux linkage is given by

$$\lambda(t) = N\phi(t) = NA_cB(t) = Li_L(t). \tag{10.94}$$

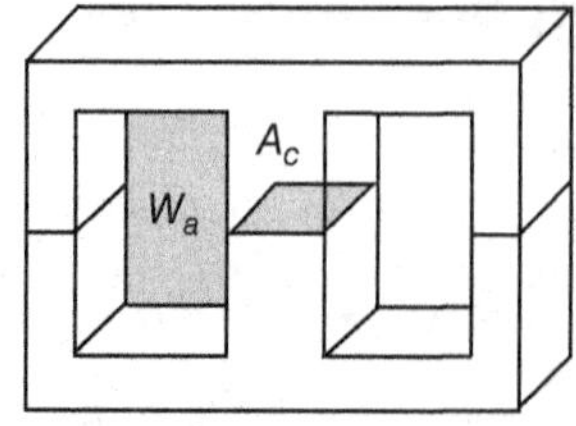

Figure 10.15 Definition of core area product $A_p = W_aA_c$ for EE core, where W_a is the core winding window area and A_c is the core cross-sectional area

For the peak values, this equation becomes

$$\lambda_{pk} = LI_{Lmax} = N\phi_{pk} = NA_cB_{pk}. \tag{10.95}$$

The cross-sectional area of the winding conductor A_w is determined by the DC and low-frequency current density

$$A_w = \frac{I_{Lmax}}{J_m} = \frac{I_{Lrms}}{J_{rms}}. \tag{10.96}$$

The number of turns N is limited by core window area W_a, cross-sectional area of the winding wire A_w, and core window utilization factor K_u

$$N = \frac{K_uW_a}{A_w} = \frac{K_uW_aJ_m}{I_{Lmax}} = \frac{K_uW_aJ_{rms}}{I_{Lrms}}. \tag{10.97}$$

The maximum energy stored in the magnetic field (in the air gap) of the inductor is

$$W_m = \frac{1}{2}LI_{Lmax}^2. \tag{10.98}$$

Substitution of (10.97) in (10.95) gives

$$LI_{Lmax} = NA_cB_{pk} = \frac{K_uW_aA_cB_{pk}}{A_w} = \frac{K_uW_aA_cJ_mB_{pk}}{I_{Lmax}} = \frac{K_uW_aA_cJ_{rms}B_{pk}}{I_{Lrms}}. \tag{10.99}$$

The *area product* of a core is defined as

$$A_p = W_aA_c = \frac{LI_{Lmax}^2}{K_uJ_mB_{pk}} = \frac{LI_{Lmax}I_{Lrms}}{K_uJ_{rms}B_{pk}} = \frac{2W_m}{K_uJ_mB_{pk}}. \tag{10.100}$$

The area product A_p gives a rough initial estimate of the core size. It is also called the *ferrite–copper area product*. The value of A_p is inversely proportional to the peak value of the flux density B_{pk} and the peak value of the current density J_m. The peak value of the flux density is limited either by core saturation or by core losses (i.e., the core maximum temperature rise). Different cores may have the same winding window area W_a, but their widths and heights may differ. Magnetic cores should have a wide and low window to minimize the proximity effect, thus reducing the winding AC loss due to lower number of layers N_l.

10.8.2 Area Product for Sinusoidal Inductor Voltage

Assume that the inductor current and voltage waveforms are sinusoidal. The DC component of the inductor current is zero. From Chapter 1, it is clear that the amplitude of a sinusoidal voltage across an inductor is given by

$$V_{Lm} = \omega NA_cB_m, \tag{10.101}$$

where B_m is the amplitude of the flux density of the core. The minimum cross-sectional area of the core for a sinusoidal inductor voltage is

$$A_c = \frac{V_{Lm}}{\omega NB_m}. \tag{10.102}$$

The definition of the inductance is

$$L = \frac{\lambda_m}{I_{Lm}} = \frac{N\phi_m}{I_{Lm}} = \frac{NA_cB_m}{I_{Lm}}, \tag{10.103}$$

resulting in the minimum required cross-sectional area of the core

$$A_c = \frac{LI_{Lm}}{NB_m}. \tag{10.104}$$

The core cross-sectional area A_c is related to the magnetic-flux-conduction capability of an inductor or a transformer.

The wire diameter is determined by the DC and low-frequency current density. The current density at DC and low frequencies is uniform. The amplitude of the current density of the winding conductor is

$$J_m = \frac{I_{Lm}}{A_w}, \tag{10.105}$$

where A_w is the cross-sectional area of the bare winding wire. For a round wire of conductor diameter d_i, the cross-sectional area of the conductor is

$$A_w = \frac{I_{Lm}}{J_m} = \frac{I_{Lrms}}{J_{rms}} = \frac{\pi d_i^2}{4}. \tag{10.106}$$

The typical values of the maximum rms uniform current density are in the range $J_{rms(\max)} =$ 6–10 A/mm^2 when the wire is short ($l_w < 1$ m) with a small number of turns N. The typical value of the maximum rms current density is $J_{rms(\max)} = 5$ A/mm^2 when the wire is long ($l_w > 1$ m) with a large number of turns N.

A core must provide sufficient space for a winding in the inductor or all windings in a transformer. The area of the opening available for a winding is called the *window area*. The diameter of the wire (or thickness of the foil) must be large enough to carry the maximum specified current density J_m. The required cross-sectional area of the conductor (copper) is given by

$$A_{Cu} = NA_w = \frac{NI_{Lm}}{J_m} = N\frac{\pi d_i^2}{4}. \tag{10.107}$$

The cross-sectional area of a core window must provide enough space for bare conductor, wire insulation, air space between turns, bobbin, insulation between layers, and shield. The *core window utilization factor* is the ratio of the total conductor (copper) cross-sectional area NA_w to the total window area W_a

$$K_u = \frac{A_{Cu}}{W_a} = \frac{NA_w}{W_a} = \frac{NI_{Lm}}{W_a J_m}, \tag{10.108}$$

resulting in the core window area

$$W_a = \frac{NA_w}{K_u} = \frac{NI_{Lm}}{K_u J_m}. \tag{10.109}$$

The product of (10.104) and (10.109) yields the *core area product*

$$A_p = W_a A_c = \frac{LI_{Lm}^2}{K_u J_m B_m} = \frac{2W_m}{K_u J_m B_m} = \frac{V_{Lm} I_{Lm}}{K_u \omega J_m B_m} = \frac{2P_Q}{K_u f J_m B_m} \ (\mathrm{m}^4), \tag{10.110}$$

where the energy stored in the inductor is

$$W_m = \frac{1}{2} LI_{Lm}^2 \tag{10.111}$$

and the apparent power of the inductor is

$$P_Q = I_{Lrms} V_{Lrms} = \frac{1}{2} I_{Lm} V_{Lm} = \frac{W_m}{T} = f W_m = \frac{\omega W_m}{2\pi}. \tag{10.112}$$

The area product A_p is quoted by many manufacturers or can be calculated from the core dimensions. The left-hand side of the expression for A_p contains the geometrical core parameters, and the right-hand side of the expressions includes the inductor magnetic and electric parameters. The core cross-sectional area A_c is related to the magnetic flux conduction capability. The window area W_a is related to the current-conduction capability.

Different cores may have the same value of area product A_p. However, there are many combinations of W_a and A_c, which have the same area product A_p. For example, core 1 has $W_a = 2$ cm^2 and $A_c = 1.5$ cm^2, and core 2 has $W_a = 1.5$ cm^2 and $A_c = 2$ cm^2. The area product for both these cores is $A_p = W_a A_c = 3$ cm^2. Magnetic cores should have a wide and low window to minimize the AC winding loss using a low number of layers N_l.

The loaded quality factor of the resonant circuit is defined as

$$Q_L = 2\pi \frac{W_m}{P_O T} = 2\pi f \frac{W_m}{P_O} = \omega \frac{W_m}{P_O}. \tag{10.113}$$

For inductors used in resonant circuits, the magnetic energy stored in the inductor is given by

$$W_m = \frac{Q_L P_O}{\omega}, \tag{10.114}$$

where Q_L is the loaded quality factor and P_O is the output power of radio frequency (RF) power amplifiers. Thus, the area product for resonant inductors is given by

$$A_p = W_a A_c = \frac{2W_m}{K_u J_m B_m} = \frac{2Q_L P_O}{K_u \omega J_m B_m} = \frac{Q_L P_O}{\pi K_u f J_m B_m}. \tag{10.115}$$

The core area product A_p is independent of the core material properties.

The number of turns of an inductor with a magnetic core is given by

$$N = \sqrt{\frac{L}{A_L}}. \tag{10.116}$$

Hence, the flux density is

$$B_m = \frac{LI_{Lm}}{NA_c} = \frac{LI_{Lm}}{A_c}\sqrt{\frac{A_L}{L}} = \frac{I_{Lm}}{A_c}\sqrt{A_L L}. \tag{10.117}$$

The core power loss density is

$$P_v = kf^a B_m^b = kf^a \left(\frac{I_{Lm}}{A_c}\sqrt{LA_L}\right)^b = kf^a \left(\frac{I_{Lm} l_c}{V_c}\sqrt{LA_L}\right)^b. \tag{10.118}$$

The core power loss is

$$P_C = P_v V_c = kf^a \left(\frac{I_{Lm}}{A_c}\sqrt{LA_L}\right)^b V_c = kf^a \left(\frac{I_{Lm} l_c}{V_c}\sqrt{LA_L}\right)^b V_c$$

$$= kf^a (I_{Lm} l_c \sqrt{LA_L})^b V_c^{1-b}. \tag{10.119}$$

For an inductor with a toroidal core, the number of turns is

$$N = \sqrt{\frac{l_c L}{\mu_{rc}\mu_0 A_c}} \tag{10.120}$$

and the flux density is

$$B_m = \frac{LI_{Lm}}{A_c}\sqrt{\frac{\mu_{rc}\mu_0 A_c}{l_c L}} = I_{Lm}\sqrt{\frac{\mu_{rc}\mu_0 L}{l_c A_c}} = I_{Lm}\sqrt{\frac{\mu_{rc}\mu_0 L}{V_c}}. \tag{10.121}$$

The core power loss density for an inductor with a toroidal core is

$$P_v = kf^a B_m^b = kf^a \left(I_{Lm}\sqrt{\frac{\mu_{rc}\mu_0 L}{V_c}}\right)^b. \tag{10.122}$$

The core power loss is

$$P_C = P_v V_c = kf^a \left(I_{Lm}\sqrt{\frac{\mu_{rc}\mu_0 L}{V_c}}\right)^b V_c = kf^a (I_{Lm}\sqrt{\mu_{rc}\mu_0 L})^b V_c^{1-\frac{b}{2}}. \tag{10.123}$$

Since $b > 2$, the core power loss P_C slightly decreases as the core volume V_c increases. For ferrite cores, $b \approx 2.5$ and P_C is proportional to $1/\sqrt[4]{V_c}$. When the frequency f increases, the power loss density P_v also increases. Therefore, the amplitude of the flux density B_m should be reduced when the frequency is increased. For example, for the Micrometals iron powder material-2, the power loss density is

$$P_v = kf^a B_m^b = 8.86 \times 10^{-10} f^{1.14} (10B_m)^{2.19} \quad (\text{mW/cm}^3) \quad \text{for} \quad 400 \text{ kHz} \le f \le 1 \text{ MHz}, \tag{10.124}$$

where f is in Hertz and B_m in milliTorr. The amplitude of the flux density is

$$B_m = \frac{1}{10}\left(\frac{P_v}{8.86 \times 10^{-10} \times f^{1.14}}\right)^{0.45562} \quad (\text{mT}). \tag{10.125}$$

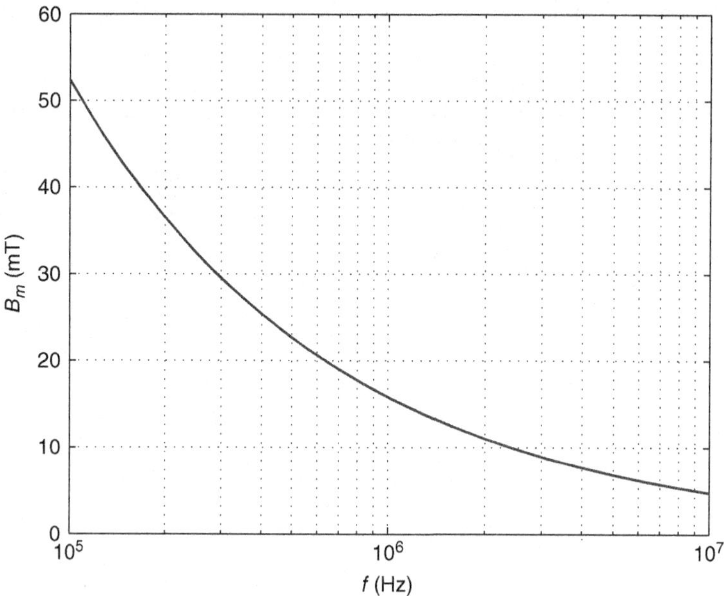

Figure 10.16 Amplitude of the flux density B_m as a function of frequency f at $P_v = 400$ mW/cm^3 for the Micrometals iron powder material-2

Figure 10.16 shows the amplitude of the flux density B_m as a function of frequency f at $P_v =$ 400 mW/cm^3 for the iron powder magnetic material-2. The minimum volume of the core at a given maximum core loss P_{Cmax} and the maximum amplitude of the inductor current I_{Lm} are given by

$$V_{cmin} = \left[\frac{P_{Cmax}}{kf^a I_{Lm} \sqrt{\mu_{rc}\mu_0 L}} \right]^{\frac{4}{2-b}}. \tag{10.126}$$

10.9 Design of AC Inductors

10.9.1 Optimum Magnetic Flux Density

Any design of magnetic components involves a trade-off between the winding and the core losses. Using fewer turns of copper allows a lower current density and lower winding loss. In contrast, using more turns of copper leads to lower magnetic flux density and lower core loss.

The amplitude of flux linkage is given by

$$\lambda_m = N\phi_m = NA_cB_m, \tag{10.127}$$

yielding the number of turns

$$N = \frac{\lambda_m}{A_cB_m}. \tag{10.128}$$

For sinusoidal waveforms,

$$\lambda_m = \frac{V_{Lm}}{\omega}, \tag{10.129}$$

resulting in

$$N = \frac{\lambda_m}{A_cB_m} = \frac{V_{Lm}}{\omega A_cB_m}. \tag{10.130}$$

Since

$$A_w = \frac{K_u W_a}{N} \tag{10.131}$$

the DC and low frequency winding resistance is

$$R_{wDC} = \frac{\rho_w l_w}{A_w} = \frac{\rho_w N l_T}{A_w} = \frac{\rho_w l_T N^2}{K_u W_a}, \tag{10.132}$$

where l_T is the MTL. The DC and low-frequency winding power loss is

$$P_{wDC} = R_{wDC} I_{Lrms}^2 = \frac{R_{wDC} I_{Lm}^2}{2} = \frac{\rho_w l_T N^2}{K_u W_a} I_{Lrms}^2 = \frac{\rho_w l_T \lambda_m^2}{K_u W_a A_c^2 B_m^2} I_{Lrms}^2$$
$$= \frac{\rho_w l_T V_{Lm}^2}{2K_u W_a A_c^2 \omega^2 B_m^2} I_{Lm}^2 = \frac{\rho_w l_T L^2 I_{Lm}^4}{2K_u W_a A_c^2 B_m^2}. \tag{10.133}$$

The copper resistivity is $\rho_{Cu} = 1.724 \times 10^{-8}$ Ωm at $T = 20$ °C. Thus, the DC and low-frequency winding loss decreases as B_m increases. The core loss is given by

$$P_C = V_c k f^a B_m^b. \tag{10.134}$$

The core loss increases with increasing B_m. Since the low-frequency winding loss P_w decreases and the core loss P_C increases when B_m increases, there is an optimum value of the magnetic flux density B_{mopt}, yielding the minimum total loss. The total power loss of an inductor is

$$P_{cwDC} = P_{wDC} + P_C = \frac{\rho_w l_T V_{Lm}^2 I_{Lm}^2}{2K_u W_a A_c^2 \omega^2 B_m^2} + V_c k f^a B_m^b = \frac{\rho_w l_T L^2 I_{Lm}^4}{2K_u W_a A_c^2 B_m^2} + V_c k f^a B_m^b. \tag{10.135}$$

The derivative of the total power loss is

$$\frac{dP_{cw}}{dB_m} = \frac{dP_w}{dB_m} + \frac{dP_C}{dB_m} = -\frac{\rho_w l_T V_{Lm}^2 I_{Lm}^2}{K_u W_a A_c^2 \omega^2 B_m^3} + bV_c k f^a B_m^{b-1} = 0. \tag{10.136}$$

Hence, the optimum value of the magnetic flux density, which yields the minimum low-frequency total power loss in an inductor, is given by

$$B_{mopt} = \left(\frac{\rho_w l_T V_{Lm}^2 I_{Lm}^2}{2bkK_u W_a A_c^2 V_c \omega^2 f^a}\right)^{\frac{1}{b+2}} = \left(\frac{\rho_w l_T V_{Lm}^2 I_{Lm}^2}{8\pi^2 bkK_u W_a A_c^2 V_c f^{a+2}}\right)^{\frac{1}{b+2}}$$
$$= \left(\frac{\rho_w l_T L^2 I_{Lm}^4}{bkK_u W_a A_c^2 V_c f^a}\right)^{\frac{1}{b+2}} = \left(\frac{\rho_w l_T W_m^2}{bkK_u W_a A_c^3 l_c f^a}\right)^{\frac{1}{b+2}}. \tag{10.137}$$

10.9.2 Examples of AC Inductor Designs

An AC inductor conducts only an AC current, and its DC component is zero. These inductors are commonly used in resonant circuits and filters. An accurate value of the desired inductance L must be obtained for most AC inductors. The following data is usually given for designing an AC inductor.

- Inductance L
- Frequency f
- Amplitude of the inductor current I_{Lm}

Example 10.5

Design an inductor for a Class E zero-voltage switching (ZVS) resonant amplifier to meet the following specifications: $L = 66.9$ μH, $Q_L = 7$, $f = 1.2$ MHz, $P_O = 80$ W, and $I_{Lm} = 1.49$ A.

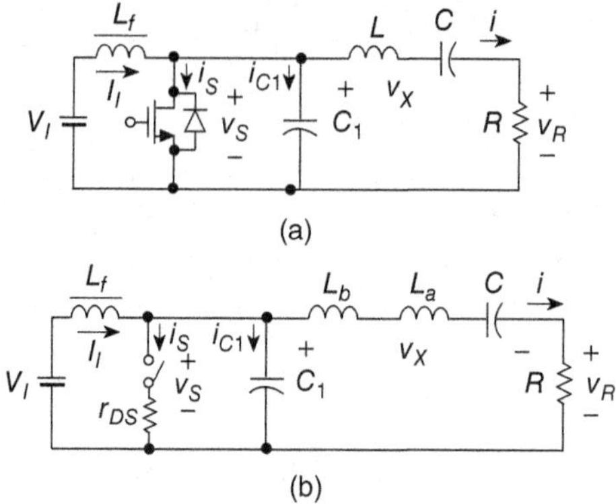

Figure 10.17 Class E ZVS power amplifier. (a) Circuit and (b) equivalent circuit

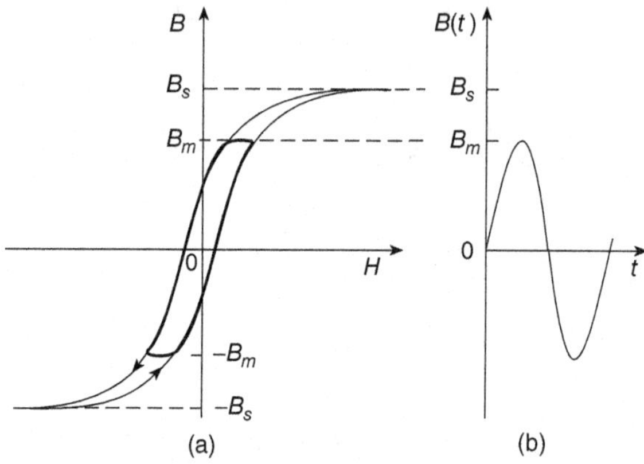

Figure 10.18 Magnetic core hysteresis and the waveform of the magnetic flux density $B(t)$ for AC inductors. (a) Hysteresis and the operating trajectory and (b) waveform of magnetic flux density $B(t)$

Solution: Figure 10.17 depicts a circuit of the Class E ZVS power amplifier. Figure 10.18 shows the core hysteresis and the magnetic flux density waveform for AC inductors. We will use a toroidal core with a single layer of the winding to avoid the proximity effect. Let us assume $K_u = 0.15$, $J_m = 4$ A/mm^2, and $B_m = 15$ mT. The required core area product is

$$A_p = W_a A_c = \frac{Q_L P_O}{\pi K_u J_m B_m f} \text{ (m}^4\text{)} = \frac{7 \times 80}{\pi \times 0.15 \times 4 \times 10^6 \times 0.015 \times 1.2 \times 10^6} \text{ (m}^4\text{)}$$
$$= 1.6505 \text{ cm}^4. \quad (10.138)$$

We need a core with a bandwidth $BW \geq 2$ MHz and $A_p \geq 2$ cm^4.

Let us select the Micrometals T130-2 ungapped toroidal carbonyl iron-powder core. The carbonyl iron powder magnetic core material No. 2 is used. The parameters of this core are as follows: $A_c = 0.698$ cm^2, $l_c = 8.28$ cm, $V_c = 5.78$ cm^3, $W_a = 3.08$ cm^2, $A_L = 11$ nH/turn, outer diameter is $D_o = 33$ mm, inner diameter is $D_i = 19.8$ mm, and the height is $H = 11.1$ mm. The core material No. 2 parameters are: $\mu_{rc} = 10$, $k = 8.86 \times 10^{-10}$, $b = 1.14$, and $c = 2.19$. $k = 8.86 \times 10^{-10}$, $a = 1.14$, and $b = 2.19$. Typically, the winding of a toroidal inductor consists of a single layer ($N_l = 1$).

The core area product is

$$A_p = A_c W_a = 0.698 \times 3.08 = 2.15 \text{ cm}^4. \quad (10.139)$$

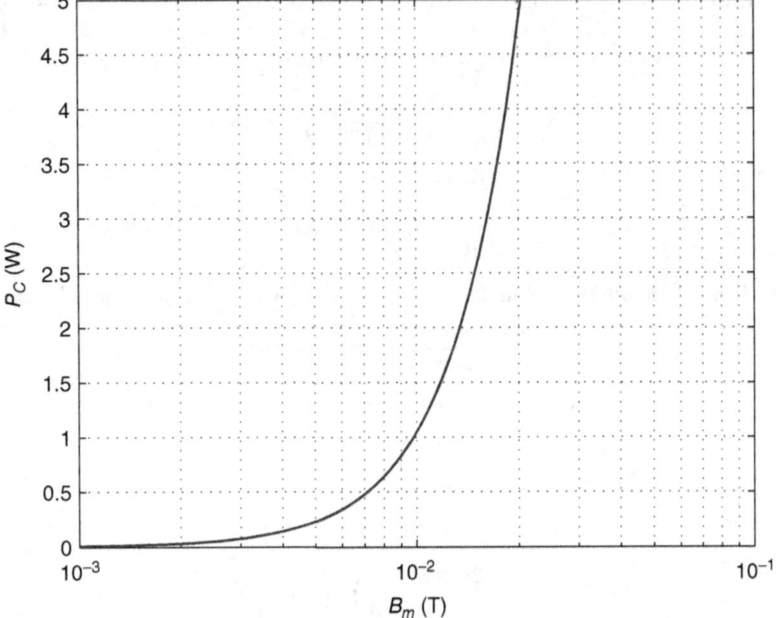

Figure 10.19 Core loss P_C as a function of the flux density B_m at $f = 1.2$ MHz for the Micrometals iron powder toroidal core T130-2 (material-2)

Figure 10.19 shows a plot of the core loss P_C as a function of B_m for the T130-2 toroidal core at $f = 1.2$ MHz. It can be seen that the core loss increases rapidly with increasing B_m. The core power loss density is given by

$$P_v = kf^a B_m^b = 8.86 \times 10^{-10} f^{1.14} (10B_m)^{2.19} \quad \text{mW/cm}^3, \tag{10.140}$$

where f is in Hertz and B_m is in milliTorr. This equation can be converted to the form

$$P_v = kf^a B_m^b = 8.86 \times 10^{-10} f^{1.14} (10^{-2}B_m)^{2.19} \quad \text{mW/cm}^3 = 0.5098 f^{1.14} (B_m)^{2.19} \quad \text{mW/cm}^3, \tag{10.141}$$

where f is in Hertz and B_m is in Torr.

The core power loss density is

$$P_v = 0.5098 f^{1.14} (B_m)^{2.19} \quad \text{mW/cm}^3 = 0.5098 \times 1\,200\,000^{1.14} \times 0.015^{2.19} = 440 \text{ mW/cm}^3. \tag{10.142}$$

The core power loss is

$$P_C = V_c P_v = 5.78 \times 0.44 = 2.5432 \text{ W}. \tag{10.143}$$

The equivalent series resistance (ESR) of the core is

$$R_{cs} = \frac{2P_C}{I_{Lm}^2} = \frac{2 \times 2.5432}{1.49^2} = 2.2911\ \Omega. \tag{10.144}$$

The number of turns is

$$N = \sqrt{\frac{L\text{nH}}{A_L\text{nH/t}}} = \sqrt{\frac{66900}{11}} = 77.986. \tag{10.145}$$

Pick $N = 78$.

The required cross-sectional area of the bare wire is

$$A_w = \frac{I_{Lm}}{J_m} = \frac{1.49}{4 \times 10^6} = 0.3725 \text{ mm}^2. \tag{10.146}$$

The nearest wire size is the AWG21 wire whose bare cross-sectional area is $A_w = 0.4116\ \text{mm}^2$, the bare wire diameter is $d_i = 0.7229$ mm, the outer diameter of the insulated wire is $d_o = 0.785$ mm, and the DC resistance per unit length is $R_{wDC}/l_w = 0.04189\ \Omega/\text{m}$. Hence, the wire insulation factor is

$$K_i = \left(\frac{d_i}{d_o}\right)^2 = \left(\frac{0.7229}{0.785}\right)^2 = 0.848. \tag{10.147}$$

The skin depth of copper at $f = 1.2$ MHz is

$$\delta_w = \frac{66.2}{\sqrt{f}}\ (\text{mm}) = \frac{66.2}{\sqrt{1.2 \times 10^6}}\ (\text{mm}) = 0.060432\ \text{mm} = 60.432\ \mu\text{m}. \tag{10.148}$$

Method I: Single-Solid-Wire Winding The ratio of the wire diameter to the skin depth is

$$\frac{d_i}{\delta_w} = \frac{0.7229}{0.060432} = 11.962 \approx 12. \tag{10.149}$$

Hence, the normalized diameter at $\eta = d_i/p = 0.8$ is

$$A = \left(\frac{\pi}{4}\right)^{\frac{3}{4}} \left(\frac{d_i}{\delta_w}\right) \sqrt{\frac{d_i}{p}} = 0.8343 \times 12 \times \sqrt{0.8} = 8.9546. \tag{10.150}$$

The AC-to-DC resistance ratio for $A \geq 3$ is given by

$$F_R = \frac{2N_l^2 + 1}{3} \left(\frac{\pi}{4}\right)^{\frac{3}{4}} \left(\frac{d_i}{\delta_w}\right) \sqrt{\frac{d_i}{p}} \tag{10.151}$$

For a single-layer inductor ($N_l = 1$) and $A \geq 3$,

$$F_R = A = 8.9546. \tag{10.152}$$

The length of a single turn of an inductor with a toroidal core is

$$l_T = D_o - D_i + 2H + 2d_o = 33 - 19.8 + 2 \times 11.1 + 2 \times 0.785 = 36.97\ \text{mm}, \tag{10.153}$$

resulting in

$$l_w = N l_T = 78 \times 36.97 = 2883.6\ \text{mm} = 2.884\ \text{m}. \tag{10.154}$$

Pick $l_w = 2.9$ m to allow for the inductor leads. The DC winding resistance is

$$R_{wDC} = \left(\frac{R_{wDC}}{l_w}\right) l_w = 0.04189 \times 2.9 = 0.1215\ \Omega. \tag{10.155}$$

The low-frequency winding power loss is

$$P_{wDC} = \frac{R_{wDC} I_{Lm}^2}{2} = \frac{0.1215 \times 1.49^2}{2} = 0.1345\ \text{W}. \tag{10.156}$$

For a single-layer inductor ($N_l = 1$), $F_R = 8.9546$ at $f = 1.2$ MHz. Hence, the ac winding resistance is

$$R_w = F_R R_{wDC} = 8.9546 \times 0.1215 = 1.088\ \Omega. \tag{10.157}$$

The AC winding power loss is

$$P_w = \frac{R_w I_{Lm}^2}{2} = \frac{1.088 \times 1.49^2}{2} = 1.2077\ \text{W}. \tag{10.158}$$

The overall series equivalent resistance of the inductor is

$$r_L = R_{cs} + R_w = 2.2911 + 1.2077 = 3.4988\Omega. \tag{10.159}$$

The power loss of the core and the winding is

$$P_{cw} = P_C + P_w = 2.5432 + 1.2077 = 3.751\ \text{W}. \tag{10.160}$$

The quality factor of the inductor at $f = 1.2$ MHz is

$$Q_{Lo} = \frac{\omega L}{r_L} = \frac{\omega L}{R_{cs} + R_w} = \frac{2\pi \times 1.2 \times 10^6 \times 66.9 \times 10^{-6}}{3.751} = 149.27. \tag{10.161}$$

The outer surface of the inductor, which radiates heat, is given by

$$A_t = \pi(D_o + D_i)\left(\frac{D_o - D_i}{2} + H + 4d_o\right)$$

$$= \pi(3.3 + 1.98)\left(\frac{3.3 - 1.98}{2} + 1.11 + 4 \times 0.0785\right) = 34.56\ \text{cm}^2. \quad (10.162)$$

The surface power loss density of the inductor is

$$\psi = \frac{P_{cw}}{A_t} = \frac{P_C + P_w}{A_t} = \frac{3.751}{34.56} = 0.1085\ \text{W/cm}^2. \quad (10.163)$$

The temperature rise of the inductor is

$$\Delta T = 450\psi^{0.826} = 450 \times (0.1085)^{0.826} = 71.84\ ^\circ\text{C}. \quad (10.164)$$

The cross-sectional area of the insulated wire is

$$A_w = \frac{\pi d_i^2}{4} = \frac{\pi \times 0.7229^2}{4} = 0.41044\ \text{mm}^2 = 0.0041044\ \text{cm}^2. \quad (10.165)$$

The core window utilization factor is

$$K_u = \frac{A_{Cu}}{W_a} = \frac{NA_w}{W_a} = \frac{78 \times 0.0041044}{3.08} = 0.1039. \quad (10.166)$$

Method II, Optimum Strand Diameter Winding This design illustrates how we can reduce the skin and proximity effects using multistrand wire with its optimum diameter. We will assume that the number of bundle layers is $N_l = 1$ and the total cross-sectional area of all strands is the same as that of the solid wire, that is, $A_{wo} = 0.4116\ \text{mm}^2$ and the bare solid wire diameter is $d_i = 0.7229\ \text{mm}^2$. The amplitude of the uniform current density in all strands is

$$J_m = \frac{I_{Lm}}{A_w} = \frac{1.49}{0.4116} = 3.62\ \text{A/mm}^2. \quad (10.167)$$

Assume the porosity factor $\eta = 0.7$. The number of strands for the optimum strand diameter is

$$k_{cr} = \frac{5N_l^2 + \sqrt{(5N_l^2)^2 - \frac{180}{\eta^2\left(\frac{\pi}{4}\right)^3}\left(\frac{\delta_w}{d_i}\right)^4}}{\frac{90}{\eta^2\left(\frac{\pi}{4}\right)^3}\left(\frac{\delta_w}{d_i}\right)^4}$$

$$= \frac{5 \times 1^2 + \sqrt{(5 \times 1^2)^2 - \frac{180}{0.7^2\left(\frac{\pi}{4}\right)^3}\left(\frac{0.060432}{0.7229}\right)^4}}{\frac{90}{0.7^2\left(\frac{\pi}{4}\right)^3}\left(\frac{0.060432}{0.7229}\right)^4} = 543.7. \quad (10.168)$$

Select $k = 544$. The optimum strand diameter is

$$d_{lopt} = \frac{d_i}{\sqrt{k_{cr}}} = \frac{0.7229}{\sqrt{544}} = 30.99\ \mu\text{m}. \quad (10.169)$$

Hence,

$$\frac{d_{lopt}}{\delta_w} = \frac{0.03099}{0.060432} = 0.5128078. \quad (10.170)$$

Select a litz wire with copper AWG48 strands. The bare strand diameter is $d_l = 31.5\ \mu\text{m}$, the insulated strand diameter is $d_{lo} = 32.3\ \mu\text{m}$, and the DC resistance per unit length is $R_{wDC}/l_w = 22.12\ \Omega/\text{m}$. The DC resistance of the single strand is

$$R_{wDCs} = \left(\frac{R_{wDC}}{l_w}\right) l_w = 22.12 \times 2.9 = 64.148\ \Omega. \quad (10.171)$$

The DC resistance of all strands is

$$R_{wDC} = \frac{R_{wDCs}}{k} = \frac{64.148}{544} = 0.1179\ \Omega. \quad (10.172)$$

The AC resistance of all strands is

$$R_w = F_{Ropt} R_{wDC} = 2 \times 0.1179 = 0.2358\ \Omega. \tag{10.173}$$

The winding AC power loss is

$$P_w = \frac{R_w I_{Lm}^2}{2} = \frac{0.1179 \times 1.5^2}{2} = 0.26179\ \text{W} \tag{10.174}$$

Thus, the winding AC power loss is reduced with respect to that of the solid wire by a factor

$$\frac{P_{w(solid)}}{P_{w(strand)}} = \frac{1.2077}{0.26179} = 4.6132. \tag{10.175}$$

The AC ESR of the inductor is

$$r_L = R_w + R_{cs} = 0.2358 + 2.2911 = 2.5269\ \Omega. \tag{10.176}$$

The total loss in the inductor is

$$P_{cw} = P_C + P_w = 2.5432 + 0.26179 = 2.80499\ \text{W}. \tag{10.177}$$

The quality factor of the inductor is

$$Q_{Lo} = \frac{\omega L}{r_L} = \frac{2\pi \times 1.2 \times 10^6 \times 66.9 \times 10^{-6}}{2.5269} = 199.6147. \tag{10.178}$$

The surface power loss density is

$$\psi = \frac{P_{cw}}{A_t} = \frac{2.80499}{34.56} = 0.0811\ \text{W/cm}^2. \tag{10.179}$$

The temperature rise of the inductor is

$$\Delta T = 450\psi^{0.826} = 450 \times 0.0811^{0.826} = 56.538\ ^\circ\text{C}. \tag{10.180}$$

The cross-sectional area of the single insulated strand is

$$A_{wso} = \frac{\pi d_l^2}{4} = \frac{\pi (31.5 \times 10^{-6})^2}{4} = 0.0007544\ \text{cm}^2. \tag{10.181}$$

The core window utilization factor is

$$K_u = \frac{NkA_{wso}}{W_a} = \frac{78 \times 544 \times 0.0007544}{3.08} = 0.10395. \tag{10.182}$$

Method III: Multiple-Strand Winding Assuming the same cross-sectional area for the solid wire and multistrand wire, the number of strands in a multiple-strand wire is

$$S_n = \frac{A_w}{A_{ws}} = \frac{\frac{\pi d_i^2}{4}}{\pi \delta_w^2} = \frac{1}{4}\left(\frac{d_i}{\delta_w}\right)^2 = \frac{1}{4}\left(\frac{0.7229}{0.060432}\right)^2 = \frac{1}{4} \times 11.9622^2 = 35.77, \tag{10.183}$$

where $A_{ws} = \pi \delta_w^2$ is the cross-sectional area of the bare strand wire. Select $S_n = 36$. The diameter of a single strand for $N_l = 1$ is

$$d_s = 2\delta_w = 2 \times 0.060432 = 0.120864\ \text{mm} \tag{10.184}$$

and the cross-sectional area of a single strand is

$$A_{ws} = \pi \delta_w^2 = \pi \times (0.060432 \times 10^{-3})^2 = 0.01147\ \text{mm}^2. \tag{10.185}$$

We will select a copper AWG36 wire with $d_{si} = 0.127$ mm, $d_{so} = 0.152$ mm, $A_{ws} = 0.01266\ \text{mm}^2$, and $R_{wDCs}/l_w = 1.3608\ \Omega$/m. The DC resistance of each strand is

$$R_{wDCs} = \left(\frac{R_{wDCs}}{l_w}\right) l_w = 1.3608 \times 2.9 = 3.946\ \Omega. \tag{10.186}$$

The equivalent winding resistance of all strands connected in parallel is

$$R_{wDC} = \frac{R_{wDCs}}{S_n} = \frac{3.946}{36} = 0.1096\Omega. \tag{10.187}$$

Since $R_w \approx R_{wDC}$ at low frequencies ($F_R \approx 1$), the winding power loss is

$$P_w = \frac{R_{wDC} I_{Lm}^2}{2} = \frac{0.1096 \times 1.49^2}{2} = 0.1216 \text{ W}. \tag{10.188}$$

The ESR of the inductor is

$$r_L = R_{cs} + R_{wDC} = 2.2911 + 0.1096 = 2.4 \ \Omega. \tag{10.189}$$

The total power loss is

$$P_{cw} = P_C + P_w = 2.5432 + 0.1216 = 2.664 \text{ W}. \tag{10.190}$$

The quality factor of the inductor is

$$Q_{Lo} = \frac{\omega L}{r_L} = \frac{2\pi \times 1.2 \times 10^6 \times 66.9 \times 10^{-6}}{2.4} = 210.06. \tag{10.191}$$

The surface power loss density of the inductor is

$$\psi = \frac{P_{cw}}{A_t} = \frac{2.664}{34.56} = 0.0771 \text{ W/cm}^2. \tag{10.192}$$

The temperature rise of the inductor is

$$\Delta T = 450\psi^{0.826} = 450 \times (0.0771)^{0.826} = 54.19\ ^\circ\text{C}. \tag{10.193}$$

The cross-sectional area of the single insulated strand is

$$A_{wso} = \frac{\pi d_{so}^2}{4} = \frac{\pi \times 0.152^2}{4} = 0.01813 \text{ mm}^2 = 0.001813 \text{ cm}^2. \tag{10.194}$$

Hence, the core window utilization factor is

$$K_u = \frac{N S_n A_{wso}}{W_a} = \frac{78 \times 36 \times 0.0001813}{3.08 \times 10^{-4}} = 0.165. \tag{10.195}$$

Example 10.6

Design an inductor for a Class E ZVS resonant power amplifier to meet the following specifications: $v_L = V_{Lm} \sin \omega t = 500 \sin 2\pi \times 100 \times 10^3 t$ (V), $i_L = I_{Lm} \sin \omega t = 1.5 \sin 2\pi \times 100 \times 10^3 t$ (A), $L = 530\ \mu\text{H}$, $J_m = 4$ A/mm^2, and $K_u = 0.4$.

Solution: Assume the amplitude of the magnetic flux density to be $B_m = 0.45$ T. The required area product of the inductor is

$$A_p = W_a A_c = \frac{L I_{Lm}^2}{K_u J_m B_m} \text{ (m}^4\text{)} = \frac{530 \times 10^{-6} \times 1.5^2}{0.4 \times 4 \times 10^6 \times 0.2} \text{ (m}^4\text{)} = 0.3727 \text{ cm}^4. \tag{10.196}$$

We shall choose a Magnetics 0F-42515EC gapped ferrite EE core with the following parameters: $A_p = 0.42$ cm^4, $A_c = 40.1$ mm^2, $l_c = 73.5$ mm, $A_L = 1500 \pm 25\%$ mH/1000 turns $=$ 15 nH/turn, $B_s = 0.5$ T, and $V_c = 2.95$ cm^3. The core is made of the F-type material whose $\mu_{rc} = 2500 \pm 20\%$, $k = 0.0573$, $a = 1.66$, and $b = 2.68$. The core winding window is

$$W_a = \frac{A_p}{A_c} = \frac{0.42}{0.401} = 1.047 \text{ cm}^2 = 104.7 \text{ mm}^2. \tag{10.197}$$

Method I, Single-Solid-Wire Winding The skin and proximity effects can be avoided if the inner wire diameter satisfies the condition $d_i \le 2\delta_w$. The skin depth of copper at $f = 100$ kHz is given by

$$\delta_w = \frac{66.2}{\sqrt{f}} \text{ (mm)} = \frac{66.2}{\sqrt{100,000}} = 0.209 \text{ mm}. \tag{10.198}$$

The cross-sectional area of the winding bare wire is

$$A_w = \frac{I_{Lm}}{J_m} = \frac{1.5}{4 \times 10^6} = 0.375 \text{ mm}^2. \tag{10.199}$$

Let us select an AWG21 copper wire with $A_w = 0.4116 \text{ mm}^2$, $d_i = 0.723$ mm, $d_o = 0.785$ mm, and the resistance per meter $R_{wDC}/l_w = 0.04189\ \Omega/\text{m}$. The amplitude of the current density is

$$J_m = \frac{I_{Lm}}{A_w} = \frac{1.5}{0.4116 \times 10^{-6}} = 3.64431 \times 10^6 \quad \text{A/m}^2. = 3.64431 \text{ A/mm}^2. \tag{10.200}$$

The cross-sectional area of the insulated wire is

$$A_{wo} = \frac{\pi d_o^2}{4} = \frac{\pi \times 0.785^2}{4} = 0.484 \text{ mm}^2. \tag{10.201}$$

The number of turns is

$$N = \frac{K_u W_a}{A_{wo}} = \frac{0.4 \times 104.7}{0.484} = 86.53. \tag{10.202}$$

Pick $N = 87$. The air-gap length is

$$l_g = \frac{\mu_0 A_c N^2}{L} - \frac{l_c}{\mu_{rc}} = \frac{4\pi \times 10^{-7} \times 40.1 \times 10^{-3} \times 87^2}{530 \times 10^{-6}} - \frac{73.5 \times 10^{-3}}{2500} = 0.7196 \text{ mm}. \tag{10.203}$$

The length of a single turn of the inductor with the EE core is

$$l_T = 2E + F + C = 2 \times 1.75 + 0.75 + 0.635 = 4.885 \text{ mm}. \tag{10.204}$$

The length of the winding wire is

$$l_w = N l_T = 87 \times 4.885 \times 10^{-2} = 4.24995 \text{ m}. \tag{10.205}$$

Pick $l_w = 4.25$ m.

Assuming the uniform current distribution in the winding wire, the DC winding resistance is

$$R_{wDC} = l_w \left(\frac{R_{wDC}}{l_w} \right) = 4.25 \times 0.04189 = 0.178\ \Omega. \tag{10.206}$$

The DC winding power loss is

$$P_{wDC} = \frac{R_{wDC} I_{Lm}^2}{2} = \frac{0.178 \times 1.5^2}{2} = 0.2 \text{ W}. \tag{10.207}$$

The ratio of the wire diameter to the skin depth is

$$\frac{d}{\delta_w} = \frac{0.723}{0.209} = 3.459. \tag{10.208}$$

Assuming that the winding consists of two identical layers ($N_l = 2$), we obtain the AC-to-DC resistance ratio from Dowell's equation

$$F_R = \frac{R_w}{R_{wDC}} = \frac{P_w}{P_{wDC}} = 7. \tag{10.209}$$

Hence, the winding AC resistance is

$$R_w = F_R R_{wDC} = 7 \times 0.178 = 1.246\ \Omega \tag{10.210}$$

and the winding loss is

$$P_w = F_R P_{wDC} = 7 \times 0.2 = 1.4 \text{ W}. \tag{10.211}$$

The amplitude of the core magnetic flux density is

$$B_m = \frac{\mu_0 N I_{Lm}}{l_g + \frac{l_c}{\mu_{rc}}} = \frac{4\pi \times 10^{-7} \times 87 \times 1.5}{0.689 \times 10^{-3} + \frac{73.5 \times 10^{-3}}{2500}} = 0.228 \text{ T}. \tag{10.212}$$

The core power loss per unit volume for the F-type material at $f = 100$ kHz and $B_m = 0.2$ T is given by

$$P_v = \frac{P_C}{V_c} = 0.0573 f^{1.66}(10B_m)^{2.68} \quad \text{mW/cm}^3$$

$$= 0.0573 \times 100^{1.66}(10 \times 0.228)^{2.68} = 1.09 \text{ W/cm}^3, \tag{10.213}$$

where f is in kilohertz and B_m is in Torr.

The total core loss is

$$P_C = V_c P_v = 2.95 \times 1.09 = 3.2155 \text{ W}. \tag{10.214}$$

Using the principle of energy conservation, we can write

$$P_C = \frac{R_{cs} I_{Lm}^2}{2}, \tag{10.215}$$

resulting in the core equivalent resistance at the operating frequency f

$$R_{cs} = \frac{2P_C}{I_{Lm}^2} = \frac{2 \times 3.2155}{1.5^2} = 2.858 \ \Omega. \tag{10.216}$$

The total power loss in the inductor is

$$P_{cw} = P_C + P_w = 3.2155 + 1.4 = 4.615 \ W. \tag{10.217}$$

The series equivalent resistance of the inductor at the operating frequency is

$$r_L = R_w + R_{cs} = 1.246 + 2.858 = 4.104 \ \Omega. \tag{10.218}$$

The quality factor of the inductor at the operating frequency is

$$Q_{Lo} = \frac{\omega L}{r_L} = \frac{2\pi \times 100 \times 10^3 \times 530 \times 10^{-6}}{4.1041} = 81.14. \tag{10.219}$$

The surface area of the inductor is

$$A_t = 4AB + 2AC + 4BC + 2(E - F)(2D + E)$$

$$= 4 \times 2.54 \times 1.59 + 2 \times 2.54 \times 0.635 + 4 \times 1.59 \times 0.635 + 2$$

$$\times (1.88 - 0.75)(2 \times 1.26 + 1.88)$$

$$= 33.36 \text{ cm}^2. \tag{10.220}$$

The surface power loss density is

$$\frac{P_{cw}}{A_t} = \frac{4.615}{37.61} = 0.1227 \text{ W/cm}^2. \tag{10.221}$$

Hence, the temperature rise of the inductor is

$$\Delta T = 450\psi^{0.826} = 450 \times 0.1227^{0.826} = 79.545\ ^\circ\text{C}. \tag{10.222}$$

The cross-sectional area of the insulated wire is

$$A_{wo} = \frac{\pi d_o^2}{4} = \frac{\pi \times 0.785^2}{4} = 0.484 \text{ mm}^2 = 0.00484 \text{ cm}^2. \tag{10.223}$$

Thus, the utilization window factor is

$$K_u = \frac{NA_w}{W_a} = \frac{87 \times 0.00484}{1.047} = 0.4022. \tag{10.224}$$

Method II, Optimum Strand Diameter Winding This design illustrates how we can reduce the skin and proximity effects using multistrand wire with its optimum diameter. We will assume that the number of bundle layers is $N_l = 2$ and the total cross-sectional area of all strands is the same as that of the solid wire, that is, $A_{wo} = 0.4116$ mm^2, and the bare solid wire diameter is $d_i = 0.7229$ mm^2. The amplitude of the uniform current density in all strands is

$$J_m = \frac{I_{Lm}}{A_w} = \frac{1.5}{0.4116} = 3.64 \text{ A/mm}^2. \tag{10.225}$$

Assume the porosity factor $\eta = 0.7$. The number of strands for the optimum strand diameter is

$$k_{cr} = \frac{5N_l^2 + \sqrt{(5N_l^2)^2 - \frac{180}{\eta^2\left(\frac{\pi}{4}\right)^3}\left(\frac{\delta_w}{d_i}\right)^4}}{\frac{90}{\eta^2\left(\frac{\pi}{4}\right)^3}\left(\frac{\delta_w}{d_i}\right)^4}$$

$$= \frac{5 \times 2^2 + \sqrt{(5 \times 2^2)^2 - \frac{180}{0.7^2\left(\frac{\pi}{4}\right)^3}\left(\frac{0.209}{0.7229}\right)^4}}{\frac{90}{0.7^2\left(\frac{\pi}{4}\right)^3}\left(\frac{0.209}{0.7229}\right)^4} = 15.059 \tag{10.226}$$

Select $k = 15$. The optimum strand diameter is

$$d_l = \frac{d_i}{\sqrt{k_{cr}}} = \frac{0.7229}{\sqrt{15}} = 186.65\ \mu\text{m}. \tag{10.227}$$

Select litz wire with copper AWG33 strands. The bare strand diameter is $d_l = 179.8\ \mu$m, the insulated strand diameter is $d_{lo} = 216\ \mu$m, and the DC resistance per unit length is $R_{wDC}/l_w = 0.6748\ \Omega$/m. The DC resistance of the single strand is

$$R_{wDCs} = \left(\frac{R_{wDC}}{l_w}\right) l_w = 0.6748 \times 4.25 = 2.8679\ \Omega. \tag{10.228}$$

The DC resistance of all strands is

$$R_{wDC} = \frac{R_{wDCs}}{k} = \frac{2.8679}{15} = 0.19119\ \Omega. \tag{10.229}$$

The AC resistance of all strands is

$$R_w = F_{Ropt} R_{wDC} = 2 \times 0.19119 = 0.382386\ \Omega. \tag{10.230}$$

The winding AC power loss is

$$P_w = \frac{R_w I_{Lm}^2}{2} = \frac{0.382386 \times 1.5^2}{2} = 0.430185 \text{ W} \tag{10.231}$$

Thus, the winding AC power loss is reduced with respect to that of the solid wire by a factor

$$\frac{P_{w(solid)}}{P_{w(strand)}} = \frac{1.4}{0.430185} = 3.2558. \tag{10.232}$$

The AC ESR of the inductor is

$$r_L = R_w + R_{cs} = 0.382386 + 2.858 = 3.24\ \Omega. \tag{10.233}$$

The total loss in the inductor is

$$P_{cw} = P_C + P_w = 3.2155 + 0.43 = 3.645685 \text{ W}. \tag{10.234}$$

The quality factor of the inductor is

$$Q_{Lo} = \frac{\omega L}{r_L} = \frac{2\pi \times 100 \times 10^3 \times 530 \times 10^{-6}}{3.24} = 102.768. \tag{10.235}$$

The surface power loss density is

$$\psi = \frac{P_{cw}}{A_t} = \frac{3.645685}{37.61} = 0.09693 \text{ W/cm}^2. \tag{10.236}$$

The temperature rise of the inductor is

$$\Delta T = 450\psi^{0.826} = 450 \times 0.09693^{0.826} = 65.469\ ^{\circ}\mathrm{C}. \tag{10.237}$$

The cross-sectional area of the single insulated strand is

$$A_{wso} = \frac{\pi d_l^2}{4} = \frac{\pi(179.8 \times 10^{-6})^2}{4} = 0.0002539\ \mathrm{cm}^2. \tag{10.238}$$

The core window utilization factor is

$$K_u = \frac{NkA_{wso}}{W_a} = \frac{87 \times 15 \times 0.0002539}{1.047 \times 10^{-4}} = 0.31647. \tag{10.239}$$

Method III, Multiple-Strand Winding The diameter of a single strand is

$$d_s = 2\delta_w = 2 \times 0.209 = 0.418\ \mathrm{mm}. \tag{10.240}$$

The cross-sectional area of the bare strand wire is

$$A_{ws} = \pi\delta_w^2 = \pi \times (0.209 \times 10^{-6})^2 = 0.1372\ \mathrm{mm}^2. \tag{10.241}$$

The number of strands is

$$S_n = \frac{A_w}{A_{ws}} = \frac{0.4116 \times 10^{-6}}{0.1372 \times 10^{-6}} = 3. \tag{10.242}$$

The cross-sectional area of each strand should be close to 0.1372 mm^2. The closest wire size is AWG26, with the bare wire diameter $d_s = 0.405$ mm, insulated wire diameter $d_{so} = 0.452$ mm, the bare wire cross-sectional area $A_{ws} = 0.128$ mm^2, and $R_{wDCs}/l_w = 0.1345$ Ω/m. The DC resistance of the single strand is

$$R_{wDCs} = l_w\left(\frac{R_{wDCs}}{l_w}\right) = 4.25 \times 0.1345 = 0.572\ \Omega. \tag{10.243}$$

The DC winding resistance is

$$R_{wDC} = \frac{R_{wDCs}}{S_n} = \frac{0.572}{3} = 0.19\ \Omega. \tag{10.244}$$

For a two-layer inductor, Dowell's equation produces $F_R = 6.45825$ and

$$R_w = F_R R_{wDC} = 6.45825 \times 0.19 = 1.230567\ \Omega. \tag{10.245}$$

The AC winding loss is

$$P_w = \frac{R_w I_{Lm}^2}{2} = \frac{1.230567 \times 1.5^2}{2} = 1.38438\ \mathrm{W}. \tag{10.246}$$

The total loss in the inductor is

$$P_{cw} = P_C + P_w = 3.2155 + 1.38438 = 4.5998\ \mathrm{W}. \tag{10.247}$$

The resistance of the inductor is

$$r_L = R_w + R_{cs} = 1.230567 + 2.8582 = 4.088767\ \Omega. \tag{10.248}$$

The quality factor of the inductor at the operating frequency is

$$Q_{Lo} = \frac{\omega L}{r_L} = \frac{2\pi \times 100 \times 10^3 \times 530 \times 10^{-6}}{4.088767} = 79.33. \tag{10.249}$$

The surface power loss density is

$$\frac{P_{cw}}{A_t} = \frac{4.5998}{37.61} = 0.1223\ \mathrm{W/cm}^2. \tag{10.250}$$

Hence, the temperature rise of the inductor is

$$\Delta T = 450\psi^{0.826} = 450 \times 0.1223^{0.826} = 79.33\ ^{\circ}\mathrm{C}. \tag{10.251}$$

The cross-sectional area of the insulated strand wire is

$$A_{wso} = \frac{\pi d_{os}^2}{4} = \frac{\pi \times 0.452^2}{4} = 0.16046 \text{ mm}^2 = 0.001646 \text{ cm}^2. \quad (10.252)$$

Thus, the core window utilization factor is

$$K_u = \frac{N S_n A_{wso}}{W_w} = \frac{87 \times 3 \times 0.0016046}{1.047} = 0.4. \quad (10.253)$$

Method IV, Multiple-Strand Winding This design illustrates how we can avoid the skin and proximity effects using multistrand wire. The diameter of a single strand is

$$d_s = \delta_w = 0.209 \text{ mm}. \quad (10.254)$$

The cross-sectional area of the bare strand wire is

$$A_{ws} = \frac{\pi d_s^2}{4} = \frac{\pi 0.209^2}{4} = 0.0343 \text{ mm}^2. \quad (10.255)$$

The number of strands is

$$S_n = \frac{A_w}{A_{ws}} = \frac{0.4116 \times 10^{-6}}{0.0343 \times 10^{-6}} = 12. \quad (10.256)$$

The cross-sectional area of each strand should be close to 0.0343 mm^2. The closest wire size is AWG32 with the bare wire diameter $d_s = 0.2019$ mm, insulated wire diameter $d_{so} = 0.241$ mm, the bare wire cross-sectional area $A_{ws} = 0.03242$ mm^2, and $R_{wDCs}/l_w = 0.53149$ Ω/m. The DC resistance of the single strand is

$$R_{wDCs} = l_w \left(\frac{R_{wDCs}}{l_w}\right) = 4.25 \times 0.53149 = 2.2588 \ \Omega. \quad (10.257)$$

The DC winding resistance is

$$R_{wDC} = \frac{R_{wDCs}}{S_n} = \frac{2.2588}{12} = 0.1882 \ \Omega. \quad (10.258)$$

From Dowell's equation, $F_R = 2.256$ and

$$R_w = F_R R_{wDC} = 2.256 \times 0.1882 = 0.42466 \ \Omega. \quad (10.259)$$

The winding loss is

$$P_w = \frac{R_w I_{Lm}^2}{2} = \frac{0.42466 \times 1.5^2}{2} = 0.4777 \text{ W}. \quad (10.260)$$

The total loss in the inductor is

$$P_{cw} = P_C + P_w = 3.2155 + 0.4777 = 3.6932 \text{ W}. \quad (10.261)$$

The resistance of the inductor is

$$r_L = R_w + R_{cs} = 0.4777 + 2.8582 = 3.2828 \ \Omega. \quad (10.262)$$

The quality factor of the inductor at the operating frequency is

$$Q_{Lo} = \frac{\omega L}{r_L} = \frac{2\pi \times 100 \times 10^3 \times 530 \times 10^{-6}}{3.2828} = 101.439. \quad (10.263)$$

The surface power loss density is

$$\frac{P_{cw}}{A_t} = \frac{3.6932}{37.61} = 0.098198 \text{ W/cm}^2. \quad (10.264)$$

Hence, the temperature rise of the inductor is

$$\Delta T = 450\psi^{0.826} = 450 \times 0.098198^{0.826} = 66.174 \ ^\circ\text{C}. \quad (10.265)$$

The cross-sectional area of the insulated strand wire is

$$A_{wso} = \frac{\pi d_{os}^2}{4} = \frac{\pi \times 0.241^2}{4} = 0.04559 \text{ mm}^2 = 0.0004559 \text{ cm}^2. \quad (10.266)$$

Thus, the core window utilization factor is

$$K_u = \frac{NS_nA_{wso}}{W_w} = \frac{87 \times 12 \times 0.0004559}{1.047} = 0.4546. \quad (10.267)$$

10.10 Inductor Design for Buck Converter in CCM

10.10.1 Derivation of Area Product A_p for Square-Wave Inductor Voltage

The window area limited by the maximum current density in the winding wire is given by

$$W_a = \frac{A_{Cu}}{K_u} = \frac{NA_w}{K_u} = \frac{NI_{Lmax}}{K_uJ_m} = \frac{NI_{Lrms}}{K_uJ_{rms}}. \quad (10.268)$$

From Chapter 1, the required core cross-sectional area for a square-wave voltage ($D = 0$) across the inductor is

$$A_c = \frac{V_{Lrms}}{4fNB_m}. \quad (10.269)$$

The maximum magnetic energy stored in an inductor is

$$W_m = \frac{1}{2}LI_{Lmax}^2. \quad (10.270)$$

Taking into account that $I_{Lmax} \approx I_L$, $I_{Lrms} \approx I_L$, and $I_{Lrms}V_{Lrms} = fW_m$, the core area product for the inductor square-wave voltage is

$$A_p = W_aA_c = \frac{NI_{Lmax}}{K_uJ_m} \times \frac{V_{Lrms}}{4fNB_m} = \frac{I_{Lmax}V_{Lrms}}{4K_ufJ_mB_m} = \frac{I_{Lrms}V_{Lrms}}{4K_ufJ_{rms}B_m} = \frac{W_m}{4K_uJ_{rms}B_m}. \quad (10.271)$$

Finally, the core area product is

$$A_p = \frac{LI_{Lmax}^2}{8K_uJ_{rms}B_m}. \quad (10.272)$$

In the pulse-width modulated (PWM) buck converter, DC and AC components flow through the inductor L. Therefore, the peak value of the magnetic flux density consists of the DC component B_{DC} and the amplitude of the AC component B_m

$$B_{pk} = B_{DC} + B_m, \quad (10.273)$$

yielding

$$A_p = W_aA_c = \frac{I_{Lrms}V_{Lrms}}{4K_ufJ_{rms}B_{pk}} = \frac{LI_{Lmax}^2}{8K_uJ_{rms}B_{pk}} = \frac{W_m}{4K_uJ_{rms}B_{pk}}. \quad (10.274)$$

10.10.2 Inductor Design for Buck Converter in CCM Using Area Product A_p Method

The inductor for PWM converters in CCM should satisfy the following conditions:

1. $L > L_{min}$ required for CCM operation.
2. $B_{pk} < B_s$ or $\lambda_{max} < \lambda_s$. For ferrite cores, $B_s < 0.5$ T at room temperature and $B_s = 0.3$ T at $T = 100\,°$C.
3. $P_L < P_{L(max)}$. Since $P_L \approx P_w = P_{rL}$, $r_L < r_{Lmax}$.
4. $J_{max} < J_{MAX}$. Typically, $J_{MAX} = 0.1$–5 A/mm^2.

The magnitude of the AC magnetic flux is caused by the peak-to-peak value Δi_L of the AC component of the inductor current, which is normally only a fraction of the DC flux. In many cases, the AC inductor current riding on top of the DC current I_L is small enough that it does not affect the overall current density of the single wire used.

Example 10.7

A buck PWM converter is operated in the continuous conduction mode (CCM) and has the following specifications: $V_I = 28 \pm 4$ V, $V_O = 12$ V, $I_O = 1-10$ A, and $f_s = 100$ kHz.

Solution: Figure 10.20 shows the PWM buck converter and its equivalent circuits for CCM operation. Figure 10.21 depicts the magnetic core hysteresis and the magnetic flux density waveform $B(t)$ for the buck PWM converter operated in CCM. In buck and boost converters, the magnetic flux excursion extends into the first quadrant only. For CCM operation, the peak value of the AC component of the magnetic flux density B_m is usually low, reducing the cores loss. The power stage of this converter is designed in [22]. The maximum output power is

$$P_{Omax} = V_O I_{Omax} = 12 \times 10 = 120 \text{ W} \tag{10.275}$$

and the minimum output power is

$$P_{Omin} = V_O I_{Omin} = 12 \times 1 = 12 \text{ W}. \tag{10.276}$$

The minimum load resistance is

$$R_{Lmin} = \frac{V_O}{I_{Omax}} = \frac{12}{10} = 1.2\ \Omega \tag{10.277}$$

and the maximum load resistance is

$$R_{Lmax} = \frac{V_O}{I_{Omin}} = \frac{12}{1} = 12\ \Omega. \tag{10.278}$$

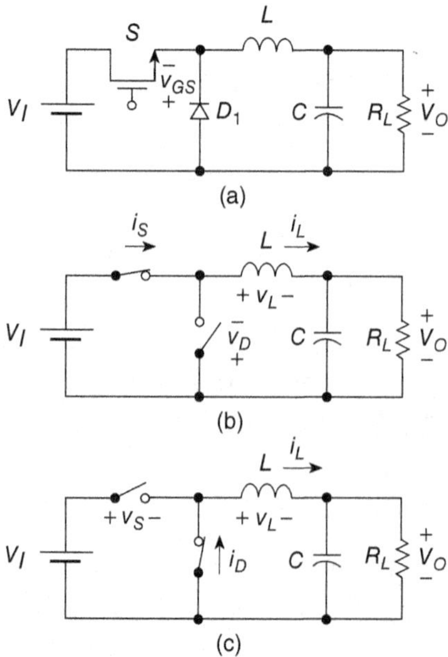

Figure 10.20 Buck PWM converter. (a) Circuit; (b) equivalent circuit when the metal-oxide-semiconductor field-effect transistor (MOSFET) is ON; and (c) equivalent circuit when the MOSFET is OFF

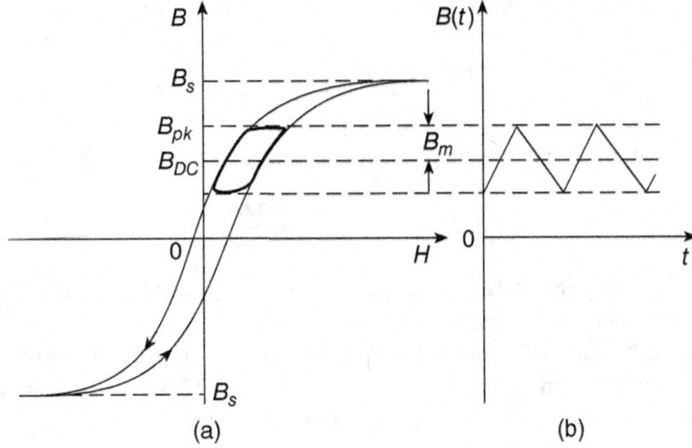

Figure 10.21 Magnetic core hysteresis and the waveform of the magnetic flux density $B(t)$ for CCM. (a) Hysteresis and the operating trajectory and (b) magnetic flux density waveform $B(t)$

The minimum DC voltage transfer function is

$$M_{VDCmin} = \frac{V_O}{V_{Imax}} = \frac{12}{32} = 0.375 \tag{10.279}$$

and the maximum DC voltage transfer function is

$$M_{VDCmax} = \frac{V_O}{V_{Imin}} = \frac{12}{24} = 0.5. \tag{10.280}$$

Assuming the converter efficiency $\eta = 0.85$, we can calculate the minimum duty cycle

$$D_{\min} = \frac{M_{VDCmin}}{\eta} = \frac{0.375}{0.85} = 0.441 \tag{10.281}$$

and the maximum duty cycle

$$D_{\max} = \frac{M_{VDCmax}}{\eta} = \frac{0.5}{0.85} = 0.5882. \tag{10.282}$$

The minimum inductance for CCM operation is

$$L_{\min} = \frac{R_{Lmax}(1 - D_{\min})}{2f_s} = \frac{12 \times (1 - 0.441)}{2 \times 10^5} = 33.54\ \mu\text{H}. \tag{10.283}$$

Pick $L = 40\ \mu$H.

The peak-to-peak inductor current ripple is

$$\Delta i_{Lmax} = \frac{V_O(1 - D_{\min})}{f_s L} = \frac{12 \times (1 - 0.441)}{10^5 \times 40 \times 10^{-6}} = 1.677\ \text{A}. \tag{10.284}$$

The buck inductor carries a large DC component of current. The peak inductor current is

$$I_{Lmax} = I_{Omax} + \frac{\Delta i_{Lmax}}{2} = 10 + \frac{1.677}{2} = 10.8385 \approx 10.839\ A. \tag{10.285}$$

The maximum energy stored in the inductor is

$$W_m = \frac{1}{2} L I_{Lmax}^2 = \frac{1}{2} \times 40 \times 10^{-6} \times 10.839^2 = 2.349\ \text{mJ}. \tag{10.286}$$

In the buck PWM converter, the inductor DC current is equal to the DC output current $I_L = I_O$. The rms value of the inductor current at $I_O = 10$ A is

$$I_{Lrms} = I_L \sqrt{1 + \frac{1}{12}\left(\frac{\Delta i_L}{I_L}\right)^2} = 10\sqrt{1 + \frac{1}{12}\left(\frac{1.677}{10}\right)^2} = 10.0117\ \text{A}. \tag{10.287}$$

The effect of the inductor current ripple Δi_L on the rms value of the inductor current is very small at full output power.

The peak-to-peak value of the AC component of the inductor current waveform is given by [22]

$$\Delta i_L = \frac{V_O(1-D)}{f_s L} = \frac{V_O\left(1 - \frac{V_O}{\eta V_I}\right)}{f_s L}. \tag{10.288}$$

The amplitudes of the fundamental component and the harmonics of the inductor current waveform are

$$I_{mn} = \frac{V_O}{f_s L}\frac{\sin \quad n\pi D}{n^2\pi^2 D}, \tag{10.289}$$

where $D = V_O/(\eta V_I)$. Figure 10.22 shows the amplitude spectrum of the AC component of the inductor current waveform for the first 10 harmonics at $V_{Imin} = 24$ V, $V_O = 12$ V, $f_s = 100$ kHz, and $L = 40$ μH. Figure 10.23 shows the amplitude spectrum of the AC component of the inductor current waveform for the first 10 harmonics at $V_{Imax} = 32$ V, $V_O = 12$ V, $f_s = 100$ kHz, and $L = 40$ μH. Figure 10.24 shows the spectrum of the AC winding resistance R_{wACn} for $R_{wDC} = 8.039$ mΩ and $d_i/\delta_w = 8.76$, computed from (5.434). Figure 10.25 shows the spectrum of the AC component of the winding power loss $P_{wn} = R_{wn}I_{mn}^2$ at $V_{Imax} = 32$ V and $I_{Omax} = 10$ A, $V_O = 12$ V, $f_s = 100$ kHz, and $L = 40$ μH. It can be seen that the winding power loss of the fundamental component is dominant in the total AC winding loss.

The *power harmonic content* is defined as the ratio of the winding power loss of the first m harmonics in the total AC winding power loss P_{wAC}

$$\gamma_m = \frac{\sum_{n=1}^{m} P_{wACn}}{\sum_{n=1}^{\infty} P_{wACn}} = \frac{\sum_{n=1}^{m} P_{wACn}}{P_{wAC}}. \tag{10.290}$$

Figure 10.26 shows the percentage power harmonic content of the winding power loss of the fundamental γ_1, the fundamental and the second harmonic γ_2, and the fundamental, the second harmonic, and the third harmonic in the total AC winding power loss P_{wAC} as functions of V_I. The power harmonic content of the fundamental γ_1 is over 95%, and the content of the first three harmonics γ_3 is over 99%.

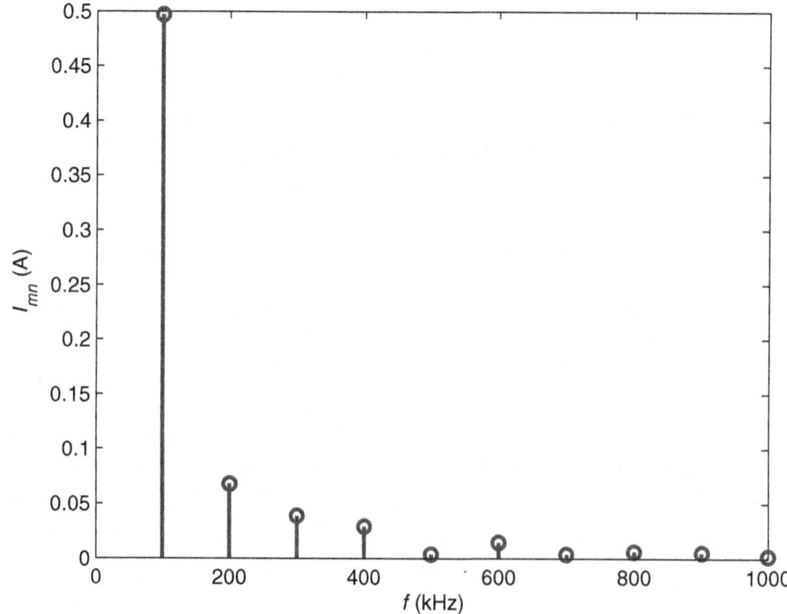

Figure 10.22 Amplitude spectrum of the inductor current waveform in the buck converter operating in CCM at $V_{Imin} = 24$ V, $V_O = 12$ V, $f_s = 100$ kHz, and $L = 40$ μH

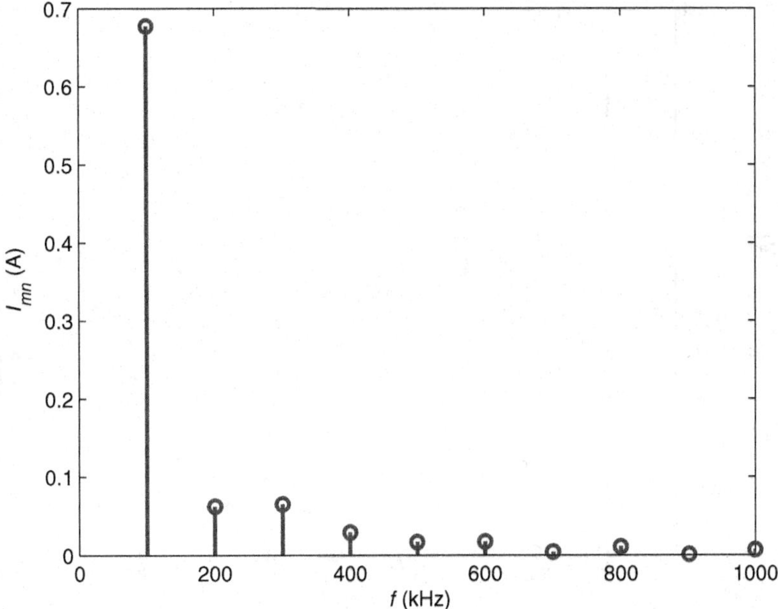

Figure 10.23 Amplitude spectrum of the inductor current waveform in the buck converter operating in CCM at $V_{Imax} = 32$ V, $V_O = 12$ V, $f_s = 100$ kHz, and $L = 40$ μH

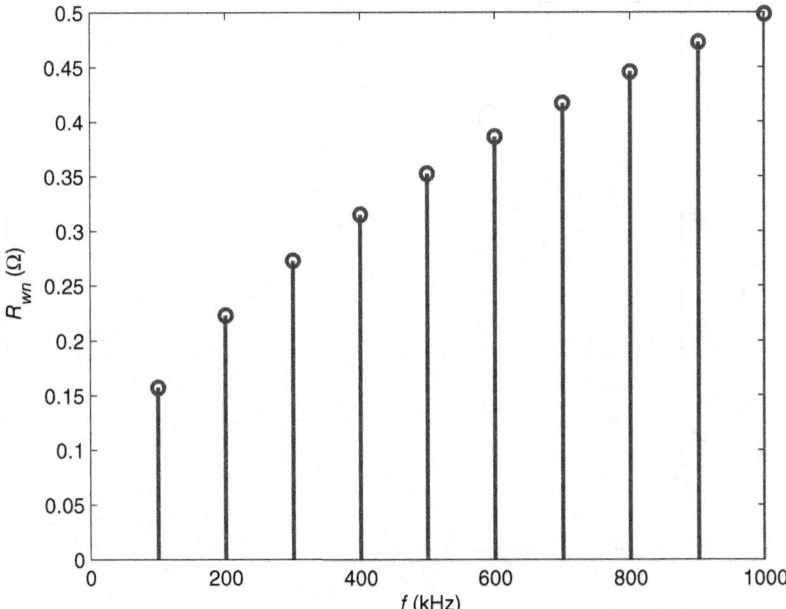

Figure 10.24 Spectrum of the AC winding resistance for $R_{DC} = 8.039$ mΩ and $d_i/\delta_w = 8.76$

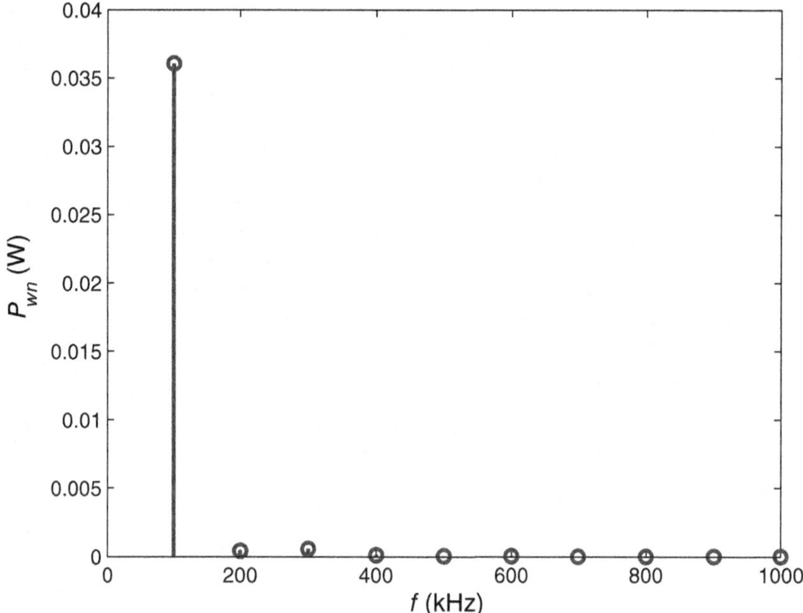

Figure 10.25 Spectrum of the AC component of the winding power loss at $V_{Imax} = 32$ V and $I_{Omax} = 10$ A

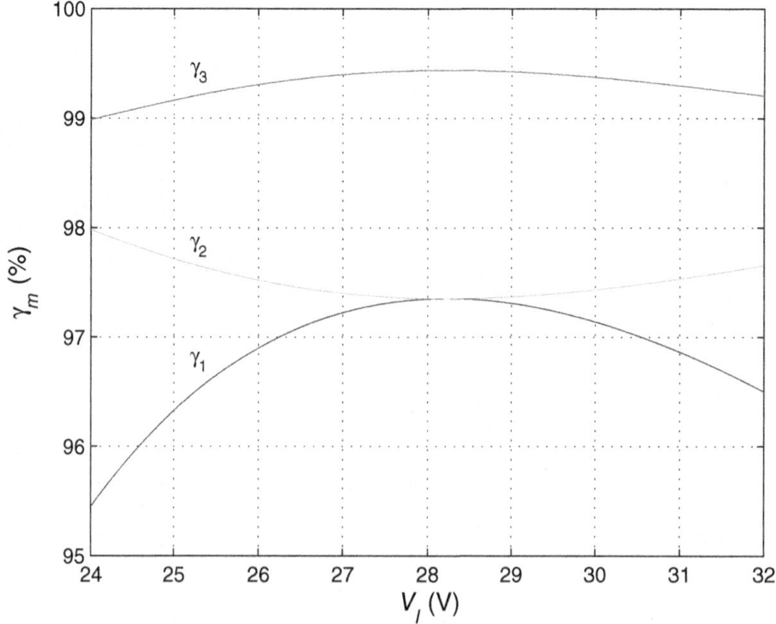

Figure 10.26 Power harmonic content in the total AC winding loss P_{wAC} as functions of the DC input voltage V_I

The winding resistance for the fundamental component can be approximated by

$$R_{w1} \approx \left(\frac{\pi}{4}\right)^{\frac{3}{4}} \left(\frac{d}{\delta_w}\right) \sqrt{\frac{d}{p}} \frac{2N_l^2 + 1}{3} R_{wDC} \quad \text{for} \quad d/\delta_w \geq 3. \tag{10.291}$$

Hence, the AC winding power loss is

$$P_{wAC} \approx \frac{1}{2} R_{w1} I_{m1}^2 = \left(\frac{\pi}{4}\right)^{\frac{3}{4}} \left(\frac{d}{\delta_w}\right) \sqrt{\frac{d}{p}} \frac{(2N_l^2 + 1) V_O^2 R_{wDC}}{6\pi^4 (f_s LD)^2} \sin^2(\pi D)$$

$$= 1.427 \times 10^{-3} \left(\frac{d}{\delta_w}\right) \sqrt{\frac{d}{p}} \frac{(2N_l^2 + 1) V_O^2 R_{wDC}}{(f_s LD)^2} \sin^2(\pi D) \quad \text{for} \quad d/\delta_w \geq 3. \tag{10.292}$$

Assume $K_u = 0.4$, $B_{pk} = 0.2$ T, and $J_m = 5$ A/mm^2. The required core area product is

$$A_p = \frac{2W_m}{K_u J_m B_{pk}} = \frac{2 \times 2.349 \times 10^{-3}}{0.4 \times 5 \times 10^6 \times 0.2} = 1.1747 \text{ cm}^4. \tag{10.293}$$

Select a Magnetics 0P-44011EC EE ungapped ferrite core with $A_p = 1.39$ cm^4, $A_c = 127$ mm^2, $l_c = 76.7$ mm, $V_c = 9.78$ cm^3, $A_L = 3260$ mH/1000 turns, $C = 10.69$ mm, $D = 10$ mm, $E = 27.6$ mm, and $F = 10.7$ mm. We select the P-type ferrite magnetic material The power loss density of this core is given by

$$P_v = 0.0434 f^{1.63} (10B_m)^{2.62} \text{ (mW/cm}^3) = 18.092 f^{1.63} B_m^{2.62} \text{ (mW/cm}^3), \tag{10.294}$$

where f is in kilohertz and B_m is in Torr. We select the Cosmo rectangular bobbin 2587-0 with the window height of $B_h = 12.59$ mm and the bobbin window width of $B_w = 7.505$ mm.

The core window area is

$$W_a = \frac{A_p}{A_c} = \frac{1.39}{1.27} = 1.094 \text{ cm}^2. \tag{10.295}$$

The cross-sectional area of the bare wire is

$$A_w = \frac{I_{Lmax}}{J_{m(\max)}} = \frac{10.8385}{5 \times 10^6} = 2.1677 \text{ mm}^2. \tag{10.296}$$

Select a round solid copper wire AWG13, which has $d_i = 1.83$ mm, $d_o = 1.9$ mm, and $R_{wDC}/l_w = 0.00656$ Ω/m. The insulated wire cross-sectional area is

$$A_{wo} = \frac{\pi d_o^2}{4} = \frac{\pi \times (1.9 \times 10^{-3})^2}{4} = 2.8353 \text{ mm}^2. \tag{10.297}$$

The number of turns is

$$N = \frac{K_u W_a}{A_{wo}} = \frac{0.4 \times 1.094 \times 10^{-4}}{2.8353 \times 10^{-6}} = 15.4409. \tag{10.298}$$

Pick $N = 16$. The desired value of the inductance can be obtained by the proper choice of the length of an air gap. The air gap length is

$$l_g = \frac{\mu_0 A_c N^2}{L} - \frac{l_c}{\mu_{rc}} = \frac{4\pi \times 10^{-7} \times 1.27 \times 10^{-4} \times 16^2}{40 \times 10^{-6}} - \frac{76.7 \times 10^{-3}}{2500} = 0.99 \text{ mm} \approx 1 \text{ mm}. \tag{10.299}$$

The peak value of the AC component of the magnetic flux density is

$$B_m = \frac{\mu_0 N}{l_g + \frac{l_c}{\mu_{rc}}} \left(\frac{\Delta i_{Lmax}}{2}\right) = \frac{4\pi \times 10^{-7} \times 16}{1 \times 10^{-3} + \frac{76.7 \times 10^{-3}}{2500}} \left(\frac{1.677}{2}\right) = 0.01636 \text{ T}. \tag{10.300}$$

The peak value of the total magnetic flux density, which consists of the DC and AC components, is given by

$$B_{pk} = \frac{\mu_0 N}{l_g + \frac{l_c}{\mu_{rc}}} \left(I_O + \frac{\Delta i_{Lmax}}{2}\right) = \frac{4\pi \times 10^{-7} \times 16}{(1 + \frac{76.7}{2500}) \times 10^{-3}} \left(10 + \frac{1.677}{2}\right)$$

$$= 0.21143 \text{ T} < B_s = 0.3 \text{ T}. \tag{10.301}$$

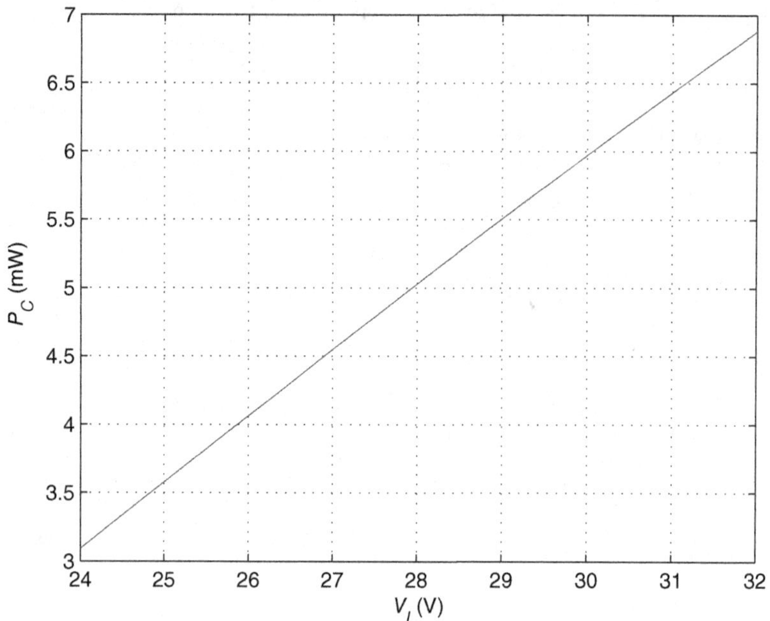

Figure 10.27 Core loss P_C as a function of the DC input voltage V_I

Thus, $B_{DC} = B_{pk} - B_m = 0.21143 - 0.01636 = 0.19507$ T. The core power loss density is

$$P_v = 0.0434 f^{1.63} (10 B_m)^{2.62} = 0.0434 \times 100^{1.63} \times (10 \times 0.01636)^{2.62} = 0.688 \text{ mW/cm}^3. \quad (10.302)$$

The core power loss is

$$P_C = V_c P_v = 9.78 \times 0.688 = 6.726 \text{ mW}. \quad (10.303)$$

Figure 10.27 shows a plot of the core loss P_C as a function of the DC input voltage V_I. The maximum magnetic flux linkage is

$$\lambda_{\max} = L I_{Lmax} = L \left(I_O + \frac{\Delta i_{Lmax}}{2} \right) = 40 \times 10^{-6} \times 10.839 = 433.56 \times 10^{-6} \quad \text{V} \cdot \text{s}. \quad (10.304)$$

The length of a single turn for the EE core is given by

$$l_T = 2(C + E) = 2(10.69 + 27.6) = 76.58 \text{ mm}, \quad (10.305)$$

where the outer width of the winding window, and C is the core thickness. The length of the winding wire is

$$l_w = N l_T = 16 \times 7.658 \times 10^{-2} = 1.225 \text{ m}. \quad (10.306)$$

The DC and low-frequency resistance of the winding is

$$R_{wDC} = l_w \left(\frac{R_{wDC}}{l_w} \right) = 1.225 \times 0.00656 = 0.008039 \ \Omega. \quad (10.307)$$

The DC and low-frequency winding power loss at $T = 20\,^{\circ}$C is

$$P_{wDC} = R_{wDC} I_L^2 = R_{wDC} I_O^2 = 0.008039 \times 10^2 = 0.8039 \ W. \quad (10.308)$$

Figure 10.28 shows the DC winding power loss P_{wDC} as a function of the DC output current I_O.

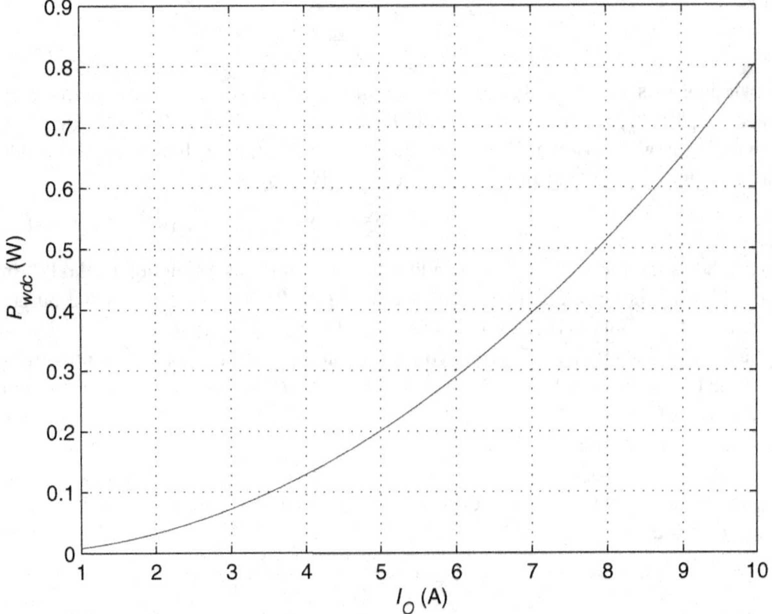

Figure 10.28 DC winding power loss P_{wDC} as a function of the output current I_O

The sum of the core loss and the DC and low-frequency winding loss is

$$P_{cwDC} = P_C + P_{wDC} = 6.729 \times 10^{-3} + 0.8039 = 0.8106 \text{ W}. \tag{10.309}$$

The height of the core winding window is

$$H_w = 2D = 2 \times 10 = 20 \text{ mm}, \tag{10.310}$$

where D is the height of one half of the core window. The maximum number of turns per layer is

$$N_{l1} = \frac{H_w}{d_o} = \frac{20}{1.9} = 10.5. \tag{10.311}$$

Pick $N_{l1} = 10$. The number of layers is

$$N_l = \frac{N}{N_{l1}} = \frac{16}{10} = 1.6. \tag{10.312}$$

Pick $N_l = 2$. Hence, the number of turns per layer is

$$N_{l1} = \frac{N}{N_l} = \frac{16}{2} = 8. \tag{10.313}$$

The ratio of the inductor peak-to-peak ripple current to the DC component of the inductor current at full power is

$$\frac{\Delta i_{Lmax}}{I_L} = \frac{1.677}{10} = 0.1677. \tag{10.314}$$

The skin depth of copper at $f_s = 100$ kHz is

$$\delta_w = \frac{66.2}{\sqrt{f_s}} \text{ (mm)} = \frac{66.2}{\sqrt{10^5}} (mm) = 0.209 \text{ mm}. \tag{10.315}$$

The ratio is

$$\frac{d_i}{\delta_w} = \frac{1.83}{0.209} = 8.74. \tag{10.316}$$

The total winding loss is higher than the DC winding loss due to skin and proximity effects. From (5.450), we compute $F_{Rh} = P_w/P_{wDC} = 1.0459$ for $N_l = 2$, $d_i/\delta_w = 8.76$, and $\Delta i_{Lmax}/I_L = 0.1677$. The total winding power loss, that is, the DC and AC winding power losses due to all the harmonics, taking into account the skin and proximity effects, is given by

$$P_w = F_{Rh} P_{wDC} = 1.0459 \times 0.8039 = 0.84 \text{ W}. \tag{10.317}$$

Figure 10.29 shows the plots of the total winding power loss P_w as a function of the DC input voltage V_I. Figure 10.30 shows the plots of the total winding loss P_w as a function of the output current I_O for fixed values of V_I. Figure 10.31 shows the ratio of the total winding loss to the DC winding loss P_w/P_{wDC} as a function of the DC input voltage V_I at I_{Omax}. Figure 10.32 shows the plots of the ratio of the total winding loss to the DC winding loss P_w/P_{wDC} as a function of the output current I_O at fixed values of V_I.

Figure 10.33 shows a plot of the AC winding loss $P_{wAC} = P_w - P_{wDC}$ as a function of the DC input voltage V_I. Figure 10.34 shows the plots of the AC winding coefficient $F_{RhAC} = P_{wAC}/P_{wDC}$ as a function of the DC current I_O for different values of the DC input voltage V_I.

The core and winding loss is

$$P_{cw} = P_C + P_w = 0.006729 + 0.84 = 0.846 \text{ W}. \tag{10.318}$$

Figure 10.35 shows the total loss P_{cw} as a function of the DC input voltage V_I at I_{Omax}. Figure 10.36 shows the total loss P_{cw} as a function of the output current I_O at fixed values of V_I. Figure 10.37 shows a plot of K_{Rh} as a function of V_I, computed from (5.526). Figure 10.38 shows the plots of K_{Rh} as a function of I_O.

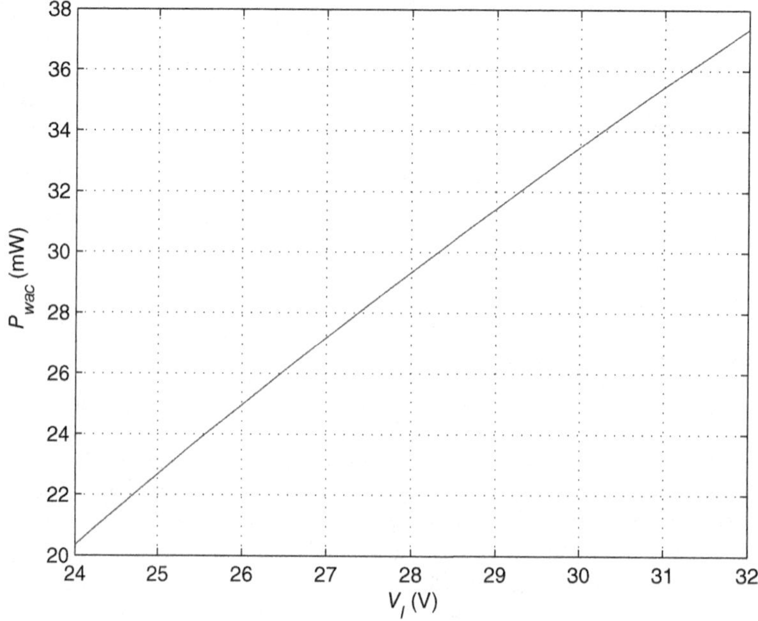

Figure 10.29 Total winding loss $P_w = P_{wDC} + P_{wAC}$ as a function of the DC input voltage V_I at I_{Omax}

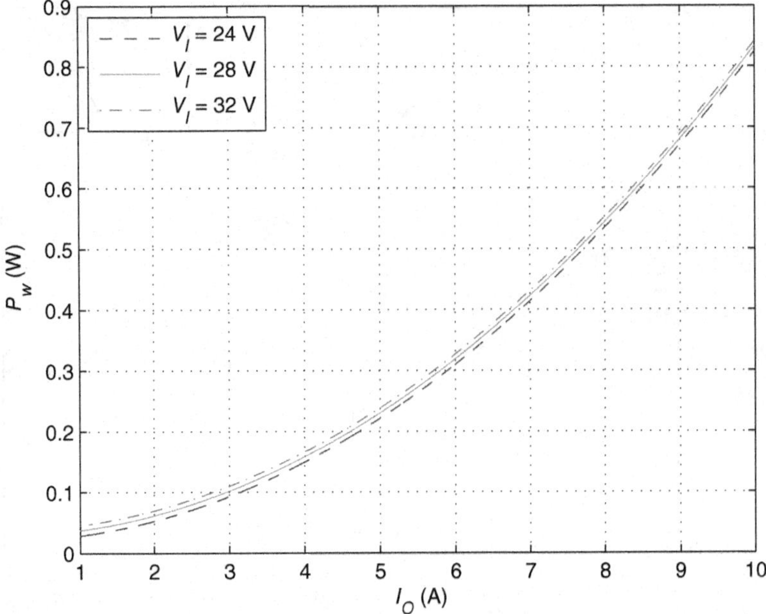

Figure 10.30 Total winding loss $P_w = P_{wDC} + P_{wAC}$ as a function of the output current I_O for fixed values of V_I

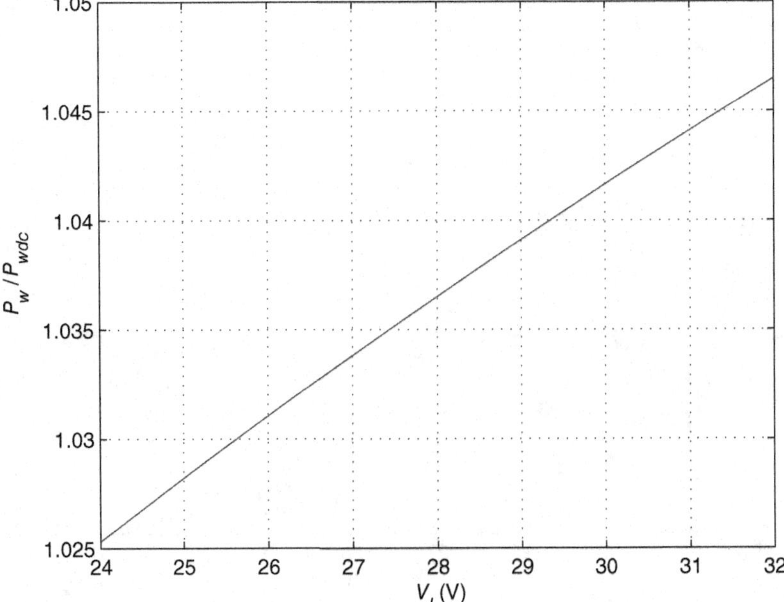

Figure 10.31 Ratio of the total winding loss to the DC winding loss P_w/P_{wDC} as a function of the DC input voltage V_I at I_{Omax}

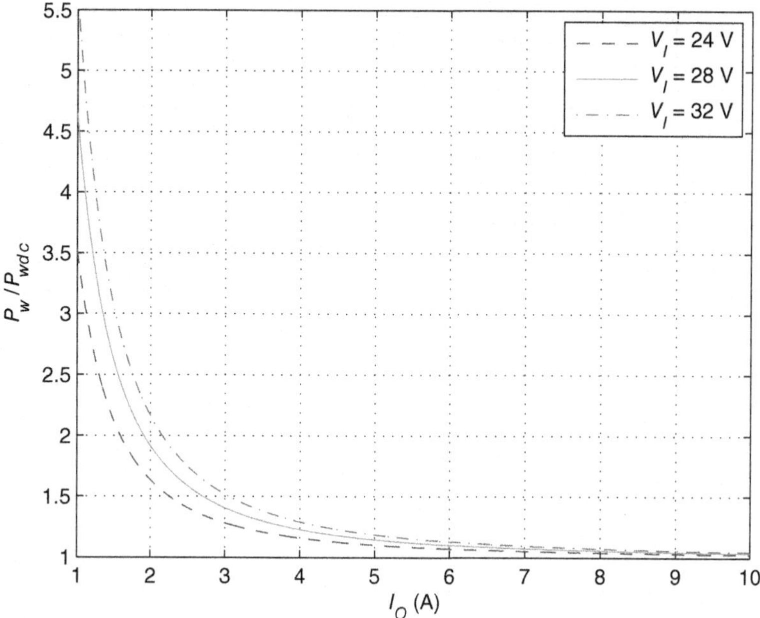

Figure 10.32 Ratio of the total winding loss to the DC winding loss P_w/P_{wDC} as a functid values of V_I

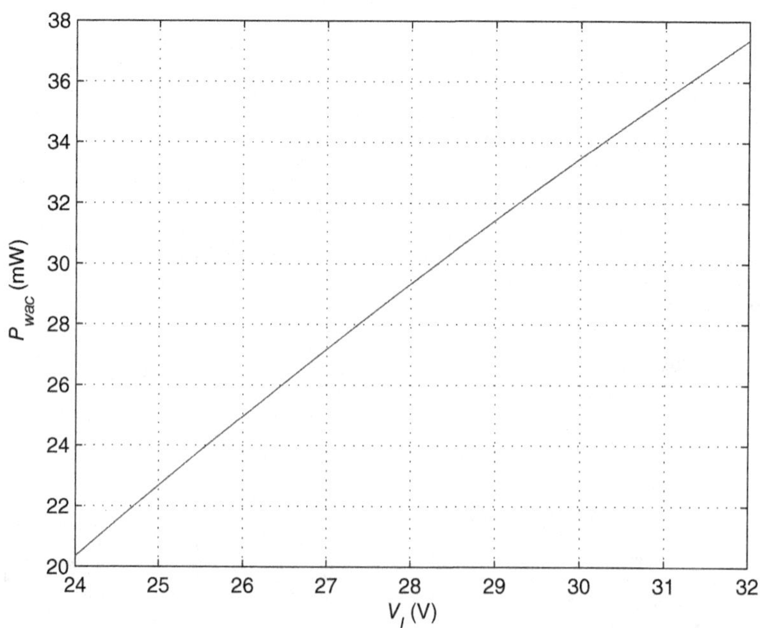

Figure 10.33 AC winding loss P_{wAC} as a function of the DC input voltage V_I

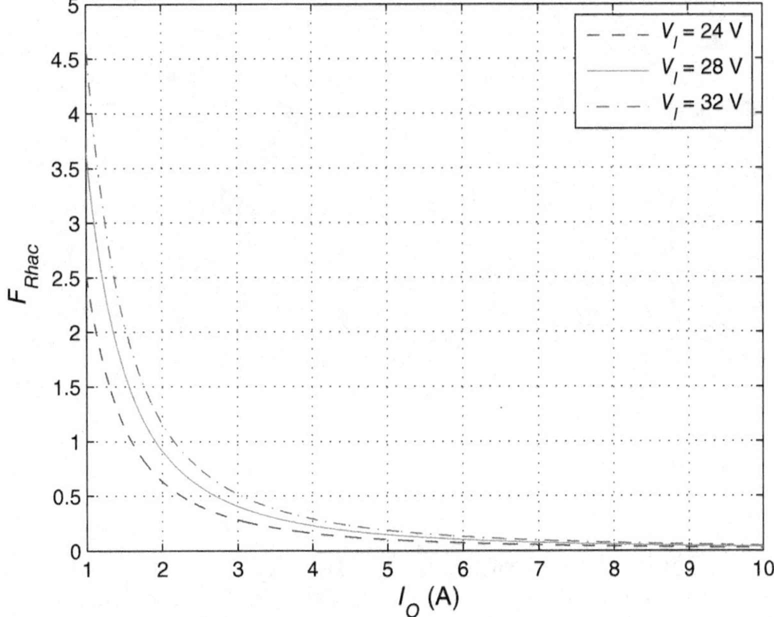

Figure 10.34 AC winding loss coefficient $F_{RhAC} = P_{wAC}/P_{wDC}$ as a function of the output current I_O

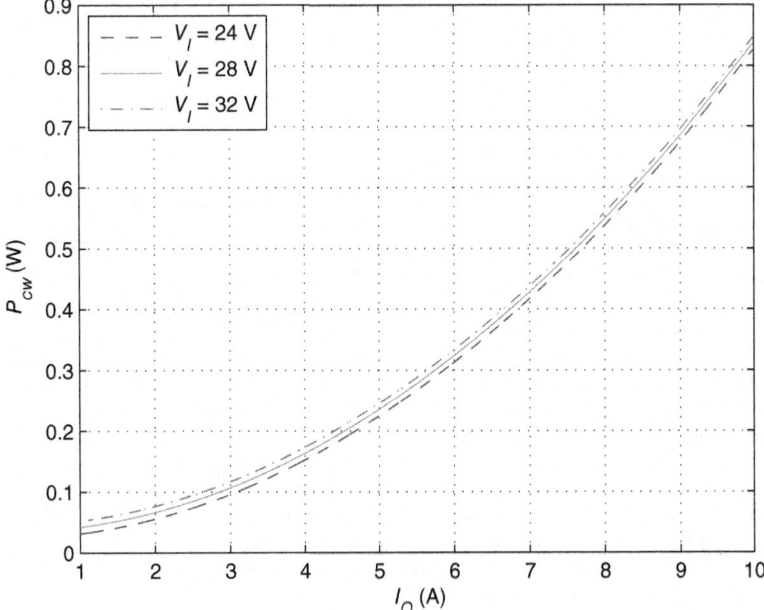

Figure 10.35 Total inductor loss $P_{cw} = P_w + P_C$ as a function of the output current I_O for fixed values of V_I

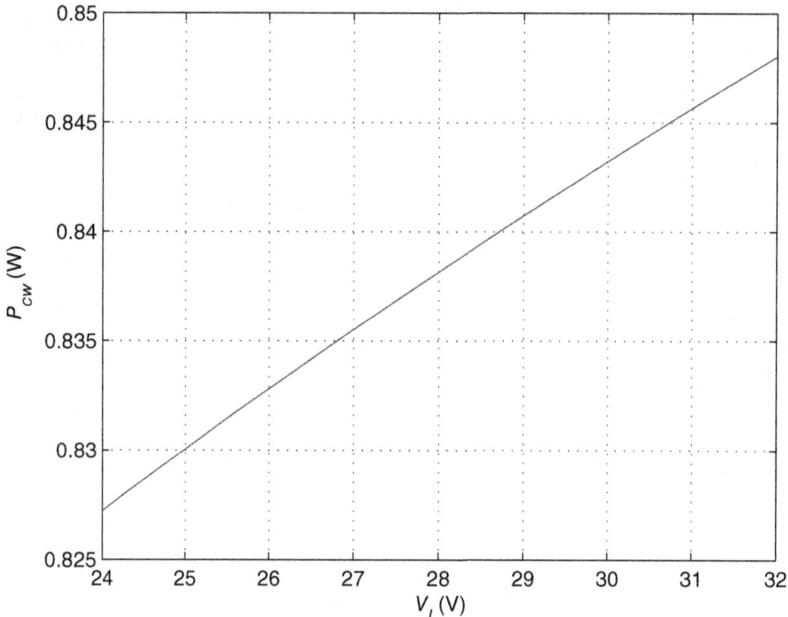

Figure 10.36 Total inductor loss $P_{cw} = P_w + P_C$ as a function of the DC input voltage V_I

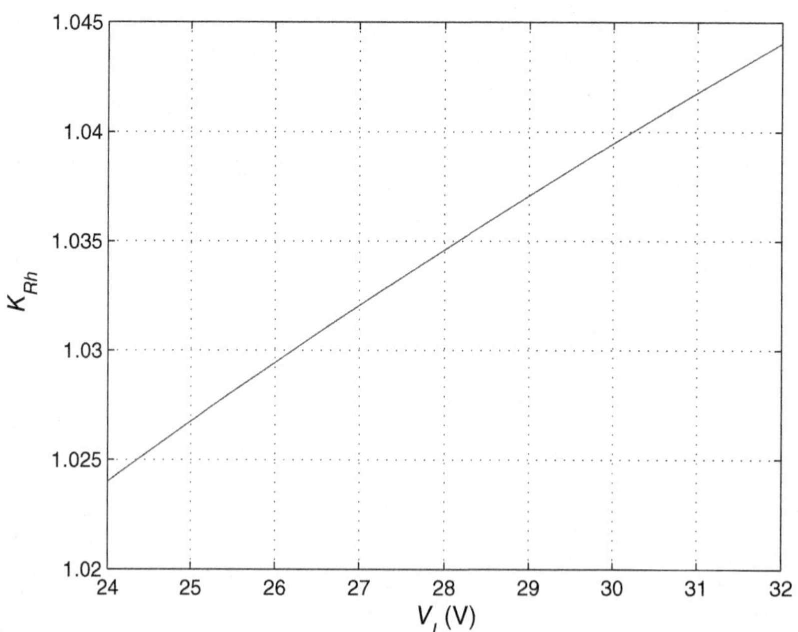

Figure 10.37 K_{Rh} as a function of the DC input voltage V_I

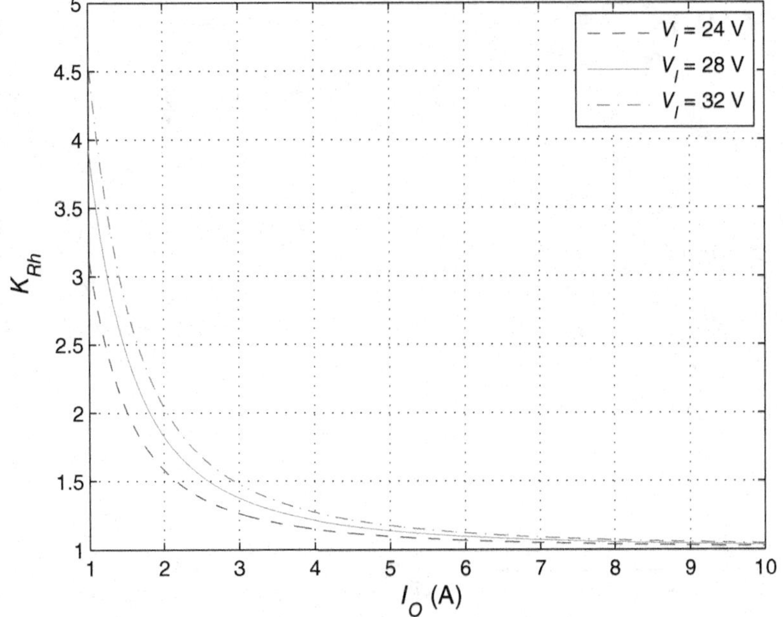

Figure 10.38 K_{Rh} as a function of the output current I_O

This loss can be expressed as

$$P_{cw} = r_{L(DC)} I_{Lrms}^2. \tag{10.319}$$

The DC and low-frequency ESR is

$$r_{L(DC)} = \frac{P_{cw}}{I_{Lrms}^2} = \frac{0.846}{10.0177^2} = 8.43\ \text{m}\Omega. \tag{10.320}$$

The outer surface of the inductor consisting of the core and the winding is estimated as

$$\begin{aligned} A_t &= 4AB + 2AC + 4BC + 2(E - F)(2D + E) \\ &= 4 \times 4 \times 1.7 + 2 \times 4 \times 1.069 + 4 \times 1.7 \times 1.069 + 2(2.76 - 1.07)(2 \times 1 + 2.76) \\ &= 59.12\ \text{cm}^2. \end{aligned} \tag{10.321}$$

The surface power loss density is

$$\psi = \frac{P_{cw}}{A_t} = \frac{0.846}{59.12} = 0.01431\ \text{W/cm}^2. \tag{10.322}$$

The temperature rise of the inductor is [1]

$$\Delta T = 450\psi^{0.826} = 450 \times 0.01431^{0.826} = 13.48^\circ\text{C}, \tag{10.323}$$

where ψ is in Watts per centimeter square. Figure 10.39 shows a plot of the temperature rise ΔT as a function of V_I. Figure 10.40 shows the plots of the temperature rise ΔT as a function of I_O.

The window utilization factor is

$$K_u = \frac{A_{Cu}}{W_a} = \frac{NA_{wo}}{W_a} = \frac{16 \times 2.8353}{1.094} = 0.4145. \tag{10.324}$$

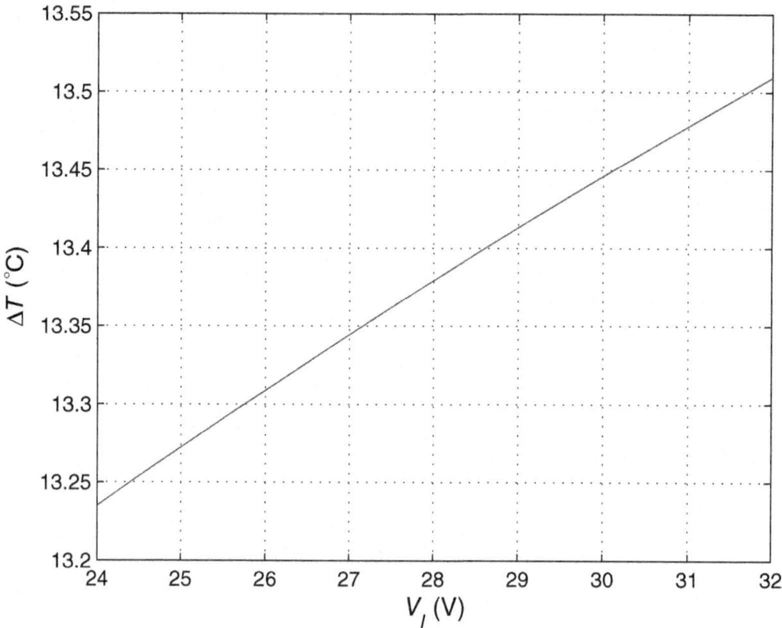

Figure 10.39 Temperature rise ΔT as a function of the DC input voltage V_I

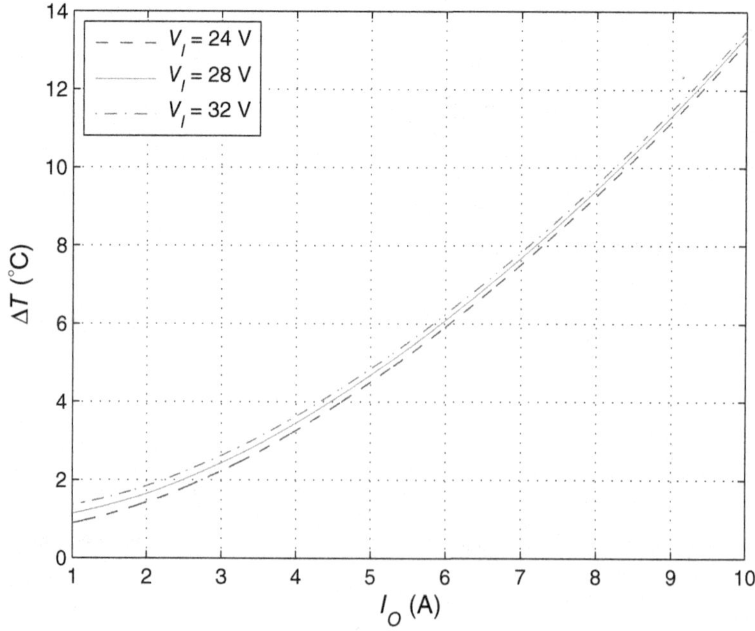

Figure 10.40 Temperature rise ΔT as a function of the output current I_O

10.11 Inductor Design for Buck Converter in DCM Using A_p Method

Example 10.8

A buck pulse-width modulated (PWM) converter is operated in the discontinuous conduction mode (DCM) and has the following specifications: $V_I = 28 \pm 4\,\text{V}$, $V_O = 12\,\text{V}$, $I_O = 0 - 10\,\text{A}$, and $f_s = 100\,\text{kHz}$.

Solution: Figure 10.41 shows the circuit of the buck converter and its equivalent circuits for DCM. Figure 10.42 shows the core hysteresis. The maximum DC output power is

$$P_{Omax} = V_O I_{Omax} = 12 \times 10 = 120\,\text{W}. \tag{10.325}$$

The minimum load resistance is

$$R_{Lmin} = \frac{V_O}{I_{Omax}} = \frac{12}{10} = 1.2\,\Omega. \tag{10.326}$$

The minimum and maximum duty cycles are

$$M_{VDCmin} = \frac{V_O}{V_{Imax}} = \frac{12}{32} = 0.375 \tag{10.327}$$

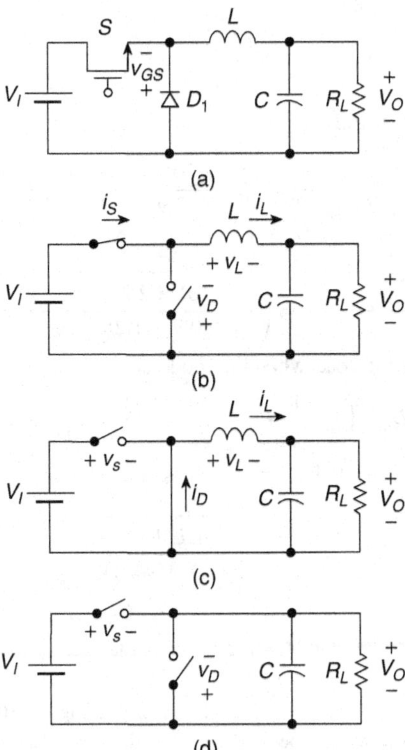

Figure 10.41 Buck PWM converter and its equivalent circuits for DCM. (a) Circuit. (b) Equivalent circuit when the metal-oxide-semiconductor field-effect transistor (MOSFET) is ON. (c) Equivalent circuit when the MOSFET is OFF. (c) Equivalent circuit when the MOSFET and diode are OFF

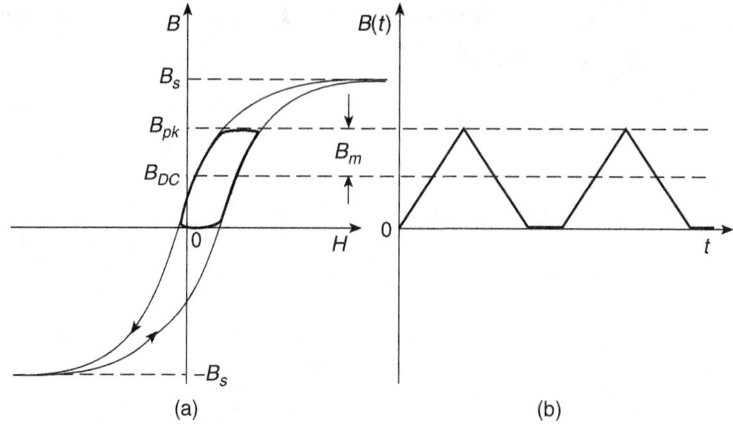

Figure 10.42 Magnetic core hysteresis and the waveform of the magnetic flux density $B(t)$ for DCM. (a) Hysteresis and operating trajectory. (b) Magnetic flux density waveform $B(t)$

and

$$M_{VDCmax} = \frac{V_O}{V_{Imin}} = \frac{12}{24} = 0.5. \quad (10.328)$$

Assume the converter efficiency $\eta = 0.9$ and the dwell duty cycle $D_w = 0.05$. The maximum inductance is

$$L_{max} = \frac{\eta R_{Lmin}(1 - M_{VDCmax})(1 - D_w)^2}{2f_s} = \frac{0.9 \times 1.2 \times (1 - 0.5)(1 - 0.05)^2}{2 \times 10^5}$$
$$= 2.4368\ \mu\text{H}. \quad (10.329)$$

Let $L = 2.2\ \mu$H.

The maximum duty cycle at full power occurs at $V_{Imin} = 24$ V and is given by

$$D_{max} = \sqrt{\frac{2f_s L M_{VDCmax}^2}{\eta R_{Lmin}(1 - M_{VDCmax})}} = \sqrt{\frac{2 \times 10^5 \times 2.2 \times 10^{-6} \times 0.5^2}{0.9 \times 1.2(1 - 0.5)}} = 0.4513. \quad (10.330)$$

and the minimum duty cycle is

$$D_{min} = \sqrt{\frac{2f_s L M_{VDCmin}^2}{\eta R_{Lmin}(1 - M_{VDCmin})}} = \sqrt{\frac{2 \times 10^5 \times 2.2 \times 10^{-6} \times 0.375^2}{0.9 \times 1.2(1 - 0.375)}} = 0.3027. \quad (10.331)$$

The minimum duty cycle at full load when the diode is ON is

$$D_{1min} = D_{max}\left(\frac{1}{M_{VDCmax}} - 1\right) = 0.4513\left(\frac{1}{0.5} - 1\right) = 0.4513. \quad (10.332)$$

Hence, $D_{max} + D_{1min} = 0.4513 + 0.4513 = 0.9026 < 1$.

The duty cycles are given by

$$D = \sqrt{\frac{2f_s L M_{VDC}^2 I_O}{(1 - M_{VDC})V_O}} \quad (10.333)$$

and

$$D_1 = \sqrt{\frac{2f_s L(1 - M_{VDC})I_O}{V_O}}. \quad (10.334)$$

Figure 10.43 shows the plots of I_{mn} as a function of I_O. Figure 10.44 shows the plots of I_{mn} as a function of D. Figure 10.45 shows the plots of D and D_1 as functions of I_O. Figure 10.46 shows the plots of $D + D_1$ as a function of the output power P_O at fixed values of the DC input voltage V_I for the buck converter operating in DCM. Figure 10.47 shows the plots of $D + D_1$ as a function of the DC input voltage V_I at fixed values of the output power P_O for the buck converter operating in DCM.

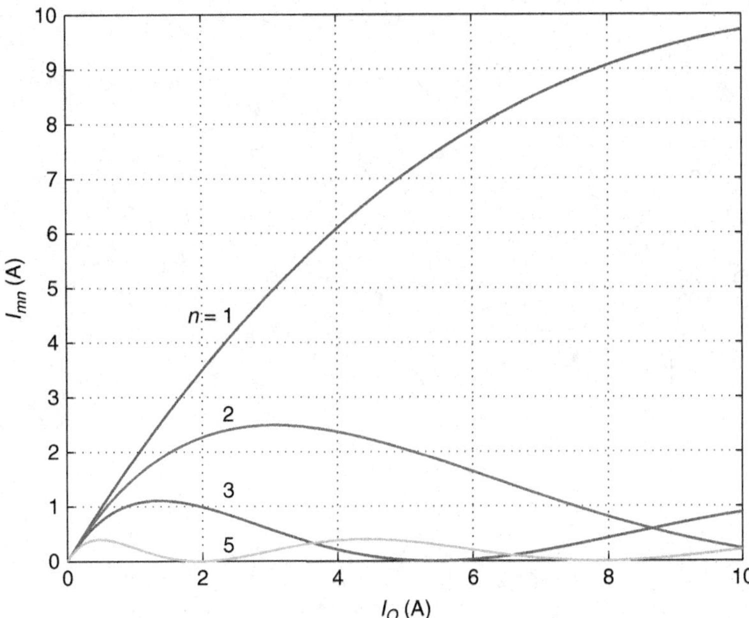

Figure 10.43 I_{mn} as a function of the output current I_O at $V_O = 12\,\text{V}$, $f_s = 100\,\text{kHz}$, $L = 2.2\,\mu\text{H}$, and $V_I = 24\,\text{V}$ for the buck converter operating in DCM

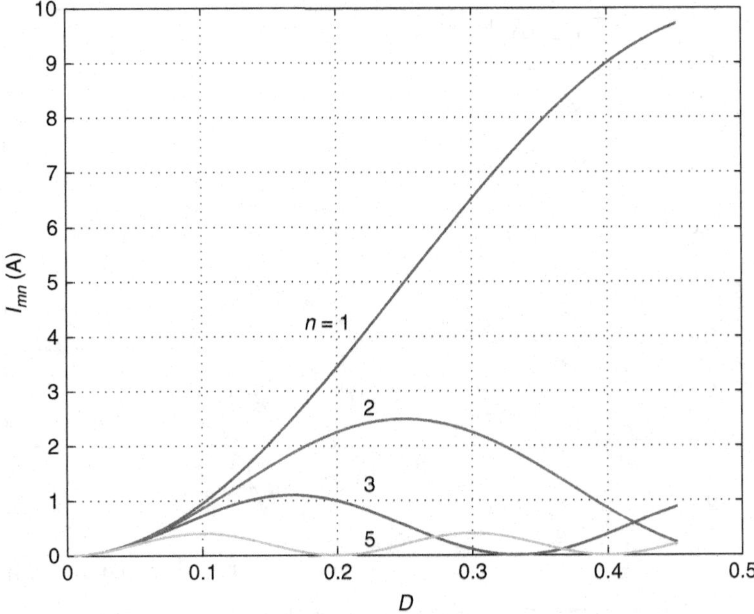

Figure 10.44 I_{mn} as a function of D at $V_O = 12\,\text{V}$, $f_s = 100\,\text{kHz}$, $L = 2.2\,\mu\text{H}$, and $V_I = 24\,\text{V}$ for the buck converter operating in DCM

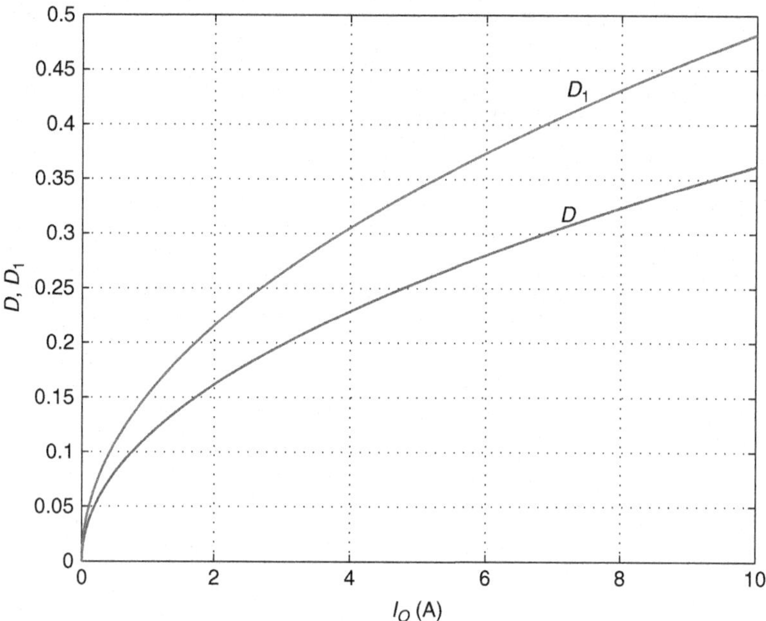

Figure 10.45 D and D_1 as functions of the output current I_O at $V_O = 12\,\text{V}, f_s = 100\,\text{kHz}, L = 2.2\,\mu\text{H}$, and $V_I = 28\,\text{V}$ for the buck converter operating in DCM

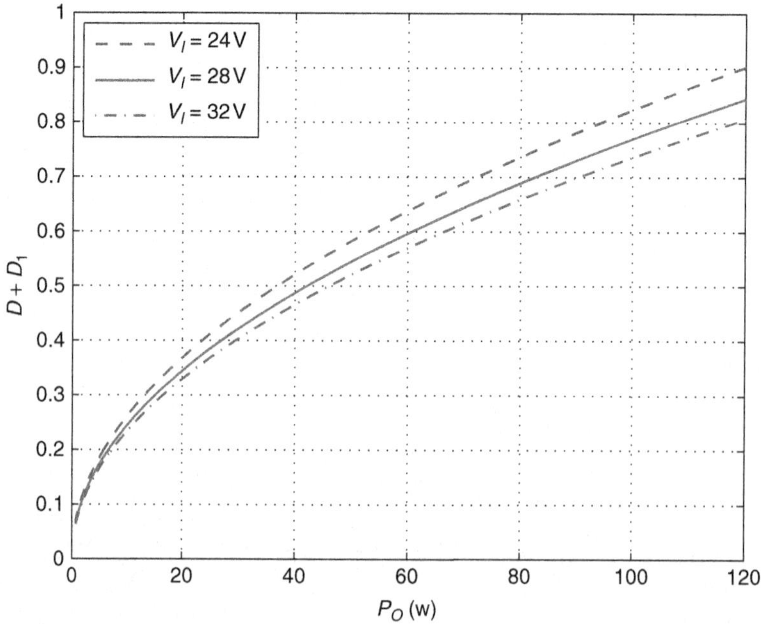

Figure 10.46 $D + D_1$ as a function of the output power P_O at fixed values of the input voltage V_I for the buck converter operating in DCM

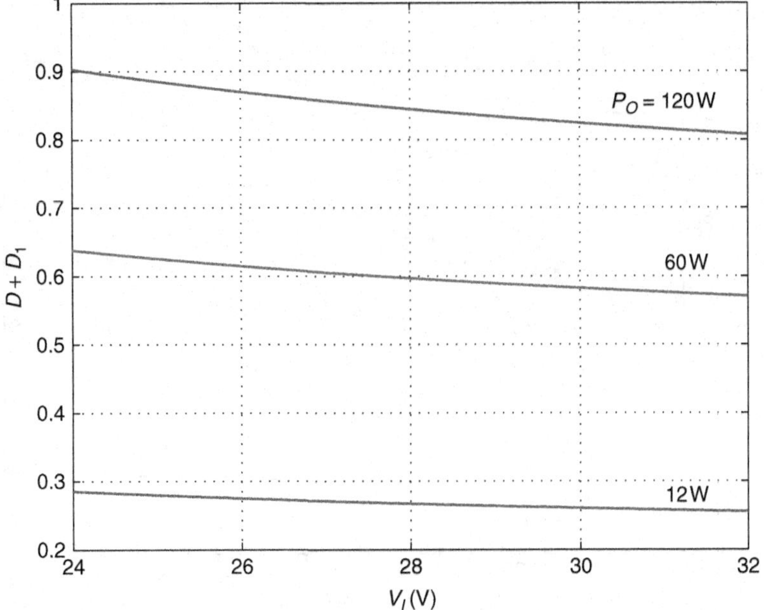

Figure 10.47 $D + D_1$ as a function of the DC input voltage V_I at fixed values of the output power P_O for the buck converter operating in DCM

The maximum inductor current is

$$\Delta i_{Lmax} = \frac{D_{min}(V_{Imax} - V_O)}{f_s L} = \frac{0.3027 \times (32 - 12)}{10^5 \times 2.2 \times 10^{-6}} = 27.518\text{ A}. \tag{10.335}$$

Figure 10.48 shows the plots of Δi_L as a function of V_I at fixed values of the output power P_O.

The maximum energy stored in the inductor is

$$W_m = \frac{1}{2} L I_{Lmax}^2 = \frac{1}{2} \times 2.2 \times 10^{-6} \times 27.518^2 = 0.832\text{ mJ}. \tag{10.336}$$

The maximum rms value of the inductor current is

$$I_{Lrms(max)} = \Delta i_{Lmax} \sqrt{\frac{D_{min} + D_{1max}}{3}} = 27.51 \sqrt{\frac{0.3027 + 0.5046}{3}} = 14.2787\text{ A}. \tag{10.337}$$

Assume $K_u = 0.4$, $B_{pk} = 0.2$ T, and $J_m = 5$ A/mm^2. The core area product is

$$A_p = \frac{2W_m}{K_u J_m B_{pk}} = \frac{2 \times 0.832 \times 10^{-3}}{0.4 \times 5 \times 10^6 \times 0.2} = 0.416\text{ cm}^4. \tag{10.338}$$

Select a Magnetics PQ42625 ferrite core. The parameters of this core are as follows: $A_p = 0.59$ cm^4, $A_c = 1.2$ cm^2, $l_c = 54.3$ mm, and $V_c = 6.53$ cm^3. The core material is F-type whose $\mu_{rc} = 3000$. We select the Magnetics PQ printed circuit bobbin PCB2625LA with the bobbin window height $B_h = 13.58$ mm and inner diameter $D_{ib} = 14.19$ mm.

The core window area is

$$W_a = \frac{A_P}{A_c} = \frac{0.59}{1.2} = 0.4917\text{ cm}^2. \tag{10.339}$$

The mean turn length (MTL) is

$$l_T = \frac{\pi(E + F)}{2} = \frac{\pi(22.5 + 12) \times 10^{-3}}{2} = 5.419\text{ cm}. \tag{10.340}$$

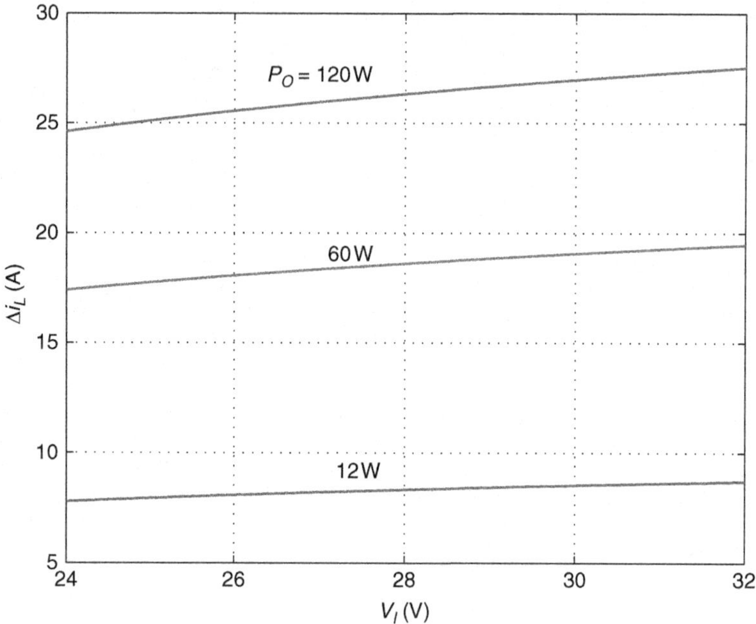

Figure 10.48 Δi_L as a function of the DC input voltage V_I at fixed values of the output power P_O for the buck converter operating in DCM

Multiple-Strand Winding. The skin depth of copper wire at $f = 100\,\text{kHz}$ is

$$\delta_w = \frac{66.2}{\sqrt{f}}(\text{mm}) = \frac{66.2}{\sqrt{10^5}}(\text{mm}) = 0.209\,\text{mm}. \tag{10.341}$$

The bare wire diameter of a strand at which the skin and proximity effects can be neglected for the fundamental component is

$$d_i = 2\delta_w = 2 \times 0.209 = 0.418\,\text{mm}. \tag{10.342}$$

A copper wire AWG26 is selected for the strands with $d_i = 0.405\,\text{mm}$, $d_o = 0.452\,\text{mm}$, $A_{wsi} = 0.1288\,\text{mm}^2$, and $R_{wDCs}/l_w = 0.1345\,\Omega/\text{m}$. The wire cross-sectional area is

$$A_w = \frac{I_{Lmax}}{J_{m(max)}} = \frac{27.518}{5 \times 10^6} = 5.5036\,\text{mm}^2. \tag{10.343}$$

The number of strands is

$$S_n = \frac{A_w}{A_{wsi}} = \frac{5.5036}{0.1288} = 42.73. \tag{10.344}$$

Select $S_n = 43$. The cross-sectional area of the insulated strand is

$$A_{wso} = \frac{\pi d_o^2}{4} = \frac{\pi \times (0.452 \times 10^{-3})^2}{4} = 0.1604\,\text{mm}^2. \tag{10.345}$$

The number of turns is

$$N = \frac{K_u W_a}{S_n A_{wso}} = \frac{0.4 \times 0.4916 \times 10^{-4}}{43 \times 0.1604 \times 10^{-6}} = 2.851. \tag{10.346}$$

Select $N = 3$.

The air gap is used to obtain the required inductance, avoid core saturation, and store magnetic energy. The air-gap length is

$$l_g = \frac{\mu_0 A_c N^2}{L} - \frac{l_c}{\mu_{rc}} = \frac{4\pi \times 10^{-7} \times 1.2 \times 10^{-4} \times 3^2}{2.2 \times 10^{-6}} - \frac{54.3 \times 10^{-3}}{3000} = 0.5984\,\text{mm}. \tag{10.347}$$

The peak value of the AC component of the magnetic flux density is

$$B_m = \frac{\mu_0 N}{l_g + \frac{l_c}{\mu_{rc}}}\left(\frac{\Delta i_{Lmax}}{2}\right) = \frac{4\pi \times 10^{-7} \times 3}{0.5984 \times 10^{-3} + \frac{54.3\times 10^{-3}}{3000}}\left(\frac{27.518}{2}\right) = 0.0841\ \text{T}. \quad (10.348)$$

The peak of the total magnetic flux density, which consists of the DC and AC components, is given by

$$B_{pk} = 2B_m = 2 \times 0.0841 = 0.1681\ \text{T}. \quad (10.349)$$

The core power loss density is

$$P_v = 0.0573 f^{1.66}(10B_m)^{2.68} = 0.0573 \times 100^{1.66} \times (10 \times 0.0841)^{2.68} = 75.26\ \text{mW/cm}^3. \quad (10.350)$$

The core loss is

$$P_C = V_c P_v = 6.53 \times 75.26 = 491.49\ \text{mW}. \quad (10.351)$$

Figure 10.49 shows the plots of the core loss P_C as a function of the output voltage P_O at fixed values of the DC input voltage V_I for the buck converter operating in DCM. Figure 10.50 shows the plots of the core loss P_C as a function of the DC input voltage V_I at fixed values of the output power P_O for the buck converter operating in DCM.

The length of the winding wire is

$$l_w = N l_T = 3 \times 5.419 = 16.257\ \text{cm}. \quad (10.352)$$

Let us assume $l_w = 18$ cm to allow for the inductor leads. The DC and low-frequency resistance of the single strand is

$$R_{wDCs} = l_w\left(\frac{R_{wDCs}}{l_w}\right) = 18 \times 10^{-2} \times 0.1345 = 0.0242196\ \Omega. \quad (10.353)$$

The DC and low-frequency resistance of the winding is

$$R_{wDC} = \frac{R_{wDCs}}{S_n} = \frac{24.22 \times 10^{-3}}{43} = 0.563\ \text{m}\Omega. \quad (10.354)$$

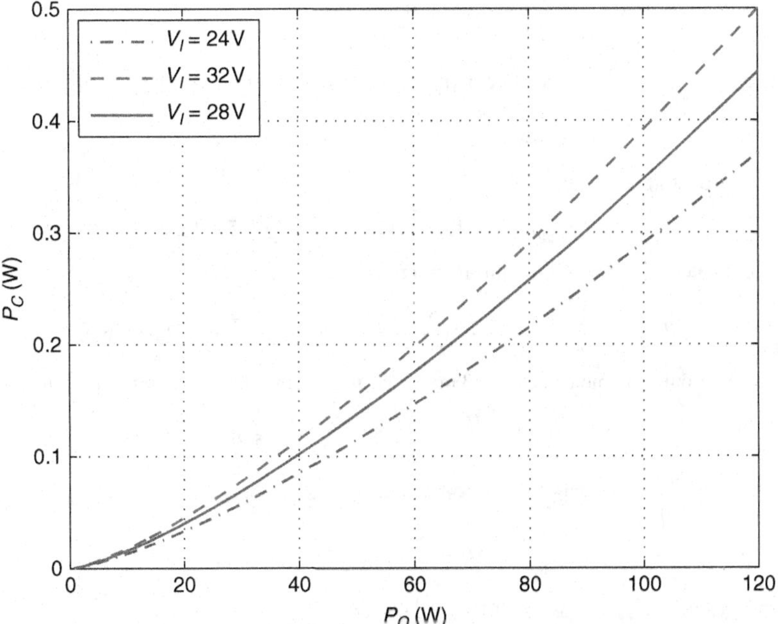

Figure 10.49 Core loss P_C as a function of the output power P_O at fixed values of the DC input voltage V_I for the buck converter operating in DCM

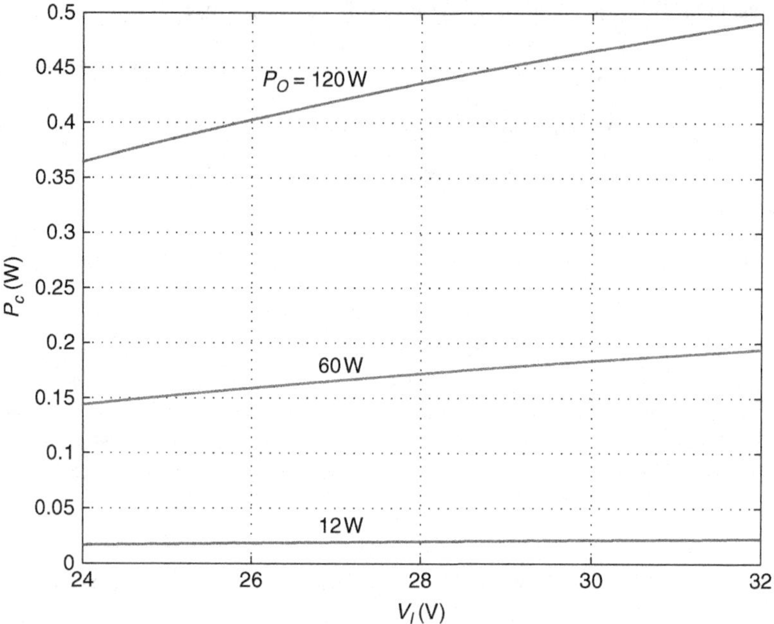

Figure 10.50 Core loss P_C as a function of the DC input voltage V_I at fixed values of the output power P_O for the buck converter operating in DCM

The DC and low-frequency winding power loss is

$$P_{wDC} = R_{wDC} I_L^2 = 0.563 \times 10^{-3} \times 10^2 = 0.0563 \text{ W}. \quad (10.355)$$

The ratio of the bare wire diameter d_i to the skin depth δ_w at $f_s = 100\,\text{kHz}$ is

$$\frac{d_i}{\delta_w} = \frac{0.405}{0.209} = 1.938. \quad (10.356)$$

The height of one-half of the core window is $D = 8.05$ mm. The height of the core winding window is

$$H_w = 2D = 2 \times 8.05 = 16.1 \text{ mm}. \quad (10.357)$$

The cross-sectional area of the bare strands is

$$A_w = S_n A_{ws} = 43 \times 0.128 \times 10^{-6} = 5.5304 \,\text{mm}^2. \quad (10.358)$$

The inner diameter of the winding wire is given by

$$d_i = \sqrt{\frac{4A_{wo}}{\pi}} = \sqrt{\frac{4 \times 5.5304}{\pi}} = 2.655 \times 10^{-3} \text{ m} = 2.654 \text{ mm}. \quad (10.359)$$

Assume that the outer diameter is $d_o = 3$ mm. The maximum number of turns per layer is

$$N_{l1} = \frac{H_w}{d_o} = \frac{16.1 \times 10^{-3}}{3 \times 10^{-3}} = 5.367. \quad (10.360)$$

Pick $N_{l1} = 5$. The number of layers is

$$N_l = \frac{N}{N_{l1}} = \frac{3}{5} = 0.6. \quad (10.361)$$

Pick $N_l = 1$. Hence, the number of turns per layer is

$$N_{l1} = \frac{N}{N_l} = \frac{3}{1} = 3. \quad (10.362)$$

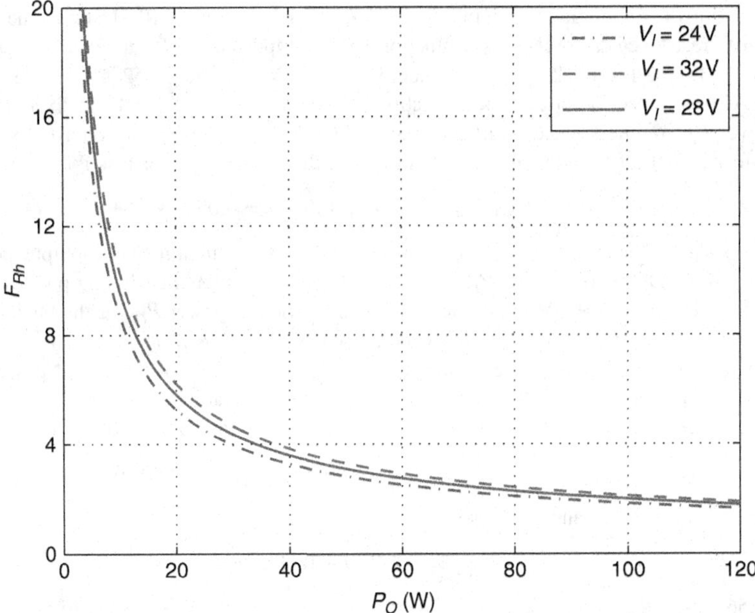

Figure 10.51 Harmonic AC resistance factor F_{Rh} as a function of the output power P_O at fixed values of the DC input voltage V_I for the buck converter in DCM

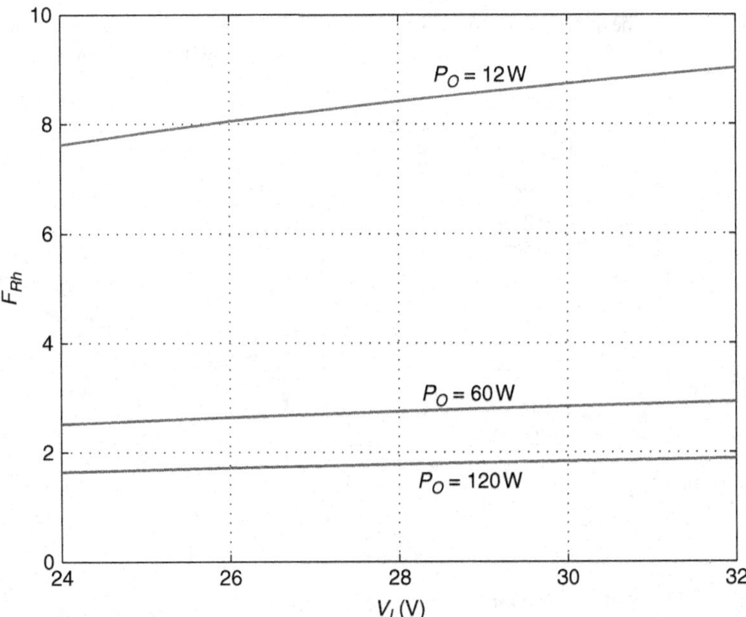

Figure 10.52 Harmonic AC resistance factor F_{Rh} as a function of the DC input voltage V_I at fixed values of the output power P_O for the buck converter in DCM

The winding consists of a single layer ($N_l = 1$) and three turns ($N = 3$). Using (5.465), the computer calculations for a single-layer winding give $F_{Rh} = 1.643$. Figure 10.51 shows the plots of the harmonic AC resistance factor F_{Rh} as a function of the output power P_O at fixed values of the DC input voltage V_I for the buck converter operating in DCM. Figure 10.52 shows the plots of the harmonic AC resistance factor F_{Rh} as a function of the DC input voltage V_I at fixed values of the output power P_O for the buck converter operating in DCM. Taking into account the skin effect of inductor current at the fundamental and its harmonics, the winding power loss is

$$P_w = F_{Rh} P_{wDC} = 1.643 \times 0.0563 = 0.0925\ \text{W}. \tag{10.363}$$

Figure 10.53 shows the plots of winding power loss P_w as a function of the output power P_O at fixed values of the DC input voltage V_I. Figure 10.54 shows the plots of winding power loss P_w as a function of the DC input voltage V_I at fixed values of the output power P_O for the buck converter in DCM. Figure 10.55 shows the plots of the winding power $P_{wx} = R_{wDC} I_{Lrms}^2$ and the winding power loss with winding resistance dependent on frequency P_w as functions of the output power P_O at the nominal DC input voltage $V_I = 28$ V for the buck converter operating in DCM.

The sum of the core loss and the DC and low-frequency winding power loss is

$$P_{cwDC} = P_C + P_{wDC} = 0.49149 + 0.0563 = 0.5478\ \text{W}. \tag{10.364}$$

The power loss in the core and in the winding is

$$P_{cw} = P_C + P_w = 0.49149 + 0.0925 = 0.584\ \text{W}. \tag{10.365}$$

Figure 10.56 shows the plots of P_{cw} as a function of P_O at fixed values of V_I. Figure 10.57 shows the plots of P_{cw} as a function of V_I at fixed values of P_O.

Neglecting all converter losses except those in the inductor, the buck converter efficiency is

$$\eta = \frac{P_O}{P_O + P_{cw}} = \frac{120}{120 + 0.584} = 99.51\%. \tag{10.366}$$

Figure 10.58 shows the plots of the buck converter efficiency as a function of P_O at fixed values of V_I, assuming that the inductor is the only lossy component. Figure 10.59 shows the plots of the

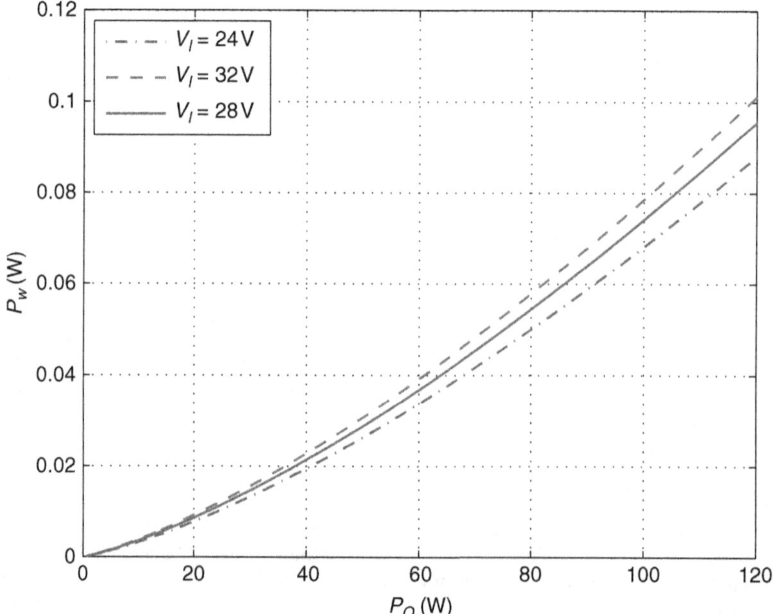

Figure 10.53 Winding power loss P_w as a function of the output power P_O at fixed values of the DC input voltage V_I

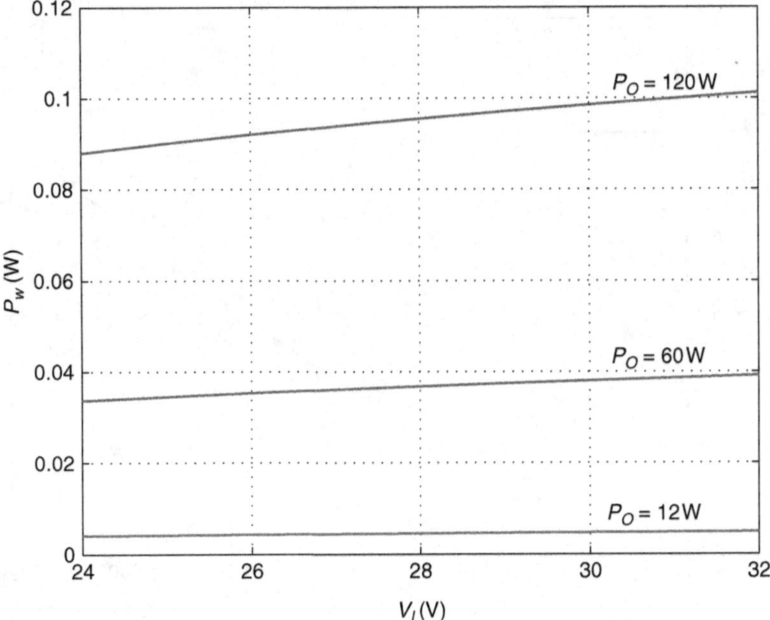

Figure 10.54 Winding power loss P_w as a function of the DC input voltage V_I at fixed values of the output power P_O for the buck converter in DCM

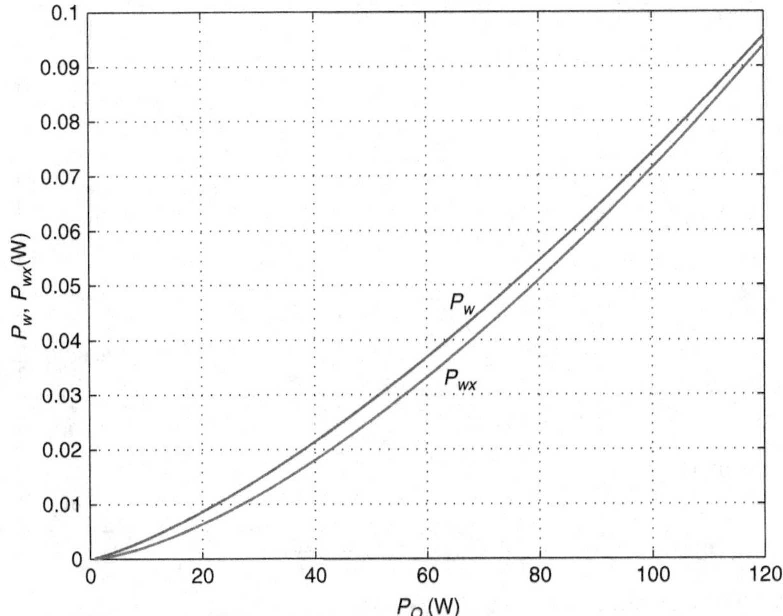

Figure 10.55 Winding loss $P_{wx} = R_{wDC} I_{Lrms}^2$ and winding loss with winding resistance dependent on frequency P_w as functions of the output power P_O at the nominal DC input voltage $V_I = 28$ V for the buck converter in DCM

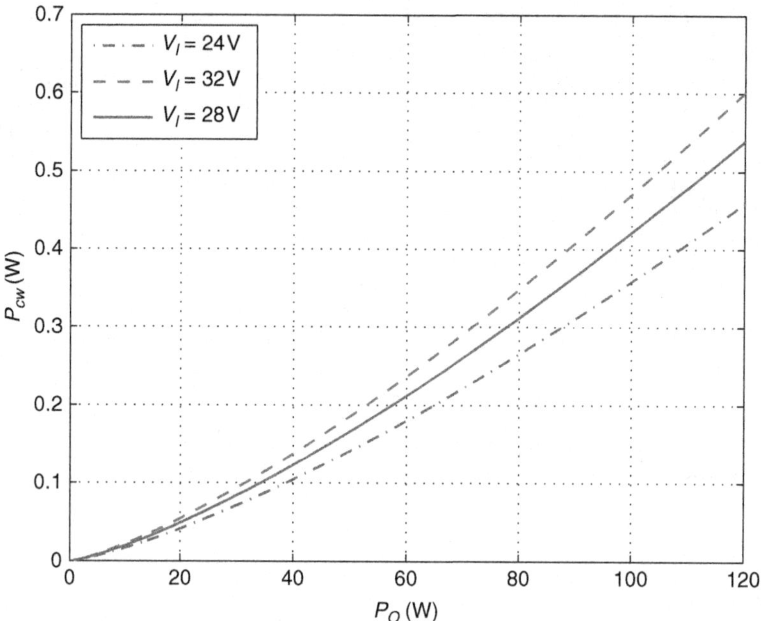

Figure 10.56 Inductor power loss P_{cw} as a function of the output power P_O at fixed values of the DC input voltage V_I for the buck converter in DCM

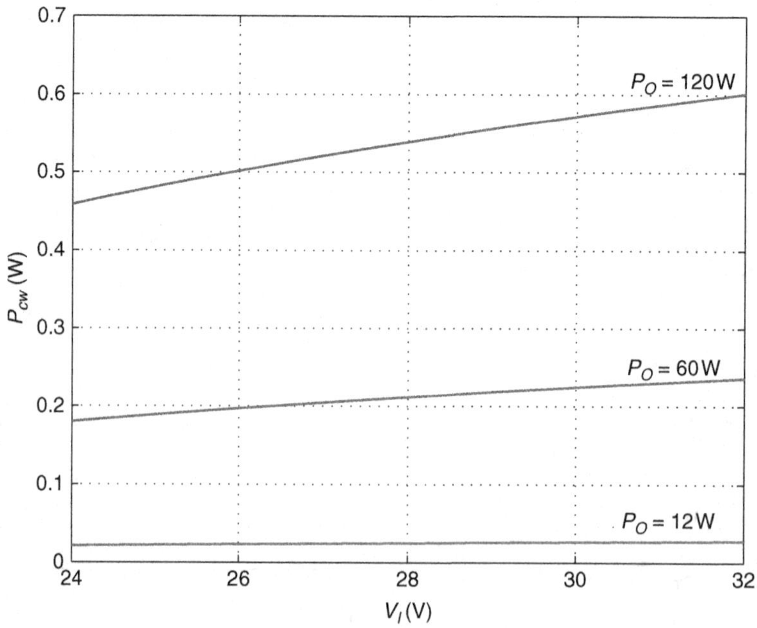

Figure 10.57 Inductor power loss P_{cw} as a function of the DC input voltage V_I at fixed values of the output power P_O for the buck converter in DCM

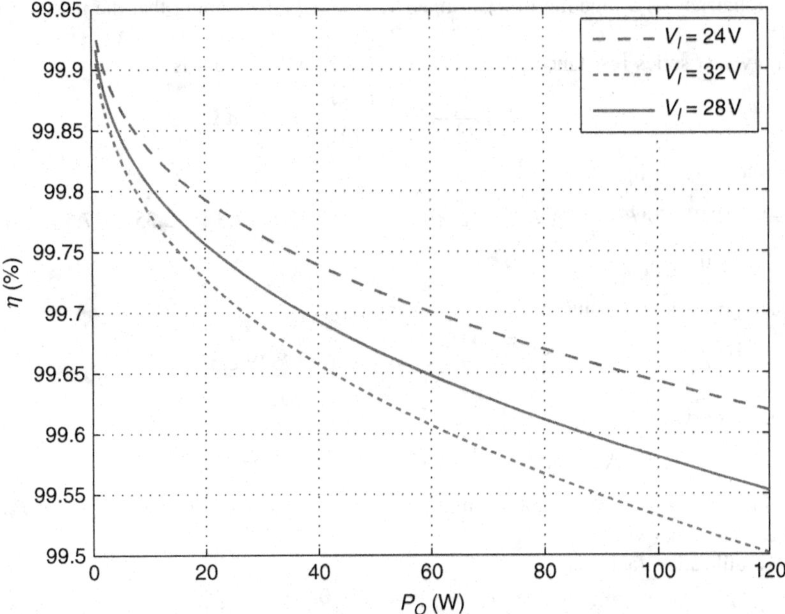

Figure 10.58 Efficiency of the buck converter as a function of the output power P_O at fixed values of the input voltage V_I for the buck converter in DCM assuming that the inductor is the only lossy component

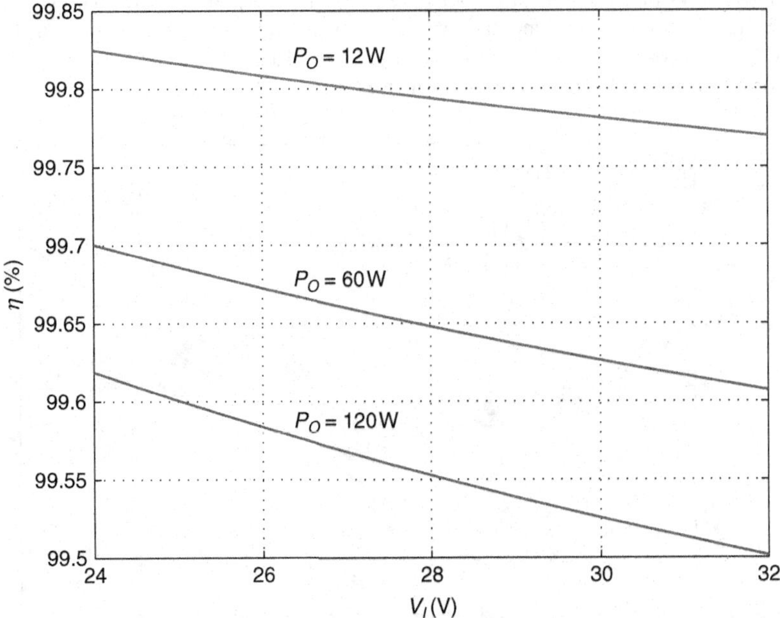

Figure 10.59 Efficiency of the buck converter as a function of the DC input voltage V_I at fixed values of the output power P_O for the buck converter in DCM, assuming that the inductor is the only lossy component

converter efficiency as a function of V_I at fixed values of P_O, assuming that the inductor is the only lossy component.

The equivalent series resistance (ESR) of the winding and the core is

$$r_{L(DC)} = \frac{P_{cw}}{I_{Lrms}^2} = \frac{0.584}{13.791^2} = 3.07\ \text{m}\Omega. \tag{10.367}$$

The surface area of the inductor is

$$A_t = 2(2BC + 2AB + AC) = 2 \times (2 \times 12.35 \times 19 + 2 \times 27.3 \times 12.35 + 27.3 \times 19)$$
$$= 33.2462\ \text{m}^2. \tag{10.368}$$

The surface power loss density is

$$\psi = \frac{P_{cw}}{A_t} = \frac{0.584}{33.2462} = 0.0175\ \text{W/cm}^2. \tag{10.369}$$

The temperature rise is

$$\Delta T = 450\psi^{0.826} = 450 \times 0.0175^{0.826} = 15.92\ ^\circ\text{C}. \tag{10.370}$$

Figure 10.60 shows a plot of the temperature rise ΔT as a function of the output power P_O at a fixed value of the DC input voltage V_I.

The core utilization factor is

$$K_u = \frac{N S_n A_{wso}}{W_a} = \frac{3 \times 43 \times 0.1604}{0.4916 \times 10^2} = 0.4209. \tag{10.371}$$

Figure 10.61 shows the plots of K_{Rh} as functions of P_O at fixed values of V_I. Figure 10.62 shows the amplitude spectrum of the inductor current waveform for P_{Omax} at V_{Imax} for the buck converter in DCM, respectively. Figure 10.63 shows the values of the winding resistance at DC and harmonics. Figure 10.64 shows the spectrum of the winding power loss P_{wn} at V_{Imax} for the buck converter in DCM. Figure 10.65 shows the amplitude spectrum of the inductor current waveform for P_{Omax} at V_{Imin} for the buck converter in DCM, respectively. Figure 10.66 shows the spectrum of the winding power loss P_{wn} at V_{Imin} for the buck converter in DCM.

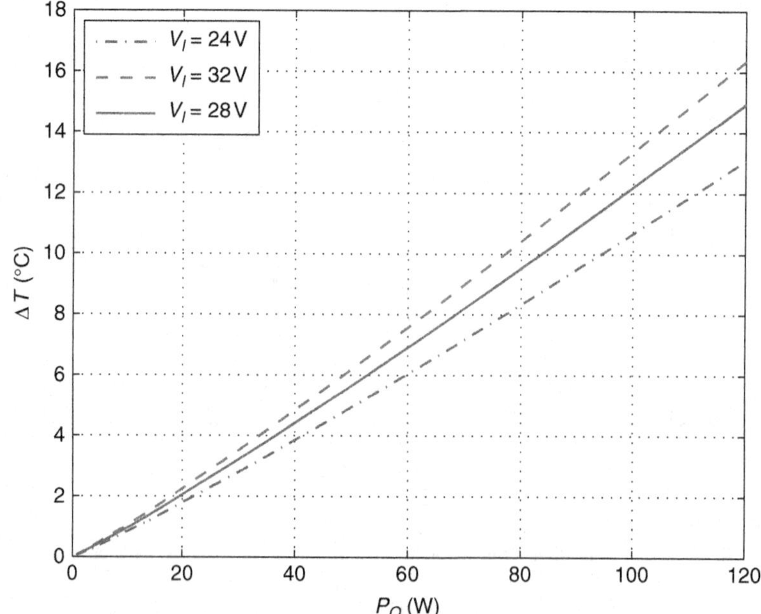

Figure 10.60 Temperature rise ΔT as a function of the output power P_O at fixed values of the output power P_O for the buck converter in DCM

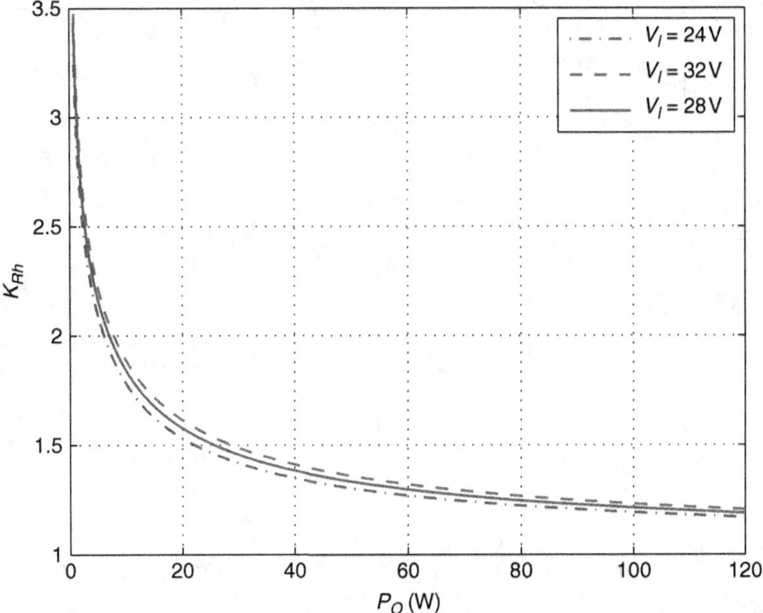

Figure 10.61 K_{Rh} as a function of the output power P_O at fixed values of the DC input voltage V_I for the buck converter in DCM

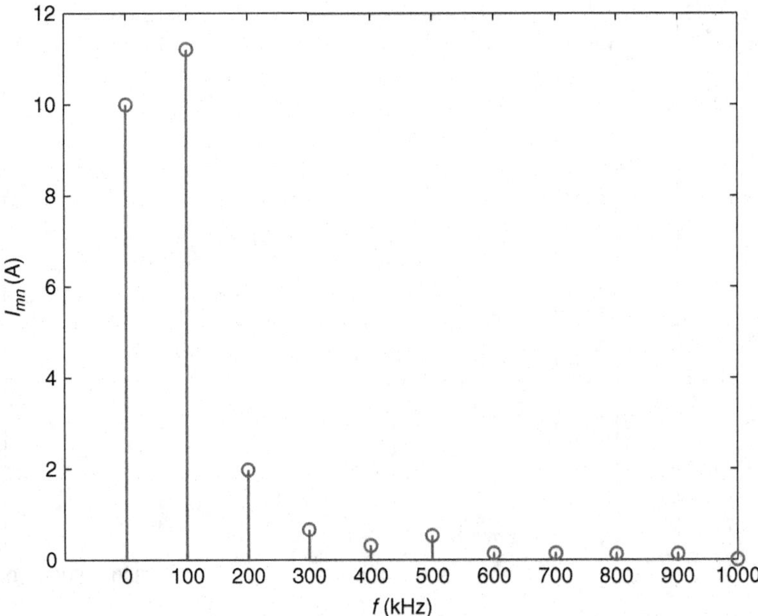

Figure 10.62 Amplitude spectrum of the inductor current I_{mn} at V_{Imax} and P_{Omax} for the buck converter in DCM

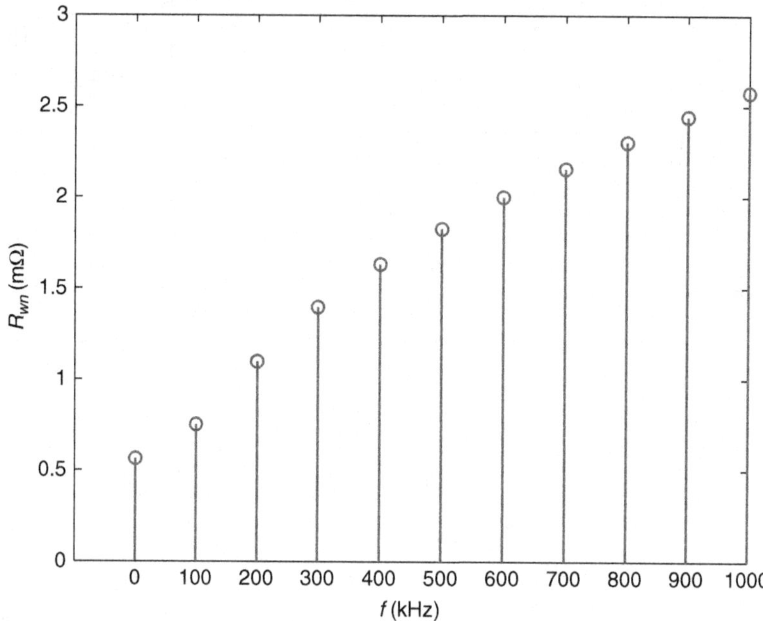

Figure 10.63 Spectrum of the winding resistance at $d_i = 0.405$ mm, $l_w = 18$ cm, $d_i/\delta_w = 1.9381$, $S_n = 43$, $N_l = 1$, and $R_{wDC}/l_w = 0.1345$

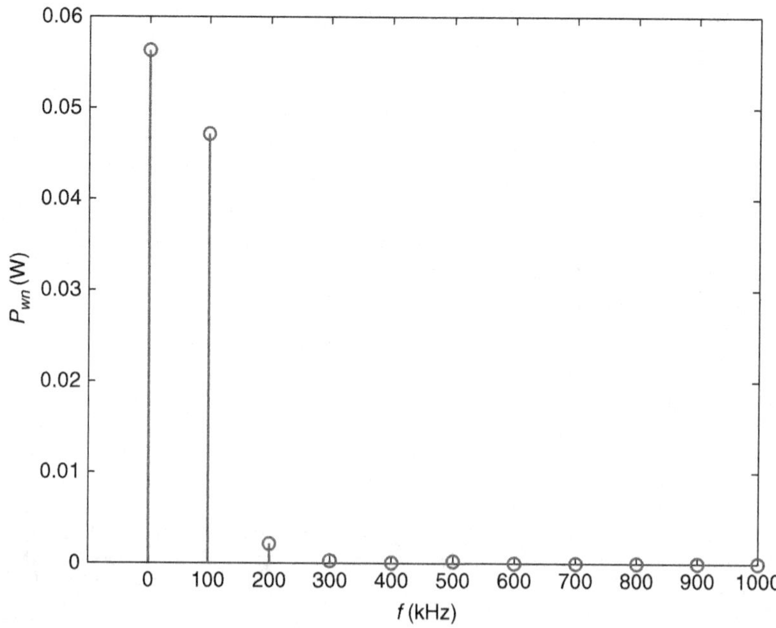

Figure 10.64 Amplitude spectrum of the winding power loss P_{wn} at V_{Imax} and P_{Omax} for the buck converter in DCM

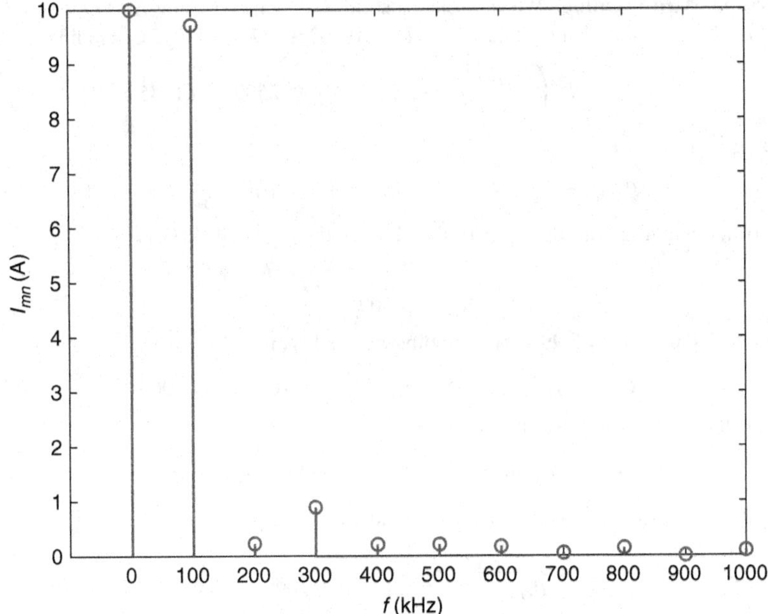

Figure 10.65 Amplitude spectrum of the inductor current I_{mn} at V_{Imax} and $0.1P_{Omax}$ for the buck converter in DCM

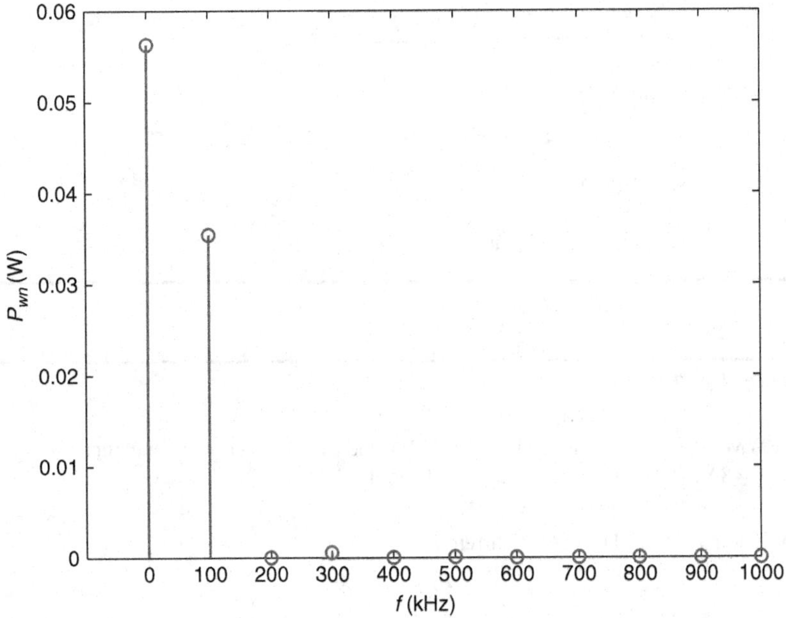

Figure 10.66 Amplitude spectrum of the winding power loss P_{wn} at V_{Imax} and $0.1P_{Omax}$ for the buck converter in DCM

Single-Solid-Wire Winding. We can select a solid-wire winding using AWG9 wire. This wire has $d_i = 2.906$ mm, $A_w = 6.632$ mm^2, and $R_{wDC}/l_w = 0.002599$ Ω/m. The DC winding resistance is

$$R_{wDC} = l_w \left(\frac{R_{wDC}}{l_w} \right) = 0.05175 \times 0.002599 = 0.1345 \text{ m}\Omega. \quad (10.372)$$

The winding loss in the DC resistance is

$$P_{wDC} = R_{wDC} I_L^2 = 0.1345 \times 10^{-3} \times 10^2 = 13.45 \text{ mW}. \quad (10.373)$$

The ratio of the winding wire diameter to the skin depth at $f = 100$ kHz is

$$\frac{d_i}{\delta_w} = \frac{2.906}{0.209} = 13.9. \quad (10.374)$$

Hence, from (5.490), $F_{Rh} = 5.993$. The winding power loss is

$$P_w = F_{Rh} P_{wDC} = 5.993 \times 13.45 \times 10^{-3} = 0.0806 \text{ W}. \quad (10.375)$$

The sum of the core loss and winding loss is

$$P_{cw} = P_C + P_w = 0.49149 + 0.0806 = 0.57209 \text{ W}. \quad (10.376)$$

Assuming that the only loss component is the inductor, the converter efficiency is

$$\eta = \frac{P_O}{P_O + P_{cw}} = \frac{120}{120 + 0.57209} = 99.52\%. \quad (10.377)$$

The equivalent DC resistance of the inductor is

$$r_{L(DC)} = \frac{P_{cw}}{I_{Lrms}^2} = \frac{0.57209}{13.791^2} = 3 \text{ m}\Omega. \quad (10.378)$$

The surface power loss density is

$$\psi = \frac{P_{cw}}{A_t} = \frac{0.57209}{33.2462} = 0.0172 \text{ W/cm}^2. \quad (10.379)$$

The temperature rise of the inductor is

$$\Delta T = 450\psi^{0.826} = 450 \times 0.0172^{0.826} = 15.69\,^{\circ}\text{C}. \quad (10.380)$$

The core window utilization is

$$K_u = \frac{N A_{wo}}{W_a} = \frac{3 \times 6.63}{0.4916 \times 10^2} = 0.4046. \quad (10.381)$$

Example 10.9

A buck PWM converter is operated in the DCM and has the following specifications: $V_I = 100 \pm 20$ V, $V_O = 48$ V, $P_O = 0 - 60$ W, and $f_s = 100$ kHz.

Solution: The maximum DC output current is

$$I_{Omax} = \frac{P_{Omax}}{V_O} = \frac{60}{48} = 1.25 \text{ A}. \quad (10.382)$$

Hence, the maximum DC inductor current for the buck converter is $I_{Lmax} = I_{Omax} = 1.25$ A. The minimum load resistance is

$$R_{Lmin} = \frac{V_O}{I_{Omax}} = \frac{48}{1.25} = 38.4\ \Omega. \quad (10.383)$$

The minimum and maximum DC voltage transfer functions are

$$M_{VDCmin} = \frac{V_O}{\eta V_{Imax}} = \frac{48}{0.9 \times 120} = 0.4444 \quad (10.384)$$

and

$$M_{VDCmax} = \frac{V_O}{\eta V_{Imin}} = \frac{48}{\eta \times 80} = 0.6667. \tag{10.385}$$

Assuming the converter efficiency $\eta = 0.9$ and the dwell duty cycle $D_w = 0.05$, the maximum inductance is

$$L_{max} = \frac{R_{Lmin}(1 - M_{VDCmax})(1 - D_w)^2}{2f_s} = \frac{38.4 \times (1 - 0.6)(1 - 0.05)^2}{2 \times 10^5} = 97.47\ \mu\text{H}. \tag{10.386}$$

Pick a standard value of the inductance $L = 56\ \mu$H.

The maximum duty cycle at full power occurs at $V_{Imin} = 80$ V and is given by

$$D_{max} = \sqrt{\frac{2f_s L M_{VDCmax}^2}{R_{Lmin}(1 - M_{VDCmax})}} = \sqrt{\frac{2 \times 10^5 \times 56 \times 10^{-6} \times 0.6^2}{38.4 \times (1 - 0.6)}} = 0.5401. \tag{10.387}$$

and the minimum duty cycle is

$$D_{min} = \sqrt{\frac{2f_s L M_{VDCmin}^2}{R_{Lmin}(1 - M_{VDCmin})}} = \sqrt{\frac{2 \times 10^5 \times 56 \times 10^{-6} \times 0.4^2}{38.4 \times (1 - 0.4)}} = 0.2939. \tag{10.388}$$

The minimum duty cycle at full load when the diode is ON is

$$D_{1min} = D_{max}\left(\frac{1}{M_{VDCmax}} - 1\right) = 0.5401\left(\frac{1}{0.6} - 1\right) = 0.36. \tag{10.389}$$

and the maximum duty cycle at full load when the diode is ON is

$$D_{1max} = D_{min}\left(\frac{1}{M_{VDCmin}} - 1\right) = 0.2939\left(\frac{1}{0.4} - 1\right) = 0.57. \tag{10.390}$$

Hence, $D_{min} + D_{1max} = 0.2939 + 0.57 = 0.8639 < 1$ and $D_{max} + D_{1min} = 0.5401 + 0.36 = 0.9001 < 1$.

Figure 10.67 shows the plots of I_{mn} as a function of I_O. Figure 10.68 shows the plots of I_{mn} as a function of D. Figure 10.69 shows the plots of D and D_1 as functions of I_O. Figure 10.70 shows

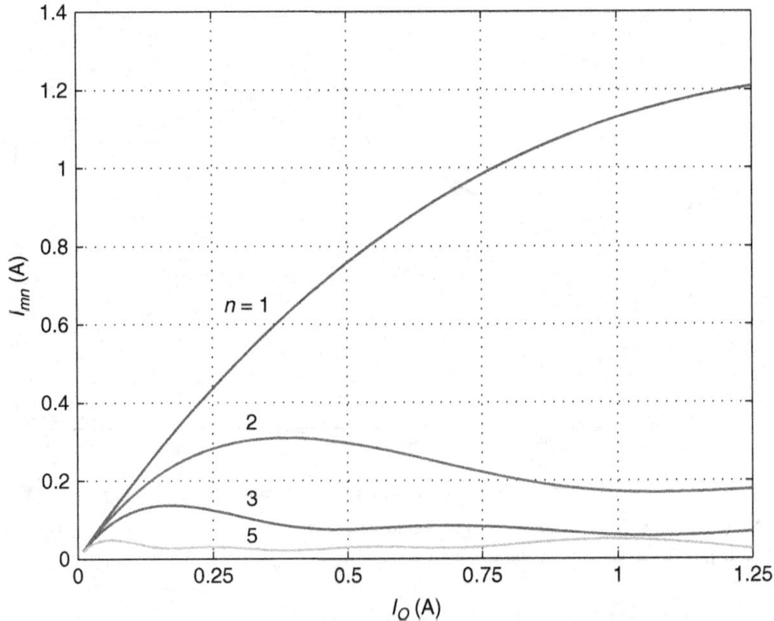

Figure 10.67 I_{mn} as a function of the output current I_O at $V_O = 48$ V, $f_s = 100$ kHz, $L = 56\ \mu$H, and $V_I = 120$ V for the buck converter operating in DCM

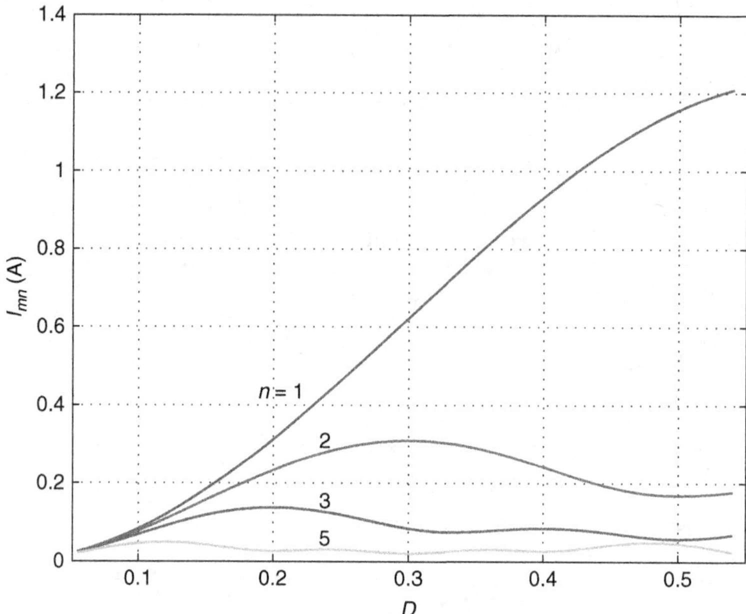

Figure 10.68 I_{mn} as a function of D at $V_O = 48$ V, $f_s = 100$ kHz, $L = 56$ μH, and $V_I = 120$ V for the buck converter operating in DCM

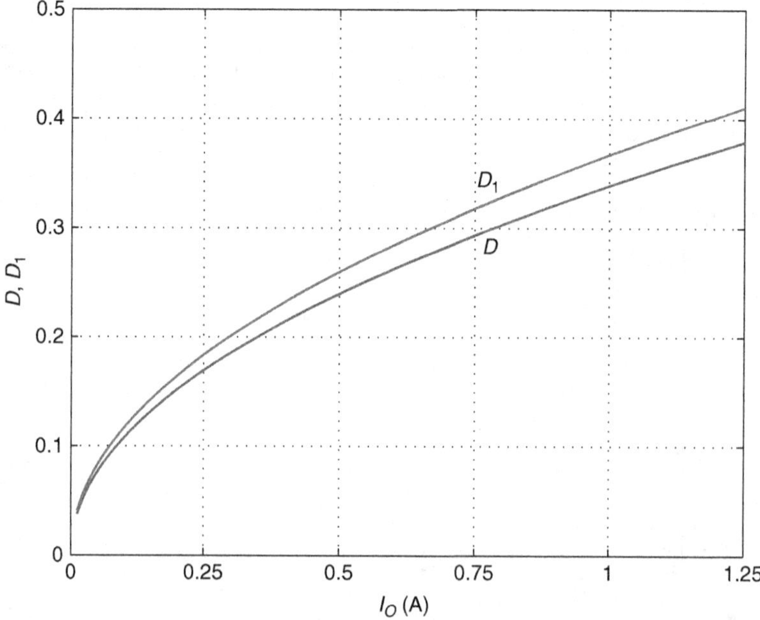

Figure 10.69 D and D_1 as functions of the output power P_O at $V_O = 48$ V, $f_s = 100$ kHz, $L = 56$ μH, and $V_I = 100$ V for the buck converter operating in DCM

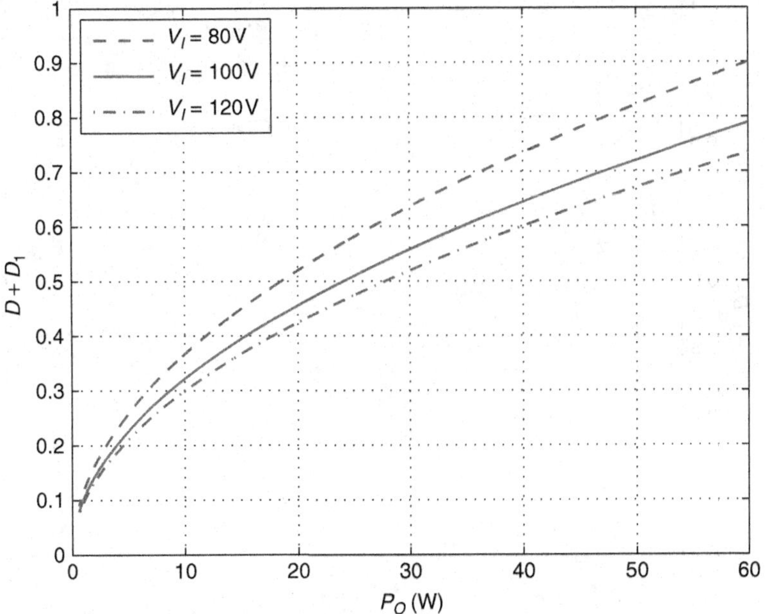

Figure 10.70 $D + D_1$ as a function of the output power P_O at fixed values of the input voltage V_I for the buck converter operating in DCM

the plots of $D + D_1$ as a function of the output power P_O at fixed values of the DC input voltage V_I for the buck converter operating in DCM. Figure 10.71 shows the plots of $D + D_1$ as a function of the DC input voltage V_I at fixed values of the output power P_O for the buck converter operating in DCM. Figure 10.72 shows plots of D and D_1 as functions of I_O at $V_O = 48$ V.

The maximum inductor current occurs at V_{Imax} and is given by

$$\Delta i_{Lmax} = \frac{D_{min}(V_{Imax} - V_O)}{f_s L} = \frac{0.294 \times (120 - 48)}{10^5 \times 56 \times 10^{-6}} = 3.78 \text{ A}. \tag{10.391}$$

Figure 10.73 shows the plots of Δi_L as a function of V_I at fixed values of P_O.

The maximum energy stored in the inductor is

$$W_m = \frac{1}{2} L I_{Lmax}^2 = \frac{1}{2} \times 56 \times 10^{-6} \times 3.78^2 = 0.4 \text{ mJ}. \tag{10.392}$$

The maximum rms value of the inductor current is

$$I_{Lrms(max)} = \Delta i_{Lmax} \sqrt{\frac{D_{min} + D_{1min}}{3}} = 3.78 \sqrt{\frac{0.294 + 0.36}{3}} = 1.76 \text{ A}. \tag{10.393}$$

Assume $K_u = 0.4$, $B_{pk} = 0.2$ T, and $J_m = 5$ A/mm^2. The core area product is

$$A_p = \frac{2W_m}{K_u J_m B_{pk}} = \frac{2 \times 0.4 \times 10^{-3}}{0.4 \times 5 \times 10^6 \times 0.2} = 0.2 \text{ cm}^4. \tag{10.394}$$

Select a Magnetics 0P-42616 ferrite pot core. The parameters of this core are as follows: $A_p = 0.39$ cm^4, $A_c = 0.939$ cm^2, $l_c = 3.76$ cm, $V_c = 3.53$ cm^3, $E = 21.6$ mm, and $F = 11.3$ mm. The diameter of the center leg is $D_c = 11.3$ mm. The core material is P-type whose $\mu_{rc} = 2500 \pm 25\%$. We select the Magnetics PQ printed circuit bobbin PCB2625LA with the bobbin window height $B_h = 13.58$ mm and inner diameter $D_{ib} = 14.19$ mm.

The core window area is

$$W_a = \frac{A_p}{A_c} = \frac{0.39}{0.939} = 0.4153 \text{ cm}^2. \tag{10.395}$$

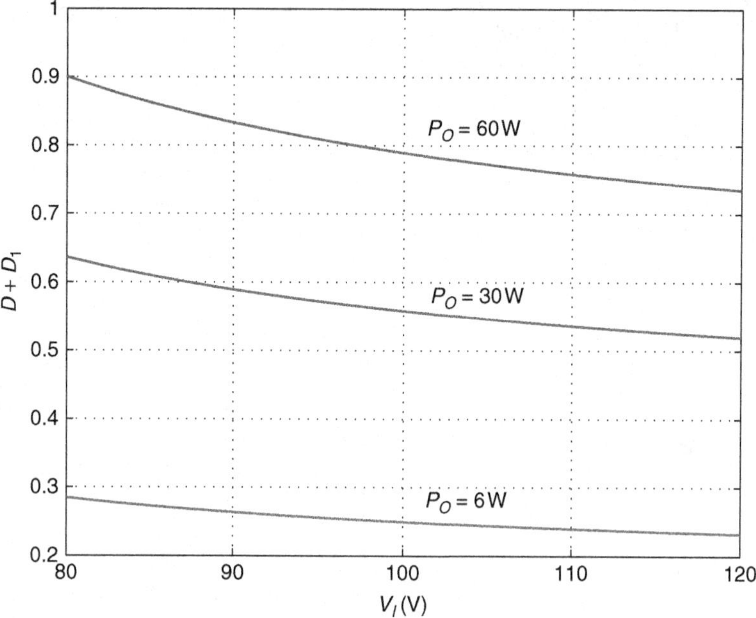

Figure 10.71 $D + D_1$ as a function of the DC input voltage V_I at fixed values of the output power P_O for the buck converter operating in DCM

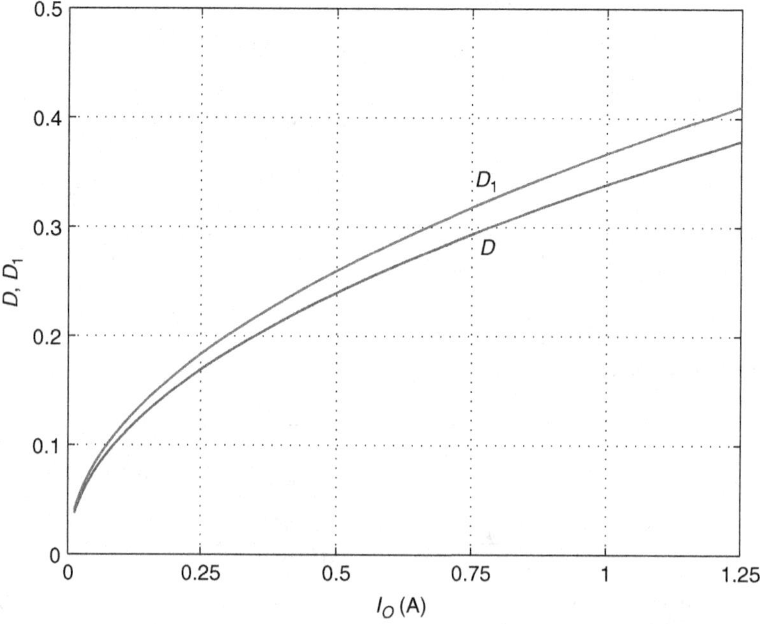

Figure 10.72 D and D_1 as functions of the output current I_O at $V_O = 48\,V$, $f_s = 100\,kHz$, $L = 56\,\mu H$, and $V_I = 120\,V$ for the buck converter operating in DCM

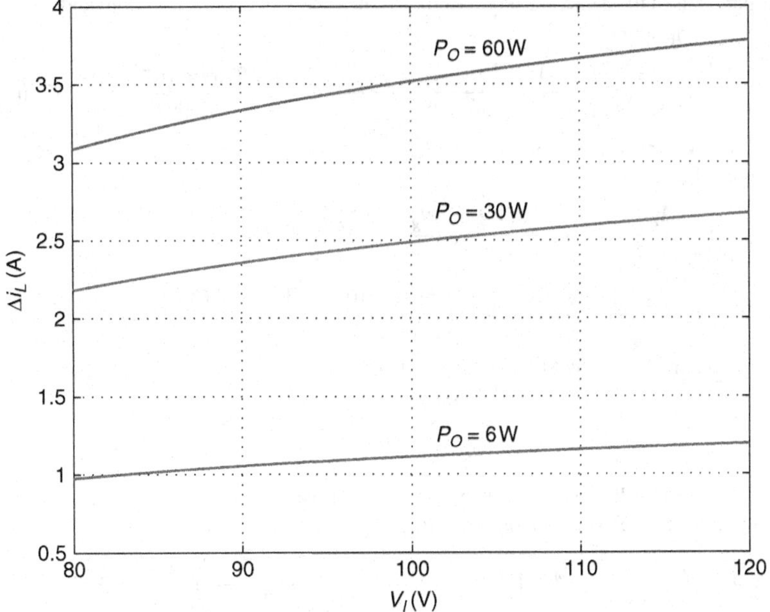

Figure 10.73 Δi_L as a function of the DC input voltage V_I at fixed values of the output power P_O for the buck converter operating in DCM

The MTL is

$$l_T = \frac{\pi(E+F)}{2} = \frac{\pi(21.6+11.3)\times 10^{-3}}{2} = 5.168\,\text{cm}. \tag{10.396}$$

Multiple-Strand Winding. The skin depth of copper wire at $f = 100\,\text{kHz}$ is

$$\delta_w = \frac{66.2}{\sqrt{f}}(\text{mm}) = \frac{66.2}{\sqrt{10^5}}(\text{mm}) = 0.209\ \text{mm}. \tag{10.397}$$

The bare wire diameter of a strand at which the skin and proximity effects can be neglected for the fundamental component is

$$d_i = 2\delta_w = 2\times 0.209 = 0.418\ \text{mm}. \tag{10.398}$$

A copper wire AWG26 is selected for the strands with $d_i = 0.405\,\text{mm}$, $d_o = 0.452\,\text{mm}$, $A_{wsi} = 0.1288\,\text{mm}^2$, and $R_{wDCs}/l_w = 0.1345\ \Omega/\text{m}$. The wire cross-sectional area is

$$A_w = \frac{I_{Lmax}}{J_{m(max)}} = \frac{3.78\ \text{A}}{5\times 10^6\ \text{A/mm}^2} = 0.756\ \text{mm}^2. \tag{10.399}$$

The number of strands is

$$S_n = \frac{A_w}{A_{wsi}} = \frac{0.756}{0.1288} = 5.8696. \tag{10.400}$$

Select $S_n = 5$. The cross-sectional area of the insulated strand is

$$A_{wso} = \frac{\pi d_o^2}{4} = \frac{\pi\times(0.452\times 10^{-3})^2}{4} = 0.1604\ \text{mm}^2. \tag{10.401}$$

The number of turns is

$$N = \frac{K_u W_a}{S_n A_{wsi}} = \frac{0.4\times 0.4153\times 10^{-4}}{5\times 0.1288\times 10^{-6}} = 25.795. \tag{10.402}$$

Select $N = 26$.

The air gap is used to obtain the required inductance, to avoid core saturation, and to store magnetic energy. The air-gap length is

$$l_g = \frac{\mu_0 A_c N^2}{L} - \frac{l_c}{\mu_{rc}} = \frac{4\pi \times 10^{-7} \times 0.939 \times 10^{-4} \times 26^2}{56 \times 10^{-6}} - \frac{37.6 \times 10^{-3}}{2500} = 0.9577\,\text{mm}. \quad (10.403)$$

Pick a standard value of $l_g = 1\,\text{mm}$.

The fringing factor is

$$F_f = 1 + \frac{A_f}{A_g}\frac{l_g}{l_f} = 1 + \frac{4\alpha l_g (D_c + \alpha l_g)}{\beta D_c^2}$$

$$= 1 + \frac{4 \times 1 \times 1 \times 10^{-3} + 1 \times 11.3 \times 10^{-3} + 1 \times 1 \times 10^{-3})}{2 \times (11.3 \times 10^{-3})} = 1.1926. \quad (10.404)$$

Hence, the number of turns with fringing effect is

$$N_f = \frac{N}{\sqrt{F_f}} = \frac{26}{\sqrt{1.1926}} = 23.08. \quad (10.405)$$

Pick $N = 21$ and three layers. Each layer has seven turns.

The peak value of the AC component of the magnetic flux density is

$$B_m = \frac{\mu_0 N}{l_g + \frac{l_c}{\mu_{rc}}}\left(\frac{\Delta i_{Lmax}}{2}\right) = \frac{4\pi \times 10^{-7} \times 21}{1 \times 10^{-3} + \frac{37.6 \times 10^{-3}}{2500}}\left(\frac{3.78}{2}\right) = 0.00234\,\text{T}. \quad (10.406)$$

The peak of the total magnetic flux density, which consists of the DC and AC components, is given by

$$B_{pk} = 2B_m = 2 \times 0.0546 = 0.1092\,\text{T}. \quad (10.407)$$

The core power loss density is

$$P_v = 0.0434 f^{1.63} (10 B_m)^{2.62} = 0.0434 \times 100^{1.63} \times (10 \times 0.0546)^{2.62} = 16.17\,\text{mW/cm}^3. \quad (10.408)$$

The core loss is

$$P_C = V_c P_v = 3.53 \times 16.17 = 57.08\,\text{mW}. \quad (10.409)$$

Figure 10.74 shows the plots of the core loss P_C as a function of the output voltage P_O at fixed values of the DC input voltage V_I for the buck converter operating in DCM. Figure 10.75 shows the plots of the core loss P_C as a function of the DC input voltage V_I at fixed values of the output power P_O for the buck converter operating in DCM.

The length of the winding wire is

$$l_w = N l_T = 21 \times 5.168 = 108.52\,\text{cm}. \quad (10.410)$$

Let us assume $l_w = 112\,\text{cm}$ to allow for the inductor leads. The DC and low-frequency resistance of the single strand is

$$R_{wDCs} = l_w \left(\frac{R_{wDCs}}{l_w}\right) = 112 \times 10^{-2} \times 0.1345 = 0.1506\,\Omega. \quad (10.411)$$

The DC and low-frequency resistance of the winding is

$$R_{wDC} = \frac{R_{wDCs}}{S_n} = \frac{0.1506 \times 10^{-3}}{5} = 0.03\,\text{m}\Omega. \quad (10.412)$$

The DC and low-frequency winding power loss is

$$P_{wDC} = R_{wDC} I_L^2 = 0.03 \times 10^{-3} \times 1.25^2 = 0.04687\,\text{W}. \quad (10.413)$$

The ratio of the bare wire diameter d_i to the skin depth δ_w at $f_s = 100\,\text{kHz}$ is

$$\frac{d_i}{\delta_w} = \frac{0.405}{0.209} = 1.938. \quad (10.414)$$

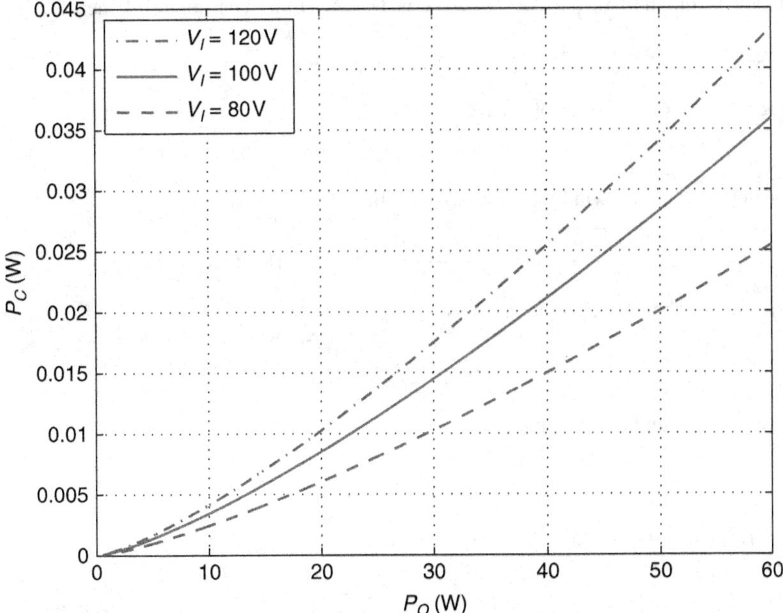

Figure 10.74 Core loss P_C as a function of the output power P_O at fixed values of the DC input voltage V_I for the buck converter operating in DCM

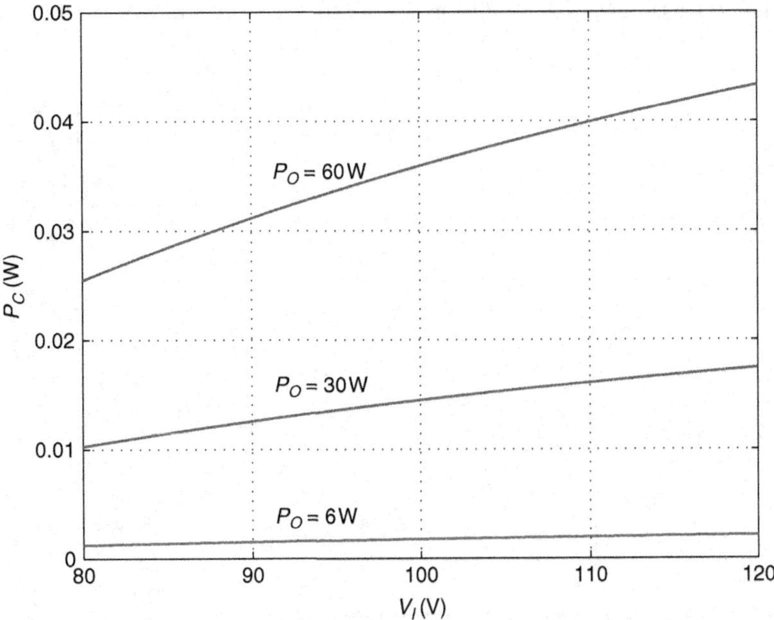

Figure 10.75 Core loss P_C as a function of the DC input voltage V_I at fixed values of the output power P_O for the buck converter operating in DCM

The height of one-half of the core window is $D = 5.5$ mm. The height of the core winding window is

$$H_w = 2D = 2 \times 5.5 = 11 \text{ mm}. \tag{10.415}$$

The copper cross-sectional area of all the strands is

$$A_w = S_n A_{wsi} = 21 \times 0.1288 \times 10^{-6} = 5.5304 \text{ mm}^2. \tag{10.416}$$

The outer diameter of the winding wire is approximately given by

$$d_o = \sqrt{\frac{4A_w}{\pi}} = \sqrt{\frac{4 \times 5.5384}{\pi}} = 2.655 \times 10^{-3} \text{ m} = 2.655 \text{ mm}. \tag{10.417}$$

The maximum number of turns per layer is

$$N_{l1} = \frac{H_w}{d_o} = \frac{16.1 \times 10^{-3}}{2.655 \times 10^{-3}} = 6.064. \tag{10.418}$$

Pick $N_{l1} = 6$. The number of layers is

$$N_l = \frac{N}{N_{l1}} = \frac{3}{6.064} = 0.4947. \tag{10.419}$$

Pick $N_l = 1$. Hence, the number of turns per layer is

$$N_{l1} = \frac{N}{N_l} = \frac{3}{1} = 3. \tag{10.420}$$

The winding consists of $N_l = 3$ and $N = 21$. Using (5.490), the computer calculations for a three-layer winding give $F_{Rh} = 4.889$. Figure 10.76 shows the plots of the harmonic AC resistance factor F_{Rh} as a function of the output power P_O at fixed values of the DC input voltage V_I for the buck converter operating in DCM. Figure 10.77 shows the plots of the harmonic AC resistance factor F_{Rh} as a function of the DC input voltage V_I at fixed values of the output power P_O for the buck converter

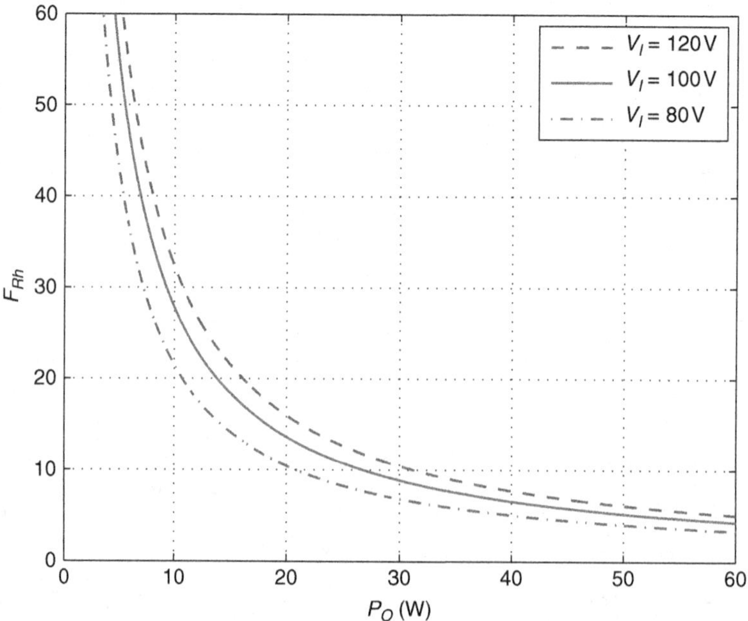

Figure 10.76 Harmonic AC resistance factor F_{Rh} as a function of the output power P_O at fixed values of the DC input voltage V_I for the buck converter in DCM

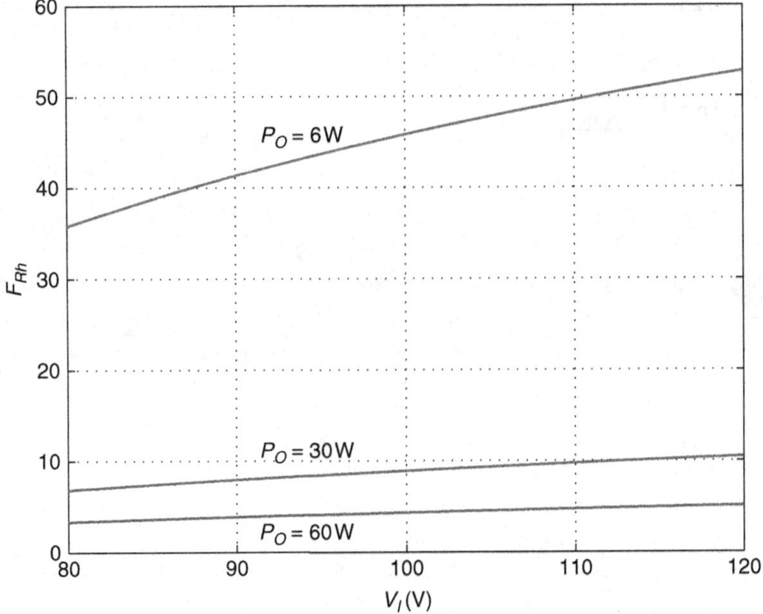

Figure 10.77 Harmonic AC resistance factor F_{Rh} as a function of the DC input voltage V_I at fixed values of the output power P_O for the buck converter in DCM

operating in DCM. Taking into account the skin effect of inductor current at the fundamental and its harmonics, the winding power loss is

$$P_w = F_{Rh} P_{wDC} = 4.889 \times 0.04687 = 0.2291 \text{ W}. \tag{10.421}$$

Figure 10.78 shows the plots of winding power loss P_w as a function of the output power P_O at fixed values of the DC input voltage V_I. Figure 10.79 shows the plots of winding power loss P_w as a function of the DC input voltage V_I at fixed values of the output power P_O for the buck converter in DCM. Figure 10.80 shows the plots of the winding power $P_{wx} = R_{wDC} I_{Lrms}^2$ and the winding power loss with winding resistance dependent on frequency P_w as functions of the output power P_O at the nominal DC input voltage $V_I = 120$ V for the buck converter operating in DCM.

The sum of the core loss and the DC and low-frequency winding power loss is

$$P_{cwDC} = P_C + P_{wDC} = 0.49149 + 0.0563 = 0.5478 \text{ W}. \tag{10.422}$$

The power loss in the core and in the winding is

$$P_{cw} = P_C + P_w = 0.05708 + 0.2291 = 0.2862 \text{ W}. \tag{10.423}$$

Figure 10.81 shows the plots of P_{cw} as functions of P_O at fixed values of V_I. Figure 10.82 shows the plots of P_{cw} as functions of V_I at fixed values of P_O.

Neglecting all converter losses except those in the inductor, the converter efficiency is

$$\eta = \frac{P_O}{P_O + P_{cw}} = \frac{60}{60 + 0.2862} = 99.51\%. \tag{10.424}$$

Figure 10.83 shows the plots of the converter efficiency as a function of P_O at fixed values of V_I, assuming that the inductor is the only lossy component. Figure 10.84 shows the plots of the converter efficiency as a function of V_I at fixed values of P_O, assuming that the inductor is the only lossy component.

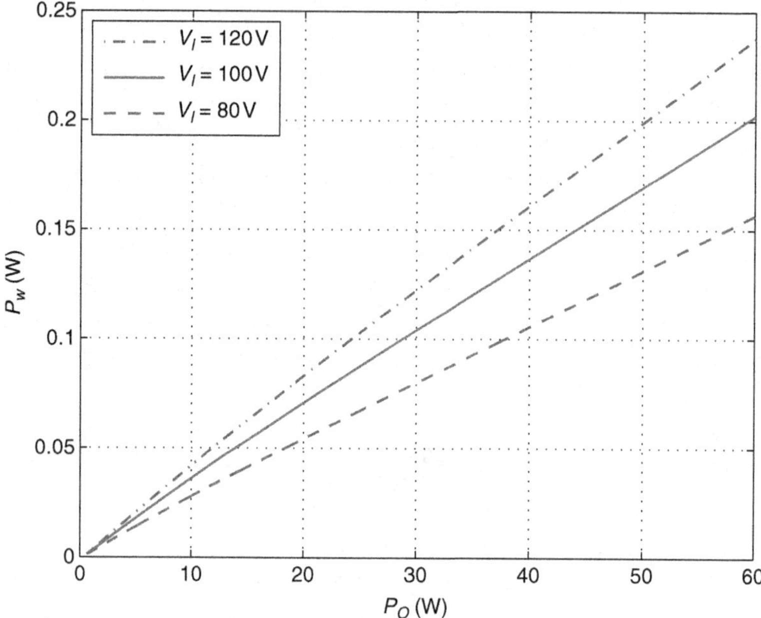

Figure 10.78 Winding power loss P_w as a function of the output power P_O at fixed values of the DC input voltage V_I

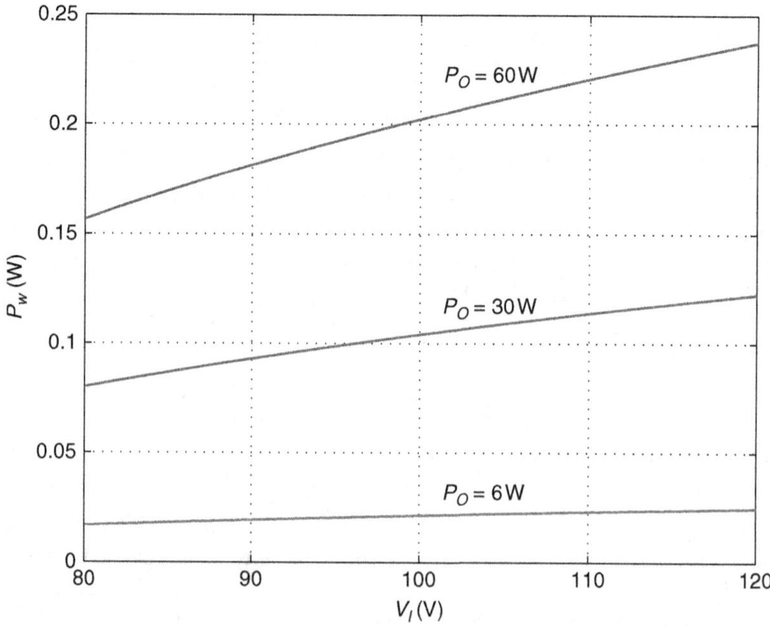

Figure 10.79 Winding power loss P_w as a function of the DC input voltage V_I at fixed values of the output power P_O

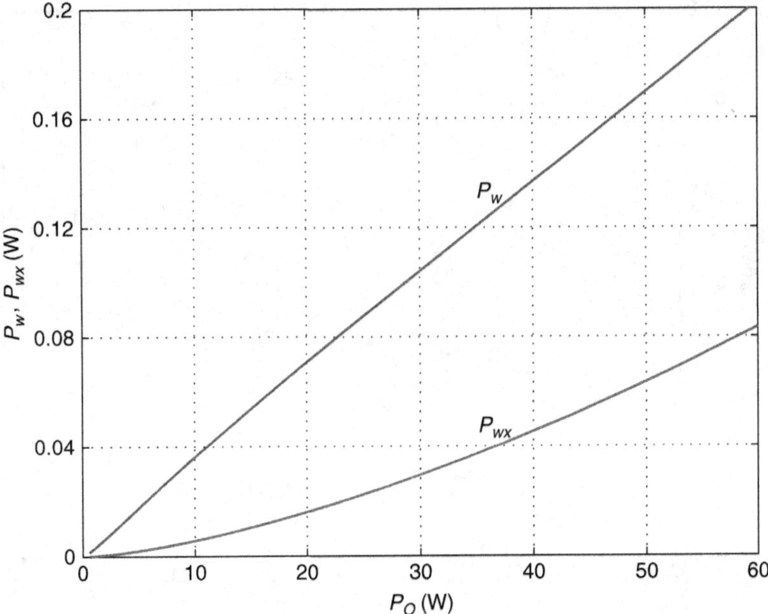

Figure 10.80 Winding loss $P_{wx} = R_{wDC} I_{Lrms}^2$ and winding loss with winding resistance dependent on frequency P_w as functions of the output power P_O at the nominal DC input voltage $V_I = 120$ V for the buck converter in DCM

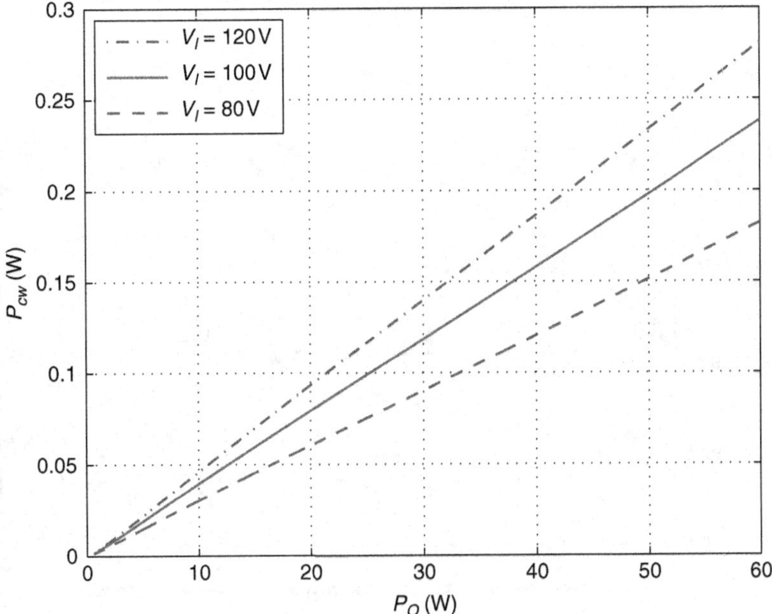

Figure 10.81 Inductor power loss P_{cw} as a function of the output power P_O at fixed values of the DC input voltage V_I for the buck converter in DCM

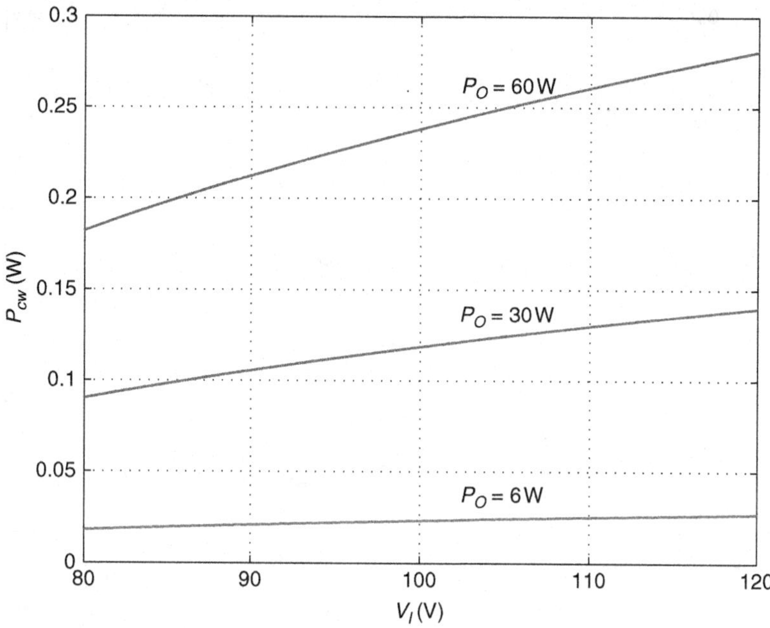

Figure 10.82 Inductor power loss P_{cw} as a function of the DC input voltage V_I at fixed values of the output power P_O for the buck converter in DCM

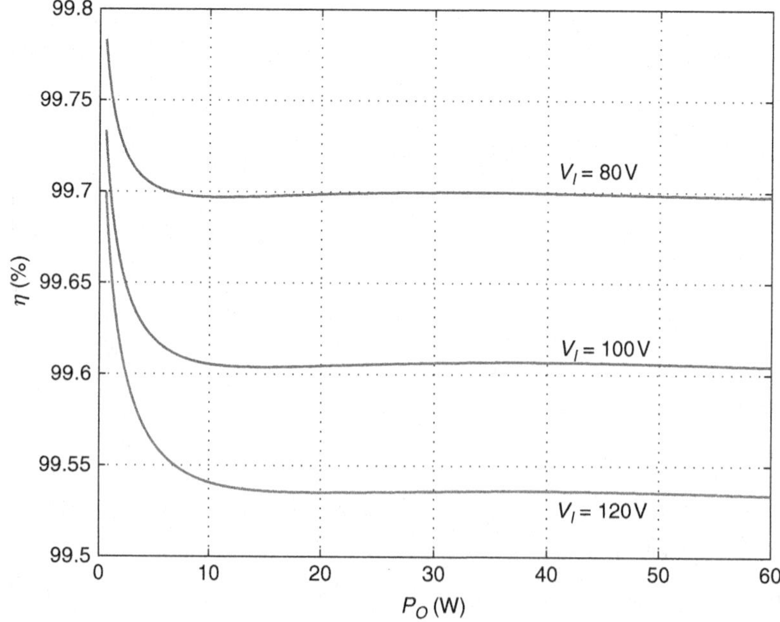

Figure 10.83 Efficiency of the buck converter as a function of the output power P_O at fixed values of the input voltage V_I for the buck converter in DCM, assuming that the inductor is the only lossy component

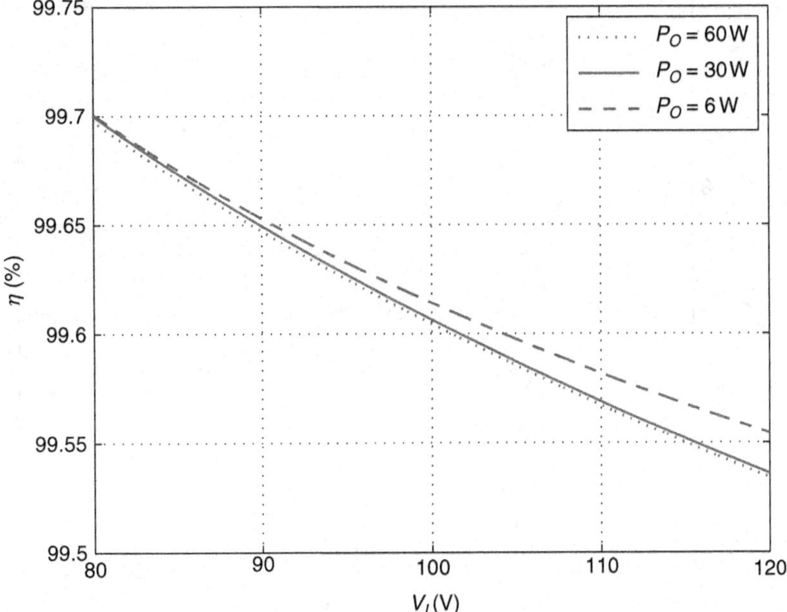

Figure 10.84 Efficiency of the buck converter as a function of the DC input voltage V_I at fixed values of the output power P_O for the buck converter in DCM, assuming that the inductor is the only lossy component

The ESR of the winding and the core is

$$r_{L(DC)} = \frac{P_{cw}}{I_{Lrms}^2} = \frac{0.2862}{3.854^2} = 0.0832\ \text{m}\Omega. \tag{10.425}$$

The surface area of the inductor is

$$A_t = 2(2BC + 2AB + AC) = 2 \times (2 \times 12.35 \times 19 + 2 \times 27.3 \times 12.35 + 27.3 \times 19)$$
$$= 23.112\ \text{m}^2. \tag{10.426}$$

The surface power loss density is

$$\psi = \frac{P_{cw}}{A_t} = \frac{0.2862}{23.112} = 0.01238\ \text{W/cm}^2. \tag{10.427}$$

The temperature rise is

$$\Delta T = 450\psi^{0.826} = 450 \times 0.01238^{0.826} = 11.97\ ^\circ\text{C}. \tag{10.428}$$

Figure 10.85 shows a plot of the temperature rise ΔT as a function of the output power P_O at a fixed value of the DC input voltage V_I.

The core utilization factor is

$$K_u = \frac{NS_nA_{wsi}}{W_a} = \frac{21 \times 5 \times 0.1288}{0.4153 \times 10^2} = 0.3256. \tag{10.429}$$

Figure 10.86 shows the plots of K_{Rh} as a function of the output power P_O at $V_I = 80$, 100, and 120 V. Figure 10.87 shows the amplitude spectrum of the inductor current waveform for P_{Omax} at V_{Imax} for the buck converter in DCM, respectively. Figure 10.88 shows the values of the winding resistance at DC and harmonics. Figure 10.89 shows the spectrum of the winding power loss P_{wn} at V_{Imax} for the buck converter in DCM. Figure 10.90 shows the amplitude spectrum of the inductor current waveform for $0.1P_{Omax}$ at V_{Imin} for the buck converter in DCM, respectively. Figure 10.91 shows the spectrum of the winding power loss P_{wn} at $0.1P_{Omax}$ and V_{Imin} for the buck converter in DCM.

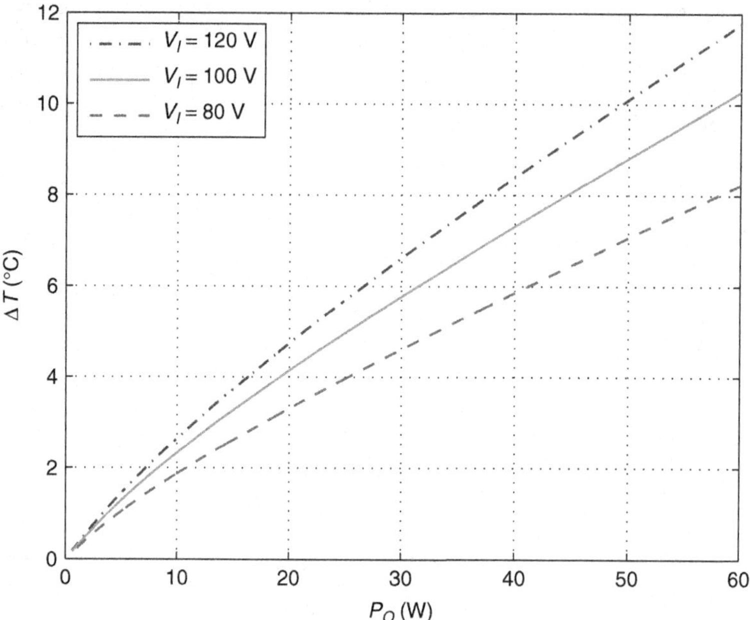

Figure 10.85 Temperature rise ΔT as a function of the output power P_O at fixed values of the output power P_O for the buck converter in DCM

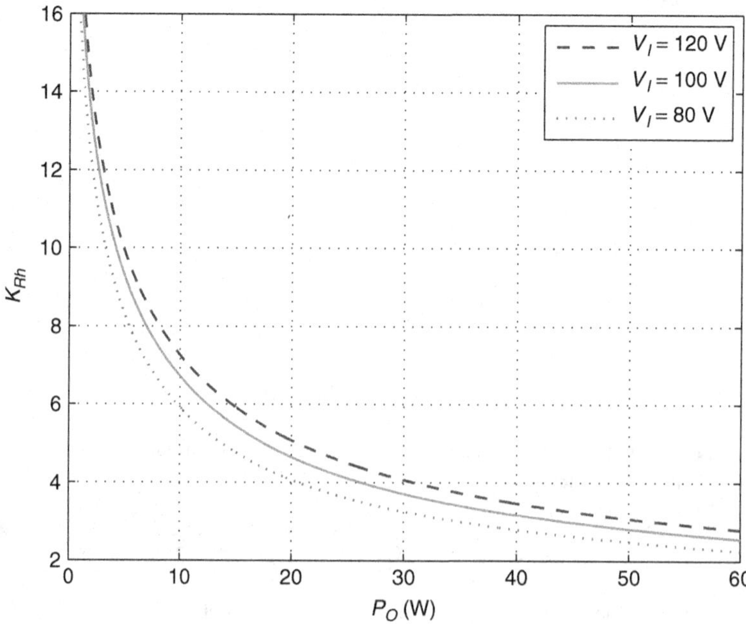

Figure 10.86 K_{Rh} as a function of the output power P_O at fixed values of the DC input voltage V_I for the buck converter in DCM

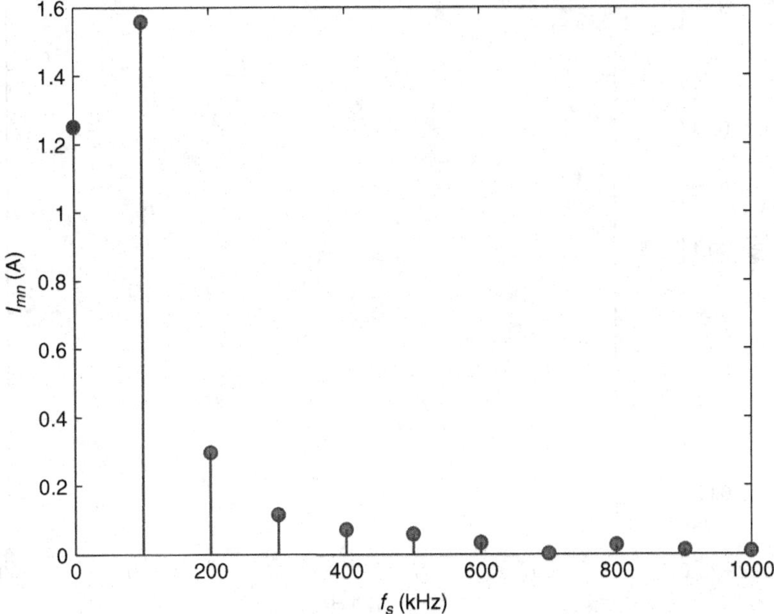

Figure 10.87 Amplitude spectrum of the inductor current I_{mn} at V_{Imax} and P_{Omax} for the buck converter in DCM

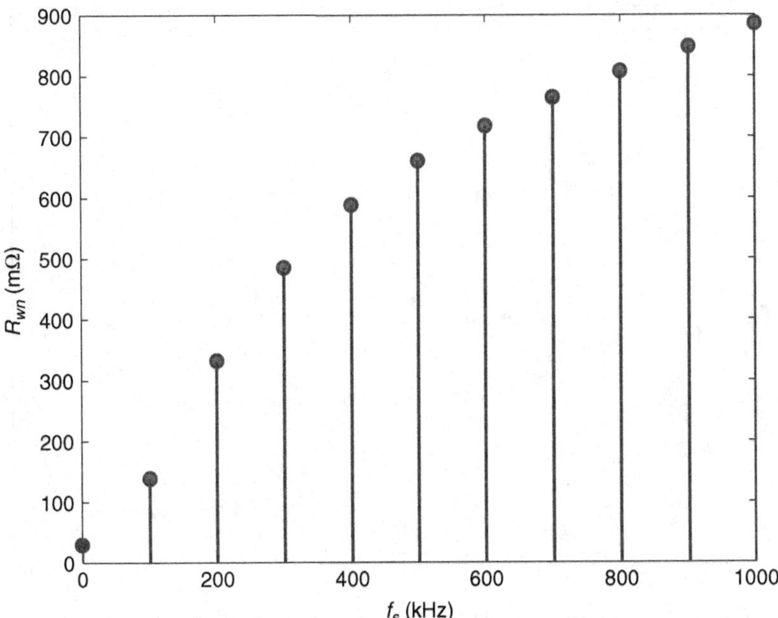

Figure 10.88 Spectrum of the winding resistance at $d_i = 0.405$ mm, $l_w = 18$ cm, $d_i/\delta_w = 1.9381$, $S_n = 43$, $N_l = 1$, and $R_{wDC}/l_w = 0.1345$

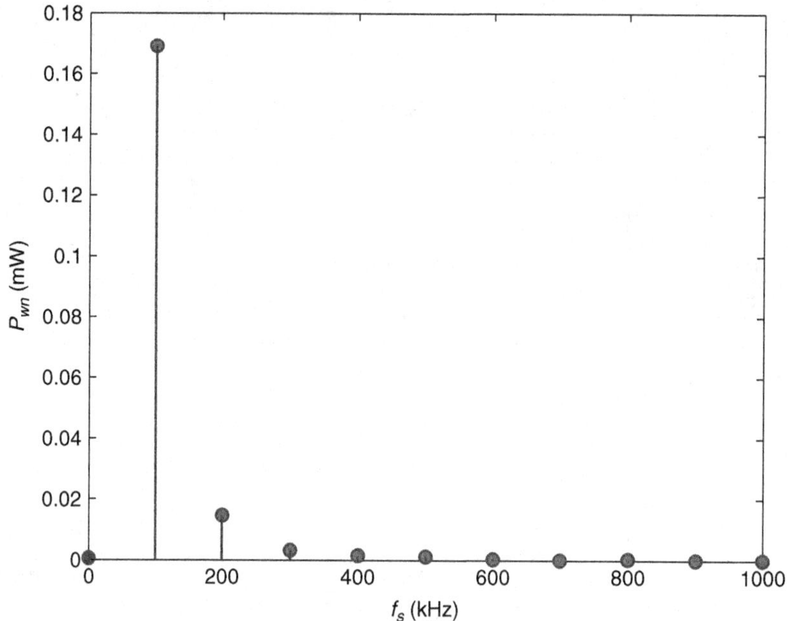

Figure 10.89 Spectrum of the winding power loss at P_{Omax} and V_{Imax}

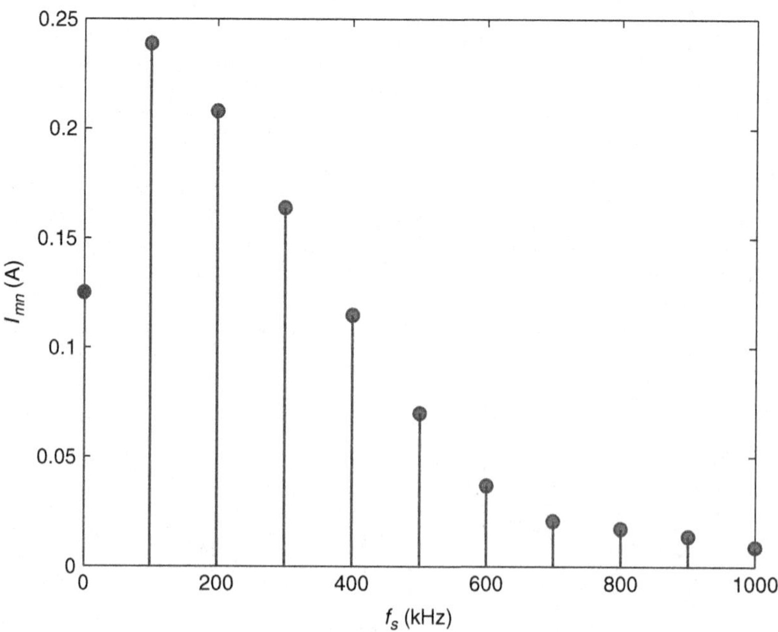

Figure 10.90 Amplitude spectrum of the inductor current I_{mn} at V_{Imax} and $0.1P_{Omax}$ for the buck converter in DCM

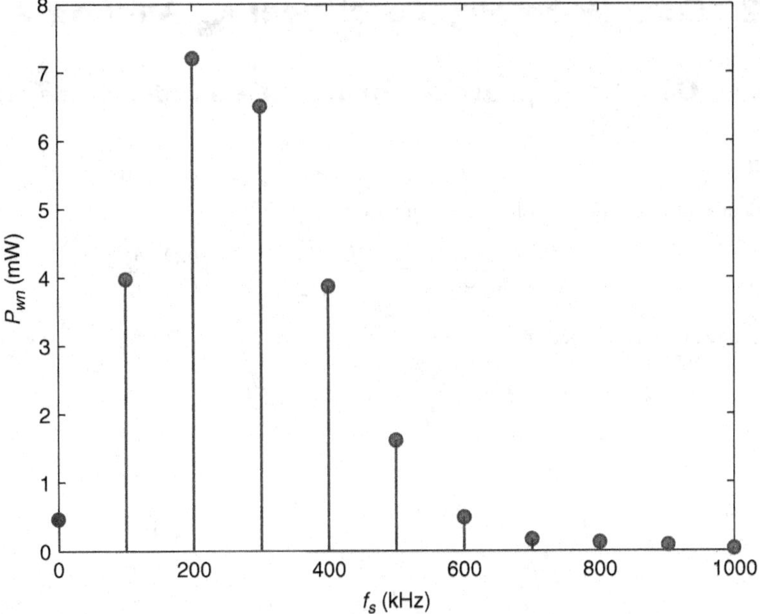

Figure 10.91 Spectrum of the winding power loss at $0.1P_{Omax}$ and V_{Imax}

Solid-Wire Winding. The required cross-sectional area of the bare conductor was calculated to be $A_w = 0.756\,\text{mm}^2$. Let us select a solid round wire AWG18 with $d_i = 1.02\,\text{mm}$, $d_o = 1.09\,\text{mm}$, and $R_{wDC}/l_w = 0.02095\,\Omega/\text{m}$. For $f_s = 100\,\text{kHz}$,

$$\frac{d_i}{\delta_w} = \frac{1.02}{0.209} = 4.8803. \tag{10.430}$$

Assuming $\eta = 0.9$,

$$A = A_o = \left(\frac{\pi}{4}\right)^{0.75} \frac{d_i}{\delta_w}\sqrt{\eta} = \left(\frac{\pi}{4}\right)^{0.75} \times 4.8803 \times \sqrt{0.9} = 3.8632. \tag{10.431}$$

From (5.490), for $N_l = 3$, we get $F_{Rh} = 27.50642$. The DC resistance is

$$R_{wDC} = l_w\left(\frac{R_{wDC}}{l_w}\right) = 1.12 \times 0.02095 = 0.023464\,\Omega. \tag{10.432}$$

The effective DC winding resistance is

$$R_{wDC} = F_{Rh}R_{wDC} = 27.50642 \times 0.023464 = 0.6454\,\Omega. \tag{10.433}$$

The DC winding power loss is

$$P_{wDC} = R_{wDC}I_L^2 = 0.023464 \times 1.25^2 = 36.662\,\text{mW}. \tag{10.434}$$

The total winding loss is

$$P_w = F_{Rh}P_{wDC} = 27.50642 \times 36.662 \times 10^{-3} = 1.008454\,\text{W}. \tag{10.435}$$

The ratio of the winding power loss with the solid wire to the winding power loss with strands is

$$\frac{P_{w(solid)}}{P_{w(strands)}} = \frac{1.008454}{0.2291} = 4.4018. \tag{10.436}$$

10.12 Core Geometry Coefficient K_g Method

10.12.1 General Expression for Core Geometry Coefficient K_g

General Case

The inductance of an inductor with an air gap is given by

$$L = \frac{N^2}{\mathcal{R}} = \frac{N^2}{\mathcal{R}_g + \mathcal{R}_c} = \frac{N^2}{\frac{l_g}{\mu_0 A_c} + \frac{l_c}{\mu_{rc}\mu_0 A_c}} = \frac{\mu_0 N^2 A_c}{l_g + \frac{l_c}{\mu_{rc}}}. \tag{10.437}$$

This equation can be simplified to the form

$$L \approx \frac{\mu_0 N^2 A_c}{l_g} \text{ for } l_g >> \frac{l_c}{\mu_{rc}}. \tag{10.438}$$

In general,

$$Ni = \phi(\mathcal{R}_g + \mathcal{R}_c) = BA_c\left(\frac{l_g}{\mu_0 A_c} + \frac{l_c}{\mu_{rc}\mu_0 A_c}\right) = B\left(\frac{l_g}{\mu_0} + \frac{l_c}{\mu_{rc}\mu_0}\right). \tag{10.439}$$

Thus,

$$NI_{Lmax} = \frac{B_{pk}}{\mu_0}\left(\frac{l_c}{\mu_{rc}} + l_g\right) \tag{10.440}$$

Hence,

$$B_{pk} = \frac{\mu_0 N I_{Lmax}}{\frac{l_c}{\mu_{rc}} + l_g} < B_s. \tag{10.441}$$

The air-gap length is

$$l_g = \frac{\mu_0 N I_{Lmax}}{B_{pk}} - \frac{l_c}{\mu_{rc}} \approx \frac{\mu_0 N I_{Lmax}}{B_{pk}}. \tag{10.442}$$

The maximum flux linkage is

$$\lambda_{max} = N\phi_{max} = NA_c B_{pk} = LI_{Lmax}. \tag{10.443}$$

To avoid core saturation, the number of turns is

$$N = \frac{LI_{Lmax}}{A_c B_{pk}}. \tag{10.444}$$

The length of the winding wire is

$$l_w = N l_T, \tag{10.445}$$

where l_T is the MTL. The cross-sectional area of the winding wire is

$$A_{Cu} = NA_w. \tag{10.446}$$

The DC winding resistance is

$$R_{wDC} = \rho_w \frac{l_w}{A_w} = \rho_w \frac{N l_T}{A_w}, \tag{10.447}$$

yielding

$$N = \frac{R_{wDC} A_w}{\rho_w l_T}. \tag{10.448}$$

Thus,

$$\frac{R_{wDC} A_w}{\rho_w l_T} = \frac{LI_{Lmax}}{A_c B_{pk}} \tag{10.449}$$

which gives the cross-sectional area of the winding wire

$$A_w = \frac{\rho_w L I_{Lmax} l_T}{R_{wDC} A_c B_{pk}}. \tag{10.450}$$

The core window utilization factor is defined as

$$K_u = \frac{A_{Cu}}{W_a} = \frac{NA_w}{W_a}. \tag{10.451}$$

Substitution of (10.444) and (10.450) in (10.451) yields the minimum required area of the core window

$$W_a = \frac{NA_w}{K_u} = \frac{\rho_w L^2 I_{Lmax}^2 l_T}{K_u R_{wDC} A_c^2 B_{pk}^2}. \tag{10.452}$$

The *core geometry coefficient* is defined as [1]

$$K_g = \frac{W_a A_c^2 K_u}{l_T} = \frac{\rho_w L^2 I_{Lmax}^2}{R_{wDC} B_{pk}^2} = \frac{2\rho_w L W_m}{R_{wDC} B_{pk}^2} \, (\text{m}^5), \tag{10.453}$$

where

$$W_m = \frac{1}{2} L I_{Lmax}^2. \tag{10.454}$$

The coefficient K_g is determined by the core geometry A_c, W_a, and l_T. Since

$$R_{wDC} = \frac{P_{wDC}}{I_{Lrms}^2} \tag{10.455}$$

(10.453) becomes

$$K_g = \frac{\rho_w L^2 I_{Lmax}^2 I_{Lrms}^2}{B_{pk}^2 P_{wDC}}. \tag{10.456}$$

The ratio of the copper DC power P_{wDC} to the output power P_O is defined as

$$\alpha = \frac{P_{wDC}}{P_O}. \tag{10.457}$$

Hence,

$$K_g = \frac{\rho_w L^2 I_{Lmax}^2 I_{Lrms}^2}{B_{pk}^2 \alpha P_O}. \tag{10.458}$$

The waveform factor of the inductor current is defined as

$$K_i = \frac{I_{Lrms}}{I_{Lmax}}, \tag{10.459}$$

yielding

$$K_g = \frac{\rho_w K_i^2 L^2 I_{Lmax}^4}{B_{pk}^2 \alpha P_O} = \frac{4\rho_w K_i^2 W_m^2}{B_{pk}^2 \alpha P_O}. \tag{10.460}$$

The core geometry coefficient K_g can be expressed in terms of the area product A_p as

$$K_g = \frac{A_p A_c K_u}{l_T} = \frac{A_c L I_{Lmax}^2}{l_T J_m B_{pk}} = \frac{2A_c W_m}{l_T J_m B_{pk}} \, (\text{m}^5). \tag{10.461}$$

10.12.2 AC Inductor with Sinusoidal Current and Voltage

For an AC inductor conducting only a sinusoidal current, the DC component of the inductor current is $I_L = 0$, $I_{Lrms} = I_{Lmax}/\sqrt{2}$, $B_{DC} = 0$, and $B_{pk} = B_m$. Hence,

$$K_g = \frac{\rho_w L^2 I_{Lmax}^4}{2B_m^2 P_{wDC}} = \frac{2\rho_w W_m^2}{B_m^2 P_{wDC}} = \frac{2\rho_w W_m^2}{B_m^2 \alpha P_O}. \tag{10.462}$$

The quality factor of resonant circuits is defined as

$$Q_L = \omega \frac{W_m}{P_O}, \tag{10.463}$$

yielding

$$K_g = \frac{2\rho_w W_m^2}{B_m^2 P_{wDC}} = \frac{2\rho_w Q_L^2 P_O^2}{\omega^2 B_m^2 P_{wDC}} = \frac{2\rho_w Q_L^2 P_O^2}{\omega^2 B_m^2 \alpha P_O} = \frac{2\rho_w Q_L^2 P_O}{\alpha \omega^2 B_m^2}. \tag{10.464}$$

10.12.3 Inductor for PWM Converter in CCM

For PWM DC–DC power converters operated in continuous conduction mode (CCM), the inductor current waveform i_L increases when the transistor is ON and it decreases when the transistor is OFF. Neglecting the inductor current ripple, the inductor current waveform can be approximated by

$$i_L \approx I_L. \tag{10.465}$$

Therefore, the peak value of the inductor current can be approximated by

$$I_{Lmax} \approx I_L \tag{10.466}$$

and the rms value of the inductor current can be approximated by

$$I_{Lrms} \approx I_L, \tag{10.467}$$

yielding

$$K_g = \frac{\rho_w L^2 I_L^4}{P_{wDC} B_{pk}^2}. \tag{10.468}$$

The energy stored in the inductor is

$$W_m = \frac{1}{2} L I_{Lmax}^2 \approx \frac{1}{2} L I_L^2, \tag{10.469}$$

resulting in the core geometry coefficient K_g describing the energy-handling capability of an inductor

$$K_g = \frac{4\rho_w W_m^2}{B_{pk}^2 P_{wDC}} = \frac{4\rho_w W_m^2}{B_{pk}^2 \alpha P_O}. \tag{10.470}$$

For copper, $\rho_w = \rho_{Cu} = 1.724 \times 10^{-8}$ Ωm at $T = 20\,°\mathrm{C}$. Hence,

$$K_g = \frac{\rho_w L^2 I_L^4}{B_{pk}^2 P_{wDC}} = \frac{4 \times 1.724 \times 10^{-8} \times W_m^2}{B_{pk}^2 P_{wDC}} = \frac{6.8966 \times 10^{-8} \times W_m^2}{B_{pk}^2 P_{wDC}}$$
$$= \frac{W_m^2 \times 10^{-8}}{0.145 B_{pk}^2 P_{wDC}} (\mathrm{m}^5) = \frac{W_m^2}{0.145 B_{pk}^2 \alpha P_O \times 10^{-2}} (\mathrm{cm}^5). \tag{10.471}$$

Assuming that $\alpha = 0.01$, we get $P_{wDC} = 0.01 P_O$. Hence, we obtain

$$K_g = \frac{W_m^2}{0.145 B_{pk}^2 P_O \times 10^{-4}} (\mathrm{cm}^5). \tag{10.472}$$

10.12.4 Inductor for PWM Converter in DCM

For DCM operation,

$$I_{Lrms} = I_{Lmax}\sqrt{\frac{D + D_1}{3}} = I_{Lmax}\sqrt{\frac{1 - D_w}{3}} \approx \frac{I_{Lmax}}{\sqrt{3}}, \tag{10.473}$$

where $D_w = 1 - D - D_1$. The winding DC resistance is

$$R_{wDC} = \frac{P_{wDC}}{I_{Lrms}^2} = \frac{3P_{wDC}}{(1 - D_w) I_{Lmax}^2}. \tag{10.474}$$

The core geometry coefficient is

$$K_g = \frac{\rho_w L^2 I_{Lmax}^2}{R_{wDC} B_{pk}^2} = \frac{\rho_w L^2 I_{Lmax}^4 (1 - D_w)}{3 B_{pk}^2 P_{wDC}} = \frac{\rho_w L^2 I_{Lmax}^4 (1 - D_w)}{3\alpha B_{pk}^2 P_O}$$
$$= \frac{4(1 - D_w)\rho_w W_m^2}{3\alpha B_{pk}^2 P_O}. \tag{10.475}$$

For copper, $\rho_w = \rho_{Cu} = 1.724 \times 10^{-8}$ Ωm at $T = 20\,°$C. Hence,

$$K_g = \frac{4(1-D_w) \times 1.724 \times 10^{-8} W_m^2}{3\alpha B_{pk}^2 P_O} = \frac{(1-D_w) W_m^2}{0.435\alpha B_{pk}^2 P_O \times 10^8}\ (\text{m}^5)$$

$$= \frac{(1-D_w) W_m^2}{0.435\alpha B_{pk}^2 P_O \times 10^{-2}}\ (\text{cm}^5). \tag{10.476}$$

Another method of deriving K_g is as follows. All the energy stored in the inductor is transferred to the load in each cycle. The energy in the inductor is

$$W_m = \frac{1}{2} L I_{Lmax}^2. \tag{10.477}$$

The output power is given by

$$P_O = \frac{W_m}{T_s} = f_s W_m, \tag{10.478}$$

resulting in

$$W_m = \frac{P_O}{f_s} = \frac{P_O}{\eta f_s}. \tag{10.479}$$

The geometry coefficient is

$$K_g = \frac{2\rho_w L W_m}{R_{wDC} B_{pk}^2} = \frac{2\rho_w L P_O}{R_{wDC} f_s B_{pk}^2}. \tag{10.480}$$

For copper, $\rho_w = \rho_{Cu} = 1.724 \times 10^{-8}$ Ωm at $T = 20\,°$C. Hence,

$$K_g = \frac{2\rho_w L P_O}{R_{wDC} f_s B_{pk}^2} = \frac{2 \times 1.724 \times 10^{-8} \times L P_O}{R_{wDC} f_s B_{pk}^2} (\text{m}^5) = \frac{L P_O}{0.29 R_{wDC} f_s B_{pk}^2 \times 10^8} (\text{m}^5)$$

$$= \frac{L P_O}{0.29 R_{wDC} f_s B_{pk}^2 \times 10^{-2}} (\text{cm}^5). \tag{10.481}$$

We can write

$$K_g = \frac{2\rho_w L P_O}{R_{wDC} f_s B_{pk}^2} = \frac{2\rho_w P_O (1-D_w) L I_{Lmax}^2}{3 P_{wDC} f_s B_{pk}^2} = \frac{4(1-D_w)\rho_w P_O W_m}{3 P_{wDC} f_s B_{pk}^2}$$

$$= \frac{4(1-D_w)\rho_w P_O^2}{3 P_{wDC} f_s^2 B_{pk}^2} = \frac{4(1-D_w)\rho_w P_O}{3\alpha f_s^2 B_{pk}^2}. \tag{10.482}$$

For copper, $\rho_w = \rho_{Cu} = 1.724 \times 10^{-8}$ Ωm at $T = 20\,°$C. Thus,

$$K_g = \frac{4(1-D_w)\rho_w P_O}{3\alpha f_s^2 B_{pk}^2} = \frac{4 \times 1.724 \times 10^{-8} (1-D_w) P_O}{3\alpha R_{wDC} f_s^2 B_{pk}^2} = \frac{(1-D_w) P_O}{0.435\alpha f_s^2 B_{pk}^2 \times 10^8}\ (\text{m}^5)$$

$$= \frac{(1-D_w) P_O}{0.435\alpha f_s^2 B_{pk}^2 \times 10^{-2}}\ (\text{cm}^5). \tag{10.483}$$

The geometry coefficient K_g describing the apparent power-handling capability is given by [1]

$$K_g = \frac{P_t}{2\alpha \times 0.145 K_f^2 f^2 B_{pk}^2 \times 10^{-2}} (\text{cm}^5), \tag{10.484}$$

where $P_t = V_{Lrms} I_{Lrms}$ is the apparent power and $K_f = 4.44$ for sinusoidal waveform of the inductor voltage and $K_f = 4$ for a square-wave inductor voltage.

10.13 Inductor Design for Buck Converter in CCM Using K_g Method

Example 10.10

A buck PWM converter is operated in the CCM and has the following specifications: $V_I = 28 \pm 4\,\mathrm{V}$, $V_O = 12\,\mathrm{V}$, $I_O = 1 - 10\,\mathrm{A}$, $f_s = 100\,\mathrm{kHz}$, and $V_r/V_O \leq 1\%$. Design an inductor.

Solutions: The maximum output power is

$$P_{Omax} = V_O I_{Omax} = 12 \times 10 = 120\,\mathrm{W} \tag{10.485}$$

and the minimum output power is

$$P_{Omin} = V_O I_{Omin} = 12 \times 1 = 12\,\mathrm{W}. \tag{10.486}$$

The minimum load resistance is

$$R_{Lmin} = \frac{V_O}{I_{Omax}} = \frac{12}{10} = 1.2\,\Omega \tag{10.487}$$

and the maximum load resistance is

$$R_{Lmax} = \frac{V_O}{I_{Omin}} = \frac{12}{1} = 12\,\Omega. \tag{10.488}$$

The minimum DC voltage transfer function is

$$M_{VDCmin} = \frac{V_O}{V_{Imax}} = \frac{12}{32} = 0.375 \tag{10.489}$$

and the maximum DC voltage transfer function is

$$M_{VDCmax} = \frac{V_O}{V_{Imin}} = \frac{12}{24} = 0.5. \tag{10.490}$$

Assuming the converter efficiency $\eta = 0.9$, we can calculate the minimum duty cycle

$$D_{min} = \frac{M_{VDCmin}}{\eta} = \frac{0.375}{0.9} = 0.4167 \tag{10.491}$$

and the maximum duty cycle

$$D_{max} = \frac{M_{VDCmax}}{\eta} = \frac{0.5}{0.9} = 0.5556. \tag{10.492}$$

The minimum inductance for CCM operation is

$$L_{min} = \frac{R_{Lmax}(1 - D_{min})}{2f_s} = \frac{12 \times (1 - 0.4167)}{2 \times 10^5} = 35\,\mu\mathrm{H}. \tag{10.493}$$

Pick $L = 40\,\mu\mathrm{H}$.

The buck inductor current contains a large DC component I_L. The peak inductor current is

$$I_{Lmax} = I_{Omax} + \frac{\Delta i_{Lmax}}{2} = 10 + \frac{1.677}{2} = 10.8385 \approx 10.839\,A. \tag{10.494}$$

The maximum energy stored in the inductor is

$$W_m = \frac{1}{2} L I_{Lmax}^2 = \frac{1}{2} \times 40 \times 10^{-6} \times 10.839^2 = 2.349\,\mathrm{mJ}. \tag{10.495}$$

Assume $B_{pk} = 0.2\,\mathrm{T}$, $\alpha = 0.01$, and $K_u = 0.4$. The core geometry coefficient K_g is given by [1]

$$K_g = \frac{W_m^2}{0.145 P_O B_{pk}^2 \times 10^{-4}}\,(\mathrm{cm}^5) = \frac{(2.349 \times 10^{-3})^2}{0.145 \times 120 \times 0.2^2 \times 10^{-4}}\,(\mathrm{cm}^5) = 0.0792\mathrm{cm}^5. \tag{10.496}$$

Select a Magnetics 0P-44011EC EE-ungapped ferrite core with $K_g = 0.08275\,\text{cm}^5$, $A_p = 1.39\,\text{cm}^4$, $A_c = 127\,\text{mm}^2$, $l_c = 76.7\,\text{mm}$, $V_c = 9.78\,\text{mm}^3$, $A_L = 3260\,\text{mH}/1000$ turns. We select the P-type ferrite magnetic material whose $\mu_{rc} = 2500 \pm 25\%$ and $B_s = 0.5\,\text{T}$. We choose the Cosmo rectangular bobbin 2587-0 with the window height $B_h = 12.59\,\text{mm}$ and the bobbin window width $B_w = 7.505\,\text{mm}$.

The core winding window area is

$$W_a = \frac{A_p}{A_c} = \frac{1.39}{1.27} = 1.095\,\text{cm}^2. \tag{10.497}$$

The maximum amplitude of current density in the winding wire is

$$J_m = \frac{2W_m}{K_u A_p B_{pk}} = \frac{2 \times 2.349 \times 10^{-3}}{0.4 \times 1.39 \times 10^{-4} \times 0.2} = 422.48\,\text{A/cm}^2 = 4.2248\,\text{A/mm}^2. \tag{10.498}$$

The cross-sectional area of the bare wire is

$$A_w = \frac{I_{Lmax}}{J_m} = \frac{10.839}{4.2256} = 2.565\,\text{mm}^2. \tag{10.499}$$

Select a round solid copper wire AWG13, which has $d_i = 1.83\,\text{mm}$, $d_o = 1.9\,\text{mm}$, and $R_{wDC}/l_w = 0.00656\,\Omega/\text{m}$. The insulated wire cross-sectional area is

$$A_{wo} = \frac{\pi d_o^2}{4} = \frac{\pi \times (1.9 \times 10^{-3})^2}{4} = 2.833\,\text{mm}^2. \tag{10.500}$$

The skin depth of copper at $f = 100\,\text{kHz}$ is

$$\delta_w = \frac{66.2}{\sqrt{f}}\,(\text{mm}) = \frac{66.2}{\sqrt{100\,000}}\,(\text{mm}) = 0.209\,(\text{mm}). \tag{10.501}$$

To avoid the skin and proximity effects at the switching frequency, the diameter of the strands is

$$d_{is} = 2\delta_w = 2 \times 0.209 = 0.418\,\text{mm}. \tag{10.502}$$

Select the AWG26 strand wire with the bare wire diameter $d_{is} = 0.405\,\text{mm}$, $d_{os} = 0.452\,\text{mm}$, the bare wire cross-sectional area $A_{ws} = 0.128\,\text{mm}^2$, and $R_{wDCs}/l_w = 0.1345\,\Omega/\text{m}$. The number of strands is

$$S_n = \frac{A_w}{A_{ws}} = \frac{2.565}{0.128} = 20. \tag{10.503}$$

Pick $S_n = 20$. The number of turns is

$$N = \frac{K_u W_a}{S_n A_{ws}} = \frac{0.4 \times 1.095}{20 \times 0.00128} = 17.109. \tag{10.504}$$

Pick $N = 18$.

The length of the air gap is

$$l_g = \frac{\mu_0 A_c N^2}{L} - \frac{l_c}{\mu_{rc}} = \frac{4\pi \times 10^{-7} \times 1.27 \times 10^{-4} \times 18^2}{40 \times 10^{-6}} - \frac{76.7 \times 10^{-3}}{2500} = 1.26\,\text{mm}. \tag{10.505}$$

The peak of the total magnetic flux density, which consists of the DC and AC components, is given by

$$B_{pk} = \frac{\mu_0 N}{l_g + \frac{l_c}{\mu_{rc}}}\left(I_O + \frac{\Delta i_{Lmax}}{2}\right) = \frac{4\pi \times 10^{-7} \times 18}{(1.26 + \frac{76.7}{2500}) \times 10^{-3}}\left(10 + \frac{1.677}{2}\right)$$
$$= 0.1899\,\text{T} < B_s = 0.5\,\text{T}. \tag{10.506}$$

The amplitude of the AC component of the magnetic flux density is given by

$$B_m = \frac{\mu_0 N}{l_g + \frac{l_c}{\mu_{rc}}}\left(\frac{\Delta i_{Lmax}}{2}\right) = \frac{4\pi \times 10^{-7} \times 18}{(1.26 + \frac{76.7}{2500}) \times 10^{-3}}\left(\frac{1.677}{2}\right) = 0.0147\,\text{T}. \tag{10.507}$$

The core power loss density is

$$P_v = 0.0434 f^{1.63} (10B_m)^{2.62} = 0.0434 \times 100^{1.63} \times (10 \times 0.0147)^{2.62} = 519.83\,\text{mW/cm}^3. \tag{10.508}$$

The core loss is

$$P_C = V_c P_v = 0.00978 \times 519.83 = 5.084\,\text{mW}. \tag{10.509}$$

The length of a single turn of the EE core is

$$l_T = 2E + F + C = 2 \times 2.76 + 1.07 + 1.069 = 7.659 \text{ cm}. \quad (10.510)$$

The winding length is

$$l_w = N l_T = 18 \times 7.659 \times 10^{-2} = 1.378 \text{ m}. \quad (10.511)$$

The DC resistance of a single strand is

$$R_{wDCs} = l_w \left(\frac{R_{wDCs}}{l_w} \right) = 1.378 \times 0.1345 = 0.1853 \, \Omega. \quad (10.512)$$

The DC winding resistance is

$$R_{wDC} = \frac{R_{wDCs}}{S_n} = \frac{0.1853}{19} = 0.009752 \, \Omega. \quad (10.513)$$

The AC power loss is approximately equal to the winding DC power loss, which is given by

$$P_{wDC} = R_{wDC} I_L^2 = 0.009752 \times 10^2 = 0.9752 \text{ W}. \quad (10.514)$$

The total power loss is

$$P_{cw} = P_C + P_{wDC} = 0.005084 + 0.9752 = 0.9802 \text{ W}. \quad (10.515)$$

The surface area of the core is $A_t = 64.16 \text{ cm}^2$. The surface power loss density is

$$\psi = \frac{P_{cw}}{A_t} = \frac{0.9802}{64.16} = 0.01527 \text{ W/cm}^2. \quad (10.516)$$

The temperature rise of the inductor is

$$\Delta T = 450 \psi^{0.826} = 450 \times 0.01527^{0.826} = 14.22 \, ^\circ\text{C}. \quad (10.517)$$

The core window utilization is

$$K_u = \frac{N S_n A_{ws}}{W_a} = \frac{18 \times 19 \times 0.00128}{1.095} = 0.3997. \quad (10.518)$$

The coefficient α is

$$\alpha = \frac{P_{wDC}}{P_O} = \frac{0.9752}{120} = 0.00812. \quad (10.519)$$

The assumed value of α is 0.01.

10.14 Inductor Design for Buck Converter in DCM Using K_g Method

Example 10.11

A buck PWM converter is operated in the DCM and has the following specifications: $V_I = 28 \pm 4$ V, $V_O = 12$ V, $I_O = 0 - 10$ A, $f_s = 100$ kHz, and $V_r/V_O \leq 1\%$. The maximum temperature rise is to be $\Delta T_{max} = 40\,^\circ$C. The maximum power loss in the winding is $P_{rw} = 1$ W.

Solution: The maximum DC output power is

$$P_{Omax} = V_O I_{Omax} = 12 \times 10 = 120 \text{ W}. \quad (10.520)$$

The minimum load resistance is

$$R_{Lmin} = \frac{V_O}{I_{Omax}} = \frac{12}{10} = 1.2 \, \Omega. \quad (10.521)$$

The minimum and maximum DC duty cycles are

$$M_{VDCmin} = \frac{V_O}{V_{Imax}} = \frac{12}{32} = 0.375 \tag{10.522}$$

and

$$M_{VDCmax} = \frac{V_O}{V_{Imin}} = \frac{12}{24} = 0.5. \tag{10.523}$$

Assume the converter efficiency $\eta = 0.9$ and the dwell duty cycle $D_w = 0.05$. The maximum inductance is

$$L_{max} = \frac{\eta R_{Lmin}(1 - M_{VDCmax})(1 - D_w)^2}{2f_s} = \frac{0.9 \times 1.2(1 - 0.5)(1 - 0.05)^2}{2 \times 10^5}$$
$$= 2.4368\ \mu\text{H}. \tag{10.524}$$

Let $L = 2.2\ \mu$H. The maximum duty cycle at full power occurs at $V_{Imin} = 24$ V and is given by

$$D_{max} = \sqrt{\frac{2f_s L M^2_{VDCmax}}{\eta R_{Lmin}(1 - M_{VDCmax})}} = \sqrt{\frac{2 \times 10^5 \times 2.2 \times 10^{-6} \times 0.5^2}{0.9 \times 1.2(1 - 0.5)}} = 0.4513. \tag{10.525}$$

and the minimum duty cycle is

$$D_{min} = \sqrt{\frac{2f_s L M^2_{VDCmin}}{\eta R_{Lmin}(1 - M_{VDCmin})}} = \sqrt{\frac{2 \times 10^5 \times 2.2 \times 10^{-6} \times 0.375^2}{0.9 \times 1.2(1 - 0.375)}} = 0.3027. \tag{10.526}$$

The minimum duty cycle at full load when the diode is ON is

$$D_{1max} = D_{max}\left(\frac{1}{M_{VDCmax}} - 1\right) = 0.4513\left(\frac{1}{0.5} - 1\right) = 0.4513. \tag{10.527}$$

The maximum inductor current is

$$I_{Lmax} = \Delta i_{Lmax} = \frac{D_{min}(V_{Imax} - V_O)}{f_s L} = \frac{0.3027(32 - 12)}{10^5 \times 2.2 \times 10^{-6}} = 27.518\ \text{A}. \tag{10.528}$$

The maximum magnetic energy stored in the inductor is

$$W_m = \frac{1}{2} L I^2_{Lmax} = \frac{1}{2} \times 2.2 \times 10^{-6} \times 27.518^2 = 0.832\ \text{mJ}. \tag{10.529}$$

In the buck PWM converter, the inductor DC current is equal to the DC output current $I_L = I_O$. The maximum rms value of the inductor current is

$$I_{Lrms(max)} = \Delta i_{Lmax}\sqrt{\frac{D_{min} + D_{1max}}{3}} = 27.51\sqrt{\frac{0.3027 + 0.4513}{3}} = 13.791\ \text{A}. \tag{10.530}$$

Assume $K_u = 0.4$, $B_{pk} = 0.2$ T, $D_w = 0.05$, and $\alpha = 0.005$. The core geometry coefficient is

$$K_g = \frac{(1 - D_w)W_m^2}{0.435\alpha B^2_{pk} P_O \times 10^{-2}}\text{cm}^5 = \frac{(1 - 0.05) \times (0.832 \times 10^{-3})^2}{0.435 \times 0.005 \times 0.2^2 \times 120 \times 10^{-2}} = 0.006293\text{cm}^5. \tag{10.531}$$

Select a Magnetics EE ferrite core 0P-43007 with $K_g = 0.00189\ \text{cm}^5$, $A_p = 0.5\ \text{cm}^4$, $l_c = 66$ mm, $A_c = 49.1\ \text{mm}^2$, $V_c = 4000\ \text{mm}^3$, and $A_L = 1680$ mH/1000 turns. The P material is used whose $\mu_{rc} = 2500 \pm 25\%$ and $B_s = 0.5$ T. We choose the Magnetics printed circuit bobbin PCB3007T1 with the bobbin window height $B_h = 17.27$ mm and the bobbin width $B_w = 5.675$ mm.

The area of the winding window is

$$W_a = \frac{A_p}{A_c} = \frac{0.5}{0.491} = 1.018\ \text{cm}^2. \tag{10.532}$$

The amplitude of the current density is

$$J_m = \frac{2W_m}{K_u A_p B_m} = \frac{2 \times 0.832 \times 10^{-3}}{0.4 \times 0.5 \times 10^{-8} \times 0.2} = 4.16 \times 10^6\ \text{A/m}^2 = 4.16\ \text{A/mm}^2. \tag{10.533}$$

The cross-sectional area of the bare wire is

$$A_w = \frac{I_{Lmax}}{J_m} = \frac{27.518}{4.16} = 6.615\ \text{mm}^2. \tag{10.534}$$

The mean length of a single turn of the EE core is

$$l_T = 2E + F + C = 2 \times 1.95 + 0.72 + 0.73 = 5.35\ \text{cm}. \tag{10.535}$$

The skin depth of copper at $f = 100\,\text{kHz}$ is

$$\delta_w = \frac{66.2}{\sqrt{f}}(\text{mm}) = \frac{66.2}{\sqrt{10^5}}(\text{mm}) = 0.209\ \text{mm}. \tag{10.536}$$

To avoid the skin and proximity effects, the bare wire diameter of a strand is

$$d_{is} = 2\delta_w = 2 \times 0.209 = 0.418\ \text{mm}. \tag{10.537}$$

We will select wire AWG26 with $d_{is} = 0.405\,\text{mm}$, $d_{os} = 0.452\,\text{mm}$, $A_{ws} = 0.128\,\text{mm}^2$, and $R_{wDC}/l_w = 0.1345\ \Omega/\text{m}$. The number of strands is

$$S_n = \frac{A_w}{A_{ws}} = \frac{6.615}{0.128} = 51.68. \tag{10.538}$$

Select $S_n = 50$. The cross-sectional area of each insulated strand is

$$A_{wso} = \frac{\pi d_{os}^2}{4} = \frac{\pi \times (0.452 \times 10^{-3})^2}{4} = 0.1604\ \text{mm}^2. \tag{10.539}$$

The number of turns is

$$N = \frac{K_u W_a}{S_n A_{wso}} = \frac{0.4 \times 1.018 \times 10^{-4}}{50 \times 0.1604 \times 10^{-6}} = 5. \tag{10.540}$$

Pick $N = 10$.

The air-gap length is

$$l_g = \frac{\mu_0 A_c N^2}{L} - \frac{l_c}{\mu_{rc}} = \frac{4\pi \times 10^{-7} \times 0.491 \times 10^{-4} \times 5^2}{2.2 \times 10^{-6}} - \frac{66 \times 10^{-3}}{2500} = 0.674\ \text{mm}. \tag{10.541}$$

The length of the winding wire is

$$l_w = N l_T = 5 \times 5.35 = 26.75\ \text{cm}. \tag{10.542}$$

Let us assume $l_w = 30\,\text{cm}$ to allow for the inductor leads. The DC and low-frequency resistance of the single strand is

$$R_{wDCs} = l_w\left(\frac{R_{wDC}}{l_w}\right) = 30 \times 10^{-2} \times 0.1345 = 40.35\ \text{m}\Omega. \tag{10.543}$$

The DC and low-frequency resistance of the winding is

$$R_{wDC} = \frac{R_{wDCs}}{S_n} = \frac{40.35 \times 10^{-3}}{50} = 0.000807\ \Omega. \tag{10.544}$$

The DC and low-frequency winding power loss is

$$P_{wDC} = R_{wDC} I_L^2 = 0.000807 \times 10^2 = 0.0807\ \text{W}. \tag{10.545}$$

The assumed value of α is 0.005. The ratio of the bare wire d_i to the skin depth at $f_s = 100\,\text{kHz}$ is

$$\frac{d_{is}}{\delta_w} = \frac{0.405}{0.209} = 1.938. \tag{10.546}$$

The winding consists of one layer ($N_l = 1$) and 10 turns $N = 5$. Using (5.194), the computer calculations for a single-layer winding yield $F_{Rh} = 1.6429$. Taking into account the skin effect of inductor current at the fundamental and its harmonics, the winding power loss is

$$P_w = F_{Rh} P_{wDC} = 1.6429 \times 0.0807 = 0.1326\ \text{W}. \tag{10.547}$$

The peak value of the AC component of the magnetic flux density is

$$B_m = \frac{\mu_0 N}{l_g + \frac{l_c}{\mu_{rc}}}\left(\frac{\Delta i_{Lmax}}{2}\right) = \frac{4\pi \times 10^{-7} \times 5}{0.674 \times 10^{-3} + \frac{66\times 10^{-3}}{2500}}\left(\frac{27.518}{2}\right) = 0.1234\,\mathrm{T}. \tag{10.548}$$

The peak of the total magnetic flux density, which consists of the DC and AC components, is given by

$$B_{pk} = 2B_m = 2 \times 0.1234 = 0.2468\,\mathrm{T}. \tag{10.549}$$

The core power loss density is

$$P_v = 0.0434 f^{1.63}(10B_m)^{2.62} = 0.0434 \times 100^{1.63} \times (10 \times 0.1234)^{2.62} = 137\,\mathrm{mW/cm^3}. \tag{10.550}$$

The core loss is

$$P_C = V_c P_v = 4 \times 137 \times 10^{-3} = 0.548\,\mathrm{W}. \tag{10.551}$$

The sum of the core loss and the DC and low-frequency winding power loss is

$$P_{cwDC} = P_C + P_{wDC} = 0.548 + 0.0807 = 0.6287\,\mathrm{W}. \tag{10.552}$$

The power loss in the core and in the winding is

$$P_{cw} = P_C + P_w = 0.548 + 0.1326 = 0.6806\,\mathrm{W}. \tag{10.553}$$

The ESR of the winding and the core is

$$r_{L(DC)} = \frac{P_{cw}}{I_{Lrms}^2} = \frac{0.6806}{13.791^2} = 3.57\,\mathrm{m\Omega}. \tag{10.554}$$

The estimated surface area of the inductor is $A_t = 36.94\,\mathrm{cm^2}$. The surface power loss density is

$$\psi = \frac{P_{cw}}{A_t} = \frac{0.6806}{36.94} = 0.01842\,\mathrm{W/cm^2}. \tag{10.555}$$

The temperature rise is

$$\Delta T = 450\psi^{0.826} = 450 \times 0.01842^{0.826} = 16.609\,^\circ\mathrm{C}. \tag{10.556}$$

The core utilization factor is

$$K_u = \frac{S_n N A_{wso}}{W_a} = \frac{50 \times 5 \times 0.1604 \times 10^{-6}}{1.018 \times 10^{-4}} = 0.3939. \tag{10.557}$$

Also,

$$\alpha = \frac{P_{wDC}}{P_O} = \frac{0.0807}{120} = 0.0006725. \tag{10.558}$$

10.15 Summary

- For AWG round magnet wire, the wire area A_w is doubled or is halved with every three wire sizes.
- The magnet wire diameter increases or decreases by 12.5% for each wire size.
- When the diameter of a wire is doubled, the AWG will decrease by a factor of 6.
- When the AWG wire size changes by a decade, the wire cross-sectional area changes by a factor of 10.
- When a round wire is changed to a square wire, the wire area increases by one wire size
- The window utilization factor K_u determines the maximum space that may be used by the conductor (copper) in the window.

- The temperature rise of magnetic components ΔT is nearly proportional to the heat radiating surface area A_t
- A core should have a long and narrow window to allow for fewer layers and therefore to minimize AC winding loss.
- As the number of turns N is increased, the flux density is reduced, and therefore core losses are decreased, but winding loss is increased.
- The core area product A_p provides a crude initial estimate of the core size.
- The core area product $A_p = W_a A_c$ is determined by the core geometry.
- Different cores may have the same value of the core area product $A_p = A_c W_a$, for example, a large winding window W_a and a small core cross-sectional area A_c, or vice versa.
- The core area product A_p design approach requires specifications of K_u, B_{pk}, and J_m.
- The core geometry coefficient $K_g = W_a A_c^2 K_u / l_T$ is determined by core geometry
- The core geometry coefficient K_g design approach requires specifications of K_u, B_{pk}, and α.
- Both the A_p and K_g methods do not require information on the core material at the stage of selecting the core.
- The amplitude of the AC component of the magnetic flux density B_m determines the core size and core losses. As the AC flux density is increased, a smaller core size is allowed, but core losses will increase.
- In AC resonant inductors used in radiofrequency (RF) power amplifiers and DC-to-AC inverters the amplitude of the inductor current I_{Lm} is large and the amplitude of the magnetic field density B_m is also large, causing large core loss. The amplitude B_m is usually limited by the core loss. Increased core losses usually require the amplitude of the flux density B_m to be derated. In addition, the core relative permeability μ_{rc} is low at high frequencies, which requires more turns N and causes large winding loss. The DC component and harmonics of the inductor current are usually zero.
- In inductors used in DC-to-DC PWM converters operating in CCM the peak-to-peak value of the inductor current is low, the peak magnetic field density B_{pk} is large and is usually limited by the core saturation. The amplitude of the AC component of the magnetic field density B_m is low, yielding low core loss. The effect of harmonics on the winding loss is usually low.
- In inductors used in DC-to-DC PWM converters operating in DCM the peak-to-peak value of the inductor current is high, the peak magnetic field density B_{pk} is large and is usually limited by the core saturation. The amplitude of the AC component of the magnetic field density B_m is large, causing large core loss. The effect of harmonics on the winding loss is significant. However, the winding loss is low because the inductance is low and requires a short winding wire.
- A core with a low aspect ratio h/w runs cooler and has a better magnetic coupling between the outside winding layers and the core.
- The maximum temperature often occurs between the winding and the core.

10.16 References

[1] C. W. T. McLyman, *Transformer and Inductor Design Handbook*, 3rd Ed., New York: Marcel Dekker, 2004.

[2] C. W. T. McLyman, *Magnetic Core Selection for Transformer and Inductor*, 2nd Ed., New York: Marcel Dekker, 1997.

[3] C. W. T. McLyman, *High Reliability Magnetic Devices, Design and Fabrication*, New York: Marcel Dekker, 2002.
[4] A. I. Pressman, *Switching Power Supplies*, New York: McGraw-Hill, 1991.
[5] A. K. Ohri, T. G. Wilson, and H. A. Owen, Jr., "Design of air-gapped magnetic-core inductors for superimposed direct and alternating currents," *IEEE Transactions on Magnetics*, vol. 12, no. 5, pp. 564–574, September 1976.
[6] W. M. Chew. P. D. Evens, and W. J. Hefferman, "High-frequency inductor design concept," 22nd Annual IEEE Power Electronics Specialists Conference Record (PESC'91), Cambridge, MA, June 24–27, 1991, pp. 673–678.
[7] K. W. E. Cheng and P. D. Evens, "Optimization of high frequency inductor design of series resonant converter," 23rd Annual IEEE Power Electronics Specialists Conference Record (PESC'92), Toledo, Spain, June 29–July 3, vol. 2, 1992, pp. 1416–1422.
[8] W.-J. Gu and R. Liu, "Study of volume and weight vs. frequency for high-frequency transformers," 24th Annual IEEE Power Electronics Specialists Conference (PESC), Seattle, WA, June 2–24, 1993, pp. 1123–1129.
[9] U. Kirchenberger, M. Marx, and D. Schreder, "A contribution to the design optimization of resonant inductors for high power resonant DC-DC converters," 27th IEEE Industry Applications Society Annual Meeting, Houston, TX, October 4–9, vol. 1, 1992, pp. 994–1001.
[10] A. Rahimi-Kian, A. Keyhani, and D. J. M. Powell, "Minimum loss design of a 100 kHz inductor with litz wire," 22nd IEEE Industry Applications Society (IAS) Annual Meeting, New Orleans, LA, October 5–9, 1997, vol. 2, pp. 1414–1420.
[11] L. Dixon, *Transformer and Inductor Design for Optimum Circuit Performance*. 2002 Power Supply Design Seminar, SEM1500, Texas Instruments, 2002.
[12] C. R. Sullivan, W. Li, S. Parbhakaran, and S. Lu, "Design and fabrication of low-loss toroidal air-core inductors," 38th Annual IEEE Power Electronics Specialists Conference (PESC) Record, 2007, pp. 1754–1759.
[13] M. Nigam and C. R. Sullivan, "Multi-layer folded high-frequency toroidal inductor windings," 23rd Annual Applied Power Electronics Conference, Austin, TX, February 24–28, 2008, pp. 682–688.
[14] B. J. Lyons, J. G. Hayes, and M. G. Egan, "Magnetic material comparisons for high-current inductors in low-medium frequency dc-dc converters," 22nd Annual IEEE Applied Power Electronics Conference, Anaheim, CA, February 25–March 1, 2007, pp. 71–77.
[15] B. J. Lyons, J. G. Hayes, and M. G. Egan, "Design algorithm for high-current gapped foil-wound inductors in low-to-medium frequency dc-dc converters," 22nd Annual IEEE Power Electronics Specialists Conference (PESC) Record, Orlando, FL, June 17-21, 2007, pp. 1760–1766.
[16] B. J. Lyons, J. G. Hayes, and M. G. Egan, "Experimental investigation of iron-based amorphous metal and 6.5% silicon steel for high-current inductors in low-medium frequency dc-dc converters," 42nd IEEE Industry Applications Meeting Conference, New Orleans, LA, September 23-27, 2007, pp. 1781–1786.
[17] M. S. Rylko, B. J. Lyons, J. G. Hayes, and M. G. Egan, "Magnetic material comparisons for high-current gapped and gapless foil wound inductors in high-frequency dc-dc converters," Power Electronics and Motion Control Conference (EPE-PEMC), Poznan, Poland, September 1-3, 2008, pp. 1249–1256.
[18] H. Njiende, R. Rrohleke, and J Bocker, "Optimized size design of integrated magnetic components using area product approach," European Power Electronics and Applications Conference, Dresden, Germany, EPE, 2005, pp. 1–10.
[19] K. Billings, *Switchmode Power Supply Handbook*, 2nd Ed., Boston, MA: McGraw-Hill, 2008.
[20] C. P. Basso, *Switch-mode Power Supplies*, New York: McGraw-Hill, 2008.
[21] A. V. Bossche and V. C. Valchev, *Inductors and Transformers for Power Electronics*, Boca Raton, FL: Taylor & Francis, 2005.
[22] M. K. Kazimierczuk, *Pulse-width Modulated DC-DC Power Converters*, New York: John Wiley & Sons, 2008.
[23] M. K. Kazimierczuk, *RF Power Amplifiers*, New York: John Wiley & Sons, 2008.
[24] M. K. Kazimierczuk and H. Sekiya, "Design of ac resonant inductors using area product method," IEEE Energy Conversion Congress and Exhibition, San Jose, CA, September 20-24, 2009, pp. 994–1001.
[25] H. Sekiya and M. K. Kazimierczuk, "Design of RF-choke inductors using core geometry coefficient," Proceedings of the Electrical Manufacturing and Coil Winding Conference, Nashville, USA, September 29-October 1, 2009.
[26] D. Murty-Bellur and M. K. Kazimierczuk, "Harmonic winding loss in buck PWM dc-dc converter in discontinuous conduction mode," *ITE Proceedings, Power Electronics*, vol. 3, no. 5, pp. 740–754, September 2010.

[27] N. Kondrath and M. K. Kazimierczuk, "Inductor winding loss with owing to skin and proximity effects including harmonics in non-isolated PWM dc-dc converters in CCM," *ITE Proceedings, Power Electronics*, vol. 3, no. 6, pp. 989–1000, November 2010.
[28] www.cosmocorp.com.
[29] www.mag-inc.com, www.micrometals.com, www.ferroxcube.com, www.ferrite.de, www.metglas.com, and www.eilor.co.il.

10.17 Review Questions

10.1. What is a magnet wire?

10.2. What is the AWG?

10.3. How does the wire diameter change when the AWG increases by a factor of 6?

10.4. How much does the wire diameter change when the AWG increases by a factor of 1?

10.5. How much does the wire cross-sectional area change when the AWG increases by a factor of 10?

10.6. What is the window utilization factor K_u?

10.7. What is the core area product A_p?

10.8. What is the core geometry coefficient K_g?

10.9. Are the AC magnetic flux excursions large in resonant inductors?

10.10. Is core loss large in resonant inductors?

10.11. How does the temperature rise ΔT depend on the heat radiating surface area?

10.18 Problems

10.1. Determine the wire insulation factor K_i for the AWG40 wire.

10.2. Determine the wire insulation factor K_i for the AWG10 wire.

10.3. Determine the fill and wire insulation factor K_{ai} for the AWG10 wire:

(a) For the triangular winding pattern.

(b) For the square winding pattern.

10.4. Determine the fill and wire insulation factor K_{ai} for the AWG40 wire.

(a) For the triangular winding pattern.

(b) For the square winding pattern.

10.5. Determine the air and wire insulation factor K_{ai} for the AWG20 wire with quad insulation:

(a) For the triangular winding pattern.

(b) For the square winding pattern.

10.6. Design an inductor for the boost PWM converter with the following parameters: $V_O = 400\,\text{V}$, $I_O = 11.25 - 225\,\text{mA}$, $V_I = 127 - 187\,\text{V}$, and $f = 500\,\text{kHz}$.

10.7. Design an inductor for the boost PWM converter with the following parameters: $V_O = 24\,\text{V}$, $V_I = 8 - 18\,\text{V}$, $P_O = 0 - 48\,\text{W}$, and $f = 100\,\text{kHz}$.

10.8. Design an inductor for the buck-boost PWM converter with the following parameters: $V_O = 12\,\text{V}$, $V_I = 28 \pm 4\,\text{V}$, $I_O = 1 - 10\,\text{A}$, and $f = 100\,\text{kHz}$.

10.9. Design an inductor for the buck-boost PWM converter with the following parameters: $V_O = 12\,\text{V}$, $V_I = 28 \pm 4\,\text{V}$, $I_O = 0 - 10\,\text{A}$, and $f = 100\,\text{kHz}$.

10.10. Design an inductor for a buck converter with $V_I = 8 - 16\,\text{V}$, $V_O = 5\,\text{V}$, $I_O = 1 - 10\,\text{A}$, and $f_s = 300\,\text{kHz}$.

10.11. A buck converter operates in CCM with $V_I = 48 \pm 8\,\text{V}$, $V_O = 28\,\text{V}$, $I_O = 2 - 5\,\text{A}$, and $f_s = 200\,\text{kHz}$. Assume a converter efficiency of 100%. Design an inductor for the buck converter. Calculate the DC winding loss and the core loss for the P material core with $\mu_{rc} = 2500$.

Design of Transformers

11.1 Introduction

This chapter is devoted to the design of high-frequency power transformers [1–27]. The area product method A_p and the core geometrical coefficient K_g method [1] of designing the transformers are presented. Expressions for the core area product A_p and the core geometrical coefficient K_g are derived. Step-by-step design procedures are given for several typical transformers. This chapter presents practical design considerations for off-line flyback converters. The design of a transformer involves the following steps:

- Core material selection
- Core shape and size selection
- Allocation of core window area for transformer windings
- Winding wire selection.

11.2 Area Product Method

11.2.1 Derivations of Core Area Product A_p

Consider a two-winding transformer. The apparent power of the primary winding is

$$P_p = I_{prms} V_{prms} \tag{11.1}$$

and the apparent power of the secondary winding is

$$P_s = I_{srms} V_{srms}. \tag{11.2}$$

Hence, the apparent power of the transformer is

$$P_t = P_p + P_s = I_{prms} V_{prms} + I_{srms} V_{srms}. \tag{11.3}$$

High-Frequency Magnetic Components, Second Edition. Marian K. Kazimierczuk.

Companion Website: www.wiley.com/go/kazimierczuk_High2e

The number of turns of the primary winding to avoid the core saturation is given by

$$N_p = \frac{V_{prms}}{K_f f A_c B_{pk}} \tag{11.4}$$

and the number of turns of the secondary winding is

$$N_s = \frac{V_{srms}}{K_f f A_c B_{pk}}, \tag{11.5}$$

where, from Chapter 1, the waveform factor for transformer sinusoidal input and output voltages is

$$K_f = \frac{2\pi}{\sqrt{2}} = \sqrt{2}\pi = 4.44, \tag{11.6}$$

the waveform factor for transformer square wave input and output voltages is

$$K_f = 4, \tag{11.7}$$

and the waveform factor for transformer rectangular wave input and output voltages is

$$K_f = \frac{2}{\sqrt{D(1-D)}}. \tag{11.8}$$

The cross-sectional areas of the primary and secondary bare wires are

$$A_{wp} = \frac{I_{prms}}{J_{prms}} \tag{11.9}$$

and

$$A_{ws} = \frac{I_{srms}}{J_{srms}}. \tag{11.10}$$

The copper cross-sectional area of all the turns is

$$A_{Cu} = K_u W_a = N_p A_{wp} + N_s A_{ws} = N_p \frac{I_{prms}}{J_{prms}} + N_s \frac{I_{srms}}{J_{srms}}$$

$$= \frac{1}{K_f f A_c B_{pk}} \left(\frac{I_{prms} V_{prms}}{J_{prms}} + \frac{I_{srms} V_{srms}}{J_{srms}} \right). \tag{11.11}$$

The core area product of a two-winding transformer is

$$A_p = A_c W_a = \frac{1}{K_f K_u f B_{pk}} \left(\frac{I_{prms} V_{prms}}{J_{prms}} + \frac{I_{srms} V_{srms}}{J_{srms}} \right). \tag{11.12}$$

If $J_{prms} = J_{srms} = J_{rms}$, the area product is expressed by

$$A_p = \frac{I_{prms} V_{prms} + I_{srms} V_{srms}}{K_f K_u f J_{rms} B_{pk}} = \frac{P_p + P_s}{K_f K_u f J_{rms} B_{pk}} = \frac{P_t}{K_f K_u f J_{rms} B_{pk}}. \tag{11.13}$$

The apparent power for a transformer with m windings is given by

$$P_t = \sum_{n=1}^{m} I_n V_n = \sum_{n=1}^{m} P_n \tag{11.14}$$

$$A_p = \frac{P_t}{K_f K_u f J_{rms} B_{pk}} = \frac{\sum_{n=1}^{m} I_n V_n}{K_f K_u f J_{rms} B_{pk}}, \tag{11.15}$$

where I_n is the rms value of the current through the nth winding, V_n is the rms value of the voltage across the nth winding, and P_n is the apparent power of the nth winding.

For a two-winding transformer with a resistive load and with sinusoidal currents and voltages, the input power is $P_p = P_I = I_{prms} V_{prms}$ and the output power is $P_s = P_O = I_{srms} V_{srms}$. Ignoring the winding and core losses in the transformer, $P_O = P_I$ and $P_t = P_p + P_s = 2P_O$. Hence, the core area product is

$$A_p = \frac{2P_O}{K_f K_u f J_{rms} B_{pk}} = \frac{\sqrt{2} P_O}{\pi K_u f J_{rms} B_{pk}} = \frac{2P_O}{\pi K_u f J_m B_{pk}}. \tag{11.16}$$

11.2.2 Core Window Area Allocation for Transformer Windings

Consider a two-winding transformer. The length of the primary winding wire is

$$l_{w1} = N_1 l_T \tag{11.17}$$

and the length of the secondary winding wire is

$$l_{w2} = N_2 l_T. \tag{11.18}$$

The cross-sectional area of the copper of the primary winding is

$$A_{Cu1} = N_1 A_{w1} \tag{11.19}$$

and the cross-sectional area of the copper of the secondary winding is

$$A_{Cu2} = N_2 A_{w2}. \tag{11.20}$$

The fraction of the total winding window area allocated to the copper of the primary winding is

$$x_1 = \frac{A_{Cu1}}{A_{Cu}} = \frac{A_{Cu1}}{K_u W_a} = \frac{N_1 A_{w1}}{K_u W_a} \tag{11.21}$$

and the fraction of the total winding window area allocated to the copper of the secondary winding is

$$x_2 = \frac{A_{Cu2}}{A_{Cu}} = \frac{A_{Cu2}}{K_u W_a} = \frac{N_2 A_{w2}}{K_u W_a} = \frac{A_{Cu} - A_{Cu1}}{A_{Cu}} = 1 - \frac{A_{Cu1}}{A_{Cu}} = 1 - x_1. \tag{11.22}$$

The cross-sectional area of the primary winding is

$$A_{w1} = \frac{K_u W_a x_1}{N_1} \tag{11.23}$$

and the cross-sectional area of the secondary winding is

$$A_{w2} = \frac{K_u W_a x_2}{N_2} = \frac{K_u W_a (1 - x_1)}{N_2}. \tag{11.24}$$

The DC and low-frequency resistance of the primary winding is

$$R_{wpDC} = \frac{\rho_w l_{w1}}{A_{w1}} = \frac{\rho_w N_1 l_T}{A_{w1}} = \frac{\rho_w l_T N_1^2}{K_u W_a x_1} \tag{11.25}$$

and the DC and low-frequency resistance of the secondary winding is

$$R_{wsDC} = \frac{\rho_w l_{w2}}{A_{w2}} = \frac{\rho_w N_2 l_T}{A_{w2}} = \frac{\rho_w l_T N_2^2}{K_u W_a x_2} = \frac{\rho_w l_T N_2^2}{K_u W_a (1 - x_1)}. \tag{11.26}$$

The DC and low-frequency resistance of the primary winding is

$$P_{wpDC} = R_{wpDC} I_1^2 = \frac{\rho_w l_T N_1^2 I_1^2}{K_u W_a x_1} \tag{11.27}$$

and the DC and low-frequency resistance of the secondary winding is

$$P_{wsDC} = R_{wsDC} I_2^2 = \frac{\rho_w l_T N_2^2 I_2^2}{K_u W_a x_2} = \frac{\rho_w l_T N_2^2 I_2^2}{K_u W_a (1 - x_1)}, \tag{11.28}$$

where I_1 and I_2 are the rms values of the primary and secondary currents, respectively. As x_1 increases from 0 to 1, P_{wpDC} decreases from infinity to a low value and P_{wsDC} increases from a low value to infinity. Therefore, an optimum value of the allocation of the primary winding copper x_1 exists at which the overall DC and low-frequency winding losses have a minimum value. The total DC winding power loss is

$$P_{wDC} = P_{wpDC} + P_{wsDC} = \frac{\rho_w l_T}{K_u W_a}\left(\frac{N_1^2 I_1^2}{x_1} + \frac{N_2^2 I_2^2}{x_2}\right) = \frac{\rho_w l_T}{K_u W_a}\left(\frac{N_1^2 I_1^2}{x_1} + \frac{N_2^2 I_2^2}{1 - x_1}\right). \tag{11.29}$$

Since

$$I_2 = \frac{N_1}{N_2} I_1, \tag{11.30}$$

we have

$$P_{wDC} = \frac{\rho_w l_T N_1^2 I_1^2}{K_u W_a}\left(\frac{1}{x_1} + \frac{1}{x_2}\right) = \frac{\rho_w l_T N_1^2 I_1^2}{K_u W_a}\left(\frac{1}{x_1} + \frac{1}{1 - x_1}\right) = \frac{\rho_w l_T N_1^2 I_1^2}{K_u W_a}\left[\frac{1}{x_1(1 - x_1)}\right]. \tag{11.31}$$

Figure 11.1 shows the plots of P_{wpDC}, P_{wsDC}, and P_{wDC} at $\rho_w l_T N_1^2 I_1^2/(K_u W_a) = 1$. The derivative of the overall winding DC power loss is

$$\frac{dP_{wDC}}{dx_1} = \frac{\rho_w l_T N_1^2 I_1^2}{K_u W_a}\left[\frac{1 - 2x_1}{x_1(1 - x_1)}\right] = 0. \tag{11.32}$$

Hence, the minimum value of the overall DC winding loss occurs for

$$x_{1opt} = \frac{1}{2} \tag{11.33}$$

and

$$x_{2opt} = 1 - x_{1opt} = 1 - \frac{1}{2} = \frac{1}{2} \tag{11.34}$$

yielding the minimum winding DC power loss for a two-winding transformer

$$P_{wDC(min)} = \frac{4\rho_w l_T N_1^2 I_1^2}{K_u W_a}. \tag{11.35}$$

Since $x_1 = x_2$,

$$A_{Cu1} = A_{Cu2} \tag{11.36}$$

which gives

$$N_1 A_{w1} = N_2 A_{w2} \tag{11.37}$$

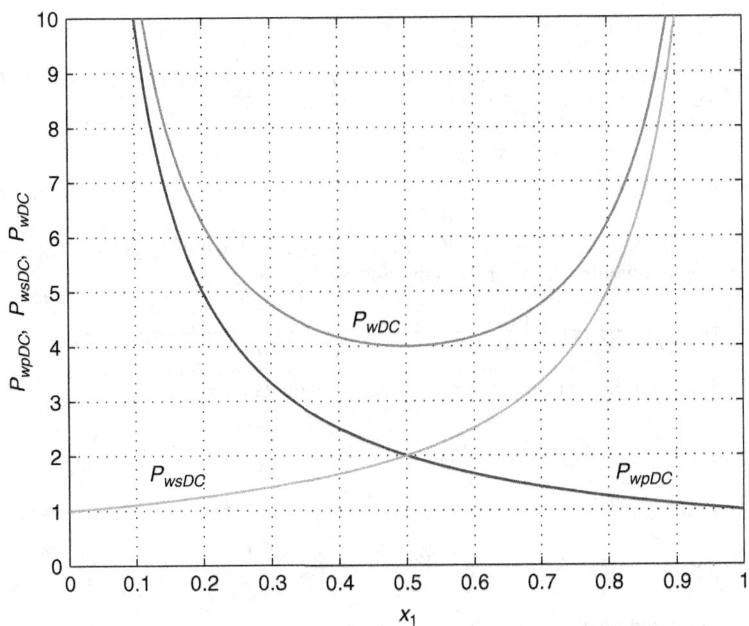

Figure 11.1 Plots of the winding DC power losses P_{wpDC}, P_{wsDC}, and P_{wDC} as functions of x_1 at $\rho_w l_T N_1^2 I_1^2/(K_u W_a) = 1$

or

$$\frac{A_{w2}}{A_{w1}} = \frac{N_1}{N_2} = n. \tag{11.38}$$

The minimum winding loss occurs when

$$x_1 = \frac{N_1 I_1}{N_1 I_1 + N_2 I_2} \tag{11.39}$$

and

$$x_2 = \frac{N_2 I_2}{N_1 I_1 + N_2 I_2}. \tag{11.40}$$

The low-frequency winding losses in a transformer are minimized when the window area is allocated to the windings in proportion to the ampere-turns. The voltage transfer function of the transformer is

$$\frac{V_2}{V_1} = \frac{N_2}{N_1} = n. \tag{11.41}$$

Substitution of (11.41) in (11.39) and (11.40) yields

$$x_1 = \frac{W_{a1}}{W_a} = \frac{\frac{V_1}{N_1} N_1 I_1}{\frac{V_1}{N_1} N_1 I_1 + \frac{V_1}{N_1} N_2 I_2} = \frac{V_1 I_1}{V_1 I_1 + V_2 I_2} \tag{11.42}$$

and

$$x_2 = \frac{W_{a2}}{W_a} = \frac{\frac{V_2}{N_2} N_2 I_2}{\frac{V_2}{N_2} N_1 I_1 + \frac{V_2}{N_2} N_2 I_2} = \frac{V_2 I_2}{V_1 I_1 + V_2 I_2}. \tag{11.43}$$

Thus, the low-frequency winding losses in a transformer are minimized when the window area is allocated to the windings in proportion to the apparent power. These principles of the window area allocation also hold true for transformers with the number of windings greater than 2.

For a multiple-output transformer, the DC and low-frequency power loss in the kth winding is

$$P_{wDCk} = R_{wDCk} I_k^2 = \frac{\rho_w l_{wk} I_k^2}{A_{wk}} = \frac{\rho_w l_T N_k^2 I_k^2}{K_u W_a x_k}. \tag{11.44}$$

The DC and low-frequency power loss in the all the windings is

$$P_{wDC} = P_{wDC1} + P_{wDC2} + \ldots + P_{wDCk} + \ldots = \frac{\rho_w l_T}{K_u W_a}\left(\frac{N_1^2 I_1^2}{x_1} + \frac{N_2^2 I_2^2}{x_2} + \ldots + \frac{N_k^2 I_k^2}{x_k} + \ldots\right)$$

$$= \frac{\rho_w l_T N_1^2 I_1^2}{K_u W_a}\left(\frac{1}{x_1} + \ldots + \frac{N_k^2 I_k^2}{N_1^2 I_1^2 x_k} + \ldots\right). \tag{11.45}$$

It is shown in [5] that the minimum DC and low-frequency winding power loss occurs for the following core window allocation for the kth winding:

$$x_k = \frac{N_k I_k}{N_1 I_1 + N_2 I_2 + \ldots + N_k I_k + \ldots} = \frac{V_k I_k}{V_1 I_1 + V_2 I_2 + \ldots + V_k I_k + \ldots}. \tag{11.46}$$

The AC resistances of the primary and secondary windings are

$$R_{wp} = F_{Rph} R_{wpDC} = F_{Rph} \frac{\rho_w l_{wp}}{A_{wp}} = F_{Rph} \frac{\rho_w l_T N_p}{A_{wp}} = F_{Rph} \frac{\rho_w l_T N_p^2}{K_u W_a x_1} \tag{11.47}$$

and

$$R_{ws} = F_{Rsh} R_{wsDC} = F_{Rsh} \frac{\rho_w l_{ws}}{A_{ws}} = F_{Rsh} \frac{\rho_w l_T N_s}{A_{ws}} = F_{Rsh} \frac{\rho_w l_T N_s^2}{K_u W_a x_2} = F_{Rsh} \frac{\rho_w l_T N_s^2}{K_u W_a (1 - x_1)} \tag{11.48}$$

yielding the total winding AC resistance

$$R_w = R_{wp} + R_{ws} = F_{Rph} R_{wpDC} + F_{Rsh} R_{wsDC} = \frac{\rho_w l_T}{K_u W_a}\left(\frac{F_{Rph} N_p^2}{x_1} + \frac{F_{Rsh} N_s^2}{1 - x_1}\right). \tag{11.49}$$

The winding AC power loss in the primary and secondary windings is

$$P_{wp} = R_{wp} I_p^2 = F_{Rph} R_{wpDC} I_p^2 = F_{Rph} \frac{\rho_w l_{wp}}{A_{Cup}} I_p^2 = F_{Rph} \frac{\rho_w l_{Tp} N_p}{A_{Cup}} I_p^2 = F_{Rph} \frac{\rho_w l_{Tp} N_p^2 I_p^2}{K_u W_a x_1} \tag{11.50}$$

and

$$P_{ws} = R_{ws} I_s^2 = F_{Rsh} R_{wsDC} I_s^2 = F_{Rsh} \frac{\rho_w l_{ws}}{A_{Cus}} I_s^2 = F_{Rsh} \frac{\rho_w l_{Ts} N_s}{A_{Cus}} I_s^2 = F_{Rsh} \frac{\rho_w l_{Ts} N_p^2 I_p^2}{K_u W_a x_2}. \tag{11.51}$$

The total winding power loss is

$$P_w = \frac{\rho_w l_T N_p^2 I_p^2}{K_u W_a} \left(\frac{F_{Rph}}{x_1} + \frac{F_{Rsh}}{x_2} \right) = \frac{\rho_w l_T N_p^2 I_p^2}{K_u W_a} \left(\frac{F_{Rph}}{x_1} + \frac{F_{Rsh}}{1 - x_1} \right). \tag{11.52}$$

Thus, the optimum allocation of the core window for the windings at the minimum total AC winding loss depends on the harmonic AC resistance factors F_{Rph} and F_{Rsh}. If $F_{Rph} = F_{Rsh}$, the optimum allocation is $x_{1opt} = x_{2opt} = 0.5$. Otherwise,

$$x_{1opt} = \frac{F_{Rph}}{2F_{Rsh}} \tag{11.53}$$

and

$$x_{2opt} = 1 - x_{1opt} = 1 - \frac{F_{Rph}}{2F_{Rsh}}. \tag{11.54}$$

The last two equations are valid if F_{Rph} and F_{Rsh} are independent of x_1 and x_2.

11.3 Optimum Flux Density

Consider a two-winding transformer. The number of turns of the primary winding is given by

$$N_1 = \frac{\lambda_{max}}{A_c B_m}, \tag{11.55}$$

where B_m is the maximum value of the AC component of the flux density and λ_{max} is the peak value of the flux linkage. The minimum winding DC power loss is

$$P_{wDC(min)} = \frac{4\rho_w l_T I_1^2 \lambda_{max}^2}{K_u W_a A_c^2 B_m^2}. \tag{11.56}$$

Thus, the winding DC power loss decreases as B_m increases. The core power loss is given by

$$P_C = V_c k f^a B_m^b. \tag{11.57}$$

The core loss increases as B_m increases. The total loss in a transformer is

$$P_{cwDC} = P_{wDC(min)} + P_C = \frac{4\rho_w l_T I_1^2 \lambda_{max}^2}{K_u W_a A_c^2 B_m^2} + V_c k f^a B_m^b. \tag{11.58}$$

Figure 11.2 shows the plots of $P_{wDC(min)}$, P_C, and P_{cwDC}. Setting the derivative of P_{cwDC} to zero, we obtain

$$\frac{P_{cwDC}}{dB_m} = \frac{dP_{wDC(min)}}{dB_m} + \frac{dP_C}{dB_m} = -\frac{8\rho_w l_T I_1^2 \lambda_{max}^2}{K_u W_a A_c^2 B_m^3} + bV_c k f^a B_m^{(b-1)} = 0. \tag{11.59}$$

Hence, the optimum value of the flux density is

$$B_{mopt} = \frac{8\rho_w l_T I_1^2 \lambda_{max}^2}{K_u W_a A_c^2 A_c^2 b V_c f^a} < B_s - B_{DC}. \tag{11.60}$$

If we take into account AC winding losses, the total loss in a transformer is

$$P_{cw} = P_{wmin} + P_C = \frac{8\rho_w l_T I_1^2 \lambda_{max}^2 F_{Rsh}}{K_u W_a A_c^2 B_m^2} \left(\frac{1}{F_{Rsh}} + \frac{1}{2F_{Rsh} - F_{Rph}} \right) + V_c k f^a B_m^b. \tag{11.61}$$

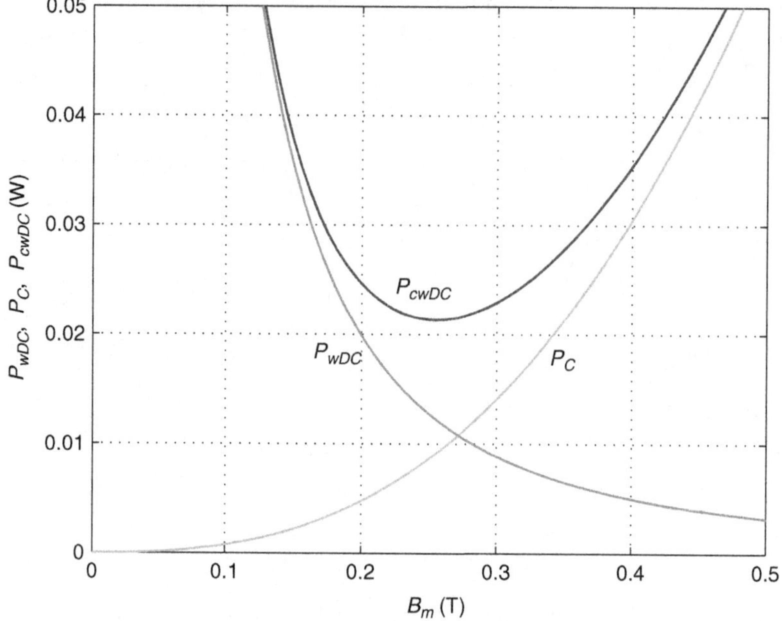

Figure 11.2 Plots of $P_{wDC(min)}$, P_C, and P_{cwDC} as functions of B_m

In this case, the optimum value of the flux density is

$$B_{mopt} = \frac{16\rho_w l_T I_1^2 \lambda_{max}^2 F_{Rsh}}{K_u W_a A_c^2 A_c^2 b V_c f^a}\left(\frac{1}{F_{Rsh}} + \frac{1}{2F_{Rsh} - F_{Rph}}\right) < B_s - B_{DC}. \tag{11.62}$$

11.4 Area Product A_p for Sinusoidal Voltages

The cross-sectional area of the core is

$$A_c = \frac{V_{1m}}{\omega N_1 B_m} = \frac{\sqrt{2}V_1}{\omega N_1 B_m}. \tag{11.63}$$

The cross-sectional area of the primary winding conductor is

$$A_{w1} = \frac{I_1}{J_{rms}} = \frac{I_{1m}}{J_m}. \tag{11.64}$$

The cross-sectional area of the secondary winding conductor is

$$A_{w2} = \frac{I_2}{J_{rms}} = \frac{I_{2m}}{J_m}. \tag{11.65}$$

The core window area is

$$W_a = \frac{A_{Cu1} + A_{Cu2}}{K_u} = \frac{N_1 A_{w1} + N_2 A_{w2}}{K_u} = \frac{N_1 I_1 + N_2 I_2}{K_u J_{rms}} = \frac{N_1 I_{1m} + N_2 I_{2m}}{K_u J_m}. \tag{11.66}$$

where $K_u = 0.4$ for solid wire and $K_u = 0.3$ for litz wire. The area product of the transformer with sinusoidal voltages is

$$A_p = W_a A_c = \frac{N_1 I_{1m} + N_2 I_{2m}}{K_u J_m} \times \frac{V_{1m}}{\omega N_1 B_m} = \frac{I_{1m} V_{1m} + I_{2m} V_{1m} N_2/N_1}{K_u \omega J_m B_m}$$
$$= \frac{I_{1m} V_{1m} + I_{2m} V_{2m}}{K_u \omega J_m B_m} = \frac{I_1 V_1 + I_2 V_2}{\pi K_u f J_m B_m}. \tag{11.67}$$

11.5 Transformer Design for Flyback Converter in CCM

11.5.1 Practical Design Considerations of Transformers

The transformer is the most important component of the flyback converter. It provides DC isolation, AC coupling, voltage level change, and magnetic energy storage. The current flows only in the primary winding when the magnetic energy in the transformer is increased and in the secondary winding while the magnetic energy is decreased. Thus, the physical transformer in the flyback converter serves as an inductor to store and transfer the magnetic energy and as a transformer to provide DC isolation, AC coupling, and voltage and current level transformation. The transformer in a flyback converter is often described as a coupled inductance.

The primary winding should start from the innermost layer of the bobbin. This reduces the length of wire, minimizing the copper loss of the primary winding. In addition, the electromagnetic interference (EMI) noise radiation is reduced because the secondary windings act as Faraday's shields. The innermost layer of the primary winding should be connected to the drain of a power metal-oxide-semiconductor field-effect transistor (MOSFET). Therefore, the shielding of the winding with the highest voltage is maximized by the secondary windings. In multiple-output transformers, the secondary winding with the highest output power should be placed closest to the primary winding to ensure good coupling and to reduce leakage inductance. If the secondary winding has small number of turns, the turns of this winding should be spread along the entire window breadth to maximize the coupling coefficient. If the secondary winding has a low number of turns and is made of strands, the strands should form one flat layer along the window breadth to increase the coupling coefficient and to reduce the proximity effect loss.

11.5.2 Area Product A_p for Transformer Square Wave Voltages

The window area limited by the maximum current density in the winding wire is given by

$$W_a = \frac{N_p A_{wp} + N_s A_{ws}}{K_u} \tag{11.68}$$

Assuming the winding allocation is such that $N_p A_{wp} = N_s A_{ws}$, the window area is

$$W_a = \frac{2N_p A_{wp}}{K_u} = \frac{2N_p I_{Lpmax}}{K_u J_{pm}} = \frac{2N_p I_{Lprms}}{K_u J_{prms}}. \tag{11.69}$$

From Chapter 1, the required core cross-sectional area is given by

$$A_c = \frac{\lambda_m}{N_p B_{pk}} = \frac{L_p I_{Lpmax}}{N_p B_{pk}}. \tag{11.70}$$

Hence, the core area product is obtained as

$$A_p = W_a A_c = \frac{2N_p I_{Lpmax}}{K_u J_{pm}} \times \frac{L_p I_{Lpmax}}{N_p B_{pk}} = \frac{2L_p I_{Lpmax}^2}{K_u J_{pm} B_{pk}} = \frac{4W_m}{K_u J_{pm} B_{pk}}. \tag{11.71}$$

The core cross-sectional area for transformer square voltage waveforms is

$$A_c = \frac{V_{Lprms}}{4f_s N_p B_{pk}}. \tag{11.72}$$

Thus, the core area product is

$$A_p = W_a A_c = \frac{2N_p I_{Lprms}}{K_u J_{prms}} \times \frac{V_{Lprms}}{4f_s N_p B_{pk}} = \frac{I_{Lprms} V_{Lprms}}{2K_u f_s J_{prms} B_{pk}}. \tag{11.73}$$

11.5.3 Area Product A_p Method

The inductor for pulse-width modulated (PWM) converters in continuous conduction mode (CCM) should satisfy the following conditions:

1. $L > L_{min}$ required for CCM operation.
2. $B_{max} < B_s$ or $\lambda_{max} < \lambda_s$. For ferrite cores, $B_s < 0.5$ T at room temperature and $B_s = 0.3$ T at $T = 100\,^{\circ}$C.
3. $P_L < P_{L(max)}$. Since $P_L \approx P_w = P_{rL}$, $r_L < r_{Lmax}$.
4. $J_{max} < J_{max}$. Typically, $J_{max} = 0.1$–5 A/mm^2.

The magnitude of the AC magnetic flux is caused by the peak-to-peak value Δi_L of the AC component of the inductor current and is normally only a fraction of the DC flux. In many cases, the AC inductor current riding on top of the DC current I_L is small enough that it does not affect the overall current density of the single wire used.

Example 11.1

A flyback PWM converter is operated in the CCM and has the following specifications: $V_I = 28 \pm 4$ V, $V_O = 5$ V, $I_O = 2$–10 A, and $f_s = 100$ kHz.

Solution: Figure 11.3 shows a circuit of the flyback converter and its models for CCM. Figure 11.4 shows waveforms in the flyback converter for CCM. The maximum and minimum output powers are

$$P_{Omax} = V_O I_{Omax} = 5 \times 10 = 50 \text{ W} \tag{11.74}$$

and

$$P_{Omin} = V_O I_{Omin} = 5 \times 2 = 10 \text{ W}. \tag{11.75}$$

The minimum and maximum load resistances are

$$R_{Lmin} = \frac{V_O}{I_{Omax}} = \frac{5}{10} = 0.5\ \Omega \tag{11.76}$$

and

$$R_{Lmax} = \frac{V_O}{I_{Omin}} = \frac{5}{2} = 2.5\ \Omega. \tag{11.77}$$

The minimum and maximum DC voltage transfer functions are

$$M_{VDCmin} = \frac{V_O}{V_{Imax}} = \frac{5}{32} = 0.15625\ \Omega \tag{11.78}$$

and

$$M_{VDCmax} = \frac{V_O}{V_{Imin}} = \frac{5}{24} = 0.2083\ \Omega. \tag{11.79}$$

Assume the converter efficiency $\eta = 0.92$ and the maximum duty cycle $D_{max} = 0.5$. The transformer turns ratio is

$$n = \frac{\eta D_{max}}{(1 - D_{max}) M_{VDCmax}} = \frac{0.92 \times 0.5}{(1 - 0.5) \times 0.2083} = 4.4167. \tag{11.80}$$

The minimum duty cycle is

$$D_{min} = \frac{n M_{VDCmin}}{n M_{VDCmin} + \eta} = \frac{4.414 \times 0.15625}{4.4167 \times 0.15625 + 0.92} = 0.4286. \tag{11.81}$$

The minimum inductance of the primary winding is determined by the operation in CCM

$$L_p \approx L_{m(min)} = \frac{n^2 R_{Lmax} (1 - D_{min})^2}{2 f_s} = \frac{4.4167^2 \times 2.5 \times (1 - 0.4286)^2}{2 \times 10^5} = 79.61\ \mu\text{H}. \tag{11.82}$$

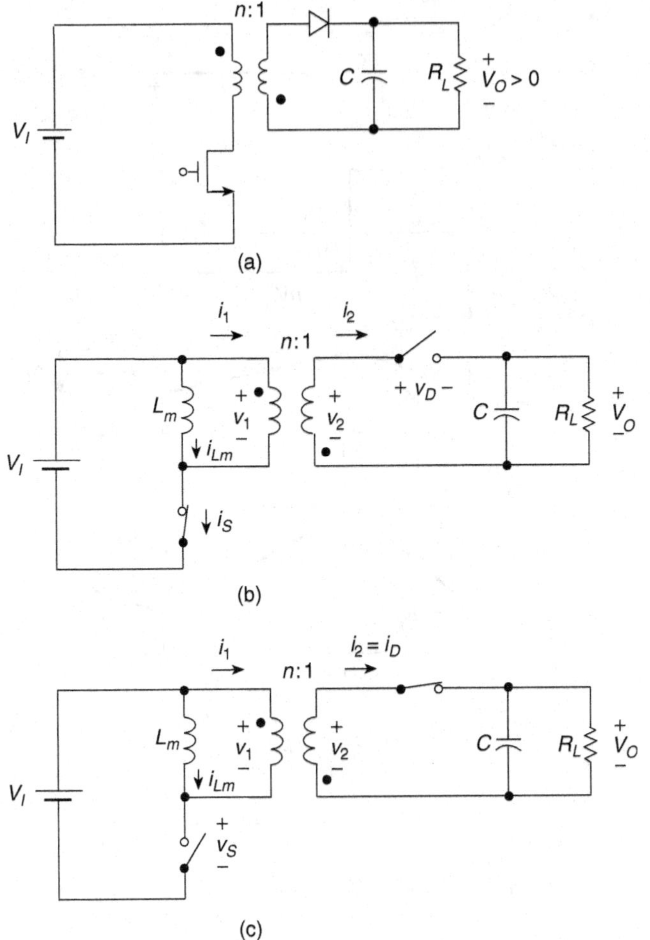

Figure 11.3 Flyback PWM converter and its models for CCM. (a) Circuit. (b) Model when MOSFET is on. (c) Model when MOSFET is OFF

Pick $L_p = 82\ \mu\text{H}$. The inductance of the secondary winding is

$$L_s = \frac{L_p}{n^2} = \frac{82 \times 10^{-6}}{4.4167^2} = 4.2036\ \mu\text{H}. \tag{11.83}$$

Pick $L_s = 4.7\ \mu\text{H}$.

The maximum peak-to-peak value of the magnetizing current and the primary current ripple is

$$\Delta i_{Lp(max)} = \Delta i_{Lm(max)} = \frac{nV_O(1 - D_{min})}{f_s L_p} = \frac{4.4167 \times 5 \times (1 - 0.4286)}{10^5 \times 82 \times 10^{-6}} = 1.5387\ \text{A}. \tag{11.84}$$

The minimum peak-to-peak value of the magnetizing current and the primary current ripple at full power is

$$\Delta i_{Lp(min)} = \Delta i_{Lm(min)} = \frac{nV_O(1 - D_{max})}{f_s L_p} = \frac{4.4167 \times 5 \times (1 - 0.5)}{10^5 \times 82 \times 10^{-6}} = 1.346\ \text{A}. \tag{11.85}$$

The maximum DC input current is

$$I_{Imax} = \frac{P_O}{\eta V_{Imin}} = \frac{M_{VDCmax} I_{Omax}}{\eta} = \frac{0.2083 \times 10}{0.92} = 2.2641\ \text{A}. \tag{11.86}$$

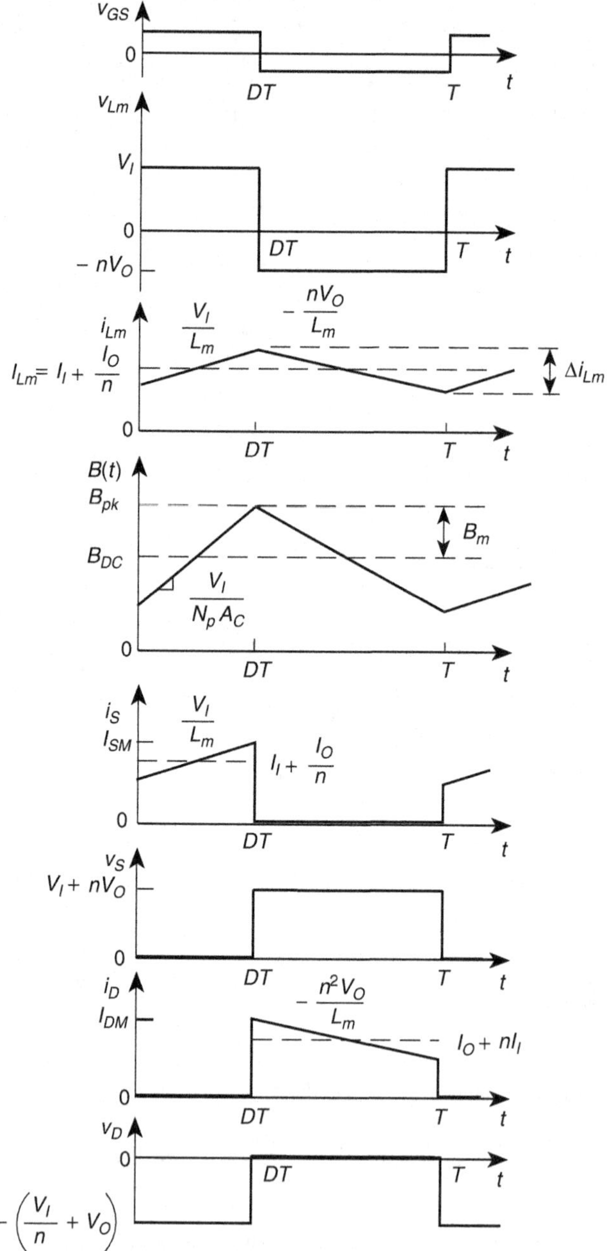

Figure 11.4 Waveforms in flyback PWM converter for CCM

The maximum peak value of the primary winding current is

$$I_{pmax} = \frac{I_{Omax}}{n(1 - D_{max})} + \frac{nV_O(1 - D_{max})}{f_s L_p}$$
$$= \frac{10}{4.4167 \times (1 - 0.5)} + \frac{4.4167 \times 5 \times (1 - 0.5)}{10^5 \times 82 \times 10^{-6}} = 5.874 \text{ A}. \quad (11.87)$$

The maximum rms value of the primary winding current is

$$I_{prms(max)} = \frac{I_{Omax}\sqrt{D_{max}}}{n(1-D_{max})} = \frac{10\sqrt{0.5}}{4.4167 \times (1-0.5)} = 3.202 \text{ A}. \tag{11.88}$$

The maximum energy stored in the magnetic field of the transformer is

$$W_m = \frac{1}{2}L_m I_{Lmmax}^2 = \frac{1}{2}L_p I_{pmax}^2 = \frac{1}{2} \times 82 \times 10^{-6} \times 5.874^2 = 1.414 \text{ mJ}. \tag{11.89}$$

Since the saturation flux density B_s decreases as the temperature increases, the high-temperature $B-H$ characteristics should be considered in the design. The typical values of the saturation flux density for ferrites at high temperatures is $B_s = 0.3\text{–}0.35$ T.

Assume the core window utilization factor $K_u = 0.3$, $J_m = 5$ A/mm^2, and $B_{pk} = 0.25$ T. The core area product is

$$A_p = \frac{4W_m}{K_u J_m B_{pk}} = \frac{4 \times 1.414 \times 10^{-3}}{0.3 \times 5 \times 10^6 \times 0.25} = 1.1312 \text{ cm}^4. \tag{11.90}$$

A Magnetics ferrite core PC 0F-43622 is selected. The dimensions of this core are $A_p = 1.53$ cm^2, $A_c = 2.02$ cm^2, $l_c = 5.32$ cm, and $V_c = 10.7$ cm^3. The F-type magnetic material is used to make this core. The core parameters are $\mu_{rc} = 3000 \pm 20\%$, $A_L = 10\,000$ mH/1000 turns, and $B_s = 0.49$ T. The coefficients of this material are $k = 0.0573$, $a = 1.66$, and $b = 2.68$.

The core window area is

$$W_a = \frac{A_p}{A_c} = \frac{1.53}{2.02} = 0.757 \text{ cm}^2. \tag{11.91}$$

The skin depth at $f = 100$ kHz is

$$\delta_w = \frac{66.2}{\sqrt{f_s}}(\text{mm}) = \frac{66.2}{\sqrt{10^5}} = 0.209 \text{ mm}. \tag{11.92}$$

To avoid the skin and proximity effects, the diameter of a bare strand should be

$$d_{is} = 2\delta_w = 2 \times 0.209 \times 10^{-3} = 0.418 \text{ mm}. \tag{11.93}$$

The primary and secondary windings will consist of many strands. Select the copper round wire AWG27 for the strands, which has $d_{os} = 0.409$ mm, $d_{is} = 0.3606$ mm, $A_{wst} = 0.1021$ mm^2, and $R_{wDC}/l_w = 0.1687$ Ω/m.

The cross-sectional area of the primary winding wire is

$$A_{wp} = \frac{I_{pmax}}{J_m} = \frac{5.874}{5} = 1.17 \text{ mm}^2. \tag{11.94}$$

The number of strands in the primary winding is

$$S_p = \frac{A_{wp}}{A_{wst}} = \frac{1.17 \times 10^{-6}}{0.1021 \times 10^{-6}} = 11.45. \tag{11.95}$$

Pick $S_p = 10$. The area allocated to the primary winding is

$$W_{ap} = \frac{W_a}{2} = \frac{0.757 \times 10^{-4}}{2} = 0.3785 \text{ cm}^2. \tag{11.96}$$

The cross-sectional area of the insulated strand wire is

$$A_{wpos} = \frac{\pi d_{os}^2}{4} = \frac{\pi \times (0.409 \times 10^{-3})^2}{4} = 0.1313 \text{ mm}^2. \tag{11.97}$$

The number of turns of the primary winding is

$$N_p = \frac{K_u W_{ap}}{S_p A_{wpos}} = \frac{0.3 \times 0.3785 \times 10^{-4}}{10 \times 0.1313 \times 10^{-6}} = 8.648. \tag{11.98}$$

Pick $N_p = 9$. The terminals of the winding in a transformer with a pot core are usually on the opposite sides of the core. The number of turns of the secondary winding is

$$N_s = \frac{N_p}{n} = \frac{9}{4.4167} = 2.03. \tag{11.99}$$

Select $N_s = 2$.

The length of the air gap is

$$l_g = \frac{\mu_0 A_c N_p^2}{L_p} - \frac{l_c}{\mu_{rc}} = \frac{4\pi \times 10^{-7} \times 2.02 \times 10^{-4} \times 9^2}{82 \times 10^{-6}} - \frac{5.32 \times 10^{-2}}{3000} = 0.233 \text{ mm}. \quad (11.100)$$

The maximum peak value of the magnetic flux density is

$$B_{pk} = \frac{\mu_0 N_p I_{pmax}}{l_g + \frac{l_c}{\mu_{rc}}} = \frac{4\pi \times 10^{-7} \times 9 \times 5.874}{0.233 \times 10^{-3} + \frac{5.32 \times 10^{-2}}{3000}} = 0.2649 \text{ T} < B_s. \quad (11.101)$$

The maximum peak value of the AC component of the magnetic flux density is

$$B_{m(max)} = \frac{\mu_0 N_p \frac{\Delta i_{Lp(max)}}{2}}{l_g + \frac{l_c}{\mu_{rc}}} = \frac{4\pi \times 10^{-7} \times 9 \times \frac{1.5387}{2}}{0.233 \times 10^{-3} + \frac{5.32 \times 10^{-2}}{3000}} = 0.0347 \text{ T}. \quad (11.102)$$

The minimum peak value of the AC component of the magnetic flux density at full power is

$$B_{m(min)} = \frac{\mu_0 N_p \frac{\Delta i_{Lp(min)}}{2}}{l_g + \frac{l_c}{\mu_{rc}}} = \frac{4\pi \times 10^{-7} \times 9 \times \frac{1.346}{2}}{0.233 \times 10^{-3} + \frac{5.32 \times 10^{-2}}{3000}} = 0.03035 \text{ T}. \quad (11.103)$$

The core power loss density is

$$P_v = 0.0573 f^{1.66} (10 B_m)^{2.68} = 0.0573 \times 100^{1.66} \times (10 \times 0.0347)^{2.68} = 7.018 \text{ mW/cm}^3. \quad (11.104)$$

The minimum core power loss density at full power is

$$P_{vmin} = 0.0573 f^{1.66} (10 B_{m(min)})^{2.68} = 0.0573 \times 100^{1.66} \times (10 \times 0.03035)^{2.68} = 4.9 \text{ mW/cm}^3. \quad (11.105)$$

The core loss is

$$P_C = V_c P_v = 10.7 \times 7.018 \times 10^{-3} = 75.09 \text{ mW}. \quad (11.106)$$

The minimum core loss at full power is

$$P_{Cmin} = V_c P_{vmin} = 10.7 \times 4.9 \times 10^{-3} = 52.43 \text{ mW}. \quad (11.107)$$

Figure 11.5 shows a plot of the core loss as a function of the DC input voltage V_I for the flyback converter operating in CCM.

The mean turn length (MTL) is

$$l_T = \frac{\pi(F + E)}{2} = \frac{\pi(15.9 + 30.4)}{2} = 72.728 \text{ mm} = 7.2728 \text{ cm}, \quad (11.108)$$

where $F = 15.9$ mm is the inner diameter of the winding window and $E = 30.4$ mm is the outer diameter of the winding window. The length of the primary winding wire is

$$l_{wp} = N_p l_T = 9 \times 7.278 = 65.502 \text{ cm}. \quad (11.109)$$

Pick $l_{wp} = 68$ cm. The DC and low-frequency resistance of each strand of the primary winding is

$$R_{wpDCs} = \left(\frac{R_{wDC}}{l_w}\right) l_{wp} = 0.1687 \times 0.68 = 0.1147 \ \Omega. \quad (11.110)$$

Hence, the DC and low-frequency resistance of the primary winding is

$$R_{wpDC} = \frac{R_{wpDCs}}{S_p} = \frac{0.1147}{10} = 0.01147 \ \Omega. \quad (11.111)$$

The DC and low-frequency power loss in the primary winding is

$$P_{wpDC} = R_{wpDC} I_{Imax}^2 = 0.01147 \times 2.2641^2 = 0.05879 \text{ W}. \quad (11.112)$$

Using (7.234), we obtain $F_{Rph} = 2.6$. Figure 11.6 shows a plot of F_{Rph} as a function of the DC input voltage V_I for the flyback converter operating in CCM.

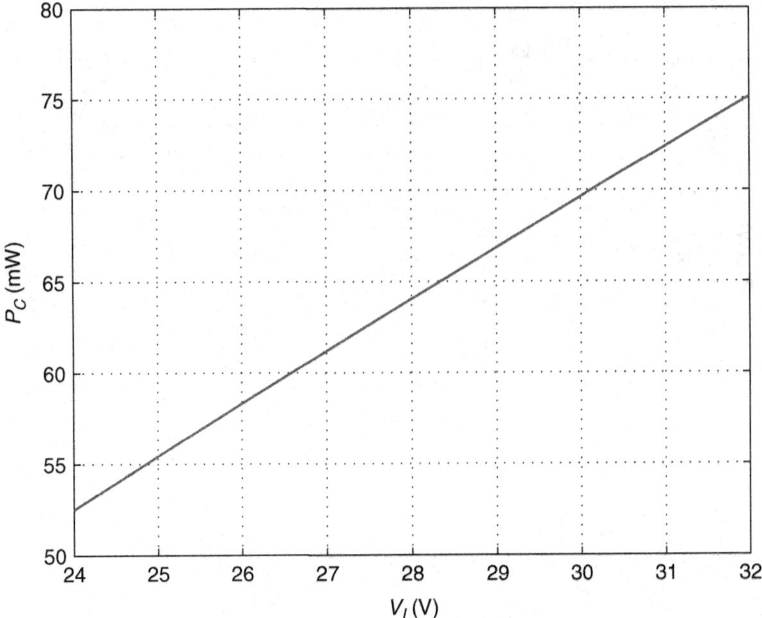

Figure 11.5 Core power loss P_C as a function of the DC input voltage V_I for the flyback converter operating in CCM

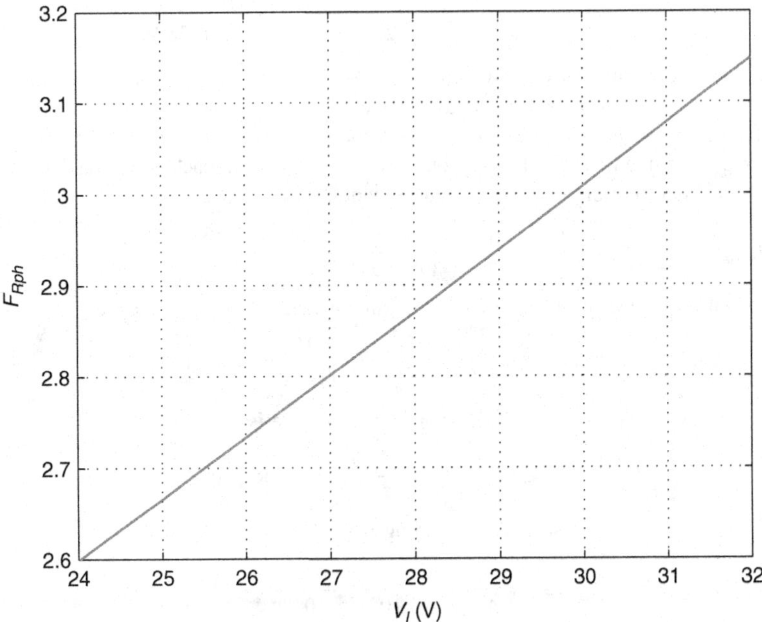

Figure 11.6 F_{Rph} as a function of the DC input voltage V_I for the flyback converter operating in CCM

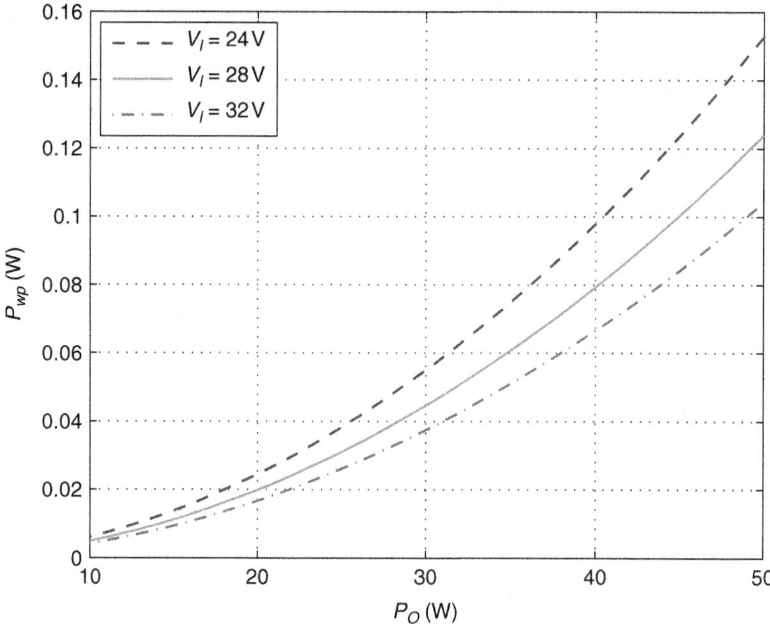

Figure 11.7 Primary winding power loss P_{wp} as a function of the output power P_O at fixed values of the DC input voltage V_I for the flyback converter operating in CCM

The primary winding power loss is

$$P_{wp} = F_{Rph} P_{wpDC} = 2.6 \times 0.05879 = 0.1528 \text{ W}. \tag{11.113}$$

Figure 11.7 shows the plots of the power loss in the primary winding P_{wp} as a function of the output power P_O at fixed values of the DC input voltage V_I for the flyback converter operating in CCM. Figure 11.8 shows the plots of the power loss in the primary winding P_{wp} as a function of the DC input voltage V_I at fixed values of the output power P_O for the flyback converter operating in CCM.

The maximum current through the secondary winding is

$$I_{smax} = \frac{I_{Omax}}{1 - D_{max}} + \frac{n \Delta i_{Lpmax}}{2} = \frac{10}{1 - 0.5} + \frac{4.4167 \times 1.5387}{2} = 23.398 \text{ A}. \tag{11.114}$$

The maximum rms value of the secondary winding current is

$$I_{srms(max)} = \frac{I_{Omax}}{\sqrt{1 - D_{max}}} = \frac{10}{\sqrt{1 - 0.5}} = 14.142 \text{ A}. \tag{11.115}$$

The cross-sectional area of the total secondary winding wire is

$$A_{ws} = \frac{I_{smax}}{J_m} = \frac{23.398}{5} = 4.68 \text{ mm}^2. \tag{11.116}$$

The number of strands of the secondary winding is

$$S_s = \frac{A_{ws}}{A_{wst}} = \frac{4.68 \times 10^{-6}}{0.1021 \times 10^{-6}} = 45.83. \tag{11.117}$$

Pick $S_s = 45$. The length of the secondary winding is

$$l_{ws} = N_s l_T = 2 \times 7.27 = 14.54 \text{ cm}. \tag{11.118}$$

Pick $l_{ws} = 16$ cm. The DC and low-frequency resistance of each strand of the secondary winding is

$$R_{wsDCs} = \left(\frac{R_{wsDCs}}{l_w}\right) l_{ws} = 0.1687 \times 0.16 = 0.027 \ \Omega. \tag{11.119}$$

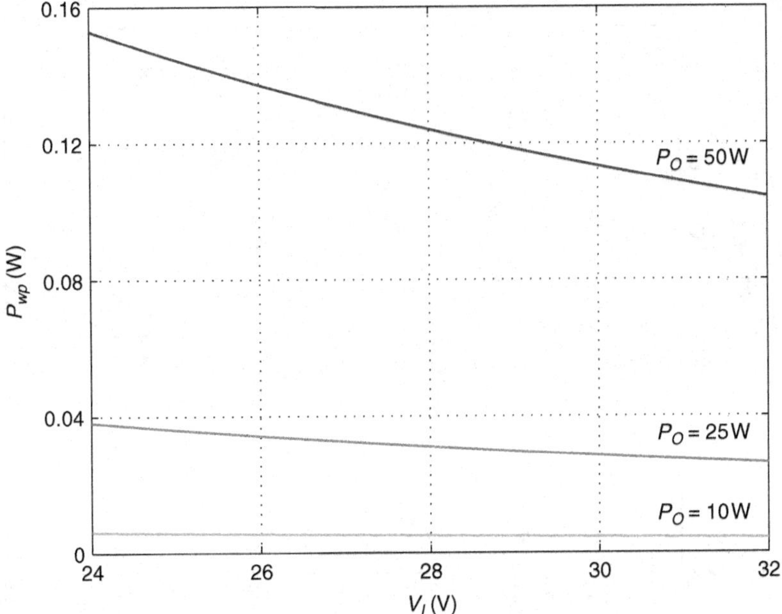

Figure 11.8 Primary winding power loss P_{wp} as a function of the DC input voltage V_I at fixed values of the output power P_O for the flyback converter operating in CCM

The DC and low-frequency resistance of all strands in the secondary winding is

$$R_{wsDC} = \frac{R_{wsDCs}}{S_s} = \frac{0.027}{45} = 0.0006\ \Omega. \tag{11.120}$$

The DC and low-frequency power loss in the secondary winding is

$$P_{wsDC} = R_{wsDC} I_{Omax}^2 = 0.0006 \times 10^2 = 0.06\ \text{W}. \tag{11.121}$$

Using (7.238), we compute $F_{Rsh} = 2.6$. Figure 11.9 shows a plot of F_{Rsh} as a function of the DC input voltage V_I for the flyback converter operating in CCM.

The secondary winding power loss is

$$P_{ws} = F_{Rsh} P_{wsDC} = 2.6 \times 0.06 = 0.156\ \text{W}. \tag{11.122}$$

Figure 11.10 shows the plots of the power loss in the secondary winding P_{ws} as a function of the output power P_O at fixed values of the DC input voltage V_I for the flyback converter operating in CCM. Figure 11.11 shows the plots of the power loss in the secondary winding P_{ws} as a function of the output power P_O at fixed values of the DC input voltage V_I for the flyback converter operating in CCM.

The DC and low-frequency power loss in both the windings is

$$P_{wDC} = P_{wpDC} + P_{wsDC} = 0.05879 + 0.06 = 0.11879\ \text{W}. \tag{11.123}$$

The sum of the power loss in both the windings is

$$P_w = P_{wp} + P_{ws} = 0.1528 + 0.156 = 0.3088\ \text{W}. \tag{11.124}$$

Figure 11.12 shows the plots of the total winding power loss P_w as a function of the output power P_O at fixed values of the DC input voltage V_I for the flyback converter operating in CCM. Figure 11.13 shows the plots of the power loss in both primary and secondary windings P_w as a function of the DC output voltage V_I at fixed values of the output power P_O for the flyback converter operating in CCM. The sum of the core loss and the resistance winding loss in the transformer is

$$P_{cw} = P_C + P_w = 0.07509 + 0.3088 = 0.3839\ \text{W}. \tag{11.125}$$

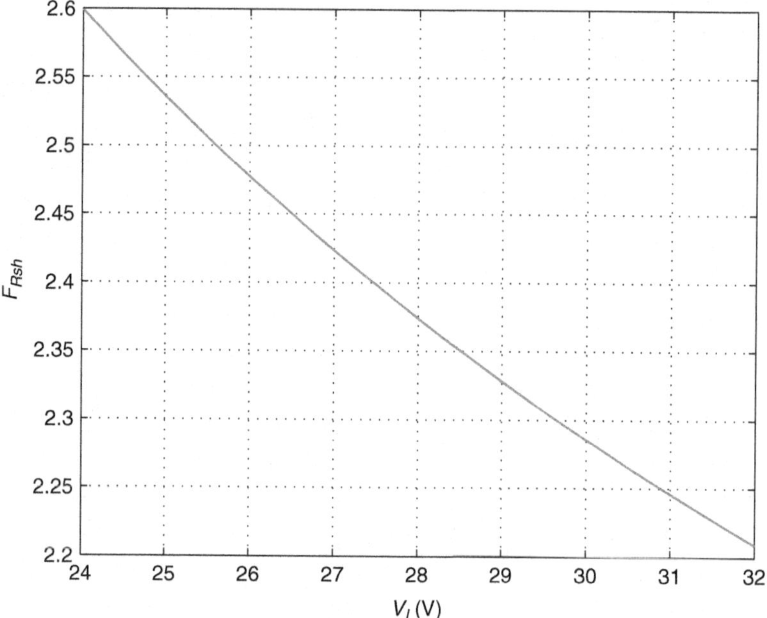

Figure 11.9 F_{Rsh} as a function of the DC input voltage V_I for the flyback converter operating in CCM

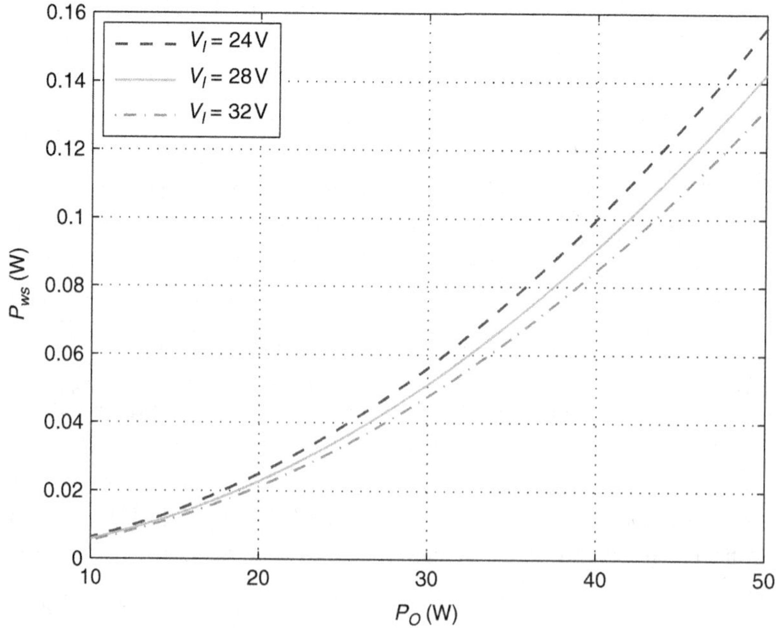

Figure 11.10 Secondary winding power loss P_{ws} as a function of the output power P_O at fixed values of the DC input voltage V_I for the flyback converter operating in CCM

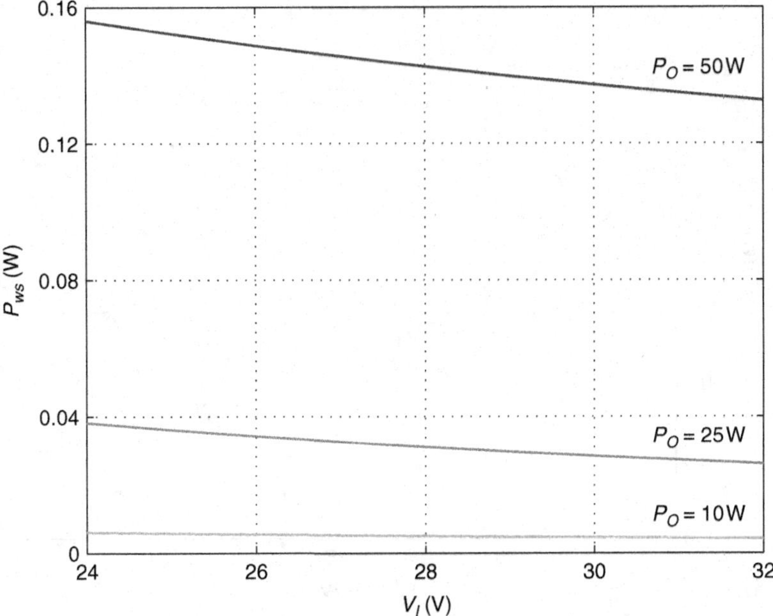

Figure 11.11 Secondary winding power loss P_{ws} as a function of the DC input voltage V_I at fixed values of the output power P_O for the flyback converter operating in CCM

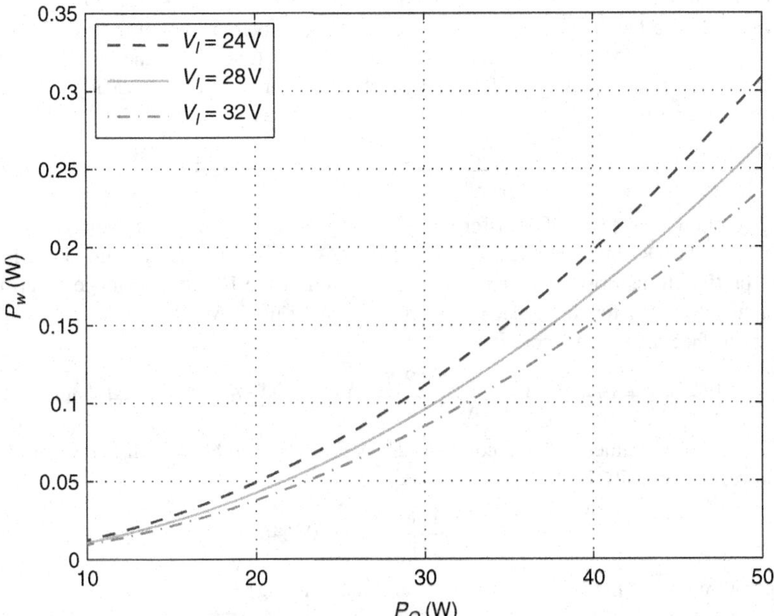

Figure 11.12 Power loss in both primary and secondary windings P_w as a function of the output power P_O at fixed values of the DC input voltage V_I for the flyback converter operating in CCM

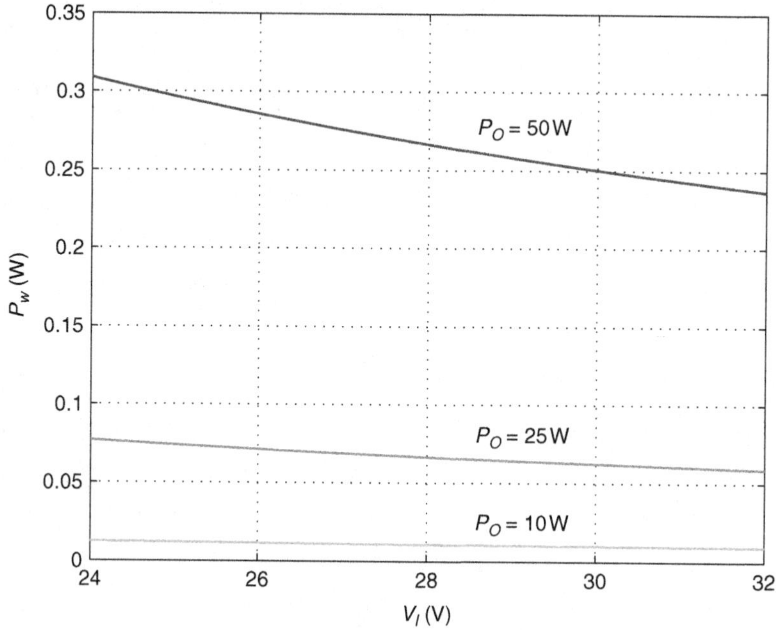

Figure 11.13 Power loss in both primary and secondary windings P_w as a function of the DC input voltage V_I at fixed values of the output power P_O for the flyback converter operating in CCM

Figure 11.14 shows the plots of core and winding power loss P_{cw} as a function of the output power P_O at fixed values of the DC input voltage V_I for the flyback converter operating in CCM. Figure 11.15 shows the plots of the core and winding power loss P_{cw} as a function of the DC input voltage V_I at fixed values of the output power P_O for the flyback converter operating in CCM.

The transformer efficiency at full power can be estimated as

$$\eta_t = \frac{P_O}{P_O + P_{cw}} = \frac{50}{50 + 0.3839} = 99.23\%. \tag{11.126}$$

Figure 11.16 shows the plots of transformer efficiency η_t as a function of the output power P_O at fixed values of the DC input voltage V_I for the flyback converter operating in CCM. Figure 11.17 shows the plots of transformer efficiency η_t as a function of the DC input voltage V_I at fixed values of the output power P_O for the flyback converter operating in CCM.

The total surface area of the core is

$$A_t = 2\left(\frac{\pi A^2}{4}\right) + (\pi A)(2B) = 2\left(\frac{\pi \times 3.56^2}{4}\right) + \pi \times 3.56 \times 2.19 = 44.4 \text{ cm}^2, \tag{11.127}$$

where A is the outer diameter of the core and $2B$ is the total height of both halves of the core. The surface power loss density is

$$\psi = \frac{P_{cw}}{A_t} = \frac{0.3839}{44.4} = 0.008646 \text{ W/cm}^2. \tag{11.128}$$

The temperature rise of the inductor (the core and the winding) is

$$\Delta T = 450\psi^{0.826} = 450 \times 0.008646^{0.826} = 8.89\ ^\circ\text{C}. \tag{11.129}$$

The core window utilization is

$$K_u = \frac{(N_p S_p + N_s S_s) A_{wpo}}{W_a} = \frac{(9 \times 10 + 2 \times 45) \times 0.1313 \times 10^{-6}}{0.757 \times 10^{-4}} = 0.3122. \tag{11.130}$$

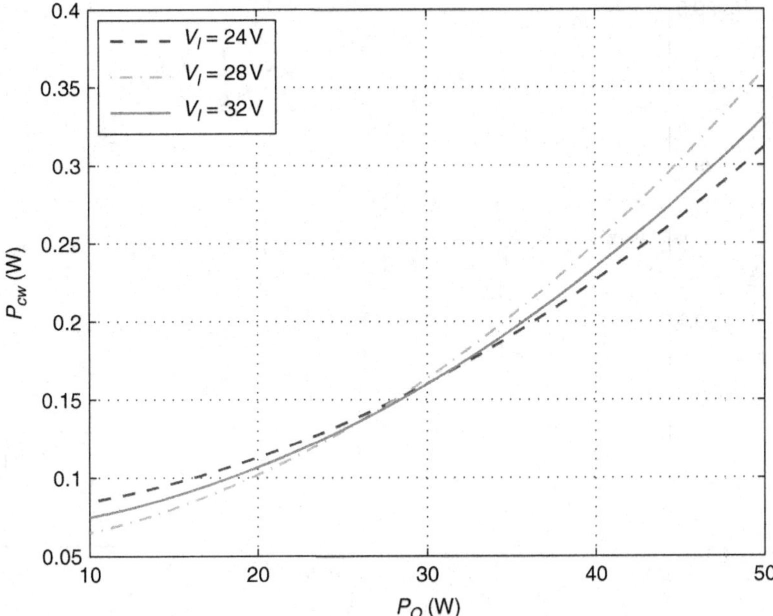

Figure 11.14 Core and winding power loss P_{cw} as a function of the output power P_O at fixed values of the DC input voltage V_I for the flyback converter operating in CCM

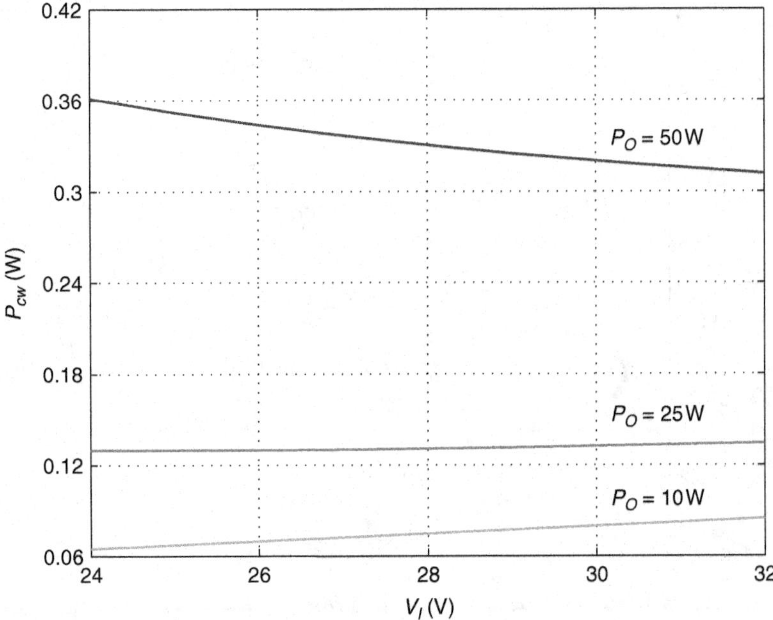

Figure 11.15 Core and winding power loss P_{cw} as a function of the DC input voltage V_I at fixed values of the output power P_O for the flyback converter operating in CCM

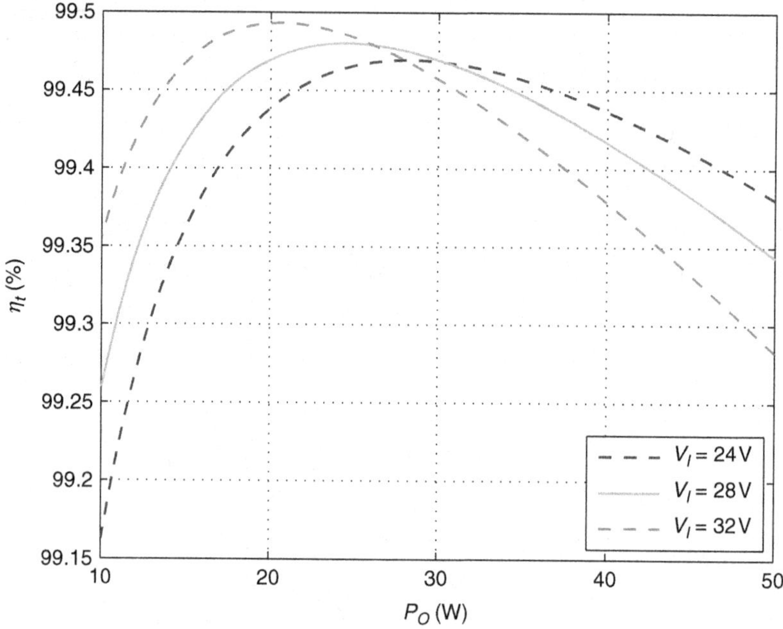

Figure 11.16 Transformer efficiency η_t as a function of the output power P_O at fixed values of the DC input voltage V_I for the flyback converter operating in CCM

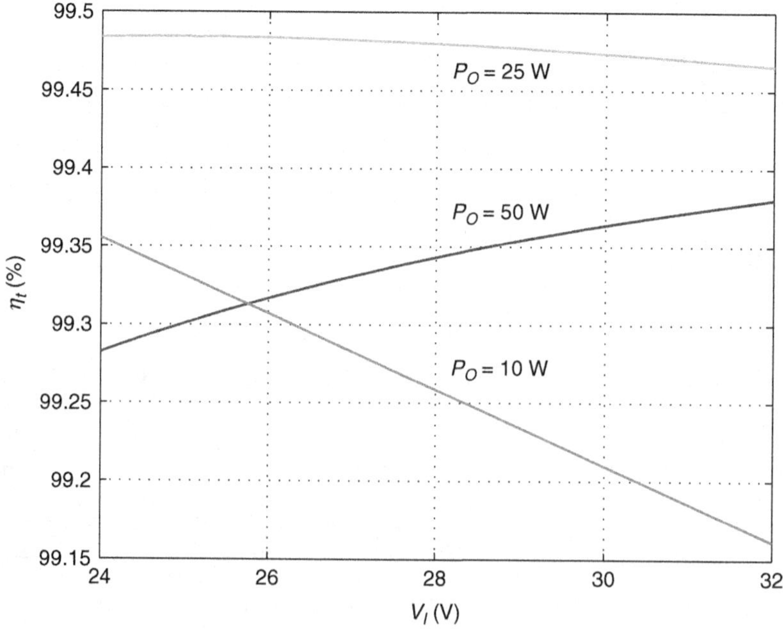

Figure 11.17 Transformer efficiency η_t as a function of the DC input voltage V_I at fixed values of the output power P_O for the flyback converter operating in CCM

11.6 Transformer Design for Flyback Converter in DCM

Example 11.2

A flyback PWM converter is operated in the discontinuous conduction mode (DCM) and has the following specifications: $V_I = 28 \pm 4$ V, $V_O = 5$ V, $I_O = 0$–10 A, and $f_s = 100$ kHz.

Solution: Figure 11.18 shows the waveforms in the flyback converter for DCM. The maximum and minimum output powers are

$$P_{Omax} = V_O I_{Omax} = 5 \times 10 = 50 \text{ W} \tag{11.131}$$

and

$$P_{Omin} = V_O I_{Omin} = 5 \times 0 = 0 \text{ W}. \tag{11.132}$$

The minimum and maximum load resistances are

$$R_{Lmin} = \frac{V_O}{I_{Omax}} = \frac{5}{10} = 0.5\ \Omega \tag{11.133}$$

and

$$R_{Lmax} = \frac{V_O}{I_{Omin}} = \frac{5}{0} = \infty. \tag{11.134}$$

The minimum and maximum DC voltage transfer functions are

$$M_{VDCmin} = \frac{V_O}{V_{Imax}} = \frac{5}{32} = 0.15625\ \Omega \tag{11.135}$$

and

$$M_{VDCmax} = \frac{V_O}{V_{Imin}} = \frac{5}{24} = 0.2083\ \Omega. \tag{11.136}$$

Assume the converter efficiency $\eta = 0.88$ and the maximum duty cycle $D_{max} = 0.5$. The transformer turns ratio is

$$n = \frac{\eta D_{max}}{(1 - D_{max}) M_{VDCmax}} = \frac{0.88 \times 0.5}{(1 - 0.5) \times 0.2083} = 4.2247. \tag{11.137}$$

The magnetizing inductance is

$$L_{m(max)} = \frac{n^2 R_{Lmin} (1 - D_{Bmax})^2}{2f_s} = \frac{4.2247^2 \times 0.5 \times (1 - 0.5)^2}{2 \times 10^5} = 11.15\ \mu\text{H}. \tag{11.138}$$

Let $L_m = 10\ \mu$H.

Assume the converter efficiency $\eta = 0.88$. The maximum duty cycle at full load is

$$D_{max} = M_{VDCmax} \sqrt{\frac{2f_s L_m}{\eta R_{Lmin}}} = 0.2083 \sqrt{\frac{2 \times 10^5 \times 10 \times 10^{-6}}{0.88 \times 0.5}} = 0.444. \tag{11.139}$$

The minimum duty cycle at full load is

$$D_{min} = M_{VDCmin} \sqrt{\frac{2f_s L_m}{\eta R_{Lmin}}} = 0.15625 \sqrt{\frac{2 \times 10^5 \times 10 \times 10^{-6}}{0.88 \times 0.5}} = 0.333. \tag{11.140}$$

The maximum duty cycle when the diode is ON at full load is

$$D_{1max} = \sqrt{\frac{2f_s L_m}{n^2 R_{Lmin}}} = \sqrt{\frac{2 \times 10^5 \times 10 \times 10^{-6}}{4.2247^2 \times 0.5}} = 0.4734. \tag{11.141}$$

Hence,

$$D_{max} + D_{1max} = 0.444 + 0.4734 = 0.9174 < 1. \tag{11.142}$$

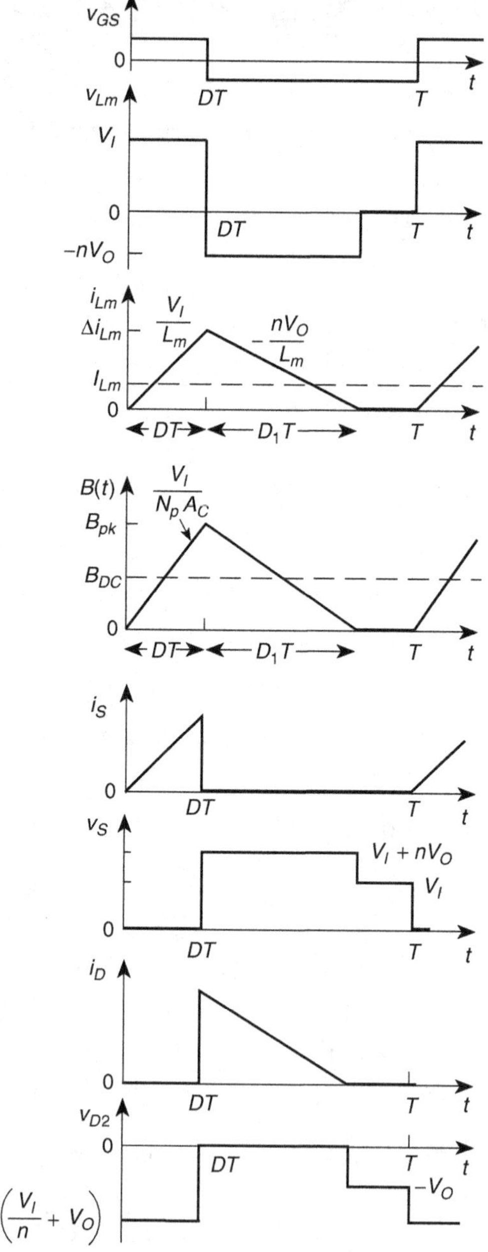

Figure 11.18 Waveforms in flyback PWM converter for DCM

Figure 11.19 shows the plots of $D + D_1$ as a function of the DC input voltage V_I at fixed values of the output power P_O for the flyback converter operating in DCM.

The maximum DC input current is

$$I_{Imax} = \frac{P_{Omax}}{\eta V_{Imax}} = \frac{M_{VDCmax} I_{Omax}}{\eta} = \frac{0.2083 \times 10}{0.88} = 2.367 \text{ A}. \tag{11.143}$$

The maximum current through the primary winding is

$$I_{pmax} = \Delta i_{Lm(max)} = \frac{D_{min} V_{Imax}}{f_s L_m} = \frac{0.333 \times 32}{10^5 \times 10 \times 10^{-6}} = 10.656 \text{ A}. \tag{11.144}$$

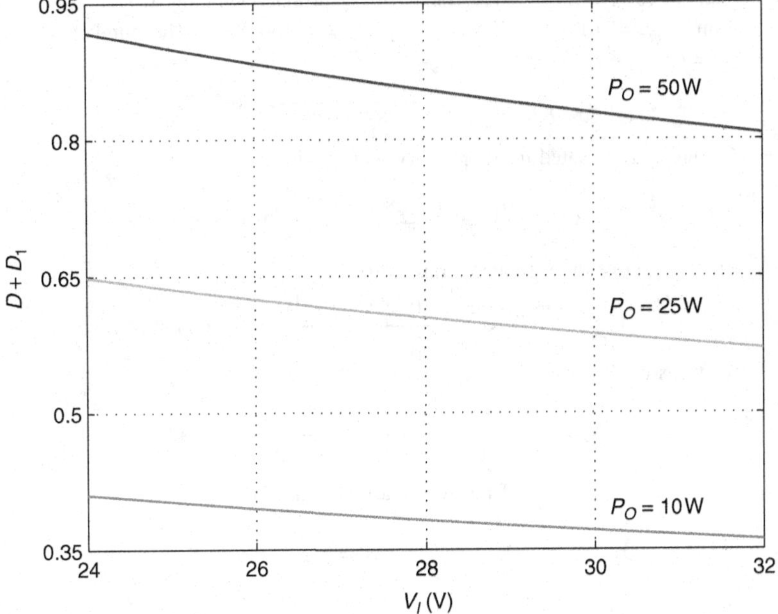

Figure 11.19 $D + D_1$ as a function of the DC input voltage V_I at fixed values of the output power P_O for the flyback converter operating in DCM

The maximum rms value of the current through the primary winding is

$$I_{prms(max)} = I_{pmax}\sqrt{\frac{D_{max}}{3}} = 10.656\sqrt{\frac{0.444}{3}} = 4.1 \text{ A}. \tag{11.145}$$

The maximum energy stored in the magnetic field of the transformer is

$$W_m = \frac{1}{2}L_m I_{pmax}^2 = \frac{1}{2} \times 10 \times 10^{-6} \times 10.656^2 = 0.5677 \text{ mJ}. \tag{11.146}$$

Assuming $K_u = 0.3$, $B_{pk} = 0.21$ T, and $J_m = 5$ A/mm^2. The core area product is

$$A_p = \frac{4W_m}{K_u J_m B_{pk}} = \frac{4 \times 0.5677 \times 10^{-3}}{0.3 \times 5 \times 10^6 \times 0.21} = 0.72088 \text{ cm}^4. \tag{11.147}$$

A Magnetics ferrite core PC 0F-43019 is selected, which has $A_p = 0.73$ cm^4, $A_c = 1.37$ cm^2, $l_c = 4.52$ cm, $V_c = 6.19$ cm^3, $A_L = 8100$ nH/turn, $\mu_{rc} = 3000 \pm 20\%$, and $B_s = 0.49$ T. The F-type magnetic material is used to make this core. The coefficients of this material are $k = 0.0573$, $a = 1.66$, and $b = 2.68$.

The core window area is

$$W_a = \frac{A_p}{A_c} = \frac{0.73}{1.37} = 0.5328 \text{ cm}^2. \tag{11.148}$$

The cross-sectional area of the primary winding wire is

$$A_{wp} = \frac{I_{pmax}}{J_m} = \frac{10.656}{5} = 2.1312 \text{ mm}^2. \tag{11.149}$$

The skin depth of copper at $f = 100$ kHz is

$$\delta_w = \frac{66.2}{\sqrt{f_s}}(\text{mm}) = \frac{66.2}{\sqrt{100\,000}} = 0.209 \text{ mm}, \tag{11.150}$$

resulting in the diameter of a strand for the primary winding at which the skin and proximity effects can be neglected

$$d_{is} = 2\delta_w = 2 \times 0.209 \times 10^{-3} = 0.418 \text{ mm}. \tag{11.151}$$

Select the copper wire AWG26 for the strands of the primary winding, which has $d_{is} = 0.405$ mm, $d_{os} = 0.452$ mm, $A_{wst} = 0.1288$ mm^2, and $R_{wDC}/l_w = 0.1345$ mm. The number of strands in the primary winding is

$$S_p = \frac{A_{wp}}{A_{wst}} = \frac{2.1312 \times 10^{-6}}{0.1288 \times 10^{-6}} = 16.547. \tag{11.152}$$

Pick $S_p = 16$. The area allocated to the primary winding is

$$W_{ap} = \frac{W_a}{2} = \frac{0.5328 \times 10^{-4}}{2} = 0.2664 \text{ cm}^2. \tag{11.153}$$

The cross-sectional area of the insulated strand wire is

$$A_{wpo} = \frac{\pi d_o^2}{4} = \frac{\pi \times (0.452 \times 10^{-3})^2}{4} = 0.1604 \text{ mm}^2. \tag{11.154}$$

The number of turns of the primary winding is

$$N_p = \frac{K_u W_{ap}}{S_p A_{wpo}} = \frac{0.3 \times 0.2664 \times 10^{-4}}{16 \times 0.1604 \times 10^{-6}} = 3.114. \tag{11.155}$$

Pick $N_p = 4$. The number of turns of the secondary winding is

$$N_s = \frac{N_p}{n} = \frac{4}{4.2247} = 0.94. \tag{11.156}$$

Select $N_s = 1$.

The length of the air gap is

$$l_g = \frac{\mu_0 A_c N_p^2}{L_p} - \frac{l_c}{\mu_{rc}} = \frac{4\pi \times 10^{-7} \times 1.37 \times 10^{-4} \times 4^2}{10 \times 10^{-6}} - \frac{4.52 \times 10^{-2}}{3000} = 0.2603 \text{ mm}. \tag{11.157}$$

The maximum peak value of the magnetic flux density is

$$B_{pk} = \frac{\mu_0 N_p I_{pmax}}{l_g + \frac{l_c}{\mu_{rc}}} = \frac{4\pi \times 10^{-7} \times 4 \times 10.656}{0.2603 \times 10^{-3} + \frac{4.52 \times 10^{-2}}{3000}} = 0.1945 \text{ T} < B_s. \tag{11.158}$$

The maximum peak value of the AC component of the magnetic flux density is

$$B_m = \frac{B_{pk}}{2} = \frac{0.1945}{2} = 0.09725 \text{ T}. \tag{11.159}$$

The core power loss density is

$$P_v = 0.0573 f^{1.66} (10 B_m)^{2.68} = 0.0573 \times 100^{1.66} \times (10 \times 0.09725)^{2.68} = 111.1 \text{ mW/cm}^3. \tag{11.160}$$

The core loss is

$$P_C = V_c P_v = 6.19 \times 111.1 = 687.7 \text{ mW}. \tag{11.161}$$

Figure 11.20 shows the plots of core loss P_C as a function of the output power P_O at fixed values of the DC input voltage V_I for the flyback converter operating in DCM.

The MTL is

$$l_T = \frac{\pi(F + E)}{2} = \frac{\pi(13.3 + 25.4)}{2} = 60.79 \text{ mm}. \tag{11.162}$$

The length of the primary winding wire is

$$l_{wp} = N_p l_T = 4 \times 6.079 = 24.316 \text{ cm}. \tag{11.163}$$

Pick $l_{wp} = 26$ cm. The DC and low-frequency resistance of each strand of the primary winding is

$$R_{wpDCs} = \left(\frac{R_{wpDCs}}{l_w}\right) l_{wp} = 0.1345 \times 0.26 = 0.035\ \Omega \tag{11.164}$$

Hence, the DC and low-frequency resistance of the primary winding is

$$R_{wpDC} = \frac{R_{wpDCs}}{S_p} = \frac{0.035}{16} = 0.002187\ \Omega. \tag{11.165}$$

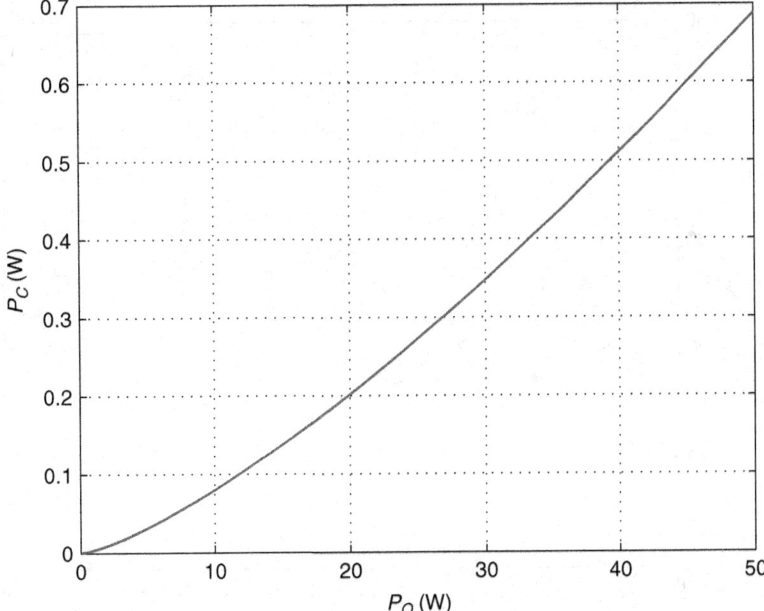

Figure 11.20 Core loss P_C as a function of the output power P_O at fixed values of the DC input voltage V_I for the flyback converter operating in DCM

The DC and low-frequency power loss in the primary winding is

$$P_{wpDC} = R_{wpDC} I_{Imax}^2 = 0.002187 \times 2.367^2 = 0.01225 \text{ W}. \quad (11.166)$$

The skin and proximity effects result in $F_{Rph} = 8.78$ for the number of layers in the primary winding equal to $N_{lp} = 1$ and $D = 0.444$. Figure 11.21 shows the plots of F_{Rph} as a function of the output power P_O at fixed values of the DC input voltage V_I for the flyback converter operating in DCM. Figure 11.22 shows the plots of F_{Rph} as a function of the DC input voltage V_I at fixed values of the output power P_O for the flyback converter operating in DCM.

The primary winding power loss is

$$P_{wp} = F_{Rph} P_{wpDC} = 8.78 \times 0.01225 = 0.10755 \text{ W}. \quad (11.167)$$

Figure 11.23 shows the plots of primary winding power loss P_{wp} as a function of the output power P_O at fixed values of the DC input voltage V_I for the flyback converter operating in DCM. Figure 11.24 shows the plots of primary winding power loss P_{wp} as a function of the DC input voltage V_I at fixed values of the output power P_O for the flyback converter operating in DCM.

The maximum current through the secondary winding is

$$I_{smax} = n \Delta i_{Lm(max)} = 4.2247 \times 10.656 = 45.018 \text{ A}. \quad (11.168)$$

The maximum rms value of the secondary winding current is

$$I_{srms(max)} = I_{smax} \sqrt{\frac{D_{1max}}{3}} = 45.018 \sqrt{\frac{0.4734}{3}} = 17.883 \text{ A}. \quad (11.169)$$

The cross-sectional area of the total secondary winding wire is

$$A_{ws} = \frac{I_{smax}}{J_m} = \frac{45.018}{5} = 9 \text{ mm}^2. \quad (11.170)$$

The number of strands of the secondary winding is

$$S_s = \frac{A_{ws}}{A_{wst}} = \frac{9 \times 10^{-6}}{0.1288 \times 10^{-6}} = 69.87. \quad (11.171)$$

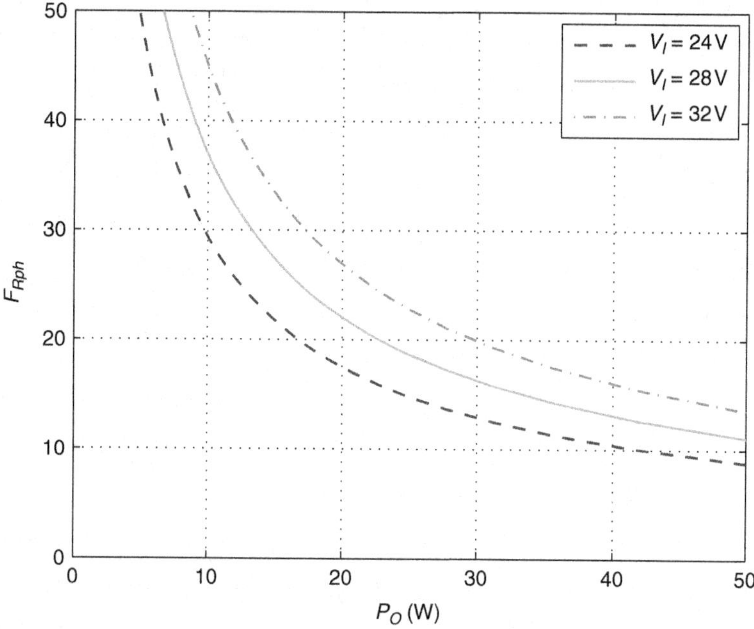

Figure 11.21 F_{Rph} as a function of the output power P_O at fixed values of the DC input voltage V_I for the flyback converter operating in DCM

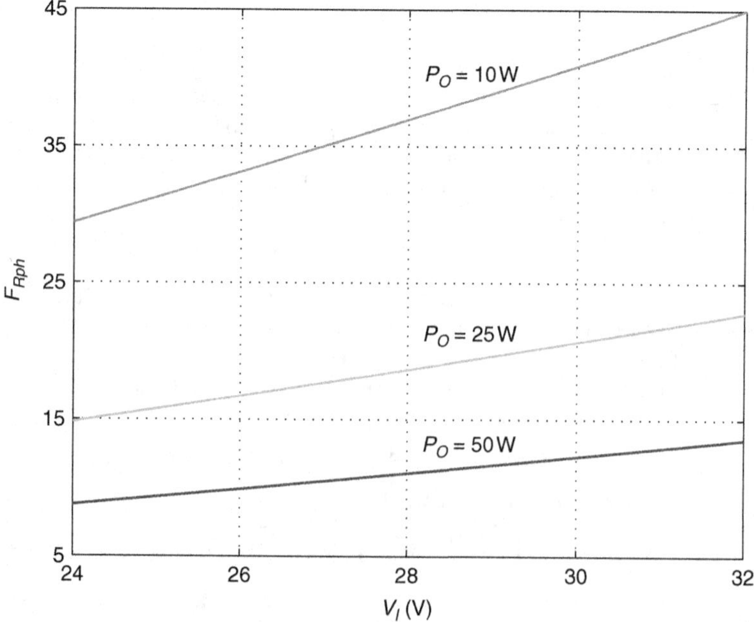

Figure 11.22 F_{Rph} as a function of the DC input voltage V_I at fixed values of the output power P_O for the flyback converter operating in DCM

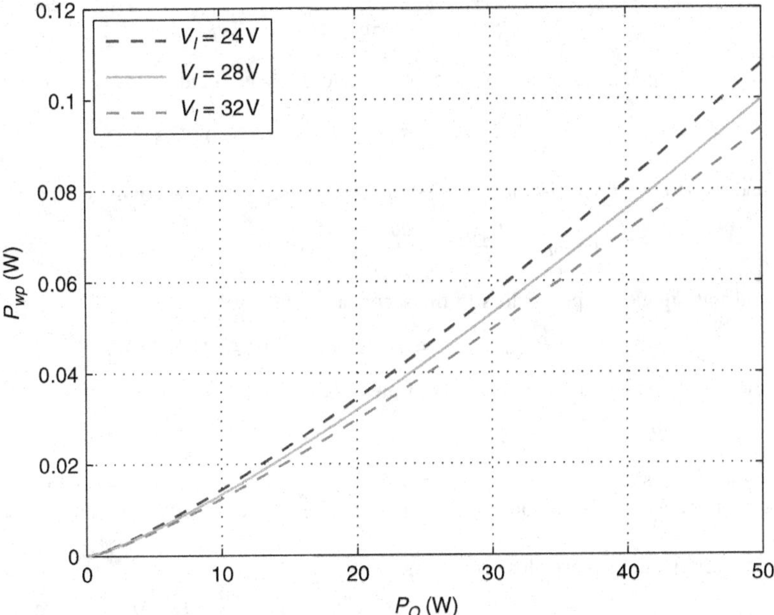

Figure 11.23 Primary winding power loss P_{wp} as a function of the output power P_O at fixed values of the DC input voltage V_I for the flyback converter operating in DCM

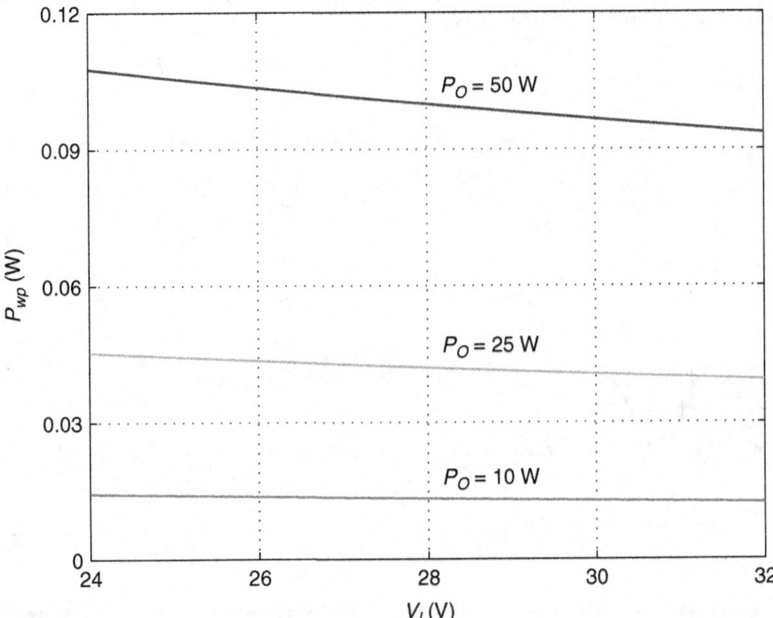

Figure 11.24 Primary winding power loss P_{wp} as a function of the DC input voltage V_I at fixed values of the output power P_O for the flyback converter operating in DCM

Pick $S_s = 70$. The length of the secondary winding is

$$l_{ws} = N_s l_T = 1 \times 6.079 \times 10^{-2} = 6.079 \text{ cm}. \tag{11.172}$$

Pick $l_{ws} = 7$ cm. The DC and low-frequency resistance of each strand of the secondary winding is

$$R_{wsDCs} = \left(\frac{R_{wsDCs}}{l_w}\right) l_{ws} = 0.1345 \times 0.07 = 0.009415 \ \Omega. \tag{11.173}$$

The DC and low-frequency resistance of all strands in the secondary winding is

$$R_{wsDC} = \frac{R_{wsDCs}}{S_s} = \frac{0.009415}{70} = 0.0001345 \ \Omega. \tag{11.174}$$

The DC and low-frequency power loss in the secondary winding is

$$P_{wsDC} = R_{wsDC} I_{Omax}^2 = 0.0001345 \times 10^2 = 0.01345 \text{ W}. \tag{11.175}$$

The DC and low-frequency power loss in both the windings is

$$P_{wDC} = P_{wpDC} + P_{wsDC} = 0.01225 + 0.01345 = 0.0257 \text{ W}. \tag{11.176}$$

The harmonic AC resistance factor is $F_{Rsh} = 2.563$ for $N_{ls} = 1$ and $D_{1max} = 0.4734$. Figure 11.25 shows the plots of F_{Rsh} as a function of the output power P_O for the flyback converter operating in DCM.

The secondary winding power loss is

$$P_{ws} = F_{Rsh} P_{wsDC} = 2.563 \times 0.01345 = 0.03447 \text{ W}. \tag{11.177}$$

Figure 11.26 shows the plots of the primary and secondary winding power loss P_w as a function of the output power P_O at fixed values of the DC input voltage V_I for the flyback converter operating in DCM. Figure 11.27 shows the plots of secondary winding power loss P_{ws} as a function of the DC input voltage V_I at fixed values of the output power P_O for the flyback converter operating in DCM.

The power loss of both the windings is

$$P_w = P_{wp} + P_{ws} = 0.1159 + 0.03447 = 0.15037 \text{ W}. \tag{11.178}$$

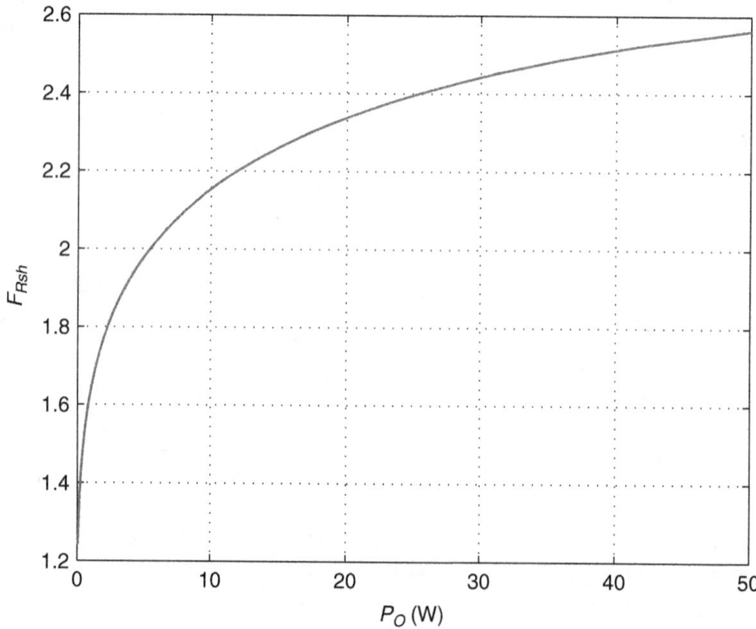

Figure 11.25 F_{Rsh} as a function of the output power P_O for the flyback converter operating in DCM

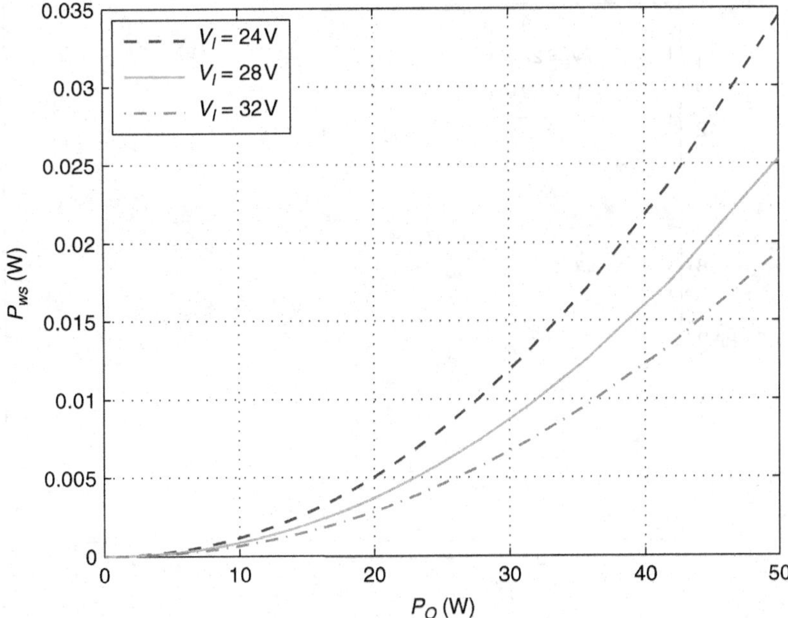

Figure 11.26 Secondary winding power loss P_{ws} as a function of the output power P_O at fixed values of the DC input voltage V_I for the flyback converter operating in DCM

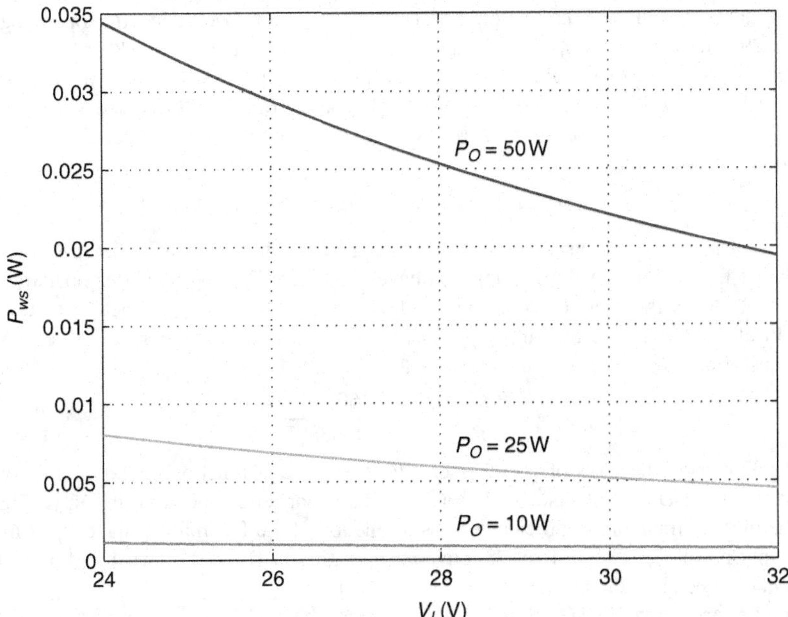

Figure 11.27 Secondary winding power loss P_{ws} as a function of the DC input voltage V_I at fixed values of the output power P_O for the flyback converter operating in DCM

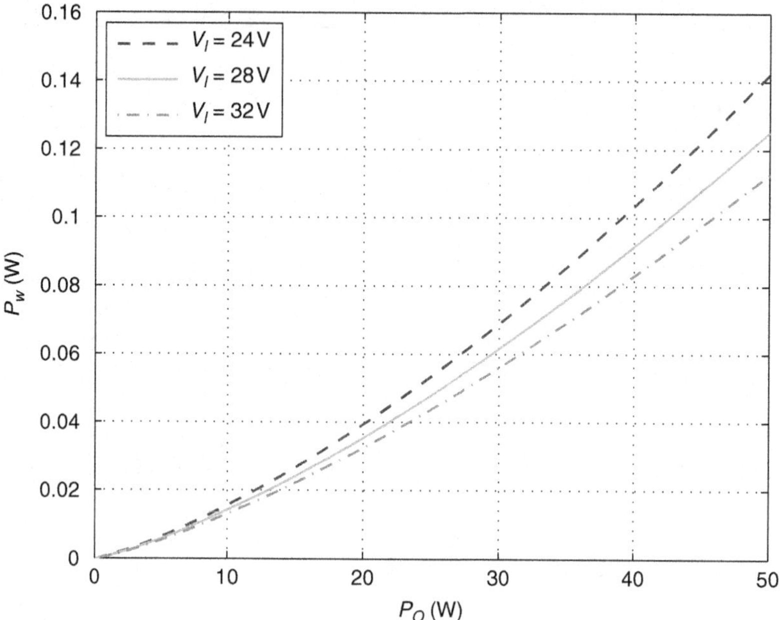

Figure 11.28 Primary and secondary winding power loss P_w as a function of the output power P_O at fixed values of the DC input voltage V_I for the flyback converter operating in DCM

Figure 11.28 shows the plots of secondary winding power loss P_{ws} as a function of the output power P_O at fixed values of the DC input voltage V_I for the flyback converter operating in DCM. Figure 11.29 shows the plots of primary and secondary winding power loss P_w as a function of the DC input voltage V_I at fixed values of the output power P_O for the flyback converter operating in DCM. Figures 11.34 and 11.35 show the amplitude spectra of the primary and secondary winding currents at P_{Omax} and V_{Imin} for the flyback converter operating in DCM, respectively.

The total power loss is

$$P_{cw} = P_C + P_w = 0.6877 + 0.15037 = 0.83807 \text{ W}. \tag{11.179}$$

Figure 11.30 depicts the plots of core and winding power loss P_{cw} as a function of the output power P_O at fixed values of the DC input voltage V_I for the flyback converter operating in DCM. Figure 11.31 shows the plots of core and winding power loss P_{cw} as a function of the DC input voltage V_I at fixed values of the output power P_O for the flyback converter operating in DCM.

The transformer efficiency is

$$\eta_t = \frac{P_O}{P_O + P_{cw}} = \frac{50}{50 + 0.83807} = 98.35\%. \tag{11.180}$$

Figure 11.32 depicts the plots of transformer efficiency η_t as a function of the output power P_O at fixed values of the DC input voltage V_I for the flyback converter operating in DCM. Figure 11.33 shows the plots of transformer efficiency η_t as a function of the DC input voltage V_I at fixed values of the output power P_O for the flyback converter operating in DCM. Figures 11.34 and 11.35 show the amplitude spectra of the primary and secondary currents at P_{Omax} and V_{Imin}.

The total surface area of the core is

$$A_t = 2\left(\frac{\pi A^2}{4}\right) + (\pi A)(2B) = 2\left(\frac{\pi \times 3^2}{4}\right) + \pi \times 3 \times 1.89 = 31.95 \text{ cm}^2, \tag{11.181}$$

where A is the outer diameter of the core and $2B$ is the total height of both halves of the core. The surface power loss density is

$$\psi = \frac{P_{cw}}{A_t} = \frac{0.83807}{31.95} = 0.02623 \text{ W/cm}^2. \tag{11.182}$$

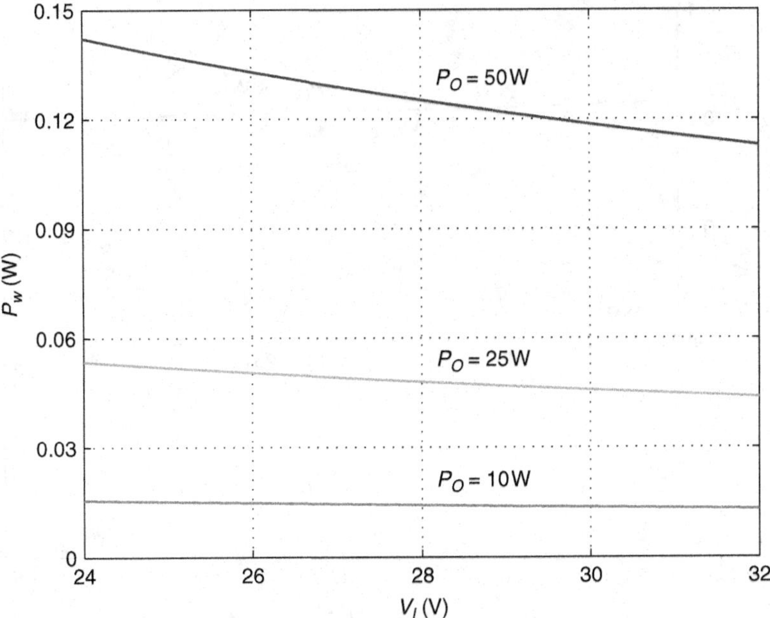

Figure 11.29 Primary and secondary winding power loss P_w as a function of the DC input voltage V_I at fixed values of the output power P_O for the flyback converter operating in DCM

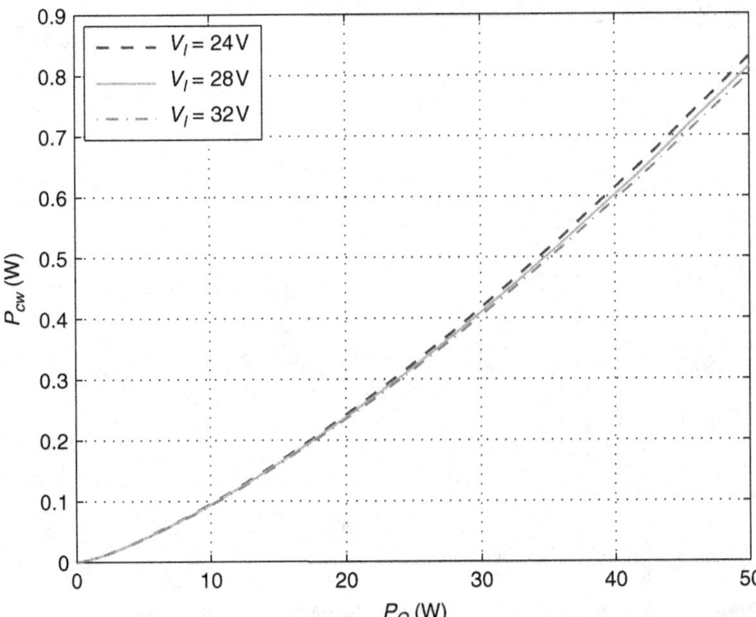

Figure 11.30 Core and winding power loss P_{cw} as a function of the output power P_O at fixed values of the DC input voltage V_I for the flyback converter operating in DCM

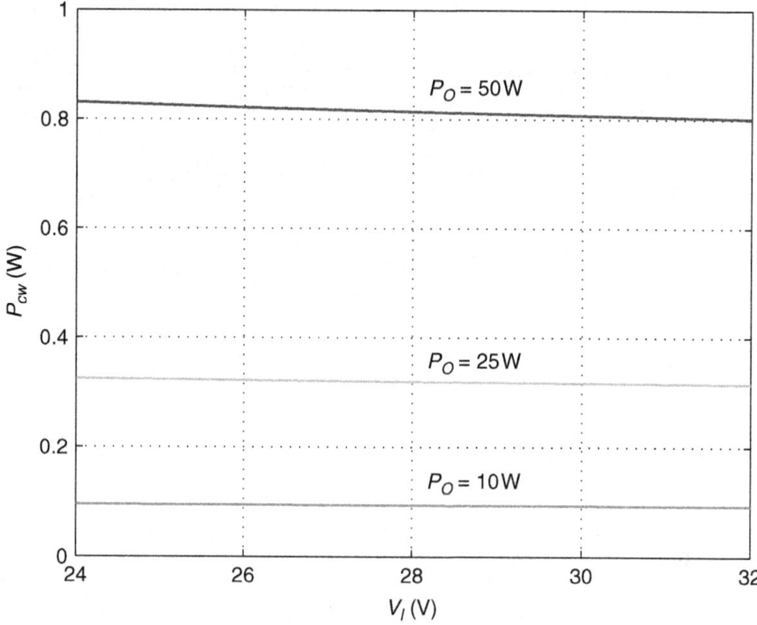

Figure 11.31 Core and winding power loss P_{cw} as a function of the DC input voltage V_I at fixed values of the output power P_O for the flyback converter operating in DCM

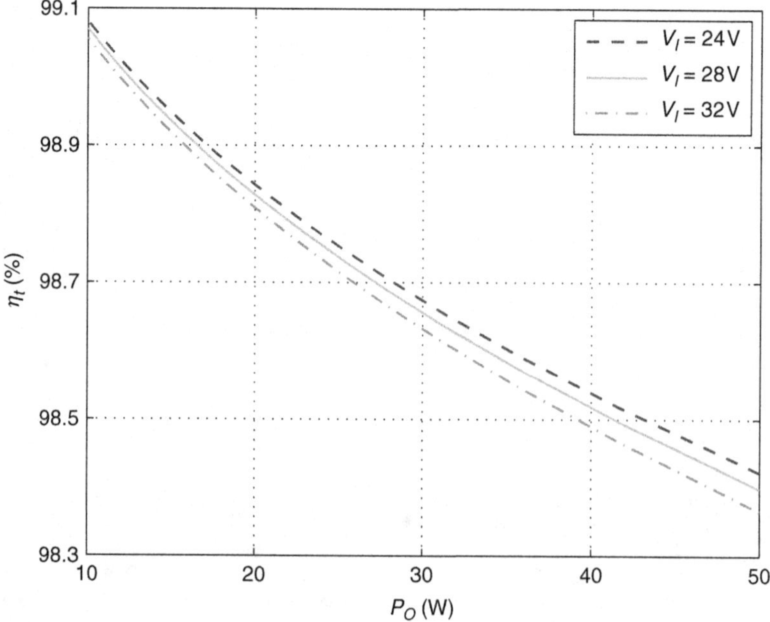

Figure 11.32 Transformer efficiency η_t as a function of the output power P_O at fixed values of the DC input voltage V_I for the flyback converter operating in DCM

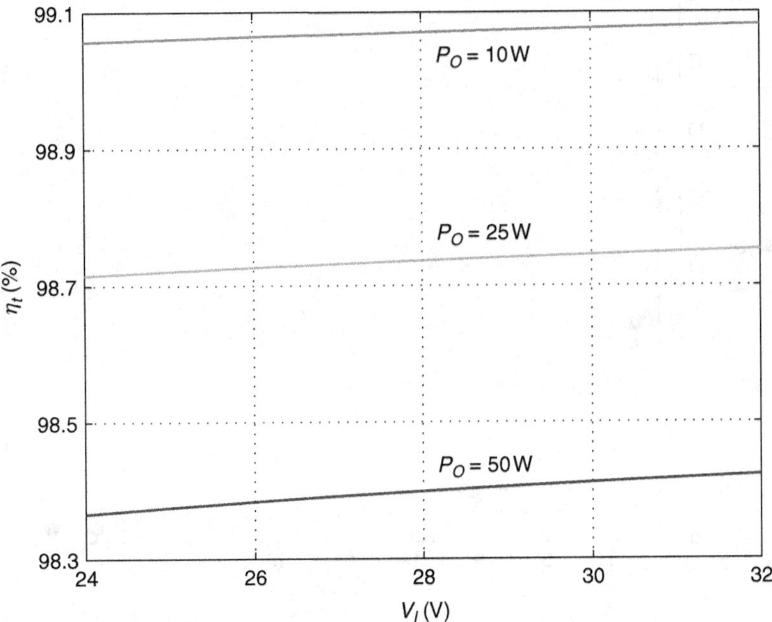

Figure 11.33 Transformer efficiency η_t as a function of the DC input voltage V_I at fixed values of the output power P_O for the flyback converter operating in DCM

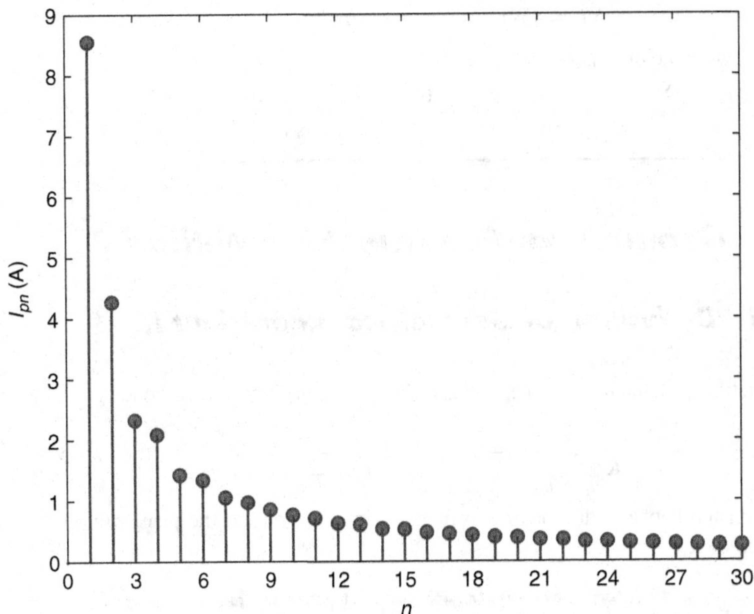

Figure 11.34 Amplitude spectrum of the primary winding current at P_{Omax} and V_{Imin} for the flyback converter operating in DCM

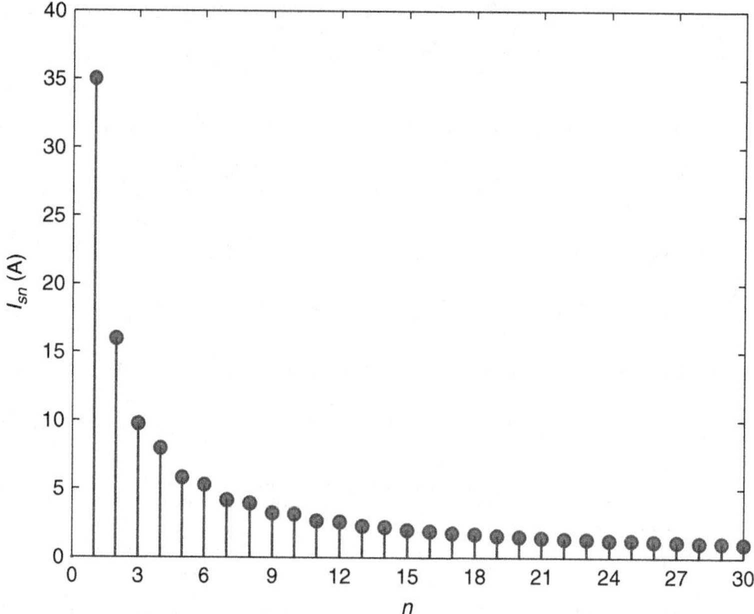

Figure 11.35 Amplitude spectrum of the secondary winding current at P_{Omax} and V_{Imin} for the flyback converter operating in DCM

The temperature rise of the inductor is

$$\Delta T = 450\psi^{0.826} = 450 \times 0.02623^{0.826} = 22.24\ ^{\circ}\mathrm{C}. \tag{11.183}$$

The core window utilization is

$$K_u = \frac{(N_p S_p + N_s S_s)A_{wpo}}{W_a} = \frac{(4 \times 16 + 1 \times 70) \times 0.1604 \times 10^{-6}}{0.5328 \times 10^{-4}} = 0.4034. \tag{11.184}$$

11.7 Geometrical Coefficient K_g Method

11.7.1 Derivation of Geometrical Coefficient K_g

The magnetizing inductance on the primary side of a transformer with an air gap is given by

$$L_m = \frac{N_p^2}{\mathcal{R}} = \frac{N_p^2}{\mathcal{R}_g + \mathcal{R}_c} = \frac{N_p^2}{\frac{l_g}{\mu_0 A_c} + \frac{l_c}{\mu_{rc}\mu_0 A_c}} = \frac{\mu_0 N_p^2 A_c}{l_g + \frac{l_c}{\mu_{rc}}} \approx \frac{\mu_0 N_p^2 A_c}{l_g} \quad \text{for} \quad l_g \gg \frac{l_c}{\mu_{rc}}. \tag{11.185}$$

The relationship between the maximum magnetizing current and the peak value of the flux density B_{pk} is expressed as

$$N_p I_{Lmmax} = \frac{B_{pk}}{\mu_0}\left(\frac{l_c}{\mu_{rc}} + l_g\right). \tag{11.186}$$

Thus,

$$B_{pk} = \frac{\mu_0 N_p I_{Lmmax}}{\frac{l_c}{\mu_{rc}} + l_g} < B_s. \tag{11.187}$$

The air gap length is

$$l_g \approx \frac{\mu_0 N_p I_{Lpmax}}{B_{pk}}. \tag{11.188}$$

The maximum flux linkage is

$$\lambda_{max} = N_p \phi_{max} = N_p A_c B_{pk} = L_p I_{Lpmax}. \tag{11.189}$$

Hence, the number of turns of the primary winding is given by

$$N_p = \frac{L_p I_{Lpmax}}{A_c B_{pk}}. \tag{11.190}$$

The length of the primary winding wire is

$$l_{wp} = N_p l_T, \tag{11.191}$$

where l_T is the mean length per turn (MLT). The cross-sectional area of the primary winding wire is

$$A_{Cup} = N_p A_{wp} \tag{11.192}$$

The DC winding resistance of the primary winding is

$$R_{wpDC} = \rho_w \frac{l_{wp}}{A_{wp}} = \rho_w \frac{N_p l_T}{A_{wp}} \tag{11.193}$$

producing

$$N_p = \frac{R_{wpDC} A_{wp}}{\rho_w l_T}. \tag{11.194}$$

Thus,

$$\frac{R_{wpDC} A_{wp}}{\rho_w l_T} = \frac{L_m I_{Lpmax}}{A_c B_{pk}}, \tag{11.195}$$

which leads to the cross-sectional area of the primary winding wire

$$A_{wp} = \frac{\rho_w L_m I_{Lpmax} l_T}{R_{wpDC} A_c B_{pk}}. \tag{11.196}$$

The core window utilization factor is defined as

$$K_u = \frac{A_{Cup} + A_{Cus}}{W_a} = \frac{N_p A_{wp} + N_s A_{ws}}{W_a}. \tag{11.197}$$

Assuming the winding allocation such that

$$N_p A_{wp} = N_s A_{ws} \tag{11.198}$$

we get

$$K_u = \frac{2A_{Cup}}{W_a} = \frac{2N_p A_{wp}}{W_a}. \tag{11.199}$$

Substituting (11.190) and (11.196) in (11.199), we obtain the minimum area of the core window

$$W_a = \frac{2N_p A_{wp}}{K_u} = \frac{2\rho_w L_m^2 I_{Lmmax}^2 l_T}{K_u R_{wpDC} A_c^2 B_{pk}^2}. \tag{11.200}$$

Hence, the *core geometry coefficient* is defined as [1]

$$K_g = \frac{W_a A_c^2 K_u}{l_T} = \frac{2\rho_w L_m^2 I_{Lmmax}^2}{R_{wpDC} B_{pk}^2} \ (\mathrm{m}^5). \tag{11.201}$$

The DC and low-frequency resistance of the primary winding is

$$R_{wpDC} = \frac{P_{wpDC}}{I_{prms}^2} \tag{11.202}$$

yielding

$$K_g = \frac{2\rho_w L_m^2 I_{Lmmax}^2 I_{prms}^2}{P_{wpDC} B_{pk}^2} \ (\mathrm{m}^5). \tag{11.203}$$

The ratio of the DC and low-frequency power loss in the primary and the secondary windings to the output power is the *transformer load voltage regulation* and is defined as

$$\alpha = \frac{P_{wDC}}{P_O} = \frac{P_{wpDC} + P_{wsDC}}{P_O} = \frac{2P_{wpDC}}{P_O}, \tag{11.204}$$

where $P_{wpDC} = P_{wsDC}$ for equal winding allocations. The efficiency of the transformer at low frequencies, where the core loss and the skin and proximity losses can be ignored, is given by

$$\eta_{tDC} = \frac{P_O}{P_O + P_{wDC}} = \frac{1}{1 + \frac{P_{wDC}}{P_O}} = \frac{1}{1+\alpha}. \tag{11.205}$$

Hence,

$$\alpha = \frac{1}{\eta_{tDC}} - 1. \tag{11.206}$$

For example, for $\eta_{tDC} = 0.99$, we get $\alpha = 1/0.99 - 1 = 0.01$. Using α, we can express K_g in terms of P_O as

$$K_g = \frac{4\rho_w L_m^2 I_{Lmmax}^2 I_{prms}^2}{\alpha P_O B_{pk}^2} \text{ (m}^5\text{)}. \tag{11.207}$$

The energy stored in the magnetizing inductance L_m of the transformer is

$$W_m = \frac{1}{2} L_m I_{Lmmax}^2. \tag{11.208}$$

The geometry coefficient K_g can be expressed in terms of the area product A_p as

$$K_g = \frac{A_p A_c K_u}{l_T} = \frac{A_c L_p I_{Lmmax}^2}{l_T J_m B_{pk}} = \frac{2A_c W_m}{l_T J_m B_{pk}} \text{ (m}^5\text{)}. \tag{11.209}$$

11.7.2 K_g for Transformer with Sinusoidal Currents and Voltages

Assume that the DC components of the currents in the primary and secondary windings are zero. Therefore, $B_{DC} = 0$, $B_{pk} = B_m$, and $I_{prms} = I_{pmax}/\sqrt{2}$. Thus,

$$K_g = \frac{\rho_w L_p^2 I_{Lpmax}^4}{P_{wpDC} B_m^2} = \frac{4\rho_w W_m^2}{P_{wpDC} B_m^2} = \frac{8\rho_w W_m^2}{\alpha P_O B_m^2} \text{ (m}^5\text{)}. \tag{11.210}$$

For a transformer in the resonant circuit, the loaded quality factor is given by

$$Q_L = \omega \frac{W_m}{P_O} \tag{11.211}$$

producing

$$K_g = \frac{4\rho_w Q_L^2 P_O^2}{\omega^2 P_{wpDC} B_m^2} = \frac{4\rho_w Q_L^2 P_O}{\alpha \omega^2 B_m^2}. \tag{11.212}$$

11.7.3 Transformer for PWM Converters in CCM

The current flowing through the magnetizing inductance is given by Kazimierczuk [23]

$$i_{Lm} = I_{Lm} = I_I + \frac{I_O}{n} = \frac{I_O}{n(1-D)} \approx I_{Lmmax}, \tag{11.213}$$

where the ripple of the magnetizing current is ignored. The primary winding current i_p is the same as the switch current i_S

$$i_S = i_p = \begin{cases} i_{Lm} = I_{Lm}, & \text{for} \quad 0 < t \leq DT \\ 0, & \text{for} \quad DT < t \leq T \end{cases} \tag{11.214}$$

The rms value of the primary current is

$$I_{prms} = \sqrt{\frac{1}{T}\int_0^{DT} i_{Lm}^2 dt} = \sqrt{\frac{1}{T}\int_0^{DT} I_{Lm}^2 dt} = \sqrt{D} I_{Lm}. \tag{11.215}$$

For full load,

$$I_{prms(max)} = \sqrt{D_{max}} I_{Lmmax}. \tag{11.216}$$

Thus,

$$K_g = \frac{2D_{max}\rho_w L_m^2 I_{Lmmax}^4}{P_{wpDC} B_{pk}^2} = \frac{8D_{max}\rho_w W_m^2}{P_{wpDC} B_{pk}^2} = \frac{16D_{max}\rho_w W_m^2}{\alpha P_{Omax} B_{pk}^2}. \tag{11.217}$$

For copper, $\rho_w = \rho_{Cu} = 1.724 \times 10^{-8}$ Ωm at $T = 20$ °C. Then,

$$K_g = \frac{16 \times 1.724 \times 10^{-8} D_{max} W_m^2}{\alpha P_{Omax} B_{pk}^2} = \frac{D_{max} W_m^2}{0.03625\alpha P_{Omax} B_{pk}^2 \times 10^{-2}}. \tag{11.218}$$

For $D_{max} = 0.5$,

$$K_g = \frac{W_m^2}{0.0725\alpha P_{Omax} B_{pk}^2}. \tag{11.219}$$

11.7.4 Transformer for PWM Converters in DCM

For the flyback converter operating in DCM,

$$I_{prms} = I_{Lmmax}\sqrt{\frac{D_{max}}{3}}. \tag{11.220}$$

Hence,

$$K_g = \frac{2\rho_w L_m^2 I_{Lmmax}^2 I_{prms}^2}{P_{wpDC} B_{pk}^2} = \frac{2D_{max}\rho_w L_m^2 I_{Lmmax}^4}{3P_{wpDC} B_{pk}^2} = \frac{8D_{max}\rho_w W_m^2}{3P_{wpDC} B_{pk}^2} = \frac{16D_{max}\rho_w W_m^2}{3\alpha P_{Omax} B_{pk}^2}. \tag{11.221}$$

For copper, $\rho_w = \rho_{Cu} = 1.724 \times 10^{-8}$ Ωm at $T = 20$ °C. In this case,

$$K_g = \frac{16 \times 1.724 \times 10^{-8} \times D_{max} W_m^2}{3\alpha P_{Omax} B_{pk}^2} = \frac{D_{max} W_m^2}{0.108\alpha P_{Omax} B_{pk}^2 \times 10^{-2}} \ (\text{cm}^5). \tag{11.222}$$

11.8 Transformer Design for Flyback Converter in CCM Using K_g Method

Example 11.3

A flyback PWM converter is operated in the CCM and has the following specifications: $L = 40$ μH, $V_I = 28 \pm 4$ V, $V_O = 5$ V, $I_O = 2$–10 A, $f_s = 100$ kHz, and $V_r/V_O \leq 1\%$.

Solution: The maximum and minimum output powers are

$$P_{Omax} = V_O I_{Omax} = 5 \times 10 = 50 \text{ W} \tag{11.223}$$

and

$$P_{Omin} = V_O I_{Omin} = 5 \times 2 = 10 \text{ W}. \tag{11.224}$$

The minimum and maximum load resistances are

$$R_{Lmin} = \frac{V_O}{I_{Omax}} = \frac{5}{10} = 0.5\ \Omega \tag{11.225}$$

and

$$R_{Lmax} = \frac{V_O}{I_{Omin}} = \frac{5}{2} = 2.5\ \Omega. \tag{11.226}$$

The minimum and maximum DC voltage transfer functions are

$$M_{VDCmin} = \frac{V_O}{V_{Imax}} = \frac{5}{32} = 0.15625\ \Omega \tag{11.227}$$

and

$$M_{VDCmax} = \frac{V_O}{V_{Imin}} = \frac{5}{24} = 0.2083\ \Omega. \tag{11.228}$$

Assume the converter efficiency $\eta = 0.92$ and the maximum duty cycle $D_{max} = 0.5$. The transformer turns ratio is

$$n = \frac{\eta D_{max}}{(1 - D_{max}) M_{VDCmax}} = \frac{0.92 \times 0.5}{(1 - 0.5) \times 0.2083} = 4.4167. \tag{11.229}$$

The minimum duty cycle is

$$D_{min} = \frac{n M_{VDCmin}}{n M_{VDCmin} + \eta} = \frac{4.4167 \times 0.15625}{4.4167 \times 0.15625 + 0.92} = 0.4286. \tag{11.230}$$

The minimum magnetizing inductance required for CCM operation is

$$L_{m(min)} = \frac{n^2 R_{Lmax} (1 - D_{min})^2}{2 f_s} = \frac{4.4167^2 \times 2.5 \times (1 - 0.4286)^2}{2 \times 10^5} = 78.78\ \mu\text{H}. \tag{11.231}$$

Pick $L_p = 82\ \mu$H. The inductance of the secondary winding is

$$L_s = \frac{L_p}{n^2} = \frac{82 \times 10^{-6}}{4.4167^2} = 4.2036\ \mu\text{H}. \tag{11.232}$$

Pick $L_s = 4.7\ \mu$H.

The maximum peak-to-peak value of the magnetizing current is

$$\Delta i_{Lm(max)} = \frac{n V_O (1 - D_{min})}{f_s L_p} = \frac{4.4167 \times 5 \times (1 - 0.4286)}{10^5 \times 82 \times 10^{-6}} = 1.5387 \text{ A}. \tag{11.233}$$

The maximum DC input current is

$$I_{Imax} = \frac{P_O}{\eta V_{Imin}} = \frac{M_{VDCmax} I_{Omax}}{\eta} = \frac{0.2083 \times 10}{0.92} = 2.2641 \text{ A}. \tag{11.234}$$

The maximum peak value of the primary winding current is

$$I_{pmax} = \frac{I_{Omax}}{n(1 - D_{max})} + \frac{n V_O (1 - D_{max})}{f_s L_p}$$
$$= \frac{10}{4.4167 \times (1 - 0.5)} + \frac{4.4167 \times 5 \times (1 - 0.5)}{10^5 \times 82 \times 10^{-6}} = 5.874 \text{ A}. \tag{11.235}$$

The maximum rms value of the primary winding current is

$$I_{prms(max)} = \frac{I_{Omax}\sqrt{D_{max}}}{n(1 - D_{max})} = \frac{10\sqrt{0.5}}{4.4167 \times (1 - 0.5)} = 3.202 \text{ A}. \tag{11.236}$$

The maximum energy stored in the magnetic field of the transformer is

$$W_m = \frac{1}{2}L_m I_{pmax}^2 = \frac{1}{2} \times 82 \times 10^{-6} \times 5.874^2 = 1.414 \text{ mJ}. \quad (11.237)$$

Assume $B_{pk} = 0.25$ T, $D_{max} = 0.5$, and $\alpha = 0.01$. The core geometry coefficient is

$$K_g = \frac{D_{max} W_m^2}{0.03625 \alpha B_{pk}^2 P_{Omax} \times 10^{-2}} (\text{cm})^5 = \frac{0.5 \times (1.414 \times 10^{-3})^2}{0.03625 \times 0.01 \times 0.25^2 \times 50 \times 10^{-2}}$$
$$= 0.08825 \text{ cm}^5. \quad (11.238)$$

We shall select a Magnetics PQ43230 core with $K_g = 0.0996$ cm^5, $A_p = 1.6$ cm^4, $A_c = 1.67$ cm^2, $l_c =$ 7.47 cm, and $V_c = 12.5$ cm^3. We will choose a ferrite P material with $B_s = 0.5$ T, $\mu_{rc} = 2500 \pm 25\%$, and $A_L = 3810$ mH/1000 turns.

The core window area is

$$W_a = \frac{A_p}{A_c} = \frac{1.6}{1.67} = 0.958 \text{ cm}^2. \quad (11.239)$$

The maximum amplitude of the current density of the primary and secondary windings is

$$J_m = \frac{4W_m}{K_u A_p B_{pk}} = \frac{4 \times 1.414 \times 10^{-3}}{0.3 \times 1.6 \times 10^{-8} \times 0.25} = 4.713 \times 10^6 \text{ A/m}^2 = 4.713 \text{ A/mm}^2. \quad (11.240)$$

The copper skin depth at $f_s = 100$ kHz is

$$\delta_w = \frac{66.2}{\sqrt{f_s}} (\text{mm}) = \frac{66.2}{\sqrt{10^5}} \text{ mm} = 0.209 \text{ mm}. \quad (11.241)$$

The diameter of the bare strands is

$$d_{is} = 2\delta_w = 2 \times 0.209 = 0.418 \text{ mm}. \quad (11.242)$$

The primary bare wire area is

$$A_{wp} = \frac{I_{pmax}}{J_m} = \frac{5.874}{4.713} = 1.24 \text{ mm}^2. \quad (11.243)$$

We shall select the AWG26 strand wire with the bare wire diameter $d_{is} = 0.405$ mm, $d_{os} = 0.452$ mm, the bare wire cross-sectional area $A_{wps} = 0.128$ mm^2, and $R_{wpDCs}/l_w = 0.1345$ Ω/m. The number of strands of the primary winding is

$$S_p = \frac{A_{wp}}{A_{wps}} = \frac{1.24}{0.128} = 9.68. \quad (11.244)$$

Pick $S_p = 9$. The window area allocated to the primary winding is

$$W_{ap} = \frac{W_a}{2} = \frac{0.958}{2} = 0.479 \text{ cm}^2. \quad (11.245)$$

The cross-sectional area of the insulated strands is

$$A_{wpos} = \frac{\pi d_{os}^2}{4} = \frac{\pi \times (0.452 \times 10^{-3})^2}{4} = 0.1604 \text{ mm}^2. \quad (11.246)$$

The number of turns of the primary winding is

$$N_p = \frac{K_u W_{ap}}{S_p A_{wpos}} = \frac{0.3 \times 0.479 \times 10^{-4}}{9 \times 0.1604 \times 10^{-6}} = 9.95. \quad (11.247)$$

Pick $N_p = 9$. The number of turns of the secondary winding is

$$N_s = \frac{N_p}{n} = \frac{9}{4.4167} = 2.03. \quad (11.248)$$

Pick $N_s = 2$. The length of the air gap is

$$l_g = \frac{\mu_0 A_c N_p^2}{L_p} - \frac{l_c}{\mu_{rc}} = \frac{4\pi \times 10^{-7} \times 1.67 \times 10^{-4} \times 9^2}{82 \times 10^{-6}} - \frac{7.47 \times 10^{-2}}{2500} = 0.177 \text{ mm}. \quad (11.249)$$

The MTL is

$$l_T = \frac{\pi(F+E)}{2} = \frac{\pi(13.5+27.5)}{2} = 64.4 \text{ mm}. \quad (11.250)$$

The length of the primary winding is

$$l_{wp} = N_p l_T = 9 \times 6.44 = 57.96 \text{ cm} = 57.96 \text{ m}. \quad (11.251)$$

Pick $l_{wp} = 0.6$ m. The DC and low-frequency resistance of each strand of the primary winding is

$$R_{wpDCs} = \left(\frac{R_{wpDCs}}{l_w}\right) l_{wp} = 0.1345 \times 0.6 = 0.0807\ \Omega. \quad (11.252)$$

Thus, the DC and the low-frequency resistance of the primary winding is

$$R_{wpDC} = \frac{R_{wpDCs}}{S_p} = \frac{0.0807}{9} = 0.00896\ \Omega. \quad (11.253)$$

The DC and low-frequency power loss in the primary is

$$P_{wpDC} = R_{wpDC} I_{Imax}^2 = 0.00896 \times 2.2641^2 = 0.0459 \text{ W}. \quad (11.254)$$

From (7.234), we compute $F_{Rph} = 2.77$. Hence, the primary winding power loss is

$$P_{wp} = F_{Rph} P_{wpDC} = 2.77 \times 0.0459 = 0.1271 \text{ W}. \quad (11.255)$$

The maximum peak value of the magnetic flux density is

$$B_{pk} = \frac{\mu_0 N_p I_{pmax}}{l_g + \frac{l_c}{\mu_{rc}}} = \frac{4\pi \times 10^{-7} \times 9 \times 5.874}{0.177 \times 10^{-3} + \frac{7.47 \times 10^{-2}}{2500}} = 0.3211 \text{ T}. \quad (11.256)$$

The maximum peak value of the magnetic flux density is

$$B_m = \frac{\mu_0 N_p \frac{\Delta i_{Lpmax}}{2}}{l_g + \frac{l_c}{\mu_{rc}}} = \frac{4\pi \times 10^{-7} \times 9 \times \frac{1.5387}{2}}{0.177 \times 10^{-3} + \frac{7.47 \times 10^{-2}}{2500}} = 0.04205 \text{ T}. \quad (11.257)$$

The core power loss density is

$$P_v = 0.0434 f^{1.63} (10 B_m)^{2.62} = 0.0434 \times 100^{1.63} \times (10 \times 0.04205)^{2.62} = 8.161 \text{ mW/cm}^3. \quad (11.258)$$

The core loss is

$$P_C = V_c P_v = 12.5 \times 8.161 \times 10^{-3} = 102.012 \text{ mW}. \quad (11.259)$$

The maximum current through the secondary winding is

$$I_{smax} = \frac{I_{Omax}}{1 - D_{max}} + \frac{n \Delta i_{Lpmax}}{2} = \frac{10}{1 - 0.5} + \frac{4.4167 \times 1.5387}{2} = 23.398 \text{ A}. \quad (11.260)$$

The maximum rms value of the secondary winding current is

$$I_{srms(max)} = \frac{I_{Omax}}{\sqrt{1 - D_{max}}} = \frac{10}{\sqrt{1 - 0.5}} = 14.142 \text{ A}. \quad (11.261)$$

The cross-sectional area of the total secondary winding wire is

$$A_{ws} = \frac{I_{smax}}{J_m} = \frac{23.398}{4.713} = 4.964 \text{ mm}^2. \quad (11.262)$$

The number of strands of the secondary winding is

$$S_s = \frac{A_{ws}}{A_{wst}} = \frac{4.964 \times 10^{-6}}{0.128 \times 10^{-6}} = 38.78. \quad (11.263)$$

Pick $S_s = 38$. The length of the secondary winding is

$$l_{ws} = N_s l_T = 2 \times 6.44 \times 10^{-2} = 0.1288 \text{ m}. \quad (11.264)$$

Pick $l_{ws} = 14$ cm. The DC and low-frequency resistance of each strand of the secondary winding is

$$R_{wsDCs} = \left(\frac{R_{wsDCs}}{l_w}\right) l_{ws} = 0.1345 \times 0.14 = 0.01883\ \Omega. \quad (11.265)$$

The DC and low-frequency resistance of all strands in the secondary winding is

$$R_{wsDC} = \frac{R_{wsDCs}}{S_s} = \frac{0.01883}{38} = 0.0004955\ \Omega. \tag{11.266}$$

The DC and low-frequency power loss in the secondary winding is

$$P_{wsDC} = R_{wsDC} I^2_{Omax} = 0.0004955 \times 10^2 = 0.04955\ \text{W}. \tag{11.267}$$

Using (7.238), we calculate $F_{Rsh} = 2.77$. The secondary winding power loss for $N_{ls} = 1$ is

$$P_{ws} = F_{Rsh} P_{wsDC} = 2.77 \times 0.09077 = 0.25\ \text{W}. = 2.77 \times 0.04955 = 0.1372\ \text{W}. \tag{11.268}$$

The DC and low-frequency power loss in both windings is

$$P_{wDC} = P_{wpDC} + P_{wsDC} = 0.0459 + 0.04955 = 0.09545\ \text{W}. \tag{11.269}$$

The power loss in both the windings is

$$P_w = P_{wp} + P_{ws} = 0.1271 + 0.1372 = 0.2643\ \text{W}. \tag{11.270}$$

The sum of the core loss and the resistance winding loss in the transformer is

$$P_{cw} = P_C + P_w = 0.102 + 0.2643 = 0.3663\ \text{W}. \tag{11.271}$$

The transformer efficiency at full power is

$$\eta_t = \frac{P_O}{P_O + P_{cw}} = \frac{50}{50 + 0.3663} = 99.27\%. \tag{11.272}$$

The total surface area of the core is

$$A_t = 2\left(\frac{\pi A^2}{4}\right) + (\pi A)(2B) = 2\left(\frac{\pi \times 3.3^2}{4}\right) + \pi \times 3.3 \times 1.515 = 32.81\ \text{cm}^2, \tag{11.273}$$

where A is the outer diameter of the core and $2B$ is the total height of both halves of the core. The surface power loss density is

$$\psi = \frac{P_{cw}}{A_t} = \frac{0.3663}{32.81} = 0.01116\ \text{W/cm}^2. \tag{11.274}$$

The temperature rise of the inductor (the core and the winding) is

$$\Delta T = 450\psi^{0.826} = 450 \times 0.01116^{0.826} = 10.97\ ^\circ\text{C}. \tag{11.275}$$

The core window utilization is

$$K_u = \frac{(N_p S_p + N_s S_s) A_{wpos}}{W_a} = \frac{(9 \times 9 + 2 \times 38) \times 0.1604 \times 10^{-6}}{0.958 \times 10^{-4}} = 0.2628. \tag{11.276}$$

The α parameter is

$$\alpha = \frac{P_{wDC}}{P_O} = \frac{0.09545}{50} = 0.0019. \tag{11.277}$$

11.9 Transformer Design for Flyback Converter in DCM Using K_g Method

---

Example 11.4

A flyback PWM converter is operated in the DCM and has the following specifications: $V_I = 28 \pm 4$ V, $V_O = 5$ V, $I_O = 0$–10 A, and $f_s = 100$ kHz.

Solution: Figure 11.18 shows the waveforms in the flyback converter for DCM. The maximum and minimum output powers are

$$P_{Omax} = V_O I_{Omax} = 5 \times 10 = 50 \text{ W} \tag{11.278}$$

and

$$P_{Omin} = V_O I_{Omin} = 5 \times 0 = 0 \text{ W}. \tag{11.279}$$

The minimum and maximum load resistances are

$$R_{Lmin} = \frac{V_O}{I_{Omax}} = \frac{5}{10} = 0.5 \ \Omega \tag{11.280}$$

and

$$R_{Lmax} = \frac{V_O}{I_{Omin}} = \frac{5}{0} = \infty. \tag{11.281}$$

The minimum and maximum DC voltage transfer functions are

$$M_{VDCmin} = \frac{V_O}{V_{Imax}} = \frac{5}{32} = 0.15625 \ \Omega \tag{11.282}$$

and

$$M_{VDCmax} = \frac{V_O}{V_{Imin}} = \frac{5}{24} = 0.2083 \ \Omega. \tag{11.283}$$

Assume the converter efficiency $\eta = 0.88$ and the maximum duty cycle $D_{max} = 0.5$. The transformer turns ratio is

$$n = \frac{\eta D_{max}}{(1 - D_{max}) M_{VDCmax}} = \frac{0.88 \times 0.5}{(1 - 0.5) \times 0.2083} = 4.2247. \tag{11.284}$$

The magnetizing inductance is

$$L_{m(max)} = \frac{n^2 R_{Lmin} (1 - D_{Bmax})^2}{2 f_s} = \frac{4.2247^2 \times 0.5 \times (1 - 0.5)^2}{2 \times 10^5} = 11.15 \ \mu\text{H}. \tag{11.285}$$

Let $L_m = 10\ \mu$H.

Assume the converter efficiency $\eta = 0.88$. The maximum duty cycle at full load is

$$D_{max} = M_{VDCmax} \sqrt{\frac{2 f_s L_m}{\eta R_{Lmin}}} = 0.2083 \sqrt{\frac{2 \times 10^5 \times 10 \times 10^{-6}}{0.88 \times 0.5}} = 0.444. \tag{11.286}$$

The minimum duty cycle at full load is

$$D_{min} = M_{VDCmin} \sqrt{\frac{2 f_s L_m}{\eta R_{Lmin}}} = 0.15625 \sqrt{\frac{2 \times 10^5 \times 10 \times 10^{-6}}{0.88 \times 0.5}} = 0.333. \tag{11.287}$$

The maximum duty cycle when the diode is ON at full load is

$$D_{1max} = \sqrt{\frac{2 f_s L_m}{n^2 R_{Lmin}}} = \sqrt{\frac{2 \times 10^5 \times 10 \times 10^{-6}}{4.2247^2 \times 0.5}} = 0.4734. \tag{11.288}$$

Hence,

$$D_{max} + D_{1max} = 0.444 + 0.4734 = 0.9174 < 1. \tag{11.289}$$

The maximum current through the primary winding is

$$I_{pmax} = \Delta i_{Lm(max)} = \frac{D_{min} V_{Imax}}{f_s L_m} = \frac{0.333 \times 32}{10^5 \times 10 \times 10^{-6}} = 10.656 \text{ A}. \tag{11.290}$$

The maximum DC input current is

$$I_{Imax} = \frac{P_{Omax}}{\eta V_{Imax}} = \frac{M_{VDCmax} I_{Omax}}{\eta} = \frac{0.2083 \times 10}{0.88} = 2.367 \text{ A}. \tag{11.291}$$

The maximum rms value of the current through the primary winding is

$$I_{prms(max)} = I_{pmax} \sqrt{\frac{D_{max}}{3}} = 10.656 \sqrt{\frac{0.444}{3}} = 4.1 \text{ A}. \tag{11.292}$$

The maximum energy stored in the magnetic field of the transformer is

$$W_m = \frac{1}{2}L_m I_{pmax}^2 = \frac{1}{2} \times 10 \times 10^{-6} \times 10.656^2 = 0.5677 \text{ mJ}. \quad (11.293)$$

Assume $B_{pk} = 0.21$ T and $\alpha = 0.002$. The core geometrical coefficient is

$$K_g = \frac{D_{max} W_m^2}{0.108\alpha P_{Omax} B_{pk}^2 \times 10^{-2}} = \frac{0.444 \times (0.5677 \times 10^{-3})^2}{0.108 \times 0.002 \times 50 \times 0.21^2 \times 10^{-2}} = 0.03 \text{ cm}^4. \quad (11.294)$$

A Magnetics ferrite core PQ 0P-42625 is selected, which has $K_g = 0.03445$ cm^5, $A_p = 0.59$ cm^4, $A_c = 1.2$ cm^2, $l_c = 5.43$ cm, $V_c = 6.53$ cm^3, $A_L = 3750$ nH/turn, $\mu_{rc} = 2500 \pm 25\%$, and $B_s = 0.5$ T. The P-type magnetic material is used to make this core. The coefficients of this material are $k = 0.0434$, $a = 1.63$, and $b = 2.62$. The core window area is

$$W_a = \frac{A_p}{A_c} = \frac{0.59}{1.2} = 0.4916 \text{ cm}^2. \quad (11.295)$$

The maximum amplitude of the current density of the primary and secondary windings is

$$J_m = \frac{4W_m}{K_u A_p B_{pk}} = \frac{4 \times 0.5677 \times 10^{-3}}{0.3 \times 0.59 \times 10^{-8} \times 0.21} = 6.11 \times 10^6 \text{ A/m}^2 = 6.11 \text{ A/mm}^2. \quad (11.296)$$

The copper skin depth at $f_s = 100$ kHz is

$$\delta_w = \frac{66.2}{\sqrt{f_s}} \text{(mm)} = \frac{66.2}{\sqrt{10^5}} = 0.209 \text{ mm}. \quad (11.297)$$

The diameter of the bare strands is

$$d_{is} = 2\delta_w = 2 \times 0.209 = 0.418 \text{ mm}. \quad (11.298)$$

The primary bare wire area is

$$A_{wp} = \frac{I_{pmax}}{J_m} = \frac{10.656}{6.11 \times 10^{-6}} = 1.744 \text{ mm}^2. \quad (11.299)$$

We shall select the AWG26 strand wire with the bare wire diameter $d_{is} = 0.405$ mm, $d_{os} = 0.452$ mm, the bare wire cross-sectional area $A_{wps} = 0.128$ mm^2, and $R_{wpDCs}/l_w = 0.1345$ Ω/m. The number of strands of the primary winding is

$$S_p = \frac{A_{wp}}{A_{wps}} = \frac{1.744}{0.128} = 13.625. \quad (11.300)$$

Pick $S_p = 13$. The window area allocated to the primary winding is

$$W_{ap} = \frac{W_a}{2} = \frac{0.4916}{2} = 0.2458 \text{ cm}^2. \quad (11.301)$$

The cross-sectional area of the insulated strands is

$$A_{wpos} = \frac{\pi d_{os}^2}{4} = \frac{\pi \times (0.452 \times 10^{-3})^2}{4} = 0.1604 \text{ mm}^2. \quad (11.302)$$

The number of turns of the primary winding is

$$N_p = \frac{K_u W_{ap}}{S_p A_{wpos}} = \frac{0.3 \times 0.2458 \times 10^{-4}}{13 \times 0.1604 \times 10^{-6}} = 3.53. \quad (11.303)$$

Pick $N_p = 4$. The number of turns of the secondary winding is

$$N_s = \frac{N_p}{n} = \frac{4}{4.2247} = 0.9468. \quad (11.304)$$

Pick $N_s = 1$. The length of the air gap is

$$l_g = \frac{\mu_0 A_c N_p^2}{L_p} - \frac{l_c}{\mu_{rc}} = \frac{4\pi \times 10^{-7} \times 1.2 \times 10^{-4} \times 4^2}{10 \times 10^{-6}} - \frac{5.43 \times 10^{-2}}{2500} = 0.219 \text{ mm}. \quad (11.305)$$

The MTL is

$$l_T = \frac{\pi(F+E)}{2} = \frac{\pi(12+22.5)}{2} = 54.19 \text{ mm}. \quad (11.306)$$

The length of the primary winding is

$$l_{wp} = N_p l_T = 4 \times 5.419 \times 10^{-2} = 0.21676 \text{ m}. \quad (11.307)$$

Pick $l_{wp} = 0.22$ m. The DC and low-frequency resistance of each strand of the primary winding is

$$R_{wpDCs} = \left(\frac{R_{wpDCs}}{l_w}\right) l_{wp} = 0.1345 \times 0.22 = 0.0296 \ \Omega. \quad (11.308)$$

The DC and the low-frequency resistance of the primary winding is

$$R_{wpDC} = \frac{R_{wpDCs}}{S_p} = \frac{0.296}{13} = 0.002277 \ \Omega. \quad (11.309)$$

The DC and low-frequency power loss in the primary is

$$P_{wpDC} = R_{wpDC} I_{Imax}^2 = 0.002277 \times 2.367^2 = 0.01275 \text{ W}. \quad (11.310)$$

The harmonic AC resistance factor is $F_{Rph} = 5.07$ for $N_{lp} = 1$, $D = 0.444$, and $D_{1max} = 0.4734$. Thus, the power loss in the primary winding is

$$P_{wp} = F_{Rph} P_{wDC} = 5.07 \times 0.01275 = 0.0646 \text{ W}. \quad (11.311)$$

The maximum peak value of the magnetic flux density is

$$B_{pk} = \frac{\mu_0 N_p I_{pmax}}{l_g + \frac{l_c}{\mu_{rc}}} = \frac{4\pi \times 10^{-7} \times 4 \times 10.656}{0.219 \times 10^{-3} + \frac{5.43 \times 10^{-2}}{2500}} = 0.2225 \text{ T}. \quad (11.312)$$

The maximum peak value of the magnetic flux density is

$$B_m = \frac{\mu_0 N_p \frac{\Delta i_{pmax}}{2}}{l_g + \frac{l_c}{\mu_{rc}}} = \frac{4\pi \times 10^{-7} \times 4 \times \frac{10.656}{2}}{0.219 \times 10^{-3} + \frac{5.43 \times 10^{-2}}{2500}} = 0.11125 \text{ T}. \quad (11.313)$$

The core power loss density is

$$P_v = 0.0434 f^{1.63} (10 B_m)^{2.62} = 0.0434 \times 100^{1.63} (10 \times 0.11125)^{2.62} = 104.42 \text{ mW/cm}^3. \quad (11.314)$$

The core loss is

$$P_C = V_c P_v = 6.53 \times 104.42 \times 10^{-3} = 681.86 \text{ mW}. \quad (11.315)$$

The maximum current through the secondary winding is

$$I_{smax} = n \Delta i_{Lmmax} = 4.2247 \times 10.656 = 45.018 \text{ A}. \quad (11.316)$$

The maximum rms value of the secondary winding current is

$$I_{srms(max)} = I_{smax} \sqrt{\frac{D_{1max}}{3}} = 45.018 \sqrt{\frac{0.4734}{3}} = 17.883 \text{ A}. \quad (11.317)$$

The cross-sectional area of the total secondary winding wire is

$$A_{ws} = \frac{I_{smax}}{J_m} = \frac{45.018}{6.11} = 7.36 \text{ mm}^2. \quad (11.318)$$

The number of strands of the secondary winding is

$$S_s = \frac{A_{ws}}{A_{wps}} = \frac{7.36 \times 10^{-6}}{0.128 \times 10^{-6}} = 57.5. \quad (11.319)$$

Pick $S_s = 60$. The length of the secondary winding is

$$l_{ws} = N_s l_T = 1 \times 5.419 \times 10^{-2} = 0.05419 \text{ m}. \quad (11.320)$$

The DC and low-frequency resistance of each strand of the secondary winding is

$$R_{wsDCs} = \left(\frac{R_{wsDCs}}{l_w}\right) l_{ws} = 0.05419 \times 0.01345 = 0.00729\ \Omega. \quad (11.321)$$

The DC and low-frequency resistance of all strands in the secondary winding is

$$R_{wsDC} = \frac{R_{wsDCs}}{S_s} = \frac{0.00729}{60} = 0.1215 \times 10^{-3}\ \Omega. \quad (11.322)$$

The DC and low-frequency power loss in the secondary winding is

$$P_{wsDC} = R_{wsDC} I_{Omax}^2 = 0.1215 \times 10^{-3} \times 10^2 = 0.01215\ \text{W}. \quad (11.323)$$

The DC and low-frequency power loss in both windings is

$$P_{wDC} = P_{wpDC} + P_{wsDC} = 0.01275 + 0.01215 = 0.0249\ \text{W}. \quad (11.324)$$

The sum of the core loss and the DC resistance winding loss in the transformer is

$$P_{cwDC} = P_C + P_{wDC} = 0.68186 + 0.0249 = 0.7065\ \text{W}. \quad (11.325)$$

The harmonic AC resistance factor is $F_{Rsh} = 5.7$ for $N_{ls} = 1$, $D = 0.444$, and $D_{1max} = 0.4734$. Hence, the AC resistance of the secondary winding is

$$P_{ws} = F_{Rsh} P_{wsDC} = 5.7 \times 0.01215 = 0.0692\ \text{W}. \quad (11.326)$$

The total AC winding power loss is

$$P_w = P_{wp} + P_{ws} = 0.0646 + 0.0692 = 0.1338\ \text{W}. \quad (11.327)$$

The total loss in the transformer is

$$P_{cw} = P_C + P_w = 0.6816 + 0.1338 = 0.8154\ \text{W}. \quad (11.328)$$

The transformer efficiency at full power is

$$\eta_t = \frac{P_O}{P_O + P_{cw}} = \frac{50}{50 + 0.8154} = 98.39\%. \quad (11.329)$$

The total surface area of the core is

$$A_t = 2\left(\frac{\pi A^2}{4}\right) + (\pi A)(2B) = 2\left(\frac{\pi \times 2.73^2}{4}\right) + \pi \times 2.73 \times 1.235 = 22.3\ \text{cm}^2, \quad (11.330)$$

where A is the outer diameter of the core and $2B$ is the total height of both halves of the core. The surface power loss density is

$$\psi = \frac{P_{cw}}{A_t} = \frac{0.8154}{22.3} = 0.0365\ \text{W/cm}^2. \quad (11.331)$$

The temperature rise of the inductor (the core and the winding) is

$$\Delta T = 450\psi^{0.826} = 450 \times 0.0365^{0.826} = 29.21\ ^\circ\text{C}. \quad (11.332)$$

The core window utilization is

$$K_u = \frac{(N_p S_p + N_s S_s) A_{wpo}}{W_a} = \frac{(4 \times 13 + 1 \times 60) \times 0.1604 \times 10^{-6}}{0.4916 \times 10^{-4}} = 0.3654. \quad (11.333)$$

The α factor is

$$\alpha = \frac{P_{wDC}}{P_O} = \frac{0.0249}{50} = 0.00498. \quad (11.334)$$

11.10 Summary

- To reduce the DC and low-frequency winding loss, the allocation of the core window to the transformer winding should be proportional to the ampere-turn N_kI_k of each winding and to the apparent power V_kI_k of each winding.
- In a two-winding transformer, 50% of the core window area should be allocated to the primary winding and 50% to the secondary winding to achieve the minimum value of the total DC and low-frequency copper loss.
- If the harmonic AC resistance factors are equal ($F_{Rph} = F_{Rsh}$) for a two-winding transformer, then the minimum value of the total DC and high-frequency winding power loss is achieved for equal allocation of core window space for the primary and secondary windings. Otherwise, this rule does not hold true.
- The core area product A_p provides crude initial estimate of core size.
- The primary winding should be placed the closest to the bobbin to reduce the EMI noise radiation.
- The innermost layer of the primary winding should be connected to the drain of a power MOSFET to maximize the shielding effect by the secondary windings.
- The maximum rms value of the current density is usually made equal in each winding to provide uniform heat generation throughout the window.
- In a multiple-output transformer, the secondary winding with the highest output power should be placed closest to the primary for the best coupling and the lowest leakage inductance.
- The secondary winding with a few turns should be spaced along the width of the bobbin to maximize the coupling to the primary winding.
- The secondary winding with only a few turns should be made of multiple parallel strands to maximize the coupling to the primary winding and to increase core window utilization.
- If the secondary winding has only a few turns and it is made of strands, the strands should form a single layer spaced along the bobbin breadth.

11.11 References

[1] C. W. T. McLyman, *Transformer and Inductor Design Handbook*, 3rd Ed., New York: Marcel Dekker, 2004.
[2] C. W. T. McLyman, *Magnetic Core Selection for Transformers and Inductors*, 2nd Ed., New York: Marcel Dekker, 1997.
[3] C. W. T. McLyman, *High Reliability Magnetic Devices, Design and Fabrication*, New York: Marcel Dekker, 2002.
[4] A. I. Pressman, *Switching Power Supplies*, New York: McGraw-Hill, 1991.
[5] R. W. Erickson and D. Maksimović, *Fundamentals of Power Electronics*, 2nd Ed., Norwell, MA: Kluwer Academic Publishers, 2001.
[6] M. Ivancovic, "Optimum SMPS transformer design," Proceedings of PCI, 1986, pp. 183–188.
[7] A. Estrov, "Transformer design for 1 MHz resonant converter," Proceedings of High Frequency Power Conversion Conference, 1986, pp. 36–54.
[8] N. R. Coonrod, "Transformer computer design aid for high frequency switching power supplies," *IEEE Transactions on Power Electronics*, vol. 2, no. 4, pp. 248–256, October 1986.
[9] J. Hendrix, "Optimising transformers for power conversion," Proceedings of SATECH, 1987, pp. 324–335.
[10] K. Sakakibara and N. Murami, "Analysis of high frequency resistance in transformers," Proceedings of 20th Annual IEEE Power Electronics Specialists Conference, vol. 2, pp. 618–624, Milwaukee, WI, June 26-29, 1989.
[11] J. Spreen, "Electrical thermal representation of conductor loss in transformers," *IEEE Transaction on Power Electronics*, vol. 5, no. 4, pp. 424–429, October 1990.

[12] R. Severns, "Additional losses in high frequency magnetics due to non-ideal field distribution," 7th Annual IEEE Applied Power Electronics Conference, Boston, MA, February 23–27, 1992, pp. 333–338.
[13] W.-J. Gu and R. Liu, "A study of volume and weight vs. frequency for high-frequency transformers," 24th Annual IEEE Power Electronics Specialists Conference, Seattle, WA, June 20–24, 1993, pp. 1123–1129.
[14] A. M. Pernia, F. Nuno, and J. M. Lopera, "1D/2D transformer electric model for simulation in power converters," 27th Annual IEEE Power Electronics Specialists Conference, Atlanta, GA, June 18–22, 1995, vol. 2, pp. 1043–1049.
[15] W. A. Roshen, R. L. Steigerwald, R. Charles, W. Earls, G. Claydon, and C. F. Saj, "High-efficiency, high-density MHz magnetic components for a low profile converter," *IEEE Transactions on Industry Applications*, vol. 31, no. 4, pp. 869–878, July/August 1995 with; also in IEEE Applied Power Electronics Conference, 1992, pp. 674–683.
[16] R. Prieto, J. A. Cobos, O. Garcia, P. Alou, and J. Uceda, "High-frequency resistance in flyback type transformer," 15th Annual IEEE Applied Power Electronics Conference, New Orleans, LA, February 6–10, 2000, vol. 2, pp. 714–719.
[17] L. Dixon, *Transformer and inductor design for optimum circuit performance*, 2002 Power Supply Design Seminar, SEM1500, Texas Instruments, 2002.
[18] D. Lavers and E. Lavers, "Waveform dependent switching losses in flyback transformer foil windings," IEEE APEC, 2002, pp. 1236–1241.
[19] Z. Y. Lu and W. Chen, "Novel winding loss analytical model of flyback transformer," Proceedings of 33rd Annual IEEE Power Electronics Specialists Conference, June 18–22, 2006, pp. 1213–1218.
[20] K. Billings, *Switchmode Power Supply Handbook*, 2nd Ed., Boston, MA: McGraw-Hill, 2008.
[21] C. P. Basso, *Switch-mode Power Supplies*, New York: McGraw-Hill, 2008.
[22] A. V. Bossche and V. C. Valchev, *Inductors and Transformers for Power Electronics*, Taylor & Francis, Boca Raton, FL: CRC, 2005.
[23] M. K. Kazimierczuk, *Pulse-width Modulated DC-DC Power Converters*, New York: John Wiley & Sons, 2008.
[24] M. K. Kazimierczuk, *RF Power Amplifiers*, New York: John Wiley & Sons, 2008.
[25] F. Forest, E. Labouré, T. Meynard, and M. Arab, "Analytic design method based on homothetic shape of magnetic core for high-frequency transformers," *IEEE Transaction on Power Electronics*, vol. 22, no. 5, pp. 2070–2080, September 2007.
[26] D. Murthy-Bellur and M. K. Kazimierczuk, "Winding loss caused by harmonics in high-frequency flyback transformer for pulse-width modulated DC-DC converters in discontinuous conduction mode," *Proceedings of IET, Power Electronics*, vol. 3, no. 5, pp. 804–817, May 2011.
[27] www.mag-inc.com, www.micrometals.com, www.ferroxcube.com, www.ferrite.de, www.metglas.com, and www.eilor.co.il.

11.12 Review Questions

11.1. Explain the principle of allocation of the transformer windings in the core window.

11.2. What is the best location of the primary winding in a transformer?

11.3. Which layer of the primary winding should be connected to the drain of the power MOSFET?

11.4. How can the coupling of the secondary winding to the primary winding be maximized when the secondary winding has only a few turns?

11.5. What is the best geometry of the secondary winding made of strands?

11.6. Is the effect of harmonics on the transformer winding resistance significant?

11.7. What is the optimum allocation of the core window area for the primary and secondary windings in a two-winding transformer?

11.13 Problems

11.1. Design a transformer for the flyback PWM DC–DC converter to meet the following specifications: $V_I = 100–200$ V, $V_O = 3.3$ V, $I_O = 0.5–1.5$ A, and $f_s = 250$ kHz.

11.2. Design a transformer for the flyback PWM DC–DC converter to meet the following specifications: $V_I = 100–200$ V, $V_O = 3.3$ V, $I_O = 0–1.5$ A, and $f_s = 250$ kHz.

11.3. Design a transformer for the flyback PWM DC–DC converter used as a universal power supply to meet the following specifications: $V_I = 120–374$ V, $V_O = 5$ V, $I_O = 1–10$ A, and $f_s = 100$ kHz.

11.4. Design a transformer for the flyback PWM DC–DC converter used as a universal power supply to meet the following specifications: $V_I = 120–374$ V, $V_O = 15$ V, $I_O = 0–2$ A, and $f_s = 100$ kHz.

Answers to Problems

CHAPTER 1

1.1 $B = 0$.

1.3 $L = 55.45$ μH.

1.4 $H = 200$ A/m, $\phi = 0.2$ mWb, $\lambda = 60$ mWb.

1.5 $B = 1$ T, $H = 994.7$ A/m, $I = 0.313$ A.

1.6 (a) $N = 100$. (b) $N = 20$. (c) $N = 40$. (d) $N = 35$.

1.7 $N = 18$.

1.8 $L = 200$ mH, $A_L = 0.8$ μH/turn, $R = 1.25\ N^2/\mu\text{H}$, $H_m = 3978.87$ A/m, $B_m = 0.9997$ T, $\phi_m = 0.399$ mWb, $\lambda_m = 0.19998$ V · s.

1.9 (a) $\mathcal{R}_g = 1.22 \times 10^6$ A · turns/Wb, $\mathcal{R}_c = 6.75 \times 10^4$ A · turns/Wb, $\mathcal{R} = 122.25 \times 10^4$ A · turns/Wb.

(b) $B_g = 50.9$ mT, $H_g = 40.5$ kA/m, $H_c = 13.5$ A/m.

(c) $I = 257.5$ mA.

1.10 $\lambda(t) = (V_{Lm}/\omega) \sin \omega t, f_{min} = 530.52$ Hz.

1.11 $L = \frac{\mu_0 l_w}{4\pi} + \frac{\mu_0 l_w}{\pi} \ln \left(\frac{d}{a}\right)$.

1.12 $L_{new} = 400$ μH.

1.13 (b) $\mathcal{R}_g = 1\,989\,436.8$ turns/H, $\mathcal{R}_1 = 65\,651$ turns/H, $\mathcal{R}_2 = 2\,121\,245.68$ turns/H, $\mathcal{R} = 2\,207\,245.68$ turns/H.

(c) $\phi_{m1} = 94.3$ μWb, $\phi_{m2} = 47.14$ μWb.

(d) $B_{m1} = 0.2357$ T, $B_{m2} = 0.1178$ T.

(e) $H_{mg} = 187\,572.56$ A/m, $H_{m1} = 62.52$ A/m, $H_{m2} = 31.247$ A/m.

(f) $L = 4.53$ mH.

1.14 (a) $\mathcal{R} = 6\,354\,651.88$ turns/H.

(b) $L = 685.48$ μH.

High-Frequency Magnetic Components, Second Edition. Marian K. Kazimierczuk.

Companion Website: www.wiley.com/go/kazimierczuk_High2e

1.15 (a) $B_c(t) = B_g(t) = 41.44 \sin \omega t (\text{mT})$.
(b) $H_c(t) = 33 \sin \omega t (\text{kA/m})$.
(e) $W_{c(max)} = 0.144\ \mu\text{J}$, $w_{c(max)} = 6.78\ \text{mJ/m}^3$.
(f) $W_{g(max)} = 0.84834\ \mu\text{J}$, $w_{g(max)} = 678.6\ \text{J/m}^3$.

1.16 (a) $F_f = 1.196$.
(b) $L = 686.6\ \mu\text{H}$.
(c) $L_f = 821.2\ \mu\text{H}$.

1.17 (a) $B_{cm} = B_{gm} = B_s = 1.5\ \text{T}$.
(b) $H_{cm} = 1.194\ \text{A/m}$. $H_{gm} = 1.194\ \text{MA/m}$.
(c) $w_{c(max)} = 0.89575\ \text{J/m}^3$. $w_{g(max)} = 0.89575\ \text{MJ/m}^3$.
(d) $W_{c(max)} = 19.1\ \mu\text{J}$. $W_{g(max)} = 618.3\ \text{J/m}^3$.
(e) $I_{Lm(max)} = 9.046\ \text{A}$.

1.18 (a) $Q_{Lo} = 418.879$.
(b) $f_r = 5.032\ \text{MHz}$.

CHAPTER 2

2.1 $\mu_{rc} = 2546.479$.

2.2 (a) $B = 2\ \text{T}$.
(b) $B = 0.5\ \text{T}$.

2.3 $\chi = 1799$.

2.4 (a) $P_v = 120\ \text{mW/cm}^3$, $P_C = 240\ \text{mW}$.
(b) $P_v = 767\ \text{mW/cm}^3$, $P_C = 1.534\ \text{W}$.
(c) $P_v = 2.27\ \text{mW/cm}^3$, $P_C = 4.54\ \text{W}$.
(d) $P_v = 4.92\ \text{mW/cm}^3$, $P_C = 9.84\ \text{W}$.
(e) $P_v = 6.74\ \text{mW/cm}^3$, $P_C = 13.48\ \text{W}$.
(f) $P_v = 8.94\ \text{mW/cm}^3$, $P_C = 17.88\ \text{W}$.

2.6 $P_e = 2.827\ \text{mW/cm}^3$.

2.7 (a) $P_h = 159\ \text{W/cm}^2$
(b) $A_{BH} = 159\ \text{J/m}^3$.
(c) $P_e = 706.6\ \mu\text{W/cm}^3$.
(d) $P_h/P_e = 225 \times 10^3$.
(e) $P_C = 626\ \text{W}$.

2.8 (a) $P_H = 4774.5\ \text{W}$.
(b) $P_E = 13.332\ \text{mW}$.
(c) $P_E = 4774.5\ \text{mW}$.

CHAPTER 3

3.1 $\delta_c = 4.11\ \text{mm}$.

3.2 (a) $\delta = 205.468$ m at $f = 60$ Hz.
(b) $\delta = 1.592$ m at $f = 1$ MHz.
(c) $\delta = 5$ cm at $f = 1$ GHz.

3.3 $\delta_c = 4.708$ mm.

3.4 $\delta_w = 1.4777$ μm, $\delta_c = 3.5588$ m.

3.5 (a) $\delta_c = 1.59$ mm at $f = 1$ MHz.
(b) $\delta_c = 0.503$ mm at $f = 10$ MHz.

3.6 $R = 1.275 \times 10^{-8}$ Ω.

3.7 $R_{wDC} = 0.775$ mΩ, $\delta_w = 2.429$ μm, $d/\delta_w = 265$, $R_w = 0.0513$ Ω, and $F_R = 66.19$.

3.8 $R_w = 74.3$ mΩ, $F_R = 6.7675$, $P_w = 2.3776$ W.

3.9 (a) $\mathbf{H}_\phi = \frac{J_o r^2}{3r_o}\mathbf{a_z}$ for $r \leq r_o$.
(b) $\mathbf{H}_\phi = \frac{J_o r_o^2}{3r}\mathbf{a_z}$ for $r > r_o$.

3.10 (a) $\mathbf{H}_\phi = \frac{J_o r^3}{3r_o^2}\mathbf{a_z}$ for $r \leq r_o$.
(b) $\mathbf{H}_\phi = \frac{J_o r_o^2}{4r}\mathbf{a_z}$ for $r > r_o$.

3.11 (a) $A_e = 0.2175 \times 10^{-6}$ m^2.
(b) $R_w = 73.36$ mΩ.
(c) $F_R = 3.864$ mΩ.
(d) $P_w = 0.917$ W.

3.12 (a) $\rho_v = -1.8126 \times 10^{10}$ C/m^3.
(b) $N_{Cu} = 1.13 \times 10^{23}$ electrons/cm^3.
(c) $J = 5.8$ A/mm^2.
(d) $I = 4.57$ A.
(e) $v_d = -0.032$ cm/s.

CHAPTER 4

4.2 $P_w = 33.667$ W.

4.3 (a) $I_{22(rms)} = 60.8276$ A.
(b) $P_{L22} = 9.25$ W.
(c) $P_w = 71.06$ W.

4.4 (a) $I_{32(rms)} = 44.55$ A.
(b) $P_{L22} = 1.985$ W.
(c) $P_w = 21.856$ W.

4.5 (a) $I_{15(rms)} = 14.08$ A.
(b) $P_{L15} = 8.42$ W.
(c) $P_w = 45.1$ W.

4.6 $P_{L25} = 336.28$ W.

CHAPTER 5

5.1 (a) $\delta_w = 0.3304$ mm, $h_{opt} = 0.1948$ mm.
(b) $h_B = 0.1213$ mm.

5.2 (a) $\delta_w = 0.191$ mm, $h_{opt} = 0.1292$ mm.
(b) $h_B = 0.0804$ mm.

5.3 (a) $\delta_w = 0.2413$ mm, $h_{opt} = 0.1922$ mm.
(b) $h_B = 0.1196$ mm.

5.4 (a) $\delta_w = 0.1478$ mm, $d_{opt} = 0.10893$ mm.
(b) $d_B = 0.0733$ mm.

5.5 $R_w = 8.2497\ \Omega$.

5.6 (a) $R_w = 0.162\ \Omega$.
(b) $R_w = 0.06869\ \Omega$.
(c) $S = 2.358$.

5.7 $F_R = 9.118$.

5.8 $R_w = 0.318\ \Omega$.

5.9 (a) $N = 20$.
(b) $h_{opt} = 0.05886$ mm.
(c) $R_{wDC} = 35.8$ mΩ.
(d) $R_w = 47.8$ mΩ.
(e) $P_w = 1.3433$ W.
(f) $J_{AV} = 12.931$ A/mm^2.

5.10 (a) $h_B = 0.036724$ mm.
(b) $R_{wDC} = 57.6$ mΩ.
(c) $R_w = 60.4$ mΩ.
(d) $P_{wB} = 1.6999$ W.
(e) $R_{wB}/R_{w(opt)} = 1.2654$.

CHAPTER 6

6.1 $w_{min} = 0.0256$ mm.

6.2 $w_{max} = 0.06366$ mm.

CHAPTER 7

7.1 $L_m = 1.26$ mH, $L_{ms} = 12.56\ \mu$H.

7.2 $n = 10$, $\mathcal{R} = 515,780$ A · turn/Wb, $\mathcal{P} = 1.937 \times 10^{-6}$ Wb/A · turn, M = 1.24 mH, $L_1 = 12{,}4$ mH, $L_2 = 0.124$ mH, $L_m = 12.276$ mH, $L_{ms} = 0.12267$ mH, $L_{11} = 0.124$ mH, $L_{12} = 620$ nH, $k = 0.9925$.

7.3 $N = 100$.

7.4 $L_m = 494.8$ mH.

7.5 $R_s = 0.219$ Ω.

7.6 $f_L = 16.08$ Hz.

CHAPTER 8

8.1 (a) $\delta_{Cu} = 2.0897$ μm,
(b) $\delta_{Al} = 2.59$ μm,
(c) $\delta_{Ag} = 2.0069$ μm,
(d) $\delta_{Au} = 2.486$ μm.

8.2 $R_{DC} = 1.325$ Ω, $R_{AC} = 2.29$ Ω, $\delta_{Al} = 0.819$ μm, $F_R = 1.728$.

8.3 $L = 0.1022$ nH.

8.4 $L = 1.134$ nH.

8.5 $L = 0.1$ nH, $\delta = 1.348$ μm, $L_{HF} = 0.0134$ nH.

8.6 $L = 1.37$ nH.

8.7 $L = 39.47$ nH.

8.8 $D = 1000$ μm, $L = 42.64$ nH.

8.9 $D = 1000$ μm, $L = 43.21$ nH.

8.10 $D = 1500$ μm, $L = 354.8$ nH.

8.11 $D = 1000$ μm, $L = 45$ nH.

8.12 $D = 1000$ μm, $l = 21\,060$ μm, $L = 59.13$ nH.

8.13 $D = 1000$ μm, $L = 41.855$ nH.

8.14 $l = 22\,140$ μm, $b = 183.33$ μm, $L = 54.59$ nH.

8.15 $L = 495.679$ nH.

8.16 $D = 1000$ μm, $L = 36.571$ nH.

8.17 $D = 520$ μm, $L = 5.75$ nH.

8.18 $D = 1000$ μm, $L = 122.38$ nH.

8.19 $l_c = 380$ μm, $L = 13.228$ nH.

CHAPTER 9

9.1 $C_s = 44.67$ pF.

9.2 $C_s = 0.8934$ pF.

9.3 $C_s = 0.6236$ pF.

9.4 (a) $C_s = 1607.59$ pF. (b) $C_s = 1827.25$ pF. (c) $C_s = 572.44$ pF.

9.5 (a) $C_s = 312.1$ pF. (b) $C_s = 354.75$ pF. (c) $C_s = 111.136$ pF.

CHAPTER 10

10.1 $K_i = 0.6855$.

10.2 $K_i = 0.941$.

10.3 (a) $K_{ai} = 0.8534$. (b) $K_{ai} = 0.739$.

10.4 (a) $K_{ai} = 0.6217$. (b) $K_{ai} = 0.5384$.

10.5 (a) $K_{ai} = 0.6456$. (b) $K_{ai} = 0.5584$.

Physical Constants

Table A.1 Physical constants

Physical constants	Symbol	Value
Boltzmann's constant	$k = \frac{E_T}{T}$	$1.3806488 \times 10^{-23}$ J/K $= 8.62 \times 10^{-5}$ eV/K
Planck's constant	h	6.62617×10^{-34} J·s $= 414 \times 10^{15}$ eV·s
Electron charge magnitude	q	1.60218×10^{-19} C
Free-space permittivity	$\epsilon_0 = \frac{10^{-9}}{36\pi}$	8.85418×10^{-12} F/m
Free-space permeability	μ_0	$4\pi \times 10^{-7}$ H/m
Speed of light in free space	$c = \frac{1}{\sqrt{\epsilon_0 \mu_0}}$	2.998×10^{8} m/s
Mass of the electron	m_e	9.1095×10^{-31} kg
Mass of the proton	m_h	1.673×10^{-27} kg
Mass of the neutron	m_h	1.649×10^{-27} kg
Thermal energy	E_T	kT

High-Frequency Magnetic Components, Second Edition. Marian K. Kazimierczuk.

Companion Website: www.wiley.com/go/kazimierczuk_High2e

Maxwell's Equations

Table B.1 Maxwell's equations

Law	Differential form	Integral form
Ampère's law	$\nabla \times \mathbf{H} = \mathbf{J} + \frac{\partial \mathbf{D}}{\partial t}$	$\oint_C \mathbf{H} \cdot d\mathbf{l} = \int \int_S \left(\mathbf{J} + \frac{\partial \mathbf{D}}{\partial t}\right) \cdot d\mathbf{S} = i_{enc}$
Faraday's law	$\nabla \times \mathbf{E} = -\frac{\partial \mathbf{B}}{\partial t}$	$\oint_C \mathbf{E} \cdot d\mathbf{l} = -\int \int_S \frac{\partial \mathbf{B}}{\partial t} \cdot d\mathbf{S} = V$
Gauss's law for electric field	$\nabla \cdot \mathbf{D} = \rho$	$\oint_S \mathbf{D} \cdot d\mathbf{S} = \int \int \int_V \rho dV = Q$
Gauss's law for magnetic field	$\nabla \cdot \mathbf{B} = 0$	$\oint_S \mathbf{B} \cdot d\mathbf{S} = 0$

Table B.2 Maxwell's equations in phasor forms

Law	Differential form	Integral form
Ampère's law	$\nabla \times \mathbf{H} = (\sigma + j\omega\epsilon)\mathbf{E}$	$\oint_C \mathbf{H} \cdot d\mathbf{l} = (\sigma + j\omega\epsilon) \int \int_S \mathbf{E} \cdot d\mathbf{S}$
Faraday's law	$\nabla \times \mathbf{E} = -j\omega\mu\mathbf{H}$	$\oint_C \mathbf{E} \cdot d\mathbf{l} = -j\omega\mu \int \int_S \mathbf{H} \cdot d\mathbf{S} = V$
Gauss's law for electric field	$\nabla \cdot \mathbf{D} = \rho$	$\oint_S \mathbf{D} \cdot d\mathbf{S} = \int \int \int_V \rho dV = Q$
Gauss's law for magnetic field	$\nabla \cdot \mathbf{B} = 0$	$\oint_S \mathbf{B} \cdot d\mathbf{S} = 0$

Constitutive Relationships

$$\mathbf{D} = \epsilon \mathbf{E}$$

$$\mathbf{B} = \mu \mathbf{H}$$

$$\mathbf{J} = \sigma \mathbf{E}$$

High-Frequency Magnetic Components, Second Edition. Marian K. Kazimierczuk.

Companion Website: www.wiley.com/go/kazimierczuk_High2e

INDEX

A

AC magnetic flux density, 4, 32, 53, 111, 115, 117, 118, 126, 129, 130, 568
AC inductors, 590, 593
AC resistance, 166, 172, 189, 191, 300, 309, 318, 326, 329
AC-to-DC resistance ratio (F_R), 32, 173, 294, 296, 299
Air factor (K_a), 576
Air gap, 51, 53, 73, 99
Air gap factor (F_g), 51
Air gap length (l_g), 51
Air and wire insulation factor (K_{ai}), 575
Aluminum, 83, 165
Alloys, 81, 91, 104
  cobalt-iron (CoFe), 81, 91, 104
  nickel-iron (FeNi), 81, 91, 104
  nickel-zinc (NiZn), 81, 91, 104
  manganese-zinc (MnZn), 81, 91, 104
American Wire Gauge (AWG), 136, 569, 570, 571
Amorphous material, 81, 104
Ampacity, 166
Ampère's law, 3, 9, 12, 19, 21, 53
Amplitude spectrum, 252, 461, 652
Angular momentum (m_a), 86
Antiferromagnetic, 94
Antiproximity effect, 244
Area product (A_p), 586, 687, 603, 619, 668
Autotransformer, 439

B

B-*H* loop, 70, 97, 98
Bandwidth (*BW*), 82, 452
Bessel functions, 173, 220
Bifilar winding, 71
Biot-Savart's law, 18, 42
Bobbin, 99, 577
Bobbin factor (K_b), 577
Buck converter, 585

C

CC core, 100
Capacitance (*C*), 69, 438, 521
Ceramics, 105
Coils, 6, 71, 417
Cobalt, 81, 82, 91, 94
Cobal-iron (CoFe) cores, 81, 105
Complex permeability (μ), 136
  parallel (μ_p), 148
  series (μ_s), 137
Complex propagation constant (γ), 22
Complex magnetic susceptibility (χ_m), 138
Contactless energy transfer, 446
Conductivity (σ), 18
Conservative field, 20
Constitutive equations, 22
Continuity equation, 22
Converter, 585
Cooling of magnetic core, 151
Copper, 23, 105, 165, 168, 353, 360, 571

High-Frequency Magnetic Components, Second Edition. Marian K. Kazimierczuk.

Companion Website: www.wiley.com/go/kazimierczuk_High2e

Copper loss, 265
Core, 32, 81
Core area product (A_p), 569
Core cross-sectional area (A_c), 6, 265, 305
Core geometry, 99
Core geometry coefficient (K_g), 654, 702, 709
Core loss (P_C), 70, 113, 129
Core material, 81
Core permeability (μ_{rc}), 94, 98
Core saturation, 4, 32, 34, 36, 38, 72
Core shape, 99
Core window area (W_a), 579
Core window utilization (K_u), 579
Coupling coefficient (k), 425
Curie temperature (T_c), 82, 90, 91
Curl, 19, 20, 21, 24, 297
Current density (J), 17, 18, 166, 175, 177, 178, 180, 183, 206, 274, 473, 570
Current probe, 446, 453
Current transformer, 446, 453

D
DC magnetic flux density (B_{DC}), 4, 5, 32, 53
DC resistance, 173, 278, 293, 296, 300, 305
Diamagnetic, 87, 95
Dipole, 83, 84
Dipole moment, 84, 85
Displacement current density (J_d), 19, 21
Divergence, 20
Domain, 88, 89, 90
Domain wall, 89
Domain motion, 89, 90
Dot convention, 428, 430
Dowell's equation, 294, 298, 375
Drift velocity (v_d), 17

E
EC core, 99
EE core, 99, 102, 580
Eddy current (i_E), 29, 164
Eddy-current loss, 70, 113, 384, 389, 393
Eddy-current core loss (P_E), 70, 113, 384, 389, 393
Edge factor (K_{ed}), 578
Effective relative permeability (μ_{re}), 52, 53
Efficiency (η), 438
Electron, 7, 26, 84, 85
EMI, 99
Energy (W_m), 25, 26, 28, 84
Energy density (w_m), 25, 26, 28
Equivalent series resistance (ESR), 521
Equivalent circuit, 6, 146, 149
Excess core loss, 129

F
Faraday's law, 13, 19, 21, 419, 420
Ferrite, 81, 105
Ferrite core, 81, 105
Ferrite material, 81, 105
Ferromagnetic, 81, 87, 93, 95, 103, 105
Ferrimagnetic, 95
Field, 2
Field density, 2, 569
Field intensity, 3
Flux, 2
Flux density, 2, 569
Flux linkage (λ), 4, 6, 16, 417, 424
Foil, 215, 294, 301, 364, 570, 581
Fourier series, 135, 455, 456, 457, 458, 459, 460
Free space permeability (μ_0), 4
Free space permittivity (ϵ_0), 532
Fringing effect, 54, 74
Fringing magnetic field intensity (H_f), 55, 56
Fringing flux (ϕ_f), 54, 55, 56
Fringing factor (F_f), 55, 57, 58, 60

G
Gap, 51, 53, 65
Gap lenght (l_g), 51, 53, 65
Gapped core, 51, 134
Gadolinium, 82
Gauss's law, 97

H
Hard magnetic material, 97
Harmonics, 135, 351, 455, 353, 355, 361
Heat conduction, 151
Helmholtz, 167, 267
Hysteresis, 70
Hysteresis curve, 70
Hysteresis loop, 70
Hysteresis loss (P_H), 70, 109

I
Ideal transformer, 413
Impedance (Z), 15, 145, 184, 211, 293, 400, 560
Incremental relative permeability (μ_{rss}), 98
Inductance (L), 40, 62, 63, 72, 145, 189, 195, 335, 477, 476
Inductance factor (A_L), 49
Inductor, 40, 47, 49, 473
Inductor design, 560
Integrated inductors, 473
Inverting transformer, 422
Insulation factor (K_i), 574
Interleaved windings, 444
Iron, 81, 87, 105

Iron cores, 103, 104
Iron powder cores, 103, 106, 107

J
Joule's law, 26

K
Kirchhoff's laws, 8, 9

L
Laminated core, 128, 283, 389
Lamination, 283
Landau-Lifshitz-Gilbert (LLG), 139
Leakage flux (ϕ_l), 9, 417, 418
Leakage inductance (L_l), 436
Leakage inductance measurement, 440
Lenz's law, 15, 18, 163
Litz wire, 338
London's penetration depth, 109
Loop, 84, 85, 97
Loss, 61, 129
Lumped-circuit model, 68, 521

M
Maclaurin, 375
Magnet wire, 569
Magnetic anisotropy, 93
Magnetic circuit, 6, 51, 55
Magnetic core, 32, 81
Magnetic domain, 83, 81
Magnetic energy (W_m), 28, 42, 64, 65, 66, 67
Magnetic energy density (w_m), 25, 64, 65, 66, 67
Magnetic dipole, 83, 84
Magnetic field, 2, 53, 204, 230
Magnetic field intensity (H), 3, 11, 12, 13, 193, 204
Magnetic laws, 9, 71
Magnetic flux (ϕ), 3, 8
Magnetic flux density (B), 3, 4, 37, 53
Magnetic flux linkage (λ), 5, 6, 32, 344, 35, 36, 37, 40
Magnetic flux leakage (ϕ_l), 417, 418
Magnetic loss tangent (tan δ_m), 141
Magnetic material, 81, 83
Magnetic moment, 84, 87
Magnetic monopole, 4
Magnetic pole, 84
Magnetic potential, 423
Magnetic spin, 84
Magnetic susceptibility (χ_m), 91, 92
Magnetization, 87, 88, 152
Magnetization curve, 97
Magnetization process, 87, 96, 97
Magnetizing current (i_{Lm}), 435
Magnetizing inductance (L_m), 433
Magnetomotive force (MMF), 2
Manganese, 81, 82
Maxwell's equations, 19, 21, 24
Mean core length (l_c), 6, 41
Mean turn length (l_T), 286, 288, 293, 585, 586
Meander inductors, 481
Meissner effect, 109
MEMS inductors, 507
Metallic glass, 81
MMF, 2, 9
Model, 68, 373, 415, 422, 437, 447, 464, 510
MPL, 6
Multiple-winding transformers, 439
Mutual flux (ϕ_m), 417, 419
Mutual inductance (M), 419, 422, 424

N
Nanocrystalline cores, 81, 107, 108
Nanocrystalline magnetic material, 81
Neumann's formula, 422
Nickel, 81, 105
Nickel-iron (NiFe) core, 81, 105
Nickel-zinc (NiZn) core, 81, 105
NiFe alloy, 81, 105
Nonideal transformer, 417
Noninverting transformer, 429, 430
Noninverleaved winding, 442
Nonlinear inductance, 15
Number of winding turns (N), 6, 32, 52, 305, 414, 578

O
Ohm's law, 6, 16, 18
Optimization, 286, 301, 303, 315, 321, 322, 331, 332, 345, 346, 365, 367
Optimum flux density, 673
Orbital moment (m_o), 84, 85
Orthogonality, 227
Orbital magnetic moment, 84, 85
Oxides, 105

P
Parallel complex permeability (μ_p), 148
Paramagnetic, 87, 91, 96
Parasitic capacitances (C_s), 69
PCB inductors, 103, 512
Penetration depth (δ), 166
Permalloy, 91, 105
Permanent magnet, 97
Permeance ($\mathcal{P}$), 6, 49, 56
Permeability (μ), 4, 7, 52, 82, 98
Permittivity (ϵ), 7, 20, 532
Phase (ϕ), 102, 179, 186

Planar core, 103
Planar inductors, 103, 492
Planar transformers, 506
Planck's constant (h), 86
Pole, 84
Power (P), 69, 200,
Power density (p), 200
Power loss, 69, 74, 208, 235, 248, 293, 351
Power loss density, 69, 74
Pot core, 49, 100, 583
Powder core, 106
Powder iron core, 106
Potential, 3, 43
Poynting vector, 24
Poynting theorem, 24
Primary winding, 445, 460
Printed circuit board (PCB), 512
Proximity effect, 227, 226, 227, 228, 230, 249, 250, 260
PWM converter, 585, 622, 623, 624, 627
PQ core, 54, 101, 584

Q
Quality factor (Q), 69, 141, 523, 526, 596, 597, 599

R
Reactance (X), 172
Relative permeability (μ_r), 4, 82, 569
Relative permittivity (ϵ_r), 532
Reluctance ($\mathcal{R}$), 6, 8, 41, 49, 53, 72, 144
Reluctor, 6
Relaxation time constant, 18
Remenace (B_r), 96
Resistivity (ρ), 164
Resonant frequency, 69, 165
RM core, 101
Rotational field, 20

S
Saturable reactor, 454
Saturation flux density (B_s), 4, 32, 94
Scalar field, 2
Secondary winding, 456, 462
Self-capacitance (C_s), 69, 521
Self-inductance (L), 40, 417, 419
Self-resonant frequency, 69, 522
Series complex permeability (μ_s), 127
Series-aiding connection, 431
Series-opposing connection, 431
Silicon, 82, 83
Silicon steel, 82
Silver, 165
Skin effect, 113, 163, 173, 204, 215, 230, 474
Skin depth (δ), 22, 23, 166, 305
Skin penetration (δ), 22, 23, 166
Small-signal inductance, 43
Small-signal relative permeability (μ_{rss}), 98
Soft magnetic material, 97
Solenoid, 45
Spectrum, 252, 461
Spin, 84, 87, 95
Steinmetz, 129
Steel, 81, 82
Stray capacitance (C_s), 69, 438, 521
Superconductors, 108
Superconductivity, 108
Supermalloy, 94
Surface area, 151, 152, 582, 583, 584, 585
Susceptibility (χ_m), 91, 92, 93, 138
Susceptance, 68

T
Tapered spiral inductor, 511
Temperature (T), 82, 90, 91, 164
Temperature rise (ΔT), 164, 465, 581
Tesla (T), 4
Thermal model, 370, 373
Thermal resistance, 373, 582
Toroidal core, 47, 99, 580, 581, 582, 583
Toroidal inductor, 13, 582
Torus core, 49
Torque (T_q), 80, 84
Transfer function, 428
Transformer, 412, 668
Transformer design, 634, 668
Transformer turns ratio (n), 414
Transformer coupling coefficient (k), 425
Transformer efficiency (η), 438
Turns ratio (n), 414, 672
Turn-to-turn capacitance (C_{tt}), 547, 548, 554

U
UU core, 100

V
Vector field, 2
Volt-second balance, 16

W
Waveform factor (K_f), 35
Weber (Wb), 3
Winding loss, 70
Winding resistance, 265, 300
Winding ac resistance (R_w), 166, 300, 309, 318, 326, 341, 367
Winding dc resistance (R_{wDC}), 164, 293
Winding patterns, 569

Winding power loss, 70, 364, 366, 455, 460
Winding dc power loss (P_{wDC}), 455, 460
Winding ac power loss (P_w), 455, 460
Winding wire, 569
Window area (W_a), 579
Window utilization factor (K_u), 574, 670
Wire, 11, 569, 570, 571
Wire diameter (d), 573
Wire insulation, 573
Wire insulation factor (K_i), 574
Wireless energy transfer, 446

Z
Zinc, 81, 82, 104

Printed in the United States
By Bookmasters